Fibre Reinforced Concrete: Improvements and Innovations II

RILEM BOOKSERIES

Volume 36

RILEM, The International Union of Laboratories and Experts in Construction Materials, Systems and Structures, founded in 1947, is a non-governmental scientific association whose goal is to contribute to progress in the construction sciences, techniques and industries, essentially by means of the communication it fosters between research and practice. RILEM's focus is on construction materials and their use in building and civil engineering structures, covering all phases of the building process from manufacture to use and recycling of materials. More information on RILEM and its previous publications can be found on www.RILEM.net.

Indexed in SCOPUS, Google Scholar and SpringerLink.

More information about this series at http://www.springer.com/series/8781

Pedro Serna · Aitor Llano-Torre ·
José R. Martí-Vargas · Juan Navarro-Gregori
Editors

Fibre Reinforced Concrete: Improvements and Innovations II

X RILEM-fib International Symposium
on Fibre Reinforced Concrete (BEFIB) 2021

 Springer

Editors

Pedro Serna ⓘ
ICITECH
Universitat Politècnica de València
Valencia, Spain

Aitor Llano-Torre ⓘ
ICITECH
Universitat Politècnica de València
Valencia, Spain

José R. Martí-Vargas ⓘ
ICITECH
Universitat Politècnica de València
Valencia, Spain

Juan Navarro-Gregori ⓘ
ICITECH
Universitat Politècnica de València
Valencia, Spain

ISSN 2211-0844   ISSN 2211-0852   (electronic)
RILEM Bookseries
ISBN 978-3-030-83721-1   ISBN 978-3-030-83719-8   (eBook)
https://doi.org/10.1007/978-3-030-83719-8

This Springer imprint is published by the registered company Springer Nature Switzerland AG
The registered company address is: Gewerbestrasse 11, 6330 Cham, Switzerland

symposium venue, but I hope you could enjoy the virtual BEFIB2021 experience, and I encourage all of you to visit Valencia at the first opportunity.

Looking forward to hosting you in Valencia.

Pedro Serna
BEFIB2021 Chairman

Organization

Committees

Organising Committee

Pedro Serna (Chairman)   Universitat Politècnica de València, Spain
Aitor Llano-Torre (Secretariat)   Universitat Politècnica de València, Spain
José R. Martí-Vargas   Universitat Politècnica de València, Spain
Juan Navarro-Gregori   Universitat Politècnica de València, Spain

Scientific Committee

M. A. Aiello, Italy
A. Aguado, Spain
C. Aldea, Canada
S. Al-Toubat, UAE
G. L. Balázs, Hungary
N. Banthia, Canada
B. Barragan, France
J. A. O. Barros, Portugal
S. Bernard, Australia
A. Bettencourt Ribeiro, Portugal
S. Billington, USA
J. Bolander, USA
P. Borges, Brazil
W. P. Boshoff, South Africa
N. Buratti, Italy
S. H. P. Cavalaro, UK
J. P. Charron, Canada
A. Conforti, Italy
E. Cuenca, Italy
F. Dehn, Germany
A. De La Fuente, Spain

E. Denarie, Switzerland
M. di Prisco, Italy
Y. Ding, China
A. Fantilli, Italy
L. Ferrara, Italy
A. Figueiredo, Brazil
S. Foster, Australia
J. Gálvez, Spain
E. Garcia-Taengua, UK
R. Gettu, India
G. M. Giaccio, Argentina
S. Grunewald, Netherlands
P. Kabele, Czech Republic
T. Kanda, Japan
T. Kanstad, Norway
I. Khan, Saudi Arabia
K. Kobayashi, Japan
M. Konsta-Gdoutos, Greece
M. Kunieda, Japan
V. Li, USA
Y. M. Lim, South Korea

Institutions

Organised by

UPV

Universitat Politècnica de València

ICITECH

Institute of Concrete Science and Technology

Supporting Associations

RILEM

International Union of Laboratories and Experts in Construction Materials, Systems and Structures

fib

The International Federation for Structural Concrete

ACI

Amercian Concrete Institute

Exhibitors

Gold Exhibitor

Silver Exhibitors

Bronze Exhibitors

Affiliate

Contents

Technological Aspects

Effects of Fly Ash Content and Curing Age on High Temperature
Residual Compressive Strength of Strain-Hardening
Cementitious Composites 3
Dhanendra Kumar, Amr Ashraf Soliman, and Ravi Ranade

Optimization of Functionally Graded Concretes Incorporating Steel
Fibres and Recycled Aggregates 13
Ricardo Chan, Charles K. S. Moy, and Isaac Galobardes

Mix Proportioning of Fiber Reinforced Self-compacting Concrete
Adopting the Compressible Packaging Method: Comparison
of Two Methods 24
Matheus G. Cardoso, Rodrigo M. Lameiras, and Ian B. Cavalcante

Influence of Steel Fibers on Damage Induced by Alkali-Silica
Reaction of Concrete with Reactive Sudbury Aggregates 35
Stefano Giuseppe Mantelli, Daman K. Panesar, and Fausto Minelli

Mechanical Properties

Effect of Alkali Treatment to Improve Fiber-Matrix Bonding
and Mechanical Behavior of Sisal Fiber Reinforced
Cementitious Composites 51
Raylane de Souza Castoldi, Lourdes Maria Silva de Souza,
and Flávio de A. Silva

Drop-Weight Impact Test for Fibre Reinforced Concrete:
Analysis of Test Configuration 61
Juan Carlos Vivas, Facundo Isla, María Celeste Torrijos,
Graciela M. Giaccio, Bibiana Luccioni, and Raúl L. Zerbino

Influence of Specimen Size for Impact Testing of Fiber Reinforced Cementitious Composite by Charpy Test 74
Gerbert P. Moreira Neto, Philipe de O. Vital, Margareth da S. Magalhães, and Reila V. Velasco

The Effect of Organic Fiber Hybridization on Fresh and Hardened Concrete Properties 86
Imane Bentegri, Tien Tung Ngo, Hamza Soualhi, Othmane Boukendakdji, and El-Hadj Kadri

The Antifragility of FRC in the Crack Pattern of Reinforced Concrete Ties 98
Alessandro P. Fantilli and Francesco Tondolo

Fatigue of Plastic Fibre Reinforced Concrete in Bending: Assessment and Prediction 109
Debora Martinello Carlesso, Sergio H. P. Cavalaro, and Albert de la Fuente

Fatigue of Cracked Steel Fibre Reinforced Concrete Subjected to Bending 121
Debora Martinello Carlesso, Albert de la Fuente, and Sergio H. P. Cavalaro

Direct Tensile Tests of Supercritical Steel Fibre Reinforced Concrete 132
Katharina Look, Peter Heek, and Peter Mark

Uniaxial Tension Tests of Steel Fibre Reinforced Concrete with AE Monitoring 143
Maure De Smedt, Eline Vandecruys, Rutger Vrijdaghs, Els Verstrynge, and Lucie Vandewalle

An Experimental Study on the Flexural Fatigue Behaviour of Pre-cracked Steel Fibre Reinforced Concrete 155
Humaira Fataar, Riaan Combrinck, and William P. Boshoff

Experimental Study on Punching Shear Strength of Fiber Reinforced Concrete Slabs 166
Salah Altoubat, Nadia Nassif, and Mohamed Maalej

Mechanical and Fracture Behaviour of an HPFRC 174
Sandra Nunes, Mário Pimentel, and Carlos Sousa

Property Assessment of Self-compacting Basalt Fiber Reinforced Concrete 186
Piotr Smarzewski

Influence of Steel Fibers and Casting Direction on the Bond Between Concrete and Reinforcement Bars 198
Pedro Henrique de Omena Jucá, Camila Vargas Cardoso,
Guilherme Durigon Cocco, Felipe Eduardo Kulzer,
Letícia Larré de Oliveira, Ederli Marangon, and Luis Eduardo Kosteski

Evaluation of Behavior of the Joint Between Two Concrete Layers During Splitting 208
Jakob Šušteršič, Rok Ercegovič, David Polanec, and Andrej Zajc

Short-Term Effects of Moderate Temperatures on the Mechanical Properties of Steel and Macrosynthetic Fiber Reinforced Concretes . . . 220
Marta Caballero-Jorna, Marta Roig-Flores, and Pedro Serna

Residual Flexural Strength of SFRC: A Multivariate Perspective 232
Emilio Garcia-Taengua, José R. Martí-Vargas, and Pedro Serna

Dynamic Behaviour of Steel Fiber Reinforced Concrete Plates Under Gun Fire and Free Fall Tests 244
Vahan Zohrabyan, Tobias Seltner, Thomas Braml, and Manfred Keuser

Spinability and Characteristics of Particle-Shell PP-bicomponent Fibers for Crack Bridging in Mineral-Bonded Composites 255
Mihaela-Monica Popa, Harald Brünig, Iurie Curosu, Viktor Mechtcherine, and Christina Scheffler

Damage of Fracture Properties of Polyolefin Fibre Reinforced Concrete Under High Temperature 265
Marcos. G. Alberti, Jaime C. Gálvez, Alejandro Enfedaque, and Ramiro Castellanos

Evaluation of the Inertia Force in Compressive Impact Loading on Steel Fiber-Reinforced Concrete 277
Mohammad Bakhshi, Isabel B. Valente, Honeyeh Ramezansefat, and Joaquim A. O. Barros

Long-Term Properties

International Round-Robin Test on Creep Behaviour of FRC - Part 2: An Overview of Results and Preliminary Conclusions 291
Aitor Llano-Torre, Pedro Serna, and Sergio H. P. Cavalaro

Evaluation of Creep and Shrinkage of Patented Mixture of UHPC with Applied Heat-Treatment 307
Vladimír Příbramský

Cracking Behaviour of FRC Members Reinforced with GFRP Bars under Sustained Loads 319
Razan H. Al Marahla and Emilio Garcia-Taengua

Experimental Investigation on the Influence of Temperature
Variations on Macro-synthetic Fibre Reinforced Concrete Short
and Long Term Behaviour.................................... 331
Clementina Del Prete, Nicola Buratti, and Claudio Mazzotti

Short and Long Term Behaviour of Polypropylene Fibre Reinforced
Concrete Beams with Minimum Steel Reinforcement.............. 342
Nikola Tošić and Albert de la Fuente

Analytical and Numerical Models

Computational Mesoscale Modeling for the Mechanical Behavior
of Fiber Reinforced Concrete................................. 357
Marcello Congro, Deane Roehl, and Eleazar C. M. Sanchez

Numerical Multi-level Model for Fibre Reinforced Concrete:
Validation and Comparison with Fib Model Code................ 365
Gerrit E. Neu, Vladislav Gudžulić, and Günther Meschke

Isotropy-Based Analytical Model to Estimate the Residual Strength
of FRC... 377
Eduardo Galeote, Ana Blanco, and Albert de la Fuente

Different Approaches for FEM Modelling of Strain-Hardening
Cementitious Composites 389
Hassan Baloch, Steffen Grünewald, and Stijn Matthys

Exploring the Performance of a Single Panel SFRC Slab Under
a Point Load with Fe Analysis 400
Olugbenga B. Soymi and Ali A. Abbas

An Analytical Study of Shear Transfer Mechanisms in Macro-
synthetic Fibre Reinforced Concrete 409
Francisco Ortiz-Navas, Juan Navarro-Gregori, and Pedro Serna

Creep of Macro Synthetic Fibre Reinforced Concrete: Experimental
Results and Numerical Model Calibration 420
Clementina Del Prete, Ioannis Boumakis, Roman Wan-Wendner,
Nicola Buratti, and Claudio Mazzotti

Statistical Modelling of Flexural Fatigue Response of Steel Fibre
Reinforced Concrete... 433
Ajeesh Koorikkattil, Sunitha K Nayar, and Veena Venudharan

Numerical Modeling of the Steel Fiber Reinforced Concrete
Behavior Under Combined Tensile and Shear Loading by a
Micromechanical Model Taking into Account Fiber Orientation 443
Duc-Tam Vu, François Toutlemonde, Benjamin Terrade, Pierre Marchand,
and Sébastien Bouteille

Predicting the Residual Flexural Strength of Concrete Reinforced
with Hooked-End Steel Fibers: New Empirical Equations 456
Enrico Faccin, Luca Facconi, Fausto Minelli, and Giovanni Plizzari

Structural Design

Design and Performance of a Precast Bridge Barrier with Ultra-
high Performance Fibre Reinforced Concrete (UHPFRC) 471
Clélia Desmettre, Jean-Philippe Charron, and Frédérick Gendron

Shear Behaviour of V-shape Webbed Steel Fibre Reinforced
Concrete Beams 483
Divan Visser and William P. Boshoff

Punching Shear Resistance of SFRC Flat Slabs with and Without
Punching Shear Reinforcement 492
Josef Landler and Oliver Fischer

Elevated Steel Fibre Reinforced Concrete Slabs and the Hybrid
Alternative: Design Approach and Parametric Study at Ultimate
Limit State 504
Stanislav Aidarov, Luca Sutera, Manuela Valerio,
and Albert de la Fuente

Shear Behavior of Hollow-Core Slabs Reinforced
by Macro-synthetic Fibers 514
Alan Piemonti, Francisco Ortiz-Navas, Antonio Conforti,
Giovanni Plizzari, Sandro Moro, Martin Hunger, Steve Schaef,
and Bruno Della Bella

How Can We Verify Structural Members Made of FRC Only? 525
Luca Facconi and Fausto Minelli

Codes and Standards

Eurocode 2 – Annex L – European Harmonized Standard for Steel
Fibre Reinforced Concrete 539
Marco di Prisco, Terje Kanstad, Giovanni Plizzari, Fausto Minelli,
and Andreas Haus

Reliability of Shear Strength Models for Fibre Reinforced Concrete
Members Without Shear Reinforcement 552
Nikola Tošić, Jesús Miguel Bairán, and Albert de la Fuente

Yield Line Design for SFRC Elevated Slabs 564
Matteo Colombo, Marco di Prisco, and Ali Pourzarabi

Review

Code Provisions for Shear Strength in Prestressed FRC Members: A Critical Review 579
Stefano Giuseppe Mantelli, Luca Facconi, Katharina Look,
Filippo Medeghini, Peter Mark, Fausto Minelli, and Giovanni Plizzari

A Short Review on the Utilization of Basalt Fibres in Concrete 591
Suman Saha

Quality Control

Using Energy Absorption Capacity to Determine Residual Resistances of FRC 601
Sergio Carmona and Climent Molins

Case Studies: Structural and Industrial Applications

Use of Steel-Fibre Reinforced Concrete to Extend Service Life of Temporary Safety Concrete Barriers 615
Clélia Desmettre and Jean-Philippe Charron

A Fiber Reinforced Concrete for a Nuclear Waste Container 628
Erik Coppens, Petra Van Itterbeeck, Bram Dooms, Thomas Richir,
and Guillaume Debournonville

Design and Execution of Floors on Ground and Industrial Pavements with Fibre Reinforced Concrete 640
Roberto Pombo, Marcelo G. Altamirano, Graciela M. Giaccio,
and Raúl L. Zerbino

Design Optimization of Fibres Reinforced Concrete Railway Tracks by Using Non-linear Finite Elements Analysis 652
Jean-Louis Tailhan and Pierre Rossi

Elevated Flat Slab of Fibre Reinforced Concrete Non-linear Simulation up to Failure 666
Alejandro Nogales and Albert de la Fuente

TBM Thrust on Fibre Reinforced Concrete Precast Segment Simulation 678
Alejandro Nogales and Albert de la Fuente

Design, Specification and Failure Investigation of Fibre Reinforced Concrete Ground Bearing Industrial Floors and Hardstandings and Pile Suspended Industrial Ground Floors 690
Chris H. Peaston

Self-compacting Steel Fibre Reinforced Concrete: Material Characterization and Real Scale Test up to Failure of a Pile Supported Flat Slab 702
Stanislav Aidarov, Francisco Mena, and Albert de la Fuente

Structural Behavior of Precast Tunnel Segments Reinforced by Macro-synthetic Fibers During Temporary Loading Phases 714
Ivan Trabucchi, Antonio Mudadu, Giuseppe Tiberti, Antonio Conforti, Giovanni Plizzari, and Ralf Winterberg

Fibre Reinforced Cement Sheaths for Zonal Isolation in Oil Wells – Quantification and Mitigation of Shrinkage-Induced Cracking 727
Pablo Alberdi-Pagola, Victor Marcos-Meson, and Gregor Fischer

Recent FRC Developments in Uruguay: Quality Control, Durability and Three Structural Applications 739
Luis Segura-Castillo, Nicolás García, Diego Figueredo, Andrés Clavijo, Enzo González, Bruna Muniz, Iliana Rodriguez, and Gemma Rodriguez

Large-Scale Pressure-Swelling Tests on Panels Made of Strain-Hardening Cement-Based Composites with Different Bedding 749
Steffen Müller and Viktor Mechtcherine

SFRC Underwater Slab for the Potsdamer Platz (PP) Berlin 761
Horst Falkner and Ulf Hinke

Polypropylene Fibre Reinforced Concrete for the Structural Panels of the Pavillions of the Motril Port (Spain) 771
Elisa Valero Ramos, Juana Sánchez Gómez, Diego Jimenez López, Aurora Montalbán, and Albert de la Fuente

Use of Macro Synthetic Fibre in Segmental Tunnels 780
Ralf Winterberg

Design and Verification of Elevated Slabs Made with Hybrid Reinforced Concrete: Case Studies 796
Luca Facconi, Fausto Minelli, and Giovanni Plizzari

A New Sustainability Assessment Method for Façade Cladding Panels: A Case Study of Fiber/Textile Reinforced Cement Sheets 809
Payam Sadrolodabaee, S. M. Amin Hosseini, Monica Ardunay, Josep Claramunt, and Albert de la Fuente

Validation Testing of Precast Tunnel Lining Segments Using Polymeric Fibers 820
Devansh Patel, Chidchanok Pleesudjai, Yiming Yao, Steve Schaef, and Barzin Mobasher

Wind Tower FRC Foundations: Research and Design 831
Marco di Prisco, Claudio di Prisco, Giancarlo Fraraccio, Bruno Dal Lago,
Paolo Martinelli, Luca Flessati, Matteo Colombo, and Giulio Zani

Textile-Reinforced Concrete

**Development of Polymeric Textile Reinforced Concrete
Structural Members** 845
Vikram Dey, Anling Li, Gozdem Dittel, Thomas Gries, Steve Schaef,
and Barzin Mobasher

Geopolymers

**Mechanical Evaluation of Na-Based Strain-Hardening Geopolymer
Composites (SHGC) Reinforced with PVA, UHMWPE,
and PBO Fibers** 857
Ana C. C. Trindade, Iurie Curosu, Marco Liebscher, Viktor Mechtcherine,
and Flávio de A. Silva

Nano-technologies Related to FRC

**Influence of Dispersion Methods of Microcrystalline Cellulose
on the Mechanical Behavior of Cement Pastes** 871
Letícia O. de Souza, Lourdes Maria Silva Souza, and Flávio de A. Silva

**Effect of Nano-SiO$_2$ Coating on the Mechanical Recovery of Debonded
Fiber-Cement Interface Under Water Curing** 879
Bo Wu and Jishen Qiu

**Piezoresistivity of Carbon Black/Cement-Based Sensor Enhanced
with Polypropylene Fibre** 889
Wengui Li, Wenkui Dong, and Surendra P. Shah

UHPFRC

**Development of Eco-Efficient UHPC and UHPFRC by Recycling
Granite Waste Powder (GWP)** 903
David Bouchard, Thomas Sanchez, Luca Sorelli, and David Conciatori

**Flexural Behaviour of Ultra High Performance Fiber
Reinforced Concrete (UHPFRC) Under Monotonic Loads
and Loading-Unloading Cycles** 915
Nicola Generosi, Jacopo Donnini, Giovanni Lancioni,
and Valeria Corinaldesi

**Application of 3D Digital Image Correlation to Capture the Impact
Beahviour of UHPFRC Plate** 925
Yuanye He, Esmaeel Esmaeeli, and Marios N. Soutsos

Experimental Characterization of the Tensile Constitutive Behaviour of Ultra-High Performance Concretes: Effect of Cement and Fibre Type 936
Francesco Lo Monte, Eduardo J. Mezquida-Alcaraz,
Juan Navarro-Gregori, Pedro Serna, and Liberato Ferrara

Mechanical and Durability Assessment of Concretes Obtained from Recycled Ultra-High Performance Concretes 947
Estefania Cuenca, Marta Roig-Flores, Roberto Garofalo,
Milena Lozano-Násner, Cecilia Ruiz-Muñoz, Fabrizio Schillani,
Ruben Paul Borg, Liberato Ferrara, and Pedro Serna

An Experimental Evaluation of Direct Tensile Strength for Ultra-high Performance Concrete 958
An Hoang Le

Finite Element Modelling of UHPFRC Tensile Bars 965
Eduardo J. Mezquida-Alcaraz, Juan Navarro-Gregori, Majid Khorami,
and Pedro Serna

**Correction to: Wind Tower FRC Foundations:
Research and Design** C1
Marco di Prisco, Claudio di Prisco, Giancarlo Fraraccio, Bruno Dal Lago,
Paolo Martinelli, Luca Flessati, Matteo Colombo, and Giulio Zani

Author Index 977

RILEM Publications

The following list is presenting the global offer of RILEM Publications, sorted by series. Each publication is available in printed version and/or in online version.

RILEM Proceedings (PRO)

PRO 1: Durability of High Performance Concrete (ISBN: 2-912143-03-9; e-ISBN: 2-351580-12-5; e-ISBN: 2351580125); *Ed. H. Sommer*

PRO 2: Chloride Penetration into Concrete (ISBN: 2-912143-00-04; e-ISBN: 2912143454); *Eds. L.-O. Nilsson and J.-P. Ollivier*

PRO 3: Evaluation and Strengthening of Existing Masonry Structures (ISBN: 2-912143-02-0; e-ISBN: 2351580141); *Eds. L. Binda and C. Modena*

PRO 4: Concrete: From Material to Structure (ISBN: 2-912143-04-7; e-ISBN: 2351580206); *Eds. J.-P. Bournazel and Y. Malier*

PRO 5: The Role of Admixtures in High Performance Concrete (ISBN: 2-912143-05-5; e-ISBN: 2351580214); *Eds. J. G. Cabrera and R. Rivera-Villarreal*

PRO 6: High Performance Fiber Reinforced Cement Composites - HPFRCC 3 (ISBN: 2-912143-06-3; e-ISBN: 2351580222); *Eds. H. W. Reinhardt and A. E. Naaman*

PRO 7: 1st International RILEM Symposium on Self-Compacting Concrete (ISBN: 2-912143-09-8; e-ISBN: 2912143721); *Eds. Å. Skarendahl and Ö. Petersson*

PRO 8: International RILEM Symposium on Timber Engineering (ISBN: 2-912143-10-1; e-ISBN: 2351580230); *Ed. L. Boström*

PRO 9: 2nd International RILEM Symposium on Adhesion between Polymers and Concrete ISAP '99 (ISBN: 2-912143-11-X; e-ISBN: 2351580249); *Eds. Y. Ohama and M. Puterman*

PRO 10: 3rd International RILEM Symposium on Durability of Building and Construction Sealants (ISBN: 2-912143-13-6; e-ISBN: 2351580257); *Ed. A. T. Wolf*

PRO 11: 4th International RILEM Conference on Reflective Cracking in Pavements (ISBN: 2-912143-14-4; e-ISBN: 2351580265); *Eds. A. O. Abd El Halim, D. A. Taylor and El H. H. Mohamed*

PRO 12: International RILEM Workshop on Historic Mortars: Characteristics and Tests (ISBN: 2-912143-15-2; e-ISBN: 2351580273); *Eds. P. Bartos, C. Groot and J. J. Hughes*

PRO 13: 2nd International RILEM Symposium on Hydration and Setting (ISBN: 2-912143-16-0; e-ISBN: 2351580281); *Ed. A. Nonat*

PRO 14: Integrated Life-Cycle Design of Materials and Structures - ILCDES 2000 (ISBN: 951-758-408-3; e-ISBN: 235158029X); (ISSN: 0356-9403); *Ed. S. Sarja*

PRO 15: Fifth RILEM Symposium on Fibre-Reinforced Concretes (FRC) - BEFIB'2000 (ISBN: 2-912143-18-7; e-ISBN: 291214373X); *Eds. P. Rossi and G. Chanvillard*

PRO 16: Life Prediction and Management of Concrete Structures (ISBN: 2-912143-19-5; e-ISBN: 2351580303); *Ed. D. Naus*

PRO 17: Shrinkage of Concrete – Shrinkage 2000 (ISBN: 2-912143-20-9; e-ISBN: 2351580311); *Eds. V. Baroghel-Bouny and P.-C. Aïtcin*

PRO 18: Measurement and Interpretation of the On-Site Corrosion Rate (ISBN: 2-912143-21-7; e-ISBN: 235158032X); *Eds. C. Andrade, C. Alonso, J. Fullea, J. Polimon and J. Rodriguez*

PRO 19: Testing and Modelling the Chloride Ingress into Concrete (ISBN: 2-912143-22-5; e-ISBN: 2351580338); *Eds. C. Andrade and J. Kropp*

PRO 20: 1st International RILEM Workshop on Microbial Impacts on Building Materials (CD 02) (e-ISBN 978-2-35158-013-4); *Ed. M. Ribas Silv*

PRO 21: International RILEM Symposium on Connections between Steel and Concrete (ISBN: 2-912143-25-X; e-ISBN: 2351580346); *Ed. R. Eligehausen*

PRO 22: International RILEM Symposium on Joints in Timber Structures (ISBN: 2-912143-28-4; e-ISBN: 2351580354); *Eds. S. Aicher and H.-W. Reinhardt*

PRO 23: International RILEM Conference on Early Age Cracking in Cementitious Systems (ISBN: 2-912143-29-2; e-ISBN: 2351580362); *Eds. K. Kovler and A. Bentur*

PRO 24: 2nd International RILEM Workshop on Frost Resistance of Concrete (ISBN: 2-912143-30-6; e-ISBN: 2351580370); *Eds. M. J. Setzer, R. Auberg and H.-J. Keck*

PRO 25: International RILEM Workshop on Frost Damage in Concrete (ISBN: 2-912143-31-4; e-ISBN: 2351580389); *Eds. D. J. Janssen, M. J. Setzer and M. B. Snyder*

PRO 26: International RILEM Workshop on On-Site Control and Evaluation of Masonry Structures (ISBN: 2-912143-34-9; e-ISBN: 2351580141); *Eds. L. Binda and R. C. de Vekey*

PRO 27: International RILEM Symposium on Building Joint Sealants (CD03; e-ISBN: 235158015X); *Ed. A. T. Wolf*

PRO 28: 6th International RILEM Symposium on Performance Testing and Evaluation of Bituminous Materials - PTEBM'03 (ISBN: 2-912143-35-7; e-ISBN: 978-2-912143-77-8); *Ed. M. N. Partl*

PRO 29: 2nd International RILEM Workshop on Life Prediction and Ageing Management of Concrete Structures (ISBN: 2-912143-36-5; e-ISBN: 2912143780); *Ed. D. J. Naus*

PRO 30: 4th International RILEM Workshop on High Performance Fiber Reinforced Cement Composites - HPFRCC 4 (ISBN: 2-912143-37-3; e-ISBN: 2912143799); *Eds. A. E. Naaman and H. W. Reinhardt*

PRO 31: International RILEM Workshop on Test and Design Methods for Steel Fibre Reinforced Concrete: Background and Experiences (ISBN: 2-912143-38-1; e-ISBN: 2351580168); *Eds. B. Schnütgen and L. Vandewalle*

PRO 32: International Conference on Advances in Concrete and Structures 2 vol. (ISBN (set): 2-912143-41-1; e-ISBN: 2351580176); *Eds. Ying-shu Yuan, Surendra P. Shah and Heng-lin Lü*

PRO 33: 3rd International Symposium on Self-Compacting Concrete (ISBN: 2-912143-42-X; e-ISBN: 2912143713); *Eds. Ó. Wallevik and I. Níelsson*

PRO 34: International RILEM Conference on Microbial Impact on Building Materials (ISBN: 2-912143-43-8; e-ISBN: 2351580184); *Ed. M. Ribas Silva*

PRO 35: International RILEM TC 186-ISA on Internal Sulfate Attack and Delayed Ettringite Formation (ISBN: 2-912143-44-6; e-ISBN: 2912143802); *Eds. K. Scrivener and J. Skalny*

PRO 36: International RILEM Symposium on Concrete Science and Engineering – A Tribute to Arnon Bentur (ISBN: 2-912143-46-2; e-ISBN: 2912143586); *Eds. K. Kovler, J. Marchand, S. Mindess and J. Weiss*

PRO 37: 5th International RILEM Conference on Cracking in Pavements – Mitigation, Risk Assessment and Prevention (ISBN: 2-912143-47-0; e-ISBN: 2912143764); *Eds. C. Petit, I. Al-Qadi and A. Millien*

PRO 38: 3rd International RILEM Workshop on Testing and Modelling the Chloride Ingress into Concrete (ISBN: 2-912143-48-9; e-ISBN: 2912143578); *Eds. C. Andrade and J. Kropp*

PRO 39: 6th International RILEM Symposium on Fibre-Reinforced Concretes - BEFIB 2004 (ISBN: 2-912143-51-9; e-ISBN: 2912143748); *Eds. M. Di Prisco, R. Felicetti and G. A. Plizzari*

PRO 40: International RILEM Conference on the Use of Recycled Materials in Buildings and Structures (ISBN: 2-912143-52-7; e-ISBN: 2912143756); *Eds. E. Vázquez, Ch. F. Hendriks and G. M. T. Janssen*

PRO 41: RILEM International Symposium on Environment-Conscious Materials and Systems for Sustainable Development (ISBN: 2-912143-55-1; e-ISBN: 2912143640); *Eds. N. Kashino and Y. Ohama*

PRO 42: SCC'2005 - China: 1st International Symposium on Design, Performance and Use of Self-Consolidating Concrete (ISBN: 2-912143-61-6; e-ISBN: 2912143624); *Eds. Zhiwu Yu, Caijun Shi, Kamal Henri Khayat and Youjun Xie*

PRO 43: International RILEM Workshop on Bonded Concrete Overlays (e-ISBN: 2-912143-83-7); *Eds. J. L. Granju and J. Silfwerbrand*

PRO 44: 2nd International RILEM Workshop on Microbial Impacts on Building Materials (CD11) (e-ISBN: 2-912143-84-5); *Ed. M. Ribas Silva*

PRO 45: 2nd International Symposium on Nanotechnology in Construction, Bilbao (ISBN: 2-912143-87-X; e-ISBN: 2912143888); *Eds. Peter J. M. Bartos, Yolanda de Miguel and Antonio Porro*

PRO 46: ConcreteLife'06 - International RILEM-JCI Seminar on Concrete Durability and Service Life Planning: Curing, Crack Control, Performance in Harsh Environments (ISBN: 2-912143-89-6; e-ISBN: 291214390X); *Ed. K. Kovler*

PRO 47: International RILEM Workshop on Performance Based Evaluation and Indicators for Concrete Durability (ISBN: 978-2-912143-95-2; e-ISBN: 9782912143969); *Eds. V. Baroghel-Bouny, C. Andrade, R. Torrent and K. Scrivener*

PRO 48: 1st International RILEM Symposium on Advances in Concrete through Science and Engineering (e-ISBN: 2-912143-92-6); *Eds. J. Weiss, K. Kovler, J. Marchand and S. Mindess*

PRO 49: International RILEM Workshop on High Performance Fiber Reinforced Cementitious Composites in Structural Applications (ISBN: 2-912143-93-4; e-ISBN: 2912143942); *Eds. G. Fischer and V. C. Li*

PRO 50: 1st International RILEM Symposium on Textile Reinforced Concrete (ISBN: 2-912143-97-7; e-ISBN: 2351580087); *Eds. Josef Hegger, Wolfgang Brameshuber and Norbert Will*

PRO 51: 2nd International Symposium on Advances in Concrete through Science and Engineering (ISBN: 2-35158-003-6; e-ISBN: 2-35158-002-8); *Eds. J. Marchand, B. Bissonnette, R. Gagné, M. Jolin and F. Paradis*

PRO 52: Volume Changes of Hardening Concrete: Testing and Mitigation (ISBN: 2-35158-004-4; e-ISBN: 2-35158-005-2); *Eds. O. M. Jensen, P. Lura and K. Kovler*

PRO 53: High Performance Fiber Reinforced Cement Composites - HPFRCC5 (ISBN: 978-2-35158-046-2; e-ISBN: 978-2-35158-089-9); *Eds. H. W. Reinhardt and A. E. Naaman*

PRO 54: 5th International RILEM Symposium on Self-Compacting Concrete (ISBN: 978-2-35158-047-9; e-ISBN: 978-2-35158-088-2); *Eds. G. De Schutter and V. Boel*

PRO 55: International RILEM Symposium Photocatalysis, Environment and Construction Materials (ISBN: 978-2-35158-056-1; e-ISBN: 978-2-35158-057-8); *Eds. P. Baglioni and L. Cassar*

PRO 56: International RILEM Workshop on Integral Service Life Modelling of Concrete Structures (ISBN 978-2-35158-058-5; e-ISBN: 978-2-35158-090-5); *Eds. R. M. Ferreira, J. Gulikers and C. Andrade*

PRO 57: RILEM Workshop on Performance of cement-based materials in aggressive aqueous environments (e-ISBN: 978-2-35158-059-2); *Ed. N. De Belie*

PRO 58: International RILEM Symposium on Concrete Modelling - CONMOD'08 (ISBN: 978-2-35158-060-8; e-ISBN: 978-2-35158-076-9); *Eds. E. Schlangen and G. De Schutter*

PRO 59: International RILEM Conference on On Site Assessment of Concrete, Masonry and Timber Structures - SACoMaTiS 2008 (ISBN set: 978-2-35158-061-5; e-ISBN: 978-2-35158-075-2); *Eds. L. Binda, M. di Prisco and R. Felicetti*

PRO 60: Seventh RILEM International Symposium on Fibre Reinforced Concrete: Design and Applications - BEFIB 2008 (ISBN: 978-2-35158-064-6; e-ISBN: 978-2-35158-086-8); *Ed. R. Gettu*

PRO 61: 1st International Conference on Microstructure Related Durability of Cementitious Composites 2 vol., (ISBN: 978-2-35158-065-3; e-ISBN: 978-2-35158-084-4); *Eds. W. Sun, K. van Breugel, C. Miao, G. Ye and H. Chen*

PRO 62: NSF/ RILEM Workshop: In-situ Evaluation of Historic Wood and Masonry Structures (e-ISBN: 978-2-35158-068-4); *Eds. B. Kasal, R. Anthony and M. Drdácký*

PRO 63: Concrete in Aggressive Aqueous Environments: Performance, Testing and Modelling, 2 vol., (ISBN: 978-2-35158-071-4; e-ISBN: 978-2-35158-082-0); *Eds. M. G. Alexander and A. Bertron*

PRO 64: Long Term Performance of Cementitious Barriers and Reinforced Concrete in Nuclear Power Plants and Waste Management - NUCPERF 2009 (ISBN: 978-2-35158-072-1; e-ISBN: 978-2-35158-087-5); *Eds. V. L'Hostis, R. Gens, C. Gallé*

PRO 65: Design Performance and Use of Self-consolidating Concrete - SCC'2009 (ISBN: 978-2-35158-073-8; e-ISBN: 978-2-35158-093-6); *Eds. C. Shi, Z. Yu, K. H. Khayat and P. Yan*

PRO 66: 2nd International RILEM Workshop on Concrete Durability and Service Life Planning - ConcreteLife'09 (ISBN: 978-2-35158-074-5; ISBN: 978-2-35158-074-5); *Ed. K. Kovler*

PRO 67: Repairs Mortars for Historic Masonry (e-ISBN: 978-2-35158-083-7); *Ed. C. Groot*

PRO 68: Proceedings of the 3rd International RILEM Symposium on 'Rheology of Cement Suspensions such as Fresh Concrete (ISBN 978-2-35158-091-2; e-ISBN: 978-2-35158-092-9); *Eds. O. H. Wallevik, S. Kubens and S. Oesterheld*

PRO 69: 3rd International PhD Student Workshop on 'Modelling the Durability of Reinforced Concrete (ISBN: 978-2-35158-095-0); *Eds. R. M. Ferreira, J. Gulikers and C. Andrade*

PRO 70: 2nd International Conference on 'Service Life Design for Infrastructure' (ISBN set: 978-2-35158-096-7, e-ISBN: 978-2-35158-097-4); *Eds. K. van Breugel, G. Ye and Y. Yuan*

PRO 71: Advances in Civil Engineering Materials - The 50-year Teaching Anniversary of Prof. Sun Wei' (ISBN: 978-2-35158-098-1; e-ISBN: 978-2-35158-099-8); *Eds. C. Miao, G. Ye and H. Chen*

PRO 72: First International Conference on 'Advances in Chemically-Activated Materials – CAM'2010' (2010), 264 pp, ISBN: 978-2-35158-101-8; e-ISBN: 978-2-35158-115-5, *Eds. Caijun Shi and Xiaodong Shen*

PRO 73: 2nd International Conference on 'Waste Engineering and Management - ICWEM 2010' (2010), 894 pp, ISBN: 978-2-35158-102-5; e-ISBN: 978-2-35158-103-2, *Eds. J. Zh. Xiao, Y. Zhang, M. S. Cheung and R. Chu*

PRO 74: International RILEM Conference on 'Use of Superabsorsorbent Polymers and Other New Addditives in Concrete' (2010) 374 pp., ISBN: 978-2-35158-104-9; e-ISBN: 978-2-35158-105-6; *Eds. O. M. Jensen, M. T. Hasholt and S. Laustsen*

PRO 75: International Conference on 'Material Science - 2nd ICTRC - Textile Reinforced Concrete - Theme 1' (2010) 436 pp., ISBN: 978-2-35158-106-3; e-ISBN: 978-2-35158-107-0; *Ed. W. Brameshuber*

PRO 76: International Conference on 'Material Science - HetMat - Modelling of Heterogeneous Materials - Theme 2' (2010) 255 pp., ISBN: 978-2-35158-108-7; e-ISBN: 978-2-35158-109-4; *Ed. W. Brameshuber*

PRO 77: International Conference on 'Material Science - AdIPoC - Additions Improving Properties of Concrete - Theme 3' (2010) 459 pp., ISBN: 978-2-35158-110-0; e-ISBN: 978-2-35158-111-7; *Ed. W. Brameshuber*

PRO 78: 2nd Historic Mortars Conference and RILEM TC 203-RHM Final Workshop – HMC2010 (2010) 1416 pp., e-ISBN: 978-2-35158-112-4; *Eds. J. Válek, C. Groot and J. J. Hughes*

PRO 79: International RILEM Conference on Advances in Construction Materials Through Science and Engineering (2011) 213 pp., ISBN: 978-2-35158-116-2, e-ISBN: 978-2-35158-117-9; *Eds. Christopher Leung and K. T. Wan*

PRO 80: 2nd International RILEM Conference on Concrete Spalling due to Fire Exposure (2011) 453 pp., ISBN: 978-2-35158-118-6, e-ISBN: 978-2-35158-119-3; *Eds. E. A. B. Koenders and F. Dehn*

PRO 81: 2nd International RILEM Conference on Strain Hardening Cementitious Composites (SHCC2-Rio) (2011) 451 pp., ISBN: 978-2-35158-120-9, e-ISBN: 978-2-35158-121-6; *Eds. R. D. Toledo Filho, F. A. Silva, E. A. B. Koenders and E. M. R. Fairbairn*

PRO 82: 2nd International RILEM Conference on Progress of Recycling in the Built Environment (2011) 507 pp., e-ISBN: 978-2-35158-122-3; *Eds. V. M. John, E. Vazquez, S. C. Angulo and C. Ulsen*

PRO 83: 2nd International Conference on Microstructural-related Durability of Cementitious Composites (2012) 250 pp., ISBN: 978-2-35158-129-2; e-ISBN: 978-2-35158-123-0; *Eds. G. Ye, K. van Breugel, W. Sun and C. Miao*

PRO 84: CONSEC13 - Seventh International Conference on Concrete under Severe Conditions – Environment and Loading (2013) 1930 pp., ISBN: 978-2-35158-124-7; e-ISBN: 978-2-35158-134-6; *Eds. Z. J. Li, W. Sun, C. W. Miao, K. Sakai, O. E. Gjorv and N. Banthia*

PRO 85: RILEM-JCI International Workshop on Crack Control of Mass Concrete and Related issues concerning Early-Age of Concrete Structures – ConCrack 3 – Control of Cracking in Concrete Structures 3 (2012) 237 pp., ISBN: 978-2-35158-125-4; e-ISBN: 978-2-35158-126-1; *Eds. F. Toutlemonde and J.-M. Torrenti*

PRO 86: International Symposium on Life Cycle Assessment and Construction (2012) 414 pp., ISBN: 978-2-35158-127-8, e-ISBN: 978-2-35158-128-5; *Eds. A. Ventura and C. de la Roche*

PRO 87: UHPFRC 2013 – RILEM-fib-AFGC International Symposium on Ultra-High Performance Fibre-Reinforced Concrete (2013), ISBN: 978-2-35158-130-8, e-ISBN: 978-2-35158-131-5; *Eds. F. Toutlemonde*

PRO 88: 8th RILEM International Symposium on Fibre Reinforced Concrete (2012) 344 pp., ISBN: 978-2-35158-132-2, e-ISBN: 978-2-35158-133-9; *Eds. Joaquim A. O. Barros*

PRO 89: RILEM International workshop on performance-based specification and control of concrete durability (2014) 678 pp, ISBN: 978-2-35158-135-3, e-ISBN: 978-2-35158-136-0; *Eds. D. Bjegović, H. Beushausen and M. Serdar*

PRO 90: 7th RILEM International Conference on Self-Compacting Concrete and of the 1st RILEM International Conference on Rheology and Processing of Construction Materials (2013) 396 pp, ISBN: 978-2-35158-137-7, e-ISBN: 978-2-35158-138-4; *Eds. Nicolas Roussel and Hela Bessaies-Bey*

PRO 91: CONMOD 2014 - RILEM International Symposium on Concrete Modelling (2014), ISBN: 978-2-35158-139-1; e-ISBN: 978-2-35158-140-7; *Eds. Kefei Li, Peiyu Yan and Rongwei Yang*

PRO 92: CAM 2014 - 2nd International Conference on advances in chemically-activated materials (2014) 392 pp., ISBN: 978-2-35158-141-4; e-ISBN: 978-2-35158-142-1; *Eds. Caijun Shi and Xiadong Shen*

PRO 93: SCC 2014 - 3rd International Symposium on Design, Performance and Use of Self-Consolidating Concrete (2014) 438 pp., ISBN: 978-2-35158-143-8; e-ISBN: 978-2-35158-144-5; *Eds. Caijun Shi, Zhihua Ou and Kamal H. Khayat*

PRO 94 (online version): HPFRCC-7 - 7th RILEM conference on High performance fiber reinforced cement composites (2015), e-ISBN: 978-2-35158-146-9; *Eds. H. W. Reinhardt, G. J. Parra-Montesinos and H. Garrecht*

PRO 95: International RILEM Conference on Application of superabsorbent polymers and other new admixtures in concrete construction (2014), ISBN: 978-2-35158-147-6; e-ISBN: 978-2-35158-148-3; *Eds. Viktor Mechtcherine and Christof Schroefl*

PRO 96 (online version): XIII DBMC: XIII International Conference on Durability of Building Materials and Components (2015), e-ISBN: 978-2-35158-149-0; *Eds. M. Quattrone and V. M. John*

PRO 97: SHCC3 – 3rd International RILEM Conference on Strain Hardening Cementitious Composites (2014), ISBN: 978-2-35158-150-6; e-ISBN: 978-2-35158-151-3; *Eds. E. Schlangen, M. G. Sierra Beltran, M. Lukovic and G. Ye*

PRO 98: ERRO-11 – 11th International Symposium on Ferrocement and 3rd ICTRC - International Conference on Textile Reinforced Concrete (2015), ISBN: 978-2-35158-152-0; e-ISBN: 978-2-35158-153-7; *Ed. W. Brameshuber*

PRO 99 (online version): ICBBM 2015 - 1st International Conference on Bio-Based Building Materials (2015), e-ISBN: 978-2-35158-154-4; *Eds. S. Amziane and M. Sonebi*

PRO 100: SCC16 - RILEM Self-Consolidating Concrete Conference (2016), ISBN: 978-2-35158-156-8; e-ISBN: 978-2-35158-157-5; *Ed. Kamal H. Kayat*

PRO 101 (online version): III Progress of Recycling in the Built Environment (2015), e-ISBN: 978-2-35158-158-2; *Eds. I. Martins, C. Ulsen and S. C. Angulo*

PRO 102 (online version): RILEM Conference on Microorganisms-Cementitious Materials Interactions (2016), e-ISBN: 978-2-35158-160-5; *Eds. Alexandra Bertron, Henk Jonkers and Virginie Wiktor*

PRO 103 (online version): ACESC'16 - Advances in Civil Engineering and Sustainable Construction (2016), e-ISBN: 978-2-35158-161-2; *Eds. T. Ch. Madhavi, G. Prabhakar, Santhosh Ram and P. M. Rameshwaran*

PRO 104 (online version): SSCS'2015 - Numerical Modeling - Strategies for Sustainable Concrete Structures (2015), e-ISBN: 978-2-35158-162-9

PRO 105: 1st International Conference on UHPC Materials and Structures (2016), ISBN: 978-2-35158-164-3, e-ISBN: 978-2-35158-165-0

PRO 106: AFGC-ACI-fib-RILEM International Conference on Ultra-High-Performance Fibre-Reinforced Concrete – UHPFRC 2017 (2017), ISBN: 978-2-35158-166-7, e-ISBN: 978-2-35158-167-4; *Eds. François Toutlemonde and Jacques Resplendino*

PRO 107 (online version): XIV DBMC – 14th International Conference on Durability of Building Materials and Components (2017), e-ISBN: 978-2-35158-159-9; *Eds. Geert De Schutter, Nele De Belie, Arnold Janssens and Nathan Van Den Bossche*

PRO 108: MSSCE 2016 -Innovation of Teaching in Materials and Structures (2016), ISBN: 978-2-35158-178-0, e-ISBN: 978-2-35158-179-7; *Ed. Per Goltermann*

PRO 109 (2 volumes): MSSCE 2016 - Service Life of Cement-Based Materials and Structures (2016), ISBN Vol. 1: 978-2-35158-170-4, Vol. 2: 978-2-35158-171-4, Set Vol. 1&2: 978-2-35158-172-8, e-ISBN : 978-2-35158-173-5; *Eds. Miguel Azenha, Ivan Gabrijel, Dirk Schlicke, Terje Kanstad and Ole Mejlhede Jensen*

PRO 110: MSSCE 2016 - Historical Masonry (2016), ISBN: 978-2-35158-178-0, e-ISBN: 978-2-35158-179-7; *Eds. Inge Rörig-Dalgaard and Ioannis Ioannou*

PRO 111: MSSCE 2016 - Electrochemistry in Civil Engineering (2016), ISBN: 978-2-35158-176-6, e-ISBN: 978-2-35158-177-3; *Ed. Lisbeth M. Ottosen*

PRO 112: MSSCE 2016 - Moisture in Materials and Structures (2016), ISBN: 978-2-35158-178-0, e-ISBN: 978-2-35158-179-7; *Eds. Kurt Kielsgaard Hansen, Carsten Rode and Lars-Olof Nilsson*

PRO 113: MSSCE 2016 - Concrete with Supplementary Cementitious Materials (2016), ISBN: 978-2-35158-178-0, e-ISBN: 978-2-35158-179-7; *Eds. Ole Mejlhede Jensen, Konstantin Kovler and Nele De Belie*

PRO 114: MSSCE 2016 - Frost Action in Concrete (2016), ISBN: 978-2-35158-182-7, e-ISBN: 978-2-35158-183-4; *Eds. Marianne Tange Hasholt, Katja Fridh and R. Doug Hooton*

PRO 115: MSSCE 2016 - Fresh Concrete (2016), ISBN: 978-2-35158-184-1, e-ISBN: 978-2-35158-185-8; *Eds. Lars N. Thrane, Claus Pade, Oldrich Svec and Nicolas Roussel*

PRO 116: BEFIB 2016 – 9th RILEM International Symposium on Fiber Reinforced Concrete (2016), ISBN: 978-2-35158-187-2, e-ISBN: 978-2-35158-186-5; *Eds. N. Banthia, M. di Prisco and S. Soleimani-Dashtaki*

PRO 117: 3rd International RILEM Conference on Microstructure Related Durability of Cementitious Composites (2016), ISBN: 978-2-35158-188-9, e-ISBN: 978-2-35158-189-6; *Eds. Changwen Miao, Wei Sun, Jiaping Liu, Huisu Chen, Guang Ye and Klaas van Breugel*

PRO 118 (4 volumes): International Conference on Advances in Construction Materials and Systems (2017), ISBN Set: 978-2-35158-190-2, Vol. 1: 978-2-35158-193-3, Vol. 2: 978-2-35158-194-0, Vol. 3: ISBN:978-2-35158-195-7, Vol. 4: ISBN:978-2-35158-196-4, e-ISBN: 978-2-35158-191-9; *Eds. Manu Santhanam, Ravindra Gettu, Radhakrishna G. Pillai and Sunitha K. Nayar*

PRO 119 (online version): ICBBM 2017 - Second International RILEM Conference on Bio-based Building Materials, (2017), e-ISBN: 978-2-35158-192-6; *Eds. Sofiane Amziane and Mohammed Sonebi*

PRO 120 (2 volumes): EAC-02 - 2nd International RILEM/COST Conference on Early Age Cracking and Serviceability in Cement-based Materials and Structures, (2017), Vol. 1: 978-2-35158-199-5, Vol. 2: 978-2-35158-200-8, Set: 978-2-35158-197-1, e-ISBN: 978-2-35158-198-8; *Eds. Stéphanie Staquet and Dimitrios Aggelis*

PRO 121 (2 volumes): SynerCrete18: Interdisciplinary Approaches for Cement-based Materials and Structural Concrete: Synergizing Expertise and Bridging Scales of Space and Time, (2018), Set: 978-2-35158-202-2, Vol.1: 978-2-35158-211-4, Vol. 2: 978-2-35158-212-1, e-ISBN: 978-2-35158-203-9; *Eds. Miguel Azenha, Dirk Schlicke, Farid Benboudjema and Agnieszka Knoppik*

PRO 122: SCC'2018 China - Fourth International Symposium on Design, Performance and Use of Self-Consolidating Concrete, (2018), ISBN:

978-2-35158-204-6, e-ISBN: 978-2-35158-205-3; *Eds. C. Shi, Z. Zhang and K. H. Khayat*

PRO 123: Final Conference of RILEM TC 253-MCI: Microorganisms-Cementitious Materials Interactions (2018), Set: 978-2-35158-207-7, Vol.1: 978-2-35158-209-1, Vol.2: 978-2-35158-210-7, e-ISBN: 978-2-35158-206-0; *Ed. Alexandra Bertron*

PRO 124 (online version): Fourth International Conference Progress of Recycling in the Built Environment (2018), e-ISBN: 978-2-35158-208-4; *Eds. Isabel M. Martins, Carina Ulsen and Yury Villagran*

PRO 125 (online version): SLD4 - 4th International Conference on Service Life Design for Infrastructures (2018), e-ISBN: 978-2-35158-213-8; *Eds. Guang Ye, Yong Yuan, Claudia Romero Rodriguez, Hongzhi Zhang and Branko Savija*

PRO 126: Workshop on Concrete Modelling and Material Behaviour in honor of Professor Klaas van Breugel (2018), ISBN: 978-2-35158-214-5, e-ISBN: 978-2-35158-215-2; *Ed. Guang Ye*

PRO 127 (online version): CONMOD2018 - Symposium on Concrete Modelling (2018), e-ISBN: 978-2-35158-216-9; *Eds. Erik Schlangen, Geert de Schutter, Branko Savija, Hongzhi Zhang and Claudia Romero Rodriguez*

PRO 128: SMSS2019 - International Conference on Sustainable Materials, Systems and Structures (2019), ISBN: 978-2-35158-217-6, e-ISBN: 978-2-35158-218-3

PRO 129: 2nd International Conference on UHPC Materials and Structures (UHPC2018-China), ISBN: 978-2-35158-219-0, e-ISBN: 978-2-35158-220-6;

PRO 130: 5th Historic Mortars Conference (2019), ISBN: 978-2-35158-221-3, e-ISBN: 978-2-35158-222-0; *Eds. José Ignacio Álvarez, José María Fernández, Íñigo Navarro, Adrián Durán and Rafael Sirera*

PRO 131 (online version): 3rd International Conference on Bio-Based Building Materials (ICBBM2019), e-ISBN: 978-2-35158-229-9; *Eds. Mohammed Sonebi, Sofiane Amziane and Jonathan Page*

PRO 132: IRWRMC'18 - International RILEM Workshop on Rheological Measurements of Cement-based Materials (2018), ISBN: 978-2-35158-230-5, e-ISBN: 978-2-35158-231-2; *Eds. Chafika Djelal and Yannick Vanhove*

PRO 133 (online version): CO2STO2019 - International Workshop CO2 Storage in Concrete (2019), e-ISBN: 978-2-35158-232-9; *Eds. Assia Djerbi, Othman Omikrine-Metalssi and Teddy Fen-Chong*

PRO 134: 3rd ACF/HNU International Conference on UHPC Materials and Structures - UHPC'2020, ISBN: 978-2-35158-233-6, e-ISBN: 978-2-35158-234-3; *Eds. Caijun Shi and Jiaping Liu*

RILEM Reports (REP)

Report 19: Considerations for Use in Managing the Aging of Nuclear Power Plant Concrete Structures (ISBN: 2-912143-07-1); *Ed. D. J. Naus*

Report 20: Engineering and Transport Properties of the Interfacial Transition Zone in Cementitious Composites (ISBN: 2-912143-08-X); *Eds. M. G. Alexander, G. Arliguie, G. Ballivy, A. Bentur and J. Marchand*

Report 21: Durability of Building Sealants (ISBN: 2-912143-12-8); *Ed. A. T. Wolf*

Report 22: Sustainable Raw Materials - Construction and Demolition Waste (ISBN: 2-912143-17-9); *Eds. C. F. Hendriks and H. S. Pietersen*

Report 23: Self-Compacting Concrete state-of-the-art report (ISBN: 2-912143-23-3); *Eds. Å. Skarendahl and Ö. Petersson*

Report 24: Workability and Rheology of Fresh Concrete: Compendium of Tests (ISBN: 2-912143-32-2); *Eds. P. J. M. Bartos, M. Sonebi and A. K. Tamimi*

Report 25: Early Age Cracking in Cementitious Systems (ISBN: 2-912143-33-0); *Ed. A. Bentur*

Report 26: Towards Sustainable Roofing (Joint Committee CIB/RILEM) (CD 07) (e-ISBN 978-2-912143-65-5); *Eds. Thomas W. Hutchinson and Keith Roberts*

Report 27: Condition Assessment of Roofs (Joint Committee CIB/RILEM) (CD 08) (e-ISBN 978-2-912143-66-2); *Ed. CIB W 83/RILEM TC166-RMS*

Report 28: Final report of RILEM TC 167-COM 'Characterisation of Old Mortars with Respect to Their Repair (ISBN: 978-2-912143-56-3); *Eds. C. Groot, G. Ashall and J. Hughes*

Report 29: Pavement Performance Prediction and Evaluation (PPPE): Interlaboratory Tests (e-ISBN: 2-912143-68-3); *Eds. M. Partl and H. Piber*

Report 30: Final Report of RILEM TC 198-URM 'Use of Recycled Materials' (ISBN: 2-912143-82-9; e-ISBN: 2-912143-69-1); *Eds. Ch. F. Hendriks, G. M. T. Janssen and E. Vázquez*

Report 31: Final Report of RILEM TC 185-ATC 'Advanced testing of cement-based materials during setting and hardening' (ISBN: 2-912143-81-0; e-ISBN: 2-912143-70-5); *Eds. H. W. Reinhardt and C. U. Grosse*

Report 32: Probabilistic Assessment of Existing Structures. A JCSS publication (ISBN 2-912143-24-1); *Ed. D. Diamantidis*

Report 33: State-of-the-Art Report of RILEM Technical Committee TC 184-IFE 'Industrial Floors' (ISBN 2-35158-006-0); *Ed. P. Seidler*

Report 34: Report of RILEM Technical Committee TC 147-FMB 'Fracture mechanics applications to anchorage and bond' Tension of Reinforced Concrete Prisms – Round Robin Analysis and Tests on Bond (e-ISBN 2-912143-91-8); *Eds. L. Elfgren and K. Noghabai*

Report 35: Final Report of RILEM Technical Committee TC 188-CSC 'Casting of Self Compacting Concrete' (ISBN 2-35158-001-X; e-ISBN: 2-912143-98-5); *Eds. Å. Skarendahl and P. Billberg*

Report 36: State-of-the-Art Report of RILEM Technical Committee TC 201-TRC 'Textile Reinforced Concrete' (ISBN 2-912143-99-3); *Ed. W. Brameshuber*

Report 37: State-of-the-Art Report of RILEM Technical Committee TC 192-ECM 'Environment-conscious construction materials and systems' (ISBN: 978-2-35158-053-0); *Eds. N. Kashino, D. Van Gemert and K. Imamoto*

Report 38: State-of-the-Art Report of RILEM Technical Committee TC 205-DSC 'Durability of Self-Compacting Concrete' (ISBN: 978-2-35158-048-6); *Eds. G. De Schutter and K. Audenaert*

Report 39: Final Report of RILEM Technical Committee TC 187-SOC 'Experimental determination of the stress-crack opening curve for concrete in tension' (ISBN 978-2-35158-049-3); *Ed. J. Planas*

Report 40: State-of-the-Art Report of RILEM Technical Committee TC 189-NEC 'Non-Destructive Evaluation of the Penetrability and Thickness of the Concrete Cover' (ISBN 978-2-35158-054-7); *Eds. R. Torrent and L. Fernández Luco*

Report 41: State-of-the-Art Report of RILEM Technical Committee TC 196-ICC 'Internal Curing of Concrete' (ISBN 978-2-35158-009-7); *Eds. K. Kovler and O. M. Jensen*

Report 42: 'Acoustic Emission and Related Non-destructive Evaluation Techniques for Crack Detection and Damage Evaluation in Concrete' - Final Report of RILEM Technical Committee 212-ACD (e-ISBN: 978-2-35158-100-1); *Ed. M. Ohtsu*

Report 45: Repair Mortars for Historic Masonry - State-of-the-Art Report of RILEM Technical Committee TC 203-RHM (e-ISBN: 978-2-35158-163-6); *Eds. Paul Maurenbrecher and Caspar Groot*

Report 46: *Surface delamination of concrete industrial floors and other durability related aspects guide - Report of RILEM Technical Committee TC 268-SIF (e-ISBN: 978-2-35158-201-5); Ed. Valerie Pollet*

Technological Aspects

Effects of Fly Ash Content and Curing Age on High Temperature Residual Compressive Strength of Strain-Hardening Cementitious Composites

Dhanendra Kumar[✉], Amr Ashraf Soliman, and Ravi Ranade

Department of Civil, Structural and Environmental Engineering (CSEE),
University at Buffalo, State University of New York,
Buffalo, NY 14260-4300, USA
dkumar1@buffalo.edu

Abstract. Class F fly ash is an integral component of most Strain Hardening Cementitious Composites (SHCCs). It reacts with the calcium hydroxide produced in the primary hydration of cement. As calcium hydroxide disintegrates at lower temperatures than other cement hydration products, reducing it through the pozzolanic reaction with fly ash has been shown to improve the thermal stability of SHCC. However, the degree (or extent) of the pozzolanic reaction increases with curing age. Therefore, the effects of fly ash content on the residual compressive strength of SHCC after exposure to elevated temperatures must be investigated at different curing ages, which is the motivation behind this study. Three different SHCCs with fly/ash to cement (FA/c) weight ratios of 1.2, 2.4, and 3.6 were tested at 28 days and 150 days under residual conditions after being subjected to high temperatures of up to 600 °C. The results show that the beneficial effects of high FA/c ratios on the residual compressive strength, associated with the accelerated pozzolanic reaction of fly ash at high temperatures, are lost at the long curing age.

Keywords: SHCC · ECC · FRC · High temperature · Fire · Fly ash · Residual strength · Curing age

1 Introduction

The high temperature behavior of concrete materials is influenced by the chemical stability of the cementitious matrix at high temperatures [1, 2]. Two major components of the hydrated cementitious matrix: calcium hydroxide (CH) or $[Ca(OH)_2]$ and calcium silicate hydrate (CSH) govern material behavior at high temperatures, particularly above 400 °C. CH dissociates into CaO and H_2O around 400 °C, leading to the coarsening of the pore structure. Furthermore, the rehydration of CaO during the cooling phase (after high temperature exposure) causes expansion and capacity reduction [1, 3]. Researchers [2, 4–7] have therefore investigated the influence of different supplementary cementitious materials (SCMs), such as slag, fly ash, micro-silica, and nano-silica, on the high-temperature performance of concrete materials, as these SCMs consume CH through the pozzolanic reaction.

© RILEM 2022
P. Serna et al. (Eds.): BEFIB 2021, RILEM Bookseries 36, pp. 3–12, 2022.
https://doi.org/10.1007/978-3-030-83719-8_1

The focus of this study is fly ash, which has been shown to improve the residual mechanical properties of concrete after high temperature exposure through the CH consumption mechanism described above. After exposure to temperatures of 200–300 °C, the 28-day residual compressive strength was observed to increase (relative to room temperature strength) with increase in fly ash to cement (FA/c) weight ratio in conventional concrete [8–12]. In an another study [13], the residual compressive strength increased (relative to room temperature strength) at 200 °C for concrete with an FA/c weight ratio of 1.5, however, the magnitude of the effect of FA/c ratio on the compressive strength of concrete (with FA/c ratios of 0.25 and 0.67) seemed to reduce after 180 days of curing. It was also stated that the change in heating and curing conditions could lead to further variation in the properties. In the above studies, the residual compressive strength improvement (relative to room temperature strength) was partially due to the reduction in CH, caused by the accelerated hydration of fly ash.

Strain Hardening Cementitious Composites (SHCC) is a particular class of fiber-reinforced concrete with ductile behavior under direct tension, achieved through micro-mechanics guided specific material tailoring [14]. The most commonly used SHCC compositions primarily consist of cement, fly ash, and fine silica sand with a small volume fraction (typically, 2%) of polyvinyl alcohol (PVA) fibers [15]. Due to the absence of coarse aggregates, SHCC uses a greater amount of cementitious materials per unit volume than conventional concrete. Class F fly ash is commonly used to improve the matrix rheology, lower matrix fracture toughness, and reduce the cement content [16, 17]. Although researchers have investigated FA/c weight ratios ranging from 0.1 to 5.6 in various SHCCs, the most commonly used FA/c weight ratio in SHCC is 1.2 [16, 17]. Due to such high content of fly ash, fly ash significantly influence the properties of an SHCC.

SHCC materials have shown superior performance than conventional concretes at high temperatures attributed to the presence of polymer fibers such as PVA which melt around 230 °C providing open space to relieve vapor pressure, absence of coarse aggregates, and use of SCMs such as fly [18, 19]. Only a few studies have investigated the effect of fly ash on the residual mechanical properties of SHCC after subjecting the materials to high temperatures. Sahmaran et al. [18] studied the 28 days residual mechanical behavior of SHCC with FA/c weight ratios of 1.2 and 2.2 after exposing these materials to temperatures of up to 800 °C. Among the two materials, the SHCC with FA/c ratio of 2.2 exhibited better residual performance after exposure to 200 °C and above, which was attributed to the hydration of unhydrated fly ash particles by the authors. Another study [19] observed that the 28 days residual compressive strength of an SHCC with a very high volume of fly ash (FA/c = 4.4) increased (relative to room temperature strength) up to 100 °C, and then declined at 200 °C.

Ke-quan et al. [20] investigated the influence of curing age on the residual per-formance of SHCC with FA/c weight ratio of 0.43. The normalized residual com-pressive strength (normalized by room temperature strength) after subjecting to 200 °C and above decreased as curing age increased from 1 day to 28 days. Relative to the room temperature compressive strength, a significant increase in compressive strength was observed at 200 °C after 1 day of casting. This observation was attributed to the strengthening of cement paste caused by increase in the van-der-Waal attraction between the CSH gel layers, which is due to evaporation of free water. In addition, the authors identified the accelerated hydration of the cementitious materials as another explanation for the above observation.

The degree of hydration for cementitious systems, blended with fly ash, is typically low at 28 days compared to that for cement paste. Many unhydrated fly ash and cement particles remain in the cementitious matrix at 28 days, which continue to hydrate and improve strength with time. However, there is a lack of studies on the effects of fly ash content and curing age (beyond 28 days) on high-temperature mechanical properties of SHCC materials.

The objective of this research was to address the above knowledge gap. A systematic experimental study was conducted on three SHCC materials with FA/c weight ratios of 1.2, 2.4, and 3.6, with each material subjected to two different curing ages of 28 days and 150 days. As the past studies [11, 13, 18] have shown that the effect of fly ash diminished after 600 °C, the maximum target temperature in this study was limited to 600 °C. All the mixtures were prepared on the same day using the same fly ash source and cured in identical conditions. The thermal conditioning of all the specimens of all SHCC materials at a particular target temperature was also done together. All these measures were taken to minimize random error. The details of the experimental program are presented in Sect. 2, followed by results and discussion in Sect. 3 and conclusions in Sect. 4.

2   Materials and Methods

2.1   Mixture Compositions

The mix proportions of the three SHCC materials investigated in this study are given in Table 1. All the materials utilized Type-I/II cement conforming to ASTM C150 [21], Class F fly ash conforming to ASTM C618 [22], fine silica sand, polycarboxylate-ether (PCE) based high range water reducing admixture (HRWRA), and viscosity modifying admixture (VMA). The specific gravity of cement, fly ash, and silica sand was 3.15, 2.40, and 2.60, respectively.

Table 1. Mixture proportions of the SHCCs.

Materials	Matrix proportions (relative to cement weight)						Fiber volume fraction (%)
	Cement	Fly ash	Sand	Water (w/cm)	HRWRA	VMA	PVA
FA-1.2	1	1.2	0.82	0.682 (0.31)	0.004	0.020	2
FA-2.4	1	2.4	1.30	1.054 (0.31)	0.008	0.031	2
FA-3.6	1	3.6	1.77	1.426 (0.31)	0.011	0.041	2

The volume fraction of the PVA fibers was kept constant in all the three materials at 2% of the composite volume. The geometry and physical properties of the PVA fibers are given in Table 2. The PVA fibers were coated (by the manufacturer) with an oiling agent (1.2% by weight), which has been shown to optimize the fiber-matrix interfacial properties to achieve robust multiple cracking [23].

Table 2. Properties of the PVA fiber.

Diameter (μm)	Length (mm)	Density (kg/m^3)	Young's Modulus (GPa)	Tensile Strength (MPa)	Elongation at break (%)	Melting Point (°C)
39	12	1300	42.8	1600	6	230

2.2 Specimen Preparation

All the mixtures were prepared in a high-shear mixer with a maximum volume of 28.4 L. Powder ingredients, i.e., cement, fly ash, and sand were first dry-mixed for three minutes. Water was then added along with HRWRA and mixed for another three minutes. A small amount of VMA was added subsequently to achieve the desired rheology of the cementitious matrix for facilitating homogeneous fiber dispersion. The final amounts of HRWRA and VMA used in various mixtures are given in Table 1. Finally, the PVA fibers were added slowly to the matrix and mixed for another three minutes.

The fresh mixture was poured in molds, and the casting process was finished within 30 min after the addition of water. Thirty cylinders (diameter 76 mm, height 152 mm) were prepared for each material. Three cylinders were tested at each of the five target temperatures (20 °C, 100 °C, 200 °C, 400 °C, and 600 °C) at two different curing ages. The molds were mildly vibrated on a vibration table to achieve the appropriate compaction of the specimens. All the specimens were demolded after 24 h of casting. Subsequently, all the specimens were cured under water for 14 days, and then kept open in air under laboratory conditions (temperature: 20 ± 2 °C and RH: 50 ± 20%) until testing. Half of the specimens were heated and tested after 28 days of curing, and the rest half after 150 days of curing.

2.3 Heating Protocol

The specimens were heated to various target temperatures of 100 °C, 200 °C, 400 °C, and 600 °C in a 42.5-L capacity air-furnace with a maximum achievable temperature of 1,100 °C. The furnace temperature was set to increase at 5 °C/min (ramp rate) until the target temperature. The furnace temperature was kept constant at the target temperature for three hours (dwell time). The furnace was then switched off, and the specimens were cooled naturally to room temperature inside the furnace with the door shut. The same heating protocol was adopted to characterize rebar-SHCC bond strength by Deshpande et al. [24], which showed the temperature profile within the specimen for this heating protocol at 100 °C and 400 °C. A schematic of the heating protocol and the placement of specimens within the furnace is shown in Fig. 1. The specimens were then tested at room temperature, i.e., in <u>residual condition</u>, within 24 h of the end of cooling.

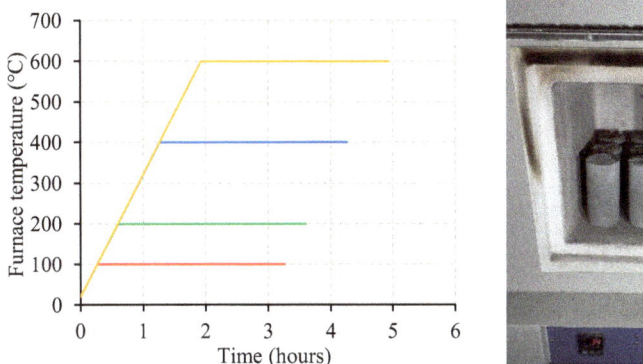

Fig. 1. Schematic of heating protocol and placement of the specimens in a furnace.

2.4   Mechanical Testing Procedure

The uniaxial compression tests on cylinder specimens were performed following the ASTM C39 [25] standard. The cylinder specimens were capped using a sulfur capping compound before testing in compression following ASTM C617 [26] to achieve smooth and flat loading surfaces. The compressive strength of each specimen was determined from the peak load reached during the test.

3   Results and Discussion

3.1   Weight Loss

The variation in weight loss with temperature, which indicates the loss of water and decomposition of hydrated cementitious products, for cylinder specimens is presented in Fig. 2. The weight loss after exposing the specimens to furnace target temperature of 100 °C at 28 days was below 3% for all the three materials. This is because the temperature within the specimens was less than the furnace target temperature in spite of a 3-h dwell time. The weight loss increased at furnace target temperatures greater than 100 °C up to 400 °C for curing age of 28 days due to the loss of free and physically bound water between 100–200 °C and decomposition of chemically bound water and PVA fibers above 200 °C. Overall, the weight loss as a function of furnace target temperature was similar for all the three materials at 28 days.

The overall trend of weight loss with temperature at 150 days was similar to that at 28 days; however, the magnitude of weight loss at each temperature decreased for all the materials at 150 days. This is attributed to the increased drying and degree of hydration after 150 days, which reduces the physically-held water in the microstructure of SHCC. The weight loss up to 600 °C did not exceed the water content (18%) in any material, which is consistent with the observations in other literature [27, 28].

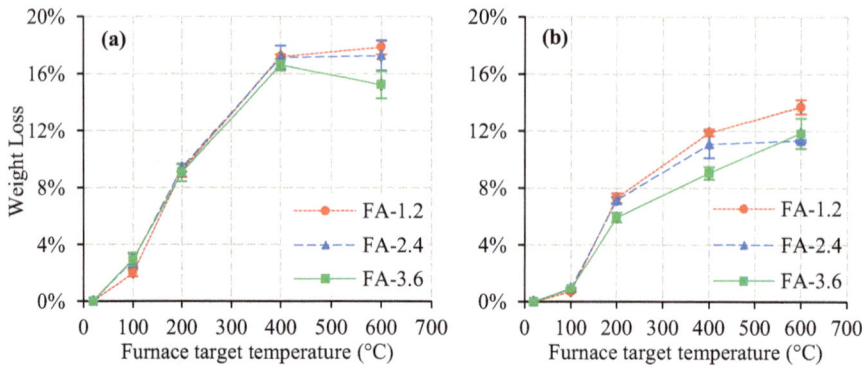

Fig. 2. Variation in weight loss with furnace target temperature at: **(a)** 28 days, and **(b)** 150 days.

3.2 Compressive Strength

The room temperature compressive strength of the three SHCCs at 28 days and at 150 days are plotted in Fig. 3. The compressive strength of the materials decreased with increase in FA/c weight ratio from 1.2 to 3.6 both at 28 and 150 days. The improvement in the room temperature compressive strength at 150 days relative to 28 days strength was 16%, 29%, and 56% for - FA-1.2, FA-2.4, and FA-3.6, respectively. It shows that the improvement in strength was greater for higher FA/c ratios. However, Yang et al. [16] observed minimal increase in strength from 28 to 90 days for SHCCs with FA/c ratios in the range of 1.2 to 5.6. While the ingredients (especially fly ash) are different in the two studies, the main reason behind the above disagreement between this study and Yang et al. [16] could be the difference in curing protocols. In the present study, all the specimens were cured under water for 14 days and then kept in air, compared to only seven days of sealed curing followed by air exposure in Yang et al. [16]. At high replacements of cement by fly ash, the relatively low water/cementitious materials (w/cm) ratios used in SHCC provide a limited amount of water for secondary reactions.

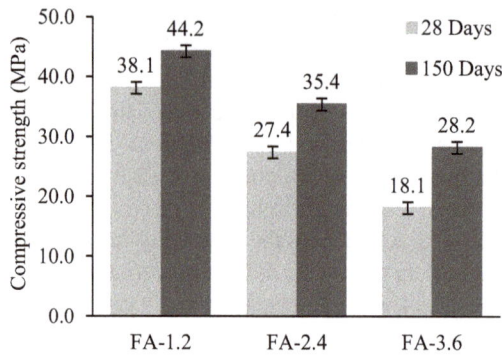

Fig. 3. Compressive strength at room temperature (20 °C).

Overall, the room temperature compressive strength of SHCCs decreased with increase in FA/c weight ratio, and it increased with curing age in this study.

The variations of *normalized residual compressive strength* (f_{cn}) with temperature for the three SHCCs investigated in this study after 28 days of curing are plotted in Fig. 4. The f_{cn} was computed by dividing the observed residual compressive strength [$f_c(T)$] of a composite after subjecting it to a temperature (T) by its compressive strength [$f_c(20\ °C)$] at room temperature (20 °C). Relative to their room temperature compressive strengths, all the SHCCs retained or showed an increase in their compressive strength up to 200 °C (i.e. f_{cn} was in the range of 0.97 to 1.15). The f_{cn} after exposure to 200 °C was dependent on the FA/c weight ratio (greatest for FA-3.6 SHCC and smallest for FA-1.2 SHCC), as larger FA/c ratio provides greater reserve of unhydrated fly ash that can be hydrated (through the pozzolanic reaction) at moderately elevated furnace temperatures (up to 200 °C in this study).

The 28-day compressive strength of all the three SHCCs showed a marginal reduction from 200 °C to 400 °C. The increase in porosity due to the melting of PVA fibers at temperatures greater than 200 °C is the likely cause behind the reduction in compressive strength. Despite the increased porosity, the average pore diameter shows little change up to 400 °C [18, 27], explaining the significant retention of compressive strength with f_{cn} greater than 0.90 for all the three SHCCs at 400 °C. In addition, the partial dehydration of CSH also takes place between 200 °C to 400 °C (and complete decomposition above 600 °C), however any detrimental effects of CSH dehydration at 28 days are negated by the accelerated pozzolanic reaction in this temperature range. After exposure to 600 °C, the f_{cn} reduces to 0.63–0.66 for the three SHCCs. The sharp reduction in f_{cn} between 400 °C and 600 °C is mainly due to the deterioration of hydration products and increased micro-cracking and coarsening of the pore structure [3, 18, 27, 29].

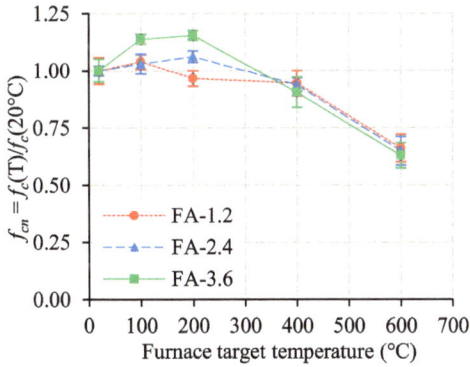

Fig. 4. Variation of normalized residual compressive strength at 28 days.

The variations of f_{cn} with temperature for the three SHCCs after 150 days of curing are plotted in Fig. 5, which are significantly different from those observed in Fig. 4 for 28-day cured specimens. The increase in compressive strength up to 200 °C with increase in the FA/c ratio observed in 28-day cured specimens was not observed in 150-day cured specimens. In fact, the order of f_{cn} at 200 °C is reversed in Fig. 5 relative to Fig. 4. The reason behind this observation is as follows. As noted above in the discussion comparing room temperature compressive strengths after 28 days and 150 days, the secondary hydration of fly ash has a significant effect on the compressive strength. At 150 days, the reserves of unhydrated fly ash and calcium hydroxide are depleted as the hydration process approaches toward completion. This reduces the possibility of further hydration at elevated temperatures up to 200 °C. As a result, f_{cn} reduced to 0.87 and 0.83 for FA-2.4 and FA-3.6 SHCCs, respectively; in contrast, FA-1.2 SHCC fully retained its room temperature compressive strength at 200 °C. Thus, for 150-day cured specimens, increase in FA/c ratios did not lead to increase in residual compressive strength at temperatures of up to 200 °C. The f_{cn} for 150-day cured specimens degraded almost linearly with temperature above 200 °C for all the three SHCCs, similar to that observed for the 28-day cured specimens.

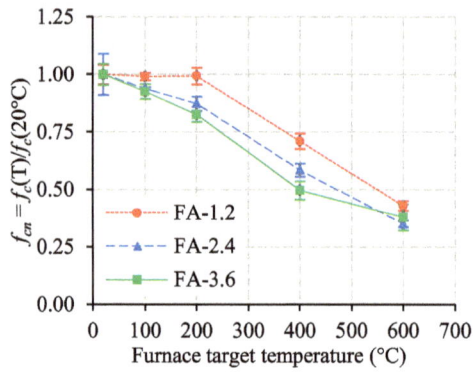

Fig. 5. Variation of normalized residual compressive strength at 150 days.

4 Conclusions

The effects of fly ash content and curing age on the residual compressive strength of SHCC after subjecting to high temperatures up to 600 °C were investigated in this study. Cylinder specimens of three SHCCs with fly ash to cement (FA/c) weight ratios of 1.2, 2.4, and 3.6 were subjected to thermal and mechanical loading at 28 days and 150 days. The following conclusions are drawn from the study:

- The room temperature compressive strength decreased with increase in FA/c weight ratio at both the curing ages of 28 days and 150 days. All the three SHCCs gained compressive strength at 150 days relative to 28 days, and the said gains increased with the FA/c ratio of the SHCCs.

- Relative to their room temperature strengths, all the SHCCs retained or showed higher residual compressive strengths up to 200 °C at 28 days. The increase in strength increased with FA/c weight ratio. Above 200 °C, all the SHCCs showed a linear reduction in the residual compressive strength.
- The above increase in compressive strength up to 200 °C (relative to room temperature strength) for SHCCs with higher FA/c ratios was not observed in 150-day cured specimens. In fact, the normalized residual compressive strength of the SHCC with the lowest FA/c ratio (1.2) investigated in this study was the highest among the three SHCCs at 200 °C.

Overall, the beneficial effects of high volumes of fly ash in SHCC at high temperatures are lost at longer curing ages. Thus, the curing age of the composites should be considered while determining the strength reduction factor at high temperatures for the cementitious composites containing high volumes of fly ash.

Acknowledgements. We are grateful to the Institute of Bridge Engineering (IBE) at the University at Buffalo (UB) for providing the financial support for this research. The material suppliers: Lafarge-Holcim, WR Grace, and Nycon are gratefully acknowledged for providing the SHCC ingredients at no cost. We also thank the staff of the Structural Engineering and Earthquake Simulation Laboratory (SEESL) at UB for their assistance with the experiments.

References

1. Khoury, G.A.: Effect of fire on concrete and concrete structures. Prog. Struc. Eng. Mater. **2**(4), 429–447 (2000)
2. Lim, S., Mondal, P.: Micro-and nano-scale characterization to study the thermal degradation of cement-based materials. Mater. Charac. **92**, 15–25 (2014)
3. Deshpande, A.A., Kumar, D., Ranade, R.: Influence of high temperatures on the residual mechanical properties of a hybrid fiber-reinforced strain-hardening cementitious composite. Constr. Build. Mater. **208**, 283–295 (2019)
4. Heikal, M., Al-Duaij, O., Ibrahim, N.: Microstructure of composite cements containing blast-furnace slag and silica nano-particles subjected to elevated thermally treatment temperature. Constr. Build. Mater. **93**, 1067–1077 (2015)
5. Seleem, H.E.D.H., Rashad, A.M., Elsokary, T.: Effect of elevated temperature on physico-mechanical properties of blended cement concrete. Constr. Build. Mater. **25**(2), 1009–1017 (2011)
6. Rashad, A.M.: An investigation on very high volume slag pastes subjected to elevated temperatures. Constr. Build. Mater. **74**, 249–258 (2015)
7. Donatello, S., Kuenzel, C., Palomo, A., Fernández-Jiménez, A.: High temperature resistance of a very high volume fly ash cement paste. Cem. Concr. Compos. **45**, 234–242 (2014)
8. Khan, M., Prasad, J., Abbas, H.: Effect of high temperature on high-volume fly ash concrete. Arabian J. Sci. Eng. **38**(6), 1369–1378 (2013)
9. Vyšvařil, M., Bayer, P., Chromá, M., Rovnaníková, P.: Physico-mechanical and microstructural properties of rehydrated blended cement pastes. Constr. Build. Mater. **54**, 413–420 (2014)

10. Sarshar, R., Khoury, G.: Material and environmental factors influencing the compressive strength of unsealed cement paste and concrete at high temperatures. Mag. Concr. Res. **45** (162), 51–61 (1993)
11. Poon, C.-S., Azhar, S., Anson, M., Wong, Y.-L.: Comparison of the strength and durability performance of normal-and high-strength pozzolanic concretes at elevated temperatures. Cem. Concr. Res. **31**(9), 1291–1300 (2001)
12. Xu, Y., Wong, Y., Poon, C., Anson, M.: Impact of high temperature on PFA concrete. Cem. Concr. Res. **31**(7), 1065–1073 (2001)
13. Nadeem, A., Memon, S.A., Lo, T.Y.: The performance of fly ash and metakaolin concrete at elevated temperatures. Constr. Build. Mater. **62**, 67–76 (2014)
14. Li, V.C.: On engineered cementitious composites (ECC). J. Adv. Concr. Tech. **1**(3), 215–230 (2003)
15. Wang, S., Li, V.C.: Polyvinyl alcohol fiber reinforced engineered cementitious composites: material design and performances. In: International RILEM Workshop on HPFRCC Structural Applications, pp. 65–73 (2005)
16. Yang, E.-H., Yang, Y., Li, V.C.: Use of high volumes of fly ash to improve ECC mechanical properties and material greenness. ACI Mater. J. **104**(6), 620 (2007)
17. Wang, S., Li, V.C.: Engineered cementitious composites with high-volume fly ash. ACI Mater. J. **104**(3), 233 (2007)
18. Şahmaran, M., Özbay, E., Yücel, H.E., Lachemi, M., Li, V.C.: Effect of fly ash and PVA fiber on microstructural damage and residual properties of engineered cementitious composites exposed to high temperatures. J. Mater. Civ. Eng. **23**(12), 1735–1745 (2011)
19. Yu, J., Lin, J., Zhang, Z., Li, V.C.: Mechanical performance of ECC with high-volume fly ash after sub-elevated temperatures. Constr. Build. Mater. **99**, 82–89 (2015)
20. Yu, K.-Q., Lu, Z.-D., Yu, J.: Residual compressive properties of strain-hardening cementitious composite with different curing ages exposed to high temperature. Constr. Build. Mater. **98**, 146–155 (2015)
21. ASTM C150/C150M-18: Standard specification for portland cement. ASTM International, West Conshohocken, PA (2018)
22. ASTM C618-17a: Standard specification for coal fly ash and raw or calcined natural pozzolan for use in concrete. ASTM International, West Conshohocken, PA (2017)
23. Li, V.C., Wu, C., Wang, S., Ogawa, A., Saito, T.: Interface tailoring for strain-hardening polyvinyl alcohol-engineered cementitious composite (PVA-ECC). Mater. J. **99**(5), 463–472 (2002)
24. Deshpande, A.A., Kumar, D., Ranade, R.: Temperature effects on the bond behavior between deformed steel reinforcing bars and hybrid fiber-reinforced strain-hardening cementitious composite. Constr. Build. Mater. **233**, 117337 (2020)
25. ASTM C39/C39M-18: Standard test method for compressive strength of cylindrical concrete specimens. ASTM International, West Conshohocken, PA (2018)
26. ASTM C617/C617M-15: Standard practice for capping cylindrical concrete specimens. ASTM International, West Conshohocken, PA (2015)
27. Sahmaran, M., Lachemi, M., Li, V.C.: Assessing mechanical properties and microstructure of fire-damaged engineered cementitious composites. ACI Mater. J. **107**(3), 297 (2010)
28. Phan, L.T., Lawson, J.R., Davis, F.L.: Effects of elevated temperature exposure on heating characteristics, spalling, and residual properties of high performance concrete. Mater. Struc. **34**(2), 83–91 (2001)
29. Kumar, D., Deshpande, A.A., Ranade, R.: Influence of fiber length on the mechanical behavior of steel-PVA hybrid fiber-reinforced strain-hardening cementitious composites at high temperatures. The Indian Concr. J. **93**(12), 30–38 (2019)

Optimization of Functionally Graded Concretes Incorporating Steel Fibres and Recycled Aggregates

Ricardo Chan[1]([⊠]), Charles K. S. Moy[1], and Isaac Galobardes[2]

[1] Department of Civil Engineering, Xi'an Jiaotong-Liverpool University, Suzhou, China
r.chan@xjtlu.edu.cn
[2] School of Architecture, Planning and Design, Mohammed VI Polytechnic University, Ben Guerir, Morocco

Abstract. Functionally graded material (FGM) refers to a class of material produced with grading composition and structure to achieve enhanced performance compared to homogeneous materials. Several studies have explored the application of the concept of FGM to enhance the flexural behaviour of concrete, producing functionally graded concrete (FGC). Previous results indicated that FGC produced with fibre reinforced recycled aggregate concrete (FRRAC) exhibited higher residual flexural strength than homogeneous FRRAC for ratios of reinforced height to total beam height (h/H) equal or higher than 0.75, demonstrating the benefits of FGC with FRRAC. Hence, this study aims to verify the optimum value of h/H to obtain the highest residual flexural performance of the FGC. To achieve this goal, an experimental program was carried out, in which, FGC fabricated with FRRAC was assessed under bending considering a content of fibre of 0.50% in volume, and values of h/H ranging from 0.70 to 1.00. The effect of h/H in fibre orientation was also evaluated using the inductive method. The results indicated that the highest residual flexural strength is obtained with $h/H = 0.90$. However, a balance between pre-cracking and post-cracking behaviour should be defined for each application, resulting in optimized values of h/H in FGC with FRRAC. Furthermore, since the fibre orientation was not affected by h/H, the same orientation factor used in the design of the fibre reinforced concrete elements can be adopted for FGC elements, increasing the potential application of FGC.

Keywords: Functionally graded concrete · Steel fibre reinforced concrete · Recycled aggregate concrete · Fibre orientation

1 Introduction

Functionally graded materials (FGM) are a class of material characterized by its grading composition and structure to achieve predefined properties [1]. The concept of FGM has been applied to enhance the flexural behaviour of concrete by adding 0.50% of steel fibres, in volume, only into the bottom layer [2, 3], resulting in functionally graded concretes (FGC). To increase the sustainability of FGC, fibre reinforced

© RILEM 2022
P. Serna et al. (Eds.): BEFIB 2021, RILEM Bookseries 36, pp. 13–23, 2022.
https://doi.org/10.1007/978-3-030-83719-8_2

recycled aggregate concrete (FRRAC) has been used in the bottom layer of two-layered FGC [2, 3]. The results indicated that FGC comprising of PCC and FRRAC could yield better mechanical performance than homogeneous FRRAC. In addition, an optimum value of h/H between 0.75 and 1.00 could result in FGC with FRRAC presenting higher residual flexural strength [3]. Thus, determining the optimum value of h/H can be very important to enhance the potential of using FGC with FRRAC in real life applications.

In general, the substitution of natural aggregates for recycled aggregates resulted in a decrease of flexural performance of concrete before the first crack [4]. On the other hand, the addition of fibres can enhance the post-cracking behaviour of concretes produced with recycled aggregates [5]. Therefore, the optimum value of h/H should reflect a balance between the flexural performance before and after cracking. As a result, FGC with FRRAC could be made viable for some applications, such as low traffic roads, car parks, cycling lines or pedestrian pavements [3].

Hence, this paper aims to verify the optimum value of h/H to obtain the highest flexural performance of FGC with FRRAC. To achieve this goal, an experimental program was carried out, in which, FGC with FRRAC was assessed under bending considering a content of fibre of 0.50% in volume, and different values of h/H ranging from 0.70 to 1.00. In addition, the influence of fibre orientation in the compressive behaviour was evaluated considering moulded and extracted cubes from beams. Finally, the effect of h/H in fibre orientation was also assessed using a non-destructive method, known as inductive method [6].

2   Experimental Program

2.1   Materials

Portland cement CEM I-42.5N [7], and tap water at room temperature (20 °C) were used in this study. Coarse and fine aggregates from natural and recycled sources were adopted. Natural aggregates were composed by limestone and river sand, while recycled aggregates were produced by crushed demolition waste. The sieving analysis of coarse and fine aggregates was done according to BS 812-103.1:1985 [8] and the results are shown in Fig. 1. Table 1 presents the main properties of the aggregates, which are the oven dry density (ρ_{rd}) and the water absorption (WA).

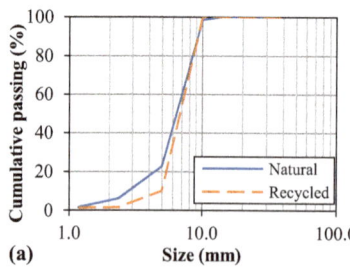
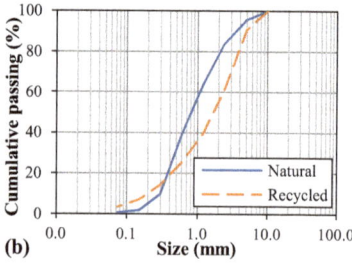

Fig. 1. Particle size distribution and grading limits for (a) coarse and (b) fine aggregates.

Table 1. Main characteristics of the aggregates.

Property	Natural aggregate		Recycled aggregate	
	Coarse	Fine	Coarse	Fine
ρ_{rd} (Mg/m³)	2.60	2.70	1.99	2.21
WA (%)	1.17	2.53	12.73	11.87

The oven-dry density of natural aggregates is higher than the recycled aggregates, while the recycled aggregates present higher water absorption than natural aggregates. This is probably due to the old mortar attached to the recycled aggregate, which decreases its density and increases the water absorption [9].

Hooked-end steel fibre that fulfilled the specifications required by BS EN 14889-1:2006 [10] was used in this study. According to the supplier, these fibres have 60 mm of length (L), 0.75 mm of diameter (D), aspect ratio of 80 (L/D), and tensile strength of 1150 MPa.

2.2 Mixes and Production

Two concrete groups are considered in this study and are represented in Fig. 2. In the first group, homogenous FRRAC was adopted (Fig. 2a) and in the second group, FGC was considered. The FGC consisted of a two-layered system, in which the top layer was produced with plain cement concrete (PCC) and the bottom layer with FRRAC (Fig. 2b). The FGC configuration is the same used in previous studies and focus on improving the material efficiency in structures subjected to bending, such as beams and slabs [2, 3]. In addition, FGC with FRRAC was studied considering h/H = 0.70, 0.80, 0.90 and 1.00, noting that h/H = 1.00 is equivalent to homogeneous FRRAC.

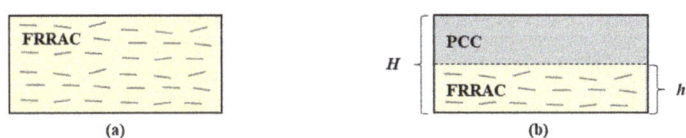

(a)   (b)

Fig. 2. Concrete families considered: (a) FRRAC and (b) PCC + FRRAC.

Reference mixes for PCC and FRRAC are presented in Table 2. A cement content of 475 kg/m³ and free water/cement ratio (w/c) of 0.45 were adopted for both mixes. The quantities of water and aggregates were adjusted due to the differences in the actual moisture content and the water absorption of the aggregates [11]. In addition, a content of fibre (cf) of 40 kg/m³ equivalent to 0.50% in volume was adopted.

Table 2. Reference mix designs (kg/m^3).

Reference mix	Cement	Water	Type of aggregate	Fine aggregate	Coarse aggregate
PCC	475	215	Natural	790	890
FRRAC			Recycled	805	715

Four concrete mixes were produced, as listed in Table 3. The mixes were identified by a number and letter. The number stands for the relation h/H and the letter R for recycled aggregates. Note that for homogeneous FRRAC, the relation h/H is equal to 1.00 since the whole volume is reinforced with steel fibres.

Prismatic and cubic specimens were produced. Beams presented 150×150 mm cross section and 550 mm length, while cubes presented 150 mm nominal size. The casting procedure described in BS EN 14651:2005 + A1:2007 [12] was followed for the production of homogeneous FRRAC. Regarding the FGC, they were produced according to the following method: (1) the concrete of the bottom layer was placed in the mould; (2) then it was compacted for 20 s by mechanical vibration and its height was carefully verified afterwards; and (3) the upper layer was cast about 30 min after finishing the previous layer, being vibrated for half the time spent for the bottom layer (10 s) in order to avoid mixing the layers to each other. This production method was used in other previous researches and assures that the specimens are well compacted and the bonding between layers is enough to guarantee a monolithic behaviour [2, 3]. A total of 16 prismatic specimens and 20 cubic specimens were produced. All specimens were demoulded 24 h after casting and cured in water at a temperature of approximately 20 °C for 28 days following the requirements of BS EN 12390-2:2009 [13].

Table 3. Concrete mixes produced to study the mechanical behaviour of FGC.

Family	Code	h/H	cf (%)
PCC + FRRAC	0.70R	0.70	0.50
	0.80R	0.80	0.50
	0.90R	0.90	0.50
FRRAC	1.00R	1.00	0.50

2.3   Test Methods

2.3.1   Bending Test

The bending test used in this study consists of the four-point bending test described in the ASTM standard C 1609/C 1609M-06 [14]. Four prismatic specimens per mix were tested at an age of 28 days. In this test, the load (F) was controlled, and the net deflection (δ) was measured using a pair of LVDTs sustained by a steel frame, obtaining the F–δ curve (Fig. 3a). These results were used to determine the flexural strength at the first crack (f_1) and the residual flexural strengths corresponding to a net deflection equal to $L/600$ (0.75 mm) and $L/150$ (3.0 mm) ($f_{0.75}$ and $f_{3.0}$, respectively).

Fig. 3. Test setups for: (a) four-point bending test; (b) compressive strength test; and (c) inductive method.

2.3.2   Compressive Test

The quality of the concretes was assessed by means of compressive strength of FRRAC and FGC. 11 cubes were tested for each concrete at the age of 28 days. Among these 11 cubes, three were produced, and the other 8 were extracted from the beams after the bending test. The extraction was done as follows: (1) two marks 50 mm distant from each end were made, then additional marks were drawn every 150 mm (Fig. 4a); (2) the prismatic specimens were cut in the marked lines using a wet-cutting diamond saw (Fig. 4b), and (3) both ends and the central part were discarded (Fig. 4c). The extraction procedure was elaborated to obtain the parts that did not suffer any plastic deformation to ensure the validity of compressive strength results. The method presented in BS EN 12390-3:2009 [15] was followed for surface preparation and determination of compressive strength. The specimens were loaded according to the casting direction, as shown in Fig. 3b.

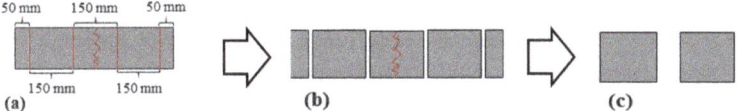

Fig. 4. Extraction of cubes from prismatic specimens: (a) marking; (b) cutting, and (c) selection.

2.3.3   Inductive Test

Before testing for compressive strength, the cubic specimens were tested using the inductive method to estimate the orientation and content of fibres [6]. In this method, the cube is placed inside a coil (see Fig. 3c) in three directions corresponding to the main axes (x in length direction, y in width direction, z in height direction), as indicated in Fig. 5. For each direction, the increase in impedance (ΔL_i) was measured using a TH2830 LCR meter (see Fig. 3c). Then, the real cf of each cube can be estimate using Eq. (1) [16], which depends on the sum of impedance increases in the main axes (ΔL_T), the volume of the specimen (V) and a constant of proportionality (ω). The constant ω is obtained through the calibration of the inductive method. In addition, the fibre orientation can be assessed using the fibre contribution in a certain direction (C_i) estimated with Eq. (2) [6]. The estimated real cf can be compared to designed cf for quality control purposes, while C_i can be used to understand the role of fibre orientation in the mechanical behaviour of concretes reinforced with fibres.

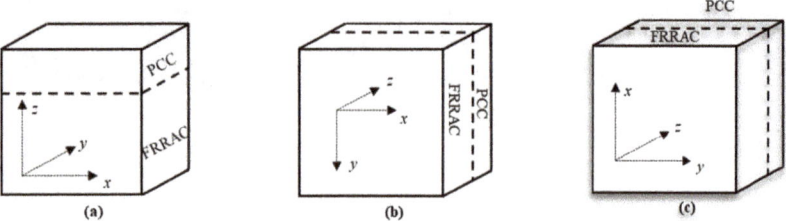

Fig. 5. Measurement directions for cubic specimens: (a) z axis; (b) y axis; and (c) x axis.

$$cf = \omega \times \sum_{i=x,y,z} \frac{\Delta L_i}{V} = \omega \times \Delta L_T \tag{1}$$

$$C_i = \frac{\Delta L_i}{\Delta L_T} \tag{2}$$

3   Results and discussion

3.1   Bending Test

Figure 6 presents the average $F–\delta$ curves obtained from the bending test. The post-cracking is enhanced with the increase of h/H up to 0.90, possibly due to the increase of cracking area containing fibres [17]. However, when h/H reaches 1.00, the post-cracking response drops to a level lower than 0.70R. The same behaviour was seen in a previous research and the difference in modulus of elasticity between layers was pointed as a possible explanation [3]. In this sense, the layer of PCC, which presents higher modulus of elasticity than FRRAC [18], absorbs part of the energy and the stresses transferred to the bottom layer of in FGC with FRRAC are reduced [19]. On the other hand, because homogeneous FRRAC presents lower modulus of elasticity throughout its volume, the concrete cracks under lower loads.

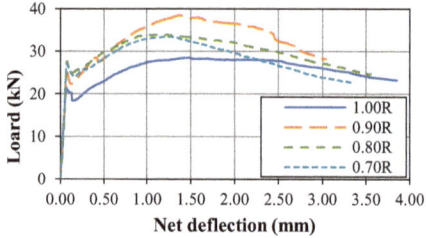

Fig. 6. $F–\delta$ curves of FGC with FRRAC with h/H ranging from 0.70 to 1.00.

Table 4 presents the results of flexural strength at first crack (f_1) and residual flexural strengths corresponding to a net deflection equal to 0.75 mm and 3.0 mm ($f_{0.75}$ and $f_{3.0}$, respectively). The average results are expressed in MPa and the coefficient of variation is added in parenthesis.

Table 4. Results obtained by means of the bending test.

Concrete	f_1 (MPa)	$f_{0.75}$ (MPa)	$f_{3.0}$ (MPa)
0.70R	3.67 (9.34%)	4.17 (13.06%)	3.19 (7.69%)
0.80R	3.70 (3.85%)	4.30 (5.14%)	3.62 (16.02%)
0.90R	3.40 (13.75%)	4.41 (16.30%)	3.80 (6.15%)
1.00R	2.93 (7.76%)	3.41 (19.32%)	3.47 (23.10%)

In general, flexural strength (f_1) results present lower coefficient of variation than results for residual flexural strengths ($f_{0.75}$ and $f_{3.0}$), as expected [17]. As observed in other study [3], f_1 decreases with the increase of h/H because of the lower strength recycled aggregates. On the other hand, $f_{0.75}$ and $f_{3.0}$ rise from $h/H = 0.70$ to 0.90, then drop when $h/H = 1.00$. The drop of $f_{0.75}$ when $h/H = 1.00$ is higher than of $f_{3.0}$. As a result, when $h/H = 1.00$, $f_{0.75}$ value is below the result obtained in 0.70R, while $f_{3.0}$ value is higher in 1.00R than 0.70R. The lower $f_{0.75}$ result in 1.00R may be explained by the lesser impact of content of fibres and the higher influence of concrete matrix in smaller deflections, such as 0.75 mm. On the contrary, in larger deflections, such as 3.0 mm, the presence of fibres throughout the volume is more influent and may compensate the weaker matrix [17].

These results indicate that the flexural performance of FGC with FRRAC can be enhanced with the addition of a top layer of PCC as thin as 10% of the total height of the specimen. However, the value of h/H needs to be properly selected, since an increase in h/H leads to lower flexural strength at first crack but higher residual flexural strength. Thus, a balance between pre-cracking and post-cracking behaviour should be obtained for each application, resulting in optimized values of h/H in FGC with FRRAC.

3.2 Compressive Test

The results obtained in the compressive test for moulded cubes and cubes extracted from beams are presented in Fig. 7. The mean compressive strength (f_{cm}) and standard deviation (presented as error bars) are calculated for each type of concrete.

The quality of the concrete produced was assured, since the variation of the results is smaller than 10%. Also, f_{cm} values tend to decrease with higher values of h/H, as expected, since the content of recycled aggregate increases with h/H. Besides, extracted cubes present slightly higher f_{cm} than moulded cubes, indicating that the fibre orientation may affect the compressive behaviour of FGC with FRRAC. Further details on the fibre orientation are presented in Sect. 3.3. As shown in Fig. 8a, moulded cubes failed with all exposed faces cracked, while only the faces parallel to the length direction cracked when the extracted cubes failed (Fig. 8b).

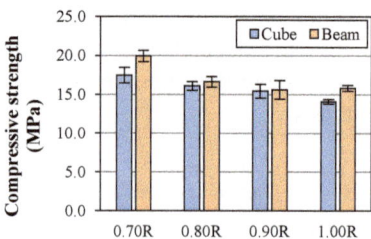

Fig. 7. Compressive strength results.

Fig. 8. Failure mode in (a) moulded cubes and (b) extracted cubes.

Furthermore, f_{cm} is commonly used in quality control of concrete structures under bending due to the well-known relationship between f_1 and f_{cm} [9]. Values ranging from 20.8–23.0% and 18.4–22.2% were found for the relation between f_1 and f_{cm} obtained in cubes and beams, respectively. Thus, a consistent relation between f_1 and f_{cm} can be experimentally stablished and used for the quality control of FGC with FRRAC.

3.3   Inductive Test

The calibration of inductive method was achieved using three expanded polystyrene specimens with different known contents of fibre. The expanded polystyrene specimens were cubes with 150 mm of nominal size. Also, the same steel fibres used to produce the concrete specimens were used in the calibration, in amounts of 18, 36 and 54 g. The calibration curve is presented in Fig. 9, and results in $\omega = 0.0275$ with R^2 of 0.99.

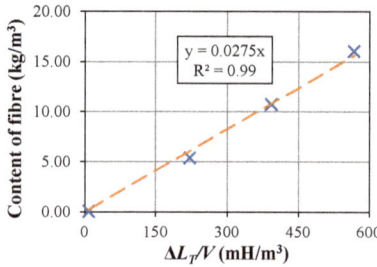

Fig. 9. Calibration of inductive method.

The real content of fibres was estimated using constant ω from Fig. 9 in Eq. (1) and the calculated results are presented in Fig. 10a. The average values are shown along with the standard deviation. Furthermore, the fibre contribution in the main axes is indicated in Fig. 10b.

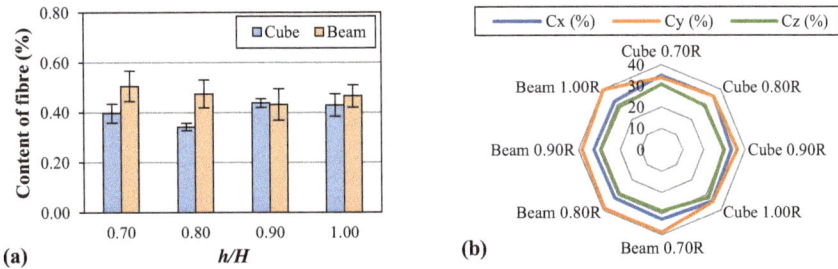

Fig. 10. Inductive method results: (a) content of fibre and (b) fibre orientation.

The average variation in cf is 7.1% and 12.2% for extracted and moulded cubes, respectively, which are expected values [6]. Also, cf in extracted cubes tend to be closer to the design value of 0.50% and higher than in moulded cubes. This difference may probably be due to the random distribution of fibres in the concrete mass after the mixing process [17]. Regarding the fibre orientation, Fig. 10b indicates that, due to the wall effect and vibration process, fibres are more orientated in the horizontal plane (x, y) [17]. In addition, in moulded cubes, C_x and C_y are approximately the same, as can be observed in Fig. 10b. On the other hand, C_y is higher than C_x in extracted cubes, as observed in other study [6] and in Fig. 10b. This happens because moulded cubes are constrained in all directions (x, y, z) while extracted cubes are constrained only in two (y, z), as indicated in Fig. 11a and Fig. 11b, respectively.

Furthermore, no clear influence of h/H ranging from 0.70 to 1.00 is verified in the fibre orientation. This indicates that small variations in reinforced layer thickness is not significant to impact the fibre orientation induced by the wall effect, in comparison with the variation in specimen length. Thus, an orientation factor related to length/width ratio plays a major role in fibre orientation, as verified in other studies [20, 21] for designing FRC slabs or panels using results from FRC beams. Considering that the orientation factor between a certain size of beam and slab is constant, the relation between the post-cracking behaviour of FGC slabs and FGC beams would be proportional to the orientation factor, regardless of h/H. Therefore, the optimum value of h/H would be the same in FGC beams and FGC slabs. Consequently, the same orientation factor used for the design of FRC slabs can be adopted to design FGC slabs, increasing the potential of FGC.

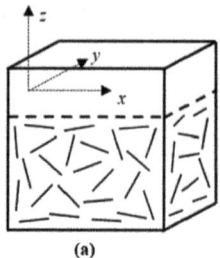 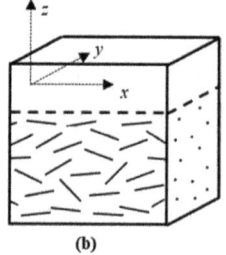

(a)   (b)

Fig. 11. Representation of fibres orientation in (a) moulded and (b) extracted cubes.

4 Conclusions

The optimum value of h/H to obtain the highest flexural performance of FGC with FRRAC was assessed through an experimental program considering different values of h/H ranging from 0.70 to 1.00. Also, the impact of h/H in the fibre orientation was verified using a non-destructive method. Overall, a balance between pre-cracking and post-cracking behaviour should be defined for each application, resulting in optimized values of h/H in FGC with FRRAC. Furthermore, the same orientation factor used for FRC elements can be adopted for FGC elements, increasing the potential of FGC. Additional conclusions were drawn and are presented as follows:

- The flexural performance of FGC with FRRAC could be enhanced with the addition of a top layer of PCC as thin as 10% of the total height of the specimen;
- The increase of h/H up to 0.90 lowered the flexural performance before cracking but improved the performance after cracking;
- The fibre orientation may affect the failure mode under compression of cubes made of FGC with FRRAC;
- The wall effect induced by the specimen geometry was evidenced in the difference in fibre orientation observed in moulded and extracted cubes; and
- The results obtained from the inductive method suggest that h/H has no influence in the fibre orientation.

Acknowledgements. The authors would like to acknowledge the Xi'an Jiaotong-Liverpool University (XJTLU) Research Development Fund for the financial support received from the project with reference RDF-16-02-42.

References

1. Kawasaki, A., Watanabe, R.: Concept and P/M fabrication of functionally gradient materials. Ceram. Int. **23**, 73–83 (1997)
2. Liu, X., Yan, M., Galobardes, I., Sikora, K.: Assessing the potential of functionally graded concrete using fibre reinforced and recycled aggregate concrete. Constr. Build. Mater. **171**, 793–801 (2018)

3. Chan, R., Liu, X., Galobardes, I.: Parametric study of functionally graded concretes incorporating steel fibres and recycled aggregates. Constr. Build. Mater. **242**, 118186 (2020)

4. de Brito, J.M.C.L., Saikia, N.: Recycled Aggregate in Concrete. Bentham Science Publishers. Springer London, London (2013). https://doi.org/10.1007/978-1-4471-4540-0

5. Chan, R., et al.: Analysis of potential use of fibre reinforced recycled aggregate concrete for sustainable pavements. J. Clean. Prod. **218**, 183–191 (2019)

6. Torrents, J.M., et al.: Inductive method for assessing the amount and orientation of steel fibers in concrete. Mater. Struct. **45**, 1577–1592 (2012)

7. BSI: BS EN 197-1:2011, Cement - Composition, specifications and conformity criteria for common cements. BSI Standards Limited, London, England (2011). https://doi.org/10.3403/30205527

8. BSI: BS 812-103.1:1985, Testing aggregates - Method for determination of particle size distribution - Sieve tests. BSI Standards Limited, London, England (1998). https://doi.org/10.3403/00139627

9. Mehta, P.K., Monteiro, P.J.M.: Concrete: Microstructure, Properties, and Materials. McGraw-Hill Education, New York (2006). https://doi.org/10.1036/0071462899

10. BSI: BS EN 14889-1:2006, Fibres for concrete - Steel fibres - Definitions, specifications and conformity. BSI Standards Limited, London, England (2006). https://doi.org/10.3403/30110332

11. Teychenné, D.C., Franklin, R.E., Erntroy, H.C.: Design of normal concrete mixes. Construction Research Communications Ltd, Watford, England (1997)

12. BSI: BS EN 14651:2005+A1:2007, Test method for metallic fibred concrete - Measuring the flexural tensile strength (limit of proportionality (LOP), residual). BSI Standards Limited, London, England (2008). https://doi.org/10.3403/30092475

13. BSI: BS EN 12390-2:2009, Testing hardened concrete - Making and curing specimens for strength tests. BSI Standards Limited, London, England (2009). https://doi.org/10.3403/30164903

14. ASTM: C 1609/C 1609M-06, Standard test method for flexural toughness and first-crack strength of fibre-reinforced concrete (Using beam with third-point loading). ASTM International, West Conshohocken, United States (2006). https://doi.org/10.1520/C1609_C1609M-06

15. BSI: BS EN 12390-3:2009, Testing hardened concrete - Compressive strength of test specimens. BSI Standards Limited, London, England (2011). https://doi.org/10.3403/30164906

16. Silva, C.L., et al.: Assessment of fibre content and orientation in SFRC with the inductive method. Part 2: application for the quality control of sprayed concrete. In: International Symposium Non-Destructive Testing in Civil Engineering (NTDTCE 2015) (2015)

17. Bentur, A., Mindess, S.: Fibre Reinforced Cementitious Composites. Taylor & Francis, London and New York (2007)

18. Silva, R.V., de Brito, J.M.C.L., Dhir, R.K.: Establishing a relationship between modulus of elasticity and compressive strength of recycled aggregate concrete. J. Clean. Prod. **112**, 2171–2186 (2016)

19. Hibbeler, R.C.: Mechanics of Materials. Pearson Education Limited, Boston, United States (2017)

20. Pujadas, P., Blanco, A., Cavalaro, S.H.P., de la Fuente, A., Aguado de Cea, A.: Fibre distribution in macro-plastic fibre reinforced concrete slab-panels. Constr. Build. Mater. **64**, 496–503 (2014)

21. Blanco, A., Pujadas, P., de la Fuente, A., Cavalaro, S.H.P., Aguado, A.: Assessment of the fibre orientation factor in SFRC slabs. Compos. Part B Eng. **68**, 343–354 (2014)

Mix Proportioning of Fiber Reinforced Self-compacting Concrete Adopting the Compressible Packaging Method: Comparison of Two Methods

Matheus G. Cardoso[✉], Rodrigo M. Lameiras, and Ian B. Cavalcante

Department of Civil and Environmental Engineering (ENC) of the University of Brasília (UnB), Brasília, Brazil
matheus-ssdo@hotmail.com

Abstract. Fiber reinforced self-compacting concrete (FRSCC) is a material that combines the advantages of self-compacting concrete and fiber-reinforced concrete, which can act on two problems of conventional concrete, improving concrete in the fresh state eliminating the need for vibrations with its high workability, in addition to increasing the ductility and toughness of the concrete due to the inclusion of the fiber. This material can be used in structures with high reinforcement rates, allowing a more efficient concreting and, at the same time, reducing the reinforcement rate. There are several ways to dosage self-compacting concrete, one of the most accurate being is the Compressible Packaging Method (CPM). This method is based on the solution of packaging dry mixtures in all components used in the concrete dosing. However, FRSCC dosage studies using CPM are still incipient. There are some ways to consider the effect of fiber on concrete, one approach assesses the effect of fibers considering a perturbed volume that they can generate in the mixture, another simulates the effect of fibers through the concept of equivalent diameter. This work examined these two methods, comparing the compactness results obtained experimentally for 3 types of steel fiber and one synthetic fiber, seeking to evaluate what is the most efficient way to consider the effect of the fibers on the CPM. The results showed that the two approaches can be used for dosage of the FRSCC, however, for larger fiber volumes (0.09% for synthetic fiber and 2% for steel fiber), the second mentioned approach presented the best results.

Keywords: Fiber reinforced self-compacting concrete · Compressible packing method · Perturbed volume · Equivalent diameter

1 Introduction

Portland cement concrete is the main engineering material used by man (Gartner 2004). Which means that more and more research is carried out in order to develop technologies and variations of this material that increase its applicability, mainly due to the fact that the concrete presents some peculiarities, such as the fragile behavior and the low deformation capacity when subject to tensile stresses (Figueiredo 2011).

P. Serna et al. (Eds.): BEFIB 2021, RILEM Bookseries 36, pp. 24–34, 2022.
https://doi.org/10.1007/978-3-030-83719-8_3

Fiber Reinforced Self-Compacting Concrete (FRSCC) is a variation of conventional concrete. It brings with it the benefits of fiber reinforced concrete such as the gain in toughness – and the ability to resist residual tensile stresses-, post-cracking behavior combined with the characteristics of self-compacting concrete, such as workability, the ability to fil forms and to overcome restrictions without segregating and eliminating the use of vibration. Figueiredo (2011) points out the main applications for FRSCC: industrial floors, tunnel lining and various precast elements. This concrete may also be used in partial replacement of the cross-cut and punch armor in beams or even in total replacement in small thickness structures (Barros et al. 2009; Lameiras et al. 2013). Many researches have been carried out on their resistance to fire (Varona et al. 2018; Park et al. 2019; Serafine et al. 2019; Wu et al. 2019; Sadrmomtazi et al. 2020) and the action of freeze-thaw (Alsaif et al. 2019; Mak and Fam 2019).

Among the various methods for FRSCC, De Larrard (1999) presents the Compressible Packaging Method (CPM). In this, De Larrard (1999) has sought the optimization of granular mixtures aiming at the maximum possible compactness coupled to a set of models of the concrete behavior in the fresh and hardened states. He determined equations that make it possible to correlate the volumetric fraction of the components with the desired rheological behavior and compressive strength of concrete.

The CPM, as well as others dosing methods, was adapted from studies on self-compacting concrete to be used in the FRSCC dosage (Grünewald 2004; Rambo 2012; Grabrois 2012). To consider the effect of the fibers on the mixture and compacity of structure, there are two approaches. One of them proposed by De Larrard (1999), who considers the effect of the fiber through a disturbed zone, where the effect of the fibers is measured as a disturbance in the compactness of the concrete aggregates. And the other, proposed by Yu et al. (1993), evaluates fiber in mixture through the concept of an equivalent diameter, where fiber is considered as a sphere that has the same surface area of fiber.

However, studies developed by Grunewald (2004) comparing the two methods, through several variations, showed the need for complementary studies that can help to define between the two methods the most accurate way of evaluating this effect on the compactness of the mixture. Specially to assess whether methodologies are also applicable for flexible synthetic fibers. Therefore, this work was developed, which aims to contribute to the choice of the best approach to consider the effect of fibers in the FRSCC dosage using the Compressible Packaging Method (CPM).

2   Materials and Methods

Three types of steel fibers were used, all with hooks at the ends, tensile strength of 1100 MPa and Elasticity Module of 210 GPa. The specific weight of these fibers is 7.85 g/cm³. A corrugated synthetic fiber from a polyolefin blend with high resistance to alkali was also used, which shows tensile strength ranging from 650 to 760 MPa, an elastic module of 4.8 GPa, specific weight of 0.96 g/cm³. The physical characteristics of the fibers studied are shown in Table 1.

Table 1. Physical characteristics of the fibers used in the research

Fiber type	Diameter (mm)	Length (mm)	Aspect ratio (length/diameter)
MaccaferriFS3N (FAC-NA-33/44)	0.75	33	44
Maccaferri FS7 (FAC-NA-33/60)	0.55	33	60
Maccaferri FF3 (FAC-NA-50/67)	0.75	50	67
Durasteel Sintética (60/60)	1.00	60	60

To evaluate the effect of adding fibers to the CPM, vibration and compression tests were performed with three types of steel fibers and one type of synthetic fiber. These are: the metallic fibers of Maccaferri FF3, FS7 and FS3N, in the contents of 1%, 2% and 4%; and Durasteel synthetic fiber, in the contents of 0.3%, 0.6% and 0.9%, mixed with two granulometric bands of gravel and one of medium sand, being the materials retained between the sieves 4.75/6.3 mm, 6.3/9.5 mm and 0.6/1.18 mm, respectively. The nomenclature adopted to identify the aggregate class in the experiments was based on the one used by De Larrard (1999), where he adopted the letter M proceeded by a number that refers to the average diameter of the grain size class in millimeters.

For the nomenclature of the fibers, the initials of the material were adopted, with ST referring to steel fibers and the initials SY to synthetic fibers, followed by the ratio between the length and the aspect ratio.

2.1 Test for Determining the Experimental Compactness of Grains

The time to perform the test was 2 min. For all mixing ratios, at least two repetitions of the test were performed. When the covariance value of the two tests deviated more than 1.25% from the average the procedure was repeated in order to minimize the test errors and to confirm the possibility of repeatability of the test. The test protocol followed the subsequent steps:

1. Weigh 2000 g of dry material to be tested;
2. Subsequently, homogenization of the binary mixtures is carried out, which was carried out in a standardized manner, for a period of 1 min;
3. With the cylinder properly fixed on the vibration table, the material is inserted into the cylinder, so that all the material has been dumped inside it;
4. Bearing in mind that the material in the cylinder is slightly uneven, it is necessary to manually spread this material just so that it becomes more leveled for later use of the weight of steel;
5. The steel weight is placed above the material in the cylinder;
6. Turn on the vibration table for a period of 2 min;
7. After switching off the vibration table, 4 measurements are taken of the difference in height of the steel weight in relation to the edge of the cylinder. It is worth mentioning that these measurements were taken at fixes points of all tests. The average of these heights is used to calculate the experimental compactness;
8. Finally, the cylinder is removed from the vibration table and the material is discarded.

After carrying out this protocol, the final height of the material was determined and with that height, using Eq. (1), the experimental compacities were calculated.

$$\phi = \frac{4M_s}{\pi \cdot D_c^2 \cdot h \cdot p_s} \tag{1}$$

Where: ϕ it the experimental compactness; p_s is the density of the material; M_s is the mass of the dry material; D_c is the internal diameter of the cylinder; h is the final height of the layer of the compacted material.

2.2 Calculation of Compactness Using the De Larrard (1999) Method

De Larrard (1999) proposed a methodology for considering the effect of this disturbance, taking into account the existence of a universal coefficient k_f, which is the ratio between the distance of propagation and the size of the particles. In the case of a sufficiently short fiber, it can fit in an interstice of coarse grains without disturbing the natural packaging. The total length of the fiber is not able to disturb the aggregate packaging.

The parameter d_F is the diameter of the cylindrical fiber, k_F is a coefficient that relates the disturbed volume of the fiber to the maximum diameter of the aggregate, d, and l_F is the fiber length. Thus, the average virtual compactness affected by the inclusion of fibers can be determined using Eq. (2):

$$\overline{\beta_i} = \left(1 - \phi_f - N_{sf} \cdot v_p\right)\beta \tag{2}$$

Where:

$\overline{\beta_i}$ is the average virtual compactness of β, in a mixture affected by the container wall or by the inclusion of fibers;

ϕ_f it the percentage of fibers in the granular skeleton;

N_{sf} is the number of fibers per unit of volume;

v_p is the disturbed volume of a fiber;

β is the undisturbed virtual compactness;

2.3 Calculation of Compactness Using the Yu et al. (1993) Method

The method proposed by Yu et al. (1993) makes it possible to include irregular particle in the calculation of the compactness of a mixture by calculating an equivalent diameter. This method takes into account several aspects such as the shape, the grain size and the compaction energy of the mixture (Yu and Zou 1998). The proposal is to find a diameter of a fictitious sphere that would represent the fibers in the mixture. The Eq. (3) was proposed by Yu et al. (1993) for a cylindrical particle such as some types of fibers:

$$d_p = \left(3.1781 - 3.6821 \cdot \frac{1}{\Psi} + 1.5040 \cdot \frac{1}{\Psi^2}\right) d_v \tag{3}$$

Onde:

Ψ is the sphericity;

d_v is the diameter of the equivalent sphere.

3   Results and Discussion

3.1   Compactness Between the Approaches Proposed by Yu et al. (1993) and by De Larrard (1999)

3.1.1   Steel Fibers

The results obtained by the two approaches for the M8 aggregate, also with the experimental results for the steel fibers are shown in Fig. 1.

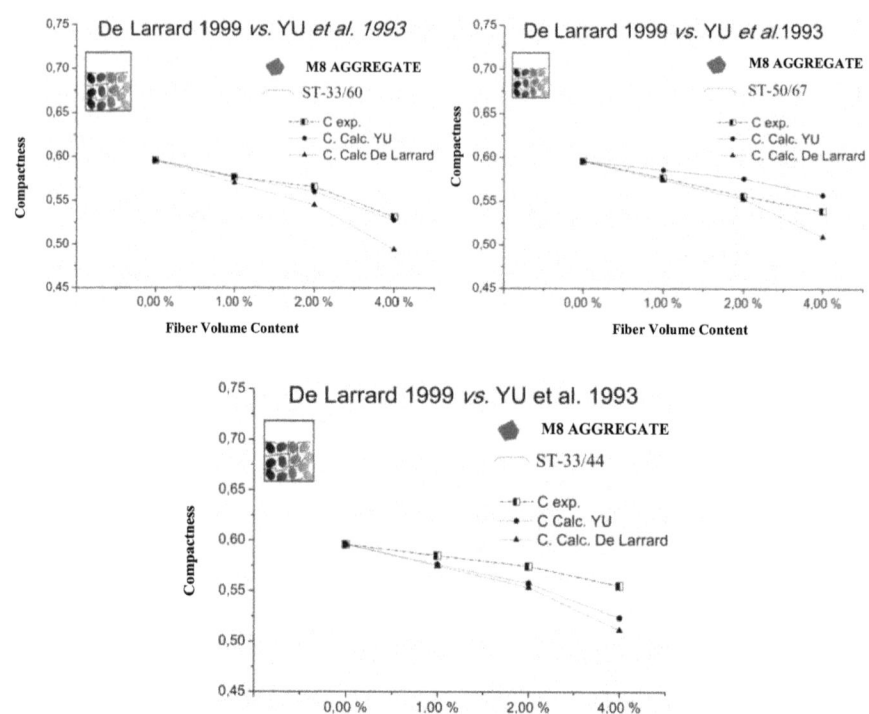

Fig. 1. Experimental Compactness and Yu et al. (1993) and De Larrard (1999) methodologies in mixtures of the grain size range M8 and the fibers ST-33/60, ST-50/67 and ST-33/44.

It is possible to observe that, for the ST-33/60 fiber and the M8 aggregate, the results obtained adopting the equivalent diameter were more satisfactory, with average errors below 2%, slightly increasing from 2% of fibers. The results obtained with the methodology proposed by De Larrard (1999) also showed small errors, specially up to

1% of fiber volume. As of 2% fiber, errors adopting the concept of disturbed volume increased, bit were still significantly small, below 3%.

Table 2. Experimental compactness and determined by the methodologies proposed by Yu et al. (1993) and De Larrard (1999), in mixtures of the grain size range M8 and the fibers ST-33/60, ST-50/67 and ST-33/44.

Fiber Volume content (%)	Experimental Compactness				De Larrard (1999) Compactness			Yu et al. (1993) Compactness	
	ST-33/60	ST-50/67	ST-33/44	ST-33/60	ST-50/67	ST-33/44	ST-33/60	ST-50/67	ST-33/44
0.00	0.5957	0.5957	0.5957	0.5957	0.5957	0.5957	0.5957	0.5957	0.5957
1.00	0.5768	0.5767	0.5847	0.5703	0.5742	0.5746	0.5774	0.5858	0.5761
2.00	0.5657	0.5563	0.5744	0.5449	0.5527	0.5535	0.5600	0.5762	0.5576
4.00	0.5315	0.5393	0.5551	0.4942	0.5096	0.5114	0.5278	0.5574	0.5235

As shown in Table 2, the results presented by the mixtures between the ST-50/67 and ST-33/44 fibers using the methodologies of Yu et al. (1993) and De Larrard (1999) were very close up until 2% fiber volume. With 4%, the proposal of De Larrard (1999) presented in its results an increase in the error having as reference the experimental data.

For mixtures comprising the M6 aggregate range and the ST-50/67 and ST-33/60 fibers, the two tested methodologies showed very close values up to the 2% fiber content, in which case with the Yu et al. (1993) approach the errors were accentuated for the fiber volume greater than this critical value, as shown in Table 3 and Fig. 2. The average errors for De Larrard (1999) and Yu et al. (1993) approaches were, respectively, 1.74% and 4.4%.

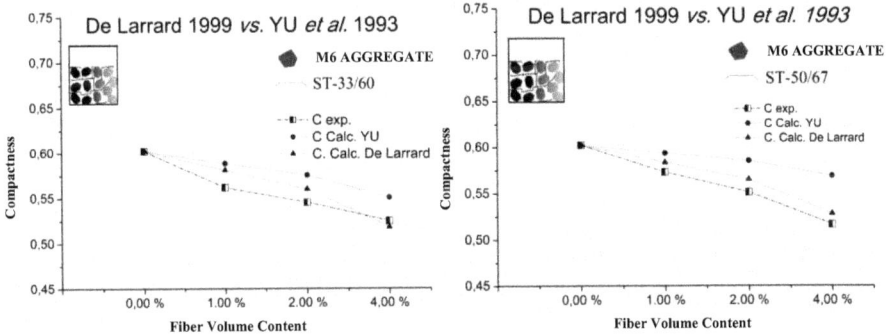

Fig. 2. Experimental compactness and Yu et al. (1993) and De Larrard (1999) methodologies in mixtures of the grain size range M6 and the fibers ST-33/60 and ST-50/67.

Table 3. Experimental compactness and determined by the methodologies proposed by Yu et al. (1993) and De Larrard (1999) in mixtures of the grain size range M6 and the fibers ST-33/60 and ST-50/67.

Fiber Volume content (%)	Experimental Compactness		De Larrard (1999) Compactness		Yu et al. (1993) Compactness	
	ST-33/60	ST-50/67	ST-33/60	ST-50/67	ST-33/60	ST-50/67
0.00	0.6021	0.6021	0.6021	0.6021	0.6021	0.6021
1.00	0.5619	0.5731	0.5811	0.5836	0.5886	0.5936
2.00	0.5456	0.5514	0.5602	0.5652	0.5756	0.5853
4.00	0.5249	0.5165	0.5183	0.5282	0.5508	0.5689

In the binary mixtures between the fine aggregate and the ST-50/67 and ST-33/60 fibers, there was an increase in the error compared to the experimental results, specially for fiber contents greater than 2%. However, the smallest errors were found adopting the concept of equivalent diameter proposed by Yu et al. (1993), as show in Table 4 and Fig. 3.

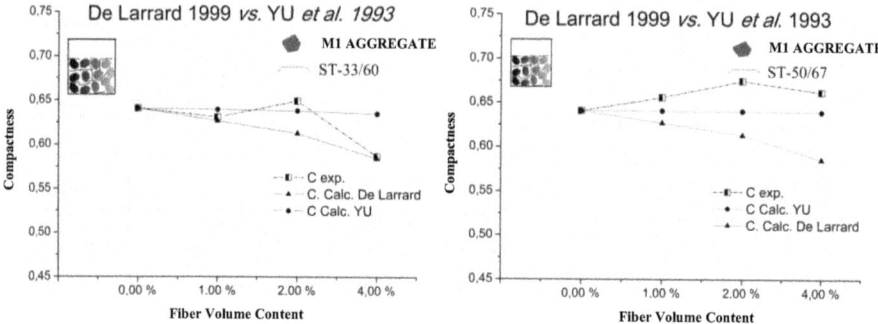

Fig. 3. Experimental compactness and Yu et al. (1993) and De Larrard (1999) methodologies in mixtures of the grain size range M1 and the fibers ST-33/60 and ST-50/67.

Table 4. Experimental compactness and determined by the methodologies proposed by Yu et al. (1993) and De Larrard (1999) in mixtures of the grain size range M1 and the fibers ST-33/60 and ST-50/67.

Fiber Content (%)	Experimental Compactness		De Larrard (1999) Compactness		Yu et al. (1993) Compactness	
	ST-33/60	ST-50/67	ST-33/60	ST-50/67	ST-33/60	ST-50/67
0.00	0.6407	0.6407	0.6407	0.6407	0.6407	0.6407
1.00	0.6307	0.6558	0.6266	0.6269	0.6393	0.6405
2.00	0.6492	0.6749	0.6126	0.6131	0.6379	0.6402
4.00	0.5866	0.6619	0.5845	0.5856	0.6347	0.6392

Based on the results found in the seven mixtures, specially taking into account the usual volumes of fiber addition in the FRSCC, where the volume of 2% fibers is generally not exceeded, the two methodologies proved to be efficient for considering the effect of fibers on compactness of binary mixtures. The results obtained through the proposal of Yu et al. (1993) and De Larrard (1999) presented values very close to the experimental values found for contents of up to 2% of fiber. Nevertheless, slightly smaller errors were found with the methodology proposed by Yu et al. (1993). With an average error below 3%, while De Larrard (1999) proposal found an average error of almost 4%, as also found by Grunewald (2004) in his research.

3.1.2   Synthetic Durasteel Fiber

It is possible to observe in Fig. 4 and Table 5 that for the SY-60/60 fiber and the M8 aggregate the two approaches showed values close to the experimental. The results obtained by adopting the equivalent diameter were more satisfactory, with an average error below 2%. The results obtained with the methodology proposed by De Larrard (1999) also showed small errors, of around 5%, underestimating the results found for compactness, but it could still be used in the dosages.

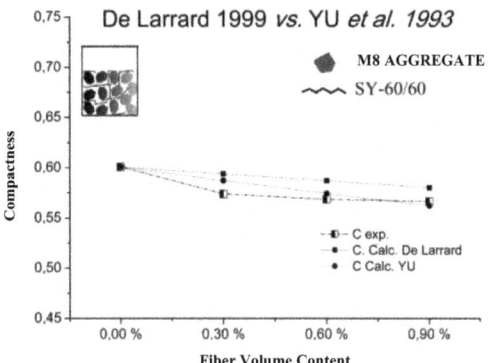

Fig. 4. Experimental compactness and Yu et al. (1993) and De Larrard (1999) methodologies in mixtures of the grain size range M8 and the fibers SY-60/60.

Table 5. Experimental compactness and determined by the methodologies proposed by Yu et al. (1993) and De Larrard (1999) in mixtures of the grain size range M8 and the fibers SY-60/60.

M8			
Fiber Volume content (%)	Experimental Compactness	De Larrard (1999) Compactness	Yu et al. (1993) Compactness
	SY-60/60	SY-60/60	SY-60/60
0.00	0.6009	0.6009	0.6009
0.30	0.5742	0.5999	0.5874
0.60	0.5686	0.5989	0.5744
0.90	0.5666	0.5979	0.5624

For the mixture of the SY-60/60 fiber with the M6 aggregate the results were very similar when comparing the experimental results and the approaches proposed by Yu et al. (1993) and De Larrard (1999), as Fig. 5 and Table 6 present. The average errors of the two approaches were below 1%.

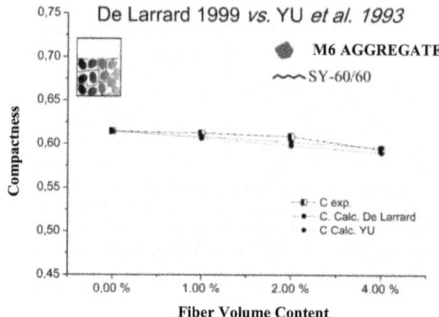

Fig. 5. Experimental compactness and Yu et al. (1993) and De Larrard (1999) methodologies in mixtures of the grain size range M6 and the fibers SY-60/60.

Table 6. Experimental compactness and determined by the methodologies proposed by YU et al. (1993) and De Larrard (1999) in mixtures of the grain size range M6 and the fibers SY-60/60.

M6			
Fiber Volume content (%)	Experimental Compactness	De Larrard (1999) Compactness	Yu et al. (1993) Compactness
	SY-60/60	SY-60/60	SY-60/60
0.00	0.6146	0.6146	0.61455
0.30	0.6125	0.6085	0.60657
0.60	0.6088	0.6023	0.59845
0.90	0.5938	0.5962	0.59001

For the mixture of Durasteel synthetic fiber with the M1 aggregate, the proposal by De Larrard (1999) presented values closer to those found experimentally, as shown in Fig. 6 and Table 7. However, the adoption of the equivalent diameter proposed by Yu et al. (1993) also proved to be applicable, given that the average errors found by the two approaches were below 1%.

For the SY-60/60 fiber, the two approaches proved to be applicable, specially in 0.6% fiber content, which is the maximum dosage indicated by the manufacturer. Even though it was stated by De Larrard (1999) that his approach would not be as accurate for flexible fibers, good results were found.

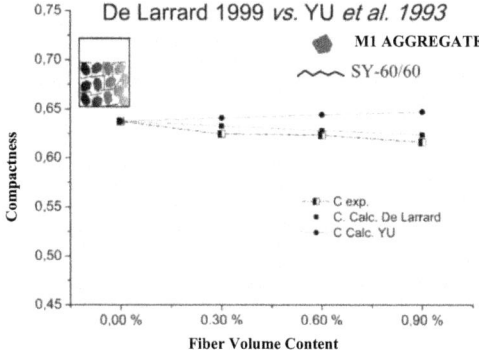

Fig. 6. Experimental compactness and Yu et al. (1993) and De Larrard (1999) methodologies in mixtures of the grain size range M1 and the fibers SY-60/60.

Table 7. Experimental compactness and determined by the methodologies proposed by Yu et al. (1993) and De Larrard (1999) in mixtures of the grain size range M1 and the fibers SY-60/60.

M1			
Fiber Volume content (%)	Experimental Compactness	De Larrard (1999) Compactness	Yu et al. (1993) Compactness
	SY-60/60	SY-60/60	SY-60/60
0.00	0.6375	0.6375	0.6375
0.30	0.6244	0.6328	0.6410
0.60	0.6233	0.6281	0.6443
0.90	0.6160	0.6234	0.6471

4 Conclusions

- It was confirmed that the two methodologies evaluated, be the fiber effect considering a perturbed volume be the fiber effect taking in account an equivalent diameter, can be used for dosages with CPM, presenting average errors below 3% when compared with the results obtained experimentally in the tests with the steel fibers.
- The two methodologies can also be applied to synthetic fibers and showed very close compactness values.
- Despite very close values, with an average error of 2.78% for the perturbed volume methodology and 2.94% for the equivalent diameter method. The first proposal showed lower error for steel fibers.
- The methodology that proposes the concept of equivalent diameter presented the lowest errors for synthetic fiber.

References

Gartner, E.M.: Industrially interesting approaches to "low-CO_2" cements. Cem. Concr. Res. **34** (9), 1489–1498 (2004)

Figueiredo, A.D.D.: Concreto reforçado com fibras. In: Tese (Livre Docência). Escola Politécninca, p. 256. Universidade de São Paulo, São Paulo (2011)

Barros, A.R., Gomes, P.C.C., Barboza, A.S.R.: Avaliação do Comportamento de Vigas de Concreto Auto-Adensável Reforçado com Fibras de Aço (2009)

Lameiras, R., Barros, J.A., Azenha, M.: Development of sandwich panels combining fibre reinforced concrete layers and fibre reinforced polymer connectors. Part I: Conception and pull-out tests. Compos. Struct. **105**, 446–459 (2013). ISSN 0263-8223

Varona, F.B., Baeza, F.J., Bru, D., Ivorra, S.: Influence of high temperature on the mechanical properties of hybrid fiber reinforced normal and high strength concrete. Constr. Build. Mater. **159**, 73–82 (2018). https://doi.org/10.1016/j.conbuildmat.2017.10.129

Park, J.-J., Yoo, D.-Y., Kim, S., Kim, S.-W.: Benefits of synthetic fibers on the residual mechanical performance of Uhpfrc after exposure to Iso standard fire. Cement Concr. Compos. **104**, 103401 (2019). https://doi.org/10.1016/j.cemconcomp.2019.103401

Serafine, R., et al.: Influence of fire on temperature gradient and physical-mechanical properties of macro-synthetic fiber reinforced concrete for tunnel linings. Constr. Build. Mater. **214**, 254–268 (2019)

Wu, L., Lu, Z., Zhuang, C., Chen, Y., Hu, R.: Mechanical properties of nano SiO2 and carbon fiber reinforced concrete after exposure to high temperatures. Materials (Basel) (2019). https://doi.org/10.3390/ma12223773

Sadrmomtazi, A., Gashti, S.H., Tahmouresi, B.: Residual strength and microstructure of fiber reinforced self-compacting concrete exposed to high temperatures. Constr. Build. Mater. **230**, 116969 (2020). https://doi.org/10.1016/j.conbuildmat.2019.116969

Alsaif, A., Bernal, S.A., Guadagninia, M., Pilakoutas, K.: Freeze-thaw resistance of steel fibre reinforced rubberised concrete. Constr. Build. Mater. **195**, 450–458 (2019)

Mak, K., Fam, A.: Freeze-thaw cycling effect on tensile properties of unidirectional flax fiber reinforced polymers. Compos. B Eng. **174**, 106960 (2019). https://doi.org/10.1016/j.compositesb.2019.106960

De Larrard, F.: Concrete Mixture Proportioning: A Scientific Approach. CRC Press, Boca Raton (1999). ISBN 1482272059

Grünewald, S.: Performance-Based Design of Self-Compacting Fibre Reinforced Concrete. Delft University of Technology, TU Delft (2004)

Rambo, D.: Concretos autoadensáveis reforçados com fibras de aço híbridas: aspectos materiais e estruturais. In: Dissertação (Mestrado). Programa de Pós-graduação em Engenharia Civil, p. 185. Universidade Federal do Rio de Janeiro, Rio de Janeiro (2012)

Grabois, T.M.: Desenvolvimento e Caracterização Experimental de Concretos Leves Autoadensáveis Reforçados com Fibras de Sisal e Aço/Thiago Melo Grabois. UFRJ/COPPE, Rio de Janeiro (2012)

Yu, A.B., Standish, N., Mclean, A.: Porosity calculation of binary mixtures of nonspherical particles. J. Am. Ceram. Soc. **76**(11), 2813–2816 (1993). ISSN 1551-2916

Yu, A., Zou, R.: Prediction of the porosity of particle mixtures. KONA Powder Part. J. **16**(0), 68–81 (1998). ISSN 0288-4534

Influence of Steel Fibers on Damage Induced by Alkali-Silica Reaction of Concrete with Reactive Sudbury Aggregates

Stefano Giuseppe Mantelli[1], Daman K. Panesar[2],
and Fausto Minelli[1](✉)

[1] Department of Civil, Environmental, Architectural Engineering
and Mathematics, University of Brescia, Brescia, Italy
fausto.minelli@unibs.it
[2] Department of Civil and Mineral Engineering,
University of Toronto, Toronto, Canada

Abstract. The main consequence of Alkali-Silica Reaction is the formation of cracks due to expansion, which can lead to a reduction in resistance capacity of the section of the structural element. The addition of steel fibers (micro fibers and Macro fibers) in the concrete, leads to considerable changes in the material properties in terms of its mechanical and durability performance, even under operating loads. This involves reducing or eliminating the formation of cracks, thus limiting the penetration of corrosive agents. In order to understand the behavior of steel fibers on concrete subjected to ASR due to the presence of highly reactive coarse aggregate (Sudbury aggregate), a 4-month accelerated laboratory testing campaign was conducted with four mix designs (plain concrete, 0.5% Macro fiber, 1.0% Macro fiber, 1.0% Hybrid: 0.5% Macro and 0.5% micro fiber). To speed up the ASR reaction process the samples were exposed to 50 °C temperature and 100% R.H. in a moisture chamber for the entire maturation period. Prisms, cubes, cylinders and beams were cast and tested for: longitudinal expansion, dynamic elastic modulus, damage rating index, compressive strength, static modulus of elasticity and flexural tensile strength. The main result was that the 1.0% Macro mix design is the one that most mitigates the formation of cracks.

Keywords: ASR · SFRC · Sudbury aggregates · Alkali-Silica Reaction · Fiber reinforced concrete · Hybrid fibers · Damage rating index

1 Introduction

Bridges, roads, dams and power plants are just a few examples of engineering works designed to last for decades while remaining exposed to weather and atmospheric agents, ensuring functionality and structural safety throughout their useful life. However, there are forms of degradation that reduce the durability of the structure. Unfortunately, some types of concrete degradation cannot yet be completely mitigated and controlled after the construction of the structure: one of these is the Alkali-Silica Reaction (ASR). The presence of the ASR phenomenon in concrete depends on the particular amorphous structure of the silica in the fine or coarse aggregates.

© RILEM 2022
P. Serna et al. (Eds.): BEFIB 2021, RILEM Bookseries 36, pp. 35–48, 2022.
https://doi.org/10.1007/978-3-030-83719-8_4

Furthermore, the silica with amorphous structure reacts chemically with the alkaline elements contained in the cement (Na, K) and the presence of moisture, leading to the formation of a hygroscopic gel which expands and generates micro-cracks in the concrete, reducing the durability of the structure [1]. In the past, for many years, concrete structures were built without consideration to the type of aggregates that would be used in the mixture. Since the mid-19th century, it became important to know the characteristics of each component of the mixture before building the structure. The retrieval of the materials, in particular the aggregates for the construction of the con- crete, is bound by the location of the construction site: in some cases, it is necessary to use reactive aggregates and, therefore, there is a need to applying preventive measures in order to reduce or even eliminate the degradation process. Regulations do in fact allow the use of reactive aggregates but with some limitations. There are methods of "chemical" intervention to mitigate the ASR (e.g., use of supplementary cementing material such as fly ash, slag cement, silica fume; as well as other additives such as lithium). The purpose of this study is to evaluate a "physical" type of intervention, which could be used in place of or in parallel with the chemical mitigation measures, by identifying the effectiveness of using macro and micro steel fibers to reduce or mitigate the expansion of concrete due to ASR.

Some studies have been carried out on the application of fibers in the mortar subjected to ASR [2, 3]. For example, Yazıcı [3] demonstrates how fibers reduce the degradation process that the ASR causes in the mechanical properties of the material (flexural strength, compressive strength and toughness). To control cracking processes, incorpo- rating fibers into the concrete leads to an increase in material toughness and structural durability. Other studies on cracked fiber reinforce concrete (FRC) under sustained loading showed that fibers modify crack patterns, with narrower and closely spaced cracks [4, 5]. It has also been found that as the content of fibers increases, the benefits to the material properties are greater [6]. Some studies have focused on the expansion due to the internal swelling caused by ASR and how steel fibers can influence this behaviour. The use of steel fibers in concrete subject to ASR may lead to less expansion and lower adverse effect on mechanical property, based on the percentage of fibers used in the mixture. Giaccio et al. [7] provided an analysis of the contribution that fibers (40 kg/m^3 macro steel fibers and synthetic 3 kg/m^3 macro and 1 kg/m^3 micro fibers) have on concrete affected by different levels of degradation by ASR. Panesar and Gautam [8] have studied triaxial expansion of ASR-affected SFRC to understand the effectiveness of steel fibers reinforcement on ASR expansion. As coarse aggregates they used Spratt ("ex- tremely reactive" aggregates [9] - Annex B.3.4). In addition, studies were conducted at a microscopic level using the damage rating index (DRI) methodology [10] to investigate in more detail the effect of steel fibers in concrete subjected to ASR. Regarding the type of fibers used, Bektaş et al. [11] propose micro fibers as a method of mitigating ASR damage because they are used primarily for controlling early age cracking [12].

Therefore, effectiveness of fibers in reducing the extent of ASR cracking in concrete has not been adequately understood. There are still many questions awaiting an answer. Several studies have been done on the effect of steel fibers (mainly micro-fibers) in the mortar subject to ASR, but with regard to concrete affected by ASR, many behaviours are still poorly understood. To date, the research carried out on the contribution that FRC can make to the concrete subjected to ASR [8, 10, 13, 14] has been evaluated mainly by testing "extremely reactive" coarse aggregates (e.g., Spratt). In order to better understand the influence of fibers on ASR concrete cracking, this study compares different amounts of steel fibers in two types (macro fibers and micro fibers). Moreover, it compares SFRCs subjected to ASR due to the presence of "highly reactive" coarse aggregates (Sudbury).

2 Materials and Methods

In this study two types of steel fibers were used: macro, (Length $[L] = 30\,\text{mm}$, Diameter $[D] = 0.35\,\text{mm}$, $L/D = 86$, *hooked shape*) and micro (Length $[L] = 13\,\text{mm}$, Diameter $[D] = 0.20\,\text{mm}$, $L/D = 65$, *straight shape*). As shown in Table 1, four mix designs were analyzed, mainly differentiated by the content and type of steel fibers. The matrix was kept the same for all four cases, except for the quantity of fibers replaced with an equivalent volume of coarse aggregates: the bulk solids volume was equal to $0.3\ \text{m}^3$ per cubic meter of concrete. The ratios of cement and fine aggregate in the mixture ensured adequate surface coating for fibers. Four types of concrete were tested: (i) reference [R] mix design without fibers (0.0% Rf), (ii) macro [M] steel fibers (0.5% Mf) (iii) macro [M] steel fibers (1.0% Mf) and, (iv) hybrid (H) dosage (1.0% Hf which contains 0.5% of macro and 0.5% of micro steel fibers). High alkali general use (GU) cement was used with a total alkali content of 0.93% Na_2O equivalent by mass of cement. The fine aggregate used was non-reactive sand from Aberfoyle Pit, CBM Aggregates, Cambridge, Ontario. The particular type of reactive coarse aggregates used in this research is called Sudbury, coming from the homonymous region of Ontario from which it is extracted. Sudbury is a greywacke-argillite gravel, crushed aggregate from a quarry in Ontario, Canada, provided by Ministry of Transportation of Ontario (MTO) for this research. The choice of the maximum aggregate size is 4.75–9.50 mm was selected to improve mixing with fibers [13]. Sudbury aggregates are classified according to Canadian legislation ([9] - Annex B.3.4) as "highly reactive" (in comparison, the Spratt aggregates are classified as "extremely reactive"); the degree of reactivity of the aggregate is determined using the concrete prism test and corresponds to: (a) 0.040% $\leq$ one year expansion < 0.120%: moderately reactive; (b) 0.120% $\leq$ one year expansion < 0.260%: highly reactive; and (c) one year expansion $\geq$ 0.260%: extremely reactive. The alkali level of the concrete mixes was increased to 5.53 kg/m^3 Na_2O equivalent of concrete by adding NaOH (sodium hydroxide) pellets to water prior to concrete mixing in order to accelerate the chemical reaction, according to ASTM C1293 [15]. The water/cement ratio used was 0.45, in accordance with ASTM C1293 [15]. To improve the workability of the concrete containing steel fibers, superplasticizer was used.

Table 1. Mix designs of concrete.

kg/m^3			Mix design			
			0.0% Rf [Plain concrete]	0.5% Mf	1.0% Mf	1.0% Hf [M0.5% + m0.5%]
Materials	Steel fibers	MACRO	–	39.25	78.50	39.25
		Micro	–	–	–	39.25
	Cement	w/c = 0.45	432.0	432.0	432.0	432.0
	Water		194.4	194.4	194.4	194.4
	Alkali pellet		1.95	1.95	1.95	1.95
	Aggregate coarse Sudbury (4.75–9.50 mm)		811.8	798.3	784.7	784.7
	Aggregate fine non-reactive sand		972.0	972.0	972.0	972.0
		TOT [kg]	2412.15	2437.90	2463.55	2463.55
Superplasticizer		l/m^3	1.0	2.0	4.0	4.0

During the casting phase, for each mix design the parameters shown in Table 2 were measured: temperature, slump, density and air content, according to the standards of the various tests required by the ASTM [16–19].

Each mix design was adjusted based on relative density (specific gravity) and absorption (ASTM C127 [20] for coarse aggregates and ASTM C128 [21] for fine aggregates) and the total evaporable moisture content of aggregate by drying, ASTM C566 [22]. Once the casting was completed and the surfaces were finished, the samples were covered with a plastic sheet, then, after four hours, they were covered with plastic and then a cloth of burlap soaked in water. To accelerate the chemical reaction process for ASR gel formation, the samples were placed inside an acceleration chamber at a constant temperature of 50 °C and 100% relative humidity, 1-day after casting, for the duration of the study. These climatic conditions increase the rate of expansion by approximately three times of that compared to 38 °C [14]. The samples were cast into molds following the procedures in ASTM C192 [23] and EN 14651 [24] based on their geometry. For each of the four mix designs the following number and geometry of specimens were cast: 12 prisms (h = 75 mm, b = 75 mm, l = 285 mm), 12 cubes (h = 150 mm, b = 150 mm, l = 150 mm), and 12 cylinders (h = 200 mm, d = 100 mm). In addition, only for mix designs with fibers, 6 beams (h = 150 mm, b = 150 mm, l = 600 mm) were prepared. Before conducting the tests, the samples were removed from the moisture chamber, placed in sealed containers and acclimatized at room temperature (23 ± 1 °C) for 16–20 h. Table 3 shows the experimental test plan for one mix design, and it indicates the code the test was conducted in accordance with, the test age and the type and number of specimens. All of the specimens for one mix design were cast on the same day. On the day of demolding, 24 h after casting, the mass and dimensions of all the samples were measured.

Table 2. Fresh property tests.

Code	TEST	–	0.0% Rf	0.5% Mf	1.0% Mf	1.0% Hf
ASTM C1064	Temperature	[°C]	31.0	29.2	32	32
ASTM C143	Slump	[mm]	40	176	151	146
ASTM C138	Air content	[%]	2.4	2.1	2.5	2.4
ASTM C231	Density	[kg/m^3]	2422.0	2380.9	2433.3	2448.2

The geometry of concrete prisms is in accordance with the 'concrete prism test' as define in ASTM C1293 [15]. The prisms were tested not only for longitudinal expansion, but also for dynamic elastic modulus and the damage rating index (DRI). The longitudinal expansion measurements [15] were taken at day 1, 7, 14, 28, 56, 90, 110, 120 and 150. The dynamic elastic modulus [25] was measured at 1, 28, 56, 90, 110, 120 and 150 days. After the two non-destructive tests (expansion and dynamic elastic modulus), the specimens were transversely cut in the middle to obtain a 2.5 cm thick slice at the ages of 28, 56, 90 and 110 days. These slices were then used to petrographically evaluate the concrete by the DRI. The methodology to evaluate the DRI was in accordance with Villeneuve et al. [26]. The purpose of the microstructural DRI examination is to identify quantitative and qualitative the type of damage that the ASR causes on the aggregate and then the paste. On the cross section of the 75 × 75 mm slice, after being polished with a manual grinder, a grid of 49 squares of 1 cm^2 each was drawn. The sample was then scanned at 3200 dpi and its computer image was analysed. For each square centimeter, seven petrographic features have been identified and weighted with the coefficients provided by Villeneuve et al. [26] shown in the Table 4.

The value of each relative petrographic feature has been summed, making an average with the value obtained from the second slice coming from the other prism and normalized on 100 cm^2. The sum of the tabulated quantity of features, will result in an overall DRI number related to a sample area of 100 cm^2.

Concrete cubes were prepared for apparent chloride diffusion test following the procedure of ASTM C1556 [27]. Three cubes for each mix design were removed from the moisture chamber at 28, 56, 90 and 110 days. Concrete cylinders were tested for compressive strength and static modulus of elasticity at 28, 56, 90 and 110 days according to ASTM C39 [28].

Beams according to EN 14651 [24] were tested under a three-point load. This is a typical test to determine the limit of proportionality (LOP) and of a set of residual flexural tensile strength values. Seven days before the test a 25 mm high notch was made at mid-span. To determine the degree of damage in the concrete caused by the ASR, observing the contribution of the presence of fibers as a function of time, it was more useful to analyze the area under the load-deflection curves, i.e. the fracture energy. Once the beams had been tested, they were separated into two halves at the notch, to count the number of fibers that worked in each section. Based on the stress distribution assumed for the calculations ([24] - Annex A), the number of counted fibers was divided into two categories: fibers that worked in the lower half of the cross mid-section and in the upper half.

Table 3. Test plan for one mix design.

Mix design × 1

0 [Cast]	1	7	14	28	56	90	110	120	150	Test	Code
12 Prisms[a]	12	12	12	12	10	8	6	4	4	Longitudinal expansion	ASTM C1293
	12	–	–	12	10	8	6	4	4	Dynamic elastic modulus	ASTM C215
	–	–	–	2	2	2	2	–	–	Damage rating index (DRI)	Villeneuve et al. (2012)
12 Cubes	–	–	–	3	3	3	3	–	–	Apparent chloride diffusion	ASTM C1556
12 Cylinders	–	–	–	3	3	3	3	–	–	Compressive strength static modulus of elasticity	ASTM C39 ASTM C469
6 Beams[b]	–	–	–	2	2	–	2	–	–	Flexural tensile strength	EN 14651

(Time [Days] spans the columns: 0 [Cast], 1, 7, 14, 28, 56, 90, 110, 120, 150)

[a]For the tests on Prisms carried out at 28, 56, 90 and 110 days before, non-destructive tests as Longitudinal Expansion and Dynamic Elastic Modulus were performed on the samples. Two samples per day were cut for DRI analysis.

[b]Beams were made only with the three mix designs that included fibers, not for Rf mix.

Table 4. List of petrographic features with the relative weighting, suggested by Villeneuve et al. [26] applied in this study.

Petrographic features	Weightings factors
Closed/tight cracks in coarse aggregate particle	0.25
Opened crack or network cracks in coarse aggregate particle	2
Crack or network cracks with reaction product in coarse aggregate particle	2
Debonded coarse aggregate	3
Disaggregate/corroded aggregate particle	2
Cracks in cement paste	3
Cracks with reaction product in cement paste	3

3   Results and Discussions

3.1   Longitudinal Expansion

The longitudinal expansion measurements (this is a non-destructive test) were conveniently collected at different ages, as shown in the Table 5.

As shown in Table 5, the number of samples reduced with time because in correspondence of the age of 28, 56, 90 and 110 days, two samples were used for DRI analysis. Figure 1 shows the variation in longitudinal expansion of the specimens as a function of age. The plotted values are the average of the test results performed at the respective ages and the error bars are the standard deviation. The expectation was that the plain concrete (0.0% Rf) had more expansion than the other mix designs which included the three types (0.5% Mf, 1.0% Mf, 1.0% Hf) of steel fibers composites.

Table 5. Number of samples tested at longitudinal expansion at age, for each mix design. (The underlined numbers indicate the ages to which the prisms must be tested according to ASTM C1293 [15]).

Age (Days)	1	7	14	28	56	90	110	120	150
N° of samples tested	12	12	12	12	10	8	6	4	4

However, in the first phase, up to 28 days, the expansion of 0.0% Rf is the lowest one. Only between the ages of 56 and 120 days is the expansion of plain concrete relatively greater compared to the mix designs containing steel fibers. At 120 days, the expansion levels of the different mix designs are similar: the longitudinal expansion of concrete with the classified "highly reactive" aggregates (Sudbury) [9] without fibers (0.0% Rf plain concrete) is only 3% higher than the longitudinal expansion measured in the mix design 1.0% Mf and 1.0% Hf, and greater than 9% compared to the mix design 0.5% Mf. However, in Gautam and Panesar [10] with "extremely reactive" aggregates (Spratt) [9], they had achieved a longitudinal expansion of the plain concrete 20% greater than the longitudinal expansion of the mix design with 0.65% of macro steel fibers at 100 days. In this study, the expansion after 150 days of observation is still below 0.04% for all mix designs (limit according to standard CSA A23.2-14 [9]). Having, therefore, limited expansions, the contribution of fibers is not clear yet. Comparing these results with those of Gautam and Panesar (2017) study [10], in which extremely reactive coarse aggregates (Spratt) were used, it is observed that at 90 days the M0.0-p (plain concrete) with Spratt aggregates has an expansion of 0.20% while the plain concrete made with Sudbury aggregate has an expansion of 0.035% after 120 days. Ramlochan et al. [29] showed a longitudinal expansion of plain concrete of the Spratt aggregate concrete prisms at 100 days of about 0.14%, while Sudbury aggregate concrete prisms, at the same age, experienced an expansion of less than 0.04%. Therefore, this comparison validates the longitudinal expansion measurements of specimens in this study. Based on the Student's t-test at a 95% confidence level the expansion for mix 1.0% Hf was statistically significantly greater than the expansion for mixes 0.5% Mf and 1.0% Mf at all ages.

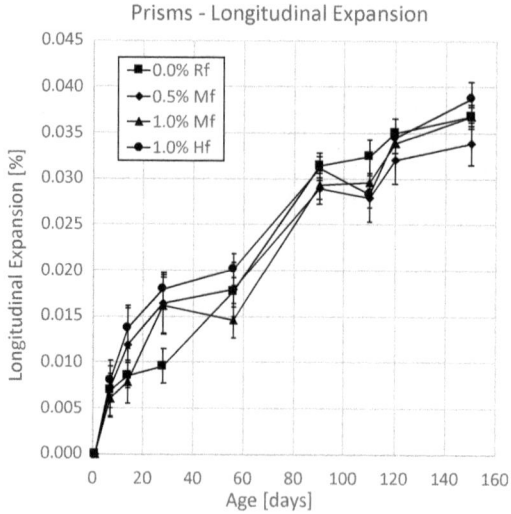

Fig. 1. Longitudinal expansion in percentage carried out on prisms at different ages.

3.2 Dynamic Elastic Modulus

The dynamic elastic modulus was determined at the age of 1, 28, 56, 90, 110, 120 and 150 days. Figure 2 shows the relative difference between the absolute values of the dynamic elastic modulus, evidencing the contribution of fibers. As ASR progresses and cracks begin to form, dynamic elastic module decreases and it is observed that:

- The mix design with 1.0% Mf shows relatively higher values than the other materials. This means that 1.0% Mf specimens have fewer cracks;
- Plain concrete samples (0.0% Rf) have the lowest dynamic elastic modulus values, so they are the most internally cracked compared to the other mix designs. The relative values of the dynamic elastic modules obtained between 28 and 150 days from 0.5% Mf and 1.0% Hf mixes are respectively 20% and 25% higher than those determined from Rf mix, while the mix design 1.0% Mf shows values 43% higher compared to plain concrete.
- The 0.5% Mf and 1.0% Hf mixes have intermediate and quite similar values, despite the different fibre dosage, which means that, probably, in the Hf mix the contribution of macro-fibers is much more significant in keeping cracks smaller than micro-fibers;
- Comparing 1.0% Mf Hf mixes, both containing 1% of fibers, it can be observed that the contribution of micro-fibers [mf] in 1.0% Hf, is insignificant in being able to eliminate or even reduce the extent of micro-cracks.

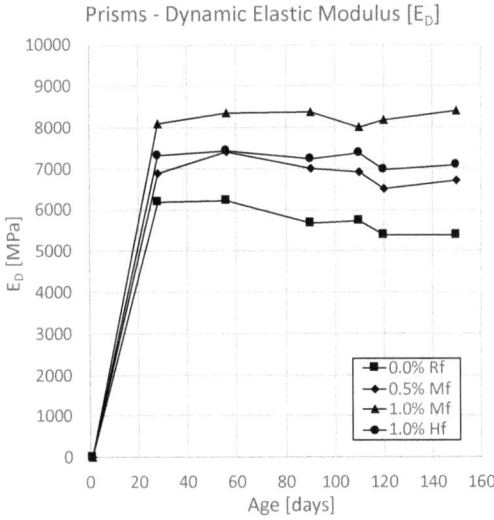

Fig. 2. Dynamic Elastic Modulus – Relative values [MPa].

3.3 Damage Rating Index (DRI)

The Damage Rating Index (DRI) is analyzed in accordance with the study of Villeneuve et al. [26]. At each age of 28, 56, 90 and 110 days, two samples were analyzed. The results represented present the average of the DRI numbers (divided into petrographic features) of two slices (49 cm^2 each one) coming from two prisms of the same age and subsequently normalized on a surface of ~ 100 cm^2. The histogram in Fig. 3 shows the values of DRI identified, for each mix design, at each age. Each bar of the histogram is divided according to the different characteristics. The values obtained at 28 and 56 days reflect the behavior found in Fig. 1 of longitudinal examination. In fact, observing the total number of DRI of each of the eight bars, it is noted that the degree of damage corresponds to the trend of the curves of longitudinal expansion. For the values obtained at 90 and 110 days a deviation is observed from the results obtained in the longitudinal expansion; however, a clear contribution of fibers can be noted. Indeed, evaluating the DRI number it is noted that the 1.0% Mf mix is the best performing to contribute to reduce damage. The portions of DRI due to the presence of cracks in the cement paste at the age of 90 and 110 days have been highlighted. The cracks in the cement paste of plain concrete in both ages (parts in purple of the bar 0.0% Rf of the histogram in Fig. 3) are greater than in the other mixes (parts in magenta of the bars of the histogram in Fig. 3). In fact, at 90 and 110 days, the presence of different fiber amounts generates a reduction from 50% to 70%, respectively, of the cracks in the cement paste. Particularly at the age of 110 days, the mix 1.0% Hf measures a 78% decrease compared to plain concrete. The cracks in the cement paste were considered the most significant petrographic features in this study since, the contribution of steel fibers acts precisely in the cement paste generating a confinement around the coarse aggregate. Furthermore, due to the durability of a structure, the cracking damage inside the cement paste is more significant than the cracking or damage of the coarse aggregate.

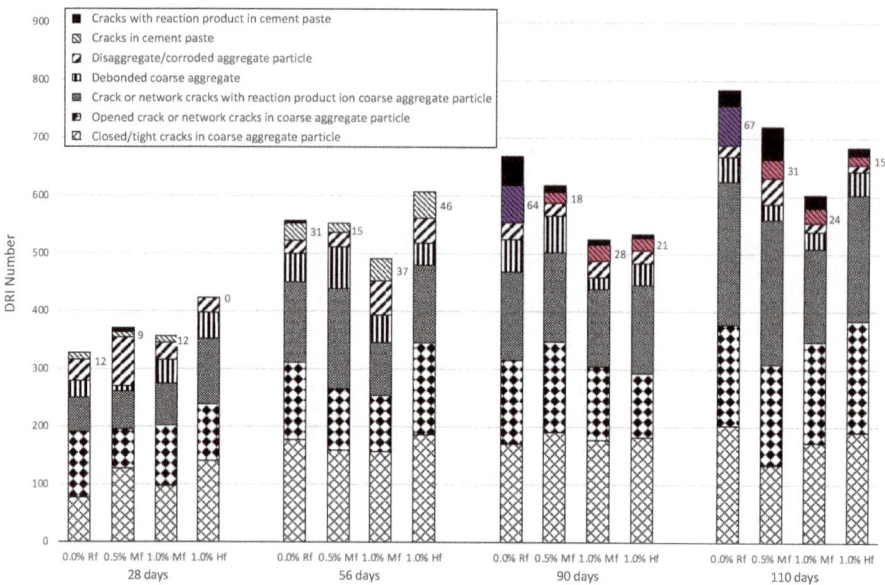

Fig. 3. Normalized DRI. The numbers next to the columns refer to the DRI number relating to the crack in cement paste.

3.4 Compressive Strength and Static Modulus of Elasticity

The compressive strength is measured in accordance with ASTM C39 [28] (Table 6) testing three samples for each age (28, 56, 90 and 110 days). In parallel, the static modulus of elasticity was also measured with a compressometer, in accordance with ASTM C469 [30]. Consequently, if the series of cylinders to be tested with compressive strength were three, to determine the static modulus of elasticity, only one sample was used for each mix design for each age; performing four cycles of loading and unloading, eliminating the first cycle, and averaging the results of the three remaining cycles. As confirmed in the literature, the compressive strength typically is insensitive to the phenomenon of ASR [31] in comparison to the static modulus of elasticity. During the compression test it was only possible to observe that the specimens with fibers reached the peak value, they do not collapse in a brittle way like the plain concrete, but owing to the confinement, the collapse is more ductile. Furthermore, the mix designs with 1.0% of fibers recorded values of compressive strength of 20% greater than the plain concrete and 0.5% Mf, thanks to the confinement of fibers.

Table 6. Compressive Strength on cylinders (d = 100mm, h = 200 mm) according to ASTM C39 [28].

Age	0.0% Rf	0.5% Mf	1.0% Mf	1.0% Hf
[days]	[MPa]	[MPa]	[MPa]	[MPa]
28	51	48	56	61
56	51	54	61	60
90	52	52	50	64
110	54	53	65	64

3.5   Flexural Tensile Strength

The flexural tensile strength (FTS) is measured in accordance with EN 14651 [24]. Two samples were tested for each mix design with fibers at each age (28, 56 and 110 days). Once the beams had been tested, they were separated into two halves at the notch, to count the number of fibers in the upper and lower half of this section. Even though the residual strengths were measured, the trend of the energy absorption as a function of time is herein considered to highlight the contribution of fibers. The energy absorption was calculated by integrating the load-deflection curve (area delimited by the curve up to a deflection value of 3.5 mm). By averaging the energy absorption values of the two beams of each mix at each age, the results shown in Fig. 4 are obtained. It should be noted that both mix designs with only macro-fibers (0.5% Mf and 1.0% Mf) have a trend that decreases in function of the age: the values obtained at 110 days are 18% for 0.5% Mf and 8% for 1.0% Mf lower than those obtained at 28 days. The hybrid mix (1.0% Hf) tends to increase, as the value at 110 days is 25% greater than that at 28 days. In addition, after a growth or decrease before the 56th day, in all mixes a plateau is reached from 56 to 110 days (in fact, the values measured at 56 days and 110 days changes only by ±2%). The following observations can be made:

- Mixes 0.5% Mf and 1.0% Mf, show a decrease in energy dissipation capacity as a function of time. This may be due to the formation of ASR, in addition to the inability of macro-fibers to mitigate damage.
- The 0.5% macro-fibers included in 1.0% Mf, compared to the 0.5% micro fibers in 1.0% Hf mix can dissipate overall 50% more energy. This is due to the presence of the hooks and the larger dimensions of the macro fibers, as expected.
- The damage of microcracks caused by the ASR over time seems to be mitigated better in 1.0% Hf compared to other mix designs. In fact, in conditions of limited expansions as verified by the longitudinal expansion test for highly reactive coarse aggregates, it is more probable that the micro fibers act before the macros.
- The mix design behavior 0.5% Mf as a function of time is similar to that recorded by 1.0% Mf. However, the energy dissipated by 0.5% Mf is close to half (60%) of that dissipated by 1.0% Mf, since the content of macro fibers present in 0.5% Mf is exactly half of that contained in 1.0% Mf.

Comparing the trend of the curves corresponding to 0.5% Mf and 1.0% Hf (differentiated only by the presence of 0.5% more micro fibers in 1.0% Hf) it can be observed how the energy difference dissipated by the two types of mix designs, is attributed to the presence of micro fibers. Furthermore, the intersection of slopes recorded between the tests carried out at 28 and 56 days, shows the progressive contribution of the micro fibers in 1.0% Hf.

Comparing the number of fibers counted in the lower half of the notched section with the dissipated energy (the peak load value), a very close relationship is observed. In particular it is noted that the contribution of micro-fibers (1.0% Hf) is not very effective with respect to the equivalent portion of macro fibers present in the design of the mixture 1.0% Mf. In fact, with the same number of fibers, 1.0% Mf dissipates 45% more energy than 1.0% Hf. This is naturally due to the different geometry and shape of the two types of fibers used.

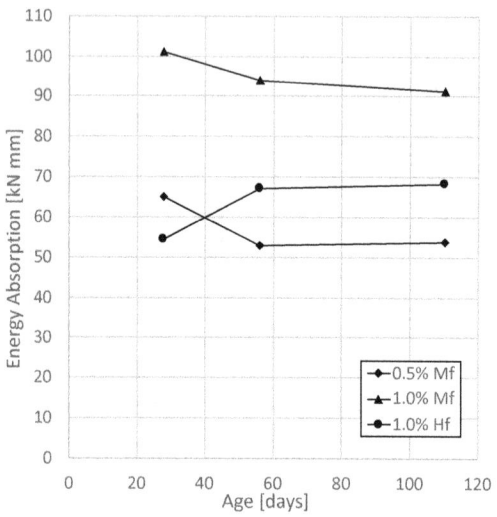

Fig. 4. Energy absorption vs. age plot.

4   Summary and Concluding Remarks

This study investigated the effects of different dosages and types of steel fibers on concrete affected by ASR, due to the presence of Sudbury highly reactive aggregates. Four mix designs of reactive concrete have been studied considering two shapes of steel fibers. At the micro-structural level, through tests such as dynamic elastic modulus and DRI, results are obtained where the 1.0% Mf mix design is more efficient in terms of crack mitigation. At a macro-structural level, with the flexural tensile strength test, even though 1.0% Mf show the highest values of energy absorption, the hybrid mix design seems to be the least affected by ASR over time. The presence of micro fibers could be, in fact, useful in the initial stages of ASR formation, when the expansion levels are still limited while the macro fibers are not engaged. Combining the results from the entire set of tests, it appears that the contribution of steel fibers starts to be significant only with dosages higher than 1.0%. Moreover, the types of coarse aggregated with high reactivity (Sudbury), compared to those with extreme reactivity (Spratt), require longer initial observation times, higher than the year, to note the contribution of fibers. Moreover, through this study it was observed that damage rating index and dynamic elastic modulus are the most sensitive tests to determine the damages due to ASR from coarse aggregates with high reactivity. As a further step, these values could be analyzed in parallel with the longitudinal expansion and the mass variation (available data but not presented herein) as a function of time to have a broader overview of the phenomenon of expansion and damage.

Acknowledgments. The authors acknowledge the support from the Ministry of Transportation Ontario (MTO), Canada for providing the highly reactive coarse Sudbury aggregates and CBM Aggregate for providing the non-reactive fine aggregates. The authors also acknowledge the staff of the University of Toronto Structural Testing Facility and Concrete Materials Laboratories for technical support throughout experimental work.

References

1. Rajabipour, F., Giannini, E., Dunant, C., Ideker, J.H., Thomas, M.D.A.: Alkali–silica reaction: current understanding of the reaction mechanisms and the knowledge gaps. Cem. Concr. Res. **76**, 130–146 (2015)
2. Turanli, L., Shomglin, K., Ostertag, C., Monteiro, P.J.: Reduction in alkali–silica expansion due to steel microfibers. Cem. Concr. Res. **31**(5), 825–827 (2001)
3. Yazıcı, H.: The effect of steel micro-fibers on ASR expansion and mechanical properties of mortars. Constr. Build. Mater. **30**, 607–615 (2012)
4. Tiberti, G., Minelli, F., Plizzari, G.: Crack control in fibrous RC elements. In: Proceedings of the 8th RILEM International Symposium on Fibre Reinforced Concrete: Challenges and Opportunities (BEFIB 2012), pp. 187–188 (2012)
5. Vasanelli, E., Micelli, F., Aiello, M., Plizzari, G.: Long term behaviour of fiber reinforced concrete beams in bending. In: Proceedings of the 8th RILEM International Symposium on Fibre Reinforced Concrete: Challenges and Opportunities (BEFIB 2012), pp. 161–162 (2012)
6. Paulík, P., Hudoba, I.: The influence of the amount of fibre reinforcement of high performance concrete. Slovak J. Civ. Eng. **2**, 1–7 (2009)
7. Giaccio, G., Bossio, M.E., Torrijos, M.C., Zerbino, R.: Contribution of fiber reinforcement in concrete affected by alkali-silica reaction. Cem. Concr. Res. **67**, 310–317 (2015)
8. Panesar, D.K., Gautam, B.P.: Triaxial expansion of plain, reinforced and fiber-reinforced ASR affected concrete. In: Proceedings of the International Conference on Structural Mechanics in Reactor Technology (SMiRT-24 BEXCO) (2017)
9. C. A23.1-14/A23.2-14: Concrete Materials and Methods of Concrete Construction/Test Methods and Standard Practices for Concrete. CSA Group (2014)
10. Gautam, B.P., Panesar, D.K.: Microscopic cracking of ASR-affected fiber-reinforced concrete. In: ACI the Concrete Convention and Exposition (2017)
11. Bektaş, F., Turanli, L., Ostertag, C.P.: New approach in mitigating damage caused by alkali–silica reaction. J. Mater. Sci. **41**(17), 5760–5763 (2006)

12. Barborak, R.: Fiber reinforced concrete (FRC) DMS-4550 tip sheet. In: Construction and Materials Tips (2011), p. 4
13. Gautam, B.P., Panesar, D.K., Sheikh, S.A., Vecchio, F.J.: Effect of coarse aggregate grading on the ASR expansion and damage of concrete. Cem. Concr. Res. **95**, 75–83 (2017)
14. Gautam, B.P., Panesar, D.K.: The effect of elevated conditioning temperature on the ASR expansion, cracking and properties of reactive Spratt aggregate concrete. Constr. Build. Mater. **140**, 310–320 (2017)
15. ASTM C1293: Standard Test Method for Determination of Length Change of Concrete Due to Alkali-Silica Reaction. ASTM International, West Conshohocken (2018)
16. ASTM C1064: Standard Test Method for Temperature of Freshly Mixed Hydraulic-Cement Concrete. ASTM International, West Conshohocken (2017)
17. ASTM C143: "Standard Test Method for Slump of Hydraulic-Cement Concrete. ASTM International, West Conshohocken (2015)
18. ASTM C138: Standard Test Method for Density (Unit Weight), Yield and Air Content (Gravimetric) of Concrete. ASTM International, West Conshohocken (2017)
19. ASTM C231: Standard Test Method for Air Content of Freshly Mixed Concrete by the Pressure Method. ASTM International, West Conshohocken (2017)
20. ASTM C127: Standard Test Method for Relative Density (Specific Gravity) and Absorption of Coarse Aggregate. ASTM International, West Conshohocken (2015)
21. ASTM C128: Standard Test Method for Relative Density (Specific Gravity) and Absorption of Fine Aggregate. ASTM International, West Conshohocken (2015)
22. ASTM C566: Standard Test Method for Total Evaporable Moisture Content of Aggregate by Drying. ASTM International, West Conshohocken (2014)
23. ASTM C192: Standard Practice for Making and Curing Concrete Test Specimens in the Laboratory. ASTM International, West Conshohocken (2016)
24. EN 14651: Test Method for Metallic Fibre Concrete – Measuring the Flexural Tensile Strength (Limit of Proportionality (LOP), Residual). European Standard (2007)
25. ASTM C215: Standard Test Method for Fundamental Transverse, Longitudinal, and Torsional Resonant Frequencies of Concrete Specimens. ASTM International, West Conshohocken (2014)
26. Villeneuve, V., Fournier, B., Duchesne, J.: Determination of the damage in concrete affected by ASR the damage rating index (DRI). In: Proceedings of the 14th International Conference on Alkali-Aggregate Reaction Concrete, p. 10 (2012)
27. ASTM C1556: Standard Test Method for Determining the Apparent Chloride Diffusion Coefficient of Cementitious Mixtures by Bulk Diffusion. ASTM International, West Conshohocken (2011)
28. ASTM C39: Standard Test Method for Compressive Strength of Cylindrical Concrete Specimens. ASTM International, West Conshohocken (2018)
29. Ramlochan, T., Thomas, M., Gruber, K.A.: The effect of metakaolin on alkali–silica reaction in concrete. Cem. Concr. Res. **30**(3), 339–344 (2000)
30. ASTM C469: Standard Test Method for Static Modulus of Elasticity and Poisson's Ratio of Concrete in Compression. ASTM International, West Conshohocken (2014)
31. Swamy, R.N.: Cement Replacement Materials. Surrey University Press, Glasgow (1986)

Mechanical Properties

Effect of Alkali Treatment to Improve Fiber-Matrix Bonding and Mechanical Behavior of Sisal Fiber Reinforced Cementitious Composites

Raylane de Souza Castoldi[1(✉)], Lourdes Maria Silva de Souza[2], and Flávio de A. Silva[1]

[1] Department of Civil and Environmental Engineering, Pontifical Catholic University (PUC-Rio), Rio de Janeiro, RJ, Brazil
[2] Tecgraf Institute, Pontifical Catholic University (PUC-Rio), Rio de Janeiro, RJ, Brazil

Abstract. The current use of dispersed fibers in cementitious matrices is focused on the enhancement in structural performance and mitigation of the effects of shrinkage, which results in microcracks in the cementitious matrix. Natural fibers appear as a low cost and eco-friendly alternative. However, these fibers are degradable in alkaline environments, resulting in changes in the mechanical performance of the composite. These fibers are also susceptible to volume variation with moisture presence, which results in interface degradation. Therefore, the main goal of this work is to evaluate the effect of alkali treatment in order to overcome these limitations and successfully utilize these materials in several applications. For this purpose, sisal fibers with 50 mm length were subjected to 1, 5 and 10 wt.% alkali solutions for chemical modification. The effect of the treatment was evaluated by pullout tests on untreated and treated fibers on a free calcium hydroxide matrix. Additionally, the mechanical performance of fiber reinforced concrete was analyzed through three-point bending tests. Treated fibers presented a brittle behavior in the pullout test. The alkali treatment did not contribute to an increase in the flexural performance of the composite. Similar values of residual strength in the post-cracking region were reached for untreated and treated fiber reinforced cementitious composites.

Keywords: Sisal fiber · Alkali treatment · Interface · Bending

1 Introduction

In the past few decades, fibers of various materials, sizes, and geometries have been added to concrete in a variety of applications. Initially, the use of dispersed fibers in cementitious matrices was focused on the mitigation of the effects of shrinkage, which results in microcracks in the cementitious matrix. It was also found that, in addition to plastic shrinkage control, fibers had the effect of improving some of the material properties of concrete such as flexural toughness, fatigue resistance, impact resistance, and post-crack strength [1–3]. Also, as a consequence of the presence of fibers, there is a reduction of cracks, improving the durability of fiber reinforced cementitious

© RILEM 2022
P. Serna et al. (Eds.): BEFIB 2021, RILEM Bookseries 36, pp. 51–60, 2022.
https://doi.org/10.1007/978-3-030-83719-8_5

elements [1]. In this context, synthetic fibers have been extensively studied and are frequently used in building elements, added to mortars and concretes.

On the other hand, natural fibers appear as a low-cost alternative, due to their natural and large-scale occurrence, mainly in developing countries and with high housing deficit [3, 4]. However, these fibers are degradable in alkaline environments [5]. Natural fibers also present high moisture absorption and subsequent swelling and interface degradation [1, 6], resulting in the formation of microcracks around the fibers [7]. Moreover, natural fibers present a weak chemical bond to cement-based matrices [8, 9]. Therefore, there is a huge concern to modify these fibers to overcome these limitations and successfully utilize these materials in several applications in which commonly synthetic fibers are used.

Several procedures to reduce natural fibers volumetric instability and improve fiber-matrix bond have been proposed, such as chemical treatments [4, 10, 11]. Chemical treatments may reduce the amount of hemicellulose, lignin, pectin and natural oils covering the surface of the fiber. Most of the chemical treatments used on natural fibers are alkali treatments [12]. The most commons chemicals used for alkalization are calcium hydroxide $Ca(OH)_2$ [10, 14, 15] and sodium hydroxide NaOH [7, 13, 16–20]. The alkalization removes some fiber constituents such as hemicellulose, lignin, and other substances that of the fiber's surface. In consequence, the fiber surface can be modified, the fibers are cleaned and most of the impurities are removed [7, 10, 16, 21]. The chemical modification provides more dimensional stability, reduces water absorption capacity and gives resistance to fiber against deterioration [12].

Despite research on natural fiber treatments, there still is a lack of information on how the use of such treatment modifies the properties of the natural fibers affecting the fiber-matrix bond with the cement-based matrix. The objective of this research is to investigate the effect of alkaline treatment on the fiber properties and interface characteristics of sisal fiber with a low alkaline cement matrix. Tensile tests and SEM analysis were carried out to study the effect of the treatment on the mechanical properties and morphology of the fiber. The potential benefits of the treatment on the fiber-matrix interface bond were investigated by pullout tests. Finally, the composite mechanical evaluation was assessed through three-point bending tests.

2   Materials and Processing

2.1   Sisal Fibers

Sisal fibers are obtained from the *Agave sisalana* plant, passing the leaves through a process called decortication [22–24]. The fibers used in this work were obtained in Bahia, Brazil. Before cutting the fibers into the desired length, a preliminary process to remove impurities was necessary. It consisted of submerging the fibers in the water at 70 ± 5 °C for one hour, and then air drying for 48 h. After that, the fibers were manually cut into segments of 50 mm and subsequently immersed in a NaOH solution at laboratory controlled temperature (of around 22 °C) for 60 min. The alkali concentrations adopted were 1%, 5% and 10%. The alkali-treated fibers were not washed to keep the excess alkali on the fiber surface. The fibers were then air dried, in a forced airflow chamber for 24 h.

Table 1. Proportions of concrete composition. All values in kg/m^3.

Cement	Metakaolin	Fly ash	Fine aggregate	Coarse aggregate	Stone dust	Water	Super-plasticizer
190	114	76	585	800	250	190	8.74

2.2   Matrix

The cementitious materials used in the matrix production were Portland cement of the grade CP II F-32 defined by ABNT NBR 16697, metakaolin and fly ash in a proportion of 50%, 30% and 20% by mass, respectively. This high dosage of pozzolanic materials was adopted in order to obtain a low alkaline matrix and minimize the degradation process of fibers. River sand was used as a fine aggregate, presenting a maximum particle size of 4.75 mm and fineness modulus of 3.02. The adopted coarse aggregate presented a maximum diameter of 12.5 mm. Stone dust was also added to the mixture. To obtain a more fluid matrix with low water content, superplasticizer Glenium 51 was included in the mix. Table 1 summarizes the matrix composition. The resulting water/cement ratio was 0.5. The compressive strength after 28 days was 35.0 ± 0.7 MPa and the slump was 230 ± 2 mm. For the pullout samples, the matrix was produced without coarse aggregate.

2.3   Specimens Preparations

To mold the pullout specimens, the mixtures were produced using a cement mixer with a capacity of 5 dm^3. The casting procedure consisted of fixing the 25 mm diameter PVC molds into an acrylic base and fill them with the matrix. After filling the mold, the upper acrylic plate was placed and then the fibers were insert in the center of small holes for better alignment. Each casting procedure resulted in ten specimens. After 24 h, the specimens were demolded and placed in a curing chamber with a temperature of around 21 °C and 100% relative humidity for 28 days.

To prepare the prismatic specimens for the bending tests, a concrete mixer with a capacity of 30 dm^3 was used. The dosage of fibers adopted was 3 kg/m^3. The specimens for the bending tests were adopted as 400 × 150 × 150 mm. After 24 h, the specimens were demolded and placed in a curing chamber with a temperature of around 21 °C and 100% relative humidity for 28 days. Before testing, notches with 15 mm of length, centered on the bottom side of the prism, were made using a 3-mm thick diamond saw. The dosage of fiber and geometry of the specimens were chosen based on the previous study of the author [6].

3   Experimental Testing Procedure

3.1   Microstructural Analysis

Fibers' diameters were measured, before and after fiber treatment, to investigate the dimensional stability of the treated fibers. The fibers were prepared by cold embedding in epoxy resin [25]. An FEI Quanta 400 scanning electron microscope was used for the microstructural investigation. The images were then evaluated in the software ImageJ, a java-based image processing program. To obtain the fiber cross-sections, a line was drawn around each image of the fiber and their areas were measured.

3.2   Direct Tensile Tests

Fiber treatments were also evaluated based on the mechanical response of untreated and treated fibers subjected to direct tensile tests. Figure 1a shows the test setup and the detail of the sample used in the test. The samples were prepared in accordance with ATSM C1557, where the fibers with a gage length of 20 mm were glued to a paper template (140 g/m^2) for the alignment inside the grips. A servo-hydraulic mechanical testing system model MTS-810 was used to perform the tests, using a displacement rate of 0.1 mm/min. A load cell of 100 N was adapted to the system to obtain the applied load. To correctly measure the vertical displacement, an LVDT was coupled to the grips. Ten samples were tested for each condition (natural, 1%, 5% and 10% of alkali concentration). To calculate the tensile stresses, the fiber cross-sectional area was estimated from the SEM images previously obtained on the microstructural analysis.

3.3   Single Fiber Pullout Tests

In order to study the fiber-matrix bond, pullout tests were performed. The pullout test setup configuration is shown in Fig. 1b. After 28 days of the casting, an MTS 810 servo-controlled hydraulic system was used for the tests and a load cell of 1 kN was used. The load was controlled by the displacement of the internal LVDT. The tests were performed using a displacement rate of 1 mm/min. Ten tests were performed for each condition (natural, 1%, 5% and 10% of alkali concentration). The embedment length of 25 mm was chosen.

3.4   Bending Tests

Three-point bending tests were performed on prismatic specimens, based on EN 14651 [26] with some modifications proposed in the previous study [6]. The details of the setup are presented in Fig. 1c. After 28 days of the casting, the specimens were tested on an MTS servo-controlled hydraulic testing machine with closed-loop control and a load cell of 100 kN. Three rollers of 37 mm of diameter will be used: two support rollers with 350 mm between them and a superior roller on the midspan for load application. A clip gage was fixed on the region of the notch to measure the crack mouth opening displacement (CMOD). The load rate adopted was 0.05 mm/min until CMOD equal to 0.1 mm and 0.2 mm/min until the test ends. For each condition (matrix, reference with natural fiber, 1%, 5% and 10% of alkaline concentration), three specimens were tested.

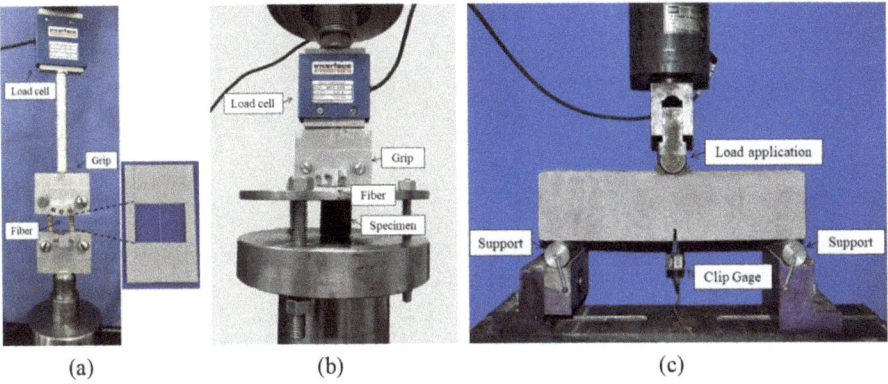

Fig. 1. Setup arrangements for (a) direct tensile tests, (b) pullout tests and (c) three-point bending tests.

4   Results and Discussion

4.1   Microstructural Analysis

Figure 2 shows the cross-section images of untreated and treated sisal fibers. An example of the contour line used to obtain the cross-sectional area is indicated in Fig. 2a. In all images is possible to observe that the fibers are formed by several individual microfibrils, linked together through the middle lamella, basically consisting of hemicellulose and lignin. Besides that, the sisal fibers present a typical non-circular shape, named as horse-shoe shape [27]. In the case of alkali treatment with 1% of NaOH, the fiber remains the same, without any visible degradation. On the other hand, high concentrations of alkali solution (5% and 10% NaOH) resulted in degradation of the fiber, by the concentration reduction of hemicellulose and lignin. In this case, the microfibrils separate from each other, which is more evident for high alkali contents. Table 2 presents some physical characteristics of the fibers obtained from the images. The equivalent diameter corresponds to the diameter of the circle having the same area as that of the average cross-sectional area of an actual fiber. It is possible to note that the cross-sectional area was little influenced due to alkali treatments.

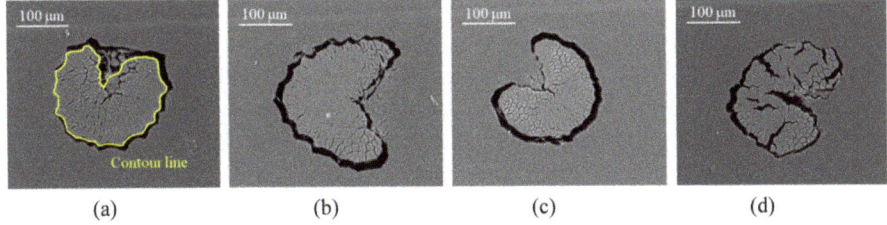

Fig. 2. Cross-section of (a) untreated sisal fiber, alkali-treated sisal fiber with (b) 1%, (c) 5% and (d) 10% by mass of NaOH.

Table 2. Morphological characteristics of untreated and treated sisal fibers.

	Natural sisal	Alkali 1%	Alkali 5%	Alkali 10%
Cross section area (mm²)	0.023 ± 0.007	0.036 ± 0.005	0.027 ± 0.007	0.027 ± 0.005
Equivalent diameter (mm)	0.171 ± 0.027	0.213 ± 0.014	0.184 ± 0.023	0.185 ± 0.019

4.2   Direct Tensile Tests

Typical tensile stress-strain curves of untreated and treated sisal fibers are shown in Fig. 3. The average results of fiber mechanical properties are presented in Table 3. The untreated fibers were characterized by a uniaxial tensile strength of 507.66 ± 74.66 MPa and a strain capacity of 0.069 ± 0.022 mm/mm, similar to previous studies [6, 22, 25, 28]. However, the tensile strength of sisal fiber decreased after all treatments. On treatment with 1% NaOH the uniaxial tensile strength was found to decrease by 32% compared to untreated fibers. For higher alkali concentrations (5% and 10% NaOH), the decrease was approximately 56%. This strength decrease related to the alkali treatment was also reported in a previous study [29] and should be caused by hemicellulose and lignin degradation and the destruction of links between the fiber cells [14]. It indicates that the immersion of these fibers in alkaline solution degraded the natural fiber, being more evident for higher alkaline concentrations.

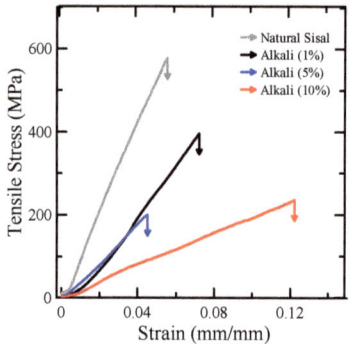

Fig. 3. Representative tensile stress-strain curves of untreated and treated fibers.

The stiffness of sisal fiber also decreased after all treatments. Considering 1 wt.% alkali concentration, Young's modulus was reduced by 25% in comparison to natural fiber. The same reduction was observed for the 5% NaOH-treated fiber. Fibers treated with 10% NaOH also presented a considerable decrease in stiffness, evidenced by a reduction of 74% on the modulus of elasticity in comparison to the natural fibers. However, the strain capacity was found to increase by 70% for the alkali concentration of 10%. The increment in the elongation capacity of the fiber can be explained by the probable removal of non-cellulosic materials and impurities by the alkali treatment [19]. An increase in cellulose to lignin ratio makes fibers less stiff and more deformable [29]. The microfibrils become more capable of rearranging along the direction of tensile deformation, resulting in a reduction in stiffness [10, 13, 29].

Table 3. Results of direct tensile tests performed on untreated and treated sisal fibers.

Fiber	Maximum load (N)	Tensile strength (MPa)	Strain capacity (mm/mm)	Modulus of elasticity (GPa)
Natural sisal	11.68 ± 1.72	507.66 ± 74.66	0.069 ± 0.022	7.45 ± 3.11
Alkali 1%	12.41 ± 2.83	344.75 ± 78.70	0.067 ± 0.012	5.57 ± 1.64
Alkali 5%	5.83 ± 0.53	215.78 ± 19.55	0.042 ± 0.007	5.56 ± 1.21
Alkali 10%	6.12 ± 1.46	226.71 ± 54.12	0.119 ± 0.005	1.92 ± 0.34

4.3   Single Fiber Pullout Tests

The representative curves of untreated and treated sisal fibers are presented in Fig. 4. The pullout curve of natural sisal is characterized by an initial point of fiber debonding, known as a slip softening behavior. After this peak load, the frictional mechanism is responsible for the pullout behavior until the complete fiber debonding. However, for the treated fibers, it was observed the fiber rupture after reach the peak load. The rupture of the fibers may occur as a result of two different mechanisms: i) the load for debonding was higher than the fiber tensile strength, leading to the fiber collapse before the debonding; ii) after debonding, the slip hardening mechanism leads to the fiber rupture during the fiber slip, due to the enhancement in the frictional bond. As was reported a considerable decrease in tensile strength in alkali treated fibers in Sect. 4.2, the first mechanism seems to be predominant. However, an increase in bond strength can be also reached, which is related to the removal of some fiber constituents by the alkali treatment [10, 16, 30]. As a result of the alkalization, the cellulose depolymerizes and the fiber fibrillation occurs, increasing the surface roughness of the fiber. Besides that, there is also an increase in roughness related to the calcium deposition on the fiber [14].

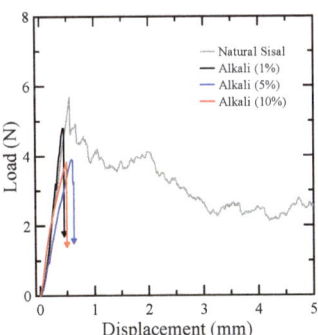

Fig. 4. Representative pullout slip curves of untreated and treated sisal.

4.4   Bending Tests

Typical Stress-CMOD curves of concretes reinforced with 3 kg/m³ of untreated and treated sisal fibers are shown in Fig. 5. The plain concrete presented a typical behavior of brittle materials: after the matrix cracking, there is a rapid stress decrease leading to the specimen fragile rupture. The presence of fibers changed the concrete flexural behavior. All the reinforced concretes presented a deflection softening behavior, characterized by a post-cracking residual stress after the matrix crack appearance. The occurrence of the first crack is related to the matrix tensile strength and was similar for all tested specimens. It was expected since the same matrix was adopted for all composites. However, the post-cracking behavior did not differ much with the fibers' treatments. From the graphs is possible to note that a higher level of residual stress was reached by natural sisal reinforced concrete. On the other hand, the stress level reached by treated fiber reinforced concretes was very similar, evidencing that the alkali treatment was not efficient considering a large-scale mechanical evaluation. Despite the small influence on the composite mechanical properties, treated fibers can have greater durability as reinforcement in cementitious matrices, as reported in previous studies [31].

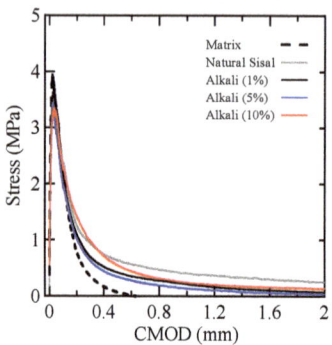

Fig. 5. Representative three-point flexural curves of concrete reinforced with 3 kg/m³ of untreated and treated fibers.

5   Conclusions

From the findings in this investigation, the following conclusions are drawn:

- A decrease of 32–56% in tensile strength was appreciated in alkali treated fibers. Regarding the stiffness, a reduction of 25–74% was reached. This fact is an indicator that there is a degradation of the fibers subjected to alkali treatment, which is evidenced by the fiber microstructural analysis.
- The alkali treatment caused brittle behavior in the pull-out test. The degradation of the fibers may be responsible for the roughness of the fiber surface, resulting in the break of the fibers in pullout tests as a result of a higher frictional bond in the fiber-matrix interface. On the other hand, the break of the fibers can be also related to the tensile strength reduction.

- The modification of sisal fibers with alkali solution was not sufficient to enhance flexural properties of the alkali treated fiber reinforced concretes in comparison to the concrete reinforced with untreated fibers.
- Even if untreated fibers are used, the residual performance of these FRC mixes did not reach the minimum requirement to be considered for structural purposes but crack propagation control instead. Higher dosages of fibers should be tested and also an optimization of the alkali treatment is necessary to guarantee the efficiency of this treatment to sisal fibers.

Acknowledgements. The authors thank Metacaulim do Brasil and Synthomer for the meta-kaolin and polymer donation, respectively. The authors are also grateful to the CETEM laboratory for the use of the scanning electron microscope. This work was supported by the Conselho Nacional de Desenvolvimento Científico e Tecnológico – CNPq and Coordenação de Aperfeiçoamento de Pessoal de Nível Superior – CAPES (Brazilian National Science Foundations).

References

1. Mukhopadhyay, S., Khatana, S.: A review on the use of fibers in reinforced cementitious concrete. J. Ind. Text. **45**(2), 239–264 (2015)
2. Shah, S.P., Kuder, K.G., Mu, B.: Fiber-reinfoced cement-based composites: a forty year odyssey. In: Proceedings of the 6th RILEM Symposium on Fibre-Reinfoced Concretes (FRC) - BEFIB - 2004, No. September (2004)
3. Zollo, R.F.: Fiber-reinforced concrete: an overview after 30 years of development. Cement Concr. Compos. **19**(2), 107–122 (1997)
4. ACI Committee 544: State-of-the-art report on fiber reinforced concrete. ACI Struct. J. **96** No. Reapproved (2002)
5. Ferreira, S.R., de Andrade Silva, F., Lima, P.R.L., Filho, R.D.T.: Effect of hornification on the structure, tensile behavior and fiber matrix bond of sisal, jute and curauá fiber cement based composite systems. Constr. Build. Mater. **139**, 551–561 (2017)
6. de Souza Castoldi, R., de Souza, L.M.S., de Andrade Silva, F.: Comparative study on the mechanical behavior and durability of polypropylene and sisal fiber reinforced concretes. Constr. Build. Mater. **211**, 617–628 (2019)
7. Mwaikambo, L.Y., Ansell, M.P.: Chemical modification of hemp, sisal, jute, and kapok fibers by alkalization. J. Appl. Polym. Sci. **84**(12), 2222–2234 (2002)
8. Ferreira, S.R., Lima, P.R.L., Silva, F.A., et al.: Effect of sisal fiber hornification on the fiber-matrix bonding characteristics and bending behavior of cement based composites. Key Eng. Mater. **600**, 421–432 (2014)
9. Ballesteros, J.E.M., dos Santos, V., Mármol, G., Frías, M., Fiorelli, J.: Potential of the hornification treatment on eucalyptus and pine fibers for fiber-cement applications. Cellulose **24**(5), 2275–2286 (2017)
10. Ferreira, S.R., Silva, F.D.A., Lima, P.R.L., et al.: Effect of fiber treatments on the sisal fiber properties and fiber-matrix bond in cement based systems. Constr. Build. Mater. **101**, 730–740 (2015)
11. Vo, L.T.T., Navard, P.: Treatments of plant biomass for cementitious building materials – a review. Constr. Build. Mater. **121**, 161–176 (2016)
12. Sood, M., Dwivedi, G.: Effect of fiber treatment on flexural properties of natural fiber reinforced composites: a review. Egypt. J. Pet. **27**(4), 775–783 (2018)

13. Kim, J.T., Netravali, A.N.: Mercerization of sisal fibers: effect of tension on mechanical properties of sisal fiber and fiber-reinforced composites. Compos. A Appl. Sci. Manuf. **41**(9), 1245–1252 (2010)
14. Zukowski, B., de Andrade Silva, F., Toledo Filho, R.D.: Design of strain hardening cement-based composites with alkali treated natural curauá fiber. Cement Concr. Compos. **89**, 150–159 (2018)
15. Zukowski, B., dos Santos, E.R.F., dos Santos Mendonça, Y.G., et al.: The durability of SHCC with alkali treated curaua fiber exposed to natural weathering. Cement Concr. Compos. **94**, 116–125 (2018)
16. Mwaikambo, L.Y., Ansell, M.P.: The effect of chemical treatment on the properties of hemp, sisal, jute and kapok for composite reinforcement. Die Angew. Makromolek. Chem. **272**(1), 108–116 (2002)
17. Mokaloba, N., Batane, R.: The effects of mercerization and acetylation treatments on the properties of sisal fiber and its interfacial adhesion characteristics on polypropylene. Int. J. Eng. Sci. Technol. **6**(4), 83 (2014)
18. Mohan, T.P., Kanny, K.: Chemical treatment of sisal fiber using alkali and clay method. Compos. A Appl. Sci. Manuf. **43**(11), 1989–1998 (2012)
19. Kundu, S.P., Chakraborty, S., Roy, A., et al.: Chemically modified jute fibre reinforced non-pressure (NP) concrete pipes with improved mechanical properties. Constr. Build. Mater. **37**, 841–850 (2012)
20. Sreekumar, P.A., Thomas, S.P., Saiter, J.M., et al.: Effect of fiber surface modification on the mechanical and water absorption characteristics of sisal/polyester composites fabricated by resin transfer molding. Compos. Part A: Appl. Sci. Manufact. **40**(11), 1777–1784 (2009)
21. Snoeck, D., Smetryns, P.A., De Belie, N.: Improved multiple cracking and autogenous healing in cementitious materials by means of chemically-treated natural fibres. Biosys. Eng. **139**(1998), 87–99 (2015)
22. Li, Y., Mai, Y.W., Ye, L.: Sisal fibre and its composites: a review of recent developments. Compos. Sci. Technol. **60**(11), 2037–2055 (2000)
23. de Silva, F., Chawla, N., de Toledo Filho, R.D.: An experimental investigation of the fatigue behavior of sisal fibers. Mater. Sci. Eng. A **516**(1–2), 90–95 (2009)
24. de Andrade Silva, F., Mobasher, B., Filho, R.D.T.: Fatigue behavior of sisal fiber reinforced cement composites. Mater. Sci. Eng., A **527**(21–22), 5507–5513 (2010)
25. Teixeira, F.P., Gomes, O.F.M., de Andrade Silva, F.: Degradation mechanisms of curaua, hemp, and sisal fibers exposed to elevated temperatures. BioResources **14**(1), 1494–1511 (2019)
26. European Standard: EN 14651: Test Method for Metallic Fibred Concrete - Measuring the Flexural Tensile Strength (Limit of Proportionality (LOP), residual) (2005)
27. de Andrade, F., Silva, B.M., Soranakom, C., Filho, R.D.T.: Effect of fiber shape and morphology on interfacial bond and cracking behaviors of sisal fiber cement based composites. Cement Concr. Compos. **33**(8), 814–823 (2011)
28. de Andrade Silva, F., Chawla, N., de Toledo Filho, R.D.: Tensile behavior of high performance natural (sisal) fibers. Compos. Sci. Technol. **68**(15–16), 3438–3443 (2008)
29. Zwane, P.E., Ndlovu, T., Mkhonta, T.T., et al.: Effects of enzymatic treatment of sisal fibres on tensile strength and morphology. Sci. Afr. **6**, e00136 (2019)
30. Ferreira, S., Lima, P., Silva, F., et al.: Effect of sisal fiber hornification on the adhesion with portland cement matrices. Rev. Matér. **17**(2), 1024–1034 (2012)
31. de Klerk, M.D., Kayondo, M., Moelich, G.M., et al.: Durability of chemically modified sisal fibre in cement-based composites. Constr. Build. Mater. **241**, 117835 (2020)

Drop-Weight Impact Test for Fibre Reinforced Concrete: Analysis of Test Configuration

Juan Carlos Vivas[1]([⊠]), Facundo Isla[2], María Celeste Torrijos[1], Graciela M. Giaccio[3], Bibiana Luccioni[2], and Raúl L. Zerbino[1]

[1] CONICET, LEMIT-CIC, Faculty of Engineering, UNLP, La Plata, Argentina
juancarlos.vivas@ext.ing.unlp.edu.ar
[2] CONICET, Institute of Structures, Faculty of Exact Sciences and Technology, National University of Tucumán, San Miguel de Tucumán, Argentina
[3] LEMIT-CIC Researcher, Faculty of Engineering, UNLP, La Plata, Argentina

Abstract. Impact resistance represents a key property of Fibre Reinforced Concrete (FRC). Recently, the authors proposed a repeated drop-weight test for FRC impact characterization that, among other advantages, is able to distinguish the contribution of different types and dosages of fibres both at cracking and in cracked state. The impacts are applied on a simply supported notched prism of $150 \times 150 \times 300$ mm, being the adopted span length 240 mm, the notch depth 25 mm and the projectile mass 5 kg. An experimental and numerical parametrical study was carried out in order to analyse the effect of the specimen geometry on impact test results. The span length (240, 350 and 500 mm), the notch depth (10, 25 and 50 mm), the specimen width (70, 100 and 150 mm), and the projectile mass (5, 10 and 20 kg) were considered as variables. The numerical model was developed to analyse the stress distribution before the first crack appears (elastic behaviour). In order to corroborate the numerical model, the impact load was measured in some tests with a dynamic load cell. Numerical results concerning the effects of the above-mentioned variables on impact response are presented.

Keywords: FRC · Impact tests · Fibres · Parametric study · Specimen geometry · Numerical modelling

1 Introduction

The assessment of the impact resistance of Fibre Reinforced Concrete (FRC) has been under study since it is one of its most outstanding properties. Industrial floors, pavements, security enclosures, tunnel lining and airports runways are some of the structures that can be exposed to this type of loads. A high complex phenomenon takes place due to impact loads and therefore, it has been difficult to develop accurate methods to characterize the material behaviour under this type of load. Instead, a wide range of different tests have been proposed.

One of the first methods consisted on repeated weight drops over the specimen, until it reaches the "no-rebound" condition [1]. The accumulated potential energy of all the impacts or directly the number of falls were used to assess the material performance. However, high variabilities among similar researches were obtained due to the

© RILEM 2022
P. Serna et al. (Eds.): BEFIB 2021, RILEM Bookseries 36, pp. 61–73, 2022.
https://doi.org/10.1007/978-3-030-83719-8_6

differences in the specimen geometries, support conditions and the way loads were applied. Therefore, the comparison of results was impossible.

Later, methods originally used for the study of metals and other materials, such as the Charpy method [2] and the Split Hopkinson Bar [3, 4] were implemented. Although, these tests allowed great advances in the comprehension of the material response, they also had limitations, such as the impossibility of evaluating specimens of representative sizes. This type of tests was only applied to fibre reinforced mortars and FRC with small maximum size aggregates.

It is in this stage that the drop tower appears as a good alternative for the evaluation of FRC impact resistance, as this machine allows evaluating impact performance of specimens with diverse sizes, subjected to different energy levels and impact velocities in the range found in many applications.

Banthia [5] presented a drop-weight test, with which impact behaviour of prismatic specimens up to $150 \times 150 \times 1525$ mm size was evaluated using accelerometers. The specimens were mainly without notch and simply supported. The supports and the projectile were also electronically instrumented to register the reaction forces (beam-supports, beam-projectile) and the impact velocity at impact time. The area under the load-deflection curve until failure was defined as "fracture energy". With a similar test, Zhang et al. [6] proposed using smaller prisms ($100 \times 100 \times 420$ mm) notched in the centre of the tensile face (50 mm). Zhu [7] also applied a drop-weight test with a drop tower to assess the impact resistance of plain concretes and FRC, but in this case using a specimen with an inverted U shape. This test, unlike the previous ones, uses repeated falls without any instrumentation. The drop number required for cracking (N_1) and for dividing the specimen (N_2) that represents the end of the test, are registered, finding a lineal relationship between N_1 and N_2. Zhang et al. [8] designed an impact test on simple supported prisms but instrumented with strain gauges and subjected to repeated impacts from fixed heights. They correlated the fall height and the projectile mass with the number of impacts required to produce failure. Although the test proposed by the ACI 544 Committee [9] does not use a fall tower, it is also a drop-weight test where a soil compacting hammer is manually used. It is a qualitative test consisting of repeated impacts. The number of impacts required for cracking and failure are registered.

This brief summary evidences that even for the same test (drop weight with fall tower) there are diverse variants such as: the number of impacts, the shape, mass and height of the projectile fall, the measured variables, the size, shape and treatment of the specimens and the support conditions, the instrumentations and data registered.

With the aim of developing a quick test, of easy instrumentation and useful for assessing impact behaviour of FRC with different residual capacities, Vivas et al. [10, 11] designed and implemented a new methodology that is still under study. One of the main advantages of this method is that it allows characterizing not only the first crack appearance but also the impact behaviour after cracking that is of particular interest depending on the application. However, as the test results are expressed in terms of energy, they depend on the specimen size and other test parameters.

The objective of this paper is to show to what extent the different test parameters (specimen geometry, span length and projectile mass) modify the results and to evaluate a correction method that allows comparing tests corresponding to different values of the mentioned variables. In addition to test results, numerical results that are used to analyse the stress distribution before cracking are presented.

2   Materials and Test Method

A FRC was prepared using cement CPF40, calcareous filler, natural siliceous sand, 12 mm maximum size granitic crushed stone, a superplasticizer and 30 kg/m^3 of steel hooked-end fibres (50 mm long, 1 mm diameter and 1100 MPa tensile strength) were incorporated. These fibres can be used in many structural applications and the dosage chosen is frequently employed in floors and pavements. Concrete slump was 140 mm, air content 3% and unit weight 2330 kg/m^3. 21 prisms of $150 \times 150 \times 600$ mm were cast for bending and impact tests and 6 cylinders of 100×200 mm were moulded for compressive strength. All the specimens were consolidated by external vibration; they were demoulded at 24 h and kept in a moist room for curing. After 28 days, the FRC had a compression strength of 54 MPa.

Characterization in bending was performed according to EN14651 [12] standard. The FRC maximum stress (f_{max}) was 5.63 MPa, the first crack stress (f_L) was 4.66 MPa and the residual stresses at CMODs equal to 500 and 2500 microns (f_{R1} and f_{R3}) were 4.68 and 5.44 MPa respectively.

The proposed impact test can be used for the evaluation of FRC cracking resistance and its post cracking capacity. Although the method, the machine, the adopted specimen and the devices have been described in previous works [10], the procedure is briefly described below. The test procedure consists of applying repeated impacts on the top of a prismatic specimen with a notch at the centre of the tensile face to localize the crack. After each impact, the Crack Opening Displacement (COD) 5 mm over the notch tip (120 mm below the top face of the prism) is measured. The impacts are applied following similar patterns for all the specimens. The drop height (h) is varied according to two phases. The objective of Phase 1 is the determination of the first-cracking resistance. It consists of an initial impact from a height (h_0) of 100 mm and then, consecutive impacts with height increments (Δh) of 50 mm between them. Phase 1 finishes when a crack is detected. The impact capacity of cracked FRC is evaluated in Phase 2. It also starts from an initial height $h_0 = 100$ mm but, unlike Phase 1, the height increments (Δh) are of 100 mm and three drops (N = 3) are executed in each height level. In phase 1 a smaller Δh is applied to determine more accurately the appearance of the first crack and then the cracking energy (E_C). The combination of Δh and N was selected to reduce the rate of energy applied in comparison to Phase 1 and avoid an excessive energy dissipation without COD growth. In this way, the effect of different fibre types and contents can be assessed and it can be verified to what extent the fibres can prevent the crack growth. The end of the test is defined when COD is greater than 3 mm.

The potential energy of each drop is determined as E = m.g.h (where m = 5 kg is the projectile mass, g represents the gravity acceleration), and the cumulated energy (E*) is the sum of the potential energy corresponding to all the impacts received by the specimen up to a certain point. The curve E* – COD is defined as the *Impact Curve*. In Fig. 1 an example of an impact curve is shown where the characteristic parameters for Phase 1 and Phase 2 can be seen. In Phase 1 E_C and COD_C are obtained, E_C is defined as the cumulated energy until the first crack and COD_C the initial opening of that crack. In Phase 2 E_P is the cumulated energy measured in the post cracking stage, it is the cumulative energy between the first crack and the COD equal to 3 mm, V_C is the COD growing rate and finally E_T is the total cumulated energy of both phases.

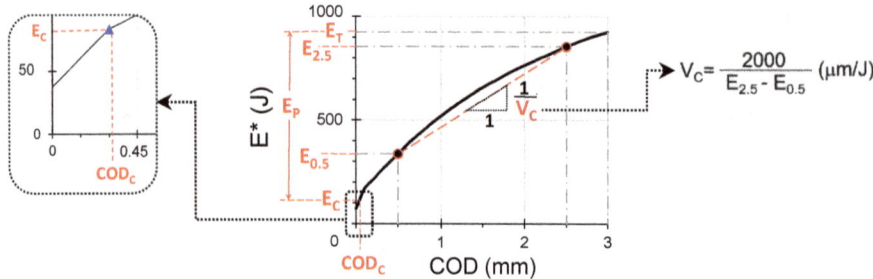

Fig. 1. Impact curve.

3   Parametric Study

The "standard specimen" (Control, C) proposed in the impact test has a square transversal section of 150×150 mm and 300 mm length. The free span between supports (L) is 240 mm, and a 25 mm notch (e) is sawn in the centre of the tensile face. The mass of the standard projectile is 5 kg.

To analyse the *individual effect* of each of the test variables, the magnitudes of: the specimen width (b), the net height (h_{sp}), that is the difference between the prism height (h_b) and the notch depth (e), the span between supports (L) and the projectile mass (m) were varied one at a time. The prisms height (h_b) remained constant. In Table 1 the adopted values for each parameter are shown. A variability study showed that with six tests, the mean values of E_T, global variable of the impact test, can be obtained with a 15% of error and a 90% of confidence level. Therefore, six tests were carried out for each series. The impact tests were performed after the bending tests EN14651 [12] using the resulting beams' halves.

Table 1. Variables in impact tests series.

Series	Id	b (mm)	$h_{sp} = h_b - e$ (mm)	L (mm)	M (kg)
Standard (Control/Reference)	**C**	**150**	**125**	**240**	**5.1**
Width	b100	**100**	125	240	5.1
	b70	**70**	125	240	5.1
Height	h100	150	**100**	240	5.1
	h140	150	**140**	240	5.1
Span	L350	150	125	**350**	5.1
	L500	150	125	**500**	5.1
Mass	m10	150	125	240	**11.2**
	m20	150	125	240	**21.2**

4   Results

The impact curve corresponding to the Reference Series C is shown in Fig. 1. The characteristic parameters for Phase 1 were E_C = 96 J and the initial opening of that crack COD_C = 38 μm; and for Phase 2, E_P = 863 J, COD growing rate V_C = 5,0 μm/J and E_T = 959 J.

The mean impact curves obtained for each Series are presented in Fig. 2a.

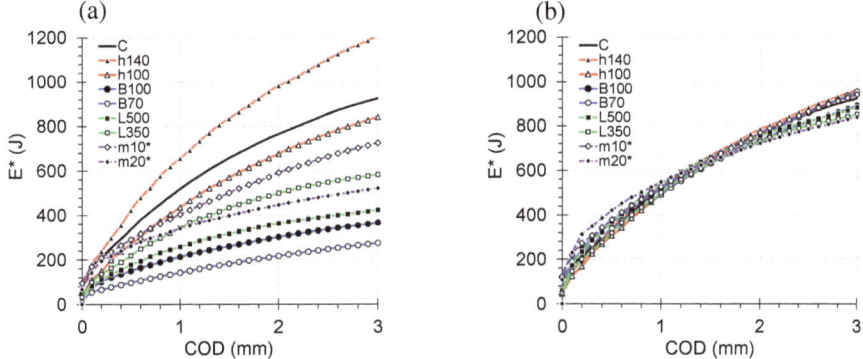

Fig. 2. (a) Mean impact curves (b) Corrected mean impact curves.

It can be seen that in most cases the shape of the curves was consistent, with exception of m20 that presented greater energy in the first 0.5 mm and a noticeable fall in the post cracking slope (that means an increase in the cracking rate). In all cases, except for h140, the total energy was reduced in relation to Series (C). The Series with the lowest E_T was b70, which suggests that, at least for this type of fibres, the width of the specimen is one of the variables that has more influence on the test results.

The results of E_C and E_P (left axis) and V_C (right log axis) are summarized in Fig. 3. The maximum and minimum values of each parameter are symbolized with vertical lines.

When comparing the different Series with reference Series C it was found that the increment in the span length reduced E_C and E_P while V_C increased 186%. All the parameters had less variability as the span length increased.

The width of the specimen also affected the crack energy (E_C, E_P) and the crack opening rate (V_C). E_C decreased with the width; in the case of b70 the decrease was almost 70%, while in b100 it was only 32%. In the case of E_P, the decrease, respect to Control Series, was near 64% in b100 and 70% in b70. The crack opening rate (V_C) was in both cases more than twice the C rate, being a 220% greater in b100 and 280% in b70. The V_C is the most affected property by the specimen width. Regarding the variability of results, there is not a clear tendency, however the CV ($\leq 30\%$) is acceptable and in accordance with the variability of the static strengths of FRCs.

The height of the specimen also modifies the results. E_C increased with the height (h_{sp}), in h140 E_C was 10% higher than the standard case (C). The increase in height from 100 to 125 mm (Series h100 and C) only increased E_P in a 7%. On the contrary, in h140 E_P increased a 32% with respect to the standard Series (C). The variability of E_P results of this Series was almost constant ($\approx$30%), being the highest value among all the test parameters and Series.

With regard to the projectile mass, there was not a clear influence on the crack energy E_C; it increased when a 10 kg projectile was used (m10 values were near 40% higher than C values with m = 5 kg) but smaller increases were found in m20. On the other hand, the energy required to reach rupture (E_P) clearly decreased as mass increased; compared with C in m10 was 20% less and 40% less with m20.

It is clear that, although the test parameters are given in terms of energy or depending on it, the results may vary with test arrangement (projectile mass, drop height or geometry).

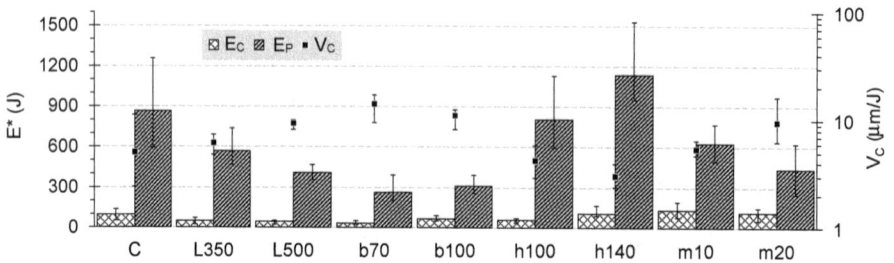

Fig. 3. Parametric study of the impact test.

5   Correction Method

A correction method that makes possible the comparison between impact tests with small geometrical or in the projectile mass differences is proposed from the results of this study. It consists of applying a dimensionless correction factor Γ, which multiplies the potential energy provided in each impact as shown in Eq. (1).

$$E_i = \Gamma \cdot m.g.h_i \tag{1}$$

Depending on the type of variable, the Γ factor will directly or inversely depend, on the relationship between the standard dimension (L, b, h_{sp}, m) and the real one (L_r, b_r, h_{spr}, m_r), affected (proportionally or potentially) by an adjustment factor (k_L, k_b, k_{hsp}, k_m). The mathematical relationships proposed for Γ are presented in Eqs. (2) to (5) where the subscripts indicate the variable to be adjusted (L, b, h_{sp}, m). The proposal contemplates the correction of only one variable per test; the combined effect will be studied in a future research. Besides, it is recommended to use the corrections as long as the magnitudes of the variables are among the range studied in this work. For different types and dosages of fibres, the ki values could possibly change.

$$\Gamma_L = k_L \frac{L_r}{L} \tag{2}$$

$$\Gamma_b = k_b \frac{b}{b_r} \tag{3}$$

$$\Gamma_{hsp} = k_{hsp} \frac{h_{sp}}{h_{sp_r}} \tag{4}$$

$$\Gamma_m = \left[\frac{m}{m_r}\right]^{k_m} \tag{5}$$

The correction factors calculated for each variable of this experience are given in Table 2. In Fig. 4, the relationship between the standard dimension (L, b, h_{sp}, m) and the real one (L_r, b_r, h_{spr}, m_r), are plotted for each case. It also includes a factor, called Measured, which is the average value of the division, in each crack opening (every 0.1 mm, from a COD of 0.5 mm), of the accumulated energy in the standard case (C) and each of the other cases (L, b, h_{sp}, m) respectively. The differences between L_r/L, h_{sp}/h_{spr}, b/b_r, m_r/m and Measured values represented in Fig. 4 justify the k_i factors; constant correction factors of each Series can be found. The corrected curves are plotted in Fig. 2b that shows the way in which the original curves fit the standard curve when the Eqs. (2) to (5) (according to each case) are applied. Figure 2b clearly shows the effectiveness of the proposed correction factors.

Table 2. Variables to consider in the impact tests in each Series.

L (mm)	L_r/L	k_L	Γ_L	h_{sp} (mm)	h_{sp} /h_{spr}	k_{hsp}	Γ_{hsp}
240	**1.00**	–	**1.00**	100	1.25	0.9	1.13
350	1.46	1	1.46	**125**	**1.00**	–	**1.00**
500	2.08	1	2.08	140	0.89	0.9	0.80
b (mm)	**b/b_r**	**k_b**	**Γ_b**	**m (kg)**	**m_r/m**	**k_m**	**Γ_m**
70	2.14	1.62	3.47	**5.1**	**1.00**	–	**1.00**
100	1.50	1.62	2.43	11.2	2.16	0.34	1.30
150	**1.00**	–	**1.00**	21.2	4.09	0.34	1.61

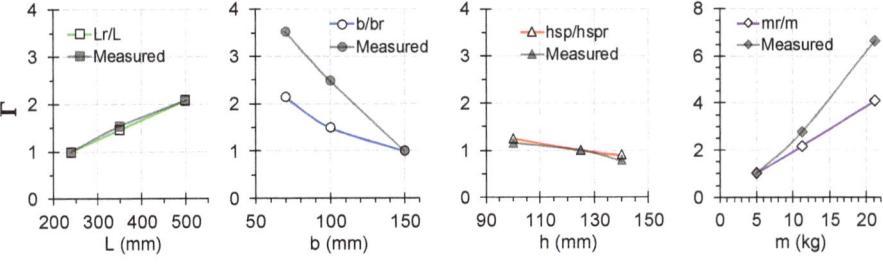

Fig. 4. Correction factors.

6 Numerical Study

6.1 Computational Model Overview

A computational model was developed to study the behavior of the specimen during the impact. The main objective was the analysis of the stress distribution in order to reinforce the hypotheses of the experimental study that the specimen works in flexion. The model was generated with LS-Dyna [13] and both the specimen and the testing machine were simulated in order to achieve representative results and minimize simplifications in boundary conditions.

The numerical model is shown in Fig. 5. The base and the fixed support are joint, while the mobile support can slide on the base with a friction coefficient of 0.17 (experimentally measured) and the movement is directed by a frictionless central guide, which also restricts the upwards movement. The bottom of the base is fixed to the global reference system. Two rollers are arranged between the specimen and the supports. Steel-steel friction (friction coefficient 0.2) is defined between the rollers and the supports, while steel-concrete friction (friction coefficient 0.3) is defined between the rollers and the specimen. Four tensors fix the specimen against the rollers with a pretension force of 1kN preventing the specimen ends from upward displacement after impact. The load cell together with two steel plates used for load distribution are also modeled on the upper face of the specimen. The projectile is modeled in detail; the one presented in Fig. 5 corresponds to a mass of 5.182 kg. Part of the projectile guide is also modeled to avoid the projectile's change of direction after the first impact; for this study, a frictionless interaction between the guide and the projectile was defined. The projectile guide is fixed to the global reference system. All the components of the testing machine are modeled with their actual mass and considering elastic material with steel mechanical properties.

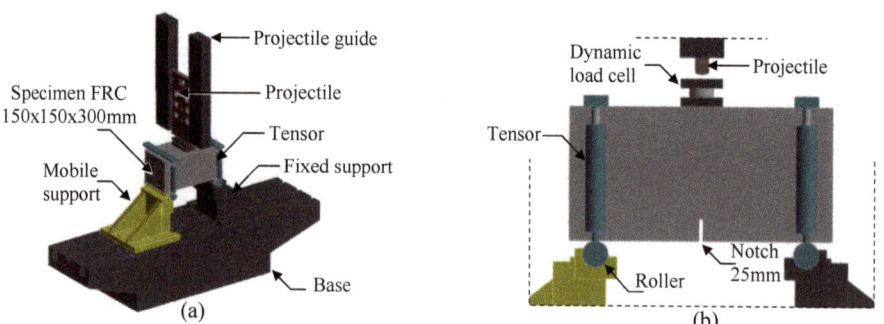

Fig. 5. Computational model. a) Perspective. b) Front view of the specimen.

The simulation begins with the pretension of the tensors and this stage takes the first 4 ms until the stabilization. The projectile is assumed to free fall with an acceleration of 9.79688 m/s². The fall heights under study vary from 100 to 500 mm, which implies a fall duration of 0.14 to 0.31 s. To shorten the calculation time, the projectile is initially positioned at a height such that it takes 4 ms to hit the loading plate and the initial velocity corresponding to free fall to this location is assigned. Once the contact has been initiated, 2 ms more are analyzed, with which the entire simulation lasts 6 ms.

Since the computational model is used to analyze the pre-cracking behavior of the specimens under the impact test, elastic behavior is assumed for FRC (Elastic modulus: 36 GPa, Poison coefficient: 0.20).

6.2 Computational Model Validation

In order to verify the finite element model developed, numerical results are compared with those obtained in the tests. The evolution of impact load in time for three fall heights (70, 300 and 500 mm) is presented in Fig. 6 where experimental and numerical results are included for comparison.

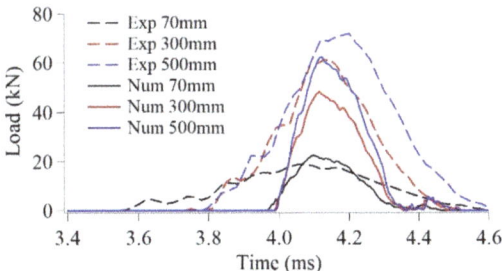

Fig. 6. Comparison of experimental and numerical load histories for different drop heights.

The distribution of principal stresses in the specimen's central plane for a fall height of 300 mm is presented in Fig. 7. The pictures show three instants of the impact corresponding to the peak load and two later instants are presented to show the evolution of stresses. The formation of compression rods can be observed, but the compression stresses do not reach high values (6 MPa), while the tensile stresses in the notch area are high (9 MPa). The maximum shear stresses are also presented. The stress distributions shown in Fig. 7 that correspond to the control specimen (C) show that this specimen can be assumed to work mainly in flexion.

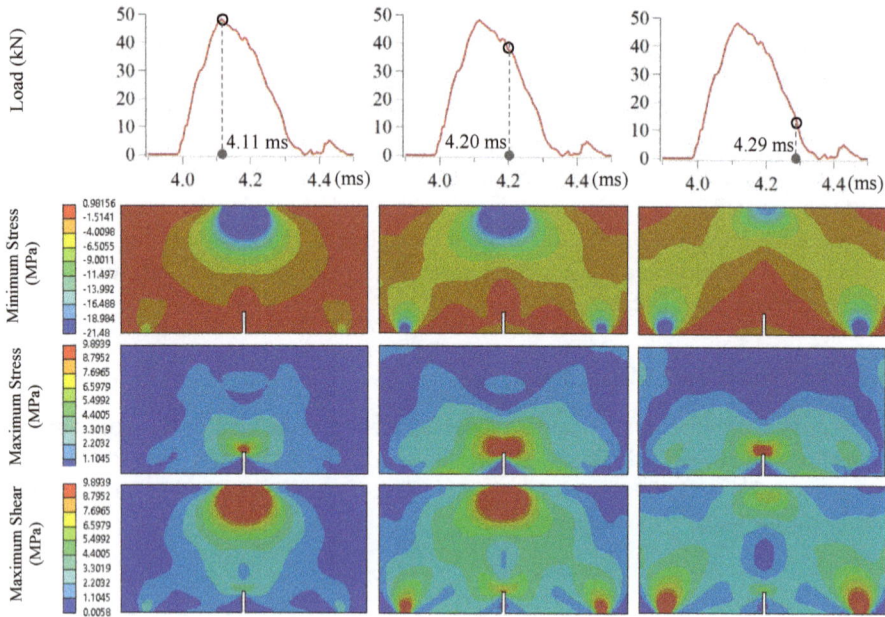

Fig. 7. Principal stress distribution.

6.3 Parametric Study Simulations

A parametric study considering the combinations presented in Table 3 is also numerically carried out. The same numerical model previously described is used for these simulations, but the load cell and distribution plates are removed to reproduce the conditions of the experimental tests. In this case, the projectile directly hits the specimen. For each Series of the parametric study, the corresponding dimension is adjusted in the model and the simulation is carried out considering a single hit. The results for a fall height of 300 mm, for which the beam behavior is in elastic range, are presented in Table 3 for all the Series studied. This fall height approximately corresponds to half the height at which the first crack appears. The results of the peak load and principal stress distribution in the specimen are included in Table 3. The stress distribution presented correspond to 0.1 ms after the peak load since it was shown that the maximum stresses are obtained for the control specimen in this instant.

Table 3 shows that when the width of the specimen is reduced or the depth of the notch is increased, the stiffness is decreased, with which the peak load is lower. On the other hand, if the depth of the notch is reduced, the stiffness increases and the peak load increases. If beam span length is increased, the stiffness is reduced, but the increase in length leads to an increase in the volume of the specimen and therefore an increase in inertia, producing higher peak loads. Finally, and as expected, the peak load increases with the mass of the projectile.

The stress distribution presented in Table 3 shows that when the thickness of the specimen is reduced, shear stresses increase. This fact could trigger a shear failure instead of a bending failure. Something similar is observed when increasing the mass of the projectile. In the remaining Series of tests, the location of tensile stresses in the notch are closer to the tensile strength of concrete, even at small notch heights. This explains why, despite the low slenderness, predominantly flexural failure occurs.

Table 3. Parametric study simulation results for 300 mm drop height

Series	Peak Load (kN)	Minimum Stress (MPa)	Maximum Stress (MPa)	Maximum Shear (MPa)	Predominant failure mechanism
C	72.5				Flexion
b100	59.7				Flexion
b70	49.1				Flexion-Shear
h100	72.3				Flexion
h140	72.6				Flexion
L350	72.8				Flexion
L500	75.8				Flexion
m10	87.5				Flexion
m20	117.1				Flexion-Shear

7   Conclusions

With the aim of analysing the influence of the specimen geometry (span length, specimen width and net height) and mass of the projectile in the proposed impact test, experimental and numerical parametric studies were done. It was found that:

- The specimen width, net height, span length and mass of the projectile affect the results, as expected, including the shape of the impact curve and the resulting impact parameters before and after cracking.
- The specimen width and the mass of the projectile are the variables that have greatest effects, the former causes the highest reduction in energy and the second modifies the shape of impact curves.
- The variability found in the impact tests results is in accordance with the typical FRC values in static tests (bending for instance), however, the increase of span length from 240 to 500 mm reduces it.
- A correction method for minor variations of each variable was proposed, based on empirical equations from the experimental results. The correction factors consider the relationship between the real variable and the proposed for the test, proportionally or potentially affected by a constant k_i.
- The numerical simulation shows that, even for beams with small notch heights, tensile stresses near the beam notch are close to concrete tensile strength. This explains why, despite the low slenderness, predominantly flexural failure occurs in most cases except when the beam width is reduced or the mass of the projectile is increased.

Acknowledgements. The authors thank to Cementos Avellaneda SA for supplying the ready-mix concrete. Funding from LEMIT-CIC and from projects CONICET PIP112-201501-00861, UNLP 11/I188, FONCyT PICT 2017 1313 and UNT PIUNT E623 is appreciated.

References

1. Edgington, J.: Steel fibre reinforced concrete. Ph.D. Thesis University of Surrey (1973)
2. Yu, R., Van Beers, L., Spiesz, P., Brouwers, H.J.H.: Impact resistance of a sustainable ultra-high performance fibre reinforced concrete (UHPFRC) under pendulum impact loadings. Constr. Build. Mater. **107**, 203–215 (2016)
3. Riisgaard, B., Ngo, T., Mendis, P., Georgakis, C.T., Stan H.: Dynamic increase factors for high performance concrete in compression using split Hopkinson pressure Bartle. Fract. Mech. Concr. Concr. Struct. (2007)
4. Sun, X., et al.: A study of strain-rate effect and fiber reinforcement effect on dynamic behavior of steel fiber-reinforced concrete. Constr. Build. Mater. **158**, 657–669 (2018)
5. Banthia, N.: Impact resistance of concrete. Ph.D. thesis, University of British Columbia (1987)
6. Zhang, X.X., Ruiz, G., Yu, R.C.: A new drop weight impact machine for studying the fracture behaviour of structural concrete. WIT Trans. Built Environ. **98**, 251–259 (2008)
7. Zhu, X.C., Zhu, H., Li, H.R.: Drop-weight impact test on U-shape concrete specimens with statistical and regression analyses. Materials **8**, 5877–5890 (2015)

8. Zhang, W., Chen, S., Liu, Y.: Effect of weight and drop height of hammer on the flexural impact performance of fiber-reinforced concrete. Constr. Build. Mater. **140**, 31–35 (2017)
9. ACI Committee 544: Measurement of Properties of Fiber Reinforced Concrete 544.2R-89 (1999)
10. Vivas, J.C., Zerbino, R., Torrijos, M.C., Giaccio, G.M.: Impact response of different classes of fibre reinforced concretes. In: RILEM-Fib X International Symposium on Fibre Reinforced Concrete, pp. 189–198 (2020)
11. Vivas, J.C., Zerbino, R., Torrijos, M.C., Giaccio, G.: Effect of the fibre type on concrete impact resistance. Constr. Build. Mater. **264**, 120200 (2020)
12. Technical Committee CEN/TC 229: EN 14651:2005 Test method for metallic fibered concrete - Measuring the flexural tensile strength (limit of proportionality, residual) Méthode (2005)
13. Livermore Software Tech. Corp. (LSTC), LS-Dyna Theory manual (2018)

Influence of Specimen Size for Impact Testing of Fiber Reinforced Cementitious Composite by Charpy Test

Gerbert P. Moreira Neto[1], Philipe de O. Vital[1],
Margareth da S. Magalhães[1,2(✉)], and Reila V. Velasco[3]

[1] Post-Graduate Program in Civil Engineering,
State University of Rio de Janeiro, Rio de Janeiro, Brazil
margareth.magalhaes@uerj.br
[2] Department of Civil Construction and Transport,
State University of Rio de Janeiro, Rio de Janeiro, Brazil
[3] Faculty of Architecture and Urbanism, Federal University of Rio de Janeiro,
Rio de Janeiro, RJ, Brazil

Abstract. To evaluate the behaviour of fiber reinforced cementitious composites under impact loads, different impact tests can be used, such as drop-weight, Charpy test, gas gun and others. The Charpy impact test was initially developed for metallic materials, however, unlike what is observed for metallic materials, there are no standards or technical procedures for cementitious materials or fiber reinforced cementitious composites. Thus, an experimental program was carried out to determine the influence of the specimen size on the energy absorption capacity measured by Charpy impact test. Composites specimens were produced with different dimensions and mixtures. The minimum dimension of the specimen was equal to twice the fibre length. The mixtures were fabricated with cement, fly ash, sand and polyvinyl alcohol (PVA) or polypropylene fibers (PP). The PP fibers used in the composite mixtures were 6 mm and 12 mm length and PVA fiber was 8 mm and 12 mm length. For comparative purposes, a reference plain mortar mixture was also produced. Mechanical behaviour of the specimens was measure by using bending and compression tests and impact behaviour by Charpy impact test. The results indicate that the bending behaviour under static loads, including the ultimate bending strength, deflection capacity and toughness, were improved by increasing the fiber length, in all the PVA and PP composites and by increasing the PP fiber content. The impact energy absorption of fiber reinforced cementitious composite depends on the fibre content and length. The result of impact energy absorption obtained by Charpy test using different specimen sizes cannot be comparable, because the impact resistance depends on the size specimen used in the Charpy test. In the specific case, when the specimen size increases, there is a decline in the energy absorption capacity for PVA fiber composite; conversely, the trend is opposite for mortar or when PP fiber is used.

Keywords: Charpy test · Impact energy absorption · Fiber reinforced cementitious composite

© RILEM 2022
P. Serna et al. (Eds.): BEFIB 2021, RILEM Bookseries 36, pp. 74–85, 2022.
https://doi.org/10.1007/978-3-030-83719-8_7

1 Introduction

Fiber reinforced cementitious composite (FRCC) are materials with the potential of exhibiting higher strength, ductility and energy absorption capacity, in comparison to plain mortar or concrete, which fail in tension immediately after the formation of a single crack. When a crack under impact loading attempts to develop in FRCC the crack bridging fibers undergo debonding and pullout, and a large amount of energy is dissipated by this mechanism. The energy absorbed by FRCC under impact is therefore many times larger than that by normal concrete [1, 2]. Impact resistance of FRCC can be measured by a number of methods, as the drop-weight, Charpy, and Izod tests, which are based on potential energy; gas gun, that is based on kinetic energy; split Hopkinson pressure bar test, based on stress wave propagation and the methods in which hydraulic machines deform specimens at a medium loading rate [3].

The Charpy impact test was first designed for testing metallic materials and has been frequently employed for characterization of cementitious composites. The Charpy impact test consists of obtaining the energy absorbed by the specimen during the fracture [4] and its procedure may be summarized as follows: the test specimen is positioned on the specimen supports against the anvils; the pendulum is released and the specimen is impacted by the striker. The Fig. 1a shows a schematic diagram of Charpy equipment and the Fig. 1b a setup used in the tests. The absorbed energy is taken as the difference between the energy in the striking member at the instant of impact with the specimen and the energy remaining after breaking the specimen, i.e., $E = m.g.(h_i - h_f)$, where m is the pendulum mass, g is the acceleration due to gravity, h_i and h_f are the initial and final height of the pendulum, respectively.

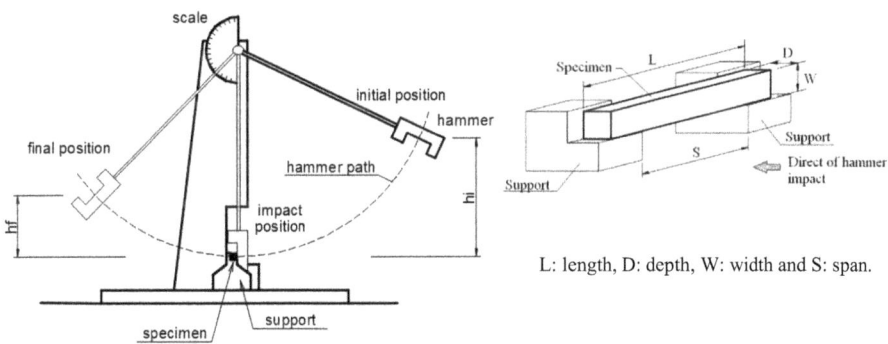

L: length, D: depth, W: width and S: span.

Fig. 1. (a) Charpy apparatus and (b) test setup.

The advantages of using Charpy apparatus for impact test of FRCC specimens are the low cost and easier machine operation. However, there are no standard testing procedures available for Charpy impact testing of FRCC [5] and a wide variety of specimen dimensions and span have been used in the investigations. For instance, researcher have used specimen with dimensions (depth × width × length) of 20 × 20 × 60 mm [6], 20 × 20 × 80 [7], 40 × 40 × 160 mm [8], 25.4 × 25.5 × 50.8 mm [9], 100 × 100

500 mm [10], 10 × 20 × 80 mm [7], 19 × 45 × 70 mm [11], 10 × 50 × 120 mm [12] and 15 × 50 × 350 mm [13], and span/depth (S/D) ratio between 1.6–4.5 [8, 9, 11], while the specimen dimension and S/D ratio used for metallic materials are 10 × 10 55 mm and 4, respectively [4]. According to Radomski [14], the minimum specimen size should not be less than twice the fiber length and five times the maximum aggregate size used for a matrix, unless the specimens are cut out from larger blocks.

According to Thomas and Sorensen [5], the complete absence of a standardized method for Charpy testing of FRCC has led to a lack of consistency in experimental parameters and has limited the basis for comparison of results between similar studies. Test parameters such as the S/D ratio is essential for the fracture mechanism to be repeated, with similar stress flows. Equally important to mention is the limited information on the literature about the specimen size-effect on the impact behaviour of FRCC [15]. In the research conducted by Banthia [15] with steel and PP fiber reinforced concrete under drop-weight impacts was reported a clear size effect on the flexural toughness under impact loads of the composites. The author concluded that the flexural toughness of the composites decreased with increasing specimen size and this effect became more pronounced at higher impact intensities.

Thus, the purpose of this work is to provide additional information about the effects of specimen size on the impact behaviour of different FRCC mixtures. In this experiment were produced specimens in the range of 12 × 12 × 72 mm to 36 × 36 216 mm of five cementitious composites reinforced with PVA or PP fibers and one plain mortar to results comparison.

2   Materials and Methods

2.1   Materials and Mixture Manufacturing

Portland cement, fly ash, sand, water and superplasticizer were used in the mixtures. The cement used is a Portland cement type IIF composed of filler (in mass: 6–10%), with a 28 days compressive strength of 32 MPa, as defined in NBR 16697 [16], density of 3.08 g/cm^3, fineness index of 4.47% and loss on Ignition (LOI) < 6.5%. The fly ash has a density of 2.63 g/cm^3, fineness index of 3.82% and LOI < 3%. This waste product is classified as type C, following the specifications from NBR 12653 [17]. Specific gravity of the binder material was determined according to NBR 16605 [18]. The major chemical compounds of fly ash is presented in Table 1. The natural sand has 300 μm maximum particles size and density of 2.33 g/cm^3. The superplasticizer used in the composites mixtures is the Adiment (manufactured by Vedacit) based on melamine-formaldehyde resin with solid content of 20.6% and density of 1.12 g/cm^3.

Table 1. Major chemical compounds in fly ash.

Compound	Al_2O_3	SiO_2	SO_3	K_2O	CaO	Fe_2O_3	TiO_2
Fly ash (% by mass)	31.82	55.20	1.35	3.00	1.54	5.66	1.15

The fibers used to produce the composites are polyvinyl alcohol (PVA) fibers (trademark REC15 and RECS15), manufactured by Kuraray Co, Japan, with lengths of 8 mm and 12 mm, and polypropylene (PP) fiber, manufactured by Maccaferri, with lengths of 6 mm and 12 mm. The physical and mechanical properties of PVA and PP fibers are listed in Table 2.

Table 2. Properties of PVA and PP fibers [19, 20].

Properties	PVA		PP	
Length (mm)	8	12	6	12
Diameter (μm)	40	40	18	18
Aspect ratio	200	300	333.3	666.7
Density (g/cm^3)	1.31	1.31	0.91	0.91
Elastic modulus (GPa)	39.9	39.9	3	3
Fiber tensile strength (MPa)	1576.2	1576.2	300	300

One unreinforced (plain) mortar and five composites were produced for the tests, which are labelled as mortar, PVA8200, PVA12200, PP6200, PP12200 and PP6286. PVA8200 was produced with 2% by volume of PVA fiber with 8 mm length, PVA12200 was produced with 2% by volume of PVA fiber with 12 mm length, PP6200 was produced with 2% by volume of PP fiber with 6 mm length, PP12200 was produced with 2% by volume of PP fiber with 12 mm length and PP6286 was produced with 2.86% by volume of PP fiber with 6 mm length. In all composite mixtures, superplasticizer was used to achieve consistent rheological properties for better fiber distribution and workability. The mixture proportions are given in Table 3.

All mixtures were produced in a mechanical mixer with 20 L capacity. To produce the mixtures (mortar and composites), all dry raw materials, except fibers, were previously mixed. Water and superplasticizer were added and the fresh mixture was stirred for approximately 5 min to allow appropriate workability of the mixture. In the composite mixture, the fibers were added manually to the cementitious matrix and the mixture was stirred for more 5 min.

Table 3. Mixtures design.

Mix	FA/C	S/C	W/B	SP/B	Fiber (%)	
					PVA	PP
Mortar	–	2.47	0.63	–	–	–
PVA08200	1.20	0.79	0.33	0.015	2.00	–
PVA12200	1.20	0.79	0.33	0.015	2.00	–
PP06200	1.20	0.79	0.33	0.015	–	2.00
PP12200	1.20	0.79	0.33	0.015	–	2.00
PP06286	1.20	0.79	0.33	0.017	–	2.86

FA: fly ash, C: cement, B: binder (C + FA), S: sand,
W: water and SP: superplasticizer.

Fresh properties of mixtures were determined through consistency and density tests. A small flow cone for conventional flow table test was used to quantify the deformability of mixtures according to NBR 13276 [21], and the bulk fresh density was measured according to EN 1015-6 [22]. The composites mixtures presented consistence between 22.9 and 30.6 cm and density in the range of 1.93–1.96 g/cm³.

The flexural and compressive specimens were cast in steel moulds. The impact specimens were cast in expanded polystyrene moulds (see Fig. 2). All specimens were demoulded 24 h after casting and cured for 28 days (27 days in a water tank with 100% relative humidity and temperature of 21 ± 1 °C and one day in laboratory environment, i.e., 60% relative humidity and temperature of 24 ± 1 °C.

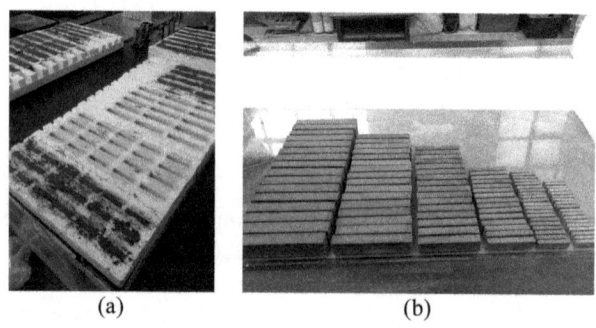

(a)   (b)

Fig. 2. Impact specimens used in the research: (a) during casting and (b) after casting.

2.2   Testing Procedure

Compressive strength and bending behaviour of composites were determined on the 28th day after casting using a Instron universal testing machine with a capacity of 100kN. Three specimens measuring 400 × 60 × 13 mm (length × width × thickness) were tested under four-point bending loads at a span of 100 mm and crosshead rate of 0.2 mm/min. Deflections at mid-span were measured using an electrical transducer. Three cylindrical specimens with dimensions of 50 × 100 mm (diameter × height) were used in compression tests. The compressive tests were performed with a cross-head rate of 0.6 mm/min.

The influence of sample dimension on the Charpy impact characteristics was evaluated for fiber reinforced cement-based composites and a mortar on the 28th day after casting. The tests were based on ASTM E23 standard [4] and was used to determine fracture energy. Impact testing machine with a maximum kinetic energy output of approximately 147 J was used in the test and the loading configuration is shown in Fig. 3a. After setting the specimen, the pendulum was released from a height h_1 and swinged through the specimen to a height h_2 with a velocity of 5.5 m/s. Assuming negligible friction and aerodynamic drag, the energy absorbed by the specimen was calculated as the height difference times the pendulum weight. Figures 3b–3d show a working scheme of Charpy impact machine and the specimen dimensions and span used in the loading for the Charpy impact test are presented in Table 4.

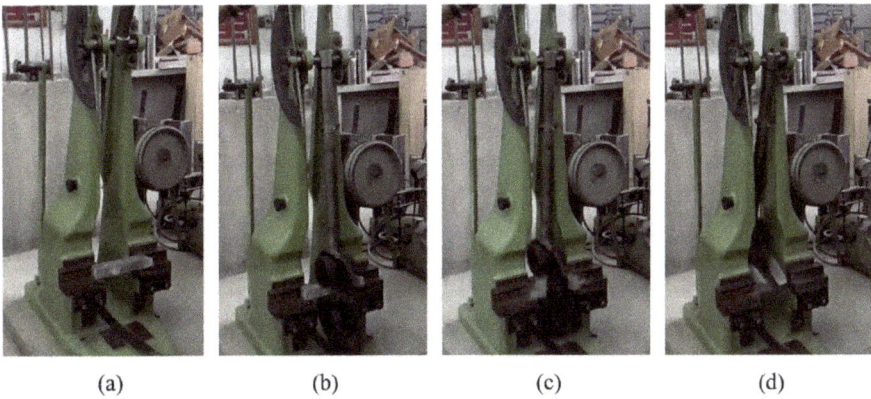

(a)   (b)   (c)   (d)

Fig. 3. (a) Setup of Charpy test and its working scheme: (b) before the impact loading, (c) hammer strikes and (d) specimen rupture.

Table 4. Details of specimen sizes and span used in the Charpy impact test.

Dimension	Mortar					PVA specimens						
						PVA8200			PVA12200			
Width (mm)	12	18	24	30	36	18	30	36	18	24	30	36
Depth (mm)	12	18	24	30	36	18	30	36	18	24	30	36
Length (mm)	72	108	144	180	216	108	180	216	108	144	180	216
Span (mm)	48	72	96	120	144	72	120	144	72	96	120	144

Dimension	PP specimens										
	PP6200				PP12200		PP6286				
Width (mm)	18	24	30	36	24	36	12	18	24	30	36
Depth (mm)	18	24	30	36	24	36	12	18	24	30	36
Length (mm)	108	144	180	216	144	216	72	108	144	180	216
Span (mm)	72	96	120	144	96	144	48	72	96	120	144

Up to twenty unnotched Charpy impact test specimens of square cross section were tested for each sample variety, at a bearing distance of 4 times specimen depth. Specimen sizes ranged from $12 \times 12 \times 72$ mm to $36 \times 36 \times 216$ mm. In this experiment the minimum width used in the impact specimen was defined based on the fiber size, as recommendation in [14] and the maximum width was based on the machine apparatus limitation, whereas, the specimen length (L) was always equal to 6 times specimen width, following [4].

A one-way analysis of variance statistical test, ANOVA, at the 5% significance level was carried out to examine the significance of the difference between various specimen-size of mortar and composites under Charpy impact test.

3 Results

3.1 Mechanical Behaviour

Typical bending stress - deflection curves obtained from the bending tests are shown in Fig. 4 and the results in relation to first cracking stress (σ_{cr}), ultimate bending strength (σ_u), ultimate deflection (δ_u) and toughness (T) are summarized in Table 5. Each result in Table 5 is the average of three specimens. In this research, the σ_{cr} is defined as the point where nonlinearity in the bending stress-deflection curve becomes evident, σ_u is related to maximum stress after the first cracking point, δ_u refers to the deflection when crack opening localization occurs, while toughness was calculated as the area under the curve up to the ultimate bending strength. In the Table 5 also is presented the compressive strength (f_c) of the composites and the standard deviation (in parenthesis).

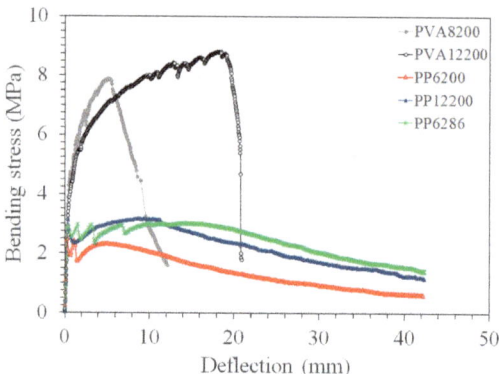

Fig. 4. Typical bending-deflection curves of the composites.

Table 5. Bending and compressive tests results of the specimens.

Mixture	σ_{cr} (MPa)	σ_u (MPa)	δ_u (mm)	T (N/mm)	f_c (MPa)
PVA8200	3.69 (0.15)	8.03 (0.94)	7.49 (2.16)	50.98 (21.02)	37.43 (1.45)
PVA12200	3.78 (1.42)	9.05 (0.23)	19.26 (0.15)	148.47 (6.28)	35.73 (0.25)
PP6200	2.68 (0.13)	2.41 (0.22)	5.41 (0.58)	10.94 (1.93)	39.27 (1.88)
PP12200	2.96 (0.59)	3.09 (0.24)	9.94 (0.99)	18.71 (7.18)	39.08 (1.02)
PP6286	3.60 (0.34)	3.43 (0.32)	13.78 (5.03)	39.63 (14.25)	34.95 (0.25)

In the graph, all composite specimens exhibit apparent multiple cracking patterns accompanying deflection-hardening behaviour with average deflection capacity ranging from 5.4 to 19.3 mm. This ductility remains much higher than that in normal concrete. It is also possible to observe that mixtures with PVA fibers achieve higher strengths, ultimate deflection and toughness due to the more elevated fiber properties (strength and elasticity modulus) when compared to PP fibers.

The bending results also show that the effect of fiber length and content is as significant as expected. From results in Table 5, it can be concluded that for a given fiber content (2%), the higher length of fibers, more elevated the composite bending performance. For instance, when it was used fiber with 12 mm length, the first cracking strength, ultimate bending strength, ultimate deflection and toughness of PVA12200 composite were increased, respectively, 2.4%, 12.7%, 157.1% and 191.2%, when compared with the values of PVA8200, while the first cracking strength, ultimate bending strength, ultimate deflection and toughness of PP12200 increased, respectively, 10.4%, 28.2%, 83.7% and 71%, when compared with the values of PP6200.

Regarding to the increase of PP fiber content from 2% to 2.86%, the effect was more significant. The first cracking strength, ultimate bending strength, ultimate deflection and toughness of PP6286 were increased, respectively, 34.3%, 42.3%, 154.7% and 262.2%, when compared with the values of PP6200.

On the other hand, there were little effect with regard to the fiber length and content on the compressive strength of composites. In details, for the same fiber content (2%), composite with a higher fiber length (12 mm) showed compressive strength up to 4.5% lower than composite with 6 mm length, regardless of fiber type. When a higher fiber content was used in the PP composite (PP6286), the reduction observed on the compressive strength was 11%, compared to PP6200 specimens. The different behaviour between compressive and bending performance happens because in the compressive specimens the fibers are randomly dispersed inside the specimen, while in the bending specimens, the fibers are arranged in a preferential orientation and perpendicular to the bending loading.

3.2   Impact Tests

The average results of impact energy absorption of the specimen are shown in Table 6, as well as the standard deviation (in parenthesis) for each measured series. The energy absorption in the different specimens is report as the energy per unit area (J/cm^2), where the area is the specimen cross-sectional area measured at the impact location. As expected, mortar specimens presented lower impact energy absorption than PP and PVA composites, due to brittle nature of plain mortar. The energy absorption capacity of PVA and PP composites is up to 5.4 and 1.3 times the energy absorption capacity of mortar, respectively, for the same specimen size. Also is seen that PVA composite presented more elevated energy absorption than PP composites due to the properties of the PVA fibers [9] and by the matrix–fibre interfacial bonding [23]. In this research, PP fiber fractures and PVA fiber pullout were seen to occur (see Fig. 5) and is well known, that fiber fractures produce a significantly more brittle impact response in FRCC as opposed to the fiber pullout mode that is highly energy absorbing [23].

Table 6. Impact energy absorption of the specimens.

Dimension (mm)	Mortar	PVA8200	PVA12200	PP6200	PP12200	PP6286
12 × 12 × 72	0.37 (0.01)	–	–	–	–	0.44 (0.03)
18 × 18 × 108	0.42 (0.01)	1.87 (0.21)	2.25 (0.29)	0.47 (0.05)	–	0.55 (0.06)
24 × 24 × 144	0.51 (0.01)	–	2.13 (0.21)	0.57 (0.02)	0.67 (0.06)	0.59 (0.02)
30 × 30 × 180	0.62 (0.03)	1.80 (0.25)	1.95 (0.23)	0.64 (0.04)	–	0.69 (0.08)
36 × 36 × 216	0.66 (0.04)	1.67 (0.20)	1.69 (0.24)	0.72 (0.03)	0.73 (0.04)	0.75 (0.04)

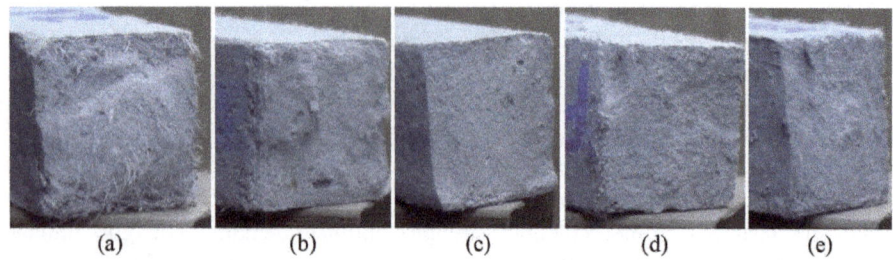

(a)   (b)   (c)   (d)   (e)

Fig. 5. Typical crack patterns for (a) PVA8200, (b) PVA12200, (c) PP6200, (d) PP12200 and (e) PP6286 specimens after impact tests.

Specimens with higher fiber length also showed higher impact energy than specimen with lower fiber length. A higher embedded length across the crack provides a higher fiber pullout resistance due to an increased bonding area between the fibers and the matrix [1]. Besides that, the effect of fiber length was more evident for specimens with lowest dimension. For instance, energy absorption capacity of PVA specimens with 12 mm length (PVA12200) were 20.3% (18 × 18), 8.2% (30 × 30) and 1.2% (36 × 36) higher than that for PVA specimens with 8 mm length (PVA8200) and energy absorption capacity of PP specimens with 12 mm length (PP12200) were 17% (24 × 24) and 1.4% (36 × 36) higher than that for PP specimens with 6 mm length (PP6200). Hence, the smaller specimen size enhances the influence of an increase in the fiber length. Energy absorption capacity is therefore highly specimen size dependent. It is also known that a smaller specimen acquires a preferential fiber orientation along the length of specimen and hence a higher ductility.

As the fiber increase in content, increase the number of bridging fibers the crack, and this can significantly increase the energy absorbed at the cracked section [15]. As a result, the energy absorption capacity of specimens with 2.86% PP fiber (PP6286) was approximately 17% (18 × 18), 4% (24 × 24), 8% (30 × 30) and 4% (36 × 36) higher compared with the energy absorption capacity of specimens with 2% PP fiber (PP6200). This result is in good agreement with other literature findings [13, 24].

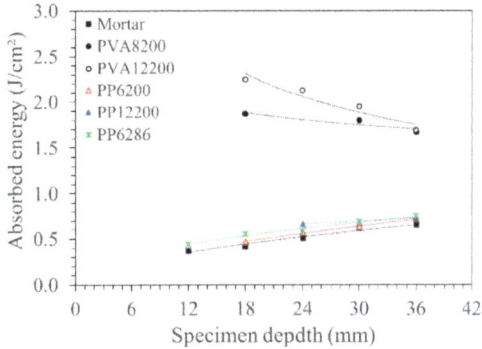

Fig. 6. Absorbed impact energy of the specimens as a function of specimen size.

The influence of specimen size in the range between $12 \times 12 \times 72$ and $36 \times 36 \times 216$ mm (expressed as $D \times W \times L$) is shown in the Fig. 6. The horizontal axis refers to specimen depth (D). Notice that for both PVA composites (PVA8200 and PVA12200), the energy absorption capacity appears to decrease as the specimen size was increased, however, statistical analysis indicates that only PVA12200 specimens with 30 mm and 36 mm have energy absorption capacity different of PVA 12200 specimens with 18 and 24 mm and energy absorption capacity of PVA8200 specimens with 36 mm depth is different de PVA8200 with 18 mm and 30 mm. This trend is coherent, because it is expected that a higher specimen depth leads to a decrease in impact energy absorption with ductile specimens [15], but it appears to be influenced by the ductility (toughness and deflection capacity) level of the composites (see Table 5 and Fig. 4) under static bending loads, as can be seen in Fig. 6. In details, the energy absorption of PVA8200 and PVA12200 specimens decreased approximately 11% and 25%, respectively, when the specimen depth increased from 18 mm to 36 mm. Instead, different trend is observed for mortar and PP composites. The impact absorption capacity of PP composites and mortar significantly increased with the increasing specimen size in all the sizes analyzed. In this case, the energy absorption of mortar, PP6200 and PP6286 specimens increased up to 57%, 53% and 36%, respectively, when the specimen depth increased from 18 mm to 36 mm, and the energy absorption of PP12200 increased 9% when the specimen depth increased from 24 mm to 36 mm.

4   Conclusions

Based on tests result the following conclusions can be drawn:

- The increase in the fiber length and content improves the bending behaviour under static loads and the energy absorption capacity under impact loads in all the specimen sizes analysed;
- The impact energy absorption is always more elevated for PVA composites than PP composites;

- The influence of fiber length on the impact energy absorption is more evidenced in specimens with lowest dimension;
- The trend of impact energy absorption (increase or decrease), measured on specimens of different dimensions, are dependent on the type of material tested.
- The results found indicate that specimens of different dimensions cannot be compared even if they are made of the same material. Hence, specimen size for using in the Charpy test must be taken into consideration to allow the experiment to be reproduced in another laboratory;
- The results obtained indicate the need to create specific standard for the use in the Charpy test for FRCC materials, especially regarding the specimen size.

Acknowledgements. The authors acknowledge the Scientific and Technological Research Agency of Rio de Janeiro State in Brazil, FAPERJ, for the financial assistance and the Kuraray Co, Japan, for supplying of the PVA fibers.

References

1. Bentur, A., Mindess, S.: Fibre Reinforced Cementitious Composites. Taylor & Francis (2007)
2. Mohammadi, Y., Carkon-Azad, R., Singh, S.P., Kaushik, S.K.: Impact resistance of steel fibrous concrete containing fibres of mixed aspect ratio. Constr. Build. Mater. **23**(1), 183–189 (2009)
3. Kim, D.J., Wille, K., El-Tawil, S., Naaman, A.E.: Testing of cementitious materials under high-strain-rate tensile loading using elastic strain energy. J. Eng. Mech. **137**, 1–8 (2011)
4. American Society for Testing and Materials ASTM E23: Standard Test Methods for Notched Bar Impact Testing of Metallic Materials (2018)
5. Thomas, R., Sorensen, A.: Charpy impact test methods for cementitious composites: review and commentary. J. Test. Eval. **46**, 2422–2430 (2018)
6. Al-Oraimi, S.K., Seibi, A.C.: Mechanical characterisation and impact behaviour of concrete reinforced with natural fibers. Compos. Struct. **32**(1), 165–171 (1995)
7. Ma, Y., Zhu, B., Tan, M.: Properties of ceramic fiber reinforced cement composites. Cem. Concr. Res. **35**(2), 296–300 (2005)
8. Alomayri, T., Shaikh, F.U.A., Low, I.M.: Synthesis and mechanical properties of cotton fabric reinforced geopolymer composites. Compos. B Eng. **60**, 36–42 (2014)
9. Lavin, T., Toutanji, H., Xu, B., Ooi, R.K., Biszick, K.R., Gilbert, J.A.: Matrix design for strategically tuned absolutely resilient structures (STARS). Presented at the SEM XI International Congress on Experimental and Applied Mechanics, Orlando, FL, 2–5 June 2008. Society for Experimental Mechanics, Bethel, CT, 12 p. (2008)
10. Erdem, S., Kağnıcı, T., Blankson, M.A.: Investigation of bond between fibre reinforce polymer (FRP) composites rebar and aramid fibre-reinforced concrete. Int. J. Compos. Mater. **5**(6), 148–154 (2015)
11. Mansur, M.A., Aziz, M.A.: Study of bamboo-mesh reinforced cement composites. Int. J. Cem. Compos. Lightweight Concrete **5**(3), 165–171 (1983)
12. Liu, Z., Cui, Q., Li, Q.: Properties of GRC modified by emulsion. Presented at the GRCA 2015 Congress, Dubai, United Arab Emirates, 19–21 April 2015. International Glassfibre Reinforced Concrete Association, Hampton, UK, 14 p. (2015)

13. Atahan, H., Tuncel, E., Pekmezci, B.: Behavior of PVA fiber-reinforced cementitious composites under static and impact flexural effects. J. Mater. Civ. Eng. **25**, 1438–1445 (2013)
14. Radomksi, W.: Application of the rotating impact machine for testing fiber-reinforced concrete. Int. J. Cem. Compos. Lightweight Concrete **3**(1), 3–12 (1981)
15. Banthia, N.: Impact resistance of HPFRCC, International RILEM Workshop on High Performance Fiber Reinforced Composites (HPFRCC) in Structural Applications, Honolulu, Hawaii, USA, pp. 479–488 (2005)
16. Brazilian Standard NBR 16696: Cimento Portland – Requisitos. Associação Brasileira de Normas Técnicas (ABNT) (2018)
17. Brazilian Standard NBR 12653: Materiais Pozolânicos – Requisitos. Associação Brasileira de Normas Técnicas (ABNT) (2014)
18. Brazilian Standard NBR 16605: Cimento Portland e outros materiais em pó - Determinação da massa específica. Associação Brasileira de Normas Técnicas (ABNT) (2017)
19. Magalhães, M.S., Filho, R.D.T., Fairbairn, E.M.R.: Durability under thermal loads of polyvinyl alcohol fibers. Matéria **18**(4), 1587–1595 (2013)
20. Fibromac polymer fibres - Technical data sheet. Maccaferri. United Kington. https://www.maccaferri.com/. Accessed 28 Jan 2021
21. Brazilian Standard NBR 13276: Argamassa para assentamento e revestimento de paredes e tetos - Determinação do índice de consistência. Associação Brasileira de Normas Técnicas (ABNT) (2016)
22. Comité Européen de Normalisation EN 1015-6: Methods of test for mortar for masonry - Part 6: Determination of bulk density of fresh mortar (1998)
23. Bindiganavile, V., Banthia, N., Aarup, B.: Impact response of ultra-high-strength fiber-reinforced cement composite. ACI Mater. J. **99**(6), 543–548 (2002)
24. Yoo, D.Y., Yoon, Y.S., Banthia, N.: Flexural response of steel-fiber-reinforced concrete beams: effects of strength, fiber content, and strain-rate. Cement Concr. Compos. **64**, 84–92 (2015)

The Effect of Organic Fiber Hybridization on Fresh and Hardened Concrete Properties

Imane Bentegri[1,2(✉)], Tien Tung Ngo[1], Hamza Soualhi[3],
Othmane Boukendakdji[2], and El-Hadj Kadri[1]

[1] CY Cergy Paris University, Laboratory L2MGC, 95000 Cergy, France
[2] University of Yahia Fares, Laboratory LME, Medea, Algeria
[3] Amar Telidji Laghouat University, Laboratory LRGC, Laghouat, Algeria

Abstract. The present study aims to evaluate the effect of two different fiber lengths (19 and 30 mm) and morphology (twist and wave) hybridization on both fresh and hardened properties of ordinary concrete. The fiber hybridization consists on the use of fiber with different form, types or morphology in the same concrete mixture. In this study, the total dosage of fiber used was fixed at 0.12% and five fiber proportion has been tested as 0/100%; 30/70%; 50/50%; 70/30% and 100/0% of fiber with 19 mm length and 30 mm respectively. The obtained results indicate that this combination of fibers leads to enhance both the plastic viscosity and the constant viscous (decrease) for all mixtures comparing to that of concrete mixtures with one fiber type. In regards to hardened properties, it has been found that the fiber hybridization with 50/50% proportion increases the mechanical strength and the elastic modulus of concrete.

Keywords: Fiber reinforced concrete · Polypropylene fibers · Fibers hybridization · Rheology · Tribology · Mechanical properties

1 Introduction

Modern concrete composites have been successfully used in the world due to its outstanding mechanical properties, such as, high flexural strength [1], good failure impact resistance [2], ductility and cracks control [3] besides the reduction of shrinkage and expansion rate of concrete [4]. However, the incorporation of fibers modifies the concrete performance at fresh and hardened states [5]. Furthermore, the nature of fiber determines its influence, for instance, polypropylene fibers (PPF) are used to control plastic shrinkage cracks and improve fire resistance, which has been largely demonstrated in previous investigations on the hardened properties of concrete.

Recently, a few studies on the fresh properties of fiber reinforced concrete have been published, using only the slump test to evaluate workability of concrete [3]. In order to control concrete implementation, however, two concrete formulation with the same slump value can have different rheological properties, that is why it is important to control not only the concrete slump but also its rheological behaviour [6]. In addition, tribological behaviour has been ignored completely in literature, despite its importance to estimate affecting pumpability parameters and to prevent the cause blockage during the pumping process.

© RILEM 2022
P. Serna et al. (Eds.): BEFIB 2021, RILEM Bookseries 36, pp. 86–97, 2022.
https://doi.org/10.1007/978-3-030-83719-8_8

Otherwise, basing on literature data, it is known that the use of two different length of fibers in concrete (hybridization) lead to enhance its mechanical properties, as the short and long fibers limit the development of microcracks and macrocracks respectively. Therefore, the present study aims to evaluate the effect of two different fiber lengths (19 and 30mm) with morphology (twist and wave respectively) hybridization on both rheological and tribological properties of ordinary concrete and mechanical strength. The total dosage of fiber was fixed at 0.12% and five fiber proportion has been tested as 0/100%; 30/70%; 50/50%; 70/30% and 100/0% of fiber 19mm and 30mm respectively.

2 Experimental Program

2.1 Materials

The cement used for all concrete mixes is an ordinary Portland cement (CEM I 52.5N), characterized by a density of 3.1 g/cm^3. The granulates characteristics of the coarse and the fine aggregates are given in Table 1 while Fig. 1 shows their grading curves.

Table 1. Aggregates properties.

Aggregates type	Size (mm)	Specific gravity (g/cm^3)	Compactness	Water absorption (%)	Finesse modulus
Gravel (G)	6.3–20	2.42	0.61	2.3	–
Sand (S)	0–4	2.79	0.64	0.9	2.32

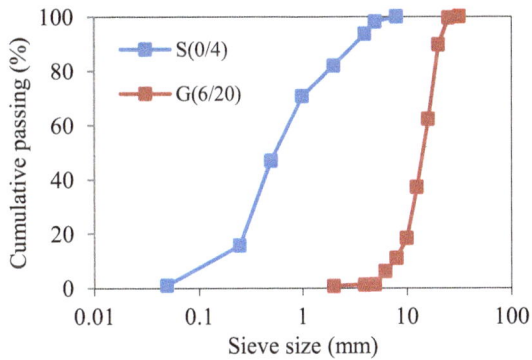

Fig. 1. Grading curves of fine and coarse aggregates.

Moreover, in the present investigation, polypropylene fiber used are produced by SikaFibre® were added to concrete mixes with 2 different lengths (19 and 30 mm) as shown in Fig. 2.

(a)   (b)

Fig. 2. Polypropylene fibers (PPF) used, PPF30 mm (a), PPF19 mm (b).

The 19 fibrillated twist PPF is intended to be used for reinforcement and precast concrete due to its high tensile strength (689 MPa). However, 30 mm PPF have a wavy shape and lower tensile strength (486 MPa). At the same time, this type of fibers is more flexible thanks to its high elastic modulus (6.9 GPa), unlike 19mm fibers (5.75 GPa). The properties of these fibers are depicted in Table 2. The difference between fibers shape can be clearly seen in Fig. 3 and Fig. 4 obtained by scanning electron microscope (SEM) analysis.

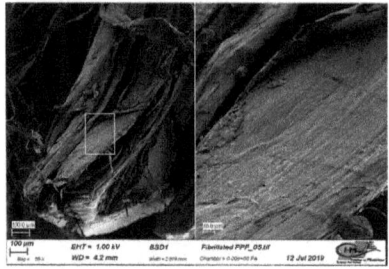
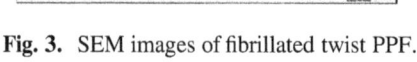
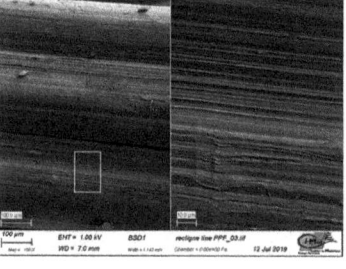

Fig. 3. SEM images of fibrillated twist PPF.   **Fig. 4.** SEM images of wave PPF.

Table 2. PPF properties.

Length (mm)	Diameter (mm)	Shape	Specific gravity	Elastic modulus (GPa)	Tensile strength (MPa)
19	0.34	Fibrillated twist	0.92	5.75	689
30	0.48	wave	0.91	6.90	486

Lastly, the superplasticizer (Sp) used was polycarboxylate based type, with a solid mass content of $33.0 \pm 1.5\%$ and a specific gravity of 1.07 ± 0.02.

2.2   Concrete Mix Design

The concrete mixes were formulated with the same optimal G/S ratio (1.44), determined by Dreux-Gorisse method [7] using the grading curves of aggregates (Fig. 1). In addition, W/C ratio has been fixed at 0.5 with a cement paste volume of 0.380 m^3. In order to control micro and macrocracks in construction sites, it is recommended to use two different lengths (short and long) in the same mix. Considering this fact, the hybridization effect has been investigated by studying five concrete mixes with different 19/30 mm fibers ratios equal to 0/100%, 30/70%, 50/50%, 70/30% and 100/0 respectively.

Noted that using fibers in concrete decreases its workability [8], researchers proposed adding superplasticizer in FRC to improve its workability and fluidity. Therefore, a superplasticizer has been added to maintain a good concrete workability behaviour with 0.2% percentage. The detailed mix proportions of all phases are shown in Table 3.

Table 3. Mix proportions.

Concrete	P.V (m3)	W/C	G/S	L Fiber (mm)	Fiber (%)	Cement (Kg)	Water (Kg)	Aggregate (Kg)	Sand (Kg)	Sp (%)
FRC19/30	0.380	0.5	1.44	19(0%)	0.000	462	231	963	669	0.2
				30(100%)	0.120					
				19(30%)	0.036					
				30(70%)	0.084					
				19(50%)	0.060					
				30(50%)	0.060					
				19(70%)	0.084					
				30(30%)	0.036					
				19(100%)	0.120					
				30(0%)	0.000					

2.3   Test Protocol and Measurements

The concrete mixes were prepared using a concrete mixer and the introduction of fibers is based on the technical sheet of fibers and according to NF EN 14845-1 standard, starting with mixing fine and coarse aggregates for 1 min, cement and fibers are then added and mixed for 2 min, and finally the water containing the superplasticizer is added. The fresh concrete was instantly subjected to the following measurements:

2.3.1   Workability

In this study concrete workability was measured using Abrams cone according to the standard NF-EN 12350-2 [9].

2.3.2 Rheological Measurement

The rheological test has been performed using a new concrete rheometer, developed by Soualhi et al. [10], to determine the rheological properties (Fig. 5). The apparatus is composed of an agitator with speed electronic control, a steel vane with a double U shape and a cylindrical container (h = 25 cm and d = 30 cm) with 18 square steel rods 9×9 mm^2 and 22 mm of length welded in the inner wall in order to optimize the adhesion of concrete to the container inner wall. It is worth noting that, the degree of accuracy is 5% and rage of confidence is 5.8% for the viscosity and 3.1% for the yield stress [10].

 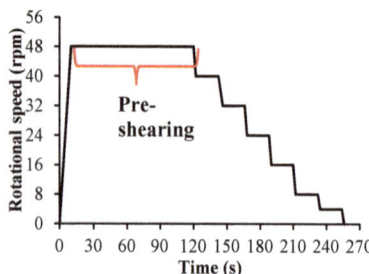

Fig. 5. Rheometer used and the imposed vane rotation speed profile [10].

The rheological test is performed according to the following steps:

- The metallic container is filled with fresh concrete in two layers; each layer is tamped 25 times with a metal rod.
- The vane is placed in the center and plunged into the concrete mix.

Thereafter, corresponding to the imposed rotation profile shown in Fig. 4, and for each speed level, the total torque is measured and recorded. The rotating vane creates two concrete layers during the rotation process. The first one is beyond the vane and moves with it while the second one still stagnant in the recipient borders. This phenomenon creates friction between the two concrete layers which allows to determine the concrete viscosity. In this study, the plastic viscosity μ was determined by assimilating concrete rheological behavior to a Bingham fluid to fit the shear stress and shear rate according to the following equation:

$$\tau = \tau_0 + \mu \dot{\gamma} \tag{1}$$

The determination of the Bingham model parameters is based on Reiner-Riwlin equation. (Eq. (2))

$$\Omega = \frac{M}{4\pi h \mu}\left(\frac{1}{R_1^2} - \frac{2\pi h \tau_0}{M}\right) - \frac{\tau_0}{2\mu}\ln\frac{M}{2\pi h\ \tau_0 R_1^2} \tag{2}$$

Where Ω, M, h, R_1, μ, and τ_0 present respectively the rotational speed (rad/s), torque (N.m), vane height (m); vane radius (m); plastic viscosity (Pa.s) and the yield stress (Pa).

Based on (Eq. 2), and using the assumed values of μ and τ_0, the mean squared error is calculated as the difference between the calculated and the measured rational speed for each measured torque point (Eq. 3).

$$mse = \frac{\sqrt{(\Omega_{calculated} - \Omega_{measured})}}{n} \tag{3}$$

Lastly, the plastic viscosity was obtained by minimizing the *mse* values [10].

2.3.3 Tribological Measurement

Given that pumpability of concrete depends on the steel-concrete friction occurred during the concrete flow through the pumping pipe, the purpose of this part is the determination of the viscous constant. Therefore, a new tribometer developed by Ngo et al. [11] was used (Fig. 6). The latter is composed of an agitator with electronic speed regulator and torque recorder, which is placed on the top of a smooth steel cylinder (h = 10 cm, d = 10.7 cm). The test consists in measuring the friction between the metal cylinder and concrete which allows to obtain the interfacial properties as the shear stress and the viscous constant of the boundary layer formed during the test as shown in Fig. 6.

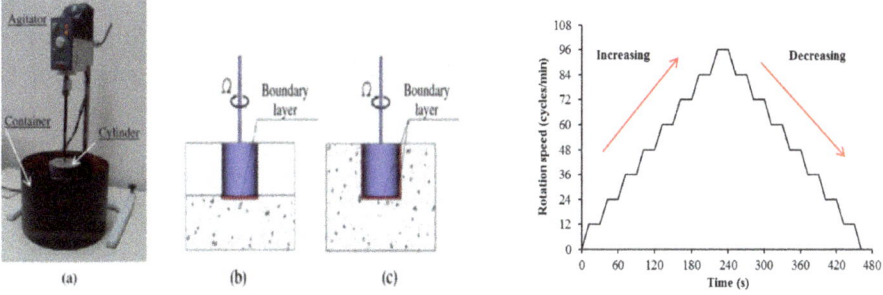

Fig. 6. Tribometer and imposed cylinder rotation speed profile used [11].

The measurement of the friction couple was performed with an imposed speed profile (Fig. 6), according to the following steps:

– As it is shown in Fig. 6 (b), the first layer of fresh concrete was filled up to the half of the container and received 25 stocks by a metal rod. Then the steel cylinder was fixed to the agitator and centered in the middle of the container.

– After saving the data from the previous phase, the second layer of concrete was filled up to the top of the container Fig. 6 (c). The same process of the previous step was followed and measurements were recorded [12].

It is worth noting that, the first step consists on determining the friction between the bottom of the cylinder and concrete, while the second one permits to measure frictions between concrete and the cylinder (bottom plus lateral surface). As only the lateral frictions are required, the final results are obtained by subtracting the first step data from the second step ones.

The viscous constant and (η) was determined from the obtained measured torques using the empirical correlation (Eq. 4) and the interface law equation (Eq. 5) to fit the test results as follows:

$$T = T_0 + kV \tag{4}$$

$$\tau = \tau_t + \eta v \tag{5}$$

Where T (N.m) and T_0 (N.m) are the torque imposed and the initial torque respectively, V (cycles/s) is the cylinder rotating speed and k (N.ms) is the linear coefficient. Besides, the interface parameters are the shear stress τ (Pa), the viscous constant η (Pa s/m), the yield stress τ_t (Pa) and the angular speed v (m/s).

In addition, taking into account the cylinder shape and dimensions, the interfacial parameters were calculated using the following equations:

$$\tau_t = \frac{T_0}{2\pi R_t^2 h_t} \tag{6}$$

$$\eta = \frac{K}{(2\pi)^2 R_t^3 h_t} \tag{7}$$

While R_t (m) and h_t (m) represent respectively the radius and the cylinder height.

2.3.4   Mechanical Strength Measurement

The compressive and splitting tensile strength testes were carried out at the age of 28 days according to the standards NF EN 12390-3 and NF EN 12390-6 respectively, using a hydraulic press machine with 3500 kN capacity. The testes were performed on concrete cylindrical specimens of 11 × 22 cm.

2.3.5   Elastic Modulus Measurement

The modulus of elasticity at the age of 28 days, was determined according to the standard NF EN 12390-13 using the hydraulic press (3R) (shown in Fig. 7(a)). The test consists of placing a cylindrical specimen (11 × 22 cm) under three loading cycles with a speed of 0.6 ± 0.2 MPa/s. The specimen is fixed inside a light metal crown (Fig. 7(b)) which contains three sensors for measuring longitudinal unit strains.

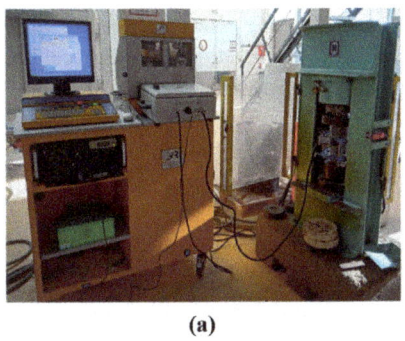

(a)   (b)

Fig. 7. The modulus of elasticity test (a), location of the specimen in the crown between the two press plates (b).

3   Results and Discussion

3.1   Effect of PPF Hybridization on Fresh Properties

The effect of fiber hybridization on slump, density and air content of different mixtures is shown in Table 4. According to the latter, we find that the combination of fiber with the superplasticizer have no effect on the concrete workability, as for all mixtures the slump value is around 22 cm. Likewise, the density and the volume of air content seemed to be unaffected by this fiber mix, this is due to the low density of the fibers used (0.9).

Table 4. The slump test, fresh density and air content results

Concrete	P.V (m³)	W/C	G/S	L Fiber (mm)	Fiber (%)	Sp (%)	Slump (cm)	Δ (Kg/m³)	Air occlus (%)
FRC19/30	0.380	0.5	1.44	19(0%)	0.000	0.2	22	2333	2,2
				30(100%)	0.120				
				19(30%)	0.036		22	2328	2.2
				30(70%)	0.084				
				19(50%)	0.060		22	2329	2.3
				30(50%)	0.060				
				19(70%)	0.084		21	2330	2.2
				30(30%)	0.036				
				19(100%)	0.120		22	2329	2,1
				30(0%)	0.000				

   In terms of rheology, Fig. 8 (b) shows the evolution of the plastic viscosity as a function of PPF19 dosages, from 0 to 100%. Up to 70% of PPF19 dosage, it appears that the plastic viscosity remains constant with the increase of PPF19 percentage at the expense of PPF30. In contrast, from 70 to 100% of PPF19, a significant increase was observed. It can be said that the combination of both shapes and lengths till 70/30 of 19 and 30 mm PPF respectively has eliminated the negative effect of the use of 19mm fiber only on the rheological behaviour of concrete. This is due to the different shapes of the fibers used and the good fibers distribution in the mix thanks to the superplasticizer introduction.

At the same time, the viscous constant slightly increased (Fig. 9 (b)) with increasing PPF19 percentage up to 70%, and decreased thereafter to become equal to the viscous constant of FRC19/30 with 30% of PPF19.

According to the data obtained, it can be concluded that the combination of two different lengths and shapes has a less significant effect on the rheological and tribological concrete behaviours than that of using only one length and shape. It is worth noting that the lowest viscosity was obtained by using 30/70% of 19 and 30 mm PPF respectively.

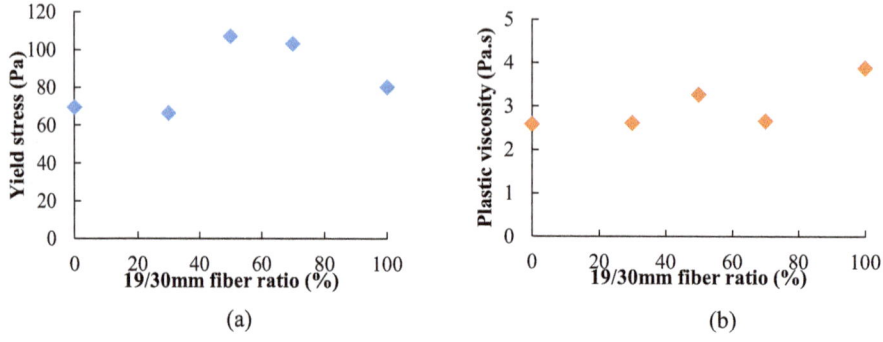

Fig. 8. Yield stress and plastic viscosity results

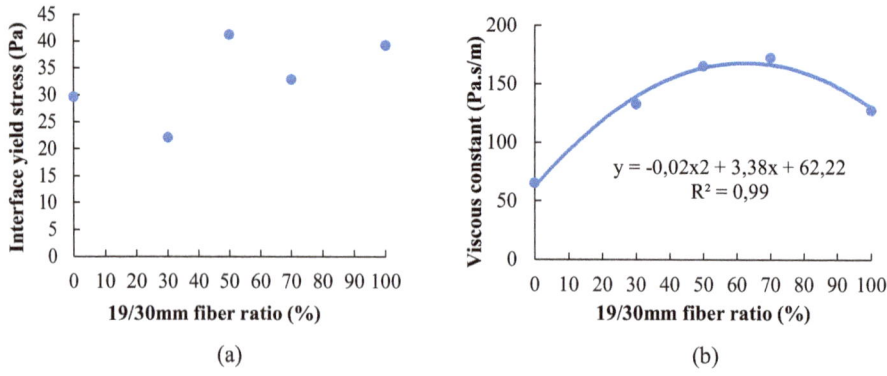

Fig. 9. Interface yield stress and viscous constant results

3.2   Effect of PPF Mixed Length on Hardened Properties

The compressive and tensile strength results of this study are shown in Fig. 10. It has been noticed that the use of multi-fibers in different proportions have not a significant effect on the concrete mechanical strength. However, with a fiber ratio of 50/50% the compressive and tensile strength increase with 10% in comparison with that of concrete made with 100% of only one type of fiber. In this regard, it has been concluded that the introduction of 50% PPF19 and 50% PPF30 permits to enhance concrete mechanical strength. Furthermore, it has been shown from Fig. 11, that the same mixture (50/50% of PPF) shows an increase of the elastic modulus with 10% in comparison with that of concrete made with 100% of PPF 19 mm. Therefore, it has been concluded that the hybridization of 50/50% of polypropylene fiber 19 and 30 mm in concrete enhance its mechanical properties.

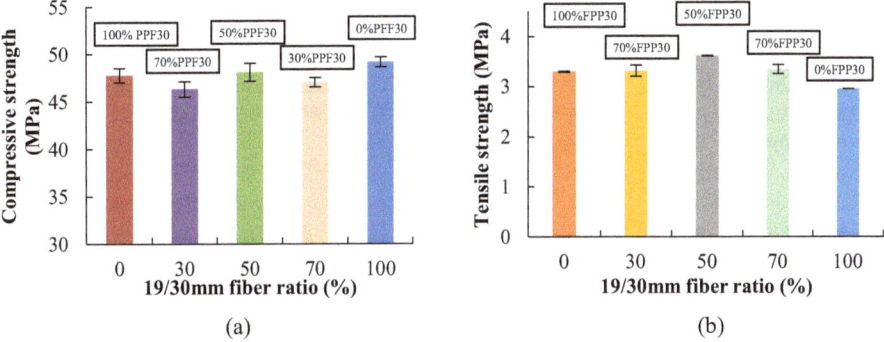

(a)   (b)

Fig. 10. Compressive and tensile strength results

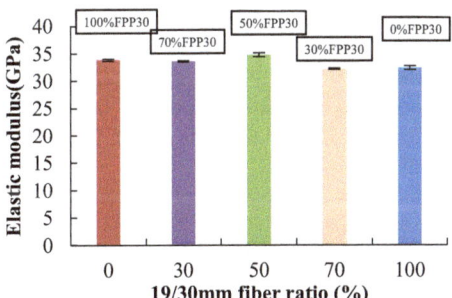

Fig. 11. Elastic modulus results

4 Conclusions

Based on the obtained results, the main conclusions can be summarized as follows:

- The introduction of fibers with a superplasticizer permit to maintain a good workability.
- The hybridization of fibers with different length and morphology enhances the concrete rheology and tribology behavior.
- The concrete formulation made with 30/70% of fiber 19 and 30 mm showed an optimum rheological and tribological parameters
- The mechanical properties have been positively affected by the introduction of fiber, and the hybridization of 50/50% showed the optimum results in regards of mechanical strength and elastic modulus.

References

1. Noushini, A., Hastings, M., Castel, A., Aslani, F.: Mechanical and flexural performance of synthetic fibre reinforced geopolymer concrete. Constr. Build. Mater. **186**, 454–475 (2018). https://doi.org/10.1016/j.conbuildmat.2018.07.110
2. jun Li, J., gang Niu, J., jun Wan, C., Jin, B., liu Yin, Y.: Investigation on mechanical properties and microstructure of high performance polypropylene fiber reinforced light-weight aggregate concrete. Constr. Build. Mater. **118**, 27–35 (2016). https://doi.org/10.1016/j.conbuildmat.2016.04.116
3. Hassanpour, M., Shafigh, P., Mahmud, H.B.: Lightweight aggregate concrete fiber reinforcement – a review. Constr. Build. Mater. **37**, 452–461 (2012). https://doi.org/10.1016/j.conbuildmat.2012.07.071
4. Kakooei, S., Akil, H.M., Jamshidi, M., Rouhi, J.: The effects of polypropylene fibers on the properties of reinforced concrete structures. Constr. Build. Mater. **27**(1), 73–77 (2012). https://doi.org/10.1016/j.conbuildmat.2011.08.015
5. Güneyisi, E., Gesoğlu, M., Akoi, A.O.M., Mermerdaş, K.: Combined effect of steel fiber and metakaolin incorporation on mechanical properties of concrete. Compos. Part B: Eng. **56**, 83–91 (2014). https://doi.org/10.1016/j.compositesb.2013.08.002
6. Chidiac, S.E., Mahmoodzadeh, F.: Plastic viscosity of fresh concrete – a critical review of predictions methods. Cement Concrete Compos. **31**(8), 535–544 (2009). https://doi.org/10.1016/j.cemconcomp.2009.02.004
7. Dreux, G., Gorisse, F., Simonnet, J.: Composition des betons : methode dreux-gorisse - bilan de cinq annees d'application en cote d'ivoire. ANN ITBTP, no 414(BETON 214), mai 1983. https://trid.trb.org/view/1039141. Accessed 09 Sept 2019
8. Zhang, K., et al.: How does adsorption behavior of polycarboxylate superplasticizer effect rheology and flowability of cement paste with polypropylene fiber? Cement Concrete Compos. **95**, 228–236 (2019). https://doi.org/10.1016/j.cemconcomp.2018.11.003
9. AFNOR et M. Hesling: NF EN 12350–2 Essais pour béton frais — Partie 2 : Essai d'affaissement, p. 3, juin 2019
10. Soualhi, H., Kadri, E.-H., Ngo, T.-T., Bouvet, A., Cussigh, F., Tahar, Z.-E.-A.: Design of portable rheometer with new vane geometry to estimate concrete rheological parameters. J. Civil Eng. Manag. **23**(3), 347–355 (2017). https://doi.org/10.3846/13923730.2015.1128481

11. Ngo, T.T., Kadri, E.H., Bennacer, R., Cussigh, F.: Use of tribometer to estimate interface friction and concrete boundary layer composition during the fluid concrete pumping. Constr. Build. Mater. **24**(7), 1253–1261 (2010). https://doi.org/10.1016/j.conbuildmat.2009.12.010
12. Ngo, T.-T., Kadri, E.-H., Cussigh, F., Bennacer, R.: Measurement and modeling of fresh concrete viscous constant to predict pumping pressures. Can. J. Civil Eng. **38**(8), 944–956 (2011). https://doi.org/10.1139/l11-058

The Antifragility of FRC in the Crack Pattern of Reinforced Concrete Ties

Alessandro P. Fantilli[✉] and Francesco Tondolo

DISEG, Politecnico di Torino, Turin, Italy
alessandro.fantilli@polito.it

Abstract. To assess the durability of Reinforced Concrete (RC) structures, a model capable of predicting the crack pattern of RC ties is herein introduced. Based on the classical tension-stiffening equations, such model provides the transfer length, which in turn depends on the bond-slip mechanism between steel and concrete. The aim is to compute the length of a tie which shows a single crack in the serviceability stage. In this particular situation, if the geometry does not change, transfer length only depends on the strength of plain or fiber-reinforced concrete (FRC). Nevertheless, the experimental investigation, performed on RC and R/FRC ties with the same geometrical and mechanical properties, reveals two different crack patterns. Specifically, RC ties show multiple cracking, whereas only one crack tends to appear in presence of FRC. This dichotomy can be ascribed to the so-called antifragility, which can be considered as the capacity of FRC to gain strength from its intrinsic disorder.

Keywords: Compressive strength · Probability distribution function · Transfer length · Serviceability stage

1 Introduction

With respect to plain concrete, greater tensile stresses can be detected in fiber-reinforced concrete (FRC) beams also in presence of large cracks. Accordingly, Model Code 2010 [1] allows the use of this concrete system with and without traditional reinforcement. At ultimate limit state, FRC structures in bending show better performances than plain concrete, especially when the amount of steel reinforcing bars is close to the minimum (e.g., in concrete slabs [2]). In service, FRC can reduce the width of cracks in the tensile zones of beams and, consequently, prevent the corrosion of rebar in hybrid reinforced concrete structures (i.e., reinforced with both fiber and steel bars [3]).

With and without the addition of fibers, reinforced concrete (RC) structures are designed through the semi-probabilistic approach, by using the same partial safety coefficients for the materials [1]. This is due to the fact that, for an ordinary amount of fibers (V_f = fiber volume fraction < 0.5%), the tensile and compressive strengths of plain concrete and FRC are practically the same. However, recent studies demonstrate that the residual tensile strengths of FRC, which are detected on the cracked surfaces, have to be reduced by safety coefficients larger than those used for the plain concrete [4]. As a result, de la Fuente et al. [4] proposed to recalibrate these coefficients by better analysing the probability distribution function of FRC.

© RILEM 2022
P. Serna et al. (Eds.): BEFIB 2021, RILEM Bookseries 36, pp. 98–108, 2022.
https://doi.org/10.1007/978-3-030-83719-8_9

In the case of material strength, the four shapes of the probability distribution functions shown in Fig. 1 can be observed [5]. Two normal/Gaussian distributions are illustrated in Fig. 1a and Fig. 1b, respectively. The first, which typically occurs to the yielding strength of steel, can be defined as robust, because small positive and negative events are expected (Fig. 1a). On the contrary, a possible distribution of the compressive strength of concrete is depicted in Fig. 1b, where the average/median value has large scatter. Such a curve, defined by Taleb [5] as fragile, is very rare, because it is quite impossible to measure symmetric probability distribution function of natural phenomena.

Asymmetric distributions, like those reported in Fig. 1c and Fig. 1d, are more frequent. The first remains fragile, because the possibility of unfavourable events is larger than those favourable, as the left tail of the distribution is thicker than the right part. On the contrary, the distribution shown in Fig. 1d can be defined antifragile, because the right fat-tail for favourable events is larger than the left part of the distribution. If the antifragility property of a complex system is to gain advantages from uncertainties and variability [5], then the larger the intrinsic disorder of a concrete system, the greater the probability of having higher strength. Accordingly, the random distribution of fibers in a cement-based matrix makes FRC more antifragile (with a distribution function of tensile strength depicted in Fig. 1d) than plain concrete, whose probability density of tensile strength remains that illustrated in Fig. 1c.

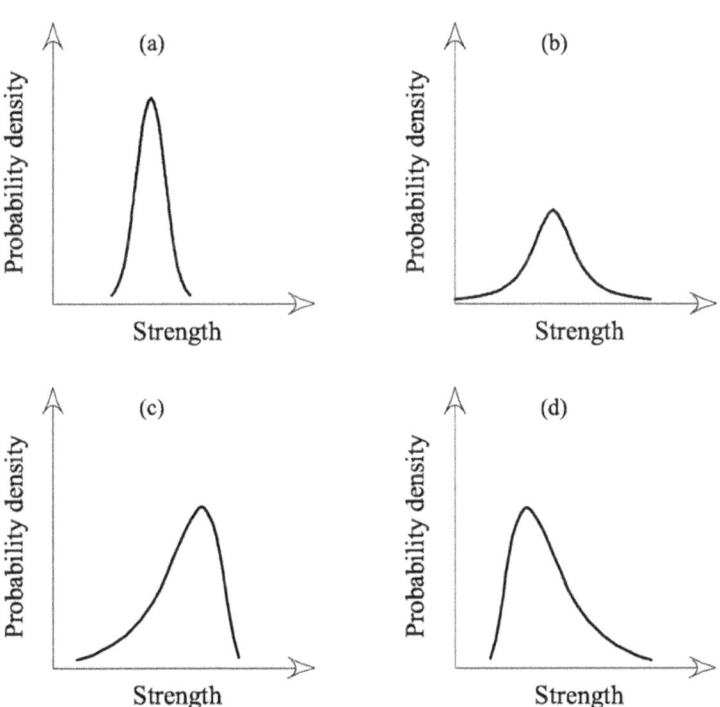

Fig. 1. Probability distribution functions of material strength: (a) symmetric – for robust materials; (b) symmetric – for fragile materials; (c) asymmetric – for fragile materials; (d) asymmetric – for antifragile materials.

Thus, the presence of fibers should reduce not only the crack width in service, but also the probability of cracking, which in turn depends on the tensile strength. This conjecture is demonstrated in the following sections by introducing a theoretical model of transfer length and by observing the crack pattern in RC and R/FRC ties of different sizes.

2   Modelling the Transfer Length in RC Ties

In 1968, Bresler and Bertero [6] carried out a series of tests on RC ties having a single crack in the midsection and subjected to cyclic loads (see Fig. 1). To avoid the formation of more than one crack in service, transfer length (or bond length) l_{tr} has to be computed in advance. For a given axial load N, l_{tr} is in turn a function of the tensile strength of concrete (f_{ct}) and of the bond-slip mechanism between the rebar and the surrounding concrete system [7]. On the other hand, transfer length also represents the domain where the classical tension-stiffening problem, defined by the following equations, has to be solved [7]:

$$\sigma_c(z)\, A_c + \sigma_s(z)\, A_s = N \tag{1}$$

$$\frac{d\sigma_s}{dz} = -\frac{p_s}{A_s}\tau(s(z)) \tag{2}$$

$$\frac{ds}{dz} = -\epsilon_s(z) + \epsilon_c(z) \tag{3}$$

where z = horizontal coordinate from the end of the tie; $\sigma_c(z)$ = stresses of concrete; $\sigma_s(z)$ = stresses of steel; A_c = cross-sectional area of concrete; and N = applied normal force; p_s = perimeter of steel bar; τ = bond stress defined as a function of the slip between steel and concrete ($s(z)$); $\varepsilon_c(z)$ = strain of concrete; and $\varepsilon_s(z)$ = strain of steel.

To calculate the distributions of $\sigma_c(z)$, $\sigma_s(z)$ and $s(z)$ (see Fig. 2), some conditions at the borders of the transfer length must be introduced. In the first stage of loading,

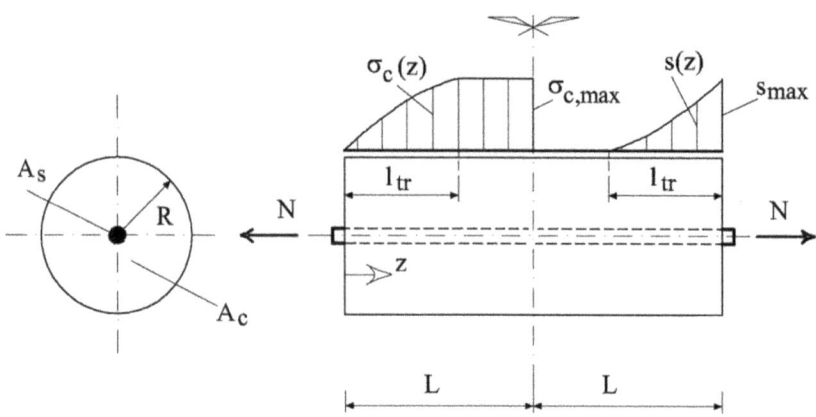

Fig. 2. Concrete tie reinforced with a single rebar.

when $N < N_{cr}$ (where N_{cr} = cracking load), the domain is a block of the tie bordered by a cross-section where $\sigma_{ct}(z = 0) = 0$ (the load is applied on the rebar), and by the so-called stage I cross-section, where no slip between steel and concrete occurs.

When $\sigma_{c,max} < f_{ct}$ ($\sigma_{c,max}$ is the stress in the midsection of the tie), perfect bond between steel and concrete (i.e., $\varepsilon_c = \varepsilon_s$) exists in the stage I cross-section. Moreover, transfer length increases with the applied normal force, as shown in Fig. 3, where the complete l_{tr}–N function is depicted. If $N \geq N_{cr}$, the domain is bordered by the cross-section without stresses in concrete, and by the cross-section at incipient cracking (i.e., $\sigma_{c,max} = f_{ct}$) [7]. In this stage of loading, during which $N_{cr} \leq N \leq N_y$ (N_y = yielding strength of steel rebar), a decrement of l_{tr} can be observed as N increases (see Fig. 3).

In addition to the boundary conditions, also the behaviour of materials has to be defined for computing l_{tr}. Both the materials are modelled with linear elastic stress-strain relationships, where E_s and E_c are the elastic moduli of steel and concrete, respectively. Whereas, for the bond-slip behaviour, the $\tau - s$ relationships proposed by Model Code 2010 [1] is adopted.

If $N_{min} \leq N \leq N_{max}$, where N_{min} ($> N_{cr}$) and N_{max} ($< N_y$) are the minimum and maximum loads applied within the serviceability stage, the RC (or R/FRC) tie depicted in Fig. 2 shows a single crack if $l_{tr,min} \leq L \leq l_{tr,max}$ (where $l_{tr,min}$ and $l_{tr,max}$ are localised on the post-cracking curve of Fig. 3 in correspondence of N_{max} and N_{min}, respectively; and L is the half-length of the tie).

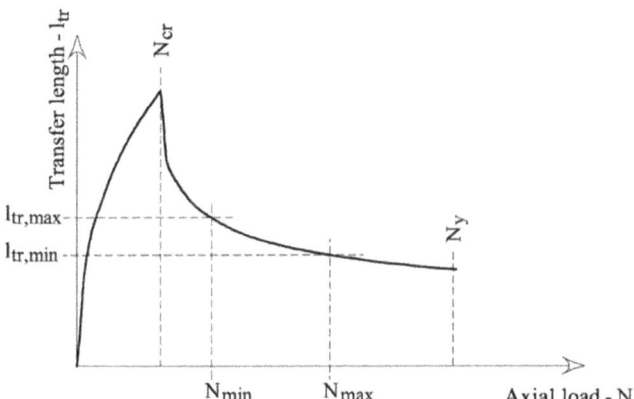

Fig. 3. Transfer length as a function of the axial load applied on RC or R/FRC ties.

3   The Experimental Campaign

A series of uniaxial tensile tests have been carried out on RC and R/FRC ties at the Politecnico di Torino. The geometrical properties of the specimens have been selected in order to generate a single crack in the serviceability stage (where $N_{min} \leq N \leq N_{max}$). Hence, the model previously described has been applied by considering the mechanical properties of the two concrete systems reported in Table 1. Except for the content of steel fibers (length = 35 mm, diameter = 0.67 mm, with hooked ends), the mixtures are the same for both plain and fiber-reinforced concrete. In the same table, the compressive strength and the standard deviation, measured on 4 cylinders, are also indicated.

Table 1. Properties of the concrete systems used in the RC and R/FRC ties.

Type of concrete	Water	Cement	Aggregate	Fiber volume fraction	f_c	Standard deviation
	kg/m^3	kg/m^3	kg/m^3	%	MPa	MPa
P	185	370	1725	0	27	14.7
F	185	370	1725	0.5		

In addition to the compressive strength measured experimentally, tensile strength, maximum bond stress (to be used in the bond-slip relationship), and the modulus of elasticity of concrete can be calculated, as a function of f_c, with the following formulae [1]:

$$f_{ct} = 0.3 \, (f_c - 8)^{2/3} \tag{4}$$

$$\tau_{max} = 1.5 \sqrt{f_c} \tag{5}$$

$$E_c = 21500 \left(\frac{f_c}{10}\right)^{1/3} \tag{6}$$

Referring to the concrete systems shown in Table 1, the proposed model provides the l_{tr}–N curves of Fig. 4. More precisely, Fig. 4a refers to a tie having the radius of concrete cover $R = 43.5$ mm and the diameter of rebar equal to 10 mm, whereas Fig. 4b illustrates the evolution of the transfer length in a tie in which the sizes are doubled. In both the ties, the geometrical reinforcement ratio is constant and equal to 1.34%.

In the smallest tie, called D10, crack appears when $N_{cr} = 17.3$ kN, and in the serviceability stage (where 22.5 kN $\leq N \leq$ 33 kN) l_{tr} varies between 164 mm and 210 mm. Thus, to maintain a single crack in service, 210 mm is assumed to be the half-length of the tie. Similarly, in the largest tie, called D20, the half-length $L = 340$ mm is selected, as the axial load varies between 90 kN and 130 (see Fig. 4b).

The geometrical properties of the ties remain the same, regardless of the content of fibers. Indeed, the transfer length of the specimen shown in Fig. 2 is only a function of concrete strength, and not of the residual tensile stresses. Thus, the presence of fibers should not modify the geometry of crack pattern, unless FRC is more antifragile than plain concrete. To assess the potential antifragility of fiber-reinforced concrete, the four series of the specimens reported in Table 2 have been subjected to uniaxial tensile loads [8].

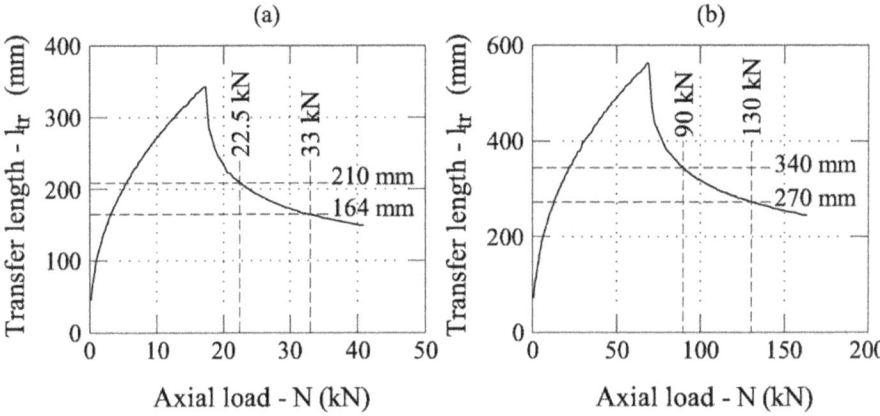

Fig. 4. Transfer length as function of the axial load: (a) N-l_{tr} curve of the tie D10; (b) N-l_{tr} curve of the tie D20.

Table 2. The specimens tested in this research project.

Series	Bar diameter	Concrete radius	Reinforcement ratio	L	Type of concrete	Nmin	Nmax
	mm	mm	%	mm		kN	kN
D10-P	10	43.5	1.34	210	P	22.5	33.0
D20-P	20	87	1.34	340	P	90.0	130
D10-F	10	43.5	1.34	210	F	22.5	33.0
D20-F	20	87	1.34	340	F	90.0	130

A single series, composed by five specimens, is labelled by DXX-Y, in which XX indicates the diameter of the rebar, and Y is P for plain concrete, and F for fiber-reinforced concrete. Each specimen has been tested by using a universal testing machine with a loading capacity of 100 kN and 250 kN for the samples D10 and D20, respectively. In all the cases, the driving parameter of the uniaxial tensile test has been the maximum stress of the reinforcing bar, which remained lower than the nominal yielding strength (i.e., 510 MPa). Before applying the cyclic loads, within the range $N_{min} \sim N_{max}$, the stabilized crack pattern has been fully created in each specimen by applying a static tensile load up to N_{max}. In each specimen, the midsection has been notched, in order to trigger the crack in this cross-section [8].

3.1 Test Results

As only the crack pattern is taken in to consideration herein, Fig. 5 illustrated the position of the cracks in the five specimens of the series D10-P. In most of the samples, three cracks have formed, despite they had been designed to show only one crack. As Fig. 6 illustrates, a different trend has been observed in the ties of the series D10-F. Indeed, the presence of three cracks dominates (80% of the cases) with respect to the formation of two cracks (20% of the cases) in the series D10-P. Vice-versa, in fiber-reinforced specimens (of the D10-F series), the presence of two cracks (60% of the cases) is more probable than that of three cracks, observed in the 40% of the cases.

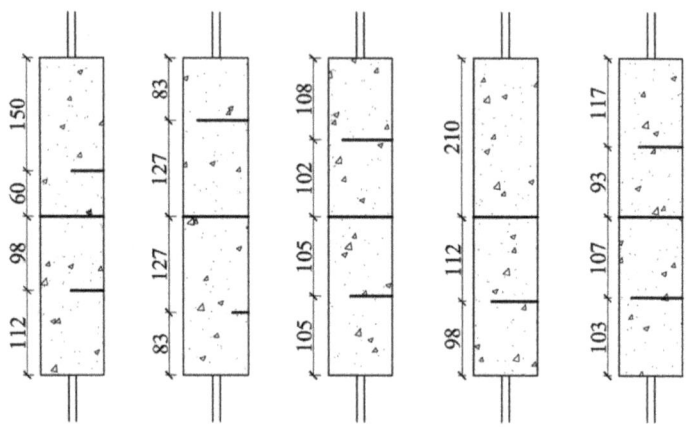

Fig. 5. Crack patterns in the D10-P ties (measures in mm).

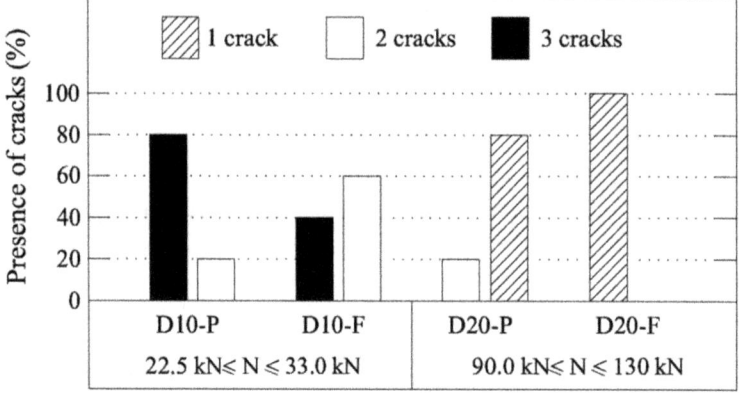

Fig. 6. The number of cracks observed in the concrete ties investigated herein.

When the size increase (i.e., in the series D20), a single crack appears in most of the specimens, with the exception of one sample of the series D20-P, in which two cracks have grown. In other words, the histogram depicted in Fig. 6 reveals that plain concrete is more prone to multiple cracking than FRC.

4   A Way to Predict the Experimental Results

Test results showed the different strengths of the concrete systems, although both the mixtures P and F (Table 1) are the same (with the exception for the fiber content). Thus, only when two different probability distribution functions of tensile strength are assumed, can the model previously described be used to predict the results of the tests. Such functions can be used to compute the whole l_{tr}-N curves, and the values of $l_{tr,min}$ as a function of concrete strength. If $l_{tr,min} < L$, three cracks appear, otherwise only one crack takes place in RC and R/FRC ties. Due to the symmetry of the element shown in Fig. 2, and to the presence of a single cracks in the midsection, the growth of two cracks cannot be predicted by the proposed model.

The numerical procedure used to compute the presence, and the probability of presence, of one or three cracks, is summarized in flow-chart depicted in Fig. 7. In this figure, the following symbols are used:

- I_{max} = number of the f_c values extracted (with a Montecarlo method) from a given statistical distribution of compressive strength.
- n_1 = number of the cases in which $l_{tr,min} > L$ (presence of a single crack).
- n_3 = number of the cases in which $l_{tr,min} \leq L$ (presence of three cracks).

To better reproduce the experimental data summarised in Fig. 6, plain concrete is assumed to be fragile, whereas FRC is antifragile, in the meaning given by Taleb [5] to these two adjectives. Therefore, the probability distribution functions are those of Fig. 1c and Fig. 1d for plain concrete and FRC, respectively. They can be numerically represented by the log-normal functions shown in Fig. 8, in which the median value (equal to f_c = 27 MPa as measured in the experimental tests - see Table 1) and the standard deviation (i.e., 14.68 MPa) are the same for the two concrete systems. In addition, to have a fat tail in the strength of FRC, the mode value is lower than that of the median. On the contrary, to make plain concrete fragile, the mode value has to be larger than the median.

The histogram depicted in Fig. 9 summarizes the shapes of the crack pattern, obtained by applying the proposed model to the four series of the RC and R/FRC ties investigated herein, and by assuming the probability distribution functions shown in Fig. 8.

When $N_{min} \leq N \leq N_{max}$, satisfactory results are obtained for the series D20, in which only one crack grows in most of the cases, especially in the presence of fibers. For the ties D10, a prevalent behaviour cannot be identified, and one or more cracks take places. However, in presence of fibers, the probability of multiple cracking is less probable than in plain concrete. This occurrence may play a significant role with regards, e.g., to durability issues of concrete structures.

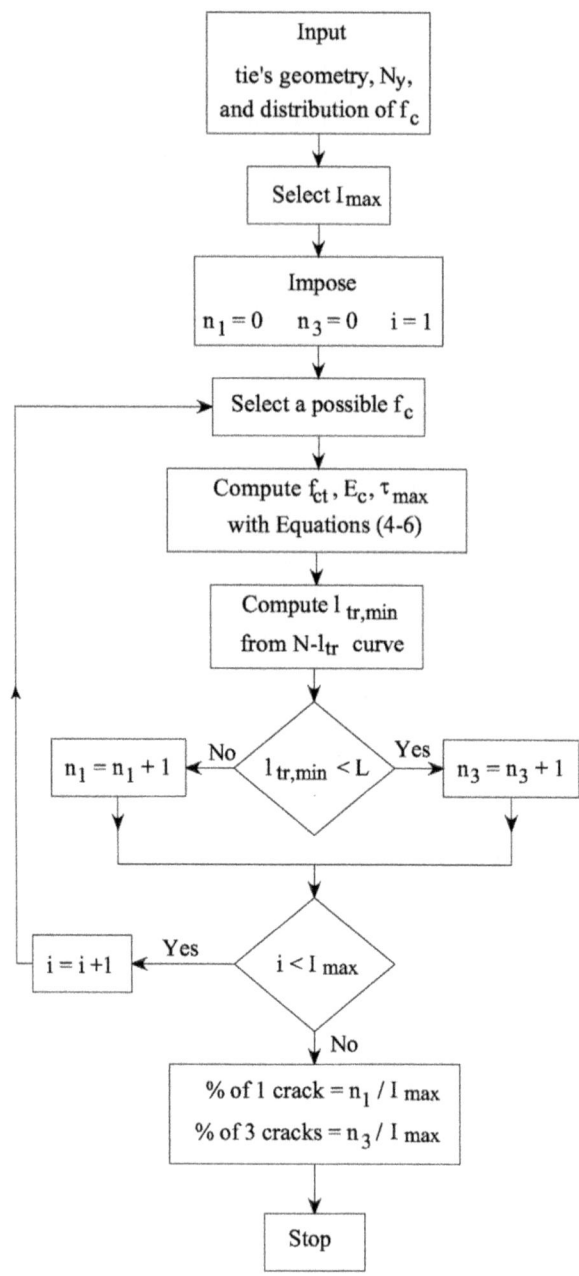

Fig. 7. Numerical procedure used to calculate the probability that 1 or 3 cracks occur in the concrete ties D10 and D20, with and without fibers.

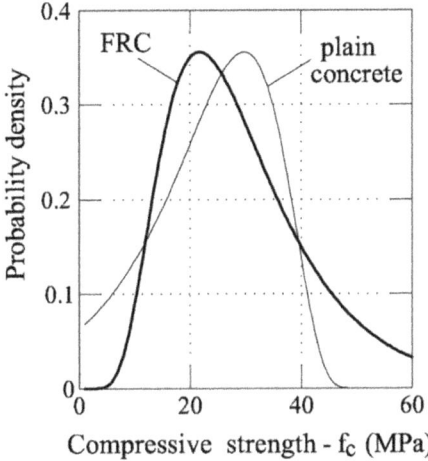

Compressive strength f_c (MPa)	plain concrete	FRC
Average	20.0	30.0
standard deviation	14.7	14.7
median	27.0	27.0
mode	30.0	21.5

Fig. 8. The log-normal functions used to model the distribution of compressive strength in plain concrete and FRC.

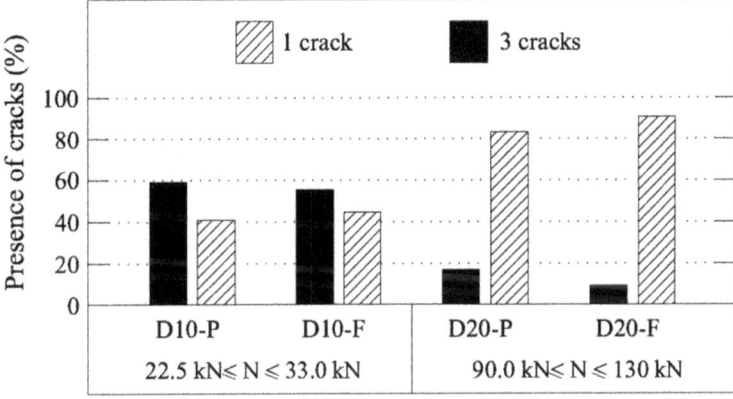

Fig. 9. The number of cracks calculated in the concrete ties by means of the numerical procedure proposed herein.

5 Conclusions

According to experimental and numerical analyses previously described, the following conclusion can be drawn:

- In the ties with the small diameter, a prevalent crack pattern cannot be identified, as in such samples one or more cracks grow. However, the presence of fibers tends to reduce the probability of having multiple cracking.

- In the ties with larger diamter, a single crack is generally present when fibers are added to the concrete mixtures. On the contrary, the growth of a single crack is less probable in plain concrete, even if one crack is more probable than multiple cracking in such elements.
- Only when plain concrete is assumed to be fragile (with the mode value of the compressive strength larger than the median) and FRC antifragile (with the median value of the compressive strength larger than the mode), can the proposed model for the computation of the transfer length be used to correctly predict the experimental data.

Further tests will be performed to better measure the probability distribution functions of plain concrete and FRC, and to correlate the fragility, or the antifragility, of concrete systems to overall structural durability.

Acknowledgements. The financial support provided by the Italian Ministry of Education, University and Research (PRIN Grant 2015HZ24KH—"Failure mechanisms caused by corrosive degrade and by lack of constructive details in the existing structures in reinforced concrete") is gratefully acknowledged. The support of the Centre SISCON (Safety of Infrastructures and Constructions) at Politecnico di Torino is also acknowledged.

References

1. *fib*: Model Code for Concrete Structures 2010, 1st edn. Ernst & Sohn, Berlin (2013)
2. ACI Committee 544: Report on Design and Construction of Steel Fiber-Reinforced Concrete Elevated Slabs - 544.6R-15. American Concrete Institute, Farmington Hill (2015)
3. Chiaia, B., Fantilli, A.P., Vallini, P.: Evaluation of crack width in FRC structures and application to tunnel linings. Mater. Struct. **42**(3), 339–351 (2009)
4. de la Fuente, A., Cugat, V., Cavalaro, S.H., Bairán, J.M.: Partial safety factor for the residual flexural strength of FRC precast concrete segments. In: Hordijk, D.A., Luković, M. (eds.) High Tech Concrete: Where Technology and Engineering Meet, pp. 1768–1775. Springer, Cham (2018). https://doi.org/10.1007/978-3-319-59471-2_203
5. Taleb, N.N.: Antifragile: Things That Gain from Disorder, 1st edn. Random House, New York (2012)
6. Bresler, B., Bertero, V.: Behavior of reinforced concrete under repeated load. ASCE-J. Struct. Div. **94**(6), 1567–1592 (1968)
7. Fantilli, A.P., Vallini, P.: Strains in steel bars of reinforced concrete elements subjected to repeated loads. J. Strain Anal. Eng. Des. **39**(5), 447–457 (2004)
8. Critelli, N., Cesetti, A., Fantilli, A.P., Tondolo, F.: Measuring crack width in RC and R/FRC ties through laser scanner. In: Pellicer, E., Adam, J. M., Yepes, V., Singh, A., Yazdani (eds.) ISEC 9, Resilient Structures and Sustainable Construction, pp. 1–6. ISEC Press (2017)

Fatigue of Plastic Fibre Reinforced Concrete in Bending: Assessment and Prediction

Debora Martinello Carlesso[1(✉)], Sergio H. P. Cavalaro[2],
and Albert de la Fuente[1]

[1] Polytechnic University of Catalonia – BarcelonaTECH, Barcelona, Spain
[2] Loughborough University, Loughborough, UK

Abstract. The present paper deals with an experimental study on the flexural fatigue behaviour of pre-cracked polypropylene fibre reinforced concrete with two different volume of fibre. Mechanical response was evaluated through compressive strength, elastic modulus and static bending test. Fatigue test considered an initial crack width accepted in the service limit state and the evolution of the crack opening displacement of the beams subjected to a prescribed number of cycles (1,000,000 or 2,000,000). After the cyclic load, the post-fatigue residual strength was evaluated and compared to the static response. Results suggest that the mechanism of crack development is independent of the adopted fibre content. The post-fatigue strength seems to be unaffected by accumulated damage due to cyclic load and the static load-crack opening displacement curve might be used as a criterion to predict the residual strength. Furthermore, a conceptual model is proposed to predict the crack opening as a function of number of cycles in view of accumulated fatigue damage. The equation was validated for different fibre content and polypropylene fibres.

Keywords: Fatigue · Fiber reinforced concrete · Cracked section · Polypropylene macrofiber

1 Introduction

The use of polypropylene fibres has increased significantly in recent years due to their contribution to post-cracking strength and their inert, non-corrosive nature, particularly for those cases in which variations in the mechanical properties depend on time are of paramount importance (e.g., sewerage buried pipelines [1, 2] and metro tunnels [3–5]). Fatigue response of polypropylene fibre reinforced concrete (PFRC) under cyclic load has been studied in terms of fatigue load versus fatigue life (S-N curves) for constant and variable loading amplitudes at different frequencies, fatigue crack evolution, strain rate and accumulated damage. Nevertheless, limited information about post-cracking and post-fatigue residual response is available. Assumptions and conclusions obtained for SFRC cannot be directly generalized for PFRC due to the differences in the properties of these fibre (e.g. both lower elastic modulus and tensile strengths). Therefore, a good understanding of the post-cracking/fatigue behaviour of PFRC is of great importance for the satisfactory design of structures.

© RILEM 2022
P. Serna et al. (Eds.): BEFIB 2021, RILEM Bookseries 36, pp. 109–120, 2022.
https://doi.org/10.1007/978-3-030-83719-8_10

Considering the surge in PFRC application with structural responsibility and the likelihood of finding elements with cracks in service, additional studies are needed to grasp the implications of the flexural fatigue of pre-cracked PFRC in terms of crack-opening evolution and residual flexural strength. Likewise, models are needed to predict the material performance in terms of the evolution of crack-opening and residual resistant capacity after the cycles. The study encompasses an extensive experimental investigation on PFRC with two types of polypropylene fibre with different fibre content and beam sizes to investigate the influence of the level of deflection-hardening on the mechanical behaviour after cycles. This research provides knowledge, unique experimental results and propose a conceptual model for the behaviour of PFRC under flexural dynamic cycles that can be used for generating specific fatigue models to be introduced in future FRC design codes and guidelines.

2   Experimental Procedures

2.1   Mix Design, Casting and Curing Procedures

Mixes contained 420 kg/m^3 Portland cement CEM I-52.5 R, limestone aggregates (860, 440 and 490 kg/m^3 of sand, coarse aggregate of 4–10 mm and from 10–200 mm, respectively), 4.2 kg/m^3 of sodium polycarboxylate ether-based superplasticiser and water/cement ratio of 0.40. Two fibre contents were used to represent different levels of residual flexural strength. The first PF1_5 (5 kg/m^3 or 0.4% by volume) is expected to achieve a limited stress recovery between f_{r1} and f_{r3} in bending according to the *fib* Model Code 2010 [6]. The second PF1_10 and PF2_10 (10 kg/m^3 or 0.8% by volume) is expected to achieve a significant stress recovery between f_{r1} and f_{r3} in bending according to the *fib* Model Code 2010 [6]. The fibre PF1 was 48-mm long and was made of virgin polypropylene with specific gravity of 0.89–0.91 g/cm^3. Fibre PF2 was 60-mm long and was made of polypropylene copolymer with specific gravity of 0.91 g/cm^3. Both had continuous embossing anchorage. PFRC compositions used in the experimental program, which should satisfy the structural requirements for typical applications in heavy-duty pavements, industrial floors, tunnel segmental lining and precast elements in general.

The mixing process took place in a vertical-axis mixer with a nominal capacity of 50 l. After homogenising all dry components (cement, sand and aggregates), water and superplasticizer were added to the mixer and, finally, fibres were added. Prismatic beams with 150 × 150 × 600 mm^3 and 75 × 75 × 275 mm^3 were cast for residual flexural strength and fatigue tests. The second beam size was chosen pondering a reduction of material and ease of handling. After casting, the moulds were covered with a thin plastic sheet and left to cure at room temperature for 24 h. Then, they were demoulded and stored in a climatic chamber at 20 °C and 95%–100% relative humidity until the date of the test.

2.2   Testing Procedures

The quasi-static flexural strength was measured following the EN 14651:2007 in three notched beams with a three-point bending test (3PBT) setup in an INSTRON hydraulic servo-controlled testing machine. A clip gauge placed at the notch controlled the CMOD during the 3PBT. All quasi-static tests were performed at 28 days.

Figure 1 shows the complete loading history of specimens subjected to the fatigue test. First, specimens were pre-cracked in 3PBT setup according to the procedure in EN 14651:2007. A constant CMOD rate of 0.05 mm/min was applied up to a total CMOD of 0.5 mm (considered the service limit value in *fib* Model Code 2010 [6]). The force corresponding to this displacement was set as the maximum load in the fatigue test (P_{upp}). The minimum load (P_{low}) during the cycles was defined by considering an amplitude of 0.3 ($R = P_{low}/P_{upp} = 0.3$). Immediately after pre-cracking, a sinusoidal cyclic load with a frequency of 6 Hz ranging from P_{upp} to P_{low} was applied, and the evolution of CMOD was recorded at every 500 cycles. N_{max} was either 1,000,000 [7–10] or 2,000,000 [11–13]. Upon reaching N_{max}, the cyclic loading was interrupted and the beams were reloaded at a constant CMOD rate of 0.2 mm/min up to failure. After failure, specimens were removed from the frame and separated in 2 halves for the manual counting of fibres. The specimens were tested in a period extending from 30 to 90 days since casting, alternating between mixes to minimise the influence of the age in the results.

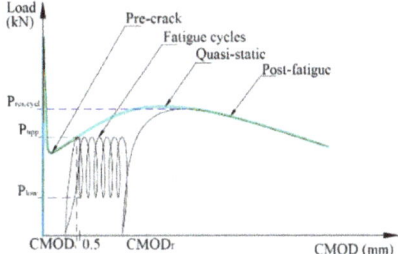

Fig. 1. Fatigue loading history

3   Results and Discussion

3.1   Flexural Response Under Monotonic Load

Table 1 presents the average residual flexural strengths (f_{R1m}, f_{R2m}, f_{R3m}, f_{R4m} corresponding to CMOD values of 0.5, 1.5, 2.5 and 3.5 mm, respectively), limit of proportionality (f_{LOPm}), maximum post-cracking stress ($f_{PC,max}$) and the respective CMOD ($CMOD_{fPC,max}$) measured in the 3PBT. The table also includes the relations between characteristic values of f_{LOP} (f_{LOPk}) and the flexural residual strengths f_{R1} (f_{R1k}) and f_{R3} (f_{R3k}) related respectively with the service and ultimate limit states [6].

Table 1. Average and characteristic 3PBT results and coefficient of variation in percentage and between parenthesis

	PF1_5-1.1C	PF1_10-1.6E	PF2_10-1.5E	PF1_5-1.0C	PF1_10-1.4E	PF2_10-1.9E
f_{LOPm} (MPa)	5.21	5.39	5.14	5.60	6.90	6.04
	(7.2)	(12.2)	(6.7)	(8.7)	(9.4)	(4.9)
$f_{PC,max}$ (MPa)	2.22	5.59	3.47	2.49	7.29	3.11
	(41.8)	(10.5)	(15.7)	(32.2)	(36.0)	(12.4)
$CMOD_{fPC,max}$ (mm)	3.06	2.92	3.99	2.29	2.75	3.84
	(10.8)	(4.9)	(0.0)	(26.6)	(25.0)	(7.0)
f_{R1m} (MPa)	1.45	3.49	2.05	1.75	4.63	1.59
	(36.2)	(9.7)	(14.8)	(23.4)	(34.5)	(20.0)
f_{R2m} (MPa)	1.92	4.93	2.73	2.41	6.53	2.28
	(42.0)	(11.9)	(16.5)	(28.6)	(35.5)	(17.2)
f_{R3m} (MPa)	2.17	5.50	3.18	2.43	7.24	2.73
	(43.4)	(10.7)	(16.7)	(33.6)	(37.6)	(15.8)
f_{R4m} (MPa)	2.17	5.42	3.40	2.36	6.96	3.06
	(41.2)	(9.0)	(16.3)	(37.4)	(36.8)	(13.9)
f_{R3k}/f_{R1k}	1.1c*	1.6e*	1.5e*	1.0c*	1.4e*	1.9e*
f_{R1k}/f_{LOPk}	0.1	0.7	0.3	0.2	0.3	0.3

*Classification of the post-cracking strength based on Model Code 2010 [6]

The f_{LOPm} is not significantly affected by the fibre content and type. Differences between same composition within specimen size can be an effect of the fibre distribution and orientation, which influence are stronger in smaller specimens. Immediately after cracking, PFRCs exhibited a sudden stress drop indicating loss of stiffness. The minimum residual stress observed just after cracking was approximately 66% (both PF1_5), 51% (both PF1_10), 57% (PF2_10-1.5E) and 48% (PF2_10-1.9E) of $f_{PC,max}$. The ductile behaviour and stress recovery are attributed to the contribution of the fibres [14, 15]. PF1_5 showed a nearly stable post-cracking response while PF1_10 and PF2_10 showed an increase in stress when reaching higher CMOD values. Notice that the last part of the name of each mix represents their classification according to the Model Code 2010 [6], which depends on the ration f_{R3k}/f_{R1k}. The further reduction of the strength after a maximum is reached arises from the progressive fibre debonding and slipping in the cross section. According to the classification proposed by the Model Code 2010 [6] for the post-cracking strength, PF1_10-1.6E can be considered as a structural material ($f_{R1k}/f_{LOPk} > 0.4$ and $f_{R3k}/f_{R1k} > 0.5$).

3.2 Remaining Residual Flexural Strength After the Fatigue Test

Table 2 summarises for each specimen the maximum number of cycles (N_{max}), P_{upp}, CMOD for P_{upp} at the first, 1,000,000 and 2,000,000 cycles ($CMOD_1$, $CMOD_{1M}$ or $CMOD_{2M}$, respectively). Table 2 also presents the maximum residual flexural strength after the fatigue cycles ($f_{res,cycl}$), the CMOD at $f_{res,cycl}$ ($CMOD_{Fres,cycl}$), the corresponding maximum measured in the control quasi-static test ($f_{PC,max}$ and $CMOD_{fPC,max}$) and the specific load level ($S'_{Fres,cycl}$) calculated as the ratio between P_{upp} and the maximum load reached after the fatigue test.

Table 2. Results of fatigue tests and post-fatigue quasi-static flexural strength

Composition	Reference	N_{max} (million cycles)	Fatigue test P_{app} (kN)	$CMOD_1$ (mm)	$CMOD_{1M}$ (mm)	$CMOD_{2M}$ (mm)	Post-fatigue quasi static $f_{res,cycl}$ (MPa)	$CMOD_{Pres,cycl}$ (mm)	$f_{res,cycl}/f_{PC,max}$	$S'_{Fres,cycl}$	Fibre/cm² (number of fibres)
PF1_5-1.1C	PF1_5-1.1C_1M-1	1	4.41	0.477	1.688	–	2.72	3.22	1.23	0.62	0.44 (83)
	PF1_5-1.1C_1M-2	1	5.37	0.466	1.385	–	3.37	2.63	1.52	0.61	0.46 (87)
	PF1_5-1.1C_1M-3	1	5.82	0.574	2.036	–	3.18	3.41	1.43	0.70	0.60 (113)
	PF1_5-1.1C_2M-4	2	4.43	0.477	1.749	1.899	2.69	3.41	1.21	0.63	0.44 (83)
	PF1_5-1.1C_2M-5	2	5.15	0.467	1.715	1.894	2.89	3.16	1.30	0.69	0.55 (104)
PF1_10-1.6E	PF1_10-1.6E_1M-1	1	8.41	0.490	2.542	–	5.08	3.55	0.91	0.64	0.83 (156)
	PF1_10-1.6E_1M-2	1	8.18	0.467	1.747	–	5.26	3.30	0.94	0.60	0.88 (165)
	PF1_10-1.6E_2M-3	2	9.60	0.491	2.403	2.718	5.56	3.82	0.99	0.66	0.95 (179)
	PF1_10-1.6E_2M-4	2	8.36	0.469	1.550	1.684	5.30	3.15	0.95	0.61	0.91 (170)
	PF1_10-1.6E_2M-5	2	4.80	0.463	1.953	2.120	3.01	3.49	0.54	0.61	0.62 (116)
PF2_10-1.5E	PF2_10-1.5E_1M-1	1	5.00	0.473	1.764	–	4.37	4.52	1.26	0.44	0.48 (91)
	PF2_10-1.5E_1M-2	1	5.00	0.475	2.947	–	3.57	5.33	1.03	0.54	0.44 (83)
	PF2_10-1.5E_1M-3	1	5.00	0.449	2.966	–	3.30	5.15	0.95	0.58	0.51 (96)
	PF2_10-1.5E_1M-4	1	4.71	0.471	2.969	–	3.08	4.97	0.89	0.59	0.42 (79)
	PF2_10-1.5E_2M-5	2	7.14	0.51	1.552	1.723	5.41	4.21	1.56	0.51	0.77 (145)
	PF2_10-1.5E_2M-6	2	3.70	0.398	1.777	1.959	2.94	5.01	0.85	0.48	0.46 (86)
	PF2_10-1.5E_2M-7	2	5.63	0.471	1.991	2.236	3.77	5.97	1.09	0.57	0.55 (104)
PF1_5-1.0C	PF1_5-1.0C_1M-1	1	0.93	0.379	1.204	–	2.60	4.11	1.04	0.50	0.45 (21)
	PF1_5-1.0C_1M-2	1	1.06	0.464	1.373	–	2.77	2.90	1.11	0.54	0.48 (23)
	PF1_5-1.0C_1M-3	1	1.98	0.521	1.365	–	4.45	2.97	1.79	0.63	0.68 (32)
	PF1_5-1.0C_2M-4	2	2.00	0.448	1.290	1.402	4.62	3.21	1.86	0.61	0.51 (24)
	PF1_5-1.0C_2M-5	2	1.32	0.443	1.253	1.382	2.91	2.54	1.17	0.64	0.35 (17)
PF1_10-1.4E	PF1_10-1.4E_1M-1	1	2.17	0.470	1.933	–	–	–	–	–	0.63 (30)
	PF1_10-1.4E_1M-2	1	2.14	0.459	3.085	–	4.43	4.06	0.61	0.68	0.73 (34)
	PF1_10-1.4E_1M-3	1	2.42	0.466	2.756	–	4.90	3.37	0.67	0.70	0.92 (43)
	PF1_10-1.4E_1M-4	1	2.94	0.477	2.275	–	6.05	3.17	0.83	0.68	0.94 (44)
	PF1_10-1.4E_2M-5	2	2.50	0.442	1.173	1.271	6.25	2.48	0.86	0.56	0.95 (45)
PF2_10-1.9E	PF2_10-1.9E_1M-1	1	1.20	0.450	1.520	–	4.02	–	1.29	0.42	0.55 (26)
	PF2_10-1.9E_1M-2	1	1.46	0.374	1.429	–	4.41	5.18	1.42	0.47	0.48 (23)
	PF2_10-1.9E_2M-3	2	1.77	0.425	2.953	3.162	4.88	6.27	1.57	0.51	0.58 (27)
	PF2_10-1.9E_2M-4	2	2.14	0.463	1.308	1.409	5.90	4.31	1.90	0.51	0.83 (39)

$CMOD_{Fres,cycl}$ is consistently larger than the measured in the quasi-static control tests ($CMOD_{fPC,max}$), suggesting that the load cycles displaced the post-fatigue peak stress towards bigger CMOD values. PF1_5-1.1C and PF1_5-1.0C exhibited bigger post-fatigue maximum flexural residual strength (average of 34% and 39%, respectively) than the equivalent control quasi-static tests. By contrast, PF1_10-1.6E and PF1_10-1.4E showed smaller values (average of 13% and 26%, respectively) in comparison to the equivalent control quasi-static tests. PF2_10-1.5E presented $f_{res,cycl}$ 9% bigger than $f_{PC,max}$ and PF2_10-1.9E presented $f_{res,cycl}$ 54% bigger than $f_{PC,max}$.

Figure 2 shows the residual flexural strength curves for the specimens subjected to the fatigue test and the results for the control quasi-static tests of PF1_5-1.1C (a), PF1_10-1.6E (b), PF2_10-1.5E (c), PF1_5-1.0C (d), PF1_10-1.4E (e) and PF2_10-1.9E (f). The thicker black line shows the 3PBT mean result and the thinner lines and upper and lower expected variabilities. The shape of the post-fatigue strength curve resembles and follows the trend found in the quasi-static control tests. This suggests that the CMOD increment observed after the load cycles led to a damage level similar to that found in the quasi-static control tests for the same CMOD increment.

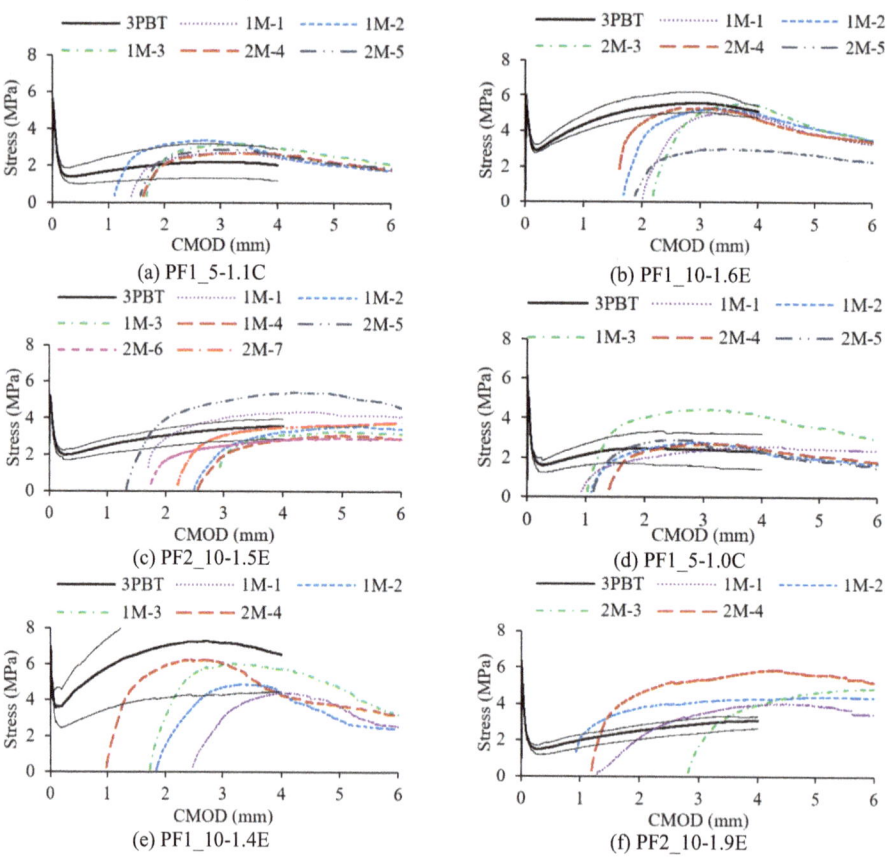

Fig. 2. Post-fatigue and control quasi-static strenth-CMOD curves of PF1_5-1.1C (a), PF1_10-1.6E (b), PF2_10-1.5E (c), PF1_5-1.0C (d), PF1_10-1.4E (e) and PF2_10-1.9E (f)

No significant difference in the post-fatigue maximum strength and the corresponding CMOD was observed between specimens subjected to a N_{max} of 1,000,000 or 2,000,000 cycles. Studies in the literature report an increase in the maximum strength of specimens after load cycles in comparison with specimens subjected only to quasi-static tests [16, 17]. This increase is associated with the application of cyclic load below the endurance limit [17–19] and depends on the stress ratio during the test (lower stress ratios promote higher post-fatigue strength) [17]. Such increase is attributed to the consolidation of microvoids at the beginning of the fatigue test [20], the relatively long duration of the tests [21] and the stochastic nature of concrete [6, 22]. The similar residual flexural strengths found in specimens characterised before and after the fatigue test do not support the findings by other studies in the literature.

Figure 3 (a) shows the relationship between f_{R1} and the corresponding maximum strength measured in the post-cracking stage ($f_{res,cycl}$ for specimens subjected to the fatigue test or $f_{PC,max}$ for specimens not subjected to the fatigue test) of 150×150 600 mm^3 and $75 \times 75 \times 275$ mm^3 PFRC beams, Notice that f_{R1} was obtained before the fatigue test in all specimens, while the others were obtained either before or after the fatigue test depending on the procedure adopted for each specimen. Should the load cycles affect the resistant capacity of the specimen, the series subjected to the fatigue test would follow a different trendline from those not subjected to the fatigue test in Fig. 3 (a). By contrast, no significant influence of the load cycle on the maximum post-fatigue resistant capacity would be expected if all specimens follow the same trendline. The analysis of Fig. 3 (a) confirms this last conjecture, as no clear difference was found regardless of the application or not of the load cycles.

Specimens PF1_10-1.6E_2M-5, PF2_10-1.5E_2M-5 and PF1_5-1.0C_2M-3 are considered outliers in their respective series. The difference in behaviour is explained by the number of fibres crossing the fracture cross-section. PF1_10-1.6E_2M-5 has 26% fewer fibres than the specimen with the second-lowest number of fibres in the same series and 30% less than the average of the other specimens in the series. The number of fibres in PF1_10-1.6E_2M-5 is closer to the average found in PF1_5-1.1C than in PF1_10-1.6E, thus explaining why its residual strength after the fatigue test approximates more the former than the latter. The opposite happens with PF2_10-1.5E_2M-5, which has 39% more fibres in the fracture surface than the specimen with the second-highest number of fibres in the same series and 61% more than the average of the other specimens in the series. In this case, the number of fibres is closer to that of PF1_10-1.6E than to that of PF2_10-1.5E, thus explaining why the residual strength of PF2-1.5E_2M-5 approximates the former. The same assumption explains PF1_5-1.0C_2M-3 response, which has 52% more fibres than the average of the respective series. Figure 3 (b) shows the relationship between the number of fibres in the cross-section and $f_{res,cycl}$ of $150 \times 150 \times 600$ mm^3 and $75 \times 75 \times 275$ mm^3 specimens for mixes with PF1 and PF2. As expected, in all cases, there is a linear trend between $f_{res,cycl}$ and the number of fibres in the cross-section.

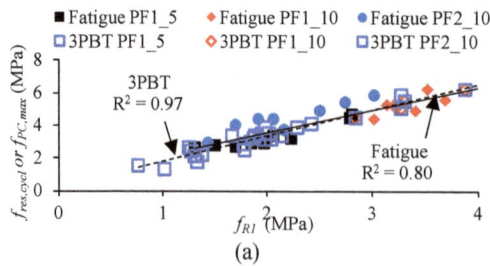 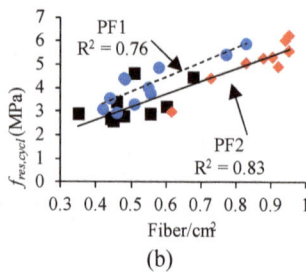

(a)   (b)

Fig. 3. Relationship between $f_{res,cycl}$ or $f_{PC,max}$ and f_{R1} of all specimens subjected to the control or fatigue tests (a) and between $f_{res,cycl}$ and fibre/cm^2 for specimens subjected to the fatigue test (b)

4   Conceptual Model for Crack Evolution

The incremental damage induced by the load cycles can be related to the logarithm of the total number of load cycles and the logarithm of CMOD variation between cycles (dCMOD/dN). This last parameter was calculated for intervals of 1000 cycles to simplify the assessment and reduce the influence of the scatter in the measurements of CMOD. The linear relationship is represented in Eq. 1, which would allow the prediction of the CMOD of the specimen subjected to the fatigue test. In this equation, v is related to the existing damage induced in the pre-cracking stage and u represents the increase in damage observed over the cycles. v and u are constants that may be determined experimentally by performing a limited number of cycles. The CMOD after N cycles (represented by $w(N)$) is calculated by integrating both sides of Eq. 1 in relation to dN, as shown in Eq. 2.

$$\log(dCMOD/dN) = u \cdot \log(N) + v \tag{1}$$

$$w(N) = \int_0^n dCMOD = \int_0^n 10^v \cdot N^u \cdot dn \tag{2}$$

The integration gives Eq. 3 for assessing the total crack opening after the fatigue test. The parameter w_0 marks the initial damage taken as a reference in the test. Since the origin of CMOD was taken before pre-cracking in this experimental programme, w_0 is 0. The parameters k_1 and k_2 are shown in Eq. 4 and 5, respectively.

$$w(N) = k_1 \cdot n^{k_2} + w_0 = k_1 \cdot n^{k_2} \tag{3}$$

$$k_2 = u + 1 \tag{4}$$

$$k_1 = 10^v / k_2 \tag{5}$$

Table 3. Minimum number of cycles needed to predict u_n and v_n using optimisation procedure, error of CMOD prediction for the CMOD at Nmax using Eq. 3, R^2 and average error of CMOD prediction for whole test

Specimen	Total number of cycles	Initial cycles used to estimate u and v		Error of prediction for maximum number of cycles (%)	Average error of prediction for whole curve (%)	R^2
		n	% of the total			
PF1_5-1.1C_1M-1	1000000	20000	2.0	13.9	10.33	0.9964
PF1_5-1.1C_1M-2	1000000	13500	1.4	19.4	8.64	0.9841
PF1_5-1.1C_1M-3	2000000	6000	0.3	4.6	2.64	0.9974
PF1_5-1.1C_2M-4	2000000	253500	12.7	11.0	6.80	0.9963
PF1_5-1.1C_2M-5	2000000	55000	2.8	8.4	6.63	0.9987
PF1_10-1.6E_1M-1	1000000	22000	2.2	10.3	2.88	0.9941
PF1_10-1.6E_1M-2	1000000	19500	2.0	15.5	8.64	0.9839
PF1_10-1.6E_2M-3	1000000	200500	20.1	12.8	2.41	0.9954
PF1_10-1.6E_2M-4	2000000	44000	2.2	8.5	2.20	0.9946
PF1_10-1.6E_2M-5	2000000	60500	3.0	15.1	2.22	0.9934
PF2_10-1.5E_1M-1	1000000	285000	28.5	8.5	4.03	0.9985
PF2_10-1.5E_1M-2	1000000	9000	0.9	7.8	7.68	0.9926
PF2_10-1.5E_1M-3	1000000	24500	2.5	8.3	7.92	0.9914
PF2_10-1.5E_1M-4	1000000	12500	1.3	14.8	11.38	0.9863
PF2_10-1.5E_2M-5	2000000	445500	22.3	9.6	16.35	0.9993
PF2_10-1.5E_2M-6	2000000	356000	17.8	8.4	5.45	0.9988
PF2_10-1.5E_2M-7	2000000	155500	7.8	6.1	4.41	0.9983
PF1_5-1.0C_1M-1	1000000	62500	6.25	0.75	5.12	0.9976
PF1_5-1.0C_1M-2	1000000	26500	2.65	3.12	3.00	0.9966
PF1_5-1.0C_1M-3	1000000	40500	4.05	6.40	5.73	0.9967
PF1_5-1.0C_2M-4	2000000	119500	5.98	4.52	7.75	0.9996
PF1_5-1.0C_2M-5	2000000	174500	8.73	8.12	8.76	0.9978
PF1_10-1.4E_1M-1	1000000	60000	6.00	12.21	2.52	0.9947
PF1_10-1.4E_1M-2	1000000	62500	6.25	12.97	5.48	0.9941
PF1_10-1.4E_1M-3	1000000	21500	2.15	3.84	3.99	0.9987
PF1_10-1.4E_1M-4	1000000	174500	17.45	0.95	4.98	0.9990
PF1_10-1.4E_2M-5	2000000	29500	1.48	2.79	5.40	0.9982
PF2_10-1.9E_1M-1	1000000	60000	6.00	36.99	2.52	0.9947
PF2_10-1.9E_1M-2	1000000	68500	6.85	5.12	1.95	0.9963
PF2_10-1.9E_2M-3	2000000	23000	1.15	26.70	6.63	0.9745
PF2_10-1.9E_2M-4	2000000	92000	4.60	3.04	9.72	0.9993

An optimisation procedure is proposed to determine when enough cycles have been applied to predict v and u so that the test can be interrupted and the values used to extrapolate the behaviour for a bigger number of cycles using Eq. 3. For a given intermediate cycle n bigger than 500 during the test of a particular specimen, it is possible to estimate u_n and v_n through the regression of the data up to n cycles using the relationship between log(dCMOD/dN) and log(N). These parameters may be used in Eq. 4 and Eq. 5 to estimate $k_{1,n}$ and $k_{2,n}$, respectively. Both are used in Eq. 6 to estimate the average prediction error of the points obtained up to n cycles. Once the error reaches an acceptable value (0.1 mm in this study) or a clear minimum value, no more cycles are needed, and the test could be interrupted. Notice that calculations were done considering 500- and 1000-cycle intervals. Since they yielded similar results, only the latter is depicted here.

$$Error(n) = \frac{\sum_{N=500}^{n} \left| CMOD_{measured} - k_{1,n} \cdot N^{k_{2,n}} \right|}{n} \tag{6}$$

To illustrate the potential of this approach to predict the flexural fatigue behaviour and shorten the fatigue test duration, Table 3 shows the minimum number of cycles (n) required to satisfy the maximum error condition in the optimisation procedure for the early interruption of the test and its proportion regarding the maximum number of cycles applied for each specimen in this experimental program (N_{max}). The table shows the error of prediction expected supposing the early interruption of the test after n cycles and the use of parameter u_n and v_n to predict the CMOD expected at N_{max}. The average error for the whole curve considering the same u_n and v_n in Eq. 3 and the R^2 are also presented in Table 3.

On average, the test can be interrupted after approximately 100000 cycles, which equates to 5.0% of N_{max}. This represents an average reduction of 10 times on the total duration of the fatigue test. Such reduction and the application of the model from Eq. 3 would entail an average error of prediction of CMOD for N_{max} of 10.0%. The average error of prediction for the whole curve would be 6.0%, which could be considered acceptable given the high scatter of the test and the significant reduction in the duration of the fatigue test.

5   Concluding Remarks

The analysis conducted in this study supports the following conclusions.

- By using the same proportion of components and fixing P_{upp} in relation to the pre-cracking load, the behaviour in terms of the evolution of the CMOD during the fatigue cycles was not affected by the fibre type and content or specimen size evaluated in this experimental program;
- The residual flexural strength of specimens subjected to the fatigue test follows the curve obtained in the quasi-static control tests of equivalent specimens not subjected to the fatigue test. Findings indicate that the damage induced by the load cycles is equivalent to that observed in the quasi-static control test for the same CMOD increment, which has important repercussions from the design standpoint;
- The numerical model to predict the fatigue behaviour is capable of accurately predicting the CMOD variation over the load cycles, thus aiding the vision described in the previous conclusion. This model depends on parameters related with the initial damage due to pre-cracking and the incremental damage induced by the cycles, which could vary depending on the loading regime, pre-cracking level and material characteristics. On the other hand, is independent on specimen size. The optimising procedure proposed here to estimate v and u using a limited number of initial load cycles can enable a significant reduction in the duration of experimental programs about the fatigue of fibre reinforced concrete.

Acknowledgments. The first author thanks the Brazilian National Council for Scientific and Technological Development for the scholarship granted (233980/2014-8). This research was enabled by funds provided by the SAES project (BIA2016-78742-C2-1-R) of Spanish Ministerio de Economía, Industria y Competitividad.

References

1. De La Fuente, A., Escariz, R.C., De Figueiredo, A.D., Aguado, A.: Design of macro-synthetic fibre reinforced concrete pipes. Constr. Build. Mater. **43**, 523–532 (2013). https://doi.org/10.1016/j.conbuildmat.2013.02.036
2. De La Fuente, A., Pons, O., Josa, A., Aguado, A.: Multi-criteria decision making in the sustainability assessment of sewerage pipe systems. J. Clean. Prod. **112**, 4762–4770 (2016). https://doi.org/10.1016/j.jclepro.2015.07.002
3. Conforti, A., Tiberti, G., Plizzari, G.A., Caratelli, A., Meda, A.: Precast tunnel segments reinforced by macro-synthetic fibers. Tunn. Undergr. Sp. Technol. **63**, 1–11 (2017). https://doi.org/10.1016/j.tust.2016.12.005
4. de la Fuente, A., Blanco, A., Armengou, J., Aguado, A.: Sustainability based-approach to determine the concrete type and reinforcement configuration of TBM tunnels linings. Case study: Extension line to Barcelona Airport T1, Tunn. Undergr. Sp. Technol. **61**, 179–188 (2017). https://doi.org/10.1016/j.tust.2016.10.008
5. Behfarnia, K., Behravan, A.: Application of high performance polypropylene fibers in concrete lining of water tunnels. Mater. Des. **55**, 274–279 (2014). https://doi.org/10.1016/j.matdes.2013.09.075
6. Fib, International Federation for Structural Concrete fib Model Code for Concrete Structures 2010, Germany (2013)
7. Breña, S.F., Benouaich, M.A., Kreger, M.E., Wood, S.L.: Fatigue tests of reinforced concrete beams strengthened using carbon fiber-reinforced polymer composites. ACI Struct. J. **102**, 305–313 (2005). https://doi.org/10.14359/14282
8. de Andrade Silva, F., Mobasher, B., Filho, R.D.T.: Fatigue behavior of sisal fiber reinforced cement composites. Mater. Sci. Eng. A. **527**, 5507–5513 (2010). https://doi.org/10.1016/j.msea.2010.05.007
9. Nanni, A.: Fatigue behaviour of steel fiber reinforced concrete. Cem. Concr. Compos. **13**, 239–245 (1991). https://doi.org/10.1016/0958-9465(91)90029-H
10. Tarifa, M., Zhang, X., Ruiz, G., Poveda, E.: Full-scale fatigue tests of precast reinforced concrete slabs for railway tracks. Eng. Struct. **100**, 610–621 (2015). https://doi.org/10.1016/j.engstruct.2015.06.016
11. Ramakrishnan, V., Wu, G.Y., Hosalli, G.: Flexural fatigue strength, endurance limit, and impact strength of fiber reinforced concretes. Transp. Res. Rec. J. Transp. Res. Board. **1226**, 17–24 (1989)
12. Zhang, H., Tian, K.: Properties and mechanism on flexural fatigue of polypropylene fiber reinforced concrete containing slag. J. Wuhan Univ. Technol. Mater. Sci. Ed. **26**, 533–540 (2011). https://doi.org/10.1007/s11595-011-0263-8
13. Johnston, C.D., Zemp, R.W.: Flexural fatigue performance of steel fiber reinforced concrete-influence of fiber content, aspect ratio, and type. ACI Mater. J. **88**, 374–383 (1991). https://doi.org/10.14359/1875
14. de Alencar Monteiro, V.M., Lima, L.R., de Andrade Silva, F.: On the mechanical behavior of polypropylene, steel and hybrid fiber reinforced self-consolidating concrete. Constr. Build. Mater. **188**, 280–291 (2018). https://doi.org/10.1016/j.conbuildmat.2018.08.103

15. Oh, B.H., Kim, J.C., Choi, Y.C.: Fracture behavior of concrete members reinforced with structural synthetic fibers. Eng. Fract. Mech. **74**, 243–257 (2007). https://doi.org/10.1016/j.engfracmech.2006.01.032

16. Nagabhushanam, M., Ramakrishnan, V., Vondran, G.: Fatigue strength of fibrillated polypropylene fiber reinforced concretes. Transp. Res. Rec. **1226**, 36–47 (1989)

17. Parant, E., Rossi, P., Boulay, C.: Fatigue behavior of a multi-scale cement composite. Cem. Concr. Res. **37**, 264–269 (2007). https://doi.org/10.1016/j.cemconres.2006.04.006

18. Naaman, A.E., Hammoud, H.: Fatigue characteristics of high performance fiber-reinforced concrete. Cem. Concr. Compos. **20**, 353–363 (1998). https://doi.org/10.1016/S0958-9465(98)00004-3

19. Ramakrishnan, V., Mayer, C., Naaman, A.E.: Cyclic behaviour, fatigue strength, endurance limit and models for fatigue behavior of FRC. Chapter **4**, 101–148 (2014)

20. Zhang, B., Wu, K.: Residual fatigue strength and stiffness of ordinary concrete under bending. Cem. Concr. Res. **27**, 115–126 (1997). https://doi.org/10.1016/S0008-8846(96)00183-4

21. fib Federation Internacionale du beton, Bond of reinforcement in concrete - State-of-the-art report (2000)

22. Lohaus, L., Oneschkow, N., Wefer, M.: Design model for the fatigue behaviour of normal-strength, high-strength and ultra-high-strength concrete. Struct. Concr. **13**, 182–192 (2012). https://doi.org/10.1002/suco.201100054

Fatigue of Cracked Steel Fibre Reinforced Concrete Subjected to Bending

Debora Martinello Carlesso[1(✉)], Albert de la Fuente[1],
and Sergio H. P. Cavalaro[2]

[1] Polytechnic University of Catalonia – BarcelonaTECH, Barcelona, Spain
[2] Loughborough University, Loughborough, UK

Abstract. This paper presents an experimental investigation on the post-crack flexural fatigue behaviour of a steel macrofiber reinforced concrete (SFRC) and a high performance steel microfiber reinforced concrete (HPFRC), on notched beams considering the crack opening for serviceability condition. Different load levels were applied by means of three-point bending tests in order to verify the fatigue life. Performance of SFRC and HPFRC was compared under cyclic dynamic tests. Higher load levels seem to lead to failure through a continuous pull-out of the fibres, generating a more ductile response. Smaller load levels can be responsible for the progressive weakening of the fibre-matrix interface through micro-cracks. The conducted probabilistic approach has demonstrated to be suitable to predict the flexural fatigue life of pre-cracked SFRC and HPFRC for a desired probability of failure. From the experimental intrinsic scatter of the fatigue phenomenon, in particular for high levels of applied fatigue load, the amount of fibres in the cracked cross section seems to play an important role in withstanding the fatigue load.

Keywords: Fatigue · Fibre reinforced concrete · Steel fibre · High performance concrete · Cracked section

1 Introduction

Cyclic load on cracked cross-sections can be crucial to some concrete structures such as pavements, wind energy towers or rail-track sleepers, where the fatigue phenomena can be a governing design parameter. Fatigue life must be investigated not only in terms of applied fatigue load, but also how the deformation process until its failure affects the load-bearing capacity of the element.

Traditionally, the fatigue of concrete has been analysed through S-N curves, which correlate the applied fatigue load and the fatigue life of concrete, allowing to predict its fatigue performance. It is known that fatigue experiments results display a considerable scatter [1, 2] and are random in nature [3]. This characteristic leads to either the formulation of models that consider the design uncertainties to ensure an adequate evaluation of failure probability [3]; or introduce high safety coefficients to assess the imprecision in fatigue prescriptions within construction codes [4].

The limited studies on flexural fatigue on pre-cracked concrete reveals that a broader understanding of the overall behaviour is necessary. Applying percentages of

© RILEM 2022
P. Serna et al. (Eds.): BEFIB 2021, RILEM Bookseries 36, pp. 121–131, 2022.
https://doi.org/10.1007/978-3-030-83719-8_11

actual resisted load of each specimen instead of using mean results from flexural test, can help reducing the scatter, providing concise information. Wider ranges of applied cyclic load, controlled pre-crack widths, tests up to 2,000,000 cycles, post-fatigue behaviour and probabilistic approach should be taken into consideration aiming at generating design-oriented constitutive models [5]. This paper presents results of an extensive experimental campaign on two types of steel fibre reinforced concrete, allowing to predict the fatigue response of these materials, a comparison between their flexural fatigue behaviour and contributing with the database.

2 Experimental Procedures

High performance fibre reinforced concrete (HPFRC) specimens were cast with 909 kg/m^3 Portland cement CEM I-52R, 1,103 kg/m^3 siliceous aggregate, 150 kg/m^3 steel microfibre, 64 kg/m^3 polycarboxylate based superplasticizer and 65 kg/m^3 nanosilica (nano-SiO2) dispersion. The effective water/cement ratio was 0.20. Steel fibre reinforced concrete (SFRC) were cast with 390 kg/m^3 Portland cement CEM I-52.5 R, 1770 kg/m^3 aggregate, 50 kg/m^3 hooked-end steel fibre and 4.8 kg/m^3 sodium polycarboxylate ether-based superplasticizer admixture. The water/cement ratio was 0.40. Microfibre has 13 mm length and diameter of 0.16 mm and corresponds to, approximately, 2.0% by volume. Macro steel fibre of SFRC has 50 mm length and diameter of 1.05 mm and corresponds to 0.65% by volume. Mixtures were chosen based on previous tests and respond to applications with structural responsibility, mainly oriented to precast concrete elements for wind towers, rail-track sleepers and industrial floors. Different type of fibres also allowed the comparison between the effect of the geometry in the overall response.

First, all dry components were mixed together (cement, sand/aggregates and calcium carbonate). In the case of HPFRC, after mixing dry components, nanosilica, superplasticizer admixture and water were added and mixed for five minutes. Subsequently, fibres were added and the concrete mixed until the total mixing time reached 18 min. For SFRC, after mixing dry components, water and superplasticizer were added to the mixture and lastly, fibres were incorporated in the mixing machine, ensuring to be well-spread. After casting, specimens were left to cure at room temperature for 24 h, covered with a thin plastic sheet; then, demoulded and stored in a humid chamber (approximately 20 °C, 95%–100% relative humidity) until the day of the test. Beam specimens were notched at midspan. Fatigue tests were performed between 30 to 120 days after cast. The flexural strength evolution in time for materials with low water/cement ratio is considered negligible within this timeframe.

Twenty-one beams were tested under fatigue loading adopting the same procedures of Standard EN 14651:2007 in an INSTRON hydraulic servo-controlled testing machine with MTS control. Fatigue tests were load-controlled with the purpose of monitoring the crack opening in the beam and its fatigue life. The crack mouth opening displacement (CMOD) was measured through a clip gauge placed on the notch at midspan. Figure 1 (a) shows the test set-up. The beam size was chosen pondering a reduction of material and ease of handling. As an attempt to reduce the scatter, it was performed a method of individual fatigue life evaluation of each beam. First, a constant deformation rate (0.05 mm/min) was imposed up to a CMOD of 0.5 mm (considered as service limit value according to fib Model Code 2010 [2]); therefore, the fatigue assessment would contemplate the fibre strength and the fibre-matrix interface within a pre-cracked cross section. Then, the corresponding load of a crack opening of 0.5 mm ($f_{R,1}$) of each specimen was obtained in the first loading stage and set as maximum load ($P_{0.5mm}$). Once $P_{0.5mm}$ is known, percentages of $P_{0.5mm}$ were chosen as cycle's upper limit of applied load (P_{upp}) being 0.65, 0.70, 0.75, 0.80, 0.85, 0.90 and 1 (S). The lower load (P_{low}) was determined as a function of the load amplitude which was kept constant ($R = P_{low}/P_{upp} = 0.3$). The cyclic load follows a sinusoidal wave with a frequency of 6 Hz. Parameters were chosen based on previous studies and on the literature [6–8]. Specimens that reached a maximum prescribed number of cycles are named "run-out". Figure 1 (b) illustrates the loading pattern.

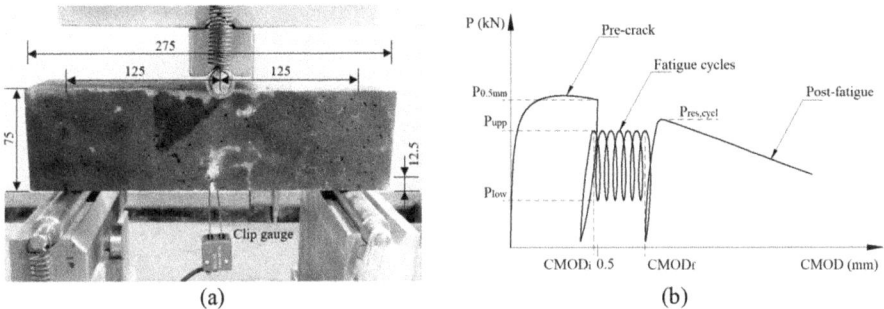

Fig. 1. Three-point bending test set-up in $75 \times 75 \times 275$ mm beam (a) and fatigue loading history of HPFRC and SFRC (b)

Fatigue life of HPFRC and SFRC was evaluated in terms of total number of cycles until rupture of specimen (N) for each S. The progressive fatigue failure process and the evolution of cracks were recorded, as well as the crack opening at the upper load of first cycle ($CMOD_i$), the crack opening of the last registered cycle ($CMOD_f$) and the crack opening range ($\Delta CMOD = CMOD_f - CMOD_i$).

The adopted criterion of incrementing the individual load was an approach to observe the tendency in a S-N relationship. This criterion also evaluates each fatigue response whilst considering a homogeneous loading criterion for all tested beams and, therefore, reducing the scatter sources. For the analysis, all specimens were included. This decision was made on the fact that specimens were pre-crack and the P_{upp} was an individual representation of each case. Omitting "run-out" specimens would underestimate the real number of cycles up to failure.

3 Results and Discussion

3.1 Fatigue Test

Table 1 summarizes the results of fatigue test on HPFRC and SFRC pre-cracked specimens under bending.

Table 1. Results from fatigue tests on HPFRC specimen

	S	$P_{0.5mm}$ (kN)	Cycles (N)	$CMOD_i$ (mm)	$CMOD_f$ (mm)	$\Delta CMOD$ (mm)
HPFRC	0.65	20.41	2,000,000[+]	0.391	0.417	0.026
	0.65	17.71	2,000,000[+]	0.402	0.495	0.093
	0.65	17.50	2,000,000[+]	0.432	0.536	0.104
	0.70	15.45	137,230	0.448	1.409	0.96
	0.70	15.88	1,000,000[+]	–	–	–
	0.70	16.33	1,581,049	0.432	1.804	1.372
	0.70	13.06	2,000,000[+]	0.429	0.663	0.234
	0.75	18.23	3,888	0.406	1.699	1.293
	0.75	17.75	4,821	0.417	2.723	2.306
	0.75	15.40	25,821	0.426	1.620	1.194
	0.80	15.88	238	0.45	5.607	5.158
	0.80	16.03	421	0.461	3.632	3.171
	0.80	14.64	1,103	0.441	–	–
	0.80	17.60	32,569	0.431	5.398	4.967
	0.85	13.96	176	0.473	3.193	2.719
	0.85	14.77	380	0.468	4.494	4.026
	0.85	14.78	448	0.473	3.497	3.024
	0.90	17.14	84	0.494	4.728	4.234
	0.90	16.08	86	0.496	4.455	3.959
	0.90	16.78	129	0.491	5.392	4.901
	1	16.87	49	0.549	5.334	4.785
SFRC	0.65	4.23	655,576	0.412	2.498	2.086
	0.65	3.87	865,807	0.427	3.298	2.871
	0.65	4.03	1,232,969	0.413	–	–
	0.65	4.67	1,250,000[+]	0.413	0.593	0.180
	0.70	5.05	284,037	0.427	2.444	2.018
	0.70	4.82	454,816	0.441	3.378	2.937
	0.70	5.17	662,702	0.436	2.969	2.533
	0.75	3.03	138,590	0.437	5.037	4.600
	0.75	4.07	799,830	0.439	1.620	1.181
	0.80	4.98	21,490	0.426	2.048	1.622
	0.80	6.18	214,800	0.466	3.936	3.470
	0.80	3.65	233,727	0.439	2.386	1.946
	0.80	4.82	450,070	0.445	3.113	2.668
	0.90	4.37	480	0.469	6.752	6.283
	0.90	5.51	710	0.505	6.347	5.842
	0.90	5.27	2,501	0.490	5.848	5.358
	0.90	5.63	3,737	0.501	4.761	4.260
	1	5.63	250	0.525	9.178	8.654
	1	3.14	874	0.471	9.715	9.244
	1	5.05	1,832	0.532	9.141	8.609
	1	4.83	4,507	0.527	6.447	5.920

[+] "Run-out"

There were relevant differences when comparing numbers of cycle to failure of specimens subjected to the same S. Both concretes showed the same average value of coefficient of variation (72%) when analysing the number of cycles to failure of each S. For HPFRC S of 0.70 series the difference of N varied from 137,230 cycles to 2,000,000 cycles (run-out), and for S of 0.80, N varied from 238 to 32,569 cycles. In the case of SFRC, specimens with applied load level of 0.65 diverged in failure after 655,576 cycles or withstood 1,2500,000 cycles with no visual sign of damage. Considering the adopted fatigue evaluation (applied dynamic load as a percentage of a specific $P_{0.5mm}$), this variation suggests that the fatigue life on pre-cracked specimens is a result of probabilistic difference in fibre orientation and distribution, imprecision of test equipment and set-up and a reflection of the fatigue scatter itself.

3.2 Cyclic Creep Curves

Figure 2 shows the average cyclic creep curves for each load level, in terms of normalized cycles (the ratio between the actual number of cycle n and the number of cycles to failure N) versus the maximum CMOD ($CMOD_{upp}$) for HPFRC (a) and SFRC (b). Since all specimens were pre-cracked, only phase II and phase III can be observed of a typical cyclic creep curve [1, 9]. Cyclic creep curve of "run-out" specimens were included for comparison.

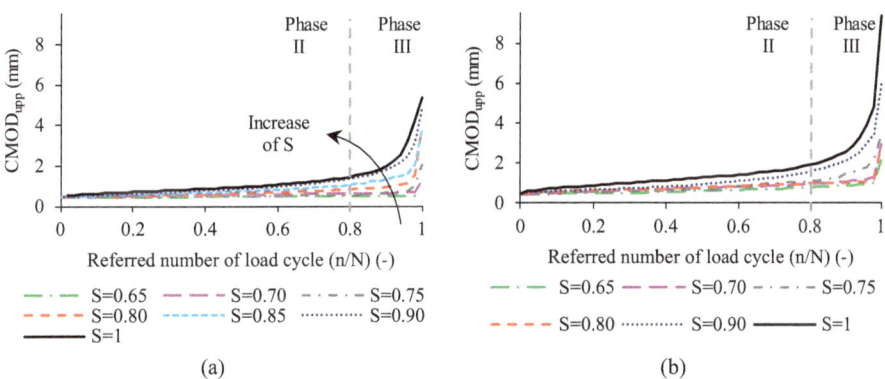

Fig. 2. Average cyclic creep curve for each load level of HPFRC (a) and SFRC (b)

In both cases, the evolution of CMOD seems to depend on the applied load level: as the load level increases, the slope of the crack increment per cycle becomes steeper and the crack opening wider. Yet, SFRC displayed higher crack opening displacements most likely due to longer fibres bridging the fractured zone and load level of 0.80 curve showed an unexpected behaviour, similar to the CMOD of S = 0.70 set.

Considering the mean fatigue life of HPFRC set for each S (1,179,570 cycles (S = 0.70); 11,510 cycles (S = 0.75); 8,583 cycles (S = 0.80); 335 cycles (S = 0.85); 100 cycles (S = 0.90); 49 cycles (S = 1)) the slope of phase II becomes steeper with smaller fatigue life. On the other hand, SFRC mean fatigue life (1,001,088 cycles

(S = 0.65); 467,185 cycles (S = 0.70); 469,210 cycles (S = 0.75); 230,022 cycles (S = 0.80); 1,857 cycles (S = 0.90); 1,866 cycles (S = 1)) between load level 0.70–0.75 and 0.90–1 showed the opposite behaviour. Since the values were similar, Welch's t-test (α equal to 0.05) was performed and revealed that there was not enough evidence to conclude that the differences between the means were statistically significant (p-value equal to 0.996 and 0.995 for S 0.70–0.75 and 0.90–1, respectively). Considering this statement, cyclic creep curves of HPFRC and SFRC suggest that the dCMOD/dn appears to be correlated to the N: lowering the secondary crack increment rate, the fatigue life increases.

Figure 3 (a) and (b) show a comparison between ΔCMOD and number of cycles to failure of HPFRC and SFRC, respectively. Higher load levels show greater upper crack opening at the last cycle (CMOD$_f$) and consequently crack opening range (ΔCMOD), than lower load cycles.

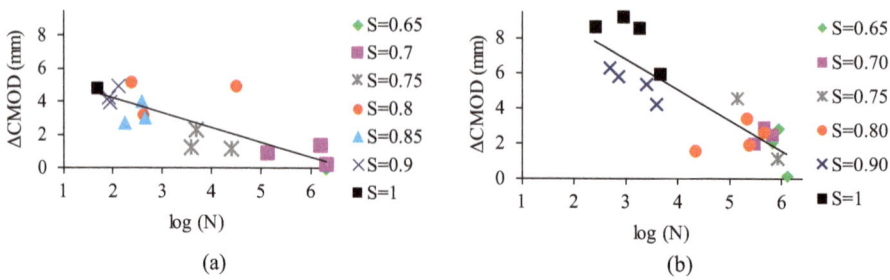

Fig. 3. Relation between crack opening range and number of cycles to failure

At higher S, the bend at phase III (Fig. 3) displays smoother shape. Also, the failure occurs at higher CMOD$_{upp}$. This suggests that lower load levels seem to produce a more brittle failure. This behaviour is in agreement with the observed in other investigations in fibre reinforced concrete in flexure [7, 10]. Higher S may failure through a continuous pull-out of the fibres, generating the ductile profile. Smaller load level can be responsible for the progressive weakening of the fibre-matrix interface through micro-cracks.

3.3   Probabilistic Approach

To estimate the probability of fatigue failure (P_f) of both investigated concretes, two different approaches were adopted. The Weibull distribution was utilized for the statistical description of fatigue data. The other approach used to describe the S–N–P_f relationship is by the mathematical model proposed by McCall [11] and slightly modified by Singh et al. [12]. The McCall model was used successfully to predict the fatigue life of various types of concretes [11–13]. To compare the investigates methods, the Wöhler curve considered the average values of N, which corresponds to a 50% of fatigue life survival [13]. Similarly, both probabilistic approaches were calculated to a probability of failure of 50%.

3.3.1 S–N Curves

Fatigue strength is obtained by the equivalent regression equation of plotting the relative load level (S) versus the logarithm of the number of cycles to failure (N). This curve is known as S–N curve, or Wöhler curve and from that, it can be obtained the fatigue strength.

The corresponding S–N equation which can be used for prediction purposes of precracked specimens of HPFRC or SFRC considered within the experimental program is given by Eq. (1) and Eq. (2), respectively.

$$S = 0.9801 - 0.0504 \log N \tag{1}$$

$$S = 1.2168 - 0.0878 \log N \tag{2}$$

Through the presented regression, both HPFRC and SFRC pre-cracked specimens seem to exhibit a fatigue endurance limit of 2,000,000 cycle of the order of 0.66 of $P_{0.5mm}$.

3.3.2 Weibull Distribution

There was a large variability in the fatigue-life data at the studied load levels and no definite trend was observed, indicating that the load levels selected for testing were probably too close together [14]. This inconveniency was mitigated by using the average value of the load levels 0.75 and 0.80 (0.78) and 0.90 and 1 (0.93) for HPFRC and the average between 0.70, 0.75 and 0.80 (0.76) and 0.9 and 1 (0.95) for SFRC. First, a graphical method was employed to verify if the fatigue-life data of the two concretes can be modelled by the two-parameter Weibull distribution. Subsequently, three different methods were used to estimate the parameters of the distribution, α and u. These methods are the graphical method (GM), method of moments (MoM) and method of maximum likelihood estimate (MLE). Results are shown in Table 2. Detailed calculation can be found elsewhere [3, 5, 15].

Table 2. Parameters α and u for fatigue-life data for all calculation methods

Type	S	Parameter	GM	MoM	MLE	Average
HPFRC	0.70	α	2.2209	3.3237	4.4104	3.3186
		u	1,779,624.93	1,701,716.77	1,683,053.13	1,721,473.90
	0.78	α	0.4816	0.7117	0.6329	0.6088
		u	7,830.17	7,893.60	7,010.31	7,578.15
	0.85	α	1.5144	2.5327	3.4926	2.5139
		u	412.21	377.06	374.27	387.86
	0.93	α	2.0540	2.8723	3.3978	2.7751
		u	102.12	97.61	97.07	98.98
SFRC	0.65	α	2.6312	3.8027	4.8085	3.7475
		u	1,145,588.46	1,107,614.88	1,098,819.32	1,117,340.89
	0.76	α	1.5802	1.8214	1.9988	1.8001
		u	472,629.75	455,470.43	459,418.75	462,506.31
	0.95	α	0.9278	1.1842	1.2211	1.1110
		u	2,096.42	1,971.94	1,990.74	2,019.70

The Kolmogorov-Smirnov test was applied as goodness-of-fit to the fatigue-life data at each load level as a function of the cumulative histogram and the hypothesized cumulative distribution function (D_n). The critical value D_c is taken from the Kolmogorov–Smirnov table for a 5% significance level. As $D_c > D_n$ the present model is accepted (Table 3).

Table 3. Kolmogorov-Smirnov test

	Load level	D_n = max $|F^* - F_N|$	Critical value D_c
HPFRC	0.70	0.1983	0.7076
	0.78	0.3269	0.4834
	0.85	0.2239	0.7076
	0.93	0.2608	0.6239
SFRC	0.65	0.2277	0.6239
	0.76	0.1822	0.4543
	0.95	0.1935	0.4543

Load level 0.78 of HPFRC revealed $\alpha < 1.0$ (Table 2), which leads to a decreasing hazard function with number of cycles. Although the graphical method as well as the goodness-of-fit test show that the Weibull distribution is a valid model in this situation, it violates the expected fatigue behaviour. For this reason, the value of $\alpha = 1.0$ can assumed [14] and the value of u recalculated to 9,837.29.

The fatigue lives corresponding to different failure probabilities P_f at different load levels can be calculated through Eq. (3) using the mean values of the parameters of the Weibull distribution.

$$\ln N = \frac{\ln\left[\ln \frac{1}{1-P_f}\right] + \alpha \ln u}{\alpha} \tag{3}$$

3.3.3 Mathematical Method: McCall Model

The McCall model [11] is based on a nonlinear relationship between S and logarithm of N given by Eq. (4).

$$L = 10^{-a(S)^b(\log N)^c} \tag{4}$$

in which $L = 1 - P_f$ is the survival probability; a, b and c are the experimental constants.

A multiple linear regression analysis was performed in order to fit the experimental data with the analytical model, resulting in Eq. (5) for HPFRC and Eq. (6) for SFRC.

$$L = 10^{-1.83 \times 10^{-2} (S)^{40.25} (\log N)^{10.56}} \tag{5}$$

$$L = 10^{-2.69 \times 10^{-5} (S)^{15.23} (\log N)^{8.53}} \tag{6}$$

3.3.4 Comparison Between Probabilistic Methods

To compare the investigates methods, the Wöhler curve considered the average values of N, which corresponds to a 50% of fatigue life survival [13]. Similarly, both probabilistic approaches were calculated to a probability of failure of 50%. Figure 4 shows the S–N–P_f curves of Weibull distribution, McCall model, Wöhler curve and experimental data points and the respective calculated R^2 for HPFRC (a) and SFRC (b).

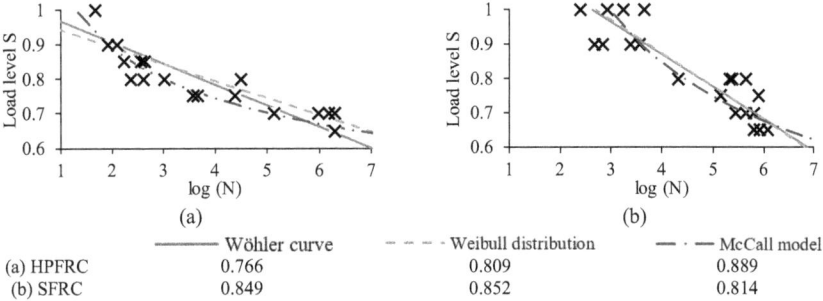

	Wöhler curve	Weibull distribution	McCall model
(a) HPFRC	0.766	0.809	0.889
(b) SFRC	0.849	0.852	0.814

Fig. 4. Comparison between methods considering a probability of failure of 50% and the experimental data and calculated values of R^2 for HPFRC (a) and SFRC (b)

Even though the scope of this research presents low number of results for each load level, the McCall mathematical method predicts reasonably well the flexural fatigue life of pre-cracked concretes for a desired probability of failure. Differences may be explained by variations in orientation and distribution of fibre which have more impact in the overall behaviour of SFRC specimen.

4 Conclusions

The main findings of this research are outlined below.

- Dynamic tests indicate higher dispersion of SFRC results compared to HPFRC most likely due to considerable lower number of fibres bridging the damaged zone, consequently minor variations in fibre orientation and distribution can have great influence in the overall behaviour;
- Applied load level plays an important role on the CMOD development through cycles and the equivalent CMOD at failure. As the load level increases, the slope of the crack increment per cycle becomes steeper and the crack opening displacement grows as well. At higher S, the bend at phase III displays smoother shape and the failure occurs at higher CMOD$_{upp}$. This suggests that lower load levels can cause a

reduction of the ductility. Higher S may lead to failure through a continuous pull-out of the fibres, this generating a more ductile response. Smaller S can be responsible for the progressive weakening of the fibre-matrix interface through micro-cracks;

- The McCall mathematical method predicts reasonably well the flexural fatigue strength of pre-cracked concrete specimens for a desired probability of failure.

Acknowledgements. The first author would like to thank the Brazilian National Council for Scientific and Technological Development for the scholarship granted (233980/2014-8). This research has been possible owe to the economic funds provided by the SAES project (BIA2016-78742-C2-1-R) of Spanish Ministerio de Economía, Industria y Competitividad.

References

1. Fib, International Federation for Structural Concrete: Constitutive modelling of high strength/high performance concrete, Germany (2008)
2. Fib, International Federation for Structural Concrete fib Model Code for Concrete Structures 2010, Germany (2013)
3. Oh, B.H.: Fatigue analysis of plain concrete in flexure. J. Struct. Eng. **112**, 273–288 (1986). https://doi.org/10.1061/(ASCE)0733-9445(1986)112:2(273)
4. Tarifa, M., Ruiz, G., Poveda, E., Zhang, X., Vicente, M.A., González, D.C.: Effect of uncertainty on load position in the fatigue life of steel-fiber reinforced concrete under compression. Mater. Struct. **51**(1), 1–11 (2018). https://doi.org/10.1617/s11527-018-1155-6
5. Carlesso, D.M., de la Fuente, A., Cavalaro, S.H.P.: Fatigue of cracked high performance fiber reinforced concrete subjected to bending. Constr. Build. Mater. **220**, 444–455 (2019). https://doi.org/10.1016/j.conbuildmat.2019.06.038
6. Kim, J.-K., Kim, Y.-Y.: Experimental study of the fatigue behavior of high strength concrete. Cem. Concr. Res. **26**, 1513–1523 (1996). https://doi.org/10.1016/0008-8846(96)00151-2
7. Lappa, E.: High strength fibre reinforced concrete static and fatigue behaviour in bending, Delft University of Technology (2007). 978-90-9021935-6
8. Huang, B.T., Li, Q.H., Xu, S.L., Liu, W., Wang, H.T.: Fatigue deformation behavior and fiber failure mechanism of ultra-high toughness cementitious composites in compression. Mater. Des. **157**, 457–468 (2018). https://doi.org/10.1016/j.matdes.2018.08.002
9. Plizzari, G.A., Cangiano, S., Alleruzzo, S.: The fatigue behaviour of cracked concrete. Fatigue Fract. Eng. Mater. Struct. **20**, 1195–1206 (1997). https://doi.org/10.1111/j.1460-2695.1997.tb00323.x
10. Leung, C.K.Y., Cheung, Y.N., Zhang, J.: Fatigue enhancement of concrete beam with ECC layer. Cem. Concr. Res. **37**, 743–750 (2007). https://doi.org/10.1016/j.cemconres.2007.01.015
11. McCall, J.T.: Probability of fatigue failure of plain concrete. J. Proc. **55**, 233–244 (1958)
12. Singh, S.P., Singh, B., Kaushik, S.K.: Probability of fatigue failure of steel fibrous concrete. Mag. Concr. Res. **57**, 65–72 (2005). https://doi.org/10.1680/macr.2005.57.2.65
13. Do, M.-T., Chaallal, O., Aïtcin, P.-C.: Fatigue behavior of high-performance concrete. J. Mater. Civ. Eng. **5**, 96–111 (1993). https://doi.org/10.1061/(ASCE)0899-1561(1993)5:1(96)

14. Singh, S.P., Kaushik, S.K.: Flexural fatigue life distributions and failure probability of steel fibrous concrete. ACI Mater. J. **97**, 658–667 (2000)
15. Mohammadi, Y., Kaushik, S.K.: Flexural fatigue-life distributions of plain and fibrous concrete at various stress levels. J. Mater. Civ. Eng. **17**, 650–658 (2005). https://doi.org/10.1061/(ASCE)0899-1561(2005)17:6(650)

Direct Tensile Tests of Supercritical Steel Fibre Reinforced Concrete

Katharina Look[1(⊠)], Peter Heek[2], and Peter Mark[1]

[1] Institute of Concrete Structures, Ruhr University Bochum, Bochum, Germany
katharina.look@rub.de
[2] FH Münster, University of Applied Science, Münster, Germany

Abstract. Steel fibre reinforced concrete (SFRC) becomes increasingly interesting for structural design and application. However, to reinforce structures just with steel fibres – not including any rebar – supercritical fibre contents are essential to ensure hardening behaviour in the post-cracking domain. Material properties are usually determined from experiments conducting three- or four-point bending tests. Specific conversion factors capture the softening behaviour and enable to transform flexural into tensile strengths. Own experiments prove that fibre contents of 1.8 Vol.-% yield flexural strengths of about 8 MPa. To get definite and reliable tensile strengths, direct tensile tests on optimised bone-shaped specimens made of supercritical SFRC are proposed here. As a specimen a slab ($w \times h \times l = 200 \times 100 \times 720$ mm^3) is casted horizontally. That way, fibre orientation and distribution representative for practically relevant slabs with 10 cm thickness are simulated. To eliminate the so-called wall-effect that occurs during casting, the edges are cut off by water jet cutting before testing. Two pairs of displacement transducers on each face of the slab record the crack opening over a measuring length of 100 mm on the top and 400 mm on the bottom face. A new test set-up is introduced. Loading is applied to the specimen by friction using pre-tensioned threaded steel rods. Coating with an epoxy resin and corundum guarantees the required coefficient of friction. Displacement transducers on the top and bottom of the specimen record the relative displacement between the specimen and the test station. Axial loading is induced by a triangular steel structure (framework). Strain gauges on the outer faces of the diagonal struts control inevitable eccentricities of the load transfer. Consequently, highly accurate measurements are recorded. During testing, the crack flanks are slowly pulled apart from another (up to 4 mm) but without complete separation. On average a maximum tensile strength of 3 MPa and a coefficient of variation of 11% for maximum force is recorded what indicates a small scatter and highly accurate strengths.

Keywords: Steel fibre reinforced concrete · Tensile test · Substitute reinforcement · Supercritical fibre content · Hardening behaviour · Bone shape

© RILEM 2022
P. Serna et al. (Eds.): BEFIB 2021, RILEM Bookseries 36, pp. 132–142, 2022.
https://doi.org/10.1007/978-3-030-83719-8_12

1 Introduction

Nowadays, the construction industry experiences a drastic change: the amount of degraded bridges and the demand for new living space is growing. Simultaneously, the need for fast, ecologic and cost-effective construction is increasing [1–3]. As steel fibre reinforced concrete (SFRC) has many advantages for concrete constructions in the fields of e.g. fire [4, 5], fatigue [6, 7], crack control [8] or creep [9], it is in the focus of current research. SFRC contributes to the reduction of construction times and helps to produce cost-effective prefabricated components as fibres are added during manufacturing and thus eliminate costly reinforcing work. At best the entire reinforcement is replaced by steel fibres [10]. Therefore, steering of fibre orientation and adaptation of the load bearing behaviour to advantageous fields of application for SFRC are fundamental [11, 12]. One approach is presented in [13]: Crosswise reinforcement in slabs intended to be used as foundations or wall-elements is replaced by SFRC. Bending beam tests on specimens with a supercritical steel fibre content of 1.8 Vol.-% have already proven highly accurate flexural tensile strengths of 8 MPa and a distinct ductile hardening behaviour in the post-cracking domain [11]. Current standards and regulations provide conversion factors to derive axial tensile strengths from flexural tensile strengths for structural design. Often based on too strong simplifications, these factors are questioned to be realistic. In general, the tensile strength is influenced by many effects such as fibre orientation and formwork geometry [14, 15]. To get well-founded and safe values for the design of SFRC slabs, tensile tests, which map the real conditions, are presented here. For these, the specimens shape is adapted to a real component's one to reproduce fibre orientation and get highly accurate and realistic tensile strengths for the substitution of reinforcing bars. Thereof, an innovative test set-up for axial tensile tests, especially designed for supercritical SFRC, is developed.

2 Specimen

2.1 Shaping

The geometry of the test specimen is designed to map the real component's conditions – e.g. fibre orientation, manufacturing process, fibre distribution – in the most appropriate way. To reproduce the expected two-dimensional fibre orientation of a flat component, the height of the test specimen is adjusted to the height of the prefabricated SFRC-slab ($h = 10$ cm) in [13] for which such an orientation has already been verified.

Even if easiest to manufacture, constant cross-sections of SFRC specimens in tensile tests are not useful since the specimens would suffer from the wall-effect. The wall-effect, which enforces one-dimensional fibre orientation in direction of the tensile stresses at a distance equal to the half of the fibre length l_f from the formwork's edge, significantly increases the tensile strength recorded in experiments [16–18]. This issue is resolved cutting some regions near the edges off the specimen before testing. However, any cutting of material induces inhomogeneity to the stress-field. To limit this effect to certain regions of minor interest shape optimization of the cut-outs has been performed with an algorithm published by *Mattheck* [17] that has successfully been applied to other engineering problems [18, 19]. Additionally, two restrictions had to be met. At centre of the specimen a length of 10 cm with a constant cross-section was maintained to monitor cracking in an undisturbed region. And at both ends of the specimen a length of 12.5 cm was foreseen to host PVC tubes to connect the specimen to the testing device. Figure 1 compares stress-fields of three alternative cut-outs qualitatively. Alternative a) has a simple quarter-circular cut-out in transition (radius $r = 5$ cm) which is enough to eliminate the wall-effect in case of fibres with $l_f = 6$ cm. The other alternatives are obtained from the algorithm with a higher degree of subdivision $n = 3$ (b) and $n = 5$ (c). While a) still shows stress peaks at the cut-outs (dark red area), the stress-fields of the two alternatives appear more and more homogeneous as n increases. Consequently, the specimen is expected to fail in the weakest region at centre.

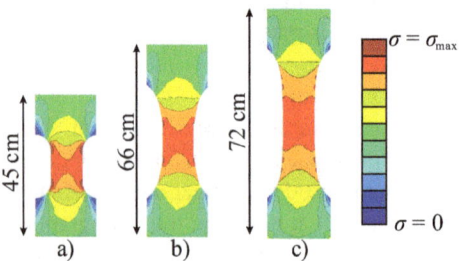

Fig. 1. Qualitative stress-distributions of the three alternatives.

Figure 2 presents the final dimensions of the formwork (dashed lines) and the shape-optimised specimen (grey shaded).

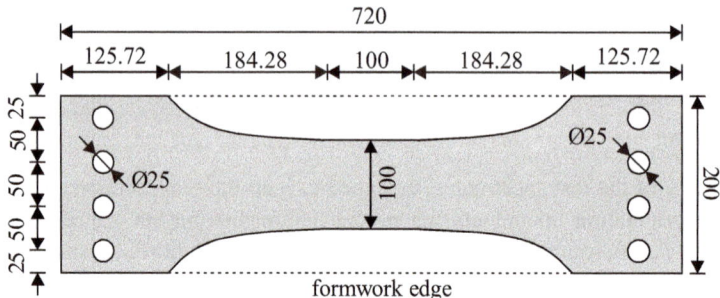

Fig. 2. Final dimensions of the formwork (dashed lines) and the final shape-optimised specimen (grey shaded).

2.2   Fabrication

Also casting and manufacturing are adapted to the required fibre orientation in pre-fabricated slabs. Besides limiting the height of the specimen, casting is done horizontally to align the fibres in the plane [11, 15]. As detailed in the next section, the load is applied to the specimen purely by friction using threaded steel rods for clamping.

Fig. 3. Water jet cutting of a specimen   **Fig. 4.** Cut of a specimen.

During casting four PVC tubes with an outer diameter of $d_o = 25$ mm are installed into the specimens at top and bottom to provide the required holes for clamping (cf. Fig. 2 and Fig. 3).

SFRC contains 1.8 Vol.-% macro steel fibres of type Dramix® 5D 65/60 BG with a length of 60 mm and a diameter of 0.9 mm. The concrete mix is detailed in [13]. Three associated cubes are casted to determine the compressive strength of the concrete. After 24 h all specimens are stripped from the formwork and stored dry at 20 °C in the lab until testing.

After 21 days the edges were cut off via water jetting. Figure 3 shows the specimen during water jetting while Fig. 4 presents the face after cutting. No outbreak of aggregates is visible which could induce cracking. But, with increasing penetration depth of the jet, the corrugation of the surface becomes coarser. Comparisons of a 3D scan with the original 3D-CAD model show minor deviations of −0.5 mm on average at the top (5 mm penetration depth) and -3.5 mm at the bottom of the specimen (95 mm penetration depth). The calculated means are below the acceptable limits for deviations of pre-casted elements according to EN 13369 [19]. Nevertheless, with increasing depth, the cross-section gets smaller than intended. Thus, the tensile strength calculated with a cross-section of 100×100 mm tends to be slightly underestimated.

In general, this method allows SFRC-specimens being shaped as desired and as can hardly be done with conventional formwork especially regarding the wall-effect. It opens new fields of application for concrete, as arbitrary shapes can be realized that were previously not possible.

3 Design of Test Set-Up

3.1 Assembly

The test set-up consists of two identical parts on bottom and top, which are locked into the two hydraulic clamping jaws of a 1,000 kN-testing machine. Axial loading of the specimen is ensured by a triangular steel construction (S355). The force is transmitted via two threaded steel rods (M36–10.9) and then introduced into the specimen purely by friction over two rigid steel plates (S235) clamped to the specimen.

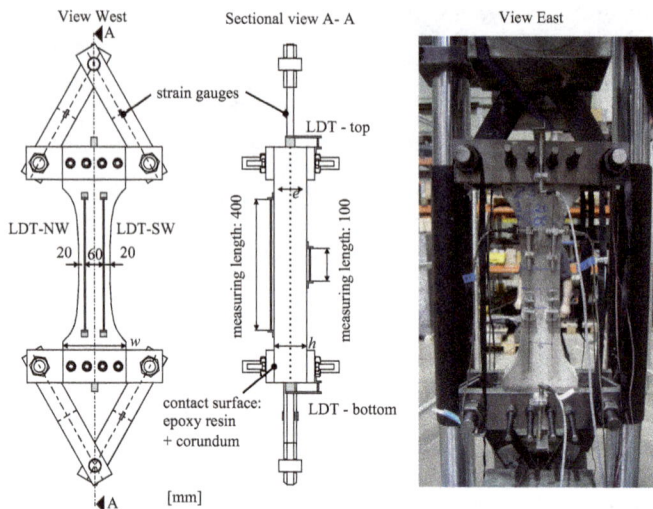

Fig. 5. Assembly of the test set-up and measurement instrumentation.

With this set-up widths of the specimen (w) from 50 to 150 mm and heights (h) from 100 to 200 mm are possible (cf. Fig. 5). The maximum length of the specimen depends on the testing machine, which here is limited to about 100 cm. The maximum force expected in the experiment is determined to 150 kN. The surface of the steel plate in contact with the specimen is coated with an epoxy resin-corundum mix to ensure enough friction. Four high-strength threaded steel rods (M16–10.9) are used for clamping. Each is alternately pre-stressed with 50 kN using a torque spanner. The test device has been designed to move freely in all three spatial directions so that the test specimen is not subjected to any constraints during testing. At the beginning of the test, the device is aligned by a laser to avoid unwanted bending.

3.2   Measurement Instrumentation

Six strain gauges applied on the diagonal struts control the centric load transfer. Two linear displacement transducers (LDT) record the total crack opening on the front face over a measuring length of 400 mm (LDT-NW and LDT-SW, Fig. 5). Two LDTs (LDT-NO and LDT-SO) record the crack opening over the specimen's central length of 100 mm (Fig. 5). Two more LDTs (LDT-top and LDT-bottom, Fig. 5) are applied on top and bottom of the set-up to record the relative displacement (slip) between specimen and test set-up. In addition, the western face of the specimen is recorded by video during the whole test. The experiment is conducted path-controlled at a speed of 0.1 mm/min. to capture the post-cracking behaviour.

4   Results

4.1   Qualification of the Test Set-Up

4.1.1   Centring

The specimen must not be bent unintentionally by skewed or eccentric loads – which are controlled by strain gauges applied on the outer faces of the diagonal struts of the steel construction. The associated strains enable to illustrate the distribution of stresses and to calculate the resulting eccentricities in the load transmission. In comparison to the known forces induced by the testing machine eccentricities of the load transfer can be detected. Figure 6 shows the calculated eccentricities when cracking occurs.

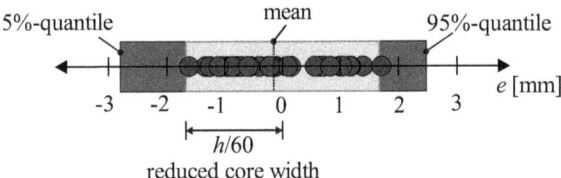

Fig. 6. Eccentricities of load application at cracking.

If the resulting longitudinal force lies inside the core width of a cross-section, only stresses of one sign occur, i.e. the whole cross-section is predominantly loaded in tension or compression. The core width of a rectangular cross-section is simply $h/6 = 100/6 \sim 17$ mm. The absolute mean of the load eccentricities at cracking is calculated to $e = 1.6$ mm and fits safely into the core (Fig. 6). This even holds true regarding the total scatter in the eccentricities. The 5%- and 95%-quantiles contain 90% of the data by definition and are computed to $\mu \pm \sigma = 1.6 \pm 0.21$ mm here. Most eccentricities even fall within a tenth of the core width ($h/60 = 100/60 \sim 1.7$ mm). It thus is concluded that the load is introduced sufficiently centric.

4.1.2   Slip

The maximum measured displacements of both LDTs on top and bottom (cf. Fig. 5) of each specimen are listed in Table 1. In most cases they are quite small. In the limit the threaded rod would get into contact with the concrete and therefore would generate constraints. From this the acceptable distance from the outer face of the threaded rod with a diameter of $d_t = 16$ mm to the concrete surface in the hole with a diameter of $d_o = 25$ mm follows from Eq. (1):

$$\frac{d_o - d_t}{2} = \frac{25 - 16}{2} = 4.5 \,\text{mm} = 4500 \,\text{µm} \tag{1}$$

Table 1. Maximum values of slip for each specimen top and bottom in [µm].

	S1	S2	S3	S4	S5	S6
LDT-top [µm]	2.19	5.99	0.09	0.16	3.97	18.39
LDT-bottom [µm]	3.22	6.50	1.31	38.63	2.11	476.91

None of the readings in Table 1 exceeds that limit. Just one displacement is conspicuous since it is significantly greater than the others (S6 - bottom). When examining the displacement versus load curves of the two LDTs of S6, it is noted that a maximum slip of 14.85 µm occurs before cracking. This corresponds to 0.5% of the crack opening width of 3 mm. Moreover, the steep rise of the slip first happens in the declining branch beyond 18 kN and thus after cracking. It has therefore no influence on the cracking-behaviour of the specimen. Furthermore, the stress-crack-opening relation of S6 shows no irregularities (cf. Fig. 7), so that even the largest value has no effect on the crack opening. The pre-tensioning force in combination with the friction coefficient of the coating is always sufficient to prevent the specimen from slipping.

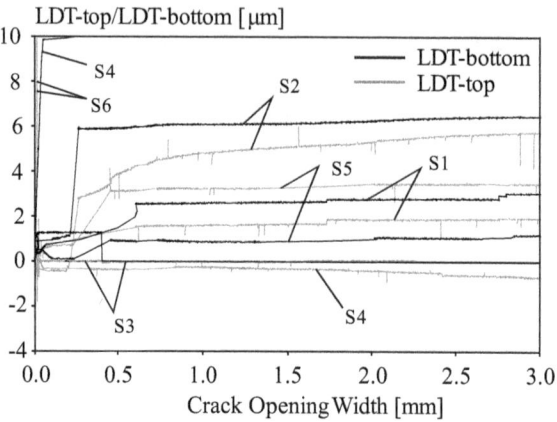

Fig. 7. Relative displacement of the specimen to the test device at top and bottom as a function of the crack opening width.

Figure 7 presents all displacements of the LDTs at top (grey) and bottom (black) in µm as a function of the crack opening width. Sudden changes in the curves characterise single crack opening events. They immediately cause load redistribution and let the specimens slip minimally. Most of the LDTs reach a constant plateau after cracking, which is seen as an indicator for the high sensitivity of the test set-up. Just three curves do not reach a constant level: bottom and top LDTs of S6 and bottom LDT of S4. As already mentioned, the LDTs of S6 are nevertheless well in an acceptable range. The LDT of S4 at bottom reaches its maximum of 9.83 µm before cracking. Here, this corresponds to 0.3% of the maximum measured crack width of 3 mm and is therefore also seen negligible.

4.2   Stress-Crack Opening

The purpose of the tensile tests is to finally obtain the stress-crack-opening relation of SFRC in a direct way and to prepare and use it for further investigations such as FEM calculations. Figure 8 presents the stress-crack-opening relations of the six specimens (grey lines) and its mean course (black line) on the primary y-axis. The secondary y-axis presents the coefficient of variation (COV) as an indicator for scatter (dotted line). During evaluation of the experimental results, it was noticed that the cracks did not develop exclusively in the central 10 cm of the specimens with the smallest cross-section width as expected. Due to the shape-optimisation to eliminate the wall-effect and the general inhomogeneity of the composite material SFRC cracks occur widely spread, on average at a distance of ±7.34 cm from the centre. Thus, most of the LDTs on the eastern face with a limited measuring length of 100 mm fail due to multiple cracks propagating in the vicinity or completely outside that measuring range. For these no stress-crack-opening relation could be measured at all. Consequently, the stress-crack-opening relations in Fig. 8 base only on the mean value of the two LDTs on the western face with a measuring length of 400 mm.

The mean compressive strength of the cubes after 28 days is 84.6 MPa and thus almost identical to the bending tests published in [13]. In contrast to these, no hardening in the post-cracking domain was achieved with the same concrete mix due to the lower ability of load redistribution of components in tension compared to components subjected to bending. On the contrary, we found a softening but still ductile behaviour after cracking. The SFRC specimen resumes the force after cracking but no sudden load decline is observed as it would be the case for plain concrete. Maximum tensile strength of 3 MPa is reached on average. Compared to the flexural tensile strength of 8 MPa, a conversation factor of 0.38 is obtained, which is close to the factors proposed

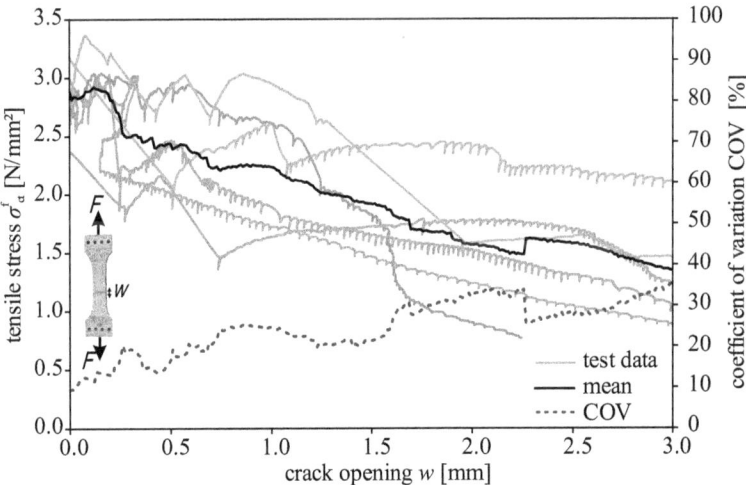

Fig. 8. Stress-crack-opening relations of the six specimens, mean course and coefficient of variation (COV).

in recent international standards, e.g. [20, 21]. Since the crack flanks are slowly pulled apart without complete separation a ductile behaviour is still observed here. Around the critical crack and occasionally in the specimen itself, small branching cracks develop in all specimens, indicating ductile failure.

Scatter increases as expected with crack propagation. For the mean load at cracking (approx. 30 kN) a COV of 11% is calculated. This is quite low especially regarding the COVs of bending tests assumed in national and international standards [21] and confirms the practicability of the developed test set-up.

5   Conclusions

In this contribution, an innovative test set-up for direct tensile tests of shape-optimized SFRC specimen with supercritical steel fibre content is developed. The aim is to provide safe tensile strength data of SFRC to support complete substitution of bar reinforcement by fibres in prefabricated slabs. The following conclusions are drawn:

- A new method to cut SFRC specimens highly accurate and freed from the wall-effect by water jetting is introduced.
- On such specimens realistic tensile strengths in real conditions, especially regarding fibre alignment, can be determined from experiments with low scatter of around 10%. Cracking occurs in a justifiable range within the specimen and thus confirms the high degree of optimisation of the specimen geometry.
- An investigation of the relative displacements between the specimen and the test rig proves negligible slip, which does not impair the measured tensile strengths.
- Small eccentricities confirm a largely axial transmission of tensile forces into the specimens. The test rig is thus capable to determine realistic tensile strengths from specimens made of supercritical SFRC.
- Measured tensile strengths of 3 MPa are stable and do not scatter much. They can therefore be used for design of SFRC components without any rebar.
- Comparing the obtained tensile to the flexural strength reported in [11], a conversion factor of 0.38 close to the ones proposed in recent national and international standards is confirmed.
- If the maximum tensile strength of 3 MPa is smeared over the cross-section of a virtual slab with a height of 10 cm, the slab would have a resistance of 300 kN/m in both directions. The same resistance would require standard reinforcement ($f_{yk} =$ 500 MPa) of 6.0 cm^2/m. This corresponds to a mesh reinforcement of about Ø10 every 12.5 or 15 cm, which is substitutable with the presented SFRC mix.

Acknowledgements. The authors would like to thank BASF SE, BauMineral GmbH, NV Bekaert SA and Dyckerhoff GmbH for the friendly provision of the test materials. Many thanks also to the members of the Structural Testing Laboratory KIBKON.

References

1. Forman, P., Gaganelis, G., Mark, P.: Optimierungsgestützt entwerfen und bemessen. Bautechnik **97**(10), 697–707 (2020). https://doi.org/10.1002/bate.202000054
2. Forman, P., Penkert, S., Mark, P., Schnell, J.: Design of modular concrete heliostats symmetry reduction methods. Civil Eng. Des. **2**(4), 92–103 (2020). https://doi.org/10.1002/cend.202000013
3. Gaganelis, G., Mark, P.: Downsizing weight while upsizing efficiency – an experimental approach to develop optimized ultra-light UHPC hybrid beams. Struct. Concr. **20**(6), 1883–1895 (2019). https://doi.org/10.1002/suco.201900215
4. Heek, P., Tkocz, J., Mark, P.: A thermo-mechanical model for SFRC beams or slabs at elevated temperatures. Mater. Struct. **51**(4), 1–16 (2018). https://doi.org/10.1617/s11527-018-1218-8
5. Heek, P., Tkocz, J., Thiele, C., Vitt, G., Mark, P.: Fasern unter Feuer. Beton- und Stahlbetonbau **110**(10), 656–671 (2015). https://doi.org/10.1002/best.201500046
6. Heek, P., Ahrens, M.A., Mark, P.: Incremental-iterative model for time-variant analysis of SFRC subjected to flexural fatigue. Mater. Struct. **50**(1), 1–15 (2017). https://doi.org/10.1617/s11527-016-0928-z. Article No. 62
7. Heek, P., Mark, P.: Zur Ermüdung von Beton und Stahlfaserbeton. Beton- und Stahlbetonbau **111**(4), 221–232 (2016). https://doi.org/10.1002/best.201500054
8. Empelmann, M., Oettel, V., Cramer, J.: 'Berechnung der Rissbreite von mit Stahlfasern und Betonstahl bewehrten Betonbauteilen'. Beton- und Stahlbetonbau **115**(2), 136–145 (2020). https://doi.org/10.1002/best.201900065
9. Plizzari, G., Serna, P.: Structural effects of FRC creep. Mater. Struct. **51**(6), 1–11 (2018). https://doi.org/10.1617/s11527-018-1290-0. Article No. 167
10. Di Prisco, M., Plizzari, G., Vandewalle, L.: Fibre reinforced concrete: new design perspectives. Mater. Struct. **42**(9), 1261–1281 (2009). https://doi.org/10.1617/s11527-009-9529-4
11. Look, K., Heek, P., Mark, P.: Stahlfaserbetonbauteile praxisgerecht berechnen, bemessen und optimieren. Beton- und Stahlbetonbau **114**(5), 296–306 (2019). https://doi.org/10.1002/best.201800097
12. Mark, P., Oettel, V., Look, K., Empelmann, M.: Neuauflage DAfStb-Richtlinie Stahlfaserbeton. Beton- und Stahlbetonbau **116**(1) (2021). https://doi.org/10.1002/best.202000065
13. Look, K., Heek, P., Mark, P.: Towards rebar substitution by fibres - tailored supercritical fibre contents. In: Serna, P., et al. (eds.) BEFIB 2020, RILEM Bookseries, vol. 30, pp. 908–919 (2020)
14. Lin, Y.-z.: DAfStb-Heft 494: Tragverhalten von Stahlfaserbeton. Institut für Massivbau und Baustofftechnologie, Universität Karlsruhe, Dissertation (1996)
15. Soroushian, P., Lee, C.-D.: Tensile strength of steel fiber reinforced concrete - correlation with some measures of fiber spacing. Mater. J. **87**(6), 542–546 (1990)
16. Soroushian, P., Lee, C.-D.: Distribution and orientation of fibers in steel fiber reinforced concrete. Mater. J. **87**(5), 433–439 (1990)
17. Stähli, P., Custer, R., van Mier, J.G.M.: On flow properties, fibre distribution, fibre orientation and flexural behaviour of FRC. Mater. Struct. **41**(1), 189–196 (2008). https://doi.org/10.1617/s11527-007-9229-x
18. Lanwer, J.-P., Oettel, V., Empelmann, M., Höper, S., Kowalsky, U., Dinkler, D.: Bond behavior of micro steel fibers embedded in ultra-high performance concrete subjected to monotonic and cyclic loading. Struct. Concr. **20**(4), 1243–1253 (2019). https://doi.org/10.1002/suco.201900030

19. Mattheck, C., Kappel, R., Sauer, A.: Shape optimization the easy way - the method of tensile triangles. Int. J. Des. Nat. Ecodyn. **2**(4), 301–309 (2007). https://doi.org/10.2495/D&N-V2-N4-301-309

20. Mattheck, C.: Design and growth rules for biological structures and their application to engineering. Fatigue Fract. Eng. Mater. Struct. **13**(5), 535–550 (1990). https://doi.org/10.1111/j.1460-2695.1990.tb00623.x

21. Mattheck, C.: Engineering components grow like trees. Materialwiss. Werkstofftech. **21**(4), 143–168 (1990). https://doi.org/10.1002/mawe.19900210403

22. EN 13369: Common rules for precast concrete products (2018)

23. International Federation for Structural Concrete: *fib* Model Code for Concrete Structures 2010. Ernst & Sohn, Berlin (2013)

24. Deutscher Ausschuss für Stahlbeton: DAfStb-Richtlinie Stahlfaserbeton, Beuth, Berlin (2012)

25. prEN 1992-1-1 – Annex L, Draft D7 to the Eurocode 2 (2020)

Uniaxial Tension Tests of Steel Fibre Reinforced Concrete with AE Monitoring

Maure De Smedt[(✉)], Eline Vandecruys, Rutger Vrijdaghs,
Els Verstrynge, and Lucie Vandewalle

Department of Civil Engineering, KU Leuven, Leuven, Belgium
maure.desmedt@kuleuven.be

Abstract. The uniaxial tensile behaviour of steel fibre reinforced concrete (SFRC) is an important material characterisation. However, uniaxial tensile tests (UTT) are difficult to perform and thus less reported in literature compared to bending tests. This paper investigates the uniaxial tensile behaviour of four SFRC mixtures. An advanced UTT setup was developed with acoustic emission (AE) monitoring. Besides monotonic loading, also progressive cyclic loading has been applied. Two hooked end steel fibre types are both used in a content of 20 and 40 kg/m^3. These fibre characteristics and their distribution and orientation have a large impact on the load-displacement behaviour. Micro-cracking is detected by means of AE activity. Failure mode analysis shows the onset of macro-cracking by the downshift of average frequency and the increase of rise angle. A larger amount of frictional damage results in more AE events during unloading and an increased AE energy amount. Localisation of AE events validated the position of the fibres in the fracture plane. During the different loading stages, the crack's initiation and development are accurately localised. Lastly, predictions of the tensile strength and behaviour according to Model Code 2010 are in good agreement with the experimental results.

Keywords: Steel fibre reinforced concrete · Uniaxial tensile behaviour · Damage · Acoustic emission

1 Introduction

The uniaxial tensile behaviour of steel fibre reinforced concrete (SFRC) is an important knowledge of the material, for example when implementing a numerical model. Performing uniaxial tensile tests (UTT) results directly in this behaviour because of the uniform tensile stress distribution [1], compared to the more complicated non-uniform stress distribution of a three-point bending test [2, 3]. However, UTT are difficult to perform due to the technical challenges such as alignment and gripping [4].

Furthermore, advanced non-destructive measurement methods are required to obtain a more profound damage assessment, instead of solely measuring load and crack opening. Acoustic emission (AE) monitoring is such a method. By recording the elastic waves generated by a strain energy release within the material, damage and cracking are continuously detected, localised and characterised. However, the interpretation must be done carefully to relate AE data to possible sources. [5–8] Previous research

© RILEM 2022
P. Serna et al. (Eds.): BEFIB 2021, RILEM Bookseries 36, pp. 143–154, 2022.
https://doi.org/10.1007/978-3-030-83719-8_13

proved the adequacy of AE monitoring on SFRC, but mostly in monotonic bending tests [9]. Research on UTT with AE is very limited [10, 11], especially including SFRC [12, 13, 18].

In conclusion, the combination of UTT on notched SFRC cylinders assessed with AE is the novelty of this research. Furthermore, not only monotonic loading, but also progressive cyclic loading is applied. The aim is to evaluate the fibre influence on the tensile behaviour, taking into account different fibre types and volumes. AE monitoring is used to examine the damage initiation and propagation.

2   Experimental Research and Methods

In total, 31 UTT were performed, namely 16 monotonic and 15 progressive cyclic. Two fibre types and two fibre volumes were examined, leading to four SFRC groups. All tests were monitored with AE.

2.1   Materials

The SFRC groups consisted of normal strength concrete (class C40/50 [9]) and two types of hooked-end steel fibres. Tested according to EN 12390–3 [14], the mean cube concrete compressive strength equalled 59 MPa. The two types of steel fibres are denoted as 3D and 5D fibres. Both have a length of 60 mm and an elastic modulus of 200 GPa. The 3D fibre has a single end-hook, a diameter of 0.75 mm and 1125 MPa tensile strength. The 5D fibre has a double end-hook, 0.90 mm diameter and 2300 MPa tensile strength. Two fibre volumes were investigated, namely 20 or 40 kg/m^3, corresponding to 0.25 or 0.50 V% respectively, in order to examine softening and hardening bending behaviour [9]. The four resulting SFRC mixtures are indicated as 3D20, 3D40, 5D20 and 5D40.

2.2   Test Specimens

SFRC cylinders with a height of 300 mm and a diameter of 100 mm were tested in the uniaxial tensile setup. At mid-height, a notch of 4 mm wide and 10 mm deep was sawn to initiate the crack position. The cylinders were cored from prisms, which were subjected to bending tests [9]. Since the prisms had dimensions of 150 × 150 660 mm, their mid-span section is at sufficient distance from the cylinders' notched section to assume no influence of the fracture plane during the bending test. Based on the previously performed three-point bending tests, the average flexural tensile strength of the prisms was equal to 4.14 MPa [9]. Furthermore, the post-cracking classes according to Model Code 2010 [15] were 1b, 1.5b, 1c and 1.5e for 3D20, 3D40, 5D20 and 5D40 respectively.

2.3   Test Setup and Loading Pattern

Figure 1 presents the applied test setup. The cylinders were glued between two steel plates. Additionally, PVC glue moulds were used at both ends to avoid failure at the glued section. The steel plates were connected to the universal testing machine with a maximum capacity of 100 kN. This machine recorded the tensile load F and displacement of the press head (stroke) with a frequency of 1 Hz. Furthermore, three clip gauges at an angle of 120° of each other measured the crack opening. The average of the three clip gauges is taken as the crack width w, as well recorded at 1 Hz.

Two loading patterns were applied on each SFRC group. For the monotonic (M) loading, the stroke increased by 0.005 mm/min until 0.1 mm and then by 0.1 mm/min until 4 mm. For the progressive cyclic (C) loading, the M pattern is extended with 12 unloading-reloading cycles at strokes equal to 0.1, 0.2, 0.3, 0.4, 0.5, 1.0, 1.5, 2.0, 2.5, 3.0, 3.5 and 4.0 mm. The unloading rate equalled 0.1 mm/min until a minimum load of 5 N. After completion of a test, the two halves of the cylinders are pulled apart to investigate the fibre distribution and position in the fracture plane.

2.4 Acoustic Emission (AE) Monitoring

Continuous AE monitoring has been applied on all UTT to examine the internal damage development. Six piezoelectric 150 kHz resonance sensors were used in a three dimensional setup, as shown in Fig. 1. At 30 mm above and below the notch, three sensors were installed with vacuum gel and tape, at an angle of 120° of each other. This configuration allows localizing damage in all dimensions, measuring at both sides of the crack and avoiding a ringing effect due to the cylindrical specimen.

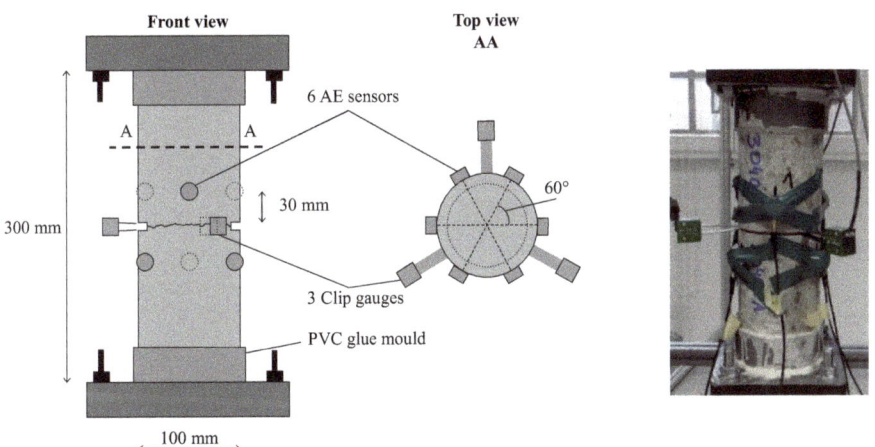

Fig. 1. Overview and picture of the test setup and AE sensors.

Each sensor was connected to a 6-channel acquisition system, using a preamplifier with 34 dB gain. Background noise is avoided by the 50–850 kHz frequency filter and the 40 dB amplitude threshold. A second amplitude filter of 50 dB is added in the post-processing phase. The sampling rate equalled 5 MHz. The maximal location uncertainty determined by pencil lead breaks equalled 20 mm.

2.5   AE Analysis

The advantage of extending the traditional UTT test setup with AE monitoring is the detection, characterisation and localisation of damage initiation and propagation. Therefore, the AE measurements are examined by a parameter-based approach. The AE activity presents the amount of recorded AE events, localised within the specimen. Based on the first hit of each event, the amplitude (dB), the energy (1 eu = 10^{-14} V^2s) and the source localisation of the signal are studied.

Furthermore, failure mode analysis is performed based on rise angle (RA, in ms/V) and average frequency (AF, in kHz). The former is rise time divided by amplitude of an AE signal; the latter is number of threshold crossing divided by signal duration. A high AF and low RA indicates mode I damage (tensile cracks), while the reverse indicates mode II damage (shear cracks) [5, 16].

3   Results and Discussion

The uniaxial tensile behaviour of the four SFRC groups is investigated in function of fibre type and volume. The influence of loading cycles is discussed as well. Thereafter, the damage imitation and localisation and the micro- and macro-cracking are examined. All results are based on both the mechanical as well as the acoustic emission characteristics. Lastly, the experimental behaviour is compared to predictions according to the Model Code 2010 (MC2010) [15].

3.1   SFRC Behaviour in Function of Fibre Characteristics

For each SFRC group (3D20, 3D40, 5D20 and 5D40), three to six specimens were tested in monotonic or cyclic loading. For 3D40, only two cyclic tests were performed successfully. Figure 2 presents the stress – crack opening diagrams for all tests. The stress is calculated with Eq. (1) [1]. F is the applied tensile load and A_s is the cross-sectional area of the notched section.

$$\sigma = F/A_s \tag{1}$$

The average cyclic envelope curve is determined based on the envelope curve of each specimen. In general, a good comparison between the monotonic and cyclic post-peak behaviour is found [18]. A larger fibre volume leads to a higher post-peak behaviour, as well as increasing the amount of end-hooks of the fibres (from 3D to 5D). However, this last effect is less pronounced in case of 20 kg/m^3.

After completion of a UTT, the fibre amount in the fracture plane is counted. These values are compared to the theoretically expected fibre amount n_f, calculated by Eq. (2) and based on the fibre volumes V_f of 20 or 40 kg/m^3. A_f is the cross-sectional area of the fibre. The orientation factor α is equal to 0.5 since no wall effect restricts the random fibre orientation in the notched section of the cylinder [17], as the notched dimeter of 80 mm is one fibre length smaller than the original prism dimension of 150 mm.

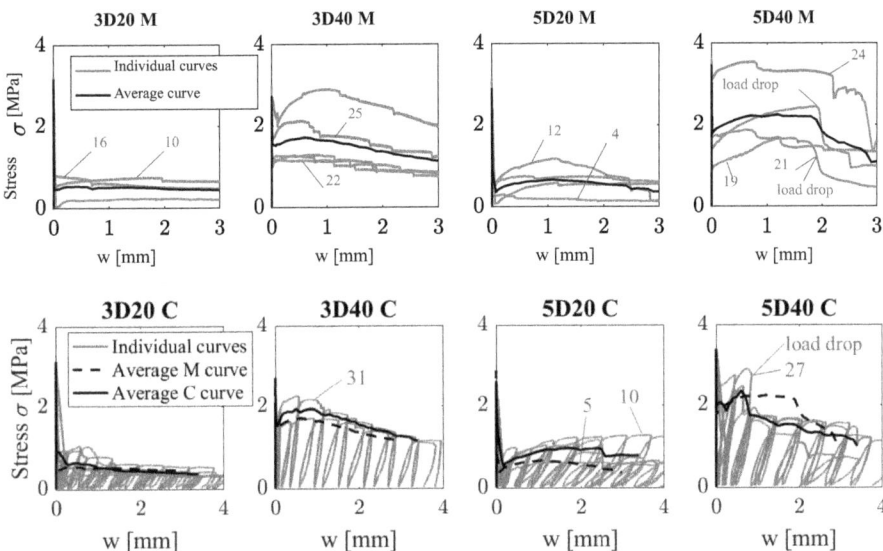

Fig. 2. Monotonic and cyclic uniaxial tensile behaviour for each SFRC group. Numbers indicate the amount of fibres in the cracked section of that specimen.

$$n_f = \alpha \cdot V_f \cdot A_s / A_f \qquad (2)$$

Table 1 compares the theoretical fibre amount to the counted one, whereby the specimens are classified according to the loading pattern (M or C). A rather large scatter between SFRC specimens is observed (COV up to 50%). For the 5D40 group, some alignment in the flow direction might have been present, given the higher counted fibre amount (22.5 and 28.0 compared to 19.8).

Table 1. Comparison of the theoretical and the counted fibre amount in the fracture plane. The number between brackets indicates the coefficient of variation (COV) of the counted fibre amount.

Fibre amount	Loading	3D20	3D40	5D20	5D40
Theoretical		14.2	28.4	9.9	19.8
Counted	Monotonic	10.7 (47.2%)	27.6 (28.0%)	7.3 (49.6%)	22.5 (13.8%)
	Cyclic	13.0 (37.7%)	29.5 (-)	9.0 (30.1%)	28.0 (12.9%)

The fibre amount directly influences the post-peak strength of the specimens. Figure 2 indicates the amount for some specimens. Nevertheless, not only the fibre amount, but also their inclination is of importance. Highly inclined fibres result in a lower post-peak strength because of the limited force resistance. An example is found in the SFRC group 3D20 M. Two specimens behave similarly (refer to Fig. 2), although the fibre amounts equal 10 and 16 fibres. After $w = 0.7$ mm, the specimen with 10 fibres even has an increased post-peak strength compared to the other one. However, investigating the cracked section in Fig. 3 shows multiple inclined fibres for the specimen with 16 fibres.

Fig. 3. Comparison of the cracked section with highly inclined fibres.

The most occurring fibre failing mechanisms for 3D fibres are pull-out and rupture. For 5D fibres, more matrix cracking and non-straightened end-hooks are observed due to the higher tensile strength and anchorage of the fibres. Figure 4 shows these differences in the fracture plane for 3D and 5D fibres. The fibre failing mechanisms lead to small load drops along the post-peak curve. The larger concrete cracking in case of 5D40 causes larger load drops. The load drops are accompanied with a change of AE rate. When (multiple) fibres are pulled-out or ruptured, the AE event rate is reduced.

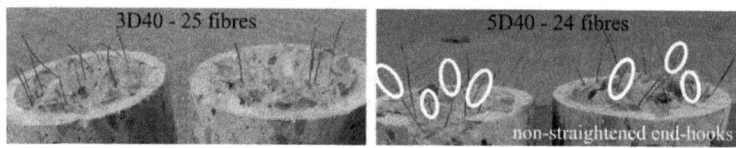

Fig. 4. Comparison of straightened 3D fibres and non-straightened 5D fibres.

3.2   AE Behaviour in Function of Fibre Characteristics

Figure 5 presents the average (envelope) curve of the total amount of localised AE events for both loading types. In general, AE activity increases with larger fibre volume. At the onset of macro-cracking, the concrete damage leads to a rapid increase of recorded AE events with higher energy and an increased amplitude. This phase is followed by local damage caused by fibre pull-out in the fracture plane.

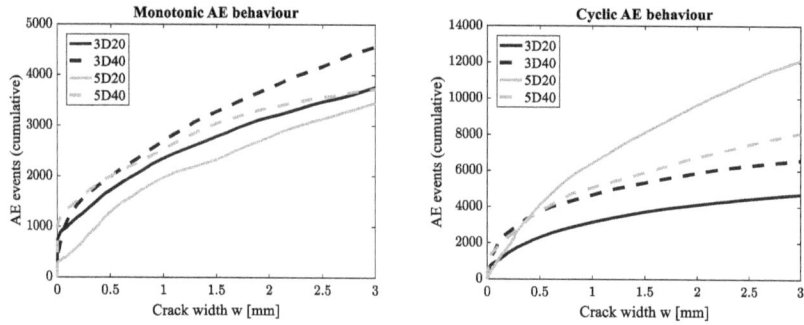

Fig. 5. Cumulative AE behaviour in for monotonic and cyclic loading.

For monotonic loading, the differences in AE behaviour are not significant due to the large scatter. The double end-hook of 5D fibres typically causes more concrete damage and thus more AE events [9]. However, their slightly larger diameter leads to fewer fibres in the cracked section for a specific fibre volume (refer to Table 1). Therefore, this effect is not pronounced.

For cyclic loading, the behaviour of 5D20 is remarkable. Due to the lower fibre amount and higher tensile strength of the fibres, the tensile stresses at the moment of cracking are transferred to less fibres in case of 5D20 compared to 5D40. Therefore, matrix damage occurs sooner. When cracks are initiated in an earlier loading cycle, more frictional AE activity is accumulated during the following loading cycles. Macro-cracking occurs on average at a stroke of 0.29 mm for 5D20 and at 0.41 mm for 5D40. The lower tensile strength of 3D fibres leads to sooner fibre rupture or plastic deformation, avoiding the effect of accumulated frictional damage.

Table 2 shows the average total AE energy at $w = 3$ mm. The AE measurements are sensitive to a large scatter. Increasing fibre type or volume leads to a larger energy release. The AE sources of 5D specimens release more energy compared to the 3D, given the ratio of AE energy to AE events. Again, the 5D20 C group has a remarkably larger total AE energy due to the accumulated frictional damage.

Table 2. Total amount of AE energy at w = 3 mm, with the coefficient of variation between brackets.

AE energy [x10^8 eu]	3D20	3D40	5D20	5D40
Monotonic loading	3.6 (41%)	4.1 (-)	4.5 (48%)	5.0 (23%)
Cyclic loading	5.9 (59%)	7.9 (37%)	10.9 (29%)	8.7 (19%)

3.3   Damage Development During Loading Cycles

Applying load cycles increases the total amount of AE activity and energy (refer to Fig. 5 and Table 2). Furthermore, the stiffness (indicated by the slope of the loading cycle) decreases for each loading cycle (refer to Fig. 2). The largest decrease occurs at the smallest crack openings. A larger fibre volume leads to a larger remaining stiffness because of the increased tensile resistance. 5D fibres at equivalent fibre volumes have a lower remaining stiffness than 3D fibres due to the increased concrete matrix damage by the double end-hook and larger fibre tensile strength.

Within one loading cycle, the majority of AE events is registered during (re)loading due to the formation of new cracks. Friction in the existing cracks causes AE events during unloading. Figure 6 presents two examples, whereby the percentages of AE events during loading relative to the whole loading cycle are indicated for each cycle. While the 3D40 specimen only presents 3–7% events during unloading, this value increases to 10–15% for the 5D20 specimen. It indicates a greater amount of frictional damage for the 5D20 group.

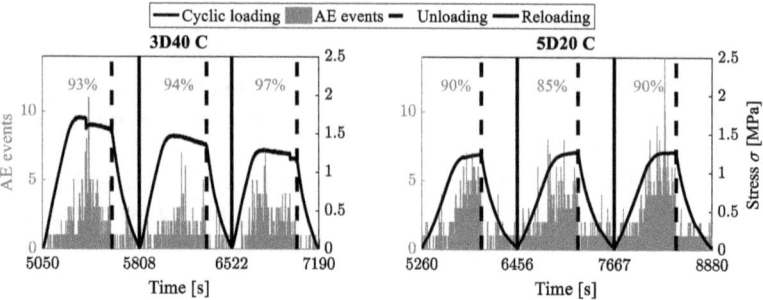

Fig. 6. AE events during representative loading cycles for 3D40 and 5D20.

3.4 Micro- and Macro-cracking Development

During the pre-peak phase, almost no AE events are registered. Right before the peak load, a small increase of AE events occurs, as shown in Fig. 7. These events are induced by developing micro-cracks with low energy. For cyclic loading, this micro-cracking occurs at the load cycle prior to peak load. Macro-cracking with higher energy occurs when the peak load is reached and the main crack develops.

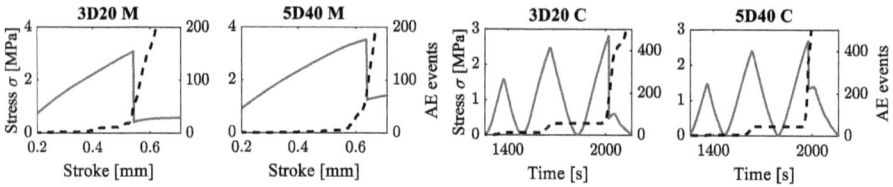

Fig. 7. Cumulative AE events (dotted line) during the pre-peak phase.

The distinction between micro- and macro-cracking is seen in the AE failure mode analysis as well. Figure 8 presents the moving average (based on 50 consecutive AE events) of AF and RA with time, for representative specimens. At the moment of cracking, a downshift of AF associated with a rising of RA is observed. The micro-cracking phase is characterised by high AF and low RA, which indicates mode I fracture. Although the damage remains attributed to mode I during the test, the change of AF and RA indicates a change in failure mode towards mode II. Based on the analysis of AF and RA, the moment of concrete cracking can be determined.

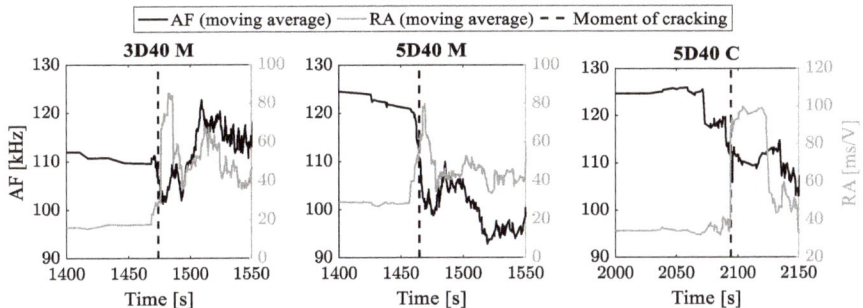

Fig. 8. AE events during representative loading cycles for 3D40 and 5D40.

3.5 Damage Initiation and Localisation in Relation to the Fibre Positions

AE monitoring with six sensors allows to localise the damage sources in the specimen. Figure 9 and 10 present an example of monotonic and cyclic loading. In both cases, the horizontal plane at the notch, i.e. the cracked section, is shown during different stages of the test, with indication of the fibre positions.

In stage I, only a few AE events are registered, representing the micro-cracks. The events mostly occur where the least amount of fibres are present, since the tensile stresses cannot be distributed to nearby fibres. Stage II include the moment of macro-cracking, resulting in the largest amount of AE events due to the high concrete damage. The events are more scattered within the whole section. During the post-peak stages III and IV, only fibres are bridging the crack opening. The pull-out damage results in AE events increasingly clustered around the remaining active fibre positions. These positions are the most clear in the stage IV since the amount of events decreases with increasing testing time.

Cyclic loading follows the same conclusions. Figure 10 is subdivided for loading cycles instead of stages. Micro-cracking before the peak load results in a low amount of events away from the fibres, especially in the load cycle prior to macro-cracking. Thereafter, an increased amount of scattered AE events is registered. Again, concentration around the active bridging fibres with decreasing AE event amount occurs during the last loading cycles.

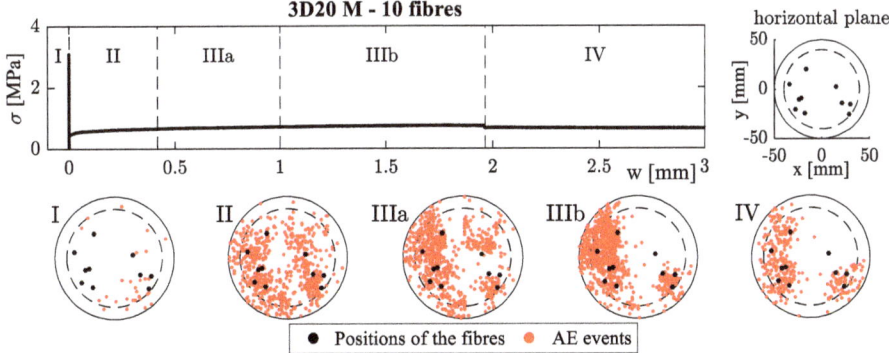

Fig. 9. AE localisation in the horizontal plane during stages of a monotonic UTT.

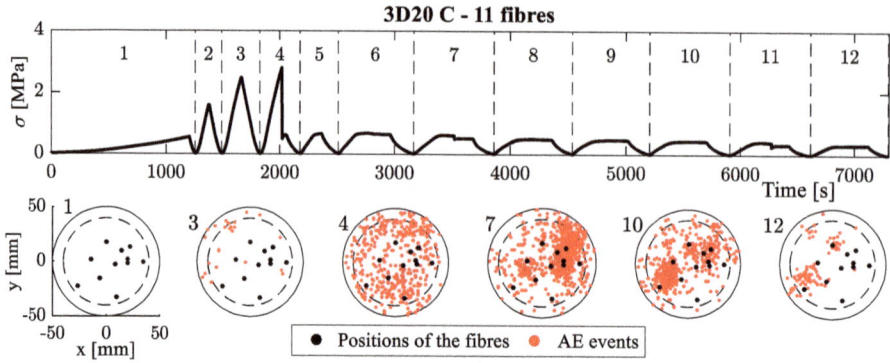

Fig. 10. AE localisation in the horizontal plane during some load cycles of a cyclic UTT.

3.6   Prediction of Tensile Behaviour and Strength

The average uniaxial tensile strength f_{ctm} for all 16 monotonically loaded specimens equals 3.03 MPa, with a standard deviation of 0.35 MPa. Since the tensile strength is mainly defined by the concrete matrix itself, it is rather independent of fibre type or volume. MC2010 [15] provides two ways to predict the uniaxial tensile strength based on the compressive strength in Eq. (3) or the flexural tensile strength in Eq. (4). The average value of these predictions equals 3.04 MPa, which corresponds very well to the experimentally observed average.

$$f_{ctm} = 0.3 \cdot \left(0.79 \cdot f_{cm,cube} - 8\right)^{2/3} = \ 0.3 \cdot (0.79 \cdot 59 - 8)^{2/3} = 3.43 \ MPa \qquad (3)$$

$$f_{ctm} = \frac{0.06 \cdot h_{sp}^{0.7}}{1 + 0.06 \cdot h_{sp}^{0.7}} \cdot f_{ctm,fl} = \frac{0.06 \cdot 125^{0.7}}{1 + 0.06 \cdot 125^{0.7}} \cdot 4.14 = 2.64 \ MPa \qquad (4)$$

Furthermore, MC2010 provides a model to predict the post-peak uniaxial tensile behaviour of FRC based on bending tests, which were performed in [9] on the prisms before coring the UTT cylinders. The results of the model are compared against the experimental data in Fig. 11. A relatively good agreement is found regarding the bi-linear post-peak behaviour. Both 3D20 and 5D20 have a lower amount of counted fibres than the theoretical fibre amount (refer to Table1). Therefore, the experimental post-peak strength is lower than the predicted strength. In case of 5D40, the counted fibre amounts exceed the theoretical amount, which leads to an opposite result.

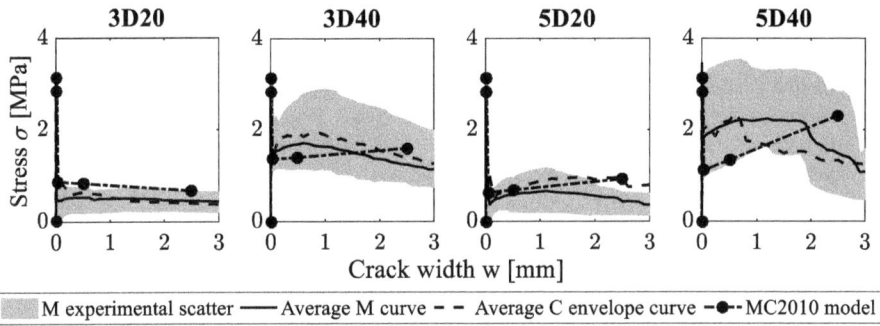

Fig. 11. Modelled uniaxial tensile behaviour of MC2010 compared to experiments.

4 Conclusions

This paper experimentally investigated monotonic and progressive cyclic behaviour of SFRC in UTT assessed by AE monitoring. The main research conclusions are summarised as follows.

- The fibre characteristics largely influence the post-cracking tensile behaviour, as well as the AE behaviour. These characteristics include fibre volume, type of end-hook, effective fibre amount and distribution in the cracked section and inclination of the fibres. Fibre counting, clip gauges and AE localisation can investigate these parameters.
- The majority of AE events during load cycles originates from the crack formation. More frictional damage results in more AE events during unloading and an increased AE energy amount.
- Micro- and macro-cracking are distinguished based on AE activity. The onset of macro-cracking is indicated by the failure mode analysis, given the increase of RA and decrease of AF.
- Localisation of AE events validates the fibre positions, damage stages and main crack position. The crack's initiation and development are accurately localised during the different loading stages. AE events are more clustered around the remaining active fibres with increasing test time.
- Predictions of the tensile strength and post-peak behaviour according to Model Code 2010 are in good agreement with the experimental results.

Acknowledgements. The authors gratefully acknowledge the financial support of Research Foundation Flanders (FWO, PhD-grant no. 1S32717N), and the supply of steel fibres used in this study by Bekaert nv.

References

1. RILEM Technical Committee: RILEM TC 162-TDF: Test and design methods for steel fibre reinforced concrete – uni-axial tension test for SFRC. Mater Struct **34** (2001)
2. Barragan, B., et al.: Uniaxial tension test for steel fibre reinforced concrete – a parameteric study. Cem. Concr. Compos. **25**, 767–777 (2003)
3. Graybeal, B., Baby, F.: Development of direct tension test method for ultra-high-performance fiber-reinforced concrete. ACI Mat. J. **110**(M17), 177–186 (2013)
4. van Mier, J., van Vliet, M.: Uniaxial tension test for the determination of fracture parameters of concrete: state of the art. Eng. Fract. Mech. **69**(2), 235–247 (2002)
5. Grosse, C., Ohtsu, M.: Acoustic Emission Testing. Springer, Heidelberg (2008). https://doi.org/10.1007/978-3-540-69972-9
6. Behnia, A., Chai, H., Shiotani, T.: Advanced structural health monitoring of concrete structures with the aid of acoustic emission. Constr. Build. Mater. **65**, 282–302 (2014)
7. Noorsuhada, M.: An overview on fatigue damage assessment of reinforced concrete structures with the aid of acoustic emission technique. Constr. Build. Mater. **112**, 424–439 (2016)
8. Wevers, M.: Listening to the sound of materials: acoustic emission for the analysis of material behaviour. NDT&E Int. **30**(2), 99–106 (1997)
9. De Smedt, M., et al.: Damage analysis in steel fibre reinforced concrete under monotonic and cyclic bending by means of AE monitoring. Cem. Concr. Compos. **114,** 103765 (2020)
10. Wang, J.-Y., Guo, J.-Y.: Damage investigation of ultra-high performance concrete under direct tensile test using acoustic emission techniques. Cem. Concr. Compos. **88**, 17–28 (2018)
11. Xiangqian, F., et al.: Acoustic emission properties of concrete on dynamic tensile test. Constr. Build. Mater. **114**, 66–75 (2016)
12. Li, B., et al.: Cyclic tensile behavior of SFRC: experimental research and analytical model. Constr. Build. Mater. **190**, 1236–1250 (2018)
13. Li, Z., Shah, S.: Localization of microcracking in concrete under uniaxial tension. ACI Mat . J. **91**(M37), 372–381 (2019)
14. CEN: NBN EN 12390-3: Testing hardened concrete - Part 3: Compressive strength of test specimens (2009)
15. Fédération Internationale du Béton (fib): fib Model Code for Concrete Structures 2010, Wilhelm Ernst und Sohn Verlag für Architektur, Berlin (2013)
16. RILEM Technical Committee: Recommendation of RILEM TC 212-ACD: Acoustic emission and related NDE techniques for crack detection and damage evaluation in concrete. Mater. Struct. **43**, 1187–1189 (2010)
17. Dupont, D., Vandewalle, L.: Distribution of steel fibres in rectangular sections. Cem. Concr. Compos. **27**, 391–398 (2005)
18. Plizzari, G.A., Cangiano, S., Cere, N.: Post-peak behavior of fiber-reinforced concrete under cyclic tensile loads. ACI Mat. J. **97**(M24), 1–11 (2000)

An Experimental Study on the Flexural Fatigue Behaviour of Pre-cracked Steel Fibre Reinforced Concrete

Humaira Fataar[1]([✉]), Riaan Combrinck[1], and William P. Boshoff[2]

[1] Unit for Construction Materials, Stellenbosch University,
Stellenbosch, South Africa
humairaf@sun.ac.za
[2] University of Pretoria, Pretoria, South Africa

Abstract. Fatigue behaviour of concrete has become an increasingly popular topic in the last century, especially with the development of railway bridges. Fatigue loading may appear in various forms, from physically applied loads, to indirect loads, including corrosion and thermal fatigue among others. These fatigue loadings may occur independently or in conjunction with the applied fatigue loadings, which could exacerbate the fatigue process and likely decrease the lifespan of the structure. The mechanisms of fatigue failure in concrete may be divided into three phases: (1) crack initiation, (2) progressive growth of micro-cracks, and (3) convergence of micro-cracks to form macro-cracks. In fibre reinforced concrete (FRC), energy is dissipated in the wake of the crack tip, which increases the load carrying capacity, thereby providing post-cracking ductility. Unlike ferrous materials, concrete was found to exhibit no fatigue limit after 2 million load cycles. However, its performance may be influenced by stress levels, load frequency, boundary conditions, matrix composition, and number of applied cycles. In this paper, the flexural fatigue behaviour of pre-cracked steel fibre reinforced concrete was investigated. Various pre-cracks and load levels were considered. X-ray Computed Tomography (CT) scans were implemented to determine the extent of damage to the fibres within the concrete matrix after fatigue loading.

Keywords: Fatigue · Pre-crack · FRC · Flexure

1 Introduction

Fatigue is defined as the process of continuous, permanent internal structural changes to a material subjected to cyclic loading. For concrete, the internal structural changes are associated with the growth of internal micro cracks [1–3]. Since concrete is known for its weak tensile strength, reinforcing steel bars or fibres are generally added to improve the tensile capacity of the composite [4, 5]. Fibres play a vital role in the suppressing the formation and propagation of cracks in concrete [6–9]. Once there is crack initiation, only then does the fibres contribute towards the load carrying capacity. The load is transferred from the concrete matrix to the fibre by shear stress at the fibre-matrix interface [8, 10, 11].

© RILEM 2022
P. Serna et al. (Eds.): BEFIB 2021, RILEM Bookseries 36, pp. 155–165, 2022.
https://doi.org/10.1007/978-3-030-83719-8_14

Fatigue loading can generally be divided into two categories: high-cycle loading and low-cycle loading. The high-cycle fatigue loading has large number of cycles at low load levels, whereas the low-cycle loading has fewer load cycles at high load levels [1]. Structures which are subjected to earthquakes may experience low-cycle fatigue loading. In contrast, high-cycle fatigue loading will be applicable to airport pavements and highway bridges [1, 4].

This research focuses on the fatigue behaviour of pre-cracked steel fibre reinforced concrete. The concrete is tested at pre-cracks of 0.6 mm, 1.2 mm, 1.8 mm and 2.5 mm. The fatigue loading is applied at 50%, 70% and 85% of the average maximum static load.

2   Experimental Framework

2.1   Material Properties and Concrete Mix

The materials used for the concrete mix, apart from the steel fibres, were locally sourced. The steel fibres used in this study were DRAMIX 3D-65/50-BG hooked-end steel fibres, supplied by BEKAERT. The concrete mix constituents and proportions are shown in Table 1 and the steel fibre properties are shown in Table 2. The cement used, supplied by Pretoria Portland Cement (PPC), was CEM II 52.5 N. The fine aggregate used was a natural pit sand (locally known as Malmesbury sand), and the coarse aggregate was a 13 mm crushed greywacke stone. Municipal water was used, as well as a high range water reducing superplasticiser called Chryso Optima 206.

Table 1. Concrete mix proportions

Material type	kg/m^3
Cement	450
Fine aggregate (Malmesbury sand)	880
Coarse aggregate (greywacke stone)	820
Water	194
Superplasticiser (1.4% by weight of binder)	6.3
Fibres (0.8% by volume)	65

Table 2. Steel fibre properties [12]

Tensile strength	1 160 MPa
Modulus of elasticity	200 GPa
Length (l)	60 mm
Diameter (d)	0.9 mm

2.2   Specimen Preparation

Once the fresh concrete was thoroughly mixed, it was cast into steel moulds with dimensions of $150 \times 150 \times 700$ mm. The mould size was then adjusted to $150 \times 150 \times 550$ mm by adding a wooden spacer to one end of the mould as shown in Fig. 1. The interior of the mould was coated in a thin film of demoulding oil, after which the concrete was cast. The moulds were filled to approximately 90% of its height before being compacted individually on the shaker table for 30 s. The remaining 10% was added as the concrete was compacted, as specified by EN 14651:2005. The specimens were demoulded after 24 h and placed in curing tanks, at a temperature of 24 ± 1 °C for an additional 27 days.

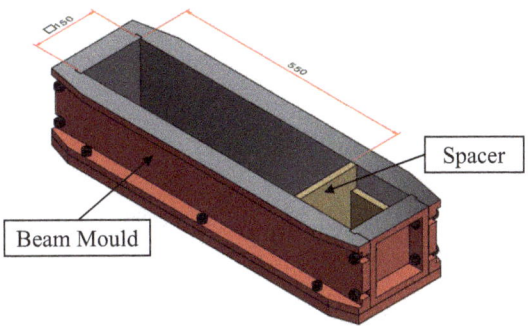

Fig. 1. Beam specimen mould (unit = mm)

After the 27 days of curing in the water tanks, the specimens were removed and prepared for testing. A notch was sawn through the width of the specimen, no less than 3 h prior to testing. Each beam specimen was rotated 90° on its longitudinal axis and notched at the midspan to a depth of 25 mm as specified by EN 14651:2005 and is illustrated in Fig. 2. The notch width did not exceed 5 mm.

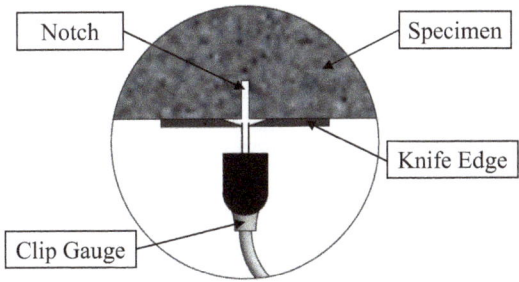

Fig. 2. Notch sawn into beam specimen

Knife edges were glued to the beam specimen using Pratley Steel Quickset® epoxy. Once the knife edges were fixed in place and the epoxy had set, the specimens were ready for testing. A clip gauge was then secured into the knife edges, which measured the crack mouth opening displacement (CMOD) as the test progressed, and is shown in Fig. 2. The static tests were conducted on the 28-day strength of the concrete specimens. However, due to the lengthy nature of the fatigue tests, the specimens tested in fatigue were tested between 28 and 40 days in order to maintain consistent results.

2.3   Test Setup

The flexural fatigue tests were performed on a servo-controlled hydraulic actuator, with a 500 kN load cell and maximum stroke length of 50 mm. The static tests, however, were performed on the 2 MN Instron Universal Material Testing Machine. The static flexural results provided insight into the maximum loads to be expected, and as a result, a 50 kN external load cell was attached to the flexural fatigue setup to provide a better resolution on the load control of the test. The flexural fatigue test setup can be seen in Fig. 3.

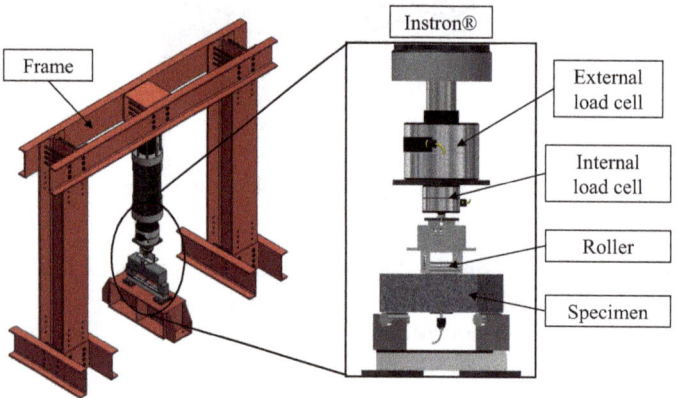

Fig. 3. Flexural fatigue test setup

The actuator was mounted on a steel frame, which was bolted to the ground. The beam specimen was then placed onto two rollers, whose frame was attached to a reinforced I-beam, which was fixed to the ground. The rollers were placed at 500 mm apart as indicated by EN 14651:2005, thereby allowing 25 mm of the beam to overhang on each side. The beam specimen was placed with the notch facing downwards, after which the clip gauge was attached to the knife edges. Once the beam was correctly mounted, the Instron head was slowly lowered to make contact with the beam without excessively loading it. Thereafter, the testing commenced.

2.4   Loading Regime

Static tests were performed in order to determine the average maximum load. These tests were position-controlled, and were based on the CMOD extension according to EN 14651:2005. The CMOD was increased at a rate of 0.05 mm/min up to 0.1 mm. Thereafter, the CMOD extension was increased to 0.2 mm/min, until the test ended at a CMOD of 3.5 mm.

The fatigue tests were performed in three-point bending, and was loaded by a compressive sinus load application as shown in Fig. 4. The maximum load (A_{max}) was taken as a percentage (50%, 70% or 85%) of the average maximum static load. The minimum load (A_{min}) was taken as a small non-zero value (usually between 3 kN and 3.5 kN). The fatigue loading was applied at a frequency of 5 Hz until a CMOD of 3.5 mm, which was deemed as failure, or up to 2 million load cycles.

Compressive Load (kN)

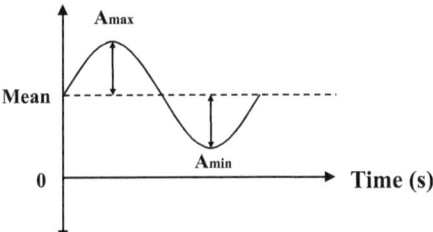

Fig. 4. Compressive loading for flexural fatigue tests

The fatigue tests were conducted in three steps: (1) pre-crack; (2) load/unload; and (3) fatigue loading. The first step was position-controlled, and the beam was cracked to a pre-specified CMOD at a rate of 0.6 mm/min. Upon reaching the required CMOD, the test changed to load-controlled in order to load/unload to the required mean value of the sinus load application. Once the mean value was reached, the final step commenced, which was the fatigue loading. The third step was load-controlled, and the load cycled between the maximum and minimum load until failure or until the 2 million load cycles were completed.

3   Test Results and Discussion

3.1   Static Tests

The static tests were performed to determine the average maximum static load. Three different fibre dosages were tested in order to find the desired dosage and concrete strength. The fibre dosages were tested at 0.6%, 0.8% and 1.0% by volume of concrete. The static test results are displayed in Fig. 5. At least five specimens were tested for each fibre dosage and the strength gain differs for each dosage. All the tested specimens displayed deflection hardening behaviour.

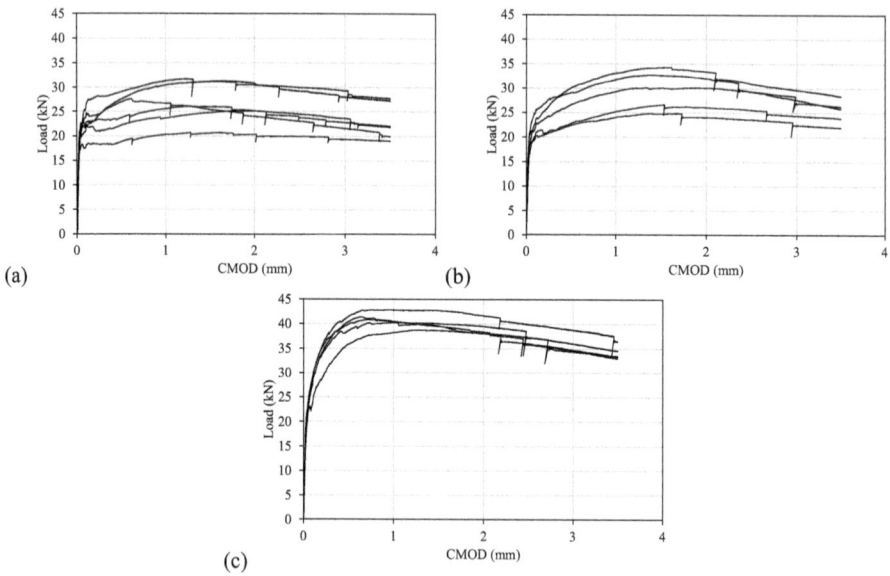

Fig. 5. Static flexural tests with fibre volume percentage (a) 0.6%; (b) 0.8% and (c) 1.0%

At the 0.6% dosage in Fig. 5 (a), there is a big variation in results when compared to the other dosages. The first crack strength ranges from a low of 17 kN, to a high of 25 kN. The strength gain after the first crack is gradual, reaching a peak of 31 kN. The test results from the 0.8% fibre dosage showed improved results over the 0.6% dosage. The first crack strength increased to a range of 20 kN to 25 kN, followed by a steady strength gain. The 1.0% fibre dosage displayed the most improved results, with the lowest variation overall. The first crack strength occurred between 23 kN and 25 kN. The load then increased rapidly to a reach an overall maximum of 43 kN. Thereafter, the load gradually decreased until the test ended at a CMOD of 3.5 mm.

As expected, the post-cracking strength increased as the fibre dosage increased, thereby providing a more ductile post-cracking behaviour. Furthermore, all the tests displayed noticeable rapid load reductions, followed by a quick recovery. The load reductions decreased in frequency as the fibre dosage increased. This was possibly due to failure of critical fibres bridging the cracks. However, further investigation is required. It was decided to use the 0.8% fibre dosage as it displayed the best overall results.

3.2 Fatigue Tests

Once the static tests were completed, the fatigue tests were performed. The tests were conducted at three different load levels (50%, 70% and 85%) and at four different pre-cracks (0.6 mm, 1.2 mm, 1.8 mm and 2.5 mm). The tests ended after the completion of 2 million load cycles, or when the CMOD reached 3.5 mm. Each test was repeated three times, unless otherwise stated. The flexural fatigue test results are displayed in Table 3.

The fatigue tests for the 70% and 85% load levels were completed for all the pre-cracks. The 50% load level at 0.6 mm pre-crack could not be completed. However, based on the 1.2 mm pre-crack at the same load level, the average number of cycles sustained for the 50% load level at 0.6 mm pre-crack was assumed to be 2 million. From the results in Table 3, a clear trend can be seen, where the average number of cycles sustained decreases as the load level and pre-crack increases.

At the 85% load level, the fatigue resistance was poor, and decreases as the pre-crack increases. This is more pronounced at the 2.5 mm pre-crack, where the average number of cycles sustained is only 1000 cycles. The coefficient of variation is also increased as the pre-cracked increased. This is due to the poor fatigue behaviour caused by the large pre-crack, which results in permanent and progressive crack growth as each cycle of the fatigue loading passes.

The 50% load level showed the best fatigue test results. Even though the 0.6 mm pre-crack tests could not be completed, based on the results from the 1.2 mm pre-crack, it was assumed that the 0.6 mm pre-crack specimens would not fail at or before 2 million load cycles. The 70% load level showed intermediate test results as anticipated. The test results showed good repeatability overall, with some variation as expected based on literature [13–15].

Table 3. Flexural fatigue test results

Pre-crack (mm)	Specimen	Number of cycles		
		Load level		
		50%	70%	85%
0.6	1	–	56 000	26 000
	2	–	75 000	38 000
	3	–	100 000	53 000
Average		**2 000 000***	**77 000**	**39 000**
Coeff. of variation (%)		**N/A**	**23.3**	**27.9**
1.2	1	1 700 000†	84 000	18 000
	2	2 000 000	208 000	20 000
	3	–	307 000	31 000
Average		**1 850 000**	**199 000**	**23 000**
Coeff. of variation (%)		**N/A**	**45.6**	**23.9**
1.8	1	746 000	43 000	6 000
	2	1 280 000	75 000	10 000
	3	–	–	29 000
Average		**1 013 000**	**59 000**	**15 000**
Coeff. of variation (%)		**26.3**	**27.3**	**63.7**
2.5	1	235 000	21 000	52
	2	888 000	26 000	651
	3	1 097 000†	37 000	2 300
Average		**740 000**	**28 000**	**1 000**
Coeff. of variation (%)		**49.6**	**23.4**	**95.9**

†Test could not be completed.
*Assumed value.

The average number of load cycles sustained for each load level and pre-crack was then plotted and is displayed in Fig. 6. The fatigue capacity of the 85% load level decreased at a steady rate between the 0.6 mm and 1.8 mm pre-crack. At the 2.5 mm pre-crack, the fatigue capacity decreased drastically. This was due to the large pre-crack coupled with the large load level applied. The 70% load level also indicated a decrease in fatigue capacity as the load level increased. However, the 1.2 mm pre-crack was the outlier since it sustained more load levels on average than the 0.6 mm pre-crack at the same load level. The 50% load level results also showed similar trends to the 70% and 85% load levels. This further demonstrates that an increase in load level and pre-crack results in decreased fatigue capacity, which is drastically exaggerated at the higher extremity.

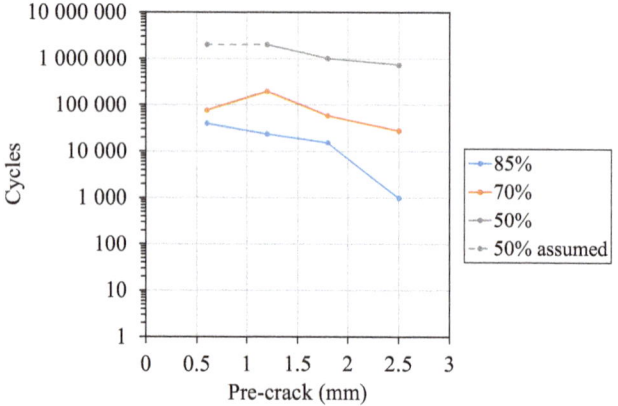

Fig. 6. Average number of cycles sustained until failure or 2 million load cycles

3.3   X-ray Computed Tomography (CT) Scans

X-ray CT scans were performed on tested specimens after pre-cracking, as well as after fatigue loading. The X-ray CT scans provided a non-destructive method of viewing the internal structure of the specimens before and after fatigue tests [16].

The specimens shown in Fig. 7 and Fig. 8 illustrates the fibre deformation before the fatigue loading for pre-cracks of 1.2 mm and 2.5 mm, respectively. The fibres closest to the notch shows more fibre deformation than fibres further away from the notch. As expected, the larger the pre-crack, the more fibre deformation, and this is seen in Fig. 8 (a), where the fibre is closest to the notch. There is more fibre end pull-out and the hooks of the fibres begin to straighten out. However, as the fibres move further away from the notch, the pull-out and fibre deformation decreases, as seen in Fig. 7 (b) and Fig. 8 (b).

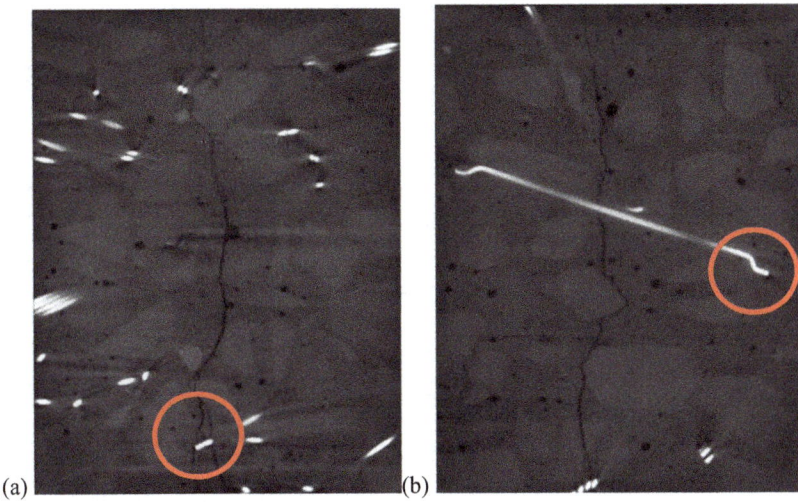

Fig. 7. X-ray CT scan of beam specimen pre-cracked to 1.2 mm (a) fibre at the notch, (b) fibre away from the notch

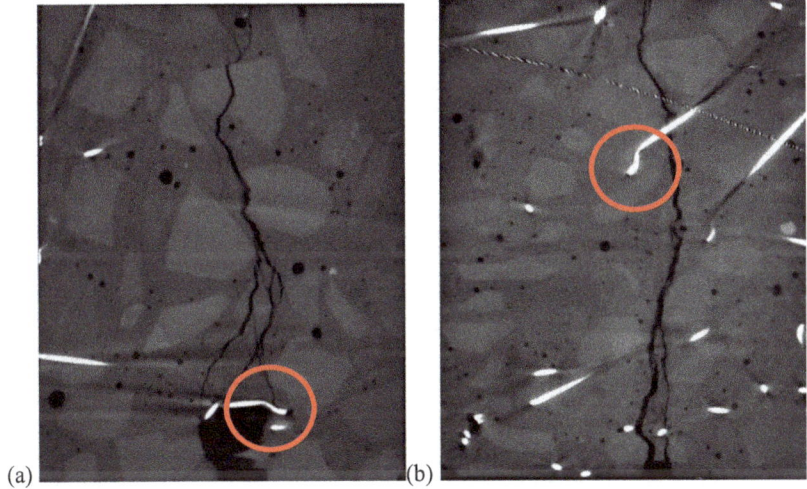

Fig. 8. X-ray CT scan of beam specimen pre-cracked to 2.5 mm (a) fibre at the notch, (b) fibre away from the notch

In Fig. 9, the specimen has undergone fatigue loading at a load level of 85% and a pre-crack of 1.2 mm. The fatigue loading resulted in multiple fibre ruptures shown in Fig. 9 (a), as well as fibre end pull-out shown in Fig. 9 (b). This fibre end pull-out and rupture is also experienced in fatigue behaviour at a single fibre level [17].

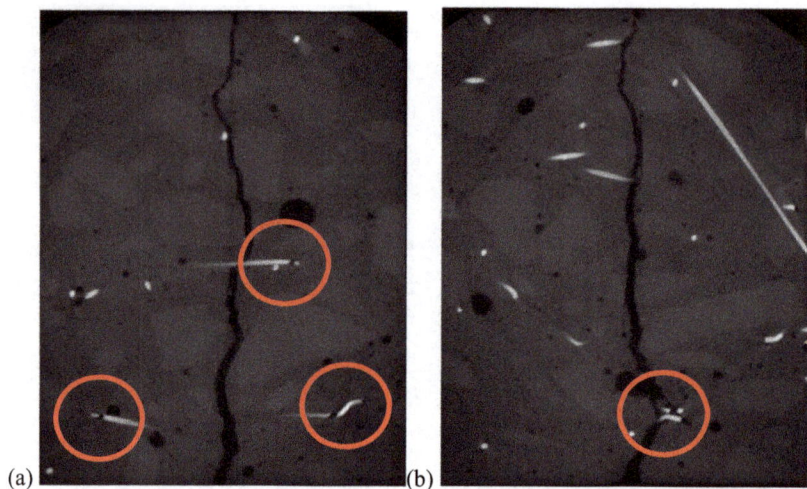

Fig. 9. Post-fatigue test X-ray CT scan of Specimen 3, tested to failure at 85% load level and 1.2 mm pre-crack (a) multiple fibre ruptures, (b) fibre pull-out

However, due to the fibre strength, length and shape of the fibre hook, as well as the final CMOD when the test ended, the fibre hooks were never able to straighten out completely. Figure 7, Fig. 8 and Fig. 9 all clearly only illustrates partial straightening of the fibre hooks.

4   Conclusion

The fatigue behaviour of pre-cracked hooked-end steel fibre reinforced concrete was tested at pre-cracks ranging between 0.6 mm and 2.5 mm, and load levels of 50%, 70% and 85%. The results showed that as the load level and pre-crack increased, the fatigue capacity decreased. This was more pronounced at the 85% load level at the 2.5 mm pre-crack, which had the weakest overall fatigue capacity. As expected, the 50% load level showed the best fatigue capacity.

X-ray CT scans were performed on specimens before and after fatigue loading. Once the specimens were pre-cracked, X-ray CT scans were taken to determine the extent of damage caused to the fibres. The fibres before fatigue loading illustrated fibre end pull-out to various degrees, based on the location of the fibre relative to the notch. The fibres closest to the notch presented with more pull-out than fibres further away from the notch. The X-ray CT scans taken after fatigue loading showed that the main fibre failure mechanism was fibre rupture.

References

1. Lee, M.K., Barr, B.I.G.: An overview of the fatigue behaviour of plain and fibre reinforced concrete. Cem. Concr. Compos. **26**(4), 299–305 (2004). https://doi.org/10.1016/S0958-9465 (02)00139-7
2. Parvez, A., Foster, S.J.: Fatigue of steel-fibre-reinforced concrete prestressed railway sleepers. Eng. Struct. **141**, 241–250 (2017). https://doi.org/10.1016/j.engstruct.2017.03.025
3. Keerthana, K., Chandra Kishen, J.M.: An experimental and analytical study on fatigue damage in concrete under variable amplitude loading. Int. J. Fatigue**111**, 278–288 (2018). https://doi.org/10.1016/j.ijfatigue.2018.02.014
4. Hsu, T.T.C.: Fatigue of plain concrete. J. Proc. **78**(4), 292–305 (1981)
5. Shah, A.A., Ribakov, Y.: Recent trends in steel fibered high-strength concrete. Mater. Des. **32**(8), 4122–4151 (2011). https://doi.org/10.1016/j.matdes.2011.03.030
6. Banthia, N., Trottier, J.-F.: Deformed steel fiber—cementitious matrix bond under impact. Cem. Concr. Res. **21**(1), 158–168 (1991)
7. Buratti, N., Mazzotti, C., Savoia, M.: Post-cracking behaviour of steel and macro-synthetic fibre-reinforced concretes. Constr. Build. Mater. **25**(5), 2713–2722 (2011)
8. Namur, G.G., Alwan, J.M., Najm, H.S.: Fiber pullout and bond slip I: Analytical study **117** (9), 2769–2790 (1992)
9. di Prisco, M., Plizzari, G., Vandewalle, L.: Fibre reinforced concrete: new design perspectives. Mater. Struct. **42**(9), 1261–1281 (2009). https://doi.org/10.1617/s11527-009-9529-4
10. Beaudoin, J.J.: Handbook of Fiber-Reinforced Concrete: Principle Properties, Developments and Applications. Noyes Publications, Park Ridge (1990)
11. Ghoddousi, P., Ahmadi, R., Sharifi, M.: Fiber pullout model for aligned hooked-end steel fiber. Can. J. Civ. Eng. **37**(9), 1179–1188 (2010). https://doi.org/10.1139/L10-053
12. Dramix® steel fiber concrete reinforcement - Bekaert.com. https://www.bekaert.com/en/ products/construction/concrete-reinforcement/dramix-steel-fiber-concrete-reinforcement. Accessed 21 Feb 2018
13. Naaman, A.E., Hammoud, H.: Fatigue characteristics of high performance fiber-reinforced concrete. Cem. Concr. Compos. **20**(5), 353–363 (1998). https://doi.org/10.1016/S0958-9465 (98)00004-3
14. Nanni, A.: Fatigue behaviour of steel fiber reinforced concrete. Cem. Concr. Compos. **13**(4), 239–245 (1991). https://doi.org/10.1016/0958-9465(91)90029-H
15. Singh, S.P., Kaushik, S.K.: Fatigue strength of steel fibre reinforced concrete in flexure. Cem. Concr. Compos. **25**(7), 779–786 (2003). https://doi.org/10.1016/S0958-9465(02) 00102-6
16. du Plessis, A., le Roux, S.G., Guelpa, A.: The CT scanner facility at Stellenbosch University: an open access X-ray computed tomography laboratory. Nucl. Instruments Methods Phys. Res. Sect. B Beam Interact. Mater. Atoms **384**, 42–49 (2016). https://doi.org/10.1016/j. nimb.2016.08.005
17. Fataar, H., Combrinck, R., Boshoff, W.P.: An Experimental study on the fatigue failure mechanisms of pre–damaged steel fibre reinforced concrete at a single fibre level. In: Serna, P., Llano-Torre, A., Martí-Vargas, J.R., Navarro-Gregori, J. (eds.) BEFIB 2020. RB, vol. 30, pp. 199–208. Springer, Cham (2021). https://doi.org/10.1007/978-3-030-58482-5_18

Experimental Study on Punching Shear Strength of Fiber Reinforced Concrete Slabs

Salah Altoubat[1,2(✉)], Nadia Nassif[1,2], and Mohamed Maalej[1,2]

[1] Department of Civil and Environmental Engineering, University of Sharjah, 27272 Sharjah, United Arab Emirates
saltoubat@sharjah.ac.ae
[2] Sustainable Construction Materials and Structural System Research Group, Research Institute of Sciences and Engineering, University of Sharjah, Sharjah, United Arab Emirates

Abstract. This paper is part of an ongoing research project on punching shear strength of reinforced concrete flat slabs incorporating macro synthetic fibers. Results obtained from testing six slabs (1500 × 1500 × 150 mm) are reported and discussed in this paper. The slabs were reinforced with flexural reinforcement ratio of 0.9% and macro synthetic fibers were added at two volume fractions of 0.5% and 1.0%. Slabs were tested in simply-supported configuration system with center point load increasing monotonically in displacement control mode. Loads, slab deflection, strain in the concrete and steel were measured in the experiment and crack patterns and failure mode were observed. The results revealed noticeable enhancement in the punching shear capacity, cracking behavior, energy absorption and ductility of the reinforced concrete slab. The addition of macro synthetic fibers increased the punching shear strength by up to 17% and the energy absorption by up to 73% for the tested FRC slabs relative to the control slab. Furthermore, the slab deflection profile showed that the addition of fibers modify the shape of slab deflection which indicates that fibers help engaging the whole slab in carrying the applied load.

Keywords: Punching shear · Fibers · Slabs · Reinforced concrete

1 Introduction

Punching shear is a failure mechanism in structural members (e.g. slabs) by shear under the action of concentrated loads. This type of failure is highly catastrophic as it happens suddenly and can trigger the progressive collapse of the whole structure. Punching shear failure can be prevented using traditional reinforcement methods such as shear stirrups, studs or drop panels [1–3]. Alternatively, integrating discrete macro fibers into the concrete mix has emerged as an effective countermeasure against punching [1–4]. The use of synthetic fibers increases the punching capacity, improves the energy absorption performance of the column–slab connection and transforms the brittle-type punching shear failure into a gradual and ductile shear failure [2, 3]. These fibers are manufactured from polymer-based materials such as nylon, polyethylene and polypropylene. Synthetic fibers are highly durable, have lesser unit weight and they are non-corrosive. Previous experiments in literature indicate that fibers contribution to punching shear

© RILEM 2022
P. Serna et al. (Eds.): BEFIB 2021, RILEM Bookseries 36, pp. 166–173, 2022.
https://doi.org/10.1007/978-3-030-83719-8_15

strength lies in their bridging effect to limit crack propagation and minimize their width at critical sections. This bridging effect considerably enhances the post-cracking tensile strength at the tensile zone. In recent years, many researchers investigated improvement in punching shear strength in flat slabs containing steel fibers [5–8].

Utilization of steel fibers to enhance punching shear strength of flat plates has been introduced by Swamy and Ali [5]. The authors reported enhancement in maximum shear load capacity between 23% to 42% corresponding to adding steel fiber volume ratio contents from 0.6% (48 kg/m^3) to 1.2% (96 kg/m^3). Harajli [6] investigated the effect of steel fiber on the punching shear resistance of flat slabs by testing 12 laboratory scale slabs. The authors reported that adding steel fibers up to 2.0% (160 kg/m^3) of the total volume ratio increased the ultimate punching capacity by about 36% compared with plain concrete slabs. Cheng and Gustavo [7] studied the behaviour of 10 fiber reinforced concrete slabs. Their results have shown that the addition of steel fibers led to an increase in punching shear capacity and deflection capacity, particularly when 1.5% of fibers were used. Moreover, the results showed that slabs with lower flexural reinforcement ratio exhibited a change in failure mode from punching shear to flexural yielding. Nguyen-Minh et al. [1] conducted 12 tests on fiber reinforced concrete slabs to study the effect of steel fiber inclusion. The authors concluded that using steel fibers increases punching shear capacity up to 39%. It also reduces the average crack width up to 40% considering serviceability limit state. Furthermore, up to 36% reduction of deflection was observed. Abdel-Rahman et al. [8] tested 14 fiber reinforced concrete slab-column connections to study the improvement of punching shear caused by addition of steel fibers. The results revealed that using steel fiber increased both the failure load and energy absorbing capacity. Increasing the steel fiber by 1.5% enhanced the ultimate capacity of the specimens by 24%.

While several research studies have been conducted on the effect of using steel fibers on the punching shear strength and deflection capacity of FRC slabs, similar research studies using synthetic fibers still remain scarce in the existing literature and thus more experimental work is needed which motivates the current study. In the present paper, the authors report the results of a laboratory investigation involving the testing under center-points loading of simply-supported slabs incorporating macro synthetic fibers. The performance of the FRC slabs has been evaluating based on measured shear strength, energy absorption and ductility as well as observed cracking behavior.

2   Experimental Program

2.1   Materials

The concrete mix was designed to achieve a 28-day compressive strength of 30 MPa. Ordinary Portland cement (OPC) of 360 kg/m^3 was used with dune sand of 372 kg/m^3 and aggregates of 601, 271 and 581 kg/m^3 for sizes of 5, 10, 20 mm, respectively. The water to cement (w/c) ratio was 0.58. A high efficiency polycarboxylate-based superplasticizer (CONCERA™) was used as an admixture in the concrete mix. The amounts of admixture used were 0.5, 3 and 3.67 kg/m^3 for concrete mixes with 0, 0.5 and 1% fiber content. Test slabs were reinforced with 16mm diameter rebars of 420 MPa yield strength used as conventional flexural reinforcement.

The macro synthetic fiber used in the testing program is shown in Fig. 1. The main components of this polymeric fiber type are polyethylene and polypropylene. The length of individual fiber is 40 mm with aspect ratio of 90. The fibers have modulus of elasticity and tensile strength of 9.5 GPa and 620 MPa, respectively.

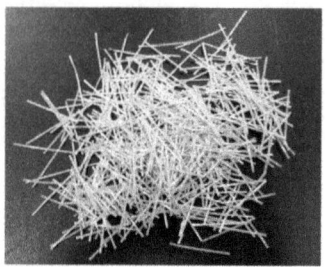

Fig. 1. Synthetic fiber used in the current study

2.2   Slab Specimens and Testing Procedure

A total of six reinforced concrete slabs were tested. The slabs dimensions were the same for all six specimens (1.5 m × 1.5 m) with thickness of 150 mm and tensile reinforcement ratio ρ = 0.9%, as shown in Fig. 2. The slabs were divided into three groups with respect to their fiber content as follows: 2 plain concrete slabs with no fiber; 2 slabs with 0.5% of macro synthetic fiber and 2 slabs with 1% of macro synthetic fiber, as shown in Table 1. Synthetic fibers were added to the concrete slabs at dosage rates of 0.5 and 1% by volume. Duplicate slabs were tested to assess the reliability and reproducibility of the test results. Slabs were cast and cured under similar conditions at 28 days. Measured slumps of plain concrete and synthetic fiber reinforced concrete were 145 mm and 160 mm, respectively.

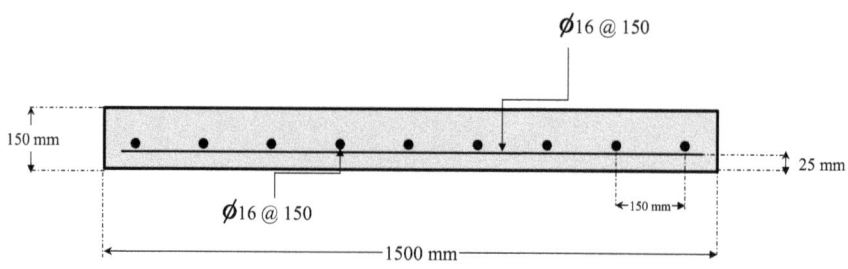

Fig. 2. Detailed cross-section of the test slabs

All slabs were tested under centre-point load using a square column stub of 100 mm. The load was applied monotonically in displacement control mode at a rate of 0.015 mm/sec. A vertically-oriented hydraulic actuator connected to a steel reaction frame was used for application of the load to the slab specimens, as shown in Fig. 3. All

Table 1. Slabs details

Slabs	Quantity	Dimension (m)	Thickness (mm)	Effective depth (mm)	Steel ratio %	Fiber %	Compressive strength (MPa)
Group 1	2	1.5 × 1.5	150	125	0.9	0	28.0
Group 2	2	1.5 × 1.5	150	125	0.9	0.5	29.0
Group 3	2	1.5 × 1.5	150	125	0.9	1	29.45

slabs tested were simply-supported along four edges. Total of 5 Linear Variable Displacement Transducers (LVDTs) were placed along the slab diagonal direction (corner to corner) to monitor the deflection profile as illustrated in Fig. 4. At each load level, vertical deflection and crack patterns on the bottom surface of all slabs were recorded.

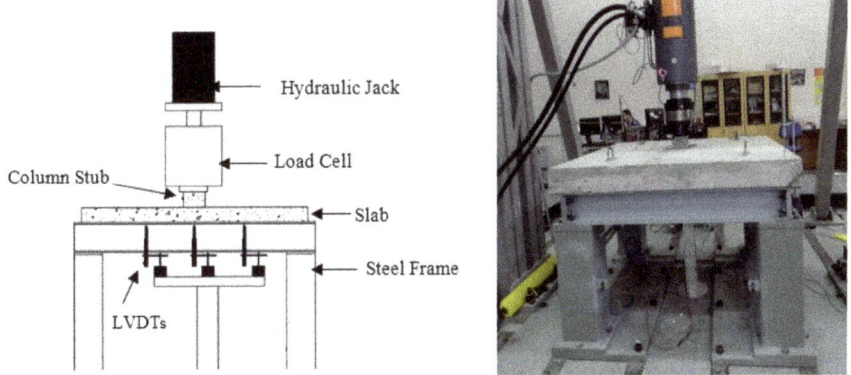

Fig. 3. Test Setup

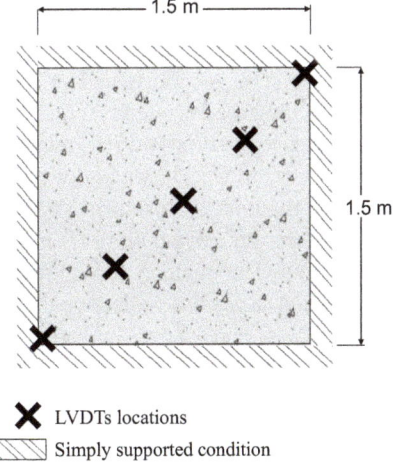

Fig. 4. LVDTs locations and boundary conditions for the slabs

3   Results and Discussion

3.1   Cracks and Failure Mode

Punching shear failure primarily occurred in all the tested slabs. This was clearly evidenced by the column stub punch through the slab top surface and defined by the cracking patterns in the bottom of the slab surface. The crack patterns for all tested slabs are shown in Fig. 5. The figure indicates that the addition of macro synthetic fibers changed the nature of failure pattern. Control slabs without fibers punched essentially instantaneously, as it was observed that some parts of the bottom concrete cover fell apart with limited cracking in the punching area (Fig. 5a).

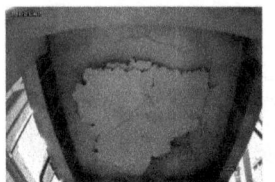 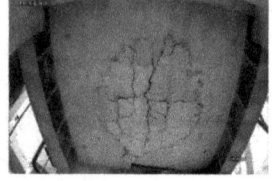 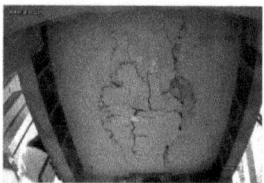

(a) Slab with 0% Fibers   (b) Slab with 0.5% Fibers   (c) Slab with 1.0% Fibers

Fig. 5.   Observed cracking patterns on the bottom surface of slabs

On the other hand, all fiber concrete slabs failed in a more gradual and ductile manner. It was noticed that the punching area became well defined and smaller than slabs with no fibers due to bridging effect of fibers (Figs. 5b and 5c). Moreover, the crack intensity increased as the percentage of fibers increased. Furthermore, slabs with 1% synthetic fibers have shown signs of post-peak flexural cracks which indicates that the synthetic fibers improved the ductility of the slab before failure. (Fig. 5c).

3.2   Load-Deflection Response

Load-deflection curves for tested slabs are presented in Fig. 6. The results show that fibers slightly increased the punching load capacity of the slab. Adding macro synthetic fibers at 1.0% increased the peak load (Pu) by 17%, relative to control slabs.

Energy absorption is defined by the total area under the load-deflection curve. The areas under the curves were computed. To preform consistent comparison, the area under the load deflection curves was calculated up to 40 mm central deflection at which the control slabs completely failed. The ductility of control slabs was much lower than the slabs with fibers which failed at 75mm deflection. The result shows that macro synthetic fibers at 0.5 and 1% fibers increased the absorbed energy (relative to control) by 47% and 73%, respectively. The results of peak loads and energy absorbed are summarized in Table 2 and Fig. 7.

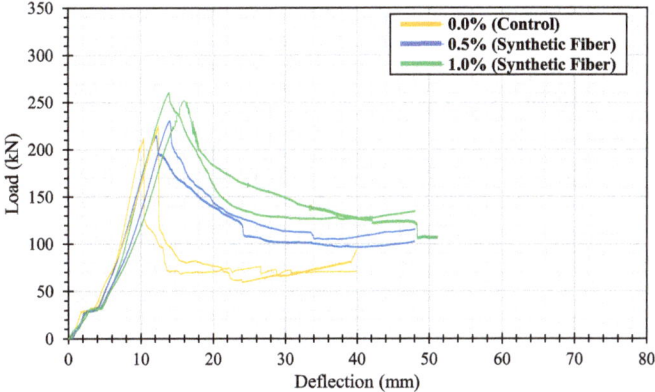

Fig. 6. Load versus deflection response for all tested slabs

Table 2. Summarized results of peak loads and energy absorbed

Fiber %	Average peak load (kN)	Percentage of increase relative to control slab	Average Energy absorbed (kJ)	Percentage of increase relative to control slab
0	219	–	3105	–
0.5	223	1.8%	4571	47%
1	256	17%	5396	73%

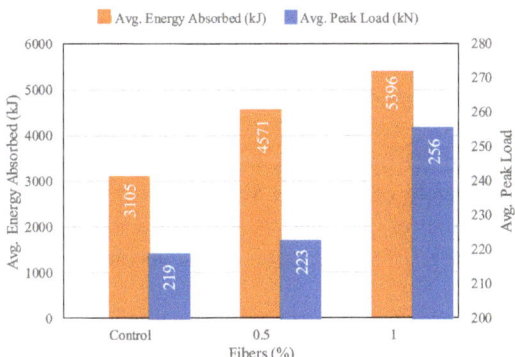

Fig. 7. Peak loads and energy absorption of tested slabs at different synthetic fiber ratios

3.3 Deflection Profile

Total of 5 LVDTs were placed along the slab diagonal (corner to corner) direction to study the surface deflection profiles of the tested slabs. Comparison of deflection profiles was done at ultimate load of (80 kN) with results shown in Fig. 8. A general observation from this deflection profile is that the addition of synthetic fibers to the slabs had reduced the overall deflections as well as the lift offs at the sides. It can be concluded that for FRC slabs, deflection beneath and adjacent to column stub showed

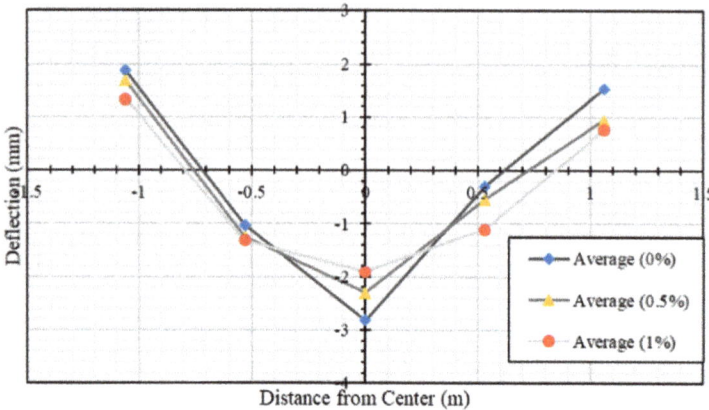

Fig. 8. Deflection profiles of tested slabs at different synthetic fibers

minor variation forming a concave profile rather than V-shape profile as in control slabs. This indicates that the fibers have distributed the loads to allow increased contribution of concrete in load carrying. All slabs have shown a decrease in the lift off values with addition of synthetic fibers.

4 Conclusions

An experimental testing program has been conducted to investigate the effect of macro synthetic fibers on the punching shear resistance and cracking behavior of reinforced concrete slabs. The experimental program included six slabs of 1500 x 1500 x 150 mm with a reinforcement steel ratio of 0.9%. Fibers were added to the concrete slabs at dosage rates of 0.5% and 1% by volume. The concrete mix was designed to achieve a 28-day compressive strength of 30 MPa. Slabs were simply supported along all four edges by a steel frame and tested under center-point load increasing monotonically in displacement control mode. The important conclusions obtained from this investigation are as follows:

- The addition of macro synthetic fibers to the concrete slabs have improved the behavior of punching shear. Control slabs behaved in a brittle manner and the punching area was large with low crack intensity. However, slabs with fibers have failed in more ductile manner and the punching area became well defined and smaller with high crack intensity due to bridging effect of fibers.
- The addition of macro synthetic fibers up to 1% slightly increased the punching shear strength of the slab by 17%, while significant increase was observed in the energy absorption by 73% for the tested FRC slabs relative to the control slab.
- The load-deflection curves of the tested slabs showed significant increase in ductility due to the addition of macro synthetic fibers.

- From the surface deflection profiles, it can be concluded that the addition of macro synthetic fibers had reduced the mid-span deflection and sides lift offs. This indicates that the fibers have distributed the loads to increase the contribution of concrete in load carrying.

References

1. Nguyen-Minh, L., Rovňák, M., Tran-Quoc, T., Nguyenkim, K.: Punching shear resistance of steel fiber reinforced concrete flat slabs. Procedia Eng. **14**, 1830–1837 (2011)
2. Roesler, J.R., Altoubat, S.A., Lange, D.A., Rieder, K.-A., Ulreich, G.R.: Effect of synthetic fibers on structural behavior of concrete slabs-on-ground. ACI Mater. J. **103**(1), 3 (2006)
3. Roesler, J.R., Lange, D.A., Altoubat, S.A., Rieder, K.-A., Ulreich, G.R.: Fracture of plain and fiber-reinforced concrete slabs under monotonic loading. J. Mater. Civ. Eng. **16**(5), 452–460 (2004)
4. Higashiyama, H., Ota, A., Mizukoshi, M.: Design equation for punching shear capacity of SFRC slabs. Int. J. Concrete Struct. Mater. **5**(1), 35–42 (2011)
5. Swamy, R.N., Ali, S.A.R.: Punching shear behavior of reinforced slab-column connections made with steel fiber concrete. J. Proc. **79**(5), 392–406 (1982)
6. Harajli, M.H., Maalouf, D., Khatib, H.: Effect of fibers on the punching shear strength of slab-column connections. Cement Concr. Compos. **17**(2), 161–170 (1995)
7. Cheng, M.-Y., Gustavo, J.P.-M.: Evaluation of steel fiber reinforcement for punching shear resistance in slab-column connections–part i: monotonically increased load. ACI Struct. J. **107**(1) (2010)
8. Abdel-Rahman, A.M., Hassan, N.Z., Soliman, A.M.: Punching shear behavior of reinforced concrete slabs using steel fibers in the mix. HBRC J. **14**(3), 272–281 (2018)

Mechanical and Fracture Behaviour of an HPFRC

Sandra Nunes[✉], Mário Pimentel, and Carlos Sousa

CONSTRUCT-LABEST, Faculty of Engineering (FEUP), University of Porto,
Porto, Portugal
snunes@fe.up.pt

Abstract. The current paper analyses the mechanical and fracture behaviour of a High-Performance Fibre Reinforced Concrete (HPFRC). An HPFRC was developed in a previous stage aiming to simultaneously, maximise aggregates content, achieve a compressive strength of 90–120 MPa and maintaining self-compactability (SF1+VS2). The benefits of fibres hybridisation (using fibres with lengths of 13, 35 and 60 mm) on flexural strength are investigated using the wedge-splitting test, in order to achieve the highest performance while keeping a relatively low fibre content. The final selected mixture was characterised in terms of workability, compressive strength and modulus of elasticity. Six notched prismatic specimens were subjected to three-point bending tests, according to EN 14651, for classification according to the MC2010. Based on the bending tests data, the simplified linear characteristic tensile stress vs. crack opening displacement relationship of the HPFRC was evaluated according to MC2010 and two other analytical approaches available in the literature.

Keywords: High performance fibre reinforced concrete (HPFRC) · Flat slabs · Compressive strength · Tensile and bending behaviour

1 Introduction

The concrete industry is continually under pressure to increase productivity and lower costs while still maintaining high quality and safety standards. This driving force for technological advancement has influenced both concrete and steel technology, resulting in the development of new types of concrete, new reinforcement solutions, and new building systems and construction processes. Within the field of concrete technology, the progress has been exceptional, notably in the improvement of concrete strength. In addition to compressive strength, there were significant improvements in other material properties, such as workability, permeability and ductility. Good examples of such advancements are self-compacting concrete (SCC) and fibre reinforced concrete (FRC). By eliminating the need for vibration, SCC is well suited for a more rational concrete production, reducing technical costs, improving the quality, durability and reliability of concrete structures, and improving health and safety in the construction site. In turn, FRC allows overcoming concrete's intrinsic brittleness and offers an opportunity to reduce one of the more labour-intensive activities necessary for concrete construction (reinforcement preparation and placement). In some structural elements (such as slabs on grade, walls and foundations) fibres can replace ordinary reinforcement completely.

© RILEM 2022
P. Serna et al. (Eds.): BEFIB 2021, RILEM Bookseries 36, pp. 174–185, 2022.
https://doi.org/10.1007/978-3-030-83719-8_16

In other elements, such as beams and suspended slabs, fibres can be used in combination with ordinary or prestressed reinforcement.

This paper describes research on high-performance fibre reinforced concrete (HPFRC). The envisaged applications are twofold: pre-cast and highly prestressed beams that are lighter, easier to transport and require less workmanship in the reinforcement preparation than their normal strength concrete counterparts; localised area at the slab-column connection zones of flat slabs (while the rest of the slab is made of normal-strength concrete), in order to improve its punching shear resistance. The potential benefits of these solutions are more industrialised construction and lower overall costs (including technical and social costs). From a structural point of view, the main reason for incorporating fibres is to enhance the fracture characteristics and structural behaviour through the fibres' ability to bridge cracks. This mechanism influences both the serviceability and ultimate limit states. Under service conditions crack propagation is more controlled, which primarily reduces the crack spacing and crack width and increases flexural stiffness. In terms of the ultimate limit state, the load resistance is increased and, for shear and punching failures, fibres also improve the ductility.

The fracture behaviour of FRC can be described by the stress-crack opening (σ-w) relationship, but appropriate test methods are necessary to determine this fundamental relationship. The uniaxial tensile test is the most accurate method for evaluating the post-cracking σ-w relationship of FRC directly. However, direct tensile tests require sophisticated testing equipment and careful preparation of the test set-up and specimens and are quite time-consuming. Therefore, extensive research efforts have been made to indirectly assess the σ-w response of FRC through inverse analysis procedures that consider the experimental results obtained by more straightforward test configurations. Distinct test methods may be employed to indirectly assess the σ-w response, such as either three- or four-point bending tests on prismatic specimens, wedge splitting tests, and round panel tests.

In the present study, the benefits of fibres hybridisation on the tensile behaviour of an HPFRC was first investigated using the wedge-splitting test (WST) [1, 2]. This test was selected since the fabrication of the WST specimens requires less than 1/4 of the volume required for the prismatic three-point bending test (3PBT) specimens according to the EN 14651:2007, which facilitated the realisation of a large experimental campaign. Besides, the WST specimen and test set-up resemble the central part of the 3PBT; and for this reason, it was selected to the detriment of other indirect compact tests. The post-cracking response of the selected HPFRC mixture was characterised by executing the 3PBT on six prismatic specimens, according to EN 14651:2007. All mixtures were characterised in terms of workability. The compressive strength and modulus of elasticity of the final selected HPFRC were also evaluated.

2   Test Program

2.1   Concrete Phase Design

In the first part of this study, the concrete matrix phase was optimised to achieve a target compressive strength of 90–120 MPa, at 28 days, and self-compacting ability, which is reported elsewhere [3]. The aggregate used for the concrete matrix was a crushed amphibolite rock and a natural siliceous sand with maximum sizes of 8 mm and 2 mm for coarse and fine aggregate, respectively. The powders fraction included cement CEM I 42.5R, limestone filler (Betocarb HP-OU) and silica fume (Elkem 940-U). A polycarboxylate-based superplasticiser (Sika Viscocrete 20HE) was also incorporated. At this stage of the study, 1% steel microfibres were included to avoid brittle failures in compression and reduce the variability of test results. The final concrete mix design is provided in Table 1.

Table 1. HPFRC mix design (kg/m^3).

Material	Cement	Limestone powder	Silica fume	Water	Superpl.	Fine aggregate	Coarse aggregate	Steel fibres
Quantity	531.86	203.72	53.19	147.85	12.55	811.82	721.43	78.50

2.2 Composite Design

In the second stage of the study, the benefits of fibres hybridisation on the post-cracking response of HPFRC were investigated. The concrete phase mix-proportions provided in Table 1 were maintained, and the total fibre content was kept equal to 1%. Three different types of steel fibres have been used as reinforcement: straight microfibres (S, $l_f = 13$ mm and $d_f = 0.2$ mm, $f_y = 2750$ MPa); double hooked-end fibres (H4D, $l_f = 35$ mm and $d_f = 0.55$ mm, $f_y = 1850$ MPa) and triple hooked-end fibres (H5D, $l_f = 60$ mm and $d_f = 0.59$ mm, $f_y = 2300$ MPa). It should be noted that all fibres have a similar aspect ratio (l_f/d_f) of around 65. The proportions of each fibre type in each mixture are summarised in Table 2. A total of ten different mixtures were assessed. Replicate runs of four mixes were included to assess the repeatability of results.

Table 2. Tested HPFRC mixtures and corresponding workability test results

Mix n°	Fibre content			Workability test results		
	S	H4D	H5D	t500 (s)	Dflow (mm)	Consistency class
1	1%	0	0	46.63	616.2	SF1
2	0.5%	0.5%	0	19.91	705.0	SF2
3	0.5%	0	0.5%	11.94	736.2	SF2
4	0	1%	0	5.38	788.3	SF3
5	0	0.5%	0.5%	6.93	773.8	SF3
6	0	0	1%	6.25	787.5	SF3
7	0.67%	0.17%	0.17%	18.13	676.2	SF2
8	0.17%	0.67%	0.17%	8.81	760.0	SF3
9	0.17%	0.17%	0.67%	8.09	760.0	SF3
10	0.33%	0.33%	0.33%	11.47	721.2	SF2
11 (= 1)	1%	0	0	42.00	598.8	SF1
12 (= 4)	0	1%	0	6.97	767.5	SF3
13 (= 6)	0	0	1%	6.72	783.8	SF3
14 (= 10)	0.33%	0.33%	0.33%	11.97	727.5	SF2

2.3   Mixing Procedure and Test Methods

The 14 mixes listed in Table 2 were prepared in 40 L batches in an open pan mixer. The sequence of mixing consisted of the following: (1) mixing fine and coarse aggregates with 25% of total mixing water during 2.5 min; (2) stop mixing and waiting 2.5 min for water absorption by aggregates; (3) adding powder materials and the remaining mixing water and mixing for 5.0 min; (4) stopping the mixer to scrape off the material adhering to the mixer walls; (5) adding fibres and mixing for additional 3.0 min. Immediately after mixing, the concrete slump flow test was carried out, in accordance with EN 12350–8, to evaluate the time necessary to reach a 500 mm diameter (t500) and the average final spread diameter (Dflow). Finally, seven 150 × 150 × 130 mm³ specimens were cast to carry out the wedge-splitting test (WST), at 28 days. All samples were covered with a plastic sheet to avoid drying and water evaporation. After 24 h, the specimens were demoulded and placed in the climatic chamber under controlled environmental conditions (temperature = 20 ± 2 °C and humidity = 95–98%) until testing age.

The WST uses cubes that have previously been prepared by casting a groove and sawing a starting-notch (53 mm depth and 4 mm width) in the central part of the specimen and two guide notches (15 mm depth) on the lateral sides of the specimen. A transversal force is applied, through the use of a wedge and rollers, resulting in an eccentric tensile axial force with respect to the crack section (see Fig. 1c). The test method is well explained, for example, by Löfgren [2]. The specimens were placed on a circular steel bar with a diameter of 8 mm, centred on the specimen and parallel to the groove's direction. Two L-shape steel devices, each equipped with two rollers, were placed on top of the specimen over both lips of the groove. In addition, two steel wedges and a profile were placed below the hydraulic actuator (see Fig. 1a). A tension-compression 300-DX Instron machine was used, with a load capacity of 300 kN. The cross-head speed was kept constant with 0.3 mm/min for all tests. A load cell with a nominal load of 100 kN and an error below 0.03% was placed above the steel profile to monitor the applied load. For the measurement of the crack mouth opening displacement (CMOD), two linear variable displacement transducers (LVDT) (see Fig. 1b) were used with a range of 5 mm and an accuracy of ± 0.0125 mm. Both the vertical load and the CMOD were monitored during the test with a frequency of 1.0 Hz. CMOD was defined as the average value of the two LVDTs on the front and rear faces. The vertical force can be separated at roller level into two forces: a vertical force (0.5 F_v) and a horizontal force or the splitting force, as shown in Fig. 1b. The splitting force (F_{sp}) can be estimated using Eq. (1), as given below:

$$F_{sp} \approx \frac{F_v}{2\tan \alpha} = 1.866 \, F_v \tag{1}$$

where F_v is the vertical force and α is the angle of the wedge device. In this study, $\alpha = 15°$.

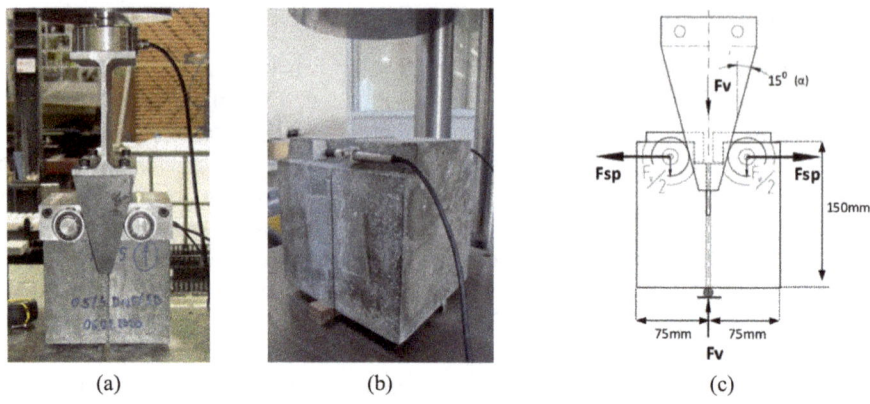

(a)	(b)	(c)

Fig. 1. (a) WST set-up; (b) LVDTs positioning; (c) forces acting in the WST

After analysing the WST results and selecting the best mixture, two larger batches (60 L) were prepared (using the same mixing procedure) to cast six prisms (150 × 150 × 600 mm^3), three cylinders (ϕ = 150 mm, h = 300 mm) and four cubes (150 × 150 × 150 mm^3) to carry out the three-point bending test (3PBT), the modulus of elasticity test and the compressive strength test, respectively, at 28 days. The curing procedure was also maintained. The compressive strength and modulus of elasticity tests were conducted according to EN 12390-3 and EN 12390-13, respectively. The 3PBT was carried out according to EN 14651:2005 + A1:2007. This test was carried out in close-loop displacement control using an LVDT installed at the specimen's mid-span. A displacement rate of 0.05 mm/min at the mid-span was adopted up to a deflection of 0.13 mm to avoid instability during the first phase of the crack formation and propagation. Then, the displacement rate was increased to 0.2 mm/min and kept constant until a CMOD of at least 4 mm was reached. The CMOD of the specimens was recorded using a clip-shape TML displacement transducer with a measurement range of 5 mm, positioned across the notch at the bottom surface of the prism (see Fig. 2).

Fig. 2. 3PBT set-up

3   Workability Results

Fresh state test results are summarised in Table 2. Comparing mixes 1 (or 11), 4 (or 12) and 6 (or 13), it can be concluded that S-type fibres have a significant negative effect on HPFRC workability, significantly reducing Dflow and increasing t500. For the same fibre volume (1%) the number of S-type fibres (microfibres) is much higher compared to the number of macrofibres (H4D or H5D) present in the mixture. Consequently, the fibres surface area increases requiring a larger volume of paste to be completely lubricated and reducing the excess of paste in the mixture, which explains the increased viscosity and lower deformability. Conversely, when increasing the proportion of the longer fibres in the mixture (H5D) Dflow increases and t500 decreases. Except for mix 1 (or 11), all mixes belong to class SF2 or SF3 according to EN 12350-8. In terms of viscosity, all mixes exhibit relatively high t500 (≥ 2 s), which can be explained by the relatively low water/binder ratio and increased density of the mixture, due to both the presence of heavier coarse aggregates (3020 kg/m^3) and the fibres. The replicate runs results show good reproducibility of fresh state test results.

4   WST Results

To enable easier comparison the valid results for each series have been combined into one average curve. The average F_{sp} vs CMOD curve for each series is shown in Fig. 3a–c. Figure 3.d shows the scatter in the test results; it can be seen the coefficient of variance (CoV) is between about 10% and 30%, which is in agreement with the scatter found in Löfgren's experiments [2]. A major factor contributing to this scatter in the test results is related to variations in the fibres distribution and orientation. Figure 4 shows the CMOD for which the maximum force was obtained for each mixture.

The area below the average F_{sp} - CMOD curve is here loosely designated as the work of fracture, W, which provides a measure of the material's energy dissipation when subjected to tensile loading [1]:

$$G_{F,WST} = \frac{W}{A_{lig}} \qquad (2)$$

In this study $A_{lig} = 7500$ mm^2. It is noted that $G_{F,WST}$ is not a material property and serves here the purpose of comparing the tensile performance of the different mixes tested under the same experimental conditions. During the experiments, some specimens reached relatively high CMOD values, and the fracture surfaces were not still separated. Thus, in order to enable the comparison of the W values of all specimens, the determination was made up to a maximum CMOD value of 3.0 mm and denoted with $G_{F,WST(3\,mm)}$. Figure 5 ranks the tested mixtures as a function of $G_{F,WST(3\,mm)}$. These results ranged from 1215 to 1791 N/m.

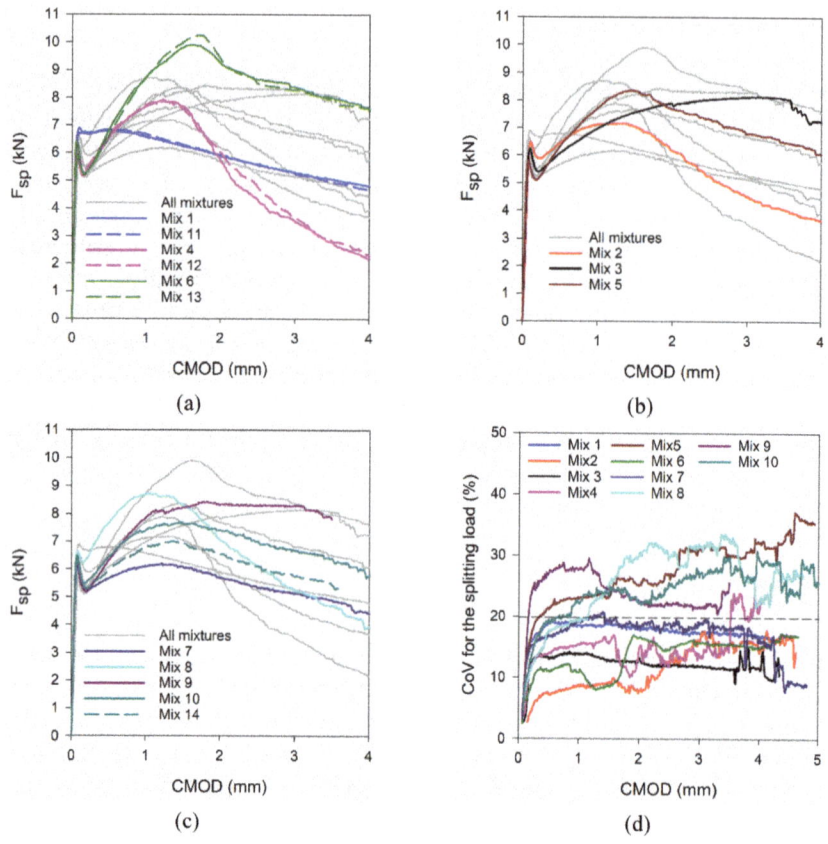

Fig. 3. Average FSP vs CMOD curves highlighting: (a) single, (b) binary, and (c) ternary mixtures; (d) comparison of scattering in the test results

4.1 Single Fibre Mixtures

Figure 3.a clearly shows the influence of the fibre type (its mechanical and geometrical properties) on the post-cracking behaviour of HPFRC. A steeper softening branch after the peak load is observed in the mixes with hooked-end macrofibres (Mixes 4 and 6) than with straight microfibers (Mix 1), where a more gradual decrease of the force with increasing CMOD is observed. The results also show that the straight microfibres are activated for smaller crack widths while the macrofibers are actived later. This is evidenced by the sharp drop of the load right after the first cracking occurred. The superior performance of fibres H5D compared to fibres H4D is also shown in Fig. 3a. Figure 5 shows that Mix 6 (or 13) with the H5D fibres exhibited the highest $G_{F(3\,mm)}$ from all the tested mixtures. Besides the lower anchorage provided by the double-hook in H4D fibres compared to the triple hook of the H5D fibres, the corresponding material has lower tensile strength leading to fibre rupture, which was confirmed by visual observation of the fracture surfaces. For such a high-strength concrete matrix, the tensile strength and ductility of H4D-type fibres was found to be insufficient to avoid fibre rupture during the fracture process.

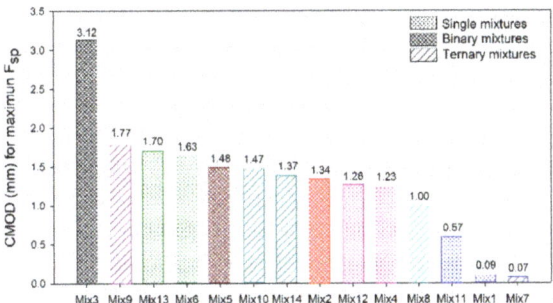

Fig. 4. CMOD corresponding to maximum Fsp

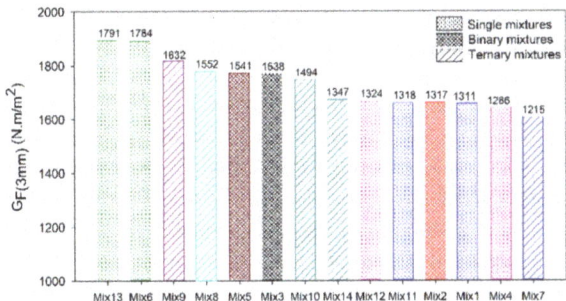

Fig. 5. $G_{F,WST(3\,mm)}$ values presented by descending order

4.2 Binary Fibre Mixtures

Several previous studies suggested that combining short and long fibres (hybrid mixtures) improves the initial tensile strength as well as the performance in the post-cracking regime. Figure 3b and Fig. 5 show the effect of binary mixtures of steel fibres while keeping a total fibre content of 1%. Excluding Mix 6 (or 13), binary mixtures 3 and 5 exhibited higher $G_{F(3\,mm)}$ than single mixtures 1 (or 11) and 4 (or 12). The presence of microfibres (S) shows an apparent beneficial effect in binary mixtures for small CMOD values (<0.5 mm) (see Fig. 3b). As the CMOD grows, the long steel fibres (H4D and H5D) become more active in crack bridging. Short fibres will then become less active, because they are being pulled out as the crack width increases. Again, the binary mixture combining 0.5%S + 0.5%H4D fibres (Mix 2) exhibited worse performance compared to 0.5%S + 0.5%H5D binary mixture (Mix 3).

4.3 Ternary Fibre Mixtures

As can be observed in Fig. 3c and Fig. 5, the ternary mixture with the best overall performance was Mix 9 (0.17%S + 0.17%H4D + 0.67%H5D). Nevertheless, the corresponding $G_{F(3\,mm)}$ does not differ significantly from the best performing binary mixtures (mixes 5 and 3). In particular, the average F_{sp} - CMOD curves of Mixes 9 and 3 are very close. The worst performing ternary mixture was Mix 7.

4.4   Selected Mixture

Based on the obtained experimental results, there are no strong reasons to select a ternary mixture. Besides, from the practical point of view, for industrial production, single/binary mixtures are preferred to ternary mixtures. Despite Mix 6 exhibited the highest $G_{F(3\,mm)}$, binary mixture (Mix 3) was selected from all the tested mixtures, since it exhibits a relatively high $G_{F(3\,mm)}$, it reaches the maximum F_{sp} at the highest CMOD from all mixtures (see Fig. 4), and potentially has less scatter due to the presence of many microfibres. Finally, it showed good self-compacting characteristics (Table 2).

5   Mechanical Behaviour of Selected Mixture

5.1   Compressive Strength and Elasticity Modulus

The average values of the secant modulus of elasticity and compressive strength of the selected HPFRC (Mix 3), at the age of 28 days, were E_{cm} = 53.5 GPa (CoV = 1.1%), f_{ccm} = 124.9 MPa (CoV = 1.6%) (cubic specimens) and f_{cm} = 113.9 MPa (CoV = 1.2%) (cylindrical specimens), respectively. Based on the obtained experimental results, the average tensile strength of the selected HPFRC was estimated as 5.3 MPa, using the following equation proposed by MC2010 [4]:

$$f_{ctm} = 2.12 \cdot \ln(1 + 0.1 \cdot f_{cm}), \text{for concrete grades} > \text{C50} \tag{3}$$

5.2   Three-Point Bending Test

Figure 6 depicts the post-cracking response of the selected HPFRC in terms of nominal flexural stress versus CMOD relationship, abbreviated hereafter by σ_N - w relationship. From these responses, the stress at the limit of proportionality, f_L, (corresponding to the maximum load reached within a CMOD of 0.05 mm) and the residual flexural tensile strengths of the HPFRC, f_{R1} to f_{R4}, corresponding to distinct values of CMOD, were obtained as indicated in Table 3. In accordance with [4], the toughness class of this concrete is "13c" (f_{R3k}/f_{R1k} = 1.08).

Table 3. Limit of proportionality and residual flexural strength of the selected HPFRC

	f_L	f_{R1} (MPa) CMOD$_1$ = 0.5	f_{R2} (MPa) CMOD$_2$ = 1.5	f_{R3} (MPa) CMOD$_3$ = 2.5	f_{R4} (MPa) CMOD$_4$ = 3.5
Average	10.0	15.4	18.0	16.4	12.9
CoV	4.4%	6.1%	4.4%	5.5%	11.7%
Characteristic	9.1	13.4	16.3	14.4	9.7

In all the 3PBT specimens, more than one crack was observed, which suggests that the post cracking tensile strength of this HPFRC is higher than its cracking strength, i.e., the material may exhibit hardening behaviour in direct tension. This has to be investigated in more detail. Crack patterns representative of those obtained in the tests are shown in Fig. 7.

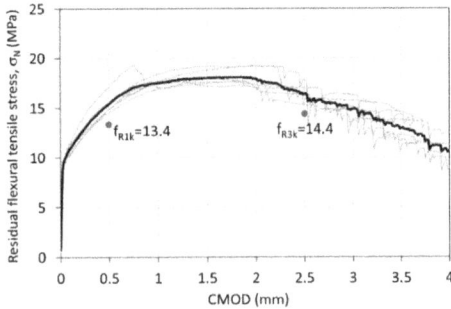

Fig. 6. σ_N vs. CMOD curves for all 6 tested specimens (mean curve in black)

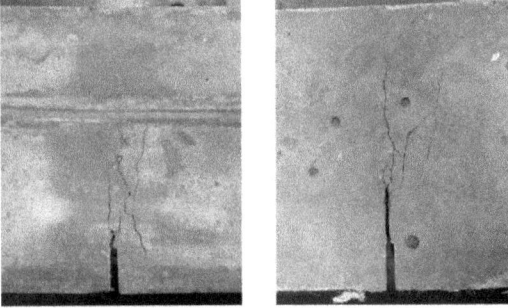

Fig. 7. Crack patterns in 3PBT showing multiple cracks above the notch

5.3   Evaluation of σ-w Relationship

Based on the experimental results presented in the previous section, the simplified linear characteristic tensile stress vs crack opening displacement relationship, σ-w, of the HPFRC was evaluated according to three analytical approaches available in the literature, namely: a) the proposal of the fib Model Code 2010 [4]; b) the proposal by di Prisco et al. [5]; and c) the formulation proposed by Amin et al. [6]. In all cases, the linear post cracking tensile stress versus crack opening displacement is derived from the 3PBT based on a simple kinematic model of the cracked section and establishing equilibrium after making suitable simplifying assumptions about the corresponding post-cracking stress distribution. The proposals a) and b) are quite similar (the derivation is given in [5]) and consider only the f_{R1} and f_{R3} values for defining the σ-w relationship. The formulation a) that ended being retained in the Model Code 2010, entails an additional consideration on the ratio f_{R3}/f_{R1}, which is not enforced in the formulation b). The procedure proposed in reference [6] uses the f_{R2} and f_{R4} values instead and includes a reduction coefficient $k_2 = 0.82$ applied to the f_{Rj}. According to the authors, this reduction is necessary to account for the overestimation of the resistance based on the results obtained from notched specimens due to the fact that the crack path is enforced by the notch and does not follow the path with the least resistance. The three linear curves corresponding to the characteristic values are presented in Fig. 8. Considerable differences can be observed between the three models. As the definition of accurate material laws is the base for proper design models for flexure, shear, punching taking advantage of the behaviour of this HPFRC, this clearly points out the need for further research on this topic.

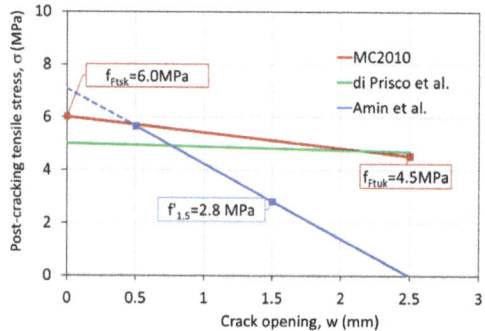

Fig. 8. Simplified linear σ-w relationships according to references [4–6]

6   Conclusions

Most relevant conclusions of the current study are the following:

- Concerning fibres mixture, no strong reasons were found to select a ternary mixture.
- Combining 0.5% straight microfibres (l_f = 13 mm) with 0.5% triple hooked-end fibres (l_f = 60 mm), a toughness class "13c" was achieved with this HPFRC, according to MC2010.
- In all the 3PBT specimens, more than one crack was observed, which suggests that the post cracking tensile strength of this HPFRC is higher than its cracking strength.
- Significant differences were found in the linear σ-w relationships of the developed HPFRC when assessed by three different analytical approaches. Further research is needed on this topic.

Acknowledgements. This work was financially supported by: Base Funding - UIDB/04708/ 2020 of the CONSTRUCT - Instituto de I&D em Estruturas e Construções - funded by national funds through the FCT/MCTES (PIDDAC), the project PTDC/ECI-EST/30511/2017 funded by national funds through the FCT/MCTES (PIDDAC), and the iPBRAIL project - POCI-01-0247-FEDER-039894 co-funded by the European Regional Development Fund (ERDF), through the

Operational Programme for Competitiveness and Internationalization (COMPETE 2020) and Lisbon Regional Operational Programme (Lisboa 2020). Collaboration and materials supply by EUROMODAL, Secil, Omya Comital, Sika and Dramix and is gratefully acknowledged.

References

1. Brühwiler, E., Wittmann, F.H.: The wedge splitting test, a new method for performing stable fracture mechanics test. Eng. Fracture Mech. **35**(1–3), 117–125 (1990)
2. Löfgren, I.: Fibre-reinforced Concrete for Industrial Construction-a fracture mechanics approach to material testing and structural analysis. Ph.D. thesis, Chalmers University of Technology (2005)
3. Blazy, J., Nunes, S., Sousa, C., Pimentel, M.: Development of an HPFRC for use in flat slabs. In: Serna, P., Llano-Torre, A., Martí-Vargas, J.R., Navarro-Gregori, J. (eds.) BEFIB 2020. RB, vol. 30, pp. 209–220. Springer, Cham (2021). https://doi.org/10.1007/978-3-030-58482-5_19
4. *fib* Model Code for Concrete Structures 2010. Ernst & Sohn, Berlim (2013)
5. di Prisco, M., Colombo, M., Dozio, D.: Fibre-reinforced concrete in *fib* Model Code 2010: principles, models and test validation. Struct. Concr. **15**(4), 342–361 (2013)
6. Amin, A., Foster, S.J., Muttoni, A.: Derivation of the σ-w relationship for SFRC from prism bending tests. Struct. Concr. **16**(1), 93–105 (2015)

Property Assessment of Self-compacting Basalt Fiber Reinforced Concrete

Piotr Smarzewski[✉]

Lublin University of Technology, Lublin, Poland
p.smarzewski@pollub.pl

Abstract. The aim of the research was to determine the suitability of chopped basalt fibers to reduce the brittleness of high performance concrete (HPC) while maintaining the requirements for self-compacting concrete (SCC). The study investigated the influence of the basalt fiber volume content on the fresh properties of self-compacting concrete and the mechanical properties of basalt fiber reinforced high performance concrete. Basalt fibers were added at the ratios of 0, 0.025, 0.05, 0.075, 0.125, and 0.25% by volume to the concrete mixtures in which 46% of the cement was replaced by ground granulated blast furnace slag (GGBS). The water to binder ratio (w/b) in the SCC mixtures was kept constant at 0.32. The influence of fiber volume content on the fresh properties of SCC, including filling ability (slump diameter, flow time) and passing ability (L-box) of the mixtures as well as mechanical properties such as the compressive strength, tensile splitting strength, flexural strength, and toughness indexes were analyzed after the moist curing of concrete at 7 and 28 days. The toughness, as the one of an important parameter for fiber reinforced concrete, was determined in accordance with ASTM C 1609. None of the basalt fiber reinforced concrete beam specimens reached the intended deflection of span/150. The load-displacement curves showed a rapid decrease in load after peak load, and the beam specimens reinforced with chopped basalt fibers showed their little potential for producing ductile HPC. Nevertheless, the higher content of basalt fibers improved the strength, and toughness of the SCC up to 0.25 vol.% fibers, and therefore the use of these fibers improved the overall SCC performance. The mechanical properties of the sustainable SCC mixtures containing the high content of GGBS as a cement substitute and a low content of basalt fibers falls within ranges suitable for structural engineering applications.

Keywords: Self-compacting concrete · Basalt fiber · Fresh properties · Strength · Toughness

1 Introduction

Fiber reinforced concrete is used in structural engineering applications due to its high tensile strength, energy absorption capacity, and ductility. Between the numerous types of fibers available, basalt fiber (BF) is considered as promising reinforcement for concrete on account of its excellent properties [1]. Basalt fiber is a new type of eco-logical, high-performance and relatively cheap to produce fiber, with excellent strength characteristics, good thermal resistance, and high resistance to an alkaline environment

© RILEM 2022
P. Serna et al. (Eds.): BEFIB 2021, RILEM Bookseries 36, pp. 186–197, 2022.
https://doi.org/10.1007/978-3-030-83719-8_17

[2–5]. Basalt fiber can improve the flexural strength, toughness and fracture energy of cementitious matrix composites [6, 7]. On the other hand, basalt fibers significantly deteriorate the workability of fresh concrete mixture compared to steel, glass and polypropylene fibers [8]. The use of basalt fibers, in contrast with the most commonly applied steel fibers, have not concerns in regard to thermal conductivity, fire resistance, and corrosion of fiber closed to the concrete matrix.

In general, replacing cement with ground granulated blast furnace slag (GGBS) reduces the permeability of concrete, increases the resistance to sulphate corrosion and increases the corrosion resistance of the reinforcing steel [9], and consequently improves the overall durability of concrete and steel reinforcement. GGBS is used in self-compacting concrete (SCC) mixtures as cement replacement due to its inherent binder properties. The addition of GGBS to SCC mixtures results in their lower permeability and greater chemical stability. This stability is attributed to the reaction of GGBS with an excess of soluble calcium hydroxide reducing its presence in SCC [10].

It is known from the previous studies that the addition of fibers to self-compacting concrete (SCC) has a positive effect on its splitting and flexural tensile strength [6, 8, 11]. Most of the previous tests of concretes with BF addition concerned the determination of their influence on the properties of normal strength concrete [2, 4, 5, 12] or high performance concrete [7, 13–16]. Some researchers reported that the optimal strength parameters of normal strength and high-performance concrete are obtained with 0.3–0.5% BF [2, 17] and 1.5% BF volume contents [7, 14], respectively. In contrast, studies on the influence of BF on the properties of SCC are limited. Algin and Ozen [18] investigated the mechanical properties of 3/6/12/24 mm length BFs with a percentage of 0/0.1/0.3/0.5% in SCC mixtures. The highest compressive strength was obtained in concretes containing 0.1% BF and lengths of 12 and 24 mm. On the other hand, the highest values of splitting tensile and flexural strength were obtained in SCCs containing 24 mm long BF with 0.5% volume content. Çelik and Bingöl [6, 19] studied the effect of the type and volume content of basalt, glass and polypropylene fibers on the rheological and mechanical properties of SCC. The results showed that each type of fiber increased the splitting strength, flexural strength, impact resistance and fracture toughness of the concrete, but the optimal SCC parameters for basalt fibers were obtained with 0.25–0.3% volume content. Cement replacement ratios by GGBS usually ranged from 5% to 80% by weight of total cement [9]. Mohamed and Najm [10] reported that replacement of cement with the optimum value of 35% GGBS produced the highest 28-day compressive strength.

The studies of the mechanical properties of HPC and SCC reinforced with basalt fiber that have been carried out so far are not sufficient. This paper presents a study on the development and evaluation of a composite designed as a combination of SCC, HPC and basalt fiber reinforced concrete with high, over 46% GGBS replacement ratio by total cement weight. The main aim of this research is to investigate the effect of BF on fresh and hardened properties of sustainable SCC. The interest of this study is to quantify and discuss the impact of basalt fibers on the rheological and mechanical properties of SCC. This research also aims to find the optimum basalt fiber content that could enhance the mechanical properties of SCC.

2   Experimental Program

2.1   Materials, Mixture Proportions and Preparation of Specimens

The raw materials used for preparing SCCs included cement (C), ground granulated blast furnace slag (GGBS), fine aggregate (FA), coarse aggregate (CA), tap water (W), and high range water reducing admixtures (HRWA). To examine the effect of basalt fiber on the strength development of sustainable SCC mixtures, a control mixture without any fibers was developed in which CEM I 42.5R ordinary Portland cement complying with PN-EN 197-1:2012 and PN-B-19707:2013 and GGBS were used as cementitious materials. Various sustainable mixtures were then created in which basalt fibers were added to the control SCC in 0.025, 0.05, 0.075, 0.125, and 0.25% volume fractions. The physical properties of cement and the specific chemical compositions and of the cementitious materials are given in Tables 1 and 2, respectively.

Table 1. Physical properties of cement.

Material characteristics	Cement
Specific surface area (cm^2/g)	4164
Water demand (%)	27
Start of setting (min)	180
End of setting (min)	215
Volume stability acc. to Le Chateliere (mm)	1
Compressive strength at 2 days (MPa)	29.6
Tensile strength at 2 days (MPa)	5.1

Table 2. Chemical constituents of cement and slag (in %).

Compound	C	GGBS
SiO_2	20.19	33.14
Al_2O_3	4.30	13.55
Fe_2O_3	3.25	1.30
CaO	64.61	43.36
MgO	1.41	6.48
SO_3	2.96	0.29
K_2O	2.59	0.31
Na_2O	0.26	0.29
Cl	0.111	0.006
Loss on ignition	3.41	0.76
Insoluble matter	0.48	0.31

By means of the chemical analysis, it was found that GGBS contains a high level of SiO_2, which is beneficial with reference to the mechanical strength of SCC.

In all SCC mixtures tested, a High Range Water Reducing Admixtures (HRWAs) based on lignosulfonates and polycarboxylic ethers were used and kept constant at 3% of the total weight of cementitious materials. The HRWAs are manufactured by CEMEX Corporation under the commercial names ISOLA BV, ISOFLECX 833, and ISOFLOW 755. The dosage of HRWAs was adjusted to provide a target flow of 720 ± 10 mm for control SCC, corresponding to a SF2 flow class. In all SCC mixtures total amount of cementitious materials was approximately 650 kg/m^3 and the W/(C +GGBS) ratio was maintained at approximately 0.32. Coarse aggregates were gravel aggregate passing 8 mm sieve size with a total amount of 400 kg/m^3. Fine aggregate consisted of 980 kg/m^3 quartz sand with a grain size of 0.5/2 mm and fineness modulus of 1.84. Fine aggregate in saturated-surface-dry condition was used. The particle size distributions of gravel coarse aggregate and quartz sand were ascertained according to PN-EN 933-1:2012. The basalt fiber bundles are flat with approximately 1 mm wide, 13 μm diameter filaments, and aspect ratio of l/d = 923. Other fiber properties were as follows: density 2700 kg/m^3, tensile strength 1700 MPa, modulus of elasticity 70,000 MPa, and elongation at break 0.25‰. Basalt fibers used in this study are shown in Fig. 1.

The mixtures proportions of basalt fiber reinforced self-compacting concretes are shown in Table 3, where SCC represents the concrete without basalt fibers and BSCC signifies the concretes with BF. The numbers immediately following the letters represent the basalt fiber volume content (%). In order to better determine the effect of BF on BSCCs, it ought to be noted that all the other mixture ingredients were kept constant.

Table 3. SCC mixture proportions (kg/m^3).

Component	SCC	BSCC-0.025	BSCC-0.05	BSCC-0.075	BSCC-0.125	BSCC-0.25
BF	0	0.675	1.35	2.025	3.375	6.75
C				350		
GGBS				300		
FA				980		
CA				400		
W				210		
HRWA				19		

Fig. 1. Basalt fibers used in this study.

The specific mixing process for BSCC is shown in Fig. 2. The feeding sequence of raw materials was as follows: aggregates (FA + CA), cementitious materials (C + GGBS), uniformly mixed W and HRWA, as well as BF. The mixing times for above mentioned raw materials were 2 min, 2 min, 4 min, and 4 min, respectively. After tests were carried out in order to determine properties of fresh concrete such as slump flow and L-box tests, the mixture was poured into the prepared moulds and compacted on a vibration table at a rate of 150 Hz. After curing for 24 h under the foil cover at a room temperature of 20 ± 2 °C and relative humidity of >95%, the specimens were demoulded and placed in a tank with water at temperature of 20 ± 2 °C for another 6 d or 27 d prior to testing at 7 and 28 curing days. Thereafter, the mechanical properties of SCCs were tested.

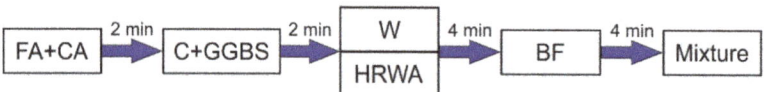

Fig. 2. Schematic diagram of BSCC mixing process.

2.2 Test Set-Up

The experimental study is intended to determine the effect of basalt fiber volume contents on the workability parameters and mechanical properties of BSCC. In order to determine properties of fresh mixtures, slump flow and L-box tests were carried out according to the recommendations of European Guidelines for Self-Compacting Concrete [20]. The slump flow test was carried out by filling the slump cone and measuring the maximum uninterrupted flow diameter in two orthogonal directions and the time (T_{500}) it takes the SCC mixture to reach a diameter of 500 mm. The L-box test was carried out by filling up the vertical part of the equipment and after 1 min, the three steel reinforcements were removed to allow the mixture to pass through them and fill the horizontal portion of the L-box. The heights between the level of the mixture and the top of the apparatus at both ends of the horizontal tank were measured and then the average depths of the mixture at both ends of the apparatus were calculated. The ratio of these depths is a measure of the passing ability (PA).

The hardened properties of SCCs were tested after 7 and 28 days of moist curing.

The compressive strength was carried out using a load-controlled universal testing machine of 3000 kN capacity according to PN-EN 12390-3, and the specimens dimensions were 100 mm × 100 mm × 100 mm. The cubes were loaded up to failure at a loading rate of 0.5 MPa/s. Six specimens were tested for each mixture at each curing period, and the average compressive strength was considered as the compressive strength of that mixture.

The splitting tensile strength of SCCs specimens was tested using the Controls universal material testing machine in accordance with PN-EN 12390–6. Each specimen was loaded at a loading rate of 0.2 mm/min until it was destroyed. Six cubical specimens of 100 mm × 100 mm × 100 mm size were tested for each mixture at each curing period, and the average value of the splitting tensile strength of all specimens was considered as the splitting tensile strength of that mixture.

The flexural strength and post-cracking behavior of BSCCs were carried out at three-point bending scheme by a servo-hydraulic testing machine according to PN-EN 12390-5. The load was applied continuously and monotonically with a rate of 0.05 mm/min. Three prismatic specimens of 100 mm × 100 mm × 500 mm size were tested for each mixture at each curing period, and the average value of the flexural strength of specimens was considered as the flexural strength of that mixture. Moreover, the toughness indexes were determined with using the ASTM C 1609 standard.

3   Results and Discussion

3.1   Fresh Properties

BSCC was assessed for fresh properties including filling ability (slump flow) and passing ability (L-box). The fresh properties test results are shown in Table 4, and typical slump flow and L-box are shown in Fig. 3.

Table 4. Fresh state properties of SCCs.

Designation	SCC	BSCC-0.025	BSCC-0.05	BSCC-0.075	BSCC-0.125	BSCC-0.25
Slump flow (mm)	722.5	680	613.5	575	552	478.5
T_{500} (s)	4.1	4.4	5	5.5	5.9	—
L-box, PA	0.94	0.89	0.87	0.84	0.81	0.61

The GBBS based SCC control mixture design achieved the slump flow spread with the value of 722.5 mm. The slump flow value was within the range of 552–722.5 mm which complied with the requirement of European guidelines [20], except the BSCC-0.25 mixture with the highest content of basalt fibers for which the slump flow diameter reached 478.5 mm. According to European guidelines [20], the SCC and BSCC-0.025 mixtures designs were classified as SF2 class as the slump flow results fell within the range of 660–750 mm. The mixtures of class SF2 are suitable for normal application including walls and columns. On the other hand the BSCC-0.05, BSCC-0.075 and BSCC-0.125 mixtures designs were classified as SF1 class as the slump flow results fell within the range of 550–650 mm. These three mixtures are appropriate for housing slabs, tunnel linings, piles and some deep foundations. Clearly, all mixtures of basalt fiber reinforced concrete produced final diameter lower than the control mixture, indicating a negative effect flow resulting from adding BF. The effect of increased BF content on deteriorated flowability is consistent with the previous findings [7, 8, 13]. The T_{500} time shown in Table 4 indicates mixtures were high viscosity. All test groups exhibited a flow time between 4.10 and 5.90 s and as per guidelines, all mixture groups can be classified as VS2 with regard to their viscosity. The L-box blocking ratio is obtained in between 0.61 and 0.94 for all groups of mixtures. The results of the L-box indicate that the basalt fibers decreased this ratio, but the mixtures showed results which were accepted according to the guide, except the BSCC-0.25 mixture.

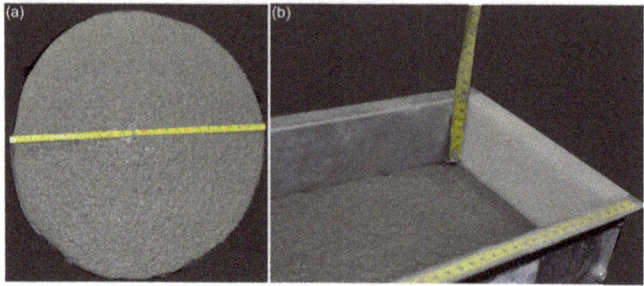

Fig. 3. Typical (a) slump flow, (b) passing ability of BSCC.

3.2   Compressive Strength

The compressive strength of the SCC mixtures containing BF was ascertained at 7 and 28 days. An average of six cubes were tested per data point. The impact of the BF content on the development of the compressive strength can be seen in Fig. 4. The lines at the top of the bars denote the standard deviation.

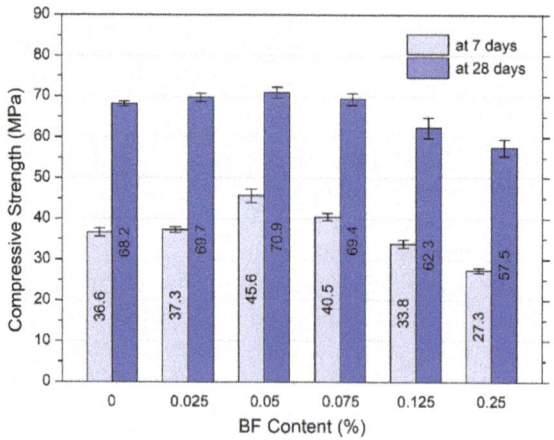

Fig. 4. Compressive strength of BSCC for different basalt fiber volume content.

The compressive strength values of BSCCs ranged between 27.3 and 37.3 MPa as well as 57.5 and 69.7 MPa after 7 and 28 days of moist curing, respectively. In BSCC series, the lowest compressive strength was observed in 0.25% basalt fiber added mixture. The highest compressive strength result was obtained in BSCC-0.05 series with an increase by 24.6% and 3.96% after 7 and 28 days compared to control mixture SCC. The BSCC mixture having 0.25% BF content had lower compressive strength in comparison with BSCCs having 0.025% BF content. These differences in BSCCs compressive strength can be rather attributed to the changes in the cement paste than the BF effect themselves. Basalt fibers adversely impacted on the workability of fresh SCC and the voids were formed in the BSCCs. In the matter of the fact, voids formation is reason for the decrease in compressive strength. In contrast, increases in compressive strength were affected from the reduction of W/B ratio due to the water absorption by the basalt fiber [5, 19].

3.3 Splitting Tensile Strength

Splitting tensile strength examinations were carried out at 7 and 28 days. Figure 5 plots the splitting tensile strength against the BF content level.

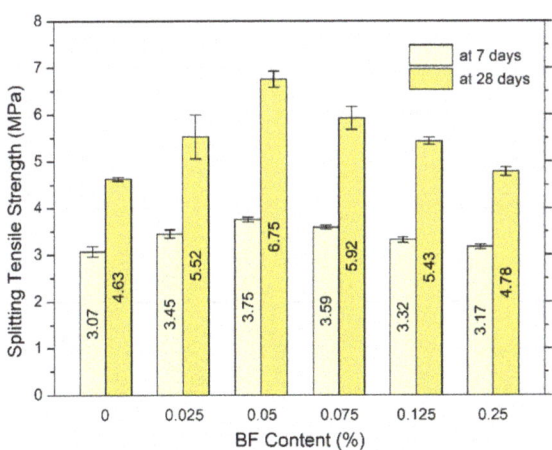

Fig. 5. Splitting tensile strength of BSCC for different basalt fiber volume content.

The results illustrate that with age the flexural strength rises. According to the splitting tensile strength test values, the strength of BSCCs ranged between 4.63 and 6.75 MPa at 28 days of moist curing. All BSCC mixtures exhibited rise in splitting tensile strength, in comparison to the control at each curing period. The highest improvement in splitting tensile strength was observed in BSCC-0.05 by 22.15% and 45.79% at 7 and 28 days, respectively. When the basalt fiber volume contents were increased to 0.075%, 0.125% and 0.25%, a gradual reduction in splitting tensile strength was observed compared to previous content.

3.4 Flexural Strength

Flexural strength tests were carried out at 7 and 28 days. Figure 6 shows the differences in the flexural strength of BSCC with the various BF contents.

These variation with any amounts of BF were quite alike to that seen as regards the splitting tensile strength. In all curing periods, a trend of a rise in flexural strength with an increase in BF up to 0.05% followed by a decrease trend was noticed. The flexural strength values for the reference mixture SCC of 3.02 MPa and 6.41 MPa at 7 and 28 days, respectively were observed. As with the splitting tensile strength, a significant rises of roughly 82.12% and 37.75% in the results for the 0.05% BF mixture after 7 and 28 days were noted. A decline in the flexural strength of BSCC for the 0.075%, 0.125% and 0.25% BF ensued. In particular, there were decreases at 7 and 28 days of 5.82–36.18% and 9.29–22.42% for the BSCC from 0.05% to 0.25% BF.

Jiang et al. [2] reported 9.58% improvement in the flexural strength of concrete containing 0.3% basalt fibers compared to the reference mixture and decrease this property for 0.5% BF compared to 0.3% fiber content. Branston et al. [4] studied concrete mixtures having 4, 8 and 12 kg/m3 basalt fiber contents and they stated that increases in flexural strength by 8.4%, 16.05%, and 21.72%, almost linearly dependent on the BF contents. Çelik and Bingöl [6, 19] investigated properties of hardened self-compacting concrete having basalt fibers and reported that the addition of 0.2% basalt fibers provided the highest improvement in flexural strength by 11.58% compared to the control specimens.

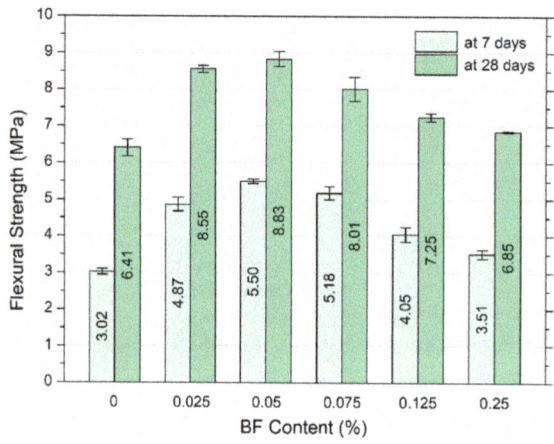

Fig. 6. Flexural strength of BSCC for different basalt fiber volume content.

Branston et al. [4] studied concrete mixtures having 4, 8 and 12 kg/m^3 basalt fiber contents and they stated that increases in flexural strength by 8.4%, 16.05%, and 21.72%, almost linearly dependent on the BF contents. Çelik and Bingöl [6, 19] investigated properties of hardened self-compacting concrete having basalt fibers and reported that the addition of 0.2% basalt fibers provided the highest improvement in flexural strength by 11.58% compared to the control specimens.

3.5 Toughness

The flexural toughness is an important parameter in assessing the influence of basalt fibers on the post-peak behaviour of SCC. The load–deflection curves as obtained in this investigation for BSCC containing different contents of BF were analysed. None of the BSCC prismatic specimens reached the intended deflection of span/150 due to specimens fail almost immediately after the formation of first crack, and therefore, the flexural toughness was obtained as total area under the load–deflection curve. Figure 7 displays the effect of the BF content on the flexural toughness development at 28 days of curing.

It can be observed that the best performance was given by BSCC containing 0.05% BF. Further addition of BF for SCC, decreases in the values of flexural toughness were observed. The highest value of flexural toughness was 3.511 kN × mm for BSCC-0.05, whereas, the lowest value of 2.254 kN × mm was obtained for control CSS without BF. It can also be concluded from Fig. 7 that self-compacting concretes containing basalt fibers gave better performance as compared to control concrete. Moreover, the flexural toughness values show more or less similar trends as presented by flexural and splitting tensile strengths.

Çelik and Bingöl [19] informed that the flexural toughness values of the self-compacting concrete mixtures with BF varied between 0.92 and 1.48 kN × mm. The results acquired in this test are somewhere about agreed with above values obtained for similar contents of BF in normal-strength SCC mixtures.

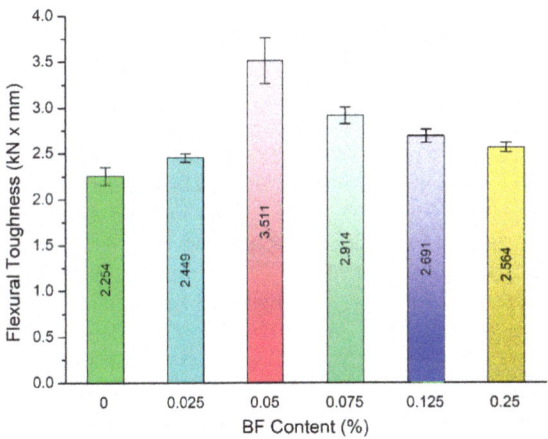

Fig. 7. Flexural toughness of BSCC for different basalt fiber volume content.

4 Conclusions

Sustainable SCC mixtures that produce high strength were developed and studied. In these mixtures up to 46% by weight of the cement was replaced with ground granulated blast furnace slag. In this study, water-to-binder ratio was maintained at 0.32, therefore, all findings pertain to this particular water-to-binder ratio. Outlined below are the major findings of the study:

- Workability of BSCCs is adversely affected with increasing basalt fiber content. Flow times to 500 mm diameter ranged from 4.1 s for SCC without fibers to 5.9 s for BSCC with 0.125% content of BF. The L-box blocking ratio was obtained in range of 0.61–0.94 and the increasing basalt fibers content decreased passing ability.
- Compressive strength increases with increasing basalt fiber content up to 0.05% after 7 and 28 days of curing. In the range of 0.075–0.25% of BF content, compressive strength decreases with increasing fiber content. The highest compressive strength values for each curing period were obtained in 0.05% basalt fiber content.
- A significant improvement in the splitting tensile strength and the flexural strength of BSCCs compared to fiber free SCC were obtained with the increase in basalt fiber content. The highest splitting tensile and flexural strength were 6.75 MPa and 8.83 MPa at 28 days of curing for BSCC-0.05 series.
- The flexural toughness values of the BSCC mixtures varied between 2.449 and 3.511 kN × mm. In addition, toughness value of BSCC-0.05 series was increased by 58% compared to SCC reference series.

Acknowledgements. This work was funded under the grant "Subvention for Science" (MEiN), project no. FN-4/2021. The author would also like to thank the CEMEX Company for donating the materials for this study.

References

1. Monaldo, E., Nerilli, F., Vairo, G.: Basalt-based fiber-reinforced materials and structural applications in civil engineering. Compos. Struct. **214**(February), 246–263 (2019)
2. Jiang, C., Fan, K., Wu, F., Chen, D.: Experimental study on the mechanical properties and microstructure of chopped basalt fibre reinforced concrete. Mater. Des. **58**, 187–193 (2014)
3. Kizilkanat, A.B., Kabay, N., Akyüncü, V., Chowdhury, S., Akça, A.H.: Mechanical properties and fracture behavior of basalt and glass fiber reinforced concrete: an experimental study. Constr. Build. Mater. **100**, 218–224 (2015)
4. Branston, J., Das, S., Kenno, S.Y., Taylor, C.: Mechanical behaviour of basalt fibre reinforced concrete. Constr. Build. Mater. **124**, 878–886 (2016)
5. Pickel, D.J., West, J.F., Alaskar, A.: Use of basalt fibres in fibre-reinforced concrete. ACI Mater. J. **115**(M79), 867–876 (2019)
6. Çelik, Z., Bingöl, A.F.: Mechanical properties and postcracking behavior of self-compacting fiber reinforced concrete. Struct. Concr. 1–10 (2019)
7. Smarzewski, P.: Flexural toughness evaluation of basalt fibre reinforced HPC beams with and without initial notch. Compos. Struct. **235**(111769), 1–12 (2020)
8. Smarzewski, P.: Comparative fracture properties of four fibre reinforced high performance cementitious composites. Materials **13**(2612), 1–16 (2020)
9. El-Chabib, H., Syed, A.: Properties of self-consolidating concrete made with high volumes of supplementary cementitious materials. J. Mat. Civ. Eng. **25**(11), 1579–1586 (2013)
10. Mohamed, O.A., Najm, O.F.: Compressive strength and stability of sustainable self-consolidating concrete containing fly ash, silica fume, and GGBS. Front. Struct. Civ. Eng. **11**(4), 406–411 (2016). https://doi.org/10.1007/s11709-016-0350-1

11. Mastali, M., Dalvand, A., Sattarifard, A.: The impact resistance and mechanical properties of the reinforced self-compacting concrete incorporating recycled CFRP fibre with different lengths and dosages. Compos. Part B **112**, 74–92 (2017)
12. Arslan, M.E.: Effects of basalt and glass chopped fibres addition on fracture energy and mechanical properties of ordinary concrete: CMOD measurement. Construct. Build. Mater. **114**, 383–391 (2016)
13. Smarzewski, P.: Flexural toughness of high-performance concrete with basalt and polypropylene short fibres. Adv. Civ. Eng. **2018**, 1–8 (2018)
14. Smarzewski, P.: Influence of basalt-polypropylene fibres on fracture properties of high performance concrete. Compos. Struct. **209**, 23–33 (2019)
15. Wang, D., Ju, Y., Shen, H., Xu, L.: Mechanical properties of high performance concrete reinforced with basalt fiber and polypropylene fiber. Constr. Build. Mater. **197**, 464–473 (2019)
16. Smarzewski, P.: Study of bond strength of steel bars in basalt fibre reinforced high performance concrete. Crystals **10**(436), 1–14 (2020)
17. Borhan, T.M.: Properties of glass concrete reinforced with short basalt fibre. Mater. Des. **42**, 265–271 (2012)
18. Algin, Z., Ozen, M.: The properties of chopped basalt fibre reinforced self-compacting concrete. Construct. Build. Mater. **186**, 678–685 (2018)
19. Çelik, Z., Bingöl, A.F.: Fracture properties and impact resistance of self-compacting fiber reinforced concrete (SCFRC). Mater. Struct. **53**(3), 1–16 (2020). https://doi.org/10.1617/s11527-020-01487-8
20. The SCC European Project Group: BIBM, CEMBUREAU, ERMCO, EFCA, EFNARC,. The European Guidelines for Self-Compacting Concrete. Specification, Production and Use (2005)

Influence of Steel Fibers and Casting Direction on the Bond Between Concrete and Reinforcement Bars

Pedro Henrique de Omena Jucá[1]([⊠]), Camila Vargas Cardoso[1],
Guilherme Durigon Cocco[1], Felipe Eduardo Kulzer[1],
Letícia Larré de Oliveira[1], Ederli Marangon[1,2],
and Luis Eduardo Kosteski[1,2]

[1] Department of Civil Engineering and MAEC Group,
Federal University of Pampa, Bagé, Brazil
[2] Engineering Post Graduated Program (PPENG) and MAEC Group,
Federal University of Pampa, Bagé, Brazil

Abstract. This research aimed to analyze the stress transfer mechanism in specimens under direct tensile stress, through the cracking process on two types of self-compacting concrete mixtures, without fibers and using steel fibers (hybrid reinforcement), which were cast horizontally and vertically. To achieve the purpose, prismatic specimens were produced and their mechanical properties evaluated through of the stiffness, load and displacement of the cracking process characterization. The specimens used dimensions of $150 \times 150 \times 750$ mm and were cast in the vertical and horizontal position with 20 mm steel bars longitudinally, centralized in the cross section. The composition of self-compacting concrete using fibers has a volumetric fraction of 1.5%. The results of experimental tests indicated different crack patterns and fracture mechanism. To mixture without fibers, the vertical direction casting group shown a cracking process synchronously on each specimen faces, demonstrating homogeneity of the mixture and less presence of voids in steel-concrete interface. Moreover, both groups without fibers, vertical and horizontal, performed a brief cracking level, quickly reaching the steel bar stiffness. On the other hand, the steel fibers presence stiffened the prismatic rods, which resulted in gradual levels of stress transfers between steel and concrete. The first crack opening in specimens of horizontal direction casting group showed increase in load and stiffening modulus, when compared to vertical group, and a steady stiffening modulus after. The vertical group performed a gradual decrease stiffening modulus in post-first crack stage. The specimens face of vertical group in the post-cracking stage showed regular and simultaneous spacing between the cracks, while the horizontal group performed an asymmetric crack pattern. Thus, the direction of horizontal casting influenced the inside distribution of fibers, resulting in a favorable mechanical orientation caused during the casting of the fresh mixture.

Keywords: Mechanical behavior · Cementitious composites · Anisotropy · Casting direction

P. Serna et al. (Eds.): BEFIB 2021, RILEM Bookseries 36, pp. 198–207, 2022.
https://doi.org/10.1007/978-3-030-83719-8_18

1   Introduction

Reinforced concrete is based on the hypothesis that both elements (steel bars/concrete) work together in the elastic regime and that changes in the strain process are inherent factors in crack control [1]. However, it is difficult to obtain a perfect relation between bar and concrete [2, 3] and improving the interface between these elements is of vital importance in conditions of maintenance, durability and safety of the structure, as they control cracking, delaying bond degradation [4]. Factors as water/cement ratio, hydration time, internal exudation, cast direction and reinforcement bars in the structural elements are relevant aspects to quantify and qualify the properties of the concrete-bar interface [5].

High performance fiber concretes, with a dense and low-porous structure, allows the element to develop more compatible levels of strain between the bar and the composite. When transmitting the efforts after the first crack through stress transfer mechanisms from fibers, the maintenance and integrity of the concrete cover around the steel bar may be ensured due to the greater confinement and multiple cracking of the matrix. Despite this, such performance is largely dependent on the control of high reinforcement strain and the shear of the bar-composite interface [6–9].

Using fibers has proven to be an essential factor in changing the mechanical properties of tension, pre-and post- first cracking of the concrete, through the distribution of concentrated stresses, maintenance of the fall of residual stresses and improvement of the steel-concrete adhesion in reinforced concrete structures [3, 10, 11]. The characteristics of the mixture flow during concreting, wall effect and fibers geometry in the composites may result in different properties of performance and heterogeneity of the mixture along the structural element [12, 13]. But, a controlled distribution of the fibers may not be guaranteed by the concreting process, implying in a different mechanical behavior among the specimens tested [14].

According to Song et al. [15], the efficiency of fibrous reinforcement in mixtures with high flow may be increased through optimized cast methods, in which the orientation and distribution of fibers are influenced by the direction and distance of casting, fiber volume and the wall effect caused by molds. The authors report that, when cast in an optimized way, the spreading of the mixture in a fresh state may divide the specimen into periods of stability or disorder, in which the fibers will have the tendency to organize themselves aligned along the flow direction or in homogeneous way, respectively.

In pullout tests, the confinement of the surrounding concrete and the failure pattern of the specimens may be modified from steel fibers addition [16, 17]. Zhang et al. [18] found that the use of oriented fibers enhanced the presence of reinforcement around the bar, which maximizes the mechanical performance of the transfer of bar-concrete stresses by 51%. According to studies by Lee [19] and Rossi et al. [20], the absorption of greater efforts in the bar-concrete adhesion and residual efforts absorbed by the fibers indicate that the use of this reinforcement may allow for the revision and, consequently, reducing the length of the transfer between bars.

Variations in the casting method also show its influence on the dispersion of fibers, on the post-cracking behavior of fiber mixtures [21, 22] and on the porosity of the reinforcement-concrete bond [23–25]. Mudadu et al. [26] studied specimens cast in two directions orthogonal to each other, analyzing the post-cracking behavior through three point flexural and uniaxial tensile tests. Their results present that, regardless of the type of test, prismatic and cylindrical specimens that were cast vertically and that they have similarities in fiber density and orientation factor resulting in similar post-cracking properties, in spite of showing lower mechanical performance than specimens molded horizontally, which were directly influenced by a favorable orientation and distribution of the fibers.

This work was proposed with the objective of experimentally evaluating the influence of casting direction on the stress transfer mechanism between steel bar and concrete composite through prismatic specimens using two different self-compacting concrete mixtures, with and without steel fibers. The mechanical properties of the rods were evaluated through of stiffness, load and displacement of the cracking process.

2   Experimental Program

2.1   Materials

The cementitious materials used were Brazilian high-early strength Portland cement (CP-V ARI RS from Votorantin) with compressive strength of 48.3 MPa, tested in according to ABNT NBR 16697 [27]; fly ash (from Presidente Médici Thermoelectric) and rice husk silica (Sílica Verde do Arroz Ltda from Pilecco Nobre Group). The aggregate mixing was composed to natural fine aggregate, coarse aggregate and silica 325 mesh (from quartz crushed). The physical properties of these materials are shown in Table 1. Powder viscosity modifier additive (Rheomac UW 410 from BASF) and polyacrylate superplasticizer with 22.7% solid content were employed to workability adjust.

Table 1. Physical properties of materials.

Constituent	Density (g/cm^3)	Maximum diameter (mm)
Coarse aggregate	2.70	9.500
Fine aggregate	2.64	1.180
Silica 325 mesh	2.68	0.045
Portland cement	2.98	
Fly ash	1.98	
Rice husk silica	2.09	

The hybrid reinforcement of the prismatic direct tension test specimens was realized with steel fibers (DRAMIX 3D 65/35 BG from ArcelorMittal) which had aspect ratio, length, diameter, and tensile strength: 65, 35 mm, 0.55 mm and 1345 MPa, respectively. The steel bars (from Gerdau) have a 20 mm diameter and strength of 500 MPa.

2.2   Mix Proportions and Mixing

For the analysis of the mechanical properties of the specimens, the experimental program described next was devise. Reference mixture (C1), without fibers, which was developed by Marangon [28] was adapted to this research using commercially available materials. Fiber volume fraction of the 1.5% was added in order to produce the second mixture (C2). The composition and characterization (for fresh and hardened state) of the mixtures are shown in Table 2 and Table 3, respectively.

Table 2. Mix composition

Constituent (kg/m^3)	C1	C2
Coarse aggregate	539.6	454.0
Fine aggregate	930.06	930.06
Cement	360	360
Fly ash	168	168
Silica 325 mesh	70	70
Rice husk silica	45	45
Superplasticizer	8.00	13.46
Water	160	150
Steel fibers	–	117
VMA	0.036	0.01217

Table 3. Rheological and mechanical properties

Properties	C1	C2
Abrams cone spread (mm) [29]	615	575
"J" ring spread (mm) [30]	570	440
"V" funnel fluidity time (s) [31]	22	59
Compressive strength (MPa)	66.33	79.89
Coefficient of variation (%)	3.31	1.44

The mixing method started with the addition and homogenization of dry constituents by a conventional mixer. Posteriorly, the water and superplasticizer addition were made and the mixed started. Finally, fibers were randomly and manually added and the mixing procedure sustained. The workability and cohesion were then evaluated and, if required, corrected by additions of superplasticizer and powder viscosity modifier additive.

2.3   Specimen Preparation

For direct tension tests with and without fibers, were produced using prismatic specimens with dimensions of $150 \times 150 \times 750$ mm (four specimens to each casting direction), which a single steel bar (20 mm diameter and 1150 mm length) was longitudinally centralized in the cross section. Regarding the analysis of the influence of the casting direction, mixtures were cast in two directions: horizontal (perpendicular casting to the steel bar) (Fig. 1.a) and vertical (parallel casting to the steel bar) (Fig. 1. b), both in central position to the open face of mold.

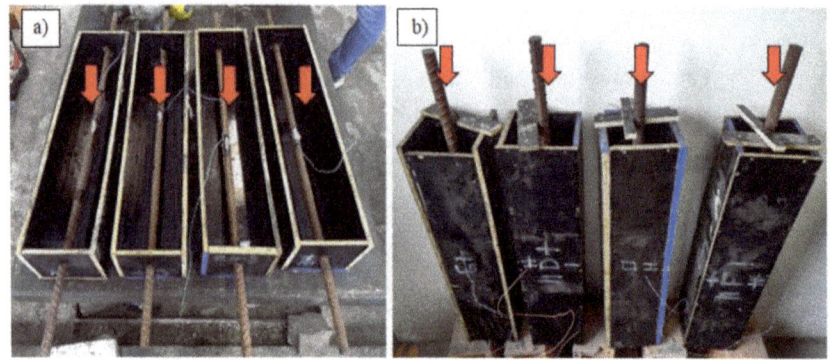

Fig. 1. Casting direction: horizontal (a) and vertical (b).

2.4   Testing

The direct tension test specimens, produced according to 2.3, were tested at a constant speed of 0.3 mm/min until the steel bar yield, and then, increased to 1 mm/min. The instrumentation set-up of the direct tension test was performed with a pair of electrical transducers to steel bar and concrete strains determination, one for each constituent, as shown in Fig. 2.

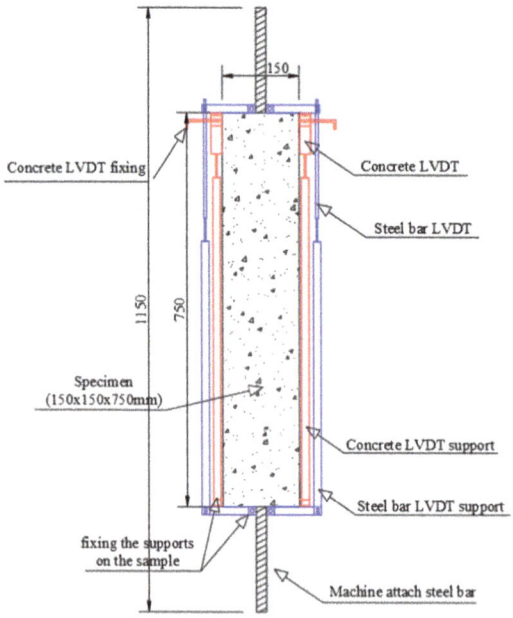

Fig. 2. Instrumentation set-up of tension test specimens.

3   Results

In this chapter are shown the results of experimental tests previously described. The direct tensile tests were performed using prismatic specimens for stiffness, load and displacement of the cracking process characterization. The analysis was carried out using four specimens in each group.

According to ABNT NBR 6118 [32], the maximum steel strain in concrete structures must be of 1%. Thus, this was adopted as a limit value in the analysis below. The load-strain curves of C1 and C2 direct tension test are presented in Figs. 3 and 4, respectively. These parameters might be evaluated through the changes that occur in the cracking process and variations in tension stiffness.

Figures 3 and 4, show blue notes which represent the first and last crack stage. Moreover, strain fields in the back (B) and front (F) faces of the specimen at these stages were included. When analyzing the results between prismatic groups, it was noticed that different cracking patterns are presented.

It may be seen through Fig. 3 that the behavior of the prismatic rods is parallel to the curve of the steel bar (in black), that is, from the cracking level the stiffness is the same as that of the bar. For both curves of mixture C1, the crack level occurs after the formation of the first crack, between the strain of 0.0039% and 0.05%. At this level, the matrix reaches the maximum tensile stress. In addition, it is worth noting that the flow level of the bar, initiated with a deformation of approximately 0.3%, and a force of approximately 175 kN.

The vertical group without fibers showed the formation of cracks that cross the specimen's section, which demonstrates greater homogeneity and mechanical capacity of the steel-concrete adhesion interface, possibly resulting in a lower presence of voids. Despite this, specimens without fibers molded horizontally present reduction in their adherence capacity and had their crack opening process occurring primarily in specimen back face.

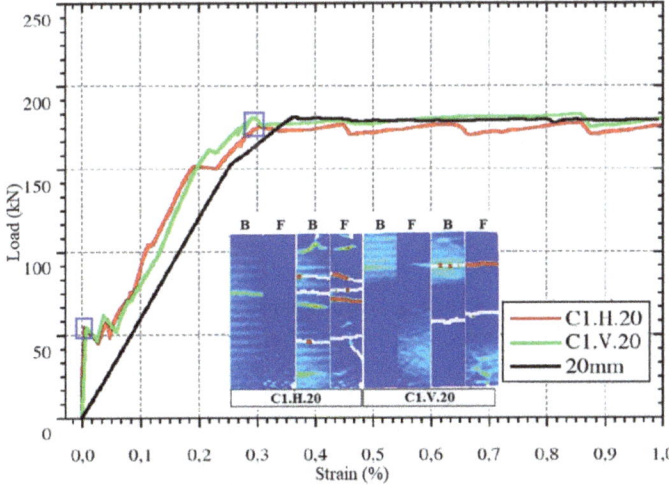

Fig. 3. Typical curves of the tensile test for self-compacting concrete without fibers.

For prismatic rods of the fiber groups (C2) (Fig. 4), the first crack opening of horizontal direction casting group showed average increase about of 75% in load and 48% in stiffening modulus when compared to vertical group. Besides, there is a smooth reduction in the stiffening modulus to the horizontal group compared to the vertical, which occurred in the first crack and remained stable after that. On the order hand, the cracking opening process of the vertical group showed a rapid decrease in stiffening modulus in the post-first crack stage.

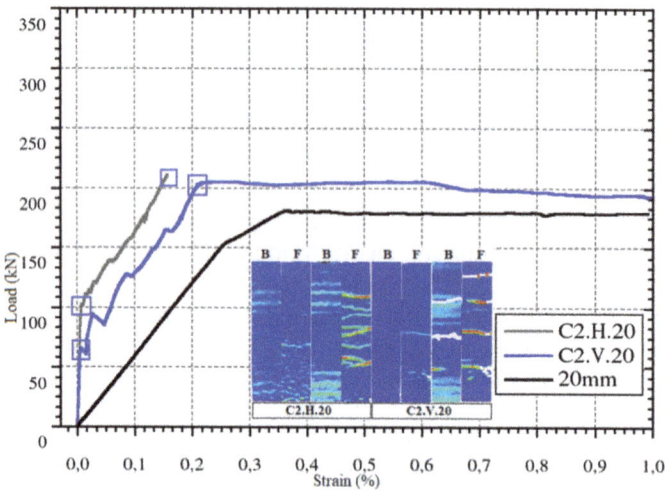

Fig. 4. Typical curves of the tensile test for self-compacting concrete with fibers.

In Fig. 4 of the self-compacting concrete C2 that specimens face of the vertical group in the post-cracking stage showed more regular and simultaneous spacing between the cracks (indicating a homogeneous distribution of fibers). On the other hand, the horizontal group performed an asymmetric behavior, with higher presence of cracks at the ends (to back face) or middle (to front face) of the specimens. Thus, the direction of casting influenced the inside distribution of fibers, caused during the spread of the fresh mixture. A favorable mechanical orientation zone (parallel direction to tests) was observed in the middle of the back face, stiffening the concrete in this range, whereas, the ends of the back face and middle of the front face showed a disorder zone. It is worth noting that the behavior found is in agreement with results already reported [15].

Due to specimen high stiffness, digital image analysis of the strain field of the C2. H.20 group was only possible from a 10x ampliation in the scale, once that the cracks are not perceptible through visual evaluation (Fig. 5.a). In the same group there was a flow of the steel bar on the outside of the prismatic rod, with its failure pattern depending on the bar's tightness (Fig. 6.a).

The confinement of the matrix caused by the insertion of fibers in C2.V.20 group caused adhesion rupture process at the ends of the prismatic rod (Fig. 6.b). This fact is related to the high absorption of strain energy around the bar, being considered a ductile failure pattern [16, 17]. Besides, its failure pattern is characterized by transverse cracks and perpendicular to the loading direction (Fig. 5.b).

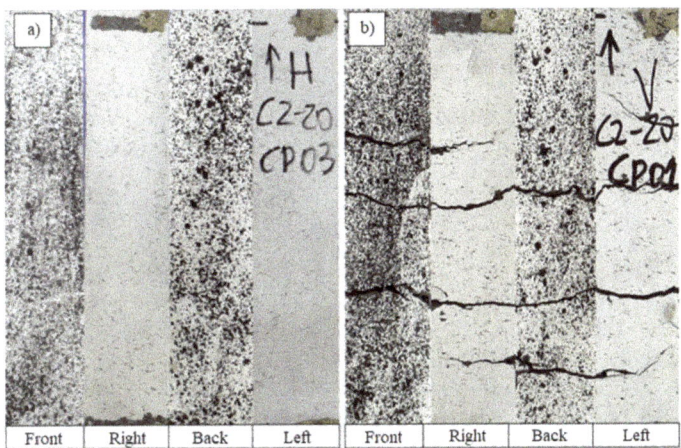

Fig. 5. Cracking map.

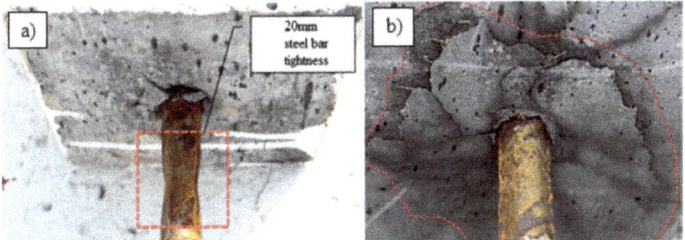

Fig. 6. Failure mode.

4 Conclusions

The influence of casting direction on the stress transfer mechanism between steel bar and concrete were investigate through direct tensile tests for two different self-compacting concrete mixtures. Based on the results obtained from the experimental tests, the following observations were obtained:

- The without fibers group performed a brief cracking level, quickly reaching the steel bar stiffness. The prismatic rods were stiffened when in steel fibers presence, which resulted in a gradual stress transfer levels in steel bar-concrete bond.
- The vertical direction casting groups shown a uniform cracking process in the cross section, with more regular and symmetric post-first crack stages. This demonstrates

that the vertical direction casting did not influence the homogeneity in both mixtures (without and with fibers).

- When cast by horizontal direction, the fiber mixture rods showed increase of load and stiffening modulus, when compared to vertical group, and an asymmetric crack pattern. This casting method resulted in a favorable orientation of fibers during the casting of the fresh mixture.

Acknowledgements. The authors would like to acknowledge the foundation of Rio Grande do Sul (FAPERGS), Coordination of Improvement of Higher-Level Personnel in Brazil (CAPES) and Nacional Council for Scientific and Technological Development (CNPq).

References

1. Dybeł, P., Kucharska, M.: Experimental assessment of the casting position factor of reinforcing bars in high performance concretes (HPC, HPSCC). Arch. Civ. Mech. Eng. **19**(1), 127–136 (2019)
2. Kim, S.W., Yun, H.D.: Evaluation of the bond behavior of steel reinforcing bars in recycled fine aggregate concrete. Cem. Concr. Compos. **46**, 8–18 (2014)
3. Huang, L., Chi, Y., Xu, L., Chen, P., Zhang, A.: Local bond performance of rebar embedded in steel-polypropylene hybrid fiber reinforced concrete under monotonic and cyclic loading. Constr. Build. Mater. **103**, 77–92 (2016)
4. Eligehausen, R., Popov, E.P., Bertero, V.V.: Local bond stress-slip relationships of deformed bars under generalized excitations (1982)
5. Horne, A.T., Richardson, I.G., Brydson, R.M.D.: Quantitative analysis of the microstructure of interfaces in steel reinforced concrete. Cem. Concr. Res. **37**(12), 1613–1623 (2007)
6. Fischer, G., Li, V.C.: Influence of matrix ductility on tension-stiffening behavior of steel reinforced engineered cementitious composites (ECC). Struct. J. **99**(1), 104–111 (2002)
7. Li, V.C.: On engineered cementitious composites (ECC). J. Adv. Concr. Technol. **1**(3), 215–230 (2003)
8. Kang, S.B., Tan, K.H., Zhou, X.H., Yang, B.: Influence of reinforcement ratio on tension stiffening of reinforced engineered cementitious composites. Eng. Struct. **141**, 251–262 (2017)
9. Nguyen, W., Bandelt, M.J., Trono, W., Billington, S.L., Ostertag, C.P.: Mechanics and failure characteristics of hybrid fiber-reinforced concrete (HyFRC) composites with longitudinal steel reinforcement. Eng. Struct. **183**, 243–254 (2019)
10. Toledo Filho, R.D., Marangon, E., de Andrade Silva, F., Mobasher, B.: Effect of steel fibers on the tensile behavior of self-consolidating reinforced concrete blocks. FRC **2014**, 609 (2014)
11. Chu, S.H., Kwan, A.K.H.: A new bond model for reinforcing bars in steel fibre reinforced concrete. Cem. Concr. Compos. **104**, 103405 (2019)
12. Zerbino, R., Tobes, J.M., Bossio, M.E., Giaccio, G.: On the orientation of fibres in structural members fabricated with self compacting fibre reinforced concrete. Cem. Concr. Compos. **34**(2), 191–200 (2012)
13. Švec, O., Žirgulis, G., Bolander, J.E., Stang, H.: Influence of formwork surface on the orientation of steel fibres within self-compacting concrete and on the mechanical properties of cast structural elements. Cem. Concr. Compos. **50**, 60–72 (2014)

14. Mínguez, J., González, D.C., Vicente, M.A.: Fiber geometrical parameters of fiber-reinforced high strength concrete and their influence on the residual post-peak flexural tensile strength. Constr. Build. Mater. **168**, 906–922 (2018)
15. Song, Q., Yu, R., Shui, Z., Wang, X., Rao, S., Lin, Z.: Optimization of fibre orientation and distribution for a sustainable Ultra-High Performance Fibre Reinforced Concrete (UHPFRC): experiments and mechanism analysis. Constr. Build. Mater. **169**, 8–19 (2018)
16. Zhao, M., Zhang, X., Yan, K., Fei, T., Zhao, S.: Bond performance of deformed rebar in steel fiber reinforced lightweight-aggregate concrete affected by multi-factors. Civ. Eng. J. **3**, 276–290 (2018)
17. Majain, N., Rahman, A.B.A., Adnan, A., Mohamed, R.N.: Pullout behaviour of ribbed bars in self-compacting concrete with steel fibers. Mater. Today: Proc. **39**, 1034–1040 (2020)
18. Zhang, W., Zou, C., Han, X., Zhang, X., Ikechukwu, O.: Bond performance between magnetized rebar and self-compacting steel fiber reinforced concrete. J. Adv. Concr. Technol. **17**(12), 686–699 (2019)
19. Lee, J.K.: Bonding behavior of lap-spliced reinforcing bars embedded in ultra-high strength concrete with steel fibers. KSCE J. Civ. Eng. **20**(1), 273–281 (2016)
20. Rossi, C.R., Oliveira, D.R., Picanço, M.S., Pompeu Neto, B.B., Oliveira, A.M.: Development length and bond behavior of steel bars in steel fiber–reinforced concrete in flexural test. J. Mater. Civ. Eng. **32**(1), 04019333 (2020)
21. Torrijos, M.C., Barragán, B.E., Zerbino, R.L.: Placing conditions, mesostructural characteristics and post-cracking response of fibre reinforced self-compacting concretes. Constr. Build. Mater. **24**(6), 1078–1085 (2010)
22. Ferrara, L., Ozyurt, N., Di Prisco, M.: High mechanical performance of fibre reinforced cementitious composites: the role of "casting-flow induced" fibre orientation. Mater. Struct. **44**(1), 109–128 (2011)
23. Jirsa, J.O., Breen, J.E.: Influence of casting position and shear on development and splice length–design recommendations (1981)
24. Dybel, P.: Effect of bond conditions on local bond-slip relationships of ribbed bars in high performance self-compacting concrete. Arch. Civ. Mech. Eng. **19**(4), 1399–1408 (2019)
25. Dybel, P., Kucharska, M.: Effect of bottom-up placing on bond properties of high-performance self-compacting concrete. Constr. Build. Mater. **243**, 118182 (2020)
26. Mudadu, A., Tiberti, G., Germano, F., Plizzari, G.A., Morbi, A.: The effect of fiber orientation on the post-cracking behavior of steel fiber reinforced concrete under bending and uniaxial tensile tests. Cem. Concr. Compos. **93**, 274–288 (2018)
27. ABNT, NBR 16697: Portland cement – Requirements. Associação Brasileira de Normas Técnicas, vol. 12 (2018)
28. Marangon, E.: Caracterização Material e Estrutural de concretos autoadensáveis reforçados com fibras de aço. Rio de Janeiro COOPE/UFRJ (2011)
29. ABNT, NBR 15823-2: Self-consolidating concrete Part 2: Slump-flow test, flow time and visual stability index - Abrams cone method. Associação Brasileira de Normas Técnicas, vol. 5 (2017)
30. ABNT, NBR 15823-3: Self-consolidating concrete Part 3: Determination of the passing ability - J-ring method. Associação Brasileira de Normas Técnicas, vol. 4 (2017)
31. ABNT, NBR 15823-5: Self-consolidating concrete Part 5: Determination of the viscosity - V-funnel test. Associação Brasileira de Normas Técnicas vol. 4 (2017)
32. ABNT, NBR 6118: Design of concrete structures—procedure. Associação Brasileira de Normas Técnicas, vol. 238 (2014)

Evaluation of Behavior of the Joint Between Two Concrete Layers During Splitting

Jakob Šušteršič[(⊠)], Rok Ercegovič, David Polanec, and Andrej Zajc

IRMA Institute for Research in Materials and Applications, Ljubljana, Slovenia
jakob.sustersic@irma.si

Abstract. The evaluation of behavior of the joint between two concrete layers during splitting was given on the basis of the results of wedge splitting tests (WST). The wedge splitting test was performed on a cube with an edge of 20 cm, which was made of two layers. There is a joint between the layers that is loaded by the splitting force. The upper (second) layer of concrete was placed after 2, 3, 4, 7 and 10 days. The WST was performed at the age of the second layer of 28 and 56 days. An important conclusion of these investigations is that the joint must be treated, with a roughness that ensures the resistance of the joint against crack propagation. The fibres have a big impact on this.

Keywords: Joint · Wedge splitting test · Roughness · Steel fibers

1 Introduction

The evaluation of behavior of the joint between two concrete layers during splitting was discussed in the research and development project [1]. One of the main purposes of the project was to find the optimal concrete mix-proportion that would be suitable for the construction of a secondary (internal) lining of an underground silo for the storage of low and intermediate level radioactive waste. Extensive investigations of various properties of concrete have been carried out in laboratories and in the test field, where various technologies for the construction of lining were also tested.

The silo is designed as a reinforced concrete cylindrical structure with a clear diameter of 27,3 m and a height (depth) of 55 m, seen from the level of the plateau to the lower elevation point of the floor calotte and floor arch, respectively. The thickness of the lining is 1,0 m [2, 3].

Due to the design of the structure, the method of construction and the required safety and durability of the LILW silo, the most critical elements are the joints between the concreting phases or rings with a height of 2,5 m. For the long-term insulating function of the lining, it is very important to make quality joints, both in terms of tightness and mechanical resistance. Due to the amount, density, and distribution of reinforcement, which has greatly increased after the inclusion of experts in seismic safety and geo-technology, we abandoned the use of the originally planned physical barriers to block water from penetrating through joints. Also, with the increase of the reinforcement, it is very difficult to prepare the surface with high-pressure washing, as the reinforcement represents an obstacle in the washing process. Therefore, additional

© RILEM 2022
P. Serna et al. (Eds.): BEFIB 2021, RILEM Bookseries 36, pp. 208–219, 2022.
https://doi.org/10.1007/978-3-030-83719-8_19

investigations of joints were carried out, both in the test field and in simulations of joints in the laboratory.

The tightness of the joint, between two layers of massive concrete in the test field, was investigated by the method of measuring watertightness on drilled samples at the joint and by a water pressure test (WPT) [1]. The starting point for the implementation as well as the WPT treatment itself is the standard EN ISO 22282-3: 2012 Geotechnical investigation and testing - Hydrogeological investigations - Part 3: Pressure test in rocks.

This paper discusses the behavior of the joint between two layers of concrete during splitting based on the results of laboratory tests. A wedge splitting test (WST) method was used [4]. The method is patented by the Austrian Patent Office [5]. This method was also used to characterize the mechanical properties of bonds between cement-like materials [6]. Old-new concrete bonds with different adhesives and with different pre-treatments of the old concrete surface have been tested. For this characterization, the load-deformation curve is used, from which the specific fracture energy G_F and the maximum load F_{max} are determined. The values of these two parameters were compared with the values of the same parameters obtained by tests of homogeneous concrete.

The WST method was also used to characterize the bond between the layers of PM-SFRC (Polymer Modified Steel Fiber Reinforced Concrete) [7]. Load - CMOD (crack mouth opening displacement) curves were measured for stable crack propagation. Strength at limit of proportion f_{LP}, maximum strength f_{max} and parameters for evaluation post-crack behavior: ductility factor $1/B$ and modulus of toughness MT were determined from these curves. They are compared with those of homogeneous PM-SFRC.

2 Experimental Program

WST was performed on a cube with an edge of 20 cm, which was made of two layers (Fig. 1).

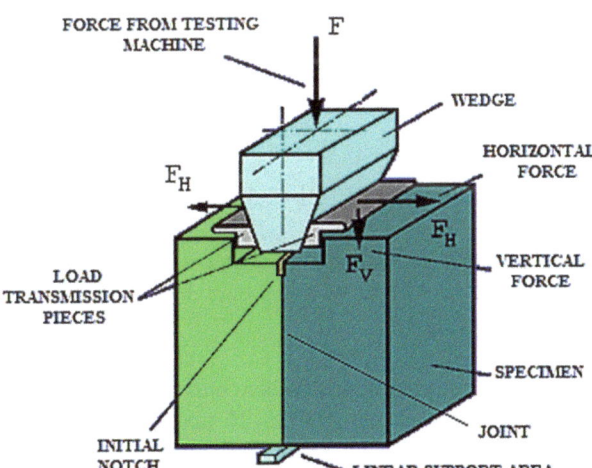

Fig. 1. Schematically view of WST of cube made of two concrete layers.

There is a joint between the layers that is loaded by the splitting force. At the top of the joint, a notch runs along its entire length. During the action of the splitting force, stresses are concentrated at the top of the notch. When these stresses are greater than the joint strength, a crack occurs, the propagation of which depends on the resistance of the joint against this propagation. In this way, this test can be used to determine how the joint treatment and the time at which the upper (second) layer of concrete was installed affect the strength and resistance of the joint to crack propagation.

During the test, a diagram of load - CMOD (Crack Mouth Opening Displacement) was recorded. From this diagram, the following properties were determined: tensile splitting strength at first crack (f_{fc}), maximum tensile splitting strength (f_{ct}) and equivalent tensile splitting strength up to crack width cw = 0,1, 0,2, 0,3 and 0,4 mm (f_{cw}). The resistance to crack propagation (RCP) could then be calculated according to the following equation:

$$RCP = \frac{f_{0,2}}{f_{ct}} \tag{1}$$

Concrete without fibers with designation PP-1 and SFRC (Steel Fiber Reinforced Concrete) with designation PP-1-JV were used to prepare samples. Their mix-proportions are given in Table 1.

Table 1. Mix-proportions (per 1 m^3 of concrete mixture).

(1)	(2)	(3)	(4)	(5)	(6)	(7)	(8)
Designation of concrete	w/b ratio	CEM	SF	Steel fibers	SPL	AFA	AGGD$_{max}$
	–	(kg)	(% of CEM)	% by volume	(% of CEM)	(% of CEM)	(mm)
PP-1	0,38	380	6,6	–	0,5	1,0	32
PP-1-JV	0,38	380	6,6	0,77	0,5	1,0	32

Materials:

(3) CEM – cement: 30% by mass CEM I 42,5 N SRO and 70% by mass CEM III/B 32,5 N – LH/SR;
(4) SF – silica fume slurry: average dry substance = 50,7% by mass;
(5) Steel Fibers - hooked steel fibres with length of 16 mm and diameter of 0,40 mm;
(6) SPL – superplasticizer;
(7) AFA – antifoam agent;
(8) AGG – crushed limestone aggregate divided into the following fractions: 0–2 mm (37%), 0–4 mm (16%), 4–8 mm (8%), 8–16 mm (19%) and 16–32 mm (20%).

Test specimens were prepared so that concrete is first placed up to half of the molds (the first layer with a thickness of 10 cm) (Fig. 2).

Fig. 2. The lower (first) layers of concrete in the molds. Their surfaces are treated (washed).

The models for the groove and notch were inserted in the molds before concrete was placing.

When the treated (rough) surface of the first layer was to be obtained, the surface was sprayed with an agent to prevent the setting of cement. The next day the cement paste was washed to get a rough surface. The roughness index was determined on each treated surface according to the method given in SIST EN 1766: 2002 (Fig. 3).

Fig. 3. Determination of the roughness index of the first layer of concrete.

The upper (second) layer of concrete was placed after 2, 3, 4, 7 and 10 days. WST was performed at the age of the second layer concrete of 28 and 56 days. For the purpose of comparison, specimens without joint were also prepared and tested.

3   Results and Discussion

3.1   Untreated Joint, Concrete Without Fibers

The dependence of the maximum splitting tensile strength f_{ct} on the time when the second (upper) layer of concrete was placed is given in Fig. 4.

It can be seen from Fig. 4 that f_{ct} of joint without treatment is decreasing with increasing the placement time of the second layer at concrete ages 28 and 56 days. f_{ct} of joints without treatment are much smaller than f_{ct} of homogenous concrete PP-1.

The dependence of the RCP of joint without treatment on the time when the second (upper) layer of concrete was placed is given in Fig. 5.

Figure 5 shows that the RCP decreases significantly with increasing time of placement of the second layer of concrete at its age of 28 and 56 days.

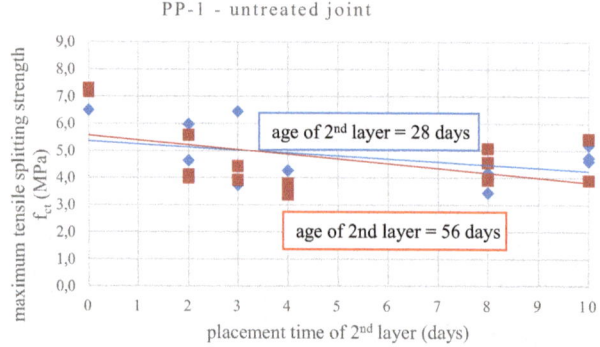

Fig. 4. Maximum splitting tensile strength f_{ct} depending on the time when the second (upper) layer of concrete was placed.

Figure 6 shows a significant reduction in the equivalent tensile splitting strengths f_{cw} of joints without treatment by increasing the crack width (cw).

The placement time of the second layer and its age at the test do not have a large effect on f_{cw}. f_{cw}, on the other hand, decrease relatively considerably with respect to the tensile splitting strength at the first crack f_{fc}. There is an even bigger difference between f_{cw} of untreated joints and f_{cw} of concrete PP-1.

Figure 7 shows a photograph of test specimens with untreated joint after the test. From the configuration of the cracks, it can be concluded that the untreated joints did not offer much resistance to crack propagation during splitting.

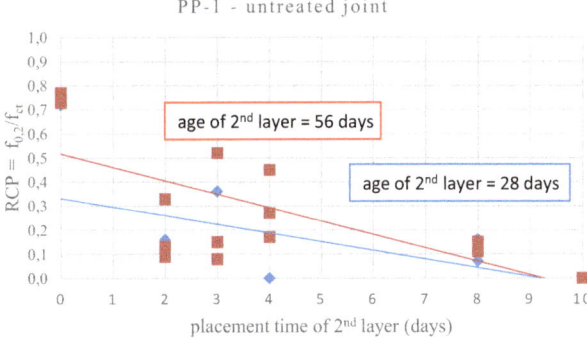

Fig. 5. RCP of untreated joint depending on the time when the second (upper) layer of concrete was placed.

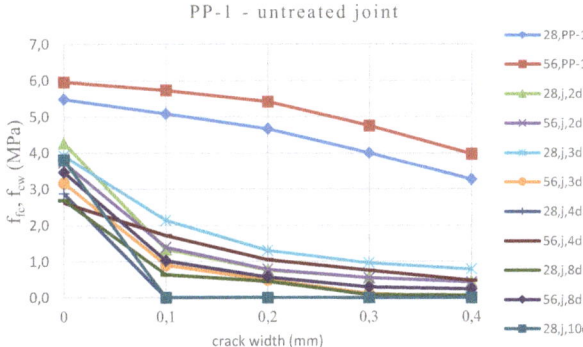

Fig. 6. Tensile splitting strength at the first crack f_{fc} and equivalent tensile splitting strength f_{cw} (cw = 0,1, 0,2, 0,3 and 0,4 mm) of untreated joints and homogenous concrete PP-1.

Fig. 7. Concrete PP-1 was placed in the second layer on second day; its age at the test was 28 days.

3.2   Treated Joint, Concrete Without Fibers

The dependence of the maximum tensile splitting strength f_{ct} on the time when the second (upper) layer of concrete was placed is given in Fig. 8.

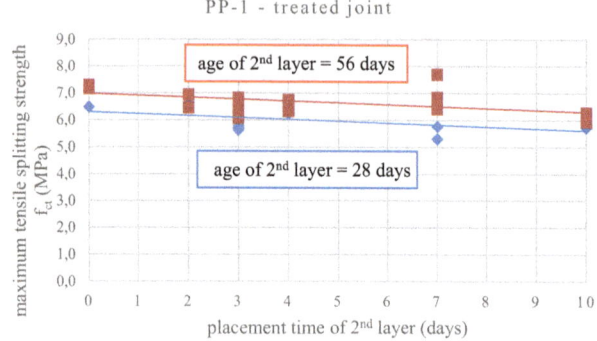

Fig. 8. Maximum splitting tensile strength f_{ct} depending on the time when the second (upper) layer of concrete was placed.

It can be seen from Fig. 8 that the f_{ct} of the treated joint in concrete PP-1 is moderately reduced by increasing the placement time of the second layer of concrete at its ages 28 and 56 days. f_{ct} of treated joints are only slightly smaller than f_{ct} of homogeneous concrete PP-1.

The dependence of the RCP of treated joint in the concrete PP-1 on the time when the second (upper) layer of concrete was placed is given in Fig. 9. The figure shows that the RCP decreases quite moderately with increasing time of placement of the second layer of concrete at its age of 28 and 56 days. This reduction in RCP is much smaller compared to untreated joint (Fig. 5). Also, the RCP of the treated joint (Fig. 9) is much higher than the RCP of the untreated joint (Fig. 5). If the concrete is placed early enough in the second layer (up to 4 days), the RCP of the treated joint is approximately equal to the RCP of the homogeneous concrete PP-1.

Figure 10 shows a moderate decrease in the equivalent tensile splitting strengths f_{cw} of treated joints by increasing the crack width (cw). The placement time of the second layer and the age of the second layer at the test did not have a large effect on the f_{cw}. However, f_{cw} decrease relatively moderately with respect to the tensile splitting strength at the first crack f_{fc}. $f_{0,1}$ is approximately equal to f_{fc}. There is very little difference between f_{cw} of treated joints and f_{cw} of homogeneous concrete PP-1.

Figure 11 shows a photograph of test specimens with treated joint after the test. From the configuration of the cracks, it can be concluded that the treated joints provided good resistance to crack propagation during splitting.

Figure 12 shows the dependence of the RCP of the treated joint on the surface roughness index. As can be seen from the figure, the RCP practically does not change in the range of roughness index from 1,6 to 3,2 mm, at the age of concrete PP-1 28 and 56 days.

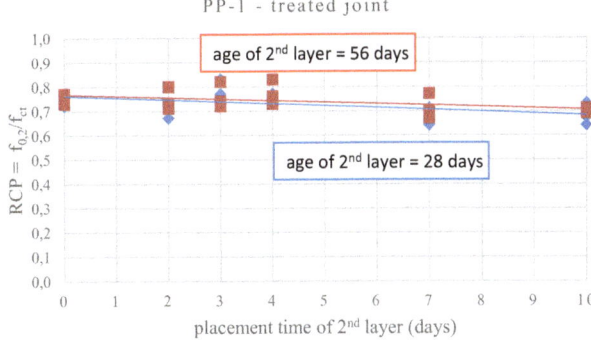

Fig. 9. RCP of treated joint depending on the time when the second (upper) layer of concrete was placed.

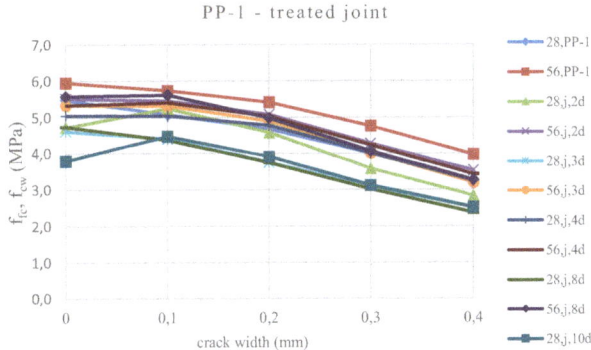

Fig. 10. Tensile splitting strength at the first crack f_{fc} and equivalent tensile splitting strength f_{cw} (cw = 0,1, 0,2, 0,3 and 0,4 mm) of treated joints and homogeneous concrete PP-1.

3.3   Treated Joint, SFRC

The dependence of the maximum tensile splitting strength f_{ct} on the time when the second (upper) layer of SFRC PP-1-JV was placed is given in Fig. 13. It can be seen from the figure that the f_{ct} of the treated joint in SFRC PP-1-JV remains almost the same at all times of placement of the second layer of SFRC at its ages 28 and 56 days. f_{ct} of treated joints are only slightly smaller than f_{ct} of homogeneous SFRC PP-1-JV.

The dependence of the RCP of treated joint in the SFRC PP-1-JV on the time when the second (upper) layer of the SFRC was placed is given in Fig. 14.

It can be seen from Fig. 14 that the RCP decreases quite moderately with increasing time of placement of the second layer of SFRC at its age of 28 and 56 days; similar to joints in the concrete PP-1 (Fig. 9), but RCP values of joints in SFRC PP-1-JV are slightly higher. If the SFRC is placed in the second layer up to 7 days, the RCP of the treated joint is approximately equal to the RCP of the homogeneous SFRC PP-1-JV.

Figure 15 shows a moderate decrease in the equivalent tensile splitting strengths f_{cw} of treated joints by increasing the crack width (cw).

Fig. 11. Concrete PP-1 was placed in the second layer on 4th day; its age at the test was 28 days.

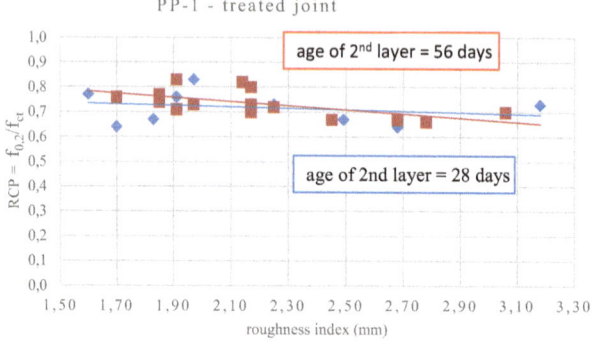

Fig. 12. RCP of the treated joint as a function of the surface roughness index.

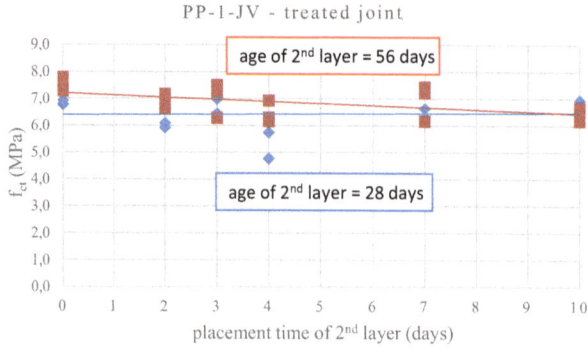

Fig. 13. Maximum splitting tensile strength f_{ct} depending on the time when the second (upper) layer of SFRC was placed.

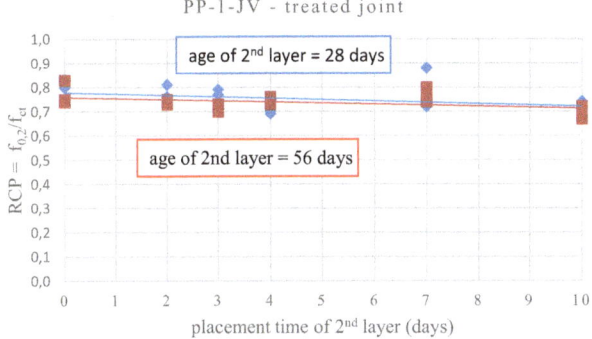

Fig. 14. RCP of treated joint depending on the time when the second (upper) layer of SFRC was placed.

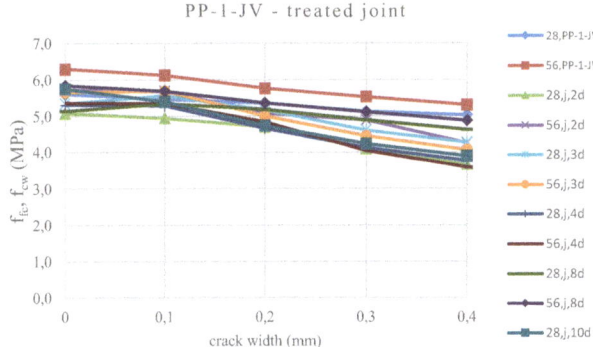

Fig. 15. Tensile splitting strength at the first crack f_{fc} and equivalent tensile splitting strengths f_{cw} (cw = 0,1, 0,2, 0,3 and 0,4 mm) of treated joints and homogeneous SFRC PP-1-JV.

This reduction is more moderate compared to f_{cw} of treated joints in the concrete PP-1 (Fig. 10). The placement time of the second layer and the age of the second layer at the test did not have a large effect on the f_{cw}. However, f_{cw} decrease relatively moderately with respect to the tensile splitting strength at the first crack f_{fc}. $f_{0,1}$ is approximately equal to f_{fc}. There is very little difference between f_{cw} of treated joints and f_{cw} of homogeneous SFRC PP-1-JV.

Figure 16 shows a photograph of test specimens with treated joint after the test. From the configuration of the cracks, it can be concluded that the treated joints provided a fairly high resistance to crack propagation during splitting.

Figure 17 shows the dependence of the RCP of the treated joint on the surface roughness index. As can be seen from the figure, the RCP does not change in the range of roughness index from 1,7 to 3,2 mm, at the age of SFRC PP-1-JV 28 and 56 days. Throughout this range, RCP values are around 0,75.

Due to the presence of fibres, the surface roughness indices of SFRC PP-1-JV are higher than the surface roughness indices of concrete PP-1 (Fig. 18), which is reflected

Fig. 16. SFRC PP-1-JV was placed in the second layer on 7th day; its age at the test was 28 days.

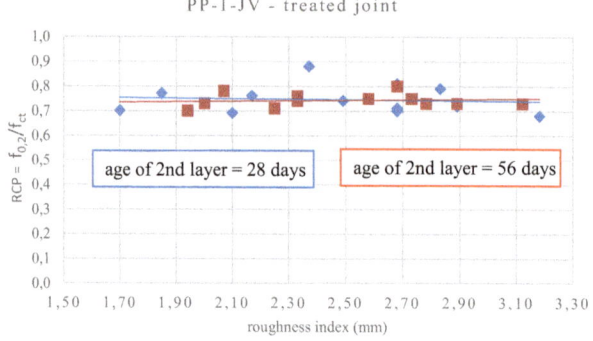

Fig. 17. RCP of the treated joint in the SFRC as a function of the surface roughness index.

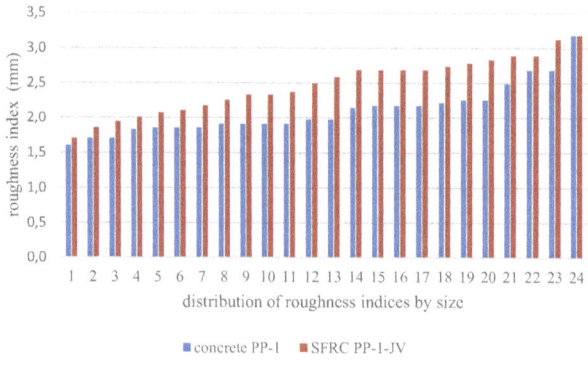

Fig. 18. Comparison of surface roughness indices of concrete PP-1 and SFRC PP-1-JV.

in the greater RCP of treated joints in the SFRC PP-1-JV during splitting. The average value of surface roughness indices of concrete PP-1 is 2,10 mm, and SFRC PP-1-JV is 2,47 mm.

4 Conclusions

An important conclusion of these investigations is that the joint must be treated, with a roughness that provides it with resistance to crack propagation (RCP). The results or findings of the investigation of the behaviour of the joint between the two layers of concrete during splitting are like those obtained by the water pressure test (WPT) in the test field.

With the roughness indices determined by the method given in SIST EN 13036-1: 2010, no correlations were found with the measured parameters of the WST. Based on the obtained results, the minimum value of the roughness index = 1.8 mm can be determined, which will ensure tightness and resistance to the crack propagation of the joint. When fibres are present on the surface of treated joint, the RCP of this joint is increased.

Acknowledgements. We would like to thank the ARAO - Agency for Radwaste Management for the successfully implemented CLS - LILW project, which fully supported its implementation. We would also like to thank the Slovenian Research Agency, which finances the implementation of the research project Massive Concrete - Technological Optimization with Advanced Experimental Methods, in which we investigate test methods used in laboratory and field tests of the CLS - LILW project.

References

1. Report of the project Study of production, placeability and characteristics of final concrete mixtures for the construction of the secondary reinforced concrete lining of the silo of the LILW repository (abbreviated name of the project: CLS-LILW), IRMA (2020)
2. Viršek, S.: Low and intermediate level waste disposal program in Slovenia. In: Proceedings of an 27th Slovenian Colloquium on Concrete, Ljubljana (IRMA, 2020), pp. 9–18 (December 2020)
3. Sinur, F., Duhovnik, B.: Low and intermediate level waste repository. In: Proceedings of an 27th Slovenian Colloquium on Concrete, Ljubljana (IRMA, 2020), pp. 41–47 (December 2020)
4. Linsbauer, H., Tschegg, E.K.: Die Bestimmung der Bruchenegie an Würfelproben (Fracture energy determination of concrete with cube-shaped specimens). Zement Beton **31**(1), 38–40 (1986)
5. Patenturkunde NR. 390328, Österreichisches Patentamt (25 April 1990)
6. Tschegg, E.K., Stanzl, S.E.: Adhesive power of bonded concrete. Fracture Processes in Concrete, Rock and Ceramics, pp. 809–818. RILEM, E. & F.N. Spon, London (1991)
7. Šušteršič, J., Šajna, A., Ukrainczyk, V., Zajc, A.: Characteristics of bond between concrete layers. In: Second International RILEM Symposium on Adhesion Between Polymers and Concrete, Dresden, pp. 83–92 (September 1999)

Short-Term Effects of Moderate Temperatures on the Mechanical Properties of Steel and Macrosynthetic Fiber Reinforced Concretes

Marta Caballero-Jorna[1]([✉]), Marta Roig-Flores[2], and Pedro Serna[1]

[1] Instituto de Ciencia y Tecnología del Hormigón,
Universitat Politècnica de València, Valencia, Spain
marcajor@upv.es
[2] Department of Mechanic Engineering and Construction,
Universitat Jaume I, Castelló de la Plana, Spain

Abstract. Fiber reinforced concrete (FRC) is a special type of concrete with improved mechanical properties due to the introduction of fibers. Macrosynthetic fibers have been recently proposed as structural reinforcement and further investigations are required to evaluate their performance in aggressive environments wherein their properties may change due to factors such as temperature.

This research examines the effect of moderate temperatures on the short-term behavior of pre-cracked and non pre-cracked FRC specimens with steel and polypropylene-based macrosynthetic fibers. A concrete type C30/37 was chosen to be as close as possible to common practice, as well as the fiber type and their content (2.1% and 0.7% respectively). The experimental campaign consists of 72 beams which were tested according to a modified procedure based on the standard EN 14651:2007 + A1:2007 in order to assess their residual flexural strengths at target temperatures (−15, 20 and 60 °C). Additionally, 18 cubes were produced to perform compression strength tests at these environmental conditions. Slump, density, air content and compression strength at standard conditions were also tested to characterize the mixes. To guarantee that the interior of the specimen has reached the target temperature during the tests, all specimens were exposed 72 h at these target temperatures before testing. During the test, the variation of internal temperatures of the specimens was greatly reduced by means of a custom insulation system. Part of the analysed beams were pre-cracked at room temperature at the age of 28 days up to a crack opening equal to 0.5 mm. These pre-cracked beams were re-loaded again at the selected moderate temperatures to investigate the effect of the temperatures on large crack mouth opening displacements in already cracked elements when compared to uncracked elements.

The results obtained show differences in the behavior depending on the type of fiber. The steel and polypropylene fiber reinforced concretes investigated in this study maintained overall good residual strength values at the temperatures selected after 3 days of exposition to moderate temperatures.

Keywords: Fiber reinforced concrete · Macrosynthetic fibers · Steel fibers · Moderate temperature · Mechanical characterization

© RILEM 2022
P. Serna et al. (Eds.): BEFIB 2021, RILEM Bookseries 36, pp. 220–231, 2022.
https://doi.org/10.1007/978-3-030-83719-8_20

1   Introduction

Over the last decades, different types of fibers have emerged as reinforcement to be introduced into the concrete matrix, resulting in fiber reinforced concretes (FRCs). Fibers often used include steel, glass, synthetic and natural fibers [1]. Their main advantage is the residual load-bearing capacity when cracked, in terms of flexural residual strength [2]. For this reason, FRCs are used for tunnel linings, bridge decks, airport pavements, slabs on grounds, industrial floors, dams, pipes, marine structures, among others.

Nowadays, a complete background about the structural contribution have been reached for steel fiber reinforced concretes (SFRC) [3], supported by standards and codes [4, 5]. However, there are many aspects of the mechanical behavior of more recent FRCs, such as macrosynthetic fiber reinforced concrete (MSFRC) that are not fully understood. In particular, the effect of moderate temperatures on the properties of MSFRC should be examined in depth to guarantee their structural contribution, both at the serviceability limit state (SLS) and ultimate limit state (ULS), in diverse conditions since the microstructure of synthetic fiber is thermally dependent and consequently, the characteristics of the concrete may change, as a result of this interaction with the environment.

Only a few investigations that covered this topic are identified in the literature. Under a hot moderate temperature range (20–60 °C), provided that the melting point of the material of the fiber is above the service temperature, deformations on fibers due to this fact might be negligible, but different experimental programs have demonstrated that MSFRCs may be affected by temperature, leading in some cases to creep failure [6].

Additionally, the variation of properties on cracked FRC with steel and macrosynthetic fibers was examined by [7]. They studied the temperature effect on the short- and long-term mechanical behavior in an extensive experimental campaign, concluding that it may play a fundamental role on the long-term behavior of MSFRC and to a lesser extent on the short-term behavior. Recently, [8] researched the flexural strength, bond strength and toughness of FRC at varying temperatures (from −20 to 60 °C). They found an increase in the performance of concrete at low temperatures as well as a minor decrease in performance at temperature of 60 °C for both steel and macrosynthetic fibers.

In the low temperature and subzero range, most of analyzed studies are carried out on conventional concrete. In [9], the authors reported that the fracture properties of a traditional concrete were improved at cryogenic temperatures, especially until −70 °C. In [10], the behavior until failure was studied under four-bending tests of large-scale reinforced concrete beams at −20 °C and room temperature. The research demonstrated that the failure load for the members tested at low temperature was around 20% higher than its coequal at room temperature and that temperature had an impact on crack widths at ultimate loads. Another research [11] studied the effect of low temperature on beams of the same dimensions of the tested in [10], but a temperature differential was applied over their depth. They showed that an increase in strength and ductility up to 13% and 34% respectively at low temperatures compared to standard temperature.

In this work, a comparative study involving MSFRC and SFRC is carried out with the aim to determine the effect of short-term moderate temperature (from −15 to 60 °C) on their mechanical properties, especially compression and residual flexural strengths, in order to determinate whether macro-synthetic fibers might be safely used in diverse environments and conditions. A markedly difference with most of the current experimental campaigns developed is that the effect of temperature is analysed under working conditions and the whole mass was at these target temperatures.

Additionally, as FRC in service could be in the cracked state and therefore, the interaction of this situation with the environment may affect the characteristics of the fibers and concretes, precracked condition is also evaluated and compared with non precracked condition.

2   Materials and Methods

2.1   Mix Design and Fiber Specifications

The concrete mix design used in the present study corresponds to a 30/37 concrete. The mix contains Portland cement type CEM I 42.5 R-SR, three coarse aggregates (4–8, 8–16 and 16–20 mm) and two types of sands (0–6 mm and 0–2 mm). Tap water was used for all mixes and two different chemical additives: superplasticizer and air-reducing admixture (0.4% vol). The percentage of volume of superplasticizer dosage was adjusted for each batch to obtain a slump of about 20 cm for all the mixes (1.2% vol. for macrosynthetic and 0.6% vol. for steel fibers). The effective w/c ratio was 0.55. Table 1 gives the details of the used concrete dosage.

Table 1. Concrete dosage used in the present study.

	Dosage (kg/m^3)
CEM I 42.5 R-SR5	280
Gravel (16–20 mm)	183
Gravel (8–16 mm)	402
Gravel (4–8 mm)	146
Sand (0–6 mm)	897
Sand (0–2 mm)	211
Water	154

Two types of fibers were used to reinforce the matrix of the concrete: polypropylene (PF) and steel (SF) fibers. To obtain similar flexural residual strengths, the fiber content was fixed in 7 kg/m^3 for MSFRCs and 21 kg/m^3 for SFRCs. Both are macro and structural fibers, according to the information provided by manufacturers. Their principal properties are collected in Table 2 and they are shown in Fig. 1.

Table 2. Fiber properties used in the present study. (Values provided by manufacturers.)

Properties	Code	
	PF	SF
Raw material	Polypropylene	Steel
Design	Embossed	Hooked end
Equivalent diameter (mm)	0.81	0.55
Length (mm)	54	35
Aspect ratio	67	65
Tensile strength (MPa)	552	1345
Modulus of elasticity (GPa)	7	210
Melting point (°C)	150–170	1375

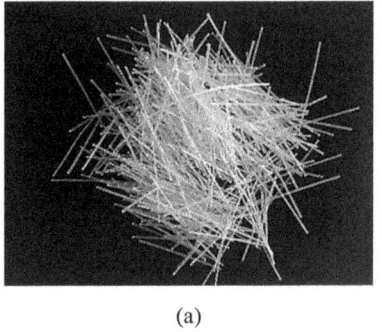   

(a)   (b)

Fig. 1. Fibers used in the present study: a) polypropylene fiber, b) steel fiber.

2.1.1   Concrete Control Properties

With the purpose of verifying the similarities of the different batches, slump, density, air content and compression strength at room temperatures were thereby tested in all the batches. The results are summarized in Table 3.

On the one hand, fresh state properties were determined. Slump was affected by the fiber type, but this was compensated with the employ of the superplasticizer. For all the cases, a high workability was achieved, being their values around 20 cm. Moreover, the results obtained from density and air content tests were similar for both types of FRCs. Values of density were around 2300–2400 kg/m^3 and the air content was <2%, in all cases. On the other hand, the mean value from three specimens and the coefficient of variation (CV) were calculated. It is important to point out that all compression strengths were uniform and around 35 MPa, except for Batch #7. This slight drop may be explained due to the weather during casting (it was an especially cold day).

Table 3. Results of the properties control of the concrete batches used.

Batch	Fiber code	Fresh state			Hardened state	
		Slump (cm)	Density (kg/m^3)	Air content (%)	Average compression strength (MPa)	CV (%)
#1	PF	22	2349.28	0.80	35.00	1.64
#2		17	2356.47	0.40	32.00	2.91
#3		21	2368.98	0.60	36.50	2.26
#4		19	2370.79	0.30	36.00	2.51
#5	SF	17	2377.19	1.10	34.00	2.11
#6		18	2373.80	0.50	35.00	5.04
#7		22	2308.35	0.40	27.00	4.69

2.2   Methods

The effect of moderate temperatures on the mechanical properties of two different groups of FRCs (specifically, reinforced with steel and macrosynthetic fibers) on short-term are investigated herein. Figure 2 sums up the implemented experimental campaign.

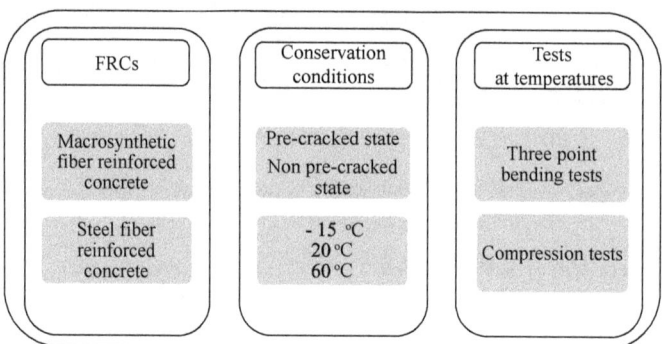

Fig. 2. Summary of the experimental campaign of the present study.

Two level of moderate temperatures were chosen: −15 and 60 °C. The performance of FRC at these temperatures will be compared with the performance at 20 °C. These temperatures were selected reaching a wider range than one defined in the Spanish regulation EHE-08 [12] for special climate conditions, in which concrete elements may be even exposed or used in specific applications. Moreover, an additional criterion in the conservation condition was considered on some specimens. Half of them were precracked at the age of 28 days at standard temperature, targeting a crack mouth opening displacement (CMOD) of 0.5 mm. Later, these specimens were stored during the same number of days than the non pre-cracked specimens at the moderate temperatures selected and all the elements, both ones and the others, were tested afterwards in the same conditions.

To evaluate the influence of the target temperatures and their interaction with the pre-cracked state after 3 days of exposure on FRCs under investigation, residual flexural strengths were determined, following EN 14651 [13]. Moreover, compression strengths tests were also performed at −15, 20 and 60 °C, following EN 12390-3 [14]. Given than the aforementioned standard methods do not contemplate the influence of the temperature factor, an insulation system, able to maintain a constant temperature during the tests, was designed and used in both cases (Fig. 3). This system is an improved version of the system used in a previous work [15].

   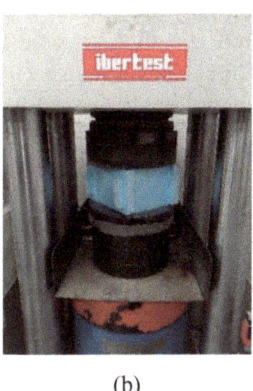

(a)   (b)

Fig. 3. Insulating thermal covering system during the tests: (a) at three point bending test and (b) at compression test.

Seven batches were produced for casting the beams and cubes specimens in a P375T08 planetary concrete mixer. Series of 6 prims of $150 \times 150 \times 600$ mm^3 and groups of 3 cubes of $150 \times 150 \times 150$ mm^3 were produced for each type of fiber, temperature and condition state, making a total of 72 prisms and 18 cubes. Three additional cubes to control and assess the compression strength at 28 days at room temperature were also produced per each batch.

Fresh-state properties tests were also determined to evaluate the quality and uniformity of the batches. After mixing procedure, slump, density, and air content were conducted, following the regulations: EN 12350-2 [16], EN 12350-6 [17] and EN 12350-7 [18] respectively.

Once manufactured the specimens, they were demoulded after 24 h and cured for 28 days in a standard humidity chamber at a temperature of 20 °C and RH 95%. After curing, the specimens from group of non-precracked specimens were kept for 72 h at their corresponding temperatures before testing. To cool the specimens until −15 °C, CH 402 T A+ horizontal freezer with automatic temperature control were used. To heat the specimens, they were submerged in water at 60 °C in thermal tanks. Specimens that were stored at 20 °C, were placed in the same chamber where they were cured. It is important to emphasise that the pre-cracking phase performed on the specimens from the group of pre-cracked specimens was carried out at 20 °C, just after curing, until reaching CMOD = 0.5 mm. After pre-cracking, these specimens were stored in the same freezer and thermal tank aforementioned to acclimate at the moderate temperatures during the same period of 3 days, before further testing at three-point bending test.

All flexural tests were performed in a universal testing machine Instron 3382 and were conducted under displacement control. To measure the CMOD, a position transducer of ±10.0 mm span range was used. Notches were cut on the middle section of all the prismatic specimens 3 days before testing. Additionally, the internal temperature of the beams was controlled. The temperature at the core of the beam was measured directly by a type K thermocouple installed at the geometric centre of the specimen, in one additional specimen from each batch. The thermocouples were embedded into the specimen during casting and continuously recorded the internal temperature via an Automatic Data Acquisition System during the tests. The temperature stability for all the specimens was ±5 °C. Other verifications were also carried out, as controlling temperature over surface before testing and over the rupture surface after testing using an infrared thermometer.

3 Results and Discussion

The mechanical properties of MSFRCs and SFRCs tested at the moderate temperatures were determined by means of mechanical tests, specifically compression and flexural tests.

3.1 Temperature Effect on Compression Strengths.

Compression strengths were evaluated at different temperatures ranging from −15 to 60 °C to show the effect of this range of temperatures on FRCs. The results are displayed in Fig. 4. A representative colour was defined for each temperature: the blue colour represents the coolest temperature (−15 °C), 20 °C is identified by the grey colour and the red colour depicts 60 °C. Polypropylene batches are those with the outlined bar. The error bars correspond to the CV. As is shown, there is a decreasing trend which is not dependent on the fiber type. When temperature increases during a short-term period of 3 days, a compression strength reduction is detected, and when temperature decreases, an increase is detected.

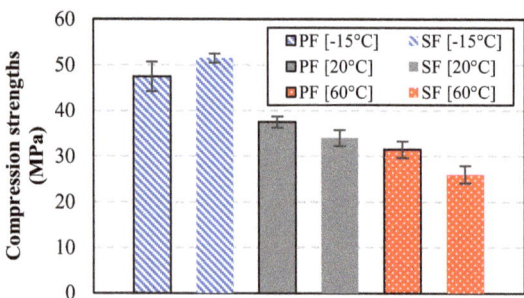

Fig. 4. Results of the compression strength tests at moderate temperatures depending on the FRCs studied.

In the literature, several authors have shown that, depending on the water content, the compressive strength at low temperatures can increase, being higher than the strength at 20 °C. These results are consistent with what was reported by [9], that explained this phenomenon through a system formed by ice and concrete. During cooling, the free water trapped in the complex network of capillary pores of the hardened cement paste solidify, gradually sealing the pores and strengthening the material. For this range of temperatures, compressive resistance can be up to three times greater than at room temperature. The results obtained in this study are in line with these results from the literature since the compression strengths experimented an increase of a 25% for MSFRC and 50% for SFRC approximately compared to the compression strengths at standard temperature. Nevertheless, for the moderate hot temperature, 60 °C, a drop of compression strengths about 15% for MSFRC and 25% for SFRC with respect these at 20 °C is detected. Further investigations are needed in this sense to clarify this fact since most of the existing literature is focused on temperatures over 100 °C.

3.2   Temperature Effect on Residual Flexural Strengths

The influence of moderate temperatures on flexural strength was analysed by comparing the peak load and the post-crack behavior through their residual strength. The mean results of the flexural strengths corresponding with the limit of proportionality (fLOP) and the residual flexural strength fR,j corresponding to CMOD = CMODj (j = 1, 3) of six specimens per group and their corresponding coefficients of variation are represented in Figs. 5, 6 and 7, respectively. The results on flexural strengths from the extra prisms with the thermocouple were not consider. The variation of the internal temperature measured was in the range of ±5 °C for all the groups, that are represented in these figures. The groups indicated as –N are those without precracking, and then, tested always at the target temperature. The groups indicated as –P are those precracked, and then, their f_{LOP} and $f_{R,1}$ were tested at room temperature and $f_{R,3}$ at the target temperature. The same colour code was assigned to depict each temperature.

Figure 5 shows the evolution of flexural strengths at the peak-load (f_{LOP}) with the temperature for both types of FRCs. From a general point of view, both MSFRC and SFRC behave in a similar manner within this range of temperature, independently on the pre-cracked and non pre-cracked states. For pre-cracked elements, the obtained values of f_{LOP} are comparable in terms of strength, but the effect of temperature is not considered in attendance since all the pre-cracked tests were performed at standard temperature. Nevertheless, an obvious change on the strengths at the limit of proportionality of non pre-cracked groups can be seen for both concretes. FRCs display a noticeable increment of f_{LOP} at low temperature (twice higher than at room temperature) and a decrease at 60 °C (over 25%), after only three days of exposure. According to [9], tensile strength effects depending on the temperature in splitting, bending or double punch tests, is similar to the effects on the compression strength, even though it can be less pronounced. On the contrary, the experimental results obtained in this work on flexural tests are as notable as on compression test, and it is logical since they are mainly dependent of the quality of the concrete matrix and hence, compression strengths.

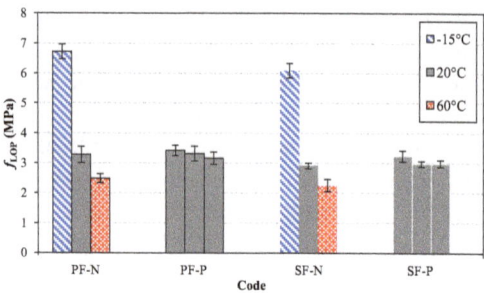

Fig. 5. Results of f_{LOP} for each condition.

Figure 6 shows the residual flexural strength at CMOD1 for the tested fibres at different target temperatures and in both pre-cracked and non pre-cracked conditions. As in the previous case, the influence of temperature is not taken into account for pre-cracked specimens due to these tests were performed at 20 °C. The outcomes of $f_{R,1}$ for the non pre-cracked SFRC and MSFRC specimens, follow the same trend than in the case of the limit of proportionality. However, it can be seen that the obtained values of $f_{R,1}$ for the non pre-cracked SFRC specimens are comparable in terms of residual flexural strengths, regardless of temperature. Regarding the behaviour at the moderate temperatures, brittle behaviour was experimented at −15 °C. Although low temperatures delay the initiation of flexural cracks at service loads because of their higher f_{LOP}, when the first crack was produced on SFRCs, specimens collapsed almost suddenly, with very small deformation and no prior warning. For MSFRCs, this behavior also occurred at low temperatures but with less intensity. This fact caused a little instability during the tests, and because of that, for following works, adjustments of the test will be investigated to ensure a better control of the process between f_{LOP} and $f_{R,1}$. At 60 °C, no significant differences are shown neither on MSFRC nor in SFRC, and the $f_{R,1}$ obtained is very similar to the $f_{R,1}$ obtained at standard temperature.

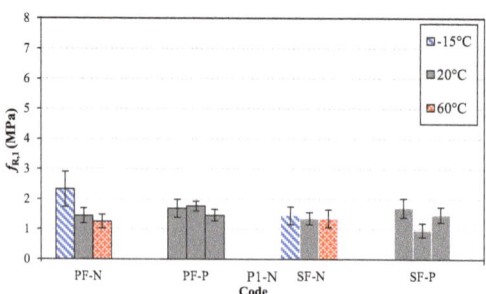

Fig. 6. Results of $f_{R,1}$ for each condition.

Figure 7 displays the residual flexural behaviour at CMOD3 at different moderate temperatures. Since this parameter has been evaluated in all cases at the moderate temperatures, it shows the difference in behaviour between pre-cracked and non pre-cracked states at the target temperatures. The residual flexural strength obtained at CMOD = 2.5 mm is similar within each type of fiber, whether the specimen was previously pre-cracked or not. Thus, no influence of pre-cracked condition is presented for large CMODs for FRC exposed during short-term periods to moderate temperatures.

Regarding temperature factor, a variation is evidenced for MSFRCs and it follows the same trend than the results at peak-load and at $CMOD_1$, obtaining the higher values at low temperature and lower at high temperature. For SFRCs, the obtained residual strengths are almost constant for all temperatures and in general, they are lower than those for MSFRCs. It may be explained because of fiber content that produces a fewer number of fibers in the cracked area of the specimen.

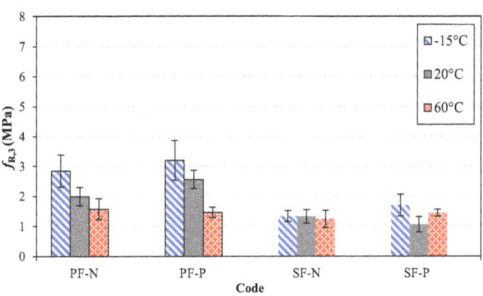

Fig. 7. Results of $f_{R,3}$ for each condition.

4   Conclusions

Reinforced concrete infrastructures are often exposed to different types of process to deterioration due to exposure conditions during their service life. Assessment of such structural damages plays a key role in safety with regards to both short- and long-term durability. This study aimed to study the effect of short-term exposure to moderate temperatures (−15 to 60 °C) on mechanical properties of macro synthetic and steel fibers reinforced concretes. The conclusions of this research are:

- An increase on the compression strengths is noted when temperature drops to −15 °C, and the opposite effect is shown when hot temperatures of 60 °C are reached, in both cases MSFRCs and SFRCs.
- Short-term exposure to the hot target temperature (60 °C) decreases, but not to a great extent, the strengths at peak-load in MSFRCs and SFRCs. The same behavior is experimented at post-crack residual strengths for both types of FRCs studied, being this more noticeable for $f_{R,3}$ values than for $f_{R,1}$ values.
- Short-term exposure to the cold target temperature (−15 °C) modifies the strengths at the peak-load in for both FRCs, being double than at obtained room temperature.

This trend is maintained for MSFRCs for small and large CMOD, showing higher residual strengths, but this increase did not happen in SFRCs. A more brittle behavior was detected at low temperatures for both types of concretes.

- The interaction of temperature with the precracked state on short-term exposure to moderate temperatures is not an influential factor for FRCs, since no significant differences are between pre-cracked and non pre-cracked elements for each type of fiber reinforced concrete.

Acknowledgements. The authors would like to express their gratitude to the Spanish Ministry of Science, Innovation and Universities for funding received under the FPU Program [FPU18/06145].

References

1. ACI Committee 544, American Concrete Institute: Report on Fiber Reinforced Concrete (2002)
2. di Prisco, M., Plizzari, G., Vandewalle, L.: Fibre reinforced concrete: new design perspectives. Mater. Struct. Constr. **42**, 1261–1281 (2009)
3. Brandt, A.M.: Fibre reinforced cement-based (FRC) composites after over 40 years of development in building and civil engineering. Compos. Struct. **86**, 3–9 (2008)
4. *fib*, Model Code for Concrete Structures 2010: Wiley-VCH Verlag GmbH & Co. KGaA, pp. 144–152 (2013)
5. Ministerio de fomento. Gobierno de España: EHE-08. Anejo 14 (2008)
6. Buratti, N., Mazzotti. C., Savoia, M., Rossi, B.: Long-term behaviour of MSFRC and SFRC. In: Fibre Concrete 2011, Prague (September 2011)
7. Buratti, N., Mazzotti, C.: Experimental tests on the effect of temperature on the long-term behaviour of macrosynthetic Fibre Reinforced Concretes. Constr. Build. Mater. **95**, 133–142 (2015)
8. Richardson, A., Ovington, R.: Temperature related steel and synthetic fibre concrete performance. Constr. Build. Mater. **153**, 616–621 (2017)
9. Rocco, C., Planas, J.: Fracture properties of concrete under cryogenic conditions. In: Borst et al. (eds) pp. 1–8 (2001)
10. Derosa, D.: Thermal effects on monitoring and performance of reinforced concrete structures. Master Thesis of Applied Science, Queen's University (October 2012)
11. Mirzazadeh, M.M., Noel, M., Green, M.F.: Effect of low temperature on the shear-fatigue performance of reinforced concrete beams. In: Proceedings, Annu. Conf. - Can. Soc. Civ. Eng., vol. 4, pp. 2682–2691 (2016)
12. Ministerio de fomento. Gobierno de España: EHE-08 Instrucción de Hormigón Estructural (2008)
13. EN 14651:2005+A1:2007. Test Method for Metallic Fibre Concrete - Measuring the Flexural tensile Strength (Limit of Proportionality (LOP), Residual), CEN - European Committee for Standardization, Brussels (2005)
14. EN 12390-3:2009, Testing hardened concrete - Part 3: Compressive strength of test specimens, CEN - European Committee for Standardization, Brussels (2009)
15. Caballero-Jorna, M., Roig-Flores, M., Serna, P.: An experimental study of the influence of moderate temperatures on the behavior of macrosynthetic fiber reinforced concrete. RILEM Bookseries **30**, 322–332 (2021)

16. EN 12350-2. Testing fresh concrete. Part 2: Slump-test, CEN - European Committee for Standardization, Brussels (2009)
17. EN 12350-6. Testing fresh concrete. Part 6: Density, CEN - European Committee for Standardization, Brussels (2019)
18. EN 12350-7. Testing fresh concrete. Part 7: Air content. Pressure methods, CEN - European Committee for Standardization, Brussels (2009)

Residual Flexural Strength of SFRC: A Multivariate Perspective

Emilio Garcia-Taengua[1(✉)], José R. Martí-Vargas[2], and Pedro Serna[2]

[1] School of Civil Engineering, University of Leeds, Leeds, UK
E.Garcia-Taengua@leeds.ac.uk
[2] ICITECH Institute of Concrete Science and Technology,
Universitat Politècnica de València, Valencia, Spain

Abstract. The main contribution of steel fibres to the hardened state performance of steel fibre-reinforced concrete (SFRC) is the residual flexural strength the material exhibits, which is commonly characterised by the residual flexural strength parameters ($f_{R1}, f_{R2}, f_{R3},$ and f_{R4}) as defined by EN 14651. A database of values of residual strength parameters corresponding to hundreds of prismatic specimens from different SFRC mix designs has been put together from previously published papers. Multiple linear regression has been applied to derive a model which relates these parameters to the steel fibres aspect ratio, length and volume fraction as well as the relative amounts of the SFRC mix constituents. The model obtained presents a very good fit to the data collected, and its relatively simple specification makes it a promising tool to optimise SFRC mix designs from the point of view of residual flexural strength. The effect of fibre dosage and dimensions and that of their interactions with other mix design parameters such as water, cement, or aggregate contents are analysed by means of response surface plots representing the average trends reproduced by the model. These modelling and analysis efforts are part of an ongoing study, and this paper focuses on the residual flexural strength parameters f_{R1} and f_{R3}. In relation to the dimensions of the fibres, the effect of fibre length on residual flexural strength has been found to be comparable to that of fibre volume fraction. This, together with the sensitivity of residual flexural strength to the fibre aspect ratio, leads to the conclusion that it is not necessary to use steel fibres in high dosages to proportion SFRC mixes with better-than-average levels of residual flexural strength. The key points emerging from the interpretation of the proposed model are presented and discussed in the context of the wide range of SFRC mixes represented by the database it is based upon.

Keywords: Data science · Mechanical properties · Residual flexural strength · Steel fibres

1 Introduction

Fibre reinforced concrete (FRC) is defined as any concrete made primarily of hydraulic cement, aggregates, and discrete reinforcing fibres [1]. Fibres are known to improve the hardened state performance of concrete, particularly in terms of mechanical properties such as tensile, flexural strength and toughness in the cracked state [2]. In fact, the residual

© RILEM 2022
P. Serna et al. (Eds.): BEFIB 2021, RILEM Bookseries 36, pp. 232–243, 2022.
https://doi.org/10.1007/978-3-030-83719-8_21

flexural strength parameters ($f_{R1}, f_{R2}, f_{R3}, f_{R4}$) and the limit of proportionality (f_L), together with compressive strength (f_c), are the basis of FRC characterisation and specification. The flexural test set-up configurations to standards EN 14651:2005 [3] and ASTM C1609/1609M [4] are shown in Fig. 1, together with an example of stress-strain curve that illustrates the limit of proportionality and the residual flexural strength parameters.

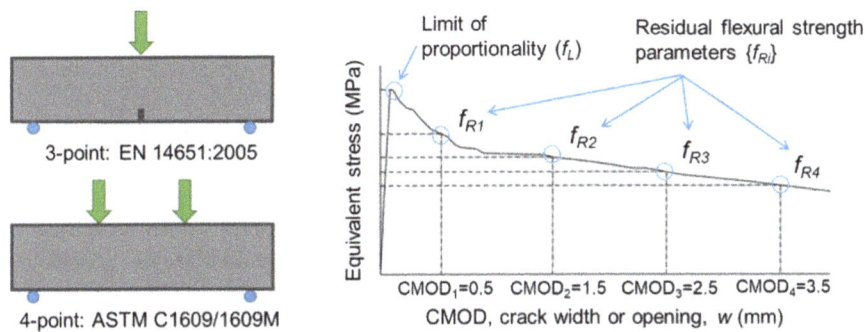

Fig. 1. Definition of residual flexural strength parameters [5].

The sensitivity of the residual flexural strength to the proportioning of the concrete mix has been rarely studied, and only in relation to specific fibres considered in the context of specific mixes. Furthermore, most studies concerned with the cracked state performance of FRC look into the effect of varying fibre contents and compare different fibre types, shapes or sizes, but the synergistic effects due to the interaction of fibres with other mix constituents are often neglected.

The two abovementioned aspects define the research gap that the work presented in this paper intends to address. It is part of an ongoing study concerned with the compilation and analysis of a database of FRC mix designs and the results of their characterisation tests [5]. In particular, this paper is concerned with steel FRC (SFRC) mixes and the analysis of the residual flexural strength parameters f_{R1} and f_{R3} in relation to the mix design. The relationships that exist between f_{R1} and f_{R3} and different variables describing the SFRC mixes (that is, relative amounts of the mix constituents and their fundamental descriptors) are analysed and quantified.

2 Dataset of SFRC Mix Designs

A dataset of steel fiber reinforced concrete (SFRC) mixtures was compiled from papers published between 2000 and 2019. The sources of information considered for this study were papers published in journals indexed in ScienceDirect® since 1999, resulting from the search with the terms "fiber-reinforced concrete" or "fibre-reinforced concrete". After an initial version of the dataset was completed, a preliminary analysis was carried out to detect and discard cases where the information was either clearly misreported or not consistent with the vast majority of the rest of the mixes. Cases where most of the information regarding the mix proportions was missing were also discarded. The resulting, final dataset comprised 765 different cases, extracted from more than 100 papers. More details on the construction of this dataset can be found elsewhere [5].

The methodological approach adopted for this analysis relied on the statistical technique known as multiple linear regression [6] in order to obtain equations that could explain f_{R1} and f_{R3} as a function of mix design variables, with two (and equally important) objectives. First, to be able to use these equations as part of the mix proportioning process. And second, to use them to produce different plots that can be used to interpret how the variation of mix design variables is associated with changes in f_{R1} and f_{R3}, and to quantify such variations.

All variables concerned with the relative proportions of the mix constituents were expressed in terms of relative weight of the constituent per unit volume of concrete, in kg/m^3. The fibre content in each mix was expressed as the volume fraction (V_f), in percentage. For each of the variables relevant to this paper, Table 1 provides the median as representative average, and the 5th and 95th percentiles as representative minimum and maximum values, respectively. A detailed descriptive analysis of the information in the SFRC dataset compiled and used in this study can be found in [5].

Table 1. Descriptive statistics of the database of SFRC mixes.

Parameter	Median	5% percentile	95% percentile
Cement (kg/m^3)	400	325	678
Additions (kg/m^3)	60	20	198
Water/cement ratio	0.45	0.22	0.60
Superplasticizer (kg/m^3)	4.0	1.3	14.0
Fibre length (mm)	45	13	60
Fibre aspect ratio	65	38	85
Fibre volume fraction (%)	0.51	0.25	2.0
Fine aggregate (kg/m^3)	835	524	1071
Coarse aggregate (kg/m^3)	880	388	1157
Max. aggregate size (mm)	15	1	20
f_{R1} (MPa)	5.3	0	21.6
f_{R3} (MPa)	4.3	0	17.6

3   Modelling of the Residual Flexural Strength

The residual flexural strength parameters f_{R1} and f_{R3} were modelled as a function of the mix design variables by means of multiple linear regression. Prior to that, the 99th percentile was calculated for each of these parameters, and the cases where they took values above this percentile were removed, thereby discarding the 1% most extreme values.

Since residual flexural strength parameters are known to be strongly correlated [5, 7], the following modelling assumption was made: the regression equations for f_{R1} and f_{R3} had to be similar and differ only in the values of their coefficients. Initial models including all pairwise interactions were considered. Statistically non-significant terms were identified and removed following the application of various model selection methods [8].

The final, refined regression equations showed very good fit to the cases in the dataset. The R-squared values were 86% and 78% for f_{R1} and f_{R3}, respectively, which are remarkably high considering that data was obtained from more than 100 different sources and that SFRC residual flexural strength parameters are known to present significant variability [9, 10].

Both regression equations had the same structure, which is shown as Eq. (1), where: f_{Ri} stands for either f_{R1} or f_{R3}, G and S are the coarse and fine aggregates contents (kg/m^3), A is the dosage of mineral additions (kg/m^3), SP is the amount of superplasticiser (kg/m^3), λ_f is the fibre aspect ratio, M is the maximum aggregate size (mm), C is the cement content (kg/m^3), L_f is the fibre length (mm), and V_f is the fibre volume fraction (percentage). The terms corresponding to the effect of fibre length, volume fraction and the total amount of aggregates were found to be dependent on other variables, which is represented by the functions noted as K_L, K_V and K_{GS} in Eq. (1). These functions are given separately in Eqs. (2) to (4). The fitted coefficients k_0 to k_6, a_0, a_1, b_0 to b_4, and c_1 to c_3 take different values for f_{R1} and f_{R3}, which are given in Table 2.

$$f_{Ri} = k_0 + k_1 \frac{G}{S} + k_2 A + k_3 SP + k_4 \lambda_f + k_5 M + k_6 C + K_L L_f + K_V V_f + K_{GS}(G+S) \quad (1)$$

$$K_L = a_0 + a_1 M \quad (2)$$

$$K_V = b_0 + b_1 C + b_2 SP + b_3 \lambda_f + b_4 L_f \quad (3)$$

$$K_{GS} = c_1 C + c_2 A + c_3 \frac{G}{S} \quad (4)$$

Table 2. Coefficients in the fitted equations.

	Coefficients in Eq. (1)						
	k_0	k_1	k_2	k_3	k_4	k_5	k_6
For $f_{Ri} = f_{R1}$	−1.61	0	0.0756	0.0963	−0.0679	0.385	−0.0112
For $f_{Ri} = f_{R3}$	1.39	−7.16	0.0207	−0.1054	−0.0668	0	0.0147
	Coefficients in Eqs. (2) and (3)						
	a_0	a_1	b_0	b_1	b_2	b_3	b_4
For $f_{Ri} = f_{R1}$	0.1842	−0.00807	−3.89	0.0056	−0.0984	0.1616	−0.0708
For $f_{Ri} = f_{R3}$	0.0894	0	−5.41	−0.0038	0.0943	0.2077	−0.0433
	Coefficients in Eq. (4)						
	c_1	c_2	c_3				
For $f_{Ri} = f_{R1}$	0.000005	−0.000044	0				
For $f_{Ri} = f_{R3}$	−0.000006	−0.000012	0.0038				

4   Analysis of the Fitted Model

The fitted model can be discussed in relation to different combinations of variables. An exhaustive examination of all possible visualisations is not attainable in one single paper. The following sections focus on the most interesting aspects, including findings in relation to variables that have usually attracted less attention such as the amount of aggregates or additions in the SFRC mix.

4.1   Fibre Content and Aspect Ratio

Figure 2 shows the response surfaces for f_{R1} and f_{R3} versus the fibre volume fraction and aspect ratio, obtained by plotting Eq. (1) and setting the rest of mix design variables to their median (Table 1).

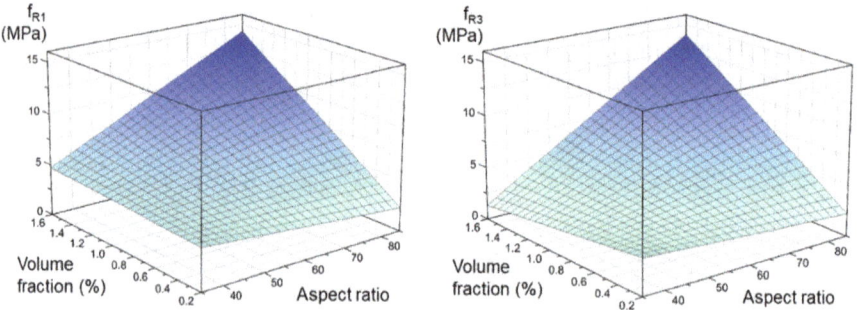

Fig. 2. Residual flexural strength vs fibre volume fraction and aspect ratio.

Both residual flexural strength parameters presented very similar trends with respect to the fibre aspect ratio and volume fraction. The model showed that the effect of fibre volume fraction on f_{R1} and f_{R3} depends on the fibre aspect ratio. Increasing the aspect ratio had an important positive effect on f_{R1} and f_{R3} for moderate to high fibre dosages, and the plots in Fig. 2 show that this effect becomes more noticeable the higher the fibre dosage is. However, this was not the case for very low volume fractions: when fibres were considered in very low contents, higher aspect ratios were not associated with better residual strength. There was, therefore, a certain volume fraction at which the trend with respect to aspect ratio changed, which was calculated by differentiating Eq. (1) with respect to the aspect ratio:

$$\frac{\partial f_{Ri}}{\partial \lambda_f} = k_4 + \frac{\partial K_V}{\partial \lambda_f} V_f = k_4 + b_3 V_f = 0 \ \rightarrow \ V_f = \frac{-k_4}{b_3} \tag{5}$$

Using the coefficient values from Table 2 in Eq. (5), the fibre dosage at which the trend with respect to the aspect ratio is reversed was 0.42% or 0.32%, for f_{R1} and f_{R3}, respectively. From this, it can be said that increasing aspect ratios were associated with increasing residual flexural strength as long as the fibre content was higher than 0.42%.

Also, for fibre contents between 0.32% and 0.42%, the effect that varying the aspect ratio has on residual flexural strength was practically negligible.

These response surfaces were also analysed in contrast with the median values of f_{R1} and f_{R3} from the database compiled for this study, which was taken as reference of the performance of an average SFRC. As Fig. 3 shows, the intersection of the response surfaces for f_{R1} and f_{R3} with their respective median planes made it possible to identify the combinations of aspect ratio and volume fraction values that were associated with better-than-average performance.

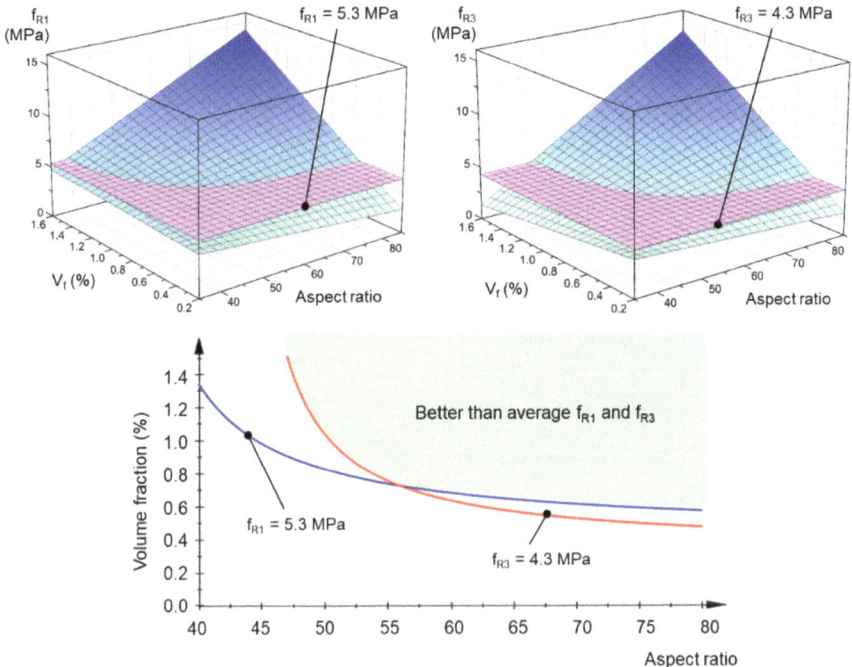

Fig. 3. Fibre dosage and aspect ratio requirements to improve residual flexural strength.

The surface plots in Fig. 3 show that improving the residual flexural strength was not necessarily linked with increasing the fibre content only. By projecting the intersections between each of these surfaces and their respective median planes unto the aspect ratio and volume fraction axes, two useful curves were obtained. For any fibre aspect ratio, these curves yield the minimum volume fraction requirement in order to achieve better-than-average performance. If fibres with an aspect ratio of 60 are considered, the fibre dosage requirement would be 0.69%, and this becomes 0.56% if the aspect ratio is 80. That is, better-than-average residual flexural strength was found to be achievable with fibre contents well below 1%. This was an interesting finding, especially bearing in mind that current trends in SFRC production indicate a preference for mixes with volume fractions not much higher than 0.5%.

4.2   Fibre Length and Maximum Aggregate Size

The surfaces for f_{R1} and f_{R3} against fibre length and maximum aggregate size shown in Fig. 4 were obtained by plotting Eq. (1) assuming the fibre volume fraction at 0.5%, 1.0% and 1.5%, and median values for the other mix design variables as per Table 1. A significant interaction between fibre length and maximum aggregate size was observed in terms of their effect on f_{R1} (Fig. 4, left). That is, the trend followed by f_{R1} with respect to the maximum aggregate size was dependent on the fibre length.

This was not the case with f_{R3} (Fig. 4, right). This was not sensitive to changes in maximum aggregate size, and higher f_{R3} values were associated with longer fibres. However, the relative effect of increasing fibre length on f_{R3} was more important at low to moderate fibre dosages. Increasing the fibre length in 10 mm was associated with an average increase of 0.68 MPa when the fibre volume fraction was 0.5%, but only 0.25 MPa when the fibre volume fraction was 1.5%.

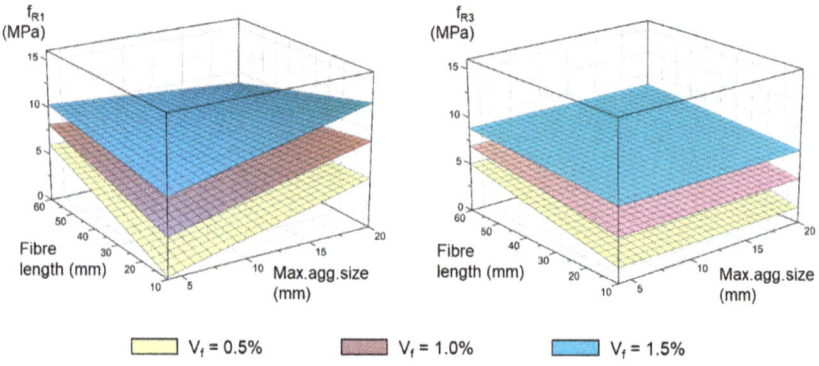

Fig. 4. Effect of maximum aggregate size and fibre length on residual flexural strength.

The discussion around f_{R1} presented more complexity (Fig. 4, left). For short fibre lengths, increasing the maximum aggregate size was associated with increasing f_{R1} values. However, with long fibres, the effect that increasing maximum aggregate sizes have on f_{R1} was the opposite. In consequence, there was a fibre length at which the trend followed by f_{R1} with respect to maximum aggregate size was reversed.

A similar observation can be made regarding the relationship between f_{R1} and the fibre length. For small values of the maximum aggregate size, increasing the fibre length was associated with an increase in f_{R1}. However, when the maximum aggregate size was higher than a certain value, longer fibres were found to reduce f_{R1} rather than improve it.

The fibre length at which the trend of f_{R1} with respect to maximum aggregate size is reversed was obtained by differentiating Eq. (1) with respect to the maximum aggregate size:

$$\frac{\partial f_{Ri}}{\partial M} = k_5 + \frac{\partial K_L}{\partial M} L_f = k_5 + a_1 L_f = 0 \ \rightarrow \ L_f = \frac{-k_5}{a_1} \tag{6}$$

Similarly, the maximum aggregate size at which the trend of f_{RI} with respect to fibre length is reversed was obtained by differentiating Eq. (1) with respect to fibre length:

$$\frac{\partial f_{Ri}}{\partial L_f} = K_L + \frac{\partial K_V}{\partial L_f} V_f = a_0 + a_1 M + b_4 V_f = 0 \ \rightarrow \ M = \frac{-a_0 - b_4 V_f}{a_1} \tag{7}$$

By using the coefficient values for $f_{Ri} = f_{RI}$ from Table 2 in Eq. (6), the fibre length at which the relationship between f_{RI} and maximum aggregate size is reversed was found to be 47.7 mm, very close to the commercially available fibre length of 45 mm. Therefore, it can be said that smaller maximum aggregate sizes were associated with higher f_{RI} values when fibres longer than 45 mm were used. Also, as the contour plots in Fig. 5 show, when the fibre length was 45 mm, f_{RI} was found to be practically insensitive to the maximum aggregate size. Interestingly, this threshold value for the fibre length was independent of the fibre content and other mix design parameters, as Eq. (6) and the contour plots in Fig. 5 show, and therefore the abovementioned considerations have general validity.

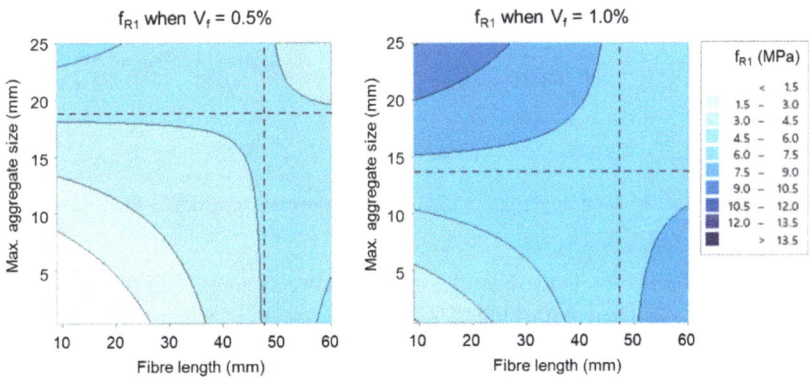

Fig. 5. Effect of maximum aggregate size, fibre length and fibre content: contour plots.

That was not the case for the maximum aggregate size that makes f_{RI} insensitive to changes in fibre length, which was a function of the fibre content as per Eq. (7). As shown in Fig. 5, for a fibre volume fraction of 0.5%, increasing the fibre length increased f_{RI} when the maximum aggregate size was not higher than 19 mm (Fig. 5, left). However, for higher fibre dosages, this limit to the maximum aggregate size decreased linearly, being 14 mm for a fibre content of 1% (Fig. 5, right), and 10 mm for a fibre content of 1.5%. Therefore, in terms of optimising f_{RI} in SFRC mixes with high fibre contents, smaller maximum aggregate sizes were found to be most effective.

4.3   Additions and Total Aggregate Content

Figure 6 shows the surfaces for f_{R1} and f_{R3} against the total contents of aggregates and additions, obtained by plotting Eq. (1) for fibre volume fractions of 0.5%, 1.0% and 1.5%, and assuming median values for the other mix design variables (Table 1). The effects of these two variables on residual flexural strength were not independent from one another, particularly in relation to f_{R1}, as the trend followed by this parameter with respect to the amount of additions varied with the total aggregate content, and vice versa. This can be observed in Fig. 6 (left). In relation to f_{R3}, however, the interaction between the total aggregate content and the amount of additions was much less relevant, as Fig. 6 (right) shows. In fact, considering the plots in Fig. 4 (right) and Fig. 6 (right), it can be said that f_{R3} was found to be practically insensitive to variations in maximum aggregate size, total aggregate content or dosage of additions, in contrast with the behaviour of f_{R1}.

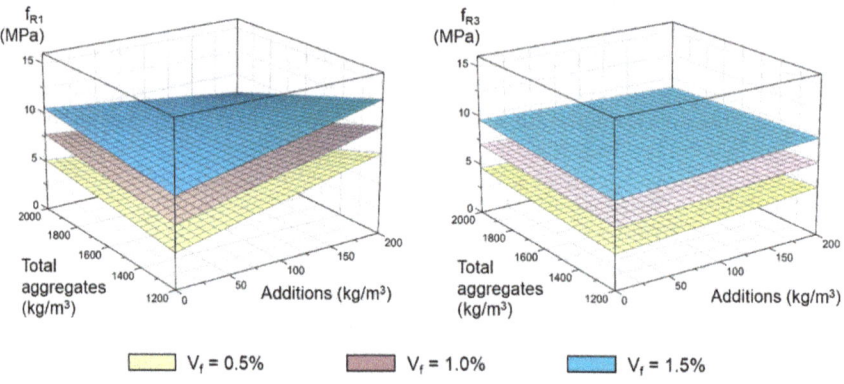

Fig. 6.   Effect of the total aggregate and additions contents on residual flexural strength.

Figure 6 (left) shows that, regardless of the fibre content, increasing dosages of additions were associated with higher f_{R1} values, especially when the total aggregate content was relatively low. However, the trend followed by f_{R1} with respect to the amount of additions was reversed when total aggregate contents on the higher end of the range were considered. There was, in consequence, a value of the total aggregate content at which the relationship between f_{R1} and dosage of additions changed. This was determined by differentiating Eq. (1) with respect to the amount of additions:

$$\frac{\partial f_{Ri}}{\partial A} = k_2 + \frac{\partial K_{GS}}{\partial A}(G+S) = k_2 + c_2(G+S) = 0 \rightarrow G+S = \frac{-k_2}{c_2}. \qquad (8)$$

Considering the coefficient values for $f_{Ri} = f_{R1}$ from Table 2 in Eq. (8), the total aggregate content at which the relationship between f_{R1} and the amount of additions is reversed was found to be 1718 kg/m^3. It is interesting to note that this value, as per Eq. (8), is independent from any other mix design parameters and, in particular, does not change with the fibre size or volume fraction.

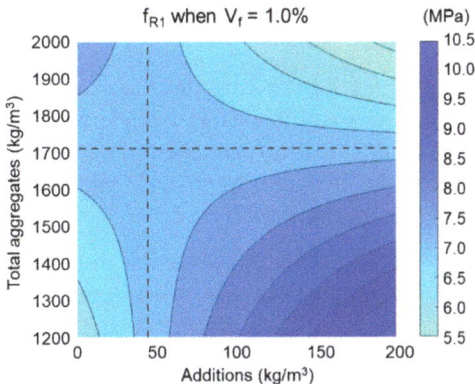

Fig. 7. Contour plot for f_{R1} versus total aggregate content and dosage of additions.

The relative effect of the aggregate content and dosage of additions on f_{R1} can be better appreciated in the contour plot shown in Fig. 7, which represents f_{R1} values as per Eq. (1) against these two parameters for a fibre volume fraction of 1.0%. The horizontal dashed line corresponds to a total aggregates content of 1718 kg/m^3. As the contour plot shows, f_{R1} was practically insensitive to the dosage of mineral additions in SFRC mixes with a total aggregates content around this value. However, in mixes with a total aggregate content lower than 1718 kg/m^3, increasing amounts of additions were associated with increasing f_{R1} values. In fact, the contour plot clearly shows that, in order to optimise f_{R1}, the introduction of mineral additions combined with a reduction in the total aggregate content is an advantageous strategy.

Conversely, the vertical dashed line in Fig. 7 corresponds to the total amount of additions which makes f_{R1} insensitive to the total aggregate content. This value was determined by differentiating Eq. (1) with respect to the total amount of aggregates, and was a function of the relative amount of cement in the SFRC mix:

$$\frac{\partial f_{R1}}{\partial (G+S)} = K_{GS} = c_1 C + c_2 A + c_3 \frac{G}{S} = c_1 C + c_2 A = 0 \quad \rightarrow \quad A = \frac{-c_1 C}{c_2} = 0.114\ C \quad (9)$$

That is, f_{R1} values were found to not be affected by changes in the total aggregate content when the amount of additions was 11.4% of the amount of cement in the SFRC mix. This corresponds to 45 kg/m^3 of additions when the cement content is considered at its median (400 kg/m^3) as per Table 1, which is the case in Fig. 7. Furthermore, as Fig. 6 (left) and Fig. 7 show, this was the minimum amount of mineral additions to be incorporated to the SFRC mix for them to have a positive effect on f_{R1} values.

In consequence, based on the dataset of SFRC mixes analysed in this study and the proposed model for residual flexural strength values, the following two general recommendations for the optimisation of f_{R1} emerged: that the total aggregate content is maintained below 1718 kg/m^3, and that mineral additions are considered in dosages of at least 11.4% of the relative amount of cement.

5 Conclusions

Based on a dataset of SFRC mixes compiled from research papers published in the last two decades, a detailed analysis and modelling is underway. This paper presents the analysis of the relationships that exist between the residual flexural parameters f_{R1}, f_{R3}, and the relative amounts of the mix constituents and their fundamental characteristics. The main conclusions can be summarised as follows:

- A regression model relating f_{R1} and f_{R3} values to the mix design parameters has been obtained. The resulting equations show a very good level of accuracy in fitting the information in the dataset (R-squared of 86% and 78%, respectively), considering the heterogeneity of the data and the variability of residual flexural strength parameters.
- The effect of fibre volume fraction on f_{R1} and f_{R3} is dependent on the fibre aspect ratio, and vice versa. For fibre contents above 0.42%, increasing aspect ratios are unequivocally associated with increasing f_{R1} and f_{R3} values.
- Fibre contents well below 1.0% should suffice to optimise the residual flexural strength and obtain f_{R1}, f_{R3} values that are better than average. The minimum dosage requirement to achieve better-than-average residual flexural strength is 0.69% for fibres with an aspect ratio of 60, and 0.56% for an aspect ratio of 80.
- When fibres with a length of 45 mm are used, f_{R1} is practically insensitive to the maximum aggregate size. However, when longer fibres are used, reducing the maximum aggregate size leads to increasing f_{R1} values. Reducing the maximum aggregate size is particularly advantageous when fibres are considered in high volume fractions.
- The total aggregate content and the amount of additions have a significant impact on f_{R1}, and their effects on this parameter are not independent from one another. On the other hand, the effect of maximum aggregate size, total aggregate content or dosage of additions on f_{R3} values is practically negligible.
- Whatever the fibre content, increasing dosages of additions are associated with higher f_{R1} values when the total aggregate content is not higher than 1718 kg/m^3. In order to optimise f_{R1}, a general recommendation derived from this study is that the total aggregate content is maintained below that threshold and the relative amount of additions is at least 11.4% the relative amount of cement.

Acknowledgements. The authors are thankful to the Concrete Research Council, the ACI Foundation and the American Concrete Institute (ACI) for the financial support awarded to the project "Optimization of Fiber-Reinforced Concrete using Data Mining" (2019–2021), with E. Garcia-Taengua as principal investigator. The support received from members of ACI Committee 544 and the industrial partners AECOM, OwensCorning and Banagher Concrete is also acknowledged.

References

1. ACI Committee 544: ACI 544.1R-96 Report on Fiber Reinforced Concrete. American Concrete Institute (1996)
2. ACI Committee 544: ACI 544.9R-17 Report on Measuring Mechanical Properties of Hardened Fiber-Reinforced Concrete. American Concrete Institute (2017)
3. European Committee for Standardization: EN 14651:2005. Test method for metallic fibre concrete. Measuring the flexural tensile strength (limit of proportionality (LOP), residual) (2005)
4. ASTM International: ASTM C1609/C1609M. Standard Test Method for Flexural Performance of Fiber-Reinforced Concrete (Using Beam with Third-Point Loading) (2012)
5. Garcia-Taengua, E.: Using Decades of Data to Rethink Proportioning and Optimisation of FRC Mixes: the OptiFRC Project, Fibre Reinforced Concrete: Improvements and Innovations. BEFIB2020. RILEM Bookseries, vol. 30, pp. 827–838 (2020)
6. James, G., Witten, D., Hastie, T., Tibshirani, R.: Linear regression. In: An Introduction to Statistical Learning. Springer Texts in Statistics, vol. 103, pp. 59–126 (2013)
7. Barros, J.A.O., Cunha, V.M.C.F., Ribeiro, A.F., Antunes, J.A.B.: Post-cracking behaviour of steel fibre reinforced concrete. Mater. Struct. **38**(1), 47–56 (2005)
8. Harrell Jr., F.E.: Multivariable modeling strategies. In: Regression Modeling Strategies, pp. 63–102. Springer (2015)
9. Cavalaro, S.H.P., Aguado, A.: Intrinsic scatter of FRC: an alternative philosophy to estimate characteristic values. Mater. Struct. **48**(11), 3537–3555 (2014). https://doi.org/10.1617/s11527-014-0420-6
10. Di Prisco, M., Plizzari, G., Vandewalle, L.: Fibre reinforced concrete: new design perspectives. Mater. Struct. **42**, 1261–1281 (2009)

Dynamic Behaviour of Steel Fiber Reinforced Concrete Plates Under Gun Fire and Free Fall Tests

Vahan Zohrabyan[✉], Tobias Seltner, Thomas Braml,
and Manfred Keuser

Chair for Concrete Construction, Bundeswehr University Munich,
Neubiberg, Germany
vahan.zohrabyan@unibw.de

Abstract. Due to an increasing threat of attacks with small arms or explosions on (government) buildings or structural facilities requiring special protection, protection against such impacts is becoming more and more important. For this reason, the Chair for Concrete Construction at the Institute of Structural Engineering at the Bundeswehr University Munich is conducting research into the improvement and development of effective structural protection systems. For this purpose, a protective layer of metal is used, which is either concreted into a steel fiber reinforced concrete slab with concrete compressive strength class of C40/50 (e.g. ring mesh) or subsequently applied to one side of the hardened slab surface (e.g. metal foam). The protective function thus achieved is validated by means of gunshot tests. Thus, in a first step, the crater ejection on the protective side, i.e. the side facing away from the bombardment, is documented. This is followed by an evaluation of the crater volume by weighing the tested specimens. The mass determined by the difference (before-after) is verified by means of a geometric control calculation.

Based on the collected results, it was shown that calibers with a dimension of 7.62 × 51 mm are absorbed by the concrete slab and thus do not pose any danger to people and material on the protective side. Despite the positive results of the protective coatings used, great research efforts are still required to optimize the protective wall panels in order to reduce the mass, size and flight distance of the concrete debris to a level that meets the safety requirements, in addition to preventing the passage of the projectile, in order to ensure the greatest possible protection.

The aim of the optimization is the best possible coordination of the material compositions, layer types, layer thicknesses and geometry of the protective layers. Based on this, the optimized protective wall panels are integrated into a completely designed and functional wall system, i.e. from the foundation to the technical control equipment for an effective and practical solution for the protection of critical infrastructures. In addition to the effectiveness of the protective wall panels, economic efficiency is a decisive criterion. Therefore, the use of materials is also to be optimized to produce the thinnest and lightest panels possible, which nevertheless fulfill all safety requirements as a result of impact loads. To investigate the behavior of the panels under impact loads, a free fall setup was developed at the institute. This allows a quantification of the panel damage due to impact loads because of differently shaped free-falling metal bodies as well as a comparison of the damage pattern between the gunshot and free fall tests.

© RILEM 2022
P. Serna et al. (Eds.): BEFIB 2021, RILEM Bookseries 36, pp. 244–254, 2022.
https://doi.org/10.1007/978-3-030-83719-8_22

Keywords: Steel fiber reinforced concrete · Crater · Gun fire tests · Protection wall · Free fall tests

1 Introduction

In general, the threat from commercial firearms, explosive blasts, and vehicle impact is increasing. This applies both domestically for public industrial facilities and on properties in foreign deployments of the German armed forces. The structural protection systems are dimensioned according to the assessed threat and the extent of the acceptable risk. They need to be constantly checked for security and, if necessary, upgraded to strengthen it. Public and industrial facilities can be, for example, embassies, government buildings, chemical and IT stations, as well as facilities for the basic supply of the population, so-called critical infrastructure [1]. To efficiently meet these growing demands, there are a variety of research areas and solution approaches.

This research work of the Chair for Concrete Structures at the Bundeswehr University Munich is part of a project that investigates different concrete compositions in combination with other materials under highly dynamic loads. The innovation here is that ring mesh or metal foam is embedded in steel fiber reinforced concrete or glued on, respectively, and thus new superstructures and compositions can be developed.

The aim is to minimize and ideally eliminate the threat posed by (secondary) debris on a wall made of this material composition to protect specifically endangered plant (parts) or technical equipment that ensures safe operation [2–4]. For this purpose, different types of concrete and compositions are tested under gun fire and exemplarily with a free fall test.

2 Test Specimens

The test series consists of a total of four different types, each with three slabs of a size of $40 \times 40 \times 6/8$ cm for testing. All twelve test specimens were produced with a concrete compressive strength of C40/50. The steel fibers added at a content of 1 Vol. % were Dramix® 4D 65/35 BG (length: 35 mm, diameter: 0.50 mm, tensile strength: 1850 N/mm^2) from Bekaert [5] or KRAMPE HAREX® DE 35/0.55 H (length: 35 mm, diameter: 0.50 mm, tensile strength: 2000 N/mm^2) [6]. The cement used for all the slabs was a CEM I 42.5 R, which was mixed with a w/c ratio of 0.5 [7]. The slab types differ with respect to their structure in that in one batch a ring mesh [8] of stainless steel was concreted horizontally in the middle of the slab, as can be seen in Fig. 1 left, during concreting. The Fig. 1 right only shows the finished concreted slab with a water stain, which results from the standard storage under water applied as for all slabs for reaching the strength class after 28 days. The aim and task of the ring mesh, which corresponds to a net, is to distribute the forces and stresses acting during an impact evenly over the surface and to retain any debris produced as far as possible by the net structure.

Fig. 1. Ring mech slab during and after concreting

The second batch is analogous to the first, but without ring mesh. Instead, after the concrete has hardened, an open-pored nickel metal foam without a PU core with dimensions of 33 × 33 × 2 cm was glued on one side with a special adhesive, Fig. 2. This side will later be the protective side during gun fire test. The metal foam, which offers plastic deformation behavior due to its structure, with simultaneous potential for absorbing large forces, is intended to hold back the resulting debris and the projectile from the concrete slab like a catch shield.

Fig. 2. Special adhesive on the plate and the finish metal foam plate before testing

Another type of plate is the construction joint. These are also slabs of 40 × 40 × 6 cm. However, first only one half of the slab was concreted with the Dramix® 4D fiber, before the second half was concreted with the KRAMPE HAREX® fiber the following day. In addition, a reinforcement grid made of 6 mm steel bars was inserted in the middle. The background to testing a transition is that if the protective wall is ultimately built around a building in the application and the construction is carried out in cast-in-place concrete and not in precast elements, there will inevitably be joints because, on the one hand, it is not possible to pour the concrete in one piece with the corresponding scope and, on the other hand, structural expansion joints have to be arranged at regular intervals. In order to simulate this, the slab was concreted in two steps, as shown in the Fig. 3 before and after production, with the intention of hitting the transition as accurately as possible during the gun fire test in order to be able to evaluate the respective behavior of the concrete sections.

Fig. 3. Plate with construction joint after the first and the second concreting

The three slabs for the first tests with the newly developed free fall test were manufactured analogously to those of the first type, ring mesh. However, the ring mesh was omitted and consisted only of the steel fiber concrete.

3 Gun Fire Test

3.1 Test Stand

The gun fire tests were carried out at the Bavarian State Office for Weights and Measures of Munich. The advantage of a gun fire channel over an open-air test is that the same ambient conditions of temperature and wind always prevail, and the applicable safety regulations are observed. The test setup, which is shown schematically in the Fig. 4, is divided into two areas. The first is the safe anteroom (1) and the channel, which is separated by a safety door. The channel has a total length of approx. 42 m. At the beginning, there is the shooting stand with suction system (2 and 3), where the ammunition is inserted. On the following approx. 10 m to the specimen there are two light barriers (4 and 5), which can measure the speed of the projectile. The plates are clamped in a frame holder (6). At the end of the channel there is an ammunition stop wall (7) which catches the projectiles if they are not held back by the test specimen.

The soft-core ammunition with a caliber of 7.62 × 51 mm used in the test has a projectile mass of 9.55 g. The impact velocity of the projectile is approx. 842 m/s.

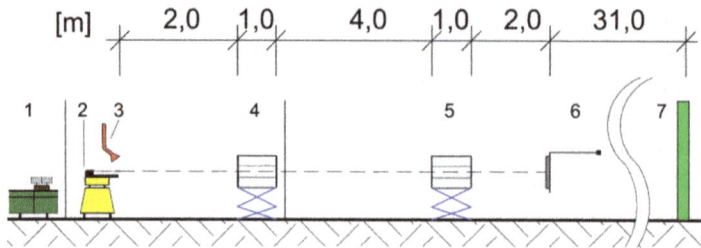

Fig. 4. Gun fire channel schematically

3.2 Evaluation and Results

The volumes were calculated, firstly, by measuring the crater (diameter or radius and depth) with subsequent calculation via a truncated cone due to the mostly existing shot through, V_k. The second volume $V_{\Delta m}$ was calculated via a differential measurement of the plate mass, which was determined before and after the gun fire respectively, and a plate density of 2.47 g/cm^3 (ring mesh) and 0.27 g/cm^3 (metal foam). Both methods have their advantages and disadvantages and especially the measurement is only an approximation, since the craters do not run ideally, which is exemplified in the Fig. 5.

Fig. 5. Not ideal circular crater

In the ring mesh type, one of the three plates were not shot through and the projectile was stopped. However, a clear crater formed on both the protection side and the load side. The volumes on both sides are approximately the same in each case, as shown in the following Table 1. The above Figure also shows a flake-like detachment from the concrete matrix, which was retained by the fibers. Due to the tensile stress given by the load on the protective side, cracks formed here in the radial direction to the perforation. These were most pronounced in the plate without perforation since more energy was dissipated here.

In the metal foam type, none of the three plates was shot through and the projectile was stopped. The craters formed only on the load side. The volumes of the craters are shown in Table 1.

The test specimens intended to simulate a construction joint were all shot through. It is noticeable that the craters with Dramix® fibers (D) on the protection side tend to be more circular. On the load side, as well as with KRAMPE HAREX® (K), the craters tend to be elongated in character, which is exemplified in the Fig. 6 left. A clear crack structure could not be determined. Mostly they ran radially with a width of up to 0.1 mm on the protective side. No cracks could be detected on the load side. The volumes are significantly larger on the protective side than on the load side, which is illustrated in the Fig. 6 right.

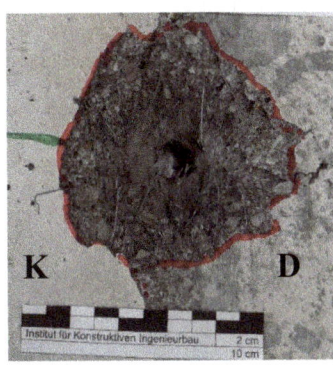
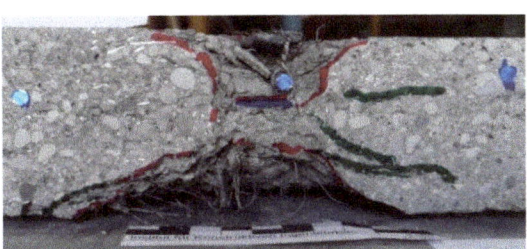

Fig. 6. Crater protection side and a cross-sectioned plate (load side is top)

The following Table 1 shows the calculation results of the volumes with both methods. In the Fig. 7 after this overview a diagram compares the volume from the protection and load side of the plates which were tested under gun fire and free fall. Here, the large difference in volume at the construction joint is particularly striking.

Table 1. Overview of the crater volume

Slab type	Volume V_k [cm^3] protection side	Volume V_k [cm^3] load side	Total V_k [cm^3]	Volume $V_{\Delta m}$ [cm^3] differential mass
Ring mesh (RM)	82.9	75.2	158.1	129.1
	84.7	94.2	178.9	185.8
	115.7	116.3	232.0	195.9
Metal foam (MF)	0.0	52.8	52.8	Not measured
	0.0	57.1	57.1	Not measured
	0.0	49.6	49.6	Not measured
Construction joint (CJ)	112.1	38.5	150.6	98.5
	108.3	66.8	175.1	42.8
	119.3	72.7	192.0	221.5
Free Fall				
Hemisphere (H)	690.1	186.1	876.2	666.1
truncated cone (TC)	753.1	224.1	977.2	956.3
Cone (C)	752.5	220.0	972.5	928.0

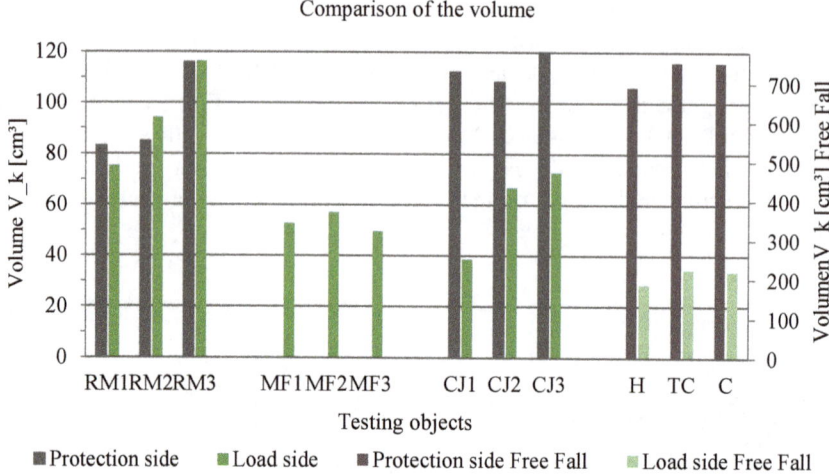

Fig. 7. Diagram of the volume comparison between load and protection side

4 Free Fall Test

4.1 Test Stand

For the newly developed test setup, shown schematically in Fig. 8 left, the potential and kinetic energy theorems were used to dimension the mass of the falling object under the geometric boundary conditions in the experimental hall. For the kinetic energy, the variables with the data of a projectile were used. This resulted in the following data, which are summarized in the Table 2. The drop body was to be made of steel and the following system of equations could be solved, where m_F (mass of drop body) is the quantity sought.

$$E_{kin} = 0.5 * m_P * v_p^2 = m_F * g * h = E_{pot} \tag{1}$$

Table 2. Overview about the data for the energy theorems

Data	Symbol	Result	Unit
Projectile mass	m_P	9.50	g
Projectile velocity	v_P	842.00	m/s
Gravitational acceleration	g	9.81	m/s²
Falling height	h	11.00	m
Steel density	ρ_S	7.85	g/cm³

This results in a theoretical mass of 31.2 kg. This was implemented using a modular cylinder (3) with three different attachments, hemisphere, cone, and truncated cone, which is shown in Fig. 8 right. The drop body was dropped through a pipe (4), which served for guidance and stability during the fall. The concrete slab was fixed to a steel frame structure (2), about 90 cm high, so that it rested only on the edge and the craters could form freely.

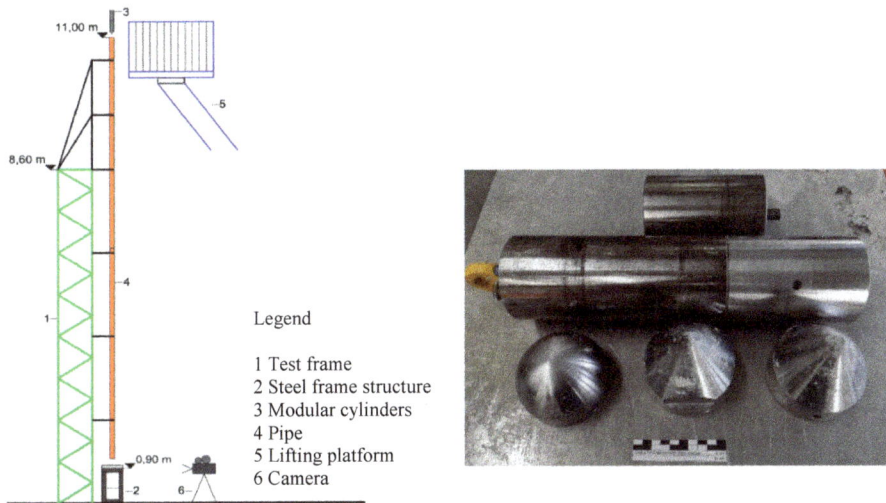

Fig. 8. Free fall test setup and steel cylinder with different attachments

4.2   Evaluation and Results

All plates showed very large destruction. On the load side, a clear punching through of the cylinder was visible. On the protective side, a clear crater formed, as well as radial cracks several millimeters wide. Due to the fastening of the plates, they cupped. One plate was completely split. The following Fig. 9 show examples of the load and protective sides of the steel fiber-reinforced concrete slab tested with a truncated cone attachment. Very large pieces of debris were detached from all slabs, as well as some debris retained by fibers in the crater. A difference can be seen between the attachments. The more pointed the impact surface, the lower the damage, since there is less surface area for force transmission at the beginning of the impact. This connection is also shown in the Table 1 and Fig. 7. The Fig. 7 shows in particular the large difference in volume between the protective side and the load side, as well as compared to the gun fire test.

 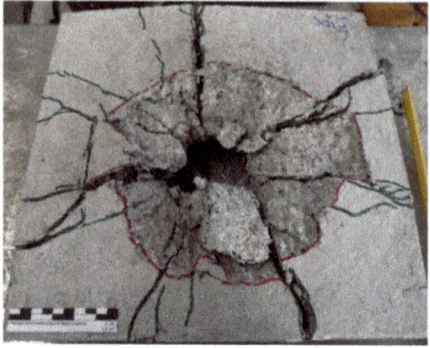

Fig. 9. Load and protective side, tested with truncated cone

5   Gun Fire and Free Fall Test in Comparison

The input variables for the free fall system are based on reference tests of gun fire or the geometric boundary conditions in the laboratory. The actual values, such as the velocity due to physical-chemical combustion processes and air resistance, vary. One of the biggest differences between these two test methods with a high influence is the velocity. She is over 57 times higher at the gunfire test. Among other things, this created a different damage pattern, and the goal is to achieve equivalent damage in the future. The correlation between different impacts on the same specimens must be found. In combination with a very small mass is the impulse smaller than at the free fall test. The masses of the individual attachments for the drop object are also not identical. The projectile comes up approximately perpendicular to the plate surface, whereas the falling object came up with up to 10° deviation from the perpendicular. This can lead to considerable deviations in the crater formation. Even the reduction of the drop height of only 0.2 m, which was finally present, would have resulted in an increase of the falling mass of 580.0 g. An essential difference is that the load direction is fundamentally different. Thus, in a gun fire test, the plate is tested perpendicular to the earth's surface, which also corresponds to the real threat. In a free fall test, the plate is necessarily horizontal. This has the disadvantage that the debris cannot fly and spread to the same extent as under gun fire. This means that no statement can be made about the flight distance, only about the number and size of the debris. The following Table 3 summarizes the essential sizes and properties of the two test methods.

Table 3. Comparison of the test methods

Characteristic	Gun fire test	Free fall tests
Projectile	soft core ammunition 7.62 × 51 mm	steel cylinder Ø 100 mm with tip, truncated cone or hemispherical attachment
Mass	9.55 g	according to energy approach: 31210 g tested with 28530 g*
Velocity	842.0 m/s ≈ 3031.0 km/h	14.7 m/s ≈ 53.0 km/h
Projectile energy	3368 J	3368 J
Distance to specimen	10.0 m	according to energy approach: 11.0 m tested with 10.8 m
Direction of bullet	horizontal	vertical
Impulse	8.0 kg*m/s	458.5 kg*m/s
*cone attachment		

The following Fig. 10 shows the debris removal from a gun fire test and a free fall test. These were take by a camera at the test stand. The left picture is a top view of the protection side and on the right a side view during the impact at the free fall test. The big debries are market with red circles and the concrete dust is marked blue. It can be clearly seen that individual larger debris and finely atomized material falls off during gun fire and a very coherent volume falls off during free fall.

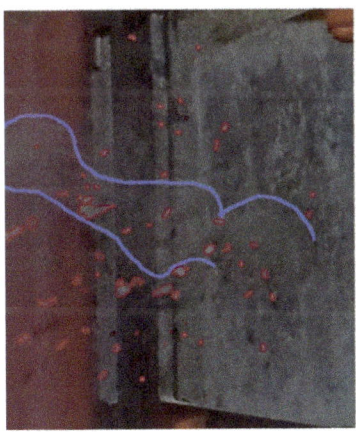 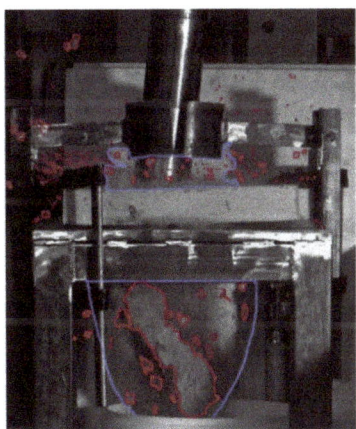

Fig. 10. Marked debris removal during impact

6   Conclusion

The aim of the research project is to develop a wall structure with a material composition that can withstand gun fire and has no or only minimal debris loss on the protective side, which was achieved.

The gun fire tests on the $40 \times 40 \times 6/8$ cm steel fiber reinforced concrete panels have shown that especially the type with bonded metal foam can absorb a projectile with a caliber of 7.62×51 mm. Furthermore, there was no debris fall on the protective side, which is relevant, so the goal is achieved. In the case of plates with ring mesh, it is basically possible to stop a projectile, but individual very large and many small debris are produced, which can fly far away. According to the results, it could be reconfirmed that steel fiber reinforced concrete improves the tensile strength and ductility of concrete elements and improves the resistance of concrete elements to dynamic loads [9]. For panels with construction joints, there is still a need for investigation. The first free fall tests show that the damage is significantly greater than with gun fire. However, it is basically possible to simulate an impact load with it.

Acknowledgement. The authors would like to thank the Munich Proof Office for the cooperation in carrying out the gun fire tests at the facility. Furthermore, the laboratory staff, in the realization of the free fall test.

References

1. Laufs, P.: Reaktorsicherheit für Leistungskernkraftwerke. Springer Verlag, Berlin-Heidelberg (2013)
2. Zohrabyan, V., Braml, Th., Zircher T., Keuser, M.: Use of steel fiber reinforced concrete for the protection of buildings against high dynamic actions. In: XV International Conference on Durability of Building Materials and Components (DBMC 2020), Barcelona, 20–23 October 2020, pp. 763–770 (2020)
3. Zircher, T., Keuser, M., Braml, Th., Berg, A., Burbach, A.: Investigations on the use of fiber concrete for infrastructure protection. In: 18th International Symposium for the Interaction of Munitions with Structures, Panama City Beach, 21–25 October 2019
4. Zohrabyan, V., Braml, Th., Zircher, T., Keuser, M.: The residual load bearing capacity of reinforced concrete as well as steel fiber reinforced concrete components after contact detonation. In: 18th International Symposium for the Interaction of Munitions with Structures, Panama City Beach, 21–25 October 2019
5. BEKAERT: Dramix® Datasheet 4D 65/35 BG (2020)
6. KRAMPEHAREX GmbH and Co. KG: Datenblatt Produkt Drahtfaser DE 35/0,55N (2017)
7. Thienel, K.-Ch., Kustermann, A.: Sonderbetone: Normalbeton, Hochfester Beton, Hochleistungsbeton, Ultrahochfester Beton, Neubiberg (2017)
8. Gerriets: Produktkatalog Bühnentextilien, Metallvorhänge. 2.7 Metall Vorhänge (G-MESH). https://www.gerriets.com/de/download-center. Accessed 29 Jan 2021s
9. Michal, M., Keuser, M., Frey, M.: Effects of a new Steel Fiber in Concrete under Small-Caliber Impact, Universität der Bundeswehr München, Neubiberg, Doktorarbeit (2014)

Spinability and Characteristics of Particle-Shell PP-bicomponent Fibers for Crack Bridging in Mineral-Bonded Composites

Mihaela-Monica Popa[1], Harald Brünig[1], Iurie Curosu[2],
Viktor Mechtcherine[2], and Christina Scheffler[1(✉)]

[1] Leibniz-Institut für Polymerforschung Dresden e. V. (IPF), Dresden, Germany
scheffler@ipfdd.de
[2] TU Dresden, Institute of Construction Materials, 01062 Dresden, Germany

Abstract. Polypropylene (PP)-fibers are one of the most widely used polymer fibers for several different applications in fiber reinforced concrete due to their availability, low price, chemical inertness and stability in high alkaline environment. In order to improve the fracture energy and toughness of fiber-reinforced mineral-based composites under impact loads, the energy absorption provided by the fiber material itself but also the failure mechanisms in the fiber-matrix interphase play a crucial role. A desirable pull-out behavior for high energy absorption is achieved for polymer fibers with high tensile strength in combination with high surface roughness. Based on this knowledge, new PP-bicomponent fibers have been developed containing different particles (e.g. Al_2O_3, $CaCO_3$) in the outer shell in order to generate a rough fiber surface. In this work the results of first spinning trials of PP-bicomponent fibers produced by a lab-scale spinning equipment are presented. The fibers` tensile strength and particle distribution along the surface was determined depending on the drawing ratio. In single-fiber pull-out tests the fibers enabled high energy absorption compared to state-of-the-art PP-fibers. Furthermore, the structure of the fibers surface before and after pull-out was analyzed by scanning electron microscopy and revealed enhanced mechanical interlocking.

Keywords: Bicomponent fibers · PP fibers · Melt-spinning · Bond strength · Loading rate

1 Introductions

Strain-hardening cement-based composites (SHCC) represent a well-known class of fiber-reinforced concrete, yielding multiple fine cracks under increasing tensile loading [1]. An important role for manufacturing or strengthening of structural elements against impact loads is played by their high energy dissipation through inelastic deformations before reaching tensile strength [2]. The reinforcement with short fibers is a feasible determinator under dynamic loading situations to improve the concrete ductility [3]. Partial fiber debonding and stretching during crack formation and growth, followed by complete debonding and pull-out with the ongoing crack opening is enhanced in a SHCC matrix, which involves a balanced bond strength [4].

© RILEM 2022
P. Serna et al. (Eds.): BEFIB 2021, RILEM Bookseries 36, pp. 255–264, 2022.
https://doi.org/10.1007/978-3-030-83719-8_23

To improve the fracture energy and toughness of fiber-reinforced mineral-based composites under impact loads, the energy absorption by the fiber material itself and failure mechanisms in the fiber-matrix interphase and interface play a crucial role. Different polymer fibers such as high-density polyethylene (HDPE), poly(p-phenylene-2,6-benzobisoxazole (PBO) or aramide fibers are known to be very effective to provide high strain capacity and ductile failure behavior, but they generate high costs. Polypropylene (PP)-fibers are among the most widely used polymer fibers for several different fibre reinforced concrete applications [5]. PP-fibers are available at a low price, and they are chemically absolutely inert and stable in a high alkaline environment. At the same time, PP-fibers reveal some weak points such as the low modulus of elasticity and tensile strength and the hydrophobic surface, which reduces the interaction with the surrounding matrix.

To overcome this drawbacks the manufacturing of bicomponent fibers [6] has been found to be a promising new approach. In general, different polymers are used to form the fiber core and the surrounding fiber shell when processed during melt spinning. First attempts to develop bicomponent-PP fibers were carried out by de Lhoeux [7] by developing fibers with pure PP in the core and a PP compound in the sheath-layer with an additional surface treatment that was applied after spinning. Also in [8] and [9] PP-based bicomponent fibers have been co-extruded to produce variable combinations of core and shell structures. The core was modified to improve the Young's modulus and tensile strength of the fibers and enhanced interfacial bond was achieved by introducing nano-particles, glass beads and fly ash in the shell. Further, it was found that the surface bond characteristics of the fiber was especially increased by the embossing of a surface structure, what was possible due to large fiber diameters in the range of 350–510 μm.

Since it is well-known, that the bond between the fibres and the cement matrix is a significant factor to adjust the performance of fibre-reinforced concrete, other surface treatments involving particles were also studied by introducing a chemical component and a mechanical/frictional bond. In this way, the fiber surface was modifed by a sol-gel coating of nano-silica particles [10] so that significant changes in the surface roughness of the fiber were produced. Further, aluminium oxide (Al_2O_3) was used for modifying the wetting characteristics of polypropylene fibers [11]. Increasing the hydrophilic rate on the fiber surface could improve the bond strength performance between fiber and cement matrix. The water dispersion technique was also performed to increase polypropylene fibers` roughness by using TiO_2 nanoparticles [12].

In our own previous studies [5], the properties of PP-fibers were systematically changed with regards to draw ratio, tensile strength and Young's modulus, filament diameter, cross-sectional shape, and fiber surface treatment using a lab-scale spinning equipment at IPF. It was shown by micromechanical testing on single fiber model composites that desirable pull-out behavior for high energy absorption is achieved for PP-fibers with high tensile strength in combination with high surface roughness. Based on these findings, strategies for a new bicomponent fiber design were derived to enhance the mechanical interlocking between fiber surface and matrix material.

In this work, the results of the first lab spinning trials of PP-bicomponent fibers are presented. Different particles are introduced into the fibres' outer shell to create a rough fibre surface during the spinning process. The fibers are characterized in terms of surface structure, fiber mechanical performance and single fiber pull-out tests.

2   Materials and Methods

2.1   Materials

A commercial available polypropylene (PP: HG 450 FB, Borealis) was used to spin the fibre core. Additionally, for spinning the fiber shell, the PP was combined with standard aluminium oxide (Al_2O_3) particles or with calcium carbonate ($CaCO_3$) particles in order to process polymer compounds. The Al_2O_3 particles presenting a diameter of 1 μm were ordered from Struers, USA. The $CaCO_3$ particles with a size ranging from 0.3 to 3 μm were delivered from Schaefer Kalk, Germany.

A cementitious matrix system designed for research purposes (see [3, 4] and Table 1) with no large aggregates was used to prepare the single-fiber model composites for pull-out tests. This model mix was used as a basis for a later improvement where the new matrices will be optimised from the application technology point of view, as well as the cement content (clinker content).

Table 1. Matrix composition

Material	kg/m^3
CEM I 52.5 R-SR3/NA, Holcim, Germany	1460
Silica fume, Elkem, Norway	292
Quartz sand 0.06–0.2 mm, Strobel Quarzsand GmbH, Germany	145
Superplasticizer, BASF SE, Germany	25
Water	315

2.2   Fiber Manufacturing

Bicomponent monofilaments were manufactured by an IPF in-house designed and built laboratory-scale piston spinning device using the melt-spinning method, see Fig. 1.

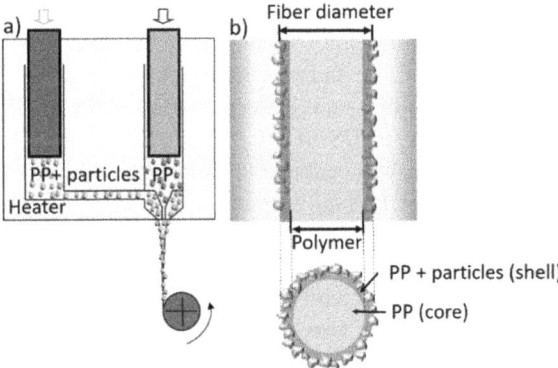

Fig. 1. Spinning of particle shell PP-bicomponent fibres, a) principle of the bicomponent melt spinning process, b) schematic drawing of a the bicomponent fiber with particle shell in concrete matrix.

Different volume percentages of Al_2O_3 or $CaCO_3$ particles were mixed with PP using a microcompounder (DSM Xplore 5&15 Micro Compounder) and used as raw material during spinning to form the outer shell of the fibers. The compounding step was required to distribute the particles uniformly in the PP and to avert clogging of the nozzles during melt-spinning. The compounded shell material was then pelletized to enable the melt-spinning process.

The fiber core was composed of the same polymer that was used in the shell. The monofilaments were molten in separate screw extruders and spun into a single-hole die with different core-shell diameters, see Table 2. Core and shell materials were brought to a temperature of 220 °C in the extruder, passed through a pump and into the spinneret. Finally, a winder with a take-up velocity of 800 m/min was used to spool the filaments on a bobbin. An offline filament drawing follows this step to initiate polymer chain orientation and therefore improve the fibers mechanical properties [3].

Fibers containing 5 vol.% of Al_2O_3 and $CaCO_3$, 10 vol.% of $CaCO_3$ and reference fibers containing PP in the core and PP in the shell were spun with different dies, see Table 2. The characterization of the fiber strength involved all of the bicomponent filaments, before and after fiber offline drawing using a ratio of three, which will be identified as 'as-spun' and 'DR = 3', respectively.

Table 2. Parameters for biomponent fiber manufacturing.

Fiber name	Die geometry [mm]	Core / shell ratio of cross sectional area [%]
PP + 5 vol. % Al_2O_3	0.3 core, 1.2 shell	50 / 50
PP + 5 vol. % $CaCO_3$	0.3 core, 1.2 shell	50 / 50
PP + 10 vol. % $CaCO_3$	0.5 core, 0.8 shell	80 / 20
PP	0.3 core, 1.2 shell	50 / 50

2.3   Characterization Methods

2.3.1   Fiber Morphology

The filaments' morphology and the particles' distribution along the fiber surface were investigated through optical and scanning electron microscopy (SEM). The scanning electron microscope ULTRA PLUS (Carl Zeiss Microscopy GmbH, Germany), equipped with an SE2 detector, was operated to obtain micrographs of the fiber surface after sputter-coating a 3 nm thick platin layer.

2.3.2   Single-fiber Tension Test

The as-spun and offline drawn bicomponent fibers' mechanical properties were investigated by single-fiber tension tests, performed with a FAVIMAT + (Textechno H. Stein GmbH & Co. KG, Germany) equipped with a 610 cN load cell and hard rubber/vulkolan clamps. Separated fibers were clamped with a gauge length of 10 mm. The linear density of each single fibre needed for the evaluation was determined before the tension test using the vibroscopic method according to ASTM D 1577. The test started with a velocity of 5 mm/min and the force-displacement curve was recorded. The stress-strain relationships were derived based on the measured diameter and defined gauge length. At least 30 samples were tested per composition.

2.3.3   Single-Fiber Pull-Out Test

Single-fiber model composites were manufactured for single-fiber pull-out tests using a device that was designed and constructed at IPF to analyze the effect of the new developed bicomponent fibres structure on the failure behavior. Therefore, pull-out tests were carried out under quasi-static and dynamic loading [13].

The cementitious matrix was prepared with a speed mixer and transferred to a sample holder. Subsequently, the fiber was embedded computer-assisted to a depth of about $l_e = 1000$ µm at room temperature and controlled climatic conditions. The specimens were stored for 28 days in a humid atmosphere until testing. The single-fiber composites were mounted in the quasi-static and dynamic SFPO devices, and the fiber end was fixed at the mandrel with minimized free fiber length. The experiments were carried out after fixing the fiber upper end to the mandrel with a cyanoacrylate adhesive. After curing the adhesive, the fibre's tensile load was adjusted to zero to eliminate the stress due to adhesive shrinkage during its curing. The tests were performed under a displacement rate of about 0.001 mm/s for the quasi-static (QSFPO) and 10 mm/s for the dynamic (DSFPO) load conditions. In the case of dynamic testing, only 300 µm displacement is taken into account to establish a constant strain-rate during the experimental procedure. The force-displacement curves were recorded for at least 20 samples for each fiber type.

3   Results and Discussion

3.1   Surface Structure of Bicomponent Fibers

Studies with SEM were carried out to investigate the particles' distribution along the fiber surface. In Fig. 2, SEM images of the fibres' surface before and after offline drawing are displayed. All fibers that are spun with particles in their shell reveal a clear increase regarding the surface roughness in comparison to the neat PP-bicomponent fibers (Fig. 2 g, h). Additionally, a monocomponent online drawn PP-fiber is shown (Fig. 2 i) that partly reveals a fibrillated surface structure.

Also, the Al2O3 as well as CaCO3-particles with varying amounts are well distributed. It can be observed that offline drawing results in strongly reduced fiber diameters (Table 3) and additional holes and stripes enhancing the surface roughness. Nevertheless, the particles are still attached to the fiber shell even though some of them could be removed.

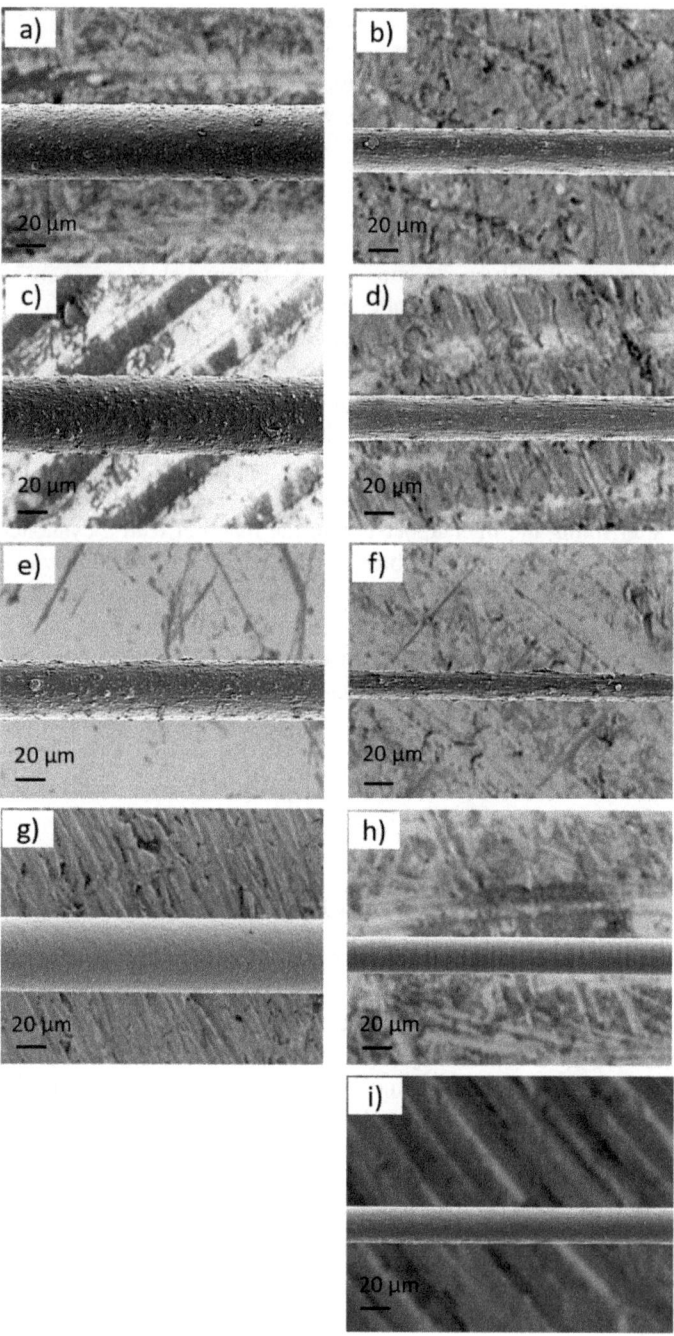

Fig. 2. SEM representative images of the fiber morphology of: as-spun (left) and offline drawn (right) bicomponent fibers containing a, b) 5 vol.% Al$_2$O$_3$, c, d) 5 vol.% CaCO$_3$, e, f) 10 vol.% CaCO$_3$ and g, h) reference bicomponennt fiber without particles and i) online drawn monocomponent PP fiber.

3.2   Mechanical Properties

For each fiber type representative tensile stress-strain curves were selected that are compared in Fig. 3 before and after drawing. In the case of as-spun fibers without drawing the polymer chains in the PP fibers are disordered, causing very high values of elongation as well as low tensile strength and Young's modulus (Fig. 3, for absolute values and standard deviations see Table 3). In the drawing process the polymer chains become aligned in fiber direction leading to a significant increase in the tensile strength and Young's modulus. This effect is disturbed by the introduction of the particles so that pure PP fibers reveal the best mechanical results and lower tensile strength are hence determined for the bicomponent fibers with particles in the shell.

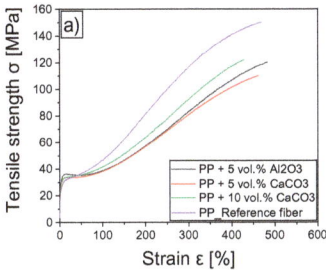 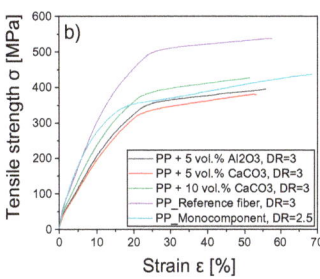

Fig. 3. Tensile stress-strain curves of a) as-spun and b) offline drawn bicomponent fibers. Note different limits of the stress and strain axes.

Table 3. Diameters and mechanical properties of bicomponent fibers.

Fiber name	Real [calculated] shell diameter [μm]	Real [calculated] core diameter [μm]	Tensile strength at break, σ [MPa]	Strain, ε [%]	Young's modulus [MPa]
PP + 5 vol.% Al$_2$O$_3$_as-spun	45.6 [50.0]	[35.0]	120 ± 3.7	483 ± 21.5	1.3 ± 0.1
PP + 5 vol.% Al$_2$O$_3$_DR = 3	25.4 [25.5]	[18.1]	396 ± 13.6	56 ± 6.4	3.7 ± 0.1
PP + 5 vol.% CaCO$_3$_as-spun	39.8 [50.0]	33.0 [35.0]	110 ± 2.6	462 ± 19.9	1.3 ± 1.2
PP + 5 vol.% CaCO$_3$_DR = 3	25.9 [25.5]	[18.1]	381 ± 3.3	53 ± 3.1	3.3 ± 0.1
PP + 10 vol.% CaCO$_3$_as-spun	40.6 [39.5]	33.2 [27.9]	122 ± 6.3	428 ± 23.7	1.4 ± 0.1
PP + 10 vol.% CaCO$_3$_DR = 3	19.5 [22.8]	[16.1]	427 ± 28.2	52 ± 7.8	4.4 ± 0.4
Reference fiber_as-spun	42.5 [50.0]	[35.0]	150 ± 7.5	470 ± 49.9	1.1 ± 0.1
Reference fiber_DR = 3	22.2 [25.5]	[18.1]	538 ± 55.9	58 ± 17.0	5.7 ± 0.5
PP monocomponent fiber_online DR = 2.5	18.1	-	437 ± 17.4	69 ± 9.1	4.5 ± 0.3

3.3   Quasi-static and Dynamic Single-Fiber Pull-Out

The force-displacement curves received during the pull-out tests of fibers with 10 vol.% CaCO$_3$ before and after drawing are shown in Fig. 4. The fibers are qualitatively compared against online drawn (DR = 2.25) monocomponent PP fibers (Fig. 2 i).

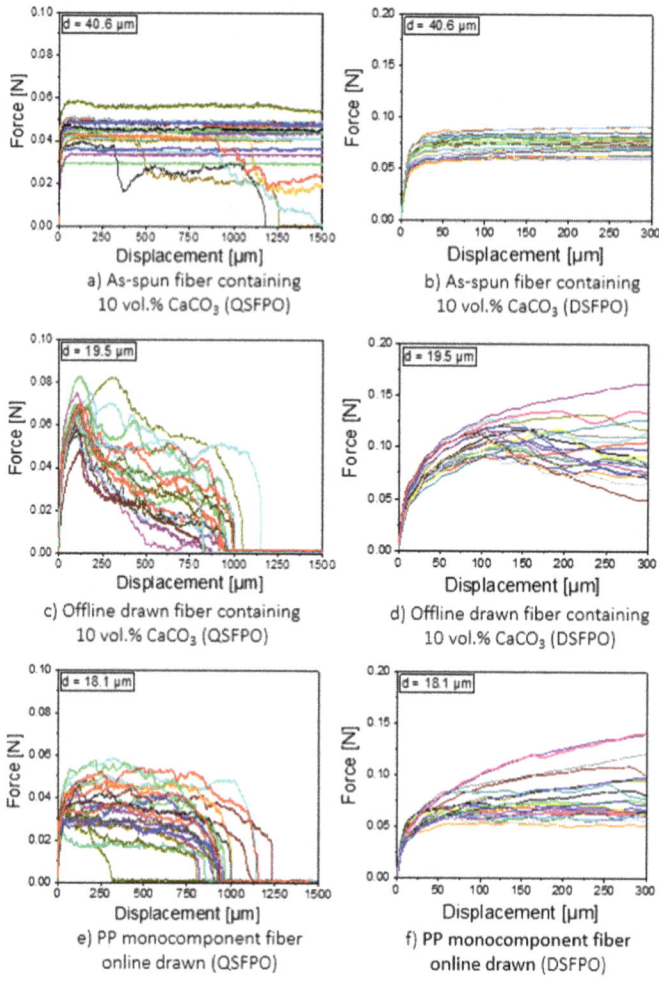

Fig. 4. Quasi-static (a, c, e) and dynamic (b, d, f) SFPO force displacement curves of as-spun and offline drawn fiber containing 10 vol.% CaCO$_3$, and online drawn (DR = 2.25) PP fiber pulled out of the high strength matrix. Note different axis limits.

A strong influence of the loading rate on the resulting pull-out force levels is revealed in general. The lowest force levels at both loading rates are observed in the case of unstretched fibers, even though they provide the largest diameter or contact area, respectively, and contain particles (Fig. 4 a,b). As can be derived from the force-displacement curves of QSFPO, nearly all fibers are stretched during testing instead of being pulled out. This is indicated by the force level that remains constant even at very high displacements that are far exceeding the embedded fiber length. This behavior was observed also for DSFPO. If the fibers reveal high tensile strength and reduced strain due to fiber drawing, a steep increase of the pull-out force at low displacements is exposed during quasi-static but also dynamic loading (Fig. 4 c,d). The rough fiber surface interacts very well with the surrounding matrix by mechanical interlocking leading to a shift of the particles along the fiber surface if the fibers are pulled (not shown in this paper). This induced plastic deformation results in increased energy absorption. Since the surface roughness of neat PP-fibers is considerably lower (Fig. 2 i), the initial pull-out forces are consequently found to be lower during Q- and DSFPO.

4 Conclusions

The work displayed in this paper aimed to introduced new developed polymeric bicomponent fibers by incorporating Al_2O_3-, and respectively $CaCO_3$-particles in the outer shell/'skin' that improve the mechanical interlocking with the concrete matrix. In this first approach, the fibers have been produced by the melt-spinning process and drawn offline to induce the orientation of the polymer chains in the axial direction for improved mechanical performance. However, the most crucial part is the improvement in the surface characteristics of the fiber. Further, the following conclusion can be drawn from this work:

- In general, the incorporation of particles hinders the orientation of the polymer chains along the whole fiber cross section so that the tensile strength pure PP-fibers is not achieved. However, the increase of the content of $CaCO_3$-particles from 5 to 10 vol.-% went along with sligthly enhanced tensile strength and Young`s modulus.
- Undrawn fibers are strongly stretched when pulled-out from the cement matrix leading to low and constant force levels.
- The addition of particles forms rough fiber surfaces that caused high energy absorption during fiber pull-out by plastic deformation of the surface layers and mechanical interlocking.

Further investigations with other particles, such as ground tire rubber (GTR) and silica (SiO_2) particles will be investigated in order to improve the strength, fracture toughness and energy absorption of fiber-reinforced composites under impact loading.

Acknowledgements. The authors greatly acknowledge the funding by the Deutsche Forschungsgemeinschaft (DFG - German Research Foundation) in the framework of the Research Training Group GRK 2250/1 "Mineral-bonded composites for enhanced structural impact safety", project number 287321140.

References

1. Curosu, I., Liebscher, M., Mechtcherine, V., Bellmann, C., Michel, S.: Tensile behavior of high-strength strain-hardening cement-based composites (HS-SHCC) made with high-performance polyethylene, aramid and PBO fibers. Cem. Concr. Res. **98**, 71–81 (2017)
2. Mechtcherine, V.: Novel cement-based composites for the strengthening and repair of concrete structures. Constr. Build. Mater. **41**, 365–373 (2013)
3. Wölfel, E.: Interphases in polypropylene and glass fiber reinforced cementitious model composites under dynamic loading. In: Proceedings of the 10th International Conference on Fracture Mechanics of Concrete and Concrete Structures (2019). (IA-FraMCoS, 2019)
4. Curosu, I., Mechtcherine, V., Millon, O.: Effect of fiber properties and matrix composition on the tensile behavior of strain-hardening cement-based composites (SHCCs) subject to impact loading. Cem. Concr. Res. **82**, 23–35 (2016)
5. Wölfel, E., Brünig, H., Curosu, I., Mechtcherine, V., Scheffler, C.: Dynamic single-fiber pull-out of polypropylene fibers produced with different mechanical and surface properties for concrete reinforcement. Materials **14**, 722 (2021)
6. Naeimirad, M., et al.: Recent advances in core/shell bicomponent fibers and nanofibers: a review. J. Appl. Polym. Sci. **135**, 46265 (2018)
7. Lhoneux, B., et al.: Development of high tenacity polypropylene fibers for cementitious composites. In: JCI International Workshop on Ductile Fiber Reinforced Cementitious Composites (2002). (DFRCC 2002)
8. Kaufmann, J., Schwitter, E.: Bi-component synthetic fibres for application in cement-bonded building materials. U.S. Patent Application **12**/067,860 (54) (2009)
9. Kaufmann, J., Lübben, J., Schwitter, E.: Mechanical reinforcement of concrete with bi-component fibers. Compos. A Appl. Sci. Manuf. **38**, 1975–1984 (2007)
10. Coppola, B., Di Maio, L., Scarfato, P., Incarnato, L.: Use of polypropylene fibers coated with nano-silica particles into a cementitious mortar. In: AIP Conference Proceedings, vol. 1695, p. 020056 (2015)
11. Ratu, R.N.: Development of polypropylene fiber as concrete reinforcing fiber (University of British Columbia (2016)
12. Szabová, R., Černáková, Ľ., Wolfová, M., Černák, M.: Coating of TiO$_2$ nanoparticles on the plasma activated polypropylene fibers. Acta Chimica Slovaca **2**, 70–76 (2009)
13. Scheffler, C., Zhandarov, S., Mäder, E.: Alkali resistant glass fiber reinforced concrete: pull-out investigation of interphase behavior under quasi-static and high rate loading. Cement Concr. Compos. **84**, 19–27 (2017)

Damage of Fracture Properties of Polyolefin Fibre Reinforced Concrete Under High Temperature

Marcos. G. Alberti, Jaime C. Gálvez[(✉)], Alejandro Enfedaque, and Ramiro Castellanos

Departamento de Ingeniería Civil: Construcción, E.T.S de Ingenieros de Caminos, Canales y Puertos, Universidad Politécnica de Madrid, C/Profesor Aranguren, 3, 28040 Madrid, Spain
jaime.galvez@upm.es

Abstract. Concrete has become the most common construction material, showing, among other advantages, good behaviour when subjected to high temperatures. Nevertheless, concrete is usually reinforced with elements of other materials such as steel in the form of rebars or fibres. Thus, the behaviour under high temperatures of these other materials can be critical for structural elements. In addition, concrete spalling occurs when concrete is subjected to high temperature due to internal pressures. Micro polypropylene fibres (PP) have shown to be effective for reducing such spalling, although this type of fibres barely improves any of the mechanical properties of the element. Hence, a combination of PP with steel rebars or fibres can be effective for the structural design of elements exposed to high temperatures. New polyolefin fibres (PF) have become an alternative to steel fibres. PF meet the requirements of the standards to consider the contributions of the fibres in the structural design. However, there is a lack of evidence about the behaviour of PF and elements made of polyolefin fibre reinforced concrete (PFRC) subjected to high temperatures. Given that these polymer fibres would be melt above 250 °C, the behaviour in the intermediate temperatures was assessed in this study. Uni-axial tests on individual fibres and three-point bending tests of PFRC specimens were performed. The results have shown that the residual load-bearing capacity of the material is gradually lost up to 200 °C, though the PFRC showed structural performance up to 185 °C.

Keywords: Fracture behaviour · Fibre reinforced concrete · High temperature · Melting point · Flexural tensile strength · Polyolefin fibres

1 Introduction

One of the most relevant advantages of concrete as a construction and building material is its behaviour when subjected to fire and high temperatures [1]. The behaviour of concrete structures subjected to high temperatures is not only related to the behaviour of concrete itself, but also to the behaviour of the reinforcing materials.

P. Serna et al. (Eds.): BEFIB 2021, RILEM Bookseries 36, pp. 265–276, 2022.
https://doi.org/10.1007/978-3-030-83719-8_24

Given that fire remains as one of the most relevant potential risks for structures, research dealing with the response of reinforced concrete and fibre reinforced concrete under this circumstance has been conducted, showing that some micro polypropylene fibres (PP) can be used to control the risk of explosive spalling [2]. Certain types of fibres have shown to enhance the response of the concrete element exposed to fire or high temperatures [3]. The damage caused by high temperatures can be limited due to the increase of pore connectivity of the PP fibres after melting. The pore structure of concrete directly influences the phenomenon of spalling occurring in a controlled or even in a forceful way (explosive spalling), if the pore structure is more closed, as in the case of HSC (high strength concrete) [4]. The use of fibres, especially mixing with PP fibres, has shown to improve significantly the properties of the concrete because the high vapor pressure due to the inner moisture in concrete is released by the micro channels left by the PP fibres when are melted [5]. The mixture between steel (SF) and PP fibres helps optimising the residual properties that the concrete exposed to high temperatures could reach [2, 6] and takes advantage of the contribution to fire resistance of PP microfibers and the structural reinforcement provided by SF. Thus, the combination of both types of fibres optimises the behaviour under high temperatures [7]. PP fibres can hardly bear any additional loadings (they are not considered as structural fibres), though they melt with high temperatures, creating a capillary network that avoids high pressures inside the element subjected to high temperature, reducing the spalling.

Several types of synthetic fibres such as PP, polyvinyl alcohol (PVA), nylon (Ny), polyethylene (PE) have shown several advantages and have been used mainly in combination with SF and specially for high strength concrete [8]. Steel fibres are the most common structural fibres. Nevertheless, research has shown that concrete reinforced only with SF, regardless of the content, exhibits explosive spalling. Recent advances have shown that polyolefin macro-fibres (PF) can also meet residual tensile strengths that can be considered in structural design to substitute steel rebars [9–11]. Such fibres have shown to be an attractive alternative to steel fibres in certain applications [12, 13]. However, there is a lack of studies dealing with this type of structural polymer fibres when exposed to high temperature. In such a sense, PF are expected to melt at medium temperatures such as 200 °C. Thus, some additional beneficial effects could appear in terms of spalling if the melting of the fibres leaves channels that help reducing the internal vapor pressure. However, PF are mainly used for structural reinforcement of concrete elements [12, 14]. Therefore, the most important characterisation is to assess the residual load bearing capacity of polyolefin fibre reinforced concrete (PFRC) elements exposed to high temperature.

The significance of this research relies on the assessment of the behaviour of PFRC exposed to high temperature. The fracture properties and the residual strengths were obtained by three-point bending tests on notched specimens previously exposed to the referred a range of temperature from 20 °C to 200 °C. The results have shown that the residual load-bearing capacity of the material is gradually lost up to 200 °C, though the PFRC showed structural performance up to 185 °C. In addition, in this research, the variation of the mechanical properties of isolated PF exposed to a range of temperature from 20 °C to 160 °C was studied. The reader is addressed to ref. [15] for detailed information of the tests performed in this research.

2 Experimental Program and Results

The experimental campaign encompassed two main research aims: characterisation of the properties of isolated fibres subjected to high temperatures and characterisation of the residual properties of FRC specimens after being exposed to high temperatures. The specimens were subjected to heat in a Memmet UFB-500 stove capable of reaching temperatures above 200 °C with an accuracy of ± 0.5 °C.

2.1 Fibre Characterisation

Isolated polyolefin-based macro fibres were introduced in an oven at various temperatures in order to assess the influence of temperature on the mechanical properties. In order to evaluate such mechanical properties, uniaxial tensile tests were performed supplying values of load and displacements. Digital image techniques (DIC) were implemented to obtain the longitudinal deformation. With such results and with a previous meticulous measurement of the fibre cross section and the initial length, it was possible to compute the ultimate residual strength for each temperature. With the stress–strain curve, the residual modulus of elasticity was calculated as the slope of the fitting curve in the quasi-straight stretch. The fibres tested were Sika-fiber T-60 [16] and their main physical and mechanical properties are presented in Table 1.

Table 1. Mechanical properties of fibres used in experimental campaign [15].

SIKA FIBER T-60	
Density	0.92 g/cm^3
Length of fibre	60 mm
Tensile strength	560 MPa
Distortion temperature	110 °C
Melting point	280 °C
Modulus of elasticity	>9 GPa

The fibres were introduced in the oven when it had reached the target temperature for an hour. Then, the fibres were kept in the oven for 24 h. This timing was chosen following the rationale found in references [2, 7]. Before testing, the fibres were released from the oven and the loss of mass was measured. The tests were performed at laboratory temperature.

The uniaxial tensile tests of the fibres were performed according to UNE-EN 6892–1: 2017 [17] assumed valid for polyolefin fibres. The testing equipment used has two jaws for fastening the fibres with a maximum load capacity of 10 kN. An actuator displacement rate of 0.169 mm/s was used. Video extensometry system with a high-definition camera was used. Four white points were painted on the fibres tested, obtaining images at one frame per second rate during the course of the test, determining the elongation of the fibre by synchronising the video with the results of the machine (load and displacement values). This setup can be seen in Fig. 1. The two points painted close to the jaws, see Fig. 1b, were used to control by DIC techniques that there were not any relative displacements between the fibre and the jaws.

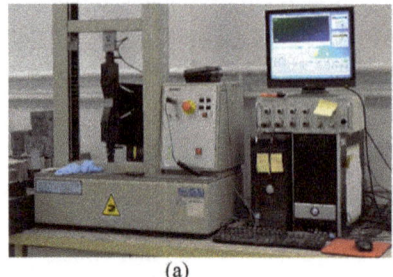

(a)   (b)

Fig. 1. (a) Testing machine for fibres characterisation; (b) fibre placed for the tensile test.

The temperatures studied were 100 °C, 125 °C, 150 °C and 160 °C. This set of temperatures permitted assessing the progressive modification of their properties. Higher temperatures would not allow performing the test because of the melting of the fibres. For each one, five specimens were tested, and stress–strain curves were recorded. A representative average curve was interpolated for each temperature. Table 2 shows these values and Fig. 2 shows an example with the five curves and the average. For the sake of clarity, the comparison showed by Fig. 3 only includes the average curves. The figure was performed considering values between 20% and 80% of the maximum fracture or sliding load. Such an interval was considered representative for obtaining the modulus of elasticity given that occasionally the fibre initially can slide and settle gradually in the jaws, shown by a non-linear stress–strain curve at the beginning of the test. The maximum stress for each temperature was obtained as the average from five tests and with the cross section of each fibre measured after being exposed to the correspondent temperature.

Figure 3 shows that the loss of mechanical properties increases with temperature. This reduction was more evident from 150 °C to 160 °C. After being exposed to those temperatures, the modulus of elasticity showed residual values of 33% and 9%, respectively, of that at room temperature as can also be seen in Table 2. It should also be worth noting that the fibres shortened with temperature and, therefore, increased the cross-section and the deformation capacity was considerably higher. Both the cross-section and the length were measured when the fibres were again at room temperature after heating.

Table 2. Physical and mechanical properties polyolefin fibres after being exposed to high temperatures (average value of five specimens).

Fibres and Temperature	Length (mm)	Width (mm)	Thickness (mm)	Max load (N)	Max stress (MPa)	Max strain (%)	Modulus of elasticity (MPa)
FPP-20 °C	60.31	1.51	0.67	264	259.44	9	2874
FPP-100 °C	57.12	1.61	0.69	243	222.79	10	2262
FPP-125 °C	54.47	1.64	0.70	256	224.63	12	2029
FPP-150 °C	40.86	1.82	0.78	286	202.28	22	943
FPP-160 °C	28.69	2.12	0.87	239	130.82	44	257

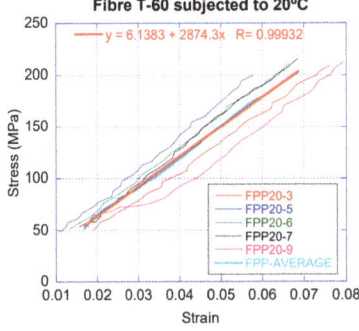

Fig. 2. Stress–strain curves of the fibres at 20 °C: individual curve for each specimen (FPP20-#), average curve and linear fitting of the average curve (shown in the legend).

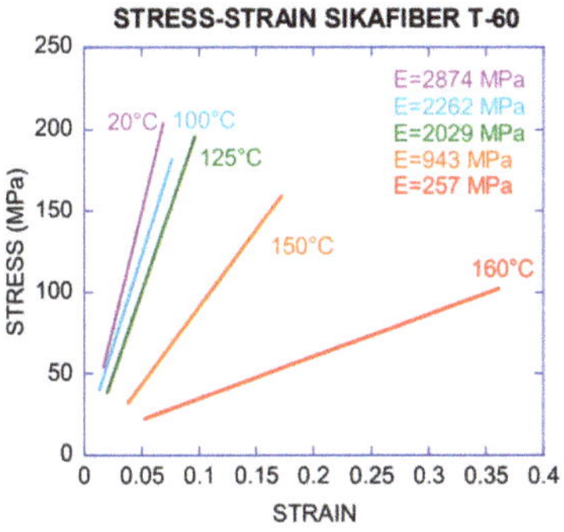

Fig. 3. Average stress–strain curves obtained with a linear fitting of five specimens tested after exposed to various temperatures.

2.2 Characterisation of Fibre Reinforced Concrete

The experimental campaign was performed with concrete elements manufactured with the concrete mix design of the reference [18]. Thus, siliceous crushed aggregates, with a maximum aggregate size of 12.7 mm, were used. Portland cement type EN-197–1 CEM I 52.5 R-SR, polycarboxylic superplasticiser Sika Viscocrete 5720, and limestone powder with a content of 98% calcium carbonate [9, 17] were used. Fibre dosages of 3 and 10 kg/m^3, with a water/cement ratio of 0.5, were employed. The mix proportioning can be seen in Table 3 and it was the same used in references [9, 18], so that the values obtained at room temperature could be compared. In addition, plain concrete specimens were also tested in order to compare the results with fibre reinforced concrete.

Table 3. Mix proportions used concrete in experimental campaign [18].

Concrete formulation	Cement (kg/m^3)	Limestone powder (kg/m^3)	Water (kg/m^3)	Sand (kg/m^3)	Gravel (kg/m^3)	Grit (kg/m^3)	Superplasticiser (% Cement Weight)	Polyolefin fibres (kg/m^3)
HF	375	100	187.5	916	300	450	0.75	–
HF3	375	100	187.5	916	300	450	0.75	3
HF10	375	100	187.5	916	300	450	0.82	10

The characterisation of the concrete elements encompassed three reference temperatures. Given that fibres were supposed to melt for higher temperatures, the highest temperature of exposure was set at 200 °C. Moreover, as the isolated fibres maintained a high degree of integrity for 150 °C, such temperature was also used for the three types of concrete (plain and with 3 and 10 kg/m^3 of polyolefin fibres). The third temperature was the room temperature in order to compare the residual properties with the original ones. After exposed to those temperatures, the properties evaluated in concrete specimens were fracture energy and residual flexural strength. These properties were assessed in prismatic specimens, cast for three-point bending tests (TPB). Given that the concrete with 10 kg/m^3 (HF10) was considered as structural material (based on the use of the same mix proportioning and fibres used in references [9, 18]), several additional temperatures were chosen between 150 °C and 200 °C for applying to the specimens made with this material. This was made in order to find the temperature at which the structural capacities were strongly affected. This can be of relevance for the structural design of PFRC elements. Thus, HF10 specimens were exposed to 150 °C, 165 °C, 175 °C, 185 °C and 200 °C.

The fracture tests were performed following RILEM TC-187-SOC [19] with specimens of 100 × 100 × 430 mm^3 (height × depth × length). Table 4 specifies the number of specimens tested in fracture for each concrete type.

Table 4. Tested specimens in concrete experimental campaign.

Nomenclature	Fibre admixture (kg/m^3)	Fibre length (mm)	Length (mm)	Width (mm)	Height (mm)	No. of specimens
HF	-	60	430	100	100	6
HF3	3	60	430	100	100	5
HF10	10	60	430	100	100	9

The heating process of specimens was carried out by means of convection heat in an oven, at an approximate heating rate of 2.80 °C/min. When the specimens reached the maximum temperature of analysis, they remained for 3 h at such temperature. The oven turned on and off using programmers during the night before the test day. The test specimens cooled on the stove for a period of 7 h. The heating time of the concrete specimens was chosen based on references [20–22] that used heated rates in the interval 1 °C/min to 10 °C/min, during 1 to 3 h at maximum temperature.

In order to obtain the residual flexural tensile strengths of PFRC, tests were carried out according to RILEM TC-187-SOC recommendation [19] and the residual strengths were computed according to UNE-EN 14561: 2007 standard [23]. Figure 4 specifies the dimensions of the specimens, being a_0 the length of the notch, h_{sp} the ligament length and P the applied load. In addition to the residual strengths, the maximum load that corresponds to the limit of proportionality (F_{LOP}), the deflection and crack opening were obtained by means of two LVDT (to measure deflection in the midspan) and one CMOD resistive transducer (to measure the crack mouth opening), respectively. The two LVDTs were placed at each side of the specimen in the midspan and the deflection was considered the average value. The CMOD was placed in the lips of the notch in order to measure the opening of the notch.

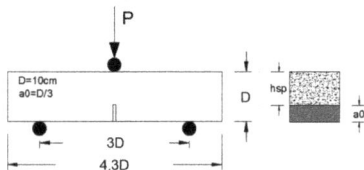

Fig. 4. Scheme of the three-point bending test [19].

With the results of the tests, the curves stress vs. crack opening (F_L-CMOD) were obtained in order to determine the residual flexural tensile strengths for crack openings of 0.5, 1.5, 2.5 and 3.5 mm. The fracture energy was calculated with the load vs. deflection curves (Load–vertical displacement of the application load point).

For the three most relevant temperatures (150 °C and 200 °C for HF3 and 150 °C, 165 °C and 200 °C for HF10) two specimens were tested. Moreover, expecting that HF10 would meet the structural requirements, some more specimens were produced. Such specimens were used in order to find the threshold at which the material behaved similar to the result at 200 °C. Thus, for 175 °C and 185 °C one specimen was tested.

2.2.1   Fracture Energy

Fracture energy was analysed for both plain and PFRC specimens. The load at the proportionality limit (F_{LOP}) was reached in all cases for CMOD below 0.1 mm. The first post-cracking branch of the curve after F_{LOP} followed the typical shape of the concrete softening behaviour up to CMOD values close to 0.5 mm (F_{MIN}) at which a new reloading branch initiated. Such reloading branch was constantly increasing up to the crack opening (CMOD) at which the fibres collapse by any of the failure mechanisms (mainly sliding or breaking). Such point of the curve was named maximum residual remaining load (F_{REM}). This can be better understood by seeing Fig. 5. The values of fracture energy (G_F) and coefficient of variation for plain concrete as a function of temperature are shown in Table 5.

The load–deflection curves of HF3 and HF10 exposed to temperatures ranging from 20 °C to 200 °C are shown in Fig. 5. This shows that these tests revealed that the degradation process of PFRC is relatively stable, with very similar behaviour for temperatures up to 185 °C. Nevertheless, the curves of the specimens subjected to 200 °C showed a clear damage although there was a remaining load-bearing capacity that could be of interest.

Table 5. Fracture energy (G_F) and coefficient of variation of HF specimens.

Fracture energy and coefficient of variation (N/m)					
Description		Specimen Dimension			LVDT = 0.4 mm
Temperature	Concrete	B (mm)	D (mm)	a_o (mm)	G_F (N/m)
20 °C	HF	100	100	33.33	88.44 (0.09)
150 °C	HF	100	100	33.33	78.86 (0.12)
200 °C	HF	100	100	33.33	115.12 (0.16)

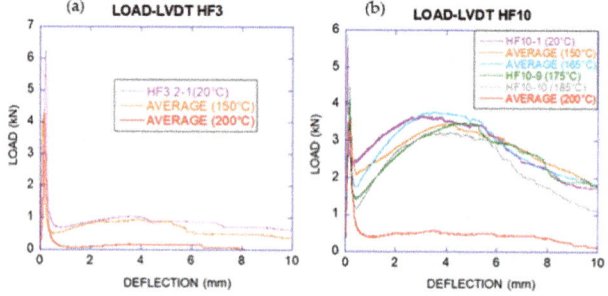

Fig. 5. Load–deflection curves of (a) HF3 and (b) HF10 at various temperatures (average at each temperature).

It should be highlighted that the results of HF10 specimens showed that the contribution of the fibres was still significant at 165 °C. As can be seen in Fig. 5, F_{REM} of the average curve at 165 °C was even higher than that of 20 °C, with a residual percentage of load-bearing capacity of 103% respect to 20 °C. The fibres, as a part of fibre reinforced concrete, reduced the degradation of its properties by temperature if this is compared with the results of isolated fibres. At 175 °C, the description of the load–deflection curve continues with typical behaviour of the polyolefin fibres, with F_{REM} remaining at 95% of the curve obtained for 20 °C. At 185 °C, F_{REM} was 88% of the curve obtained for 20 °C, still being significant.

Conversely, load–deflection curve was different for 200 °C. Such a curve showed residual values for F_{LOP}, F_{MIN} and F_{REM} of 61%, 17% and 16% respectively compared with the tests of the specimens at 20 °C. The load-bearing capacity decreased sharply and the loss of mechanical properties of the effective fibres in the section was evident. When F_{MIN} was surpassed, it reduced the load value close to zero for a 10 mm deflection. The fracture energy (G_F) assessed for HF3 and HF10 can be seen in Table 6 and Table 7 as well as the coefficient of variation.

Table 6. Fracture energy (G_F) of HF3 specimens.

Fracture energy and coefficient of variation (N/m)					
Description	Deflection (LVDT Measurement)				
	0.5 mm	1 mm	2.5 mm	5 mm	Final
Temperature	G_F (N/m)	G_F (N/m)	G_F (N/m)	G_F (N/m)	G_F (N/m)
20 °C	175	229	415	776	1352
	-	-	-	-	-
150 °C	112	152	323	659	1035
	(0.19)	(0.23)	(0.27)	(0.30)	(0.47)
200 °C	131	141	153	188	230
	(0.13)	(0.10)	(0.02)	(0.16)	(0.19)

Table 7. Fracture energy (G_F) of HF10 specimens.

Fracture energy and coefficient of variation (N/m)					
Description	Deflection (LVDT Measurement)				
	0.5 mm	1 mm	2.5 mm	5 mm	Final
Temperature	G_F (N/m)	G_F (N/m)	G_F (N/m)	G_F (N/m)	G_F (N/m)
20 °C	207	407	1140	2479	4308
	-	-	-	-	-
150 °C	183	351	954	2196	3876
	(0.12)	(0.17)	(0.16)	(0.15)	(0.19)
165 °C	169	325	993	2377	4251
	(0.12)	(0.09)	(0.08)	(0.00)	(0.05)
175 °C	152	274	810	2037	3908
	-	-	-	-	-
185 °C	142	254	803	1981	3488
	-	-	-	-	-
200 °C	131	165	268	457	727
	(0.14)	(0.17)	(0.12)	(0.12)	(0.10)

2.2.2   Residual Flexural Tensile Strengths

The structural capacity of PFRC was assessed according to UNE-EN 14561:2007 [23], EHE-08 and fib Model Code. In such standards, it is set that in order to consider the contribution of the fibres in the structural design, the residual flexural tensile strengths f_{R1} (strength at CMOD 0.5 mm) should be superior to 40% of f_{LOP} and f_{R3} (strength at CMOD 2.5 mm) shall not be less than 20% of f_{LOP}. Thus, the analysis of residual flexural tensile strengths $f_{R1}, f_{R2}, f_{R3}, f_{R4}$ were considered with respect to the increase in temperature.

Given the low fibre dosage of HF3, the specimens did not meet the requirements of the standards even when exposed to ambient temperature. At least until temperatures of 150 °C the specimens behaved similarly to those only exposed to ambient temperature.

In the case of the HF3 specimens exposed to 200 °C, the fibres that seemed to be closest to the surface reaching their melting point and the residual properties were strongly affected. The residual flexural tensile strengths of HF3 exposed to various temperatures are shown in Table 8.

Table 8. Residual flexural tensile strengths of HF3 exposed to various temperatures.

Residual Flexural Tensile Strengths (MPa) and Coefficient of Variation									
Description		f_{LOP}							
CMOD = 0.5 mm									
CMOD = 1.5 mm		CMOD = 2.5 mm							
CMOD = 3.5 mm									
Temperature	Concrete	f_{R1}	%	f_{R2}	%	f_{R3}	%	f_{R4}	%
20 °C	HF3	6.32	0.94	15%	0.74	12%	0.85	14%	0.97	15%
150 °C	HF3	4.42	0.61	14%	0.62	14%	0.80	18%	0.91	21%
		(0.01)	(0.23)		(0.36)		(0.31)		(0.31)	
200 °C	HF3	4.62	0.43	9%	0.04	1%	0.05	1%	0.08	2%
		(0.09)	(0.37)		(1.10)		(1.20)		(0.86)	

The tests conducted on HF10 specimens showed that as the temperature of exposure increased, f_{LOP} decreased with a value of 5.94 MPa in the control test to 4.22 MPa at 150 °C, a decrease of 29%. Regarding f_{R1} and f_{R3} the specimens met the structural requirements of the standards up to those exposed to 165 °C, with values 42% and 71% for f_{R1} and f_{R3} respectively. From 175 °C to 200 °C, only f_{R3} met the requirements. This is worth mentioning because such is the requirement for ultimate limit state in design. Table 9 presents the residual flexural tensile strengths of HF10 exposed to various temperature values.

Table 9. Residual flexural tensile strengths HF10 exposed to various temperatures.

Residual Flexural Tensile Strengths (MPa) and Coefficient of Variation									
Description		f_{LOP}	CMOD 0.5 mm		CMOD 1.5 mm				CMOD 2.5 mm	CMOD 3.5 mm
Temperature	Concrete		f_{R1}	%	f_{R2}	%	f_{R3}	%	f_{R4}	%
20 °C	HF10	5.94	2.39	40%	2.87	48%	3.34	56%	3.64	61%
150 °C	HF10	4.22	2.18	52%	2.43	58%	2.79	66%	3.13	74%
		(0.05)	(0.23)		(0.19)		(0.15)		(0.13)	
165 °C	HF10	4.27	1.78	42%	2.35	55%	3.03	71%	3.55	83%
		(0.04)	(0.05)		(0.02)		(0.02)		(0.04)	
175 °C	HF10	4.51	1.44	32%	2.00	44%	2.55	57%	2.99	66%
185 °C	HF10	4.63	1.23	27%	1.96	42%	2.66	57%	3.01	65%
200 °C	HF10	4.42	0.71	16%	0.40	9%	0.48	11%	0.51	12%
		(0.08)	(0.23)		(0.14)		(0.02)		(0.09)	

3 Conclusions

The use of polyolefin fibre reinforced concrete has become an alternative to steel fibres. However, the main possible drawback of this type of fibres could be their behaviour when exposed to high temperature given their low melting temperature. The main aim of this research was to assess the behaviour of PFRC specimens as well as PF when they are exposed to temperatures ranging from 100 °C to 200 °C. Among the results, the variations of the residual strengths of structural PFRC specimens subjected to elevated temperatures may be useful for designing structures with PFRC that can be exposed to such conditions.

The use of polymer structural fibres can be more adequate than other structural ones for reinforcing the concrete lining of tunnels, protection walls in industrial plants, structures subjected to marine environments or port pavements. The use in such applications requires assessing their structural behaviour when subjected to high temperatures.

This research shows that up to 150 °C the structural behaviour and the fracture energy of PFRC specimens were not affected. Between 150 °C and 200 °C, some properties were deteriorated and at 200 °C a significant percentage of the fibres were molten.

PFRC specimens with 10 kg/m^3 of PF showed that even after exposure up to 165 °C they met the required structural contributions considered by the standards (a remaining residual strength of 103% stood out with respect to the result for specimens only exposed to room temperature). Conversely, specimens exposed to 200 °C showed absence of any structural capacity, though the specimens still did not collapse or show brittle failure.

Acknowledgements. The authors offer their gratitude to the Ministry of Economy, Industry and Competitiveness of Spain for supporting this research by means of grant PID2019-108978RB-C31.

References

1. Bazant, Z.P., Kaplan, M.F.: Concrete at High Temperatures: Material Properties and Mathematical Models. Addison-Wesley Longman, Longman Harlow, United Kingdom (1996)
2. Varona, F.B., Baeza, F.J., Bru, D., Ivorra, S.: Evolution of the bond strength between reinforcing steel and fibre reinforced concrete after high temperature exposure. Constr. Build. Mater. **176**, 359–370 (2018)
3. Liu, X., Ye, G., De Schutter, G., Yuan, Y., Taerwe, L.: On the mechanism of polypropylene fibres in preventing fire spalling in self-compacting and high-performance cement paste. Cem. Concr. Res. **38**, 487–499 (2008)
4. Sanjayan, G., Stocks, L.J.: Spalling of high-strength silica fume concrete in fire. ACI Mater. J. **90**, 170–173 (1993)
5. Chen, B., Liu, J.: Residual strength of hybrid-fiber-reinforced high-strength concrete after exposure to high temperatures. Cem. Concr. Res. **34**, 1065–1069 (2004)

6. Yermak, N., Pliya, P., Beaucour, A.L., Simon, A., Noumowé, A.: Influence of steel and/or polypropylene fibres on the behaviour of concrete at high temperature: spalling, transfer and mechanical properties. Constr. Build. Mater. **132**, 240–250 (2017)
7. Varona, F., Baeza, F., Bru, D., Ivorra, S.: Influence of high temperature on the mechanical properties of hibryd fibre reinforced normal and high strength concrete. Constr. Build. Mater. **159**, 73–82 (2018)
8. Park, J.J., Yoo, D.Y., Kim, S., Kim, S.W.: Benefits of synthetic fibers on the residual mechanical performance of UHPFRC after exposure to ISO standard fire. Cem. Concr. Compos. **104**, 103401 (2019)
9. Alberti, M.G., Enfedaque, A., Gálvez, J.C.: On the mechanical properties and fracture behavior of polyolefin fiber-reinforced self-compacting concrete. Constr. Build. Mater. **55**, 274–288 (2014)
10. Picazo, A., Gálvez, J.C., Alberti, M.G., Enfedaque, A.: Assessment of the shear behaviour of polyolefin fibre reinforced concrete and verification by means of digital image correlation. Constr. Build. Mater. **181**, 565–578 (2018)
11. A. Blanco, P., A. de la Fuente, P., Cavalaro, S., Aguado, A.: Application of constitutive models in European codes to RC–FRC. Constr. Build. Mater. **40**, 246–259 (2013)
12. Alberti, M.G., Enfedaque, A., Gálvez, J.C., Pinillos, L.: Structural cast-in-place application of polyolefin fiber–reinforced concrete in a water pipeline supporting elements. J. Pipeline Syst. Eng. **8**, 05017002–1–05017002–11 (2017)
13. Alberti, M.G., Enfedaque, A., Gálvez, J.C.: Improving the reinforcement of polyolefin fiber reinforced concrete for infrastructure applications. Fibers **3**, 504–522 (2015)
14. Behfarnia, K., Behravan, A.: Application of high performance polypropylene fibers in concrete lining of water tunnels. Mater. Des. **55**, 274–279 (2014)
15. Alberti, M.G., Gálvez, J.C., Enfedaque, A., Castellanos, R.: Influence of high temperature on the fracture properties of polyolefin fibre reinforced concrete. Materials **14**, 601 (2021)
16. Sika, Sika-fiber T-60. Macrofibras sintéticas con carácter estructural para el refuerzo de hormigones, Sika: Madrid, Spain (2017)
17. UNE-EN 6892–1:2017, Metallic materials—Tensile testing—Part 1: Method of test at room temperature (ISO 6892–1:2016), AENOR: Madrid, Spain (2017)
18. Alberti, M.G.: Polyolefin fibre-reinforced concrete: from material behaviour to numerical and design considerations. Doctoral Thesis, Universidad Politécnica de Madrid, Madrid, Spain (2015)
19. Planas, J., Guinea, G., Gálvez, J., Sanz, B., Fathy, A.: Indirect test for stress-crack opening curve, de Experimental Determination of the Stress-Crack Opening Curve for Concrete in Tension—Final report of RILEM Technical Committee TC 187-SOC, RILEM Publications SARL: Paris, France, pp. 13–29 (2007)
20. Sideris, K.K.: Performance of thermally damaged fibre reinforced concretes. Constr. Build. Mater. **23**, 1232–1239 (2009)
21. Chen, B., Wu, K., Yao, W.: Conductivity of carbon fiber reinforced cement-based composites. Cem. Concr. Compos. **26**, 291–297 (2004)
22. Novak, J.: Fire response of hybrid fiber reinforced concrete to high temperature. Procedia Eng. **172**, 784–790 (2017)
23. UNE-EN-14651:2007+A1. Test method for metallic fibre concrete—Measuring the flexural tensile strength (limit of proportionality (LOP), residual), AENOR: Madrid, Spain (2008)

Evaluation of the Inertia Force in Compressive Impact Loading on Steel Fiber-Reinforced Concrete

Mohammad Bakhshi$^{(\boxtimes)}$, Isabel B. Valente, Honeyeh Ramezansefat, and Joaquim A. O. Barros

Department of Civil Engineering, Faculty of Engineering, University of Minho, Guimaraes, Portugal

Abstract. Steel-fibre-reinforced concrete (SFRC) is a strain rate sensitive material and, therefore, its dynamic and static compressive behaviour can be significantly different. In the present study, the effect of loading rate on the compressive behaviour of SFRC with 1% hooked end steel fibres is experimentally investigated. During impact loading, an inertia force is created due to acceleration along the specimen, whose effect in the range of impact is studied for a comprehensive assessment of the dynamic analysis of SFRC structures. For the evaluation of the inertia force, an instrumented drop-weight test setup is used, which includes two fast response loadcells with capacities of 1000 and 2000 kN on top (impact force) and bottom (reaction force) of specimen. The drop-weight impact tests were performed with three different drop heights, corresponding to maximum strain rates that ranged from 1 to 50 s^{-1}. Two high-capacity accelerometers (5000 g) were mounted in the middle of the cylindrical specimens to obtain the cylinder acceleration response. The results show that, by increasing the strain rates, compressive strength, maximum acceleration at the middle of cylinder, and inertia force are increased. The results in terms of the ratio between inertia and impact load of specimens are presented and discussed.

Keywords: Steel fiber reinforced concrete · Drop-weight test · Compressive strength · Modulus of elasticity · Inertia force · Strain rate

1 Introduction

Fibre reinforced concrete (FRC) has been widely used in the construction of high-rise buildings and critical protective structures, because adding fibres into the concrete matrix effectively increases the ductility and the impact energy absorption capacity of the structural elements and significantly improves the concrete's cracking performance [1]. Steel fibre-reinforced concrete (SFRC) is more ductile and tough than plain concrete [2]. Steel fibres crossing cracks limit their propagation due to the mobilization of fibre pull-out mechanisms. Despite fibre reinforcement is mainly used to increase the post-cracking tensile behaviour of cement-based materials, SFRC structural elements subjected to impact have significant zones under pronounced compressive stress field, whose sensitivity to high strain rates is important to access for a comprehensive modelling if its behaviour. Bischoff and Perry [3] investigated the effect of strain rate

© RILEM 2022
P. Serna et al. (Eds.): BEFIB 2021, RILEM Bookseries 36, pp. 277–288, 2022.
https://doi.org/10.1007/978-3-030-83719-8_25

on the compressive behaviour of plain concrete by performing different test procedures to determine its dynamic characteristics and concluded that the properties of concrete under dynamic and static loading are different.

The majority of the research developed on the dynamic material properties of SFRC is based on experimental tests and several experimental techniques have been developed for investigating the mechanical properties of concrete under impact loading. Figure 1 suggests the test method to be adopted for the desired loading rate (with a correspondence on the strain rate). The impact loadings can be applied by non-instrumented multiple drop weight, weighted pendulum (Charpy), instrumented drop weight, Split-Hopkinson Pressure Bar (SPHB), and projectile, [4].

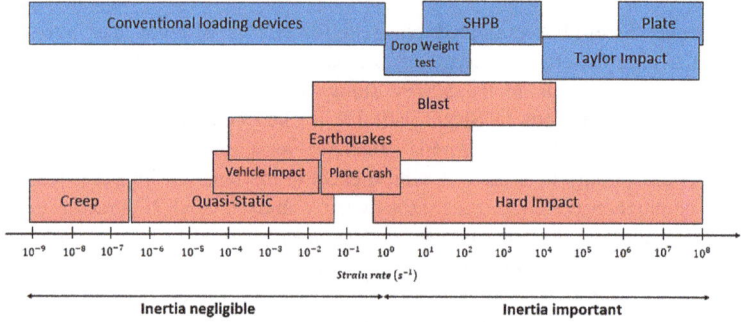

Fig. 1. Categorize of loadings based on the strain rate regime, [5].

Several factors affect the strain rate sensitivity of SFRC, including compressive strength, type of loading (i.e., compressive, tensile, and flexural), temperature, and type of steel fibres [6, 7]. Generally, the effect of strain rate on the mechanical properties of concrete is represented by the dynamic increase factor (DIF), which is defined as the ratio of dynamic to static values of the certain property. The effect of strain rate on the mechanical behaviour of SFRC has been investigated by different researchers [3, 8, 9]. Wang *et al.* [9] performed some compressive tests on plain concrete and SFRC specimens with three different volume fractions of very short straight steel fibres (1.5%, 3% and 6%). The results showed that the compressive strength of SFRC increased with both the strain rate and the volume fraction of steel fibres. Lok and Zhao [8] investigated the dynamic compressive behaviour of SFRC with hooked end fibres, using the SHPB test. They found that the compressive strength of SFRC increased with the strain-rate. Although there is no standard test for SFRC cylinders under compressive impact loading, instrumented drop weight tests were carried out by Xu *et al.* [6] on standard SFRC cylinders to study their compressive behaviour under impact loading. These authors used synthetic fibres, undulated, cold rolled, flattened, hooked end, and two new spiral shape steel fibres in their experimental study. They showed that the rate sensitivity and material properties of FRC under compressive impact loading are dependent on the material type and on the shape of fibres. Regarding the dynamic compressive behaviour of SFRC, many researchers consider the compressive dynamic increase factor (DIF) as a principal parameter [2, 7, 9]. Banthia [2] studied the effect of inertia force in SFRC beams under impact loading using an instrumented drop weight test. He reported that, for evaluating the flexural behaviour of the beam, due to inertia force in the impact loading, the reaction forces must be used instead of the contact load between the hammer and the beam. The recorded impact load of hammer can be corrected if the acceleration distribution along the length of the beam is known. Three accelerometers were used in his study for this purpose.

In the present study, the variation of the inertia force, at different strain rates, was investigated. For that purpose, acceleration was measured at the middle of SFRC cylindrical specimens. The relationship between inertia force, impact load and strain rate are determined and discussed. Moreover, the effect of strain rate on compressive strength and modulus of elasticity of SFRC is investigated. The developed SFRC mixture includes 1% volume fraction of hooked end steel fibres with length of 30 mm and aspect ratio of 80. This hooked end fibre was used due to its relatively high reinforcement performance, demonstrated in several experimental works. An instrumented drop-weight impact test setup is designed for the experimental analysis. Drop-weight impact tests are performed with four different drop heights, corresponding to maximum strain rates ranging from 1 to 50 s^{-1}. At least three cylindrical specimens are tested for each rate of loading and each height of the impactor.

2 Experimental Procedure

2.1 Materials

A single type of SFRC was used throughout the experiments, made with EN 197–1 [10] type I cement, 42.5R with a specific density of 3.15 gr/cm^3, and fly ash Type II according to EN 450–1 [11] requirements. Normal-weight natural river fine and coarse aggregates were used, with saturated-surface-dry (SSD) densities of 2.60 and 2.64 g/cm^3, respectively, and maximum size of 12 mm. The superplasticizer SIKA ViscoCrete 5920 was used. The content of each ingredient of the SFRC mixture presented in Table 1 was obtained by applying the volume method recommended by ACI-544-1R [12]. The geometric and mechanical properties of the used hooked-end steel fibres are indicated in Table 2.

Table 1. Mix proportion of SFRC in one m^3.

Cement [kg]	Fly ash [kg]	Water-cement content ratio	Fine aggregate [kg]	Coarse aggregate [kg]	Superplasticizer [kg]	Steel fiber [kg]
400	200	0.3	942	628	4.8	75.8

Table 2. Properties and amount of utilized steel fibre.

Volume [%]	Mass [kg/m³]	Diameter [mm]	Length [mm]	l/d	Modulus of elasticity [GPa]	Tensile strength [MPa]
1	75.8	0.38	30	80	210	2300

2.2 Static Tests

Quasi-static loading tests were conducted in cylindrical specimens of 100 mm diameter and 200 mm height to evaluate the compressive strength and modulus of elasticity according to EN 12390–2 [13]. These tests were executed in a servo-hydraulic testing machine with a maximum load capacity of 3000 kN, and three strain gauges were used to measure the compressive strain.

2.3 Impact Tests

The instrumented drop-weight system used in the compressive tests was developed to apply compressive impact loads with varying heights and masses, as shown in Fig. 2a. It has the capacity to drop a maximum mass of 290 kg up to a height of 9 m, which corresponds to a theoretical maximum impact velocity of 13.3 m/s. The inertia force is calculated from the difference between the forces measured on top and bottom of the specimens. The two accelerometers are installed with 90-degree radial distance, in the backside of the cylinders in the view of camera position, Fig. 2b. The recording rate capacity of this control card is 50 samples per millisecond, when multiple channels are used simultaneously, and results show that the recording rate was sufficient. The failure process, crack velocity, and deformation of specimens were measured with a high-speed video camera. These measurements were used to calculate strain values and strain rates in the tested specimens, during the impact loading process. The strain and the strain rates were also directly measured by three strain gauges installed at the specimen's mid height.

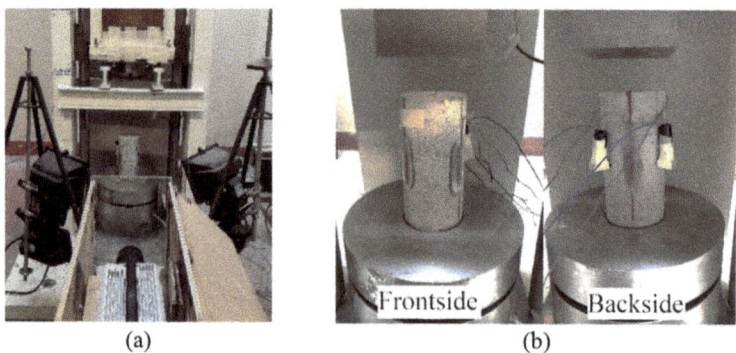

(a)   (b)

Fig. 2. (a) The drop weight impact machine and (b) Cylinders under impact test.

Two approaches were considered for measuring the deformation of the specimens. In the first approach, a PHOTRON FastCam APX-RS (PHOTRON, Japan) with the capability of recording up to 50,000 frames per second (fps) to measure surface deformations of cylindrical specimens ignoring the effect of curvature, was used, together with halogen lights that were installed to provide the lighting required for the high-speed videos. In the present experimental study, the high-speed camera was able to record the impact process with a rate of 15,000 frames per second and a resolution of 128 × 256 pixels, Fig. 3. The recorded videos were analysed by GOM Correlate software for digital image correlation (DIC).

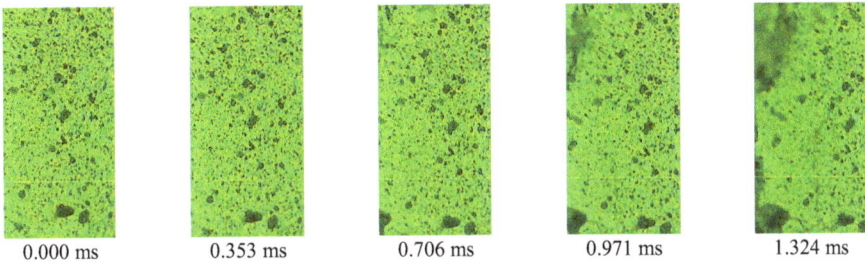

| 0.000 ms | 0.353 ms | 0.706 ms | 0.971 ms | 1.324 ms |

Fig. 3. Typical test sequence for SFRC recorded with High-speed camera (specimen HS2-2).

In the second approach, the hammer velocity during the impact period ($\dot{\delta}(t)$) was obtained through the initial impact velocity and the integral of the acceleration recorded by the accelerometer ($\ddot{\delta}(t)$). Subsequently, average compressive deformability ($\delta(t)$) was determined by the integral of the velocity.

$$\dot{\delta}(t) = \int \ddot{\delta}(t)dt \tag{1}$$

$$\delta(t) = \int \dot{\delta}(t)dt \tag{2}$$

deformability at the middle of cylinder height, respectively. In total, 16 cylinders were tested with four different heights of impactor of 1000 (HS1), 1500 (HS2), 2000 (HS3), and 2500 mm (HS4), corresponding to contact velocities of 4.42, 5.42, 6.26, and 7.00 m/s, respectively. These different heights were considered in order to determine the effect of the strain rate on the compressive strength and modulus of elasticity, and also to evaluate the variation obtained on inertia force and maximum acceleration.

The deformation along the height of cylinder is obtained with the high-speed camera that records the deformation in the middle of the cylinder surface, in a frame with 70 mm height and 5 mm width. The data collected also shows that the deformation distribution can be assumed linear along the height of cylinder. Similar results were reported by other researchers [14]. To evaluate the variation of acceleration along the height of cylinders, eight different points were defined. Afterwards, the acceleration was consecutively measured in these eight points, with two accelerometers, in specimens tested with an impactor of 20 kg, dropping from a height of 10 cm. The results show a non-uniform axial acceleration along the height of the cylinder, as presented in Fig. 4. Considering the trend presented in Fig. 4, it was assumed that the acceleration distribution along the axial direction of the specimen decreases linearly from the top to the bottom, as is also shown in Fig. 4. Based on the deformation and acceleration distribution along the height of the cylinder, maximum acceleration at the middle of cylinder is selected as an average acceleration. Regarding the maximum acceleration values measured in the specimen, during impact loading, two values of acceleration are considered: (a) the maximum acceleration at the middle of cylinder height measured by two accelerometers ($\ddot{\delta}_{mid}$), and (b) the maximum acceleration in the contact point between impactor and cylinder at the highest point of specimen ($\ddot{\delta}_{top}$) that is two times of its middle value, if a linear axial distribution is assumed.

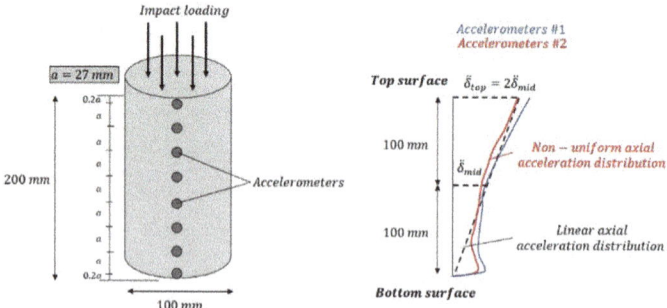

Fig. 4. Assumed acceleration distribution.

A linear variation of deformation from the top to the bottom of the specimen, and also a linear variation on the values of velocity ($\dot{\delta}$) and acceleration ($\ddot{\delta}$) is assumed, Fig. 5. It means that the acceleration distribution follows the deformation distribution along the height of specimen [2] in the time step, for a given time increment of Δt, ($\Delta t = t_2 - t_1$).

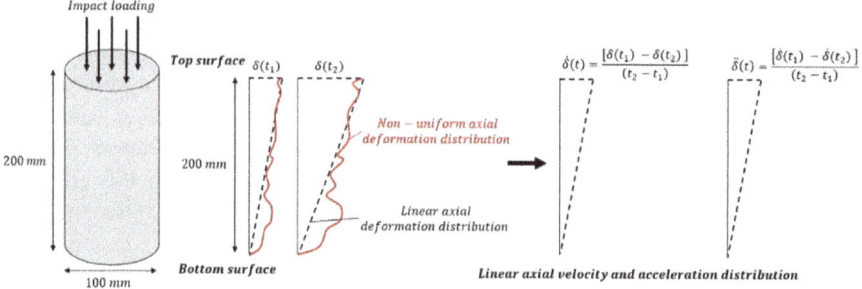

Fig. 5. Assumed deformation, velocity, and acceleration linear distribution.

3   Results and Discussion

3.1   Failure Patterns and Fracture Surfaces

In the specimens tested under static loading, micro cracks propagate mainly parallel to the direction of loading. For this loading situation, cracks progressed roughly in the axial direction of the specimen and a cone-shape failure was observed (Fig. 6a).

Under impact loading, a large number of cracks were formed and then propagated along the cylinders. By increasing the height of the impactor, an increase in the number of fractured pieces was observed in the tested specimens, showing the effect of strain rate on the failure of SFRC cylinders (Fig. 6b to Fig. 6e). The increase in the number of fractured pieces is an indication of a higher energy dissipation capacity of the specimen. Moreover, pull-out was the dominant mechanism for hooked end fibres in SFRC cylinders tested in this series. The cone-shape failure was also observed under impact loading, with more fractured pieces than under static loading. By increasing the height of impactor, the number of fractured pieces has increased. It means that by increasing the height of impactor, the energy absorption capacity of cylinder under impact loading increases.

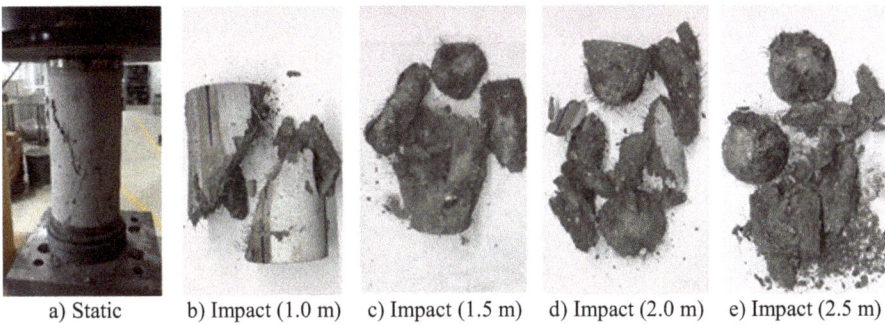

a) Static   b) Impact (1.0 m)   c) Impact (1.5 m)   d) Impact (2.0 m)   e) Impact (2.5 m)

Fig. 6. Failure modes of SFRC cylinders under static and impact loading.

3.2   Mechanical Properties

The mechanical properties of specimens tested under static and impact loading are shown in Table 3. The strain rate for the static compressive test is 1.33E-05. For all cylindrical specimens, the strain at compressive strength is almost 2.4‰. In normal concrete, this value ranges between 2.0‰ to 2.3‰. In impact loading, as expected, the strain rate influences the compressive strength and peak strain, and modulus of elasticity of SFRC. Both the compressive strength and modulus of elasticity have increased with the strain rate. The log-linear graph is used to represent the relationship between DIF and strain rate. It was observed that the values of DIF-E (modulus of elasticity) are higher than the values of DIF-f_c (compressive strength).

The results obtained are also compared with the models proposed by CEB-FIP [15, 16] to predict the modulus of elasticity and the compressive strength as a function of the strain rate. In the SFRC, the DIF-f_c has ranged between 1.04 and 1.35, while the DIF-E has varied between 1.22 and 1.58. For both compressive strength and modulus of elasticity, the CEB-FIP model overestimates the DIF up to strain values around $30s^{-1}$. For strain rates higher than $30s^{-1}$, the obtained compressive strengths are in good agreement with the values predicted by the CEB-FIP models. When the strain rate is higher than 30 s^{-1}, the CEB-FIP model underestimates the value of DIF-E, Fig. 7.

Table 3. Influence of strain rate on the mechanical properties of SFRC.

Type ID	Strain rate [s^{-1}]	Top load [kN]	Bottom load [kN]	Modulus of elasticity (E) [GPa]	DIF-E	Compressive strength (f_c) [MPa]	DIF-f_c
S-1	1.33E-05	585.0	nr	41.72	1.000	73.07	1.000
S-2	1.33E-05	601.9	nr	42.40		76.72	
S-3	1.33E-05	529.0	nr	40.55		67.09	
HS1-1	10.80	634.1	585.3	50.69	1.220	74.56	1.023
HS1-2	8.39	636.6	589.4	49.34	1.187	75.08	1.031
HS1-3	9.41	636.9	602.3	53.71	1.293	76.73	1.053
HS1-4	10.15	681.8	619.3	54.13	1.303	78.89	1.083
HS2-1	18.65	nr	610.7	61.40	1.478	77.80	1.068
HS2-2	25.18	700.2	626.8	59.63	1.435	79.85	1.096
HS2-3	19.50	685.7	632.6	52.40	1.261	80.59	1.106
HS2-4	21.22	673.4	613.1	56.54	1.361	78.10	1.072
HS3-1	29.28	795.8	640.2	59.26	1.426	81.55	1.119
HS3-2	36.17	861.4	702.4	63.29	1.523	89.48	1.228
HS3-3	34.60	852.9	649.3	65.46	1.575	82.71	1.135
HS3-4	24.65	nr	688.7	64.12	1.543	87.73	1.204
HS4-1	27.27	1052.0	716.6	60.91	1.466	91.29	1.253
HS4-2*	41.34	nr	1123.2	67.77	1.631	143.08	1.964
HS4-3	37.94	nr	755.7	61.47	1.479	96.27	1.321
HS4-4	35.83	nr	742.6	63.26	1.523	94.60	1.299
HS4-5	39.21	nr	769.1	60.53	1.457	97.97	1.345

nr – not recorded

*The results obtained with specimen HS4-2 are considered as outliers

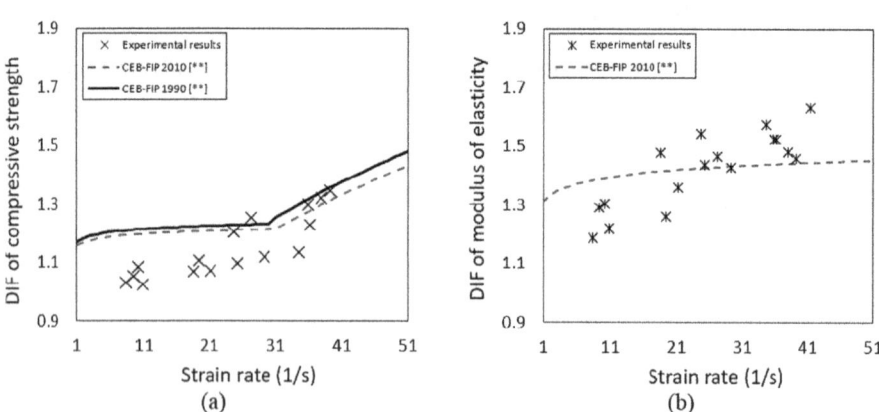

Fig. 7. DIF of SFRC for (a) compressive strength and (b) modulus of elasticity.

3.3   Effect of Inertia in Impact Loading

Figure 8 shows the values of impact and reaction forces along time (top load in three of the tested specimens could not be measured due to load cell limitations). The time intervals between the starting moment of the impact force and the reaction force are 134, 121, 99 and 80 μs, corresponding to drop heights of 1000, 1500, 2000, and 2500 mm, respectively. Therefore, increasing the height of impactor provokes an increase in impact and reaction forces, while the time gap between impact and reaction decreases. The time delay between top and bottom load cells and strain gauges installed at the mid height of cylinders can be attributed to stress wave propagation along the specimens. Based on these results, the forces recorded by top and bottom load cells show different values under impact loading, which means that the inertia force is significant. As it is known, the value of inertia force is a function of the acceleration along the specimen. The test results indicate that the ratio between inertia force and impact force (recorded by top load cell) significantly depends on the acceleration. This observation is also reported by Xu et al. [6]. The accelerations are measured by two accelerometers positioned at the mid height of specimens. The results indicate that by increasing the height of impactor, the acceleration increases and, accordingly, the value of inertia force also increases. For impactor heights of 1000, 1500, 2000, and 2500 mm, the maximum acceleration values are 1013 g, 1487 g, 1611 g, and 3173 g, respectively, and the ratio of inertia force to impact force (ω) is 9%, 10%, 24%, and 32%, respectively, as presented in Table 4. In the present study, to eliminate the inertia effect on the material properties, the reaction force measured by the bottom load cell is considered for calculating the compressive strength of SFRC under impact loading.

Table 4. Effect of strain rate on maximum acceleration and inertia force.

Type ID	Strain rate $[s^{-1}]$	m [kg]	Impact force [kN]	Inertia force [kN]	Inertia to impact force ratio (ω)	$\ddot{\delta}_{max,mid}$ [g*]	$\ddot{\delta}_{max,top}$ [g*]	$m\ddot{\delta}_{max,mid}$ [kN]	$m\ddot{\delta}_{max,top}$ [kN]
HS1-1	10.80	3.63	634.1	48.8	0.077	831	1662	30.1	60.3
HS1-2	8.39	3.75	636.6	47.2	0.074	987	1974	37.0	74.0
HS1-3	9.41	3.71	636.9	34.6	0.054	741	1482	27.5	55.0
HS1-4	10.15	3.61	681.8	62.5	0.092	1013	2026	36.6	73.2
HS2-1	18.65	3.75	nr	nr	nr	nr	nr	nr	nr
HS2-2	25.18	3.68	700.2	73.4	0.105	1487	2974	54.7	109.4
HS2-3	19.50	3.72	685.7	53.1	0.077	1339	2678	49.8	99.7
HS2-4	21.22	3.75	673.4	60.3	0.090	1270	2540	47.7	95.3
HS3-1	29.28	3.61	795.8	155.6	0.196	1574	3148	56.8	113.7
HS3-2	36.17	3.72	861.4	159	0.185	1509	3018	56.1	112.2
HS3-3	34.60	3.73	852.9	203.6	0.239	1611	3222	60.1	120.1
HS3-4	24.65	3.72	nr	nr	nr	1690	3380	62.8	125.7
HS4-1	27.27	3.74	1052.0	335.4	0.319	3173	6346	118.6	237.1
HS4-3	37.94	3.74	nr	nr	nr	2064	4128	77.2	154.3
HS4-4	35.83	3.64	nr	nr	nr	2878	5756	104.7	209.3
HS4-5	39.21	3.66	nr	nr	nr	3118	6236	11.4	22.9

nr – not recorded

*g = 9.81 m.s^{-2}

The inertia force can be analytically calculated by the equation of motion. According to structural dynamic theory, when an element moves relative to a reference, the inertia forces are proportional to the value of its acceleration in the same direction [17]. Consequently, by increasing the amount of created acceleration, the inertia force is increased. The motion equation can be written for a cylinder under impact loading. Based on this equation, the inertia force is a function of the specimen's mass and acceleration, as defined in Eq. (3),

$$\sum F = F_i = F_t - F_b = m\ddot{\delta}_{max} \tag{3}$$

where F_i, F_t, F_b, m, and $\ddot{\delta}_{max}$ are the inertia force, impact load (top), reaction load (bottom), mass of specimen, and maximum acceleration, respectively. In the compressive impact test (instrumented drop weight test), the specimen (motion of particles) is constrained to deform in the loading direction, and therefore, only the inertia load in the loading direction is calculated. For evaluating the relationship between maximum acceleration and inertia force, both the maximum acceleration at the middle of the cylinder height ($\ddot{\delta}_{max,mid}$) and the maximum acceleration in the contact point of impactor and cylinder ($\ddot{\delta}_{max,top} = 2\ddot{\delta}_{max,mid}$) are considered, as an input to the equation of motion, Fig. 8a.

The analytical inertia force is calculated based on the assumption of linearity distribution of acceleration along the height of cylinder. When the calculated analytical inertia force is compared with the experimental inertia force, it can be observed that when the loading rate (or height of impactor) increases, the analytical equation further underestimates the inertia force, Fig. 8a. This may be attributed to the real non-linear distribution of acceleration along the height of the cylinder. By increasing the contact velocity (or height of the impactor), the non-linearity of the acceleration distribution increases. It means that in the higher contact velocity, the maximum acceleration is dependent not only on the contact velocity but also on the height of the cylinder. Afterward, when the non-linearity of the acceleration distribution increases, the assumption of linearity for acceleration distribution could not be completely satisfied. Therefore, the value of the inertia force obtained by the equation of motion must be modified, assuming a non-linear distribution along the height of the cylinder. Further experimental tests are needed to confirm this hypothesis. In addition, the inertia to impact force ratio also increases when the maximum acceleration is increased, in an approximately linear trend, Fig. 8b. By increasing the height of the impactor, the effect of inertia is more significant. Similar observations have also been reported in [2] and [6].

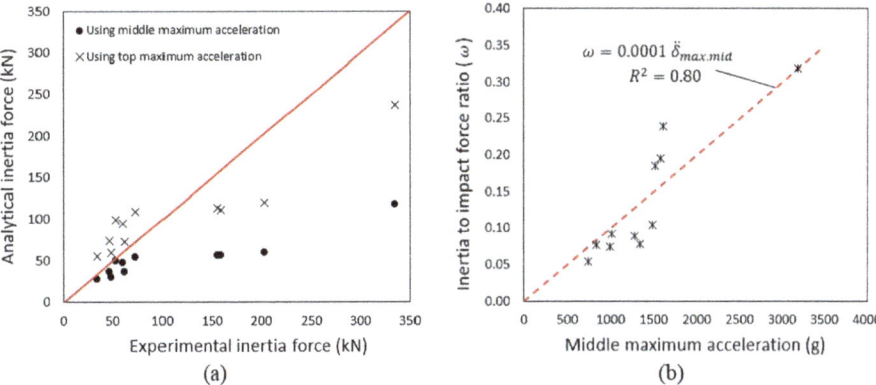

Fig. 8. Comparison between: (a) experimental and analytical inertia force, (b) inertia to impact force ratio and maximum acceleration in the middle of the specimen.

4 Conclusion

This paper investigates the effect of strain rate on the compressive behaviour of SFRC and analyses the parameters of compressive strength and modulus of elasticity. It also investigates the effect of inertia force in the compressive impact loading of hooked end SFRC and its relationship with maximum acceleration measured in cylindrical specimens. The following conclusions are reached:

1- Compressive strength and modulus of elasticity of SFRC increase with the strain rate. For the highest strain rate tested, maximum DIF values of 1.35 and 1.58 were obtained for compressive strength and modulus of elasticity, respectively.

2- The maximum acceleration value at the middle of the cylinder has increased with the strain rate. Consequently, the inertia force has also increased. The difference between analytical and experimental inertia forces has increased with the strain rate. It is concluded that a non-linear distribution of acceleration is likely, and the linear distribution assumption is not valid for all ranges of strain rate.

Acknowledgements. The study reported in this paper is part of the project "PufProtec - Prefabricated Urban Furniture Made by Advanced Materials for Protecting Public Built" with the reference of (POCI-01-0145-FEDER-028256) supported by FEDER and FCT funds. The first author gratefully acknowledges the financial support of FCT-Fundação para a Ciência e Tecnologia for the Ph.D. Grant SFRH/BD/149246/2019.

References

1. Barros, J.A., Cunha, V.M., Ribeiro, A.F., Antunes, J.A.B.: Post-cracking behaviour of steel fibre reinforced concrete. Mater. Struct. **38**(1), 47–56 (2005)
2. Banthia, N., Mindess, S.: Impact resistance of steel fiber reinforced concrete. Mater. J. **93**(5), 472–479 (1996)

3. Bischoff, P.H., Perry, S.H.: Compressive behaviour of concrete at high strain rates. Mater. Struct. **24**(6), 425–450 (1991)
4. Wu, H., Fang, Q., Gong, J., Liu, J.Z., Zhang, J.H., Gong, Z.M.: Projectile impact resistance of corundum aggregated UHP-SFRC. Int. J. Impact Eng. **84**, 38–53 (2015)
5. Nyström, U., Gylltoft, K.: Numerical studies of the combined effects of blast and fragment loading. Int. J. Impact Eng. **36**(8), 995–1005 (2009)
6. Xu, Z., Hao, H., Li, H.N.: Experimental study of dynamic compressive properties of fibre reinforced concrete material with different fibres. Mater. Des. **33**, 42–55 (2012)
7. Othman, H., Marzouk, H., Sherif, M.: Effects of variations in compressive strength and fibre content on dynamic properties of ultra-high-performance fibre-reinforced concrete. Constr. Build. Mater. **195**, 547–556 (2019)
8. Lok, T.S., Zhao, P.J.: Impact response of steel fiber-reinforced concrete using a split Hopkinson pressure bar. J. Mater. Civ. Eng. **16**(1), 54–59 (2004)
9. Wang, S., Zhang, M.H., Quek, S.T.: Effect of high strain rate loading on compressive behaviour of fibre-reinforced high-strength concrete. Mag. Concr. Res. **63**(11), 813–827 (2011)
10. British Standard Institution, BS EN 197–1, 'Cement Composition, specifications and conformity criteria for common cements', BSI London (2011)
11. British Standard Institution, BS EN 450–1, 'Fly ash for concrete Definition, specifications and conformity criteria', BSI London (2012)
12. ACI 544.1 R-96, 'State-of-the-Art Report on Fiber Reinforced Concrete', ACI Committee (1996)
13. British Standards Institution, BS EN 12390–2, 'Testing hardened concrete Making and curing specimens for strength tests', BSI London (2009)
14. Mindess, S., Zhang, L.: Impact resistance of fibre-reinforced concrete. Proc. Inst. Civ. Eng.-Struct. Build. **162**(1), 69–76 (2009)
15. CEB-FIP, C.: 'Model code 2010', Comite Euro-International du beton (2010)
16. CEB-FIP, C.: 'Design code', Comite Euro International du beton, pp. 51–59 (1990)
17. Craig Jr, R., Kurdila, A.J.: Fundamentals of Structural Dynamics, John Wiley & Sons, New Jersey (2006)

Long-Term Properties

International Round-Robin Test on Creep Behaviour of FRC - Part 2: An Overview of Results and Preliminary Conclusions

Aitor Llano-Torre[1(✉)] [iD], Pedro Serna[1] [iD], and Sergio H. P. Cavalaro[2]

[1] ICITECH Institute of Science and Technology of Concrete,
Universitat Politècnica de València, Valencia, Spain
aillator@upv.es
[2] Loughborough University, Loughborough, UK

Abstract. The International Round Robin Test (RRT) on the creep behaviour of Fibre Reinforce Concrete (FRC) cracked specimens organised by the RILEM Technical Committee 261-CCF was outlined in the BEFIB2016 Proceedings. The objective of this paper is to present an overview of the results and preliminary conclusions derived from the RRT four years later. A total of 124 specimens with either steel or macro-synthetic fibres were tested in 16 different laboratories spanning across 5 continents and following four methodologies: flexural creep of small-scale prisms, direct tension creep and flexural creep of both square and round panels. Shrinkage and creep in compression were also assessed. Specimens were subjected to sustained load for 360 days. Then, they were unloaded and left to rest for 30 days to assess the creep recovery. Finally, specimens were tested to failure to assess the residual behaviour after one-year creep test. Although a general guideline was defined for the testing procedure, each laboratory had slightly different equipment and methodology. RRT results supported the identification of the main parameters affecting the creep results. Different variables, delayed deformations, methodologies and procedures, equipment and some parameters calculated from RRT results were analysed. The preliminary conclusions from the RRT are summarised hereafter.

Keywords: Round-Robin Test · Fibre Reinforced Concrete · Creep · Cracked sections · Steel fibres · Macro-synthetic fibres

1 Introduction

The growing interest on the long-term behaviour of the Fibre Reinforced Concrete (FRC) has motivated the creation of the RILEM Technical Committee 261-CCF in 2014 to address the creep behaviour of Fibre Reinforced Concrete (FRC) in cracked state under sustained load. The absence of a standardised methodology has led to several test methods to characterize the creep behaviour of FRC. These methodologies present significant differences in terms of type and size of specimen, loading configuration, test parameters and type of equipment used. For example, in flexural creep tests, some of the differences can be observed in the load configuration, which can be either three-point bending test (3PBT) [18] or four-point bending test (4PBT) [8].

© RILEM 2022
P. Serna et al. (Eds.): BEFIB 2021, RILEM Bookseries 36, pp. 291–306, 2022.
https://doi.org/10.1007/978-3-030-83719-8_26

Specimens may also be notched or not notched before the creep test and tested in either a single setup [9] or multi-specimen setup in column of two or three specimens [1]. Direct Tension creep tests performed on cast prismatic specimens [3] or cylindrical cored specimens [17] have been also reported in the literature. As a result of research on tunnels, creep tests in panels were usually performed in square [10] or in round panels [4], following different reference standards. Moreover, some laboratories conduct creep test with sustained load on FRC structural elements [16]. These differences may be quite significant for creep results and should be deeply studied focusing on a more homogeneous creep procedure.

Therefore, in the context of the RILEM TC 261-CCF, a comprehensive Round-Robin Test (RRT) on the creep behaviour in cracked section of FRC was proposed and conducted to assess the existing methodologies with the participation of 19 laboratories from 22 institutions recognised by their excellence in the area. The RRT participants represent the technical and scientific state of the art on this topic. The RRT started in 2015 once the criteria and the program were defined. The experimental program, concrete mixes and materials, FRC characterization and criteria applied for the test parameters' definition as well as general procedures were described in previous publication [12].

The second phase of the RRT focusing on the creep test was concluded by the end of 2016 with the successful participation of 16 international laboratories. A significant database of 124 specimens tested in creep in the cracked state has been created. The resulting database reports an average of 125 data from each specimen, leading to more than 15000 data points.

This paper presents preliminary findings from such analysis. The analysis of the RRT is still ongoing and the conclusions are being drawn by the TC. Final conclusions and detailed information about the RRT will be published by the TC in the incoming RILEM State-of-the-Art Report series [14].

2   Round-Robin Test Introduction and Overview

As explained in previous publications [12], a single FRC concrete matrix C30/37 was designed with steel and macro-synthetic fibres. More than 14 tons of FRC concrete in 451 specimens were cast in the same location (Valencia, Spain) and delivered to the participants to avoid variations due to the fabrication process. The compressive and flexural residual strength characterization tests of the FRC mixes were performed by the organising laboratory LAB-01. Some of the creep test parameters were agreed and defined by the TC members considering the limitations and availability of the participants.

This paper continues the RRT description given in the previous publication [12] and completes the information. The proposed general creep test procedure and the specific procedures followed by the participants in the four different creep test methodologies are described. The data collection convention for the exchange of the RRT results is also explained. The delayed deformation curves versus time obtained as result of the creep tests are presented as well as a summary of the different analysis performed to the RRT results and the preliminary conclusions.

3   Creep Test Procedure and Methodologies

The creep test procedure followed in the RRT consists of three main phases regardless of the methodology adopted by each laboratory: *pre-cracking stage, creep stage* and *post-creep bending test stage*. This division on three phases (Fig. 1) is commonly agreed in the literature related to FRC creep in the cracked state. In the pre-cracking stage (OD), the specimens were pre-cracked up to the desired crack opening and then unloaded. In the creep stage (DF), the specimens were loaded up to a specific stress (generally expressed as a percentage of the flexure residual strength capacity) that is sustained for 360 days. Once the specified duration of sustained load was reached, the specimens were unloaded (F-H) and were left 30 additional days without load registering the delayed recovery deformation. Finally, in the post-creep stage (HK), the specimens were tested again in flexure until failure. This simplified diagram of the general procedure is defined for flexure tests in Crack Mouth Opening Displacement (CMOD), but it can be adapted for Crack Opening Displacement (COD) or deflection (δ).

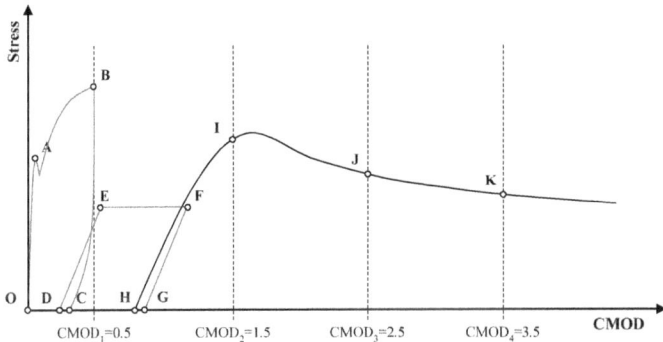

Fig. 1. Complete diagram of main phases of a creep test.

Four main methodologies were conducted following this general procedure: *flexure on prismatic specimens, direct tension,* and *flexure on square* and *round panels.* A total of 124 specimens were tested in creep. The number of specimens tested per participant and methodology are summarised in Table 1. Figure 2 shows samples of the creep frames used per methodology as a summary.

Table 1. Number of tested specimens arranged by methodology and laboratory

Creep test	Participant																Total	
	1	2	3	4	5	6	7	8	10	11	12	13	15	16	17	18	N°	%
Flexural	12	12	6	6	12	6	6	6	8	6	4	2	–	–	–	–	86	69.4
Direct tension	–	–	–	–	–	–	–	–	–	6	–	–	–	4	–	–	10	8.1
Square panel	–	–	–	–	–	–	–	–	–	–	4	–	12	–	–	8	24	19.4
Round panel	–	–	–	–	–	–	–	–	–	–	–	–	–	–	4	–	4	3.2

Considering the significant number of laboratories in the RRT and the goal of assessing the influence of the methodology in the creep results, the TC decided to allow each laboratory to maintain their respective procedure. The analysis of the results should detect the significance of those differences on both methodology and procedure.

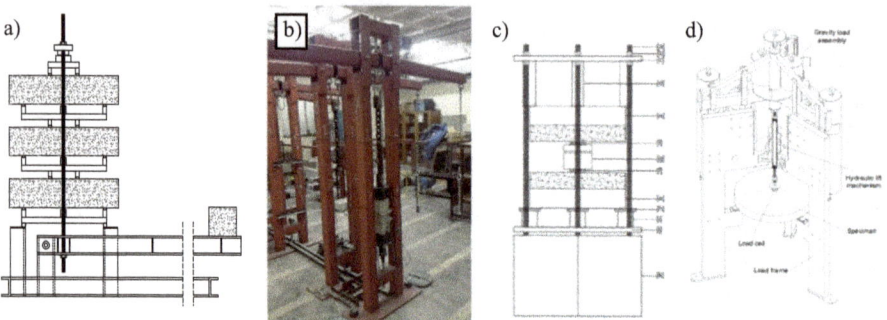

Fig. 2. Different creep test methodologies: a) flexure in prismatic specimens, b) direct tension, c) flexure in square panels and d) flexure in round panels.

4   Data Collection Definition

As a result of the RRT, a large database of creep test results was created having a similar structure as in previous database proposals [11]. The RRT database [13] compiles more than 150 parameters and variables registered for each specimen tested in the RRT. The RRT creep database assessment was used by the RILEM Technical Committee 261-CCF for the development of a new creep test procedure recommendation [15]. The RRT Creep Database [13] is available for the research community to improve the global knowledge of the long-term behaviour of cracked FRC.

In order to avoid misunderstanding between laboratories, a general convention was established with the participants regarding the data to be collected in the three stages of the creep test. In this paper, only the parameters related to the creep and post-creep bending test stages will be presented, since the parameters related to the previous stages were already discussed [12].

4.1   Creep Stage

The idealised stress-displacement curve during the creep test stage is presented in Fig. 3. Regarding the loading step (DE), the parameters defined were the stress applied during the creep stage ($f_{R,c}$) and the instantaneous CMOD immediately after reaching the reference load ($CMOD_{ci}$). The parameters covering the delayed creep stage (EF) were the total CMOD after j days in the creep test ($CMOD_{ct}{}^{j}$) and delayed CMOD after j days in the creep test ($CMOD_{cd}{}^{j}$) as the subtraction of instantaneous deformations from $CMOD_{ct}{}^{j}$. After 360 days under sustained load, the specimens were unloaded (but not removed from the creep frames) and the instantaneous recovery ($CMOD_{cri}$) was registered. The delayed recovery deformations ($CMOD_{crd}$) were registered 30 days after unloading the creep test to assess the recovery capability of FRC after the delayed stage.

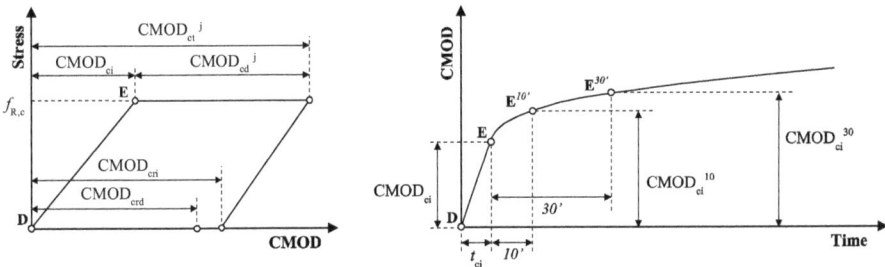

Fig. 3. Delayed creep stage parameters (left); short-term and instantaneous deformation (right).

Due to the challenge of reaching a consensus about an accurate definition of instantaneous deformation $CMOD_{ci}$ after loading, the additional reference values $CMOD_{ci}{}^{10'}$ and $CMOD_{ci}{}^{30'}$ were registered 10' and 30' minutes after reaching the desired load to assess the influence of the definition criterion in the creep coefficients, as shown in Fig. 3b. In addition to the instantaneous deformations, an additional parameter regarding the duration of the loading process in the creep test (t_{ci}) was registered to assess the influence of the different procedures in the elastic deformation.

4.2   Post-creep Stage

The post-creep bending tests were performed following the same procedure as in the pre-cracking test until specimen failure to assess the flexural behaviour after the delayed stage. Only in the case of flexural testing of prismatic specimens, one additional hysteresis loop at $f_{R,c}$ was performed to check the energy capacity of the specimens. Once the curves from the 3 stages were joined, the post–creep residual strength parameters were obtained at $CMOD_2$, $CMOD_3$ and $CMOD_4$ as depicted in Fig. 1.

5   Round-Robin Test Results

The results presented here represent only a part of those from the database of the RRT. The total $CMOD_{ct}$, COD_{ct} or δ_{ct} evolution, comprising both instantaneous and delayed deformations during the creep tests are presented for both steel and macro-synthetic FRC (SFRC and SyFRC hereafter) mixes as an example. The results were organized into four groups depending on the creep test methodology adopted. Note that the residual deformations from the pre-cracking test ($CMOD_{pr}$) are not included.

5.1   Flexural Creep Tests Procedure on Prismatic Specimens

Flexural creep testing on prismatic specimens was performed by 12 laboratories comprising 86 prismatic specimens tested. The delayed CMOD curves for some laboratories are depicted in Fig. 4 as an example. Note that two laboratories measured deflection (δ) instead of crack opening and thus, the results were converted into CMOD to obtain comparable results. Despite the high inter-laboratory variability, significant differences between FRC mixes were observed in nearly all cases.

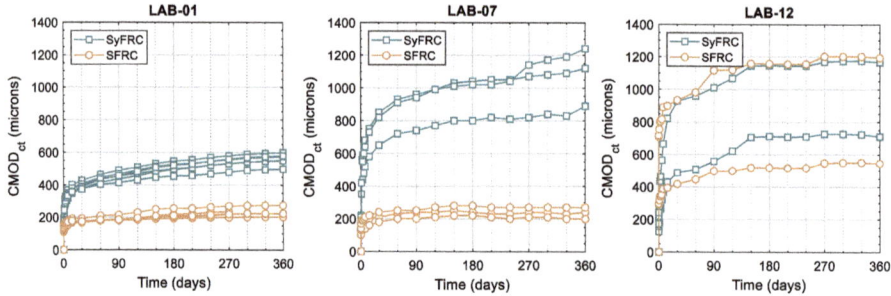

Fig. 4. Delayed CMOD obtained from flexural creep tests.

Most participants that performed creep testing in a multi-specimen setup (LAB-01, LAB-02, LAB-04, LAB-05, LAB-06, LAB-08 and LAB-11) obtained similar results. On the contrary, both LAB-12 and LAB-13 which also performed multi-specimen creep testing found significant differences regarding the long-term behaviour. In those cases where a single specimen setup was used (LAB-07 and LAB-10) similar delayed behaviour was observed. On the other hand, LAB-03 results differ from the rest of laboratories that used a single specimen setup. Such differences highlight the influence of the procedures and the equipment used in the creep tests.

5.2   Direct Tension Creep Tests Procedure

Direct tension creep tests were performed by two laboratories comprising 10 specimens in total. Similar results were obtained by both laboratories despite the different shape of the specimens and casting procedure. Delayed crack opening displacement (COD) versus time is presented in Fig. 5.

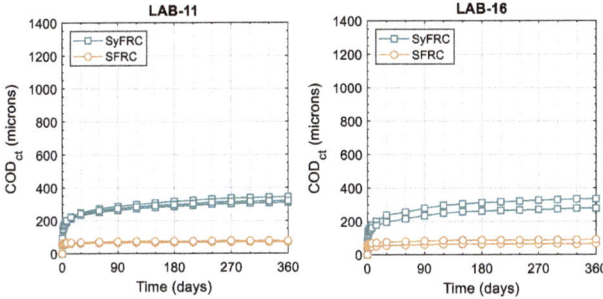

Fig. 5. Delayed COD obtained from direct tension creep tests.

5.3   Square Panel Creep Tests Procedure

The delayed deflection curves obtained from creep tests performed in 24 square panels are shown in Fig. 6. A high variability between the three participant laboratories can be observed. Significant procedure differences in terms of pre-crack level, creep index or procedures may explain the disparity in the results.

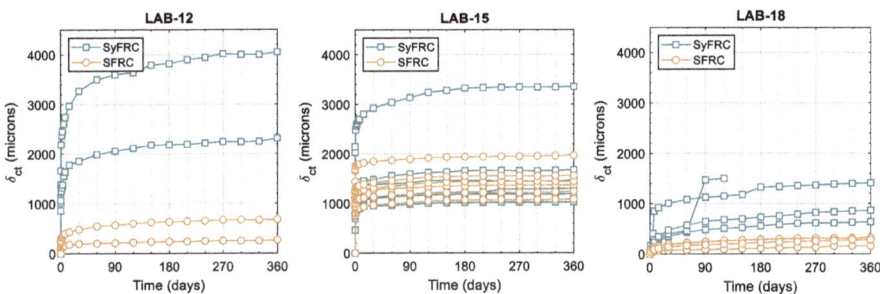

Fig. 6. Delayed deflection obtained from flexural creep tests on square panels.

5.4   Round Panel Creep Tests Procedure

Only one participant performed creep tests on 4 round panels. The results are depicted in Fig. 7. Despite using single specimen testing procedure, the results show a significant scatter. Notice that this participant could only perform the creep tests for 100 days.

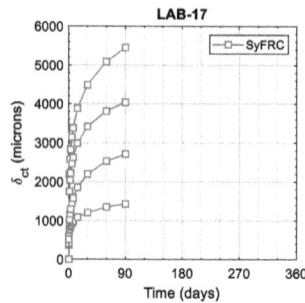

Fig. 7. Delayed deflection obtained from flexural creep tests on round panels.

6 Parameters Analysed

The conclusions from the analysis of the RRT results are limited to the studied ranges of variables and parameters considered. Therefore, extrapolation or prediction beyond the defined ranges is not advised.

The analysis of the database was divided into areas that represent the group of variables, delayed deformations, procedures and equipment aspects, methodologies and some parameters calculated from RRT results. Due to the length restrictions of this paper, only a few of these are presented hereafter. More detailed description of the analysis is available in the RILEM State-of-the-Art Report series publication about the RRT [14].

6.1 Main Variables of the RRT

The residual strength of the different FRC mixes, environmental conditions during creep tests, pre-crack level and creep index were analysed. Only a short description of most significant variables is presented in this publication.

The pre-crack level achieved in the flexural creep tests was close to the target value of 500 microns. A slightly higher variation of the pre-crack level ($COD_{pn} = 200$ microns) was observed in the direct tension test due to the difficulty in the control of the crack opening in the pre-cracking stage. In the case of panel creep test, LAB-12 and LAB-17 achieved the target deflection of 2 mm. By contrast, LAB-18 achieved 3 mm deflection and LAB-15 reported variable pre-crack deflections since the pre-cracking tests were interrupted upon reaching the limit of proportionality (LOP).

Regarding the creep index I_c, significant differences and scatter were found depending on the methodology and procedure such as those cases where multi-specimen setup was adopted for flexural creep tests or different procedures were deployed in the square panel tests, as observed in Fig. 8. Considering the histogram of creep index for flexure creep testing, the confidence interval of this RRT was assessed between 45 and 55%.

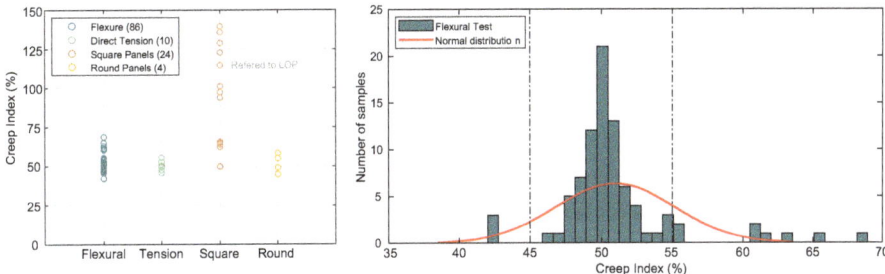

Fig. 8. Applied creep index values and histogram of creep index for flexure tests.

6.2 Analysis of Each Methodology

6.2.1 Flexural Creep Tests on Prismatic Specimens

In the case of flexure creep tests on prismatic specimens, a preliminary model for the delayed CMOD as a function of time was obtained including variables such as creep index, stress level, fibre material or laboratory. The main goal of this regression analysis procedure was to identify which variables were significant and to detect the laboratory and procedure dependence in the creep test results. This analysis showed significant differences between the results from a group of 7 laboratories named cluster A (laboratories 01, 02, 04, 05, 06, 08 and 11) and those obtained by a group of 2 laboratories named cluster B (laboratories 07 and 10). Three laboratories presented significant differences with respect to the previous ones and were placed in clusters C, D and E, respectively. As observed in Fig. 9, cluster A represent 70% of the specimens tested in flexure.

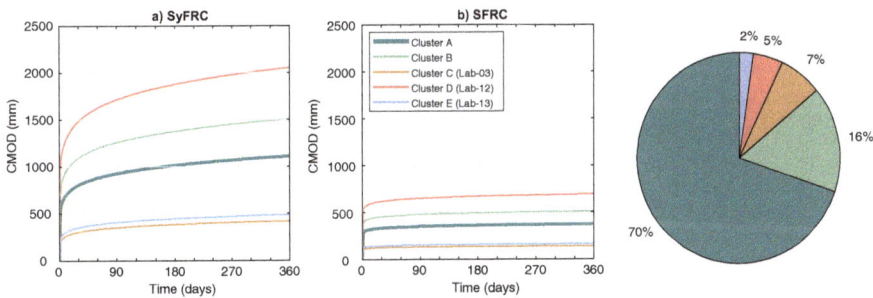

Fig. 9. Significance of different clusters in tested specimens and their basic creep curves.

6.2.2 Direct Tension Creep Tests

The limited number of specimens hindered further analysis of differences. The influence of parameters like creep index (I_c), residual strength ($f_{R,1}$) or testing procedure was not detected since only two procedures were implemented and the procedure monitorisation is more rigorous.

6.2.3 Flexural Creep Tests on Both Square and Round Panels

Due to the high scatter amongst the individual test results, the creep deflection was not as statistically relevant as the pre-cracking level. Considering the creep coefficient at 180 days, higher and more scattered creep coefficients were obtained when higher pre-crack deformations were arranged.

6.3 Procedure and Equipment Variability in the Methodologies

The flexure creep test procedure on prismatic specimens was the most common procedure, which was conducted by 12 participant laboratories. Most participants developed their own methodology from national reference standards [5, 7]. Therefore, some of the procedures and equipment differences between participants were due to the standard adaptations such as the assessment of deformation, existence or absence of notch and the load configuration during the pre-cracking or creep tests (3PBT or 4PBT). Although most participants (>80%) employed 4PBT in the creep test, up to four different load configurations were identified and classified including the use of notch from 4PBT-1 to 4PBT-4, as summarised in Table 2.

Table 2. Differences between laboratories in flexure creep test methodology.

Laboratory	Notch	Load setup		Measure	Multi	Load transfer	Air Control
		Pre-crack	Creep				
LAB-01	Yes	3 PBT	4 PBT-1	CMOD	Yes (x3)	Lever arm	Yes
LAB-02	Yes	3 PBT	4 PBT-1	CMOD	Yes (x3)	Lever arm	No
LAB-03	No	4 PBT-2	4 PBT-2	δ	No	Lever arm	No
LAB-04	Yes	3 PBT	4 PBT-1	CMOD	Yes (x3)	Lever arm	Yes
LAB-05	Yes	3 PBT	4 PBT-1	CMOD	Yes (x3)	Lever arm	Yes
LAB-06	Yes	3 PBT	4 PBT-3	CMOD	Yes (x3)	Lever arm	Yes
LAB-07	Yes	3 PBT	3 PBT	δ	No	Lever arm	No
LAB-08	Yes	4 PBT-1	4 PBT-1	CMOD	Yes (x2)	Screw bars	Yes
LAB-10	Yes	3 PBT	3 PBT	CMOD	No	Hydraulic jack	Yes
LAB-11	Yes	3 PBT	4 PBT-1	CMOD	Yes (x3)	Lever arm	Yes
LAB-12	Yes	4 PBT-1	4 PBT-1	CMOD	Yes (x2)	Hydraulic jack	Yes
LAB-13	Yes	4 PBT-4	4 PBT-4	CMOD/δ	Yes (x2)	Hydraulic jack	Yes

Additional differences arise when comparing the design and construction of the creep frames such as the use of single- or multi-specimen setup, the load transfer system, the support boundary conditions or the control of the environmental conditions. Moreover, all the support and loading rollers were analysed and classified in five different groups depending on their degrees of freedom and rotation capability. This classification revealed a significant influence of the boundary condition on the delayed performance of specimens under sustained load due to frictions or restrictions in the free deformation.

In the case of the direct tension creep test, similar procedures were followed by the two laboratories regarding the creep frame constructions and the creep load application. On the contrary, the specimen preparation differed significantly, since LAB-11 drilled cylindrical cores sized Ø94 × 150 mm from prisms while LAB-16 used 100 × 100 × 500 mm cast prisms. All specimens were notched before testing. The main methodological aspects adopted by both laboratories are summarised in Table 3.

Table 3. Differences between laboratories in direct tension creep test methodology.

Laboratory	Notch	Transducer	Measure	Multi	Load transfer	Climate Control
LAB-11	Yes	Electronic	COD	Yes (×3)	Lever arm	Yes
LAB-16	Yes	Electronic	COD	No	Lever arm	Yes

Although the square panel creep tests based on EN 14488-5 [6] were performed by three laboratories, there were mainly two different creep frames and procedures applied (see Table 4). On the one hand, the LAB-12 performed creep test on square panels following similar methodology than for flexure creep test and using a multi-specimen setup with hydraulic actuator. On the other hand, both LAB-15 and LAB-18 performed creep tests in similar frames but with some additional differences regarding the procedure, such as the pre-crack level criterion and the applied creep index. Moreover, only one participant performed round panel creep tests based on ASTM C1550 [2]. All creep tests on panels were performed on un-notched specimens.

Table 4. Differences between laboratories in square panel creep test methodology.

Laboratory	Panel	Measure	Pre-crack	Multi	Load transfer	Climate Control
LAB-12	Square	δ	2 mm	Yes (x2)	Hydraulic jack	Yes
LAB-15	Square	δ	LOP	No	Lever arm	No
LAB-18	Square	δ	3 mm	No	Lever arm	Yes
LAB-17	Round	δ	2 mm	No	Dead load	Yes

6.4 Procedure Aspects Influencing All Methodologies

Finally, procedure-related differences affecting all methodologies were detected. The time in which the load is applied by all the laboratories was analysed and significant variations from 2 to 4475 s were observed. A classification of the procedures was proposed in terms of t_{ci} in four groups as seen in Fig. 10a. A short t_{ci} of less than 60 s was employed in 29.0% of the specimens. 27.4% of the specimens were loaded in the medium t_{ci} from 60 to 360 s. The long t_{ci} classification ranges from 360 to 900 s and was performed in 30.6% of the specimens. Only 6.5% of the specimens (two laboratories) took more than 900 s to apply the load. The time in which load is applied (t_{ci}) plays a significant role in the creep coefficient calculation since this parameter has direct influence on the instantaneous deformations.

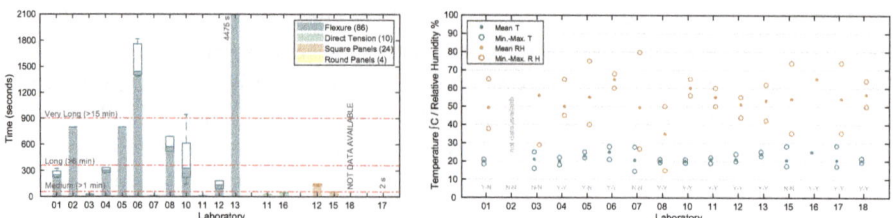

Fig. 10. Classification of t_{ci} and environmental conditions in different laboratories.

Considering the environmental conditions, 83% of participants reported controlled temperature and 75% reported controlled humidity. Although the highest variations in both temperature and humidity were reported by laboratories that did not control ambient conditions, some participants that reported environmental conditions as controlled reported a significant scatter in the environmental conditions. Such scattered values revealed that environmental conditions were not appropriately controlled, and in some cases were only restricted instead of controlled. The mean reported values as well as maximum and minimum of the reported range are depicted in Fig. 10b for each laboratory.

6.5 Creep Coefficients

The results of the flexural creep tests on beams were analysed in terms of creep coefficient. The creep coefficient φ_c evolution for both SyFRC and SFRC mixes obtained for flexure creep tests and classified by clusters are depicted in Fig. 11. This parameter is a clear indicator of the long-term deformation relative to the initial elastic deformation, as defined in the following equation:

$$\varphi^j_{w,c} = \left(CMOD_{ct}^{\;j} - CMOD_{ci}^{\;j}\right)/CMOD_{ci} \tag{1}$$

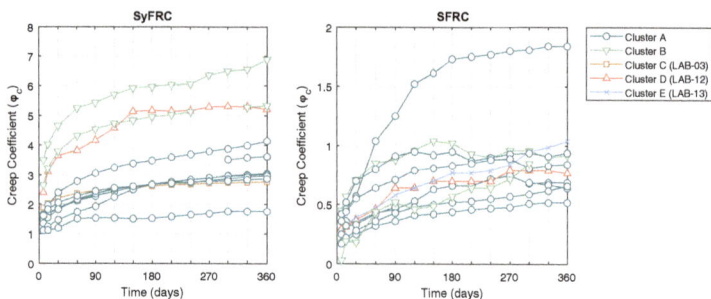

Fig. 11. Average creep coefficients φ_c evolution for both SyFRC and SFRC mixes obtained for flexure creep tests and classified by clusters.

In those cases where COD or deflection were adopted as reference measurement, the creep coefficients were obtained by using δ or COD instead of CMOD values. Moreover, the influences of the different criteria for elastic deformation ($CMOD_{ci}^{10'}$ and $CMOD_{ci}^{30'}$) were also evaluated. Results reveal that that the creep coefficients are more homogeneous and robust when $t_0 + 30'$ is used as $CMOD_{ci}$.

6.6   Crack Opening Rate

The Crack Opening Rates (COR) for three different times (COR^{30-90}, COR^{90-180} and $COR^{180-360}$) were obtained by means of the following equation:

$$COR^{j-k} = \left(CMOD_{cd}^{k} - CMOD_{cd}^{j}\right)/((k-j)/365) \qquad (2)$$

The COR from the flexure creep test decrease over time for both fibre materials as seen in Fig. 12. Moreover, COR is sensitive to the procedure adopted, which leads to high scatter between laboratories.

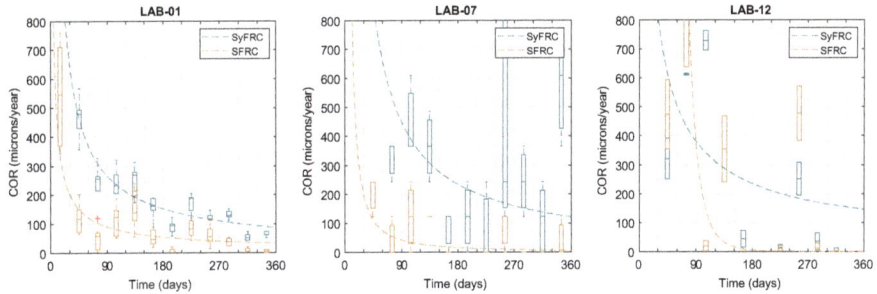

Fig. 12. COR evolution obtained for different participant laboratories.

7   Preliminary Conclusions

The RRT within the RILEM TC 261-CCF activity meant a significant effort to the organisers and participants due to the high number of specimens, methodologies and variables involved. A total of 124 specimens were tested in creep for 360 days in most cases: 86 in flexure, 10 in direct tension, 24 square panels and 4 round panels. The preliminary conclusions from the RRT are summarised below:

- The most applied test setup in the RRT is flexure in beams (69% of the specimens tested), followed by square panel testing (19%), direct tension testing (8%) and round panel testing (3%).
- There was a significant variety of the environmental conditions observed in the different laboratories. Twelve laboratories reported that creep tests were performed

in temperature-controlled conditions but regarding the relative humidity, only ten laboratories reported a control of the humidity.

- Depending on the methodology, different applied creep index deviations were observed. In the case of flexural creep testing, larger deviations were found in those laboratories which performed multi-specimen setup. In the case of square panel creep tests, scattered creep index values were observed due to significant differences in the procedures adopted.
- Despite the extended use of the flexural creep test procedure, the results of this RRT show a high variability in terms of delayed CMOD between the different laboratories. Five significant groups of methodologies were identified.
- Since the number of participants and specimens tested in direct tension creep tests is limited, the influence of parameters like creep index, concrete strength or testing procedure could not be assessed.
- Significant differences regarding the pre-crack level, creep index and applied load level were found between the three participants that performed square panel tests, which compromised the comparative analysis.
- Round panel creep tests were performed in four SyFRC specimens but unfortunately, the short duration of creep tests (around 100 days) made any comparison more difficult.

The following findings were identified regarding the differences in the flexure in beams creep test procedures by each laboratory:

- Five different load configurations were detected for the flexure creep test methodology. Despite this variability, most of the participants performed the same load configuration combination: 3PBT for pre-cracking tests and 4PBT for creep test in multi-specimen setup.
- The multi-specimen setup is an effective way of increasing the number of specimens tested in creep reducing time and costs but, in some cases, it could lead to higher scatter in the creep index.
- The different support boundary conditions were classified depending on the degrees of freedom. Five different support rollers or loading rollers were detected in the flexure creep test set-ups. This is a relevant variability that may affect the creep test results.

The following findings arise from the consideration of all the methodologies included in the RRT:

- The lever arm is the most extended way to apply the load to the creep frames and it was used by twelve laboratories. Only four laboratories used a hydraulic jack whereas both screwed bars and dead load systems were used by only one laboratory each.
- Significant differences regarding the time in which load is applied were detected due to differences in the testing procedure.
- Regarding the environmental conditions, 83% of the participants controlled the temperature while 75% of the participants controlled the humidity. Notwithstanding, even in some laboratories that controlled the environmental conditions, significant scatter of the results was reported.

- The influence on the creep coefficients of three different reference instantaneous deformations $CMOD_{ci}$ was analysed at t_0, $t_0 + 10'$ and $t_0 + 30'$. Results suggest that the creep coefficients are more homogeneous and robust when $t_0 + 30'$ is used as $CMOD_{ci}$.
- COR decreases over time regardless of the reference time considered here (COR^{30-90}, COR^{90-180} and $COR^{180-360}$).

Acknowledgements. This Round-Robin Test has been financed by the fibre manufacturers BASF, BEKAERT and ArcelorMittal Fibres. The RRT organizers wish to thank their financial support in this research project.

References

1. Arango, S., Serna, P., Martí-Vargas, J.R., García-Taengua, E.: A test method to characterize flexural creep behaviour of pre-cracked FRC specimens. Exp. Mech. **52**, 1067–1078 (2012)
2. ASTM C1550–12a, Standard Test Method for Flexural Toughness of Fiber Reinforced Concrete (Using Centrally Loaded Round Panel), ASTM International, West Conshohocken, PA (2012). https://doi.org/10.1520/C1550-12A, www.astm.org
3. Babafemi, A.J., Boshoff, W.P.: Tensile creep of macro-synthetic fibre reinforced concrete (MSFRC) under uni-axial tensile loading. Cement Concr. Compos. **55**, 62–69 (2015)
4. Bernard, E.S.: Creep of Cracked Fibre Reinforced Shotcrete Panels, Shotcrete: More Engineering Developments, Bernard ed., pp. 47–57 (2004)
5. Deutscher Ausschuss für Stahlbeton (DAfStb): DAfStb-Richtlinie Stahlfaserbeton (2012)
6. EN 14488-5:2006. Testing sprayed concrete - Part 5: Determination of energy absorption capacity of fibre reinforced slab specimens, Residual, CEN - European Committee for Standardization, Brussels (2006)
7. EN 14651:2005+A1:2007. Test Method for Metallic Fibered Concrete - Measuring the Flexural tensile Strength (Limit of Proportionality (LOP), Residual), CEN - European Committee for Standardization, Brussels (2005)
8. García-Taengua, E., Arango, S., Martí-Vargas, J.R., Serna, P.: Flexural creep of steel fiber reinforced concrete in the cracked state. Constr. Build. Mater. **65**, 321–329 (2014). https://doi.org/10.1016/j.conbuildmat.2014.04.139
9. Kusterle, W.: Viscous material behavior of solids-creep of polymer fiber reinforced concrete. In: 5th Central European Congress on Concrete Engineering, Baden (2009)
10. Larive, C., Rogat, D., Chamoley, D., Welby, N., Regnard, A.: Creep behaviour of fibre reinforced sprayed concrete, SEE Tunnel: Promoting Tunneling in SEE Region, ITA WTC 2015 Congress and 41st General Assembly, Dubrovnik, Croatia (2015)
11. Llano-Torre, A., Garcia-Taengua, E., Marti-Vargas, J.R., Serna, P.: Compilation and study of a database of tests and results on flexural creep behaviour of fibre reinforced concrete specimens. In: Proceedings of the FIB Symposium Concrete Innovation and Design, Copenhagen (2015)
12. Llano-Torre, A., Serna, P., Cavalaro, S.H.P.: International Round Robin Test on creep behavior of FRC supported by the RILEM TC 261-CCF. In: Proceedings of the BEFIB 2016, 9th RILEM International Symposium on Fiber Reinforced Concrete, Vancouver, Canada, 19–21 September 2016, pp. 127–140 (2016)

13. Llano-Torre, A., et al.: Database of the Round-Robin Test on Creep Behaviour in Cracked Sections of Fibre Reinforced Concrete organised by the RILEM Technical Committee 261-CCF (2021). https://doi.org/10.4995/Dataset/10251/163221
14. Llano-Torre, A., Serna, P. (eds.): Round-Robin test on creep behaviour in cracked sections of FRC: experimental program, results and database analysis. RILEM State-of-the-Art Reports. Springer (2021).https://doi.org/10.1007/978-3-030-72736-9
15. Llano-Torre, A., Serna, P.: Recommendation of RILEM TC 261-CCF: test method to determine the flexural creep of fibre reinforced concrete in the cracked state. Mater. Struct. **54**(3), 1–20 (2021). https://doi.org/10.1617/s11527-021-01675-0
16. Vasanelli, E., Micelli, F., Aiello, M.A., Plizzari, G.: Long term behavior of FRC flexural beams under sustained load. Eng. Struct. **56**, 1858–1867 (2013)
17. Vrijdaghs, R., di Prisco, M., Vandewalle, L.: Uniaxial tensile creep of a cracked polypropylene fiber reinforced concrete. Mater. Struct. **51**(1), 1–12 (2018). https://doi.org/10.1617/s11527-017-1132-5
18. Zerbino, R., Monetti, D.H., Giaccio, G.: Creep behaviour of cracked steel and macro-synthetic fibre reinforced concrete. Mater. Struct. **49**(8), 3397–3410 (2015). https://doi.org/10.1617/s11527-015-0727-y

Evaluation of Creep and Shrinkage of Patented Mixture of UHPC with Applied Heat-Treatment

Vladimír Příbramský[(✉)]

Department of Concrete and Masonry Structures, Faculty of Civil Engineering,
Czech Technical University in Prague, Thákurova 7/2077, 166 29 Praha 6,
Prague, Czech Republic
vladimir.pribramsky@fsv.cvut.cz

Abstract. In this paper the description and evaluation of results of experimental verification of rheological properties of patented mixture of ultra-high-performance concrete (UHPC) is presented. Specimens were cured with various curing regimes including curing by an increased temperature and in a water saturated environment. For the evaluation of the results an adapted model B4 is used, which is considered the most advanced rheological material model based on great consistency with large set of experimental results. It seems to be viable for use for prediction of creep and shrinkage of UHPC as it predicts long-term strains by incorporating effect of volume of additives and admixtures used in the fresh concrete. Model B4 also takes into effect thermal treatment of fresh concrete which accelerates cement hydration in early age. Current model B4 has several limitations that are often exceeded by characteristics of UHPC. In this paper, these limitations are identified and viable adaptation of model B4 is presented.

Keywords: UHPC · Creep · Shrinkage · Model B4 · Heat treatment

1 Rheological Properties of UHPC

Currently there is no complex mathematical model of creep and shrinkage of concrete that would be applicable for UHPC as well. Models or recommendations available in German, French or Japanese standards for concrete design use adapted models for concrete of ordinary grade. Their accuracy is limited by a number of factors that have much smaller effect on prediction of creep and shrinkage of concrete of ordinary grade than in the case of UHPC. For shrinkage of UHPC is apparent dominant effect of autogenous shrinkage over the drying shrinkage [1, 2]. The speed of hydration (and the possibility of significant speed up by curing with application of heat-treatment) has tremendous effect on autogenous shrinkage build up speed, elimination of drying shrinkage and mitigation of later creep strains [3, 5]. Additional strain mitigating factor is usage of high dosages of high-strength steel fibre reinforcement.

© RILEM 2022
P. Serna et al. (Eds.): BEFIB 2021, RILEM Bookseries 36, pp. 307–318, 2022.
https://doi.org/10.1007/978-3-030-83719-8_27

1.1 Curing UHPC with Heat-Treatment

Early age concrete curing is in the case of UHPC even more important than curing of concrete of ordinary grade. Applying steam on the surface or placing the specimens in environment with relative humidity close to 100% leads up to four times smaller long-term creep strains when compared to untreated specimen placed in environment with 65% relative humidity shortly after casting the specimens [5]. Currently frequently researched topic is applying heat-treatment to the UHPC members. Heat-treatment is a procedure of placing the specimens in the environment with high relative humidity and temperature between 60 °C – 90 °C. These conditions are ideal for swift hydration build-up which is connected to fast strength and stiffness build-up. Strength build-up is accompanied by progressing autogenous shrinkage and mitigation of long-term creep strains.

The heat-treatment is also connected to often discussed phenomenon of delayed ettringite formation (DEF). DEF is topic linked with rapid deterioration of precast and prestressed concrete and was investigated recently as a cause for massive damage on heat-treated sleepers in India [6]. The UHPC studies show that as DEF is primarily linked with water available within the concrete matrix. This is a factor that is greatly mitigated in UHPC [7]. The relative humidity in UHPC microstructure is further discussed in chapter 1.4.

1.2 Experimental Measurement of Patented UHPC Mixture

A set of experiments was performed with the help of the Klokner institute to verify the behaviour and properties of UHPC. Four sets of specimens of patented UHPC mixture with characteristic 28-day compressive strength of 140 MPa were cured in different curing regimes. One set was left untreated and three sets were cured after 24 h from casting for 24 h in a water pool with specific curing temperature. After the curing was completed mechanical propertied of the control specimens (cubes and cylinders) were measured. Mechanical properties (compressive strength and E modulus) were measured in 4 different ages: 1 day (right after demoulding and before the curing was applied), 2 days (after curing), 7 days and 28 days after casting. Heat-treatment was applied on specimens by placing them in a water pools in which the temperature was controlled and kept by system of thermometers and multiple immersion heaters placed in the pool. Four different sets of specimens were measured – each with different curing regime:

- Untreated specimens placed after demoulding in ambient temperature. These specimens are in figures below marked as "Air 20"
- Cured specimens placed in water pool with ambient temperature for 24 h ("Water 20")
- Cured specimens placed in heated water pool with temperature 70 °C for 24 h ("Water 70")
- Cured specimens placed in heated water pool with temperature 90 °C for 24 h ("Water 90")

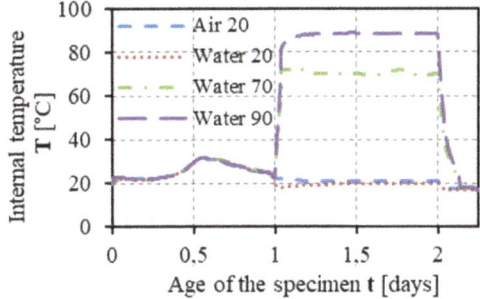

Fig. 1. Temperature of specimens during curing regime.

The exact temperature of the specimens before and during the curing process was recorded and are visible on Fig. 1 above. The average curing temperature was set for "Water 70" 70,6 °C and for "Water 90" the temperature was 88,0 °C.

For measurement of creep and shrinkage strains were in each set 3 prism specimens with dimensions 70 × 70 × 300 mm. The specimens contained vibrating wire strain gauges with thermal sensor, which enabled measurement of strains and temperature from the exact time of casting. For the water treated specimens the measurement was paused for the time when curing was applied. The temperature was measured during that time directly in the water pools (Fig. 2).

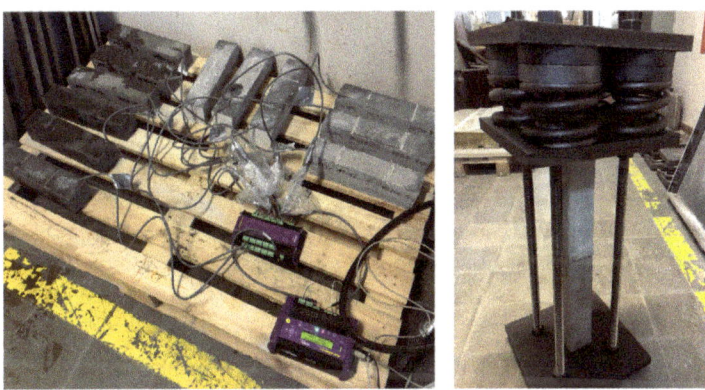

Fig. 2. Measurement of shrinkage and creep strains after the curing and applied heat-treatment.

Shrinkage strains were measured on a single specimen from each set and creep strains were measured on 2 specimens from each set, that were placed together in a stand and loaded by hydraulic press to value 150 kN right after their temperature decreased to ambient temperature. Applied load represented 37% of average compressive strength of uncured specimens ("Air 20") measured right before loading the specimens.

Heat-treatment in water pool accelerates the hydration and leads to high early-age strength as well as autogenous shrinkage of the cured specimens. The exact effect will be presented in chapter Results.

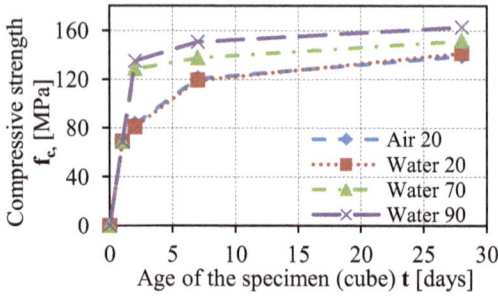

Fig. 3. Compressive strength of UHPC for different curing regimes.

The average compressive strength of cured specimens ("Water 90") was right after the curing (48 h after casting) 135,1 MPa. This value is very close to average 28-day compressive strength of the untreated specimens, which was measured 139,2 MPa. The rise of average compressive strength in time for measured sets is apparent on Fig. 3 above. From the figure is also apparent, that curing at ambient temperature ("Water 20") has no effect on strength of the UHPC.

1.3   Model B4

The most modern and most widely used model for prediction of rheological behaviour of concrete is model B4, which was published recently [4] as a successor to model B3 II. Model was developed for accurate creep and shrinkage strains prediction of large-scale structures (bridges with long span, skyscrapers, nuclear power plants, dams, cooling towers…). For such structures of high strategic value is important to predict their long-term behaviour accurately to keep them in flawless shape during their long lifespan.

Model B4 is based on extensive number of experimental results and tests both recent and published in the past. This model uses complex approach and incorporates important effects, such as concrete mixture components dosages and effects of admixtures and determines the concrete member behaviour based on its shape. The main advantage of the B4 model is its continuous development based on newly conducted experiments and the possibility of its adaptation for new modern mixtures of high-performance concrete, UHPC and fibre reinforced concrete based on parametric study of specific usage by means of behaviour and admixture coefficients.

1.4   Model B4 Limitation and Proposed Adaptations

Model B4 was developed for description of concrete of ordinary grades and it contains boundaries of the model applicability which are exceeded by mixture of UHPC. For UHPC are most relevant following boundaries [4]:

- Concrete compressive strength (influences calculation of E modulus)
- Curing temperature (influences strain build up speed)
- Water to cement ratio w/c (influences final creep and shrinkage strain magnitudes)

1.4.1   Concrete Compressive Strength

E modulus of UHPC is in the model B4 determined from average compressive strength of concrete at age 28 days. Upper boundary condition is set in the model on average compressive strength of 70 MPa. For concrete with greater compressive strength is the magnitude of E modulus overestimated by the model by approximately 25%. More suitable approach for UHPC seems to be from Model code 2010 [10]. In MC10 is presented more complex approach to determine the E modulus based also on type of aggregate. For comparison in Fig. 4 below were used E moduli values from experiments [5, 6, 8]. Alternative approach is to use Eq. (1) suggested by Alsalman [9] which is based on large number of published papers containing E moduli of UHPC in past decade (cumulatively containing measured data of 240 specimens). Good consistency of formula by Alsalman [9] is also achieved for tested specimens (in Fig. 4 marked as "CTU 2020") and for many tests published in the USA [8].

$$E_{28} = 8,01 \cdot f_c^{0,36} \tag{1}$$

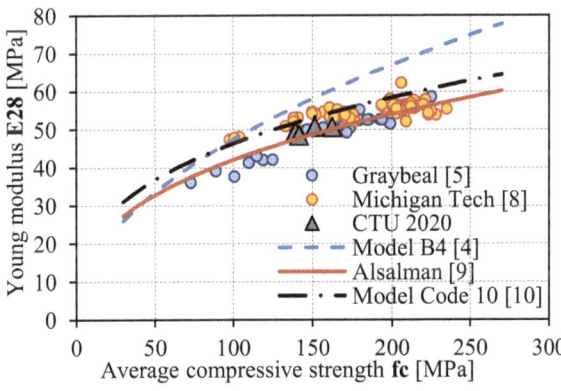

Fig. 4. Relation between E modulus and average 28-day compressive strength.

1.4.2   Curing Temperature

Model B4 describes the effect of environment temperature during curing of the concrete on speed of hydration by parameter β_{Th} [4]. Parameter β_{Th} has a linear effect on calculated effective age of the concrete after the curing process. Even though the relation of β_{Th} to curing temperature is exponential, studies and performed experiments show that if extrapolated it significantly underestimates effect of temperatures above 50 °C during curing regime. This is understandable as the upper boundary of model B4 for applicability of β_{Th} is in environment with temperature under 30 °C. A modified approach Eq. (2) with additional exponential member, which causes massive growth of parameter β_{Th} with environment temperature above 50 °C is necessary.

$$\beta_{Th,adapted} = \beta_{Th} + \exp\left(\frac{3 \cdot U_h}{R} \cdot \left(\frac{1}{293} - \frac{1}{T_{cur} + 231}\right)\right) \tag{2}$$

1.4.3 Water to Cement Ratio

Water to cement ratio "w/c" has significant effect on autogenous shrinkage and creep strains. UHPC is in this manner a specific material with very low w/c approximately 0,15–0,20. The appropriate consistency and workability of the concrete mixture is achieved by adding substantial dosages of superplasticizers. In original model B4 using such low w/c ratio leads to major inconsistency with experimentally obtained results of total creep and shrinkage strains. It is caused by extreme deficiency of water in the concrete mixture that leaves a portion of cement unhydrated. This unhydrated cement then acts as a fine aggregate in the concrete matrix and is not active in hydration.

Based on hydration equations of cement substances and their molar weights may be derived a boundary w/c ratio that would theoretically lead to full hydration of all provided water. If we mark this boundary value as w/c_{min}, its magnitude is for cement types I-III between 0,23–0,26. In practice, using water to cement ratio of w/c_{min} cannot lead to full hydration of all present cement due to drying and so bigger cement grains cannot be fully hydrated.

Heat-treatment causes rapid cement hydration in environment saturated by hot steam or liquid water. From experimental results it is apparent that additional water is transported into the concrete matrix where it promotes further hydration. This behaviour is possible to incorporate into model B4 for UHPC by adapting calculation of the w/c ratio by reduction based on parameter β_{Th} as shown in Eq. (3).

$$w/c = w/c_{min} - \frac{1}{250}\sqrt{\beta_{Th}} \tag{3}$$

This hypothesis is supported by measurement of concrete compressive strength on specimens with the same mixture recipe but different curing regimes. Additional proposed adjustment of model B4 is in aggregate to cement ratio "a/c", where portion of cement that cannot hydrate due to lack of available water is counted as part of the aggregate (acts as a filling).

$$a/c = \frac{a + c - \frac{w}{w/c}}{\frac{w}{w/c}} \tag{4}$$

As aggregate "a" is in Eq. (4) assumed combined mass of all mixture component with exception of water, cement active in hydration and superplasticizers. In model B4 acts aggregate content as mitigating factor of development of autogenous shrinkage and creep strains.

1.4.4 Adjustment of Drying Shrinkage Calculation

From majority of experiments [1, 3, 5] used for model B4 calibration is apparent that in the case of UHPC there is very low magnitude of drying shrinkage and drying creep strains or none at all. In model B4 is the magnitude of final drying shrinkage strains determined (besides mixture components) on relative humidity (RH) of surrounding environment. This effect is described by parameter r_h in Eq. (5). When relative humidity exceeds 98%, drying shrinkage is replaced by drying swell.

$$r_h = 1 - h^3 \text{ for } h \leq 0,98 \qquad (5)$$

In the case of UHPC (especially when heat-treatment is applied) occurs nearly full hydration of cement by free water in the concrete mixture and water from curing. Therefore, in the concrete matrix there is significant reduction of free water that could escape by drying shrinkage due to rapid cement hydration. In the model B4, this effect may be incorporated by reduction of relative humidity in the concrete leading to smaller RH difference between the concrete and the environment which would mitigate or eliminate drying shrinkage. Proposed modification of model B4 is by adding inner concrete relative humidity parameter h_c, which modifies calculation of r_h as apparent in Eq. (6). Measured internal humidity of UHPC based on w/c ratio was published in parametric study of internal humidity in 10 different HPC mixtures measured 1 year after casting [11]. The w/c ratio in the study was varying between 0,25 – 0,4 and it may be simply approximated by a linear function. It may be also seen, that for the UHPC with substantial silica fume content, the values of relative humidity are furthermore reduced. On the Fig. 5 are extrapolated the available data [11] for UHPC with w/c around 0,15.

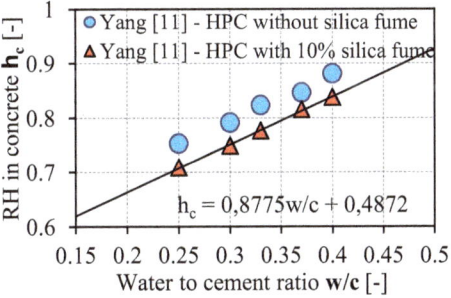

Fig. 5. Dependency of w/c parameter on internal relative humidity in concrete [11].

In this area and especially in the case of UHPC there seem not to be enough data and additional experimental verification of this phenomenon is required. Most studies simply conclude that UHPC does not develop strains due to drying shrinkage. Exception would be French study [2], where some amount of drying shrinkage was recorded.

$$rh = 1 - (h + (1 - hc))^3 \text{ and } r_h \geq 0 \tag{6}$$

Characteristic feature of UHPC that mitigates drying is also its homogeneity and resistance to environment effect due to low permeability of the surface.

1.5   Model B4 Calibration

Model B4 includes the effect of admixtures to predict creep and shrinkage strains by sets of empirical coefficients based on the admixture type. For UHPC are relevant two separate sets of admixture coefficients – superplasticizer content and silica fume content. Model B4, however, does not offer a set of coefficients for UHPC containing both superplasticizer and silica fume and sets of coefficients greatly differ for either of these admixtures. On the tested specimen, neither of the empirical sets provided is fitting to the measured values. In order to fit the measured values, fitting coefficients were iteratively acquired. Original sets and the adapted used set of coefficients are shown on Table 1 below.

Table 1. Empirical coefficients for shrinkage prediction [4].

Admixture class (% from c)	$\times p_2$	$\times p_3$	$\times p_4$	$\times p_5$
Superpl. ($\geq 0\%$)	0,72	2,19	1,72	0,48
Silica fume ($\geq 0\%$)	1,12	3,11	0,51	0,61
Used to fit experiment (CTU 2020)	**2,25**	**2,30**	**2,40**	**0,48**

However, different UHPC recipes show significant variation in autogenous shrinkage development speed when specimens are heat-treated. In the study by Flietstra [3] the speed of autogenous shrinkage development is slow, while in other studies [2, 5] and on measured specimens, the effect is substantial even during the curing of the specimens. This observation is the main obstacle in attempt to create a universal calibrated model B4 for UHPC. To calibrate the model B4 for UHPC parameter r_t, which has direct effect on the speed of shrinkage strain development, was changed from value −4,5 to −1,2 in order to fit the measured data.

Model B4 uses for the prediction of creep strains the cumulative effect of both admixture of superplasticizer and silica fume (coefficients are listed in Table 2 below). However, in the case of used UHPC recipe, the amount of used superplasticizer is exceeding the highest superplasticizer content threshold nearly four times when used together with silica fume. As in the case of coefficients for shrinkage prediction, a separate set of coefficients was used in order to fit the measured results.

Table 2. Empirical coefficients for creep prediction [4].

Admixture class (% from c)	$\times \tau_{cem}$	$\times \varepsilon_{au}$	$\times r_{ew}$	$\times r_\alpha$
Superpl. ($\leq 5\%$), Silica fume ($\leq 8\%$)	6,00	2,80	0,29	0,21
Superpl. ($\leq 5\%$), Silica fume ($\geq 8\%$)	3,00	0,96	0,26	0,71
Superpl. ($\geq 5\%$), Silica fume ($\leq 8\%$)	8,00	1,95	0,00	1,00
Silica fume (>8%, $\leq 18\%$)	2,60	0,82	0,00	1,20
Silica fume (>18%)	1,00	1,50	5,00	1,00
Used to fit experiment (CTU 2020)	**6,00**	**2,28**	**0,00**	**1,00**

The noncontinuous or discrete character of sets of coefficients for admixture level content as well as high dosages of silica fume in the UHPC recipe makes using these coefficients very problematic without proper calibration. However, the used set for creep prediction is closer to provided sets than in the case of coefficients for shrinkage. This obstacle may be overcome by a more extensive parametric experimental study of different admixture types and their isolated effect on the UHPC rheological properties.

2   Results

When previously presented adaptations and calibrations of model B4 are taken into account, the experimentally measured strains of creep and shrinkage show very good consistency with the prediction of model B4.

2.1   Shrinkage

Predicted and measured values of shrinkage strains are shown on Fig. 6 below. Interesting result is, that cured specimens show substantial decrease of the magnitude of total shrinkage strain after heat-treatment was applied. Additionally, heat-treated specimens show very low level of later shrinkage development after the curing. Values of measured total shrinkage strains show good consistency with the values obtained by the calibrated model B4.

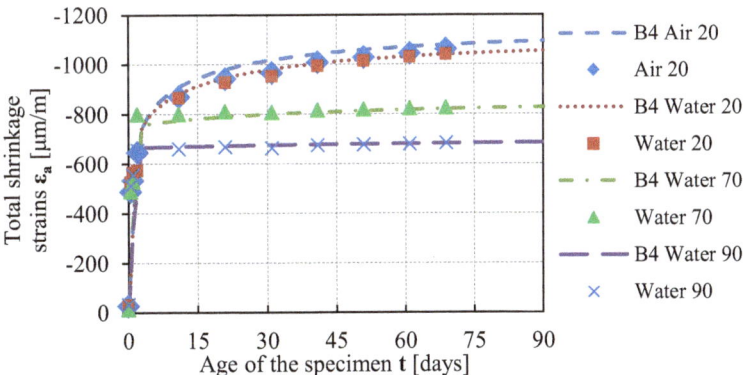

Fig. 6. Total shrinkage strain development of UHPC measured experimentally and compared with predicted values by model B4.

Even though the separate values of autogenous and drying shrinkage were not measured experimentally, they may be separated and explained using model B4. The autogenous shrinkage strain is the major part of the total shrinkage strain and in time correlates with the increase of the compressive strength of the UHPC. Heat-treated specimens show both increased compressive strengths, but also greater magnitude of autogenous shrinkage strains when compared to untreated specimens as is apparent on Fig. 7 (left).

Drying shrinkage is even more affected by the curing regime and the applied heat-treatment. As was explained earlier, heat-treatment substantially accelerates the hydration. While being cured, the specimens were placed in a water pool and swelling was observed during the curing process. As the maturing of heat-treated specimens is accelerated, when the curing stops, the swelling effect may be measured indirectly as the magnitude of total shrinkage strain of heat-treated specimens is much lower than of specimens, that were cured at ambient temperature. The heat-treated specimens after curing show also expected lower rate of drying shrinkage and therefore the specimens remain swelled even months after the curing took place as is apparent on Fig. 7 (right).

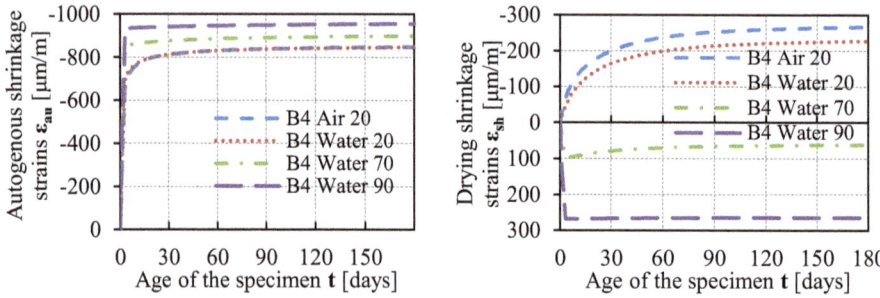

Fig. 7. Autogenous shrinkage and drying shrinkage strain development of UHPC measured experimentally and compared with predicted values by model B4.

2.2 Creep

Good consistency of measured data and calibrated model B4 was also achieved as is visible on Fig. 8 below. Adapted and calibrated model B4 show good consistency also with experiments and consistent are both magnitudes of the creep strains and also speed of creep strain development for uncured and heat-treated specimens.

As the experiment started, a failure occurred on multiple strain gauges on specimens "Water 20" and "Air 20". Valid data were acquired for these specimens only for the first 8–10 days, after loading the specimens in the stands. These specimens will be in recent future casted again and experiment will be repeated in order to measure the long-term creep strain development of loaded specimens. Despite this setback, the available data seem sufficient to show trend and consistency with the B4 model prediction.

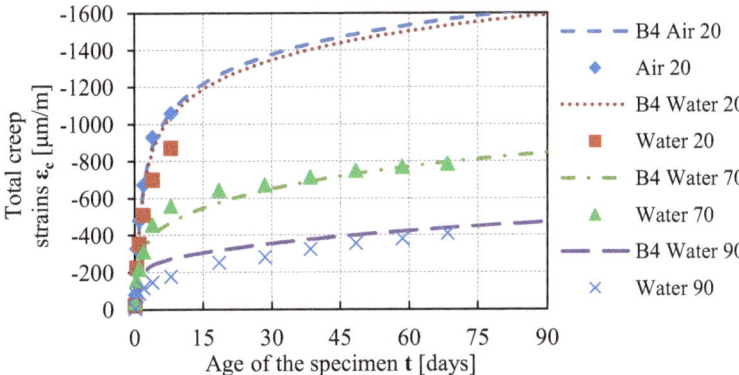

Fig. 8. Creep strain development of UHPC measured experimentally and compared with predicted values by model B4.

3 Conclusions

In this paper were presented few aspects of UHPC long-term behaviour. Curing regimes and heat-treatment was elaborated and its effect on creep and shrinkage behaviour of UHPC was measured experimentally on UHPC prism specimens. For prediction of rheological behaviour of UHPC a modified model B4 was used. Model B4 in general seems very suitable for the prediction of UHPC, as it includes coefficients for calibration of the model based on different admixtures and it also describes differences in behaviour based on curing temperature. In this paper the limits of current model B4 for prediction of UHPC behaviour were identified and proper adaptations were recommended. For the future research was identified as most vital the relation between UHPC creep and shrinkage strains and application of heat-treatment and further calibration of admixture content, especially of silica fume, superplasticizers and their synergistic effects. The obstacle for creation of general application of model B4 for UHPC recipes is the unavailability of sufficient data of creep and shrinkage of UHPC as the recipe for used UHPC mixture is in many cases kept as a business secret of the manufacturer and therefore is not published.

Acknowledgements. This work was supported by SGS grant of CTU in Prague, Czech Republic, grant No. SGS20/042/OHK1/1T/11. Experiments were carried out at the Klokner Institute in Prague.

References

1. Burkart, I., Mueller, H.S.: Creep and shrinkage characteristics of ultra high strength concrete (UHPC). In: Proceedings of 2nd International Symposium on Ultra High Performance Concrete (2008)

2. Francisco, P., Benboudjema, F., Rougeau, P., Torrenti, J.M.: Creep and shrinkage prediction for a heat-treated Ultra High Performance Fibre-Reinforced Concrete. In: Proceedings of 3rd International Symposium on UHPC and Nanotechnology for High Performance Construction materials, 3/2012 (2012)

3. Flietstra, J.C., Ahlborn, T.M., Harris, D.K., Silva, H.M.: Creep behavior of UHPC under compressive loading with varying curing regimes. In: Proceedings of 3rd International Symposium on UHPC and Nanotechnology for High Performance Construction materials (2012)

4. Bažant, Z.P.: RILEM Technical Committee TC-242-MDC. Model B4 for creep, drying shrinkage and autogenous shrinkage of normal and high-strength concretes with multi-decade applicability, Materials and Structures (2014)

5. Graybeal, B.A.: Material Property Characterization of Ultra-High Performance Concrete, Final report; Office of Infrastructure Research and Development, Federal Highway Administration (2006)

6. Awasthi, A., Matsumoto, K., Nagai, K., Asamoto, S., Goto, S.: Investigation on possible causes of expansion damages in concrete – a case study of sleepers in Indian Railways. J. Asian Concrete Fed. 3(1), 49–66 (2017)

7. Heinz, D., Ludwig, H.M.: Heat treatment and the risk of DEF delayed ettringite formation in UHPC. In: Proceedings of the International Symposium on Ultra High Performance Concrete, Kassel, Germany September 13–15, 2004, pp. 717–730 (2004)

8. Center for Structural Durability Michigan Tech Transportation Institute. Ultra High Performance Concrete for Michigan Bridges Material Performance – Phase I Final Report (2008)

9. Alsalman, A., Dang, C.N., Prinz, G.S., Hale, W.M.: Evaluation of modulus of elasticity of ultra-high performance concrete. Constr. Build. Mater. 153, 918–928 (2017)

10. Fédération internationale du béton. Bulletin No 65. Model Code 2010 - Final draft, vol. 1 (2012)

11. Yang, Q., Zhang, S.: Self-desiccation mechanism of high-performance concrete. J. Zhejiang Univ. Sci. 5(12), 1517–1523 (2005)

Cracking Behaviour of FRC Members Reinforced with GFRP Bars under Sustained Loads

Razan H. Al Marahla and Emilio Garcia-Taengua[✉]

School of Civil Engineering, University of Leeds, Leeds, UK
E.Garcia-Taengua@leeds.ac.uk

Abstract. GFRP bars are regarded as an alternative to steel reinforcement in marine and aggressive environments. However, there are some shortfalls to the use of GFRP reinforced members in flexure, which the addition of fibres can redress. This paper is concerned with the effect of synthetic fibres on the cracking behaviour of GFRP reinforced members. A number of FRC beams reinforced with GFRP bars were tested in flexure, considering different synthetic fibre contents and GFRP bar diameters. The flexural loads applied were representative of service conditions and were sustained for 90 days. The short- and long-term cracking behaviour was analysed in terms of crack spacing, distribution and development in pure bending sections. It was concluded that synthetic fibres increased the cracking moment capacity by up to 20% and reduced the crack width and crack spacing by up to 63% and 31%, respectively.

The accuracy of the models available in current codes to predict crack width and crack spacing was assessed by comparing the experimental results to theoretical predicted values. The accuracy of crack spacing, and crack width predictions was found to vary with fibre content, and higher discrepancies were associated with higher fibre contents. This study shows that current prediction models for crack width and spacing need updating to make them better suited to elements reinforced with GFRP bars adequately considering the contribution of synthetic fibres.

Keywords: Crack width · Crack spacing · FRP · Sustained load · Synthetic fibres

1 Introduction

Fibre-reinforced polymer (FRP) bars are increasingly used as reinforcement in concrete structures where the corrosion of steel reinforcement can be critical to their serviceability, being a competitive alternative to conventional steel in harsh environments [1, 2]. Serviceability conditions govern the design of FRP-reinforced members, and therefore the study of cracking and deformation is of fundamental importance. In service conditions, structural concrete elements reinforced with FRP bars operate at between 20 and 40% of their flexural capacity [3–6]. Due to its comparatively low modulus of elasticity, FRP reinforcement causes concrete members to exhibit larger deformations and bigger crack widths than their steel-reinforced equivalents [4, 7]. In

© RILEM 2022
P. Serna et al. (Eds.): BEFIB 2021, RILEM Bookseries 36, pp. 319–330, 2022.
https://doi.org/10.1007/978-3-030-83719-8_28

consequence, cracks need to be controlled to avoid undesirable behaviour due to the degradation of bond between FRP bars and concrete over time. FRP bars are available in different types based on the material embedded in the resin matrix. This study considers glass fibre reinforced polymer (GFRP).

It is well known that fibre reinforced concrete (FRC) typically presents more distributed crack patterns and smaller crack widths [8, 9]. However, although previous studies report on ductility and toughness improvements achieved with different types of fibres, the majority of literature is concerned with the contribution of steel fibres and has paid little attention to non-metallic fibres.

There is abundant literature on the short-term cracking of structural concrete members reinforced with FRP bars [7, 10, 11]. However, there is limited research published on long-term cracking of FRC beams with FRP bars. This study is part of an experimental programme aimed at investigating the influence of synthetic fibres on the long-term flexural performance of GFRP reinforced beams. This paper presents the experimental investigation of the influence of synthetic fibres on the short- and long-term cracking of GFRP beams under sustained loads, and compares the experimental observations with the theoretical values based on the prediction models proposed by the Eurocode 2 [12], ACI 440.1R-15 [1], Model code 2010 [13] and RILEM TC 162-TDF [8].

2   Experimental Programme

2.1   Materials and Mix Designs

The same reference concrete mix, with a water-to-cement ratio of 0.29 and a mean compressive strength of 60 MPa at 28 days, was considered for all beams and control specimens. The type of cement used was CEM I 52.5N, and the reference concrete mix was designed so that it could accommodate different synthetic fibre contents by only adjusting the superplasticiser dosage (SikaViscoCrete 25MP was used). Synthetic macro-fibres with a length of 54 mm, a diameter of 0.34 mm and a tensile strength of 600 MPa were used. They were considered at three different dosages: 0, 5, and 10 kg/m^3, corresponding to volume fractions of 0%, 0.55% and 1.1%, and referred to as mix 1, mix 2 and mix 3, respectively in Table 1.

Pultruded GFRP bars with diameters of 16 mm and 20 mm were used as flexural reinforcement for the beams produced in this study. These bars had helical surface and were made with vinyl ester resin with a 50 year service life. Their tensile strength and modulus of elasticity were determined by testing five bars of each diameter to ACI 440.3R-04 [14]. The stress-strain curves obtained were linear up to the rupture stress and showed no significant differences between diameters. The average tensile strength and elastic modulus were 694 MPa and 40 GPa respectively, and the ultimate strain was 1.76%.

Table 1. Synthetic FRC mix designs.

Constituents	Quantity (kg/m^3)		
	Mix 1	Mix 2	Mix 3
Water	147	147	147
Cement	510	510	510
Fine aggregate	950	950	950
Coarse aggregate (20 mm)	300	300	300
Coarse aggregate (10 mm)	580	580	580
Synthetic fibres	0	5	10
Superplasticiser	9.4	12	14

2.2 Flexural Tests

A total of eight simply supported beams were produced. All beams had the same dimensions: the rectangular cross-section was 200 mm wide and 260 mm deep, and the clear span between supports was 3050 mm. These beams were tested in flexure following a four-point configuration, as shown in Fig. 1. Steel stirrups were equally spaced to provide sufficient shear reinforcement in all sections of the beam apart from the central third of the span, in order to prevent shear failure.

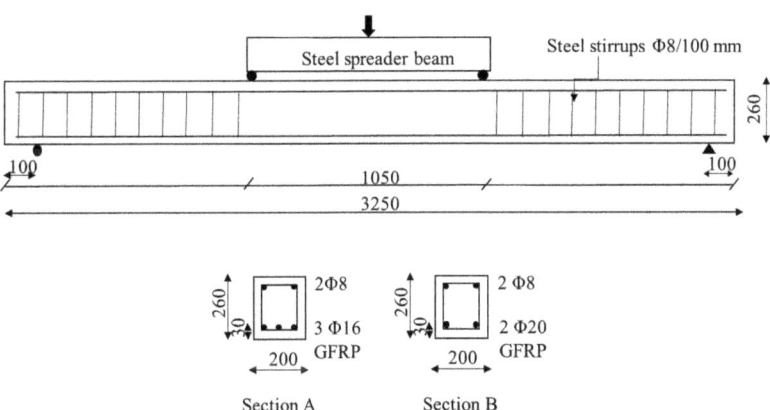

Fig. 1. Beam dimensions and reinforcement details.

Two series of beams (A and B) were produced: four beams were reinforced with 16 mm GFRP bars, and four were reinforced with 20 mm GFRP. The number of GFRP bars in each case was adjusted so that all eight beams had practically the same reinforcement ratio. As shown in Table 2, each series comprised 4 beams: one was the reference beam without fibres; two identical beams with 5 kg/m^3 fibres content, and one beam with 10kg/m^3 of fibres. Beams with moderate synthetic fibre content

(5kg/m^3) were replicated to get an indication of variability in the results. They were designated as F#D#, where F denotes the fibre content in kg/m^3, and D denotes the bar diameter in mm.

Table 2. Beams details.

Beam designation	Bar diameter (mm)	Fibre content (kg/m^3)	Section
F0D16	16	0	A
F5D16	16	5	A
F5D16* (replicate)	16	5	A
F10D16	16	10	A
F0D20	20	0	B
F5D20	20	5	B
F5D20* (replicate)	20	5	B
F10D20	20	10	B

2.3   Testing Procedure

As illustrated in Fig. 1, beams were supported by rollers and loaded at two points, each being 525 mm away from midspan. The load was applied at a constant rate of 2 kN/min until it reached the value of 37 kN, which was then sustained for 90 days. This load corresponded to 40% of the design flexural capacity of the section and was considered as representative of serviceability conditions.

Midspan deflection was monitored by means of an LVDT, and concrete surface strains were measured using DEMEC points installed in the constant bending moment zone, with a spacing of 150 mm between them. Strains on the GFRP bars were measured with strain gauges placed on their surface. Crack width, depth and spacing were measured manually and recorded immediately after the application of the load and regularly thereafter for 90 days. This paper focuses on the values observed right after the application of the load and after 90 days of sustained load.

3   Experimental Results and Discussion

3.1   Characterisation Test Results

The compressive strength, tensile strength, and modulus of elasticity of the concrete mixes were determined by testing cylindrical (150 × 300 mm) and cubic (100-mm side) specimens. From each batch of concrete produced, 3 cylinders and 3 cubic specimens were produced and tested. This information was used to monitor consistency between batches. The average values are given in Table 3.

No significant differences were observed in relation to compressive strength or elastic modulus. Tensile strength improved with the addition of synthetic fibres,

Table 3. Concrete mechanical properties.

Properties	Mechanical properties (MPa)		
	Mix 1	Mix 2	Mix 3
Compressive strength	60.3	60.1	59.7
Tensile strength	3.90	4.4	4.6
Modulus of elasticity	39.6	40.9	39.8

increasing by 13% when 5 kg/m^3 fibres were used, and by 4.5% when the fibre content was increased from 5 to 10 kg/m^3. This was consistent with studies reporting improvements of up to 19% in tensile strength when using synthetic fibres [15, 16].

In terms of flexural strength, three notched prismatic specimens per batch were tested to EN 14651. Specimens with synthetic fibres exhibited softening behaviour, and the average ratio of the residual flexural strength f_{R1} to the limit of proportionality (f_{R1}/f_L) was 0.42 for 5 kg/m^3 synthetic fibres and 0.61 for 10 kg/m^3 synthetic fibres.

3.2 Influence of Fibres on the Cracking Load and Crack Width

During the application of the load, flexural cracks appeared in the constant bending moment zone when the cracking load was reached. With the increasing load, new vertical cracks formed. The load levels at which cracks stabilised ranged between 1.5–2 times the cracking load, which was consistent with studies reporting that this occurs at load levels between 11% and 41% of the flexural capacity [17, 18].

The cracking load, number of cracks, crack width, depth and spacing were recorded. Crack width, depth and spacing were averaged taking into account the measurements from all cracks. Values at 90 days are given in Table 4, together with the average crack width right after the load was applied.

Table 4. Crack measurements recorded after 90 days sustained loading.

Beam designation	Cracking load (kN)	No. of cracks	Average spacing (mm)	Average depth (mm)	Average crack width (mm)	
					0 day	90 days
F0D16	21.1	10	201	212	0.53	0.72
F5D16	22.6	13	160	186	0.34	0.44
F5D16*	26.2	13	157	178	0.31	0.42
F10D16	25.2	15	125	147	0.18	0.31
F0D20	20.7	10	230	214	0.55	0.79
F5D20	22.3	11	180	188	0.35	0.51
F5D20*	24.0	11	183	180	0.34	0.53
F10D20	24.8	13	121	145	0.22	0.34

FRC beams exhibited higher cracking loads than the control beams. Fibres were found to improve the cracking load by up to 14% when dosed at 5 kg/m^3. Cracking loads were further increased when the fibre content increased from 5 kg/m^3 to 10 kg/m^3, but only by 4%. FRC beams developed more cracks and presented lower crack widths and depths than their control counterparts (Fig. 2). The average short-term crack width was 0.34 mm and 0.20 mm in FRC beams with 5 and 10 kg/m^3 of synthetic fibres respectively, as opposed to 0.54 mm in the reference beams, without fibres.

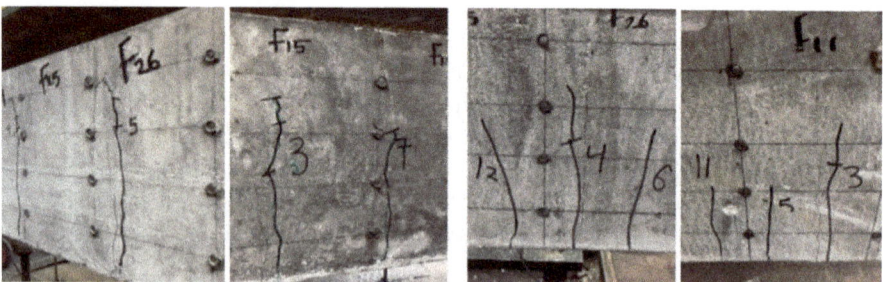

Fig. 2. Crack detail in beams without fibres (left) and with synthetic fibres (right).

The average crack widths observed in the reference beams were higher than the limit values of 0.5–0.7 mm recommended by the ACI Committee 440. It is not uncommon that the crack widths observed in beams with GFRP bars exceed recommended limits [19]. However, the addition of 5 kg/m^3 of synthetic fibres sufficed to bring crack widths down to acceptable levels.

The reference beams without fibres developed new cracks along the span within the first week of sustained loading, which was consistent with previous reports [20]. The FRC beams, on the other hand, did not develop new visible cracks during the sustained loading period. Furthermore, the average crack depth was reduced by up to 14% and 31% with fibre dosages at 5 and 10 kg/m^3, respectively.

The crack width values measured after the 90 days of sustained loading were between 40–60% higher than the values registered right after the application of the flexural load. However, the crack widths significantly decreased with increasing fibre contents consistently at all ages, as shown in Fig. 3. Reductions between 34% and 40% were observed for a synthetic fibre content of 5 kg/m^3, and between 57% and 66% for a synthetic fibre content of 10 kg/m^3.

The theoretical crack width values obtained from different models under short and long-term conditions are given in Table 5, whilst Table 6 presents the experimentally observed versus predicted ratios. Predicted values as per Eurocode 2 [12], CSA [21] and some other models [22–24] closely matched the experimental short-term crack widths of reference beams. The ACI 440.1R and Eurocode 2 models yielded good predictions for reference beams without fibres but overpredicted the long-term crack width values by 26% and 52% in FRC beams with fibre contents of 5 and 10 kg/m^3, respectively. This was attributed to these models considering higher bond coefficients

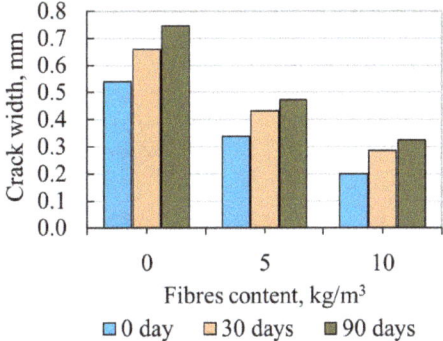

Fig. 3. Average crack widths observed for different dosages of fibres.

and not adequately accounting for the fibres contribution in increasing the area of concrete active in the tension zone.

The model proposed by RILEM TC162-TDF, on the other hand, underpredicted long-term crack widths. Although the model proposed by the Model Code 2010 accounts for the post-cracking tensile strength of FRC, it underpredicted crack width values by 45% and 34% for fibre contents of 0 and 5 kg/m^3, respectively, but the difference with the experimental values decreased to 9% for a fibre content of 10 kg/m^3. This was attributed to this model having been developed for steel FRC with steel reinforcement, not synthetic FRC with GFRP bars.

Table 5. Theoretical crack width values according to different models.

Models and provisions	Crack width (mm)	
	Bar 16 mm (short/long)	Bar 20 mm (short/long)
CSA [21]	0.56	0.58
ISIS Manual [23]	0.82	0.95
ACI 440.1R-015 [1]	0.71	0.71
Toutanji and Saafi [22]	0.55	0.66
Salib and Abdel-Sayed [24]	0.53	0.62
Wang and Belarbi [9]	0.53	0.61
EC2 [12]	0.56/0.69	0.62/0.77
RILEM TC 162-TDF [8]*	0.09/0.11	0.10/0.13
CEB-FIP Code [13]	0.28/0.37	0.35/0.46
CEB-FIP Code [13]*	0.20/0.30	0.26/0.35
CEB-FIP Code [13]**	0.18/0.26	0.24/0.32

*FRC with 5 kg/m^3
**FRC with 10 kg/m^3

Table 6. Long-term crack widths compared to theoretical values.

Beam	Exp.	EC2			MC2010		ACI 440.1R	
	$W_{exp.}$	$W_{pred.}$	$W_{pred.}/W_{exp.}$		$W_{pred.}$	$W_{pred.}/W_{exp.}$	$W_{pred.}$	$W_{pred.}/W_{exp.}$
F0D16	0.72	0.69	0.96		0.37	0.51	0.71	0.98
F5D16	0.44	0.69	1.74		0.28	0.68	0.71	1.61
F5D16*	0.42	0.69	1.65		0.28	0.71	0.71	1.68
F10D16	0.31	0.69	2.24		0.25	0.83	0.71	2.28
F0D20	0.79	0.77	0.88		0.46	0.59	0.71	0.89
F5D20	0.51	0.77	1.50		0.35	0.68	0.71	1.39
F5D20*	0.53	0.77	1.44		0.35	0.65	0.71	1.33
F10D20	0.34	0.77	2.25		0.32	0.95	0.71	2.08

The crack width values obtained experimentally were used to obtain an adjusted value for the bond coefficient (k_1 or k_b) for the different fibre contents considered. Table 7 shows the adjusted bond coefficient values and the corresponding long-term crack widths.

The plot in Fig. 4 shows the ratio of theoretical to experimental crack width versus fibre contents using original and adjusted bond coefficients in the different models. It can be observed that the ACI, ISIS and CSA models, when used with the adjusted bond coefficients, led to predicted crack width values that were quite close to those experimentally observed for plain and FRC beams. On the other hand, predicted values using EC2 were higher, which was attributed to the lower bond coefficient originally proposed in the EC2 model (0.8).

Table 7. Crack width values using adjusted bond coefficients.

Fibre content (kg/m^3)	Adjusted k_b	Crack width (mm)			
		ACI	EC2	ISIS	CSA
0	1.26	0.75	0.81	0.85	0.76
5	0.77	0.45	0.69	0.54	0.46
10	0.56	0.33	0.63	0.39	0.34

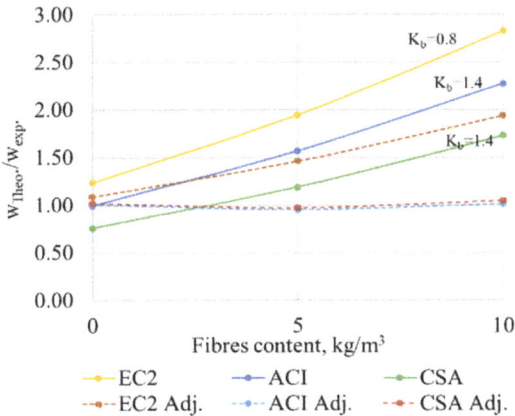

Fig. 4. Ratio of theoretical to experimental crack widths versus fibre contents.

3.3 Influence of Fibres on the Crack Spacing

The spacing between flexural cracks was measured in all tested beams. Table 8 presents the maximum, minimum and average values of crack spacing measured after 90 days of sustained load, together with the ratios of average to minimum and maximum to minimum crack spacing.

Table 8. Long-term maximum and average crack spacing.

Beam designation	Crack spacing (mm)			Ratios	
	$S_{max.}$	$S_{Ave.}$	$S_{Min.}$	$S_{ave.}/S_{Min.}$	$S_{max.}/S_{Min.}$
F0D16	250	201	170	1.18	1.47
F5D16	218	160	115	1.39	1.89
F5D16*	210	157	121	1.29	1.74
F10D16	156	125	78	1.60	2.00
F0D20	248	230	180	1.28	1.38
F5D20	220	180	123	1.46	1.79
F5D20*	222	183	113	1.62	1.96
F10D20	160	121	85	1.43	1.88

Using the expressions in Eurocode 2 for steel-reinforced beams, the maximum and average crack spacing would be 232 mm and 177 mm for a rebar diameter of 16 mm, and 255 mm and 201 mm for a diameter of 20 mm. Comparing these values to those in Table 8, the maximum crack spacing values predicted by the Eurocode were similar to those observed in the GFRP-reinforced beams without fibres. However, it overpredicted the crack spacings observed in FRC beams by up to 35%. This indicated that crack spacing estimates to Eurocode 2 reproduced well the observations with GFRP bars but failed to account for the effect of synthetic fibres.

The synthetic FRC beams presented significant reductions in crack spacing when compared to their counterparts without fibres, the maximum cracks spacing being reduced up to 37% for a synthetic fibre content of 10 kg/m³. The ratio of maximum to average crack spacing was 1.16 for the beams without fibres, and 1.29 for FRC beams. Observed values of the ratios of maximum to average crack spacing and minimum to average crack spacing are plotted against average crack spacing in Fig. 5. The average crack spacing was 1.39 times the minimum crack spacing, whilst the maximum crack spacing ranged between 1.35 and 2 times the minimum crack spacing. These observations were found to be in good agreement with the values reported in previous studies [25].

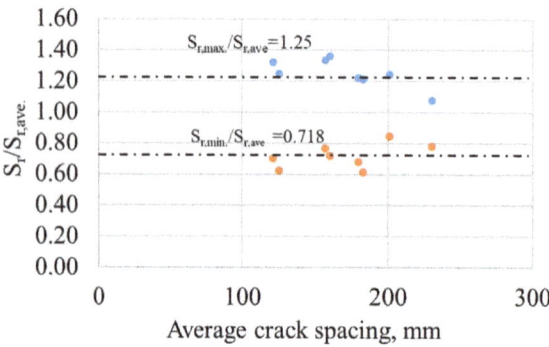

Fig. 5. Relative maximum and minimum crack spacing with respect to average crack spacing.

According to the Model Code 2010, the maximum crack spacing corresponds to twice the length over which slip occurs between concrete and reinforcement. For 16 and 20 mm bar diameters, these values would be 111 mm and 139 mm, respectively, and as a result the maximum crack spacing is significantly underpredicted in GFRP-reinforced beams with synthetic fibres. Some researchers have proposed models to calculate the crack spacing in steel FRC, e.g. Moffatt [26] accounts for the residual flexural strength of the material. However, such models do not seem to work well for synthetic FRC beams with GFRP bars, significantly underestimating the crack spacing in GFRP reinforced beams, yielding values of 77 mm and 84 mm for 16- and 20 mm bar diameters, respectively.

4   Summary and Conclusions

This paper reports on the experimental results on the short- and long-term cracking behaviour of synthetic FRC beams reinforced with GFRP bars subject to sustained flexural load for a period of 90 days. The relative effect of the synthetic fibre content and bar diameter on crack width and spacing are discussed. Experimentally obtained values of crack width and crack spacing were compared to the theoretical values obtained by means of predictive models in a number of codes and reports. The main conclusions of this study can be summarised as follows:

- The incorporation of synthetic fibres to concrete at the volume fractions considered in this study resulted in an increased cracking capacity of GFRP beams. The observed cracking loads were increased by up to 14% and 20% in beams with 5 and 10 kg/m^3 of synthetic fibres, respectively.
- The average crack width in GFRP reinforced beams with 5 kg/m^3 of synthetic fibres were up to 37% lower than those corresponding to the reference beams without fibres. Higher reductions, up to 63%, were observed in beams with 10 kg/m^3 of synthetic fibres.
- The spacing between flexural cracks was reduced by 13% and 31% when synthetic fibres were dosed at 5 and 10kg/m^3, respectively. Similar reductions were observed in the crack depth values: crack depth observed in beams with 5 and 10kg/m^3 of synthetic fibres was 14% and 31% lower, respectively.
- The comparison of experimental results for crack width, depth and spacing against the values predicted by different models confirmed that these do not accurately represent the cracking of synthetic FRC beams reinforced with GFRP bars. This was attributed to the fact that existing models mostly assume steel reinforcement and steel fibres when the contribution of fibres is accounted for.

Acknowledgements. The authors are thankful to Oscrete Construction Products (Christeyns UK Ltd) and Sika Ltd, which very kindly provided the materials used in this study. The support provided by the technical team of the laboratories in the School of Civil Engineering, University of Leeds, is also acknowledged. The authors are thankful to Al-Zaytoonah University of Jordan for the financial support granted to Ms Al Marahla in undertaking her PhD studies at the University of Leeds.

References

1. ACI 440.1R-15, ACI Committee 440: Guide for the Design and Construction of Structural Concrete Reinforced with FRP Bars. American Concrete Institute, Farmington Hills, p. 44 (2015)
2. Nanni, A., De Luca, A., Jawaheri Zadeh, H.: Reinforced Concrete with FRP Bars: Mechanics and Design. Taylor & Francis Group, CRC Press, Boca Raton (2014)
3. Pilakoutas, K., Neocleous, K., Guadagnini, M.: Design philosophy issues of fiber reinforced polymer reinforced concrete structures. J. Compos. Constr. **6**(3), 154–161 (2002)
4. Nanni, A.: North American design guidelines for concrete reinforcement and strengthening using FRP: principles, applications and unresolved issues. Constr. Build. Mater. **17**(6–7), 439–446 (2003)
5. Rafi, M.M., Nadjai, A.: Evaluation of ACI 440 deflection model for fiber-reinforced polymer reinforced concrete beams and suggested modification. ACI Struct. J. **106**(6), 4–6 (2009)
6. El-Nemr, A., Ahmed, E.A., Benmokrane, B.: Flexural behavior and serviceability of normal- and high-strength concrete beams reinforced with glass fiber-reinforced polymer bars. ACI Struct. J. **110**(6), 1077 (2013)

7. Bischoff, P.H.: Reevaluation of deflection prediction for concrete beams reinforced with steel and fiber reinforced polymer bars. J. Struct. Eng. **131**(5), 752–767 (2005)
8. RILEM TC 162-TDF, 162-TDF: Test and design methods for steel fibre reinforced concrete. Mater. Struct. **35**(9), 579–582 (2002)
9. Wang, H., Belarbi, A.: Flexural behavior of fiber-reinforced-concrete beams reinforced with FRP rebars. In: Proceedings of the 7th Symposium on FRP in Reinforced Concrete Structures—FRPRCS7, pp. 895–914 (2005)
10. Barris, C., Torres, L., Vilanova, I., Mias, C., Llorens, M.: Experimental study on crack width and crack spacing for Glass-FRP reinforced concrete beams. Eng. Struct. **131**, 231–242 (2017)
11. Goldston, M.W., Remennikov, A., Sheikh, M.N.: Flexural behaviour of GFRP reinforced high strength and ultra high strength concrete beams. Constr. Build. Mater. **131**, 606–617 (2017)
12. BS EN 1992–1–1: Eurocode 2: Design of concrete structures–Part 1–1: General rules and rules for buildings. European Committee for Standardization (2004)
13. CEB-FIP Code, M.: International Federation for Structural Concrete (fib). Federal Institute of Technology Lausanne–EPFL, Section Génie Civil, pp. 3–978 (2010)
14. ACI 440.3R-04: ACI Committee 440: Guide Test Methods for Fiber-Reinforced Polymers for Reinforcing or Strengthening Concrete Structures. American Concrete Institute, Farmington Hills (2004)
15. Song, P., Hwang, S., Sheu, B.: Strength properties of nylon-and polypropylene-fiber-reinforced concretes. Cem. Concr. Res. **35**(8), 1546–1550 (2005)
16. Fallah, S., Nematzadeh, M.: Mechanical properties and durability of high-strength concrete containing macro-polymeric and polypropylene fibers with nano-silica and silica fume. Constr. Build. Mater. **132**, 170–187 (2017)
17. Barris, C., Torres, L., Comas, J., Miàs, C.: Cracking and deflections in GFRP RC beams: an experimental study. Compos. B Eng. **55**, 580–590 (2013)
18. Mias, C., Torres, L., Turon, A., Barris, C.: Experimental study of immediate and time-dependent deflections of GFRP reinforced concrete beams. Compos. Struct. **96**, 279–285 (2013)
19. Wu, T., Sun, Y., Liu, X., Wei, H.: Flexural behavior of steel fiber–reinforced lightweight aggregate concrete beams reinforced with glass fiber–reinforced polymer bars. J. Compos. Constr. **23**(2), 04018081 (2019)
20. Gross, S. P., Yost, J.R., Kevgas, G.J.: Time-dependent behavior of normal and high strength concrete beams reinforced with GFRP bars under sustained loads. In: High Performance Materials in Bridges, pp. 451–462 (2003)
21. CSA: Canadian Standard Association, CSA-S806–02CSA-S806–02, Design and Construction of Building Components with Fibre-Reinforced Polymers. American Concrete Institute, Toronto, pp. S806-S812 (2012)
22. Toutanji, H.A., Saafi, M.: Flexural behavior of concrete beams reinforced with glass fiber-reinforced polymer (GFRP) bars. ACI Struct. J. **97**(5), 712–719 (2000)
23. ISIS Manual: Manual no.3. Strengthening Reinforced Concrete Structures with Externally Bonded Fibre Reinforced Polymers. The Canadian Network of Centers of Excellence on Intelligent Sensing for Innovative Structures, ISIS Canada, University of Winnipeg, Manitoba, p. 151 (2001)
24. Salib, S.R., Abdel-Sayed, G.: Prediction of crack width for fiber-reinforced polymer-reinforced concrete beams. Struct. J. **101**(4), 532–536 (2004)
25. Borosnyoi, A.: Serviceability of CFRP Prestressed Concrete Beams. PhD Thesis, Budapest University of Technology and Economics (2002)
26. Moffatt, K.: Analyse de dalles de pont avec armature réduite et béton de fibres métalliques. Montréal, Canada: École polytechnique de Montréal, p. 248 (2001)

Experimental Investigation on the Influence of Temperature Variations on Macro-synthetic Fibre Reinforced Concrete Short and Long Term Behaviour

Clementina Del Prete[✉], Nicola Buratti, and Claudio Mazzotti

DICAM – Structural Engineering, University of Bologna, Bologna, Italy
clementina.delprete2@unibo.it

Abstract. In the last decades, fibre reinforced concretes became widely adopted in structural applications. Nevertheless, some elements of their mechanical behaviour are not fully understood. For instance, the effect of environmental conditions on the short and long-term behaviour of these materials has been studied to a limited extent only.

In this perspective, the present paper presents the results of a large experimental campaign involving flexural tests on Macro Synthetic Fibre Reinforced Concrete (MSFRC) specimens under short- and long-term loads. Two different polypropylene Fibres were used, with dosages of 8 kg/m^3 and 10 kg/m^3. The effect of temperature on the short-term behaviour of these materials was investigated by performing three-point bending tests at 20 °C and 40 °C in cracked and uncracked conditions. The effect of temperature variations on long-term deformations was studied by means of four-point bending tests on pre-cracked notched beams at increasing temperatures, from 20 °C to 40 °C. The paper presents the test results as well as analyses the effective number of fibres crossing the cracks.

Keywords: FRC · Testing · Sustained load · Creep · Shrinkage · Temperature

1 Introduction

The mechanical characterization of the behaviour of FRC elements in cracked conditions, and in particular under long-term loads, is a topic of great interest. This is mostly motivated by the absence of official recommendations in terms not only of design criteria but also of experimental procedures [1, 2]. Furthermore, being FRCs composite materials, the identification of the contributions of their constituents to time–dependent deformations is not yet fully understood.

While for short-term loads FRCs containing either synthetic or steel fibres can feature similar performances [3, 4], many studies in the literature indicate that there are differences under long-term loads [1, 5], mostly due to the creep of fibres [2, 6, 7]. Kurtz et al. [8] tested the long-term performance of cracked beams made FRCs with polypropylene or nylon short fibres. Creep failure occurred when the stress level was higher than a certain percentage of the failure load under short-term monotonic testing.

© RILEM 2022
P. Serna et al. (Eds.): BEFIB 2021, RILEM Bookseries 36, pp. 331–341, 2022.
https://doi.org/10.1007/978-3-030-83719-8_29

MacKay et al. [9] described the results of experimental tests comparing the behaviour of one SFRC and one MSFRC under long terms loads, observing that cracked MSFRCs were associated to larger values of the creep coefficient. Kusterle [10] tested one SFRC and three different MSFRCs and concluded that MSFRCs had large long-term deformations and that a maximum creep load ratio of 50% seemed to be the maximum for obtaining good long-term performance. Zhao et al. [11, 12] carried out an experimental program to investigate the long-term behaviour of SFRCs under uniaxial tensile loads by testing cylindrical specimens. The time-dependent crack opening observed was almost at the same level of instantaneous crack opening after 3 months loading at around 30% of cracking strength. They also concluded that the damage due to debonding at the fibre/matrix interface was not increasing with creep deformation at the loading level of 30%, even though the irreversible part almost doubled during the creep loading. Babafemi and Boshoff [6] investigated the time-dependent behaviour of a MSFRC under long-term uniaxial tensile loading and observed significant crack widening over time under sustained uniaxial tensile loads. Even at loads as a low as 30% of the post-peak resistance, the time-dependent crack widening did not stabilize after 8 months. Tensile creep failure occurred within 10 days for specimens loaded at 60% of the post-peak resistance and within less than a day for a 70% loaded specimen. Average fibre counts on the cracked face of MSFRC were found to influence the time-dependent behaviour. Babafemi and Boshoff [6] also performed single fibre long-term pull-out tests observing that specimens loaded at 50% of the quasi-static capacity pulled out over time. Buratti and Mazzotti found that temperature influences the creep deformation rates in particular on MSFRCs [5].

The present paper combines these aspects, characterizing the mechanical performance of a concrete reinforced with polypropylene fibres, in particular two different types are considered. They present different geometry and mechanical properties and the FRC so produced is tested under short- and long-term loading conditions. The experiments presented have been performed with two different curing temperature at same relative humidity, at 20 °C and 40 °C with 55% RH, and in cracked and un-cracked states. In addition, the influence of the temperature variation on the long-term state is detected by increasing the rate during the flexural creep test, from 20 °C to 30 °C and 40 °C.

2   Description of the Experimental Campaign

The experimental campaign presented here had two main objectives; studying of the effect of limited. temperature variations on the i) short- and ii) long-term behaviour of macro-synthetic FRC elements.

2.1   Materials

All the specimens tested were cast with the concrete mix reported in Table 1. Concrete was produced with a concrete mixer with vertical axis. All aggregates were first mixed for 5 min with an amount of water equal to half of the total, cement and remaining water were then added and finally fibres were inserted continuing mixing for 3 min. Superplasticizer was added in order to achieve a slump of at least 170 mm.

Two types of macro-synthetic fibres were used, both of them are made of polypropylene and feature a crimped shape; their main geometrical and mechanical parameters are reported in Table 2. In the following the fibre type is identified with the codes F1 and F2. For both the fibres a dosage of 8 kg/m^3 was used.

Table 1. Specifications of the concrete mix design.

Component	Quantity
CEM I 52.5R (kg/m^3)	400
Sand 0–1 mm (kg/m^3)	886
Sand 0–5 mm (kg/m^3)	438
Gravel 5–15 mm (kg/m^3)	444
Water/Cement ratio (-)	0.44

Table 2. Geometry and mechanical properties of the Fibres used, as declared by the producer.

Properties	F1	F2
Material	Polypropylene	Polypropylene
Diameter (mm)	0.75	0.81
Length (mm)	40	54
Aspect ratio (-)	53	67
Density (g/cm^3)	0.91	0.91
Tensile strength (MPa)	420–480	510–590
Elastic Modulus (MPa)	3300–3900	5700–6300

2.2 Short-Term Tests

The analysis of the effect of temperature variations on the short-term behaviour was carried out by testing 150 × 150 × 600 mm notched specimens in three point bending after curing in four different conditions:

A-20   20 °C;

B-20   pre-cracking at 20 °C up to a CMOD of 0.15 mm, then unloading and curing at 20 °C and 55% RH for 48 hours, and finally three-point bending test to failure;

A-40   40 °C for 48 h;

B-40   pre-cracking at 20 °C up to a CMOD of 0.15 mm, then unloading and curing at 40 °C and 55% RH for 48 h, and finally three-point bending test to failure;

The FRC with the fibres F1 were subjected to the four curing conditions described above, while other FRC, only to A-20 and A-40 conditions. This difference is justified by the results obtained for the fibres F1 (Sect. 3.2). Besides, cubic specimens were produced as well in order to characterize the compressive strength of the MSFRCs at 20 °C and 40 °C.

Short-term tests were carried-out in compliance to EN 14651 [13] (Fig. 1), according which tests are carried out in CMOD control in three point bending flexural configuration. The tests were performed using a servo-hydraulic machine. The crack mouth opening was measured using a COD transducer.

Fig. 1. Three point bending test set-up according to EN 14651.

2.3 Long-Term Tests

The investigation of temperature effect on long term deformations was carried out by performing flexural creep tests on $150 \times 150 \times 600$ mm notched specimens, in cracked conditions, for a period of 280 days. A group of 8 prismatic specimens was first pre-cracked in three point bending until a CMOD of 0.5 mm and then unloaded. The three of them exhibiting the highest residual strength (f_{R1}) were tested under sustained load. During creep tests the relative humidity was set to 55% while the temperature was increased as follows:

a. 20 °C for 90 days (MSFRC with F1) and 50 days (MSFRC with F2);
b. 30 °C for 40 days (MSFRC with F1) and 128 days (MSFRC with F2);
c. 40 °C until the end of the test (approximately 6 months of total duration).

The remaining five prisms were not used for the creep tests. They were stored, not loaded, in the same environmental conditions. After that, all specimens were re-loaded, in three-point bending, until failure in order to characterize the deformation in service conditions.

The experimental procedure used for the creep tests is not standardized, but consistent with other tests in the literature [14, 15]. A class 2 lever was used to apply a constant load to a stack of three specimens. A four-point bending scheme was adopted in order to guarantee stability of the stack. The load applied was set in order to produce a target nominal flexural tensile stress in the cracked section (i.e. a bending moment) equal to 50% of the residual strength measured during pre-cracking at CMOD = 0.5 mm.

During the creep test, the CMOD on each specimen was recorded using LVDTs. At the end of creep tests, specimens were unloaded, recording CMOD for further 30 days, and then tested to failure using the same setup adopted for pre-cracking, i.e. three-point bending (Fig. 2).

Fig. 2. Creep loading frames adopted for the tests.

3   Results

3.1   Compressive Tests

Cubic specimens were cured in two different conditions, half of them for 28 days at 20 °C while the others at 20 °C for 26 days and then at 40 °C and 55% relative humidity for 48 h. The mean compressive strength obtained is reported in Table 3. In general, the FRC with fibres F2 tends to have a lower compressive strength that the FRC containing F1. This result might be attributed to the decrement of workability of the concrete admixture due to the different geometry of F2, and the consequent increment of the entrapped air. For both the FRCs specimens tested after curing at 40 °C had slightly lower strengths than those maintained at 20 °C.

Table 3. Mean compressive strength obtained testing $150 \times 150 \times 150$ mm cubes.

ID specimens group	Temperature °C	R_{cm} (MPa)
F1 – Group 1	20	65.57
F1 – Group 2	40	57.47
F2 – Group 1	20	50.61
F2 – Group 2	40	48.91

3.2   Short Term Flexural Tests

Figure 3 shows mean nominal residual flexural tensile strength versus CMOD curves for the specimens containing the fibres F1 (a) and F2 (b), for the different curing conditions. Concerning the specimens with the fibres F1 it is possible to notice that pre-cracking had no significant effect, in fact, the curves corresponding to specimens cured in cracked-conditions (B-20) and (B-40) are consistent with the curves for uncracked specimens. For this reason, fibres F2 were tested only in conditions A-20 and A-40. It should be noticed that the portion of the B-20 and B-40 curves for 0 mm $\leq$ CMOD $\leq$ 0.15 mm, correspond to pre-cracking, which was carried-out in all cases at 20 °C. Figure 3a suggests a strength decrease of 16% associated to the temperature increase for the FRC with Fibres F1. Figure 3b, confirms a similar behaviour for the concrete with fibres F2; in this case, increasing the temperature form 20 °C to 40 °C leads to a strength decrement of 25% on average. However, it is important to notice that these results may be affected by the number of fibres crossing the crack.

Therefore, to better understand the effect of temperature, the specimens were split-into two parts at the end of the tests, and the number of fibres crossing the cracks was counted. While counting, the crack surfaces were subdivided into three horizontal strips of equal height, and the number of fibres in each of them reported. Figure 4 and Fig. 5, show, for the FRC containing F1 and F2, respectively, the relationship between the number of fibres in the two thirds of the crack surfaces closer to the notch and nominal residual flexural strengths f_{R1} and f_{R3}. Comparing Fig. 4 and 5, contrarily to what was suggest by Fig. 3, it is clear that the effect of temperature on the two fibres is very similar. A comparison of Fig. 4b with Fig. 5b suggests that with reference to f_{R3} fibres F1 are slightly more sensitive to temperature than F2.

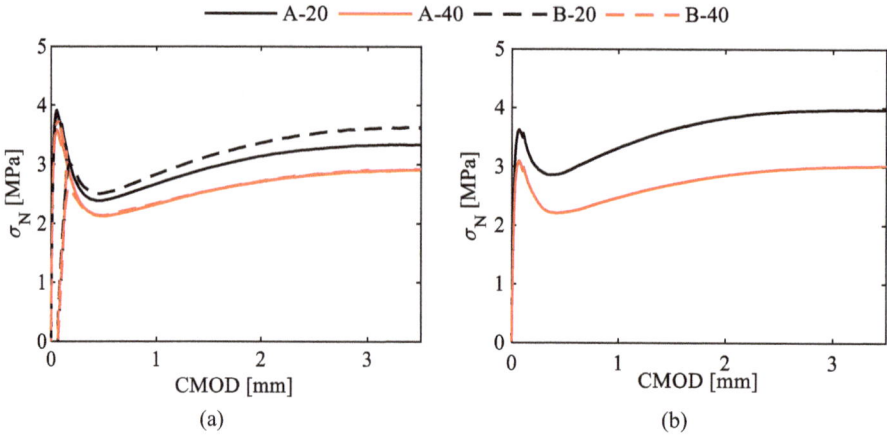

Fig. 3. Comparison of the mean nominal flexural residual tensile strength – CMOD curves for the four curing conditions (Sect. 3): (a) Fibre type F1; (b) fibre type F2.

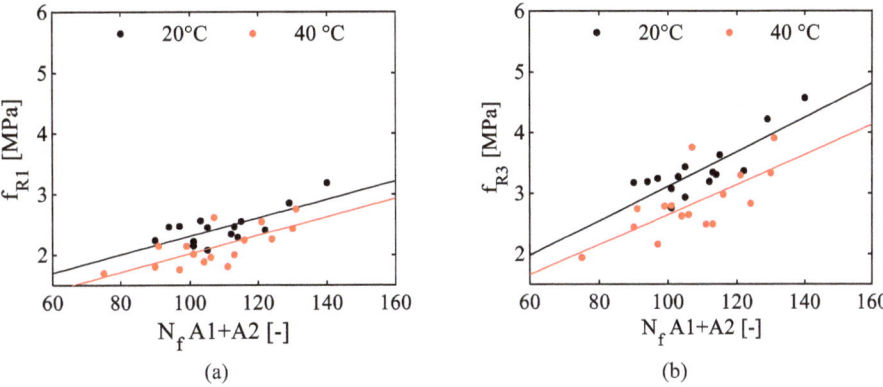

Fig. 4. Fibre type F1: linear regression of the residual flexural strength (a) f_{R1} and (b) f_{R3} with the number of Fibres for 20 °C and 40 °C curing temperature.

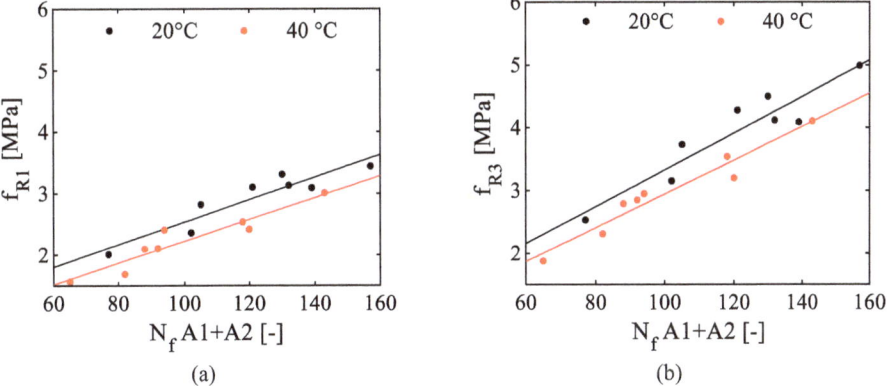

Fig. 5. Fibre type F2: linear regression of the residual flexural strength (a) f_{R1} and (b) f_{R3} with the number of Fibres for 20 °C and 40 °C curing temperature.

3.3 Long Term Flexural Tests

As discussed in Sect. 2.1, specimens were pre-cracked to CMOD = 0.5 mm before the creep test. Table 4, reports the peak nominal residual flexural tensile strength f_L and f_{R1} values measured during the pre-cracking tests. The loads applied during the creep tests were defined in order to produce a target nominal tensile stress in the cracked section $\sigma_{creep} = 0.5\ f_{R1}$. The actual load applied, is given in Table 4 in terms of creep load rate, defined as σ_{creep}/f_{R1}. It can be noticed that the deviations from the target value are minor. Figure 6 shows the results of creep tests in terms of creep coefficient, defined as:

$$\varphi(t) = (CMOD(t) - CMOD(t_0))/CMOD(t_0) \tag{1}$$

where t_0 indicates the time at which the full creep load was applied. Clearly, the duration of the isothermal conditions was different for the two types of fibres tested. The temperature was increased when the creep coefficient rate (Fig. 7) became less than 10^{-2} 1/days. The creep coefficient rate represents the slope of the creep coefficient curves and was computed as a secant, considering time steps of two days (Fig. 7). The shape of the creep coefficient curves is similar for the two FRCs with the two different fibres, even if fibres F2 are associated to lower values of creep. In both cases, it can be noticed that increasing temperature increases the creep rate.

At the end of creep tests, the specimens were unloaded, CMOD recovery was measured for at least 30 days, and then they were tested to failure in three-point bending. At the end of these tests the number of fibres was counted. Figure 8, shows the correlation between the number of fibres in the two-thirds of the crack surfaces closer to the notch, and the value of the creep coefficient at t = 50 days (both the FRCs are at 20 °C). Although data are limited, it is possible to observe that, the creep coefficient value for the F2 fibres is much lower than that for the F1. The better performance attributed to the second fibre type, F2, might be connected to their different material properties. In particular, considering also data form the literature [5] it is possible to conclude that fibres with higher elastic modulus tend to have less

Table 4. Nominal residual flexural tensile strengths from pre-cracking tests and creep load rate

Specimen	f_L (MPa)	f_{R1} (MPa)	Creep load rate
F1-S1	3.94	2.37	0.50
F1-S2	4.02	2.63	0.51
F1-S3	4.16	2.80	0.49
F2-S1	5.38	3.01	0.51
F2-S2	4.97	2.63	0.51
F2-S3	5.05	2.99	0.49

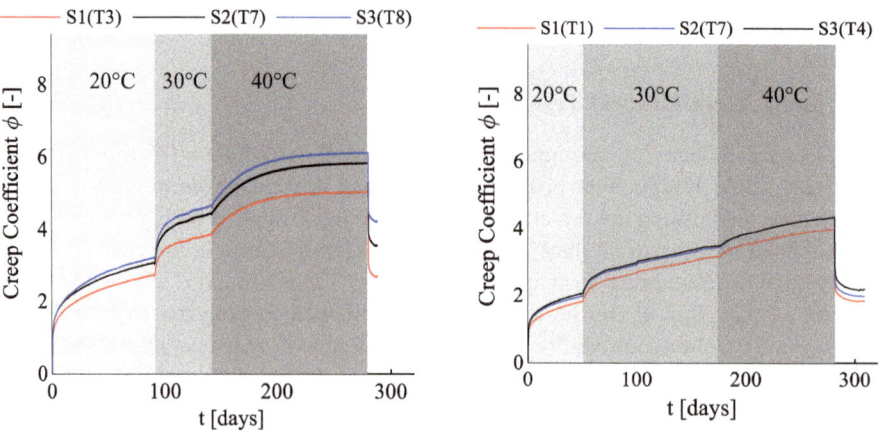

Fig. 6. Creep coefficient – time curves during the creep test for (a) Fibre type F1 and (b) Fibre type F2.

creep. This Figure also suggests that, as observed elsewhere, there is a maximum inverse correlation between the creep coefficient and the number of fibres [2].

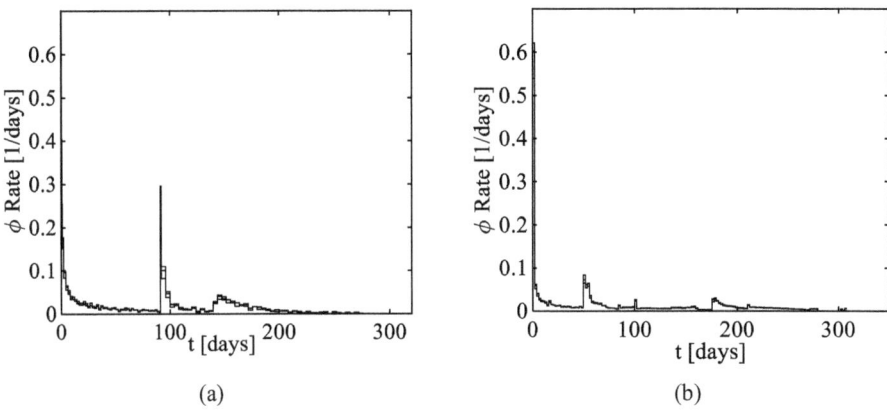

(a)   (b)

Fig. 7. Creep coefficient rate – time curves of the creep test for (a) Fibre type F1 and (b) Fibre type F2.

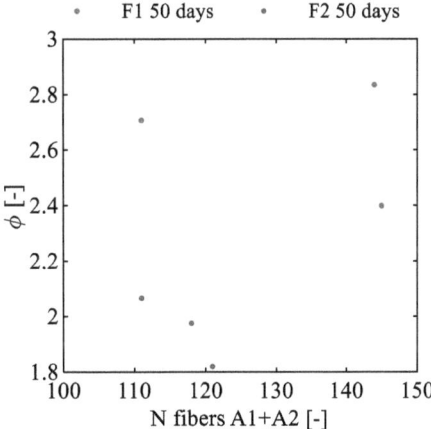

Fig. 8. Creep coefficient at 50 days – number of Fibres for Fibre type F1 and Fibre type F2.

4 Conclusions

The experimental research carried on and here described makes possible to draw some conclusions related to the behaviour of macro-synthetic reinforced concretes under short- and long-term conditions:

- small temperature variations, similar to those that can occur in serviceability conditions, can affect both the short- and long-term behaviour of FRCs with polypropylene fibres;
- considering the results of short-term tests in terms of nominal flexural residual tensile strength without taking into account the actual number of fibres crossing crack may lead to misleading conclusions on the effect of temperature on fibres. Thanking that into account the effect of temperature on the two fibres was very similar, even if F2 fibres had slightly better performances;
- temperature increases produce an increment in creep deformation rate;
- the results obtained indicate that the FRC with fibres F2 had less creep deformations than the FRC containing F1;
- the data presented here, together with other results from the literature [5] indicate that fibres with higher elastic modulus are less prone to creep;
- the data presented support the conclusion that there is an inverse correlation between the number of fibres crossing the crack and creep [2].

References

1. Pujadas, P., Blanco, A., Cavalaro, S., de la Fuente, A., Aguado, A.: The need to consider flexural post-cracking creep behavior of macrosynthetic fiber reinforced concrete. Constr. Build. Mater. **149**, 790–800 (2017)
2. Buratti, N., Mazzotti, C.: Creep Testing Methodologies and Results Interpretation, vol. 14 Springer, Dordrecht (2017). https://doi.org/10.1007/978-94-024-1001-3_2
3. Buratti, N., Mazzotti, C., Savoia, M.: Post-cracking behaviour of steel and macro-synthetic fibre-reinforced concretes. Constr. Build. Mater. **25**(5), 2713–2722 (2011)
4. Buratti, N., Incerti, A., Tilocca, A.R., Mazzotti, C., Paparella, M., Draconte, M.: Energy absorption tests on fibre-reinforced-shotcrete round and square panels. In: Tunnels and Underground Cities: Engineering and Innovation meet Archaeology, Architecture and Art-Proceedings of the WTC 2019 ITA-AITES World Tunnel Congress, pp. 1842–1851 (2019)
5. Buratti, N., Mazzotti, C.: Experimental tests on the effect of temperature on the long-term behaviour of macrosynthetic Fibre Reinforced Concretes. Constr. Build. Mater. **95**, 133–142 (2015)
6. Babafemi, A.J., du Plessis, A., Boshoff, W.P.: Pull-out creep mechanism of synthetic macro fibres under a sustained load. Constr. Build. Mater. **174**, 466–473 (2018)
7. Vrijdaghs, R., di Prisco, M., Vandewalle, L.: Creep deformations of structural polymeric macrofibers. In: Creep Behaviour in Cracked Sections of Fibre Reinforced Concrete, RILEM Bookseries, vol. 14. Springer, Dordrecht (2017). https://doi.org/10.1007/978-94-024-1001-3_5
8. Kurtz, S., Balaguru, P.: Postcrack creep of polymeric fiber-reinforced concrete in flexure. Cem. Concr. Res. **30**(2), 183–190 (2000)
9. MacKay, J., Trottier, J.F.: Post-crack creep behavior of steel and synthetic FRC under flexural loading. In: Shotcrete: More Engineering Developments, pp. 183–192. Taylor & Francis (2004)
10. Kusterle, W.: Flexural creep tests on beams—8 years of experience with steel and synthetic fibres. In: Creep Behaviour in Cracked Sections of Fibre Reinforced Concrete, RILEM Bookseries, vol. 14. Springer, Dordrecht (2017)

11. Zhao, G., di Prisco, M., Vandewalle, L.: Experimental research and numerical simulation of post-crack creep behavior of SFRC loaded in tension. In: Mechanics and Physics of Creep, Shrinkage, and Durability of Concrete - Proceedings of the Ninth International Conference on Creep, Shrinkage, and Durability Mechanics (CONCREEP-9), pp. 340–347 (2013)

12. Zhao, G., di Prisco, M., Vandewalle, L.: Experimental investigation on uniaxial tensile creep behavior of cracked steel fiber reinforced concrete. Mater. Struct. **48**(10), 3173–3185 (2014). https://doi.org/10.1617/s11527-014-0389-1

13. BS EN 14651–2005: Test method for metallic fibred concrete — Measuring the flexural tensile strength (limit of proportionality (LOP), residual). Br. Stand. Inst. **3**, 1–17 (2005)

14. Babafemi, A.J.: Tensile Creep of Cracked Macro Synthetic Fibre Reinforced Concrete. Stellenbosch University (2015)

15. Arango, S., Taungua, E.G., Vargas, J.R.M., Serna Ros, P.: A comprehensive study on the effect of fibers and loading on flexural creep of SFRC. In: BEFIB2012 - Fibre reinforced concrete, pp. 704–715 (2012)

Short and Long Term Behaviour of Polypropylene Fibre Reinforced Concrete Beams with Minimum Steel Reinforcement

Nikola Tošić[(⊠)] and Albert de la Fuente

Civil and Environmental Engineering Department, Universitat Politècnica
de Catalunya, Barcelona, Spain
nikola.tosic@upc.edu

Abstract. Fibre reinforced concrete (FRC) has quickly become an attractive solution for increasing both the mechanical performance and sustainability of concrete structures. In particular, polymeric fibres are increasingly recognized as offering significant benefits for FRC structural applications, especially in areas such as durability. Nonetheless, the behaviour of such FRC remains to be fully understood, especially from the perspective of long-term effects such as shrinkage and creep. Therefore, in this study, a comprehensive experimental programme is carried out for short- and long-term characterization of polypropylene FRC (PPFRC). The experimental program consisted of producing concretes C40/50 with 0, 3 and 9 kg/m^3 of polypropylene fibres. Besides specimens for testing mechanical properties, shrinkage and creep in compression and under bending were tested. Finally, full-scale 3–m span beams with minimum steel reinforcement (0.18%) were tested until failure and under sustained loads. The results are analysed and the contribution of polypropylene fibres to reducing deflections and crack widths is assessed, with no evidence of local flexural failures due to tertiary creep phenomena. The results of this study can provide a contribution towards a fuller understanding of PPFRC structural behaviour and its future incorporation into design codes.

Keywords: Flexure · Serviceability · Shrinkage · Creep · Deflections · Structural behaviour

1 Introduction

Over previous years, fibre reinforced concrete (FRC) has become a very dynamic area of both research and practical application [1]. As such, it has been investigated and successfully applied in various types of structures and structural members such as pavements [2], ground-supported slabs [3], tunnel linings [4], pipes [5], and flat slabs [6, 7], also demonstrating the sustainability benefits it can bring [8].

The experience gained and results collected have enabled the gradual development of FRC material and structural resistance models. However, current achievements are not final and advancements are still possible in several regards. Firstly, current design guidelines are mostly based on the experience with steel fibre reinforced concrete (SFRC). Besides durability, polypropylene fibre reinforced concrete (PPFRC) is

© RILEM 2022
P. Serna et al. (Eds.): BEFIB 2021, RILEM Bookseries 36, pp. 342–353, 2022.
https://doi.org/10.1007/978-3-030-83719-8_30

constantly proving its technical and economic potential for several structural applications and the next important step is its gradual acceptance in design guidelines. This is further supported by the fact that countries that have already introduced national regulation covering PPFRC, such as Spain [9], have seen a significant increase in polypropylene fibre use for structural application, meaning that there is a clear market demand for such products and solutions.

One important aspect of PPFRC considered as needing further research has been its time-dependent behaviour under sustained load, i.e. creep. In a systematic review of literature, Tošić et al. [10] investigate and synthesise literature on the creep of synthetic fibre reinforced concrete at the fibre, material, and structural levels. Several groups of authors found that hybrid-PPFRC, i.e. structural members with steel reinforcement and PPFRC is a practically feasible solution that can be successfully applied in members under long-term loading, but further comprehensive studies in this direction are welcome [10, 11].

In this regard, for FRC in general, the behaviour of hybrid-reinforced FRC members containing minimum reinforcement is also highly important and requires further studies. This is primarily due to potentially detrimental effects that such reinforcement configurations can have on the cross-section ductility [12].

In order to study both aspects, a comprehensive experimental programme at the Universitat Politècnica de Catalunya was developed and initiated, consisting of short- and long-term tests of hybrid-PPFRC full-scale beams, alongside accompanying short- and long-term tests on concrete specimens. The objective of the programme is (1) to determine the polypropylene fibre contribution to ultimate (early and long-term) bending strength of hybrid-PPFRC beams with minimum reinforcement and (2) to characterize the time-dependent serviceability behaviour of hybrid-PPFRC beams under sustained load.

2   Experimental Program

The experimental programme was designed as a three-year research programme on short- and long-term behaviour of PPFRC and hybrid-PPFRC members divided into three one-year stages. As the programme began in August 2020, only results that could be collected up to March 2021 are presented herein. Within the first stage of the programme, three self-compacting concretes (SCCs) were considered with a target compressive strength class of C40/50: concrete RC20 without fibres and concretes B1-3_20 and B1-9_20 with 3 and 9 kg/m^3 of polypropylene fibres, respectively (i.e. 0.33% and 1% by volume, respectively). The fibres were supplied by MBCC Group as embossed monofilament polypropylene fibres with a length of 48 mm and a diameter of 0.85 mm. The concretes were produced by PROMSA (Barcelona, Spain) with mix designs and fresh-state properties presented in Table 1. The mixes were produced in the order of RC20, B1-3_20 and B1-9_20 in September, October and December 2020, respectively.

Table 1. Concrete mix designs and fresh-state properties.

Component	RC20	B1-3_20	B1-9_20
CEM II/A-L 42.5N (kg/m^3)	500	500	500
Fine aggregate 0/2 mm (kg/m^3)	879	879	879
Fine aggregate 0/4 mm (kg/m^3)	360	360	360
Coarse aggregate 4/10 mm (kg/m^3)	150	150	150
Coarse aggregate 10/20 mm (kg/m^3)	300	300	300
Fibres (kg/m^3)	–	3	9
Plasticizer (%)[*]	0.48%	0.70%	0.70%
Superplasticizer (%)[*]	1.07%	1.73%	1.73%
Water-cement ratio (–)	0.46	0.46	0.46
Diameter of slump flow (mm)	675	625	430
Fresh-state density (kg/m^3)	2330	2290	2280

[*]expressed as percentage of cement weight

As can be seen from Table 1, the mixes B1-3_20 and B1-9_20 do not fully comply with SCC requirements in terms of slump flow diameter; nonetheless, the mixes are highly flowable. The reason for this was that the same mix design was maintained for mixes B1-3_20 and B1-9_20 (in terms of admixture content) in order to ensure satisfactory compressive strength of all mixes (>C40/50).

2.1 Short-Term Tests

Short-term tests were performed for both material and structural characterization of the concretes.

At the material level, compressive strength (f_{cm}), modulus of elasticity (E_{cm}) and splitting tensile strength ($f_{ct,sp}$) were determined at 7 and 28 days using Ø150/300 cylinders (3 specimens for each property). The Barcelona test [13] was performed on 6 Ø150/150 cylinders for each concrete, whereas residual tensile strength according to the EN 14651 test [14] was assessed for concretes B1-3_20 and B1-9_20 (on 9 150/150/600 mm prisms).

For structural level testing, for each concrete, two full-scale beams were tested until failure in three-point bending, according to Fig. 1 (P is the externally applied load and g_{sw} the beam self-weight).

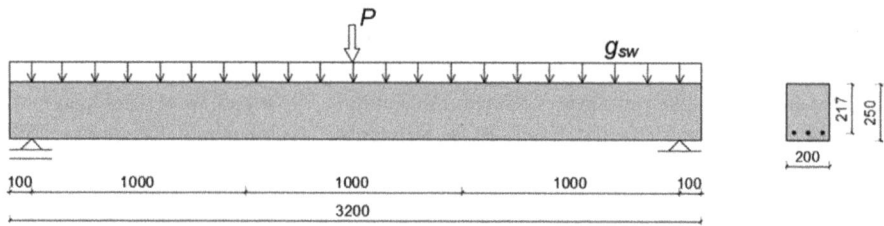

Fig. 1. Full-scale beams for ultimate bending strength testing (dimensions in mm).

The beams had a clear span of 3000 mm and were designed without considering the structural contribution of the fibres and so that the cracking moment is equal to the yielding moment, i.e. $M_{cr} = M_y$. Considering concrete C40/50 and reinforcement B500, this resulted in a reinforcement ratio of $\rho = 0.20\%$ consisting of 3Ø6 mm bars; no shear reinforcement was used or required.

2.2 Long-Term Tests

For long-term characterization, at the material level, for each concrete, 3 Ø150/300 cylinders were used for testing drying shrinkage and compressive creep (at a load level of approximately $0.25f_{cm}$).

For concretes B1-3_20 and B1-9_20 two 150/150/600 mm notched prisms were first pre-cracked in three-point bending to a crack-tip opening displacement (CTOD) of 0.3 mm (considered as a limiting crack width for durability) and then loaded in four-point bending under the full load corresponding to 0.3 mm crack width during pre-cracking ($P_{0.3}$), as shown in Fig. 2, recalculating the load from the three-point bending to the four-point bending configuration, to ensure equal bending moments. Although this load magnitude is larger than typically applied in literature [10], it was selected as this load level approximates the quasi-permanent load in the hybrid-reinforced full-scale beams.

On the structural level, two additional full-scale beams, identical to the ones presented in Fig. 1, were produced for long-term characterisation. One of the beams from each concrete was pre-cracked in three-point bending (similar to the procedure followed for prisms) until a principal crack reached a width of 0.3 mm (measured by three LVDTs in the beam midspan, placed on each side of the beam). Then, the beams were loaded in four-point bending with sustained load to the load corresponding to the 0.3 mm crack width ($P_{0.3}$), as shown in Fig. 3. For these beams, deflections were measured with dial indicators above the supports, next to load application points and in the midspan. Additionally, one LVDT was placed on the bottom side of the beam crossing the principal crack identified during pre-cracking. Another beam was produced for each concrete and left unloaded so as to be tested after 365 days alongside the sustained load beam, after it has been unloaded (to determine the effects of sustained load exposure on ultimate bending strength).

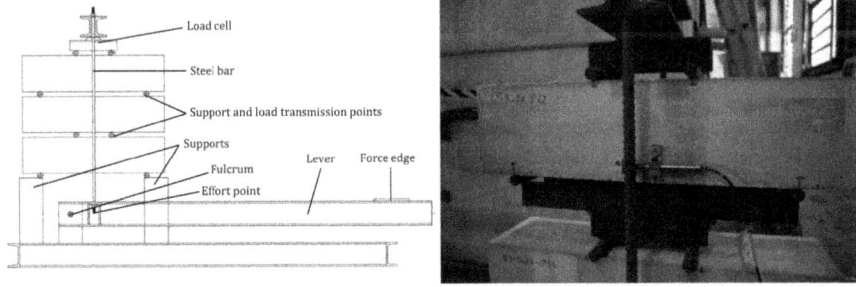

Fig. 2. Schematic of long-term FRC prism characterization (left) [15] and close-up of concrete B1-3_20 (right) and CTOD measurement by an LVDT.

Fig. 3. Pre-cracking in three-point bending (left) and sustained load testing in four-point bending (right).

3   Results and Discussion

3.1   Short-Term Testing Results

The results of mechanical properties testing of the concretes are presented in Table 2. Very similar properties of all the concrete can be seen. The presented results are averages of three measurements and the scatter of the results is similar for all concretes as well; e.g. for f_{cm} the coefficient of variation does not surpass 3%, meaning that the characteristic compressive strength f_{ck} of all three concretes is greater than 40 MPa, satisfying the C40/50 concrete class target.

Table 2. Mechanical properties of the tested concretes.

Concrete	f_{cm} (7 d) (MPa)	f_{cm} (28 d) (MPa)	$f_{ct,sp}$ (7 d) (MPa)	$f_{ct,sp}$ (28 d) (MPa)	E_{cm} (7 d) (MPa)	E_{cm} (28 d) (MPa)	γ_c (28 d) (kg/m^3)
RC20	40.7	46.0	2.85	3.40	28415	30320	2330
B1-3_20	34.5	42.9	2.95	3.13	27725	29405	2300
B1-9_20	40.0	47.4	3.70	2.82	27930	28290	2305

In terms of residual tensile strength, the results of tests on notched prisms according to EN 14651 [14] are shown in Fig. 4 in the form of an average curve (obtained by averaging the curves of the nine tested prisms) and the 5% and 95% fractile lines obtained using the scatter obtained on the nine tested prisms and assuming a normal distribution. The average values of the limit of proportionality f_L and residual strengths f_{R1}, f_{R2}, f_{R3} and f_{R4} (corresponding to crack mouth opening displacements, CMODs, of 0.5, 1.5, 2.5 and 3.5 mm, respectively) for B1-3_20 are 4.23, 0.74, 0.91, 1.00 and 1.04 MPa, respectively, whereas for B1-9_20 they are 4.37, 2.50, 3.57, 4.08 and 4.28 MPa, respectively. As such, per the *fib* Model Code 2010 [16], mix B1-3_20 could be classified as 0.5d and mix B1-9_20 as 1.5e, i.e. both mixes exhibit hardening behaviour with the f_{R3}/f_{R1} ratio 1.20 and 1.60 for B1-3_20 and B1-9_20, respectively. Furthermore, according to the *fib* Model Code 2010, FRC can be used to substitute reinforcement when $f_{R3}/f_{R1} > 0.5$ and $f_{R1}/f_L > 0.4$. The former condition is satisfied for both mixes, but the second is only satisfied for B1-9_20.

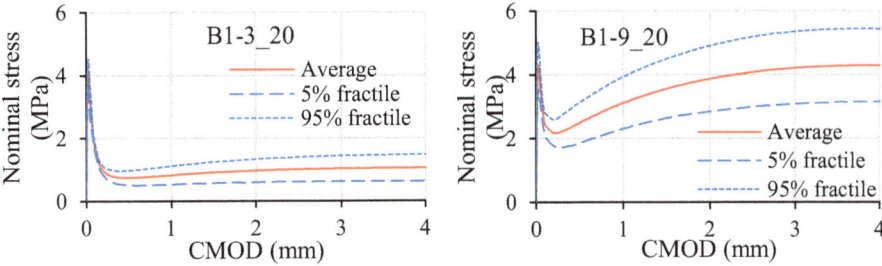

Fig. 4. Average EN 14651 results with confidence intervals for B1-3_20 (left) and B1-9_20 (right).

The results of the Barcelona test [13] are shown in Fig. 5 in the form of an average curve (obtained by averaging the curves of the nine tested prisms) and the 5% and 95% fractile lines obtained using the scatter obtained on the nine tested prisms and assuming a normal distribution. The average values of the tensile strength f_{ct}, and residual tensile strengths $f_{0.5}$, $f_{1.5}$, $f_{2.5}$ and $f_{3.5}$, corresponding to the vertical displacement of the hydraulic press clip of 0.5, 1.5, 2.5 and 3.5 mm after cracking, for concrete B1-3_20, were 3.43, 0.70, 0.55, 0.49 and 0.43 MPa, respectively and 3.14, 1.64, 1.42, 1.29 and 1.21 MPa, respectively for B1-9_20, whereas for RC20 only f_{ct} could be measured with an average value of 3.62 MPa.

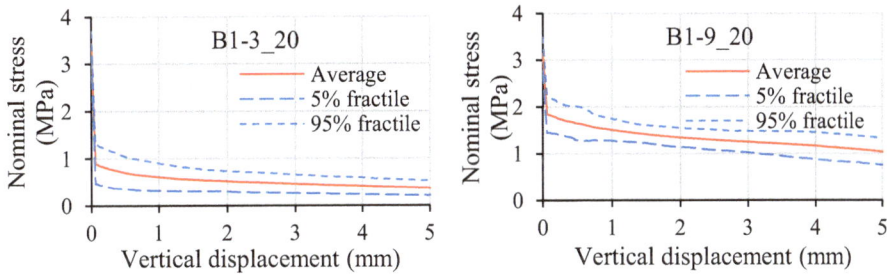

Fig. 5. Average Barcelona test results with confidence intervals for B1-3_20 (left) and B1-9_20 (right).

In summary, in terms of material properties, uniform mechanical properties were achieved among the concretes RC20, B1-3_20 and B1-9_20 and good residual behaviour was obtained both with 3 and 9 kg/m³ of polypropylene fibres.

At the structural level, prior to testing full-scale beams up to bending failure, steel reinforcement was characterized. The Ø6 mm bars were assessed as having yield strength of 525 MPa, an ultimate strength of 650 MPa, modulus of elasticity of 195 GPa and a strain at failure of 16.5%.

Two beams of each concrete were tested between the ages of 28 and 30 days under monotonic three-point bending using displacement control. Additionally, in order to monitor the evolution of the principal crack (i.e. section in which failure occurs), LVDTs were mounted on both sides of each beam as shown in Fig. 6.

The applied moment–midspan deflection graphs for all the beams are shown in Fig. 7 (wherein the moment was calculated as $M = (P \times L)/4$). In the figure, the circles, squares and triangles represent cracking moments, yielding moments and maximum moments, respectively). It should be noted that moments due to self-weight (ca. 1.30 kNm) are not included in the figure.

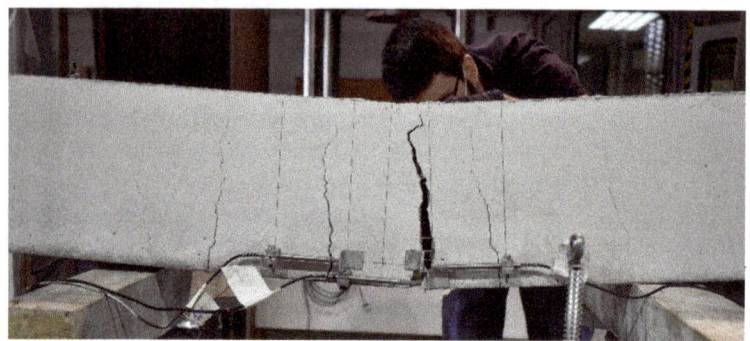

Fig. 6. Example of beam B1-3_20–2 after failure.

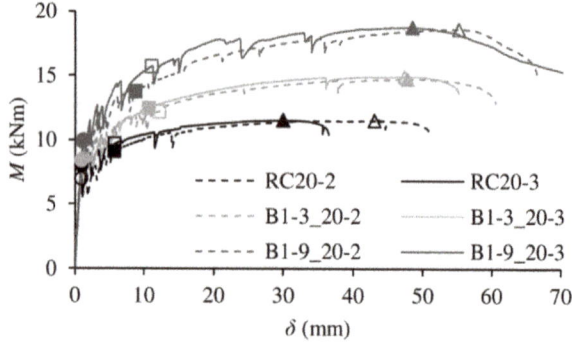

Fig. 7. Applied moment–midspan deflection diagrams for the tested beams.

If the moment due to self-weight is included, the average values of cracking moments M_{cr} for beams RC20, B1-3_20 and B1-9_20 was assessed as 8.78, 9.74 and 11.22 kNm, respectively; yielding moment M_y as 10.70, 13.79 and 16.02 kNm, respectively; and the maximum moment M_{max} as 12.83, 16.13 and 19.98 kNm, respectively. Additionally, an increase in deflections associated with each of these bending moments is noticed. In other words, a significant improvement due to fibre inclusion can be seen over the entire M–δ curve, both on moments and deflections and already at the 3 kg/m^3 level of fibre content, e.g. the increase in yielding moment over RC20 is 28.9% for B1-3_20 and 49.7% for B1-9_20 (also at larger displacements). This means that, in this case, polypropylene fibres can be considered as providing a significant contribution to the bearing capacity of the beams and enhancing their deformation capacity.

3.2   Long-Term Testing Results

The long-term tests were performed at the LATEM laboratory of the Universitat Politècnica de Catalunya in Barcelona. The tests were carried out without strict control of ambient conditions; nonetheless, over the period October 2020–February 2021 the average temperature and relative humidity were 15.1 °C and 58.9%, respectively, with CoVs of 22.3% and 21.5%, respectively. Temperature and relative humidity were recorded using two electronic loggers at 2 h intervals.

The evolution of the drying shrinkage ε_{cs} and compressive creep coefficient φ_{cc} (ratio of creep strain to initial strain) of the concretes in logarithmic time scale is presented in Fig. 8.

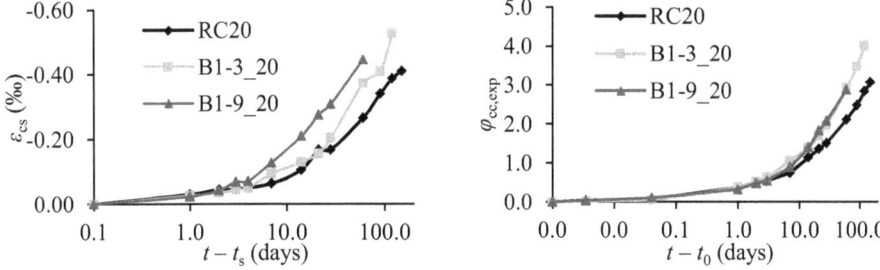

Fig. 8. Evolution of drying shrinkage strain (left) and creep coefficient (right).

The evolution over time of both properties can be seen to be quite similar, at least considering the inherent scatter associated with both shrinkage and creep.

As for long-term testing of FRC prisms, as explained in the previous section, the prisms were first tested in the configuration of EN 14651 (three-point bending of notched specimens), until a CTOD of 0.3 mm was reached as measured by an LVDT (Fig. 2). The prisms were then unloaded and transferred to a loading frame (Fig. 2) and loaded with the full load corresponding to the 0.3 mm crack width in the short-term test setup. Two beams of each of the concretes B1-3_20 and B1-9_20 were tested in this way, and using an LVDT at the level of the crack tip, the evolution of the crack width was monitored.

For the two prisms of concrete B1-3_20, after short-term testing until a CTOD of 0.3 mm and a nominal stress of 0.72 MPa, a "residual" CTOD of 0.15 mm was observed in both prisms. Immediately after loading in the long-term test setup, the CTOD increased to 0.22 and 0.29 mm. For the two prisms of concrete B1-9_20, after short-term testing until a CTOD of 0.3 mm and a nominal stress of 2.30 MPa, the "residual" CTOD was 0.14 mm, and immediately rose to 0.26 and 0.33 mm after loading. The time evolution of the CTODs is shown in Fig. 9.

In the case of prisms B1-3_20, after 116 days under sustained load, the CTODs were 0.48 and 0.58 mm, i.e. the "CTOD creep coefficient" (ratio of time-dependent increase in CTOD to initial CTOD after loading) was 1.18 and 1.00. In the case of concrete B1-9_20, after 69 days under sustained load, the CTODs were 0.46 and 0.54 mm, respectively and their "CTOD creep coefficients" were 0.77 and 0.64.

Previous studies tended to test cracked FRC specimens under sustained load only at a percentage (generally 50%) of the load corresponding to pre-cracking until 0.2–2.5 mm [10] and several reported creep failures of such specimens (sometimes within weeks), the behaviour of B1-3_20 and B1-9_20 points to good time-dependent behaviour with no indications of tertiary creep, although crack width stabilization has still not occurred.

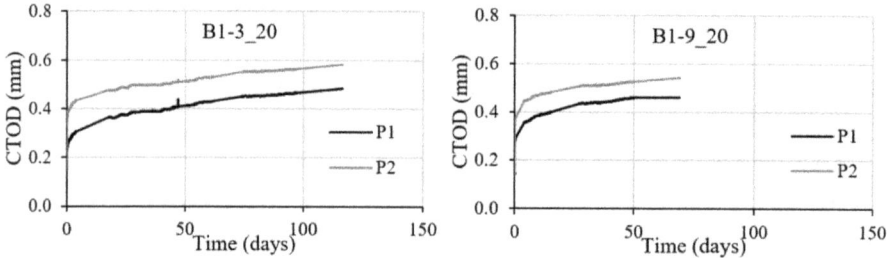

Fig. 9. Evolution of crack widths in specimens B1-3_20 (left) and B1-9_20 (right).

The long-term testing of full-scale beams from each concrete was performed following a procedure similar to prism testing: at the age of 28 days, each beam was first pre-cracked in three-point bending so that the principal/dominant crack had a crack width of 0.3 mm. Then, the beams were transferred to steel frame supports and loaded in four-point bending with the load corresponding to the principal crack width of 0.3 mm. The deflections, strain distribution and crack width of the principal crack were measured. Following the pre-cracking procedure, the beams RC20–1, B1-3_20–1 and B1-9_20–1 were loaded to a moment–yielding moment (M/M_y) ratio of 0.61, 0.53 and 0.52, respectively (although due to increasing M_y, the absolute values of M were 6.54, 7.35 and 8.31 kNm, respectively).

For beams RC20–1, B1-3_20–1 and B1-9_20–1, initial deflections were 2.88, 1.86 and 3.37 mm, respectively; after 60 days, the deflections were 5.78, 6.99 and 7.88 mm, respectively (increase by 2.00, 3.76 and 2.34, respectively). The larger increase for beam B1-3_20–1 over time is due to the fact that beams RC20–1 and B1-9_20–1 had 6 and 9 cracks, respectively after pre-cracking, whereas beams B1-3_20–1 had only 3. Over time, the number of cracks increased to 11 and 13 for beams RC20–1 and B1-9_20–1, respectively, whereas it rose to 9 for beam B1-3_20–1 causing a large increase in deflections relative to the initial state. This is also evident from the crack width evolution of the principal crack for each beam, shown in Fig. 10. The initial value for beams RC20–1, B1-3_20–1 and B1-9_20–1 were 0.32, 0.34 and 0.29 mm, respectively; after 60 days the values were 0.35, 0.43 and 0.36 mm, respectively.

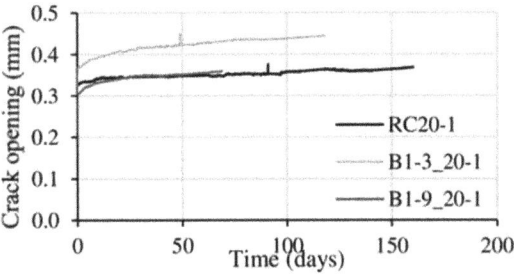

Fig. 10. Evolution of crack widths in full-scale beams.

It can be seen that the evolution of the crack width of beam B1-9_20–1 follows the evolution for beam RC20–1, whereas an increase over time can be seen for beam B1-3_20–1. However, the slope of increase for beam B1-3_20–1 follows that of the other two beams and the increase is located early after loading when additional cracks were opened. Therefore, with 9 kg/m^3 of polypropylene fibres an excellent behaviour in terms of resistance to crack width increase can be observed as the beams is loaded to a higher moment than RC20 (although they are nominally designed for the same resistance), This is in line with the finding that B1-9_20 has characteristics allowing for reinforcement substitution. With 3 kg/m^3 there is also a beneficial effect of resistance increase but accompanied by some crack width increase, even though a different loading procedure might have induced more cracks during pre-cracking, avoiding their appearance over time.

4   Conclusions

In this study, first results of a comprehensive experimental programme on short- and long-term material and structural characterisation of PPFRC are presented.

The tests consisted of mechanical properties testing, time-dependent behaviour testing, ultimate moment capacity assessment and behaviour under sustained load. In particular, for structural tests, PPFRC members with minimum steel reinforcement were tested, being an important field of FRC application that requires further study.

Considering the obtained results, the following conclusions are drawn:

- Using 3 and 9 kg/m^3 of a specific type of polypropylene fibres in an SCC concrete with compressive strength class C40/50 ensures satisfactory residual behaviour with the FRCs classified as 0.5d and 1.5e, respectively.
- In terms of short-term flexural behaviour, the addition of fibres brings significant increases in cracking, yielding and maximum moments and the associated deflections. The increases are evident already at the fibre content of 3 kg/m^3 with the yielding moment increasing by 28.9% and 49.7% relative to reinforced concrete for PPFRC with 3 and 9 kg/m^3 of polypropylene fibres, respectively.

- In terms of time-dependent behaviour of PPFRC, after at least 60 days under a load corresponding to a CMOD of 0.3 mm, no tertiary creep is observed, in neither of the PPFRC mixes, whereas crack width stabilization has still not been reached.
- In terms of behaviour under sustained load of PPFRC full-scale beams with minimum reinforcement, a content of 9 kg/m^3 of polypropylene fibres allows the application of a higher load with no increase of crack width relative to the behaviour of reinforced concrete. At the fibre content of 3 kg/m^3 a certain increase in crack width can be observed over time; however, the beam is nonetheless exposed to a higher load than the reinforced concrete beam.

The results of this study are just the first reporting of a multi-year experimental programme so that further results will be gatherer, allowing for a more in-depth analysis of the observed phenomena. Nonetheless, the results obtained thus far demonstrate that there is an effective contribution of certain polypropylene fibres to the flexural behaviour of hybrid-PPFRC structural members. Furthermore, the possibility of their safe use in members with minimum reinforcement under sustained loads is clearly shown.

Acknowledgements. The authors want also express their gratitude to the Spanish Ministry of Science and Innovation for the financial support received under the scope of the project CREEF (Creep and Fatigue of fibre reinforced concrete elements) (PID2019-108978RB-C32). This study has also received funding from MBCC Group. This support is gratefully acknowledged. Any opinions, findings, conclusions, and/or recommendations in the paper are those of the authors and do not necessarily represent the views of the individuals or organizations acknowledged.

References

1. di Prisco, M., Plizzari, G., Vandewalle, L.: Fibre reinforced concrete: new design perspectives. Mater. Struct. **42**, 1261–1281 (2009). https://doi.org/10.1617/s11527-009-9529-4
2. Meda, A., Plizzari, G.A.: New design approach for steel fiber-reinforced concrete slabs-on-ground based on fracture mechanics. ACI Struct. J. **101**, 298–303 (2004). https://doi.org/10.14359/13089
3. Roesler, J.R., Altoubat, S.A., Lange, D.A., Rieder, K.A., Ulreich, G.R.: Effect of synthetic fibers on structural behavior of concrete slabs-on-ground. ACI Mater. J. **103**, 3–10 (2006). https://doi.org/10.14359/15121
4. FIB Bulletin 83. Precast tunnel segments in fibre-reinforced concrete. Int. Fed. Struct. Concr. (fib), Lausanne (2018)
5. de la Fuente, A., Escariz, R.C., De Figueiredo, A.D., Aguado, A.: Design of macro-synthetic fibre reinforced concrete pipes. Constr. Build. Mater. **43**, 523–532 (2013). https://doi.org/10.1016/j.conbuildmat.2013.02.036
6. Aidarov, S., de la Fuente, A., Mena, F., Ángel, S.: Campaña experimental de un forjado de hormigón reforzado con fibras a escala real. ACE, pp. 1–10 (2019)
7. Gossla, U.: Development of SFRC free suspended elevated flat slabs. Aachen Univ. Appl. Sci. (2005)

8. Josa, I., de la Fuente, A., Casanovas-Rubio M del, M., Armengou, J., Aguado, A.: Sustainability-oriented model to decide on concrete pipeline reinforcement. Sustainability, **13**, 3026 (2021). https://doi.org/10.3390/su13063026

9. EHE. Instrucción de Hormigón Estructural (EHE-08). (2008). https://doi.org/10.1017/CBO9781107415324.004

10. Tošić, N., Aidarov, S., de la Fuente, A.: Systematic Review on the Creep of Fiber-Reinforced Concrete. Mater. (Basel) **13**, 5098 (2020)

11. Plizzari, G., Serna, P.: Structural effects of FRC creep. Mater. Struct. **51**(6), 1–11 (2018). https://doi.org/10.1617/s11527-018-1290-0

12. Markić, T., Amin, A., Kaufmann, W., Pfyl, T.: Discussion on Assessing the influence of fibers on the flexural behavior of reinforced concrete beams with different longitudinal reinforcement ratios by Conforti et al. [structural concrete, 2020]. Struct. Concr. (2021). https://doi.org/10.1002/suco.202000488

13. UNE 83515. Hormigones con fibras.Determinacion de la resistencia a fisuracion, tenacidad y resistencia residual a traccion. Metodo Barcelona. Madrid: AENOR. (2010)

14. EN 14651. Test method for metallic fibred concrete. Measuring the flexural tensile strength (limit of proportionality (LOP), residual). Br. Stand. Inst. (2005). 9780580610523

15. Blanco, A.: Characterization and modelling of SFRC elements. Universitat Politecnica de Catalunya (UPC) (2013)

16. FIB. fib Model code for concrete structures 2010. Int. Fed.Struct.Concr. (fib), Lausanne (2013). https://doi.org/10.1002/9783433604090

Analytical and Numerical Models

Computational Mesoscale Modeling for the Mechanical Behavior of Fiber Reinforced Concrete

Marcello Congro[1,2(✉)], Deane Roehl[1,2], and Eleazar C. M. Sanchez[2]

[1] Department of Civil and Environmental Engineering, Pontifical Catholic University of Rio de Janeiro, Rio de Janeiro, Brazil
marcellocongro@tecgraf.puc-rio.br
[2] Modeling and Multiphysics Simulation Laboratory, Tecgraf Institute, Pontifical Catholic University of Rio de Janeiro, Rio de Janeiro, Brazil

Abstract. In the last decades, fiber-reinforced concrete (FRC) has emerged in the civil engineering industry. Due to its excellent mechanical properties and functionality as crack propagation control, several numerical investigations have been carried out to study these composites. However, only few technical standards are established for these materials. However, the increase in computational cost in order to explicitly represent the fibrous reinforcement in FE models and the strategies to simulate the interfacial relations between each phase are among the greatest challenges of mesoscale modeling. This paper proposes a numerical methodology to simulate the mechanical behavior of fiber-reinforced concrete in a mesoscale approach with the Finite Element Method (FEM). The mesoscale level is the fiber scale; in this sense, the cementitious matrix, the discrete and random fiber reinforcement, and the interfacial transition zone (ITZ) are explicitly represented in the computational models. Finally, the formulation is validated against experimental data available in the literature. Moreover, the simulations considering this new mesoscale element formulation present reliable results close to the experimental responses.

Keywords: Mesoscale · Finite Element Method (FEM) · Cement composite materials · Numerical analysis

1 Introduction

Concrete is one of the most used materials worldwide in the construction industry, especially due to its versatility and adaptability. In recent years, several researchers and structural engineers have studied numerical techniques for representing and understanding the mechanical behavior of structures and their materials. From a macroscopic and structural scale, fiber reinforced concrete (FRC) is often considered a homogeneous and isotropic material. However, when observing this material at a closer level, there are several material constituents, since the cement composite materials are composed by cement paste, fibers, fine aggregates, voids and capillary pores. Given the presence of the fibrous reinforcement in the cementitious matrix, it is necessary to consider the effects of heterogeneities on the nonlinear behavior of the composite. The heterogeneities affect not

© RILEM 2022
P. Serna et al. (Eds.): BEFIB 2021, RILEM Bookseries 36, pp. 357–364, 2022.
https://doi.org/10.1007/978-3-030-83719-8_31

only the mechanical behavior, but also the damage mechanisms, which occur differently in these materials if compared to conventional concretes [1].

It is also relevant to emphasize the importance of multiscale modeling of these advanced materials for structural applications. In this way, numerical methodologies can be useful not only for the interaction investigation of concrete components, but also for several applications involving structural elements (such as slabs, larger beams, among others). Therefore, multiscale methods emerge as important techniques for Structural Engineering, since technical standards for structural design are not well established and full scale experimental tests are no feasible. Multiscale modeling is also important for structural applications [7, 8]. Several authors have studied fiber-reinforced concrete at the mesoscale, proposing distinct numerical methodologies to simulate material behavior [9–11]. Numerical analyses can be carried out to the design of real scale FRC structures in conjunction with the technical standards for structural design, especially regarding bending tests. In this sense, the mathematical models can consider distinct design variables and material working conditions before material prototyping [6].

This paper proposes a numerical model for the simulation of the mechanical behavior of FRC in direct tensile tests in a mesoscale approach. A new finite element is developed in order to capture the fiber normal effects by imposing constraints methods for models with multiple degrees of freedom. In this sense, a composite element is developed from the information obtained at the material scale. This methodology is also able to reproduce the global load carrying behavior of the composite. Computational procedures are carried out based on the Finite Element Method.

It is important to point out that this approach, denoted here by "composite model", internally considers in its formulation not only the effects of stiffness of the plane stress elements that represent the cementitious matrix, but also incorporates the stiffness of each fiber in the finite elements crossed by the fibrous reinforcement. The nonlinear behavior of the material in its post-cracking stage and the influence of the presence of fibers in the displacement field will be evaluated, observing peak loads, deflection and system stiffness, as well as the fracture propagation patterns. Finally, the composite post-peak behavior is then compared with the experimental reference data available in the literature to verify the efficiency of the numerical model.

2   Methodology for 2D Mesoscale Model

Several Authors Present Different Approaches Regarding the Material Representation at the Mesoscale, Especially for Fiber Reinforced Concrete. Several Mathematical Models, Depending on Problem Complexity, Can Perform the Material Modeling at an Intermediate Scale. A Common Approach in Computer Simulations Involving These Composites is Carried Out at the Fiber Scale, Where Concrete is Considered Biphasic, Composed of the Cementitious Matrix and the Discrete Fibers, [2, 3] as Indicates Fig. 1.

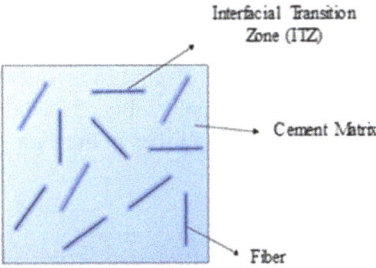

Fig. 1. 2D mesoscale model commonly adopted for FRC numerical simulations.

The presented methodology for fiber reinforced concrete mesoscale modeling involves the development and implementation of a new finite element capable of capturing the normal strength and fiber stiffness effects incorporated into the cementitious matrix when subjected to tension loads. It is necessary to consider in the mathematical formulation of the element not only the effects of stiffness caused by the cementitious matrix, but also those generated by the fiber, since this contribution is relevant for the material behavior analyses. The general idea of the method is to couple the stiffness matrix of a bar element with the stiffness matrix of a plane stress element to carry out a static condensation process of the fiber element degrees of freedom. This association between the dissimilar elements with multifreedom constraint methods is performed to simulate the compatibility between the fiber and the cementitious matrix. Figure 2 presents a schematic representation of the composite finite element for the mesoscale numerical simulations.

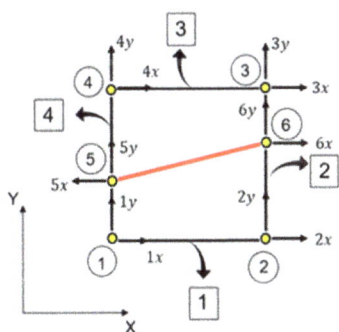

Fig. 2. Plane stress quadrangular element crossed by a fiber and its degrees of freedom.

In This Way, the Global Stiffness of the Composite Element Considering the Condensation Effects of the Cementitious Matrix and the Fibrous Reinforcement is Given by the Sum of Each Fiber/matrix $[K]_{mf}$ and Interface/spring $[K]_{is}$ Stiffness Contribution, as Shown in Eq. 1. To Introduce the Fibers Bridging Effect in the Cohesive Element, a Spring with Fiber Elastic Stiffness [k]s is Incorporated Within the

Cohesive Interface Element. Therefore, It is Possible to Capture the Stiffness Effects Incorporated by the Fiber to the Composite, as Well as the Normal and Tangential Effects Brought About by the Addition of Cohesive Interface Elements in Mode I Crack Opening Displacement. Additionally, Fig. 3 Schematically Represents the Stiffness Effects of the Cement Matrix and Fiber in the Formulation of the Element for the 2D Mesoscale Model.

$$[K]_{global} = [K]_{mf} + [K]_{is} \tag{1}$$

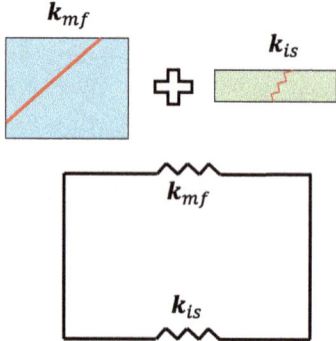

Fig. 3. Schematic representation of fiber/matrix and interface element/spring stiffness contribution to the composite finite element.

Furthermore, it is necessary to calculate the amount of fibers according to the volumetric fiber content of the experiment. This process is done automatically due to the data structure developed to check which plane elements are crossed by fibers. In this way, the process is more efficient, with no limitations on the amount of fibers to be included in the finite element model. Additionally, the quality of the finite element mesh is ensured, avoiding elements with a poor aspect ratio.

The cohesive interface elements also consider the contact between each fiber segment and the cement matrix. A constitutive damage model with exponential softening is adopted for modeling the post-peak behavior of concrete.

3   Results

The numerical analysis developed in this paper is based on the direct tensile test of a self-compacting concrete reinforced with 1% hooked-end steel fibers reported in the literature [4, 5]. The steel fibers are 60 mm long with an aspect ratio of 80, and have tensile strength of 1238 ± 25 MPa and Young modulus of 202 ± 4 GPa. The concrete samples were cast with dimensions of 100x100x400 mm with reduction of section height in the center to 75mm. For more information regarding the experimental test, please refer to [5].

The amount of fibers to be inserted in the numerical model is calculated through a quotient between the fiber volumetric content in the center region of the model specimen and the volume of each fiber. The fiber length and diameter are L = 60 mm and Φ = 0.75 mm, respectively. This approximation for the experimental test leads to the generation of 283 fibers randomly distributed in the central region of a prismatic specimen, as presented in Fig. 4(a). In addition, Fig. 4(b) indicates the dimensions of the part.

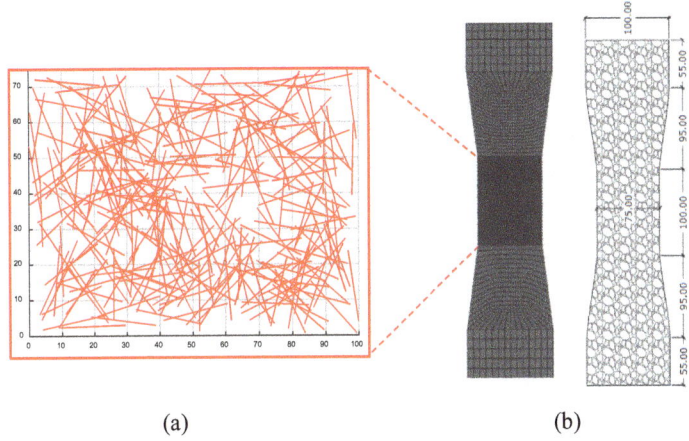

(a)   (b)

Fig. 4. (a) Random fiber distribution for the mesoscale model of [4]; (b) Prismatic specimen dimensions (in mm).

Table 1 displays the mechanical properties for the cementitious matrix and the fibers during the numerical simulations for the plane stress mesoscale model considering the experimental direct tensile test information. Note that the subscript 'm' refers to a property of the cementitious matrix, while 'f' refers to a fiber parameter.

Table 1. Mechanical properties for the mesoscale numerical model.

Mesoscale Model (based on [4])	
E_m(GPa)	33.00
v_m(-)	0.25
f_t^m(MPa)	4.63
K_n, K_t(MPa/m)	33.0 $*10^3$
E_f(GPa)	200.0
v_f(-)	0.30

Figure 5 presents the crack propagation process in the numerical model using (a) the composite formulation and (b) the experimental fracture patterns obtained by [4]. It is possible to observe that, for the proposed mesoscale methodology, the crack patterns obtained by the numerical simulations are similar to the real model tested in the laboratory. Figure 6 presents the load-displacement curves for the post-peak behavior of the composite. These results represent the experimental data, the numerical model curve considering the homogenous properties of the cement matrix (also given by experimental tests), and the computational model for the mesoscale considering the composite formulation.

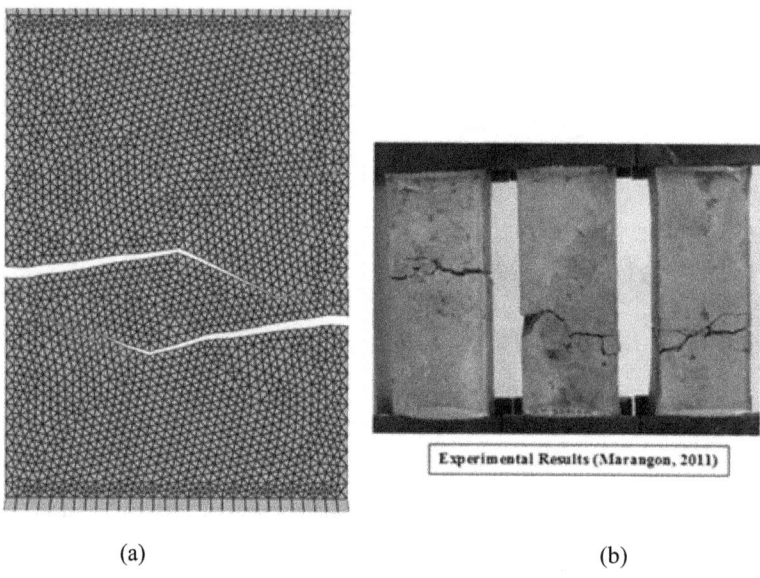

(a)   (b)

Fig. 5. Crack propagation patterns considering: (a) the numerical formulation for mesoscale; (b) experimental results of [4].

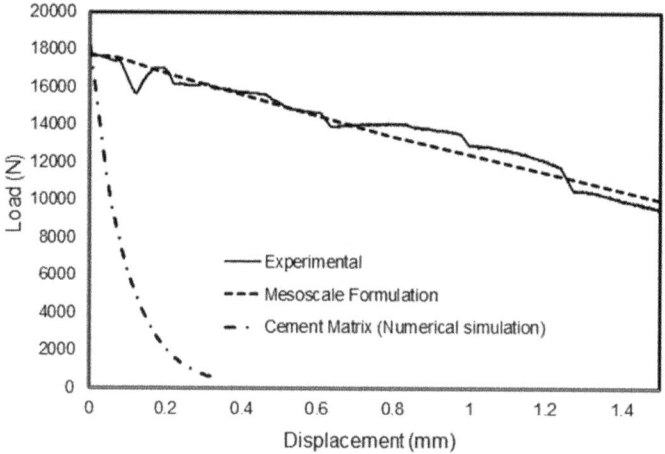

Fig. 6. Load-displacement curves for the global behavior of the composite for the direct tensile test of [4].

According to Fig. 6, it is possible to infer that the mesoscale methodology presents good agreement with the reference experimental data. It is also important to observe the increase in the composite load capacity, resulting in distinct post-peak responses when compared to the results of the homogeneous cement matrix. Due to model symmetry, fibers are inserted in the central region of the part. In this sense, the numerical results from 2D mesoscale formulation in the post-peak phase demonstrate good fit with the experimental reference.

Finally, Fig. 7 presents the damage evolution at each Gauss point of the central region of the model for the composite numerical formulation. The damage evolution is defined as an irreversible process that takes place at the microscale, firstly developing several microcracks. After the coalescence process, major fractures arise in the concrete, leading to the reduction of the material's load capacity.

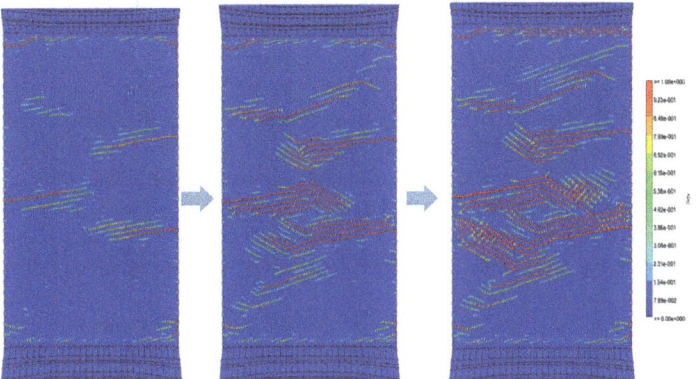

Fig. 7. Load-displacement curves for the global behavior of the composite for the direct tensile test of [4] in multiple simulation steps.

4   Conclusions

- Mesoscale is the observation level with challenging representation. The increase in computational cost for the explicit representation of fibers in the finite element model and the strategies to represent the interfacial mechanisms between materials are among the greatest challenges of this work, since several mesoscale methodologies are presented in the literature;
- This formulation considers the fiber normal stiffness to the composite system, which is neglected in other works in the literature;
- There is no limitation for the number of fibers to be inserted in the finite element models, and the mesh generation process is done easily, without elements with a poor aspect ratio;
- The proposed methodology shows how the numerical model can help to design custom-made materials according to their specific applications, including structural FRC members;

- An experimental tensile test from the literature is chosen for the validation of the numerical model using the composite element. The global behavior of the composite presents a good agreement with the experimental reference;
- Fracture propagation patterns are also very close to the experimental response.

Acknowledgements. This study was financed in part by the Coordenação de Aperfeiçoamento de Pessoal de Nível Superior – Brasil (CAPES) – Finance Code 001 and Conselho Nacional de Pesquisa (CNPq) – Grant 308547/2016–0.

References

1. Suryanto, B., Nagai, K., Maekawa, K.: Influence of damage on cracking behavior of Ductile Fiber Reinforced Cementitious Composite (DFRCC). In: Proceedings of 8th International Conference on Creep, Shrinkage and Durability of Concrete and Concrete Structures (CONCREEP8). CRC Press, Ise-Shima, Japan (2008)
2. Friedrich, L.F., Wang, C.: Continuous Modeling Technique of Fiber Pullout from a Cement Matrix with Different Interface Mechanical Properties Using Finite Element Program. Latin Am. J. Solid. Struct. **13**, 1937–1953 (2016)
3. Zhan, Y., Meschke, G.: Multilevel computational model for failure analysis of steel-fiber–reinforced concrete structures. J. Eng. Mech. **142**, 11 (2016)
4. Marangon, E.: Caracterização material e estrutural de concretos autoadensáveis reforçados com fibras de aço. 322 p. PhD Thesis Federal University of Rio de Janeiro, Rio de Janeiro, Brazil (2011)
5. Congro, M., Sanchez, E.C.M., Roehl, D., Marangon, E.: Fracture modeling of fiber reinforced concrete in a multiscale approach. Compos. Part B: Eng. **174**, 106958 (2019)
6. Congro, M., Roehl, D., Mejia, C.: Mesoscale computational modeling of the mechanical behavior of cement composite materials. Compos. Struct. **257**, 113137 (2021)
7. Chaudhuri, P.: Multiscale modeling of fracture in concrete composites. Compos. Part B: Eng. **47**, 204–215 (2013)
8. Caporale, A., Feo, L., Luciano, R.: Damage mechanics of cement concrete modeled as a four-phase composite. Compos. Part B: Eng. **65**, 124–130 (2014)
9. Liang, X., Wu, C.: Meso-scale modelling of steel fibre reinforced concrete with high strength. Constr. Build. Mater. **165**, 187–190 (2018)
10. Ogura, H., Kunieda, M., Ueda, N., Nakamura, H.: MESO-scale modeling for fiber reinforced concrete under mixed mode fracture. In: Proceedings of the 8th International Conference on Fracture Mechanics of Concrete and Concrete Structures, FraMCoS (2013)
11. Su, Y., Li, J., Wu, C., Wu, P., Tao, M., Li, X.: Mesoscale study of steel fibre-reinforced ultra-high performance concrete under static and dynamic loads. Mater. Des. **116**, 340–351 (2017)

Numerical Multi-level Model for Fibre Reinforced Concrete: Validation and Comparison with Fib Model Code

Gerrit E. Neu$^{(\boxtimes)}$, Vladislav Gudžulić, and Günther Meschke

Institute for Structural Mechanics, Ruhr University Bochum, Bochum, Germany
`gerrit.neu@rub.de`

Abstract. In this contribution, a Finite Element modelling scheme for steel-fibre reinforced concrete (SFRC) is proposed with which the post-cracking response of fibre reinforced structural members can be predicted. In contrast to the common guidelines, the post-cracking response of SFRC is derived from the actual fibre properties instead of indirectly from bending tests. The numerical model is designed to directly track the influence of design parameters such as fibre type, fibre orientation, fibre content and concrete strength on the structural response. For this purpose, sub models on the single fibre level are combined into a crack bridging model, considering the fibre orientation and the fibre content, and are integrated into a finite element model for the purpose of numerical structural analysis. The predictive capability of the proposed numerical multi-level model for SFRC is systematically validated by means of test series performed on the fibre, crack and the structural level. The experimental study comprises pull-out tests of Dramix 3D fibres, uniaxial tension tests involving different fibre contents and fibre types as well as three-point bending tests on notched beams with 23 and 57 kg/m^3 Dramix 3D fibres. Furthermore, the results are compared to the modelling approach presented in the fib model code 2010 and an inverse analysis approach.

Keywords: Steel-fibre reinforced concrete (SFRC) · FEM · Interface elements · Fibre pull-out · Uniaxial tension tests · 3-point bending

1 Introduction

Segmental tunnel linings form the final load-bearing and sealing structure of a mechanized driven tunnel. The state-of-the-art in many countries is to design these structures using reinforced concrete (RC) segments. Recent trends in the tunnelling industry are changing this paradigm, as steel fibre reinforced concrete (SFRC) is becoming more widely used worldwide. Steel fibres are capable to effectively reduce crack widths, thereby increasing the durability of the structure and providing important reinforcement at segment extremities, i.e., in areas in which traditional reinforcement bars cannot be placed due to minimum concrete cover requirements. Unfortunately, most design codes are tailored to traditional RC design, and not to SFRC design. With traditional RC most structural checks or proofs assume that stresses in concrete increase linearly until the tensile strength is reached, and that upon cracking, all tensile loads

© RILEM 2022
P. Serna et al. (Eds.): BEFIB 2021, RILEM Bookseries 36, pp. 365–376, 2022.
https://doi.org/10.1007/978-3-030-83719-8_32

within concrete members are carried by the steel reinforcement. SFRC, on the other hand, has a comparatively complex post-cracking response. The size, shape, concentration, and orientation of the fibres all affect the non-linear stress distribution across the cross section of a cracked concrete member. For design purposes, this behaviour requires several assumptions to be made of the stress distribution in a section of concrete member after cracking, or it requires a complex non-linear analysis using advanced material models by means of numerical methods. Often, numerical models for plain concrete are used, where the post-cracking response or rather the fracture energy are modified in order to capture the crack bridging capabilities of SFRC. These models consider the fibre reinforcement in a phenomenological way and are not suitable for the model-based design of SFRC structures.

In this contribution a multi-level model for the analysis of SFRC structures is presented which can assess the influence of a chosen fibre type and content on the structural response. Based on the explicit fibre geometry, the interface conditions, the material properties and an assumed fibre orientation, the model is used to determine an equivalent uniaxial traction-separation law. In order to evaluate the post-cracking response of SFRC structures by FE-analysis, a discrete crack approach using interface-elements, whose post-cracking behaviour is governed by the derived traction-separation law, is used. First, the model is briefly introduced (Sect. 2.1), validations of all sub-models on each considered scale are performed involving single fibre pull-out tests (Sect. 3.1), uniaxial tension tests (Sect. 3.2) and Sect. 3-point bending tests on notched beams (Sect. 3.3). Furthermore, the proposed Multi-Level FRC model is compared to the approach given in the fib model code 2010 [1].

2 Modeling of Steel Fibre Reinforced concrete

2.1 Multi-level FRC Model

The design of SFRC structures is not as straightforward as the design of traditional RC structures. Steel fibres provide a residual strength after onset of cracking depending on the type, content and orientation of the fibres. Available guidelines (e.g. [1, 2]) characterize the residual strength of SFRC based on bending tests and derive uniaxial stress-strain relationships for Ultimate Limit State (ULS) and Service Limit State (SLS) design. As an alternative, a multi-level FRC model [3] was developed, which allows to directly assess the influence of the individual fibre type and the fibre cocktail on the structural behaviour. As illustrated in Fig. 1, the multilevel modelling framework consists of sub-models related to three different scales involved in the numerical analyses of FRC structures:

Fibre Scale: At the level of the interaction between individual fibres and the concrete, the pull-out behaviour of a single fibre is controlled by the interface conditions, the fibre shape and the inclination θ with respect to a crack plane. Semi-Analytical models for predicting the single fibre pull-out behaviour has been developed in [4], accounting for the explicit geometry of the fibre (with and without hooked ends), the concrete strength, the inclination and embedment length of the fibre (Fig. 1, top). The plastification of the hooked-end and thereby its straightening during pull-out is captured by multiple characteristic key-states (Fig. 1, top right). Furthermore, the spalling of concrete and fibre rupture during pull-out is considered based on [5].

Crack Scale: At the level of an opening crack, the post-cracking response of SFRC is modelled by a uniaxial traction-separation (t–w) law (Fig. 1, middle), which superpose the response of plain concrete $t^{concrete}$ and the fibre bridging effect t^{fiber}:

$$t(w) = t^{concrete}(w) + t^{fiber}(w). \tag{1}$$

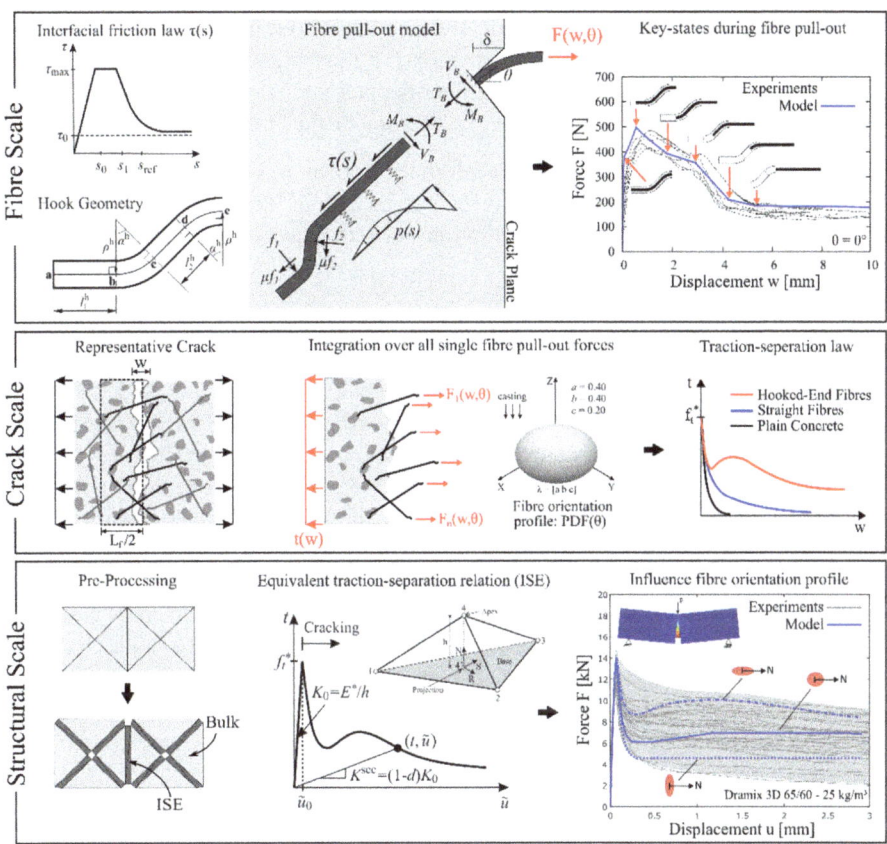

Fig. 1. Multi-Level model for SFRC, incorporating the pull-out properties of the individual fibre (top), the crack bridging effect of multiple fibres crossing a crack (middle) and the effect of a specific fibre reinforcement on the structural behaviour (bottom).

An exponential softening law for the plain concrete can be used whereas the fibre bridging effect is obtained via the integration of the single fibre pull-out responses of all effective fibres intercepting a representative crack, taking the orientation of fibres due to casting into consideration [6]:

$$f^{fiber}(w) = \frac{V_f}{A_f} \int_{\tilde{x}=0}^{L_f/2} \left[\int_{\theta=0}^{\cos^{-1}\left(2\tilde{x}/L_f\right)} F(\tilde{x}, \theta, w) \, p(\theta) d\theta \right] p(\tilde{x}) \, d\tilde{x}, \qquad (2)$$

where V_f is the volume fraction of the fibres, A_f and L_f are the cross-section area and length of one fibre, respectively. The single fibre pull-out force $F(\tilde{x}, \theta, w)$, calculated by the semi-analytical models on the fibre scale, is dependent on the embedment length $\tilde{x}$ and the inclination of the fibre θ. The spatial dispersion characteristics of the fibres in the composite is represented by the probability densities $p(\theta)$ and $p(\tilde{x})$ as functions of the inclination angle θ and the embedment length $\tilde{x}$ of the fibre, where $p(\theta)$ is characterized by an orientation tensor which defines a fibre orientation profile. A detailed description is given in [3]. Furthermore, the orientation tensor can also be obtained by simulating the fiber flow during casting with Smooth Particle Hydrodynamics [7].

Structural Scale: For capturing the development (opening, spacing) of cracks in finite element (FE) models on the structural level, interface-solid-elements (ISEs) [3] are used and equipped with the traction-separation law determined on the crack scale (Fig. 1, bottom).

2.2   Fib Model Code 2010 Approach

Generally, the design procedure for SFRC structures in accordance to guidelines begin with some sort of bending test, whose results are used to determine the residual strength of an SFRC. In the fib model code 2010 [1], a notched, 3-point bending test is called for, whereas in [2] a 4-point bending test is called for. The 3-point bending test is used to derive a Force-Crack mouth opening displacement (F-CMOD) relationship, whereas the 4-point bending test is used to derive a force-displacement relationship (F–U). The 4-point bending test provides an area of constant stress between the two load application points, which "allows" beam failure to begin at the weakest point within the constant-stress section and therefore leads to a more conservative strength characterization. In contrast, the F–CMOD relationship can more easily be used to derive a stress-crack opening relationship for fibres bridging a crack, since, because of the notch, the beam is forced to fail at a certain point due to the stress concentration at the notched section.

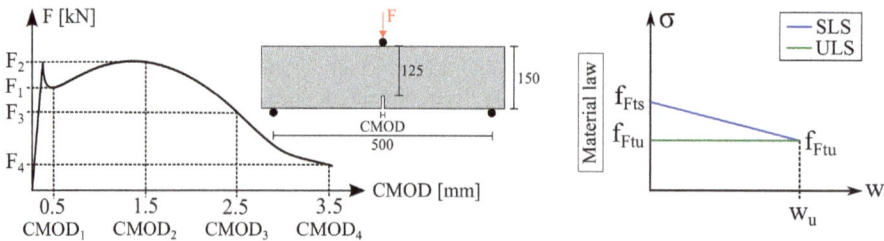

Fig. 2. Characterization of the residual strength of SFRC in accordance to [1]. The experimental results of 3-point bending tests (left) are converted in uniaxial material laws for the design of structural members (right).

In order to compare the post-cracking response of SFRC predicted by the Multi-Level FRC model (Sect. 3.1) with a guideline-based approach, the fib model code 2010 [1] approach was chosen due to the availability of stress-crack opening based material laws. After completion of the bending tests, certain discrete points taken from the experimentally obtained curves are used as a basis for characterization of the SFRC and to derive a stress-strain ($\sigma - \varepsilon$) or a stress-crack opening ($\sigma - w$) relationship, respectively. The fib model code 2010 [1] evaluates the residual flexural tensile strength at CMOD values of 0.5 mm ($CMOD_1$) and 2.5 mm ($CMOD_3$) and uses these values to determine the material laws for SLS and ULS. In general, the residual flexural tensile strength for a certain CMOD can be determined by means of the following expression:

$$f_{R,i} = \frac{3 \, F_{R,i} \, L}{2 \, b \, h_{sp}^2},$$

(3)

where $F_{R,i}$ is the measured force at the corresponding CMOD, h_{sp} is the distance between the tip of the notch and the top of the cross section (=125 mm), L is the span (=500 mm) and b is the width of the specimen (=150 mm). For ULS calculations, uniaxial residual strength is assumed to be constant and equal to the ultimate residual strength f_{Ftu} up to the ultimate crack opening w_u (Fig. 2, bottom left). The ultimate residual strength f_{Ftu} can be obtained by:

$$f_{Ftu} = \frac{f_{R3}}{3},$$

(4)

where f_{R3} is the residual flexural tensile strength at $CMOD_3$. For SLS calculations, a linear model can be used, which can be defined through residual values of flexural strength by using the following equations:

$$f_{Fts} = 0.45 \, f_{R1},$$

(5)

$$f_{Ftu} = f_{Fts} - \frac{w_u}{CMOD_3} \, (f_{Fts} - 0.5 f_{R3} + 0.2 f_{R1}) \geq 0.$$

(6)

In Fig. 2 (bottom), the corresponding material laws for SLS and ULS are shown.

Furthermore, the fib model code 2010 [1] suggests, as a more refined approach, to derive a stress-crack opening relationship based on an inverse analysis of the bending test results. In this contribution the inverse analysis approach proposed by [8] will be employed, where the total crack bridging stress, similar to Eq. 1, is the superposition of the response of plain concrete and the fibre bridging effect. An exponential softening law is used for the plain concrete component whereas the crack bridging stress of the fibres at a specific crack width $f(w)$ is calculated by:

$$f(w) = \frac{k_1 k_2 F(w) L/2}{h_{sp}^2 b}.$$

(7)

Here, $F(w)$ is the applied force at the evaluated crack width in the bending tests, k_1 was chosen to 1.25 and k_2 to 0.82. The corresponding crack width w is obtained by the measured CMOD:

$$w = \frac{CMOD \times 0.35h_{sp}}{h - 0.3h_{sp}}. \tag{8}$$

3 Experimental Validation

3.1 Fibre Scale: Single Fibre Pull-Out

On the scale of a single fibre embedded in the concrete matrix, the multi-level FRC model predicts the pull-out behaviour of a single fibre by using semi-analytical models (see Fig. 1, top). In order to calibrate and subsequently validate the model, pull-out experiments on Dramix 3D 65/60 hooked-end steel fibres for different inclinations θ with respect to the crack plane with an embedment length of 30 mm were performed [9]. These fibres are characterized by a length L_f of 60 mm, a diameter D_f of 0.92 mm and a tensile strength σ_f of 1160 N/mm^2. The geometry of the hooked ends was measured with a microscope ($\alpha_h = 43°$, $\rho_h = 2.53$ mm, $l_1{}^h = 0.24$ mm, $l_2{}^h = 0.88$ mm). For the concrete, a Young's modulus E_c of 39976 N/mm^2, a tensile strength f_t of 6.0 N/mm^2 and a mode I fracture energy G_f of 0.136 N/mm was determined. In order to estimate the bond characteristics between the fibre and the concrete matrix (Fig. 1, top left), pull-out tests on straight fibres (end hook of the Dramix 3D 65/60 fibre was removed) without inclination were performed. In Fig. 3 (top left) the experimental results and the corresponding result of the calibrated semi-analytical fibre pull-out model are shown. The parameters describing the bond characteristic were chosen to $\tau_{max} = 3.27$ [MPa], $\tau_0 = 0.5$ [MPa], $s_{ref} = 0.2$ [mm] in conjunction with a friction coefficient of $\mu = 0.37$.

The validation of the semi-analytical fibre pull-out model for Dramix 3D 65/60 hooked-end fibres is performed for inclinations θ of 0°, 30° and 60° and uses all the before mentioned parameters as a direct input. The comparison between the experimental results and the predictions of the Multi-Level FRC model are presented in Fig. 3.

In general, a good agreement between the experimental results and the model predictions can be observed. The characteristic mechanisms and interactions between the fibre and the concrete matrix during the pull-out procedure can be captured by the model. For an inclination of $\theta = 0°$ (Fig. 3, top right) as well as for an inclination of $\theta = 30°$ (Fig. 3, bottom left), the beginning of the sliding (in the model referred as "key-state 1"; see also Fig. 1, top) and consequently reaching the maximum pull-out force ("key-state 2") are captured well. Afterwards, the straightening of the hooked-end through plastification („key-states 3–5") results in a reduction of the pull-out force until the hook is completely straightened and a residual frictional resistance remains ("key-state 6"). For an inclination of $\theta = 60°$ (Fig. 3, bottom right), the maximum pull-out force and the overall pull-out response, including the rupture of the fibre at a displacement of approximately 3.8 mm, are captured well.

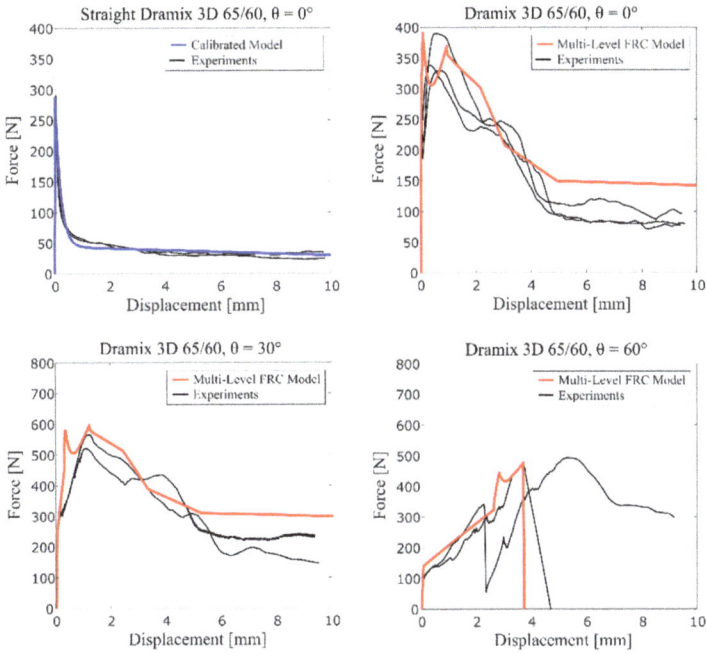

Fig. 3. Calibration (blue) and validation (red) of the Multi-Level FRC model for Dramix 3D 65/60 hooked-end steel fibres for different inclinations θ [8].

3.2   Crack Scale: Uniaxial Tension Tests

On the crack scale, the Multi-Level FRC model is used to derive an equivalent uniaxial traction-separation law, where an accurate prediction of the number of fibres crossing the representative crack and the assumed fibre orientation are of particular importance (Fig. 1, middle). In [10, 11] a systematic experimental campaign was carried out to investigate the post-cracking behaviour of SFRC and in particular, to analyse the influence of fibre type, fibre content and the fibre orientation on the post-cracking strength. Therefore, 3-point bending tests on notched beams, as required from the fib model code 2010 [1] (see Sect. 2.2), uniaxial tension tests (UTT) on small (S) and large (L) notched cylinders (Fig. 5, top left) were carried out. The experiments were performed on SFRC specimen with different fibre contents and hooked-end fibres with different lengths L_f, diameters D_f and tensile strengths σ_f. After testing, cross-sections close to the notch were cut out of all specimen and the number of fibres was detected through an image analysis of the cross-section.

For the validation regarding the assumed number of fibres crossing the representative crack in the Multi-Level FRC model, the measurements from [11] on notched beams, containing different fibre contents and different fibres, were used. The number of fibres is not only depending on the fibre content itself, but also on the geometry, especially length and diameter, of the fibres used. In the Multi-Level FRC model, due to the input of the explicit fibre geometry, the volume fraction of fibres V_f in Eq. (1) is calculated by estimating the number of fibres per m^3, depending on the fibre content along with the mass of a fibre based on its geometry, divided by the size of the representative crack. Here, the size of the representative crack was chosen equal to the cut cross-Sect. (150×150 mm) with a depth of half the fibre length (Fig. 1, middle left). A homogenous distribution of the fibres in the specimen and the same hooked-end geometry (α_h = 45°, ρ_h = 1.4 mm, l_1^h = 1.8 mm, l_2^h = 1.3 mm) was assumed for all validations. The experimentally detected number of fibres (blue and grey) and the predicted number of fibres by the Multi-Level FRC model (red) are presented in Fig. 4.

For a specific fibre type (L_f = 35 mm, D_f = 0.55 mm), a good agreement between the predicted and measured values for all investigated fibre contents can be observed (Fig. 4, left). For a constant fibre content of 39.5 kg/m^3 a relatively accurate prediction for different types of fibres can be achieved (Fig. 4, right). For the fibres with a length of 30 mm and a diameter of 0.6 mm an overestimation of the number of fibres and for the fibres with a length of 60 mm and a diameter of 0.8 mm a slight underestimation can be observed.

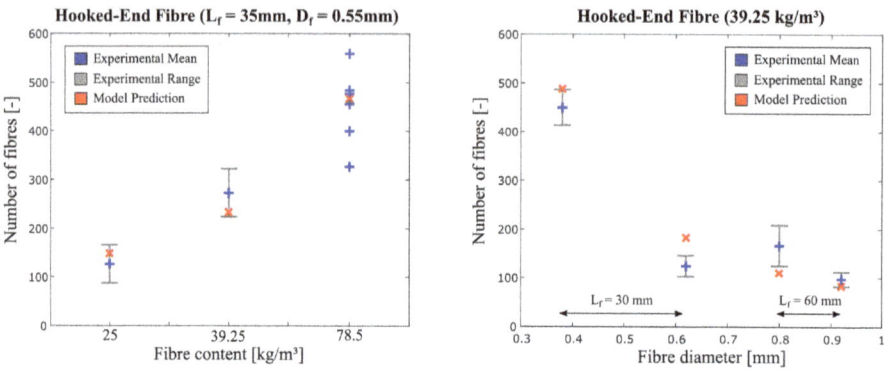

Fig. 4. Comparison of the detected fibres on a cross section through image analysis (grey and blue) and the predicted number of fibres by the Multi-Level FRC model (red) for different fibre contents (left) and different fibre types (right).

In order to validate the Multi-Level FRC model on the crack scale, the herewith obtained traction-separation laws are compared with the experimental results from UTTs performed by [10]. All available information regarding the fibre type, content and the concrete properties were used as an input of the model. No information regarding the hook-end geometry of the fibres and their pull-out behaviour was provided, therefore the hooked-end geometry from the previous investigation on the number of fibres was used and the interface parameters were chosen in accordance to Sect. 3.1. It is assumed that due to the vertical casting of the specimen and the small formwork dimensions, the fibres tend to align with the vertical plane resulting in a slightly favourable fibre orientation (fibre orientation profile λ = [0.3 0.3 0.4], see Fig. 1 middle). The resulting traction-separation relationships for different fibre types and fibre contents generated by the Multi-Level FRC model (red) are compared to the experimental results (grey and black) in Fig. 5. In addition, each diagram contains the linear uniaxial material law for SLS calculations (blue) from the fib model code 2010 [1] and the uniaxial stress crack-opening relationship obtained by inverse analysis (green) (Sect. 2.2) which were derived by the F-CMOD curves of corresponding bending tests [10].

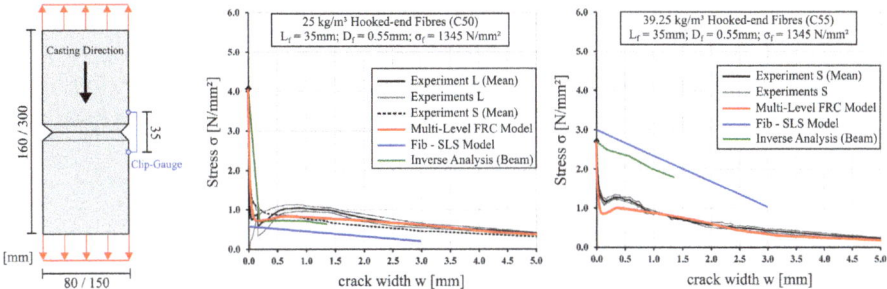

Fig. 5. Comparison between the experimental results of the UTTs (black and grey) and the uniaxial stress-crack opening relationships obtained by the Multi-Level FRC model (red) as well as the fib SLS model (blue) and the applied inverse analysis procedure (green) for different fibre contents.

Overall, a good agreement between the experimentally obtained stress-crack width curves and the predictions by the Multi-Level FRC model can be observed for all investigated cases. When comparing the linear uniaxial material laws for SLS calculations from [1] with the experimental results, the post-cracking characteristics are not well captured and the residual strength of the SFRC tends to be overestimated. The $\sigma - w$ laws obtained by inverse analysis [8] show a good agreement for a fibre content of 25 kg/m^3 whereas a poor prediction for a fibre content of 39.25 kg/m^3 can be observed. This deviation could be explained by a noticeable difference in fibre distribution and orientation between the bending tests and UTTs as well as the general tendency that bending tests estimate a higher residual strength than UTTs.

3.3 Structural Scale: 3-Point-Bending Tests on Notched Beams with Different Fibre Contents

In order to validate the Multi-Level FRC modelling framework on the structural scale, 3-point bending tests on notched beams, as required from the fib model code 2010 [1] (see Sect. 2.2), were performed. The specimens are reinforced with 23 and 57 kg/m^3 Dramix 3D 65/60 hooked-end steel fibres, respectively. The fibre as well as the concrete properties are identical to those stated in Sect. 3.1, because the single fibre pull-out tests and these bending tests were part of the same experimental campaign [9]. Based on the validated single fibre pull-out model for this fibre type (Sect. 3.1), the traction-separation laws for both fibre contents were obtained by assuming that due to the vertical casting of the specimens, the fibres tend to align with the horizontal plane (fibre orientation profile $\lambda = [0.4 \ 0.4 \ 0.2]$, see Fig. 1 middle). The resulting stress-crack opening laws predicted by the Multi-Level FRC model are shown in Fig. 6 (left) for a fibre content of 23 kg/m³ (red dotted line) and 57 kg/m³ Dramix 3D 65/60 fibres (red line). In addition, the linear stress-crack opening relationships for SLS calculations in accordance to the fib model code 2010 were derived from the experimental results of the bending tests (23 kg/m³: f_{R1} = 5.34 N/mm², f_{R3} = 3.49 N/mm²; 57 kg/m³: f_{R1} = 10.18 N/mm², f_{R3} = 6.51 N/mm²) and is also shown in Fig. 6 (blue lines). Furthermore, the stress-crack opening relationships based on an inverse analysis [8], that uses the experimental bending test data for evaluating Eqn. 7 and Eq. 8, are included in Fig. 6 (green lines). In order to simulate the bending tests on the structural scale, all the before mentioned traction-separation laws are incorporated in the numerical FE model using interface elements (Fig. 1, bottom). The resulting $F–CMOD$ curves are compared to the experimental results in Fig. 6 (right).

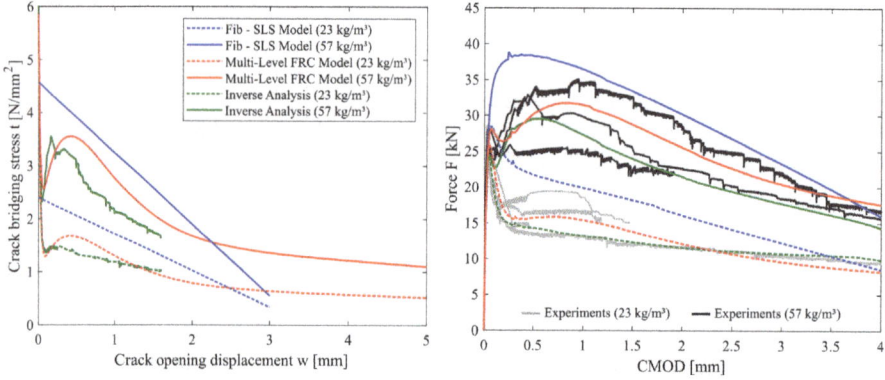

Fig. 6. Left: Traction-separation relationships for 23 and 57 kg/m³ Dramix 3D 65/60 fibres resulting from the Multi-Level FRC model (red) the linear SLS model given in fib model code 2010 (blue) and an inverse analysis (green); Right: Comparison of the experimentally obtained F-CMOD curves (grey) with the numerical results using the Multi-Level FRC model (red),the linear SLS model given in fib model code 2010 (blue) and the σ − w law derived by an inverse analysis (green).

For the Multi-Level FRC model, a good agreement between the numerical and experimental results for both fiber contents can be observed. The peak load as well as the post cracking characteristic can be captured. The same is true for using the σ − w laws derived by the inverse analysis procedure. In contrast, the linear SLS model from the fib model code 2010 [1] (Sect. 2.2) overestimates the post-cracking strength and poorly predicts the post-cracking characteristics for both fiber contents.

4   Conclusions

In this contribution a multi-level modelling framework for the numerical analysis of SFRC structures was presented and validated by experiments on all involved scales. On the fibre scale, the semi-analytical fibre pull-out model is capable to predict the characteristic behaviour of hooked-end steel fibres during the pull-out procedure for different inclinations of the fibre with respect to the crack plane. Further, on the crack scale, the presented Multi-Level FRC model is used to derive realistic uniaxial stress-crack opening relationships for different fibre types and contents. Finally, on the structural scale, the discrete crack model based on interface elements, whose behaviour is governed by the traction-separation law derived on the crack scale, is capable to reproduce the structural response for different fibre contents. It can be concluded that the Multi-Level FRC model is well suited for the model-based design of SFRC, due to its ability to directly assess the influence of major design parameters (fibre type, fibre content, fibre orientation and concrete strength) on the structural response. In contrast, the linear uniaxial material law for SLS, proposed by the fib model code 2010, is, in the investigated cases, not accurate enough to reflect the post-cracking response of SFRC beams when used as traction-separation law in interface elements and therefore should be used with caution in non-linear structural analysis involving crack width predictions. If instead an inverse analysis procedure is used to derive a multi-linear uniaxial material law from the bending test results, good predictions of the post-cracking response of SFRC in non-linear structural analysis can be obtained.

Acknowledgements. Financial support was provided by the German Research Foundation (DFG) in the framework of project B2 of the Collaborative Research Center SFB 837 *Interaction modeling in mechanized tunnelling* and the research project SPP 2020 (ME 1848/9–1). This support is gratefully acknowledged.

References

1. International Federation for Structural Concrete (fib), fib Model Code for Concrete Structures 2010 (Ernst Sohn, 2013).
2. Deutscher Ausschuss für Stahlbetonbau (DAfStb), ,Richtlinie Stahlfaserbeton (Deutscher Ausschuss für Stahlbetonbau e.V. 2012).
3. Zhan, Y. and Meschke, G., Multilevel computational model for failure analysis of steel-fiber–reinforced concrete structures, J. Eng. Mech. (ASCE) **142**, H. 11, S. 04016090 (2016)
4. Zhan, Y., Meschke, G.: Analytical model for the pullout behavior of straight and hooked-end steel fibers, J. Eng. Mech. (ASCE) **140**, H. 12, S. 04014091 (2014)
5. Laranjeira, F., Molins, C., Aguado, A.: Predicting the pullout response of inclined hooked steel fibers. Cem. Concr. Res. **40**(10), 1471–1487 (2010)
6. Wang, Y., Becker, S., Li, V.: A statistical tensile model of fiber reinforced cementitious composites. Composites **20**, 265–274 (1989)
7. Gudžulić, V., Dang, T.S., Meschke, G.: Computational modeling of fiber flow during casting of fresh concrete. Comput. Mech. **63**(6), 1111–1129 (2018). https://doi.org/10.1007/s00466-018-1639-9
8. Amin, A., Foster, S.J., Muttoni, A.: Derivation of the σ-w relationship for SFRC from prism bending tests, Struct. Concr. 16(1), 93–105 (2015)
9. Gudžulić, V., Neu, G.E., Gebuhr, G. Anders, S., Meschke, G.: Numerical multi-level model for fiber-reinforced concrete. Multi-level validation based on an experimental study on high-strength concrete (in German), vol. 273, pp. 107506, Beton- und Stahlbetonbau, Ernst, Sohn (2019)

10. Mudadu, A., Tiberti, G., Germano, F., Plizarri, G.A.: The effect of fiber orientation on the post-cracking behaviour of steel fiber reinforced concrete under bending and uniaxial tensile tests. Cement Concr. Compos. **93**, 274–288 (2018)
11. Tiberti, G., Germano, F., Mudadu, A., Plizarri, G.A.: An overview of the flexural post-cracking behaviour of steel fiber reinforced concrete. Struct. Concr. **19**, 695–718 (2018)

Isotropy-Based Analytical Model to Estimate the Residual Strength of FRC

Eduardo Galeote[1(✉)], Ana Blanco[2], and Albert de la Fuente[1]

[1] Universitat Politècnica de Catalunya, Barcelona, Spain
eduardo.galeote@upc.edu
[2] Loughborough University, Loughborough, UK

Abstract. This study describes an analytical model to estimate the post-cracking strength of FRC under flexural stresses in a three-point bending configuration. Considering the content and the number of fibres within concrete, statistical distributions are implemented to assign randomly the position and orientation of each fibre. Different degrees of isotropy can be defined depending on the properties of concrete, this leading to different distributions and orientations of the fibres according to the type of concrete. Pull-out laws are also implemented in the model to define the pull-out characteristics of the fibres at the crack section and determine the residual strength of FRC based on the contribution of each fibre to resisting the flexural load. Therefore, and assuming that only fibres contribute to the tensile strength after cracking, the pull-out load of all the fibres combined with the sectional equilibrium can be used to determine the post-cracking strength of FRC. Assembling the contribution of all the fibres provides a representative curve of the post-cracking strength of the analysed element. The estimations of the post-cracking strength curves are able to reflect the influence of the content and the orientation of the fibres, as well as the effect of the specimen dimension and the type of concrete, defined by the degree of isotropy.

Keywords: Fibre reinforced concrete · Residual strength · Analytical model · Orientation · Distribution

1 Introduction

Fibre orientation is particularly relevant given its ability to substantially modify the behaviour of fibre reinforced concrete (FRC) at the post-cracking stage [1–3]. Plenty of research has focused on the influence of the type or geometry of the fibres [4–6], however, it is the distribution and orientation of the fibres that remain unresolved yet. It is, therefore, important to account with numerical tools that take the orientation and distribution of fibres into consideration to simulate the behaviour of FRC under a three-point configuration aimed for pre-design purposes.

Despite the factors influencing the orientation of the fibres [7], constitutive models for FRC usually ignore the effect of fibre orientation on the mechanical performance. The MC2010 [8] proposes an orientation factor (K) to account for the orientation of the fibres on design, although no specific values are suggested and the parameter needs to

© RILEM 2022
P. Serna et al. (Eds.): BEFIB 2021, RILEM Bookseries 36, pp. 377–388, 2022.
https://doi.org/10.1007/978-3-030-83719-8_33

be adjusted according to the criteria of the designer. In this line, also the AFGC recommendations [9] address the use of an orientation factor K to take into account the influence of the fibre orientation, with some authors [10, 11] proposing methods that deal with fibres orientation and its effect on the material response. Accordingly, estimating the residual strength of FRC considering the distribution and orientation of the fibres becomes a matter of great interest.

The main objective of this study is to present a model to estimate the behaviour of FRC under flexural stresses. Based on the properties of the fibre, the concrete matrix and the interaction between these two elements, the model estimates the flexural behaviour of specimens subjected to a three-point bending configuration. The analytical method distributes the fibres within the cross-section of an element following a predefined orientation. Each fibre is assumed to contribute independently to the overall flexural strength according to its specific pull-out properties. Assembling the contribution of all the fibres provides a representative curve of the post-cracking strength of the analysed element.

The results show that the orientation of the fibres strongly depends on the dimensions of the cross-section and that the model is able to capture the influence of the size effect, the content of fibres and the degree of isotropy, which is influenced by the type of concrete. Consequently, the model could be used as a tool to estimate the contribution of fibres to the residual strength of FRC at the design stage and compare the influence of different factors on the post-cracking performance.

2   Description of the Model

2.1   General Overview

The model is based on the procedure for FRC described by Robins et al. [12] and subsequently used in additional research by the same authors [13, 14]. This model takes into account only the fibres that contribute to resisting the load to which the element is subjected and is built on the idea that the behaviour of a cracked section of FRC is dominated by the pull-out strength of each fibre bridging the crack.

The position, the embedded length or the orientation angle of the fibre with respect to the crack surface are defined according to a probabilistic procedure [15] that randomly calculates these three variables for all the fibres at the crack surface. The pull-out mechanisms of fibres, as described in models presented by different authors [16–18], are also implemented to assess the pull-out constitutive law of the fibres.

To determine the tensile block produced by the pull-out of the fibres on the sectional diagram for a given crack opening, each fibre is addressed individually. The model is analysed in terms of crack opening at the front tip (w_1), although each fibre has a specific crack opening (w_2) that depends on its position with regard to the depth of the specimen. The crack opening associated with each fibre is used to determine the pull-out load.

Given that the contribution of concrete on the tensile stress of FRC can be considered negligible in the post-cracked regime, fibres are assumed to be the only elements contributing to resist the flexural stresses the beam undergoes. The failure mechanism of a beam under flexure is, therefore, represented considering the upper point as the hinge bonding the two halves of the element when the crack appears (Fig. 1). This assumption is an approximation of the real failure mechanism and simplifies the assessment of the internal forces equilibrium for a certain crack opening. As shown in Fig. 1, such simplification leads to disregard the depth of the compressed block considering that the sectional compression forces act on the hinge, leading to determine the flexural moment equilibrium taking into account only the pull-out forces of the fibres under tensile stresses.

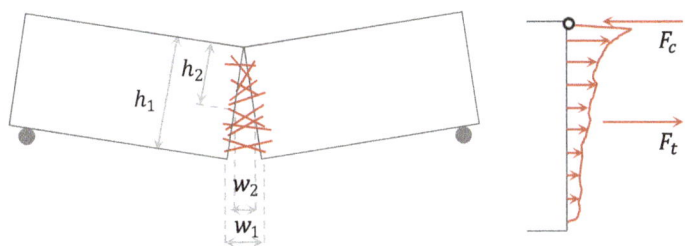

Fig. 1. Failure mechanism and sectional forces.

2.2 Position and Orientation of the Fibres

The fibre distribution is generated in the model based on the orientation defined in the three main directions. Based on previous research [19–21], the distribution of the fibres within the cross-section is generally uniform although the position and the orientation follow a probability density function that fits a normal distribution. In this study, the Gaussian law has been taken as the fundamental hypothesis mainly due to the linear relation between the orientation number and the number of fibres crossing the cracked section [22].

The number of fibres in a specific volume of concrete is determined according to the content of fibres, which are distributed and oriented following the procedure shown in Fig. 2. For this, the coordinates of the middle point of each fibre (P_X, P_Y, P_Z) are determined, followed by determining the orientation of the fibres based on the unitary vector of the direction of the fibre $\vec{D} = (D_X, D_Y, D_Z)$. Both the position and the orientation, along with the fibre length, allows determining the position of the edges of the fibre, that need to remain inside the boundaries of the specimen.

Each component of the coordinates of the position (P_X, P_Y, P_Z) and the unitary vector of the direction $\vec{D} = (D_X, D_Y, D_Z)$ are calculated according to an independent probabilistic law associated with each of the components in axes X, Y and Z. In this model, a uniform distribution of the fibres is considered given that fibres are assumed to present the same probability of being in any position. Such a distribution can be represented as a particular case of a standard distribution. For this, the standard deviation (σ) and the mean value (μ) that define the standard distribution need to present a relation $\sigma \gg \mu$. This leads to a flatter shape of the probability density function that entails a cumulative probability $\phi_{\mu,\sigma}(X) \approx 0.5$ for values $X \approx \mu$. Therefore, a random value $X \in [X_1, X_2]$ with $X_1 < \mu < X_2$ can be assessed with a probability $\phi_{\mu,\sigma}(X) \approx \phi_{\mu,\sigma}(X_1) \approx \phi_{\mu,\sigma}(X_2) \approx 0.5$.

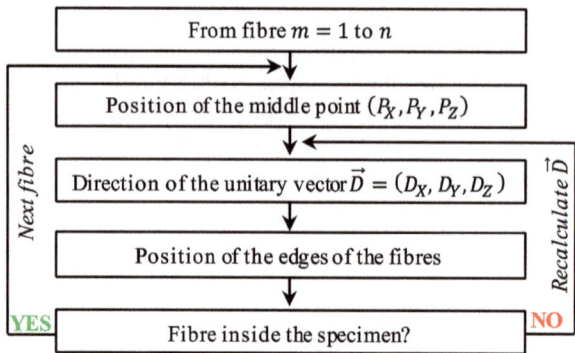

Fig. 2. Procedure for distributing the fibres in the concrete matrix.

Considering the above, the three coordinates of the position of the central point of a fibre (P_X, P_Y, P_Z) can be randomly evaluated assuming a uniform distribution along any of the axes X, Y or Z. The procedure to determine the position in a general axis i (P_i) between a maximum and a minimum position $P_{i,max}$ and $P_{i,min}$ is described in the following steps:

1. Determine the mean value μ and define a standard deviation $\sigma \gg \mu$ so the probabilistic law follows a uniform distribution.
2. Assess the maximum and the minimum position of the central point of the fibre so that $P_{i,min} < \mu < P_{i,max}$ and $P_{i,min} \approx \mu \approx P_{i,max}$ given that $\sigma \gg \mu$.
3. Calculate the difference between the cumulative normal distributions of the maximum and the minimum position of the fibre $\Delta\left(\phi_{\mu,\sigma}\left(P_{i,max}\right), \phi_{\mu,\sigma}\left(P_{i,min}\right)\right)$.
4. Generate a random number $k \in [0, 1]$ and obtain the cumulative normal distribution of a random position $\phi_{\mu,\sigma}(P_i) = \phi_{\mu,\sigma}\left(P_{i,min}\right) + k \cdot \Delta\left(\phi_{\mu,\sigma}\left(P_{i,max}\right), \phi_{\mu,\sigma}\left(P_{i,min}\right)\right)$.
5. Interpolate $\phi_{\mu,\sigma}(P_i)$ to obtain $P_i = \frac{\phi_{\mu,\sigma}(P_i) - \phi_{\mu,\sigma}\left(P_{i,min}\right)}{\phi_{\mu,\sigma}\left(P_{i,max}\right) - \phi_{\mu,\sigma}\left(P_{i,min}\right)} \cdot \left(P_{i,max} - P_{i,min}\right) + P_{i,min}$.

A similar procedure is used to determine the three components of the directional vector $\overrightarrow{D_t} = \left(D_{tx}, D_{ty}, D_{tz}\right)$ with the interpolation described in Step 5 adapted as shown in Eq. (1). These three components of $\overrightarrow{D_t}$ are subsequently normalized as shown in Eq. (2) to assess the unitary vector of the direction of the fibres $\vec{D} = (D_X, D_Y, D_Z)$.

$$D_{ti} = \frac{\phi_{\mu,\sigma}(D_{ti}) - \phi_{\mu,\sigma}\left(D_{ti,min}\right)}{\phi_{\mu,\sigma}\left(D_{ti,max}\right) - \phi_{\mu,\sigma}\left(D_{ti,min}\right)} \cdot \left(D_{ti,max} - D_{ti,min}\right) + D_{ti,min} \tag{1}$$

$$D_i = \frac{D_{ti}}{\sqrt{D_{tx}^2 + D_{ty}^2 + D_{tz}^2}} \tag{2}$$

2.3 Direction of the Fibres and Degrees of Isotropy

The components of the directional vector (D_{tx}, D_{ty}, D_{tz}) are defined by the degree of isotropy of FRC. In case of a perfect isotropic FRC these components reach values $D_{ti,max} = 1.0$ and $D_{ti,min} = -1.0$, whereas a preferential alignment of the fibres entails a certain degree of anisotropy leading $D_{ti,max}$ and $D_{ti,min}$ to present absolute values below 1 and reach 0 in case of perfect anisotropy [23, 24]. A summary of four combinations of isotropy degrees is shown in Table 1. A preferential orientation of the fibres as in self-compacting concrete (SCC) would lead to a degree of isotropy such as $(D_{tx}, D_{ty}, D_{tz}) = (1.0, 0.5, 1.0)$ according to results of previous research [15, 25]. Two additional degrees of isotropy have been considered: the first one states for an intermediate anisotropy $(D_{tx}, D_{ty}, D_{tz}) = (1.0, 0.8, 1.0)$ between full isotropic distribution and SCC. The second describes a case of concrete with a certain anisotropy on the Z-axis with a directional vector $(D_{tx}, D_{ty}, D_{tz}) = (1.0, 0.5, 0.8)$.

Table 1. Combinations of degrees of isotropy.

Degrees of isotropy	D_{tx}	D_{ty}	D_{tz}	Notation
Isotropic concrete	1.0	1.0	1.0	ID1
Quasi-isotropic concrete	1.0	0.8	1.0	ID2
Self-compacting concrete	1.0	0.5	1.0	ID3
Anisotropic concrete	1.0	0.5	0.8	ID4

2.4 Pull-out Behaviour

Given the properties of the fibres used in the experimental programme, the pull-out model proposed by Laranjeira [16] for straight fibres has been used to reproduce the pull-out behaviour. The input parameters for the pull-out model are based on those adopted in previous studies [15, 16]. A friction coefficient $\mu = 0.6$ between the steel fibre and the concrete is assumed, while the ultimate tangential debonding stress was estimated at 51% of the tensile strength of concrete. The tensile strength of concrete was calculated according to the specifications of MC2010.

3   Analysis of the Post-cracking Strength

3.1 Experimental Program

The composition of the high-performance concrete mixes used in the experimental program is detailed in Table 2. Two water-to-cement ratios ($w/c = 0.16$ and 0.23, A and B respectively) and two contents (90 and 190 kg/m^3) of steel straight fibres ($l = 13$ mm; $\emptyset = 0.2$ mm; $f_{uf} = 2750$ MPa) were used to manufacture four FRC mixes (M90A, M90B, M190A and M190B).

Table 2. Concrete mixes [kg/m³].

Materials	M90A	M190A	M90B	M190B
CEM I 52.5R	800	800	800	800
Silica sand 3 – 4 mm	1131	1098	1129	1098
Filler (CaCO₃)	200	200	200	200
Water	129	129	185	185
Nanosilica (o.c.w.)	5%	5%	5%	5%
Superplasticizer (o.c.w.)	4%	4%	2%	2%
Steel fibres	90	190	90	190

The experimental campaign is based on three-point bending tests conducted on prismatic FRC specimens of 40×40×160, 100×100×400 and 150×10×600 mm. The results of the bending tests are shown in terms of residual strength f_R - CMOD in Fig. 3. These results present the post-cracking strength of the four mixes of FRC determined using specimens of dimensions 150×150×600 mm and the results of mix M190B for the three dimensions of specimens tested.

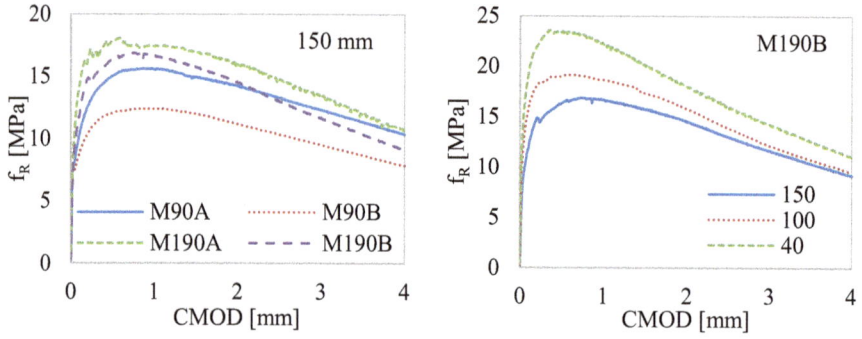

Fig. 3. f_R - CMOD of the four mixes and 150, 100 and 40 mm of M190B.

The results exhibit clear trends in the results presented in terms of the influence of the content of fibres, w/c ratio and dimension of the samples. In this line, the increasing content of fibres reveals greater strengths, while reducing the w/c ratio also leads to improved results given the greater strength of the concrete matrix and an improved adherence between the fibre and the matrix. Regarding the effect of the dimension of the specimen, a generalised size effect could be identified in all mixes, with smaller samples presenting greater residual strengths.

3.2   Results of the Analytical Model

3.2.1   Influence of the Content of Fibres

A comparison of the residual strength f_R - CMOD curves obtained by the model for the analyses of mixes (M90A, M190A, M90B and M190B) is shown in Fig. 4. The results are presented for prismatic specimens of dimensions 150, 100 and 40 mm with a defined degree of isotropy ID4.

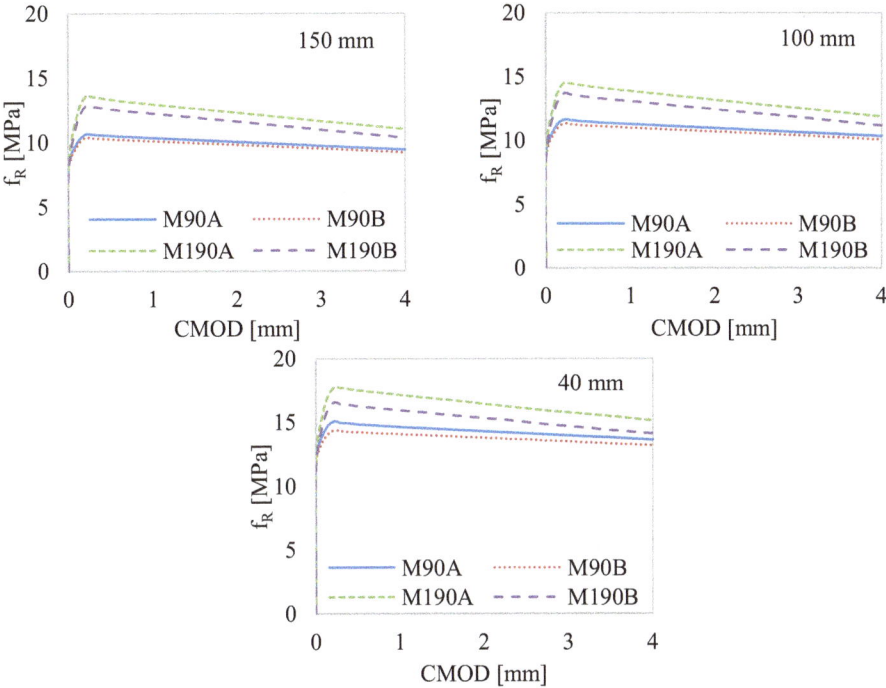

Fig. 4. f_R - CMOD curves for specimens of a) 150, b) 100 and c) 40 mm.

The results of the model reveal how increasing the fibre content leads to an enhanced post-cracking performance. In line with the results of other authors [5, 26], the effect of the matrix strength due to the w/c ratio represents an additional factor influencing the residual strength of FRC. In this case, both M90A and M190A with a $w/c = 0.16$ result in a stronger matrix leading to higher residual strengths than M90B and M190B ($w/c = 0.23$). The w/c ratio is not a defining parameter of the model, although its influence on the compressive, flexural and residual strength—inputs of the model—affect the matrix properties, the pull-out response of the fibres and, consequently, the residual strength curves.

3.2.2 Influence of the Geometry

Figure 5 shows the results of the model when using the input data of mixes M90A, M190A, M90B and M190B for three specimen dimensions and a degree of isotropy ID4. The results show a clear influence of the element size on post-cracking strength, which decreases with the size of the element.

The results of the model exhibit an evident size effect that recreates the variation of the strength with regard to the specimen dimension. The cross-section of a 100×100

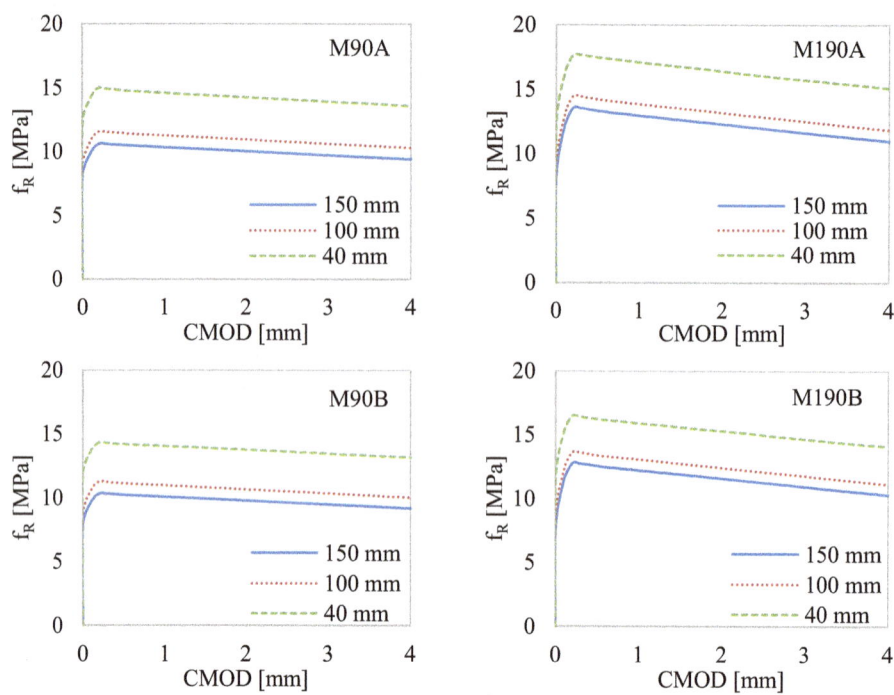

Fig. 5. Comparison of f_R of three specimen dimensions.

mm beam is 56% smaller than that of a 150×150 mm beam, such reduction entailing an average increase of the strength of 8.2%. The reduction from 100×100 mm to 40×40 mm is 84% and represents an average increase of strength of 27%. This may be attributed to the strong dependence of the flexural behaviour on the number of fibres with a greater orientation number in a perpendicular direction to the cracked surface when reducing the cross-section [27]. In this line, reducing the cross-section increases the anisotropy of the material leading to a greater orientation of the fibres in a perpendicular direction to the crack surface. Accordingly, some studies attribute the increasing strength of FRC when reducing the specimen dimension to the wall-effect and the orientation of the fibres rather than to the size effect itself [5].

Given that the flexural residual strengths f_{R1} and f_{R3} are used by many codes and guidelines to estimate the constitutive equation of FRC, the evolution of these two parameters with the increasing dimensions of the elements is shown in Fig. 6. The values presented correspond to the four mixes described under a degree of isotropy ID4. The results clearly represent the influence of the geometry of the elements as well as the effect of the content of fibre and the matrix strength. In this case, the differences in the residual strength f_{R1} and f_{R3} calculated through the model between mixes with the same content of fibres are attributed to the input properties of concrete.

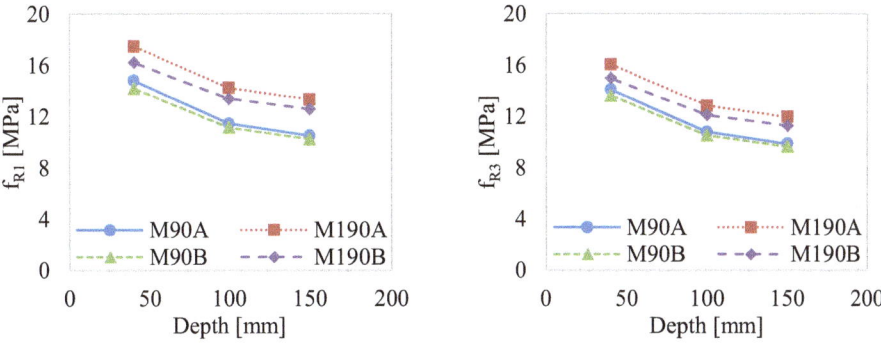

Fig. 6. f_{R1} and f_{R3} according to the specimen dimension.

3.2.3   Influence of the Degree of Isotropy

The results of Fig. 7 show the influence of the degrees of isotropy that state for an isotropic (ID1) and an anisotropic concrete (ID4) on the residual strengths f_{R1} and f_{R3}. The results also depict the evolution of the residual strengths obtained for different specimen sizes with cross-sectional dimensions of 150×150, 100×100 and 40×40 mm. The results are shown for straight fibres in a content of 90 kg/m^3.

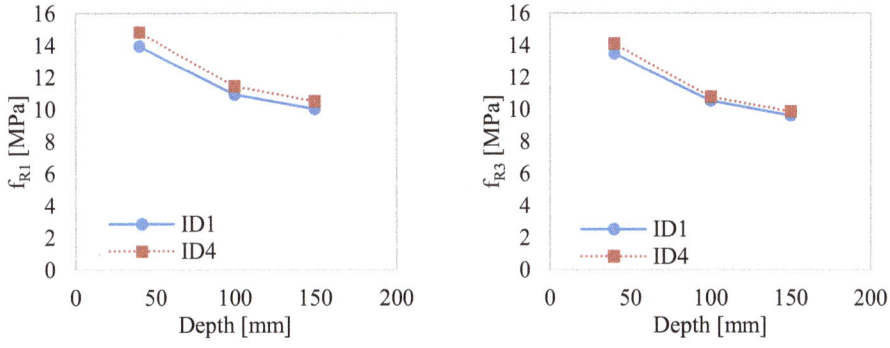

Fig. 7.   Influence of the degree of isotropy on a) f_{R1} and b) f_{R3}.

Even though ID4 entails a higher number of the fibres in the direction perpendicular to the failure plane, Fig. 7 reveals that the variation of the isotropy degree leads to almost negligible differences on the residual strength. In this line, even the scatter of the bending tests (in some cases around 20% [28]) might be higher than the differences on the residual strength exhibited between the two analysed degrees of isotropy. This highlights that, in some cases, even though casting and vibration procedures may lead to different orientations, the impact of these two factors may be lower than expected.

4   Conclusions

An analytical model based on the distribution and the pull-out of the fibres to estimate the flexural performance of FRC has been presented. One the interests of the model is that, each time the model runs, the position and the orientation of the fibres are slightly different given the probabilistic distribution used to calculate these parameters. When compared to the experimental curves, the results of the model are able to recreate the trends of the post-cracking strength according with the different properties that have been analysed and define the behaviour of FRC. Based on the experimental results presented and the results yielded by the model, several conclusions may be drawn:

- The model takes into account the influence of the content of fibres on the residual strength with mixes blended with greater contents exhibiting higher residual strengths.
- The model reflects the size effect, with specimens of greater dimensions exhibiting lower residual strengths. This is attributed to the distribution and orientation of fibres, since smaller dimensions present a greater number of fibres oriented perpendicular to the crack section, thus enhancing the residual strength.
- Isotropic concretes (ID1) present lower residual strengths than more anisotropic concretes (ID4). This effect is explained given that fibres following a degree of isotropy ID4 present a preferential orientation towards a perpendicular direction to the cracked surface and have a greater contribution to post-cracking strength.

This model is presented as a tool to estimate the contribution of fibres to the residual strength of FRC at the design stage, and can be adapted to use different types of fibres with different geometries and properties if implementing the pertinent pull-out law. Additionally, the type of concrete can be introduced by means of the parameter accounting for the degree of isotropy (ID) that takes into account de flowability of the concrete and its ability to orientate the fibres towards a specific direction. In this line, and even though the model may present some limitations adjusting real flexural results, it reflects the general influence of the fibres on the post-cracking strength of FRC.

Acknowledgements. The first author acknowledges the Spanish Ministry of Science, Innovation and Universities for providing support through the FPU fellowship. This research has been possible due to the economic funds provided by the SAES project (BIA2016–78742-C2–1-R) of the Spanish Ministry of Economy, Industry and Competitiveness. The authors also want to express their gratitude to Escofet S.A. for the collaboration during the experimental program.

References

1 Abrishambaf, A., Pimentel, M., Nunes, S.: Influence of fibre orientation on the tensile behaviour of ultra-high performance fibre reinforced cementitious composites. Cem. Concr. Res. **97**, 28–40 (2017)
2 Zhou, B., Uchida, Y.: Influence of flowability, casting time and formwork geometry on fiber orientation and mechanical properties of UHPFRC. Cem. Concr. Res. **95**, 164–177 (2017)

3 Zhou, B., Uchida, Y.: Relationship between fiber orientation/distribution and post-cracking behaviour in ultra-high-performance fiber-reinforced concrete (UHPFRC). Cem. Concr. Compos. **83**, 66–75 (2017)

4 Lawler, J.S., Zampini, D., Shah, S.P.: Microfiber and macrofiber hybrid fiber-reinforced concrete. J. Mater. Civ. Eng. **17**, 595–604 (2005)

5 Abu-Lebdeh, T., Hamoush, S., Heard, W., Zornig, B.: Effect of matrix strength on pullout behavior of steel fiber reinforced very-high strength concrete composites. Constr. Build. Mater. **25**, 39–46 (2011)

6 Wu, Z., Shi, C., He, W., Wu, L.: Effects of steel fiber content and shape on mechanical properties of ultra high performance concrete. Constr. Build. Mater. **103**, 8–14 (2016)

7 Martinie, L., Roussel, N.: Simple tools for fiber orientation prediction in industrial practice. Cem. Concr. Res. **41**, 993–1000 (2011)

8 International Federation for Structural Concrete, Fib Model Code for Concrete Structures 2010 (2010). www.fib-international.org

9 AFGC, Ultra high performance fibre-reinforced concrete. Recommendations, Association Française de Génie Civil (2013)

10 Simon, A., Corvez, D., Marchand, P.: Feedback of a ten years assessment distribution using K factor concept. In: RILEM-Fib-AFGC International Sympposium. Ultra-High Perform. Fibre-Reinforced Concrete UHPFRC 2013, pp. 669–678 (2013)

11 Guénet, T., et al.: Numerical modeling of UHPFRC tensile behavior by a micromechanics FEM model taking into account fiber orientation. In: Proceedings 9th International Conference Fracture Mechanics of Concrete and Concrete Structures IA-FraMCoS (2016)

12 Robins, P., Austin, S., Chandler, J., Jones, P.: Flexural strain and crack width measurement of steel-fibre-reinforced concrete by optical grid and electrical gauge methods. Cem. Concr. Res. **31**, 719–729 (2001)

13 Prudencio, L., Austin, S., Jones, P., Armelin, H., Robins, P.: Prediction of steel fibre reinforced concrete under flexure from an inferred fibre pull-out response. Mater. Struct. **39**, 601–610 (2007)

14 Jones, P.A., Austin, S.A., Robins, P.J.: Predicting the flexural load–deflection response of steel fibre reinforced concrete from strain, crack-width, fibre pull-out and distribution data. Mater. Struct. **41**, 449–463 (2008)

15 Cavalaro, S.H.P., Aguado, A.: Intrinsic scatter of FRC: an alternative philosophy to estimate characteristic values. Mater. Struct. **48**, 3537–3555 (2015)

16 Laranjeira, F., Aguado, A., Molins, C.: Predicting the pullout response of inclined straight steel fibers. Mater. Struct. **43**, 875–895 (2010)

17 Laranjeira, F., Molins, C., Aguado, A.: Predicting the pullout response of inclined hooked steel fibers. Cem. Concr. Res. **40**, 1471–1487 (2010)

18 Zhan, Y., Meschke, G.: Analytical model for the pullout behavior of straight and hooked-end steel fibers. J. Eng. Mech. **140**, 04014091 (2014)

19 Abrishambaf, A., Barros, J.A.O., Cunha, V.M.C.F.: Relation between fibre distribution and post-cracking behaviour in steel fibre reinforced self-compacting concrete panels. Cem. Concr. Res. **51**, 57–66 (2013). https://doi.org/10.1016/j.cemconres.2013.04.009

20 Nguyen, D.L., Ryu, G.S., Koh, K.T., Kim, D.J.: Size and geometry dependent tensile behavior of ultra-high-performance fiber-reinforced concrete. Compos. Part B Eng. **58**, 279–292 (2014)

21 Carmona, S., Molins, C., Aguado, A., Mora, F.: Distribution of fibers in SFRC segments for tunnel linings. Tunn. Undergr. Sp. Technol. **51**, 238–249 (2016)

22 Krenchel, H.: Fibre spacing and specific fibre surface (1975)

23 Cavalaro, S.H.P., López, R., Torrents, J.M., Aguado, A.: Improved assessment of fibre content and orientation with inductive method in SFRC. Mater. Struct. **48**, 1859–1873 (2015)

24 Cavalaro, S.H.P., López-Carreño, R., Torrents, J.M., Aguado, A., Juan-García, P.: Assessment of fibre content and 3D profile in cylindrical SFRC specimens. Mater. Struct. **49**, 577–595 (2016)

25 Laranjeira, F., Aguado, A., Molins, C., Grünewald, S., Walraven, J., Cavalaro, S.: Framework to predict the orientation of fibers in FRC: a novel philosophy. Cem. Concr. Res. **42**, 752–768 (2012)

26 Yoo, D.-Y., Park, J.-J., Kim, S.-W.: Fiber pullout behavior of HPFRCC: effects of matrix strength and fiber type. Compos. Struct. **174**, 263–276 (2017)

27 Zerbino, R., Tobes, J.M., Bossio, M.E., Giaccio, G.: On the orientation of fibres in structural members fabricated with self compacting fibre reinforced concrete. Cem. Concr. Compos. **34**, 191–200 (2012)

28 Molins, C., Aguado, A., Saludes, S.: Double punch test to control the energy dissipation in tension of FRC (Barcelona test). Mater. Struct. **42**, 415–425 (2009)

Different Approaches for FEM Modelling of Strain-Hardening Cementitious Composites

Hassan Baloch[(✉)], Steffen Grünewald, and Stijn Matthys

Magnel-Vandepitte Laboratory, Department of Structural Engineering
and Building Materials, Ghent University, Ghent, Belgium
Hassan.baloch@ugent.be

Abstract. Strain-hardening cementitious composites (SHCCs) are a special class of fibre-reinforced concretes which develop multiple, fine cracks when subjected to an increasing tensile load. This ensures remarkably high strain capacities of up to several percent, making SHCC an advantageous material both for new structures subjected to high mechanical loading or in severe environmental conditions and for the retrofitting/strengthening of existing structures by application of thin SHCC layers.

Finite element modelling (FEM) can be an efficient tool to predict the behaviour of SHCC under different loading types. This paper lists some recent advances in finite element modelling of strain hardening fibre reinforced cementitious composites. Based on a comprehensive literature review two main approaches are highlighted, the continuum model approach and the advanced lattice modelling approach. This paper provides an overview of both approaches and their functionalities with respect to each other, based on results from literature. The work reported in this paper is part of a larger study into the use of SHCC for shear strengthening of existing concrete structures.

Keywords: Fibres · Strain-hardening cementitious composites · Finite element modelling

1 Introduction

Strain-hardening cementitious composites (SHCCs) are a special class of high-performance fibre-reinforced cementitious composites which under tensile stress show multiple cracking and a very high ductility, [1]. This material, also commonly referred to in literature as engineering cementitious composites (ECC), are typically reinforced with high strength short plastic fibres such as poly-vinyl alcohol (PVA) or high-density polyethylene (HDPE). Through well considered fibre types and tailored matrix design, SHCC can show strain hardening behaviour having tensile strain capacities of up to 8% [2–4]. This remarkable deformation is caused by the fine distribution of cracks in the material, which are bridged by short fibres. These properties make SHCC an interesting choice not only in construction but also as a repair material to strengthen existing structures.

In order to successfully implement SHCCs in the construction industry, there is a need for reliable modelling techniques, both at the level of material performance and

© RILEM 2022
P. Serna et al. (Eds.): BEFIB 2021, RILEM Bookseries 36, pp. 389–399, 2022.
https://doi.org/10.1007/978-3-030-83719-8_34

structural behaviour. During the last two decades, researchers have tried to develop an accurate model representation of the complex SHCC material behaviour. Traditionally, micromechanics-based analytical methods are used to tailor the mix composition of SHCC and to ensure multiple cracking. According to this methodology, a fibre bridging law is applied by satisfying both energy and strength criteria which eventually leads to steady-state crack propagation. These criteria are explained in detail in the literature [5, 6].

The modelling of SHCC members is characterised by a few differences when compared to conventional concrete. The additional complexity arises with the continuous formation of multiple cracks. In order to simplify the process, continuum-based finite element models have been developed in which the composite is modelled as a single continuum instead of analysing the interaction between fibres and the matrix. Continuum models can follow various constitutive laws which are found in literature and can be broadly classified as homogenization-based constitutive laws [7, 8] and improved individual-based constitutive models [9–11]. Continuum models, however, come with limitations as they are usually not able to predict the sequential formation of cracks and the crack spacing. Considering these limitations, few advanced models are also studied which consider the effect of each fibre in the matrix and attempts to model the post multiple cracking behaviour. Most frequently studied are rigid body spring models [13, 14] and lattice models [15, 16]. In contrast to the technical work done in this field, an overview of FEM modelling approaches for SHCC materials is lacking.

In this paper, an overview of FEM models for SHCC members is presented. Firstly, different variants of continuum-based models are highlighted. After that, some recent advanced approaches are discussed including lattice and rigid based models. As this paper deals with FEM modelling of SHCC, specifically, experimental works and other types of fibre reinforced composites are excluded.

2   Continuum-Based Modelling

SHCC comprises of a cementitious matrix and embedded short fibres. Associated to this combination, the fibre-matrix interface is of governing importance. While complex micro-models can accurately depict the response by incorporating different heterogeneities in SHCC, it can require higher computational power. In this respect, continuum macro-models can provide a practical solution to simulate the structural behaviour of SHCC elements [18]. These models are easy to use and are often already incorporated in proprietary non-linear FEM packages (such as Diana, Abaqus, Atena, etc.). Various constitutive models are used by researchers in the continuum-based approach which can be broadly classified as homogenization-based constitutive models [7] and individual-based continuum models [9].

2.1   Homogenization-Based Approach

Kabele [19] proposed in 2002 a homogenization-based constitutive law for SHCC materials (Fig. 2 which is derived as a relation between total stress and total strain of a representative volume element. This volume element comprises of up to two mutually perpendicular sets of parallel distributed cracks. These cracks are assumed to form when the maximum principal stress reaches the value of first cracking strength. The overall response of the continuum is then governed by a linear hardening relationship obtained from the uniaxial tensile test. Once formed, the crack direction is assumed to remain fixed. However, the crack can slide in case of a change in direction of principal stresses. Crack sliding is assumed to be resisted by bridging fibres.

This constitutive law is used in [20] to simulate the experiments on SHCC beams which were designed to fail in shear, also referred as Ohno beams (Fig. 1a) [18]. The authors studied the effect of bond-slip behaviour, the influence of tensile properties and fibre stiffness. The authors found that only a fraction of SHCC's uniaxial capacity was utilized, possibly because of fibre damage due to relative crack slipping. To match experimental results, the authors recommended using lower fibre shear modulus values. The results in Fig. 1b shows that although the analysis predicted peak loads fairly reasonable, the peak load was higher and the post-peak response was stiffer in the simulations and the cracks were not correctly represented. The model predicted cracks at 45° which was a higher angle than that in the experimental results.

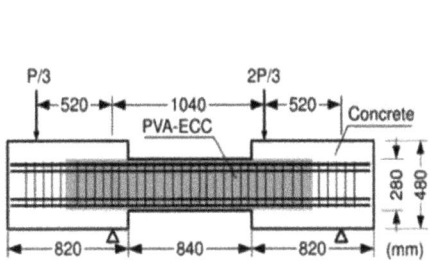

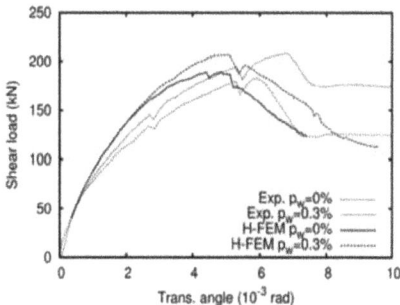

Fig. 1a. Ohno Shear beam configuration [18]

Fig. 1b. Analytical (H-FEM) and experimental results of beams with stirrup reinforcement ratios $p_w = 0$ and $p_w = 0.3\%$ [18]

2.2 Individual-Based Approach

Considering the limitations of homogenization-based models, Kabele [10, 21] and Yang and Fischer [9] proposed the individual-crack-based approach shown in Fig. 2b. In this modelling technique the crack opening vs bridging stress relationship is first derived by experimental testing of notched tensile specimens. For a particular element, the post-peak response is then calculated by dividing the crack opening by the element dimension perpendicular to the crack. As shown in Fig. 2b there is a drop in stress with crack opening and then a further increase in stress as the load is transferred to the fibres. According to Kabele [21] this approach can improve the results under non-proportional stresses as compared with homogenized-based models. However, in the earlier model of Kabele [21] one crack will form in each element as soon as the tensile strength is

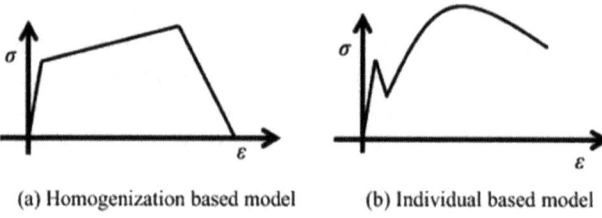

(a) Homogenization based model   (b) Individual based model

Fig. 2. Continuum-based models for SHCC

reached, so the mesh size becomes an important factor to determine crack spacing. For an accurate analysis of crack spacing, a priori information on the crack pattern is required to select the mesh size. Yang and Fischer [9] and Kabele [10] used an alternative approach where they considered variation of crack bridging behaviour among different sections based on experimental results and they imposed a minimum crack spacing criteria so that cracks are formed sequentially under increasing stresses.

3 Lattice Models

Initially lattice models found their application in the concrete industry to simulate chloride diffusion and moisture transport [22]. However, recently there is an increasing interest in their usage for structural analysis especially for high-performance cementitious composites [23, 24]. In the following section, different types of lattice models developed to model SHCC are reviewed.

3.1 Lattice Equivalent Continuum Model (LECM)

Zhang et al. [25] developed a shear lattice system for analysing the crack surface of SHCC based on a lattice equivalent continuum model (LECM) [26], which consists of a combination of compression, tension and shear lattices. LECM involves the use of discrete lattice modelling of the cement matrix. Even though the continuum model retains its uniaxiality of stress-bearing materials, a combination of these approaches leads to a change of characteristics from an orthotropic model to a more anisotropic model. Conceptually, this model starts with a lattice approach but ultimately governing equations represent the equivalent continuum element of that lattice system.

The stress-strain relationship of SHCC under tension is modelled by experimental data from uniaxial tests while the compression behaviour can be modelled by a suitable compression softening model. In LECM, the shear transfer model is formulated considering the shear stress transfer mechanism such as aggregate interlocking and shear deformation at the crack surface. The crack surface is simulated as a geometry comprising a series of triangles. In this, shear lattice contact stress in the contact area is compressive stress (S1 in Fig. 3) perpendicular to the inclined crack surface while for the crack area having no contact there is a fibre bridging stress (S2 in Fig. 3). Furthermore, an additional coefficient S is also added that takes into account the contact rate related to crack width.

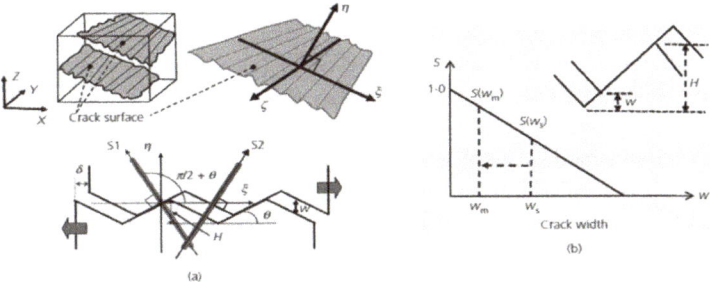

Fig. 3. Shear lattice expressing crack surface of SHCC (a) assumed crack surface, (b) contact area [27]

In literature, this particular model has been successfully applied to predict both the shear behaviour of SHCC beams [25] and also to simulate SHCC as a shear repair material [27]. Figure 4 shows the geometry of a SHCC repaired RC beam and corresponding LECM model. As demonstrated in Fig. 3, this model takes into account of contributions from both fibre bridging effect on the crack surface and the contact effect of the matrix. Figure 5 shows the experimental and numerical load-displacement curves of the repaired RC beams. The results show good agreement and validate the proposed shear transfer model, for the configurations used in that study. Numerical evaluation of the SHCC strengthening layer also revealed fine multiple cracking and final failure due to crack localization which shows the contribution of fibre bridging effect [27] (Fig. 4b).

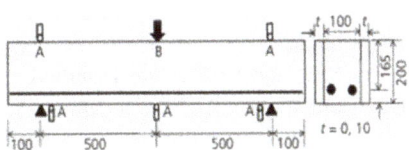

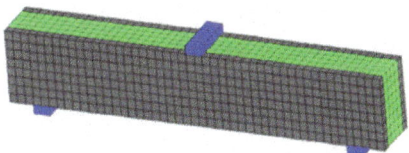

Fig. 4a. SHCC layer geometry for RC shear strengthening [18, 27]

Fig. 4b. Model of SHCC repaired RC beam [27]

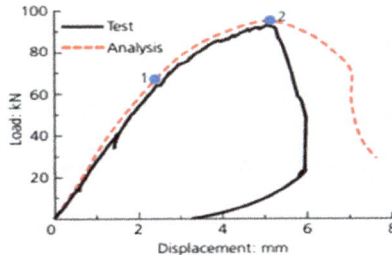

Fig. 5. Load displacement curves (experiment and simulation) of SHCC repaired RC beams [27]

3.2 Discrete Lattice Model

Recently. Lukovic et al. [23] simulated the behaviour of normal concrete and SHCC beams using a 3D lattice mesh model. Contrary to the previously discussed models, no continuum concept was applied in this model. Instead, it considers different phases of the composite and models them as truss or beam elements (Fig. 6). These elements form a lattice mesh and show usually linear elastic behaviour. Loading is then applied in steps and after each step elements that exceed the defined limit are removed till failure is reached. A 20 mm cubical grid was made with each cube having a lattice node at a random location. Four of the closest nodes were connected with concrete beams using Delaunay triangulation. These concrete beams were assumed to have linear elastic behaviour before failure and ascribed material properties such as tensile and compressive strength determined by experiments.

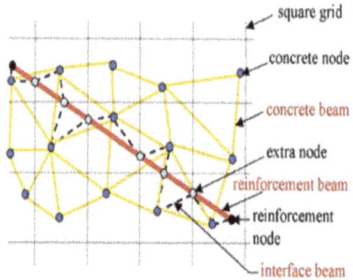

Fig. 6. SHCC layer geometry for RC shear strengthening [18]

Lukovic et al. [23] performed simulations on beams reinforced with 0.69% longitudinal reinforcement and a/d = 3.3. SHCC elements were simulated using ideally plastic material law with 5% strain capacity instead of linear elastic law which was used to simulate concrete. The tensile strength of both concrete and SHCC was assumed to be the same. Figure 7 shows the results of normal concrete and SHCC beams. The capacity of the SHCC beam was found to be almost twice the value of the normal concrete while the failure mechanism shifted from shear to bending in the SHCC beams. It was also noted that SHCC failed in a brittle manner and under smaller deflection compared to normal concrete beams.

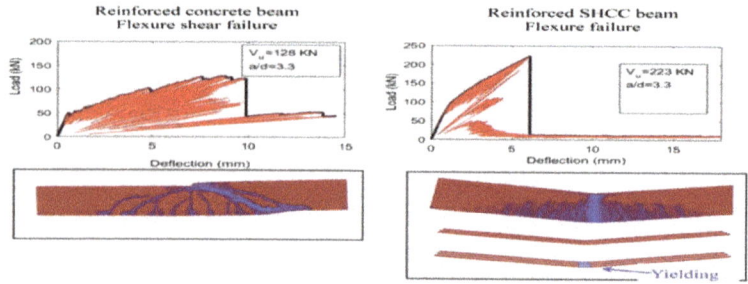

Fig. 7. Behaviour of reinforced SHCC and reinforced concrete beams [27]

3.3   Rigid-Body Lattice Model

RBSM is a discreet modelling approach based on the bond stress-slip behaviour of single fibres, as outlined for SHCC materials by Kunieda et al. [28]. Similar to other lattice models, matrix and fibres are modelled separately in RBSM with the location of each fibre identified usually by conducting a parametric analysis. Bolander and Saito [29] performed 2D parametric analysis in which fibres were modelled as beam elements to predict tensile fracture of fibre reinforced concrete [28].

Figure 8 shows a typical example of 3D polyhedral (Voronoi) cells which are assumed to be rigid bodies and assumed to have 6 degrees of freedom [30]. Neighbouring cells are connected with six springs across their common facet. Each facet comprises two tangential, one normal and three rotational springs representing 6 degrees of freedom. The mechanical properties of normal and tangential springs are determined by the properties of the matrix (tensile strength, fracture energy, elastic modulus), while the properties of rotational springs are defined by facet geometry [30]. Short fibres are placed at a random location as shown in Fig. 8b. A zero sized spring is placed at each point where a fibre crosses a facet between two cells. The bridging force provided by the spring is determined by integration of the bond stress-slip relationship. Embedment length and orientation angle are also calculated for every single fibre.

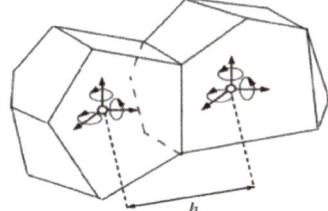
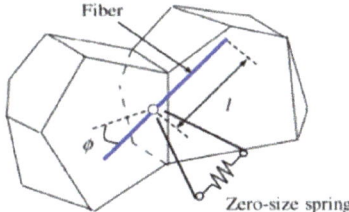

Fig. 8a. Rigid cells and defined DOF [14]   **Fig. 8b.** Fibre model and zero-size spring [14]

RBSM is shown to be applicable to predict the behavior of SHCC [9, 14] but studies are scarce, and the size of the assessed members is of small scale. Kunieda et al. [14] performed tests on rectangular beams 30×13 mm^2 having 100 mm span length. In order to simplify the model and to make the computation more efficient, the authors used cells of rectangular (30×13 mm^2) cross-sections at 1mm spacing. Fibres were randomly distributed in the matrix as shown in Fig. 9.

Figure 10 compares the experimental stress-strain relationship with that of the model. The value of tensile strength used in the analysis was 7.5 MPa which was determined through flexural testing of the plain matrix. Flexural and tensile strengths were assumed to be equal. The model predicts a softening behaviour in 1.0% fibre volume whereas experimental results show a strain-hardening up to 1.0% strain. For the 1.5% fibre volume case, the analysis predicted repeated gain and losses in stresses after the first crack. The model exhibits strain-hardening of about 5.2% while the experimental results showed an ultimate tensile strain capacity ranging from 2.0% to 2.5%. More work is needed to achieve more accurate modelling.

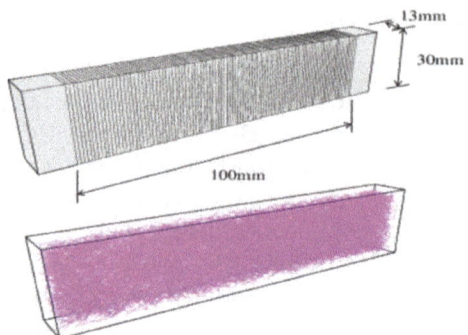

Fig. 9. Discretized RBSM SHCC beam model [27]

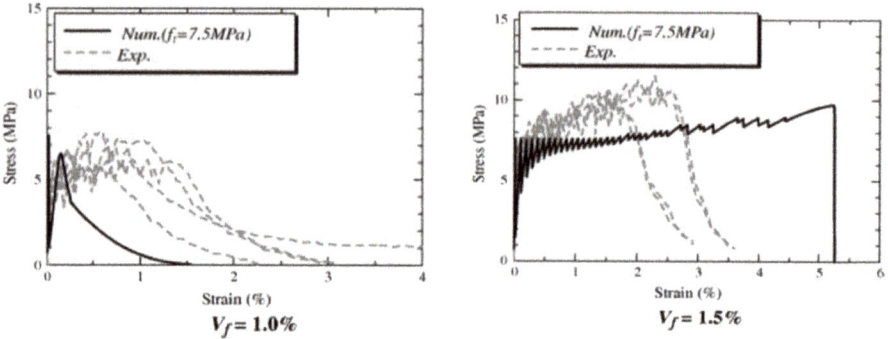

Fig. 10. Experimental and numerical load displacement curves [27]

4 Concluding Remarks

This paper gives an overview of FEM modelling techniques for simulating fracture in SHCC members. The complex fracture process in a heterogeneous material such as SHCC is not easy to predict, especially due to fibre-matrix interaction and potential multiple cracking. The physical and chemical properties of the applied fibres, and the matrix composition play a significant role in the nucleation and propagation of cracks. Most common used SHCC modelling approaches in the literature include analytical micromechanical models for micro or meso scale material modelling. For macro scale FEM modelling, different types of continuum models and lattice models have been applied. In basic FEM models, SHCC is assumed as a continuum element, whereas more advanced FEM models consider the interaction between SHCC phases more explicitly, to predict its macroscopic behavior. It is difficult to conclude which technique is the most suitable since each approach has its merits and demerits.

Conventional continuum models can provide a practical solution by idealizing the cracking state as an equivalent homogenization continuum. However, there are studied cases when this approach leads to an overestimation of load capacity of SHCC members especially when the member is loaded in non-proportional stresses such as shear [21]. The application of lattice models has proved to have potential of being more reliable and effective to simulate the shear behavior. The lattice equivalent continuum model is one of the models which has shown promising results in modelling SHCC as a shear repair material.

Acknowledgements. The authors would like to acknowledge the financial support of the Higher Education Commission, Pakistan [HRDI-UESTP (BATCH-V)].

References

1. Kong, H.J., Bike, S.G., Li, V.C.: Constitutive rheological control to develop a self-consolidating engineered cementitious composite reinforced with hydrophilic poly(vinyl alcohol) fibers. Cem. Concr. Compos. (2003). https://doi.org/10.1016/S0958-9465(02)00056-2
2. Baloch, H., Grünewald, S., Lesage, K., Matthys, S.: Influence of different fibre types on the rheology of strain hardening cementitious composites. In: Serna, P., Llano-Torre, A., Martí-Vargas, J.R., Navarro-Gregori, J. (eds.) BEFIB 2020. RB, vol. 30, pp. 3–11. Springer, Cham (2021). https://doi.org/10.1007/978-3-030-58482-5_1
3. Ranade, R., Li, V.C., Stults, M.D., Rushing, T.S., Roth, J., Heard, W.F.: Micromechanics of high-strength, high-ductility concrete. ACI Mater. J. (2013). https://doi.org/10.14359/51685784
4. Yu, K., Wang, Y., Yu, J., Xu, S.: A strain-hardening cementitious composites with the tensile capacity up to 8%. Constr. Build. Mater. (2017). https://doi.org/10.1016/j.conbuildmat.2017.01.060
5. Li, V.C., Leung, C.K.Y.: Steady-state and multiple cracking of short random fiber composites. J. Eng. Mech. **118**(11), 2246–2264 (1992). https://doi.org/10.1061/(asce)0733-9399(1992)118:11(2246)
6. Li, V.C.: On engineered cementitious composites (ECC). J. Adv. Concr. Technol. **1**(3), 215–230 (2003). https://doi.org/10.3151/jact.1.215
7. Boshoff, W.P., van Zijl, G.: A computational model for strain-hardening fibre-reinforced cement-based composites. J. S. Afr. Inst. Civ. Eng. (2007). https://www.researchgate.net/publication/228702758_A_computational_model_for_strain-hardening_fibre-reinforced_cement-based_composites
8. Kabele, P., Horii, H.: Analytical model for fracture behavior of pseudo strain-hardening cementitious composites. In JSCE NO. 29 (1997). https://www.researchgate.net/publication/261147726_Analytical_model_for_fracture_behavior_of_pseudo_strain-hardening_cementitious_composites
9. Yang, J., Fischer, G.: Simulation of the tensile stress-strain ehavior of strain hardening cementitious composites. In: Konsta-Gdoutos, M.S. (ed.) Measuring, Monitoring and Modeling Concrete Properties, pp. 25–31. Springer Netherlands, Dordrecht (2006). https://doi.org/10.1007/978-1-4020-5104-3_3
10. Kabele, P.: Stochastic finite element modeling of multiple cracking in fiber reinforced cementitious composites. In Fracture and Damage of Advanced Fibre-reinforced Cement-based Materials, Proceedings (2010). https://www.researchgate.net/publication/261147290_Stochastic_finite_element_modeling_of_multiple_cracking_in_fiber_reinforced_cementitious_composites

11. Kabele, P.: Finite element fracture analysis of reinforced SHCC members. In Advances in Cement-Based Materials - Proceedings of the International Conference on Advanced Concrete Materials, pp. 237–244. CRC Press (2010). https://doi.org/10.1201/b10162-39

12. Kabele, P.: Finite element modeling of strain-hardening fiber-reinforced cementitious composites (SHCC) (2011)

13. Kang, J., Bolander, J.: Spatial representation of fiber bridging forces in strain-hardening cement composites. In: International Association for Fracture Mechanics of Concrete and Concrete Structures (2016). https://doi.org/10.21012/fc9.299

14. Kunieda, M., Ogura, H., Ueda, N., Nakamura, H.: Tensile fracture process of strain hardening cementitious composites by means of three-dimensional meso-scale analysis. In Cement and Concrete Composites, vol. 33, pp. 956–965. Elsevier (2011). https://doi.org/10.1016/j.cemconcomp.2011.05.010

15. Zhang, Y., Ueda, N., Nakamura, H., Kunieda, M.: Numerical approach for evaluating shear failure behavior of strain hardening cementitious composite member. Eng. Fract. Mech. **156**, 41–51 (2016). https://doi.org/10.1016/j.engfracmech.2016.02.008

16. Luković, M., Šavija, B., Schlangen, E., Ye, G., van Breugel, K.: A 3D lattice modelling study of drying shrinkage damage in concrete repair systems. Materials **9**(7), 575 (2016). https://doi.org/10.3390/MA9070575

17. Spagnoli, A.: A micromechanical lattice model to describe the fracture behaviour of engineered cementitious composites. Comput. Mater. Sci. **46**(1), 7–14 (2009). https://doi.org/10.1016/j.commatsci.2009.01.021

18. Kabele, P., Kanakubo, T.: Experimental and numerical investigation of shear behavior of PVA-ECC in structural elements (2007)

19. Kabele, P.: Equivalent continuum model of multiple cracking. Eng. Mech. (2002)

20. Cervenka, V., Cervenka, J., Pukl, R.: ATENA - a tool for engineering analysis of fracture in concrete. Sadhana Acad. Proc. Eng. Sci. (2002). https://doi.org/10.1007/BF02706996

21. Kabele, P.: Finite element fracture analysis of reinforced SHCC members. In: Advances in Cement-Based Materials - Proceedings of the International Conference on Advanced Concrete Materials (2009). https://doi.org/10.1201/b10162-36

22. Recent approaches to shear design of structural concrete. J. Struct. Eng. (1998). https://doi.org/10.1061/(asce)0733-9445(1998)124:12(1375)

23. Lukovic, M., Yang, Y., Schlangen, E., Hordijk, D.: On the potential of lattice type model for predicting shear capacity of reinforced concrete and shcc structures. In: Hordijk, D.A., Luković, M. (eds.) High Tech Concrete: Where Technology and Engineering Meet, pp. 804–813. Springer, Cham (2018). https://doi.org/10.1007/978-3-319-59471-2_94

24. Pan, Y., Prado, A., Porras, R., Hafez, O., Bolander, J.: Lattice modeling of early-age behavior of structural concrete. Materials **10**(3), 231 (2017). https://doi.org/10.3390/ma10030231

25. Zhang, Y.X., Ueda, N., Umeda, Y., Nakamura, H., Kunieda, M.: Evaluation of shear failure of strain hardening cementitious composite beams. Procedia Eng. (2011). https://doi.org/10.1016/j.proeng.2011.07.257

26. Ahmad, S.I., Tanabe, T.A.: Three-dimensional FE analysis of reinforced concrete structures using the lattice equivalent continuum method. Struct. Concr. (2013). https://doi.org/10.1002/suco.201100009

27. Zhang, Y., Peng, H., Lv, W.: Shear stress transfer model for evaluating the fracture behaviour of SHCCs for RC shear strengthening. Mag. Concr. Res. (2018). https://doi.org/10.1680/jmacr.17.00104

28. Kunieda, M., Ogura, H., Ueda, N., Nakamura, H.: Tensile fracture process of strain hardening cementitious composites by means of three-dimensional meso-scale analysis. Cement Concr. Compos. **33**(9), 956–965 (2011). https://doi.org/10.1016/j.cemconcomp.2011.05.010

29. Bolander, J.E., Saito, S.: Discrete modeling of short-fiber reinforcement in cementitious composites. Adv. Cem. Based Mater. **6**(3–4), 76–86 (1997). https://doi.org/10.1016/S1065-7355(97)90014-6

30. Berton, S., Bolander, J.E.: Crack band model of fracture in irregular lattices. Comput. Methods Appl. Mech. Eng. **195**(52), 7172–7181 (2006). https://doi.org/10.1016/j.cma.2005.04.020

Exploring the Performance of a Single Panel SFRC Slab Under a Point Load with Fe Analysis

Olugbenga B. Soymi[1] and Ali A. Abbas[2(✉)]

[1] Department of Civil Engineering, The Federal Polytechnic Ilaro, Ilaro, Nigeria
[2] School of Architecture, Computing and Engineering,
University of East London, London, UK
abbas@uel.ac.uk

Abstract. This study explores the performances of a single round panel without conventional longitudinal reinforcement using a nonlinear finite element analysis (FEA) software, ABAQUS. Experimental data from an established work on a full-scale steel fibre-reinforced concrete (SFRC) panel with 1.0% fibre volume were validated using FEA. A constitutive model was used to interpret the tensile behaviour of SFRC panel, and concrete damaged plasticity (CDP) of ABAQUS deployed for both tensile and compressive behaviour of the panel. The model was relatively stiffer than the experiment. The experiment result has a failure load of 38.0 kN while the FEA failure load was 37.9kN, thou stiffer than that of the experiment. This shows the FEA simulate the behaviour of the round panel under central loading. The parametric studies were carried to investigate the impact of steel fibre on the tensile behaviour of SFRC round panel by varying the fibre volume (1%, 1.25%, 1.50%, 1.75%, 2.00% and 2.5%) and characteristic strength (30MPa, 40MPa, 50MPa and 60MPa). The results of the central displacements, yield and peak loads, strengths, ductility, strain, and crack patterns. The FEA results show that the more the fibre volume, the more the strength. The FEA was able to provide an essential performance profile of SFRC round-panel.

Keywords: Concrete · Steel-fibre · Point-load · Displacement · Round-panel

1 Introduction

Concrete has emerged over the years as the most used building material. It is very strong in compression and has about one-eighth of this strength in tension. The structural usage of it will require the inclusion of a material that has significant strength in tension [1–7]. The inclusion of steel bars in the tension zone brought an improved solution which has led to the significant development in the construction industry and its acceptance in different structures elements including suspended slabs. Over the past years, these slabs were constructed by inserting rebar or welded mesh fabric in the concrete. An increasing trend to this construction method, is the replacement of some or all of the reinforcement with fibres [1, 2, 8–17, 19, 19–21]. The replacement of rebars with fibres has proven to save construction time, lessen labour cost, improving

© RILEM 2022
P. Serna et al. (Eds.): BEFIB 2021, RILEM Bookseries 36, pp. 400–408, 2022.
https://doi.org/10.1007/978-3-030-83719-8_35

safety on site by reducing chances of getting injured and ultimately leads to cheaper slabs when compared to the traditional reinforced concrete slabs. Various types of fibres (steel, glass, polyethene etc.) are available in the market with steel-fibres emerging as most used. When plain concrete is subjected to loading, cracks are propagated on the tension surface. This damage first starts as a single crack when the tensile strength of the concrete has been exceeded and are thus mitigated by the introduction of fibres in the concrete matrix. With the improvement in the knowledge and understanding of steel-fibre usage in concrete in recent time, there has been a swift expansion in the construction of steel-fibre-reinforced concrete pile-supported slabs [23–25].

Steel-fibre in concrete started in 1874, when Berard patented an idea in which left-overs (waste) grains of steel are mixed to concrete which led to a more ductile material [22]. In the present world, there are many types of fibres being used in the construction industry for variety of works. Examples are natural fibres and artificial fibres (steel, carbon, polyethene, and so on), but the most used and researched on is the steel fibre. The steel fibres come in various shapes, sizes and strength [19]. The benefits of adding of steel-fibres in concrete in structural members as reported in earlier researches include (i) enhancing the post-cracking behaviour of the concrete, (ii) the significant reduction in construction time, (iii) improve tensile strength and energy absorption and, (iv) reduction in crack-width giving rise to enhanced damage tolerance [26, 27]. However, provisions were made for additional reinforcement to resist peak moments over piles or supports, particularly in the corner panel [28]. Also, recent experimental works on real-life full-size specimen justify the possibility of fibres-only reinforced concrete [13, 19, 29, 30].

The absence of a generally accepted code or standard method for the design and analysis of SFRC slabs has led to development of design guidance based on elastic and yield-line analyses for suspended SFRC slabs [30–32] and some national guidelines [3, 23]. Likewise, manufacturers of steel-fibres came up with their design guidance as provided by Bekaert and ArcelorMittal, which are the principal suppliers of steel fibres (Dramix® and TAB-Structural system) in the UK. The Bekaert design method [33] uses the elastic moment coefficient of the Dutch Code NEN 6720 and is thus similar to TR 63 guidelines [32]. The ArcelorMittal proprietary method, on the other hand, uses steel fibres to completely replace all traditional reinforcement [34]. Concrete is a composite material with pores and micro-cracks caused by shrinkage, expansion and thermal strains, which have been restrained by coarse aggregates and boundary conditions. During loading, the concrete matrix transfers part of the load to the fibres before any macro-crack is formed. With continuous loading, the micro-cracks will then start to expand and ultimately lead to a macro-crack which covers numerous micro-cracks [11, 15, 35].

This study aims at exploring the performance of a single panel steel-fibre reinforced concrete (SFRC) slabs under point-load with finite element analysis. The experimental work of [36] Round Plate was explored. Parametric studies were carried out and their results compared to the numerical investigations and existing experimental data with current design guidance. This research work is limited to a single suspended slab over three supports at angle 120° apart.

2   Methodology

The post-cracking behaviour of SFRC structural elements was the thrust of their research works. The experimental work involves simply supported beams and round-plates consisting of six specimens for each SFRC mix with varying amount of steel fibres (0.75–1.25%). The mechanical properties of the SFRC round-plate (labelled F35–1.0) chosen for this study in compression made was made from 1% volume fraction of fibres are compressive strength 46.9MPa, modulus of elasticity 33.5GPa and Poisson ratio of 0.23 while the post-cracking tensile strength of 2.6MPa was gotten from the uniaxial test. The hooked-end steel-fibre used in the concrete matrix has an aspect ratio of 67.2 and a length of 35mm. The strength of the steel fibre is 1200MPa.

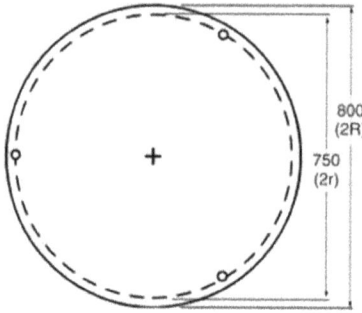

Fig. 1. Dimensions of round-plate adapted from (de Montaignac *et al.* [36]

NLFEA were carried out in 3D using the ABAQUS and Lok and Xiao constitutive models. Taking advantage of the high-performance computer in University of East London, UK, the full size of the round-plate was modelled (i.e. 800 mm in diameter, 750 mm clear distance and 80 mm thick). Round steel sections were used to mimic the supports to prevent localised failure. A depiction of the plate model is shown in Fig. 1. Calibration was done to validate the numerical model using the respective experimental data. The parametric study involves both the fibre volume (V_f) and concrete strength (f_{ck}) being changed. The experiment was conducted with 1% V_f. The slab was modelled using V_f values of 1%, 1.25%, 1.50%, 1.75%, 2.00% and 2.5%. Also, f_{ck} used in the experiment was 46.9MPa and for the parametric study f_{ck} used are 30MPa, 40MPa, 50MPa and 60MPa.

As earlier stated, the displacement-based loading was applied at the centre of the round plate. To guide against localised failure, a steel plate, 10mm thick was modelled for the loading rig. Based on the sensitivity analysis earlier done, a mesh elements size of 25 mm width was adopted, which produces the most suitable agreement with experimental data. To avoid distortion in the meshing, the wedge was used to restrain an uneven mesh pattern. The loading was applied gradually until failure using a displacement-based loading control. The following NLFEA results are at this moment discussed following sections.

3 Results and Discussion

3.1 Load-Displacement Diagrams

Figure 2 shows the Load-displacement curve for various values of V_f given that the f_{ck} for the experimental work was used. It clearly shows that as the fibre volume ratio increases, the peak load also increases. To further examined the response of the round-plate to apply loads, different amount of characteristic strength of concrete are considered. This is to determine the dependency of peak load with characteristic strength of concrete.

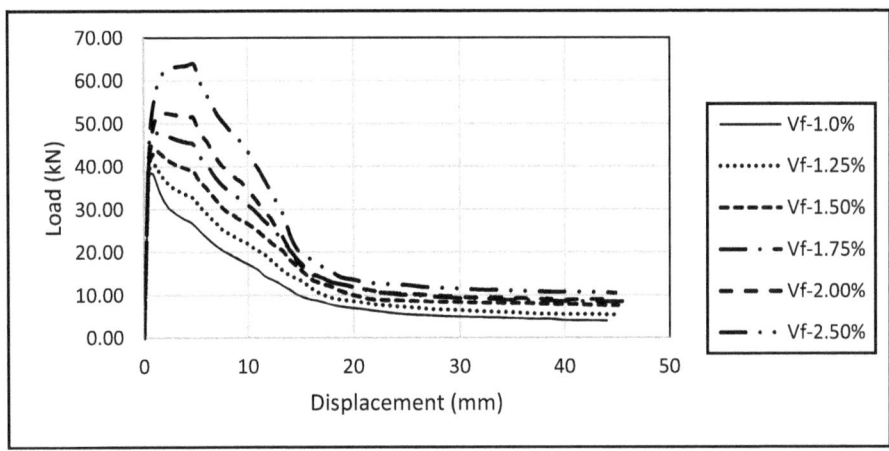

Fig. 2. Load-displacement curve for a range of V_f

The result shows that round-panel with 1% fibre volume ratio yielded 38.21kN, while that of 2.50% fibre volume is 57.11kN in control. As the fibre volume is increased, there is a gradual increase in the yield load. There is also an increase in the stiffness as the fibre volume increases. The strength of the slab was also enhanced as the fibre volume increases. An 18% and 22% increment in strength was obtained in the slab when the fibre was increased to 1.25% and 1.5% respectively. Incidentally, all the FEA results for the displacement at the yield point give 0.51mm, which is 67% lower than the experimental result.

The load-carrying capacity of the slab was enhanced as the fibre volume is increased as shown in the load-displacement curves. This simply translates into, the higher the fibre volume, the higher the strength of the round plate to resist applied load. Both the experiment and FEA with 1% volume fibre failed at 38kN. As the fibre increases, the maximum load also increases. The FEA produces a slab that is less stiff as that of the experiment. This can be attributed to the stiffness in the FEA results. The strength and ductility ratios were also measured. There is a significant increase in the strength ratio and a decrease in the ductility ratio. Also, as the fibre volume fraction content increases (from 1.75%–2.50%) a gentle plateau was observed that is relative to the peak load. The "plateau" was more obvious in beams with higher fibre volume ratio than in ones with small fibre dosage. This is due to the influence of the amount of fibre in the post-cracking phase of the slab. The yield and peak loads increase gradually as the f_{ck} increases.

3.2   Strength

The results show the panels after cracking occurred to demonstrate a sharp but controlled load reduction followed immediately by a post-cracking hardening phase which then changed gradually to smooth load reduction. After the matrix has cracked, a hardening response occurred when the peak load was approaching, followed by a gradual reduction in load as the crack width increases. The fibre volume ratios influence can be observed clearly as the fibre content increases. The Fig. 3 shows the responses of the SFRC round panel at yield and peak loads in respect of different f_{ck}. The yield and peak loads increase gradually as the f_{ck} increases. This is due to the influence of the amount of fibre in the post-cracking phase of the slab.

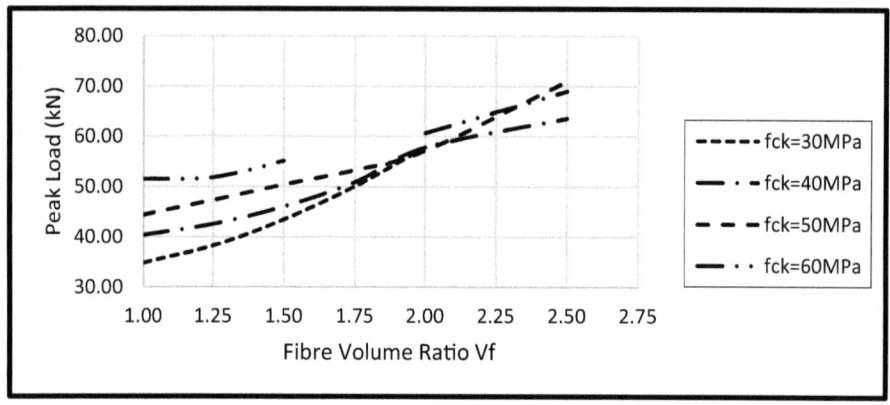

Fig. 3. Peak load and fibre volume ratio graphs for f_{ck}

The displacements at yield loads are about the same for all the fibre volume ratios of different characteristic strengths are seen in Fig. 4. This can be attributed to the fact that the linear part of the load-displacement graph (from zero to the point of yield load) is guided by the material properties of the SFRC matrix. In this part, the steel fibres place little or no role in the resistance of the load. They are resisted by the concrete matrix.

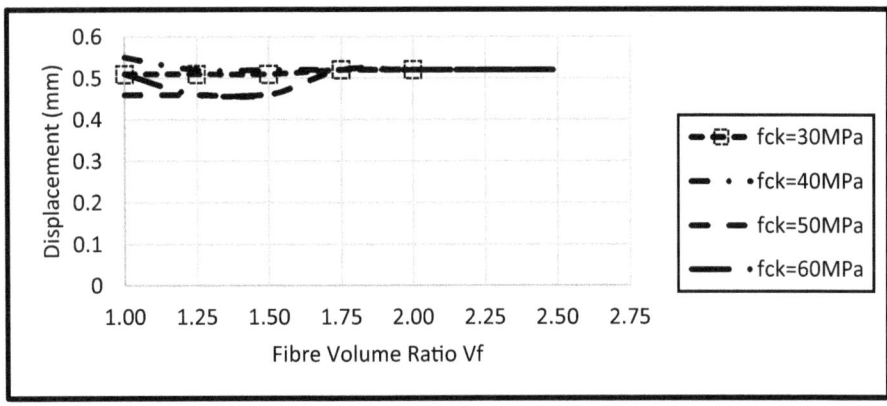

Fig. 4. Displacement at yield load and fibre volume ratio graphs for f_{ck}

At the peak load, the steel-fibre improves the resistance to load (Fig. 5). There was a jump in the displacement from 1.50% to 2.00% in f_{ck} less than 50MPa. The peak load in 60MPa strength was a gentle slope from 1.00% to 2.50% fiber volume ratio.

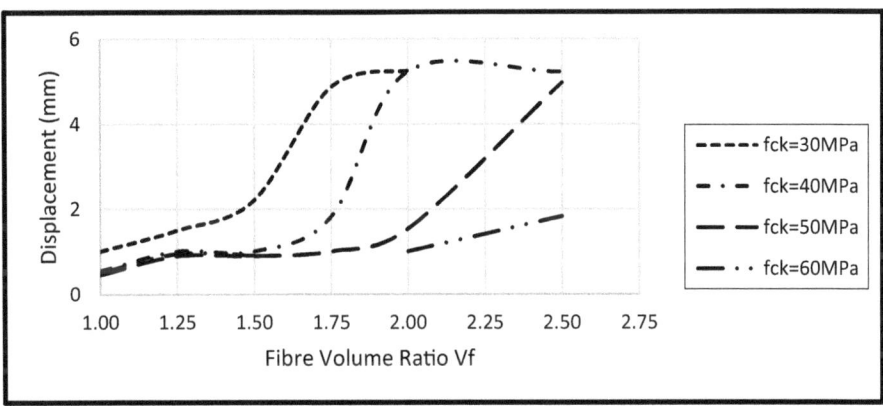

Fig. 5. Displacement at Peak load and Fibre Volume Ratio Graphs for f_{ck}

3.3   Ductility

The ultimate displacement (δ_u) taken at the ultimate load (P_u) is used to measure the ductility of the Round-plate panels. The ultimate load (P_u) represents the (post-peak) residual strength at failure. A softening behaviour of the load-deflection response accompanies the rise in ductility. The residual strength was found to be sufficiently low (<85%) in all cases studied, suggesting that the softening is very significant. Consequently, to ensure practicality of these ductility levels, the residual strength was limited to a minimum of 85% of the load-carrying capacity (P_{max}). To truly utilise the resulting increase in ductility for the design purposes, it is vital to maintain the residual strength at a tolerable level. The ductility of the Round-Plate panels can be determined based on the ductility ratio $\mu = \delta_u/\delta_y$ which shows that the addition of fibres leads to enhanced ductility ratio of the SFRC RP panels significantly. However, as the fibre volume fraction exceeds a crucial value, the addition of fibres beyond this value does not increase the ductility of the RP panel.

3.4 Deflected Shapes

A graphic presentation of the deflected shapes for the round-plate analysed at failure are given in Fig. 6. The results are in agreement and have similar patterns with a substantial deflection at the round-plates' mid-span and a collapse along the weakest plane.

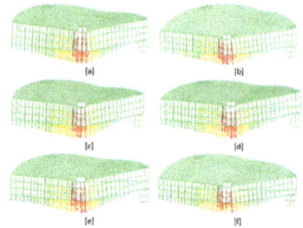

Fig. 6. Deflected shapes (y-y) for the round-plate with V_f = (a) 1.00% (b) 1.25% (c) 1.50% (d) 1.75% (e) 2.00% and (f) 2.50%

4   Conclusions

The exploration was on round-plates with three (3) isolated supports under monotonic loads. The flexural behaviour of the round-panels was studied with fibre volume ratio V_f =1.00%. It comprises of change in fibre volume ratios and the characteristic strength of the SFRC. It was observed that the characteristic strength f_{ck} has no effect on the flexural strength f_{tu} of the matrix. This is because the flexural strength is based on certain parameters (fibre volume ratio, bond stress, length and equivalent diameter of fibre). The exploration shows that the load-carrying capacity of the round-plates was enhanced as the volume of steel fibres increases. Likewise, the strength increases by $\sim 7\%$ over the previous result of the round-panel with isolated supports and $\sim 11\%$ for the round-plates with continuous support under monotonic loading. Also, additional fibres lead to increase stiffness which resulted in a reduction to deflections of the round-panels. The benefits of adding steel fibres are seen both at the serviceability and ultimate limit states. Reduction in crack formation was also achieved. Structurally, increase in fibres changes the failure mode to ductile failure from brittle failure, which is required in the structural design. As the fibres volume increases, ductility was improved. Interestingly, the different characteristic strengths produce an increase in ductility and strength, which shows that the higher the characteristic strength of the

SFRC matrix, the higher the ductility and strength. Three cracks formed along the weakest plane, from the point of load application, which are the spaces in-between two adjacent supports. Conclusively, it can be established that the load-carrying capacity of the plates are enhanced by the addition of steel fibres and ensures a more ductile structural response (thus avoiding a brittle failure mode), which is preferred in design. This also leads to a rise in ductility and strength enhancement.

References

1. Abbas, A., Syed Mohsin, S., Cotsovos, D.: Seismic response of steel fibre reinforced concrete beam-column joints. Eng. Struct. **59**, 261–283 (2014)
2. Abbas, A.A., Syed Mohsin, S.M., Cotsovos, D.M., Ruiz-Teran, A.M.: Shear behaviour of steel-fibre-reinforced concrete simply supported beams. Proc. ICE Struct. Buildings **167**(9), 544–558 (2014). https://doi.org/10.1680/stbu.12.00068
3. American Concrete Institute: State-of-the-Art Report on Fibre Reinforced Concrete: American Concrete Institute (2002)
4. Barros, J.A.O., Cruz, J.S.: Fracture energy of SFRC. J. Mech. Compos. Mater. Struct. **8**(1), 29–45 (2001)
5. British Standard Institute: BS EN 1992, Eurocode 2, Design of concrete structures, Part 1-1: General rules and rules for buildings. London: British Standard Institute (2004)
6. Crowther, D.: Fibre-reinforced concrete - origin of the species. Concrete **43**, 22–23 (2009)
7. Lok, T.S., Xiao, J.R.: Flexural strength assessment of steel fiber reinforced concrete. J. Mater. Civ. Eng. ASCE **11**(3), 0188–0196 (1999)
8. Ahmed, A.: Modeling of a reinforced concrete beam subjected to impact vibration using ABAQUS. Int. J. Civ. Struct. Eng. **4**(3), 227–236 (2014). https://doi.org/10.6088/ijcser.201304010023
9. American Concrete Institute: Measurement of properties of fibre reinforced concrete. Naples FL: American Concrete Institute (1999)
10. ArcelorMittal: Steel fibre reinforced concrete (SFRC) for industrial floors especially slabs without joints and slabs on piles (2011). www.arcelormittal.com
11. Barros, J.A.O.: Post-cracking behaviour of steel fibre reinforced concrete. Mater. Struct. **38** (275), 47–56 (2004). https://doi.org/10.1617/14058
12. Barros, J.A.O., Cruz, J.S.: Energy absorption capacity of SFRC. In: IRF'99 : Integrity, Reliability and Failure, Porto (1999). http://hdl.handle.net/1822/12840
13. Barros, J.A.O., Salehian, H., Pires, N.M.M.A., Gonçalves, D.M.F.: Design and testing elevated steel fibre reinforced self-compacting concrete slabs. BEFIB (2012)
14. Beddar, M.: Fibre-reinforced concrete: past, present and future. Concrete **38**, 47–49 (2004)
15. Bernardi, P., Cerioni, R., Michelini, E.: Analysis of post-cracking stage in SFRC elements through a non-linear numerical approach. Eng. Fract. Mech. **108**, 238–250 (2013). https://doi.org/10.1016/j.engfracmech.2013.02.024
16. Blanco, A., Pujadas, P., Cavalaro, S., de la Fuente, A., Aguado, A.: Constitutive model for fibre reinforced concrete based on the Barcelona test. Cem. Concr. Compos. **53**, 327–340 (2014). https://doi.org/10.1016/j.cemconcomp.2014.07.017
17. Destrée, X.: Steel fibre reinforcement for suspended slabs. Concrete **35**, 58–59 (2001)
18. Destrée, X.: Steel-fibre-reinforced pile-supported slabs. Concrete **41**, 26–27 (2007)
19. Rilem, T.: RILEM TC 162-TDF - Test and design methods for steel fibre reinforced concrete - bending test - final recommendation. Mater. Struct. **35**, 528–579 (2002)

20. Singh, H.: Steel fibers as the only reinforcement in concrete slabs: flexural response and design chart. Struct. Eng. Int. **25**(4), 432–441 (2015)
21. Swamy, R.N.: The technology of steel fibre reinforced concrete for practical applications. Proc. Inst. Civ. Eng. (1974)
22. Berard, A.: Improvement in artificial stone, Google Patents (1874)
23. British Standard Institute: BS EN 14889-1-2006 - Fibres for concrete - Steel fibres. London: British Standard Institution (2006)
24. Campione, G.: The effects of fibers on the confinement models for concrete columns. Can. J. Civ. Eng. **29**(5), 742–750 (2002)
25. Labib, W., Eden, N.: An investigation into the use of fibres in concrete industrial ground-floor slabs. Saetaequina (2004)
26. Rilem, T.: RILEM TC 162-TDF - Test and design methods for SFRC - principles and applications. Mater. Struct. **35**, 262–278 (2002)
27. Lok, T.-S., Pei, J.-S.: Flexural behavior of steel fibre reinforced concrete. J. Mater. Civ. Eng. ASCE **10**(2), 0086–0097 (1998)
28. IstructE.: Interim Guidance on the Design of Reinforced Concrete Structures using Fibre Composite Reinforcement. The Institution of Structural Engineers, London (1999)
29. Hedebratt, J., Silfwerbrand, J.: Full-scale test of a pile supported steel fibre concrete slab. Mater. Struct. **47**(4), 647–666 (2013). https://doi.org/10.1617/s11527-013-0086-5
30. Vollum, R.: Design of steel-fibre-reinforced pile-supported slabs. Concrete **41**, 12–14 (2007)
31. Eddy, D.: Fibre-only suspended ground-floor slabs - do they work. Concrete **42**, 14–15 (2008)
32. The Concrete Society TR63: TR63 - Guidance for the design of steel fibre reinforced concrete, vol. 63, pp. 1–109. The Concrete Society, UK (2007)
33. Viney, T.: Piled floor design - the dutch method. Concrete (2007)
34. Destrée, X., Jürgen, M.: Steel fibre only reinforced concrete in free suspended elevated slabs: case studies, design assisted by testing route, comparison to the latest SFRC standard documents. Tailor Made Concrete Structures, pp. 437–445 (2008)
35. Tlemat, H., Pilakoutas, K., Neocleous, K.: Stress-strain characteristic of SFRC using recycled fibres. Mater. Struct. **39**, 365–377 (2006). https://doi.org/DOI10.1617/s11527-005-9009-4
36. de Montaignac, R., Massicotte, B., Charron, J.-P., Nour, A.: Design of SFRC structural elements: post-cracking tensile strength measurement. Mater. Struct. **45**(4), 609–622 (2011). https://doi.org/10.1617/s11527-011-9784-z

An Analytical Study of Shear Transfer Mechanisms in Macro-synthetic Fibre Reinforced Concrete

Francisco Ortiz-Navas[✉], Juan Navarro-Gregori, and Pedro Serna

Instituto de Ciencia y Tecnología del Hormigón (ICITECH),
Universitat Politècnica de València, València, Spain
fraorna@doctor.upv.es

Abstract. Several researches have demonstrated the effectiveness of fibres to provide post-cracking strength to concrete elements. This is due to the potential of fibres to bridge the crack faces and continue transferring stresses along the shear crack by different shear transfer mechanisms. Several models have been stablished in order to explain the mechanical action of fibres in a shear crack, most of them explained as a function of parameters such as the fibre type, dispersion, inclination, aspect ratio or pull-out stress of the fibre. However, these models have been stablished only for steel fibres, which limits their use to different fibre's materials. Within this framework, the present research first investigates the different shear transfer mechanisms acting in a shear crack, and evaluate the possibility of characterize the fibre shear transfer mechanism by means of the residual flexural tensile stresses (RFTS). For this, an analytical model that involves the transfer mechanism acting in a shear crack is developed. Analytical results are compared against experimental results obtained after testing 20 pre-cracked push-off specimens of plain concrete and macro-synthetic fibre reinforced concrete. Results evidence the great differences among aggregate interlock models available in the literature against the existing differences among the fibre ones. Finally, the feasibility of using RFTS to characterize the contribution of fibres to transfer shear stresses is observed.

Keywords: Shear transfer · PFRC · Aggregate interlock · Polypropylene fibre reinforced concrete

1 Introduction

Shear transfer in a concrete crack constitutes a complex interaction of several mechanisms acting together and influencing to each other. These mechanisms have been studied by different authors for over 50 years. However, there is still no consensus about the contribution of each mechanism to the element's shear strength so far. One of the main shear transfer mechanisms in reinforced concrete elements is the aggregate interlock, also called interface shear transfer. This mechanism acts in normal-strength concrete through a crack that usually surrounds aggregates, while in lightweight and high-performance concretes a crack normally passes across them. Hence, the aggregate interlock depends mainly on aggregate types, and on the bond between aggregates and

© RILEM 2022
P. Serna et al. (Eds.): BEFIB 2021, RILEM Bookseries 36, pp. 409–419, 2022.
https://doi.org/10.1007/978-3-030-83719-8_36

the matrix. Several models [1–4] have been developed to explain this mechanism, which was first introduced in 1968 [5].

When new materials such as fibres are introduced to the concrete matrix, some of the shear transfer mechanisms could be affected. In fact, it has been evidenced the effectiveness of fibres, dosed in correct amounts, to improve the shear strength and reduce the brittleness of concrete elements. These improvements are mainly due to the bridging effect provided by fibres given their ability to sew crack faces and restrain them from opening. Due to the bridging effect of fibres, the aggregate interlock "can be improved" in fibre reinforced concrete (FRC). When comparing reinforced concrete (RC) and FRC elements in the same load stages, fibres reduce both crack slip and crack opening by theoretically enhancing the aggregate interlock effect for longer periods. Nevertheless, Kaufmann et al. [6] theoretically studied this interaction between fibres and aggregate interlock and proposed a reduction factor that affects the aggregate interlock.

Moreover, the bridging effect lets fibres to transmit tensile forces when the shear crack appears. This behaviour is influenced mainly by the type and volume of fibres. Several models [7, 8] have estimated the contribution of fibres to shear according to the volume of fibres and their bond strength, however, mostly using hooked-end or double-hooked-end steel fibres.

In the last decade, macro-synthetic polypropylene fibres have been incorporated into structural applications [9, 10]. In fact, macro-synthetic polypropylene fibres are capable of providing similar toughness and ductility to the concrete matrix to steel fibres, but with certain limitations. Some researchers have reported the success of macro-synthetic fibres to be used as shear reinforcement in structural elements [9, 11, 12]. Nevertheless, most of the shear transfer models available to explain the transfer mechanism in FRC have been developed based on steel fibres exclusively. Within this framework, the present paper explores the shear transfer mechanism acting in a crack of push-off specimens manufactured with plain concrete (PC) and macro-synthetic FRC (MSFRC). For this, an analytical model that involves several aggregate interlock and fibre models is developed. In addition, it is also evaluated the possibility of characterizing the fibre shear transfer mechanism by means of RFTS. Analytical results are compared against experimental results obtained after testing 20 pre-cracked push-off specimens of plain concrete and macro-synthetic fibre reinforced concrete.

2 Experimental Campaign

Twenty pre-cracked PC and PFRC push-off specimens were tested by a direct shear loading scheme. The geometry and details of reinforcement can be seen in Fig. 1. The total procedure to test the specimens consisted in four stages: frame assembly, pre-cracking, handling and push-off test. In the frame assembly stage, an external steel frame that confines the specimen is fixed in order to better control the crack kinematics. During the pre-cracking process, the crack is formed by splitting the shear plane of the specimen using a hydraulic jack (100 kN). In the handling stage, the cracked specimen is lifted up to the vertical position and placed under the second hydraulic jack. Before the push of test, the initial condition of the crack opening, and confinement of the specimens are set up. In fact, three possible confinement scenarios before the push-off test may occur: totally confined, partially confined or totally released. After setting the initial conditions, load was applied with a servo-hydraulic jack (500 kN) at a constant piston displacement rate of 0.015 mm/s. In this stage, slip displacement (s), crack opening (w) and strains of external steel bars used to determine confinement were recorded. In total during the four stages, 10 potentiometric displacement transducers (*PTs*) were employed to measure the crack kinematics as well as undesired displacements of the external frame. Figure 2 shows the test set up, components and instrumentation used during the pre-cracking process and the push-off test. More information about the test methodology can be found in [13].

Portland cement type CEM I 42.5N, two gravels, two sand types, limestone filler and superplasticiser were used for PC and PFRC specimens as indicated in Table 1. Macro-synthetic fibres (48 mm length) with a nominal aspect (length/diameter) ratio of 57, density of 0.91 g/cm^3, 400 MPa tensile strength and 4.7 GPa of modulus of elasticity were used in PFRC specimens dosed in volume fraction of 0.88% (8 kg/m^3) and 1.31% (12 kg/m^3). Concrete compression strength (f_c) and residual flexural tensile strength (f_{Rj}) were determined according to EN 12390-3 and EN 1465 respectively on cylindrical (150 × 300) and notched prismatic (150 × 150 × 600). The mean values of f_c and f_{Rj} (coefficient of variation in brackets) are summarised in Table 2.

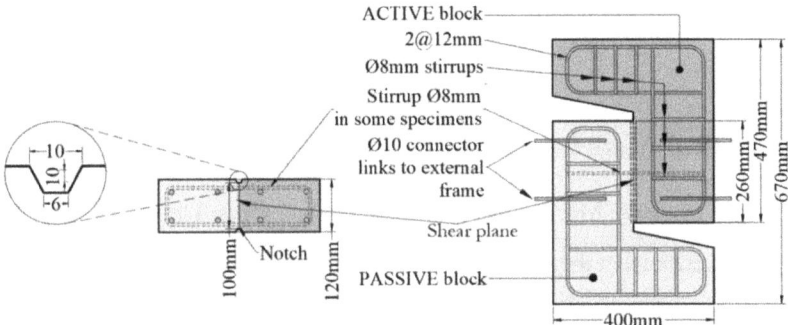

Fig. 1. Geometry and reinforcement of push-off specimens

Table 1. Mix design of PC and PFRC.

Material	PC	PFRC 8	PFRC 12
Cement CEM I 42.5 N [kg/m^3]	350	350	350
Crushed sand [kg/m^3]	950	950	1045
Limestone filler [kg/m^3]	60	60	66
Coarse gravel 7/14 mm [kg/m^3]	600	600	540
Coarse gravel 4/7 mm [kg/m^3]	300	300	270
Water [lt/m^3]	190	190	190
Fibres [kg/m^3]	0	8 ($V_f = 0.88\%$)	12 ($V_f = 1.31\%$)
Superplasticiser [lt/m^3]	3.5	3.5	3.5

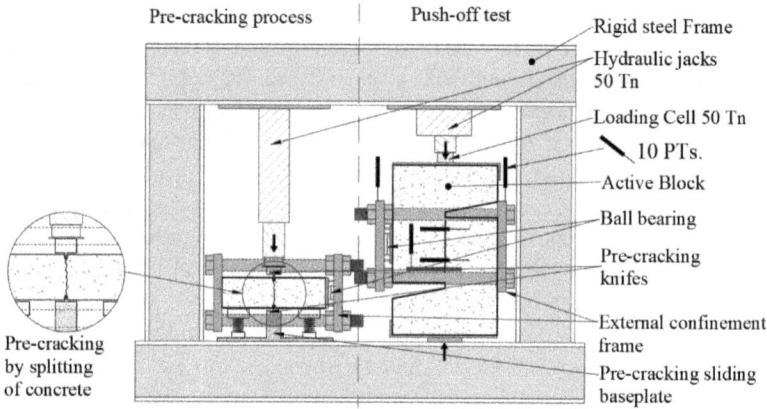

Fig. 2. Test set up.

Table 2. Main mechanical properties of PC and PFRC

Parameter	RC	PFRC8	PFRC12
f_c [MPa]	43.62 (0.06)	43.50 (0.07)	45.89 (0.09)
f_L [MPa]	4.0 (0.04)	4.39 (0.05)	4.49 (0.08)
$f_{R,1}$ [MPa]	–	1.52 (0.18)	2.27 (0.19)
$f_{R,3}$ [MPa]	–	2.09 (0.23)	3.29 (0.19)

3   Analytical Model

3.1   Shear Transfer Mechanism Acting in a FRC Open Crack

Feenstra [14] stated that the formulation of a crack constitutive relation should consider three possible states that arise at discrete cracks: before cracking, at the initiation of discrete crack and at open crack state, the latter being the only one considered in the present analytical model. The state of an open crack can be defined by its global displacements (ε), constituted by crack opening (w) and slip displacement (s), as well as its global stresses (σ), which includes normal stress (σ) and shear stress (τ) as shown in Fig. 3.

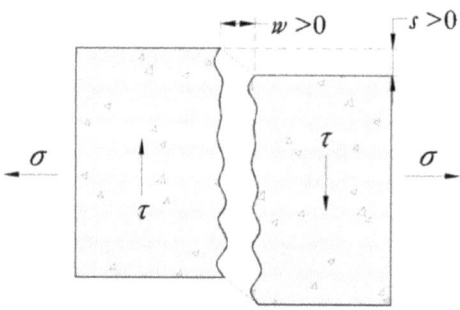

Fig. 3. Crack opening state.

The formulation to study a shear crack in the present analysis is based on an additive model which takes the following mechanisms into account: the initial confinement effect (pre-compression) (σ_0), the confinement effect by an active or passive mechanism (σ_c), the aggregate interlock (σ_a) and fibres (σ_F). Hence Eq. (1) can be expressed as follows and is represented in Fig. 4.

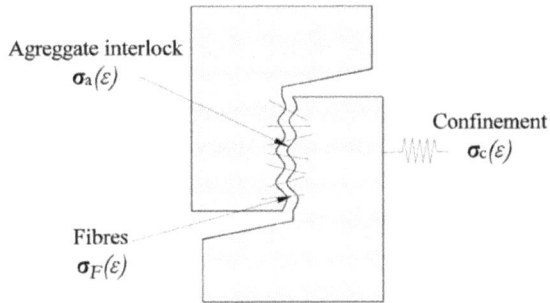

Fig. 4. Transferring mechanisms that act on the pre-cracked push-off test

$$\sigma = \sigma_0 + \sigma_c(\varepsilon) + \sigma_a(\varepsilon) + \sigma_F(\varepsilon) \tag{1}$$

The effect of pre-compression σ_0 results during the stages frame assembly, pre-cracking and handling, and is provided by the bars constituting the external rigid frame. This initial confinement causes normal stress in concrete at the start of the push-off test stage. Therefore, the effect of pre-compression could produce shear stresses due to friion when crack faces start sliding. The effect of initial confinement was taken into account in the present analysis by Eq. (2) which is based on the Coulomb friction law where μ is the characteristic friction coefficient of the crack interface and σ_0 is the confinement present on each specimen (measured on external steel bars using strain gauges) prior the push-off test. In this model, μ has been determined experimentally and is considered as 0.75.

$$\sigma_0 = \begin{bmatrix} \sigma_0 \\ \mu\sigma_0 \end{bmatrix} \tag{2}$$

Concerning to the confinement effect by an active or passive mechanism, this mechanism can be formulated by normal stress, which depends only on crack opening; therefore, it can be expressed as shown in Eq. (3). The confinement expression $\sigma_c(w)$ can be expressed as a linear function as shown in Eq. (4) where k_c is external frame stiffness and b is an adjustment coefficient, both determined by a regression analysis of the experimental normal stresses obtained in the push-off tests. In this model, as the external frame was the same in all the push-off test, the k_c value considered was 3.28 [N/mm^3].

$$\sigma_c = \begin{bmatrix} \sigma_c(w) \\ 0 \end{bmatrix} \tag{3}$$

$$\sigma_c(w) = k_c(w) + b \tag{4}$$

Regarding the aggregate interlock mechanism (σ_a), the following constitutive models are considered:

- The empirical rough crack model by Bazant and Karakoç (1983) [2] (Eqs. (7) and (8))
- The simplified model by Walraven and Reinhardt (1981) [4] (Eqs. (7) and (8))
- The simplified method by Vecchio and Collins (1986) [15] Eqs. (9) and (10))

$$\sigma_a = \frac{-0.000534}{w}(145|\tau|)^{1.3}\left(1 - \frac{0.231}{1 + 0.185w + 5.63w^2}\right) \tag{5}$$

$$\tau_a = \frac{0.195f_c 0.01D_{max}^2}{0.01D_{max}^2 + w^2} \frac{\frac{2.45}{0.195f_c} + 2.44\left(1 - \frac{4}{0.195f_c}\right)\left|\frac{s}{w}\right|^3}{1 + 2.44\left(1 - \frac{4}{0.195f_c}\right)\frac{s^4}{w}} \frac{s}{w} \tag{6}$$

$$\sigma_a = -\left(-\frac{f_{c,c}}{20} + 1.35w^{-.63} + (.191w^{-.552} - .15)f_{c,c}\right) \tag{7}$$

$$\tau_a = -\left(-\frac{f_{c,c}}{30} + 1.8w^{-.8} + (.234w^{-.707} - .20)f_{c,c}\right) \tag{8}$$

$$\tau_a = 0.18\tau_{max} + \frac{1.64\sigma_a^2}{\tau_{max}} \tag{9}$$

$$\tau_{max} = \frac{\sqrt{f_{cm}}}{0.21 + \frac{24w}{D_{max} + 16}} \tag{10}$$

Finally, the fibre contribution mechanism (σ_f) is considered using the following three models:

- Pfyl's model [16] (2003) (Eqs. (11) and (12))
- Kaufmann et al. model (2019) [6] (Eqs. (13) and (14))
- Tensile behaviour of fibres obtained by an inverse analysis (hereafter inverse model) of the small prismatic beams tested according to EN 14654, where the average residual flexural tensile strengths were determined for each PFRC type. The average tensile behaviour of each PFRC type used in the present analysis are presented in Fig. 5. The tensile behaviour of each PFRC was obtained based on MC2010 [17].

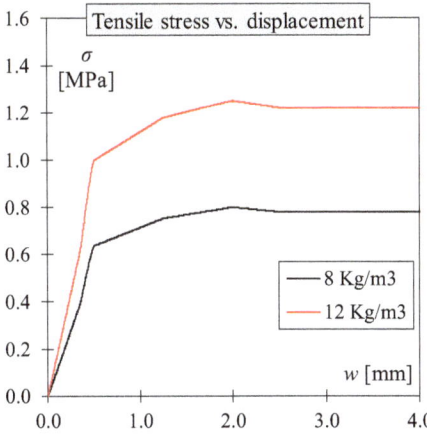

Fig. 5. Tensile behaviour of PFRC obtained by inverse analysis.

$$\sigma_{cf} = \sigma_{cfo}\left(2\sqrt{\frac{u}{u_0}} - \frac{u}{u_0}\right), \ 0 \leq u \leq u_0 \tag{11}$$

$$\sigma_{cf} = \sigma_{cfo}\left(1 - \frac{2u}{l_f}\right)^2, \ u_0 < u \leq \frac{l_f}{2} \tag{12}$$

$$\sigma_{cf\alpha} = \frac{1}{\sin \alpha_r}\sigma_{cfo}\left(2\sqrt{\frac{\delta}{u_0}} - \frac{\delta}{u_0}\right), \ 0 \leq \delta \leq u_0 \tag{13}$$

$$\sigma_{cf\alpha} = \frac{1}{\sin \alpha_r}\sigma_{cfo}\left(1 - \frac{2u}{l_f}\right)^2, \ u_0 < \delta \leq \frac{l_f}{2} \tag{14}$$

3.2   Numerical Implementation of the Model

The numerical implementation of the model was done to reproduce the test under controlled crack opening conditions. Each increase in the slip displacement in the crack led to incremented shear stresses (Δ_τ). The increase in normal stresses (Δ_σ) had to be zero because no external horizontal forces influenced the specimen as seen in Eqs. (15) and (16) (Fig. 6).

$$\Delta\sigma = \Delta\sigma_a + \Delta\sigma_c + \Delta\sigma_F + \sigma_0 = 0 \tag{15}$$

$$\Delta\sigma_a + \Delta\sigma_F + \sigma_0 = -\Delta\sigma_c \tag{16}$$

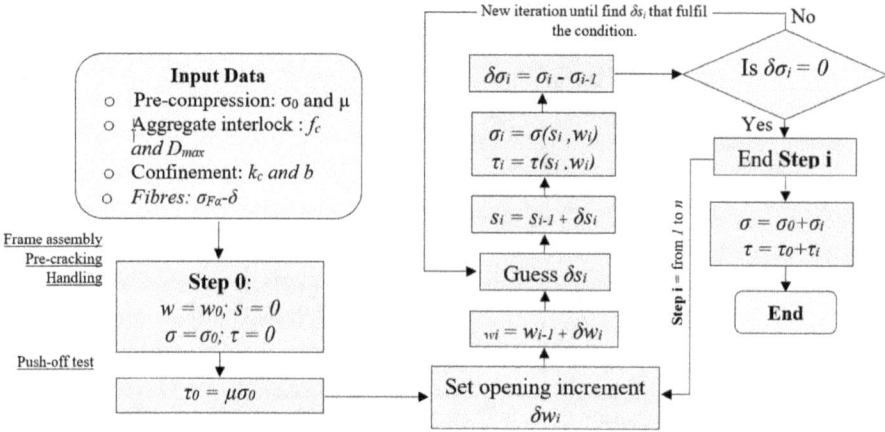

Fig. 6. Numerical procedure of the push-off test under crack opening control

It was necessary to establish an iterative numerical procedure that solves all the steps considered in the crack opening control. To solve the model, the non-linear optimisation algorithms included in Visual Basic to solve a non-linear system of equations was used. The step size was considered constant and equalled 0.05 mm, which seemed a quasi-continuous numerical simulation.

4   Results

This section presents part of the experimental results and their comparison with the analytical results obtained from the proposed model. The PC results of the numerical analysis are in Fig. 7a, while Fig. 7b and c show results of PFRC specimens with 8 kg/m^3 and 12 kg/m^3 of fibres, respectively. PC experimental results were compared against the simplified models proposed by Walraven and Reinhardt (W&R), Gambarova and Karakoç (G&K), and the simplified formula set forward by Vecchio and Collins (V&C). For each specimen, the experimental response of crack slip vs. shear stress and crack opening vs. shear stress was compared with the aforementioned aggregate interlock models. In these comparisons, slip displacements up to 2.5 mm and crack openings up to 1.2 mm were taken into account.

Similarly to PC, the PFRC experimental results were compared to the aggregate interlock models: the simplified model by Walraven and Reinhardt (W&R) and the model of Gambarova and Karakoç (G&K), by combining each one with three different fibre models.

As seen in Fig. 7a, all the aggregate interlock models quite accurately described crack opening vs. shear behaviour of push-off tests. In fact the models that better fitted the experimental results were those developed by Vecchio and Shim based on a regression analysis of Walraven's tests.

As observed from Fig. 7b and c, all the aggregate interlock models combined with the fibre model provided a good prediction of the shear stresses transmitted according

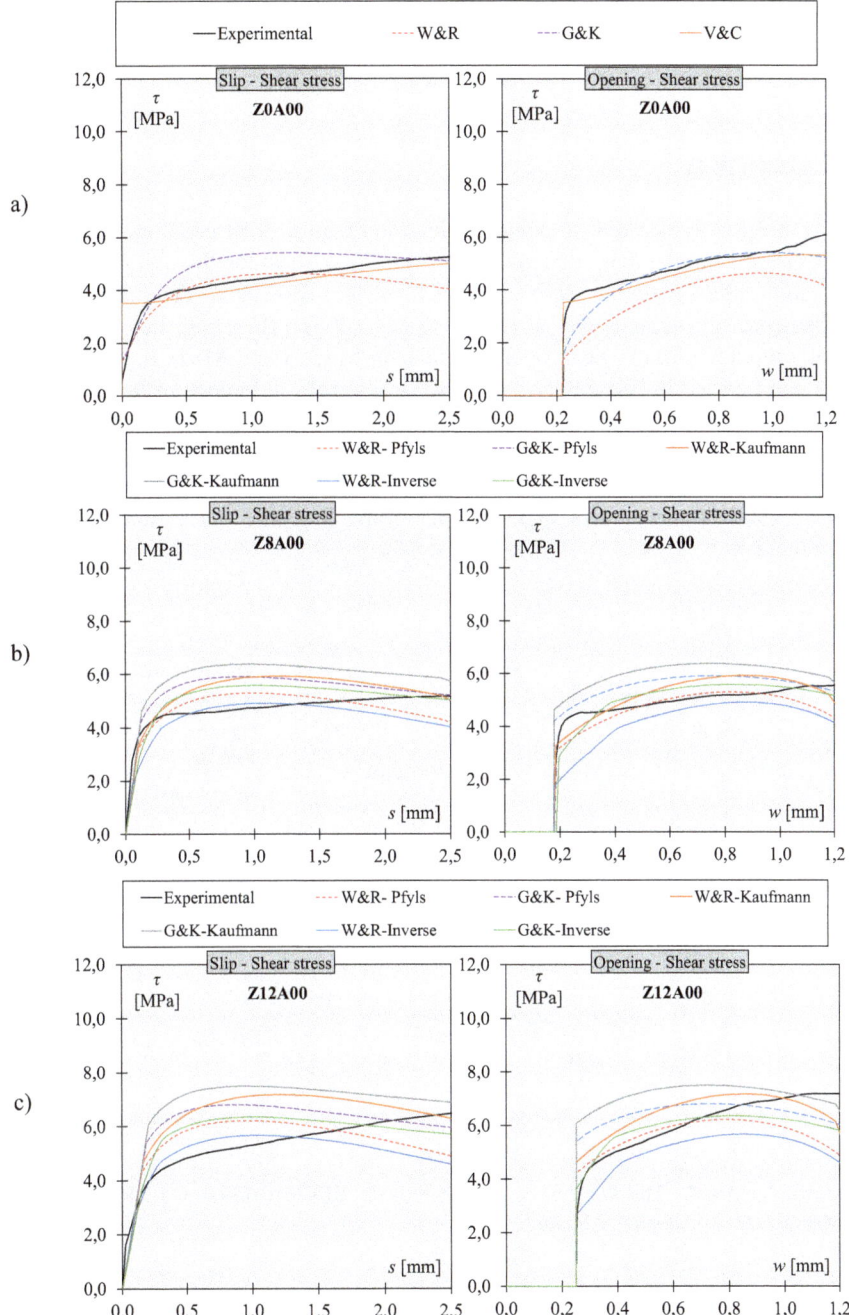

Fig. 7. Experimental and analytical results of specimen: Z0A00 (a), Z8A00 (b) and Z12A00 (c)

to crack opening. As with the PC specimens, the aggregate interlock model that better predicted the experimental results was that developed by Walraven and Reinhardt, followed by Gambarova and Karakoç's model.

5 Conclusions

- The stresses transferred by polypropylene fibres in a shear crack could be directly estimated from the tensile behaviour of PFRC; i.e. from the flexural residual tensile strength stress material properties which, in this case, were determined from the prismatic specimens tested according to EN 14651. Therefore, the work done by polypropylene fibres was probably similar in Mode I and Mode II.
- As can be seen in Fig. 8 there exists a considerable difference between the aggregate interlock models. When comparing these differences to those among the fibre models, the fibre differences were noticeably lesser than the aggregate interlock ones. These results evidence the need to firstly improve the understanding of the shear transfer mechanism in PC to determine the fibre interaction with another mechanism.

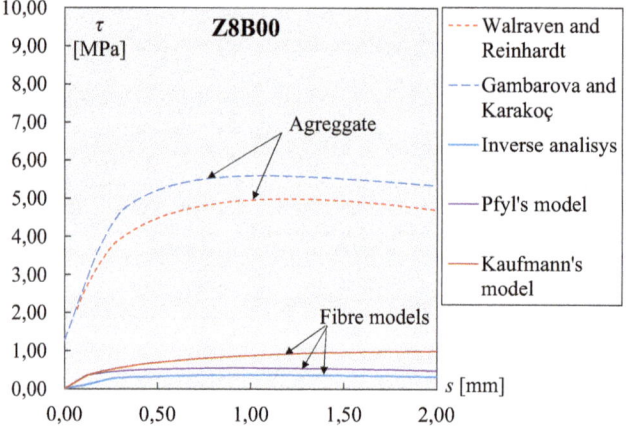

Fig. 8. Comparison between aggregate interlock and fibre contribution

Acknowledgements. This work forms part of Project "BIA2016-78460-C3-1-R" supported by the State Research Agency of Spain.

References

1. Bazant, Z.P., Gambarova, P.: Rough cracks in reinforced concrete. J. Struct. Div. **106**, 819–842 (1980). https://cedb.asce.org/CEDBsearch/record.jsp?dockey=0009420

2. Gambarova, P.G., Karakoc, C.: A new approach to the analysis of the confinement role in regularly cracked concrete elements. North-Holland, Netherlands (1983). http://inis.iaea.org/search/search.aspx?orig_q=RN:15048850

3. Li, B., Maekawa, K.: Contact density model for cracks in concrete. Comput. Mech. Concr. Struct. Adv. Appl. (1987)

4. Walraven, J.C.: Aggregate interlock: a theoretical and experimental analysis. Delft University (1980)

5. Fenwick, R., Pauley, T.: Mechanism of shear resistance of concrete beams. J. Struct. Div. **94**, 25 (1968)

6. Kaufmann, W., Amin, A., Beck, A., Lee, M.: Shear transfer across cracks in steel fibre reinforced concrete. Eng. Struct. **186**, 508–524 (2019). https://doi.org/10.1016/j.engstruct.2019.02.027

7. Narayanan, R., Darwish, I.Y.S.: Use of steel fibers as shear reinforcement. ACI Struct. J. **84**, 216–227 (1987). https://doi.org/10.14359/2654

8. Aoude, H., Belghiti, M., Cook, W.D., Mitchell, D.: Response of steel fiber-reinforced concrete beams with and without stirrups. ACI Struct. J. **109**, 359–367 (2012). https://doi.org/10.14359/51683749

9. Conforti, A., Minelli, F., Plizzari, G.A.: Shear behaviour of prestressed double tees in self-compacting polypropylene fibre reinforced concrete. Eng. Struct. **146**, 93–104 (2017). https://doi.org/10.1016/j.engstruct.2017.05.014

10. Ortiz Navas, F., Navarro-Gregori, J., Leiva Herdocia, G., Serna, P., Cuenca, E.: An experimental study on the shear behaviour of reinforced concrete beams with macrosynthetic fibres. Constr. Build. Mater. **169**, 888–899 (2018). https://doi.org/10.1016/j.conbuildmat.2018.02.023

11. Conforti, A., Minelli, F., Tinini, A., Plizzari, G.A.: Influence of polypropylene fibre reinforcement and width-to-effective depth ratio in wide-shallow beams. Eng. Struct. **88**, 12–21 (2015). https://doi.org/10.1016/j.engstruct.2015.01.037

12. Arslan, G., Ozturk, M., Secer, R., Keskin, O.: Shear behaviour of polypropylene fibre-reinforced-concrete beams without stirrups (2016)

13. Echegaray-Oviedo, J.: Upgrading the push off test to analyze the contribution of steel fiber on shear transfer mechanisms (2014). https://doi.org/10.4995/Thesis/10251/43723

14. Feenstra, P.H.: Computational aspects of biaxial stress in plain and reinforced concrete. Delf university (1993). http://resolver.tudelft.nl/uuid:faf2fd16-1c43-4711-b783-9e8e00d10c21

15. Vecchio, F.J., Collins, M.P.: The modified compression-field theory for reinforced concrete elements subjected to shear. ACI J. Proc. **83**, 219–231 (1986). https://doi.org/10.14359/10416

16. Pfyl, T.: Tragverhalten von Stahlfaserbeton (Structural Behavior of Steel Fiber Concrete). Swiss Federal Institute of Technology Zurich (20030. https://doi.org/10.3929/ethz-a-004501155

17. International Federation for Structural Concrete (fib): Model Code 2010, final drafts. Wilhelm Ernst & Sohn (2013)

Creep of Macro Synthetic Fibre Reinforced Concrete: Experimental Results and Numerical Model Calibration

Clementina Del Prete[1]($\boxtimes$), Ioannis Boumakis[2],
Roman Wan-Wendner[2,3], Nicola Buratti[1], and Claudio Mazzotti[1]

[1] DICAM – Structural Engineering, University of Bologna, Bologna, Italy
clementina.delprete2@unibo.it
[2] Department of Civil Engineering and Natural Hazards, University of Natural
Resources and Life Sciences, Wien, Austria
[3] Department of Structural Engineering, Ghent University, Ghent, Belgium

Abstract. The mechanical performance of fibre reinforced concrete presents aspects still under investigation, mostly those regarding the long-term behaviour. Even if creep and shrinkage are two well-known phenomena that characterize concrete, in case of FRCs, and in particular of macro-synthetic Fiber reinforced concretes (MSFRCs), there are no reliable models for predicting their long term-behaviour, because of the interaction between concrete, fibre creep and bond. In addition, temperature is a further variable to control since it affects the material performance.

In this perspective, the present paper shows the experimental results of a large campaign of creep tests performed on macro synthetic fibre reinforced concrete specimens. The material tested had a compressive strength of about 55 MPa and it is reinforced with 8 kg/m^3 of polypropylene crimped fibres. The experimental investigation is carried out by performing creep compression tests on cylinders and direct tensile test on notched cylinders. In addition, the tensile behaviour of the single fibre under sustained load is analysed. The tests were conducted in a humidity and temperature controlled chamber. Furthermore, the temperature was increased from 20 °C to 30 °C after a time of 50 days of testing in order to understand how this condition modifies the creep deformations evolution of the material.

The paper shows also the initial calibration of a numerical model based on the Lattice Discrete Particle Model (LDPM) theory. The LDPM is one of the most validated theories able to reflect the actual coarse aggregate distribution of a quasi-brittle material, i.e. concrete. Currently this theory has been extended to include the fibres reinforcement. The aim of the big study presented would be to elaborate a predictive model for the MSFRCs accounting also for concrete and polymers long term behaviour.

Keywords: FRC · Testing · Sustained load · Creep · Shrinkage · Temperature · LDPM

© RILEM 2022
P. Serna et al. (Eds.): BEFIB 2021, RILEM Bookseries 36, pp. 420–432, 2022.
https://doi.org/10.1007/978-3-030-83719-8_37

1 Introduction

The assessment of the durability and serviceability of FRCs is becoming a topic of great interest, in particular with reference to cracked elements. Creep and shrinkage effects characterize the concrete mechanical behaviour affecting its durability [1] In the case of FRCs, in particular those reinforced with polymeric fibres, fibres have creep deformations as well. As the material starts cracking, the creep deformations of FRC elements depend on concrete, fibres and the their interaction, i.e. bond [2].

In the literature there are many experimental tests concerning the flexural behaviour of cracked elements [3–5]. Nevertheless, the results obtained from bending tests also depend on creep in compression and on the behaviour of cracked sections. For this reason some researchers [6] proposed tests under simpler loading conditions, in particular direct tension tests, that allow an easier analysis of creep in tension. Vrijdaghs et al. [7] proposed a test procedure where notched cylinders were manually pre-cracked and then tested under sustained tensile load. A variation to the set-up adopted is proposed by Sorelli et al. [8] who made use of prismatic specimens to be tested under tensile load [9].

From a numerical modelling point of view, one of the important aspects of creep is to account for the composite nature of fibre reinforced concretes so as to describe the interaction between the reinforcement and the matrix. The lattice discrete particle model theory (LDPM) is a constitutive model widely validated to fully describe the plain concrete behaviour [10, 11]. Recently, the LDPM has been formulated so to include the fibre reinforcement with a discrete approach named LDPM-F, so threating the material as formed by two constituents with no homogenization techniques [12, 13]. Besides, the aging viscoelastic nature of concrete can be also described by the LDPM constitutive law so extended and named M-LDPM [14].

This paper can be divided into two parts, the first discusses the experimental results of creep tests, which are used in the second part to calibrate a numerical model. The experimental campaign was aimed at characterizing creep deformations of a type of MSFRC, involving different tests: compression creep and shkinkage of concrete, tension creep in cracked cylinders and tension creep of the fibre alone [15]. The numerical model adopted is based on LDPM, and made use of the software MARS (Modeling and Analysis of the Response of Structures).

2 Experimental Campaign

The experimental campaign carried-out includes different tests, aimed at quantifying the main contribution to creep in FRC elements. In particular, shrinkage, creep compression and tensile tests, on FRC cylinders and single fibres, were performed. All tests were developed in a climate chamber where the environmental conditions were controlled: the relative humidity was set at 55% and the temperature is increased of 10 °C from 20 °C to 40 °C and kept constant for respectively, 50, 128 and 104 days.

2.1   Material Properties

All the concrete specimens were cast using the admixture given in Table 1. A dosage of 8 kg/m³ of fibres was used, their properties are reported in Table 2. Their tensile strength end elastic modulus were obtained by means of tension tests.

Table 1. Concrete mix used.

Component	Quantity
CEM I 52.5R (kg/m³)	400
Sand 0–1 mm (kg/m³)	172
Sand 0–5 mm (kg/m³)	730
Gravel 5–15 mm (kg/m³)	837
Water (l/m³)	184
Water/Cement ratio (–)	0.46

Table 2. Fibre properties.

Properties	
Material	Polypropylene
Diameter (mm)	0.81
Length (mm)	54
Aspect ratio (–)	67
Tensile strength (MPa)	473
Elastic modulus (MPa)	2600

2.2   Set-Up Adopted

Compression Creep and Shrinkage
Compression creep and shrinkage tests were carried out on 150 mm × 300 mm cylinders (Fig. 1). During the tests strains were recorded using two strain gauges per cylinder (on opposite sides). Cylinders for creep tests were rectified at the top and bottom faces. A hydraulic loading frame, with an oil accumulator, was used in order to impose a constant compression load on the specimens. No wrapping was used neither on the specimens for creep tests nor on those for shrinkage tests.

(a)   (b)

Fig. 1. (a) Cylinders used for shrinkage and creep compression tests; (b) Creep compression set-up.

Uniaxial Tension Tests on FRC

There is no standard testing protocol for the characterization of FRCs creep deformations. The procedure here proposed is based on several literature proposals with improvements and modifications.

The specimens used for uniaxial tension tests are 150 mm high and 125 mm diameter cylinders, that were cored from 150 × 150 × 600 mm beams. In particular, three of them were obtained for each beam, and were cored in such a way that their axis corresponds to the longitudinal axis of the beams. This procedure guarantees that the fibre orientation is the same as one would observe in bending tests. The tests comprise three phases: pre-cracking, creep test, re-loading to failure. After coring, the top and bottom surfaces of the cylinders were rectified. Steel plates were then epoxied on these surfaces, which were then used to attach the specimens to the testing machines used. Furthermore, a circumferential notch was cut on the cylinders in order to control the crack-location (Fig. 2a). Pre-cracking was performed using a servo-hydraulic machine, operated in crack opening displacement control. To this aim three COD transducers were installed at 120° around the notched section of the cylinders. Pre-cracking tests were interrupted when an average crack opening of 0.3 mm was measured. During the test, the end of the cylinders are allowed to rotate, for consistency with the setup used for creep tests, where a chain of three specimens was loaded by means of a lever loading frame (Fig. 2b).

The load applied in the creep tests was 50% of the force measured at an average crack opening of 0.3 mm during the pre-cracking tests. Crack openings were measured on each cylinder using three LVDTs at 120° around the circumference. At the end of creep tests, the specimens were unloaded, deformations were measured for 30 days, and then the cylinders were tested to failure, using the same setup adopted for pre-cracking.

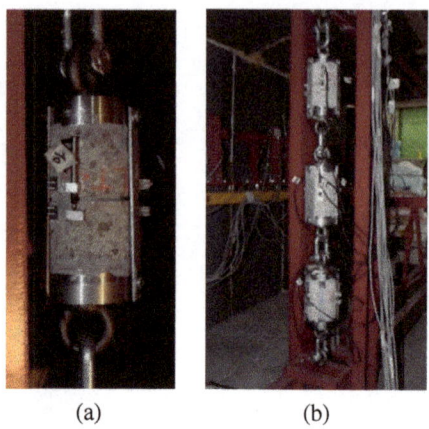

(a)   (b)

Fig. 2. (a) Notched cylinder used for creep tests; (b) Tensile creep test on MSFRC specimens.

Tension Test on Fibres

A lever system (Fig. 3) was used to apply a constant tensile load to polymeric fibres: they are gripped at both ends using steel notched cylinders with sand paper glued on the notch to create a mechanical interlock between the fiber and the system.

In this case, the load applied was 20% of the maximum strength of the fibres, this load percentage was defined based on the results of preliminary direct tension tests on notched concrete cylinders, in which the number of fibres crossing the crack surface was counted. During the test the elongation of the fibres was measured using two LVDTs.

Fig. 3. Fibre tensile creep test.

2.3 Experimental Results

Figure 4 reports the average strains measured during shrinkage and compression creep tests (average shrinkage strains have been removed from these latter). The temperature increments are marked by the discontinuities along the curves and indicated by the dashed lines in figures. The origin is defined as the test beginning.

As discussed, the uniaxial tension creep tests were executed on pre-cracked cylinders thus activating the fibre reinforcement contribution. Figure 5a shows force versus average crack opening curves for obtained from the pre-cracking test of the cylinders. Figure 5b show the results of creep tests in terms of creep coefficient, computed as:

$$\varphi(t) = (COD(t) - COD(t_0))/COD(t_0) \tag{1}$$

where $COD(t)$ and $COD(t_0)$ indicate the average crack opening displacement measured at the general time t and at time t_0 at which the full creep load was applied and the creep deformations start.

The contributions to creep deformations are due to three possible mechanisms taking place: i) concrete creep in tension; ii) fibre creep; iii) creep of the bond or pull-out. With the aim of decoupling the effect of fibre elongation, creep tensile tests on the single fibres were performed. Figure 6 shows the results of these tests in terms of creep coefficient for the two fibers tested, named FibA and FibB. It is possible to notice that the effect of the temperature variation featured by the fibre is quite comparable to that exhibited when the tensile load is applied on the cylinders.

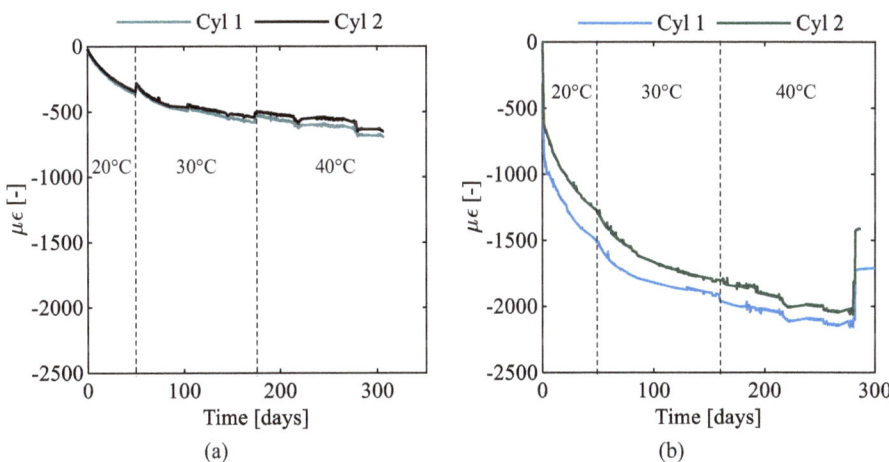

Fig. 4. (a) Shrinkage test curves; (b) Creep compressive test curves.

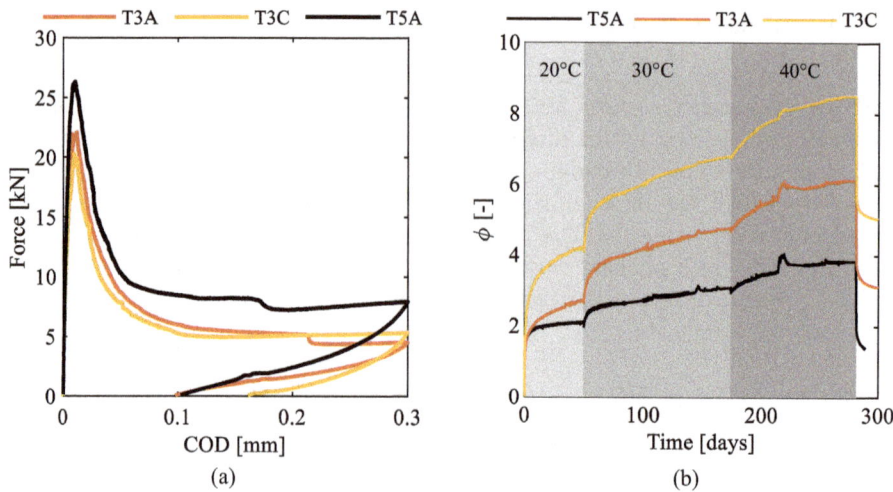

Fig. 5. Creep tensile test: (a) Pre-cracking curves; (b) Creep test curves.

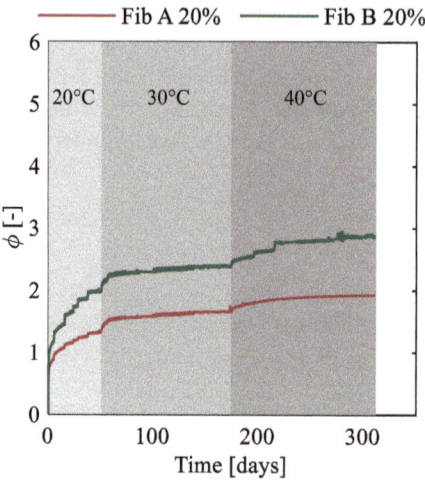

Fig. 6. Creep coefficient – time curves for fibre creep test.

3   Numerical Approach

As the mechanical behaviour of the material has been experimentally investigated in different loading conditions, it is possible to separate the contributions, in terms of deformations, occurring during the flexural cracked state of the MSFRC. The extended Lattice Discrete Particle Model (LDPM), coupled with the Hygro-thermo-chemical model and the Microprestress Solidification theory, is used to simulate the deformations induced by the shrinkage phenomenon and the creep compressive deformations.

The innovative aspect lies in the inclusion of the fibre viscoelasticity described by an analytical model derived from the Boltzmann and Volterra principle.

The procedure adopted had the following phases:

a. calibration and validation of the mechanical parameters for simulating FRC under instantaneous loads. Three-point bending and compression tests are used;
b. calibration of the parameters characterizing the HTC model (in this study they are assumed from the literature since specific experimental tests are not available);
c. calibration and validation of the parameters describing the viscoelastic behaviour of concrete, not considering the contribution of fibres, using shrinkage and creep compression tests (some of the parameters are assumed, details are given below);
d. calibration of the viscoelastic behaviour of fibres using experimental creep data;
e. validation of the different contributions, separately calibrated, on the creep flexural behaviour.

The phases here discussed are mainly *c* and *d* while the validation step *e* represents the final goal of the project, and represents a future development.

3.1   LDPM and LDPM-F Calibration

The approach adopted to simulate the mechanical behaviour of the macro-synthetic fibre reinforced concrete at issue is the LDPM, a mesoscale theory able to describe the aggregate particles interaction through a lattice system. The constitutive formulation, on which the mechanical behaviour of concrete is based, needs the calibration of a set of parameters for the elastic and inelastic behaviour of concrete [10, 11]. The concrete internal structure is generated assigning the mix properties (i.e. aggregate grading curve), the fibres are inserted according to their geometry, i.e. length and diameter. Aggregates and fibres are randomly distributed thus giving to LDPM the capability of simulating the typical scatter observed in FRC experimental tests. The mechanical contribution of fibres is mechanically identified by a set of parameters describing the interaction with the matrix, i.e. the gradual slippage from the matrix, and the mechanical properties of the fibre itself, tensile strength and elastic modulus [12, 13]. For the present study the mechanical parameters were calibrated on the experimental results of three point bending and cube compressive tests (Table 3), as discussed in [16].

Table 3. Calibration of LDPM-F parameters.

Parameters (Units)	Value	Parameters (Units)	Value
E_0 (GPa)	45	G_d (N/m)	1.0
α^* (–)	0.25	τ_0 (MPa)	4.0
σ_t (MPa)	2.5	β (–)	0.5
l_t (mm)	800	k_{sp}^* (–)	6.2
σ_s/σ_t (–)	3.0	k_{sn}^* (–)	1.0
n_t (–)	0.5	σ_{uf} (MPa)	473
σ_{c0}^* (MPa)	190	k_{rup}^* (–)	0.0

(continued)

Table 3. (*continued*)

Parameters (Units)	Value	Parameters (Units)	Value
H_{c0}/E_0^* (–)	0.4	E_f (GPa)	3.3
k_{c0}^* (–)	2	V_f (%)	0.9
k_{c1}^* (–)	1	Tortuosity	0.6
k_{c2}^* (–)	5	Shape	Crimped
μ_0^* (–)	0.35	ρ_f [kg/m^3]	910
μ_∞^* (–)	0	l_f (mm)	54
σ_{N0}^* (MPa)	600	d_f (mm)	0.81

*Assumed from literature

3.2   M-LDPM Calibration

The M-LDPM formulation accounts for creep, shrinkage and thermal effects in concrete by coupling the LDPM approach with a hygro-thermo-chemical model and the Microprestress Solidification theory [14]. By means of this present theory it is possible to consider the aging properties of concrete as connected to the growth of the volume fraction over time, caused by the processes of hydration and polymerization of cement. The total creep strain is obtained as the sum of two contributions: the viscoelastic and viscous strain, ε_v and ε_f whose increments over time are calculated approximating the creep behaviour of concrete to a series of N Kelvin units in series [17].

The HTC model describes the multi-physics of transport in concrete defining the fields of temperature, T, humidity, h, and binder reaction degree α_b. In Table 4 are collected the mechanical parameters calibrated on the shrinkage and creep compressive curves, i.e. phase 'b' of the procedure.

Table 4. Calibration of M-LDPM parameters.

Parameters (Units)	Value	Parameters (Units)	Value
α_h (x10^{-3})	1.19	ξ_4 (MPa^{-1} × 10^{-6})	2
α_T (x10^{-6})	5	n_α^* (–)	1.3
ξ_1 (MPa^{-1} × 10^{-6})	1.3	K_0 (MPa^{-2} s^{-1} × 10^{-2})	2.34
ξ_2 (MPa^{-1} × 10^{-6})	0.42	K_1 (MPa/Kel × 10^4)	0.001

*Assumed from literature

3.3   Fibre Viscoelasticity Calibration

The innovative aspect of the present formulation is the consideration of the viscoelastic behaviour of the plastic fibre as it is not possible to neglect it when dealing with time dependent phenomena: plastic behaviour is basically characterized by creep, stress relaxation and constant rate stressing [18]. The approach here proposed is based on the exponential Kernel functions and is easier than that used for concrete creep as the non-aging nature of this material. It is based on the model developed by Golub, Fernati and Lyashenko [18] and has the following expression:

$$\varepsilon(t) = \frac{\sigma_k}{E} h(t) \left\{ 1 + \frac{\lambda}{\beta} \left[1 - \exp\left(-(1+\alpha)^{1+\alpha} \beta t^{1+\alpha} \right) \right] \right\} \tag{2}$$

where α and β are the fractional exponential kernel parameters, λ connects the elastic modulus of the fibre assumed as half of the instantaneous [19], $\sigma(t) = \sigma_k h(t)$ represents the levels of stresses for which the Kernel parameters are determined, in particular σ_k is constant and $h(t)$ is the Heaviside function.

The expression (2) is used to describe the creep deformations of the single fibre and the unknown parameters are calibrated (Table 5) by fitting the experimental curves with the analytical model (Fig. 7). The experimental data used for the calibration cover a range of 50 days during which the temperature, during the experiments, has been kept constant as the analytical approach does not model the temperature variation.

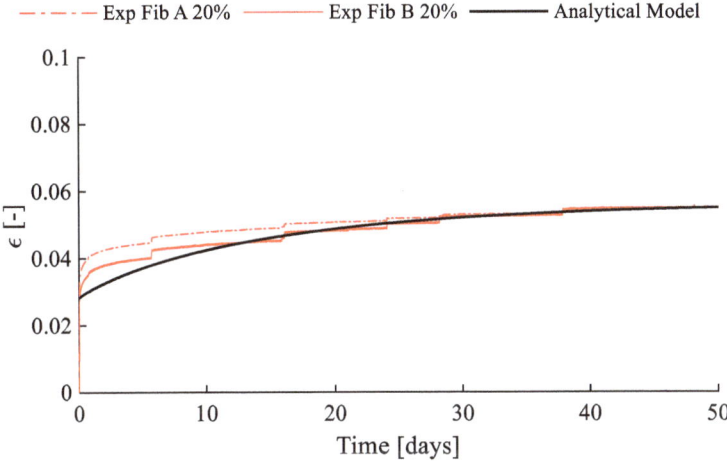

Fig. 7. Experimental creep curve and analytical creep curve according to the formulation (2).

Table 5. Values of the parameters fitting the experimental creep curves.

Parameters (Units)	Value
α	−0.0098
β	0.0998
λ	0.0998

3.4   Calibration of the Numerical Model: Simulation of Uniaxial Tensile Tests

The creep deformations exhibited by the notched specimens under tensile load are mostly due to the fibre contribution, in particular the elongation and the slippage at interface. For this reason, the inclusion of the viscoelasticity of the reinforcement, might help to describe the MSFRC behaviour. First of all the pre-cracking stage of the cylinders is simulated (Fig. 8) reproducing the experimental tests: the specimens are pre-cracked in displacement control allowing the rotations at the end as in the experiments. The notched cylinder is modelled according to the dimensions of the samples used in the experiments: the discretization is handled by setting the minimum and maximum aggregate size experimentally used. This is reflected in the irregular opening of the crack that makes the model close to the reality of the campaign performed. In Fig. 8 there are three curves representing the numerical behaviour simulated with three different arrangements of the fibres so as to catch the irregularity in the distribution. The comparison suggests that the numerical model is able to reproduce the real behaviour experimentally characterized, although the initial slope is not completely catched by the model.

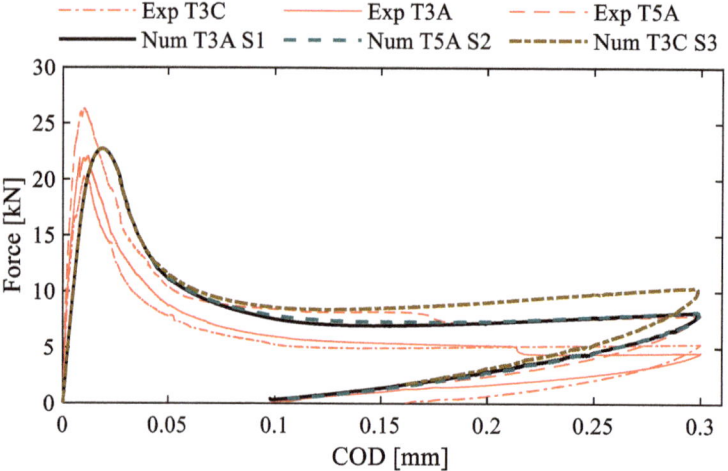

Fig. 8. Numerical simulation of the pre-cracking stage of the creep tensile test.

4   Conclusions

The first part of the present paper described the experimental characterization of the creep deformations of a macro-synthetic fibre reinforced concrete. Afterwards, the first results of the calibration of a numerical predictive model for the creep deformations, are shown. Based to the results described, the following conclusions and observations can be drawn:

- a innovative setup for creep test of fibres was proposed;
- comparing the behaviour of the MSFRC and the fibre under tensile load, one more contribution besides fibre elongation determines the crack-opening increment of the composite material, most likely bond creep or pull-out;
- the creep deformations in cracked conditions of MSFRC elements depend on different phenomena, namely the creep of fibres, the creep at the fibre-to-concrete

interface, and the creep of concrete. In bending tests, which are the most common in the literature, it is not yet clear how to separate these contributions. In the present paper direct tension tests on cracked FRC cylinders and tension tests of fibres were carried out in order to obtain data for calibrating numerical models that might contribute in quantifying the significance of the aforementioned contributions;

- On the basis of the experimental results, a numerical model based on the LDPM theory was proposed. The first phase of the procedure concerned the calibration and validation of the model for simulating the short term behaviour of FRC; consequently creep and shrinkage deformations were simulated considering only the concrete matrix deformations. A model for simulating the viscoelastic behaviour of polymeric fibres in the framework of LDPM was proposed and fitted on experimental data;

Regarding the numerical approach, the paper illustrates the first steps towards the numerical prediction of the MSFRC viscoelastic deformations. Future developments will concern the inclusion in the model of all the contributions to creep in cracked conditions and the simulation of elements in bending.

References

1. Jiràsek, M., Bazant, Z.P.: Inelastic Analysis of Structures. Wiley, Hoboken (2002)
2. Babafemi, A.J., du Plessis, A., Boshoff, W.P.: Pull-out creep mechanism of synthetic macro fibres under a sustained load. Constr. Build. Mater. **174**, 466–473 (2018)
3. Pujadas, P., Blanco, A., Cavalaro, S., de la Fuente, A., Aguado, A.: The need to consider flexural post-cracking creep behavior of macrosynthetic fiber reinforced concrete. Constr. Build. Mater. **149**, 790–800 (2017)
4. Buratti, N., Mazzotti, C.: Temperature effect on the long term behaviour of macro-synthetic-and-steel–fibre reinforced concrete. In: 8th RILEM International Symposium on Fibre Reinforced Concrete: Challenges and Opportunities (2012)
5. Buratti, N., Mazzotti, C.: Effects of different types and dosages of fibres on the long-term behaviour of fibre-reinforced self-compacting concrete. In: 8th RILEM International Symposium on Fibre Reinforced Concrete, RILEM PRO88, pp. 715–725 (2012)
6. Babafemi, A.J.: Tensile creep of cracked macro synthetic fibre reinforced concrete. Stellenbosch University (2015)
7. Vrijdaghs, R., Di Prisco, M., Vandewalle, L.: Creep of cracked polymer fiber reinforced concrete under sustained tensile loading. In: 9th International Conference on Fracture Mechanics of Concrete and Concrete Structures FraMCoS-9, pp. 1–9 (2016)
8. Sorelli, L.G., Meda, A., Plizzari, G.A.: Bending and uniaxial tensile tests on concrete reinforced with hybrid steel fibers. J. Mater. Civ. Eng. **17**(5), 519–527 (2005)
9. Buratti, N., Mazzotti, C.: Experimental tests on the long-term behaviour of SFRC and MSFRC in bending and direct tension. In: Proceedings of the 9th Rilem International Symposium on Fiber Reinforced Concrete (BEFIB) (2016)
10. Cusatis, G., Pelessone, D., Mencarelli, A.: Lattice Discrete Particle Model (LDPM) for failure behavior of concrete. I: theory. Cem. Concr. Compos. **33**(9), 881–890 (2011)
11. Cusatis, G., Mencarelli, A., Pelessone, D., Baylot, J.: Lattice Discrete Particle Model (LDPM) for failure behavior of concrete. II: calibration and validation. Cem. Concr. Compos. **33**(9), 891–905 (2011)

12. Jin, C., Buratti, N., Stacchini, M., Savoia, M., Cusatis, G.: Lattice discrete particle modeling of fiber reinforced concrete: experiments and simulations. Eur. J. Mech. A/Solids **57**, 85–107 (2016)
13. Schauffert, E.A., Cusatis, G., Pelessone, D., O'Daniel, J.L., Baylot, J.T.: Lattice discrete particle model for fiber-reinforced concrete. II: tensile fracture and multiaxial loading behavior. J. Eng. Mech. **138**(7), 834–841 (2011)
14. Abdellatef, M., Boumakis, I., Wan-Wendner, R., Alnaggar, M.: Lattice discrete particle modeling of concrete coupled creep and shrinkage behavior: a comprehensive calibration and validation study. Constr. Build. Mater. **211**, 629–645 (2019)
15. Serna, P., Llano-Torre, A., Cavalaro, S.H.P. (eds.): Creep Behaviour in Cracked Sections of Fibre Reinforced Concrete. RB, vol. 14. Springer, Dordrecht (2017). https://doi.org/10.1007/978-94-024-1001-3
16. Del Prete, C., Wan-Wendner, R., Buratti, N., Mazzotti, C.: Lattice discrete particle modeling of MSFRC. In: SSCS19 Numerical Modeling Strategies for Sustainable Concrete Structures, pp. 27–36 (2019)
17. Bažant, P.Z., Xi, J.: Continuous retardation spectrum for solidification theory of concrete creep. J. Eng. Mech. **121**(2), 281–288 (1995)
18. Findley, W.N., Lai, J.S., Onaran, K.: Creep and Relaxation of Nonlinear Viscoelastic Materials (1976)
19. Sorzia, A.: Modelling of creep and stress relaxation test of a polypropylene microfibre by using fraction-exponential kernel. Model. Simul. Eng. **2016**, 1–7 (2016)

Statistical Modelling of Flexural Fatigue Response of Steel Fibre Reinforced Concrete

Ajeesh Koorikkattil[(✉)], Sunitha K Nayar, and Veena Venudharan

Department of Civil Engineering, Indian Institute of Technology, Palakkad, India
101914001@smail.iitpkd.ac.in

Abstract. Steel fibre reinforced concrete (SFRC) is widely recommended where fatigue is a predominant mode of failure, such as concrete bridges, offshore structures, and concrete pavements. From a detailed literature review, it was observed that the different fibre parameters like fibre type, fibre volume, aspect ratio, etc., influence the crack resistance performance and thus, the fatigue endurance limit. This paper attempts to understand the effect of various test parameters (stress ratio, frequency) and fibre parameters (length, diameter, volume of fibres, aspect ratio, and reinforcing index) on the fatigue life of SFRC based on available literature data. Studies involving SFRC with volume fractions of 0.13–2.0% are included in the analysis. The objective of the current study is to propose a generalized fatigue model, based on statistical analysis of data available from literature, for pre-cracked SFRC with the consideration of the above variables. A multiple linear regression (MLR) analysis using SPSS software was used to perform analysis of variance (ANOVA). The most significant parameters, obtained from the analysis, for the prediction of fatigue life of pre-cracked SFRC are stress ratio, length of fibre and reinforcing index. The generalized expression can be used for the prediction of the post cracking fatigue life of SFRC for volume fractions within the range of 0.13–2.0%. It is envisaged that these generalized models could be integrated into any SFRC design methodology requiring fatigue response prediction, leading to better-optimized designs. Overall, this study expands the state-of-the-art on fatigue behaviour of SFRC and its analysis.

Keywords: Fatigue failure · SFRC · Post-cracking · MLR · ANOVA

1 Introduction

Fatigue cracking is a major distress found in rigid pavements due to progressive, permanent structural changes in the material subjected to repetitive loads [1–4]. It is noteworthy that the stress required to fracture a material under fatigue, due to repeated loading is less than that required for monotonic loading [5, 6]. Steel fibre reinforced concrete (SFRC) is widely recommended for applications where fatigue is a predominant mode of failure, such as concrete bridges, offshore structures, and concrete pavements due to the crack resistance imparted by the presence of fibres [6–10].

Many researchers have carried out investigations on the fatigue response of SFRC mainly to determine the fatigue endurance limit of different concrete grades with

© RILEM 2022
P. Serna et al. (Eds.): BEFIB 2021, RILEM Bookseries 36, pp. 433–442, 2022.
https://doi.org/10.1007/978-3-030-83719-8_38

different fibre parameters like type, volume fraction, and aspect ratio [11–13]. Though different test configurations are globally available to characterize the fatigue property, flexural fatigue testing is one of the commonly used methods, where the fatigue performance is measured in terms of the number of cycles to flexural fatigue failure [14, 15]. It is important to note that for the same testing configuration, the fatigue behaviour of an FRC material is independent of the specimen shape, concrete strength, curing condition, age, moisture condition at loading, etc. [15, 16].

As the fibres in concrete have proved to predominantly improve the post cracking performance, in recent years, many fatigue studies have investigated the response of both concrete in the uncracked and pre-cracked state. Most of the studies have been carried out on uncracked specimens (unnotched and notched) under flexure loading with both four-point loading and three-point loading configurations [17–19]. Researchers have developed fatigue models to predict the fatigue response of SFRC involving different types of steel fibres like hooked-end, straight, corrugated, etc. [11, 20, 21]. The models generally represent the fatigue life of SFRC expressed in terms of *S-N* curves (representing the relation between applied stress level, *S*, and the number of cycles to failure, *N*) [17, 22, 23]. These studies on uncracked specimens indicate the effectiveness of incorporation of fibres just after the initiation of the first crack, which helps to resist the fatigue stress, thus preventing, or delaying the crack growth by fibre bridging and pull-out behaviour [24, 25]. It is also observed that the fatigue life of SFRC increases with an increase in fibre dosage and higher aspect ratio [21, 26–28]. Steel fibres significantly enhance the fatigue performance of concrete compared to other fibre materials like polymer or glass. This is because of the higher stiffness of steel fibres [25]. Concrete reinforced with hooked- end fibres possess higher fatigue endurance than straight or corrugated steel fibres due to its anchorage [2, 25, 29].

From the review of literature, it is also obvious that the flexural fatigue performance of SFRC is more predominant at higher stress levels [15, 20, 30]. It has been observed that short fibres present in the matrix can prevent the micro-crack formation more efficiently for a particular volume fraction compared to long fibres but during the crack growth period, long fibres arrest the propagation of macro-cracks effectively [24]. A detailed review on the fatigue behaviour of both plain cement concrete (PCC) and SFRC has been done by Lee and Barr (2004) with the incorporation of experimental fatigue data from various literature for formulating a statistical model to predict the fatigue life. Coefficient of determination (R^2) was used to check the statistical significance of the model. The developed generalized *S-N* relation for uncracked SFRC specimens incorporates the effect of two fibre volume fractions of 0.5 and 1.0% in the prediction of fatigue life. Higher volume fractions like 1.5%, 2.0% and various fibre parameters like length, diameter, aspect ratio, etc., and the effect on pre-cracked concrete specimens were not included in the analysis. Banjara and Ramanjaneyulu (2018) carried out experimental investigations on both PCC and SFRC for the assessment of the fatigue behaviour with different fibre dosages. They have developed a model (*S-N* expression) based on their experimental study for the prediction of fatigue life for uncracked concrete by incorporating the effect of fibre content for a volume fraction ranging from 0.5–2.0%.

As discussed earlier, in addition to the fatigue studies available for uncracked concrete, there are some studies on the fatigue response of pre-cracked FRC as well. Granju et al. (2000) conducted three-point bending fatigue tests on pre-cracked concrete specimens using two different steel fibres (hooked-end and double-headed) having different fibre lengths with a crack opening of 300 μm (by monotonic loading) to determine the effect of fatigue and creep of FRC. The residual load carrying capacity ($F_{R,300}$) was taken as the reference load for the cyclic test. They found that 60 mm hooked-end fibre have maximum post-cracking residual load carrying capacity as compared to other lengths. They performed tests with different stress levels ranging from 80 to 50% and 10% of the post-cracking load-carrying capacity for maximum and minimum loading in each cycle. They found that the stress level at 80% leads to failure immediately and below 50% the specimen could withstand one million cycles of load repetitions (predefined endurance limit in this case). Germano and Plizzari (2012) performed an experimental program on notched beams under three-point bending to assess the post-peak cyclic behaviour of SFRC (hooked-end) with 0.5% and 1.0% volume fractions but with short steel fibres of length 35 mm. They performed cyclic tests by adopting a pre-cracking of 0.1 mm on the beam specimen. The tests were carried out under three different load cycles of a constant amplitude of 50%, varied between 25% and 75% in the 1st case, 35% and 85% in the 2nd case, 15% and 65% in the 3rd case. The load levels were taken as percentage of P_{max} obtained during monotonic testing. The study demonstrated that crack opening range and crack opening rate have a great influence on the fatigue life in the post-peak region (SFRC exhibit high crack opening range and crack opening rate and better performance at high stress levels as compared to PCC). They found a linear relation between fatigue life and crack opening rate. They also observed that the effect of addition of 1.0% volume fraction of fibres was less effective on high cycle fatigue as compared to 0.5% and attributed this counter-intuitive behaviour mainly to the additional flaws created by an increased fibre content.

Carlesso et al. (2019) performed an experimental study (three-point bending fatigue tests on cracked high-performance fibre reinforced concrete, HPFRC) using steel micro-fibres having a length of 13 mm and an aspect ratio of 82. A higher volume fraction of 2.0 vol% (150 kg/m^3) was chosen and the pre-cracking was done up to 0.5 mm under monotonic loading. The flexural fatigue testing was done under seven stress levels of 65, 70, 75, 80, 85, 90 and 100% of applied load. They continued the test till either the *CMOD* reaches 4 mm, or the specimen can withstand two million of load cycles, whichever happens earlier. It was observed that below 66% of load-carrying capacity the specimen exhibits fatigue endurance, i.e., below this stress level the specimen will not fail. Stephen and Gettu (2020) followed fatigue testing procedure of Granju et al. (2000) with pre-cracking till a *CMOD* of 0.5 mm. They incorporated hooked- end steel fibres of 60 mm length and different dosages to obtain the post-peak response of cracked specimens. The residual load carrying capacity ($F_{R,1}$) was taken as the reference and the maximum load level in each cycle varied from 50 to 90% $F_{R,1}$ and the lower loading level was kept constant: 10% of $F_{R,1}$. They found that when the load-*CMOD* behaviour approaches the envelope of monotonic curve then the failure will happen due to the sudden increase of *CMOD* (critical crack mouth opening). They also identified three stages of failure developments, in the first stage the enhancement of damage is steady; the second stage exhibits progressive damage, shown by a linear *CMOD*-log *N* response; and finally, the uninhibited crack propagation and lowering of load-*CMOD* stiffness. The study also noted that the benefits of incorporation of higher dosage of fibres was obtained more predominantly at a higher stress level, and there is no marked significance at lower stress levels.

The *S-N* relations for various fibre dosages as reported by several researchers from flexural fatigue testing on pre-cracked specimens are presented in Table 1. The test and fibre parameters influencing the prediction of the fatigue life are also shown in Table 1.

Table 1. Fatigue models

V_f (vol %)	Authors	Loading configur-ation	Test Parameters		Fibre Parameters			Fatigue Model Equation
			SR	f (Hz)	L (mm)	D (mm)	Aspect Ratio (AR)	
0.13	Stephen and Gettu, 2020	Three point	0.90 - 0.50	5	60	0.75	80	$\log N = 30.7 - 14.5 \log(\%F_{R,1})$
0.5	Germano and Plizzari, 2012	Three point	0.85 - 0.65	3	35	0.54	65	$\log N = 12.55 - 11.40\,SR$
0.8	Granju et al., 2000	Three point	0.80 - 0.50	5	60	0.75	80	$\log N = 20.1 - 8.1 \log(\%F_{R,1})$
1.0	Germano and Plizzari, 2012	Three point	0.85 - 0.65	3	35	0.54	65	$\log N = 13.27 - 13.04\,SR$
2.0	Carlesso et al., 2019	Three point	1.0 - 0.65	6	13	0.16	80	$\log N = 19.44 - 19.84\,SR$

Note: V_f- volume of fibres, *SR* - stress ratio, *f* - frequency, *L* - length of fibre, *D* - diameter of fibre, *AR* - aspect ratio

Figure 1 shows all models plotted with the same stress level for comparing the fatigue performance of SFRC with different fibre dosages. It may be noticed from Fig. 1 that models with the same fibre dosage show some variation in fatigue behaviour of SFRC.

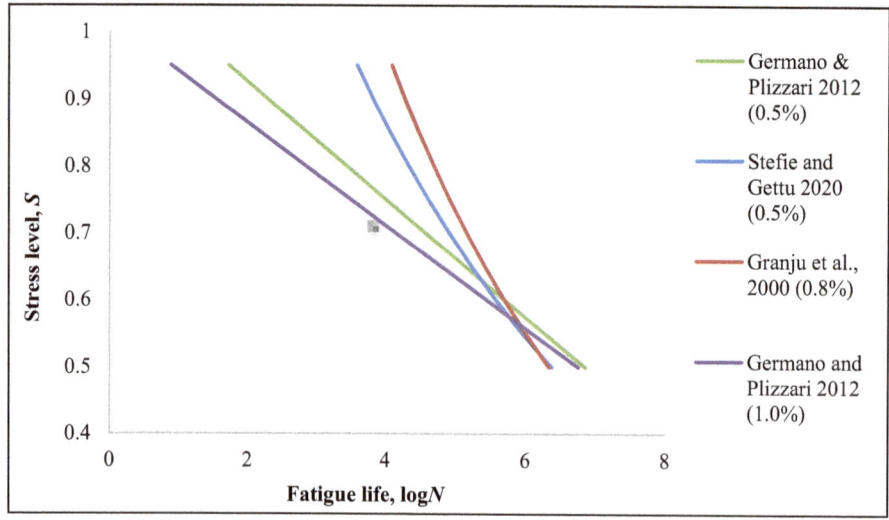

Fig. 1. S-N curves for SFRC (0.5, 0.8, 1.0 & 2.0 vol%)

Based on the review of literature, the following inferences were identified:

- It is observed that the different fatigue test parameters like applied stress level, frequency of loading, and fibre parameters like fibre type, length, diameter, volume of fibres, aspect ratio, reinforcing index, etc. influence the crack resistance performance and thereby, the fatigue endurance limit [15, 21, 27, 28, 31]
- The test configurations used for fatigue tests are different in terms of loading rate, specimen size, frequency, assumed fatigue endurance limit, etc. [16, 23, 32, 33]. Therefore, a direct comparison of one fatigue model with other models is not possible. In this regard, a generalized fatigue model with the incorporation of both testing parameters and fibre parameters is needed.
- The lack of generalized models for predicting the fatigue life of pre-cracked concrete was observed.
- For formulating a generalized fatigue model for predicting the fatigue response of SFRC, both material level and testing level parameters should be taken into consideration.

2 Objective and Scope

This work attempts to propose a generalized fatigue model based on statistical analysis encompassing the effect of various test parameters and fibre parameters for pre-cracked SFRC. The scope of the current investigation is listed below:

- Analyse and identify the various testing and fibre parameters which influence the fatigue life of pre-cracked SFRC.
- Develop a generalized statistical model to predict the fatigue life with the most significant variables.
- Recommend the proposed generalized model for a wide range of fibre dosages.

3 Methodology

3.1 Data Composition

For evaluating the effect of several test and fibre parameters on the prediction of fatigue performance of SFRC, fatigue test data from literature studies [12, 23, 25, 34] on pre-cracked specimens under flexural loading with central point loading configurations were included in this study. Various dosages of steel fibres ranging from 0.13–2.0% were included in the current study. The details of the independent variables used in this study and their range details are presented in Table 2.

Table 2. Range of independent variables

Independent variable	Range
Stress ratio	1.0-0.50
Frequency of loading (Hz)	2-5
Length of fibre (mm)	13-60
Diameter of fibre (mm)	0.16-0.75
Volume fraction of fibre (vol%)	0.13-2.0

3.2   Statistical Analysis

A statistical assessment was carried out based on the test data obtained from the experimental programs of pre-cracked SFRC specimens presented in literature. The relationship between the dependent variable log N and independent variables like stress ratio (SR), frequency (f in Hz), length (L in mm), diameter (D in mm), aspect ratio ($AR = \frac{L}{D}$), dosage of fibres (V in $\frac{kg}{m^3}$), and reinforcing index ($RI = volume\ fraction\ (vol\ \%) \times AR$), crack mouth opening displacement ($CMOD$), etc., were found by generating a correlation matrix using SPSS software. The correlation matrix of both dependent and independent variables related to pre-cracked SFRC is presented in Table 3. The interrelation between each independent variable was also assessed by the correlation matrix.

The fatigue model was developed considering the above test and fibre parameters using multiple linear regression analysis with IBM SPSS software which was used to perform the analysis of variance (ANOVA). Different combinations of fatigue models were engendered and investigated, among those models, the model with the highest R^2 value was selected. The statistical significance of every model was evaluated by t-test, F-test, and p-values.

Table 3. Correlation matrix of pre-cracked SFRC

	log N	SR	f	$SR \times f$	L	D	AR	RI	V	$CMOD$
log N	1									
SR	-0.87	1								
f	0.19	-0.09	1							
$SR \times f$	-0.36	0.61	0.78	1						
L	0.02	-0.11	0.09	0.11	1					
D	-0.02	-0.02	-0.12	-0.05	0.98	1				
AR	0.18	-0.11	0.95	0.75	0.41	0.21	1			
RI	0.13	-0.11	0.42	0.32	-0.78	-0.87	0.01	1		
V	0.10	-0.08	0.31	0.24	-0.83	-0.89	0.02	0.99	1	
$CMOD$	-0.01	0.07	0.35	0.22	0.69	0.62	0.55	-0.57	-0.62	1

4   Results and Discussion

SR, *L*, and *RI* are the most influential parameters attained from the analysis for the prediction of fatigue life of pre-cracked SFRC. The model which showed high goodness-of-fit with the R^2 being 0.805 for a degree of freedom of 3 was finalized. The model summary and ANOVA results are shown in Fig. 2 and Table 4.

Thus, the generalized model for estimating the fatigue life of pre-cracked SFRC is found as:

$$\log N = 9.543 - 9.932 \times SR + 0.03 \times L + 0.014 \times RI \qquad (1)$$

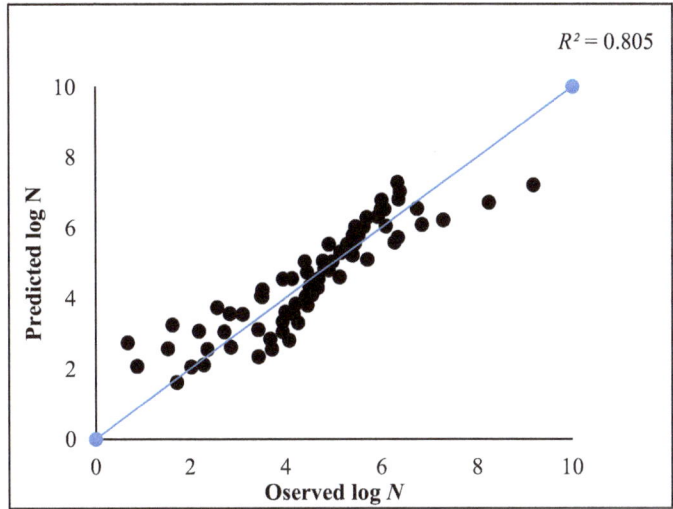

Fig. 2. Comparison of Observed log N (experimental data) vs Predicted log N (proposed fatigue model) for pre-cracked SFRC

Table 4. ANOVA outcomes and calculations

	Sum of Squares	*df*	Mean Square	*F*	Significance
Regression	152.959	3	50.986	90.583	0.000
Residual	37.150	66	0.563		
Total	190.109	69			

The *t*-test parameters and the associated *p*-values work under the assumption that samples are approximately normally distributed. The *t*-test parameters for the Constant,

SR, L, and *RI* were found as 13.45, −15.91, 3.63, and 4.31 which shows the ability of each independent variable to predict the fatigue life of pre-cracked SFRC. The estimated *t*-values are considerably higher than the critical values. Here the *p*-value associated with the *t*-test was found below 0.05 which shows the statistical significance of the model (i.e., can reject the null hypothesis).

The *F*-value designates the reliability of each independent variable (*SR, L*, and *RI*) incorporated in the multiple linear regression analysis for estimating the fatigue life using the dependent variable (log *N*). The higher *F*-value (91) specifies that the group of independent variables incorporated in the analysis has a substantial effect on the prediction of fatigue life. This test provides a comprehensive understanding of the entire variables considered in the multiple linear regression analysis.

Thus, it could be considered that the proposed generalized fatigue model as given in Eq. (1) can be used for the prediction of the post-cracking fatigue life of SFRC for volume fractions within the range of 0.5–2.0%. Different shapes of steel fibres with hooked-ends, straight, and corrugated are considered in this study, but no variable was allocated to interpret the shape factor in the model.

5   Research Significance

The main contribution of the current research investigation is the development of a generalized fatigue model for the prediction of fatigue life of pre-cracked SFRC, which could be used for a wide range of fibre dosages with the influence of various test and fibre parameters. This statistical model could be utilized to predict the fatigue life where there is a difficulty in conducting the actual fatigue test and can be included in any SFRC design methodology where fatigue is a principal mode of failure.

6   Conclusions

The objective of the current study was to propose a generalized fatigue model, based on statistical analysis, for pre-cracked SFRC with the consideration of various test parameters (applied stress level and frequency of loading cycle) and fibre parameters (diameter, volume of fibres) on the fatigue life of pre-cracked SFRC. The following conclusions were derived from the study:

- Different combinations of test parameters like applied stress level, rate of loading, test configuration, and fibre parameters like length, diameter, aspect ratio, reinforcing index, etc. can affect the fatigue performance of pre-cracked SFRC under repetitive loads.
- The test and fibre parameters, like stress ratio, The most significant parameters, pre-cracked SFRC are stress ratio, length of fibre and reinforcing index show a significant influence on the fatigue life of pre-cracked SFRC.
- A general expression for estimating fatigue life of pre-cracked SFRC was generated using experimental fatigue data available in the literature and the developed model show a very good to excellent fit with $R^2 = 0.805$.

- The proposed fatigue model can be used for the prediction of fatigue life of pre-cracked SFRC for volume fractions within the range of 0.13–2.0 vol%.
- It is contemplated that the generalized model could be integrated into any SFRC design methodology requiring fatigue response prediction, leading to better and optimized designs.

References

1. Zhang, J., Stang, H., Li, V.C.: Fatigue life prediction of fiber reinforced concrete under flexural load. Int. J. Fatigue **21**(10), 1033–1049 (1999)
2. Cachim, P.B., Figueiras, J.A., Pereira, P.A.: Numerical modelling of fibre-reinforced concrete fatigue in bending. Int. J. Fatigue **24**(2–4), 381–387 (2002)
3. Kim, J.K., Kim, Y.Y.: Experimental study of the fatigue behavior of high strength concrete. Cem. Concr. Res. **26**(10), 1513–1523 (1996)
4. Banjara, N.K., Ramanjaneyulu, K., Sasmal, S., Srinivas, V.: Flexural fatigue performance of plain and fibre reinforced concrete. Trans. Indian Inst. Met. **69**(2), 373–377 (2016)
5. Raithby, K.D.: Flexural fatigue behaviour of plain concrete. Fatigue Eng. Mater. Struct. **2**, 269–278 (1979)
6. Nayar, S.K., Gettu, R.: Mechanistic-empirical design of Fibre Reinforced Concrete (FRC) pavements using inelastic analysis. Sādhanā **45**(1), 1–7 (2020). https://doi.org/10.1007/s12046-019-1255-1
7. Li, V.C., Matsumoto, T.: Fatigue crack growth analysis of fiber reinforced concrete with effect of interfacial bond degradation. Cem. Concr. Compos. **20**(5), 339–351 (1998)
8. Cachim, P.B.: Experimental and numerical analysis of the behaviour of structural concrete under fatigue loading with applications to concrete pavements. Ph.D. thesis, Faculty of Engineering of the University of Porto (1999)
9. Singh, S.P.: Fatigue strength of hybrid steel-polypropylene fibrous concrete beams in flexure. Procedia Eng. **14**, 2446–2452 (2011)
10. Chen, M., Zhong, H., Zhang, M.: Flexural fatigue behaviour of recycled tyre polymer fibre reinforced concrete. Cem. Concr. Compos. **105**, 103441 (2020)
11. Singh, S.P., Goel, S., Lal, R., Kaushik, S.K.: Prediction of mean and design fatigue lives of steel fibrous concrete using SN relationships. Asian J. Civ. Eng. (Build. Hous.) **5**, 175–190 (2004)
12. Germano, F., Plizzari, G.A.: Fatigue behaviour of SFRC under bending. In: Barros, J. et al. (eds.) Proceedings of Eighth RILEM International Conference on Fibre Reinforced Concrete, 2012_02_503, Portugal, 12 p. (2013)
13. Parvez, A., Foster, S.J.: Fatigue behavior of steel-fiber-reinforced concrete beams. J. Struct. Eng. **141**(4), 04014117 (2015)
14. Zhang, J., Stang, H., Li, V.C.: Experimental study on crack bridging in FRC under uniaxial fatigue tension. J. Mater. Civ. Eng. **12**(1), 66–73 (2000)
15. Lee, M.K., Barr, B.I.: An overview of the fatigue behaviour of plain and fibre reinforced concrete. Cem. Concr. Compos. **26**(4), 299–305 (2004)
16. Nayar, S.K.: Design of fibre reinforced concrete slabs-on grade and pavements. Doctoral Thesis, Department of Civil Engineering, Indian Institute of Technology Madras, Chennai, India (2016)
17. Nanni, A.: Fatigue behaviour of steel fiber reinforced concrete. Cem. Concr. Compos. **13**(4), 239–245 (1991)

18. Mohammadi, Y., Kaushik, S.K.: Flexural fatigue-life distributions of plain and fibrous concrete at various stress levels. J. Mater. Civ. Eng. **17**(6), 650–658 (2005)
19. Singh, S.P., Kaushik, S.K.: Fatigue strength of steel fibre reinforced concrete in flexure. Cem. Concr. Compos. **25**(7), 779–786 (2003)
20. Ramakrishnan, V., Wu, G.Y., Hosalli, G.: Flexural fatigue strength, endurance limit and impact strength of fiber reinforced concretes. Transp. Res. Rec. **1226**, 17–24 (1989)
21. Johnston, C.D., Zemp, R.W.: Flexural fatigue performance of steel fiber reinforced concrete–influence of fiber content, aspect ratio, and type. ACI Mater. J. **88**(4), 374–383 (1991)
22. Blasón, S., Poveda, E., Ruiz, G., Cifuentes, H., Canteli, A.F.: Twofold normalization of the cyclic creep curve of plain and steel-fiber reinforced concrete and its application to predict fatigue failure. Int. J. Fatigue **120**, 215–227 (2019)
23. Carlesso, D.M., De la Fuente, A., Cavalaro, S.H.: Fatigue of cracked high performance fiber reinforced concrete subjected to bending. Constr. Build. Mater. **220**, 444–455 (2019)
24. Singh, S.P., Mohammadi, Y., Madan, S.K.: Flexural fatigue strength of steel fibrous concrete containing mixed steel fibres. J. Zhejiang Univ. Sci. A **7**(8), 1329–1335 (2006). https://doi.org/10.1631/jzus.2006.A1329
25. Stephen, S.J., Gettu, R.: Fatigue fracture of fibre reinforced concrete in flexure. Mater. Struct. **53**(3), 1–11 (2020). https://doi.org/10.1617/s11527-020-01488-7
26. Chang, D.I., Chai, W.K.: Flexural fracture and fatigue behaviour of steel-fibre-reinforced concrete structures. Nucl. Eng. Des. **156**(1), 201–207 (1996)
27. Goel, S., Singh, S.P., Singh, P.: Flexural fatigue strength and failure probability of self compacting fibre reinforced concrete beams. Eng. Struct. **40**, 131–140 (2012)
28. Tiberti, G., Germano, F., Mudadu, A., Plizzari, G.A.: An overview of the flexural post-cracking behavior of steel fiber reinforced concrete. Struct. Concr. **19**(3), 695–718 (2018)
29. Banjara, N.K., Ramanjaneyulu, K.: Experimental investigations and numerical simulations on the flexural fatigue behavior of plain and fiber-reinforced concrete. J. Mater. Civ. Eng. **30**(8), 1–15 (2018)
30. Nayar, S.K., Gettu, R.: A comprehensive methodology for the design of fibre reinforced concrete pavements. Fibre Reinforced Concrete: From Design to Structural Applications, vol. 79, pp. 321–330. Fib Bulletin (2016)
31. Wei, S., Jianming, G., Yun, Y.: Study of the fatigue performance and damage mechanism of steel fiber reinforced concrete. Mater. J. **93**(3), 206–212 (1996)
32. Singh, S.P., Kaushik, S.K.: Flexural fatigue analysis of steel fiber-reinforced concrete. Mater. J. **98**(4), 306–312 (2001)
33. Germano, F., Tiberti, G., Plizzari, G.: Post-peak fatigue performance of steel fiber reinforced concrete under flexure. Mater. Struct. **49**(10), 4229–4245 (2015). https://doi.org/10.1617/s11527-015-0783-3
34. Granju, J.L., Rossi, P., Rivillon, P., Chanvillard, G., Mesureur, B.: Delayed behaviour of cracked SFRC beams. In: Fifth RILEM Symposium on Fibre-Reinforced Concrete (FRC), Lyon, pp. 511–520, September, 2000

Numerical Modeling of the Steel Fiber Reinforced Concrete Behavior Under Combined Tensile and Shear Loading by a Micromechanical Model Taking into Account Fiber Orientation

Duc-Tam Vu[1(✉)], François Toutlemonde[1], Benjamin Terrade[1], Pierre Marchand[2], and Sébastien Bouteille[3]

[1] Materials and Structures Department, Gustave Eiffel University, Champs-sur-Marne, France
duc-tam.vu@univ-eiffel.fr
[2] CEREMA/DTecITM/DTOA/GITEX, Fontenay-sous-Bois, France
[3] Centre D'Étude des Tunnels (CETU), Bron, France

Abstract. The mechanical behavior of steel fiber reinforced concrete (SFRC) is strongly dependent on the cracks bridging brought by fibers. Thus, the fibers orientation in SFRC is one key factor. This research extends a micromechanics-based model to describe the shear transfer in addition to tensile mechanisms at the crack surface and to establish a base at micro-scale for the further homogenization at the macro-scale. The shear effect is described as a function of the fiber pullout process, the stress across the cracks is then derived from the integration of the product of the fiber pullout function by the probability of the fiber location and orientation. The simulation results provide the SFRC behavior under mixed-mode displacement of crack (slip and opening). The influence of material parameters is investigated.

Keywords: Steel fiber reinforced concrete · Fiber orientation · Shear behavior · Micromechanics

1 Introduction

Research and application of steel fiber reinforced concrete (SFRC) for over 30 years have shown the beneficial effect of fibers which improve the tensile and shear strength and the crack growth control. By bridging the cracks, the fibers enhance the aggregate interlock and participate in the transfer of the shear stress [1–3]. Thus, when a sufficient volume of fibers is added, it may transform the brittle failure mode of the member under shear force into a ductile failure mode. This property of SFRC allows increasing the structural strength under multiaxial loading.

Many studies have investigated the SFRC behavior under shear loading and a number of models have been proposed. However, these models, either propose only the method to quantify the ultimate shear strength [4] or, empirically modify the models for

© RILEM 2022
P. Serna et al. (Eds.): BEFIB 2021, RILEM Bookseries 36, pp. 443–455, 2022.
https://doi.org/10.1007/978-3-030-83719-8_39

plain concrete according to experimental data [5, 6]. The existing models do not consider the interface properties, fiber geometry and one of the most important factors – the fibers orientation.

Namely, a fiber inclined with respect to the cracking plane normal direction requires a higher release energy and hence induces a higher pullout load but also a longer plateau during the pullout process [7, 8]. Observation of oriented fiber reinforced concrete under direct shear tests [9, 10] has shown a significant difference in the fiber pullout process depending on the fiber inclination angle.

One of the effective approaches to analyze the shear behavior of SFRC considering the effect of fibers orientation is the micromechanical theory. Namely, Foster and Htut [11] have developed the Variable Engagement Model (VEM) to investigate the behavior of SFRC under mixed modes crack displacement by averaging the pullout force of all fibers across the crack, the fibers orientation effect being introduced by an empirically-determined coefficient. The SFRC shear strength so determined could not capture the inclination effect of the fibers which is unsatisfactory for small crack displacement. Conversely, Li et al. in [12, 13] have used a micromechanical approach for cementitious composites reinforced by synthetic fibers under pure tension. The strain-hardening behavior due to the fiber/matrix interaction and the snubbing effect due to the inclined fibers are considered. The effect of fiber orientation is introduced by a probability distribution function. Such a model has been exploited in [14] for UHPFRC under pure tension. The isotropic probability distribution function has been replaced by a Gaussian-like bivariate probability function, allowing to better evaluate the effect of fiber orientation and derive the orientation coefficient (α) defined in SFRC design codes such as [15–17].

In this research, the approach described in [14] has been extended to evaluate the SFRC behavior under mixed-mode: slip and opening crack displacement. Using the same probability distribution function, the effect of fibers orientation under shear loading on the pullout process has been introduced considering the bending effect by implementing the method presented in [18]. The sensitivity of the model to the properties of SFRC mix and constituents is investigated.

2   Micromechanical Model of a Single Fiber Under Combined Opening and Sliding

The micromechanical model proposed in [13] based on the average stress taken by crack bridging fibers under pure tension can be expressed as follows

$$\sigma_c = \frac{v_f}{A_f} \iiint F(\delta, L).p(L, \theta, \varphi) \sin \varphi \, \mathrm{d}L\mathrm{d}\varphi\mathrm{d}\theta \tag{1}$$

Where σ_c is the stress transferred along the crack by fibers; $F(\delta, L)$ is a function of crack displacement δ and fiber embedded length L that is responsible for the bridging force of a single fiber; $p(L, \theta, \varphi)$ is the probability for a single fiber to have an embedded length L, inclined with respect to the normal of crack with an angle φ and to the shear direction with an angle θ; v_f and A_f are the fiber content and fiber cross-sectional area, respectively.

Considering an inclined fiber in the pullout process with the corresponding crack opening w and crack sliding s as shown in Fig. 1, the bridging force can be derived as combining an axial component (N) which can be considered as opposing the pullout force of an aligned fiber and, a tangent component (P) which can be considered as opposing to the sliding displacement. Hence, the normal component i.e., the pullout force of an aligned fiber is investigated first, then the tangent component derived and finally the distribution of fibers orientation is considered. Hereafter, in the first approach focusing on the fiber influence, the shear stress is transferred through the crack only by the fibers and the contribution of the aggregate interlock is not considered, this will be further considered as an additional isotropic term.

2.1   Pullout Mechanism of a Single Fiber

The pullout tests on a single either straight or hooked-end fiber [9–21] indicate that, during the pullout process, fiber undergoes three main stages: debonding stage, straightened stage, and frictional slipping stage.

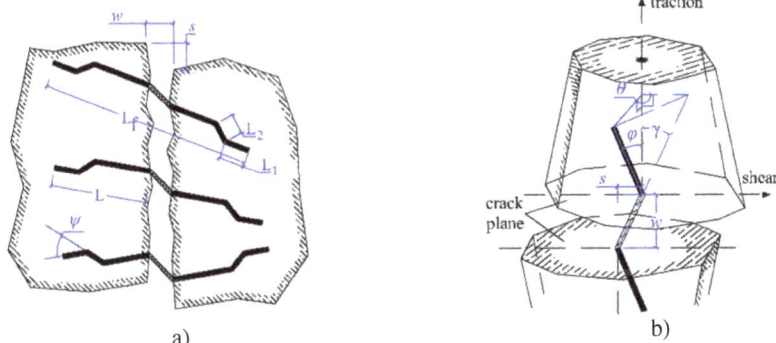

Fig. 1. Fiber bridging a crack: a) SFRC crack under opening and sliding displacement, b) A single fiber across the crack in 3D

The relationship between the debonding force (N_d), the straightening force (N_r) and the frictional pullout force (N_e) with respect to the fiber pullout displacement is derived as below:

$$N_d(\delta) = \pi\sqrt{\frac{1}{2}E_f\tau_0 d_f^3(1+\eta)}.\sqrt{\delta} \qquad\qquad \forall 0 \le \delta \le \delta_0 \qquad (2)$$

$$N_r(\delta) = \frac{Q[L_1 - R(L_1+L_2)]}{L_1L_2(L_1+L_2)}(\delta-\delta_0)^2 + \frac{Q\left[(L_1+L_2)^2R - L_1^2\right]}{L_1L_2(L_1+L_2)}(\delta-\delta_0) + N_d(\delta_0) \quad \forall \delta_0 < \delta \le \delta_e$$

$$(3)$$

$$N_e(\delta) = \pi d_f \tau_e (1 + \eta)(L - \delta + \delta_e) \qquad\qquad \forall \delta_e < \delta \leq L \quad (4)$$

$$\text{with}: \quad \delta_0 = \frac{2L^2(1+\eta)\tau_0}{E_f d_f} \; ; \quad \delta_e = \delta_0 + \sum L_i \; ; \quad \tau_e = \tau_0 \left(1 + \frac{Q}{N_d(\delta_0)}\right) \quad (5)$$

$$Q = \frac{\sigma_f \pi d_f^2}{6 \sin\psi.\left(1 - \mu \sin\frac{\psi}{2}\right)} \; ; \quad R = \frac{2}{1 - \mu \sin\frac{\psi}{2}} \; ; \quad \eta = \frac{E_f}{E_m}.\frac{v_f}{1 - v_f} \quad (6)$$

where d_f: fiber diameter; E_f: fiber elastic modulus; E_m: matrix elastic modulus; τ_0: bond strength of fiber-matrix interface; σ_f: elastic limit of steel; μ: coefficient of friction between fiber and matrix; L_f: fiber length; L: embedded length; $L_{1,2}$: length of hook and ψ: hook angle. It should be noted that $L_{1,2} = 0$ or $\psi = 0$ corresponds to the case of straight fibers.

During the pullout process, physically, the fiber can be extracted from both sides (two-sides pullout), however, for randomly distributed fibers, the embedded lengths of fibers on both sides of the crack are generally different. If the fiber bridging undergoes a slip hardening, after the shorter embedded side has reached its peak-stress, the stress in the opposite side will decrease and thus, the longer part might not exhibit the following stage and only the shorter side is pulled out (one-side pullout). Moreover, in general the pullout displacement that corresponds to the peak-load is relatively small with respect to the size of admissible crack openings in SFRC [12]. Thus, herein, only the one-side pullout case has been investigated, the solution for two-side pullout can be found with another approach in [13].

2.2   Consideration of Shear Sliding

When the crack surfaces begin to slide, the fibers that intercept the crack will be pulled out further and the extracted part of fiber will undergo a rotation with an angle γ as shown in Fig. 2a, this angle γ and the corresponding pullout length can be defined by the following relation:

$$\gamma = \arctan\left(\frac{s}{w}\right); \qquad \delta = \sqrt{w^2 + s^2} \qquad (7)$$

The fiber section at the exit point bends with an angle β, this angle is the intersection angle between the embedded and extracted part. Let φ and θ be the initial elevation and initial azimuth angle of a 3D randomly oriented fiber with respect to the normal axis to the crack plane. By the geometric relationship, the intersection angle β can be defined according to the following formula:

$$\beta = \arccos(\cos\varphi.\cos\gamma + \sin\varphi.\sin\gamma.\cos\theta) \qquad (8)$$

In case of 2D oriented fiber where the azimuthal angle $\theta = 0$, the relation (8) is simplified into:

$$\beta = |\gamma - \varphi| \tag{9}$$

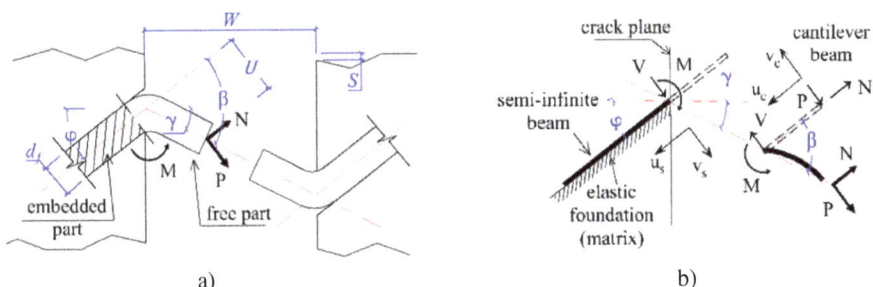

Fig. 2. One side of fiber under shear displacement: a) Notation for the model, b) The equivalent beam model

The shorter side of the fiber can be split into two elastic bodies: i) The extracted part considered as a cantilever beam subjected to an axial load N and a shear load P (referring to the exit section); ii) The embedded part considered as a continuously supported beam in an elastic foundation subjected to a moment M and a shear load V at the exit end (which are transferred from the cantilever part) as presented in Fig. 2b. By using Timoshenko's beam theory [22] for small deflection and perfectly elastic behavior of steel fiber, the deflection functions of the embedded part $v_s(u_s)$ and extracted part $v_c(u_c)$ are expressed as follows:

$$v_s = \frac{2\zeta}{k_m} e^{-\zeta u_s} [(V + \zeta M)\cos(\zeta u_s) - M\zeta \sin(\zeta u_s)]; \qquad v_c = 2C \sinh mu_c + \frac{P}{N} u_c \tag{10}$$

With $\zeta = \sqrt[4]{k_m/(4E_f I_f)}$; $m = \sqrt{N/(E_f I_f)}$ where k_m is the matrix foundation stiffness, E_f the modulus of elasticity and I_f the moment inertia of the fiber, C is a constant due to the differential equation and can be identified by continuity consideration.

Let l be the length and U the deflection at the free end of the cantilever part, the symmetry of fiber bending yields:

$$l = \frac{1}{2}(\delta + d_f); \qquad U = \frac{1}{2}l\sin\beta \tag{11}$$

By using the boundary condition:

$$v_c(u_c = l) + v_s(u_s = 0) = U; \quad v_c'(u_c = l) = v_s'(u_s = 0); \quad \frac{V}{EI} = -v_c'''(u_c = l) \quad (12)$$

the relationship between the bending force and the axial force is derived as follow:

$$P = \frac{U}{K+l}.N \quad \text{with:} \quad K = \frac{2\zeta.N(m\cosh ml + \zeta\sinh ml) - k_m\sinh ml}{2\zeta^2 N(m\cosh ml + 2\zeta\sinh ml) + k_m m\cosh ml} \quad (13)$$

The bridging force F of a single fiber can then be expressed as:

$$F(\delta, L) = N\cos\beta + P\sin\beta = N(\delta, L)\left(\cos\beta + \frac{l}{K+l}\sin^2\beta\right) \quad (14)$$

with δ, β, l, K defined in Eqs. (7), (8), (11) and (13). The normal component $N(\delta, L)$ takes the value of $N_d(\delta, L), N_r(\delta, L)$ or $N_e(\delta, L)$ for each relevant stage of the pullout process.

As mentioned in Eq. (10), the foundation stiffness k_m directly relates to the matrix rigidity. Leung and Li [23] have presented a method using finite element analysis to find its value for circular cross-section fibers, the authors proposed the ratio k_m/E_m to range from 0.2 to 0.25 for a cylindrical fiber in a regular cementitious matrix. Hereafter, the ratio $k_m/E_m = 0.25$ has been used.

2.3   Distribution of Fiber Orientations

The anisotropic orientation of fiber was firstly investigated in [13] where the contribution of all fibers bridging the crack is averaged at any possible location along its length and orientation. The averaged contribution is materialized by a product of 2 probability functions: i) $p(L)$ which accounts for the dispersion of the fiber embedded length L and ii) fiber orientation distribution $p(\varphi, \theta)$ defined by 2 variables: azimuth angle θ and elevation angle φ. A uniform distribution of fiber location in the representative element volume is described by $p(L) = 2/L_f$. For the fiber orientation distribution $p(\varphi, \theta)$, instead of a uniform law, Guenet [14], proposed an estimate based on the feedback of tomographic measurements using a π-periodic Gaussian-like distribution function expressed as:

$$p_\varphi(\varphi, \theta) = \frac{k_g}{2\pi\sinh k_g}\cosh[k_g(\cos\varphi\cos\varphi_0 - \sin\varphi\sin\theta\sin\varphi_0)] \quad (15)$$

where k_g is a parameter describing the width of the distribution: the larger k_g-value is, the more fibers are oriented toward the direction φ_0. Such an orientation distribution is presented in Fig. 3 where this probability density function is related to the orientation factor currently used in SFRC design [17].

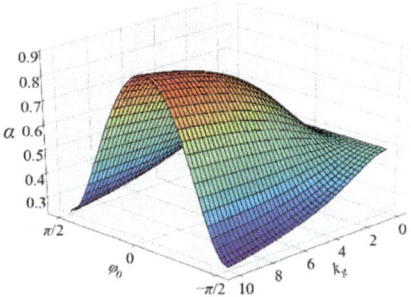

Fig. 3. The relationship between the fiber orientation factor and the probability density function

3   Shear/Normal Bridging Stress Under Opening/Sliding Relations

From Eq. (5), the crack displacements where all the fibers are fully debonded (δ^*) and when all the fiber are straightened (δ^{**}) may be expressed as follows:

$$\delta^* = \frac{L_f^2(1+\eta)\tau_0}{2E_f d_f}; \qquad \delta^{**} = \delta^* + L_1 + L_2 \tag{16}$$

During the pullout process, the fibers with different embedded lengths undergo different stages. To account for all fibers that intercept the crack, the limits for fiber embedded length are set as below:

$L_d = \sqrt{\frac{\delta d_f E_f}{2(1+\eta)\tau_0}}$: A fiber with embedded length overpassing this limit is under debonding stage.

$L_r = \sqrt{\frac{(\delta-L_1-L_2)d_f E_f}{2(1+\eta)\tau_e}}$: A fiber with embedded length under this limit is under frictional pullout stage.

From Eqs. (1), (14) and the condition derived above, the bridging stress (σ_c) is expressed as follows:

- For $\delta \in [0, \delta^*]$, several fibers are under debonding stage, the others are under straightening stage

$$
\begin{aligned}
\sigma_{c1} = \frac{4\pi v_f}{d_f^2} \Bigg[& \int_{\theta=0}^{2\pi} \int_{\varphi=0}^{\frac{\pi}{2}} \int_{L=L_d}^{L_f/2} F(N_d(\delta,L)).p(L).p(\theta,\varphi)\sin\varphi L d L d\varphi d\theta \\
& + \int_{\theta=0}^{2\pi} \int_{\varphi=0}^{\frac{\pi}{2}} \int_{L=\delta}^{L_d} F(N_r(\delta,L)).p(L).p(\theta,\varphi)\sin\varphi L d L d\varphi d\theta \Bigg]
\end{aligned}
\tag{17}
$$

- For $\delta \in (\delta^*, L_1 + L_2]$: all fibers are under straightening stage

$$\sigma_{c2} = \frac{4\pi v_f}{d_f^2} \int_{\theta=0}^{2\pi} \int_{\varphi=0}^{\frac{\pi}{2}} \int_{L=\delta}^{L_f/2} F(N_r(\delta, L)).p(L).p(\theta, \varphi) \sin \varphi dLd\varphi d\theta \qquad (18)$$

- For $\delta \in (L_1 + L_2, \; \delta^{**}]$: several fibers are under straightening stage, the others are under frictional pullout stage:

$$
\sigma_{c3} = \frac{4\pi v_f}{d_f^2} \left[\int_{\theta=0}^{2\pi} \int_{\varphi=0}^{\frac{\pi}{2}} \int_{L=L_r}^{L_f/2} F(N_r(\delta, L)).p(L).p(\theta, \varphi) \sin \varphi dLd\varphi d\theta \right.
$$
$$
\left. + \int_{\theta=0}^{2\pi} \int_{\varphi=0}^{\frac{\pi}{2}} \int_{L=\delta}^{L_r} F(N_e(\delta, L)).p(L).p(\theta, \varphi) \sin \varphi dLd\varphi d\theta \right] \qquad (19)
$$

- For $\delta \in (\delta^{**}, L_f/2]$: all fibers are under frictional pullout stage

$$\sigma_{c4} = \frac{4\pi v_f}{d_f^2} \int_{\theta=0}^{2\pi} \int_{\varphi=0}^{\frac{\pi}{2}} \int_{L=\delta}^{L_f/2} F(N_e(\delta, L)).p(L).p(\theta, \varphi) \sin \varphi dLd\varphi d\theta \qquad (20)$$

The shear and normal stress across the crack then can be derived as:

$$\tau = \sigma_c. \sin \gamma; \qquad \sigma = \sigma_c. \cos \gamma \qquad (21)$$

Note: The model can be used for straight fibers by eliminating all terms which contain $N_r(\delta, L)$, in that case, the formula of the bridging stress contains only the 1st term of σ_{c1} in Eq. (17) and σ_{c4} in Eq. (20).

As an illustration, for the properties listed in Table 1, two 3D-curve surfaces present the shear stress resulting from the model for all the values (w, s) in the range [0, 8 mm] (although the upper value may exceed the validity of fibers elastic behavior) for both straight and hooked-end fiber (Fig. 4).

Table 1. Input value of micromechanics parameters used in the model

Geometrical properties		Material properties		Interface properties	
d_f(mm)	0,75	E_m(MPa)	30000	μ	0.15
L_f(hooked) (mm)	60	E_f(MPa)	210000	τ_0(MPa)	6
L_f(straight) (mm)	48	σ_f(MPa)	1050	k_m(MPa)	7500
L_i(mm)	1.5	k_g	8		
ψ(rad)	$\pi/4$	φ_0	$\pi/15$		

From the surface shape on Fig. 4, the shear displacement s exhibits an important effect on the response, both in terms of shear strength and SFRC post-cracking stage.

The hooks have also a great contribution since their yielding corresponds to the phase where the peak capacity is reached. In the following, we focus the sensitivity study on the response of SFRC with hooked-end fibers, most frequently used in conventional SFRCs, although, the model also applies to straight fibers.

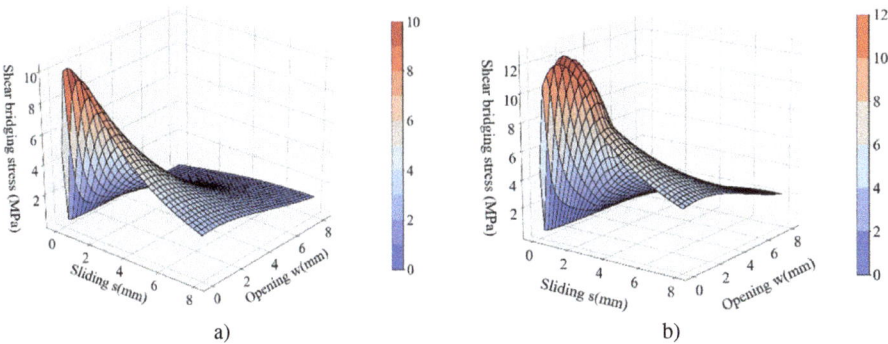

Fig. 4. Shear stress curves of the proposed model: a) Straight fiber, b) Hooked-end fiber

4   Shear Transfer along the Crack Surface

In order to analyze the behavior of a cracked SFRC member under combined shear and tensile load, the shear transfer along a crack surface is evaluated by imposing a series of crack opening displacement (COD) to the model at initial state. Using the parameters listed in Table 1, the shear stress *vs.* sliding and normal stress *vs.* sliding responses are plotted in Fig. 5.

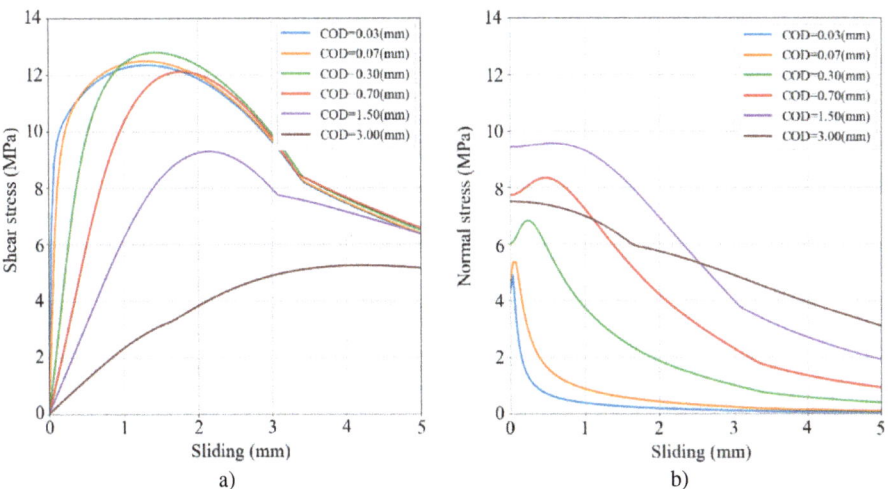

Fig. 5. Stress transmission under prefixed COD curves for hooked-end fiber: a) Shear stress vs sliding; b) Normal stress vs sliding

An increasing COD is shown to induce an increase in the displacement at peak-shear and at peak-normal stress. Moreover, increasing COD induces a lower slope of the initial stage, which can be explained by the contribution of bending effect: when shear is applied on the crack plane the extracted part of the fiber between the crack faces rotates, the higher COD is equivalent to a longer extracted part which leads to a smaller rotation angle.

Evolution of the peak shear stress with increasing COD is not monotonous: Fig. 5a shows a decrease of peak shear stress for large CODs ranging from 0.3 mm to 3.0 mm. As detailed hereabove: the smaller rotation angle leads to the lower stress transferred. Conversely, for the relatively small CODs ranging from 0.03 mm to 0.30 mm, the peak-shear stress increases with increasing COD. The hook contribution may explain this: For a small COD, many fibers are not straightened or not even debonded. Thus, all fibers that intercept the crack are in the hardening phase and many fibers are plasticized or even broken during sliding. The hook contribution also affects the normal stress as shown in Fig. 5b, the initial normal stress increases rapidly until the COD equals to 3.0 mm, *i.e.*, the crack opening where all fibers are straightened.

The shear transmission is also evaluated with a varying direction of the crack displacement or the bending angle γ. Assuming that the crack undergoes a proportional loading *i.e.*, the increment ratio $\Delta s/\Delta w$ is constant, three loading paths with $\Delta s/\Delta w$ equal to 0.5, 1, and 2 respectively are considered. The response of the model under these loading paths is shown in Fig. 6. It can be seen that the crack under the load path 1 corresponding to $\Delta s/\Delta w = 0.5$ can transfer the lowest shear stress and the crack displacement related to the peak-shear stress is also the lowest one. On the contrary, the lower the shear stress transmission through the crack, the higher the normal stress transferred.

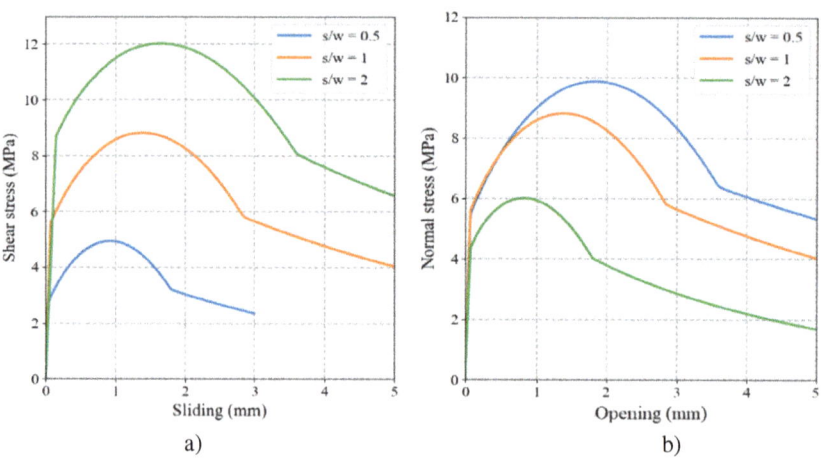

Fig. 6. Bridging stress vs crack displacement under proportional loading: a) Shear stress vs sliding, b) Normal stress vs opening

5 Effect of Micromechanical Parameters

The behavior of the model is clearly influenced by the properties of the materials *e.g.,* fiber diameter, elastic modulus, etc. A sensitivity study has been conducted varying the value of some significant parameters, namely, the fiber aspect ratio: L_f/d_f and the fiber volume fraction v_f. The result is shown in Fig. 7. All other parameters remain the same as listed in Table 1 and the ratio $\Delta s/\Delta w$ has been kept equal to 1. As shown in Fig. 7a, the fiber length, the fiber diameter and also the ratio between them have a significant influence on the shear stress, the higher length and larger diameter improve the shear stress transmission while the ratio L_f/d_f affects the shear strength at early stage of sliding.

The influence of the fiber volume fraction is shown in Fig. 7b, exhibiting significant improvement of the shear strength but also changes of the shape of curves or the global behavior of SFRC. The fiber volume fraction not only impacts the strength of SFRC but also the workability of the mix and thus, indirectly, it affects the fiber orientation. For a higher fiber content, the hooks contribution to the peak-shear stress increases. Expressing the peak shear stress as a function of the fiber content, a quasi-linear relation is obtained, which is consistent with several previous experimental studies [4].

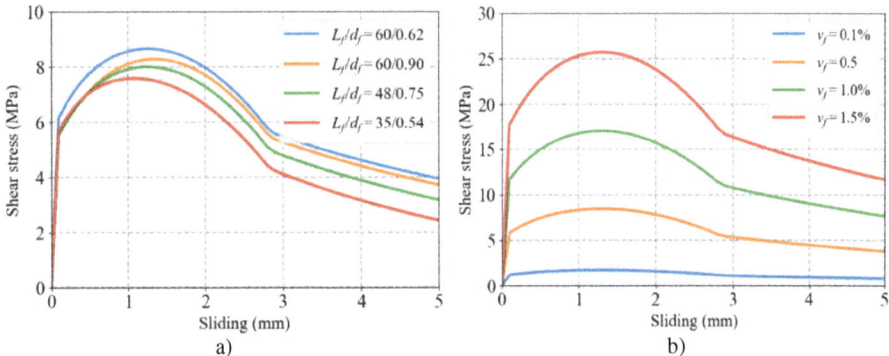

Fig. 7. Effect of fiber's parameters on the shear stress: a) Length/diameter of fiber, b) Fiber content

6 Discussion and Conclusion

In the present model, the fibers bridging mechanism during the pullout process, plays a key role in the stress transmission along the crack, therefore, the ratio between crack sliding and crack opening strongly affects the stress transfer. This will keep critical when integrating this micromechanical analysis into a constitutive law at the macro-scale. Meanwhile, the model developed is based on the hypothesis of small deflection and perfectly elastic behavior of steel fibers. In practice, under a large crack sliding, the deflection of fibers is large and fibers section at the exit point may yield. In comparison, the shear stress estimated in the present model is about 20% higher than the one that appears in the pulley approach [13] where the fiber's behavior is supposed to be perfectly plastic. This aspect will be considered in a further refined approach to avoid overestimation of the shear strength. However, the solution that accounts for large deflection or steel elastoplastic behavior (fiber yielding) may hardly give a closed-form function, thus, the present model has been deemed convenient in the first stage for further multi-scale analysis. The effect of aggregate interlock was first not included in the present model; however, the fiber contribution is considered separately from the matrix contribution. Thus, the aggregate interlock will be added by superposing algebraically this model with a widely accepted model in this field such as the "Contact Density" model presented in [24].

To summarize this research, the contribution of fibers to SFRC behavior, including normal stress and shear stress under combined opening and shear sliding, has been derived from a micromechanical model based on the pullout process of a single fiber and averaging the distribution of fiber orientation. The bending effect due to the shear sliding is related to the pullout process by an analytical function while the fiber orientation is accounted for by a Gaussian-like probability function that represents the anisotropic distribution of the fiber orientation with respect to a prefixed direction. The model provides a rational quantitative interpretation of the fiber contribution to shear transfer and of influence of fiber properties. Benchmarking data from pure shear tests on SFRC, including UHPFRC [6], are currently collected to calibrate and validate the model to make it a tool for SFRC structural optimization.

References

1. Resplendino, J., Toutlemonde, F.: Designing and Building with UHPFRC. Wiley, Hoboken (2011)
2. Rossi, P.: Les bétons de fibres métalliques. Presses De L'ENPC (1998)
3. Casanova, P.: Bétons renforcés de fibres métalliques: du matériau à la structure. Etude expérimentale et analyse du comportement de poutres soumises à la flexion et à l'effort tranchant. Ph.D. thesis, Ecole Nationale des Ponts et Chaussées (1995)
4. Khaloo, A.R., Kim, N.: Influence of concrete and fiber characteristics on behavior of steel fiber reinforced concrete under direct shear. Mater. J. **94**(6), 592–601 (1997)
5. Kaufmann, W., Amin, A., Beck, A., Lee, M.: Shear transfer across cracks in steel fibre reinforced concrete. Eng. Struct. **186**, 508–524 (2019)
6. Herrera, A.: Fonctionnement des jonctions âmes-membrures en Béton Fibrés à Ultra-Hautes Performances (BFUP). Ph.D. thesis, Université Paris-Est (2017)
7. Morton, J., Groves, G.W.: The cracking of composites consisting of discontinuous ductile fibres in a brittle matrix—effect of fibre orientation. J. Mater. Sci. **9**(9), 1436–1445 (1974)
8. Chanvillard, G.: Analyse expérimentale et modélisation micromécanique du comportement des fibres d'acier tréfilées, ancrées dans une matrice cimentaire, vol. OA12. LCPC (1992)
9. Lee, G., Foster, S.: Behaviour of Steel Fibre Reinforced Mortar in Shear I: Direct Shear Testing, University of New South Wales Concrete Testing Laboratory, Scientific UNICIV No. R-444, October 2006

10. Foster, S.J., Lee, G.G., Htut, T.N.S.: Radiographic imaging for the observation of Modes I and II fracture in Fibre Reinforced Concrete. In: Proceedings of the FraMCos6, vol. 3, pp. 1457–1465, June 2007
11. Htut, T.N.S., Foster, S.J.: Unified model for mixed mode fracture of steel fibre reinforced concrete, p. 9 (2010)
12. Li, V.C., Wang, Y., Backer, S.: A micromechanical model of tension-softening and bridging toughening of short random fiber reinforced brittle matrix composites. J. Mech. Phys. Solids **39**(5), 607–625 (1991)
13. Wu, C., Leung, C.K.Y., Li, V.C.: Derivation of crack bridging stresses in engineered cementitious composites under combined opening and shear displacements. Cem. Concr. Res. **107**, 253–263 (2018)
14. Guenet, T.: Modélisation du comportement des bétons fibrés à ultra-hautes performances par la micromécanique: effet de l'orientation des fibres à l'échelle de la structure. Ph.D. thesis, Université Paris-Est, Université Laval (2016)
15. NF P18-710. Complément national à l'Eurocode 2 - Calcul des structures en béton : règles spécifiques pour les bétons fibrés à ultra-hautes performances (BFUP). AFNOR (2016)
16. NF P18-470. Bétons fibrés à ultra hautes performances - Spécification, performance, production et conformité. AFNOR (2016)
17. fib Model Code for Concrete Structures 2010. Ernst & Sohn (2013)
18. Katz, A., Li, V.C.: Inclination angle effect of carbon fibers in cementitious composites. J. Eng. Mech. **121**(12), 1340–1348 (1995)
19. Naaman, A.E., Shah, S.P.: Pull-out mechanism in steel fiber-reinforced concrete, vol. 102. American Society of Civil Engineers ASCE (1976)
20. Alwan, J., Naaman, A.E., Guerrero, P.: Effect of mechanical clamping on the pull-out response of hooked steel fibers embedded in cementitious matrices. Concr. Sci. Eng. **1**(1), 15–25 (1999)
21. Robins, P., Austin, S., Jones, P.: Pull-out behaviour of hooked steel fibres. Mater. Struct. **35**(7), 434–442 (2002)
22. Timoshenko, S.: Strength of Materials. Part 1: Elementary Theory and Problems. D. Van Nostrand Company, Incorporated (1940)
23. Leung, C.K.Y., Li, V.C.: Effect of fiber inclination on crack bridging stress in brittle fiber reinforced brittle matrix composites. J. Mech. Phys. Solids **40**(6), 1333–1362 (1992)
24. Li, B., Maekawa, K., Okamura, H.: Contact density model for stress transfer across cracks in concrete. J. Fac. Eng. Univ. Tokyo **40**, 9–52 (1989)

Predicting the Residual Flexural Strength of Concrete Reinforced with Hooked-End Steel Fibers: New Empirical Equations

Enrico Faccin, Luca Facconi, Fausto Minelli[✉],
and Giovanni Plizzari

Department of Civil, Environmental, Architectural Engineering
and Mathematics, University of Brescia, Brescia, Italy
fausto.minelli@unibs.it

Abstract. To characterize the tensile behavior of Steel Fiber Reinforced Concrete (SFRC), international codes generally adopt performance-based approaches that require to perform either indirect or direct tensile tests on concrete samples. The fib Model Code 2010 recommends to assess the tensile performance of SFRC by a Three Point Bending test on a notched beam able to provide a series of residual strengths corresponding to different crack mouth openings detected at midspan. Therefore, when designing SFRC structures, the tensile properties considered in the safety verifications must be checked by laboratory tests involving a suitable number of beam samples. On the contrary, especially in case of preliminary structure sizing, designers need simple tools for estimating the potential tensile performance of concrete by starting from its basic properties. The present paper proposes two equations for predicting the residual strengths f_{R1} and f_{R3} included in most of the equations reported by the fib Model Code 2010 for safety verification of SFRC members. The concrete compressive strength, the fiber aspect ratio and volume fraction are the only three parameters included in the formulations. The assessment of the model effectiveness has been based on a statistical analysis including the adoption of a modified Demerit Point classification method. The good predicted performance of the proposed equations has been also proved by comparison with other similar models reported by literature.

Keywords: Steel fiber reinforced concrete · Residual strengths · Fib model code 2010 · Prediction · Design · Modified demerit point classification

1 Introduction

Steel Fiber Reinforced Concrete (SFRC) is a composite material that is generally characterized by a softening behavior under uniaxial tension. The tensile behavior of the material can be assessed experimentally either by direct tests or by indirect tests, which are often more reliable and simpler to carry out. The fib Model Code 2010 (MC2010) [1] recommends determining the post-cracking performance of fiber reinforced concrete (FRC) by testing notched prisms under three point bending according EN14651. The latter provides four residual flexural strengths, namely f_{R1}, f_{R2}, f_{R3} and

© RILEM 2022
P. Serna et al. (Eds.): BEFIB 2021, RILEM Bookseries 36, pp. 456–468, 2022.
https://doi.org/10.1007/978-3-030-83719-8_40

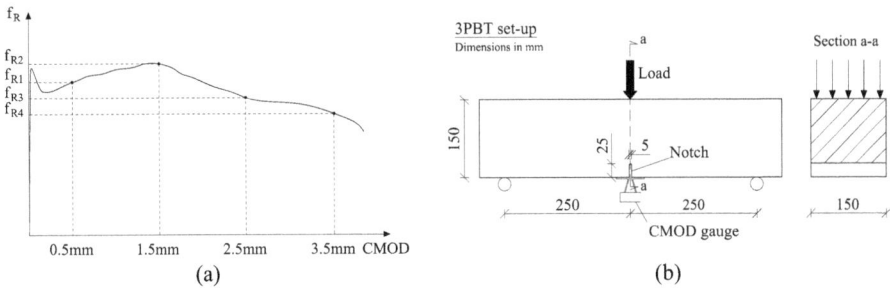

Fig. 1. (a) Typical residual strength – CMOD curve; (b) 3PBT set-up according MC2010 (dimensions in mm).

f_{R4}, which correspond respectively to Crack Mouth Opening Displacements (CMOD) of 0.5 mm, 1.5 mm, 2.5 mm and 3.5 mm (Fig. 1a).

Based on the force (F_i) (Fig. 1b) corresponding to the i-th value of the $CMOD_i$, the residual strength f_{Ri} is calculated as follows:

$$f_{Ri} \cdot = \frac{3 \cdot F_i \cdot L}{2 \cdot b \cdot h_{sp}^2} \tag{1}$$

where b = 150 mm, L = 500 mm is the beam span and h_{sp} = 125 mm is the height of the beam section over the notch. The MC2010 provides simplified stress-crack width relationships that allow to relate the uniaxial tensile behavior of the material to the flexural strength resulting from the 3PBT.

When the results of proper characterization tests are not available, the possibility of using reliable models for predicting the tensile properties of the material is much appreciated especially by designers. Few authors dealt with the development of analytical models able to predict the residual flexural strength of SFRC by starting from the basic properties of the composite materials. In 2014, Moraes-Neto et al. [2] proposed empirical equations that relate the residual strengths defined by the MC2010 to both the fiber aspect-ratio ($L_f/\varnothing_f$ = fiber length/fiber diameter) and the total volume fraction (V_f) of fibers. The model validation processes, which involved a database containing 69 results gathered from 3 Point Bending Tests (3PBT) on notched beams, highlighted the ability of the model to provide conservative predictions having a low-moderate dispersion (CV = 27–36%). Venkateshwaran et al. [3] have recently proposed an alternative empirical model including several parameters such as the fiber volume fraction, the fiber aspect ratio, the cylindrical compressive strength of concrete and the number (N) of hook-ends forming the single fiber. The model revealed to be quite accurate (CV < 25%) even if the validation was based on a limited database of 3PBT including the results obtained from 69 notched prisms tested by the authors and additional 39 tests collected from the literature.

The study described in the following sections proposes a new empirical model for predicting the residual strengths f_{R1} and f_{R3} frequently included in the MC2010's formulations for safety verification of SFRC members (e.g., verification of SFRC elements under flexure (MC2010 – clause 7.7.3.1) and/or shear actions (MC2010 – clause 7.7.3.2)). The equations include the most significant parameters governing the tensile behavior of SFRC, namely the fiber volume fraction, the fiber aspect ratio and the mean cylindrical compressive strength of the concrete matrix. To check the model accuracy and extend the applicability of the model to a wide range of material typologies, the model validation was carried out against a total of 155 3PBTs series, including both normal strength and high performance concretes, all containing hooked-end steel fibers. To further prove the model effectiveness, a comparison with the other models founds in the literature and described above was also reported at the end of the paper.

2 Description of the Database

To assess the performance of the predictive equations described in Sect. 3, an experimental database formed by 155 3PBT data was collected from different works reported by the literature [2–13]. As shown in Table 1, the main parameters included in the database are the mean cylindrical compressive strength (f_{cm}), the fiber aspect ratio ($L_f/Ø_f$) the volume fraction of fibers (V_f) and the fiber length (L_f). The residual strengths (f_{R1}, f_{R3}) are plotted in the diagrams of Fig. 1 against the three variables mentioned before. As expected, the post-cracking strengths tend to increase as the compressive strength, the aspect ratio and the fiber content increase as well. However, a linear trend was not observed and the coefficient of determinations (R^2) resulting from the linear regression of each data-set were generally lower than 0.4. Moreover, it is seen that the experimental data present a quite high dispersion because the Coefficient of Variation (CV) related to f_{R1} and f_{R3} was equal to 50% and 53%, respectively.

   To highlight the distribution of the variables that compose the database, the frequency and the related cumulative percentage of each variable have been reported in the Pareto charts of Fig. 3. The latter allow to appreciate the frequency of each variable as well as their cumulative impact on the residual strength values. About the compressive strength (Fig. 3a), 75% of the collected data fall within the range 34–63 MPa typical of normal-strength concrete. The remaining 25% of the data include tests performed on high-strength materials having compressive strengths ranging from 63 MPa to a maximum of 100 MPa. The 60% of the samples included in the database adopts hooked-end steel fibers having an aspect ratio higher than 61 and lower than 72. This cumulative percentage increases to about 77% if all the fibers with aspect ratios ranging from 50 to 78 are taken into account. The fiber volume fractions (Fig. 3c) cover a wide range of values frequently used in practical applications. In fact, about 55% of the data consist of concrete reinforced with fiber contents ranging from 20 kg/m³ (i.e., 0.25% by volume) to 50 kg/m³ (i.e., 0.6% by volume). Moreover, a significant number of data (i.e., about 40%) is represented by samples containing high volume of fibers (i.e., 0.6–1.1%) frequently used in high performance materials such as those employed for retrofitting/repairing existing masonry buildings or concrete bridges.

Table 1. Summary of 3PBTs data collected from literature.

Author	Test series ID	f_{cm}	V_f	$L_f/\varnothing_f$	L_f	N*	fR1	f_{R3}
		[MPa]	[%]	[−]	[mm]	[−]	[MPa]	[MPa]
Tiberti et al. [4]	S1	42.1	0.38	56	50	1	2.68	2.84
	S2	41.3	0.57	56	50	1	4.64	5.19
	S3	49.8	0.76	56	50	1	5.38	5.95
	S4	62.4	0.38	56	50	1	4.00	4.24
	S5	60.0	0.57	56	50	1	6.06	6.67
	S6	64.2	0.76	56	50	1	8.12	8.33
	S7	44.6	0.38	44	33	1	2.38	2.21
	S8	43.3	0.57	44	33	1	3.75	3.63
	S9	42.4	0.76	44	33	1	4.64	4.54
	S10	62.8	0.38	44	33	1	3.46	3.00
	S11	63.3	0.57	44	33	1	5.08	4.56
	S12	62.3	0.76	44	33	1	7.26	6.25
	S13	44.2	0.38	67	50	1	2.93	2.95
	S14	41.5	0.57	67	50	1	4.72	5.22
	S15	49.5	0.76	67	50	1	5.98	6.43
	S16	61.5	0.38	67	50	1	4.73	4.44
	S17	63.4	0.57	67	50	1	6.10	6.33
	S18	59.6	0.76	67	50	1	7.99	8.23
	S19	42.1	0.38	56	50	1	2.62	2.75
	S20	41.3	0.57	56	50	1	4.90	5.57
	S21	49.8	0.76	56	50	1	6.36	6.52
	S22	63.5	0.38	56	50	1	4.51	4.83
	S23	62.4	0.57	56	50	1	6.42	6.97
	S25	43.3	0.38	67	50	1	9.20	4.45
	S26	42.4	0.57	67	50	1	3.88	6.63
	S27	43.2	0.76	67	50	1	6.27	8.02
	S28	63.4	0.38	67	50	1	7.32	6.12
	S29	65.2	0.57	67	50	1	5.35	7.97
	S30	62.3	0.76	67	50	1	6.89	9.67
	S31	41.2	0.38	50	37	1	8.81	2.25
	S32	42.0	0.57	50	37	1	2.53	3.62
	S33	44.5	0.76	50	37	1	3.48	4.54
	S34	61.2	0.38	50	37	1	5.74	3.97
	S35	63.9	0.57	50	37	1	4.12	5.22
	S36	62.3	0.76	50	37	1	5.13	6.14
	S37	59.3	0.38	56	50	1	6.33	5.83
	S38	62.5	0.57	56	50	1	4.36	7.80
	S39	63.2	0.76	56	50	1	6.86	9.51
	S40	58.8	0.38	67	50	1	5.72	6.55
	S42	61.8	0.76	67	50	1	7.79	10.96
	S43	38.8	0.32	50	50	1	9.05	1.23

(continued)

Table 1. (*continued*)

Author	Test series ID	f_{cm}	V_f	$L_f/Ø_f$	L_f	N*	fR1	f_{R3}
	S44	34.3	0.38	63	50	1	1.71	2.09
	S45	33.5	0.38	50	50	1	1.94	2.05
	S46	35.4	0.38	60	60	1	2.14	1.85
	S47	36.1	0.38	45	44	1	2.11	1.59
	S48	57.0	0.50	65	60	2	2.40	7.70
	S50	73.3	0.50	80	30	1	5.76	7.75
	S51	73.0	1.00	65	60	2	7.21	11.49
	S52	74.5	1.00	80	30	1	7.26	11.78
	S54	39.7	0.50	48	30	1	10.84	4.04
	S55	25.4	1.00	48	30	1	12.39	4.40
	S57	40.8	0.50	48	30	1	5.68	3.35
	S58	27.4	1.00	48	30	1	5.00	4.36
	S59	34.9	1.00	64	35	1	5.79	6.16
	S60	48.7	0.50	64	35	1	5.09	3.50
	S61	26.1	1.00	64	35	1	4.12	7.22
	S62	45.8	0.50	64	35	1	5.43	5.91
	S63	38.6	1.00	64	35	1	6.61	8.12
	S64	40.8	0.50	64	35	1	7.27	5.42
	S65	46.1	1.00	64	35	1	9.49	9.16
	S66	39.0	0.50	64	35	1	5.35	5.19
	S67	36.8	1.00	64	35	1	10.46	5.86
	S68	42.5	0.50	64	35	1	4.83	4.77
	S69	37.6	1.00	64	35	1	4.57	6.85
	S70	42.5	0.32	64	35	1	7.38	2.05
	S71	42.5	0.32	64	35	1	2.04	1.53
	S72	46.8	1.00	64	35	1	1.44	6.97
	S74	52.3	0.50	75	60	1	7.58	8.37
	S76	50.7	0.50	100	80	1	7.75	9.87
	S78	44.1	0.50	60	30	1	8.31	6.86
	S79	43.6	0.76	60	30	1	7.64	7.19
	S80	39.3	0.32	55	33	1	2.62	2.09
	S81	38.8	0.32	75	60	1	2.77	3.59
Present Study	S83	98.2	0.96	79	30	1	12.54	11.46
	S84	91.8	0.96	79	30	1	11.90	12.04
	S85	85.7	0.96	79	30	1	13.36	13.04
	S86	87.9	0.96	79	30	1	13.18	13.35
Facconi et al. [5]	S87	50.4	0.32	67	32	1	2.81	4.88
	S88	41.5	0.25	80	32	1	4.29	4.32
	S89	45.0	0.32	80	32	1	5.94	5.54
	S90	43.7	0.32	86	32	1	4.10	4.20

(*continued*)

Table 1. (*continued*)

Author	Test series ID	f_{cm}	V_f	$L_f/\varnothing_f$	L_f	N*	fR1	f_{R3}
	S91	43.4	0.63	86	32	1	7.20	6.60
	S92	42.8	0.32	65	32	1	3.37	5.03
	S93	42.6	0.64	65	32	1	5.99	6.69
Venkateshwaran et al. [3]	M32	38.6	0.26	67	60	1	2.38	2.76
	M34	37.3	0.50	67	60	1	5.29	7.00
	M36	36.0	0.75	67	60	1	7.29	8.04
	M38	41.1	0.99	67	60	1	6.82	9.51
	M42	38.9	0.26	67	60	1.5	3.26	4.26
	M44	39.7	0.50	67	60	1.5	4.89	6.47
	M46	38.1	0.75	67	60	1.5	6.16	7.21
	M48	39.7	0.99	67	60	1.5	9.08	8.94
	M52	41.9	0.26	67	60	2	3.32	5.68
	M54	42.5	0.50	67	60	2	5.42	7.60
	M56	40.4	0.75	67	60	2	6.20	8.25
	M58	39.7	0.99	67	60	2	9.35	12.20
	L54	28.1	0.50	67	60	2	4.62	6.57
	H54	56.1	0.50	67	60	2	5.51	8.61
	H32	46.1	0.26	67	60	1	3.31	3.78
	H34	46.7	0.50	67	60	1	6.21	7.54
	H36	48.0	0.75	67	60	1	4.61	8.22
	H38	40.5	0.99	67	60	1	8.29	9.76
	H34A	47.6	0.51	80	60	1	6.23	7.88
	H34S	45.4	0.52	64	35	1	6.06	5.16
	H3'12'A	48.3	1.49	67	60	1	12.72	15.35
	H3'16'A	46.0	1.98	67	60	1	13.58	17.36
Soetens et al. [6]	3P-SH-20–1	59.6	0.25	79	30	1	2.81	4.95
	3P-SH-40–1	56.8	0.51	79	30	1	3.98	7.06
	3P-SH-60–1	54.0	0.76	79	30	1	5.41	10.26
	3P-LH-20–1	57.1	0.25	85	60	1	5.81	5.75
	3P-LH-40–1	60.8	0.51	85	60	1	10.25	8.73
	3P-LH-60–1	53.7	0.76	85	60	1	9.01	10.71
Bencardino et al. [7]	DS1%_1	66.2	1.00	80	50	1	13.80	11.60
	DS1%_2	64.6	1.00	80	50	1	11.80	11.40
	DS1%_3	62.6	1.00	80	50	1	8.00	7.10
	DS2%_1	62.7	2.00	80	50	1	14.10	15.40
	DS2%_2	60.8	2.00	80	50	1	13.70	14.30
	DS2%_3	64.2	2.00	80	50	1	13.50	13.00
Lee et al. [8]	3D-37	49.9	0.37	67	60	1	5.53	6.25
	3D-60	49.3	0.60	67	60	1	6.37	6.49
	3D-100	49.8	1.00	67	60	1	12.00	13.31

(*continued*)

Table 1. (*continued*)

Author	Test series ID	f_{cm}	V_f	$L_f/\varnothing_f$	L_f	N*	fR1	f_{R3}
	4D-37	49.8	0.37	67	60		4.76	5.95
	4D-60	51.4	0.60	67	60	1.5	6.23	7.49
	4D-100	51.7	1.00	67	60	1.5	11.26	13.57
	5D-37	48.0	0.37	67	60	2	4.48	7.49
	5D-60	52.3	0.60	67	60	2	5.62	9.74
	5D-100	49.8	1.00	67	60	2	10.41	12.03
Gouveia et al. [9]	ND1	33.8	0.50	64	35	1	3.68	3.72
	ND2	31.8	0.75	64	35	1	4.54	5.35
	ND3	46.2	0.75	64	35	1	6.97	4.99
	ND4	45.8	1.00	64	35	1	6.93	6.58
	ND5	44.5	1.25	64	35	1	9.60	9.24
Barros et al. [10]	Cf60fc50	51.9	0.76	67	37	1	5.71	2.77
	Cf60fc70	63.7	0.76	67	37	1	9.06	8.08
	Cf75fc50	55.7	0.96	67	37	1	6.32	3.25
	Cf75fc70	70.0	0.96	67	37	1	12.30	11.55
	Cf90fc50	56.4	1.15	67	37	1	11.02	12.61
	Cf90fc70	57.6	1.15	67	37	1	10.51	11.99
Chanthabouala et al. [11]	F09–03	89.0	0.30	67	60	2	4.20	6.50
	F09–06	87.0	0.60	67	60	2	8.90	10.50
	F09–09	90.0	0.90	67	60	2	16.00	16.30
	F09–12	100.0	1.20	67	60	2	17.50	22.20
	F14–03	89.0	0.30	67	60	2	4.20	6.50
	F14–06	87.0	0.60	67	60	2	8.90	10.50
	F14–09	90.0	0.90	67	60	2	16.00	16.30
	F14–12	100.0	1.20	67	60	2	17.50	22.20
Landler and Fischer [12]	M1–25	48.1	0.50	67	60	1.5	4.00	4.30
	M2–25	39.8	0.50	67	60	2	4.50	5.20
	M3–25	48.1	1.00	67	60	2	10.00	9.20
	M1–30	48.9	0.50	67	60	1.5	4.20	5.20
	M2–30	40.3	0.50	67	60	2	3.60	5.20
	M3–30	47.0	1.00	67	60	2	8.80	7.30
Gouveia et al. [13]	F0.5_R1	69.7	0.50	67	60	1.5	5.45	3.67
	F0.75_R1	67.6	0.75	67	60	1.5	5.55	5.93
	F1_R1	66.0	1.00	67	60	1.5	6.21	9.63

* N = number of hooks at the fiber end.

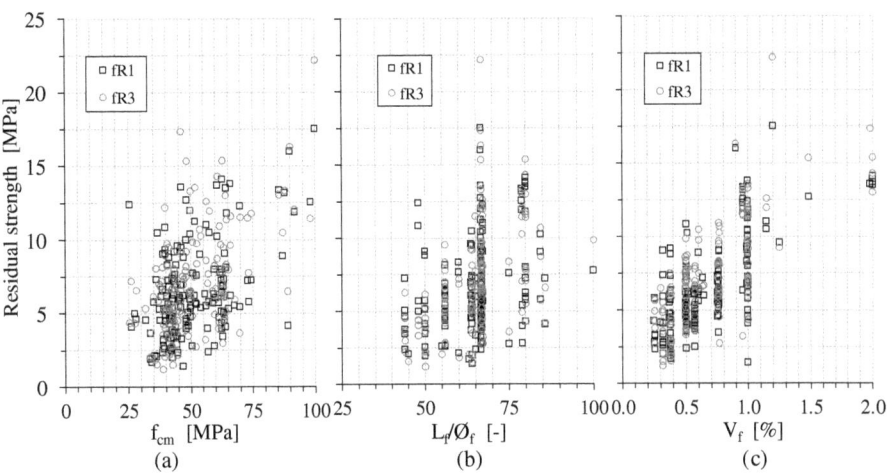

Fig. 2. Mean residual strengths (f_{R1}, f_{R3}) plotted against (a) the mean compressive strength (f_{cm}), (b) the fiber aspect ratio ($L_f/\varnothing_f$) and (c) the fibre volume fraction (V_f).

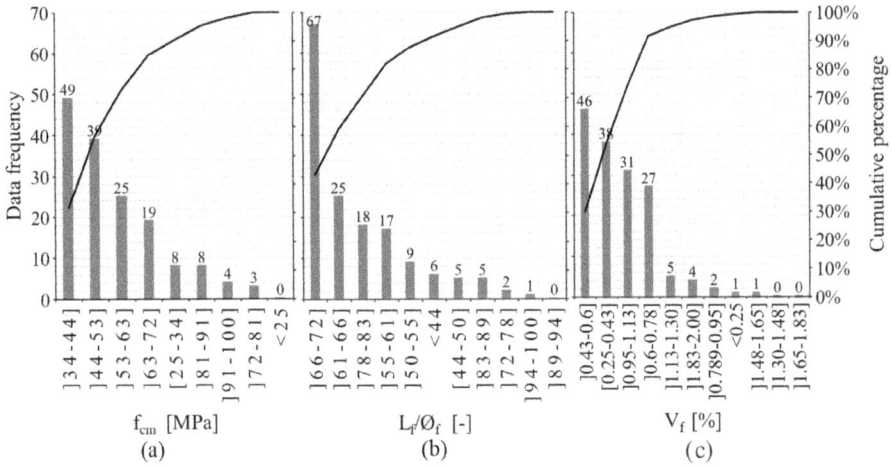

Fig. 3. Pareto charts showing the distribution of the variables forming the database: (a) mean compressive strength (f_{cm}); (b) fiber aspect ratio ($L_f/\varnothing_f$); (c) fibre volume fraction (V_f).

In view of the previous considerations one may conclude that the intervals of the variables included in the database are representative of most of the typical applications (e.g., slab on grade; elevated slabs; pre-cast elements; overlays) of SFRC. However, based on the lack of information concerning SFRCs characterized by significantly high compressive strengths and/or fiber contents, the authors recommend the following limits of applicability of the equations reported in Sect. 3: 20 MPa $\leq f_{cm} \leq$ 85 MPa; 0.25% $\leq V_f \leq$ 1.1%.

3   Proposed Predictive Equations and Effectiveness Assessment

The model described herein aims at providing a simple and practically oriented method for estimating the residual flexural strength of concrete reinforced with hooked-end steel fibers. To best fit the experimental data described in the previous section, the residual strengths were expressed by equations having the general form $f_{Ri} = a_i \cdot (L_f/\varnothing_f)^{b_i} \cdot (V_f)^{c_i} \cdot (f_{cm})^{d_i}$. A regression analysis was carried out to determine the four coefficients (i.e., a_i, b_i, c_i, d_i) able to minimize the mean absolute deviation (MAPE – Mean Absolute Percentage Error) of the residual strength predicted by the previous equation from the real values collected in the database. Moreover, the Moraes Neto et al.'s [2] modified version of the "Demerit Point Classification" (DPC) model proposed by Collins was applied to further assess the predictive performance of the proposed equations as well as to compare them with other predictive equation reported by the literature.

The two equations resulting from the best-fitting procedure can be expressed as follows:

$$f_{R1} = 2.7 \cdot \left(\frac{L_f}{\varnothing_f}\right)^{0.8} \cdot (V_f)^{0.9} \cdot (f_{cm})^{1/2} \text{ [MPa]} \tag{2}$$

$$f_{R3} = 0.5 \cdot \left(\frac{L_f}{\varnothing_f}\right) \cdot (V_f)^{0.85} \cdot (f_{cm})^{2/3} \text{ [MPa]} \tag{3}$$

where the terms V_f and f_{cm} are expressed in [−] and [MPa], respectively. The diagrams of Fig. 4 report the values of the residual strengths ($f_{Ri,pred}$) predicted by the previous equations against the corresponding values ($f_{Ri,exp}$) summarized in the database (Table 1). The accuracy of the predictions was reasonably good as the MAPE was approximately equal to 25%. In addition, the predictions appear to be on the safe side because the actual values of the residual strengths are in most of the cases higher than the corresponding predicted ones.

To better assess the actual predictive performance of the proposed equations, the factor $\beta = f_{Ri,exp}/f_{Ri,pred}$ was calculated for each prediction and then reported in a Box Plot (Fig. 5). The Box Plot chart includes five numbers representing respectively the minimum of the data set, the first quartile (Q_1 – first horizontal line in the box), the median (i.e., the horizontal line in the middle of the box), the third quartile (Q_3 – third horizontal line in the box) and the maximum of the data set. As shown in Fig. 5, the predicted values of both residual strengths presented a quite high variability (i.e., CV = 44% and 27%) that is partly dependent on the dispersion of the database. As one may note, the data falling within the Q1-Q3 range have β values varying from a minimum of about 0.9 to a maximum of 1.3, with a median of 1.07 and 1.14 respectively for f_{R1} and f_{R3}. Therefore, the interquartile range (IQR = Q3−Q2) is equal approximately to 0.38 both for f_{R1} and f_{R3}, which corresponds to a scatter of about ± 0.19 with respect to the median of β. This scatter is relatively low as it is 0.17 times the median of β reported above.

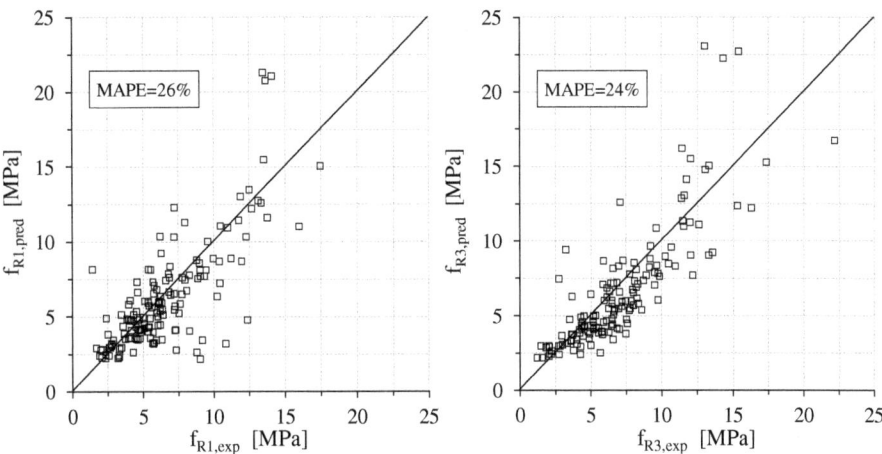

Fig. 4. Predicted vs. experimental values of the residual strengths f_{R1} and f_{R3}.

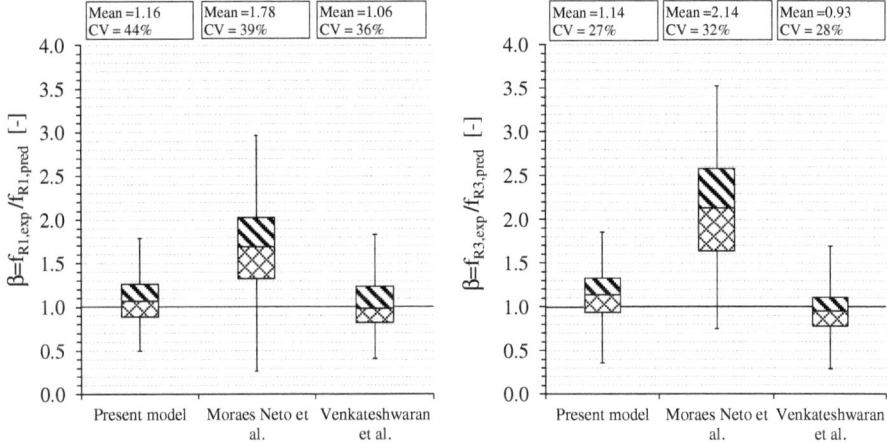

Fig. 5. Dispersion of the residual strengths predicted by different models.

The effectiveness of the proposed equations can be also checked by comparison with other equations reported by the literature. According to Moraes Neto et al. [2], the residual strengths can be calculated as follows:

$$f_{R1} = 7.5 \cdot (RI)^{0.8} [MPa] \tag{4}$$

$$f_{R3} = 6.0 \cdot (RI)^{0.7} [MPa] \tag{5}$$

where RI = $V_f \cdot L_f / \emptyset_f$. Unlike the equations proposed herein and those suggested by Moraes Neto et al., Venkateshwaran et al. [3] have recently formulated the following relations depending also on the number of hooks (N) forming the end of the single steel fiber:

$$f_{R1} = \psi \cdot \left(0.32 f_{cm}^{0.5} + 6.214 \cdot RI + 0.034 N^2\right) [MPa] \tag{6}$$

$$f_{R3} = \psi \cdot \left(0.30 f_{cm}^{0.5} + 7.629 \cdot RI + 0.373 N^2\right) [MPa] \tag{7}$$

where the factor Ψ is taken as $(1 + L_f/100)^{0.5}$. These equations were used to predict the residual strengths of the database described in this paper. The variability of the predicted strengths is highlighted by the charts depicted in Fig. 5. The Moraes Neto et al.'s model resulted to be very conservative as the median of predictions was at least 1.7 times higher than the corresponding actual values reported in the database. Moreover, the values of β values appear more dispersed than those predicted by the other two models. On the contrary, the predictions of the Venkateshwaran et al.'s equations were much more consistent with those resulting from the model proposed herein, in terms of both data variability and accuracy in predicting the actual experimental value. However, it must be remarked that the model reported by Venkateshwaran et al. proved to be the least conservative of the three analyzed models, especially while predicting f_{R3}.

As suggested by Moraes Neto et al. [2], an additional criterion to assess the performance of the model consists in assigning different penalty values (PN) to the β values group together into the ranges defined according to Table 2. Table 3 summarizes the total penalties related to the predictions of the residual strengths resulting from the implementation of the three models compared in this work.

Table 2. Modified demerit point classification (DPC) according Moraes Neto et al. [2].

β = $f_{Ri,exp}/f_{Ri,pred}$	Classification	Penalty (PN)
<0.50	Extremely dangerous	10
[0.50–0.85[	Dangerous	5
[0.85–1.15[	Safety	0
[1.15–2.00[	Conservative	1
≥ 2.00	Extremely conservative	2

It appears that the lowest and the highest penalty scores were attained by the Moraes Neto et al.'s and the Venkateshwaran et al.'s model, respectively. The former exhibited the best performance because the majority of its prediction laid within the "conservative" and the "extremely conservative" ranges. The latter was much less conservative as more than 30% of total data presented either dangerous or extremely dangerous predictions. Unlike the previous model, the one proposed herein seems to be the best balance between prediction accuracy and safety. In fact, besides the quite good accuracy of predictions described above, the proposed equations attained a total penalty score very close to that presented by the Moraes Neto et al.'s model. This good

Table 3. Summary of penalty scores.

$\beta = f_{Ri,exp}/f_{Ri,pred}$		Present model		Moraes Neto et al.		Venkateshwaran et al.	
		N. data	PN	N. data	PN	N. data	PN
f_{R1}	< 0.50	2	20	1	10	3	30
	[0.50–0.85[	27	135	3	15	42	210
	[0.85–1.15[	63	0	13	0	61	0
	[1.15–2.00[	56	56	96	96	44	44
	≥ 2.00	7	14	42	84	5	10
	TOTAL	**155**	**225**	**155**	**205**	**155**	**294**
f_{R3}	< 0.50	2	20	0	0	35	350
	[0.50–0.85[	30	150	2	10	36	180
	[0.85–1.15[	56	0	16	0	66	0
	[1.15–2.00[	66	66	47	47	18	18
	≥ 2.00	1	2	90	180	0	0
	TOTAL	**155**	**238**	**155**	**237**	**155**	**548**

performance resulted from the model calibration process, which led to the adoption of model coefficients able to provide prediction falling within the "safety" and "conservative" ranges.

4 Concluding Remarks

The present work deals with the assessment of the predictive performance of the new empirical equations proposed to predict the residual flexural mean strengths f_{R1} and f_{R3} of SFRC containing hooked-end fibers. The equations are formulated so that the residual strengths depend only on the concrete compressive strength, the fiber volume fraction and the fiber aspect ratio. To calibrate the model parameters, a database reporting 155 data of 3PBTs collected from the literature was considered. The effectiveness and reliability of the model was proved by a statistical analysis estimated the performance of prediction by the factor β, which was calculated as the ratio between the experimental and predicted values of the residual strengths. The safety level of model predictions was checked by using a modified version of the Demerit Point Classification (DPC).

The results described in the paper showed the ability of the proposed equations to provide quite accurate predictions characterized by an average β of about 1.15 and CVs equal to 44% and 27% for f_{R1} and f_{R3}, respectively. The latter are not so high considering the significant dispersion of the data forming the adopted database. As compared to other empirical models found in the literature, the one proposed ensured a good accuracy with the highest number of predictions falling within the "safe" and "conservative" ranges considered by the DPC. This makes the proposed equations suitable for a safe preliminary structural design.

References

1. fib Model Code for Concrete Structures 2010. Fédération Internationale du Béton. Lausanne: Ernst & Sohn (2013)
2. Moraes Neto, B.N., Barros, J.A., Melo, G.S.: Model to simulate the contribution of fiber reinforcement for the punching resistance of RC slabs. J. Mater. Civ. Eng. **26**(7), 04014020 (2014)
3. Venkateshwaran, A., Tan, K.H., Li, Y.: Residual flexural strengths of steel fiber reinforced concrete with multiple hooked-end fibers. Struct. Concr. **19**(2), 352–365 (2018)
4. Tiberti, G., Germano, F., Mudadu, A., Plizzari, G.A.: An overview of the flexural post-cracking behavior of steel fiber reinforced concrete. Struct. Concr. **19**(3), 695–718 (2018)
5. Facconi, L., Minelli, F., Plizzari, G.: Steel fiber reinforced self-compacting concrete thin slabs–experimental study and verification against model code 2010 provisions. Eng. Struct. **122**, 226–237 (2016)
6. Soetens, T., Matthys, S.: Different methods to model the post-cracking behaviour of hooked-end steel fibre reinforced concrete. Constr. Build. Mater. **73**, 458–471 (2014)
7. Bencardino, F., Rizzuti, L., Spadea, G., Swamy, R.N.: Implications of test methodology on post-cracking and fracture behaviour of steel fibre reinforced concrete. Compos. B Eng. **46**, 31–38 (2013)
8. Lee, S.J., Yoo, D.Y., Moon, D.Y.: Effects of hooked-end steel fiber geometry and volume fraction on the flexural behavior of concrete pedestrian decks. Appl. Sci. **9**(6), 1241 (2019)
9. Gouveia, N.D., Fernandes, N.A., Faria, D.M., Ramos, A.M., Lúcio, V.J.: SFRC flat slabs punching behaviour–experimental research. Compos. Eng. **63**, 161–171 (2014)
10. Barros, J.A., Neto, B.N.M., Melo, G.S., Frazão, C.M.: Assessment of the effectiveness of steel fibre reinforcement for the punching resistance of flat slabs by experimental research and design approach. Compos. Eng. **78**, 8–25 (2015)
11. Chanthabouala, K., Teng, S., Chandra, J., Hai, T.K., Ostertag, C.: Punching Tests of Double-Hooked-End Fiber Reinforced Concrete Slabs (Doctoral dissertation, Petra Christian University) (2018)
12. Landler, J., Fischer, O.: Punching shear capacity of steel fiber reinforced concrete slab-column connections. In: Proceedings IABSE Congress, New York (2019)
13. Gouveia, N.D., Faria, D.M., Ramos, A.P.: Assessment of SFRC flat slab punching behaviour–part I: monotonic vertical loading. Mag. Concr. Res. **71**(11), 587–598 (2019)
14. Collins, M.P.: Evaluation of shear design procedures for concrete structures. A Report prepared for the CSA technical committee on reinforced concrete design (2001)

Structural Design

Design and Performance of a Precast Bridge Barrier with Ultra-high Performance Fibre Reinforced Concrete (UHPFRC)

Clélia Desmettre, Jean-Philippe Charron[(⊠)], and Frédérick Gendron

Department of Civil, Geological and Mining Engineering,
Polytechnique Montreal, Montreal, Canada
jean-philippe.charron@polymtl.ca

Abstract. The use of precast concrete barriers instead of cast-in-place barriers avoids early-age cracking due to restrained shrinkage and thermal effects and accelerates the construction process. For these reasons, Polytechnique Montreal has developed since 2009 several durable precast concrete barriers with fiber reinforced concretes and various connection types to the bridge slab. Due to the excellent mechanical and durability properties of ultra-high performance fibre reinforced concrete (UHPFRC), it was integrated recently in a precast barrier concept developed for the Quebec Ministry of Transportation (QMT). This paper presents the concept and the mechanical behaviour under eccentric loading of the TL-5 precast hybrid barrier composed of a NSC core and a UHPFRC shell and connected to the bridge slab through a reinforced UHPFRC connection. An experimental test was performed on a full scale 2 m precast hybrid barrier and was reproduced numerically. The validated model was then used to reproduce the Canadian Highway Bridge Design Code (CSA-S6) loading conditions and validate the performance of the hybrid barrier. The model showed that the hybrid barrier capacity is 69 % and 43 % higher than the ultimate design load required in CSA-S6 when submitted to centered and eccentric loadings, respectively.

Keywords: Precast bridge barrier · Ultra-High Performance Fiber Reinforced Concrete (UHPFRC) · Eccentric loading · Centered loading · Laboratory test · Numerical modeling

1 Introduction

Cast-in-place concrete bridge barriers require long construction time and frequently show early-age cracking resulting from restrained shrinkage and thermal strains that impact their durability. Precast bridge barriers are an alternative to avoid early cracking. Different kinds of precast barriers were developed in the last decades with different types of barrier-slab connection such as anchorage by post-tension [1, 2], with bolts [3] and with recess of reinforced concrete filled on-site [4, 5]. Despite filling a recess on-site requires more construction time, this kind of connection presents more benefits than post-tensioned and bolted connections when concerning the ease of installation (alignment) and the durability (holes partially empty in the structure).

© RILEM 2022
P. Serna et al. (Eds.): BEFIB 2021, RILEM Bookseries 36, pp. 471–482, 2022.
https://doi.org/10.1007/978-3-030-83719-8_41

Some precast barrier concepts using reinforced concrete barrier-slab connections were developed at Polytechnique Montreal since 2009, with several industrial partners [4, 5] and for different barriers test levels (TL-4 and TL-5 as defined in CSA-S6 [6]). The most recent concept developed in partnership with the Quebec Ministry of Transportation (QMT) is a TL-5 precast hybrid barrier composed of a NSC core and a UHPFRC shell including a reinforced UHPFRC barrier-slab connection. The mechanical behaviour of this barrier was evaluated in two phases, Phase 1 was dedicated to the performance under centered loading while Phase 2 focused on the performance under eccentric loading. This paper presents Phase 2 of the project and in some case compare performances under the two types of loadings.

The behaviour of the precast hybrid barrier under eccentric loading is of prime importance as concrete barrier loaded at its extremity presents a lower capacity and probability of occurrence is much greater than for a cast-in-place barrier. Literature shows capacity decreases between 20 and 52% for barrier lengths between 3 to 7.3 m [3, 7, 8]. This reduction can be attributed to a smaller barrier area/volume contributing to the resistance. To counter the capacity reductions under eccentric loading, the number of stirrups is generally doubled in the first meter from the barrier extremity [9]. This practice implies a major increase of reinforcement for precast barrier with limited length, typically of 4 m.

This paper first presents the TL-5 precast hybrid barrier concept and design and then describes its experimental and numerical behaviour under an eccentric quasi-static loading.

2 Methodology

2.1 Design and Concept of the Hybrid Precast Barrier

Concrete barriers have to resist to specific longitudinal, transverse and vertical static loads, applied at a specific height and on a specific length, as specified in CSA-S6 [6]. These loads and the load position depend on the test level category of the barrier and are equivalent to the maximum crash test forces. For the TL-5 barrier category, the specified transverse, longitudinal and vertical unfactored loads specified in CSA-S6 [6] are 210, 70 and 90 kN respectively. These loads have to be applied at a height of 900 mm from the top of the driving surface and on a length of 2400 mm for the transverse load. Namy [10] showed that neglecting the longitudinal and vertical static loads which stabilized the barrier is more critical for the barrier damage and capacity and that only the transverse load can be considered to design the barrier. The minimum transverse ultimate resistance is obtained by increasing the unfactored transverse load of 210 kN by a dynamic amplification factor of 1.7 and in divided the result by the global material performance factor of 0.75 for reinforced concrete considered in this project. Therefore, the minimum transverse ultimate resistance of the precast hybrid barrier becomes 476 kN. The concept of the precast hybrid barrier was optimized by finite element models considering this ultimate capacity, the models being detailed in Sect. 2.2.

Figure 1 illustrates the cross-section of the TL-5 precast hybrid barrier (Fig. 1a) and the TL-5 cast-in-place barrier in normal strength concrete (NSC) used by QMT (Fig. 1b) for comparison purpose. Both barriers have the same height and width. The hybrid barrier is composed of a NSC core coated with a 30 to 85 mm-width UHPFRC shell containing 4% in volume of steel microfibres. The shell width and the fibres dosage were chosen for mechanical and durability considerations. Due to the excellent mechanical performance of UHPFRC, only the 20M (diameter of 19.5 mm) reinforcing bars required for the barrier-slab connection were kept in the barrier. This reinforcement is overlapped (spacing of 100 mm c/c) in the connection recess with the 20M reinforcing bars of the slab. "Terminators" couplers [11] are installed at the extremities of the overlapped reinforcing bars for a better stress transfer. The recess is then filled with UHPFRC containing 3% in volume of steel microfibres. The fibre dosage in the recess was chosen considering mechanical, durability and economical aspects. The bottom part of the connection recess penetrates the upper part of the slab to minimize the recess height in the barrier, to benefit from slab confinement and to prevent water and aggressive agent infiltration under the recess.

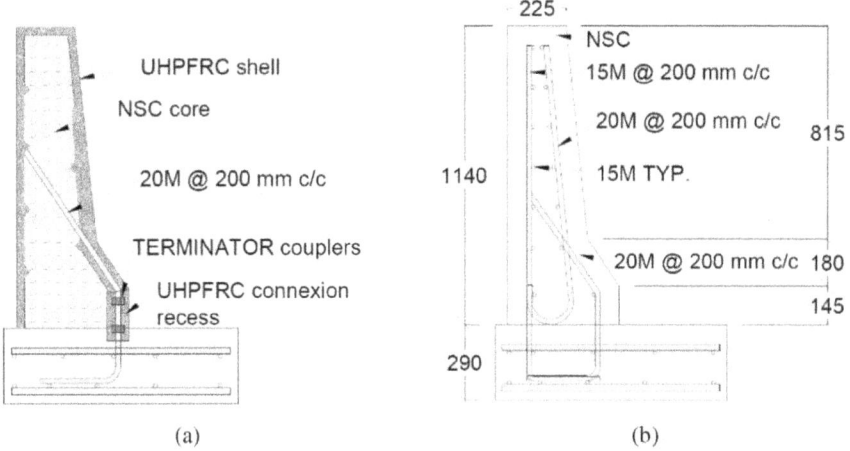

(a)   (b)

Fig. 1. Cross-sections of the TL-5 category barriers (a) Developed precast hybrid barrier with UHPFRC, (b) Reference cast-in-place barrier.

2.2   Finite Element Modelling

Finite element analyses were performed with the software ATENA (Advance Tool for Engineering Nonlinear Analysis) developed by Cervenka Consulting for reinforced concrete structures [12]. The 3D numerical model was used to optimize the precast hybrid barrier design. The model also aimed to simulate the experimental test that was performed in this project. The model took into account all the elements of the specimen such as the barrier with its connection recess, the bridge slab, interfaces between materials with contact elements, the loading plate that will be used for the experimental test and the bridge slab anchorage to the strong laboratory slab (Fig. 2a). ATENA allows the user to define a UHPFRC tensile law taking into account its hardening pre-peak behaviour and its softening post-peak behaviour. Compressive and tensile laws obtained from the characterization tests performed on the NSC and UHPFRC (see Sect. 2.3) were implemented in the model. The steel reinforcement was modelled by truss elements (Fig. 2b) and their properties implemented were also based on tensile characterization tests performed on rebar samples. 3D tetrahedral elements with one integration point were used with mesh size varying depending on the element as illustrated in Fig. 2a. The transversal loading was applied at the center of the loading plate at a rate of 0.2 mm per step of analysis. Gauss integration with linear interpolation was used to integrate the elements and the response was solved with the Newton-Raphson iterative method. More detailed information about the numerical modelling is also available in Gendron [13].

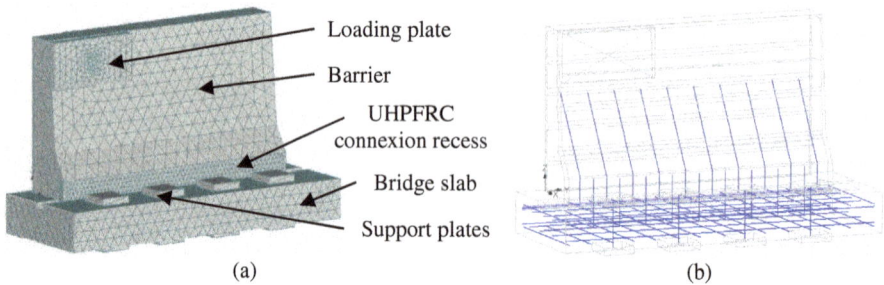

(a)   (b)

Fig. 2. Finite element model of the 2 m hybrid precast barrier and bridge slab (a) Meshes and (b) Reinforcement

2.3   Experimental Program

As the behaviour of the precast hybrid barrier under centered load was studied in Phase 1 of the project [5], the experimental part of this Phase 2 focuses on its behaviour under an eccentric load. This kind of loading is critical in the barrier design process as it decreases its capacity.

An experimental test was performed on a precast hybrid barrier of 2 m loaded at its extremity by a 700 mm-long plate (length × width × thickness = 700 × 350 × 57 mm³). CSA-S6 [6] specifies to load the barrier at a height of 900 mm from the top of the driving surface. Considering that the bridge slab is usually covered with a 65 mm-thick bituminous asphalt, the load was applied at a height of 965 mm from the top of the concrete slab during the test. It can be noticed that the experimental loading length (700 mm) differs from those of 2400 mm prescribed in CSA-S6 [6]. This choice was made for practical aspects (reduced cost of the 2 m specimen vs 4 m, availability of equipment, etc.) and also considering that shorter barrier length and loading plate length reduce the precast barrier ultimate capacity [13]. However, the experimental test will be reproduced numerically and, once validated, the numerical model will be adapted to validate the 4 m-length barrier behaviour when loaded with a 2400 mm plate as specified in the CSA-S6 [6].

The testing specimen was produced in a precast concrete factory to closely reproduce the production mode expected for such precast barriers. The NSC core (Fig. 3a) was first cast horizontally to ensure a good filling of the formwork. A concrete surface retarder was applied on the core surfaces that were planned to be in contact with the UHPFRC shell or recess. This retarder was applied directly on the formwork surfaces for the molded surfaces and directly on concrete after pouring for the other surfaces. All these surfaces were then cleaned with a high pressure water jet after unmolding to obtain exposed aggregate surfaces (Fig. 3b) to provide high quality tensile and cohesive properties of the NSC-UHPFRC interfaces. A few days later, the UHPFRC-4% shell (Fig. 3c) was cast from the top of the barrier placed in its vertical position. The bridge slab, on which the barrier will be installed during the test, was also produced in the same precast concrete factory. Its upper section on which the barrier will be installed was also prepared with an exposed aggregate surface (Fig. 3d). Once the barrier and slab production was finished, the specimens were delivered to the laboratory of Polytechnique Montreal. The barrier was installed in the bridge slab and the recess was filled with UHPFRC-3% from the front of the barrier using conventional formwork (Fig. 3e). This casting method allows a visual inspection of the good recess filling after demolding.

(a)   (b)   (c)

(d)   (e)   (f)

Fig. 3. Major steps of the specimen production (a) NSC core production, (b) exposed aggregate surface preparation, (c) UHPFRC shell production, (d) bridge slab production, (e) UHPFRC recess filling and (f) barrier-bridge slab assembly

A few days before the test, the barrier-bridge slab assembly (Fig. 3f) was installed in the experimental set-up. This setup is detailed in Fig. 4. The bridge slab was fixed to the laboratory strong slab through post-tensioned high-strength steel bars. The lateral displacement of the bridge slab during the test is prevented by another slab placed behind and also post-tensioned in the laboratory slab. This experimental set-up aims to evaluate the barrier behaviour and not the bridge slab behaviour. More information about the bridge slab behaviour can be found in Namy et al. [14]. The transverse loading was applied at the center of the loading plate through a 1000 kN capacity hydraulic actuator and at a displacement rate of 0.12 mm per minute. The applied load was measured by the actuator load cell. A system of braced column, cable, pulleys and actuator counterweight was developed to transfer only the transverse loading to the tested barrier.

The project focused on the mechanical behaviour of the hybrid barrier under quasi-static loadings instead of impact loadings. Indeed, due to the over-resistance of concrete and steel at high strain rates [15, 16], a quasi-static loading is more critical (induced greater damage at a same loading level and provide lower ultimate capacity) than a dynamic loading as demonstrated by Charron et al. [2] and Duchesneau [17] in previous projects.

Evaluation of the hybrid barrier mechanical behaviour during the test was done with several instruments. A LVDT was placed behind the barrier relative to the loading point and several LVDTs were placed at specific locations to measure crack openings of expected cracks in the barrier or relative displacement at barrier-slab and slab-strong floor interfaces. Some anchoring steel rebar in the UHPFRC connection recess were instrumented with strain gages. A digital image correlation (DIC) system was also used in areas of interest to obtain detailed information about the surface deformations.

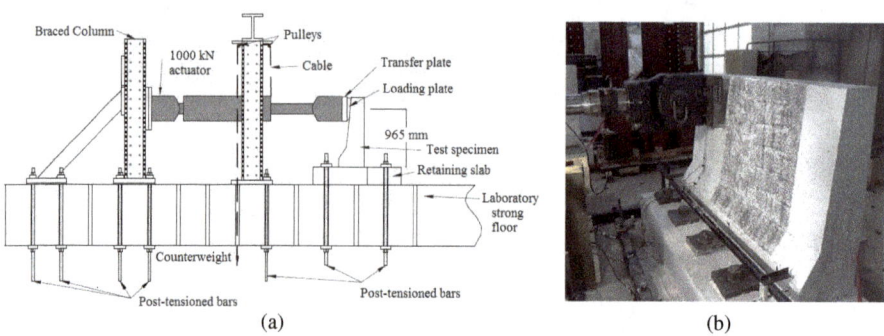

(a)   (b)

Fig. 4. Experimental setup (a) Sketch of the experimental setup and (b) Photo of the applied loading on a test specimen

3   Results and Discussion

3.1   Characterization Specimens

Characterization tests were performed to determine the compressive and tensile behaviours of the concretes (NSC and UHPFRC) used in the precast hybrid barrier and in the bridge slab. The compressive strengths (f'_c) and Young's modulus (E_c) were determined in accordance with CSA A23.2-9C and ASTM C469 respectively. The tensile strength (f'_t) of NSC was obtained from splitting tests and the tensile behaviour (f'_t and stress-strain or displacement tensile curves) of UHPFRC was obtained from direct tensile tests performed on dogbones [18]. Table 1 summarizes the average values of f'_c E_c and f'_t, whereas Fig. 5 illustrates the global tensile behaviour of the UHPFRC of the hybrid barrier shell and connection recess. Figure 5a represents the pre-peak tensile behaviour in strain, whereas Fig. 5b represents the post-peak tensile behaviour in crack width. During Phase 1 of the project, concrete cores were extracted from the UHPFRC shell to evaluate the fibre orientation in comparison to the orientation observed in the direct tensile characterization specimens. As expected, the orientation in the barrier shell was less preferential than in the characterization specimen because of their difference of size and casting method. To take into account this effect in the numerical simulation of the hybrid barrier behaviour, reduced and conservative tensile laws (dotted line in Fig. 5) were implemented in the model in comparison to the tensile laws obtained from the characterization specimens (in continuous line in Fig. 5). The tensile behaviour of the Grade 400W 20M reinforcing bars was also characterized and their yielding strength was equal to 424 MPa.

Table 1. Material properties

Element (material)	f'_c(MPa)	E_c(GPa)	f'_t(MPa)
Bridge slab (NSC)	34.3	24.4	3.23
Hybrid barrier core (NSC)	46.3	29.1	4.42
Hybrid barrier shell (UHPFRC 4% vol.)	148.3	36.4	13.5
Recess & Shear key (UHPFRC 3% vol.)	148.4	36.8	10.5

3.2   Experimental and Numerical Results

The experimental load-lateral displacement curve measured during the testing of the precast hybrid barrier of 2 m under eccentric loading is presented in continuous blue line in Fig. 6a. This figure also illustrates in dotted blue line the numerical global behaviour obtained with the same testing conditions (barrier of 2m and the L700 loading plate) and with the numerical parameters detailed in Sect. 2.2. Pictures of the experimental and numerical cracking patterns at failure are shown in Fig. 7a and b respectively. Then the global behaviour of the hybrid precast barrier under eccentric loading is compared to the global behaviour of the same barrier under centered load (green curves) in Fig. 6a and to the global behaviour of the cast-in-place NSC barrier

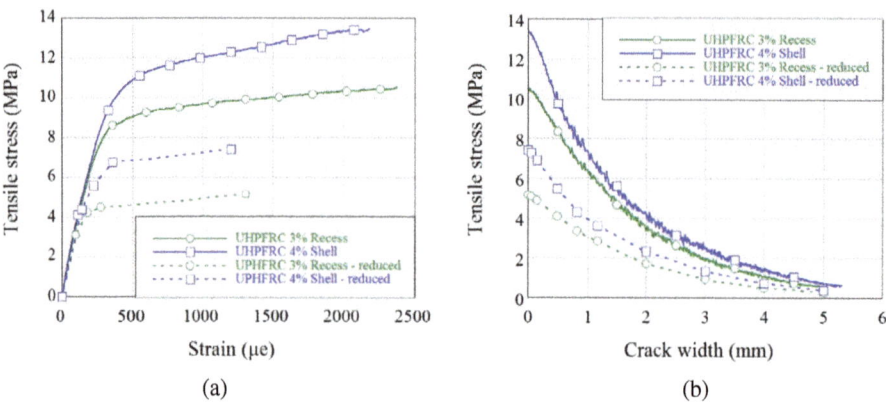

Fig. 5. Tensile behaviour of the UHPFRC used in the hybrid barrier and the connections (a) Pre-peak behaviour and (b) Post-peak behaviour

under centered and eccentric loading (respectively in orange and black lines) in Fig. 6b. These additional results come from Phase 1 of the project.

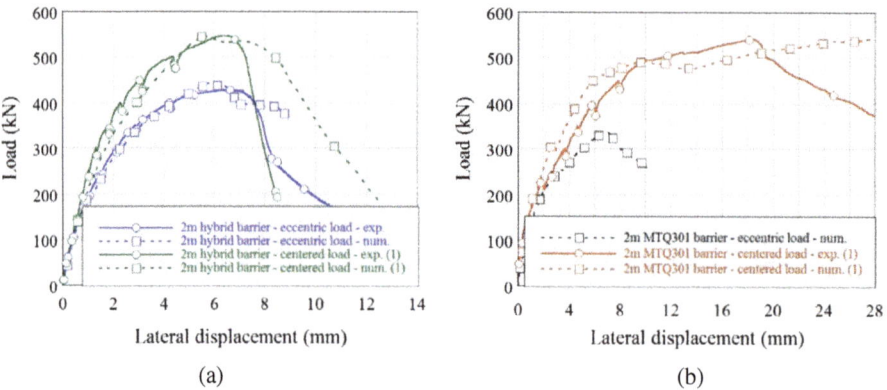

Fig. 6. Load-displacement behaviour of 2 m barriers submitted to centered or eccentric loadings with the L700 loading plate (a) Precast hybrid barrier and (b) Reference cast-in-place MTQ301 barrier

The load-displacement curve of the 2 m precast hybrid barrier tested under eccentric loading (continuous blue line in Fig. 6a) shows a linear and elastic behaviour up to 175 kN. Then, concrete cracking occurs and the curve becomes non-linear. The maximum capacity of the hybrid barrier under eccentric loading is equal to 428 kN for a transverse displacement of 6.3 mm. The post-peak resistance then drops relatively quickly. The major crack controlling the failure is a shear crack occurring around the loading plate as also observed in the reference cast-in-place barrier [5]. This crack

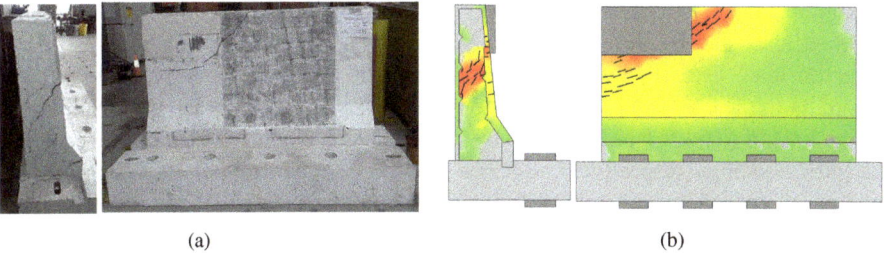

(a)   (b)

Fig. 7. Crack pattern at failure for the 2 m hybrid precast barrier (a) Experimental and (b) Numerical cracking pattern

deflects down the barrier on the barrier side (Fig. 7a). The anchoring reinforcing bars in the UHPFRC connection recess remained elastic during the whole test. The numerical pre-peak load-displacement curve (dotted blue line in Fig. 6a) well represents the experimental pre-peak behaviour of the barrier. The numerical peak is very close to the experimental one with a difference of 3% in load (442 kN versus 428 kN) and 10% in displacement (5.7 mm versus 6.3 mm). The failure mode and cracking pattern are also well reproduced as illustrated in Fig. 7b.

The comparison of the blue and green continuous lines in Fig. 6a illustrates that loading the hybrid barrier at its extremity instead at its center decreased its capacity by 22% (426 kN versus 546 kN), which is the same reduction range observed in literature (see Sect. 2.3). Finally, the comparison of the behaviour of the 2 m hybrid precast barrier (Fig. 6a) and a conventional 2 m NSC cast-in-place barrier (Fig. 6b) shows that the hybrid barrier has a similar capacity under centered loading (546 kN and 540 kN for the hybrid and NSC barriers respectively) and a higher capacity under eccentric loading (426 kN and 337 kN for the hybrid and NSC barriers respectively). In all cases, the numerical results (dotted lines in Fig. 6) reproduce with accuracy the associated experimental results (continuous lines with the same color in Fig. 6) obtained during Phase 1 or 2 of the project, which confirms the validity of the numerical models to extend the utilization to other conditions.

3.3 Validation of the Behaviour of the Precast Hybrid Barrier According to the CSA-S6

As mentioned in Sect. 2.3, the loading length of 700 mm used during the experimental test differs from the length of 2400 mm specified for TL-5 concrete barrier in CSA-S6 [6]. The validated numerical model was thus used to evaluate the behaviour of the precast hybrid barrier under the CSA-S6 conditions. A 4 m barrier length that should be used in construction project and a loading plate of 2400 mm were modelled as specified in CSA-S6. The load was applied at the center of a 2400 mm-long I-beam (W150 x 22 steel profile) that could realistically be used during experimental tests. The numerical results obtained for centered and eccentric loadings are respectively shown in green and blue in Fig. 8. It can first be noticed that increasing the loading length from 700 to 2400 mm and the barrier length from 2 to 4 m increases the barrier maximum capacity by 48% (from 546 kN in Fig. 6a to 807 kN in Fig. 8) in case of centered load, and by 60% (from 426 kN in Fig. 6a to 682 kN in Fig. 8) in case of eccentric load. Overall, the maximum capacity of the hybrid precast barrier loaded according to the CSA-S6 [6] specifications surpasses significantly the CSA-S6 ultimate design load of 476 kN (see Sect. 2.1) by 69% and 43% when submitted to centered and eccentric loadings respectively. Besides, the precast hybrid barrier under centered loading showed a bending-shear failure, while a shear failure was observed under eccentric loading. The mechanical performance of the hybrid barrier clearly demonstrates that it can be used in replacement of the reference cast-in-place barrier of QMT.

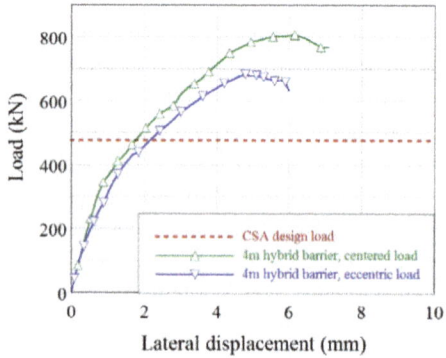

Fig. 8. Load-displacement behaviour of 4 m precast hybrid barriers submitted to centered or eccentric loadings with the 2400 mm loading plate (according to CSA-S6 specifications)

4 Conclusions

A new concept of precast hybrid barrier including a UHPFRC shell and a UHPFRC connection was developed at Polytechnique Montreal, in partnership with Quebec Ministry of Transportation (QMT). This paper focused on its structural behaviour under an eccentric load and established comparison with the performance under centered load. An experimental test was performed on a full scale barrier and the same barrier was modelled with finite element analyses. Once validated, the numerical model was used to simulate the barrier behaviour under the CSA-S6 [6] specifications. The following conclusions can be drawn:

- The tested 2 m precast hybrid barrier submitted to an eccentric load through a 700 mm-long plate showed a maximum capacity of 426 kN and a shear failure around the loading plate. This capacity represents a reduction of 22% in comparison with a centered loading measured in Phase 1 of the project.
- When comparing the precast hybrid barrier and the reference cast-in-place barrier loaded in the same conditions (2 m and loading length of 700 mm), they present a similar capacity (around 540 kN) under centered load. However, the capacity under eccentric load of the hybrid barrier is improved by 26%, in comparison to the reference cast-in-place barrier due to the UHPFRC shell contribution.

- The numerical model reproduced accurately the experimental results and was thus used to reproduced to loading conditions prescribed in CSA-S6 [6].
- Under the CSA-S6 [6] loading conditions (4 m barrier length and loading length of 2400 mm), the precast hybrid barrier capacity surpasses by 69% and 43% the CSA-S6 ultimate design load of 476 kN, respectively for a centered and eccentric loading. The hybrid barrier can thus be used in replacement of the reference cast-in-place barrier.

Acknowledgements. This research project was financially supported by Quebec Ministry of Transportation (Grant n° R686.2 - Design of precast barriers of extended durability with UHPFRC). The technical advices of Mr. Boulet and Mr. Magramane are acknowledged. Authors would like to acknowledge King Packaging Materials (a Sika company) for graciously providing the UHPFRC and the technical staff of Polytechnique Montreal for their contribution to this project.

References

1. Mitchell, G., Tolnai, M., Gokani, V., Picón, P., Yang, S., Klingner, R. E., Williamson, E. B.: Design of Retrofit Vehicular Barriers Using Mechanical Anchors, Center for Transportation Research - The University of Texas at Austin, Austin, TX FHWA/TX-07/0–4823-CT-1 (2006)
2. Charron, J.P., Niamba, E., Massicotte, B.: Static and dynamic behavior of high- and ultrahigh-performance fiber-reinforced concrete precast bridge parapets. J. Bridg. Eng. **16**, 413–421 (2011)
3. Ngan, C.L.Y.: Experimental investigations of anchorage capacity of precast concrete bridge barrier for Performance Level 2. Civil Engineering, University of British Columbia, Vancouver, BC, Mémoire de maîtrise (2008)
4. Duchesneau, F., Charron, J.P., Massicotte, B.: Monolithic and hybrid precast bridge parapets in high and ultra-high performance fibre reinforced concretes. Can. J. Civ. Eng. **38**, 859–869 (2011)
5. Charron, J.P., Damry, R., Desmettre, C., Massicotte, B., Structural use of UHFRC for precast barriers - in French, Group of Research in Structural Engineering - École Polytechnique Montréal, Montréal, QC 1220–00-BD03 (2013)
6. CSA-S6, Canadian Highway Bridge Design Code, ed. Mississauga, Ontario, Canada: Canadian Standards Association (2014)
7. Alberson, D.C., Williams, W.F., Menges, W.L., Haug, R.R.: Testing and Evaluation of the Florida Jersey Safety Shaped Bridge Rail, Texas Transportation Institute - The Texas A&M University System, College Station, TX FHWA/TX-04/9–8132–1 (2004)
8. Patel, G.: Development Of Precast Barrier Wall System For Bridge Decks, Master Thesis, Civil Engineering, Ryerson University, Toronto, ON (2008)
9. QMT, Structural Design Manual - in French. Québec, QC: Quebec Ministry of Transportation (2017)
10. Namy, M.: Structural behaviour of cast-in-place and precast concrete barriers anchored to bridge deck overhangs and subjected to transverse static loading, Master Thesis, Ecole Polytechnique de Montréal, Montreal, QC (2012)
11. nVent. nVent LENTON Terminator For Rebar Anchorage (2019). https://www.erico.com/catalog/literature/CP7E-WWEN.pdf

12. Červenka, V., Jendele, L., Červenka, J.: ATENA Program Documentation - Part 1 -Theory. République tchèque, Prague (2016)
13. Gendron, F.: Design and analysis of the mechanical behavior of precast barriers for bridges with connection joints using ultra-high performance fiber-reinforced concrete - in French, Master Thesis, École Polytechnique de Montréal, Montréal, QC (2019)
14. Namy, M., Charron, J.-P., Massicotte, B.: Structural behavior of bridge decks with cast-in-place and precast concrete barriers: numerical modelling. J. Bridge Eng. **20** (2015)
15. Rossi, P., Toutlemonde, F.: Effect of loading rate on the tensile behavior of concrete : description of the physical mechanisms. Mater. Struct. **29**, 116–118 (1996)
16. Bischoff, P.H., Perry, S.H.: Impact behavior of plain concrete loaded in uniaxial compression. J. Eng. Mech. **121**, 685–693 (1995)
17. Duchesneau, F.: Design of precast hybrid and monolithic barriers using high and ultra-high performance concretes - in French, Master Thesis, Ecole Polytechnique de Montréal, Montreal, QC, (2010)
18. Beaurivage, F.: Study of the influence of structural parameters on the constitutive laws of fiber-reinforced concrete for the design of structures - in French, Master Thesis, Ecole Polytechnique de Montréal, Montréal, QC (2009)

Shear Behaviour of V-shape Webbed Steel Fibre Reinforced Concrete Beams

Divan Visser[1(✉)] and William P. Boshoff[2]

[1] Stellenbosch University, Stellenbosch, South Africa
[2] Pretoria University, Pretoria, South Africa

Abstract. Asymmetrical three-point bending tests were performed on beams with four different types of V-shaped/tapered webs (V-beams) containing various percentages of steel fibres by volume (0, 0.6 and 1%) as well as longitudinal reinforcement in the form of two 16 mm diameter steel bars. The test results revealed that the ultimate shear capacity increases significantly with increasing fibre content and larger web cross-sectional areas. The normalised shear stresses at ultimate resistance are relatively constant for the various types of V-beams indicating that shear capacity is highly dependent on the entire cross-sectional area of the web. An average increase in shear strength of 49% for beams containing 0.6% fibres and 74% for beams containing 1% steel fibres was observed for V-beams when compared to their reference beams containing no fibres. Furthermore, the results show that the widely used assumption where shear is resisted entirely by an effective rectangular cross-sectional area calculated as the product of the narrowest part of the beam (b_w) and the effective depth (d) can result in over-conservative shear predictions for V-beams.

Keywords: Shear capacity · Steel fibres · Fibre reinforced concrete · Self compacting concrete

1 Introduction

The phenomenon of shear is complicated as it comprises of numerous interdependent mechanisms. Shear failure in beams result from a combination of shear forces and bending moments [1] and occur when the principle tensile stresses in the shear span reaches the ultimate tensile strength of the concrete [2]. Such failures are generally brittle failures and happen with little or no warning which may lead undesirable consequences. Since fibres significantly enhance the tensile capacity of concrete [3] and the shear capacity is highly dependent on tensile properties, fibres significantly improve shear strength [4]. Fibres typically only get activated after cracking when the tensile stresses are transferred from the concrete to the fibres and carried entirely by the fibres bridging the cracks [5]. Furthermore, fibres can potentially replace or reduce the amount of vertical reinforcement used in beams [4] and it has been proven that the addition of 0.75% of steel fibres by volume can be used as minimum shear reinforcement [6, 7]. Fibres not only enhance the shear capacity of beams but also allow for beams to resist loads at larger cracks widths and reach ultimate resistance at a significantly later stage compared to conventional concrete [8]. Additionally, to the

© RILEM 2022
P. Serna et al. (Eds.): BEFIB 2021, RILEM Bookseries 36, pp. 483–491, 2022.
https://doi.org/10.1007/978-3-030-83719-8_42

physical contribution of fibres to shear resistance, fibres also improve other shear mechanisms in beams such as enhancing the doweling action of longitudinal reinforcement and improving aggregate interlock by confining cracks [8] which also has a significant impact on improved durability of the concrete member [9].

Shear improvement due to fibres greatly depends on the physical properties of the fibres used, the fibre orientation in the concrete specimen and on the fibre-matrix interfacial bond strength. Self-compacting concrete (SCC) has shown significant potential to improve shear capacity of concrete through optimal fibre orientation and interfacial bond strength. The benefits of using SCC to pour specimens consisting of fibrous concrete is that as the concrete flows in the moulds, fibres align in the flow direction [10, 11]. This can be used to manipulate the fibre alignment to coincide with the principal stress directions in concrete members. Moreover, the denser microstructure of SCC caused by the larger percentage of fines compared to conventional FRC results in higher fibre-matrix interfacial bond strength and it has been shown that fibre-reinforced SCC FRC performs better in shear compared to FRC [6].

A majority of design codes used for designing fibrous beams in shear, including the Eurocode [13] and ACI Building Code [14], are based on the assumption that shear forces are entirely resisted by the rectangular portion of the web (i.e. the area is the narrowest width of the beam times the height). Also, it has been proven that the flange in a beam has a significant influence on the shear capacity [15, 16] and it can be argued that when not considering the entire web area and the flange, it may lead to over-conservative designs.

In this paper, the results of experimental investigations into the shear behaviour of V-beams are presented by comparing shear strengths and shear cracking for various beam-types and fibre volumes. The experimental results provide and indication of the influence the shape of the web has on the shear capacity of beams.

2   Experimental Framework

Three sets of beams consisting of four various steel fibre reinforced concrete V-beams, were tested in asymmetrical bending tests to investigate the shear strength of these V-beams. A total of 24 tests were performed after 14 days of curing on V-beams consisting of concrete with cylinder compressive strength near 28 MPa (14-day strength) and either 0, 0.6 or 1% steel fibres.

2.1   Materials

A steel fibre reinforced self-compacting concrete (SFR-SCC) mix was developed with adequate workability and passing ability to ensure that all gaps and corners of the complex moulds used to cast the beams are filled with concrete with minimal effort required for compaction. Table 1 lists the various mixture designs for the three beams sets and results of tests performed on the fresh concrete. The concrete consisted of CEM II/A-L 42.5 cement and a Class F fly ash. A natural pit sand, locally known as Coarse Malmesbury sand, and crushed coarse aggregate with nominal diameter of 9 mm were used. A superplasticiser named MAPEI Dynamon SP1 was used to ensure the workability of the concrete. Hooked-end steel fibres, supplied by Bekaert, with a length of 50 mm, 1.05 mm in diameter and tensile strength of 1115 MPa were used in the concrete.

Table 1. Concrete mixture and fresh concrete test results.

Constituents (kg/m³)		Control mix	Mix 1 (0.6%)	Mix 2 (1%)
Cement		238	237	236
Fly ash		159	158	157
Water		185	184	183
Sand		1075	1068	1064
Aggregate		710	706	703
Superplasticiser		3.14	3.17	3.39
Fibres		0	47.1	78.5
Slump flow (mm)		620	655	625
J-ring (mm)	Spread	570	530	590
	Blocking	24	29	18

2.2 Mechanical Properties

Flexural notched beam tests according to EN 14651-A1 [16] were performed on prismatic SFR-SCC members to evaluate the flexural performance of the various concrete mixtures that were developed. The average residual flexural tensile strength values for the two concrete mixtures containing 0.6 and 1% steel fibres are listed in Table 2 and coefficients of variations are shown in brackets to give an indication of variability of the test results.

Table 2. Average residual flexural tensile strength values for concrete mixtures with fibres.

Concrete mixture	fR1	f_{R2}	f_{R3}	fR4
Mix 1 (0.6% fibres)	4.30 (0.18)	4.54 (0.17)	4.45 (0.17)	4.20 (0.17)
Mix 2 (1% fibres)	6.10 (0.27)	5.81 (0.18)	5.26 (0.16)	4.73 (0.17)

2.3 Beam Details and Test Setup

Four beam-types were tested in this investigation with cross-sections and dimensions as seen in Fig. 1. The width of the top of the web increase from beam-type A to D with a constant bottom width of 50 mm. All beams were carefully designed for shear failure to govern by over-reinforcing the beams longitudinally with 402 mm² reinforcement area consisting of two 16 mm high yield steel bars stacked vertically with 20 mm cover to the bottom bar.

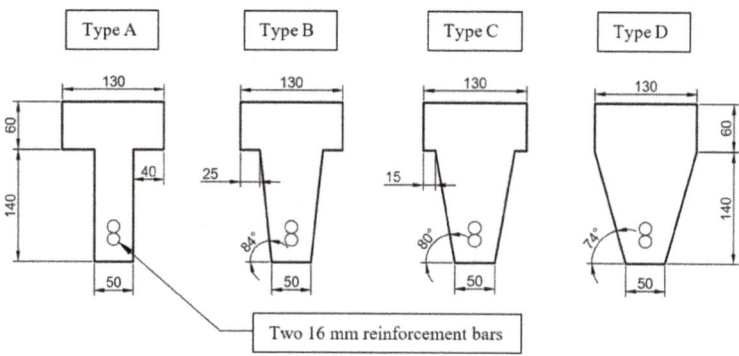

Fig. 1. Cross-sections of various beam-types and dimensions.

The test setup configuration is presented in Fig. 2, which shows that every beam was tested twice (test setup 1 in blue and test setup 2 in red), provided the beam failed in the desired shear failure mode. Vertical supports were spaced 1500 mm apart and the point load was applied 430 mm from the nearest support. A shear span-to-effective depth ratio (a/d) of 2.6 was selected to reduce the effect of arch action on shear strength. Linear Variable Differential Transformers (LVDTs) were connected to both sides of the beams in shorter shear span at 35° to the horizontal at a predicted shear crack location. The readings from the LVDTs were used to calculate shear crack widths during testing. LVDTs were used to measure vertical deflections at midspan and over supports, and net vertical deflections were determined.

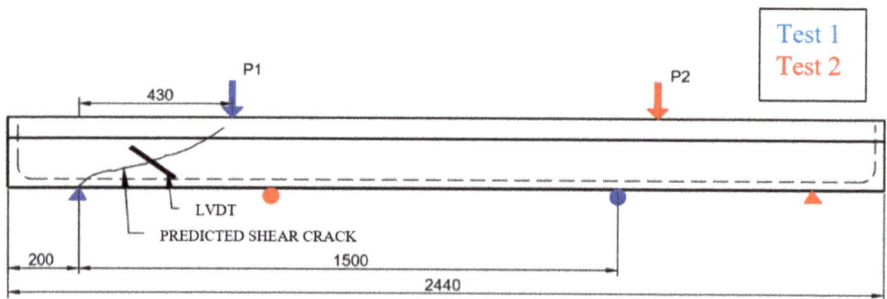

Fig. 2. Test setup configuration.

A hydraulic 5 tonne Instron actuator was used to apply load to the beams at a rate of 5 mm/min with stoppages every 5 kN. During stoppage intervals the actuator was kept in a constant position for 3 min while the beams were inspected for shear cracking and cracks were marked off. Photos were taken of the beams in stoppage intervals which were later used for studying the shear cracking patterns.

3   Results and Discussion

The experimental results are summarised in Table 3, where beams are grouped in sets according to fibre volume. Four typical failure modes were observed during testing, diagonal-tension failure (DT), flexural-shear failure (FS), shear-compression failure (SC) and rebar debonding failure (RD), however, the dominant failure mode observed in all beams is DT. DT is indicated by diagonal shear cracks spanning almost directly from the support to the point of load application. The introduction of fibres in the concrete mixture allowed for a semi-brittle failure in some beams while all the beams with no fibres displayed a more brittle failure. However, beams containing fibres that failed at relatively high loads (usually beam type C & D), brittle failures still occurred.

Table 3. Experimental results.

Beam	V_f (%)	f'_c (MPa)	V_u (kN)	v_u (MPa)	$v_u/f'^{0.5}_c$	α_{avg} (°)	Failure mode
CS-A.1-FRC0	0	27.0	23.47	1.59	0.27	34	DT
CS-A.2-FRC0	0	27.0	21.21	1.43	0.24	32	DT
CS-B.1-FRC0	0	27.7	23.95	1.42	0.25	31	DT + SC
CS-B.2-FRC0	0	27.7	24.07	1.42	0.25	30	DT + FS + RD
CS-C.1-FRC0	0	26.1	31.79	1.74	0.30	35	DT + SC
CS-C.2-FRC0	0	26.1	23.93	1.31	0.23	32	DT + SC
CS-D.1-FRC0	0	27.8	36.84	1.81	0.27	35	DT + FS
CS-D.2-FRC0	0	27.8	21.17	1.04	0.15	33	DT + FS
1-A.1-FRC0.6	0.6	26.6	26.35	1.78	0.30	23	DT + FS
1-A.2-FRC0.6	0.6	26.6	23.24	1.57	0.27	42	DT
1-B.1-FRC0.6	0.6	29.8	32.90	1.95	0.33	34	DT + FS
1-B.2-FRC0.6	0.6	29.8	38.92	2.30	0.39	32	RD + DT
1-C.1-FRC0.6	0.6	28.5	37.42	2.04	0.34	33	DT
1-C.2-FRC0.6	0.6	28.5	34.17	1.87	0.31	32	DT + SC
1-D.1-FRC0.6	0.6	26.6	41.07	2.01	0.31	34	DT + SC + FS
1-D.2-FRC0.6	0.6	26.6	45.68	2.24	0.34	35	DT + SC
2-A.1-FRC1	1	28.5	37.68	2.55	0.42	32	DT
2-A.2-FRC1	1	28.5	33.51	2.26	0.37	37	DT
2-B.1-FRC1	1	29.9	41.64	2.46	0.42	34	DT + SC + FS
2-B.2-FRC1	1	29.9	41.10	2.43	0.41	34	DT + SC + RD
2-C.1-FRC1	1	29.0	42.06	2.30	0.38	33	DT + FS
2-C.2-FRC1	1	29.0	42.89	2.34	0.39	32	DT + SC
2-D.1-FRC1	1	29.2	50.03	2.45	0.36	32	DT + FS
2-D.2-FRC1	1	29.2	42.23	2.07	0.30	33	DT + SC + FS

3.1   Shear Strengths

The average ultimate shear capacity increased significantly from beam-type A to D for all beam sets indicating that the web cross-sectional area influences the shear capacity of a beam significantly. Figure 3 shows that the average normalised shear stresses are relatively constant for the various beam-types and increase with fibre content. This indicates that shear capacity is highly dependent on the cross-sectional area of the beam, hence a much larger part of the web contributes to the shear capacity of a beam than typically stipulated in design codes. The average normalised shear stresses at ultimate resistance for the control set, Set 1 (0.6% fibres) and Set 2 (1% fibres) are 0.25, 0.37 and 0.44, respectively, which is an average increase in shear stresses of 49% for 0.6% fibres and 74% for 1% fibres when compared to the beams in the control set (no fibres).

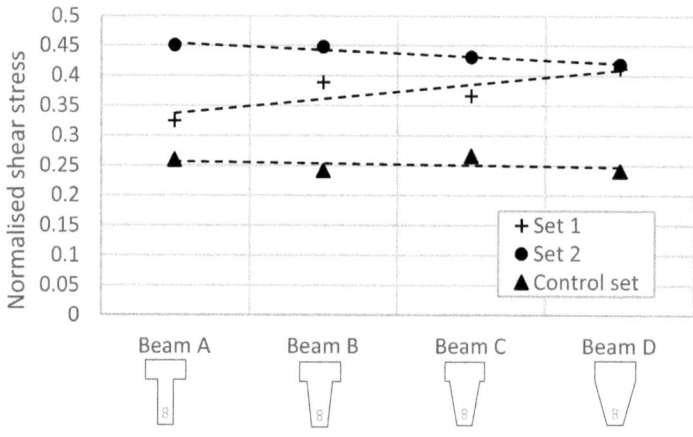

Fig. 3. Normalised shear stresses with linear trendlines indicated in dashed lines.

Regression analyses based on the average normalised shear stresses for the various fibre volumes resulted in an R2-value of 0.996 indicating that the relationship between average normalised shear stresses and fibre content is linear in this investigation. However, more research is required to investigate this relationship since it might be highly influenced by fibre properties such aspect ratio and tensile strength.

3.2   Shear Cracking

The first diagonal cracks started forming when the load reached roughly 25 ± 5 kN which is when localised tensile stresses reached the tensile capacity of the concrete. The beams containing fibres exhibited a relatively constant crack growth rate throughout the entire test until the ultimate resistance was reached. Conversely, in beams with no fibres unstable crack formation was observed with sudden and large increases in shear crack widths. This indicates that fibres allow for a steady stress transfer across cracks and maintain material integrity, resulting in post-cracking tensile strength and enhanced carrying capacity and ductility after shear cracking. Beams with no fibres failed shortly after the first cracks had formed.

Shear crack widths were significantly increased by fibres in the concrete mixture. This is because fibres allow for higher loading capacity after cracking, hence, ultimate loads are reached at wider crack widths compared to beams in the control set. Figure 4 presents the average crack width at ultimate resistance for the various beams-types and fibre volumes. The results show an increase in crack widths of approximately 40% for beam-type A and 90% for beam-type D compared to the beams containing no fibres. This shows that fibres not only increase shear resistance but also enhance the carrying capacity at larger crack widths. However, no significant difference was seen for beams with 0.6% fibres compared 1% fibres. Furthermore, there is a reduction in crack width from beam-type A to D. It is argued that as the web cross-sectional area increases more fibres bridge the crack which confines the widening of the crack keeping crack widths smaller until the ultimate load is reached.

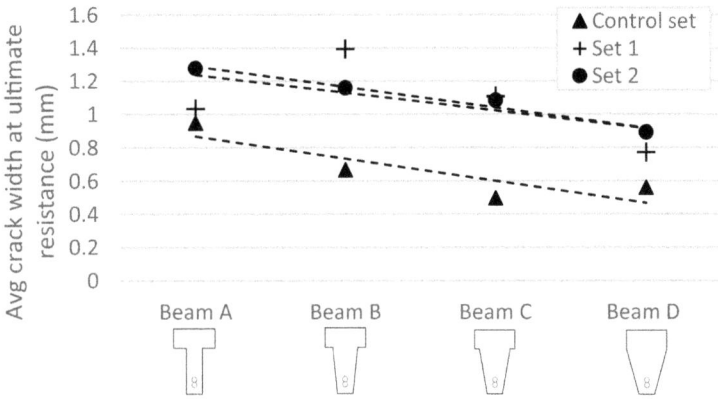

Fig. 4. Average crack width at ultimate resistance with linear trendlines indicated in dashed lines.

No differences in the diagonal shear crack angles were observed for the various beam-types and fibre volumes. An average crack angle of 33° was measured for the various beam sets. Furthermore, it was found that shear cracking was less developed and narrower in beams with fibres at service loads compared to beams with no fibres which is a significant aspect from a durability point of view. In terms of the amount of shear cracks, it was found that fibres increase the number of shear cracks that form while beams with no fibres typically only one critical shear crack formed during testing which developed and resulted in failure.

4   Conclusion

This research was done to investigate the shear behaviour of irregular shaped beams with varying web widths over the depth of the beam. The following conclusions were made from this investigation:

- Steel fibres significantly enhance the shear capacity of beams, up to 49% and 74% increase in normalised shear stresses was obtained for 0.6% and 1% fibres, respectively.
- Ultimate loads are reached at larger crack widths indicating that steel fibres ensure high load resistance at greater crack widths.
- Shear capacity is highly dependent on the entire cross-sectional area of the beam as results have shown that normalised shear stresses are relatively constant for all beam-types.
- Larger web areas significantly increase the shear strength of a beam.
- An increase in web area significantly reduces the crack width at ultimate resistance as more fibres bridge the cracks and restrict crack widening.
- Steel fibres promote a steady crack growth rate and have the potential of changing a brittle failure mode to a semi-ductile failure.
- Steel fibres have the ability to confine cracks at service loads which can lead to enhanced durability.

Acknowledgements. The authors highly appreciate the funding and facilities provided by the University of Stellenbosch which made this research possible.

References

1. Kong, F.K., Evans, R.H.: Reinforced and Pre-stressed Concrete, 3rd edn. Wokingham, England (1987)
2. Li, V.C., Ward, R., Hamza, A.M.: Steel and synthetic fibres as shear reinforcement. ACI Struct. J. **89**(5), 499–508 (1992)
3. Mohammadi, Y., Singh, S.P., Kaushik, S.K.: Properties of steel fibrous concrete containing mixed fibres in fresh and hardened state. Constr. Build. Mater. **22**, 956–965 (2008)
4. Singh, B., Jain, K.: Appraisal of steel fibers as minimum shear reinforcement in concrete beams. ACI Struct. J. **111**(5), 1191–1202 (2014)
5. FIB Bulletin 65: Model Code 2010. Laussanne, Switzerland (2010)
6. Greenough, T., Nehdi, M.: Shear behavior of fiber-reinforced self-consolidating concrete slender beams. ACI Struct. J. **105**, 468–477 (2009)
7. Parra-Montesinos, G.J.: Shear strength of beams with deformed steel fibers - evaluating an alternative to minimum transverse reinforcement. Concrete International, pp. 57–66, November 2006
8. Tung, N.D., Tue, N.V.: A new approach to shear design of slender reinforced concrete members without transverse reinforcement. Eng. Struct. **107**, 180–194 (2016)
9. Minelli, F., Conforti, A., Cuenca, E., Plizzari, G.: Are steel fibres able to mitigate or eliminate size effect in shear? Mater. Struct. **47**(3), 459–473 (2013). https://doi.org/10.1617/s11527-013-0072-y
10. Vandewalle, L., Heirman, G., Van Rickstal, F.: Fibre orientationin self-compacting fibre reinforced concrete. In: Fibre Reinforced Concrete: Design and Application, pp. 719–728. RILEM Publications SARL, Bagneux (2008)
11. Ponikiewski, T., Golaszewski, J.: Properties of steel fibre reinforced self-compacting concrete for optimal rheological and mechanical properties in precast beams. Procedia Eng. **65**, 290–295 (2013)

12. EN 1992-1-1: Eurocode 2: Design of concrete structures. Part 1-1: General rules and rules for buildings. European Committee for Standardization, Brussels (2004)
13. ACI Committee 318: Building code and commentary (ACI 318-08/318R-08). American Concrete Institute, Farmington Hills (2008)
14. Zararis, I.P., Karaveziroglou, M.K., Zararis, P.D.: Shear strength of reinforced concrete T-beams. ACI Struct. J. **103**, 693–700 (2006)
15. Thamrin, R., Tanjung, J., Aryanti, R., Nur, O.F., Devinus, A.: Shear strength of reinforced concrete T-beams without stirrups. J. Eng. Sci. Tech. **11**(4), 548–562 (2016)
16. EN 14651-A1: Test method for metallic fibered concrete - Measuring the flexural tensile strength (limit of proportionality (LOP), residual). BSI Stand. Pub. (2007)

Punching Shear Resistance of SFRC Flat Slabs with and Without Punching Shear Reinforcement

Josef Landler[(✉)] and Oliver Fischer

Department of Civil, Geo and Environmental Engineering, Technical University of Munich, Theresienstraße 90, 80333 Munich, Germany
josef.landler@tum.de

Abstract. The punching shear capacity of steel fibre reinforced concrete (SFRC) flat slabs without shear reinforcement has been the subject of extensive investigations over the past 40 years. Past investigations have shown an increase in the failure load of around 30 to 60 % and a change to a very ductile failure mode. However, most of these tests were carried out on slabs with a thickness of less than 150 mm, and are thus of only limited practical relevance. In addition, only twelve tests are available with a combination of SFRC and conventional shear reinforcement.

To partly make up for this lack of experimental data, a total of ten punching tests were carried out at the Technical University of Munich over the last three years. The experimental program included three reference tests without steel fibres and seven SFRC tests. All specimens were manufactured with conventional flexural reinforcement, a slab thickness of 250 mm and four specimens had stirrups or shear studs as punching shear reinforcement. The parameters varied were the fibre type and content as well as the flexural reinforcement ratio. All of the SFRC specimens showed a significant increase in the bearing and deformation capacity compared to reference specimens.

Keywords: Punching shear · Steel fiber reinforced concrete · Steel fibre · Shear reinforcement · Flat slabs · Stirrups · Shear studs · Punching shear test · Macro steel fibre

1 Introduction

The punching shear resistance of the slab is one of the main factors when designing flat slabs that are supported directly on columns. The combined effect of a high shear force and bending moment leads to a highly stressed area. This can result in the very brittle and abrupt failure mode "punching shear" with no warning signs and can lead to a progressive collapse of the entire structure. In the past, alternatives such as stirrups, bent-up bars or shear studs were introduced to increase the punching shear resistance. However, all these methods have disadvantages, such as special design rules, and often cause problems and deficiencies in construction. In addition, for each of these types of shear reinforcement there exists an upper limit, beyond which the punching shear capacity remains the same, regardless of an increase in the amount of shear

P. Serna et al. (Eds.): BEFIB 2021, RILEM Bookseries 36, pp. 492–503, 2022.
https://doi.org/10.1007/978-3-030-83719-8_43

reinforcement. This has come to be known as the level of maximum punching shear capacity. The use of steel fibre reinforced concrete (SFRC) in the area of a slab-column connection also has great potential to increase the punching shear capacity and therefore minimize or avoid conventional shear reinforcement. In current practical punching shear design in Germany, the required degree of flexural reinforcement ratio is often artificially increased compared to the value required for bending. Therefore, the flexural reinforcement could also be reduced by using SFRC in the slab-column connection. The labourious installation of the reinforcement as well as more difficult placing and compacting of the concrete leads to increased labour and construction costs. The use of SFRC can therefore reduce costs and improve the construction process as well as the quality of construction.

The punching shear behaviour of steel fibre reinforced concrete flat slabs without shear reinforcement has been the subject of extensive investigations by various researchers over the past 40 years (cf. [1–5]). As a result, a database assembled by the authors for punching tests on SFRC flat slabs without shear reinforcement currently contains around 400 tests from nearly 35 sources [6]. However, most of these tests were carried out on slabs with a thickness of less than 150 mm and are thus of only limited relevance for engineering practice. Additionally, novel developments in the steel strength and the anchoring behaviour of steel fibres over the past years have led to an improvement in the steel fibre performance. Compared to tests on SFRC without shear reinforcement, there are hardly any tests on SFRC flat slabs with shear reinforcement. Only twelve tests [1, 7–9] could be found in national and international literature. Due to the lack of experience with and knowledge of the interaction between SFRC and shear reinforcement, this combination is explicitly excluded in the design approaches of the German DAfStb Guideline "Steel Fibre Reinforced Concrete" [10].

The present study deals with two systematic test series on steel fibre reinforced concrete flat slabs with and without shear reinforcement. The slab thickness chosen was 250 mm in line with common applications in engineering practice. In the first test series with six specimens, the steel fibre type and content as well as the flexural reinforcement ratio were varied. The aim of the second test series, with a total of four specimens, was to investigate punching shear behaviour of SFRC slabs with stirrups or shear studs as shear reinforcement. This paper presents and discusses the results of the punching tests on SFRC flat slabs with and without shear reinforcement. Parts of this research have already been published in [5, 11].

2   Experimental Program

2.1   General

To evaluate the structural behaviour of SFRC flat slabs with and without shear reinforcement, ten punching tests were carried out on slab-column connections at an interior column. Six specimens were made without shear reinforcement and four with shear reinforcement, two each with conventionally closed stirrups and double headed studs as shear reinforcement. Three reference specimens with and without shear reinforcement were produced and tested as a direct comparison with the results of the

SFRC specimens without shear reinforcement. The test parameters investigated were the fibre type and content, the flexural reinforcement ratio and the combination of SFRC with stirrups and shear studs as shear reinforcement. All specimens with shear reinforcement were designed to fail at the level of maximum punching shear capacity. Table 1 provides an overview of the test program chosen along with the geometrical and material properties.

2.2   Test Specimens and Reinforcement Layout

A uniform notation Mx-yy-z.zz was used for all specimens (see Table 1), whereby Mx denotes the concrete mix (M0, M1, M2, M3) with varying fibre type and content, yy indicates the slab thickness in [cm] and z.zz the flexural reinforcement ratio in [%] of each test specimen. The notation used for the four test specimens with shear rein-forcement also contains information on the type of shear reinforcement. Stirrups and shear studs were used as shear reinforcement, B indicates stirrups and D shear studs. The number at the end of the notation indicates the diameter of the shear reinforcement in [mm].

All ten specimens consisted of an octagonal, reinforced concrete slab with a monolithically formed stub with identical dimensions (see Fig. 1). The inscribed diameter of the octagonal slab was $l = 2,800$ mm and the slab thickness $h = 250$ mm. The column stub at the centre of the flat slab had a side dimension of $c = 300$ mm. The specimens were restrained at 12 points radially located 1,200 mm from the centre of the column stub, thus resulting in a shear span of $a_\lambda = 1,050$ mm. The chosen concrete cover of 25 mm for the SFRC flat slabs without shear reinforcement and 35 mm with shear reinforcement led to a mean effective depth of approximately 205 mm and

Table 1. Test parameters for all slab specimens.

Specimen	c	d_m	$\varnothing_{sl}$	ρ_l	Shear reinforce-ment	ρ_{sw}	$f_{cm,cyl}$	V_f	Fibre-type	Residual tensile strength	
	[mm]	[mm]	[mm]	[%]		[%]	[MPa]	[Vol.-%]		f_{R1} [MPa]	f_{R3} [MPa]
M0-25-1.23		204	20.0	1.23			47.8		No steel fibers		
M1-25-1.23		195	20.0	1.29			48.1	0,5	4D	4.01	4.27
M2-25-1.23	300	202	20.0	1.24			39.8	0,5	5D	4.48	5.16
M3-25-0.75		205	18.0	0.77	--	--	41.2	1,0	5D	8.91	8.42
M3-25-1.23		196	20.0	1.28			44.2	1,0	5D	9.96	9.17
M3-25-1.75		200	25.0	1.80			40.9	1,0	5D	8.91	8.42
M0-25-1.75-B10		190	25.0	1.78	stirrups	0.75	39.6		No steel fibers		
M3-25-1.75-B10	300	186	25.0	1.82	stirrups	0.75	41.3	1,0	5D	8.41	7.69
M0-25-1.75-D20		186	26.5	1.85	studs	1.51	40.2		No steel fibers		
M3-25-1.75-D20		185	26.5	1.86	studs	1.51	36.3	1,0	5D	10.08	9.13

Key: c: width of square column section; d_m: average effective depth; $\varnothing_{sl}$: diameter of flexural reinforcement; ρ_l: average reinforcement ratio $(\rho_x + \rho_y)/2$; ρ_{sw}: shear reinforcement ratio $\rho_{sw} = A_{sw}/(2 \cdot s_r \cdot u_{0,5d})$; $f_{cm,cyl}$: mean cylinder compressive strength of concrete; V_f: steel fibre content; f_{R1}/f_{R3}: residual tensile strength corresponding to *CMOD* = 0.5 mm/2.5 mm

195 mm respectively. This results in a shear span-depth ratio a_λ/d of approximately 5.20.

Specimens M0-25-1.23, M1-25-1.23, M2-25-1.23 and M3-25-1.23 were designed with identical reinforcement layout to investigate the influence of the steel fibre type and content. This consisted of an upper layer of 20 mm rebar in each orthogonale direction with a characteristic yield strength of f_y = 523 MPa at a reinforcement ratio of about 1.23 %. The effect of the flexural reinforcement ratio was analysed in specimens M3-25-0.75 and M3-25-1.75 with reinforcement ratios of 0.77 % and 1.80 %. Reinforcing steel with f_y = 552 MPa and a diameter of $\varnothing_{sl}$ = 25 mm was used for the specimen M3-25-1.75. To avoid premature bending failure, high-strength reinforcing steel or post-tensioning steel was used for specimen M3-25-0.75 and for all specimens with shear reinforcement. Specimen M3-25-0.75 was designed with $\varnothing_{sl}$ = 18 mm post-tensioning bars with f_y = 986 MPa. High-strength reinforcing steel with a diameter $\varnothing_{sl}$ = 25 mm and yield strength of f_y = 764 MPa was chosen for all specimens with stirrups, whereas post-tensioning steel with f_y = 968 MPa and diameter of $\varnothing_{sl}$ = 26.5 mm was used for the specimens with shear studs. The bottom layer of reinforcement for all specimens consisted of 10 mm rebar with a characteristic yield strength of f_{yk} = 500 MPa.

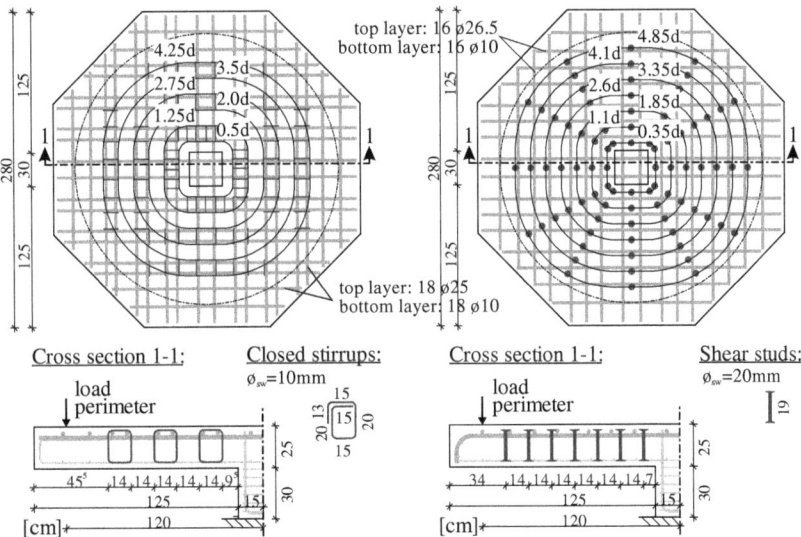

Fig. 1. Layout of flexural and shear reinforcement for test specimens with stirrups (left) and shear studs (right).

The stirrups and shear studs were made of conventional reinforcing steel with yield strengths of f_y = 529 MPa and 586 MPa, respectively, and diameters of $\emptyset_{sw}$ = 10 mm (stirrups) and 20 mm (shear studs). To ensure a failure at the level of maximum punching shear capacity, a very high shear reinforcement ratio was selected beyond the upper limit for each reinforcement type, 0.75 % and 1.51 % for the stirrups and shear studs, respectively. The layout of the stirrups was designed according to the detailing provisions of the German Annex of Eurocode 2 [12, 13]. The distance between the first perimeter of shear reinforcement and the column face was s_0 = 0.5d. The spacing between the other perimeters in a radial direction was defined as s_r = 0.75d. The selected stirrup geometry (closed stirrups) ensured sufficient anchoring quality by enclosing both the top and bottom flexural reinforcement layer. The shear studs were designed and arranged according to EOTA Technical Report TR60 [14]. The first perimeter was located at a distance of s_0 = 0.35d from the column face. All of the other perimeters had a uniform spacing of s_r = 0.75d in a radial direction. Figure 1 shows the reinforcement layout of the specimens with stirrups and shear studs in detail.

2.3   Material Properties

In addition to a plain concrete mix (M0) as a reference, three steel fibre reinforced concrete mixtures (M1, M2, M3) were used with two types of macro steel fibres from Bekaert GmbH. The maximum coarse aggregate size was 16 mm and the concrete mix was designed to reach a cylinder strength (ø150/300 mm) $f_{cm,cyl}$ of around 40 MPa at an age of 28 days. A high-strength, concrete without fibres with a compressive strength of around $f_{cm,cu}$ = 85 MPa determined using 150 mm cubes was used for the column stubs.

The type of steel fibres used were Dramix 4D 65/60 BG and Dramix 5D 65/60 BG (see Fig. 2). Dramix 4D was used in M1 with a fibre content of 0.5 Vol.-%, whereas Dramix 5D was used in M2 and M3 with a fibre content of 0.5 Vol.-% and 1.0 Vol.-% respectively. Both fibre types are hooked-end steel fibres with a length of 60 mm and a diameter of 0.9 mm; the aspect ratio is about 65. The main differences between the two fibre types are the material used and the geometry of the hooked end. Dramix 4D is made from steel with a nominal tensile strength of f_t = 1,600 N/mm^2 and a strain at ultimate strength of 0.8 %. Dramix 5D is made from a high-strength steel with a nominal tensile strength of f_t = 2,300 N/mm^2 and a strain at ultimate strength of 6.0 %. Dramix 5D has an optimised hooked-end geometry compared to Dramix 4D that prevents it being pulled out (see Fig. 2). Thanks to the geometry and material characteristics of the steel fibres, the fibre strength can be fully activated to achieve ductile post-peak behaviour.

 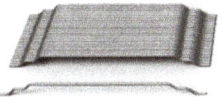

Fig. 2. Fibre types used: Dramix 4D (left) and Dramix 5D (right) [NV Bekaert SA]

The residual tensile strengths of the various concrete mixtures were determined by means of a three-point flexural test on notched specimens according to DIN EN 14651 [15]. Six specimens per SFRC mix and testing day were used for the flexural tests. The average results for the residual flexural tensile strengths f_{R1} and f_{R3} for a crack mouth opening displacement (CMOD) of 0.5 mm and 2.5 mm are summarized in Table 1. Figure 3 shows the average load-CMOD curves of all tested specimens.

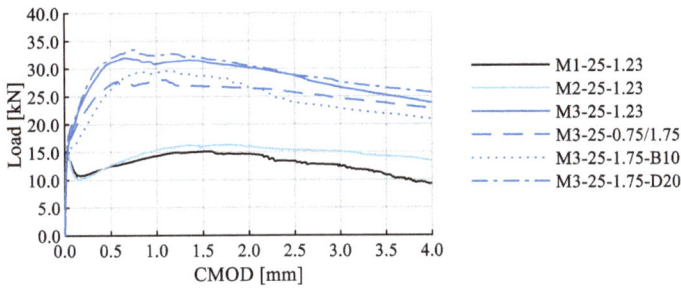

Fig. 3. Load-CMOD curves from the 3-point bending tests of all SFRC mixtures [5, 11]

2.4   Test Setup and Measurements

Figure 4 shows the test setup as a schematic. The load was applied by a central hydraulic cylinder (blue). A circular array of twelve tie rods were used to restrain the specimen and ensure the balance of vertical forces. The tie rods are anchored to an abutment plate by hollow plunger cylinders (green). All hollow plunger cylinders were linked to a common manifold and absorbed the same load independent of its individual displacement to minimize eccentricities in the application of the load.

The load was applied under load control in increments of around 150 kN up to the service load level V_{SLS}. The incremental load application allowed systematic documentation of the crack development and crack pattern on the top of the specimen. The load was cycled ten times between V_{SLS} and half its value to simulate lifetime loading. V_{SLS} was determined based on the expected failure load and the product of the partial safety factors for the material (γ_M) and the applied loads (γ_F), resulting in $\gamma = \gamma_M \cdot \gamma_F \approx 1.5 \cdot 1.4 \approx 2.1$. Following the load cycles, two or three further load steps were applied before subsequently loading the specimens continuously until failure using displacement control.

To investigate the punching shear behaviour of SFRC flat slabs in more detail, a large number of measurements was performed during the testing procedure. The vertical displacement of the specimens was recorded along the two principal axes and at the tie-rod perimeter using linear variable differential transformers (LVDT's). The increase in slab thickness and the relative displacement between the column stub and the slab was monitored at several points to investigate the development of the inner shear crack. Strain gauges were used to measure the strains in the flexural and shear reinforcement. Strain gauges also recorded the concrete strains on the compression face of the slabs near the column stub. Four inclinometers were arranged in the main axes to study the slab rotation. Further information on the test setup and the measurements performed can be found in [5, 11].

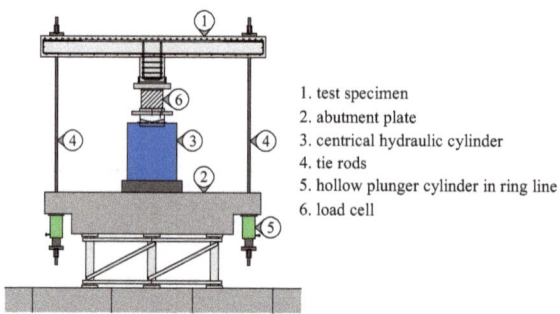

1. test specimen
2. abutment plate
3. centrical hydraulic cylinder
4. tie rods
5. hollow plunger cylinder in ring line
6. load cell

Fig. 4. Schematic of test setup [5].

3   Results of the Experiments

3.1   General

Punching shear failure was achieved in all of the ten specimens tested. Table 2 summarises the failure loads V_{Test} and the related shear forces V_{flex} that produces flexural failure according to the yield-line theory. A comparison of V_{Test} and V_{flex}, as well as an evaluation of the measured strains of the flexural reinforcement shows that the specimens had not reached their flexural capacity when punching shear failure occurred. Characteristic for all specimens was the increasing slab thickness and the penetration of the column stub into the slab before failure occurred. The cone typical for a punching shear failure became clearly visible when the specimens were later cut in two.

Table 2. Test results for all slab specimens.

Specimen	V_{flex} [kN]	V_{Test} [kN]	$V_{Test}/V_{Rk,c;EC2}$ [-]
M0-25-1.23	1859	1170	1.13
M1-25-1.23	1990	1394	1.41
M2-25-1.23	2057	1345	1.40
M3-25-0.75	2186	1692	2.00
M3-25-1.23	2248	1740	1.81
M3-25-1.75	2874	1906	1.86
M0-25-1.75-B10	3171	1986	2.03
M3-25-1.75-B10	3226	2186	2.26
M0-25-1.75-D20	3530	2153	2.24
M3-25-1.75-D20	3283	2866	3.12

V_{flex}: flexural capacity; V_{Test} failure load; $V_{Rk,c;EC2}$: punching shear resistance without shear reinforcement according to DIN EN 1992-1-1/NA(D) $V_{Rk,c;EC2} = 0{,}18/\gamma_c \cdot k \cdot (100 \cdot \rho_l \cdot f_{ck})^{1/3}$

3.2 Load-Deflection Behaviour

Figure 5 presents the load-deflection curves from the ten tests. The "deflection" plotted on the abscissa describes the deflection in the centre of the slab adjusted for the vertical deflection in the tie rod restraints. A stiff initial response followed by a decrease in stiffness due to the formation of flexural cracks can be observed in all ten test specimens. The gradients on both branches before and after flexural cracking are virtually independent of the addition of steel fibres or shear reinforcement.

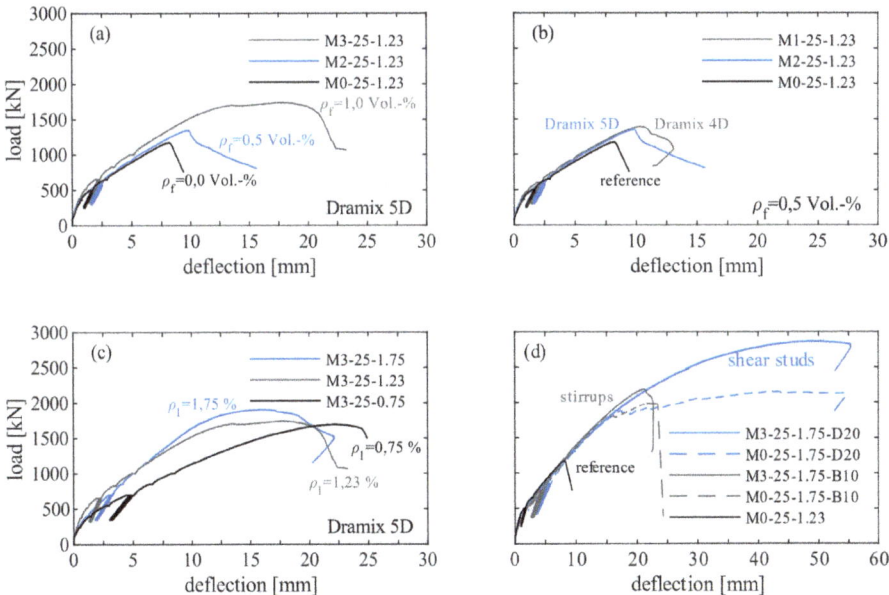

Fig. 5. Comparison of load-deflection curves with varied (a) fibre content, (b) fibre type, (c) flexural reinforcement ratio and (d) shear reinforcement.

With increasing steel fiber content, both the maximum load and the deformation increase, and the load-deflection curves merely continue with the same stiffness (Fig. 5 (a)). In addition, the increasing steel fibre content enables the complete compensation of the tensile forces released during crack formation. With a fibre content of 1.0 Vol.-%, no brittle decrease in the load-deflection curve is visible after reaching the maximum load. In Fig. 5 (b), the steel fibre type shows no significant influence on the achievable maximum load. Compared to specimen M1-25-1.23 with the Dramix 4D fibres, the improved anchorage behaviour of the Dramix 5D steel fibres in specimen M2-25-1.23 allows a more ductile post-cracking behaviour at large deflection. Varying the flexural reinforcement ratio, a change in stiffness after flexural cracking was observed, as seen in Fig. 5 (c). This tendency has already been observed in previous experimental tests on reinforced fibre-free concrete flat slabs [16]. Within typical test scatter, the present tests do not indicate a direct influence of the flexural reinforcement ratio on the fiber contribution. The arrangement of shear reinforcement in a reinforced concrete flat slab without steel fibres leads to a continuation of the load-deflection curve with a nearly constant stiffness and a growth in the failure load in a similar way (Fig. 5 (d)). The addition of steel fibres enables a further increase in the load capacity and deformation. The compression zone is strengthened due to the steel fibres and the punching shear reinforcement is subject to lower strains at the same load, which leads to the higher ultimate loads.

In general, the absolute value of the failure load and the value of the related deflection increases in a comparable manner with increasing fibre content or the addition of shear reinforcement. Since flexural cracks formed during all tests at roughly similar loads, an increase in the steel fibre content or the addition of punching shear reinforcement resulted in a continuation of the load-deflection curves with an almost constant stiffness. The punching shear strength of reinforced concrete flat slabs without fibres and shear reinforcement is governed by the width and the roughness of the shear cracks. With increasing load and slab rotation, the shear crack width increases continuously and the contributions of tensile stresses and aggregate interlock stresses in the shear crack decrease. The flat slab will fail after these contributions are lost and the critical shear crack cuts through the compression zone [17, 18]. Adding steel fibres to the concrete can delay or limit this progressive crack opening. The steel fibres connect the crack edges to each other and enable the transmission of tensile stresses after crack formation. The limited crack opening also means that the fibre-independent shear transfer mechanisms are even effective at higher loads. The combined effect of the classical punching shear transfer actions and the fibre bridging-stress of the SFRC permit a significant increase in the failure load. This correlation is graphically shown in Fig. 6 by the change in slab thickness plotted over time. When the critical shear crack opening is reached, the slab thickness of the reference specimen M0-25-1.23 increases suddenly. As a result of the addition of steel fibres, the gradient of the change in slab thickness of both specimens with steel fibres M2-25-1.23 and M3-25-1.23 decreases significantly after the maximum load is exceeded and ductile failure is observed.

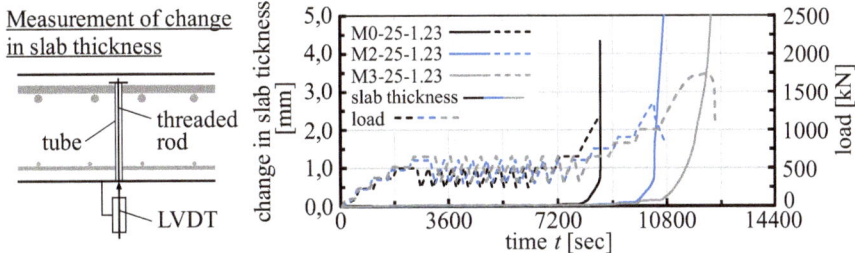

Fig. 6. Measured changes in thickness over time for test specimens M0-25-1.23, M1-25-1.23 and M3-25.123.

3.3   Cracking Pattern

Saw-cuts were made in all specimens after testing to examine the inner shear crack pattern. Figure 7 shows exemplarily photographs of the saw-cuts M0-25-1.23, M3-25-1.23, M0-25-1.75-D20 and M3-25-1.75-D20 as representative of all specimens. The saw-cut of the reference specimen M0-25-1.23 without shear reinforcement shows a fine shear crack. In contrast, test M3-25-1.23 with a steel fibre content of 1.0 Vol.-% shows a finely structured crack band due to multiple crack formation and a crack inclination similar to that of M0-25-1.23. Considering the usual test scatter, no direct influence of the steel fibres on the crack inclination can be detected.

The crack pattern of the reference specimen with punching shear reinforcement M0-25-175-D20 extends over the entire area of punching shear reinforcement and shows a separation of the compression zone. The addition of steel fibres reduces the cracking to the area around the first shear stud and therefore results in a very steep shear crack. In addition, the compression zone is not separated over a longer distance but rather is heavily damaged in the column region.

a) M0-25-1.23

b) M3-25-1.23

c) M0-25-1.75-D20

d) M3-25-1.75-D20

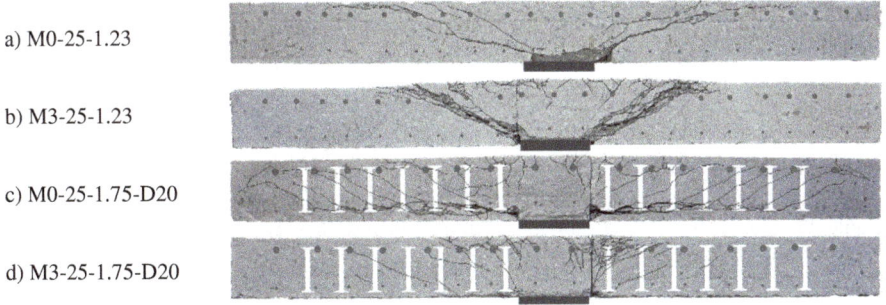

Fig. 7. Saw cuts of specimens M0-25-1.23, M3-25-1.23, M0-25-1.75-D20 and M3-25-1.75-D20.

4   Summary and conclusion

In summary, a punching shear failure could be confirmed for all specimens with and without shear reinforcement based on the evaluation of the crack patterns, the changes in slab thickness and the tensile strains in the flexural reinforcement. The specimens with shear reinforcement show a very steep failure crack between the first row of shear reinforcement and the column face and a significant concrete degradation in the compression zone. Because the strains in the shear reinforcement did not exceed the yield strength, the capacity of the shear reinforcement had not yet been reached when failure occurred. The failure mode can therefore be classified as failure at the level of the maximum punching shear capacity for all four specimens with shear reinforcement.

The results of the experimental investigations on steel fibre reinforced flat slabs with and without conventional punching shear reinforcement allow the following conclusions to be drawn:

- The punching shear capacity of slab-column connections without shear reinforcement can be significantly increased by adding steel fibres. Depending on the fibre content, an increase of 40 % to 80 % is possible compared to the punching shear capacity of flat slabs without steel fibres and shear reinforcement according to DIN EN 1992-1-1 (EC2).
- The combined use of SFRC with shear reinforcement in flat slabs enables a further significant increase in the punching shear capacity. The very steep failure crack and the strains in the shear reinforcement showed that both of the specimens that were investigated failed at the level of the maximum punching shear capacity. The failure load for the specimens with stirrups could be increased due to the addition of steel fibres by around 10 % compared to the reference specimen. The specimen with shear studs even allowed the failure load to be increased by around 33 %.
- The steel fibre types Bekaert Dramix 4D and Dramix 5D which were investigated have no notable influence on the failure load that is achieved. However, due to the better end anchorage, the post-cracking behaviour becomes more ductile with the high-strength, steel fibre type Dramix 5D.
- Increasing the fibre content leads to an increasing inner shear crack width. The fine, discrete shear crack in the saw-cut of the reference specimen changes into a coarse crack band due to multiple crack formation.

Acknowledgements. The authors want to express their gratitude to the research initiative "Zukunft Bau" of the German Federal Institute for Research on Building, Urban Affairs and Spatial Development (BBSR) and the "Stiftung Bayerisches Baugewerbe" for funding the research project this paper is based on.

References

1. Swamy, R.N., Ali, S.A.R.: Punching shear behavior of reinforced slab-column connections made with steel fiber concrete. ACI J. **79**(5), 392–406 (1982)

2. Cheng, M.Y., Parra-Montesinos, G.: Evaluation of steel fiber reinforcement for punching shear resistance in slab-column connections – Part I: monotonically increased load. ACI Struct. J. **107**(1), 101–109 (2010)
3. Gouveia, N.D., Lapi, M., Orlando, M., Faria, D.M.V., Ramos, A.M.P.: Experimental and theoretical evaluation of punching strength of steel fiber reinforced concrete slabs. Struct. Concr. **19**(1), 1–12 (2018)
4. Chanthabouala, K., Teng, S., Chandra, J., Tan, K.-H., Ostertag, C.P.: Punching tests of double-hooked-end fiber-reinforced concrete slabs. ACI Struct. J. **115**(11), 1777–1789 (2018)
5. Landler, J., Fischer, O.: Steigerung der Durchstanztragfähigkeit und Duktilität durch die Zugabe moderner Hochleistungsstahlfasern. Beton- und Stahlbetonbau **114**(9), 663–673 (2019)
6. Landler, J., Fischer, O.: Datenbank zum Durchstanzen stahlfaserverstärkter Flachdecken ohne Durchstanzbewehrung. Beton- und Stahlbetonbau **116**(special Issue S1), 59–69 (2021)
7. Zambrana Vargas, E.N.: Punção em lajes-cogumelo de concreto de alta resistência reforçado com fibras de aço. Ph.D.-thesis, Universidade de Sao Paulo (1997)
8. Azevedo, A.: Resistência e ductilidade das ligações laje-pilar em lajes-cogumelo de concreto de alta reistência armado com fibras de aço e armadura transversal de pinos. Ph.D.-thesis, Universidade de Sao Paulo (1999)
9. Musse, T.H., Liberati, E.A.P., Trautwein, L.M., Gomes, R.B., Guimaraes, G.N.: Punching shear in concrete reinforced flat slabs with steel fibers and shear reinforcement. Ibracon **11**(5), 1110–1121 (2018)
10. German Committee for Structural Concrete, DAfStb Guideline "Steel Fibre Reinforced Concrete". Beuth Verlag (2012)
11. Landler, J., Fischer, O.: Durchstanzen stahlfaserverstärkter Flachdecken mit Durchstanzbewehrung. Beton- und Stahlbetonbau **116**(5) (2021, accepted for publication)
12. DIN EN 1992-1-1:2011-01: Eurocode 2: Design of concrete structures – Part 1-1: General rules and rules for buildings; German version EN 1992-1-1:2004 + AC2010
13. DIN EN 1992-1-1/NA:2013-04: National Annex – Nationally determined parameters – Eurocode 2: Design of concrete structures – Part 1-1: General rules and rules for buildings
14. EOTA Technical Report: TR 060 – Increase of punching shear resistance of flat slabs or footings and ground slabs – double headed studs – Calculation Methods (2017)
15. DIN EN 14651:2007-12: Test method for metallic fibre concrete – Measuring the flexural tensile strength (limit or proportionality (LOP), residual)
16. Guandalini, S., Burdet, O., Muttoni, A.: Punching tests of slabs with low reinforcement ratios. ACI Struct. J. **106**(1), 87–95 (2009)
17. Muttoni, A.: Punching shear strength of reinforced concrete slabs without transverse reinforcement. ACI Struct. J. **105**(4), 440–450 (2008)
18. Schmidt, P., Kueres, D., Hegger, J.: Punching shear behaviour of reinforced concrete flat slabs with a varying amount of shear reinforcement. Struct. Concrete 1–12 (2019)

Elevated Steel Fibre Reinforced Concrete Slabs and the Hybrid Alternative: Design Approach and Parametric Study at Ultimate Limit State

Stanislav Aidarov[1,2]([✉]), Luca Sutera[3], Manuela Valerio[3], and Albert de la Fuente[2]

[1] Smart Engineering Ltd., UPC Spin-Off, Barcelona, Spain
stanislav.aidarov@upc.edu
[2] Polytechnic University of Catalonia, Barcelona, Spain
[3] Polytechnic University of Turin, Turin, Italy

Abstract. Application of Fibre Reinforced Concrete (FRC) in pile supported flat slabs is definitely challenge in term of a structural application of this material. Possible substitution of traditional reinforcement by steel fibres in the concrete mix drew attention of researchers due to clear benefits that could be provided by this technological alternative. This statement has been already proven by existing examples, in which during the construction, the optimization of resources, reduction of execution time and required labor force was highlighted.

Nevertheless, considering the knowledge base associated with structural behaviour of FRC, it has been confirmed that hybrid solutions consisting in FRC with a moderate residual flexural tensile strength combined with conventional reinforcement placed in those zones where the higher bending moments are expected can be even more attractive from the technical point of view.

Taking into account the abovementioned, the potential application of Hybrid Reinforced Concrete (HRC) for elevated slabs was studied. By modifying the ratio fibre/rebar content for different structural geometries, a parametric study was carried out in order to evaluate various solutions in terms of structural capacity in accordance with the requirements of Ultimate Limit State.

Keywords: Fibre Reinforced Concrete · Elevated slabs · Hybrid Reinforced Concrete

1 Introduction

Construction of elevated slabs gained a competitive solution since Fibre Reinforced Concrete (FRC) was accepted as a structural material by relevant codes and recommendations [1–4]. This material reinforced by high content of fibres (over 60 kg/m^3) provides a significant residual flexural tensile strength, thus allowing substitution of traditional reinforcement in the structures with relatively high stresses [5–7]. The presence of fibres in concrete composition, moreover, enhances punching shear strength and ductility [8, 9] what is of paramount importance in design of pile supported flat slabs.

© RILEM 2022
P. Serna et al. (Eds.): BEFIB 2021, RILEM Bookseries 36, pp. 504–513, 2022.
https://doi.org/10.1007/978-3-030-83719-8_44

The viability of this approach has been already proven by several real scale tests [5, 7] and existing structures where partial/total substitution of conventional reinforcement was implemented. All examples of FRC application in elevated flat slabs highlighted the significant reduction of execution time and required labour force. In case of Spanish experience [10], the economic study was realized in order to justify tangible outcomes by using FRC in the abovementioned structural element. The provided analysis demonstrated the financial advantage of novel technology: the cost savings were around 12% in comparison with the traditional solution.

However, bending and shear forces can differ significantly throughout the slab, this requiring different residual flexural strengths (f_R) of the FRC (different amounts of fibres therefore); in this sense, as FRC has to be designed for guaranteeing the most demanding f_R requirement, over-reinforced areas can result from this strategy. Considering this aspect, hybrid reinforced concrete solutions (fibres + traditional reinforcement, HRC) might be even more attractive from technical and economic points of view since a moderate values of f_R can be fixed while most demanding bending and punching forces can be resisted by HRC.

In this context, the potential application of HRC for elevated slab was analysed by means of the yield line method for designing pile supported flat slabs at Ultimate Limit State (ULS). Numerous fibre/rebar ratios were studied for 6 × 6 m panel flat slab considering different thicknesses; the procedure and results are found to be of interest for similar future designs.

2   Flexural Strength of SFRC and HRC

The significant amount of reinforcement (nearly all) in pile supported flat slabs is intended to provide the required flexural strength of the structure. Therefore, in both studied alternatives (FRC and HRC solutions), the design procedure focuses on determination of the adequate flexural reinforcement under established load conditions, having in mind the irrefutable need of further investigation of punching shear and long-term responses of the elevated FRC/HRC slabs. For this purpose, the simplified constitutive laws for concrete behaviour in both compression and tension were assumed, i.e. the rigid-plastic block of stresses (Fig. 1).

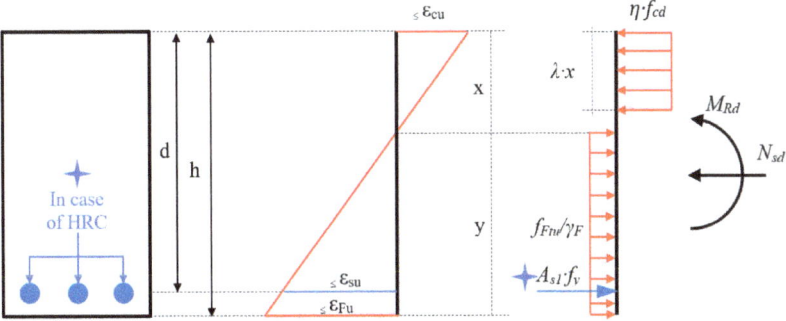

Fig. 1. Sectional model to assess flexural strength of SFRC and HRC

The simplified rigid plastic model for the assessment of the ULS behaviour of FRC in tension can be identified by a unique value of $f_{Ftu,d}$. This value only depends on the characteristic residual tensile strength at the crack mouth opening displacement (CMOD) of 2.5 mm (f_{R3k}) in accordance with EN 14651 [11] and can be calculated as $f_{Ftu,d} = f_{R3k}/(3 \cdot \gamma_F)$, where $\gamma_F = 1.50$ is the partial safety factor in tension for FRC. The structural response of FRC in compression, in turn, is to be traditionally represented by the compressive block $\eta \cdot f_{cd} \cdot \lambda \cdot x$, where f_{cd} is the design compressive strength (f_{ck}/γ_F) and η along with λ are defining the effective height of the compression zone and the effective strength, respectively [12].

Taking into consideration the above adopted distribution of concrete stresses at ULS and in accordance with fundamental propositions of concrete design, which demand the equilibrium of internal forces and external loads, Navier-Bernoulli hypothesis, equality of embedded rebar and the surrounding concrete strains, the position of neutral axis (x) and the design resisting moment (M_{Rd}) are to be calculated for both solutions by following equations:

$$x = \frac{f_{Ftu,d} \cdot h \cdot b + A_{s1} \cdot f_{yd}}{(0.8 \cdot f_{cd} + f_{Ftu,d}) \cdot b} \tag{1}$$

$$M_{Rd} = 0.48 \cdot f_{cd} \cdot x^2 \cdot b + f_{Ftu,d} \cdot \frac{(h-x)^2}{2} \cdot b + A_{s1} \cdot f_{yd} \cdot (d-x) \tag{2}$$

The geometry of the section in these equations is characterized by the established width (b) and depth (h). Additionally, it is worth highlighting that flat slabs typically have low ratio of tensile reinforcement what leads to ductile collapse, i.e. the failure occurs due to excess of ultimate strain in tension whereas the maximum compressive strain is far from reaching its ultimate value. Therefore, the contribution of steel bars in the sectional analysis can be evaluated as $A_{s1} \cdot f_{yd}$, where A_{s1} is the cross sectional area of reinforcement and f_{yd} – design yield strength of reinforcement.

Finally, applying Eqs. 1–2, the design resisting moment can be obtained for different fibre/rebar contents. The design objective is to guarantee that $M_{Rd} \geq M_{Ed}$, M_{Ed} being the design bending moment produced by the factored load combinations; M_{Ed}, in turn, is to be assessed by means of Yield Line Theory as it is described in the next section.

3 Yield Line Analysis for Slabs

Yield Line Method for the evaluation of load-carrying capacity of elevated slabs is one of the most versatile analytical tools due to absence of certain restrictions for the analysis as with various elasticity-based methods: there is no restrictions regarding the boundary conditions, dimensions of the openings, geometry of the element or load type. This approach is based on Johansen's theory [13, 14] and claims that the ultimate load of a slab can be obtained by postulating a collapse mechanism (generated by yield lines) which is compatible with the boundary conditions. The important requirement is that the slab sections are assumed to be sufficiently ductile to allow plastic rotation to occur at critical section along yield lines which is usually fulfilled by the amount of tensile reinforcement in standard elevated slabs [15] and, a fortiori, in case of SFRC/HRC pile supported slabs.

The application of Yield Line Method is comprised of three essential steps: a) selection of relevant yield line pattern, i.e. collapse mechanism b) in accordance with the selected collapse mechanism the ultimate moment/failure load ratio is to be estimated c) based on the achieved data, the section can be calculated and, as the outcome, the required reinforcement is to be found. The first from the above listed steps is of paramount importance because the error will lead to wrong evaluation of failure load and, as a consequence, to unsafe design since the presented method gives upper bound solutions. However, nowadays, yield line patterns for common geometries are developed and deeply studied, therefore, margin of error approaches zero.

The second step that deals with the relation between applied load and produced moments can be assessed by means of the method of segment equilibrium or method of virtual work. Nevertheless, for structures like pile-supported slabs of common shape, the standard formulae are already presented and there is no need to provide relatively laborious calculations in order to estimate the above named relation. For instance, the simplified case of the elevated slab without any opening under uniformly distributed load (q_d), which will be studied in this paper, has three different established load – ultimate moment ratios (Fig. 2). Two of them refers to, so-called, global failure of internal (Eq. 3) and corner panels (Eq. 4) of the slab and the third case concerns to local failure (Eq. 5) – the situation where the ultimate load (P_{col}) is transferred to column from the slab tributary area [16]. Due to this transfer of load, around each column negative radial lines emanate from the centre and a positive circumferential yield line forms at the bottom of the slab (Fig. 2). The above described equations depend on the ratio of negative to positive flexural capacities of the studied element ($\oslash_h$); this value is usually considered as 1.0 for FRC/HRC solutions, although this assumption could be controversial in terms of real fibre distribution along the depth of an element [5, 17].

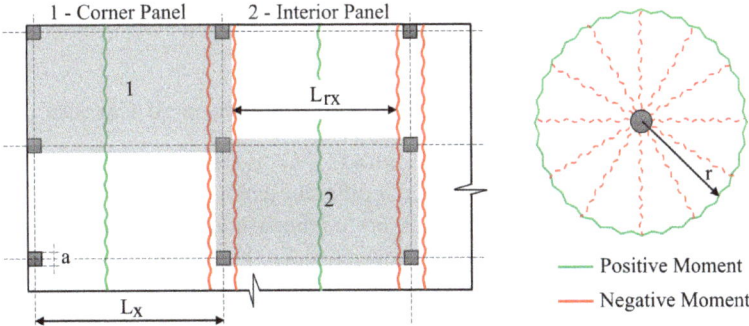

Fig. 2. Yield Line Patterns for elevated flat slabs: global and local failures

$$M_{Ed}^+ = \frac{q_d \cdot L_{rx}^2}{8 \cdot (1 + \oslash_h)} \tag{3}$$

$$M_{Ed}^+ = \frac{q_d \cdot L_{rx}^2}{2 \cdot (\sqrt{(1 + \oslash_h)} + 1)^2} \tag{4}$$

$$M_{Ed}^+ = \frac{P_{col}}{1 + \oslash_h} \cdot \frac{1}{6.2 \cdot (1 + \frac{4 \cdot a}{L_x})} \tag{5}$$

$$M_{Ed}^- = \oslash_h \cdot M_{Ed}^+ \tag{6}$$

Finally, the required flexural reinforcement can be found based on the obtained design moment and selected sectional model for the design of structural element in question (Fig. 1). In case of the presented study, design was performed in compliance with the listed phases including following details:

- Design for the selected geometry of elevated slab was carried out varying the residual tensile flexural strength of the material (f_{R3}) from 0 to 10 MPa in increment of 1 MPa.
- The variation of f_{R3} led to re-design of required traditional flexural reinforcement. However, the traditional reinforcement layout does not change throughout the study; the variable parameter is the required mm^2/m of reinforcement.
- The behaviour of FRC/HRC elevated slab under the load of 14 kN/m^2 at ultimate limit conditions (typical load magnitude for residential and office buildings in accordance with the Spanish Building Code) was studied for three different thicknesses: 190, 200 and 220 mm.

4   Design of FRC and HRC Flat Slab at ULS

4.1   Geometry of Structure and Top/bottom Reinforcement Layouts

The span length in both orthogonal directions (L_x, L_y) is set at 6 m while the slab depth is ranged between 190 and 220 mm for 3 different cases as it was stated previously. Thus, the obtained span to depth ratios are commensurate with the maximum ones related to previously constructed FRC elevated slabs (Fig. 3).

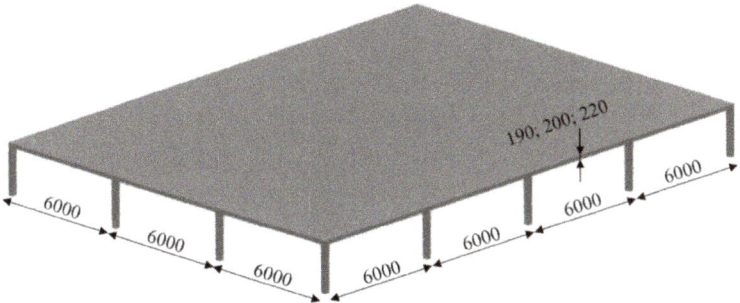

Fig. 3. Geometry of SFRC/HRC flat slab (dimensions are presented in mm)

The position of reinforcement areas is predefined: bottom reinforcement is placed across whole bays without any curtailment due to existence of several yield line patterns that demand similar load of failure (slightly less) and can be developed along the lines where the reinforcement is reduced. The top reinforcement is concentrated in vicinity of columns in accordance with Fig. 4:

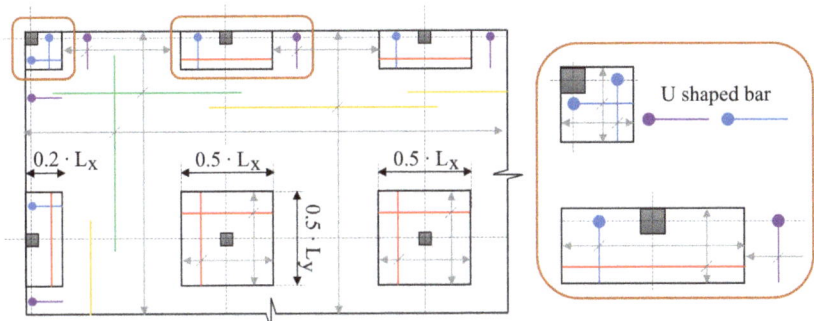

Fig. 4. Arrangement of Top & Bottom Reinforcement

Presented arrangement of the top reinforcement has certain advantages and shortcomings. The concentration of reinforcement over the columns certainly enhances the shear resistance of the slab and contributes in dealing with the peaking of the moments at service loads. From the other part, omitting top reinforcement between concentrations over column heads can lead to some incidental cracking in these areas [18]. However, the indicated aspects are out of the scope of the paper in question and should be analysed in further studies.

In case of flat slabs, the curtailment length also should be checked against local failure patterns (Fig. 2). For this purpose, the radius of circular yield line should be calculated by means of Eqs. 7–8 and the required anchorage length of the reinforcement is to be provided. Analysing the presented geometry and load conditions, the top reinforcement of $0.5 \cdot L_x / 0.5 \cdot L_y$ for internal columns fulfilled the requirements of the minimum anchorage length. The same statement is also related to the top reinforcement areas of edge and corner pillars.

$$r = c \cdot \sqrt[3]{\frac{P}{q \cdot A}} \tag{7}$$

$$c = \sqrt{\frac{a \cdot b}{\pi}} \tag{8}$$

Where: r – the radius of the positive circumferential yield line (Fig. 2); c – the radius of an equivalent circular column; a, b – dimensions of the column; q – ultimate uniformly distributed load; A – area of column cross section and P – ultimate load transferred to column from the slab tributary area.

4.2   Parametric Study of SFRC/HRC Solutions

The post-cracking residual tensile strength of the FRC slightly modifies the calculation of the ultimate bending moments along the yield lines in comparison with conventionally reinforced concrete solutions (Fig. 5a), because the zones without any top reinforcement still provide the certain flexural strength of the element and, as a consequence, the pick moments in the vicinity of the columns are to be reduced (Fig. 5b). Also, the possible variation in top reinforcement arrangement should be pointed out: the designer can keep to the (elastic) code recommendations of dividing the total negative moment in the proportions 75% and 25% in column and middle strips respectively in order to avoid uncontrollable cracks in the zones without any reinforcement [18]. However, the design process will follow the same procedure described in this paper.

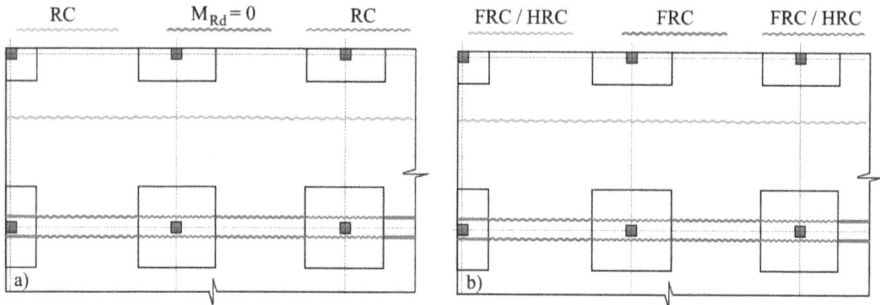

Fig. 5. Moment proportioning in: a) Traditionally reinforced concrete solution; b) FRC/HRC approach.

Considering this fact and according to the presented equations, required amount of reinforcing steel bars was calculated for each solution, varying the characteristic value of residual tensile strength for a crack mouth opening displacement of 2.5 mm (f_{R3k}) of the FRC. The results showed similar tendency: the increment of f_{R3k} linearly reduces the amount of conventional reinforcement per cubic meter (ρ_s) up to a certain f_{R3k} ranged between 6 and 8 MPa depending on the slab thickness (Fig. 6).

Finally, the Serviceability Limit States in terms of deformations and cracking adjusted the relationships presented in Fig. 6. In this regard, Fig. 7 depicts the relationship $\rho_s - f_{R3k}$ for the quasi-permanent combination of loads $q_{cp} = 9$ kN/m^2 and a maximum crack with of $w_{max} = 0.20$ mm. It must be noted that the SLS of cracking governs the amount of reinforcement.

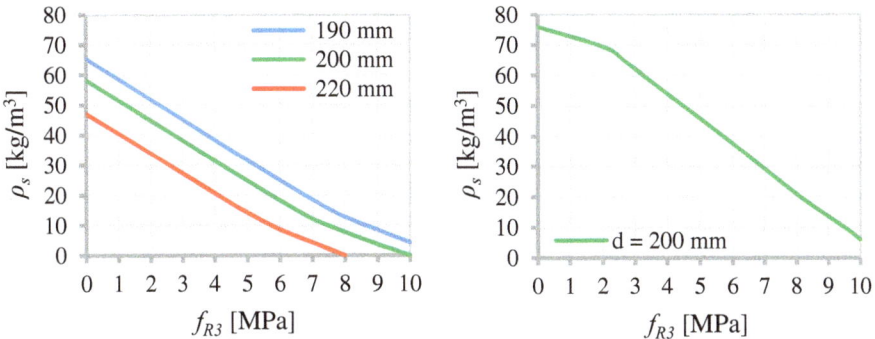

Fig. 6. Relationship $f_{R3k} - \rho_s$ at UDL for $q_d = 14$ kN/m^2.

Fig. 7. Relationship $f_{R3k} - \rho_s$ at SLS for $q_{cp} = 9$ kN/m^2 ($w_k = 0.20$ mm)

The decision (the combination of f_{R3k} and ρ_s to be considered) should be made based onto the sustainability concept. In this regard, the economic aspect can be evaluated by the consideration of direct/indirect costs and the established period of construction procedure whereas the analysis of environmental and social aspects can involve the study of the amount of emission, occupational risk during the construction and third-party effects such as noise pollution and other inconveniences [19]. Additionally, the necessity of carrying out a rigorous control of FRC elaboration is to be pointed out, especially for the concrete mixes with a relatively high amount of fibres; the mixing procedure, transportation, and pouring approach should be taken into account in order to provide a suitable workability of the material during the construction process.

5   Conclusions

The application of FRC/HRC for elevated slabs is already an attractive solution nowadays. However, the analytical approaches for the design is not suitably covered, especially in case of HRC. This paper describes the use of Yield Line Method for estimation of various HRC solutions at Ultimate Limit State using rigid-plastic model for FRC in accordance with the *fib* Model Code 2010. Based on the obtained results, the following conclusions can be derived:

- Hybrid Reinforced Concrete solution for elevated slabs can be even more attractive solution in comparison with solely FRC since over-reinforced zones result from the high demand of residual flexural strengths in certain areas (ex., negative bending moment around the columns).
- The cases analyzed in the parametric study confirm that SLS of cracking appears to be the most demanding in terms of concrete reinforcement requirement.

The results presented herein are partial since these are part of an ongoing PhD thesis developed by the first author. The numerical-aided design method proposed for this type of FRC structural elements as well as analyses and provisions for both the crack width and deflections are being developed and these will be published.

Acknowledgements. The authors wish to express their gratitude to the Spanish Ministry of Economy and Competitiveness for the financial support in the scope of the project eFIB (reference: RTC-2016-5236-5) which was carried out along with SACYR Ingeniería e Infraestruturas. The first author also thanks the Department of Enterprise and Education of Catalan Government for providing support through the PhD Industrial Fellowship (2018 DI 77) in collaboration with Smart Engineering Ltd.

References

1. fib 2010. fib Model Code for Concrete Structures (2010)
2. Ministerio de Fomento 2008 Instrucción de Hormigón Estructural (EHE-08)
3. RILEM TC 162-TDF 2002 Recommendations of RILEM TC 162-TDF: Test and design methods for steel fibre reinforced concrete: bending test. Mater. Struct. **35**, 579–582
4. Italian National Research Council 2006 CNR. CNR-DT 204 /2006 Guide for the Design and Construction of Fiber-Reinforced Concrete Structures (Rome)
5. Aidarov, S., Mena Sebastiá, F., de la Fuente, A.: Structural response of a fibre reinforced concrete pile-supported flat slab: full-scale test. Eng. Struct. (2021)
6. Destrée, X., Mandl, J.: Steel fibre only reinforced concrete in free suspended elevated slabs: Case studies, design assisted by testing route, comparison to the latest SFRC standard documents. Tailor Made Concr. Struct. 437–443 (2008)
7. Gossla, U.: Development of SFRC Free Suspended Elevated Flat Slabs (2005)
8. Gouveia, N.D., Lapi, M., Orlando, M., Faria, D.M.V.: Experimental and theoretical evaluation of punching strength of steel fiber reinforced concrete slabs 13 (2017)
9. Gouveia, N.D., Fernandes, N.A.G., Faria, D.M.V., Ramos, A.M.P., Lúcio, V.J.G.: SFRC flat slabs punching behaviour – experimental research . Compos. PART B **63**, 161–171 (2014)
10. Maturana Orellana, A.: Estudio teórico-experimental de la aplicabilidad del hormigón reforzado con fibras de acero a losas de forjado multidireccionales (2013)
11. CEN 2007 EN 14651:2007. Test method for metallic fibre concrete. Measuring the flexural tensile strength (limit of proportionality (LOP), residual)
12. EN 1992-1-1 2004 Eurocode 2: Design of concrete structures: General rules and rules for buildings. CEN, Brussels
13. Johansen, K.W.: Yield-line theory (Cement and Concrete Association) (1962)

14. Johansen, K.W.: Yield-line formulae for slabs (Cement and Concrete Association) (1972)
15. Nilson, A.H., Darwin, D., Dolan, C.W., Charles, W.: Design of concrete structures
16. ACI Committee 544. and American Concrete Institute. 2015 Report on the Design and Construction of Steel Fiber-Reinforced Concrete Elevated Slabs (American Concrete Institute)
17. Barros, J., Salehian, H., Pires, M., Gonçalves, D.: Design and testing elevated steel fibre reinforced self-compacting concrete slab. Fibre Reinf. Concr. 1–12 (2012)
18. Kennedy, G., Goodchild, C.H.: Practical Yield Line Design (2004)
19. de la Fuente, A., Casanovas-Rubio, M.D.M., Pons, O., Armengou, J.: Sustainability of column-supported RC slabs: fiber reinforcement as an alternative. J. Constr. Eng. Manag. **145**, 1–12 (2019)

Shear Behavior of Hollow-Core Slabs Reinforced by Macro-synthetic Fibers

Alan Piemonti[1], Francisco Ortiz-Navas[2], Antonio Conforti[1(✉)],
Giovanni Plizzari[1], Sandro Moro[3], Martin Hunger[4], Steve Schaef[5],
and Bruno Della Bella[6]

[1] Department of Civil, Environmental, Architectural Engineering
and Mathematics (DICATAM), University of Brescia, Brescia, Italy
antonio.conforti@unibs.it
[2] Concrete Science and Technology Institute,
Universitat Politecnica de Valencia, Valencia, Spain
[3] Master Builders Solution Italia Spa, Treviso, Italy
[4] Master Builders Solution Deutschland GmbH, Mannheim, Germany
[5] Master Builders Solution Beachwood, Beachwood, USA
[6] Gruppo Centro Nord, Belfiore, Italy

Abstract. Several researches demonstrated that the addition of fibers in correct proportions can significantly increase the shear behavior of reinforced concrete elements, allowing to partially or totally replace the traditional web reinforcement. However, despite of this increase of knowledge about fiber reinforced concrete, there is still a gap in the applicability of fibers as shear reinforcement in certain precast and prestressed structural elements where the use of conventional transverse reinforcement is difficult due to their manufacturing process. In this context, the present paper presents the experimental results of full-scale tests on Hollow-Core Slabs (HCS) made with Polypropylene Fiber Reinforced Concrete (PFRC). In order to analyze the HCS end zones (critical zones in shear), several tests were performed on slabs adopting a shear-span-to-effective depth ratio equal to 3.5. Results show that macro-synthetic fibers are able to improve the shear strength of hollow-core slabs. The experimental results obtained were compared against the predictions of four international codes standards (Eurocode 2, ACI 318-11, Model Code 2010 and EN1168) as well.

Keywords: Fiber reinforced concrete · Hollow-core slabs · Macro-synthetic fibers · Prestressed elements · Shear strength

1 Introduction

Hollow-core slabs (HCS) are precast elements widely used in residential, parking and industrial buildings due to their high quality control, easy installation and reduced construction time. These slabs are commonly manufactured by extrusion or slip-formworks using concrete with very low workability. Initially, their sections were provided with circular voids, but in the last decades, they have been improved using kind of ellipsoidal voids and prestressed concrete. The manufacturing process of HCS entail some restrictions on placing conventional reinforcement: HCS are generally made

© RILEM 2022
P. Serna et al. (Eds.): BEFIB 2021, RILEM Bookseries 36, pp. 514–524, 2022.
https://doi.org/10.1007/978-3-030-83719-8_45

without any web reinforcement and without any reinforcing bar for tendon anchorage. These elements are usually simply supported at their end zones (support width of about 10–15 cm), so their end zones (for a length of about 60–70 times tendon diameter, i.e. transfer length) are very critical regions. In fact, end zones are disturbed regions mainly stressed in tension by shear forces (in a zone where the beneficial effects of prestressing are not totally active) and splitting actions (due to the transferring of prestressing force from steel to concrete). In addition, since tendons have a negligible anchorage, tendon slip is also possible. All this makes end zones critical and underlines the importance of studying them under shear loading for finding new reinforcement solutions.

Fibre Reinforced Concrete (FRC) was proved to be very effective in enhancing the shear strength of Reinforcement Concrete (RC) structures and prestressed elements. Fibres can generally replace the conventional web reinforcement required in these elements for both minimum shear reinforcement and equilibrium. Several researches were conducted for analysing the influence of steel fibres (added in different percentage in the concrete mixture) on the shear behaviour of these HCS elements (Paine et al. [1, 2] in 1998, Cuenca et al. [3] in 2012, Simasathien and Chao [4] in 2015 and Dudnik et al. [5] in 2017). These authors found that steel fibres increased the shear capacity of the element and improved the post peak ductility as compared to the sample with conventional reinforcement. Dudnik et al. [5] also found that a steel fibre volume fraction higher than 0.50% may lead to concrete compaction problem in HCS, with potential detrimental effects on slabs shear strength.

Concerning the use of macro-synthetic fibres in HCS, which showed to be very effective in shear in the last decade, only a very limited study is present in the literature [6]. Two 356 mm deep HCS were tested for evaluating the feasibility of using copolymer/polypropylene fibres (twisted-bundle nonfibrillating monofilament and fibrillating network fibre) with a volume fraction of 0.67% and 1%. Specimens were tested at their end zones with a/d = 3.0. The authors underlined that polypropylene fibres could increase the shear strength of HCS, even if the shear strength increment was not clearly quantified. Therefore, there is still a lack in the literature concerning the shear behaviour of HCS reinforced with macro-synthetic fibres. Within this framework, the research work presented herein aims at investigating the possibility of using macro-synthetic fibres as reinforcement of HCS end zones.

2   Experimental Program

2.1   Specimen Geometry and Reinforcement Details

Specimens were manufactured indoor in a precast concrete plant (Gruppo Centro Nord S.p.A.) by using an extrusion procedure and dry concrete. Three days after casting, slabs were cut with a diamond-tipped saw to the selected length of 6000 mm and stored. One reference slab was made in conventional reinforcement (without any web reinforcement, RC specimen), while four slabs were produced with 10.5 kg/m^3 of macro-synthetic fibres (PFRC specimens).

The cross section geometry of the adopted HCS is given in Fig. 1, while Table 1 summarizes the main characteristics of the elements. This geometry was chosen since

HCS with a depth greater than 400 mm could be subjected to an applied shear force greater than the shear resistance of concrete, while thinner HCS are generally no critical in shear. All specimens contained the same longitudinal reinforcement. Ten 6/10″ and two 3/8″ seven-wire strands were placed as bottom longitudinal reinforcement; the top reinforcement consisted of six three-wire strands (wire diameter 3 mm). The designation of bottom tendons is also reported in Fig. 1 (from (1) to (12)). Tendons were characterized by having a pretension of 1200 MPa, which is equivalent to a mean compressive stress on the concrete cross section of 7.96 MPa. External webs were wider than internal ones in order to withstand actions at temporary loading conditions (lifting by an overhead crane with clamps and transportation), as well as in the external webs a low amount of longitudinal reinforcement was provided due to the lower amount of concrete confining tendons (splitting forces).

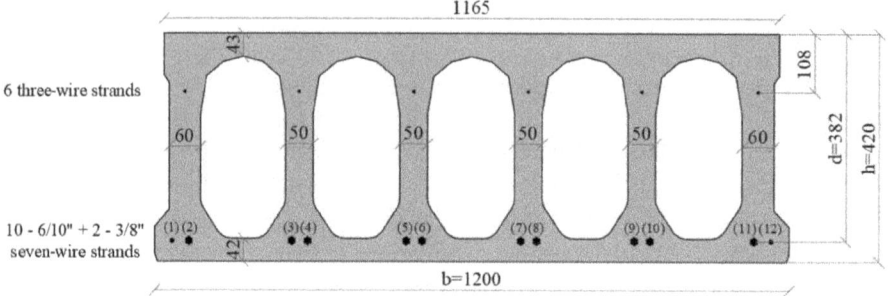

Fig. 1. Nominal cross section of HCS with reinforcement details (measures in mm) and tendon designation.

Table 1. Main characteristics of HCS.

Length (L)	[mm]	6000
Width (b)	[mm]	1200
Web width (b_w)	[mm]	2 x 60 + 4 x 50
Total depth (h)	[mm]	420
Effective depth (d)	[mm]	382
Top reinforcement	[-]	6 three-wire strands (wire diameter 3 mm)
Bottom reinforcement	[-]	10 – 6/10″ + 2 – 3/8″ seven-wire strands
Top reinforcement area	[mm²]	127
Bottom reinforcement area	[mm²]	1494
Longitudinal reinforcement ratio (ρ)	[%]	1.05
Initial prestress	[MPa]	1200

Usually these elements have a span of about 12.1 m; a reduced total length (6 m) was adopted just to have laboratory samples that allow reliable shear tests.

With the objective of analysing the shear behaviour of HCS end zones, the loading scenario is shown in Fig. 2. Each specimen was tested adopting an *a/d* ratio equal to *3.5* (in order to avoid as much as possible the arch effect on the shear behaviour of these elements).

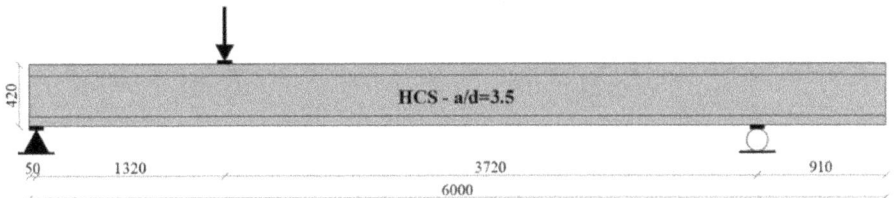

Fig. 2. Third point loading scenario for the end zone: load scheme for *a/d* = 3.5.

2.2 Material Properties

Table 2 lists the mix design of dry concrete adopted for RC and PFRC samples. In the PFRC elements, the amount and type of superplasticizer was specifically chosen in order to have both a good distribution of fibres and an optimum consistency of dry concrete. Only one type, geometry and dosage of polypropylene (PP) fibres was employed. Crimped polypropylene fibres, 40 mm long with a nominal aspect ratio (length/diameter) equal to 53 and density of 910 kg/m^3, were embedded in concrete in the amount of 10.5 kg/m^3 in PFRC slabs. The tensile strength and modulus of elasticity were 400 MPa and 3.64 GPa respectively. The fibre type and amount was chosen after pilot experiments carried out to obtain a structural PFRC according to Model Code 2010 [7] which guarantees casting feasibility (dry concrete).Tendon properties were evaluated according to EN15630-1 [8]. The yielding stress f_{ym} and ultimate stress f_{um} of the longitudinal reinforcement were 1819 MPa (CV = 0.01) and 1986 MPa (CV = 0.01) for 6/10″ seven-wire strands; 1734 MPa (CV = 0.01) and 1933 MPa (CV = 0.01) for 3/8″ seven-wire strands; 1854 MPa (CV = 0.01) and 2021 MPa (CV = 0.01) for three-wire strands.

Table 2. Concrete mix design for RC and PFRC.

River sand	[kg/m^3]	330
Crushed sand	[kg/m^3]	410
Crushed gravel 2/7 mm	[kg/m^3]	551
Crushed gravel 8/15 mm	[kg/m^3]	650
Cement I 52.5R	[kg/m^3]	120
Cement II/A-LL 42.5R	[kg/m^3]	240
Water/cement ratio	[-]	0.43
Superplasticiser	[l/m^3]	0 (RC); 0.7 (PFRC)
Polypropylene fibres	[kg/m^3]	0 (RC); 10.5 (PFRC)

Table 3 lists the mechanical properties of concretes (coefficient of variation CV in bracket) at the age of testing, i.e. 28 days. Six cubic specimens of $150 \times 150 \times 150$ mm^3 and five $150 \times 150 \times 600$ mm^3 small beams were used to determine compressive strength ($f_{c,cube}$) and post-cracking residual strengths according to EN14651 [9] (limit of proportionality f_L, residual flexural tensile strengths $f_{R,1}$, $f_{R,2}$, $f_{R,3}$, $f_{R,4}$, corresponding to a Crack Mouth Opening Displacement (CMOD) equal to 0.5, 1.5, 2.5 and 3.5 mm), respectively.

Table 3. Mechanical properties of HCS (CV in brackets).

HCS		RC	PFRC
$f_{c,cube}$	[MPa]	54.3 (0.04)	48.9 (0.05)
f_c	[MPa]	45.1	40.6
f_L	[MPa]	5.67 (0.02)	5.92 (0.03)
$f_{R,1}$	[MPa]	–	2.66 (0.17)
$f_{R,2}$	[MPa]	–	3.00 (0.19)
$f_{R,3}$	[MPa]	–	3.43 (0.19)
$f_{R,4}$	[MPa]	–	3.39 (0.17)

Concerning the five post-cracking performances of PFRC, Fig. 3 shows the flexural tensile stress vs. CMOD curves of seven PFRC and two plain concrete small beams.

It can be underlined that polypropylene fibres showed the typical post-cracking behaviour, characterized by a load drop after the peak load followed by a stable and pretty constant behaviour up to a CMOD around 3.5 mm. Moreover, fibres provide significant post-cracking performances after only three days. Finally, the obtained results, fulfil the requirements of MC2010 for the use of fibres in structural elements since $f_{R,1}/f_L > 0.40$ and $f_{R,3}/f_{R,1} > 0.50$ in terms of mean values.

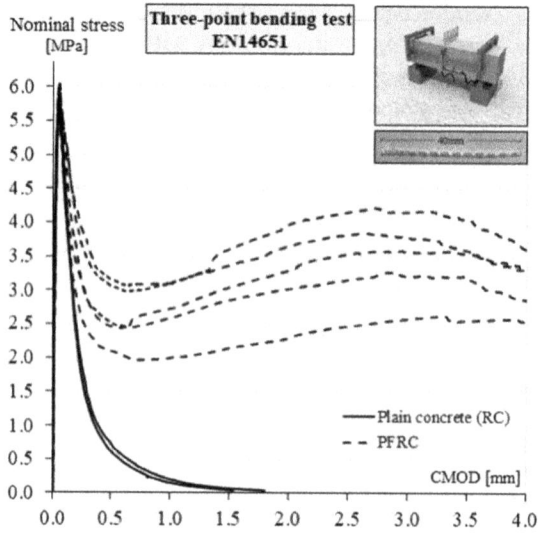

Fig. 3. Nominal stress vs. CMOD curves according to EN14651 [9].

2.3   Test Set-Up and Instrumentation

HCS were subjected to a three point loading test following the recommendations indicated in EN1168 [10], but with a shear span of $a/d = 3.5$ instead of 2.8. Tests were carried out under displacement control by using an electromechanical screw jack having a capacity of 500 kN. A lateral and front view of the loading system, as well as the pinned support detail are shown in Fig. 4. This picture summarizes also the instrumentation adopted: six Linear Variable Differential Transducers (LVDT) were used for measuring deflection and support displacements, while eight Potentiometric Displacement Transducers (PT) were adopted to capture the inclined shear crack opening at both external webs of specimens. Flexural crack opening and compression chord shortening were also evaluated under the point load by using 3 and 2 PT, respectively. Furthermore, in order to capture the tendon slip (as the difference between tendon and concrete displacement), 8 PT were placed on the sides of a sample (Fig. 4) for measuring the displacement of both concrete (in two points) and tendons at any HCS webs. More details about the instrumentation can be found in Conforti et al. [11].

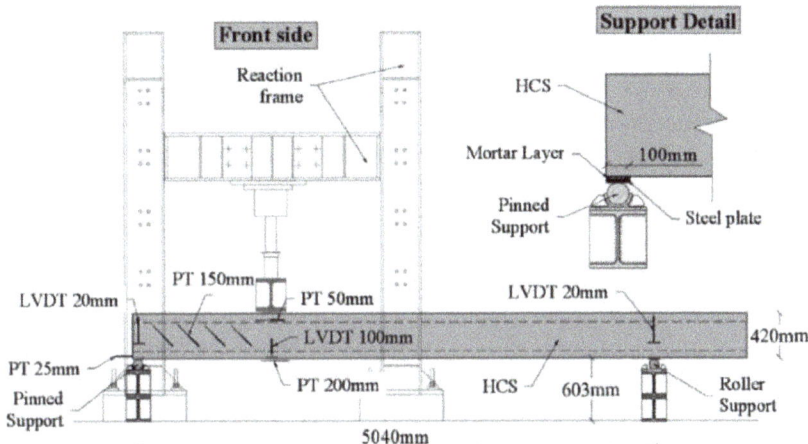

Fig. 4. Set-up and instrumentation details.

3   Experimental Results and Discussion

All samples showed a web-shear cracking (cracking due to diagonal tension before the appearance of any flexural crack) with loss of bond between concrete and tendons. Table 4 summarizes the main experimental results in terms of maximum applied load (*Pmax*), point load deflection at *Pmax (δmax)*, shear strength (*Vu*) (which was evaluated considering *Pmax* as a function of the static scheme (Fig. 2), self-weight of sample and bearing system), nominal shear strength ($vu = Vu/(\Sigma bw \cdot d)$) and normalized shear strength ($vu/(fc)0.5$).

Figure 5 shows the load-deflection response, the load vs. critical shear crack curves of all samples and the tendon-to-concrete slip (in particular, Fig. 5b shows the load vs. slip of the tendon in the first web failing in shear (the number displayed in parenthesis after sample designation shows the considered tendon)). The experimental response of HCS was characterized by four main consecutive phases:

- Phase I (up to 200 kN): uncracked stage with no significant tendon slip (Fig. 5b);
- Phase II (between 200 kN to *Pmax*): increment of tendon slip (Fig. 5b) with consequent progressive reduction of the effectiveness of prestressing actions on the shear strength (the increment of tendon slip resulted higher for tendon (1), (2), (11) and (12) as compared to the other ones, due to the lower amount of concrete confining tendons). During this phase, a slight change in slope can be observed in the load-deflection response of samples (Fig. 5a) as compared to Phase I;
- Phase III (at *Pmax*): an external web fails in shear (Fig. 5c), leading to a quite sudden collapse of all other webs. This is reasonable since, even if the external webs are thicker, they are both less prestressed at the end zone and characterized by tendons less confined by surrounding concrete. The weakness of external webs is confirmed by the higher tendon slip captured at P_{max}. It should be also noticed that the failure occurred in all samples at a load level significantly lower than their flexural bearing capacity as well as no flexural cracks were detected anywhere;
- Phase IV (after *Pmax*): the RC sample showed a very brittle failure with sudden increase of shear crack opening and tendon slip, while PFRC samples showed a post-cracking resistance with stable softening (Fig. 5a).

One should also note that HCS including polypropylene fibres reached a greater load capacity as compared to the RC of about 25–30% (Table 4). In fact, the web shear failure occurred after reaching an average shear stress of about 0.28 *fc* in the RC sample, while in case of PFRC specimens it was around 0.37 *fc*. This increment of shear strength is mainly due to the ability of polypropylene fibres in better controlling the tendon slip as compared to a RC sample (Phase II). In fact, in Fig. 5b, it can be observed that the presence of fibres reduced the tendon slip for a given load (allowing a second change of slope in the load vs. tendon slip response at a load of about the maximum reached by the RC sample). Therefore, polypropylene fibres were able to improve the bond between tendons and concrete, which allows a better control of the development of potential splitting cracks, which guarantee the effectiveness of prestressing actions on the shear strength even for higher loads, allowing PFRC samples to reach a higher shear strength as compared to RC one. This experimental evidence is in

Table 4. Main experimental results of tests (CV in brackets).

HCS	Specimen	P_{max}	δ_{max}	V_u	v_u	$v_u/(f_c)^{0.5}$
		[kN]	[mm]	[kN]	[MPa]	[-]
RC	42-RC-3.5	346.6	4.31	273.4	1.89	0.28
PFRC	42-PFRC-3.5-I	489.0	6.75	378.7	2.60	0.41
	42-PFRC-3.5-II	413.9	4.94	323.4	2.11	0.33
	42-PFRC-3.5-III	465.8	5.38	361.3	2.44	0.38
	42-PFRC-3.5-IV	392.8	5.55	307.9	2.14	0.34
	Mean PFRC	440.4 (0.10)	5.65 (0.14)	342.8	2.32	0.37

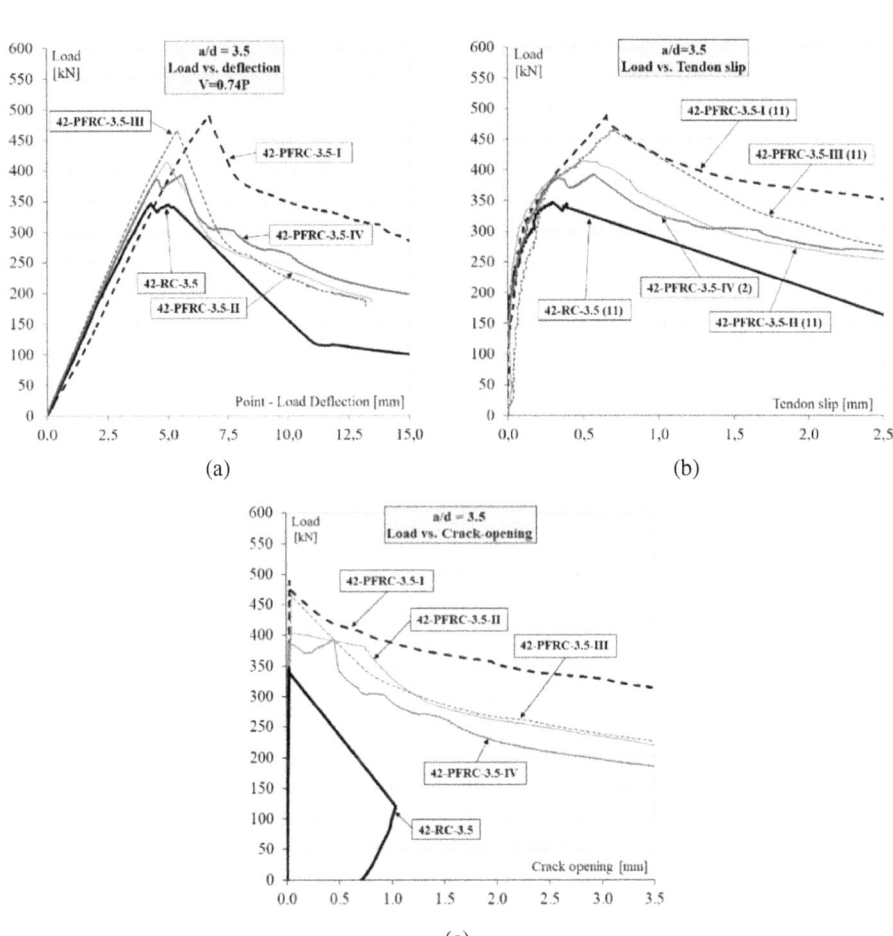

Fig. 5. Load vs. deflection curves (a), Load vs. tendon slip curves (b), Load vs. Crack opening curves (c).

accordance with the results of Palmer and Schultz [6] who observed that, at greater end slips of strands, less web shear strength is obtained.

Concerning the web-shear cracking, when the first crack appears on an external web (principal tensile stress reaches the tensile strength of concrete), polypropylene fibres were only able to guarantee a stable softening behaviour characterized by significant post-cracking resistance to HCS, since a slightly increase of shear strength was only observed in the sample 42-PFRC-3.5-IV. This softening behaviour is due to the progressive tendon slip and web-shear crack opening. Therefore, polypropylene fibres at the given dosage were not able to redistribute the loads to the different webs and to provide a further increase of shear strength. Figure 6 exhibits an example of the final crack patterns at the front and backside of a sample, showing the critical shear crack on both external webs.

(a)   (b)

Fig. 6. Example of a final crack pattern on front (a) and back side (b) (42-PFRC-3.5-II specimen).

4   Comparison Against Code Predictions

The prediction of the shear strength of HCS without fibres were evaluated by using the analytic expressions provided by ACI 318-14 [12], Eurocode 2 (EC2) [13], MC2010 [7] and EN1168 [10]. In the case of HCS with fibres, only a formulation proposed by MC2010 was evaluated; more details about that can be found in [11]. In order to make this comparison significant, the shear strength was calculated by assuming safety factors equal to 1 (γ_c for EC2, MC2010 and EN1168, ϕ for ACI 318-14), and mean values of the material mechanical properties (Table 3). Prestress losses were estimated in 20% of tendons prestress [14]. Figure 7 shows the comparison between the experimental and the predicted shear strength ($V_u/V_{u,code}$): for RC slabs, all codes resulted in unconservative predictions: the level II of EN1168 was the most accurate ($V_u/V_{u,code} = 0.97$), followed by level I of MC2010 (0.93). It should be noted that, in both cases, formulations include a calibration factor of 0.8, which reduces by 20% the shear strength of the element. EC2 overestimates the shear capacity by about 25%, similar as ACI 318-14. Finally, in HCS with fibres, MC2010 formulation significantly overestimate the shear capacity. However, it should be underlined that this equation was not specifically developed for this type of elements.

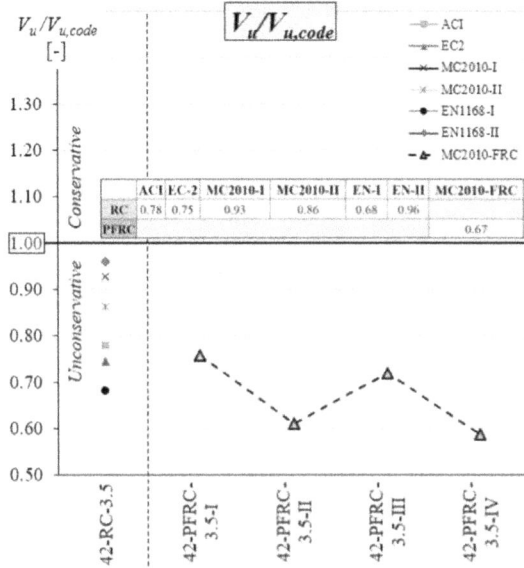

Fig. 7. Comparison against shear strength predictions of different standards.

5   Conclusion Remarks

Based on this experimental study, the following conclusions can be drawn:

- Polypropylene fibre reinforced concrete can be used to enhance the shear strength of hollow-core slab end zones;
- PFRC enhances the shear strength of hollow-core slab end zones mainly by improving the bond between tendons and concrete (better controlling the development of potential splitting cracks) with consequence reduction of tendon slip. The latter better guarantees the effectiveness of prestressing actions on the shear strength even for higher loads, allowing PFRC samples to reach greater shear strength as compared to the RC hollow-core slabs;
- The formulations adopted by codes for estimating the shear strength of RC hollow-core slabs seems to be unconservative. In particular, Eurocode 2, ACI 318 and EN1168-level I led to the worst predictions, while EN1168-level II was the most accurate. For PFRC hollow-core slabs, the prediction of Model Code 2010 resulted in even more unconservative, underlining the need of developing proper analytical models.

References

1. Paine, KA.: Steel fibre reinforced concrete for prestressed hollow core slabs, Ph.D. Thesis (University of Nottingham, 1998).

2. Peaston, C.H., Elliott, K.S., Paine, K.A.: Steel fiber reinforcement for extruded prestressed hollow core slabs. ACI Spec. Publ. **182**, 87–107 (1999)
3. Cuenca, E., Serna, P.: Failure modes and shear design of prestressed hollow core slabs made of fiber-reinforced concrete. Compos. Part B Eng. **45**, 952–964 (2013)
4. Simasathien, S., Chao, S.-H.: Shear strength of steel-fiber-reinforced deep hollow-core slabs. PCI J. **60**, 85–101 (2015)
5. Dudnik, V.S., Milliman, L.R., Parra-Montesinos, G.J.: Shear behavior of prestressed steelfiber- reinforced concrete hollow-core slabs. PCI J. **62**, 58–72 (2017)
6. Palmer, K.D., Schultz, A.E.: Experimental investigation of the web-shear strength of deep hollow-core units. PCI J. **56**, 83–104 (2011)
7. International Federation for Structural Concrete (fib): Model Code 2010, final drafts. vol. 1 & 2. Wilhelm Ernst & Sohn (2013)
8. European Committee for Standardization: EN ISO 15630–1: Steel for the reinforcement and prestressing of concrete. Test methods. Part 1: Reinforcing bars, wire rod and wire (ISO 15630–1:2010), **28** (2010)
9. European Committee for Standardization: EN 14651: Test method for metallic fibres concrete. Measuring the flexural tensile strength (2005)
10. European Committee for Standardization: EN 1168:2005+A3: Precast concrete products. Hollow core slabs **82** (2012)
11. Conforti, A., Ortiz-Navas, F., Piemonti, A., Plizzari, G.A.: Enhancing the shear strength of hollow-core slabs by using polypropylene fibres. Eng. Struct. **207**, 110172 (2020)
12. ACI: Building Code Requirements for Structural Concrete (ACI 318–14) and Commentary (ACI 318R-14) (2014)
13. European Committee for Standardization: Eurocode 2: Design of concrete structures, Brussels, Belgium (2004)
14. Naaman, A.E.: Prestressed concrete analysis and design. Fundamentals, 3rd Edn. Ann Arbor, MI (2012)

How Can We Verify Structural Members Made of FRC Only?

Luca Facconi[✉] and Fausto Minelli

Department of Civil, Environmental, Architectural Engineering
and Mathematics, University of Brescia, Brescia, Italy
luca.facconi@unibs.it

Abstract. The fib Model Code for Concrete Structures 2010 (MC2010) recognizes fiber reinforced concrete (FRC) as a new structural material, favoring its usage in innovative structural applications. According to MC2010, FRC can be adopted as the only reinforcement in structures provided that both strength and ductility requirements are fulfilled. This paper presents a discussion on the MC2010 requirements for FRC structures applied to both experimental and numerically simulated structural elements, proposing a refined set of equations for the ULS and SLS verification of FRC structures (not containing any classical reinforcement).

Keywords: Fiber Reinforced Concrete · Verification · Model Code 2010 · Ductility · Service load · Low reinforcement ratio

1 Introduction

The improved tensile strength and toughness resulting from the addition of fibers to concrete make Fiber Reinforced Concrete (FRC) an effective solution for reducing or, with special caution, totally replacing flexural reinforcement in many structural applications. As shown by different studies, fibers may be used in place of main and secondary reinforcement in statically indeterminate structures such as two-way slabs and tunnel linings [1, 2].

After cracking, the ability of the structure to increase its capacity depends on both the tensile resistance of the cross-section, the residual strength provided by fibers and the degree of static indeterminacy of the structural system. However, when considering the load-deformation response of a FRC structure containing either very low amount or no conventional reinforcement, the post-peak response generally consists of a softening curve whose degrading behavior is strongly related to the tensile performance of FRC. The residual capacity exhibited after the attainment of the maximum resistance is of primary importance to fulfill the structural code ductility requirements that prevent from potential brittle failure mechanisms.

The fib Model Code 2010 [3] (MC2010) provides a series of design recommendations specifically developed for concrete structures reinforced with fibers only. As compared to reinforced concrete structure, the determination of minimum ductility of FRC members not containing rebars represents an important issue. According to MC2010 (see clause 7.7.2), a FRC structure with a load (P) – deformation (δ) response

© RILEM 2022
P. Serna et al. (Eds.): BEFIB 2021, RILEM Bookseries 36, pp. 525–535, 2022.
https://doi.org/10.1007/978-3-030-83719-8_46

similar to that depicted Fig. 1 has to comply with at least one of the two following conditions:

$$\delta_{u,MC} \geq 20 \cdot \delta_{SLS,MC} \qquad (1)$$

$$\delta_{peak} \geq 5 \cdot \delta_{SLS,MC} \qquad (2)$$

where $\delta_{u,MC}$ is the ultimate deformation, $P_{u,MC}$ is ultimate capacity, δ_{peak} is the deformation at the maximum load (P_{max}) and $\delta_{SLS,MC}$ is the deformation at the service load determined by performing a linear elastic analysis. Besides the two minimum deformation requirements above, the ultimate load must fulfill both following conditions:

$$P_{u,MC} \geq P_{SLS} \qquad (3)$$

$$P_{u,MC} \geq P_{cr} \qquad (4)$$

which include the maximum service load (P_{SLS}) as well as the load at first cracking (P_{cr}).

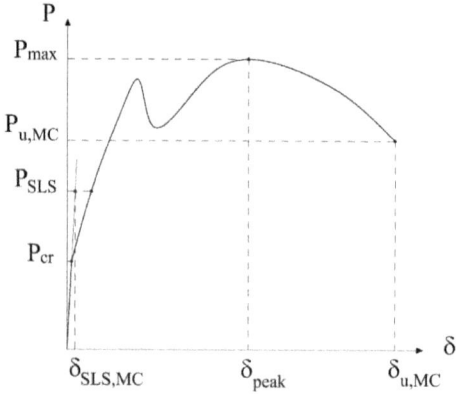

Fig. 1. Load (P) – deflection (δ) response of a FRC structure according to MC2010 (adapted from [3]).

From a conceptual point of view, the rationale behind the MC2010's approach is clear and suitable for many of the FRC members typically used in real applications. Anyway, one should observe that the code does not set any limiting condition for selecting the ultimate deformation ($\delta_{u,MC}$) which, on the contrary, is simply defined as the deformation corresponding to the ultimate capacity. Moreover, as the ultimate deformation of the structure lays on the post-peak branch of the global response, it can be generally assessed by performing experimental tests or simulations under deflection control. Unlike real structures, whose behavior is naturally "load-controlled", the deflection control allows governing the softening response leading to a significant reduction of the structure resistance and an unrealistic overestimation of the ductility. To address these deficiencies, the authors of this paper recently proposed an alternative verification approach [4] considering a bi-linear curve equivalent to the global load-deformation response of the structure.

After a brief description of the aforementioned new verification approach, the present work reports a series of verification examples aiming at extending the database originally collected by the authors [4]. Finally, as a worked example consisting in an analysis of a FRC cross-section under pure bending is reported and discussed in order to find the minimum tensile performance of FRC able to fulfill the new ductility requirement.

2 Verification Procedure

2.1 Method Description

Consider the typical load-deflection response of a FRC structure such as that depicted in Fig. 2. To estimate the structure ductility, the actual response of the structure can be simplified by a bi-linear curve consisting of a first linear branch with constant slope ($K_{1,id}$) followed by a constant plateau. The bilinearization of the structure response have been proposed by different authors [5] and standards [6] to perform push-over analyses frequently used in the seismic verification of existing structures. Here, the same approach has been proposed and adapted for the assessment of FRC structures containing little or no conventional reinforcement.

To define the bi-linear curve, the conventional ultimate displacement (δ_u) must be determined so that the corresponding ultimate capacity (P_u) of the structure is higher than 0.85 times the maximum load (P_{max}). The total dissipated energy can be then determined by calculating the area (A_{eff}) delimited by the actual curve up to the displacement δ_u. The conventional yielding capacity P_y of the idealized bi-linear relationship reads as follows:

$$P_y = K_{1,id} \cdot \left(\delta_u - \sqrt{\delta_u^2 - \frac{2 \cdot A_{eff}}{K_{1,id}}} \right) \tag{5}$$

where $K_{1,id} = 0.6 \, P_{max}/\delta_{0.6}$ and $\delta_{0.6}$ is the displacement of the structure corresponding to the load $0.6 \cdot P_{max}$.

Once the yielding capacity is known, a conventional ductility index (μ_δ) can be calculated as follows:

$$\mu_\delta = \frac{\delta_u}{\delta_1} \tag{6}$$

Where δ_1 is the deflection corresponding to the elastic limit on the idealized curve.

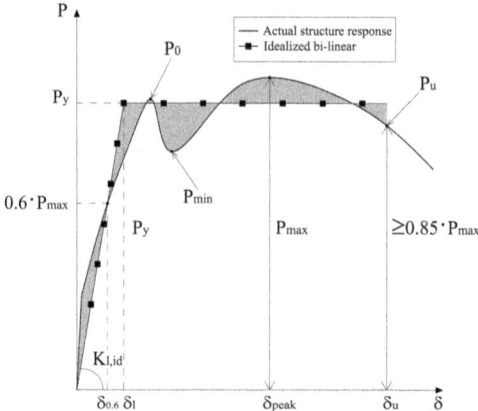

Fig. 2. Bilinearization of the actual response of a FRC structure. (adapted from [4])

Based on the validation study carried out by Facconi and Minelli [4], the minimum ductility conditions reported below must be fulfilled to ensure a good structural behavior at Ultimate Limit State (ULS) conditions:

$$\mu_\delta > 2 \quad \text{if} \quad \frac{P_{max}}{P_{cr}} > 3 \tag{7}$$

$$\mu_\delta > 11 - 13 \cdot \frac{P_{max}}{P_{cr}} \quad \text{if} \quad 2 < \frac{P_{max}}{P_{cr}} \leq 3 \tag{8}$$

$$\mu_\delta > 5 \quad \text{if} \quad \frac{P_{max}}{P_{cr}} \leq 2 \tag{9}$$

The check of the previous three ductility requirements can be avoided when the following condition occurs:

$$\frac{P_{cr}}{P_{SLS}} \geq 3 \tag{10}$$

One should note that the lower (i.e., $\mu_\delta = 2$) and the upper (i.e., $\mu_\delta = 5$) bound of the ductility domain represented by the Eqs. 7, 8 and 9 are similar to the limits recommended by FEMA 306 [7] for reinforced concrete structures. A critical point of the curve shown in Fig. 2 is the minimum load P_{min} that may be observed after the first peak load P_0. A significant drop of resistance after the first peak could lead to a brittle failure of the structure. To prevent this, one should carefully assume that P_{min} is not lower than $0.6 \cdot P_0$.

Regarding the structural behavior at Serviceability Limit State (SLS), the maximum allowable deflection ($\delta_{SLS,max}$) must fall within the elastic branch of the bi-linear equivalent response. It is here assumed that the actual deflection exhibited by the structure at SLS, i.e. δ_{SLS}, has to comply with the following requirement:

$$\delta_{SLS} \leq \delta_{SLS,max} \leq \frac{\delta_1}{3} \qquad (11)$$

In addition to the previous condition, the following recommendation related to the maximum allowable service load ($P_{SLS,max}$) has to be fulfilled:

$$P_{SLS} \leq P_{SLS,max} = \frac{P_{max}}{\gamma_F \cdot k_{SLS}} \qquad (12)$$

where γ_F is the material partial safety factor for FRC, which is typically considered equal to 1.3 or 1.5 for members constructed with or without an improved production control procedure, respectively (see MC2010 – clause 5.6.6). The factor k_{SLS} represents the structural partial safety factor relating the mean failure load to the maximum service load. According to the authors' experience, the values of the safety factors mentioned above should be chosen so that $2.6 \leq \gamma_F \cdot k_{SLS} \leq 3.5$. The values of the maximum allowable service load ($P_{SLS,max}$) and deflection ($\delta_{SLS,max}$) resulting from the Eqs. 10 and 11 may help the designer in structural sizing.

2.2 Verification of Different FRC Members

The proposed verification approach has been used to verify different FRC members whose global behavior has been reported and described by some research works found in the literature. Table 1 reports a summary of the works considered and of the outcomes of the verification procedure. Different structural typologies have been analyzed, including static determinate structures (e.g. beams, tunnels segments) as well as structures with a high degree of redundancy, such as slabs on grade and elevated slabs. All the members are reinforced with either fibers only or fibers combined with very low amount of conventional reinforcement.

Table 1. Verification of different FRC members by the proposed method.

Ref.	Structure type	Fiber content [kg/m³]	P_{cr} [kN]	P_y [kN]	δ_{SLS} [mm]	δ_1 [mm]	δ_u [mm]	μ_δ [-]	P_{cr}/P_{SLS} [-]	P_{max}/P_{cr} [-]	Check
[8]	Beam	100*	2.3	7.6	0.09	0.46	4.08	8.9	0.8	3.6	OK
[9]	Slab on grade	30*	125	227	0.4	1.1	3.2	2.9	1.4	1.8	NO
		45*	135	258	0.5	1.4	4.0	2.7	1.1	2.2	NO
[10]	One-way slab	3.3**	11	17	0.02	0.1	0.5	5.8	2.6	1.7	OK
		6.7**	12	20	0.04	0.2	0.7	4	2.9	1.8	NO
		10**	12	22	0.04	0.3	10	40	2.9	1.9	OK
[11]	One-way slab	29.6*	7 (kPa)	17 (kPa)	0.9	2.6	48	19	1.2	3	OK
		14.8*	7 (kPa)	14 (kPa)	0.4	2.2	38	17	1.8	2.1	OK
[12]	Elevated slab	80*	50	272	4.6	17	40	2.3	0.6	6	OK
		40*	60	120	1.2	5.7	47	8.2	1.7	2.1	OK
[13]	Elevated slab	70*	85	315	3.3	17	86	5.0	0.9	3.9	OK
[14]	Two-way slab	9†		55	169	0.1	3.4	72	21	1.1	3.2	OK

(continued)

Table 1. (*continued*)

Ref.	Structure type	Fiber content [kg/m³]	P_{cr} [kN]	P_y [kN]	δ_{SLS} [mm]	δ_1 [mm]	δ_u [mm]	μ_δ [-]	P_{cr}/P_{SLS} [-]	P_{max}/P_{cr} [-]	Check
		9[†]	55	188	0.2	1.2	43	36	0.9	3.7	OK
[15]	Two-way slab	25*	30	55	0.8	2.3	57	24	1.6	1.9	OK
		25*	40	59	0.8	2.5	49	20	1.8	1.6	OK
		25*	42	66	0.9	2.5	68	27	1.8	1.7	OK
[2]	Tunnel segments	40*	154	181	0.07	0.6	5.1	8.5	8.6	1.3	OK
		40*	164	156	0.09	0.7	1.9	2.7	9.1	1	OK
		40*	165	179	0.09	0.9	3.6	4.0	9.2	1.1	OK
[16]	Four-point supported plate	35*	113	217	0.60	1.94	11.7	6.0	1.5	2.1	OK
		35*	105	224	0.80	2.51	16.6	6.6	1.3	2.3	OK
[17]	Composite one-way two-span slab	50 (topping)* 40 (substrate)*	70	123	0.80	2.42	17.4	7.2	1.1	1.9	OK
		50 (topping)* 40 (substrate)*	60	120	0.4	1.21	11.3	9.3	0.8	2.2	OK
[18]	Pipe	20*	50	52 (kN/m²)	0.13	0.39	1.83	4.7	3.0	1.1	NO
		35*	50	56 (kN/m²)	0.10	0.30	10.0	33.3	2.2	1.3	OK

* Steel fibers
** Synthetic fibers.
† Polymeric fibers.

To fully perform the element verification, the service load applied to the structure must be known. However, the authors of the research works mentioned in Table 1 do not generally report the value of P_{SLS}. To overcome this lack of information, it has been assumed, according to Eq. 12, that the service load is $P_{max}/3$.

The verification check shows that most of the structures fulfill the requirements of the proposed method. On the contrary, some of the members has not passed the final check because of their deficiency in ductility.

For the sake of example, Fig. 3a reports the response of two slabs on grade, namely S4 and S11, tested by Sorelli et al. [9]. Each specimen has dimensions of 1.5 × 1.5 × 0.15 (thickness) m and is reinforced only with normal strength double-hooked end steel fibers having a length of 50 mm and a diameter of 1 mm. Total fiber amounts of 30 and 45 kg/m³ were used to cast the slab S4 and S11, respectively. Both specimens have been loaded by a single vertical force applied in the middle of the member. As shown in Fig. 3a, the post-peak response of the specimens does not appear in the diagram as the tests have been stopped at peak load. The lack of the post-peak response limited the ductility index of both slabs, which do not meet the ductility requirements of the Eqs. 8 and 9 (see Table 1). However, it is worth remarking that a ductile behavior is generally not required for slabs on ground, such as industrial pavements, which are typically subjected to static loads significantly lower than the maximum capacity. Therefore, in the real case, in order to avoid the ductility check, a service load P_{SLS} equal or lower than 42 kN and 45 kN (see Eq. 10) should be assumed respectively for the slab S4 and S11. As compared to the verification method adopted herein, the MC2010's provisions appears less straightforward. In fact, if one assumes $\delta_u = \delta_{peak,MC} = \delta_{u,MC}$, the maximum service load corresponding to the deflection limit sets by the Eq. 2 results to be even higher than the maximum capacity of the slabs S4 and S11. However, considering Eq. 1, the maximum service load of the slab S4 (P_{SLS} = 36 kN) and S11 (P_{SLS} = 45 kN) are very consistent with those calculated by the proposed procedure.

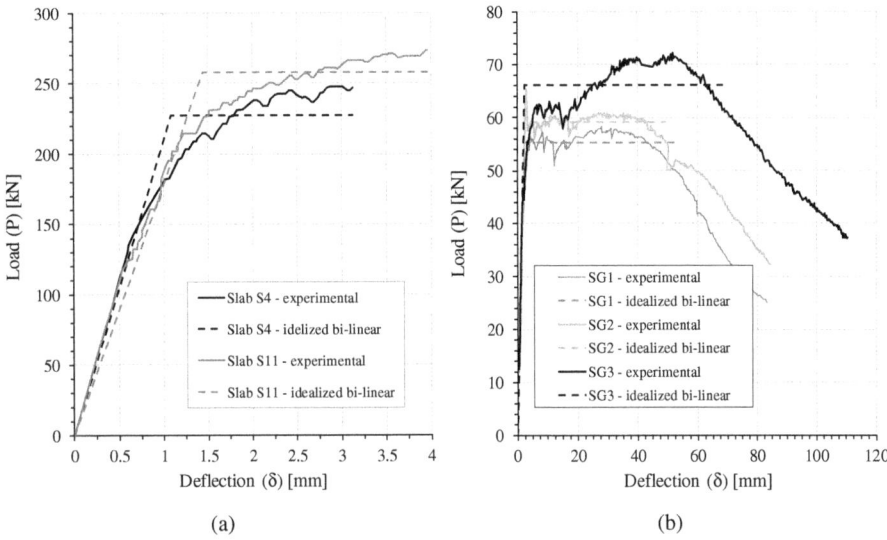

(a)   (b)

Fig. 3. Bi-linearization of the load-deflection curves adapted from the originals reported by literature: (a) slabs on grade [9]; (b) two-way slabs [15].

Similar considerations apply also to the pipe containing 20 kg/m^3 (see Table 1 – tests by de la Fuente et al. [18]) that fails the ductility check required by the Eq. 9. Likewise the slab S4 and S11, the ductility check can be avoided if the service load is not higher than 1/3 times the cracking load (i.e., $P_{SLS} \leq P_{cr}/3 = 17$ kN/m^2).

Figure 3b reports the load-deflection curves reported by Facconi et al. [15], who tested three rectangular FRC slabs provided with a continuous support along the four sides. The three specimens have dimensions of 2.5 × 4.2 × (thickness) 0.08 m and are reinforced only with 25 kg/m^3 of high-strength double-hooked end steel fibers having a length of 32 mm and a diameter of 0.4 mm. As shown in Table 1, the cracking load ranges from a minimum of 30 kN (i.e., specimen SG1) to a maximum of 42 kN (i.e., specimen SG3). As the tests have been performed under deflection control, the maximum deflection attained by the members is considerably higher than both the deflection at first cracking and the deflection at peak load. If one assumes $P_{U,MC}$ = P_{cr} = P_{SLS}, the ultimate deflections $\delta_{u,MC}$ of the slabs SG2 and SG3 result 71 mm and 101 mm, respectively. The latter easily fulfill Eq. 1 as they are more than 40 times higher than the deflection ($\delta_{SLS,MC}$ = 1.8 mm) corresponding to the service load. However, a real structure may hardly withstand a reduction of the post-peak bearing capacity comparable to that observed in the deflection-controlled tests. A more reliable approach is provided by the verification procedure herein recalled, which provides lower ultimate deflections and, consequently, ultimate loads (P_y) closer to the maximum capacities attained by the specimens. As shown in Table 1, all the slabs fulfill the ductility limit required from the Eq. 9.

3   Worked Example: Minimum Performance of FRC in a Flexural Member

When designing FRC sections subjected to flexure, the verification procedure described in the previous paragraph can be used to select the best FRC performance able to ensure the minimum ductility requirement represented by the Eqs. 7–9. A simple worked example is briefly reported and discussed in the following to show a possible design procedure aiming at assessing the effect of the FRC properties (i.e., the tensile residual strengths) on ductility.

Assuming a simply supported beam, subjected to a given load distribution, which has a total depth (h) of 150 mm and a width of 1000 mm. The bending resistance of the section can be calculated by the equilibrium of the stress distribution depicted in the schematic of Fig. 4a, which reports the characteristic values of concrete properties. A parabolic distribution consistent with that suggested by the Eurocode 2 (clause 3.1.5) is assumed for concrete in compression. The mean cylindrical compressive strength (f_{cm}), the elastic modulus (E_c) and the strain at peak strength are equal to 38 MPa, 33000 MPa and 2×10^3, respectively. About tension, the linear stress-strain distribution recommended by the MC2010 [3] (clause 5.6.5) has been considered. The latter consists of a first elastic branch, having a constant slope equal to E_c, that is followed by a bi-linear curve representing the behavior of concrete after the attainment of the uniaxial tensile strength $f_{ctm} = 2.9$ MPa. The post-cracking characteristic strength $f_{t(w),k}$ is related to the crack width (w) by the following equation:

$$f_{t(w),k} = f_{ftu,k} + (f_{Ftu,k} - f_{Fts,k}) \cdot \frac{w - 2.5}{2} \tag{13}$$

where w is the crack width (in mm); $f_{Fts,k} = 0.45f_{R1k}$; $f_{Ftu,k} = 0.5f_{R3k} - 0.2f_{R1k}$; f_{R1k} and f_{R3k} are the characteristic residual strengths corresponding to Crack Mouth Opening Displacement (CMOD) values of 0.5 mm and 2.5 mm, respectively. The residual strengths f_{R1k} and f_{R3k} can be found by testing notched prisms according EN14651 [19]. Crack widths (w) are divided by the characteristic length ($l_{cs} = h - x$) to get the corresponding strain defining the post-cracking stress-strain response (Fig. 4). Note that the maximum allowable tensile strain is equal to 2%. As the simulation has to be performed by implementing the mean properties of materials, the value of $f_{t(w),k}$ has been multiplied by 1.3 to get a reasonable estimation of the corresponding mean value.

A layered approach has been here implemented to integrate the variable stress-strain distribution acting along the cross section to find the total resisting moment (M_R) against the section curvature (χ). To assess the effect of the FRC performance on the bending behavior, different values of the residual strengths f_{R1k} and f_{R3k} have been considered. In more detail, each sectional analysis has been carried out by selecting

f_{R1k} and the ratio $f_{R3,k}/f_{R1,k}$, being the latter higher than the minimum value, i.e. $f_{R3,k}/f_{R1,k} > 0.5$, required by the MC2010 for structures reinforced with fibers only. Figure 4b reports some typical moment-curvature responses resulting from the parametric study together with the corresponding idealized bi-linear curves calculated according to the verification procedure above described.

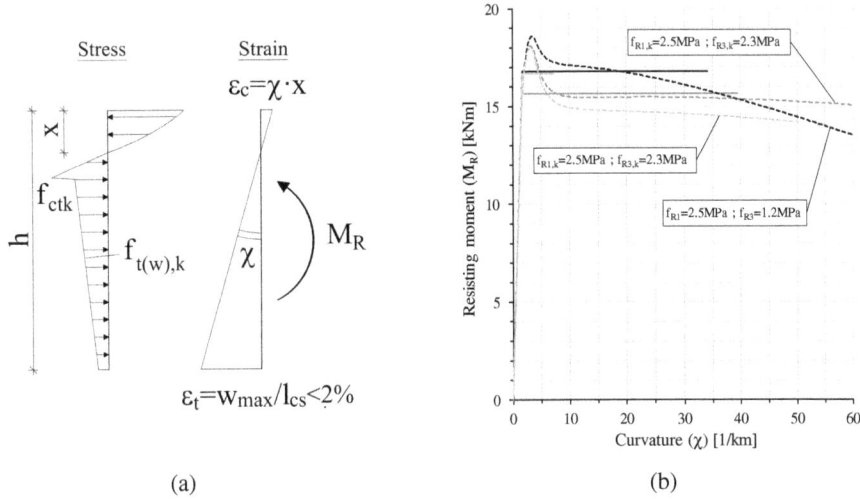

Fig. 4. (a) Stress and strain distribution acting along the FRC; (b) typical moment (M_R) – curvature (χ) obtained from sectional analysis.

The cracking moment of the cross section is approximately equal to 15.7 kNm whereas the maximum resisting moment ranges from a minimum of 17.5 kNm to a maximum of about 18.5 kNm. Therefore, as the maximum to cracking moment ratio is lower than 2, the condition that governs the verification is provided by the Eq. 9, namely $\mu_\delta \geq 5$.

Figure 5 shows the most significant values of the ductility index resulting from the bi-linearization of the moment-curvature response. Each curve is plotted assuming a constant $f_{R1,k}$. The horizontal axis has been subdivided into four ranges (i.e., a, b, c, d) corresponding to the FRC classes reported by the MC2010 [3] (clause 5.6.3). The results prove that the ductility requirement is fully verified if $f_{R1,k} \geq 2.5$ MPa, irrespective of the value of $f_{R3,k}$. On the contrary, if $f_{R1,k} < 2.5$ MPa, the ductility tends to reduce as $f_{R3,k}$ increases. Such a reduction can be explained considering that the increment of the ratio $f_{R3,k}/f_{R1,k}$ leads to a reduction of both the slope of the post-cracking curve (see Eq. 13) and the strength $f_{t(w),k}$ at w = 0. This results in a higher drop in the decrease of the resisting moment after the peak and, as a consequence, in a reduced ductility. For FRC having $f_{R1,k} < 2.2$ MPa, the ductility requirement is totally unfulfilled.

According to Eq. 12, the maximum moment for service conditions can be assumed equal to 1/3 of the maximum resistance. As the maximum resisting moment is approximately equal to 17–19 kNm, the service load applied to the beam should be chosen so that the corresponding acting moment be lower than 5.7–6.6 kNm.

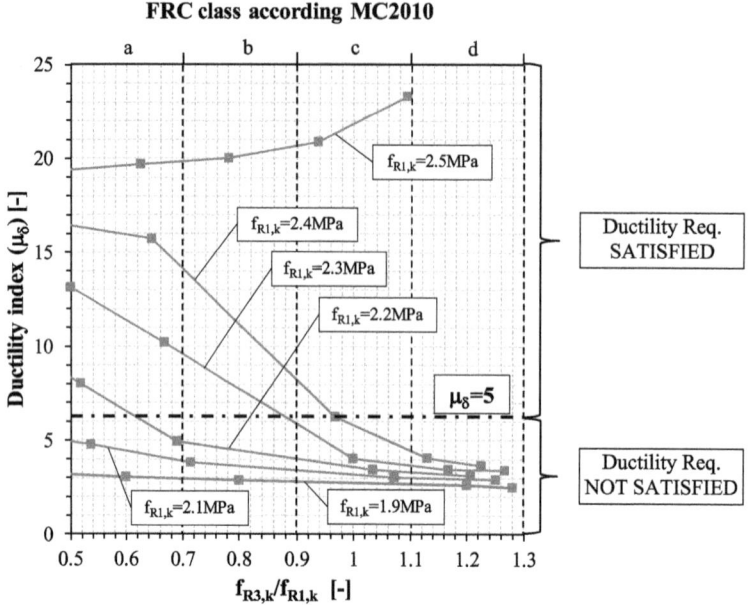

Fig. 5. Flexural verification of a 150 mm depth section: ductility index vs. $f_{R3,k}/f_{R1,k}$.

4   Concluding Remarks

A critical discussion on the actual MC2010 requirements for FRC structures not including any classical reinforcing bars was herein resented, as well as a new and more refined verification tool, comprising both requirements at ULS (strength and ductility) and at SLS (deflection). The ductility requirements are coupled with the maximum load over the load at first cracking ratio, being more sever where this ratio is lower. A worked example on a simple supported FRC beam was also developed for different classes of FRC emphasizing the crucial role of the f_{R3}/f_{R1} ratio in providing ductility to the FRC structure considered.

References

1. Pujadas, P., Blanco, A., Cavalaro, S., Aguado, A.: Plastic fibres as the only reinforcement for flat suspended slabs: experimental investigation and numerical simulation. Constr. Build. Mater. **57**, 92–104 (2014)
2. Caratelli, A., Meda, A., Rinaldi, Z.: Design according to MC2010 of a fibre-reinforced concrete tunnel in Monte Lirio. Panama. Struct. Concr. **13**(3), 166–173 (2012)
3. fib Model Code for Concrete Structures 2010, October 2013, 434 p. ISBN 978-3-433-03061-5

4. Facconi, L., Minelli, F.: Verification of structural elements made of FRC only: a critical discussion and proposal of a novel analytical method. Eng. Struct. **131**, 530–541 (2017)
5. Tomaževič, M.: Earthquake-Resistant Design of Masonry Buildings. p. 281. Imperial College Press, London (1999). ISBN: 1-86094-0668
6. D.M. 17/01/2018 (NTC2018). Aggiornamento delle Norme tecniche per le costruzioni. (in Italian)
7. FEMA 306: Evaluation of Earthquake Damaged Concrete and Masonry Wall Buildings - Basic Procedures Manual, prepared by the Applied Technology Council (ATC-43) (1999)
8. Ferrara, L., Ozyurt, N., di Prisco, M.: High mechanical performance of fibre reinforced cementitious composites: the role of '"casting-flow induced"' fibre orientation. Mater. Struct. **44**, 109–128 (2011)
9. Sorelli, L., Meda, A., Plizzari, G.: Steel fiber concrete slabs on ground: a structural matter. ACI Struct. J. **103**(4), 551–558 (1997)
10. Fantilli, A., Gorino, A., Chiaia, B.: Precast plates made with lightweight fiber-reinforced concrete. In: FRC 2014 Joint ACI-fib International Workshop, Proceedings of FRC 2014 Joint ACI-fib International Workshop on Fibre Reinforced Concrete: from Design to Structural Applications, Montreal, Canada, 24–25 July 2014, pp. 224–234 (2014)
11. Roberts-Wollmann, C.L., Guirola, M., Easterling, W.S.: Strength and performance of fiber-reinforced concrete composite slabs. J. Struct. Eng. **130**(3), 520–528 (2004)
12. Hedebratt, J., Silfwerbrand, J.: Full-scale test of a pile supported steel fibre concrete slab. Mater. Struct. **47**(4), 647–666 (2013). https://doi.org/10.1617/s11527-013-0086-5
13. Parmentier, B., Van Itterbeeck, P., Skowron, A.: The behaviour of SFRC flat slabs: The Limelette full-scale experiments for supporting design model codes. In: FRC 2014 Joint ACI-fib International Workshop, Proceedings of FRC 2014 Joint ACI-fib International Workshop on Fibre Reinforced Concrete: from design to Structural Applications, Montreal, Canada, 24–25 July 2014, pp. 394–405 (2014)
14. Pujadas, P., Blanco, A., Cavalaro, S., Aguado, A.: Plastic fibres as the only reinforcement for flat suspended slabs: parametric study and design consideration. Constr. Build. Mater. **70**, 88–96 (2014)
15. Facconi, L., Minelli, F., Plizzari, G.: Steel fiber reinforced self-compacting concrete thin slabs–experimental study and verification against Model Code 2010 provisions. Eng. Struct. **122**, 226–237 (2016)
16. di Prisco, M., Colombo, M., Pourzarabi, A.: Biaxial bending of SFRC slabs: is conventional reinforcement necessary? Mater. Struct. **52**(1), 1–15 (2018). https://doi.org/10.1617/s11527-018-1302-0
17. di Prisco, M., Colombo, M., De Wilder, K., Vandewalle, L.: A new FRC solution for a partially prefabricated industrial deck. In: Proceedings of the 9th RILEM International Symposium on Fiber Reinforced Concrete (BEFIB 2016), Vancouver, BC, Canada, pp. 19–21 (2016)
18. de la Fuente, A., Escariz, R.C., de Figueiredo, A.D., Molins, C., Aguado, A.: A new design method for steel fibre reinforced concrete pipes. Constr. Build. Mater. **30**, 547–555 (2012)
19. EN 14651-5. Precast concrete products – test method for metallic fibre concrete – Measuring the flexural tensile strength, European Standard (2005)

Codes and Standards

Eurocode 2 – Annex L – European Harmonized Standard for Steel Fibre Reinforced Concrete

Marco di Prisco[1], Terje Kanstad[2], Giovanni Plizzari[3], Fausto Minelli[3], and Andreas Haus[4(✉)]

[1] Politecnico di Milano, Milan, Italy
[2] Norwegian University of Science and Technology, Trondheim, Norway
[3] University of Brescia, Brescia, Italy
[4] Bekaert GmbH, Neu-Anspach, Germany
andreas.haus@bekaert.com

Abstract. The current Eurocodes are under revision and estimated to be available in 2023. For the first time in the European history, Eurocode 2: "Design of concrete structures" will be extended with a European-wide harmonized annex covering steel fibre reinforced concrete. The work on Annex L – Steel Fibre Reinforced Concrete has already started in 2012 and significantly benefitted from the work carried out for the fib Model Code for Concrete Structures 2010. The use of performance classes of Model Code 2010 as well as parts of the design approach were the basis for the new steel fibre reinforced concrete annex. In addition, the latest state of science has been used to prepare a powerful but, in the same way, easy-to-use design document for structural engineers, covering both ultimate and serviceability limit states for steel fibre reinforced structures, with or without reinforcement.

Keywords: Steelfibre · Eurocode 2 · Concrete · Reinforcement · Ductility · Ultimate limit states · Serviceability limit states

1 Introduction and Organization of the Work

The first Eurocode for design of concrete structures, EN 1992-1-1 (Eurocode 2) was published in 2004. When the current revision work started in 2012, it was decided to also include Steel Fibre Reinforced Concrete, Stainless Steel, Carbon Fibre Reinforced Polymers, and Existing Structures in separate Annexes. While the overall responsibility for the revision is placed on the committee CEN TC250/SC2, the work with the Steel Fibre Reinforced Concrete (SFRC) is carried out by members of the committee CEN TC250/SC2/WG1/TG2 established in 2012, and the Project Team CEN TC 250/SC2/PT3 (2018–2020) being responsible for the final text. PT3 has completed the last draft on April 30th this year, while the entire new Eurocode 2-draft will be submitted to CEN for the final processing by November 2020. The process will continue for some years, and a realistic estimate is that translation into the various languages will be available around 2024.

© RILEM 2022
P. Serna et al. (Eds.): BEFIB 2021, RILEM Bookseries 36, pp. 539–551, 2022.
https://doi.org/10.1007/978-3-030-83719-8_47

2 Scope of Eurocode 2 – Annex L

The main aim of the Annex L is to assess the suitability of steel fibres for use in design of concrete structures in the context of EC2. At the beginning, also alternative fibres were meant to be considered if agreed by WG1 and SC2, but this agreement was not reached due to the lack of knowledge on many crucial doubts related mainly to creep and durability at environmental conditions. The TG2 Committee had to: propose the Annex L text in the Normative as "motherhood statements" allowing the use of steel fibres for the detailed design in SLS and ULS using fibres for reinforcement in concrete structures, as "compact" as possible [asked 5–10 pages; written 26 pages!]; propose text for detailing rules and to ensure robustness in tying systems etc.; propose text necessary for inclusion in Part 1–2 concerning fibres in fire design; prepare the necessary background document(s); identify requirements to materials and execution that should be covered by other standards as conditions for the design rules incorporated in EC2.

3 Performance Classes

The performance classes introduced in the Annex L follows the same idea introduced in the Model Code 2010 [1]. It should be highlighted that the range 1–10 MPa on f_{R1k} as the characteristic residual flexural strength at $CMOD_1 = 0,5$ mm clearly shows the aim to include the mechanical performance of high performance SFRC. In fact, even if small cross section specimens made of high performance SFRC can show nominal bending strength values up to 20–30 MPa for a crack opening displacement of 0.5 mm, when testing the material according to EN 14651 standard, the performance class is around 10c (ratio $f_{R,3k}/f_{R,1k}$). For common use, it should be underlined that, even if maximum aggregate size, water/cement ratio and compressive strength usually qualify concrete matrixes, often they are not enough for fully qualifying the interaction between fibres and concrete mix. Fibre type, as well as aspect ratio, steel mechanical performance and its shape combined with grading curve, w/c ratio and filler type can significantly contribute to the final SFRC performance. The knowledge on this topic requires further research efforts, because it could allow the producer to dynamically recalibrate the final performance of the composite without facing with a new qualification procedure every time that one of the parameters has to be changed. Several experimental results have confirmed how SFRC performance can be predicted by considering the change of the fibre content (with the same concrete matrix) and assuming a linear relationship (between material properties and fibre content) in favour of safety.

According to the German guidelines, in order to favour the control on site, an upper bound of $0.6f_{R,im}$ for the characteristic value of $f_{R,ik}$ is assumed. This condition finds a justification in the difference between the on-site empirical evaluations as compared with the lab-measured strengths. The Committee has also discussed a recently presented classification on the toughness in uniaxial compression presented by Gonzalo-Ruiz and co-workers [2–4], that is able to modify the softening in uniaxial compression on the basis of a parabola softening law, requiring only one parameter affected by f_{R1k} and f_{ck}.

4   Constitutive Laws

The constitutive law introduced is rather close to that proposed by Model Code 2010 [1]. The only difference consists in the introduction of the maximum characteristic value introduced as 60% of the mean value to take into account the distance between real execution on site and lab preparation of the specimens. Furthermore, the correlation between the residual strengths $f_{R,ik}$ and the uniaxial tensile strength was assumed according to di Prisco et al. [5]. In particular, the residual tensile strength for crack widths at the serviceability limit state f_{Fts} is regarded as a fixed point at a crack width $w_1 = 0.5$ mm, and the two points of the linear pull-out branch are referred to as follows (Fig. 1):

$$f_{Ft,1ef} = 0,37\ f_{R,1k} \tag{1}$$

$$f_{Ft,3ef} = 0,57\ f_{R,3k} - 0,26\ f_{R,1k} \tag{2}$$

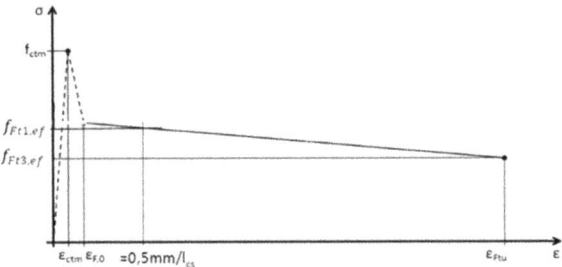

Fig. 1. Constitutive law in uniaxial tension according to Annex L

with the ultimate limit tensile strain for SFRC ε_{Ftu} and the structural length used to convert the stress-crack width relationship of SFRC to a stress-strain relationship compatible with concrete design l_{cs}:

$$\varepsilon_{Ftu} = 2,5mm/l_{cs};\ l_{cs} = \min\{h; s_{rm}\} \tag{3}$$

By examining 13 different materials having a quite wide range of performance classes (0.5c, 0.5e, 1b, 1e, 2c, 2.5c, 3e, 5c, 6c), it is possible to compare the experimental results obtained by means of EN 14651 [6] with those computed by means of the plane section approach, by adopting the design values identified according to Annex L rules ($\kappa_0 = \kappa_G = 1$; as factors for fibre orientation and volume effects). The ratio that corresponds to the safety coefficient computed in relation to the ultimate limit state and the maximum bending moment is shown in Fig. 2: the reliability of the safety coefficient of 1.5 selected is always guaranteed.

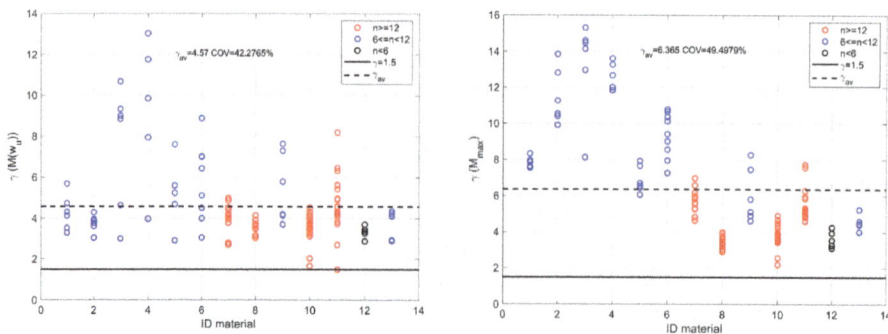

Fig. 2. Safety coefficients guaranteed by the uniaxial tensile constitutive law proposed in Annex L in relation to the pure bending behaviour with reference to respectively (a) the Ultimate Limit State $M(w_u)$ and (b) the Maximum bending moment M_{max}.

5   Orientation and Size Factors

Fibre Reinforced Concrete has the peculiarity to be reinforced by a random distribution of fibres. The latter can be regarded as a strength as well as a weakness. It is a strength because the single fibre randomly oriented can work for a very wide range of directions. In fact, if we consider scantly effective the fibre contribution in a double cone characterized by an angle $\alpha = 30°$ astride its axis, the solid angle covered by a single fibre corresponds to a solid angle ω:

$$\omega = 4\pi - 2\pi(1 - \cos(\alpha/2)) \cong 3,86\,\pi \qquad (4)$$

and this observation clarifies why FRC is very competitive when compared to common reinforcement when the stress state is 3D oriented. This peculiarity is also a weakness, because the designer cannot know precisely its distribution and orientation that is affected by casting procedure and its boundary conditions. To overpass this crucial point, in the Annex FRC is always assumed as 3D randomly oriented. The constitutive relationship is identified starting from the bending test according to EN 14651 [6] and it is well known, from a theoretical point of view, that the 3-point bending test has an average orientation factor α which is higher ($\cong 0.65$) than a theoretical 3D orientation factor ($\cong 0.45$). However, the tensile behaviour of the fibre is not so much affected by a small deviation of its axis due to kinking reorientation and, therefore, the tensile response is usually much more affected by fibre number, rather than by fibre orientation. For this reason in the Annex L, as well in the Model Code, a single parameter κ_0 is introduced and its value is always 1 for common bending situation and is set equal to 0.5 for local check, where a random casting procedure could locally introduce an ineffective fibre orientation [7–9], and for ties and walls vertically cast, when the bending moment acting on the horizontal surface is considered. In the Annex, as in the Model Code 2010 [1], also a favourable distribution can be considered: it is the case of self-compacting concrete with a casting procedure that favours a very efficient fibre distribution with a relatively small dispersion. The factor κ_0 in this case cannot exceed the double ($\leq$ 2); it has to be checked by means of design-by-testing and involves always an orthotropic behaviour of concrete with a coefficient κ_0 that orthogonally becomes lower than unity.

Another concept already exploited in Model Code 2010 [1] is the size-factor, named K_{Rd}, that takes into account the reduction of the standard deviation of the residual strengths f_{Ri} and, therefore, the distance between the characteristic value $f_{Ri,k}$ and the mean value $f_{Ri,m}$, when a large portion of the structure can contribute in the failure mechanism to its global bearing capacity. This factor depends on the structure volume involved in the failure mechanism, but also by the stability of the failure response. In the Annex it is denominated κ_G and depends only on the volume, but more precise approaches can be used according to [10–12]. The Annex introduces both the coefficients, κ_0 and κ_G, as multiplier coefficients of the design strengths f_{Ftid}.

6   Durability of Steel Fibre Reinforced Concrete

In addition to several decades with practical experience, the background of the durability-clauses in Annex L is based on literature reviews and on the experimental research carried out by Berrocal [13] and Marcos [14].

The main content is briefly described by the following aspects:

- for SFRC, the concrete cover requirements $c_{min,dur}$ shall only apply to the embedded reinforcement, not to the steel fibres. There is no documentation that steel fibres influence the corrosion protection of ordinary reinforcing bars.
- For design of SFRC in exposure classes XS2, XD2, XS3 and XD3, the residual tensile strength in the outer most tension fibres shall be disregarded within a layer of $c_{f,dur}$. This statement will be adapted to the new durability system with Exposure Resistance Classes (ERC), and the values of $c_{f,dur}$ are still under debate. Amongst others it is discussed whether the value of $c_{f,dur}$ also shall be related to the calculated crack widths versus the crack width limits.
- If protective measures are applied to the steel fibres to protect them for the design life of the structure, the residual tensile strength of the outermost tension fibres within depth $c_{f,dur}$ may be used.
- The residual tensile strength of fibres in the layer $c_{f,dur}$ may be used in temporary situations immediately after curing such as during the construction phase.

7   Ductility of Steel Fibre Reinforced Concrete and Structural Design

The ductility of SFRC structure response was long debated by the Committee. This is due to the softening that characterizes the uniaxial tension behaviour. Model Code 2010 clarified that the key point is the overall structural behaviour rather than the local material one. Many accepted strengthening techniques are characterized by brittle behaviour: the focus is that the final structure response should be ductile and robust, according to the capacity design approach. An important doubt concerns the 2D structures like slabs. In fact, 1D structures (linear elements) have always a conventional reinforcement even if made of FRC, because the dominant one-direction behaviour determines the need and the convenience to maintain a classical continuous main longitudinal reinforcement. On the contrary, in slabs (due to their high level of internal redundancy) the conventional reinforcement made of rebars could be strongly reduced by using steel fibres, which become much more convenient for their ability to satisfy, at the same time, bending moment, shear, torsion and punching, thus exploiting their full capabilities [15].

A recent paper [16] evidenced the different behaviour of conventional reinforcement and FRC in biaxial bending: it should be underlined that this difference is measured for relative small amount of fibre contents and, therefore, a new research investigation is promoted in the TG2 Committee to investigate the behaviour in case of higher fibre contents. Anyway, in [17] it is demonstrated that, even in case of small fibre amounts, the limit analysis approach indicated in the Model Code 2010 (recalled in the Annex L) can be regarded as reliable, reaching safety coefficients much larger than those guaranteed by conventional reinforcement. The only doubts are oriented to robustness and this topic becomes the frontier of our knowledge that, up to now, can be faced by prescribing a certain amount of conventional reinforcement able to link the supporting columns, in case of elevated slabs (this extra reinforcement is well known in literature as anti-collapse reinforcement).

In favour of safety, by the way, in the Annex plastic analysis of SFRC structures without any direct check of rotation capacity may be used for the ultimate limit state analysis of the following structures:

(1) Foundations and slabs supported directly on ground.
(2) Statically redundant rafts and slabs on piles, if the ductility class is at least c; if the member is needed for structural stability, $f_{R,3k}f_{ct,flm} \geq 1,0$.
(3) Statically redundant elevated slabs if the ductility class is at least c and $f_{R,3k}f_{ct,flm} \geq 1,0$.
    For members not fulfilling (1), (2) or (3), methods based on plastic analysis, or linear analysis with limited redistribution, shall only be applied where the deformation capacity of the critical sections, derived from the predicted failure mechanisms, is demonstrated to be sufficient.
(4) Due to possible localization effects occurring at yielding of longitudinal reinforcement, minimum reinforcement due to robustness, as prescribed above, is not sufficient to generally allow plastic analysis without verification of the deformation capacity.
(5) Methods to be used for verification of plastic deformation capacity shall take into account local variations in residual tensile strength.

In order to guarantee robustness, in members reinforced with steel fibres without or with axial force, a general rule on the minimum reinforcement is provided so that:

$$M_{R,min}(N_{ED}) \geq k \cdot M_{cr}(N_{ED}) \tag{5}$$

where:

$M_{R,min}$ is the bending strength of the section with $A_{s,min}$ in presence of an axial force N_{ED}, and the effects of the fibres included by the effective residual tensile strength, $f_{Ftu,ef}$ in agreement with the stress distributions.

The reduction in $A_{s,min}$ due to the fibre-contribution has the following limits: in members subjected to axial force, $A_{s,min}$ shall meet the following requirement

$$N_{R,min} \geq k \, N_{cr} \tag{6}$$

where the effects of fibres are included in terms of residual tensile strength, and the reduction $A_{s,min}$ due to fibre contribution depends on the specific type of structure; for members reinforced with steel fibres requiring shear or torsion reinforcement, the minimum shear reinforcement area $\rho_{w,min}$ may be taken as in the Model Code 2010, that means:

$$\rho_{Fw,min} = \rho_{w,min} - f_{Ftu,ef}/f_{yk} \geq 0 \tag{7}$$

provided that $f_{Ftu,ef} \geq 0,08 \cdot \sqrt{f_{ck}}$.

8   Ultimate Limit States

As already mentioned above, SFRC may significantly enhance structural behaviour at Ultimate Limit State (ULS); however, depending on the degree of redundancy of the structure and on SFRC mechanical properties, some conventional rebars may be necessary.

The Annex includes the classical ULSs, starting from bending, where the contribution of SFRC is usually not particularly significant because most of the bending stresses are better resisted by conventional rebars (especially in linear and beam-like structural elements). In fact, a little increase of ultimate bending resistance can be observed in SFRC beams with longitudinal reinforcement. The latter contribution may be more significant in elevated slabs where a lower reinforcement ratio is generally adopted due to the higher possibility of stress redistribution. However, depending on the longitudinal reinforcement ratio and the SFRC performance class, an increased bond strength of the longitudinal rebars may reduce the yielding penetration and, as a consequence, structural ductility [18].

SFRC may be particularly useful for the ULS of anchorage since the post-cracking resistance provides confinement around the anchored bar that contrasts the splitting crack development, especially when stirrups are not present and low concrete covers are adopted. It should be mentioned that the SFRC confinement may become particularly important in internal laps where the confinement efficiency of external stirrups is significantly reduced.

In general, a stress-block or linear state of stress in tension due to steel fibres can be taken into account for the determination of the resisting design moment. In the first case, a simplified rigid plastic approach for the residual tensile strength may be used for ductility classes *a*, *b* and *c*. For classes *d* and *e* this approach may only be used to determine the ultimate bending moment at the ultimate strain. The following tensile strain limit shall be used: $\varepsilon_{Ftu} = \varepsilon_{Ftud} = 0{,}02$.

A more refined bi-linear residual tensile stress distribution can be adopted by considering design strengths $f_{Ft1d} = f_{Ft1,ef}/\gamma_{SF}$ and $f_{t3d} = f_{Ft3,ef}/\gamma_{SF}$. For design of statically redundant slabs at the ultimate limit states, the design strengths, f_{Ftud}, f_{Ft1d} and f_{Ft3d}, may be increased by multiplying the latters with the abovementioned factor κ_G.

The other Ultimate Limit States considered are discussed in the following.

8.1 The Influence of Fibres on Shear

Fibres are particularly effective in shear critical beams due to the real possibility of reducing or avoiding the transverse reinforcement [19]. In fact, due to its post-cracking strength, the different shear resistant mechanisms may be enhanced by the use of SFRC: among these, the aggregate interlock plays a major contribution as fibres act in reducing shear crack width with a more diffused and steady shear crack pattern. SFRC is particularly efficient in shallow beams and prestressed beams (except in the diffusion zone of prestressing) where the stirrups may be totally removed [19]. However, the contribution of SFRC to shear resistance of stirrups, even though considered as an additional term, should properly consider the compatibility of deformations when activating the shear resistance provided by stirrups and SFRC.

Based on the work reported in [20], the following equations, consistent with the one included in the current draft of EC2 for RC elements, were derived and proposed. For members not requiring shear reinforcement, having classical longitudinal reinforcement, the design shear strength can be computed as follows:

$$\tau_{Rd,cF} = \eta \frac{0,6}{\cdot\gamma_c}\left(100\rho_l f_{ck}\frac{d_{dg}}{d}\right)^{\frac{1}{3}} + f_{Ftud} \geq \eta\tau_{Rdcmin} + f_{Ftud} \tag{8}$$

where:

$$\eta = max(1/(1+0.43f_{Ftuk}^{2.85}); 0.4) \tag{9}$$

As general statements, shear reinforcement is not required in regions where $\tau_{Ed} \leq \tau_{RD,cF}$. Moreover, steel fibres cannot be considered for members resisting shear in combination with axial tension.

For members requiring shear reinforcement (again, having longitudinal classical reinforcement), it is possible to combine shear reinforcement (truss-analogy) with fibres as it reads:

$$\tau_{Rd,sF} = \left(\eta_s \frac{A_{sv}}{s}\frac{z}{bd}f_{yd} + \eta_F f_{Ftud}\right) \tag{10}$$

where $\eta_s = 0.75$ and $\eta_F = 1.0$.

Finally, for SFRC containing neither longitudinal bars nor prestressing steel in the tensile zone, the resisting design shear strength should be computed as:

$$\tau_{Rd,cF} = f_{Ftud} \geq \tau_{Rd,cmin} \tag{11}$$

8.2 The Influence of Fibres on Torsion

Torsion determines a diffused stress state in the beam, requiring both longitudinal and transverse reinforcement, according to the well-known 3D spatial truss analogy. In this context, SFRC can hardly substitute longitudinal rebars but could be particularly efficient in the partial or complete substitution of stirrups. A case of practical interest is represented by beams under a limited torsion requiring a heavy minimum amount of transverse reinforcement [21]. The latter could be avoided when using SFRC by considering its post-cracking performance. However, although some results are available into the literature [22], this topic requires further research. The influence of fibres in pure torsion, as well as in combination with flexure and shear, should be clearly studied in the next future.

The Annex L provides the following equations for the determination of torsional strength of members having classical longitudinal reinforcement. The design resisting torque in the longitudinal and transverse direction are defines as it reads:

$$T_{Rd,slF} = \eta_s T_{Rd,sl} + \eta_F T_{Rd,clF} \tag{12}$$

$$T_{Rd,swF} = \eta_s T_{Rd,sw} + \eta_F T_{Rd,cwF} \tag{13}$$

where $\eta_F = 1.0$ and $\eta_s = 0.75$ (likewise in the shear provisions) and factors $T_{Rd,sl}$ and $T_{Rd,sw}$ are the classical formulation for RC members (3D space truss analogy). Moreover, the fibre contributions to torsion are defined as follows:

$$T_{Rd,clF} = f_{Ftud} \cdot t_{ef,i} \cdot 2 \cdot A_k \tag{14}$$

and

$$T_{Rd,cwF} = f_{Ftud} \cdot t_{ef,i} \cdot 2 \cdot A_k \tag{15}$$

8.3 The Influence of Fibres on Punching

As for shear, SFRC is particularly efficient to punching resistance due to its post-cracking performance. The latter enhances the punching-resistant mechanisms especially when using concretes having higher strength since a slab with a reduced thickness may be adopted. Based on the work of Muttoni [23], for slabs not requiring punching reinforcement, not subjected to significant tensile actions, the design punching shear strength is defined as follows:

$$\tau_{Rd,cF} = \eta_C \tau_{Rd,c} + \eta_F f_{Ftud} \geq \eta \tau_{Rdcmin} + f_{Ftud} \tag{16}$$

where $\eta_c = \tau_{Rd,c}/\tau_{Ed} \leq 1,0$ and $\eta_F = 1.0$. Note that $\tau_{Rd,c}$ is the formulation for classical RC members.

Where shear reinforcement is required in SFRC slabs with flexural reinforcement,

$$\tau_{Rd,cs} = \eta_C \tau_{Rd,c} + \eta_s \rho_w f_{ywd} + \eta_F f_{Ftud} \geq \rho_w f_{ywd} + \eta_F f_{Ftud} \tag{17}$$

For the calculation of the control perimeter at which shear reinforcement is not required ($b_{0,out}$ as in the draft for RC), an enhanced expression can be utilized by replacing the one for RC elements:

$$b_{0,out} = b_0 \cdot \left(\frac{d_v}{d_{v,out}} \cdot \left(\frac{1}{\eta_c} - \frac{\eta_F \cdot f_{Ftud}}{\tau_{Rd,c}} \right) \right)^2 \tag{18}$$

where all parameters involved are identified in the EC2 draft for RC except for:

$$\eta_s = \left(1.5 \frac{d_{dg}}{d} \right)^{1/2} \left(\frac{1}{\eta_c \cdot k_{pb}} \right)^{3/2} \leq 0,8 \tag{19}$$

For SFRC members with no longitudinal reinforcement, the following design punching shear strength can be adopted:

$$\tau_{Rd,cF} = f_{Ftud} \geq \tau_{Rdcmin} \tag{20}$$

9   Serviceability Limit State

SFRC is particularly efficient at Serviceability Limit State (SLS), especially for its significant contribution to crack opening reduction. In fact, due to the post-cracking resistance at the crack face as well as the enhanced bond strength, the addition of fibres was seen to promote a crack pattern with narrower and more closely space cracks, with a reduction of crack spacing up to 50% [24]. This may be particularly helpful for the protection of rebars in aggressive environments. SFRC also reduces structural deformability due to the enhanced tension stiffening. The Annex adopts the following formulation for the maximum crack spacing in members reinforced with steel fibres and longitudinal rebars:

$$s_{r,max,cal} = \left(2c + 0.35 \cdot k_b \frac{\phi}{\rho_{b,ef}} \right) \cdot \left(1 - \frac{f_{Fts,ef}}{f_{ctm}} \right) \tag{21}$$

10 Requirements for Special Concrete Structures

In the Annex, also SFRC characterized by a performance class 10e are considered. A class 10e means that the characteristic uniaxial tensile strength at a crack opening of 0.5 mm is close to 3.7 MPa and, therefore, it is equal to the $f_{ctk,min}$ of a concrete C100 according to Model Code 2010, while the ultimate tensile strength is even higher than 4.81 MPa. In case of a structure with a κ_G larger than 1, a hardening material in uniaxial tension is expected. The expressions proposed in [1] were suitably checked for conventional SFRC and, therefore, in these cases a uniaxial tensile test is strongly recommended to guarantee the uniaxial tensile behaviour.

Finally, in §14 the Annex introduces special rules that can satisfy robustness following specific requirements based on performance classes: 1b for SFRC columns footing on rocks, 2c for foundations on piles, 4c for segmental linings. These requirements are based on the accumulated experience and on the main idea that soil interaction and a generalized compressive stress in the structure can prevent any risk of lack of robustness.

11 Conclusion

A very significant effort made by TG2 Committee with 27 meetings in 8 years of around 25 experts of 13 different countries was aimed at the preparation of the Annex L, which represents a significant step forward in relation to Model Code 2010. New equations related to the fibre contribution in compression toughness, on SLS and ULS, on ductility and robustness, on the execution rules and fire resistance (not discussed specifically in the paper, but sent to the specific Committees) have been introduced. About SLS and ULS, a strong effort was accomplished to provide practitioners with design rules consistent with those developed for RC members not containing fibres: by neglecting the performance of fibres, all equations related to ULS and SLS are identical to the corresponding of RC elements. Specific topics like fatigue were left to design-by-testing approaches, while creep was assumed not too critical in common situations, because although it could potentially cause a crack opening increase with the time, it is not able to affect crack spacing and usually it favours stress redistribution.

Many subjects will require further research, but the knowledge on this material is enough to introduce it in the range of the possible choices for a common designer, because crack-bridging action offered by fibres, randomly distributed in the concrete, will favour durability and consequently sustainability, thus reducing the resources required to achieve the safety requirements specified in the Codes.

Acknowledgements. The authors are in debt with the members of CEN TC250/SC2/WG1/TG2 for their contributions and the passionate, effective and wise discussions carried out during the 27 meetings in the last eight years.

References

1. fib Model Code for Concrete Structures 2010. Ed. Fib – International Federation for Structural Concrete, Ernst & Sohn, Berlin (2013)
2. De La Rosa, Á., Ruiz, G., Poveda, E.: Study of the compression behavior of steel-fiber reinforced concrete by means of the Response Surface Methodology. Appl. Sci. **9**(24), 5330 (2019)
3. Ruiz, G., De La Rosa, Á., Poveda, E.: Relationship between residual flexural strength and compression strength in steel fiber reinforced concrete within the new Eurocode 2 regulatory framework. Theoret. Appl. Fract. Mech. 103 (2019)
4. Ruiz, G., De La Rosa, Á., Wolf, S., Poveda, E.: Model for the compressive stress–strain relationship of steel fiber–reinforced concrete for non–linear structural analysis. Hormigón y Acero **69**(Suppl. 1), 75–80 (2018)
5. di Prisco, M., Colombo, M., Dozio, D.: Fibre-reinforced concrete in fib Model Code 2010: Principles, models and test validation. Struct. Concr. **14**(4), 342–361 (2013)
6. EN 14651: Test method for metallic fibered concrete – Measuring the flexural tensile strength (limit of proportionality (LOP), residual). Brussels (2004)
7. Martinelli, P., Colombo, M., Pujadas, P., de la Fuente, A., Cavalaro, S., di Prisco, M.: Characterization tests for SFRC elevated slabs: identification of fibre distribution and orientation effects, submitted to Materials & Structures (2020)
8. Laranjeira, F., Grünewald, S., Walraven, J., et al.: Characterization of the orientation profile of steel fiber reinforced concrete. Mater. Struct. Constr. **44**, 1093–1111 (2011)
9. Soroushian, P., Lee, C.-D.: Distribution and orientation of fibers in steel fiber reinforced concrete. ACI Mater. J. **87**, 433–439 (1990)
10. Colombo, M., Martinelli, P., di Prisco, M.: On the evaluation of the structural redistribution factor in FRC design: a yield line approach. Mater. Struct. **50**(1), 1–18 (2016). https://doi.org/10.1617/s11527-016-0969-3
11. di Prisco, M., Martinelli, P., Dozio, D.: The structural redistribution coefficient K_{Rd}: a numerical approach to its evaluation. Struct. Concr. **17**, 390–407 (2016)
12. Pourzarabi, A., Colombo, M., Di Prisco, M.: On the mechanical response of a fibre reinforced concrete redundant structure; the redistribution factor. In: Proceedings of the 12th fib International Ph.D. Symposium in Civil Engineering, pp. 625–632 (2018)
13. Berrocal, C.G.: Corrosion of steel bars in fibre reinforced concrete: corrosion mechanisms and structural performance, TU Chalmers, Gothenburg, Sweden (2017)
14. Meson, V.M.: Durability of steel fibre reinforced concrete in corrosive environments, Department of Civil Engineering, Technical University of Denmark (2019)
15. Facconi, L., Plizzari, G., Minelli, F.: Elevated slabs made of hybrid reinforced concrete: proposal of a new design approach in flexure. Struct. Concr. **20**(1), 52–67 (2019)
16. di Prisco, M., Colombo, M., Pourzarabi, A.: Biaxial bending of SFRC slabs: is conventional reinforcement necessary? Mater. Struct. **52**(1) (2019). art. no. 1
17. di Prisco, M., Colombo, M., Pourzarabi, A.: Yield line design for SFRC elevated slabs. In: Proceedings of the Befib International Conference, Valencia (2020, accepted)
18. Meda, A., Minelli, F., Plizzari, G.A.: Flexural behaviour of RC beams in fibre reinforced concrete. Compos. B Eng. **43**(8), 2930–2937 (2012)
19. Minelli, F., Plizzari, G.A.: On the effectiveness of steel fibers as shear reinforcement. ACI Struct. J. **110**(3), 379–389 (2013). ISSN 0889-3241
20. Minelli, F., Plizzari, G.A.: Shear strength of FRC members with little or no shear reinforcement: a new analytical model. In: Minelli, F., Plizzari, G.A.: fib Bulletin 57: Shear and punching shear in RC and FRC elements, Workshop 15–16 October 2010, Salò, Italy, vol. unico, pp. 211–225 (2010). ISSN 1562-3610, ISBN 978-2-88394-097-0

21. Chalioris, C.E., Karayannis, C.G.: Effectiveness of the use of steel fibres on the torsional behaviour of flanged concrete beams. Cement Concr. Compos. **31**(5), 331–341 (2009)

22. Facconi, L., Minelli, F., Plizzari, G., Ceresa, P.: Experimental study on steel fiber reinforced concrete beams in pure torsion. In: Proceedings of the fib Symposium 2019: Concrete - Innovations in Materials, Design and Structures, pp. 1811–1818 (2019)

23. Muttoni, A., Ruiz, M.F., Simões, J.T.: Recent improvements of the critical shear crack theory for punching shear design and its simplification for code provisions (2019) FIB 2018 - Proceedings for the 2018 fib Congress: Better, Smarter, Stronger, pp. 116–129.

24. Giuseppe, T., Fausto, M., Giovanni, P.: Cracking behavior in reinforced concrete members with steel fibers: a comprehensive experimental study. Cem. Concr. Res. **68**, 24–34 (2015)

Reliability of Shear Strength Models for Fibre Reinforced Concrete Members Without Shear Reinforcement

Nikola Tošić$^{(\boxtimes)}$, Jesús Miguel Bairán, and Albert de la Fuente

Civil and Environmental Engineering Department,
Universitat Politècnica de Catalunya, Barcelona, Spain
nikola.tosic@upc.edu

Abstract. The scope and amount of fibre reinforced concrete (FRC) structural applications have seen significant increases. This means that safe and reliable ultimate limit state (ULS) models are necessary for FRC structural members. Among these, shear strength of FRC members without shear reinforcement is highly important due to the brittleness of shear failure. Because of this, the *fib* Model Code 2010 introduced two shear strength models: an empirical model based on Eurocode 2 and a physical model based on the Modified Compression Field Theory. However, a comprehensive reliability assessment of these models has been lacking. Therefore, in this study, the safety format of these models is assessed and the partial safety factors for FRC in shear, γ_c and γ_F are updated. As a first step, a large database of experimental results on FRC beams is used to determine model uncertainties. Following this, a comprehensive parametric probabilistic analysis is performed using the First Order Reliability Method to determine the adequate values of γ_c for different target reliability indices β. The results of this study show that in order to reach typical reliability indices used in ULS design, γ_c and γ_F values need to be increased for FRC members without shear reinforcement for both models proposed by the *fib* Model Code 2010.

Keywords: Fiber reinforced concrete · Safety · Partial safety factor · Beams · Database · Model Code

1 Introduction

Fibre reinforced concrete (FRC) is already recognized as a feasible and more sustainable, alternative to traditional reinforced concrete (RC) solutions in many structural applications [1–4]. One attractive and advantageous use of FRC is in members exposed to shear where its use can permit the elimination or reduction of shear reinforcement. Such a use of FRC can bring significant savings in material costs and time of construction. However, to design such members, reliable ultimate limit state (ULS) models for FRC are necessary. For the case of flexural strength of FRC, comprehensive reliability-based assessments of flexural design models have been performed [5]. At the same time, such work for shear strength, and in particular, for shear strength of members without shear reinforcement, has been lacking.

© RILEM 2022
P. Serna et al. (Eds.): BEFIB 2021, RILEM Bookseries 36, pp. 552–563, 2022.
https://doi.org/10.1007/978-3-030-83719-8_48

Shear itself has been a topic that has attracted interest of practitioners and researchers for decades [6] with numerous empirical, semi-empirical and theoretical models having been proposed. Still, the behaviour of members without shear reinforcement has remained a challenging modelling task [7], even for RC members, e.g. the current version of Eurocode 2 [8] contains an empirical model for the shear strength of RC members without shear reinforcement. As for the *fib* Model Code 2010 [9] it presents two models for the shear strength of FRC members without shear reinforcement. The first model—herein referred to as MC2010-A—is the officially adopted model that arose from the work of Minelli et al. [10–12], based on the Eurocode 2 empirical shear resistance model. Although fibres provide several contributions to shear strength—toughness, aggregate interlock, improved bending strength of struts, increased dowel action of the longitudinal reinforcement—the MC2010-A model considers only the contribution of the fibres through the pull-out mechanism and as a type of "distributed reinforcement" [13]. The second model—herein referred to as MC2010-B—is an alternative model presented in the *fib* Model Code 2010, based on the Modified Compression Field Theory [14] and is more consistent with the *fib* Model Code 2010 shear strength model for RC members [9].

So far, research has mostly focused on assessing the "model error", δ, of these models, i.e. the ratio between actual behaviour/shear strength measured in experiments and shear strength predicted by models. Even for RC members, the scatter of model errors for shear strength of members without shear reinforcement is quite high with coefficients of variation (CoV) routinely exceeding 20% [15, 16]. For FRC, Marí et al. [17] calculated the model error on a database of experimental results on steel fibre reinforced concrete (SFRC) beams for models MC2010-A and MC2010-B and found average values of 1.04 and 0.99, respectively, and CoV values of 23% and 24%. Even though these results are commensurable with model uncertainties for RC members without shear reinforcement, they are inconclusive about the reliability of FRC shear design since the probability of failure remains unquantified. Both models MC2010-A and MC2010-B are based on an FRC partial safety factor $\gamma_c = 1.50$, accepted in order to maintain continuity with RC but, apparently, without developing a full probabilistic analysis (at least it is not reported in the *fib* Model Code 2010 background documentation).

However, considering the high scatter associated with FRC tensile/flexural residual strength, the reliability index β achieved by these models is not clear in advance. Therefore, the goal of this study is to perform a probabilistic analysis of both *fib* Model Code 2010 models for the shear strength of FRC members without shear reinforcement and calibrate the partial safety factor γ_c required for achieving code-prescribed failure probabilities P_f according to different consequence classes. To achieved this, first, the model error was determined on a database of experimental results. Then, a parametric study was performed using the First Order Reliability Method (FORM) considering different probability distributions and parameters of input variables. Finally, based on the results, the FRC partial safety factor was calibrated based on the target reliability index and failure probability. The analysed models are based on SFRC; nonetheless, the models have been reported to be compatible with polymeric fibre reinforced concrete as well [18, 19]. The results and conclusions presented herein have the potential to become a reference for future revisions of the national and international design codes for FRC members.

2 Description of the Models and Assessment of Model Error

2.1 Model Code 2010 Models for Shear Strength of FRC Members Without Shear Reinforcement

2.1.1 Model MC2010-A

According to the MC2010-A model, shear strength of FRC members "with conventional longitudinal reinforcement and without shear reinforcement" is given by the following expression:

$$V_{Rd,F} = \left\{ \frac{0.18}{\gamma_c} \cdot k \cdot \left[100 \cdot \rho_1 \cdot \left(1 + 7.5 \cdot \frac{f_{Ftuk}}{f_{ctk}} \right) \cdot f_{ck} \right]^{1/3} + 0.15 \cdot \sigma_{cp} \right\} \cdot b_w \cdot d \qquad (1)$$

where:

$V_{Rd,F}$ is the shear strength in [N];

γ_c is the partial safety factor for concrete;

k is a factor considering the size effect, determined as

$$k = 1 + \sqrt{\frac{200}{d}} \leq 2.0 \qquad (2)$$

d is the effective depth in [mm];

ρ_1 is the longitudinal reinforcement ratio defined as

$$\rho_1 = \frac{A_{sl}}{b_w \cdot d} \qquad (3)$$

b_w is the smallest width of the cross-section in the tensile zone in [mm];

f_{Ftuk} is the characteristic value of the ultimate residual tensile strength of FRC, considering a crack width $w_u = 1.5$ mm, according to Eq. (5.6–6) of the *fib* Model Code 2010:

$$f_{Ftu} = f_{Fts} - \frac{w_u}{CMOD_3} \cdot (f_{Fts} - 0.5 \cdot f_{R3} + 0.2 \cdot f_{R1}) \geq 0 \qquad (4)$$

CMOD3 is the crack mouth opening displacement (CMOD) of 2.5 mm per EN 14651 [20];

f_{Ftu} is the mean value of the ultimate residual tensile strength of FRC,

f_{Fts} is the mean FRC serviceability residual strength equal to 0.45·f_{R1k};

f_{R1} is the mean FRC residual strength corresponding to CMOD = 0.5 mm;

f_{R3} is the mean FRC residual strength corresponding to CMOD = 2.5 mm;

f_{ctk} is the characteristic value of the tensile strength of concrete in [MPa];

f_{ck} is the characteristic value of the compressive strength of concrete in [MPa] defined as $f_{cm} - 8$ MPa, where f_{cm} is the mean compressive strength;

σ_{cp} is the average stress acting on the concrete cross-section A_c [mm2] due to an axial force N_{Ed} [N] caused by loading or prestress (N_{Ed} is positive for compression), i.e. $\sigma_{cp} = N_{Ed}/A_c < 0.2 \cdot f_{cd}$, where f_{cd} is the design compressive strength.

As in Eurocode 2, the shear resistance $V_{Rd,F}$ cannot be smaller than a minimum value $V_{Rd,F,min}$:

$$V_{Rd,F,min} = \left(0.035 \cdot k^{3/2} \cdot f_{ck}^{1/2} + 0.15 \cdot \sigma_{cp}\right) \cdot b_w \cdot d \tag{5}$$

It should be noted that Eq. (1) uses the partial safety factor γ_c for "concrete without fibres" [9].

2.1.2   Model MC2010–B

The MC2010-B model defines $V_{Rd,F}$ as an expansion of $V_{Rd,c}$ defined for RC members:

$$V_{Rd,F} = \frac{1}{\gamma_F}\left(k_v \cdot \sqrt{f_{ck}} + k_v \cdot f_{Ftuk} \cdot \cot\theta\right)z \cdot b_w \tag{6}$$

where:

$k_f = 0.8$

f_{Ftuk} is the characteristic value of the FRC ultimate residual tensile strength of FRC at w_u.

k_v for members without shear reinforcement:

$$k_v = \frac{0.4}{1 + 1500 \cdot \varepsilon_x} \cdot \frac{1300}{1000 + k_{dg} \cdot z} \tag{7}$$

k_{dg} is the aggregate size parameter defined as $32/(16 + dg) \geq 0.75$ where d_g is the maximum aggregate size (if smaller than 16 mm, k_{dg} can be taken as 1.0).

ε_x is the longitudinal strain at the mid-depth of the effective shear depth z

$$\varepsilon_x = \frac{1}{2 \cdot E_s \cdot A_{sl}}\left(\frac{M_{Ed}}{z} + V_{Ed} + N_{Ed} \cdot \left(\frac{1}{2} \mp \frac{\Delta e}{z}\right)\right) \tag{8}$$

M_{Ed}, V_{Ed}, N_{Ed} are the design bending moment, shear force, and axial force at the shear control section, respectively and Δe is the eccentricity of the design axial force.

The limits of the compressive stress field angle θ are between a minimum value θ_{min} and 45° with θ_{min} defined as

$$\theta_{min} = 29° + 7000 \cdot \varepsilon_x \tag{9}$$

Finally, the crack width at the ultimate limit state w_u is defined as

$$w_u = 0.2 + 1000 \cdot \varepsilon_x \geq 0.125 \text{mm} \tag{10}$$

Importantly, unlike the MC2010-A model, the MC2010-B model in Eq. (6) prescribes the use of the γ_F partial safety factor for FRC. Nonetheless, the partial safety factor is also defined as 1.50 [9].

2.2 Database of Experimental Results for Model Error Assessment

In order to adequately assess the error model δ a comprehensive database of experimental results is needed. In this case, the database of experiments on SFRC beams without shear reinforcement, compiled by Lantsoght and made available online [21, 22] was used. The database is described in detail by Lantsoght [21]. The database consisted of 488 results on SFRC beams with longitudinal reinforcement and without shear reinforcement, collected from 65 individual studies. The parameter range of the original database is shown in Table 1 under the "Original database" column. All results reported in the database were from simply supported beams tested in either three- or four-point bending; the majority of had rectangular cross-sections. Significantly, residual strength was not a parameter reported in the database, because not all studies reported these values, for a variety of reasons. However, a large number of fibre properties is reported, as well as a large variety of steel fibre types [21].

Since the range of parameters was very wide, even outside of the scope of the models in some cases, three filtering criteria were imposed:

1. Concrete classes between C12 and C120 were considered (mean compressive strengths between 20 and 128 MPa);
2. Only beams with a longitudinal reinforcement ratio smaller than 4% were considered;
3. Only beams with a clear shear span-to-effective depth ratio larger than 2.0 were considered.

The first criterion was applied as these classes are the lower and upper compressive strength class, respectively, as defined by the *fib* Model Code 2010. This reduced the number of results to 477.

The second criterion was applied as some of the beams had very large reinforcement ratios unrepresentative of practical applications. This further reduced the number of results to 443.

Finally, the third criterion was applied because a different resisting mechanism is activated in these cases, consisting in the direct transfer of load to the support for values of the shear span-to-effective depth ratio smaller than 2.0. This reduced the number of results to 332. The ranges of parameter values for the filtered database are provide in Table 1 under the "Filtered database" heading.

Around 90% of the results were on beams with $f_{cm} < 70$ MPa, effective depth between 100 and 500 mm and with a longitudinal reinforcement ratio between 1.0% and 3.5%. The fibre volume fraction of 329 out of the 332 beams was below 2.0% (160 kg/m^3) and for 296 beams it was below 1.5% (120 kg/m^3).

Table 1. Example of construction of a table.

Parameter	Original database [22]		Filtered database	
	n = 488		n = 332	
	Min	Max	Min	Max
b_w (mm)	50	610	50	610
h (mm)	100	1,220	100	1,220
d (mm)	85	1,118	85	1,118
l_{span} (mm)	204	7,823	459	7,823
a/d (–)	0.46	6.00	2.22	6.00
a_v/d (–)	0.20	5.95	2.00	5.95
ρ_l (%)	0.37%	5.72%	0.37%	3.70%
f_y (MPa)	276	900	276	610
f_{cm} (MPa)	9.8	215.0	20.2	111.5
V_f (%)	0.2%	4.5%	0.2%	4.5%
λ (–)	25	191	25	191
f_{uf} (MPa)	260	4,913	260	4,913

l_{span} – clear span of the beam; a/d – shear span-to-effective depth ratio measured from left side of loading plate to left side of support;

a_v/d – clear shear span-to-effective depth ratio measured from face of loading plate to face of support; f_y – yield strength of steel reinforcement;

V_f – fibre volume fraction; λ – fibre aspect ratio (ratio of fibre length to diameter);

f_{uf} – tensile strength of fibres

2.3   Calculation of Model Errors

As reported above, residual tensile strengths for SFRC in the database of experimental results were not reported. Hence, the f_{Ri} values had to be estimated. This was done using the regressions presented in Figs. 1a (for f_{R3}) and 1b (for f_{R1}), derived from a statistical analysis of experimental results using the EN 14651 standard [20] on notched $150 \times 150 \times 600$ mm beams, as reported by Venkateshwaran et al. [23], Tiberti et al. [24], Galeote et al. [25], as well as other experimental programs conducted at the Structures and Materials Technology Laboratory (LATEM) of the Polytechnic University of Catalonia (UPC). The database included a large variety of concrete mixes, with the compressive strength range of 15–117 MPa, volume fraction of fibres 0.33–2.52%, fibre aspect ratios 35–110, fibre tensile strength 1000 to 3000 MPa and fibre modulus of elasticity (E_f) 190000–210000 MPa. Figure 1 demonstrates a good fit between the proposed linear regressions and the observed data, with the values of the coefficients of determination (R^2) of 0.90 and 0.75 for f_{R1} and f_{R3} predictions, respectively.

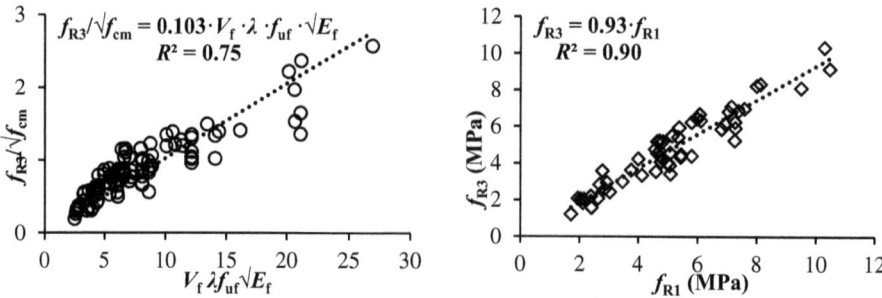

Fig. 1. Correlations used to assess f_{R3} (left) and f_{R1} (bottom) of the SFRC.

Then, the model error δ could be determined as $\delta = V_{\text{experiment}}/V_{\text{model}}$. For this purpose, the partial safety factor γ_c and γ_F were eliminated from Eqs. (1) and (6), respectively, and mean values of material properties were used. Values of mean compressive strength were used, f_{R1} and f_{R3} were predicted based on the above-presented regression and the tensile strength f_{ct} was calculated based on *fib* Model Code 2010 expressions [9].

For MC2010-A, the upper limit of 2% was applied to the longitudinal reinforcement ratio. For MC2010-B the compression field angle was adopted as $\theta = 45°$ as it provides optimal results. The descriptive statistics of the model errors on the basis of 332 results are reported for models MC2010-A and MC2010-B in Table 2. A box-and-whiskers plot was used to eliminate outliers, i.e., values of δ smaller than $Q_1 - 1.5 \cdot IQR$ and greater than $Q_3 + 1.5 \cdot IQR$ were excluded (where Q_1 and Q_3 are the first and third quartile, respectively, and IQR is the "interquartile range", i.e. $Q_3 - Q_1$). The obtained results can be seen to be comparable to previous analyses of the same models as well as analyses of RC model errors.

Table 2. Summary statistics of the model error δ for the MC2010-A and MC2010-B models.

	MC2010-A $n = 327$	MC2010-B $n = 328$
Mean, μ	1.075	0.912
Standard deviation, σ	0.245	0.266
CoV	22.8%	29.1%

3 Partial Safety Factor Calibration

3.1 Design Set and Probability Analysis

To assess the reliability of the studied models, a set of design cases was defined, as presented in Table 3. A typical range of thicknesses of building and bridge-deck slabs, beams, footings, and mat foundations was selected as corresponding to the range of 200–1000 mm. The cross-section width was considered as constant and equal to

300 mm since shear strength depends linearly on it. The effective depth was determined as $d = h - d_s = h - 50$ mm. The variables in Table 3 produce 140 combinations of geometry, longitudinal reinforcement and concrete class.

The process is initiated by computing the reference minimum ($V_{Rd,min}$) and maximum ($V_{Rd,max}$) shear capacities for every combination of the design variables of Table 3. For this purpose, current code value of the resistance safety factors $\gamma_c = _F = 1.50$ were retained. FRC residual flexural capacities were limited to $f_{R3k,min}$ and $f_{R3k,max}$ of 3 and 10 MPa, respectively. The range was further divided in quarters, so that five design loads were obtained for each case. Considering the 5 design loads, a total of 700 design cases (140 × 5) were generated.

Table 3. Range of variables in the design set.

Parameter	Values of parameters in the design set						
b (mm)	300						
h (mm)	200	400	600	800	1000		
ρ_l (–)	0.002	0.005	0.010	0.015	0.020	0.025	0.030
f_{ck} (MPa)	30	50	70	90			

The reliability of each design case was assessed by the reliability index β, related to the probability of failure (P_f) by $\beta = -\Phi^{-1}(P_f)$, where Φ is the cumulative standard normal distribution. FORM [26, 27] was used to estimate the reliability index.

A design failure is identified when a negative value is found in the limit state function $G = V_R - V_S = \delta \cdot V_{R,model} - V_S$, where $V_{R,model}$ is the shear resistance predicted by the model and V_S is the shear load. Considering G a function of random variables, the probability of failure is computed as the probability of obtaining a negative value of G, i.e.

$$P_f = P(G < 0) = P(\delta \cdot V_R - V_S < 0) \tag{11}$$

To calculate $V_{R,model}$, models MC2010-A and MC2010-B were as described in Sect. 2.1, without the safety factor and using the observed values of the materials and geometry variables. However, it must be noted that in Eq. (8), the value of V has to be multiplied by γ_F. The set of random variables and the corresponding distribution functions used are summarized in Table 4. The model error was selected as lognormally distributed according to the recommendations of the Joint Committee on Structural Safety (JCSS) [28].

Table 4. Definition and distribution of random variables.

Variable	Description	Statistical model	Mean value (μ)	CoV
Δ	Model error	Lognormal	1.075 (A); 0.912 (B)	0.228 (A); 0.291 (B)
f_c	Compression strength	Lognormal	$f_{ck} + 8MPa$	0.050–0.128
f_{ct}	Tensile strength	Lognormal	$f_{ck} \leq 50MPa \rightarrow 0.3f_{ck}^{\frac{2}{3}}$ $f_{ck} > 50MPa \rightarrow$ $2.12 \cdot \ln\left(1 + \frac{f_{cm}}{10}\right)$	0.182
f_{Ftu}	Residual strength at w_u	Lognormal	$1.412 \cdot f_{Ftuk}$	0.2
Δb	Geometrical error in section width	Normal	$0.003 \cdot b \leq 3$ mm	$\frac{4 + 0.006 \cdot b \leq 10mm}{\mu_b}$
Δd	Geometrical error in effective depth	Normal	10 mm	1

3.2 Calibration of the Partial Safety Factor γ_c in Model MC2010-A and γ_F in Mode MC2010-B

To establish the relationship between γ_c and γ_F to β, the required f_{Ftu} has to be designed for each element belonging to the design set of Sect. 3.1, for different values of γ_c and γ_F. For this purpose, $V_{Rd} = V_{Sd}$ is imposed and the reliability index is computed for each case. The design shear load V_{Sd} was assumed to be deterministic; therefore, the computed reliability index refers to the probability of reaching a shear strength (V_R) smaller than the design resistance (V_{Rd}): $P(V \leq V_{Rd}) = \Phi(\beta_R)$ where β_R is the resistance reliability index.

The reliability indices associated to the models MC2010-A and MC2010-B for estimating the shear strength capacity of FRC members without shear reinforcement (Eqs. (1) and (6)) were assessed for a range of safety factors γ_c and γ_F varying between 1.50 and 2.50, Fig. 2.

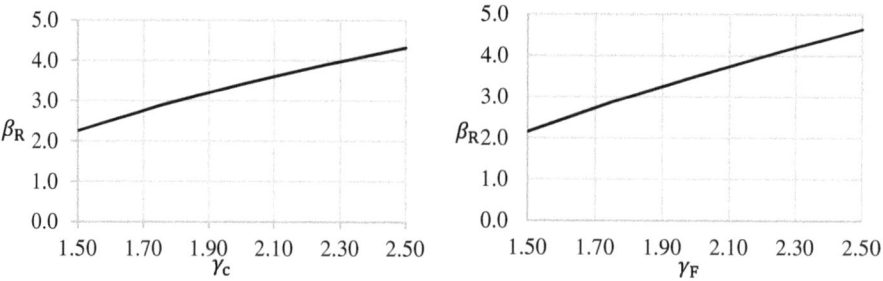

Fig. 2. Variation of the resistance reliability index with respect to γ_c for model MC2010-A (left) and γ_F for model MC2010-B (right).

As can be seen, the obtained values of β_R are 2.15 for both $\gamma_c = 1.50$ in model MC2010-A and for $\gamma_F = 1.50$ in model MC2010-B, with values going up to 4.32 and 4.64 for γ_c and γ_F equal to 2.50, respectively.

Generally, the target reliability index can be established as a result derived from an analysis coupling economic costs and failure consequences for human beings. As reference, a target reliability index for ULS verifications for a period of 50 years and medium consequences of failure of $\beta_{target} = 3.8$ is suggested in the *fib* Model Code 2010. Additionally, β_{target} of 3.1 and 4.3 are suggested in the same code for low and high consequences of failure, respectively, for a period of 50 years [9]. It should be noted that β_{target} accounts for uncertainties associated with both resistance and loads, as random variables, whereas $\beta_{R,target}$ includes only those associated with the resistance. However, it may be assumed that the β_{target} and $\beta_{R,target}$ are linearly related through the resistance sensitivity coefficient (α_R) as $\beta_{R,target} = \alpha_R \cdot \beta_{target}$, where α_R can be considered as 0.8 [29].

Table 5 gathers the β_{target} values for the three reference consequences of failure defined in the *fib* Model Code 2010, together with the computed γ_c and γ_F required to guarantee these β_{target} values in the design set.

Table 5. Safety factors γ_c and γ_F for the target reliability of different consequences of failure.

Consequence of failure	β_{target} For 50 years	$\alpha R \cdot \beta target$ (taking $\alpha_R = 0.8$)	γ_c	γ_F
Low	3.1	2.48	1.59	1.61
Moderate	3.8	3.04	1.82	1.81
High	4.3	3.44	2.01	1.97

As can be seen, for the reference consequence of failure ($\beta_{target} = 3.8$), the required γ_c is 1.82 for model MC2010-A, which is 21% larger than that currently proposed in the code ($\gamma_c = 1.50$). A similar value of was found for γ_F (1.81) for model MC2010-B.

4   Conclusions

This paper presents reliability-based calibration of partial safety factors γ_c and γ_F for the shear design of FRC members without shear reinforcement according to two *fib* Model Code 2010 models. For this purpose, a database of experimental results was used for assessing model errors, after which FORM analyses were performed to calibrate the partial safety factors. Based on the obtained results, the following conclusions can be drawn:

- The model errors of the MC2010-A and MC2010-B models were determined to be lognormally distributed with mean of 1.075 and 0.912, respectively, and CoV values of 22.8% and 29.1%, respectively, based on 332 experimental results.
- Based on a probability analysis and a parametric study of 700 individual cases, the relationship between the target failure probability β_R and partial safety factors γ_c

and γ_F was determined. It was found that β_R values associated with the value of $\gamma_c = \gamma_F = 1.50$—currently adopted in the *fib* Model Code 2010—are 2.25 and 1.88, for models MC2010-A and MC2010-B, respectively, i.e. below the target values of $\alpha_R \cdot 3.1 = 2.48$; $\alpha_R \cdot 3.8 = 3.04$ and $\alpha_R \cdot 4.3 = 3.44$ associated with low, moderate and high consequences of failure, respectively, and a service life of 50 years, considering the resistance sensitivity coefficient as $\alpha_R = 0.8$.

The results of this study allow confirming that target failure probabilities could not be achieved by the MC2010-A and MC2010-B models for the shear resistance of FRC members without shear reinforcement with $\gamma_c = \gamma_F = 1.50$. Therefore, calibration of the partial safety factor is required for future code revisions. Although the results of the study are dependent on the range of parameters considered in the experimental database used for assessing the model error and on the choice of parameter ranges in the probability analysis, they can still be considered as robust enough for drawing general conclusions.

Acknowledgements. This study has received funding from the European Union's Horizon 2020 research and innovation programme under the Marie Sklodowska-Curie grant agreement No 836270. This support is gratefully acknowledged. The authors also wish to express their acknowledgement to the Ministry of Economy, Industry and Competitiveness of Spain for the financial support received under the scope of the projects PID2019-108978RB-C32. Any opinions, findings, conclusions, and/or recommendations in the paper are those of the authors and do not necessarily represent the views of the individuals or organizations acknowledged.

References

1. de la Fuente, A., Blanco, A., Armengou, J., Aguado, A.: Sustainability based-approach to determine the concrete type and reinforcement configuration of TBM tunnels linings. Case study: Extension line to Barcelona Airport T1. Tunn. Undergr. Sp. Technol. **61**, 179–88 (2017). https://doi.org/10.1016/j.tust.2016.10.008.
2. de La Fuente, A., Casanovas-Rubio, M.D.M., Pons, O., Armengou, J.: Sustainability of column-supported RC Slabs: fiber reinforcement as an alternative. J. Constr. Eng. Manag. **145**, 1–12 (2019). https://doi.org/10.1061/(ASCE)CO.1943-7862.0001667
3. Winkler, A., Edvardsen, C., Kasper, T.: Examples of bridge, tunnel lining and foundation design with Steel-fibre-reinforced concrete. Am. Concr. Institute, ACI Spec. Publ., 451–60 (2014). https://doi.org/10.35789/fib.bull.0079.ch42
4. Parra-Montesinos, G.J., et al.: Earthquake-resistant fibre-reinforced concrete coupling beams without diagonal bars. Am. Concr. Institute, ACI Spec. Publ. 461–70 (2014). https://doi.org/10.35789/fib.bull.0079.ch43
5. Cugat, V., Cavalaro, S.H.P., Bairán, J.M., de la Fuente, A.: Safety format for the flexural design of tunnel fibre reinforced concrete precast segmental linings. Tunn Undergr Sp Technol. **103**, 103500 (2020). https://doi.org/10.1016/j.tust.2020.103500
6. Balász, G.: A historical review of shear. Shear punching Shear RC FRC Elem., Salò: International Federation for Structural Concrete (fib), pp. 1–14 (2010)
7. FIB Bulletin 85: Towards a rational understanding of shear in beams and slabs. Lausanne (2018)

8. EN 1992-1-1: Eurocode 2: Design of concrete structures - Part 1-1: General rules and rules for buildings. CEN, Brussels (2004)
9. FIB: fib Model Code for Concrete Structures 2010. Lausanne: International Federation for Structural Concrete (fib) (2013). https://doi.org/10.1002/9783433604090
10. Meda, A., Minelli, F., Plizzari, G.A., Riva, P.: Shear behaviour of steel fibre reinforced concrete beams. Mater. Struct. Constr. **38**, 343–351 (2005). https://doi.org/10.1617/14112
11. Minelli, F., Plizzari, G.A.: On the effectiveness of steel fibers as shear reinforcement. ACI Struct. J. **110**, 2013 (2013). https://doi.org/10.14359/51685596
12. Minelli, F.: Plain and fiber reinforced concrete beams under shear loading: structural behavior and design aspects. University of Brescia (2005)
13. di Prisco, M., Plizzari, G., Vandewalle, L.: MC2010: overview on the shear provisions for FRC. Shear punching Shear RC FRC Elem., Salò: International Federation for Structural Concrete (fib); pp. 61–76 (2010)
14. Vecchio, F.J., Collins, M.P.: The modified compression-field theory for reinforced concrete elements subjected to shear. J. Am. Concr. Inst. **83**, 219–31 (1986). https://doi.org/10.14359/10416
15. Sykora, M., Holicky, M., Prieto, M., Tanner, P.: Uncertainties in resistance models for sound and corrosion-damaged RC structures according to EN 1992-1-1. Mater. Struct. **48**(10), 3415–3430 (2014). https://doi.org/10.1617/s11527-014-0409-1
16. Reineck, K.H., Kuchma, D.A., Kim, K.S., Marx, S.: Shear database for reinforced concrete members without shear reinforcement. ACI Struct. J. **100**, 240–249 (2003)
17. Marì Bernat, A., Spinella, N., Recupero, A., Cladera, A.: Mechanical model for the shear strength of steel fiber reinforced concrete (SFRC) beams without stirrups. Mater. Struct. **53** (2), 1–20 (2020). https://doi.org/10.1617/s11527-020-01461-4
18. Conforti, A., Minelli, F., Plizzari, G.A.: Shear behaviour of prestressed double tees in self-compacting polypropylene fibre reinforced concrete. Eng. Struct. **146**, 93–104 (2017). https://doi.org/10.1016/j.engstruct.2017.05.014
19. Conforti, A., Minelli, F., Tinini, A., Plizzari, G.A.: Influence of polypropylene fibre reinforcement and width-to-effective depth ratio in wide-shallow beams. Eng. Struct. **88**, 12–21 (2015). https://doi.org/10.1016/j.engstruct.2015.01.037
20. EN 14651: Test method for metallic fibred concrete — Measuring the flexural tensile strength (limit of proportionality (LOP), residual). Br Stand Inst (2005). 9780580610523
21. Lantsoght, E.O.L.: Database of shear experiments on steel fiber reinforced concrete beams without stirrups. Materials (Basel) **12**, 917 (2019). https://doi.org/10.3390/ma12060917
22. Lantsoght, E.: Database of experiments on SFRC beams without stirrups failing in shear (Version 1.0). Zenodo (2019). https://doi.org/10.5281/ZENODO.2578061
23. Venkateshwaran, A., Tan, K.H., Li, Y.: Residual flexural strengths of steel fiber reinforced concrete with multiple hooked-end fibers. Struct. Concr. **19**, 352–365 (2018). https://doi.org/10.1002/suco.201700030
24. Tiberti, G., Germano, F., Mudadu, A., Plizzari, G.A.: An overview of the flexural post-cracking behavior of steel fiber reinforced concrete. Struct. Concr. **19**, 695–718 (2018). https://doi.org/10.1002/suco.201700068
25. Galeote, E., Blanco, A., Cavalaro, S.H.P., de la Fuente, A.: Correlation between the Barcelona test and the bending test in fibre reinforced concrete. Constr. Build. Mater. **152**, 529–538 (2017). https://doi.org/10.1016/j.conbuildmat.2017.07.028
26. Melchers, R.E.: Structural Reliability Analysis and Prediction. Ellis Horwood (1987)
27. Madsen, H.O., Krenk, S., Lind, N.C.: Methods of structural safety. Dover, New York (2006)
28. JCSS. Probabilistic Model Code (2001)
29. EN 1990. Eurocode - Basis of structural design. Brussels: CEN (2002)

Yield Line Design for SFRC Elevated Slabs

Matteo Colombo$^{(\boxtimes)}$, Marco di Prisco, and Ali Pourzarabi

Department of Civil and Environmental Engineering,
Politecnico di Milano, Milan, Italy
matteo.colombo@polimi.it

Abstract. The paper aims at discussing the application of Yield Line methods for the design of elevated slabs made of Steel Fibre Reinforced concrete. The paper considers three different case studies of Fibre reinforced concrete slabs ranging from laboratory small case test to real scale application. In particular, three different slabs are considered: circular simply supported slab (diameter 60 cm and thickness 2 cm), two different configurations of centrally loaded slabs with four point supports with 2 m span and a real scale elevated slab on columns with clear span of 6 m and nominal thickness of 20 cm. The paper discusses the approach that is going to be proposed by the Annex L of Eurocode 2 for the design of FRC structures and particularly compare the results of such a design approach and the experimental results.

Keywords: Fibre reinforced concrete · Elevated salbs · Yield line design

1 Introduction

The advent of design guidelines and codes for structural applications of fibre reinforced concrete is opening the door for engineers and designers to exploit the enhanced properties of fibre reinforced concrete (FRC), both in serviceability and ultimate limit state. The Model Code 2010 [1] has proposed two stress-crack opening laws in uniaxial tension defined for the post-cracking behaviour of FRC. Similar approach is going to be presented also by Annex L of Eurocode 2. The development of these tensile post-cracking relationships is derived from a standard three-point bending test (3PB) on notched specimens which results are expressed in stress-Crack Mouth Opening Displacement (CMOD). The residual stress values at a CMOD of 0.5 mm (corresponding to SLS condition) and 2.5 mm (corresponding to ULS condition), named respectively f_{R1} and f_{R3}, are related to uniaxial post-cracking tensile behaviour through a simplified inverse analysis based on equilibrium [2].

While a 3PB test according to EN 14651 [3] is an accepted method for the characterization of FRC, the results obtained from such a test, suffer from a high scatter, which is mainly due to the small fracture plane, the low number of fibres in the ligament area [4, 5], and the fibre distribution [6]. The high dispersion of the post-peak tensile properties obtained from bending tests leads to smaller characteristic values of residual strength parameter, directly affecting the design strength parameters.

© RILEM 2022
P. Serna et al. (Eds.): BEFIB 2021, RILEM Bookseries 36, pp. 564–575, 2022.
https://doi.org/10.1007/978-3-030-83719-8_49

When referring to elevated slabs, designers often refer to Yield Line analysis [7] as a quick and reliable method. The application of this method is generically accepted for R/C structures because of the ductility of the sectional response (moment-curvature diagram) despite the brittleness of concrete. Some doubts may arise when this approach is applied to elevated slabs just made of Fibre Reinforced Concrete where the cross-sectional response can present a post-peak behaviour with limited ductility or even with softening.

The paper adopts the constitutive law proposed by Annex L of Eurocode 2 to apply the Yield Line Design approach to four different case studies:

1) A centrally loaded slab 2 m × 2 m (thickness 15 cm) simply supported at the four corners: both SFRC and hybrid (SFRC + traditional reinforcement) solutions are considered.
2) A centrally loaded slab 2 m × 2 m (thickness 15 cm) simply supported at the midspan of each side: both SFRC and hybrid (SFRC + traditional reinforcement) solutions are considered.
3) A real scale elevated slab on columns with net span of 6 m and a thickness of 21 cm loaded with a concentrated load: only SFRC solution is available.
4) A simply supported circular slab (diameter 56 cm and thickness 2 cm) loaded with a central point load: only SFRC solution is available.

2 Eurocode Design Rules

The paper refers to the design approach proposed by the upcoming Annex L of Eurocode 2. The approach proposed is similar to those proposed by Model Code 2010 and, in particular, the material characterization for FRC tensile behaviour refers to EN14651 bending test on notched specimens. One of the main novelties of the Eurocode approach consist in the definition of the characteristic value of the material that introduce an upper bound $\kappa_{k,max}$ of the characteristic values that should be assumed to adjust the performance taking into account the variability of the casting procedure of the structure compared to the preparation of the beam samples in EN 14651. In particular the definition of the residual strengths $f_{Rj,k}$ (with j = 1, 3) is proposed as follow:

$$f_{Rj,k} = \min(f_{Rj,k}^{exp}; \kappa_{k,max} \cdot f_{Rj,m})$$

(1)

The proposed value for $\kappa_{k,max}$ is fixed by the annex to 0.6.

The uniaxial tensile constitutive law proposed to compute resistant bending moment at the ULS with a plane section approach is presented in Fig. 1.

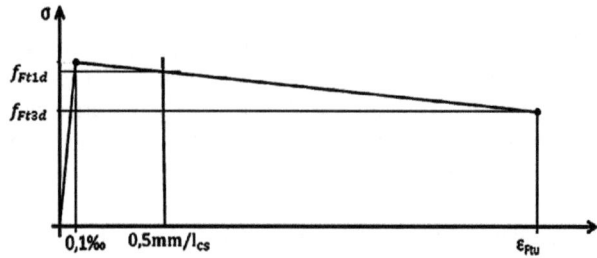

Fig. 1. Uniaxial tension constitutive law adopted.

In the presented uniaxial model, the different parameters are defined as follow:

$$f_{Ft1d} = \kappa_G \cdot \kappa_O \cdot f_{Ft1,ef} / \gamma_{SF} \tag{2}$$

$$f_{Ft3d} = \kappa_G \cdot \kappa_O \cdot f_{Ft3,ef} / \gamma_{SF} \tag{3}$$

$$f_{Ft1,ef} = 0.37 \cdot f_{R1,k} \tag{4}$$

$$f_{Ft3,ef} = 0.57 \cdot f_{R3,k} - 0.26 \cdot f_{R1,k} \tag{5}$$

where γ_{SF} is the material safety coefficient that can be taken equal to 1.5, κ_O is the coefficient taking into account fibre orientation and distribution and is set equal to 1 for plate bending, while κ_G is a coefficient that takes into account the capability of the structure to redistribute stresses. As a matter of fact, it has been observed that for structures capable of redistributing stresses, and in which large crack surfaces dominate the failure process, even if the material topological variability in the structure remains high [8, 9] the structural behaviour shows a very low dispersion [10]. Accordingly, design of FRC structures with characteristic values of parameters obtained from a highly scattered testing modality leads to over-conservative design solutions and may downplay the cost-competitiveness of FRC as a structural material. The coefficient κ_G, in the Annex L, is going to be defined as follow:

$$\kappa_G = 1 + A_{ct} \cdot 0.5 \leq \kappa_{G,max} \tag{6}$$

$$\kappa_{G,max} = \min \left(\frac{0.9}{\kappa_{k,max}} = 1.5; 0.9 \cdot \frac{f_{Rj,m}}{f_{Rj,k}} \right) (\text{with } j = 1, 3) \tag{7}$$

In the following calculation the coefficient κ_G was take equal to $\kappa_{G,max}$ for sake of simplicity.

Finally, it worth to underline that the use of plastic analysis, such as Yield Line approach, is allowed by the code for statically indeterminate elevated slabs subject to the following conditions:

- ductility class is at least c ($f_{R3,k}/f_{R1,k} > 0.9$)
- $f_{R3,k} \geq f_{ct,fl,m}$ being $f_{ct,fl,m}$ the mean value to the tensile strength in bending

3   Elevated Slabs Case Studies

As already discussed, the Yield Line approach is going to be applied to different case studies to assess the prediction capability of the method and to verify if it is able to guaranty a proper safety level to the structural design. The material mechanical properties for all the cases are reported in Table 1, while the average value of maximum load reached during the experimental test is reported in Table 2, 3, 4 and 5 for each case.

3.1   CASE A: Centrally Loaded Slab with Supports in the Corners

The first case study refers to an experimental campaign [11, 12] that consists in 9 standard notched specimens tested in a three-point bending setup, and 6 concrete slabs with dimensions of $2 \times 2 \times 0.15$ m tested under a concentrated central load and supported on four supports at each corner. Three different slab solutions were studied: SFRC without any reinforcement (SFRC), plain concrete with steel rebars in two directions (R/C) and a Hybrid solution in which fibre reinforcement was used together with traditional steel rebars (Hybrid). Two nominally identical tests were performed for each case under study.

Both in the case of R/C and Hybrid solutions 12 Φ 1 2 mm steel rebars have been placed parallel to the slab sides in each direction at the bottom of the slab with a cover of 30 mm. All the rebars were grade B450 and uniaxial tensile tests provide an average yielding strength of 527 MPa and an average ultimate strength of 647 MPa.

Figure 2 shows a picture with the general set-up of the slab, while the crack pattern is shown in Fig. 3a.

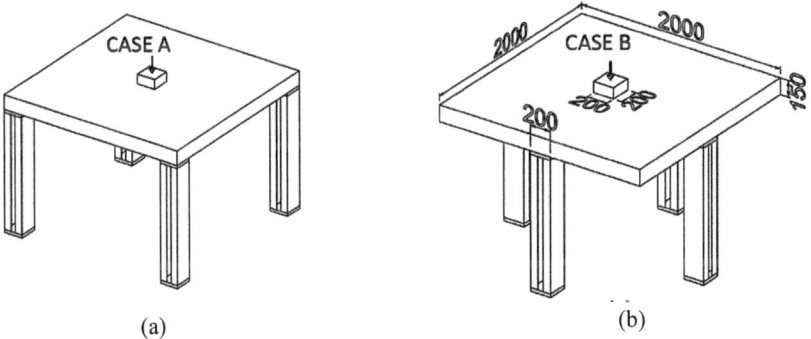

(a)   (b)

Fig. 2. Test configuration of test for Case A and B [12]

3.2   CASE B: Centrally Loaded Slab with Supports at Mid-Span of Each Side

The second case study refers to the same experimental campaign of case A and always consider concrete slabs with dimensions of 2 × 2 × 0.15 m tested under a concentrated central load, but supported at the mid-span of each side.

The material adopted is the same of case A, because all the slabs of both the cases have been cast during the same batch. R/C, SFRC and Hybrid solutions were also in this case considered. The traditional reinforcement disposal and properties are the same of case A.

Figure 3b shows the set-up of the tests.

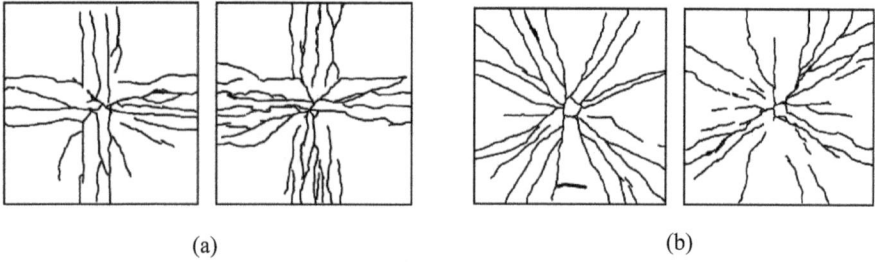

(a)   (b)

Fig. 3.   Experimental crack pattern for (a) case A and (b) case B [12]

3.3   CASE C: Real Scale Elevated Slab on Columns

A real scale elevated slab was taken into account: the SFRC elevated flat slab was erected and tested at the Belgian Building Research Institute (BBRI) site in Limelette, Belgium [9]. The slab consisted of nine bays (panels) measuring 6 × 6 m supported by sixteen 0.3 m diameter circular concrete columns in C30/37 (see Fig. 4). The nominal thickness of the slab was 0.2 m (0.21 the measured one) and its overall size was 18.3 × 18.3 m. The columns were reinforced with six 12 mm dia. longitudinal reinforcing bars together with circular stirrups and anchored in 1.5 × 1.5 × 0.35 m concrete footings. The longitudinal bars in the columns extended to mid-depth of the slab without any other additional anchoring length. The centre-to-centre distance between the columns was 6 m in both orthogonal directions. The flexural behaviour of the slab was checked against its serviceability limit state (SLS) and its post-cracking behaviour for the ultimate limit state (ULS) and, in particular, its ultimate bending capacity by applying a uniformly distributed load and later a concentrated load in the centre of specific panels.

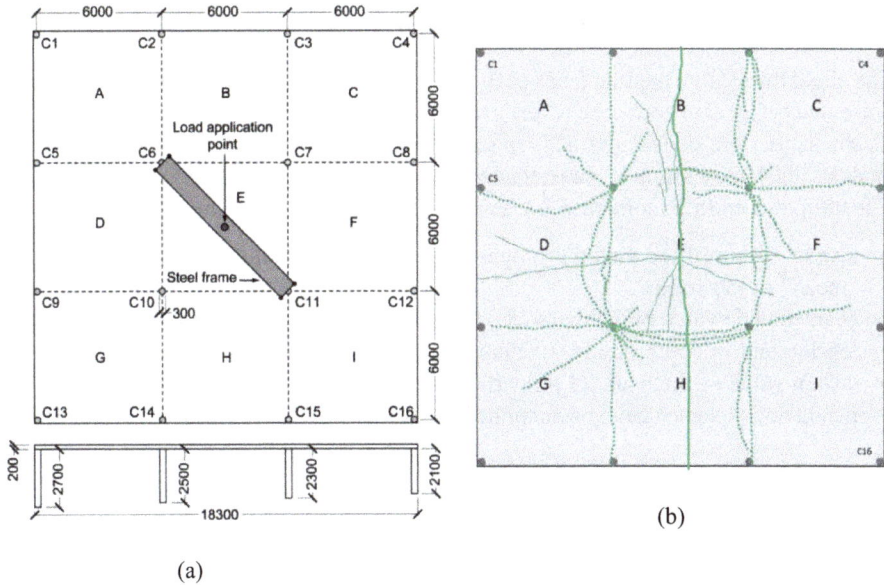

(a)

(b)

Fig. 4. CASE D: a) slab test setup (b) experimental crack pattern [9]

3.4   CASE D: Simply Supported Circular Slab Centrally Loaded

A circular slab 20 mm thick and 56 cm in diameter was tested under a central point load; the slab was simply supported around the edge by a steel ring with a thickness of 20 mm [13]. A piece of neoprene was placed under the loading head to avoid stress concentrations given by the uneven top surface of the plates. Figure 5a shows the testing setup. The material adopted is a material classified as 9b by means of EN14651 bending test, but that experience a hardening behaviour in bending up to 3 mm of crack opening when a structural specimen with the same thickness of the slab is tested.

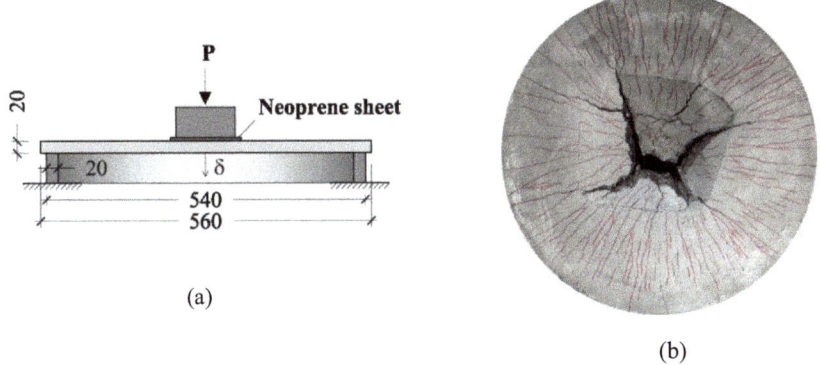

(a)

(b)

Fig. 5. CASE D: a) slab test setup (b) crack pattern [13]

4 Yield Line Design Approach

The Yield line design approach has been adopted to predict the ultimate load of the cases presented before. For each case investigated (with the only exception of case D), different mechanisms were considered and, as prescribed by the lower bound theorem of limit analysis, the lowest ultimate load obtained was considered as the prediction ultimate load. The ultimate bending moment was computed according to different situations:

- average value of the material properties ($f_{c,m}$, $f_{R1,m}$ and $f_{R3,m}$) – situation define as "mean" in the follow.
- characteristic value of the material properties as defined by the material class both in tension and in compression – situation defined as "class" in the follow.
- design value of the material properties for the definition of which both safety coefficient and κ_G factor have been applied – situation defined as "design" in the follow

The described situations have been adopted to compute the resistant bending moment according to a plane section approach and the design prescription previously discussed in paragraph 2. For each situation the moment curvature diagram of the cross section has been built with a multilayer approach assuming the parabola rectangular constitutive law in compression and the previously described linear constitutive law in tension. From each moment – curvature diagram, two different resistant bending moment were defined:

- one corresponding to the peak of the diagram (m_{max})
- one corresponding to the reaching of the ultimate crack opening (w_u = min (0.02l_{cs}; 2.5 mm)) at the bottom of the cross section (m_{wu})

It is worth to note that the structural characteristic length (l_{cs}) was assumed equal to the cross section thickness (h) when only SFRC was available and according to the following Eq. 8 when also traditional reinforcement is available.

$$l_{cs} = \min(h; s_{rm}) \tag{8}$$

being s_{rm} the average crack spacing.

For the traditional reinforcement an infinite elastic-plastic model has been adopted.

The different yield line mechanisms adopted for each case are reported in Figs. 6, 7, 8 and 9.

The yield line predictions (P_u) for each of the case considered are summarized in Tables 2–5 together with the maximum experimental load (P_{exp}) and the related safety coefficient ($\gamma = P_{exp}/P_u$).

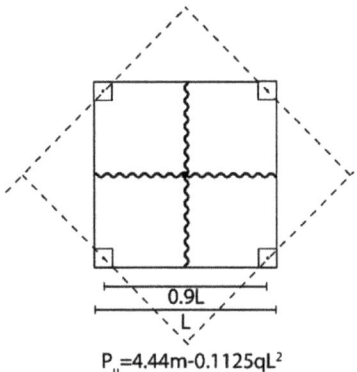

$P_u = 4.44m - 0.1125qL^2$

Fig. 6. CASE A: Yield line mechanism

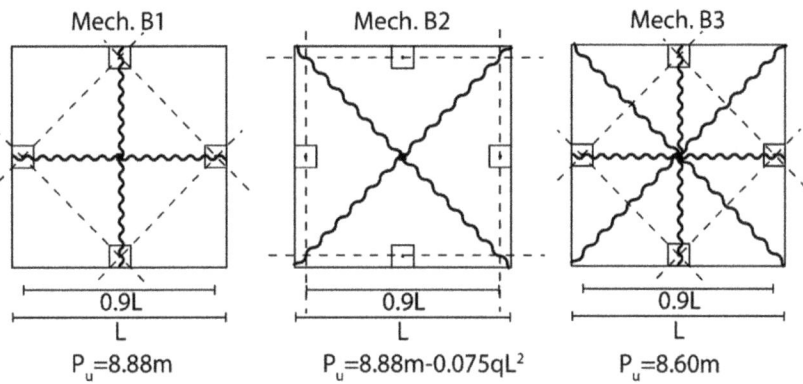

Mech. B1 Mech. B2 Mech. B3

$P_u = 8.88m$ $P_u = 8.88m - 0.075qL^2$ $P_u = 8.60m$

Fig. 7. CASE B: Yield line mechanisms

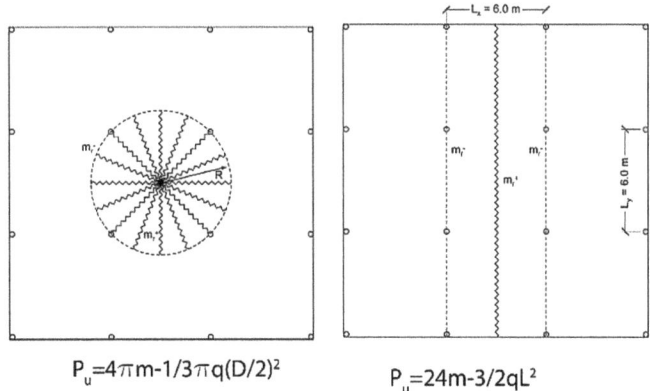

$P_u = 4\pi m - 1/3\pi q(D/2)^2$ $P_u = 24m - 3/2qL^2$

Fig. 8. CASE C: Yield line mechanisms [9]

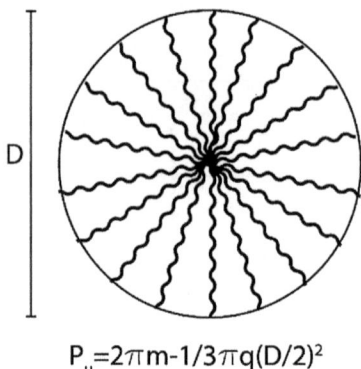

$$P_u = 2\pi m - 1/3\pi q (D/2)^2$$

Fig. 9. CASE D: Yield line mechanism

Table 1. Materials mechanical properties for all the cases investigated.

Case	Compr class	$f_{R1,av}$	$f_{R3,av}$	n.	$f_{R1,k}^{EXP}$	$f_{R3,k}^{EXP}$	Tough. class	$f_{R1,k}$	$f_{R3,k}$	$f_{ct,fl,m}$
A	C40	5.64	4.92	5	3.82	3.10	3c	3	2.7	5.50
B	C40	5.64	4.92	5	3.82	3.10	3c	3	2.7	5.50
C	C30	7.37	6.85	20	3.90	4.20	3c	3	2.7	4.54
D	C110	15.46	12.87	3	14.17	8.87	9b	9	6.3	8.47

Table 2. CASE A – Yield line predictions.

Material prop.	$P_{u,exp}$ [kN]	m type	m [kNm/m]	P_u [kN]	γ
Case A – SFRC					
Mean	162	m_{max}	21.65	94.53	1.71
		m_{wu}	17.69	76.93	2.11
Class		m_{max}	11.74	50.49	3.21
		m_{wu}	9.78	41.78	3.88
Design		m_{max}	11.5	49.42	3.28
		m_{wu}	9.69	41.38	**3.92**
Case A – Hybrid					
Mean	295	m_{max}	57.77	243.25	1.21
		m_{wu}	60.90	257.16	1.15
Class		m_{max}	45.34	188.01	1.57
		m_{wu}	44.28	183.30	1.61
Design		m_{max}	40.15	164.94	1.79
		m_{wu}	39.00	159.83	**1.85**

Table 3. CASE B – Yield line predictions.

Material prop.	$P_{u,exp}$ [kN]	m type	m [kNm/m]	Mec. B1	Mec. B2	Mec. B3	P_u [kN]	γ
Case B – SFRC								
Mean	236	m_{max}	21.65	192.44	191.32	186.19	186.19	1.27
		m_{wu}	17.69	157.24	156.12	152.13	152.13	1.55
Class		m_{max}	11.74	104.36	103.23	100.96	100.96	2.34
		m_{wu}	9.78	86.93	85.81	84.11	84.11	2.81
Design		m_{max}	11.50	102.22	101.10	98.90	98.90	2.39
		m_{wu}	9.69	86.13	85.01	83.33	83.33	**2.83**
Case B – Hybrid								
Mean	495	m_{max}	57.77	513.51	512.39	496.82	496.82	1.00
		m_{wu}	60.9	541.33	540.21	523.74	523.74	0.94
Class		m_{max}	45.34	403.02	401.90	389.92	389.92	1.27
		m_{wu}	44.28	393.60	392.48	380.81	380.81	1.30
Design		m_{max}	40.15	356.89	355.76	345.29	345.29	1.43
		m_{wu}	39.00	345.67	345.54	335.40	335.40	**1.47**

Table 4. CASE C – Yield line predictions.

Material prop.	$P_{u,exp}$ [kN]	m type	m [kNm/m]	Mec. C1	Mec. C2	P_u [kN]	γ
Mean	328	m_{max}	52.40	568.00	998.40	568.00	0.58
		m_{wu}	46.90	498.88	866.40	498.88	0.66
Class		m_{max}	22.31	189.88	276.24	189.88	1.73
		m_{wu}	18.90	147.03	194.40	147.03	2.23
Design		m_{max}	21.74	182.72	262.56	182.72	1.80
		m_{wu}	18.50	142.00	184.80	142.00	**2.31**

Table 5. CASE D – Yield line predictions.

Material prop.	$P_{u,exp}$ [kN]	m type	m [kNm/m]	P_u [kN]	γ
Mean	17.5	m_{max}	1.37	8.56	2.05
		m_{wu}	1.32	8.24	2.12
Class		m_{max}	0.83	5.16	3.39
		m_{wu}	0.78	4.85	3.61
Design		m_{max}	0.83	5.16	3.39
		m_{wu}	0.78	4.85	**3.61**
Mean struct. specimen(*)			2.04	12.73	1.37

(*) In this case the resistance bending moment is obtained by the mean results of bending tests performed on structural specimens that were characterized by the same thickness of the slab.

5 Conclusions

Four cases of FRC and Hybrid plates are investigated to compute the safety coefficient guaranteed by the approach suggested in the Eurocode 2 Annex L, mainly inspired by Model Code 2010.

The main conclusions are:

- even though the ductility class was only b in case D and the residual flexural strength $f_{R3,k}$ is always lower than $f_{ct,flm}$, with the only exception of case D, the design predictions on FRC plates without any reinforcement made by using limit analysis according to the rules indicated by the current draft of Annex L - EC2 show always a safety factor which is higher than 1. The minimum value is 2.31, which is larger than 1.5 (the safety coefficient γ_{SF} adopted for fibre reinforced concrete), found in the full-size test (case C), while the highest safety coefficient is 3.92 in a reduced scale case (case A);
- the hybrid cases show a larger ductility in terms of vertical displacements than that shown by FRC cases, but a lower safety coefficient even though larger than 1.15, that adopted for conventional reinforcement, ranging between 1.47 and 1.85.
- The ideal material safety coefficient could be computed as a weighted average by considering the reinforcement ratios of conventional reinforcement and of FRC material, that gives a value close to 1.325 for solutions like those investigated characterized by comparable reinforcement ratios.
- The factor $\kappa_{k,max}$ reduces the class only of the material D.

Acknowledgements. The authors are in debt with the CEN TC250/WG1/SC2/TG2 members for the profitable discussions on this topic.

References

1. fib, International Federation for Structural Concrete. Model code 2010, Final Complete Draft, Bulletins 65 and 66, March 2012 (2012). ISBN 978-2-88394-105-2 and April 2012. ISBN 978-2-88394-106-9
2. di Prisco, M., Colombo, M., Dozio, D.: Fibre-reinforced concrete in fib model code 2010: principles, models and test validation. Struct. Concr. **14**(4), 342–361 (2013)
3. EN 14651. CEN, Test method for metallic fibre concrete. Measuring the flexural tensile strength (limit of proportionality (LOP), residual, CEN, p. 20 (2005)
4. Bernard, E.S., Xu, G.G.: Influence of fibre count on variability in post-crack performance of fibre reinforced concrete. Mater. Struct. **50**(3), 1–9 (2017). https://doi.org/10.1617/s11527-017-1035-5
5. Minelli, F., Plizzari, G.: A new round panel test for the characterization of fiber reinforced concrete: a broad experimental study. J. Test. Eval. **39**(5), 889–897 (2011)
6. Colombo, M., Martinelli, P., di Prisco, M.: On the evaluation of the structural redistribution factor in FRC design: a yield line approach. Mater. Struct. **50**(1), 100 (2017)
7. Park, R., Gamble, W.L.: Reinforced Concrete Slabs. Wiley , Hoboken (2000)
8. di Prisco, M., Martinelli, P., Dozio, D.: The structural redistribution coefficient KRd: a numerical approach to its evaluation. Struct. Concr. **17**(3), 390–407 (2016)

9. di Prisco, M., Martinelli, P., Parmentier, B.: On the reliability of design approach for FRC structures according to MC2010: the case of elevated slabs. Struct. Concr. **4**, 588–602 (2016)
10. Sorelli, L.G., Meda, A., Plizzari, G.A.: Steel fiber concrete slabs on ground: a structural matter. ACI Struct. J. **103**(4), 551 (2006)
11. di Prisco, M., Colombo, M., Pourzarabi, A.: Biaxial bending of SFRC slabs: Is conventional reinforcement necessary? Mater. Struct. **52**(1) (2019). Art. no. 1
12. Pourzarabi, A., di Prisco, M., Colombo, M.: Deflection and cracking behavior of SFRC plates and the prediction of the load bearing capacity. In: Proceedings of fib Italy YMG Symposium, Parma, Italy (2019)
13. Zani, G., Pourzarabi, A., Colombo, M., di Prisco, M.: Fire resistance of VHPFRC thin plates, In: Proceedings of Conference PROTECT 2019, Whistler Canada (2019)

Review

Code Provisions for Shear Strength in Prestressed FRC Members: A Critical Review

Stefano Giuseppe Mantelli[1], Luca Facconi[1(✉)], Katharina Look[2],
Filippo Medeghini[2], Peter Mark[2], Fausto Minelli[1],
and Giovanni Plizzari[1]

[1] DICATAM - Department of Civil, Environmental, Architectural Engineering
and Mathematics, University of Brescia, Brescia, Italy
`luca.facconi@unibs.it`
[2] Institute of Concrete Structures, Ruhr University, Bochum, Germany

Abstract. Since its introduction, FRC has gained quite high attention from precast industry for the potential advantages resulting from partial or total removal of conventional reinforcement, especially related to shear. As compared to non-prestressed elements, the use of precompression may allow to significantly reduce the required amount of transversal reinforcement, taking advantage of the shear force carried by prestressing tendons/cable. As proved by several research studies, the adoption of fibers may further reduce shear reinforcement, which could be only limited to support areas or to those regions where the prestressing action is not fully developed.

The current structural codes are generally very conservative in predicting the shear strength of these elements. Therefore, new research studies are needed to improve the accuracy of models for calculating the shear strength. Special focus could be also done to existing prestressed structures, which might benefit from more refined models in terms of reduction of structural repairing costs.

This paper presents a database of shear tests of prestressed FRC members, emphasizing the rather conservative predictions of the prEN 1992-1-1 (2020) and fib Model Code 2020 drafts. A critical discussion will follow.

Keywords: Shear · Prestressed members · Fiber Reinforced Concrete · Database · prEN 1992-1-1 · Fib Model Code 2020

1 Introduction

Fiber Reinforced Concrete (FRC) has been extensively used by precast industry for producing structural members (e.g., roof elements, beams/columns, foundation elements, slabs, etc.) typically used for several structural applications such as commercial and industrial buildings or bridges. Despite steel fibres are relatively more expensive than traditional stirrups or links, the good tensile performance resulting from the addition of steel fibers to concrete allows reducing or totally replacing conventional shear reinforcement, often resulting in faster production process as well as in the reduction of labor cost.

© RILEM 2022
P. Serna et al. (Eds.): BEFIB 2021, RILEM Bookseries 36, pp. 579–589, 2022.
https://doi.org/10.1007/978-3-030-83719-8_50

Despite the vast amount of experiments that have been carried out to assess the shear capacity of structural concrete members (collected in database as in [1]), the behavior and design of reinforced concrete (RC) elements under shear remains an area of much concern. Design codes are continually changing and generally becoming more stringent so that structures that were designed several decades ago typically do not comply with the requirements of current codes.

The fib Model Code 2010 (MC2010) [2] first introduced design recommendations for FRC structures, including a method for the assessment of the shear capacity (MC2010 – clause 7.7.3.2). In the last few years significant efforts have been made to further improve to the original MC2010's formulations, leading to the new proposals recently reported in the Annex L to prEN 1992-1-1 (2020) [3], hereafter named as prEN 1992, as well as in the draft of the fib Model Code 2020 [4], here named as MC2020.

The study described in the following sections aims at assessing the reliability of the formulations proposed by the aforementioned draft codes in predicting the shear capacity of prestressed FRC beams not containing conventional shear reinforcement. Despite the limited number of experimental results available into the literature, a database of reliable results has been collected by including only prestressed FRC beams with steel fibers and bonded tendons. After a brief description of the code formulations, which will include some minor modifications proposed by the Authors, the analytical prediction of the shear capacity is compared with the experimental results. Finally, some considerations on the accuracy of the code predictions will be stated and discussed.

2   EC2 Annex L and Model Code 2020 Provisions for Shear Strength

The latest version of the EC2 [5], which was released in 2004, did not report any recommendations concerning the design of FRC members. The draft document of the new Eurocode 2 (prEN 1992) [3], aims at filling this gap with an annex (i.e., ANNEX L) to the code specifically developed to include the most recent advancements in the field of FRC structural design. About FRC elements subjected to shear, the design approach introduced by the new draft is based on the equations originally reported in the MC2010 [2] (see Table 1). In more detail, when considering members without shear reinforcement, the prediction of the design shear resistance (V_{Rd}) depends on the shear strength $\tau_{Rd,cF}$, which in turn results from the combination of two contributions representing respectively the behavior of plain concrete ($\tau_{Rd,c}$) and the effect of fibers (f_{Ftud}). The former is calculated by the formulation adopted by the EC2 (clause 6.2.2) for calculating the strength of plain concrete beams not requiring shear reinforcement, whereas the latter corresponds to the ultimate residual strength of FRC reported by the MC2010 (clause 5.6.4). It is worth noting that the prEN 1992 introduces a factor η in order to reduce the contribution of plain concrete to the total shear resistance as the tensile residual strength of FRC increases (corresponding to greater values of shear crack width, for which the aggregate interlock will significantly diminish. To simplify the equation reported by the prEN 1992 for calculating η, the following tri-linear function is proposed and adopted in the calculations hereafter performed:

$$\eta = max[1.2 - 0.5 \cdot f_{Ftum}; 0.4] \leq 1 \tag{1}$$

where f_{Ftum} is the mean value of the residual strength of FRC determined according MC2010 [2] at ultimate limit states. A comprehensive summary of the equations reported by the prEN 1992 is shown in Table 1. The approach proposed by the most recent draft of the MC2020 to calculate the shear resistance is developed from the Simplified Modified Compression Field Theory [6] and is consistent with the level-of-approximation (LoA) method [2] originally formulated for traditional concrete members subjected to either shear or punching actions. The equations described in Table 1 show that, likewise the prEN 1992, the MC2020 estimates the shear resistance by combining the shear contributions of plain concrete ($V_{Rd,c}$) and fibers ($V_{Rd,FRC}$), with consideration of a correlation coefficient that act in diminishing the plain concrete contribution k_v, through coefficient ζ, as the fiber contribution increases. The model is based on the mid-depth longitudinal strain (ε_x) calculation, used to determine the inclination of the strut (θ_F) as well as the crack width (w), the latter necessary to determine the residual strength of FRC within the LoA III approximation. The longitudinal strain is dependent on the design values of the bending moment (M_{Ed}), shear (V_{Ed}) and axial force (N_{Ed}) acting on the beam cross section; the latter taking positive in case of tension. The internal forces resulting from prestress and the corresponding eccentricities with respect to the beam axis are depicted in the schematic of Fig. 1. Note that Table 1 does not report equations of LoA I, which is not meant for modeling purpose but just for a preliminary safe and easy design tool.

Both the prEN 1992 and the MC2020 adopt the approach suggested by the MC2010 [2] to determine the post-cracking uniaxial tensile strength (f_{Ftw}) of FRC. The latter is related to the flexural residual strengths obtained from 3-Point Bending Tests (3PBTs) on notched beams (150 × 150 × 600 mm) performed according EN14651 [7]. Four residual strengths, named as f_{R1}, f_{R2}, f_{R3} and f_{R4}, are derived from the bending test at different values of the Crack Mouth Opening Displacements (CMOD). The residual strengths f_{R1} and f_{R3}, which correspond respectively to CMODs of 0.5 mm and 2.5 mm, are those considered by the codes in the assessment of the design shear resistance (Table 1). Unfortunately, a large number of shear tests on FRC beams reported in literature does not provide information about the actual post-cracking flexural strength defined by the MC2010. In these cases, the implementation of the equations reported in Table 1 becomes difficult to undertake. The literature reports different methods [8, 9] to predict the residual strength of FRC by starting from the properties of fibers and of the concrete matrix. Here, new empirical formulations will be used to determine the mean values of the residual strengths f_{R1} and f_{R3} of FRC when the actual values are not available, as follows:

$$f_{R1} = 2.7 \cdot \left(\frac{L_f}{\phi_f}\right)^{0.8} \cdot \left(V_f\right)^{0.9} \cdot \left(f_{cm}\right)^{1/2} \quad [MPa] \tag{2}$$

$$f_{R3} = 0.5 \cdot \left(\frac{L_f}{\phi_f}\right)^{0.85} \cdot (V_f)^{0.85} \cdot (f_{cm})^{2/3} \quad [MPa] \tag{3}$$

Table 1. Code equations for predicting the shear strength of prestressed FRC beams without stirrups.

prEN 1992 – Annex L (2020)	fib Model Code 2020
$V_{Rd} = \tau_{Rd,cF} \cdot b_w \cdot d$ where: -V_{Rd} = design shear resistance -$\tau_{Rd,cF}$ = deign shear strength of FRC member -d =effective depth to tension reinforcement [mm] -b_w =smallest web width	$V_{Rd} = V_{Rd,c} + V_{Rd,FRC}$ where: -V_{Rd} = design shear resistance -$V_{Rd,c}$ = design shear resistance of plain concrete member -$V_{Rd,FRC}$ = FRC contribution to shear resistance
$\tau_{Rd,cF} = \eta \cdot \tau_{Rd,c} + f_{Ftud} \geq \eta \cdot \tau_{Rdc,min} + f_{Ftud}$ where: -$f_{Ftud} = \frac{f_{R3d}}{3} = \frac{f_{R3k}}{3 \cdot \gamma_c}$ -f_{R3d} = design residual strength of FRC at CMOD$_3$=2.5 mm -f_{R3k} = characteristic residual strength of FRC at CMOD$_3$=2.5 mm -$\eta = MAX\left\{\frac{1}{1+0.43 \cdot f_{Ftuk}^{2.85}} ; 0.4\right\}$ -$\tau_{Rdc,min} = \frac{10}{\gamma_c} \cdot \sqrt{\frac{f_{ck}}{f_{yd}}} \cdot \frac{d_{dg}}{d}$ minimum shear strength of concrete -γ_c = partial safety factor of concrete -d_{dg} = size parameter describing the roughness of the failure zone: $d_{dg} = 16 + D_{lower} \leq 40$ [mm], $f_{ck} \leq 60$ MPa $d_{dg} = 16 + D_{lower}\left(\frac{60}{f_{ck}}\right)^2 \leq 40$ [mm], $f_{ck} > 60$MPa -D_{lower} is the smallest value of the maximum aggregate size (D_{max}) for the coarsest fraction of aggregates in the concrete permitted by the specification of concrete -f_{ck} = characteristic compressive strength of concrete -$f_{yd} = f_{pd} - \sigma_{p\infty}$ design yield strength of prestressing tendons -f_{pd} = design yield strength of prestressing tendons - $\sigma_{p\infty}$ = long term stress level in prestressing tendons -$\tau_{Rd,c} = \frac{0.6}{\gamma_c} \cdot \left(100 \cdot \rho_l \cdot f_{ck} \cdot \frac{d_{dg}}{d}\right)^{1/3}$ design shear strength of concrete	$V_{Rd,c} = k_{v,FRC} \cdot \frac{\sqrt{f_{ck}}}{\gamma_c} \cdot z_v \cdot b_w$ where: -$k_{v,FRC} = \frac{0.4}{(1+1500 \cdot \epsilon_x) + \zeta} \cdot \frac{1300}{1000 + k_{dg} \cdot z_v}$ -$k_{dg} = \frac{32}{16+d_g} \geq 0.75$ - $\zeta = (20-14000 \cdot \epsilon_x) \cdot \frac{f_{Ftuk}}{f_{ck}}$ -f_{ck} = characteristic concrete compressive strength [MPa] -z_v=0.9d (0.72h) internal lever arm [mm] -d = effective depth to tension reinforcement [mm] -b_w = smallest web width [mm] -d_g = maximum aggregate size [mm] -γ_c = partial safety factor of concrete -ϵ_x = longitudinal strain at section mid-depth
	$\theta_F = 29° + 7000 \cdot \epsilon_x \leq 45°$ where θ_F is the strut slope to the horizontal
	$V_{Ed} = V_{Ed0} - F_p \cdot \sin(\delta_p)$ $M_{Ed} = M_{Ed0}$ $N_{Ed} = N_{Ed0} - F_p \cdot \cos(\delta_p)$ -M_{Ed}=design bending moment -V_{Ed}=design shear action -N_{Ed}=design axial force Design actions according to Figure 1
	$\epsilon_x = \dfrac{\left(\dfrac{M_{Ed}}{z_v} + V_{Ed} + N_{Ed} \cdot \dfrac{(z_p - e_p)}{z}\right)}{2 \cdot \left(\dfrac{z_s}{z_v} \cdot E_s \cdot A_s + \dfrac{z_p}{z_v} \cdot E_p \cdot A_p\right)} \geq 0$

(continued)

Table 1. (*continued*)

$-d=\frac{d_s^2 \cdot A_s + d_p^2 \cdot A_p}{d_s \cdot A_s + d_p \cdot A_p}$ = effective depth to tension reinforcement [mm] $-d_s$ = effective depth to tension reinforcement $-d_p$ = effective depth to prestressed tendons $-A_s$ = total area of tension reinforcement $-A_p$ = total area of tendons $-\rho_l = \frac{d_s \cdot A_s + d_p \cdot A_p}{b_w \cdot d^2}$ equivalent reinforcement ratio	where: $-E_s$ = Young modulus of steel reinforcement $-E_p$ = Young modulus of tendons For beams with prestressed tendons: $z_v = \frac{z_s^2 \cdot A_s + z_p^2 \cdot A_p}{z_s \cdot A_s + z_p \cdot A_p} \geq 0.72 \cdot h$ $-z_p = 0.9d_p$ $-z_s = 0.9d_s$ $-d_s$ = effective depth to tension reinforcement $-d_p$ = effective depth to prestressed tendons $-A_s$ = total area of tension reinforcement $-A_p$ = total area of tendons
	LoA II approximation: $V_{Rd,FRC} = \dfrac{f_{Ftuk}}{\gamma_F} \cdot \cot\theta_F \cdot z_v \cdot b_w$ where: $-f_{Ftuk} = \frac{f_{R3k}}{3}$ for h $\leq$ 800 mm $-f_{Ftuk} = 0$ for h > 800 mm $-\zeta = (20 - 14000 \cdot \varepsilon_x) \cdot \frac{f_{Ftuk}}{f_{ck}} \geq 0$
	LoA III approximation: $V_{Rd,FRC} = \dfrac{f_{Ftwk}}{\gamma_F} \cdot \cot\theta_F \cdot z_v \cdot b_w$ where: $-f_{Ftwk} = \left[f_{Ftsk} + \frac{f_{Ftuk} - f_{Ftuk}}{w_{SLS} - w_u} \cdot (w - 0.5) \right] \geq 0$ $-f_{Ftsk} = 0.37 \cdot f_{R1k}$ $-f_{Ftuk} = f_{Ftsk} - 0.5 \cdot (0.63 f_{R1k} - 0.57 f_{R3k})$ $-w_{SLS} = 0.5$ mm $-w_u = 1.25$ mm $-\zeta = (20 - 14000 \cdot \varepsilon_x) \cdot \frac{f_{Ftwk}}{f_{ck}} \geq 0$ $-w = 0.2 \left(1 + 60 \frac{f_{Ftuk}}{f_{ck}} \right) + 1000\varepsilon_x \left(1 - 3 \frac{f_{Ftuk}}{f_{ck}} \right) \geq 0.5$mm

where V_f [−] is the volume fraction of fibers; f_{cm} [MPa] is the men compressive strength of concrete; L_f and $\varnothing_f$ are the length and the diameters of fibers, respectively. To calibrate the two equations above, a database containing 155 3PBTs collected from literature was considered. Further details about the model and its validation can be found in [10]. Note that Eqs. (2) and (3) apply only for 20 MPa $\leq f_{cm} \leq$ 85 MPa and 0.25% $\leq V_f \leq$ 1.1%. Where f_{cm} and V_f do not meet these limits, the FRC performance obtained in [11] was utilized with $f_{R3} = 3 f_{tf}$, where the latter is the tensile strength provided by the fibers, according to the VEM theory [12]. While using the latter approach, it was assumed for simplicity that $f_{R3} = f_{R1}$.

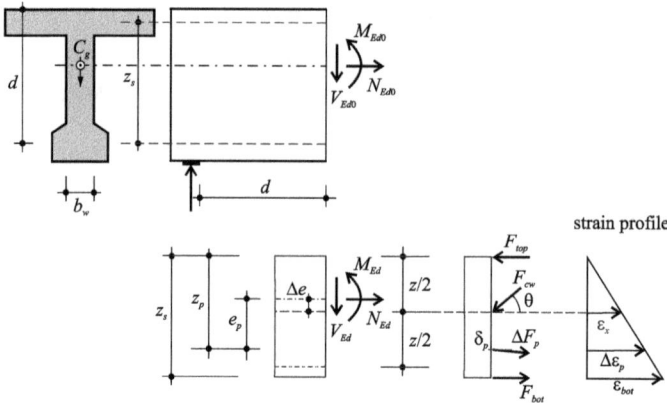

Fig. 1. Main variable considered by the MC2020.

3   Database of Shear Tests

To assess the performance of the code equations reported in Table 1, a database (see Table 2) of 29 shear tests performed on steel fiber reinforced concrete (SFRC) prestressed beams with bonded tendons was collected from different works reported by the literature [11, 13–17]. As shown in Table 2, the prestressed members consist of I- and T-cross section beams with a shear span (a/d) of 2.5–5.2 and a total height (h) ranging from a minimum of 400 mm to 800 mm.

All the main parameters required by the prEN 1992 and MC2020 formulations are highlighted in Table 2, including the mean compressive strength (f_{cm}) as well as the residual strengths f_{R1} and f_{R3} of FRC. Except for the few experimental data marked with an asterisk (*), most of the values of the residual strengths were estimated by the Eqs. (2) and (3). When not provided, the elastic moduli of ordinary reinforcement (E_s) and prestressing tendons (E_p) were both assumed equal to 200 GPa. For all the experiments, low relaxation strands having a mean value of yielding strength f_{py} of 1690 MPa were assumed, except for [13], which explicitly reported a yielding strength of 1900 MPa.

4   Shear Formulations Assessment

To assess the performance of the prEN 1992 and MC2010 shear formulations, the equations reported in Table 1 were used to calculate the shear capacity of all beams belonging to the shear database collected herein.

In order to obtain predictions consistent with the experimental results, the partial safety factor γ_c and γ_F were set equal to 1 and the values of the design or characteristic mechanical properties of concrete and FRC were assumed equal to the corresponding mean values.

Table 2. Summary of shear tests on prestressed FRC beams.

#	References	Specimen ID	Section Type	b_w mm	h mm	a/d -	D_{lower} d_g mm	d_s mm	A_s mm²	A_p mm²	d_p mm	Δe mm	e_p mm	σ_{pi} MPa	$\Delta\sigma_p$ %	σ_{pe} MPa	F_p kN	f_{cm} MPa	V_f %	l_f mm	$\varnothing_f$ mm	f_{ct} MPa	f_{cs} MPa
1		TB22	T	75	250	2.5	9.5	185	402	104	185	69	146	1330	15**	1131	118	34.7	0.50	30	0.5	3.57	3.53
2		TB23	T	75	250	2.5	9.5	185	402	104	185	69	146	1330	15**	1131	118	40.5	1.00	30	0.5	7.20	7.06
3	[13]	TB25	T	75	250	2.5	9.5	185	402	104	185	69	146	1330	15**	1131	118	38.2	1.00	30	0.5	7.00	6.79
4		TB26	T	75	250	2.5	9.5	185	603	52	185	69	146	1330	15**	1131	59	39.1	1.00	30	0.5	7.08	6.90
5	[14]	TB28	T	75	250	2.5	9.5	0	0	208	185	69	146	1330	15**	1131	235	34.0	1.00	30	0.5	6.60	6.28
6		Beam 1	I	70	300	3.7	0.0	0	0	1200	260	14	131	1000	15**	850	1020	177.8	2.50	13	0.2	19.68	19.68
7		B3-DZ	I	120	700	3.7	15.0	600	2281	2562	627	74	362	1400	15	1190	3049	65.8	0.64	30	0.6	5.18	2.18*
8		B4-DZ	I	120	700	3.7	15.0	600	2281	2562	627	74	362	1400	15	1190	3049	74.2	0.64	30	0.4	8.13	8.61*
9	[15]	B5-DZ-I	I	120	700	3.7	15.0	600	2281	2562	627	74	362	1400	15	1190	3049	66.9	0.64	30	0.6	5.22	2.18*
10		B5-DZ-II	I	120	700	3.7	15.0	600	2281	2562	627	74	362	1400	15	1190	3049	66.9	0.64	30	0.6	5.22	2.18*
11		B6-DZ-I	I	120	700	3.7	15.0	600	2281	2562	627	74	362	1400	15	1190	3049	67.4	0.64	30	0.4	7.74	8.61*
12		B6-DZ-II	I	120	700	3.7	15.0	600	2281	2562	627	74	362	1400	15	1190	3049	67.4	0.64	30	0.4	7.74	8.61*
13		T1a	I	60	400	3.8	0.0	0	0	693	316	43	183	1263	15**	1074	744	132.9	0.90	18	0.2	9.09	9.09
14	[16]	T1b	I	60	400	3.8	0.0	0	0	891	316	43	183	1263	15**	1074	957	153.1	0.90	18	0.2	9.75	9.75
15		T4a	I	60	400	3.8	0.0	0	0	891	316	43	183	1263	15**	1074	957	154.9	0.90	18	0.2	9.81	9.81
16		T4b	I	60	400	4.4	0.0	0	0	891	316	43	183	1263	15**	1074	957	161.0	0.90	18	0.2	9.99	9.99
17		X-B1	I	50	650	3.2	1.0	0	0	1668	620	0	0	1400	20	1120	1868	125.0	1.00	15	0.2	6.15	6.15
18		X-B2	I	50	650	3.2	1.0	0	0	1668	620	0	0	1400	20	1120	1868	126.0	1.00	15	0.2	6.18	6.18
19		X-B3	I	50	650	3.2	1.0	0	0	1668	620	0	0	1400	20	1120	1868	135.0	1.00	15	0.2	6.39	6.39
20	[11]	X-B4	I	50	650	2.5	1.0	0	0	1668	620	0	0	1400	20	1120	1868	122.0	1.00	15	0.2	6.09	6.09
21		X-B5	I	50	650	3.5	1.0	0	0	1668	620	0	0	1400	20	1120	1868	140.0	1.00	25	0.2	11.37	11.37
22		X-B6	I	50	650	4.5	1.0	0	0	1668	620	0	0	1400	20	1120	1868	140.0	1.00	25	0.2	11.37	11.37
23		X-B7	I	50	650	2.5	1.0	0	0	1668	620	0	0	1400	20	1120	1868	122.0	1.50	20	0.2	12.51	12.51
24		HF400h/6	I	100	800	2.8	12.0	0	0	1540	739	59	392	1354	26.2	999	1539	59.5	0.75	40	0.6	8.96*	5.96*
25		HF600/4	I	100	750	3.0	12.0	0	0	1540	689	82	392	1354	26.2	999	1539	65.4	0.75	40	0.6	10.46*	6.24*
26	[17]	HF600/5	I	100	750	3.0	12.0	0	0	1540	689	82	392	1354	26.2	999	1539	65.9	0.75	40	0.6	8.55*	5.55*
27		HF400/7	I	100	750	3.0	12.0	0	0	1540	689	82	392	1354	26.2	999	1539	63.5	0.75	40	0.6	6.64*	4.77*
28		HF400/8	I	100	750	3.0	12.0	0	0	1540	689	82	392	1354	26.2	999	1539	70.0	0.75	40	0.6	8.10*	4.68*
29		HF260/9	I	100	750	3.0	12.0	0	0	1540	689	82	392	1354	26.2	999	1539	65.0	0.75	40	0.6	6.45*	4.38*

* Experimental data provided by literature.

** Total prestress losses not specified. In these cases, a total loss of 15% was assumed.

Table 3. Summary of the predicted shear capacities and comparison with test results.

#	Ref.	Specimen ID	V_{exp} kN	prEN 1992 – Annex L		MC2020 LoA II		MC2020 LoA III	
				V_{ana} kN	V_{exp}/V_{ana} -	V_{ana} kN	V_{exp}/V_{ana} -	V_{ana} kN	V_{exp}/V_{ana} -
1		TB22	87.50	30.62	2.86	48.77	1.79	51.01	1.72
2		TB23	113.00	42.49	2.66	74.16	1.52	77.69	1.45
3	[13]	TB25	80.00	40.76	1.96	71.32	1.12	74.86	1.07
4		TB26	110.10	42.49	2.59	72.44	1.52	75.98	1.45
5		TB28	93.50	35.96	2.60	65.96	1.42	69.48	1.35
6	[14]	Beam 1	273.00	136.81	2.00	238.36	1.15	254.48	1.07
7		B3-DZ	587.00	152.41	3.85	232.64	2.52	341.32	1.72
8		B4-DZ	722.00	258.82	2.79	449.48	1.61	461.30	1.57
9	[15]	B5-DZ-I	579.00	152.48	3.80	234.20	2.47	344.34	1.68
10		B5-DZ-II	555.00	152.48	3.64	234.20	2.37	344.34	1.61
11		B6-DZ-I	586.00	258.53	2.27	440.10	1.33	440.05	1.33
12		B6-DZ-II	620.00	258.53	2.40	440.10	1.41	440.05	1.41
13		T1a	234.00	70.80	3.30	139.60	1.68	148.48	1.58
14	[16]	T1b	267.00	76.85	3.47	150.84	1.77	160.42	1.66
15		T4a	344.00	77.28	4.45	151.83	2.27	161.46	2.13
16		T4b	291.00	78.24	3.72	154.99	1.88	164.82	1.77
17		X-B1	314.65	79.48	3.96	162.75	1.93	172.53	1.82
18		X-B2	338.52	79.83	4.24	163.54	2.07	173.37	1.95
19		X-B3	345.65	82.37	4.20	169.70	2.04	179.89	1.92
20	[11]	X-B4	434.62	79.42	5.47	160.82	2.70	170.49	2.55
21		X-B5	403.00	133.79	3.01	245.08	1.64	263.39	1.53
22		X-B6	372.62	133.12	2.80	245.08	1.52	263.39	1.41
23		X-B7	497.55	145.76	3.41	256.31	1.94	274.68	1.81
24		HF400h/6	420.03	176.31	2.38	329.25	1.28	446.45	0.94
25		HF600/4	392.44	172.03	2.28	325.97	1.20	475.73	0.82
26	[17]	HF600/5	347.17	156.20	2.22	305.07	1.14	421.19	0.82
27		HF400/7	389.95	138.57	2.81	278.07	1.40	357.17	1.09
28		HF400/8	428.31	137.81	3.11	283.73	1.51	419.52	1.02
29		HF260/9	325.58	134.33	2.42	268.31	1.21	354.70	0.92
	Mean		-	-	3.1	-	1.7	-	1.5
	CV [%]		-	-	26.4	-	25.8	-	27.1

The shear capacities predicted by the codes (V_{ana}) are summarized in Table 3 together with the corresponding experimental values (V_{exp}). The code formulations proved to be very conservative as compared to the actual shear capacities observed in the experimental tests. As shown in Table 3, the predictions of the prEN 1992 were about 3.1 times higher than the experimental values, with peak values of more than 5. The predictions of MC2020 LoA II and LoA III were less conservative. They exhibited mean values of V_{exp}/V_{ana} respectively equal to 1.7 and 1.5. In addition, the coefficient of variation (CV), which ranges from 27.1% in MC2020 (LoA III) to a minimum of 25.8% in MC2020 (LoA II), seems to be rather high. The pretty conservative trend of the models is well highlighted by the comparison diagrams of Fig. 2, which report also the MAPE (mean absolute percentage error) related to each predicting formulation.

It emerges a pressing need to further study and evaluate the shear behavior of prestressed members. Even applying the most up to date models available at this stage, the draft of MC2020 above all [4], it seems that further shear fundamental resisting mechanisms, peculiar of p/s members, are currently neglected. Moreover, the influence of fibers can be further enhanced by the presence of prestressing in a synergetic combined action: this should be experimentally verified and, eventually, analytically modeled.

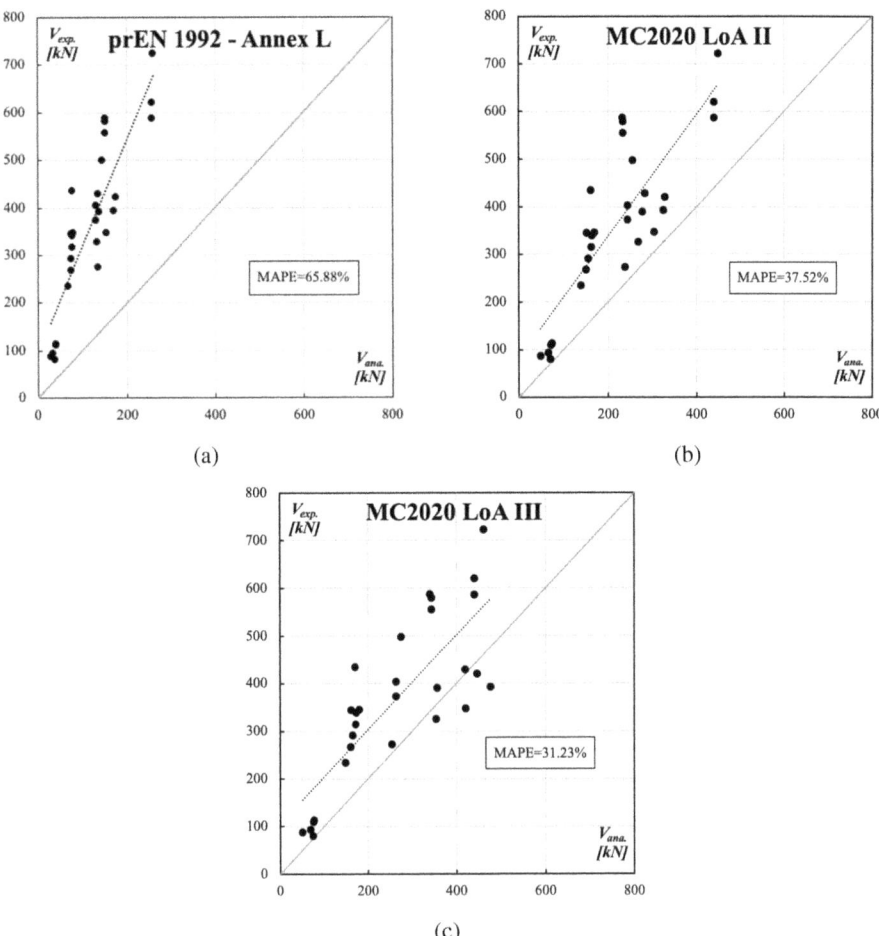

Fig. 2. Experimental vs. analytical prediction of the shear resistance: (a) prEN 1992 – Annex L; (b) MC2020 – LoA II; (c) MC2020 – LoA III.

Last but not least, the inability of the available analytical models to properly predict the actual shear -strength of prestressed members can have a strong impact on the shear assessment of existing structures, many of those are bridge girders or beams, very often prestressed with bonded tendons. Analytical models very conservative, as herein shown, could determine retrofitting or strengthening interventions heavier than needed, with a direct impact on cost, resulting in a lower number of interventions for a specific budget given by road administrations to structural strengthening.

Figure 3 reports the normalized experimental shear strength (normalized to the square root of the compressive strength) as a function of the residual strength f_{R3}. The latter is the most important parameter describing the influence of fibers in the shear response. Note that also the level of prestressing should be quite influential with this respect, but all specimens herein considered were described by having a prestressing level, at the time of testing ($\sigma_{p\infty}$), in the range of 850–1190 MPa, taking into account both short term and long term losses. The influence of the different level of prestressing was not herein investigated, considering the limited set of data and some uncertainties about the effective measure of losses by the authors (as above mentioned, some assumptions were taken about this point). However, at a first glance, this plot evidences a clear ascending trend of the normalized shear strength vs. the FRC residual strength. This is in accordance to other studies on classical non prestressed members, even though the prestressing action helps in providing higher normalized shear strength.

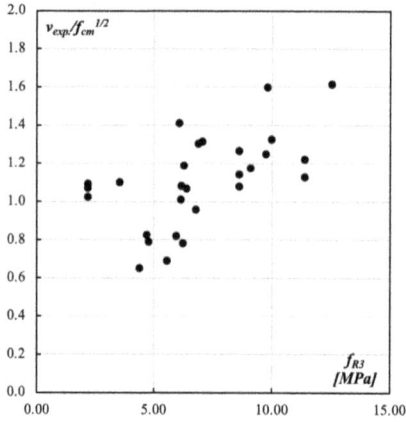

Fig. 3. Normalized experimental shear stress $v_{exp}/(f_{cm})^{1/2}$ vs. f_{R3}.

5   Conclusions

A database of shear experiments on prestressed (bonded tendons only) I or T shaped beams was presented herein. The predictions of the prEN 1992 – Annex L as well as that of current draft of MC2020 were discussed. It was found that, as expected, the most up to date analytical models are still very conservative in predicting the shear strength of these members, differently from classical RC members, in which the same models perform much better. Especially prEN 1992 – Annex L results overconservative. The COV values were also rather high, indicating that further peculiar shear resisting mechanisms should be investigated. Attention should be especially devoted towards the shear assessment of existing prestressed elements, in order to avoid structural retrofitting interventions heavier than needed. Finally, this database should be extended to further check and improve the formulations herein discussed.

References

1. Heek, P., Look, K., Minelli, F., Mark, P., Plizzari, G.: Datenbank fur querkraftbeanspruchte Stahlfaserbetonbauteile: bewertung der Bemessungsansätze nach DAfStb-Richtlinie und fib Model Code 2010. BETON- UND STAHLBETONBAU **112**, 144–154 (2017). https://doi.org/10.1002/best.201600075. ISSN: 1437–1006
2. fib Model Code for Concrete Structures 2010: Fédération Internationale du Béton. Ernst & Sohn, Lausanne (2013)
3. Annex L to prEN 1992–1–1–2020. CEN TC 250/SC2/WG1/TG2 D4
4. fib Model Code for Concrete Structures 2020 – Draft Volume 1V, 2020
5. EN 1992–2: Eurocode 2. Design of Concrete Structures. Part 2: Concrete bridges: design and detailing rules (2005)
6. Bentz, E.C., Vecchio, F.J., Collins, M.P.: The simplified MCFT for calculating the shear strength of reinforced concrete elements. ACI Struct. J. **103**(4), 614–624 (2006)
7. EN 14651: Test Method for Metallic Fibre Concrete – Measuring the Flexural Tensile Strength (Limit of Proportionality (LOP), Residual). European Standard (2007)
8. Moraes Neto, B.N., Barros, J.A., Melo, G.S.: Model to simulate the contribution of fiber reinforcement for the punching resistance of RC slabs. J. Mater. Civ. Eng. **26**(7), 04014020 (2014)
9. Venkateshwaran, A., Tan, K.H., Li, Y.: Residual flexural strengths of steel fiber reinforced concrete with multiple hooked-end fibers. Struct. Concr. **19**(2), 352–365 (2018)
10. Faccin, E., Facconi, L., Minelli, F., Plizzari, G.: Predicting the residual strength of concrete reinforced with hooked-end steel fibers: new empirical equations. In: Proceedings of the RILEM-fib X International Symposium on Fibre Reinforced Concrete BEFIB2021, Valencia (Spain), 20 to 22 September 2021 (2021)
11. Voo, Y.L., Foster, S.J.: Shear strength of steel fiber reinforced ultra-high performance concrete beams without stirrups. In: Proceedings of the 5th International Specialty Conference on Fibre Reinforced Materials, vol. 136, no. 11, pp. 177–184 (2009)
12. Voo, J.Y.L., Foster, S.J.: Tensile fracture of fiber reinforced concrete: variable engagement model. In: Proceedings of the 6th Rilem Symposium on Fiber Reinforced Concrete (FRC), RILEM, Bagneux, France, pp. 875– 884 (2004)
13. Tan, K.H., Paramasivam, P., Murugappan, K.: Steel fibers as shear reinforcement in partially prestressed beams. ACI Struct. J. **92**(6), 643–652 (1995)
14. Hegger, J., Tuchlinsky, D., Kommer, B.: Bond anchorage behavior and shear capacity of ultra high performance concrete beams. In: Proceedings of the International Symposium on Ultra High Performance Concrete, pp. 351–360 (2004)
15. Meda, A., Minelli, F., Plizzari, G.A., Riva, P.: Shear behavior of steel fibre reinforced concrete beams. Mater. Struct. **38**, 343–351 (2005). ISSN 1359–5997
16. Hegger, J., Bertram, G.: Shear carrying capacity of Ultra-High Performance Concrete beams. In: Tailor Made Concrete Structures – Walraven & Stoelhorst (eds), Taylor & Francis Group, London, pp. 341–347 (2008). ISBN 978-0-415-47535-8
17. Cuenca, E., Serna, P.: Shear behavior of prestressed precast beams made of self-compacting fiber reinforced concrete. Constr. Build. Mater. **45**, 145–156 (2013)

A Short Review on the Utilization of Basalt Fibres in Concrete

Suman Saha[(✉)]

Department of Civil Engineering, National Institute of Technology Calicut,
Kozhikode 673601, Kerala, India
sumansaha@nitc.ac.in

Abstract. In recent days, fibre reinforced concrete already gained significant momentum to be used for the structural applications due to its better mechanical properties than the normal concrete. Steel fibres, synthetics fibres, glass fibres, carbon fibres, natural fibres etc. are few commonly used fibres in concrete mixes to improve the mechanical properties, post cracking behaviours, durability properties etc. for using in structural applications. In recent days, basalt fibres are also gaining interest to be used in concrete mixes due to its better physio-mechanical properties and economic production processes. This paper intends to provide a systematic short review of the recent developments, experimental test results of the basalt fibre reinforced concrete desired to utilize in the structural applications. Promising advantages of the utilization of basalt fibres are highlighted based on the published works, and problems are also highlighted.

Keywords: Fibre reinforced concrete · Basalt fibre · Mechanical properties

1 Introduction

Concrete is the most widely consumed construction material worldwide due to its own versatile nature. Concrete performs way better under compression and it is brittle in nature as well. Although it is having so many advantageous, there is always requirement to improve the other characteristics, specifically in which it performs relatively poor. With that point of view, to enhance the ductile behavior, shrinkage resisting capacity and to improve post cracking tensile behavior of hardened concrete, several types of fibres of different dimension are incorporated in the concrete mix. Metallic, synthetic, mineral and vegetable fibres are generally used. Steel, glass, polyethylene (PE), polypropylene (PP), acrylics (PAN), polyvinyl acetate (PVA), polyamides (PA), aramid, polyester (PES), and carbon are the few examples of the used fibres in concrete mixes in order to enhance its hardened properties. However, some of these fibres may not be available at all places and again some of these fibres' production require high amount of energy. Therefore, it is always recommended to find any suitable fibres, which will be having lesser negative impact on the environment in its production process. Fiore et al. (2015) indicated basalt fibres as sustainable alternative for other types of fibres such as carbon. Sim et al. (2005) stated that the production process of basalt fibres is cheaper than that of other types of fibres. In this article, it is aimed to explain briefly about the basalt fibres, its utilization in concrete and influence of it on

© RILEM 2022
P. Serna et al. (Eds.): BEFIB 2021, RILEM Bookseries 36, pp. 590–597, 2022.
https://doi.org/10.1007/978-3-030-83719-8_51

the characteristics of the concrete in different following sections to have better understanding about the basalt fibre reinforced concrete.

2 Manufacturing Process

Volcanic basalt rocks are the primary sources of the basalt fibres (Fiore et al. 2011). Basalt is an igneous rock found in abundance throughout the world. Basalt rock is crushed, loaded into a furnace and liquefied. Next, basalt filaments are drawn through platinum-rhodium bushings. Basalt of high acidity (over 46% silica content) and low iron content is considered desirable for fiber production. Unlike with other composites, such as glass fiber, essentially no materials are added during its production. The basalt is simply washed and then melted. The manufacture of basalt fiber requires the melting of the crushed and washed basalt rock at about 1500 °C (2,730 °F). The molten rock is then extruded through small nozzles to produce continuous filaments of basalt fiber. The basalt fibers typically have a filament diameter of between 10 μm and 20 μm which is far enough above the respiratory limit of 5 μm to make basalt fiber a suitable replacement for asbestos. Usually, basalt fibres also have a high elastic modulus, resulting in high specific strength - three times that of steel. Thin fiber is usually used for textile applications mainly for production of woven fabric. Thicker fiber is used in filament winding, for example, for production of Compressed Natural Gas (CNG) cylinders or pipes. The thickest fiber is used for pultrusion, geogrid, unidirectional fabric, multi-axial fabric production and in form of chopped strand for concrete reinforcement. One of the most prospective applications for continuous basalt fiber is the production of basalt rebar that substitutes more traditional steel rebar on construction market, and this is the most modern trend at the moment. Figure 1 shows some common basalt fibre products developed for reinforcing concrete. The chemical composition of the basalt fibres is shown in Table 1.

Table 1. Chemical composition of the basalt fibres

Author	SiO_2	Al_2O_3	CaO	MgO	Na_2O	K_2O	Fe_2O_3	TiO_2
Jamshaid and Mishra (2016)	52–58	17–18	5–8	1–4	3–6	1–5	4–10	–
Militký et al. (2002)	43.3–47	11–13	10–12	8–11	<5	<5	<5	<5
Deák and Czigány (2009)	42.43–55.69	14.21–17.97	7.43–8.88	4.06–9.45	2.38–3.79	1.06–2.33	10.80–11.68	1.10–2.55
Adesina (2021)	42–58	11–18	5–12	1–11	2–6	0.8–5	4–12	–
Ayub et al. (2014)	51.6–59.3	14.6–18.3	5.9–9.4	3–5.3	0.8–2.25		9–14	0.8–2.25

(a) Rebar   (b) Mesh   (c) Chopped Fibre

Fig. 1. Common basalt fibre

Basalt fibres can be manufactured directly from a single raw material (basalt rock) without the need for additives, making the process simpler. As a result, the fibres can be manufactured with conventional processes and equipment, and less energy, which offers an economic advantage. Moreover, the fibres are considered 100% natural, have no toxic reaction with air or water, and the fiberization process is said to be more environmentally friendly than that of glass fibre. In terms of mechanical and physical properties, basalt fibre has gathered attention due to its high elastic modulus, high strength, corrosion resistance, high temperature resistance, and light-weight. Table 2 provides the properties of basalt fibres reported in the bibliography.

Table 2. Properties of basalt fibres

Properties	Adesina (2021)	Fiore et al. (2011)	Lyu et al. (2021)	Jiang et al. (2014)	Ayub et al. (2014)	Wlodarczyk, and Jedrzejewskia (2016)
Density (g/cm^3)	1.85–2.75	2.8	2.61	2.65	2.63–2.80	2.63
Elastic modulus (GPa)	80–115	89	93–110	93–110	93.1–110	75–90
Tensile strength (GPa)	2.6–4.84	2.8	2.5–3.4	4.1–4.8	4.1–4.8	3–3.2
Elongation at fracture (%)	2.4–3.15	3.15	2.5–3.1	3.1–3.2	3.1	3.1

3   Past Studies on the Usage of Basalt Fibres in Concrete

Utilization of basalt fibres in the concrete mix to enhance its properties is getting increased day by day. Adesina (2021) pointed out the increasing the number of articles published globally on the topic of basalt fibres. This indicates the growing interest to utilize the basalt fibres in concrete mixes. In this section, different characteristics e.g. workability, compressive strength, flexural strength, splitting tensile strength etc. of concrete mixes, which are observed by few past studies, has been explained and summarized briefly.

3.1 Workability

Most of the studies observed the reduction in workability of the concrete mixes containing basalt fibres. It can be attributed by the basalt fibres' water absorption capacity. It leads to lower the amount of water in the mixes and as results reduction in water absorption is usually observed. However, researchers pointed out that the length of the fibres (Jiang et al. 2014), dosage and use of chemical admixtures also play important role on the workability of the concrete mixes. Table 3 represents the observation made by few studies on workability of concrete containing basalt fibres.

Table 3. Observation on workability of concrete containing basalt fibres

Authors	Observation
Niu et al. (2020)	26%, 51%, 58%, and 67% lower workability of concrete mixes containing 0.05%, 0.1%, 0.15%, and 0.2% basalt fibres (volumetric fraction) respectively than the concrete mixtures without basalt fibres
Jiang et al. (2014)	7%, 15%, 53%, and 65% lower workability when basalt fibres (12 mm length) were incorporated at 0.05%, 0.1%, 0.3% and 0.5% (volumetric fraction) respectively
Sadrmomtazi et al. (2018)	Balling effect was observed with the incorporation of higher percentage of basalt fibres
Kabay (2014)	No significant variation in workability found. It's because of the use of superplasticizers efficiently
Loh et al. (2019)	About 29.8–40.8% decrease in workability was observed with the increase of basalt fibres by volume

3.2 Mechanical Properties of Concrete

Researchers conducted significant amount of investigation to determine the mechanical properties of concrete e.g. compressive strength, splitting tensile strength, flexural strength etc. when the basalt fibres were incorporated into the mixes. From the observation of different studies, it's difficult to conclude a single point how the incorporation of basalt fibres affects the compressive strength of concrete. Few studies (Niaki et al. 2018, Algin and Ozen 2018, Loh et al. 2019, Guo et al. 2019) reported that incorporation of the basalt fibres enhanced the compressive strength of the concrete mixes. But, few studies (Kabay 2014, Sadrmomtazi et al. 2018, Guo et al. 2019) reported the negative impacts on the compressive strength of the concrete mixes when basalt fibres were incorporated. Du et al. (2020) observed no significant variation of compressive strength when basalt fibres were used. Table 4 provides the optimum dosage of the basalt fibres to achieve best mechanical properties of the concrete mixes by few past studies.

Table 4. Optimum content (volumetric fraction) of basalt fibres based on mechanical properties

Authors	Optimum dosage of basalt fibres (by volume)
Niu et al. (2020)	0.05% dosage of 18 mm length of basalt fibres exhibited highest compressive and splitting tensile strength
Jiang et al. (2014)	0.1%, 0.5% and 0.3% dosage of basalt fibres (12 mm and 22 mm length) showed highest compressive strength, splitting tensile strength and flexural strength respectively
Sadrmomtazi et al. (2018)	1.5% dosage of basalt fibres (6mm length) showed highest compressive strength, and flexural strength respectively
Kabay (2014)	0.07%, and 0.14% dosage of basalt fibres (12 mm and 24 mm length) showed highest compressive strength, and flexural strength respectively

Borhan (2013) conducted experimental investigations to determine the compressive, split-tensile, and flexural strength of basalt fibre reinforced concrete (BFRC). In this research, the volume fractions of basalt fibre were varied as 0.1%, 0.2%, 0.3%, and 0.5% by total mix volume. Results indicated decrease in slump of concrete mixes with increasing the volume fraction of basalt fibre. A slight increase in the splitting tensile strength was observed with increase the volume fraction of fibre till 0.3% and then decreased with 0.5% basalt fibre. The compressive strength was also increased with the increase in fibre content till 0.3% and a reduction of 12% when 0.5% fibre content was used in concrete mixes. Ayub et al. (2014) determined compressive strength, elastic modulus and tensile strength of concrete mixes with the incorporation of 1%, 2% and 3% basalt fibre by volume fraction. Apart from this, 10% silica fume and metakaolin were also used as replacement of cement content. Experimental results showed the improvement of compressive strength of concrete mixes with the addition of basalt fibres up to 2% for all combination of mixes. At 3% basalt fibre content, around the range from 2.37% to 15.1% decrease in compressive strength was observed for the concrete mixes when metakaolin and silica fume were used. Li and Xu (2009) found basalt fibres can significantly increase the energy absorption capacity of geopolymer concrete under impact loading by using a Split-Hopkinson pressure bar system. However, the performance of basalt fibre reinforced concrete mixes under impact in general is still largely unknown. Dias and Thaumaturgo (2005) investigated the influence of the volumetric fraction of the basalt fibers on the fracture toughness of geopolymer concrete. Volumetric fractions of basalt fibers were varied as 0%, 0.5% and 1% by volume. 26.4% and 12% reductions in the compressive and splitting tensile strengths were observed for ordinary Portland cement concretes with the addition of 1.0% of basalt fibers. No significant changes in compressive strength and 25% increase in splitting tensile strength were found for geopolymer concrete with 1% basalt fibres content. Higher ultimate load and larger displacement before failure were observed for the geopolymer concrete beams than normal concrete beams. Geopolymer concretes with fibers exhibited larger values of Crack Mouth Opening Displacement (CMOD) than normal concrete beams.

3.3 Durability Properties of Concrete

Past studies also explored the durability properties of concrete mixes containing basalt fibres. From their observations, it's again difficult to make certain concluding remarks how incorporation of basalt fibres affects the durability characteristics of the concrete mixes. Kabay (2014) concluded from their investigation that there is no significant variation of porosity when different length of basalt fibres were used at same dosage. But, the increasing the dosage of basalt fibres lead to a slight increment of the porosity of the concrete mixes. Their study observed 10% and 11.3% of void content when basalt fibres of 24 mm length were incorporated at 0.07% and 0.14% respectively. Guo et al. (2019) found increased porosity with the incorporation of basalt fibres in concrete mixes. Increment in water absorption capacity of the concrete mixes with the increase of the dosage of basalt fibres in concrete mixes was found by the study of Kabay (2014). Sadrmomtazi et al. (2018) also concluded the similar remarks on water absorption capacity of concrete mixes with basalt fibres. In chloride ion penetration tests, Guo et al. (2019) found the cumulative current passed more through the concrete mixes when basalt fibres were incorporated. Lee et al. (2014) and Rabinovich et al. (2001) have shown basalt fibre loses tensile strength over time in a calcium hydroxide solution intended to replicate hydrating cement. In these cases, basalt fibre generally retained more strength than E-glass fibre. However, the improvement may be considered trivial when after 90 days; the basalt fibres still lost more than 50% of its tensile strength.

4   Conclusions

In this article, an overview on the properties of basalt fibres, and their influence, as a concrete reinforcement, on the behavior of the resulting basalt fibre reinforced concrete at fresh and hardened state is assessed. Based on the analysis with the help of few published articles, it can be concluded that production and utilization of basalt fibres is an environment-friendly approach and economic as well. The length and dosage of the basalt fibres play very significant role on the characteristics of the concrete mixes. It is evident that the characteristics of concrete mixes (fresh, mechanical and durability) can be improved with the utilization of shorter length of the basalt fibres. However, there is a requirement to conduct detailed investigations to understand the effect of basalt fibre on the characteristics of basalt fibre reinforced concrete mixes on large scale applications.

References

Fiore, V., Scalici, T., Di Bella, G., Valenza, A.: A review on basalt fibre and its composites. Compos. B Eng. **74**, 74–94 (2015)

Sim, J., Park, C.: Characteristics of basalt fiber as a strengthening material for concrete structures. Compos. B Eng. **36**(6–7), 504–512 (2005)

Fiore, V., Di Bella, G., Valenza, A.: Glass–basalt/epoxy hybrid composites for marine applications. Mater. Des. **32**(4), 2091–2099 (2011)

Jamshaid, H., Mishra, R.: A green material from rock: basalt fiber–a review. J. Text. Inst. **107**(7), 923–937 (2016)

Militký, J., Kovačič, V., Rubnerova, J.: Influence of thermal treatment on tensile failure of basalt fibers. Eng. Fract. Mech. **69**(9), 1025–1033 (2002)

Deák, T., Czigány, T.: Chemical composition and mechanical properties of basalt and glass fibers: a comparison. Text. Res. J. **79**(7), 645–651 (2009)

Adesina, A.: Performance of cementitious composites reinforced with chopped basalt fibres–an overview. Constr. Build. Mater. **266**, 120970 (2021)

Ayub, T., Shafiq, N., Nuruddin, M.F.: Mechanical properties of high-performance concrete reinforced with basalt fibers. Procedia Eng. **77**, 131–139 (2014)

Lyu, Z., Shen, A., Meng, W.: Properties, mechanism, and optimization of superabsorbent polymers and basalt fibers modified cementitious composite. Constr. Build. Mater. **276**, 122212 (2021)

Jiang, C., Fan, K., Wu, F., Chen, D.: Experimental study on the mechanical properties and microstructure of chopped basalt fibre reinforced concrete. Mater. Des. **58**, 187–193 (2014)

Wlodarczyk, M., Jedrzejewski, I.: Concrete slabs strengthened with basalt fibres–experimental tests results. Procedia Eng. **153**, 866–873 (2016)

Niu, D., Su, L., Luo, Y., Huang, D., Luo, D.: Experimental study on mechanical properties and durability of basalt fiber reinforced coral aggregate concrete. Constr. Build. Mater. **237**, 117628 (2020)

Sadrmomtazi, A., Tahmouresi, B., Saradar, A.: Effects of silica fume on mechanical strength and microstructure of basalt fiber reinforced cementitious composites (BFRCC). Constr. Build. Mater. **162**, 321–333 (2018)

Kabay, N.: Abrasion resistance and fracture energy of concretes with basalt fiber. Constr. Build. Mater. **50**, 95–101 (2014)

Niaki, M.H., Fereidoon, A., Ahangari, M.G.: Experimental study on the mechanical and thermal properties of basalt fiber and nanoclay reinforced polymer concrete. Compos. Struct. **191**, 231–238 (2018)

Algin, Z., Ozen, M.: The properties of chopped basalt fibre reinforced self-compacting concrete. Constr. Build. Mater. **186**, 678–685 (2018)

Loh, Z.P., Mo, K.H., Tan, C.G., Yeo, S.H.: Mechanical characteristics and flexural behaviour of fibre-reinforced cementitious composite containing PVA and basalt fibres. Sādhanā **44**(4), 1–9 (2019). https://doi.org/10.1007/s12046-019-1072-6

Guo, Y., Hu, X., Lv, J.: Experimental study on the resistance of basalt fibre-reinforced concrete to chloride penetration. Constr. Build. Mater. **223**, 142–155 (2019)

Du, Q., Cai, C., Lv, J., Wu, J., Pan, T., Zhou, J.: Experimental investigation on the mechanical properties and microstructure of basalt fiber reinforced engineered cementitious composite. Materials **13**(17), 3796 (2020)

Borhan, T.M.: Thermal and mechanical properties of basalt fibre reinforced concrete. In: Proceedings of World Academy of Science, Engineering and Technology, no. 76, p. 313. World Academy of Science, Engineering and Technology (WASET), April 2013

Li, W., Xu, J.: Mechanical properties of basalt fiber reinforced geopolymeric concrete under impact loading. Mater. Sci. Eng., A **505**(1–2), 178–186 (2009)

Dias, D.P., Thaumaturgo, C.: Fracture toughness of geopolymeric concretes reinforced with basalt fibers. Cem. Concr. Compos. **27**(1), 49–54 (2005)

Lee, J.J., Song, J., Kim, H.: Chemical stability of basalt fiber in alkaline solution. Fibers Polym. **15**(11), 2329–2334 (2014)

Rabinovich, F.N., Zueva, V.N., Makeeva, L.V.: Stability of basalt fibers in a medium of hydrating cement. Glass Ceram. **58**(11), 431–434 (2001)

Quality Control

Using Energy Absorption Capacity to Determine Residual Resistances of FRC

Sergio Carmona[1([⊠])] and Climent Molins[2]

[1] Department of Civil Engineering, Universidad Técnica Federico Santa María,
Valparaíso, Chile
`sergio.carmona@usm.cl`
[2] Department of Civil and Environmental Engineering, UPC - Barcelona Tech,
Barcelona, Spain

Abstract. In this paper, the residual strengths and flexural toughness of fibre reinforced concretes (FRC), determined from the load - CMOD response obtained by mean of flexural test conducted according to EN 14651, are related experimentally through non-linear fits, which depend on the fibres content and crack opening. On the other hand, it is experimentally verified that there is a linear relationship between the flexural toughness and the energy absorption capacity of FRCs determined by testing square panels following EN 14488 – 5.

Using the proposed relationships, the residual strengths of FRCs have been estimated from the energy absorption capacity, with differences between experimental and estimated values less than 3%.

Keywords: Flexural toughness · Equivalent strengths · Square panel test · Energy absorption capacity

1 Introduction

The constitutive equations for fibre reinforced concrete structures, given in the CEB Model Code (MC – 2010) [1], use the residual strengths determined through three point bending test performed on notched beam (3PB), in accordance with the EN 14651 standard [2]. However, a very rigid frame with a closed-loop control system is needed to perform this test, which is not normally available on-site quality control laboratories. Then, the energy absorption capacity, determined through the square panel test, performed as EN 14488–5 [3] standard or EFNARC recommendation [4] is used for controlling fibre reinforced shotcretes, because they can be performed in a conventional framework and even using non-standardized equipment.

With the aim to obtain a relationship between different tests, many authors proposed correlations based on crack width [5–9]. Nevertheless, crack width is not measured in square panel test as in the 3PB test, and the panels exhibit many cracks in the final state after testing, which is governed by fibre content and friction at the supports [10].

Even though the failure mechanisms between both tests might be different, this should not pose a problem to obtain a correlation between both tests. In fact, several codes and studies from the literature propose correlations between the results of test

© RILEM 2022
P. Serna et al. (Eds.): BEFIB 2021, RILEM Bookseries 36, pp. 601–611, 2022.
https://doi.org/10.1007/978-3-030-83719-8_52

methods with completely different cracking mechanisms [9, 10]. Using this idea, Carmona and Molins [10, 11] proposed a code-type expression to correlate the energy absorption capacity determined by testing square panel with the dissipated energy in cylinders subjected to double punching test or Barcelona test.

Considering this result, the 3PB test and the square panel test are correlated in terms of energy, which is also related to the residual strengths used by MC – 2010 to classify the fibre reinforced concretes and in its constitutive equations. However, in standardized 3PB test there is no definition of toughness at all. Then, flexural toughness and parameters which relate it to residual strengths have been defined, establishing experimental relationships between them.

2   Flexural Toughness and Equivalent Strength

According to EN 14651 standard [2], the characterization of the properties of the FRCs is carried out by testing a beam with a 25 mm deep central notch, loaded at the midspan (3PB), as shown in Fig. 1. The test should be carried out in a closed loop control system under crack mouth opening displacement ($CMOD$) control with a high stiffness loading system to avoid instability in the transition between the pre- and post-cracking stage. During the test, the load and the $CMOD$ must be recorded continuously. Using the $P - CMOD$ response obtained during test, the EN 14651 standard defines the residual strengths, $f_{R,j}$, as:

$$f_{R,j} = \frac{3 \, F_j \, l}{2 \, b \, h_{sp}^2} \tag{1}$$

where l, b and h_{sp} are the span, width and height of the specimen in the notched section, respectively, shown in Fig. 1; and F_j, with $j = 1, 2, 3, 4$, is load at $CMOD_1 = 0.5$ mm, $CMOD_2 = 1.5$ mm, $CMOD_3 = 2.5$ mm and $CMOD_4 = 3.5$ mm, respectively.

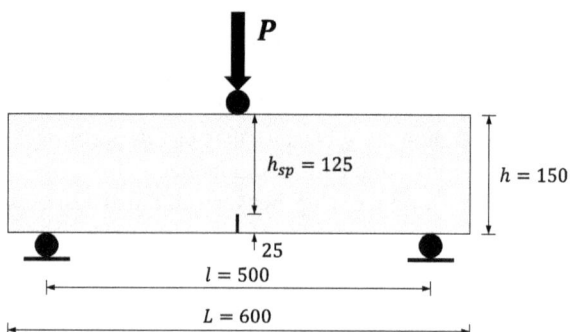

Fig. 1. Test setup as EN – 14651 standard [2] (dimensions in mm).

Based on the flexural toughness defined by ASTM C–1609 standard [12] and considering the linear relationship between the $CMOD$ and the deflection of the beam, Carmona and Molins [13] defined flexural toughness, T_j, as the area under the curve $P - CMOD$ up to certain value of $CMOD_j$, calculated as:

$$T_j = \int_0^j P(CMOD)dCMOD \qquad (2)$$

where $j = 1, 2, 3, 4$, is the area until $CMOD_1 = 0.5$ mm, $CMOD_2 = 1.5$ mm, $CMOD_3 = 2.5$ mm and $CMOD_4 = 3.5$ mm, respectively. The flexural toughness is used in this research to define the equivalent load, $P_{e,j}$, as:

$$P_{e,j} = \frac{T_j}{CMOD_j} \qquad (3)$$

Then, substituting $P_{e,j}$ in Eq. (1), the equivalent strengths, $f_{e,j}$, can be calculated using expression (4):

$$f_{e,j} = \frac{3}{2} \frac{T_j}{CMOD_i} \frac{l}{bh_{sp}^2} \qquad (4)$$

Finally, using the Eqs. (1) and (4), the residual equivalent strength ratio can be defined as:

$$R_{e,j} = \frac{f_{e,j}}{f_{Rj}} \qquad (5)$$

This ratio is used in this research to establish a relationship between residual strengths and flexural toughness.

3 Energy Absorption Capacity

The energy absorption capacity, $E(\delta)$, is defined in the EN 14488 – 5 standard and the EFNARC recommendation, as the area under load – deflection curve until a deflection $\delta = 25$ mm, and can be calculated using the expression (6):

$$E_{25} = \int_0^{25} P(\delta)d\delta \qquad (6)$$

where $P(\delta)$ is load as function of deflection. This test is conducted on a square panel of 600 mm side and 100 mm thick supported on its four edges, which is loaded at the center by a contact surface of 100×100 mm, as can be seen in Fig. 2. Load is applied under deformation control at a rate of midspan deflection of 1.0 mm/min [3] or 1.5 mm/min [4] until a displacement of 30 mm is achieved at the centre point of the panel.

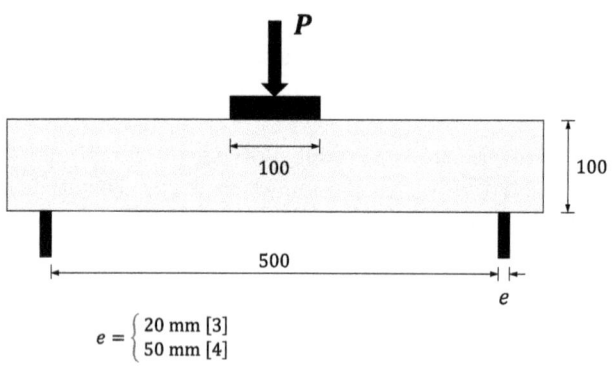

Fig. 2. Square panel test setup (dimension in mm).

4   Experimental Details

4.1   Properties of Tested FRC

The concretes tested in this research were designed by a mixed concrete supplier to be sprayed, and were made using a Chilean pozzolanic cement, Type IP [14], and crushed river sand. The mixes proportions are presented in Table 1. Concretes were reinforced with macro–synthetic fibres, which properties and features are given in Table 2, where l_f is the length of the fibre, d_f the quivalent diameter determined using manufacturer's information, f_{st} the tensile strength and E the Young modulus.

Table 1. Mix properties used in this research.

Material	Dose (kg/m^3)
Cement IP	420
Sand 1	1655
Sand 2	331
Plasticizer admixture	2.10
Superfluidiying admixture	2.10
Silica coloidal	2.94
Total water	156
Water/cement ratio	0.371

Concretes were cast in the Laboratory at Federico Santa Maria Technical University in Valparaiso, Chile, using a conventional paddle mixer of 200 L' capacity. For each FRC 600 × 600 × 100 mm square panels, standardized beams and three compression cylindrical specimens were also cast. Fibre content, volumetric substitution (V_f), slump, compressive strength (f_c) and the number of testing specimens are given in Table 3.

Table 2. Synthetic fibres properties (manufacturer's data).

Designation	l_f (mm)	d_f (mm)	λ_f l_f/c	f_{st} (MPa)	E_f (GPa)	Fibres/kg (Nr)
BC–54	54	0.84	64	640	12	37,000

Table 3. Properties of tested concretes and number of testing specimens.

Concrete	Fibre content (kg/m³)	V_f (%)	Slump (mm)	f_c (MPa)	Nr of specimens	
					Beams	Panels
FRC – 4	4.0	0.44	235	39.5	12	10
FRC – 8	8.0	0.88	200	40.9	12	10
FRC – 12	12.0	1.32	190	42.3	14	10

4.2 Three-Point Bending Tests Results

The 3PB tests were performed on a closed-loop control system of 100 kN capacity, under *CMOD* control according to the standard [2], which was measured using a clip–gauge extensometer with a total range of 5.0 mm, at a rate of 0.05 mm/min until reaching a *CMOD* = 0.1 mm. Then, the rate was increased to 0.2 mm/min until the end of test at a *CMOD* = 4.0 mm. Before testing, a 25 mm deep notch was cut at midspan of each beam. During the tests, the load and *CMOD* were recorded continuously at a rate of 5 data/s.

The mean curves obtained with each tested concrete are shown in Fig. 3. In these curves, different behaviors in the post-cracking regime can be observed, which vary from softening to hardening, depending on the fiber content. The concrete with 4 kg/m³ of fibers exhibits a strong drop after reaching the peak load, then the curves remain almost constant with low load capacity until the end of tests. Concrete reinforced with 8 kg/m³ of fibers or more, the deep of drop after first peak decreases on fibers content. After that, the responses show hardening, reaching a second peak at a *CMOD* = 3 mm, approximately. The second peak load of concrete reinforced with 12 kg/m³ is more than 30% higher than the first peak load.

The mean values of residual strengths and flexural toughness calculated for each tested concrete using the Eqs. (2) y (3), for $CMOD_1$ and $CMOD_3$, are presented in Table 4.

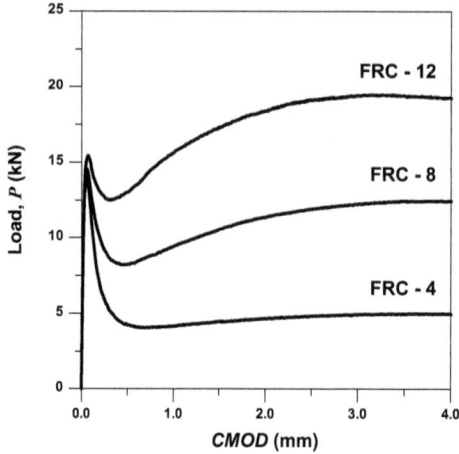

Fig. 3. Mean P-CMOD curves obtained with 3PB tests.

Table 4. Residual strengths, flexural toughness, and energy absorption capacity of tested FRC.

Concrete	Fiber content (kg/m^3)	$f_{R,j}$ (MPa)		T_j (J)		E_{25} (J)
		$f_{R,1}$	$f_{R,3}$	T_1	T_3	
BC–54–4.0	4	1.45	1.72	3.70	13.48	804
BC–54–8.0	8	2.66	3.87	5.01	26.07	1064
BC–54–12.0	12	3.82	5.36	5.94	35.83	1252

4.3   Square Panel Test Results

The square panel tests were also conducted in a hydraulic closed-loop control system of 100 kN capacity, under deformation control. In these tests, the midspan deflection was measured with a LVDT of 50 mm range, placed in the center bottom face of the specimen. During tests, load and deflection were recorded continually at a rate of 3 data/s and were used to obtain the $P - \delta$ responses plotted in Fig. 4. The energy absorption capacity, $E(\delta)$, of each FRC tested was calculated as the area under $P - \delta$ curve using Eq. (6) until a midspan deflection $\delta = 25$ mm. The curves $E(\delta) - \delta$ are shown in the Fig. 4 and the values of the energy absorption capacity at a $\delta = 25$ mm, E_{25}, are also given in Table 4.

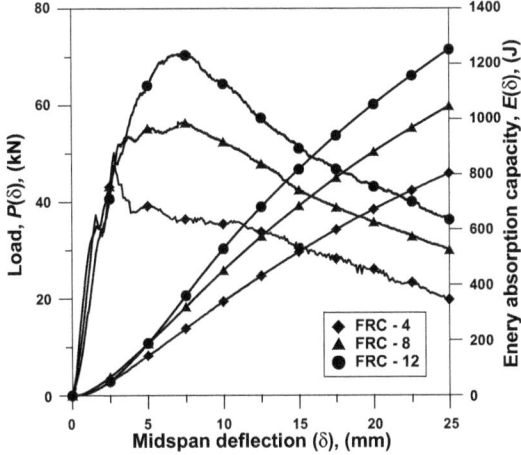

Fig. 4. Mean square panel tests results.

5 Relationship Between f_{Rj} and E_{25}

Considering that the MC - 2010 uses the values of $f_{R,1}$ and $f_{R,3}$ in the constitutive equations that describe the post-cracking behavior of FRCs, this research will be limited only to establishing correlations that allow establishing these values from the energy absorption capacity.

Carmona and Molins [13] obtained an experimental linear relationship between the energy absorption capacity and the flexural toughness of the FRCs, therefore, if a relationship is established between the residual strengths and the flexural toughness, then, it will be possible to relate the capacity absorption of energy with residual strengths.

Using the results of the bending tests, $f_{e,j}$ and $R_{e,j}$ are calculated with the Eq. (5) and Eq. (6), obtaining the values given in Table 5. As can be seen in Fig. 5a, the highest values of $R_{e,j}$ are reached for the lowest fiber contents and for the small crack openings ($CMOD_1 = 0.5$ mm), where the toughness is strongly influenced by the first peak loads of the $P - CMOD$ response, which is governed by the tensile strength of the concrete and, then, they exhibit a strong drop in load (Fig. 4), until the fibers control the crack opening. Therefore, the equivalent strengths $f_{e,1}$ are considerably higher than the residual strength $f_{R,1}$. The values of $R_{e,j}$ decrease when increasing fiber content, which show lower drops in load after the first peak is reached, reaching values of $R_{e,1} < 1$ in concrete reinforced with high fiber contents, which exhibit $P - CMOD$ response with hardening, which causes high values of residual strengths $f_{R,1}$, at low levels of crack opening.

Table 5. Values of $f_{e,j}$ and $R_{e,j}$ obtained using experimental results.

Concrete	Fiber content (kg/m³)	V_f (%)	$f_{e,j}$(MPa)		$R_{e,j}$	
			$j = 1$	$j = 3$	$j = 1$	$j = 3$
FRC – 4	4	0.44	2.37	1.73	1.633	1.003
FRC – 8	8	0.88	3.21	3.34	1.205	0.862
FRC – 12	12	1.32	3.80	4.59	0.995	0.856

From the results presented in Fig. 5a, it can be seen that $R_{e,j}$ depends on the fiber volumetric substitution, $V_{f,i}$, where $i = 4$, 8 or 12 kg/m³, and on the $CMOD_j$. Then, normalizing $R_{e,j}$ by $V_{f,i}$, the curves shown in Fig. 5b are obtained, which are adjusted to expressions of the form:

$$\left(\frac{R_{e,j}}{V_{f,i}}\right) = a \cdot V_f^b \tag{7}$$

Where the experimental parameters a and b depend on $V_{f,i}$ and $CMOD_j$. Using the experimental results, by means of a non–linear regression analysis, carried out with the XLSTAT© application for Excel ©, these parameters have been determined for each value of j considered in this research, which are given in Table 6. As can be seen in Fig. 5b, Eq. (7) fits well with the experimental data, with values of the correlation coefficient, $r^2 \geq 0.998$.

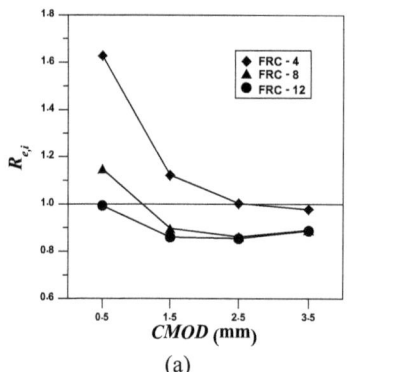

(a)

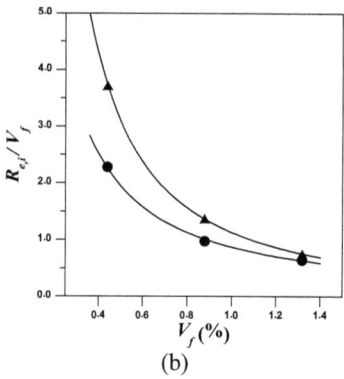

(b)

Fig. 5. (a) $R_{e,j} - CMOD$, and (b) $\frac{R_{e,j}}{V_{f,i}} - V_f$ relationships.

Table 6. Experimental parameters of Eq. (7) as function of $CMOD_j$.

j	$CMOD_j$ (mm)	a	b	r^2
1	0.5	1.131	−1.449	1.000
3	2.5	0.874	−1.153	0.998

On the other hand, Carmona and Molins [13] proved that there is a linear relationship between E_{25} and T_j, as can be seen in Fig. 6, so that:

$$T_j = m_j \cdot E_{25} + c_j \tag{8}$$

By means of a linear regression analysis, the parameters m_j and c_j of Eq. (8) have been determined using the experimental results, which are given in Table 7, where slope m_3 is 10 times higher than m_1 showing a proportionality between the energy absorption capacity, E_{25}, and the toughness increasing when $CMOD$ increases.

Table 7. Values of experimental parameters of Eq. (8).

j	$CMOD_j$ (mm)	m_j	c_j	r^2
1	0.5	0.0005	−0.329	0.9999
3	2.5	0.0498	−26.668	0.9996

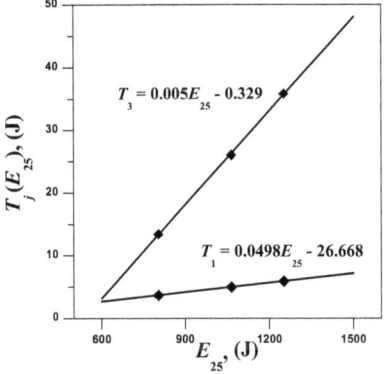

Fig. 6. Relationships between energy absorption capacity and flexural toughness for each j–value studied.

Using the correlations developed above, the residual strengths can be estimated from the energy absorption capacity. In this way, once $E_{25,j}$ is known, the T_i can be determined with Eq. (8) and the values of m_j and c_j given in Table 7. By replacing T_i in Eq. (4), the $f_{e,j}(T_j)$ can be estimated. On the other hand, for the corresponding $V_{f,i}$, the $R_{e,j}$ is determined with Eq. (7) and the values of a and b from Table 6. Finally, with $f_{e,j}(T_i)$ and $R_{e,j}$, the residual strengths $f_{R,j}(E_{25}, R_{e,j})$ can be computed with the following Eq. (9):

$$f_{Rj}(E_{25}, V_{f,i}) = \frac{f_{e,j}}{R_{e,j}} \tag{9}$$

Following the procedure described above, the residual strengths have been estimated reaching differences with the experimental values less than −2.33%, as can be seen in Table 8.

Table 8. Residual strengths estimated using the proposed procedure.

$V_{f,i}$ (%)	E_{25} (J)	$T_j(E_{25})$ (J) Eq. (8)		$R_{e,j}$ Eq. (7)$\times V_{f,i}$		$f_{Rj}(E_{25}, V_{f,i})$ (MPa)		Difference (%)	
		$j=1$	$j=3$	$j=1$	$j=3$	$j=1$	$j=3$	$j=1$	$j=3$
0.44	804	3.69	13.37	1.635	0.991	1.44	1.73	−0.36	0.41
0.88	1064	4.99	26.32	1.198	0.891	2.67	3.78	0.25	−2.33
1.32	1252	5.93	35.68	0.998	0.838	3.80	5.45	−0.48	1.72

6 Conclusions

The most relevant conclusions of this research are the following:

- Using $P - CMOD$ response obtained with Three-point bending test on a notched beam, the flexural toughness, equivalent strengths, and equivalent strengths ratio have been defined.
- Using experimental results, a non – linear relationship between residual strengths and the equivalent strengths ratio have been established.
- A linear relationship between energy absorption capacity of panels and flexural toughness of 3PB beams has been determined.
- The residual strengths can be estimated from energy absorption capacity by means of the proposed correlations with a difference less than 3%.

Acknowlegements. This research was supported by Fondecyt Project "Use of the Generalized Barcelona Test for Characterization and Quality Control of Fibre Reinforced Shotcretes in Underground Mining Works", N°1150881.

References

1. CEB-FIP: Model code – first complete draft. FIB Bull. **55**, 1–318 (2010)
2. European Committee for Standardization: EN 14651: test method for metallic fibre concrete–measuring the flexural tensile strength (Limit of Proportionality (LOP), residual) (2007)
3. European Committee for Standardization: EN 14488-5:2006 Testing sprayed concrete. Determination of energy absorption capacity of fibre reinforced slab specimens (2006)
4. European Federation of National Associations of Specialist Contractors and Material Suppliers for the Construction Industry: European Specification for Sprayed Concrete (1996)
5. Alberti, M.G., Enfedaque, A., Gálvez, J.C.: On the mechanical properties and fracture behavior of polyolefinfiber-reinforced self-compacting concrete. Constr. Build. Mater. **55**, 274–288 (2014)
6. Conforti, A., Minelli, F., Plizzari, G.A., Tiberti, G.: Comparing test methods for the mechanical characterization of fiber reinforced concrete. Struct. Concr. **19**, 656–669 (2018). https://doi.org/10.1002/suco.201700057
7. Carmona, S., Molins, C., Aguado, A.: Correlation between bending test and Barcelona tests to determine FRC properties. Constr. Build. Mater. **181**, 673–686 (2018)

8. Carmona, S., Molins, C.: Use of BCN test for controlling tension capacity of fibre reinforced shotcrete in mining works. Constr. Build. Mater. **198**, 399–410 (2019)
9. Galeote, E., Blanco, A., Cavalaro, S.H.P., De la Fuente, A.: Correlation between the Barcelona test and the bending test in fibre reinforced concrete. Constr. Build. Mater. **152**, 529–538 (2017)
10. Carmona, S., Molins, C., García, S.: Application of Barcelona test for controlling energy absorption capacity of FRS in underground mining works. Constr. Build. Mater. **246**, 118458 (2020). https://doi.org/10.1016/j.conbuildmat.2020.118458
11. Carmona, S., Molins, C.: Application of BCN test for controlling fibre reinforced shotcrete in tunnelling works in Chile. In: IOP Conference Series: Materials Science and Engineering, vol. 246, p. 012010 (2017)
12. ASTM: C1609/C1609M-19, Standard Test Method for Flexural Performance of Fibre-Reinforced Concrete (Using Beam With Third-Point Loading). ASTM International, West Conshohocken (2019). www.astm.org.
13. Carmona, S., Molins, C.: Equivalence between flexural toughness and energy absorption capacity of FRC. In: BEFIB 2020, RILEM Bookseries, vol. 30, pp. 253–261 (2021)
14. ASTM: C595/C595M-18, Standard Specification for Blended Hydraulic Cements. ASTM International, West Conshohocken, PA (2018). www.astm.org

Case Studies: Structural and Industrial Applications

Use of Steel-Fibre Reinforced Concrete to Extend Service Life of Temporary Safety Concrete Barriers

Clélia Desmettre and Jean-Philippe Charron[(⊠)]

Department of Civil, Geological and Mining Engineering,
Polytechnique Montréal, Montréal, Canada
jean-philippe.charron@polymtl.ca

Abstract. Temporary concrete safety barriers used by the Quebec Ministry of Transportation (QMT) have very short service life, mainly due to repetitive damages resulting from low velocity impacts occurring during their handling procedure. Utilization of steel fibre reinforced concrete (SFRC) in barrier represents an economical alternative to delay crack initiation and propagation. The quasi-static and pseudo-dynamic flexural and shear behaviours of concretes containing 0, 0.5 and 1% in volume of steel macrofibres were evaluated to select an appropriate SFRC to enhance the barriers service life. The maximal flexural and shear quasi-static strengths of SFRC increased by more than 40% in comparison to normal strength concrete (NSC) and the flexural and shear failure energies exploited in service condition also increased by more than 170%. Increasing the loading rate enhanced the maximal strengths by more than 12% for all the tested concrete, while keeping similar strength improvement between SFRC and NSC. Satisfactory results obtained at materials scale convinced QMT to complete a pilot application. 50 SFRC temporary safety barriers containing 0.5% of steel fibres were produced in a precast plant and installed on construction sites to establish their performance (cracking, spalling) during 1.5 years in comparison to 50 NSC barriers. Toughness of SFRC led to significant decrease of the damages observed in-field in the SFRC barriers in comparison to the NSC barriers.

Keywords: Steel Fibre Reinforced Concretes (SFRC) · Flexural strength · Shear strength · Static loading · Dynamic loading · Concrete barriers · Field performance

1 Introduction

The Quebec Ministry of Transportation (QMT) uses temporary concrete safety barriers to delineate construction zones of traffic areas. These barriers have short service life (around 3–4 years), mainly due to important damages resulting from repetitive low velocity impacts of the handling equipment. These impacts are mainly closed to lifting zones and at connection zones with other barriers (Fig. 1).

Use of fibres into concrete improves the cracking control and thus the flexural ductility. Hooked-end steel fibre reinforced concrete with fibre dosages superior to

© RILEM 2022
P. Serna et al. (Eds.): BEFIB 2021, RILEM Bookseries 36, pp. 615–627, 2022.
https://doi.org/10.1007/978-3-030-83719-8_53

around 0.5% in volume can also present a hardening phase after the first cracking and before the softening phase [1, 2], which increases the maximal strength. The increase of the fibre length and the fibre aspect ratio also increases the flexural behaviour (strength and ductility) [1–3]. Under shear loading, presence of fibres provides to FRC additional pull-out and dowel action in the shear crack and leads to better performance (higher shear stress and ductility) [4]. For steel fibre dosages superior to around 0.5% in volume for NSC and to around 0.25–0.3% for high-performance concrete (HPC), the maximal shear strength increases almost linearly with the fibre dosage [5–8]. The shear ductility also increases with the fibre dosage [5, 9]. Increasing the loading rate enhances the compressive and tensile strengths of concrete [10, 11] and, as a consequence, also increases the flexural and shear strengths. The strength improvement with higher loading rates seems quite similar regardless the concrete type or whether fibres are present or not [10, 12]. However, less crack branching was sometimes observed [13, 14] on steel fibre reinforced concrete (SFRC) prisms when increasing the loading rate.

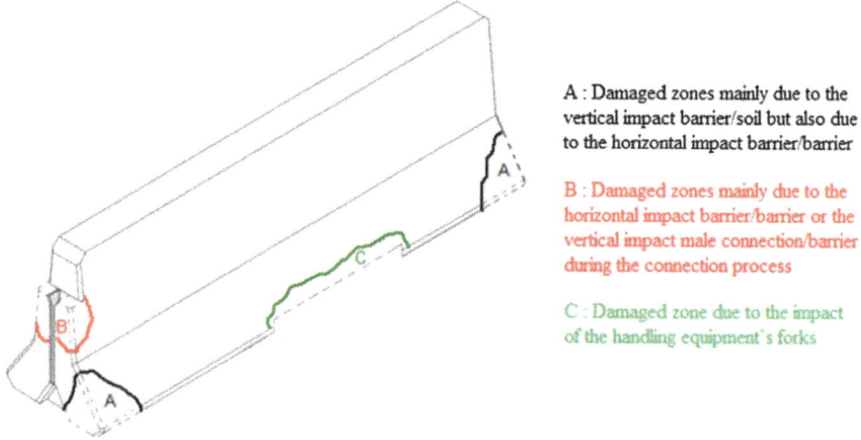

A : Damaged zones mainly due to the vertical impact barrier/soil but also due to the horizontal impact barrier/barrier

B : Damaged zones mainly due to the horizontal impact barrier/barrier or the vertical impact male connection/barrier during the connection process

C : Damaged zone due to the impact of the handling equipment's forks

Fig. 1. Damaged zones caused by the concrete safety barriers handling.

Mechanical benefits of including fibres into concrete led Polytechnique Montreal to propose at QMT to induce fibre reinforcement in temporary safety concrete barriers to increase their toughness under impact loading and extend their service life. The research project was divided in three phases. Phase A consisted in evaluating, in laboratory, the quasi-static and pseudo-dynamic flexural and shear behaviours of fibre reinforced concretes prisms. Phase B aimed to validate the industrial production of a FRC and to produce FRC safety barriers to establish their in-field performance in comparison to the conventional NSC barriers. Phase C will consist in submitted full-scale FRC and NSC barriers to laboratory impact tests representative of handling operations to better quantify the gain brought by SFRC in such elements.

2   Methodology

2.1   Experimental Program

Phase A of this research project consisted in developing in laboratory 35 MPa air-entrained NSC and fibre reinforced concretes with 0.5 and 1% in volume of synthetic and steel fibres (SyFRC and SFRC respectively). The five materials were submitted to quasi-static flexural and shear tests. Strength and ductility improvements provided by the fibres were evaluated for each testing condition. SFRC-0.5% was judged optimal for the targeted application and selected to be submitted to pseudo-dynamic flexural and shear tests. This paper only presents the results obtained with NSC and SFRC (SFRC-0.5% and SFRC-1%). Detailed information about the behaviour of the SyFRC is available in [15].

In Phase B, the FRC with 0.5% vol. of steel macrofibres (SFRC-0.5%-barrier) was produced with industrial equipment at a precast plant. Its quasi-static flexural and shear behaviours were evaluated and compared to the behaviour of the SFRC-0.5% developed in Phase A. The industrial SFRC mix was then used to build 50 SFRC temporary concrete safety barriers of 4-m length. These barriers were submitted to field conditions and the damage evolution, mainly due to handling operations, was followed during 1.5 years and compared to the damage evolution observed on NSC barriers built at the same period.

2.2   Material Properties

Industrial production of SFRC in Phase B was submitted to a public tender, specific material suppliers could not be specified to obtain the same as in Phase A. The air-entrained SFRC were based on V-P QMT concrete mixtures of 35 MPa generally used for construction of temporary concrete safety barriers in Quebec [16]. Compositions of the concretes are summarized in Table 1. The binary cement used in the project is a Canadian GUb-SF type composed of 91% of general use (GU) type cement and 9% of silica fume. Minimal fibre aspect ratio (length/diameter, l_f/d_f) of 65 was specify in the tender (Phase B) instead of the ratio of 80 used during the laboratory development (Phase A) to allow more suppliers to participate. The physical and mechanical properties of the selected fibres during Phase A (SFRC-0.5% and 1%) and Phase B (SFRC-0.5%-barrier) are presented in Table 2.

Table 1. Concrete mix designs.

Concrete	NSC-0%	SFRC-0.5%	SFRC-0.5%-barrier	SFRC-1%
GUb-SF cement (kg/m^3)	460	460	458	460
Water (kg/m^3)	206	205	202	203
Sand (kg/m^3)	868	868	783	867
Coarse aggregate (kg/m^3)	789	789	891	761
Steel fibres (kg/m^3)	0	39	39	78
Superplasticizers (lit/m^3)	1.15	2.8	0.95	5.2
Air entrainer (lit/m^3)	0.115	1.4	0.12	0.069
Water/binder ratio	0.45	0.45	0.44	0.45

Table 2. Main properties of the steel fibres.

Specimen	Length l_f	l_f/d_f	Elastic modulus	Tensile strength
SFRC-0.5% and 1%	60 mm	80	210 GPa	1050 MPa
SFRC-0.5%-barrier	60 mm	65	200 GPa	1500 MPa

2.3   Experimental Procedures

The fresh and hardened states properties were determined using the standards presented in Table 3. For the flexural tests, ASTM C1018 was preferred to ASTM C1609 as the set-up is more common in Canadian private laboratories. For the shear tests, the set-up was inspired from the standard JSCE SF6. However, as the JSCE SF6 set-up causes rotation of the ends of the specimen as well as cracking at the supports, the set-up modifications proposed in [8,9] were adopted to achieve consistent results. Sketches of the flexural and shear tests set-up are illustrated in Fig. 2a and 1b respectively. The flexural and shear prisms were cast in two layers following the casting procedure proposed in RILEM EN14651.

Flexural and shear quasi-static tests were performed on all the studied concretes (NSC-0%, SFRC-0.5%, SFRC-0.5%-barrier and SFRC-1%), whereas the pseudo-dynamic flexural and shear tests were only performed on NSC-0% and SFRC-0.5%. Three specimens were tested for each condition.

Table 3. Standards for determination of fresh and hardened states properties.

	Properties	Standards	Specimens
Fresh state	Temperature	ASTM C1064	As indicated in standard
	Mass density	ASTM C138	As indicated in standard
	Air content	ASTM C231	As indicated in standard
	Slump	ASTM C143	As indicated in standard
Hardened state	Compressive strength	ASTM C39	Cylinders (d = 100 mm, h = 200 mm)
	Flexural strength	ASTM C1018	Prisms 150 mm × 150 mm × 600 mm
	Shear strength	JSCE SF6 modified	Prisms 150 mm × 150 mm × 600 mm

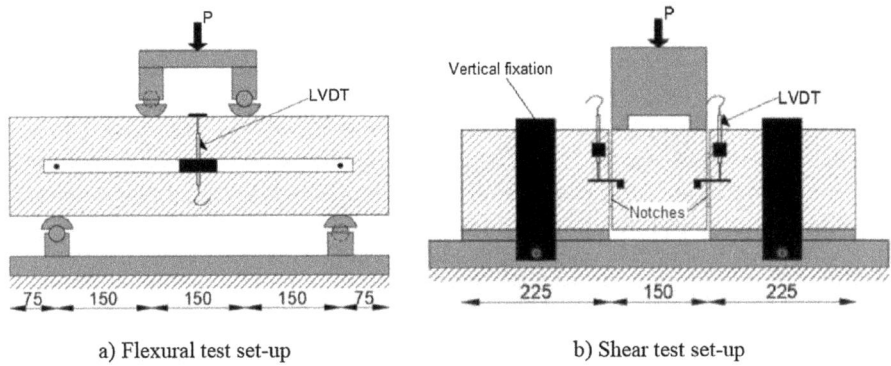

a) Flexural test set-up   b) Shear test set-up

Fig. 2. Sketch of the flexural and shear set-up.

Table 4 summarizes the applied flexural and shear quasi-static loading rates. For the flexural tests, the quasi-static loading rate for NSC-0% was based on ASTM C78 and the same initial loading rate was applied to SFRC for comparison purpose. Then the loading rate was accelerated for the SFRC specimens when reaching 90% of the ultimate force in the post-peak region to reach loading rate proposed in ASTM C1018 and C1609 and perform tests in a reasonable time. For the shear tests, the initial quasi-static loading rate followed the rate range of 0.06 to 0.1 MPa/s specified in JSCE SF6 and also mentioned in other references using the same test set-up [8, 9]. The loading rate was then accelerated when reaching 75% of the ultimate force in the post-peak region to be able to complete the test in a reasonable duration.

The pseudo-dynamic loading rates (300 mm/min and 200 mm/min for the flexural and shear tests respectively) were the same for all concretes. They were chosen to have a significant rate increase while considering the limitations of the load frame and control system. Pseudo-dynamic tests involved strain rate around 4.9×10^{-4} s^{-1}, which corresponds to low velocity vehicle impacts.

Table 4. Quasi-static loading rates.

Flexural test	Specimen	Until 90% of $F_{ult\ post-peak}$	From 90% to 70% of $F_{ult\ post-peak}$	From 70% of $F_{ult\ post-peak}$ to the end of the test
	NSC	0.01 mm/min	0.01 mm/min	0.01 mm/min
	SFRC	0.01 mm/min	0.05 mm/min	1 mm/min
Shear test	Specimen	Until 75% of $F_{ult\ post-peak}$	From 75% to 30% of $F_{ult\ post-peak}$	From 30% of $F_{ult\ post-peak}$ to the end of the test
	NSC/SFRC	0.05 mm/min	0.5 mm/min	1 mm/min

2.4   Method for On-Site Evaluation of Full-Scale Concrete Safety Barriers

50 temporary concrete safety barriers were industrially produced with the mix SFRC-0.5%-barrier. At the same time, 50 standard barriers built with NSC were also produced. A testing section composed of 45 SFRC and 45 NSC barriers (total length of 90 × 4 m = 360 m) was installed in real conditions around construction sites to establish the damage evolution in both kind of temporary concrete barriers when submitted to normal handling operations. The 5 remaining SFRC barriers were kept for Phase C of the project involving impact tests in laboratory.

The commissioning of the testing section was in August 2017. Then, all the barriers of the testing section were subjected to 13 handling operations (loading and unloading on the delivery truck, installations, transversal displacement, etc.) including 5 operations that consisted in connecting the male and female connections of adjacent barriers. Global and detailed inspections of barriers were completed in November and December 2018 respectively. The global inspection consisted in observing one side of 45 barriers of each type; the other side being exposed to vehicle circulation. This global inspection allowed comparison of proportion of damaged barriers of each type. 10 typical SFRC and NSC damaged barriers of the testing section were then selected for the detailed inspection that consisted in thoroughly analyzing both sides of the barriers to compare their damage characteristics.

During the inspections, the areas of concrete loss (spalling) were reported and calculated. The damages were classified in 5 categories associated to concrete loss area observed on temporary concrete barriers in several construction sites of the region of Montreal (Quebec,Canada) (Table 5). Category 1 corresponds to very minor damage resulting of concrete loss of less than 10 000 mm^2, while Category 5 represents the largest observed damage with concrete loss over 200 000 mm^2.

Table 5. Damages categories with examples.

Category 1 0-10000 mm^2	Category 2 10000 - 40000 mm^2	Category 3 40000 - 100000 mm^2	Category 4 100000 - 200000 mm^2	Category 5 200000 - 400000 mm^2
550 mm^2 9750 mm^2	11500 mm^2 36500 mm^2	42000 mm^2 64500 mm^2	138000 mm^2	221000 mm^2

3 Experimental Results and Discussion

3.1 Fresh State and Compressive Properties

The tested concretes had a slump of 150 ± 30 mm and an air content of 5 to 8%, as required for V-P QMT concrete mixtures used in temporary concrete safety barriers. Furthermore, all concretes presented compressive strengths at 28 days superior or really close to the minimum requirement of 35 MPa (Table 6).

Table 6. Compressive properties (at 28 days) of the studied concretes.

Properties	NSC-0%	SFRC-0.5%	SFRC-0.5%-barrier	SFRC-1%
Compressive strength, f_c (MPa)	42	41.6	34.6	54.1
Young modulus, E_c (GPa)	32.8	32.2	$-^*$	30.8

* *Young's modulus was not measured for the industrial mixture*

3.2 Flexural and Shear Quasi-static Behaviours

Variability of test results on three specimens of a same testing condition was low during the flexural and shear tests, thus the average results are used in Sects. 3.2 and 3.3. All results are available in [15]. The loading rate for quasi-static tests are detailed in Table 4.

Figure 3 illustrates the flexural and shear quasi-static stress-displacement average curves of concretes and Table 7 summarizes the maximal flexural and shear stresses (*MOR* and τ_{max} respectively) as well as the energy up to 1 mm of measured displacement, E_{1mm}. The energy represents the area under the load-displacement curve and the criterion E_{1mm} is presented instead of the total energy because it is more representative of the service conditions where small displacement within the barrier are expected during their handling.

The use of 0.5 and 1% in volume of steel macrofibres in specimens submitted to flexural or shear quasi-static loading led to a hardening phase after the first cracking, which increased their maximal flexural and shear strengths (*MOR* and τ_{max} respectively) in comparison to NSC-0%. The presence of this hardening phase is coherent with literature that often shows flexural and shear hardening phases when the fibre dosage is higher to 0.5–0.6% in volume [1, 3, 5–8, 17]. The *MOR* increased by 40 and 107% for SFRC-0.5% and SFRC-1% respectively, in comparison to NSC-0%. The τ_{max} increased by 101 and 229% for SFRC-0.5% and SFRC-1% respectively, in comparison to NSC-0% (Table 7). As already observed in literature for steel fibre reinforced concrete with a hardening behaviour [5–8], the *MOR* and τ_{max} increased almost linearly with fibre dosage in this range.

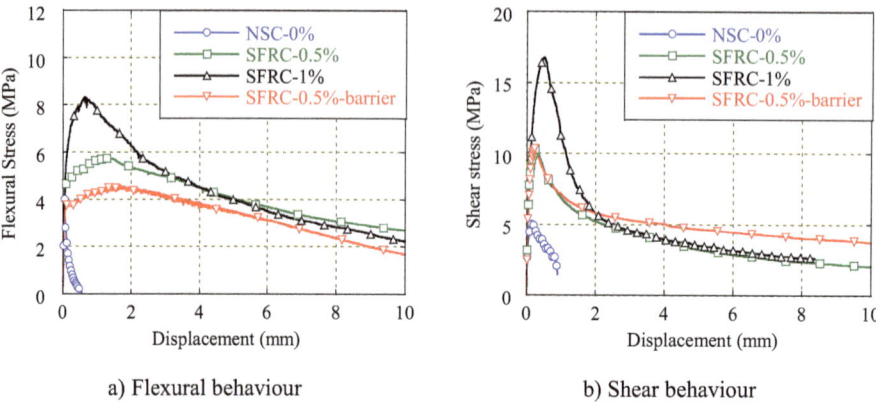

a) Flexural behaviour   b) Shear behaviour

Fig. 3. Average quasi-static flexural and shear stress-displacement curves of the studied concretes.

Table 7. Effect of the fibres on the quasi-static flexural and shear behaviors.

	NSC-0%	SFRC-0.5% (% vs NSC)	SFRC-1% (% vs NSC)
MOR_{Stat}	4.10 MPa	5.75 MPa (+40%)	8.5 MPa (+107%)
$\tau_{max\ stat}$	5.11 MPa	10.29 MPa (+101%)	16.8 MPa (+229%)
$E_{f\ 1mm\ Stat}$	4.54 J	41.5 J (+814%)	57.1 J (+1158%)
$E_{\tau 1mm\ Stat}$	39.6 J	107.0 J (+170%)	173.5 J (+338%)

While SFRC-0.5% and SFRC-1% used the same type of fibres, SFRC-0.5%-barrier included fibres with the same length (60 mm) but a larger diameter (0.92 mm versus 0.75 mm) and thus a lower aspect ratio (65 versus 80, Table 2). Due to this lower aspect ratio, the *MOR* increase of the SFRC-0.5%-barrier (+12% in comparison to NSC-0%) was less important than the laboratory concrete SFRC-0.5% (+40% in comparison to NSC-0%) containing the same fibre dosage. This impact of the aspect ratio on the flexural performance was already shown in literature [1–3]. However, the difference in the aspect ratio did not change the τ_{max} value between these two concretes and the shear behaviour of both concretes (SFRC-0.5%-barrier and SFRC-0.5%) was quite similar. This could be explained by a higher impact of the fibre diameter than the aspect ratio on the shear behaviour. This hypothesis should be more investigated.

Although the SFRC strength improvements were higher under shear loading than under flexural loading, the trend is reversed when considering E_{1mm} with increases of +814 and +1158% under flexural loads and of +170 and +338% under shear loads, for SFRC-0.5% and SFRC-1% respectively in comparison to NSC-0%. The increase of $E_{\tau\ 1mm}$ (shear energy) is quasi-linear with the increase of the fibre dosage.

3.3 Flexural and Shear Pseudo-dynamic Behaviours

Quasi-static flexural and shear loadings showed that the improvements in strength and failure energy obtained with 0.5% of steel macrofibres were adequate. Therefore, only the SFRC-0.5% mix was tested under pseudo-dynamic loading and compared to NSC-0%. The pseudo-dynamic loading rates for flexural and shear tests are 200 mm/s and 300 mm/s, respectively. Figure 4 compares the flexural and shear pseudo-dynamic stress-displacement average curves of both concretes and Table 8 summarizes the differences of *MOR* and τ_{max} as well as E_{1mm} between the pseudo-dynamic and quasi-static loadings.

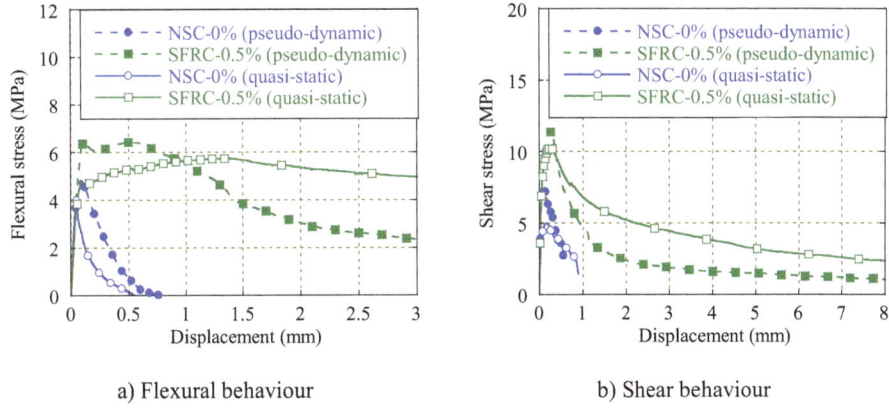

a) Flexural behaviour   b) Shear behaviour

Fig. 4. Average stress-displacement curves of the studied concretes.

Table 8. Comparison of the quasi-static and pseudo-dynamic behaviors.

Concrete	Flexural behavior			Shear behavior		
	MOR_{Stat}	MOR_{Dyn}	$MOR_{Dyn/Stat}$	τ_{stat}	τ_{Dyn}	$\tau_{Dyn/Stat}$
NSC-0%	4.10 MPa	4.91 MPa	+20%	5.11 MPa	7.37 MPa	+44%
SFRC-0.5%	5.75 MPa	6.88 MPa	+20%	10.29 MPa	11.48 MPa	+12%
Concrete	$E_{f\,1mm\,Stat}$	$E_{f\,1mm\,Dyn}$	$E_{f1mm\,Dyn/Stat}$	$E_{\tau1mm\,Stat}$	$E_{\tau1mm\,Dyn}$	$E_{\tau1mm\,Dyn/Stat}$
NSC-0%	4.54 J	10.5 J	+131%	39.6 J	35.3 J	−11%
SFRC-0.5%	41.5 J	44.1 J	+6%	107.0 J	99.8 J	−7%

Increasing the loading rates in the pseudo-dynamic tests led to *MOR* increases of around +20% for both NSC and SFRC and to τ_{max} increases of +44% and +12% for NSC and SFRC respectively (Table 8). The *MOR* increase is coherent with the values of around +25% found in some references [12, 18] for strain rates between 2.9×10^{-7} and 4.9×10^{-4} s^{-1}, which are the rates used in the flexural quasi-static and pseudo-dynamic tests respectively. It was impossible to make adequate comparison with literature for shear values, however τ_{max} increases stays in the same range as the flexural gains. Table 8 also shows quite similar strength improvements when increasing the loading rate whether or not fibres are included in the mix. In this way, quasi-static tests seem to bring good information on the expected strength difference between NSC and SFRC in realistic loading rates.

However, except for NSC-0% submitted to flexural loading, the strength gain with higher loading rate resulted in steeper stress decreases in the post-peak region. This more fragile post-peak behavior under pseudo-dynamic loading was already observed in [13]. Consequently, the flexural and shear E_{1mm} were quite similar or a little inferior ($\pm 10\%$) to the corresponding quasi-static value. The failure energy loss is more marked at higher displacements (Fig. 4).

3.4   On-Site Evaluation of Full-Scale Temporary Safety Concrete Barrier

The aim of the on-site evaluation of the SFRC temporary safety concrete barrier was to evaluate how much the strength and failure energy improvement obtained at the material scale could improve the performance of the full-scale concrete barrier when submitted to low velocity impact during handling operations.

Two inspections of the testing section took place sixteen months after commissioning of the testing section, a global one on 45 barriers of each type (NSC and SFRC) and a detailed one on 10 representative (typical) barriers of each type. During the 16-month period, 13 handling operations of the barrier occurred, including 5 operations that consisted in connecting the male and female connections of adjacent barriers. The global inspection showed that damage progress was particularly visible on NSC barriers. When excluding the very minor damages, this inspection indicated that 44% of the NSC barriers presented concrete loss in comparison to 11% of the SFRC barriers, which represents a decrease of 75% of the damaged barriers when using SFRC instead of NSC. The concrete loss were mainly located in the lifting area, at barrier inferior corners or at the top of the male and female connections.

The detailed inspection of 10 barriers of each type allowed to better quantify the differences between the damaged SFRC and NSC barriers. Figure 5 summarizes the concrete loss (total number, number per damaged areas categories and average concrete loss area) found in the 10 NSC and FRC inspected barriers. Concrete loss on the damaged SFRC barriers were less numerous and of smaller areas than on the damaged NSC barriers. Total number of damages on the 10 inspected NSC and SFRC barriers was 49 and 26 respectively, thus SFRC allowed a decrease of 47%. Moreover, the average area of concrete loss was reduced with SFRC. The damaged NSC barriers presented concrete loss of Categories 1 to 5 (Fig. 5) with an average area of 20 700 mm^2, whereas the damaged SFRC barriers presented concrete loss of Categories 1 to 3 with an average area of 12 150 mm^2, which represents a decrease of 41% of the average damaged area with SFRC.

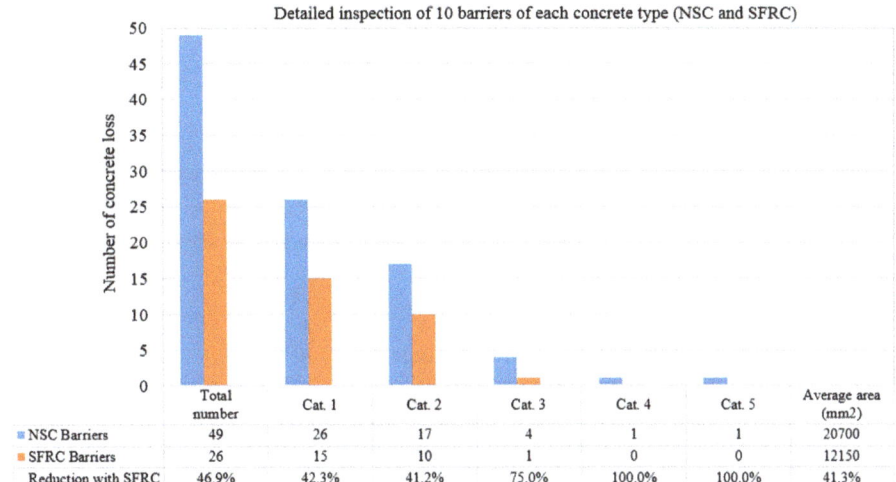

Fig. 5. Summarize of the concrete loss observed on 10 barriers of each type (NSC and SFRC)

4   Conclusions

Temporary concrete safety barriers used by the Quebec Ministry of Transportation (QMT) have very short service life (around 3–4 years), mainly due to important damages resulting from low velocity impacts occurring during their handling procedure. This project aimed to develop fibre reinforced concrete (FRC) to improve the barriers mechanical performance and thus their service life when submitting to handling operations. FRC containing 0.5 and 1% in volume of steel or synthetic macrofibres were first developed in laboratory where their quasi-static and pseudo-dynamic flexural and shear behaviours were evaluated. This paper only presented the results for the FRC with steel macrofibres (SFRC). Following the FRC laboratory development, SFRC with 0.5% vol. of fibres was selected for an industrial production of SFRC temporary concrete safety barriers. The produced barriers were put into service and followed during around 1.5 years to compare their damage evolution to conventional temporary concrete safety barriers made of normal-strength concrete (NSC). Based on the results of this project, the following conclusions can be drawn:

- The inclusion of 0.5 and 1% vol. of hooked-end steel macrofibres with an aspect ratio (length/diameter) of 80 in the laboratory mixes increased the maximal flexural strength (MOR) by 40 and 107% respectively and the maximal shear strength (τ_{max}) by 101 and 229% respectively. The flexural and shear energies for a displacement of 1 mm (representative of service condition) were also increased for flexural gains of 814 and 1158% and shear gains of 170 and 338%, for SFRC-0.5% and SFRC-1% respectively.
- Due to the lower aspect ratio (65) of its fibres, the industrially produced SFRC-0.5% (SFRC-0.5%-barrier) presented smaller MOR gains (+12%) than the laboratory produced SFRC-0.5%.

- Increasing the loading rate enhanced the *MOR* by around 20% for both NSC and SFRC-0.5% and τ_{max} by 44% and 12% for NSC and SFRC-0.5% respectively. The presence of fibres did not change significantly the strength gain when increasing the loading rate. In return for the strength gain observed when increasing the loading rate, the concrete ductility decreased for high displacements.
- The SFRC-0.5% temporary concrete safety barriers used in field during around 1.5 years performed better than the NSC barriers. The number of damaged barriers decreased by 75% when using SFRC and these damaged barriers presented 47% less damaged zones and with areas 41% inferior.

Phase C of the project will possibly take place to submit 5 SFRC and 5 NSC barriers, kept for this purpose, to laboratory impact tests to quantify more precisely the damage evolution differences between these two types of barriers.

Acknowledgements. This research project was financially supported by Quebec Ministry of Transportation and the technical advices of Mr. Desmarchais, Mrs. Baljic and Mrs. Durand are acknowledged. Authors would like to acknowledge CRH Canada, Euclid Canada and Bekaert for graciously providing materials and the technical staff of Polytechnique Montreal for their contribution to this project.

References

1. De Montaignac, R., Massicotte, B., Charron, J.-P., Nour, A.: Design of SFRC structural elements: post-cracking tensile strength measurement. Mater. Struct. **45**, 609–622 (2012)
2. Soutsos, M., Le, T., Lampropoulos, A.: Flexural performance of fibre reinforced concrete made with steel and synthetic fibres. Constr. Build. Mater. **36**, 704–710 (2012)
3. Balaguru, P., Narahari, R., Patel, M.: Flexural toughness of steel fiber reinforced concrete, ACI Mater. J. **89**(6) (1992)
4. Barragan, B., Gettu, R., Agullo, L., Zerbino, R.: Shear failure of steel fiber-reinforced concrete based on push-off tests. ACI Mater. J. **103**(4), 205 (2006)
5. Boulekbache, B., Hamrat, M., Chemrouk, M., Amziane, S.: Influence of yield stress and compressive strength on direct shear behaviour of steel fibre-reinforced concrete. Constr. Build. Mater. **27**, 6–14 (2012)
6. Khaloo, A.R., Kim, N.: Influence of concrete and fiber characteristics on behavior of steel fiber reinforced concrete under direct shear. ACI Mater. J. **94**(6), 592-601 (1997)
7. Khanlou, A., MacRae, G., Scott, A., Hicks, S., Clifton, G.: Shear performance of steel fibre-reinforced concrete. In: Australasian Structural Engineering Conference 2012: The past, present and future of Structural Engineering, p. 400 (2012)
8. Majdzadeh, F., Soleimani, S.M., Banthia, N.: Shear strength of reinforced concrete beams with a fiber concrete matrix. Can. J. Civ. Eng. **33**, 726–734 (2006)
9. Mirsayah, A.A., Banthia, N.: Shear strength of steel fiber-reinforced concrete. ACI Mater. J. **99**(5), 473-479 (2002)
10. Pająk, M.: The influence of the strain rate on the strength of concrete taking into account the experimental techniques. Arch. Civ. Eng. Environ. **3**, 77–86 (2011)
11. Tedesco, J.W., Ross, C.A.: Experimental and numerical analysis of high strain rate splitting-tensile tests. ACI Mater. J. **90**(2), 162-169 (1993)
12. Suaris, W., Shah, S.P.: Strain-rate effects in fibre-reinforced concrete subjected to impact and impulsive loading. Composites **13**, 153–159 (1982)

13. Naaman, A.E., Gopalaratnam, V.: Impact properties of steel fibre reinforced concrete in bending. Int. J. Cem. Compos. Lightweight Concrete **5**, 225–233 (1983)
14. Zhang, X., Elazim, A.A., Ruiz, G., Yu, R.: Fracture behaviour of steel fibre-reinforced concrete at a wide range of loading rates. Int. J. Impact Eng **71**, 89–96 (2014)
15. Charron, J.-P., Desmettre, C., Androuët, C.: Flexural and shear behaviors of steel and synthetic fiber reinforced concretes under quasi-static and pseudo-dynamic loadings. Constr. Build. Mater. **238**, 117659 (2020)
16. Norme du ministère des Transports du Québec (MTQ) [Tome VIII : Dispositif de retenue] (2015)
17. Buratti, N., Mazzotti, C., Savoia, M.: Post-cracking behaviour of steel and macro-synthetic fibre-reinforced concretes. Constr. Build. Mater. **25**, 2713–2722 (2011)
18. Suaris, W., Shah, S.P.: Constitutive model for dynamic loading of concrete. J. Struct. Eng. **111**, 563–576 (1985)

A Fiber Reinforced Concrete for a Nuclear Waste Container

Erik Coppens[1($\boxtimes$)], Petra Van Itterbeeck[2], Bram Dooms[2],
Thomas Richir[3], and Guillaume Debournonville[3]

[1] ONDRAF/NIRAS, The Belgian Agency for Radioactive Waste and Enriched
Fissile Materials, Kunstlaan 14, 1210 Brussels, Belgium
e.coppens@nirond.be

[2] WTCB-CSTC-BBRI, The Belgian Building Research Institute,
Lombardstraat 42, 1000 Brussels, Belgium

[3] Tractebel ENGIE, Boulevard Simón Bolívar 34-36, 1000 Brussels, Belgium

Abstract. For the production of caissons, which are cubic containers for nuclear waste conditioning, the use of a fiber reinforced concrete in alternative to a classic reinforced concrete is studied. A numerical analysis approximating the behaviour of a caisson under load is presented in this paper. This numerical analysis allows for identifying the concrete properties that lead to satisfying structural performance of the caisson under load. This information is on his turn used in a second numerical analysis of laboratory bending experiments on notched beams according to the EN 14651 and round panel testing according to the ASTM C1550. The result of this second numerical analysis is a limit curve which identifies the mechanical performance that should be attained in these tests with 95% confidence. It allows thus for easy discriminating between an adequate concrete composition and a non-adequate one. The experimental work focused first on identifying a suitable fiber among three possible commercial candidates. Next a fiber dosage was defined and the concrete composition was submitted to small variations while verifying his performance according to EN 14651. It could be demonstrated however small changes in the concrete in terms of e.g. a variation in the water-to-cement ratio for 0,45 to 0,50 do not influence the performance according to EN 14651 in a significant way. Also it is demonstrated in this paper a concrete with Bekaert's 3D 80/30 hooked end steel fibers at a dosage of 55 kg/m^3 does not perform sufficiently well for the production of these caissons. The development of an adequate concrete would probably demand for higher fiber loads beyond 55 kg/m^3 or a significant change in other concrete constituents beyond the current investigated ranges.

Keywords: Fiber reinforced concrete · EN 14651 · ASTM C1550 · Numerical analysis · Statistical analysis

1 Introduction

ONDRAF/NIRAS, the Belgian Agency for Radioactive Waste and Enriched Fissile Materials, is preparing the construction of a surface repository for low and intermediate short-lived radioactive waste. The disposal concept consists of emplacing the waste in

© RILEM 2022
P. Serna et al. (Eds.): BEFIB 2021, RILEM Bookseries 36, pp. 628–639, 2022.
https://doi.org/10.1007/978-3-030-83719-8_54

reinforced concrete containers or 'caissons'. The caisson is essentially a cubic box and is filled with the waste and a cement-based grout. This provides a monolithic element known as a 'monolith' (maximum mass: 20 tons). An artist impression of such a caisson with four coli radioactive waste can be found in Fig. 1 (right hand side, mark the four lifting anchors). The monoliths are placed in disposal modules, which are made of reinforced concrete and which are covered by a multi-layer earth cover. The long-term safety functions of the monoliths include restricting radionuclide release, limiting water infiltration, and providing chemical retention (e.g. sorption) properties that retard radionuclide migration.

The use of a fiber reinforced concrete (FRC) – without traditional rebars – is considered for the caissons, due to the excellent durability of FRC towards chloride and carbonation. Indeed it is not only expected for fibers to be less sensitive to corrosion [1] also, it is expected once the corrosion of fibers starts, this will in turn result in less additional damage to the concrete as is the case for a classic reinforced concrete [2–6]. Due to this outstanding durability towards chlorides and carbonation the research program was first focused on engineering a FRC tailored to this application.

The FRC requirements from a mechanical point of view were determined based on the *fib* Model Code 2010. Different fiber concrete mixtures were tested on lab-scale and compared with the required mechanical performance as calculated with a numerical analysis. The findings and the methodology used in this study are presented in this paper.

2 Methodology

2.1 Numerical Analysis

Two numerical analysis are executed in this study. The first analysis, based on the *fib* Model Code 2010 is executed to approximate the performance the FRC should attain to be suitable for the foreseen application. The information gained from the first model i.e. the concrete properties that result in sufficient mechanical performance, is used in a second numerical analysis. This second analysis is a numerical reproduction of the test EN 14651 and ASTM C1550. Therefore the results from this second analysis allows for easy comparison with laboratory test results. Indeed this second numerical analysis allows for identifying for both EN 14651 and ASTM C1550 what level of performance the concrete samples should attain with 95% confidence for these tests. If a concrete compositions does not achieve this performance, it is not adequate for the foreseen application.

2.1.1 Numerical Analysis of the Caisson
According to the Model Code 2010 (Sect. 7.7.2) a FRC can be considered as appropriate, in terms of ductility and resistance if,

- the ultimate displacement is at least 20 times the displacement at service load or,
- the displacement at peak load is at least 5 times the displacement at service load.

and the maximum load is at least equal to 1,5 times the service load (based on Sect. 5.6.6).

The Menetrey-William concrete model is used to simulate within ANSYS finite element software the behaviour of a caisson out of FRC with varying types of concrete characteristics. The most important values of the parameters of the Menetrey-William concrete model which are used in the present study are given in Table 1. Three values (0.4; 0.6 and 0.8) of the residual tensile relative strength parameter are considered in order to simulate different concretes. Other parameters are kept constant.

In the analysis only one fourth of a caisson is modelled, taking into account the symmetries (see Fig. 1). The loading conditions of the caisson are relatively severe. In the analysis the vertical translation is blocked at the top of one handling device in order to simulate the fact that only two opposite handling devices are active. Conservatively waste and liquid mortar are modelled with a hydrostatic pressure, corresponding to a liquid with a mass density such to give a correct total vertical load (4 300 kg/m³). Lateral pressure is therefore overestimated. A vertical acceleration is applied statically and gradually to the caisson so that the evolution of the maximum displacement with the vertical load can be determined.

Table 1. Used parameters for the Menetrey-William concrete model in ANSYS

Parameters	Value	Parameters	Value
E-modulus	35 GPa	Poisson's ratio	0.2
Compr. strength	40 MPa	Res. compr. strength	0.2
Tensile strength	2.5 MPa	Dilatancy angle	35°
Res. stress at start of non-lin. hard.	0.8	Pl. st. at inaxial compr. strength	0.23%
Plastic strain limit in tension	0.2%	Ultim. eff. plastic strain in compr.	0.25%
Res. tensile relative strength	0.4; 0.6; 0.8	Plastic strain limit in tension	0.2%

For each of these three concretes – with varying residual tensile relative strength – we calculated the total vertical load in function of the maximum displacement of the caisson (Fig. 1). It is to be noted that the load (not displacement) is applied to the model, as a result the decreasing part of the load-displacement curve could not be simulated. The service load and service displacement correspond with the dotted lines in Fig. 1.

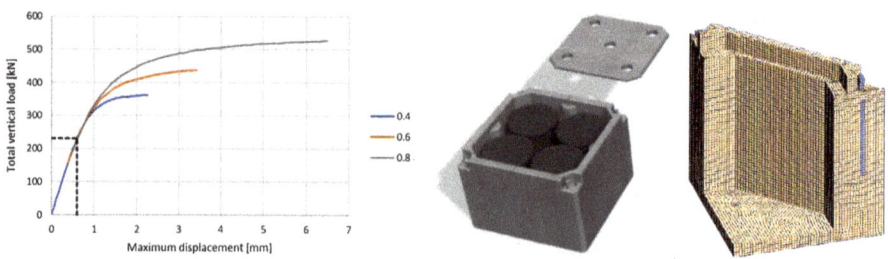

Fig. 1. Numerical analysis with three different fiber reinforced concretes (left) of a caisson (middle) and the modelled quarter (right – mark the lifting anchor in blue).

It can be observed that among the simulated concretes, the one with a residual tensile relative strength of 0.6 can be considered adequate for the foreseen application since the peak load is at least 1.5 times the service load and the maximum displacement is at least 5 times the service displacement.

2.1.2 Numerical Analysis of EN 14651 and ASTM C1550

In a consequent step the performance under standard laboratory trials of this same concrete with a residual tensile relative strength of 0.6 was numerically simulated. More precisely were the classic three-point bending test on notched beams according to the EN 14651 and a centrally loaded round panel on three point supports according to the ASTM C1550 modelled. The resulting curves (displacement versus load) are so-called limit curves. These limit curves are used to evaluate the mechanical performance between different concrete compositions. Indeed they represent the 95% confidence limit of what should be the minimum mechanical performance in these tests.

2.2 Experimental Program

The experimental investigation is two-fold. Firstly, a suitable fiber for the foreseen application is selected and secondly, the concrete composition submitted to variations and verified for the application. Based on a market survey three different fibers were subject of the current investigation: Saint-Gobain Fibraflex FF30L6, Bekaert's 3D 80/30 SL hooked end steel fibers, and Bekaert's 5D 65/60 BG hooked end steel fibers. Table 2 describes the main characteristics of these fibers, as communicated by their producer.

Table 2. Description of fibers used in the study

	Fibraflex FF30L6	Bekaert 3D 80/30 SL	Bekaert 5D 65/60 BG
Material	Slightly alloyed steel	Stainless steel	Steel
Length	30 mm	30 mm	50 mm
Diameter/thickness	1.6 mm × 29 μm	Diameter 0.38 mm	Diameter 0.90 mm
E-modulus	Unknown	200 kN/mm^2	300 kN/mm^2
Strength	Unknown	2000 N/mm^2	2300 N/mm^2

In the first experimental phase, all laboratory concrete is made with one and the same sulphate resistant Portland cement, CEM I 52.5 N, Ultimat, from Vicat France. The gravel and sand are both calcareous and in some cases calcareous filler is also used. The overall concrete composition used in this phase can be found in Table 3.

Since the discrimination between adequate or non-adequate FRC is done by considering the conclusions from the numerical analysis as described in (Sect. 2.1.2), bending experiments on notched beams according to the EN 14651 and round panel testing according to the ASTM C1550 are executed.

Also, it was estimated, based on these same numerical analysis a fiber dosage of at least 30 kg/m^3 and up to 50 kg/m^3 would be needed to give sufficient residual strength and ductility to the FRC for the foreseen application. Therefore all three fibers were tested in the concrete composition according to Table 3 at either 50 kg/m^3 or 30 kg/m^3, with as main evaluation parameter being the workability of the concrete in the laboratory.

In a second experimental phase the concrete composition was further optimized, taken into account the results of the first phase and additional design requirements related to the caisson manufacturing plant which was under construction at the moment of conducting the research. Indeed, since in the caisson manufacturing plant besides a FRC also a conventional reinforced concrete is foreseen to be produced, the final FRC-composition should preferably use the same (or less) raw materials; Portland cement, calcareous filler, calcareous calibres, 0/4, 2/6 and 6/14, but no 6/20. These conditions have to be taken into account and forced us to reduce the D$_{max}$ from 20 mm tot 14 mm in the second experimental phase.

Table 3. Reference concrete composition

Material	kg/m^3
Cement	350
Calcareous filler	50
0/4 calcareous sand	868.9
6/14 calcareous gravel	595.9
6/20 calcareous gravel	392.5
Fibers	30 or 50
Water	168.9
Water-to-cement	0.47

While taking this constraint into consideration it was foreseen to identify the optimum concrete composition for a fixed amount and type of fiber by applying small variations in the concrete composition. For all concrete compositions samples were manufactured for bending experiments according to EN 14651 and all datasets were analysed in a statistical manner.

3 Results and Discussion

3.1 Numerical Analysis

Bending experiments on notched beams according to the EN 14651 and round panel testing according to the ASTM C1550 were modelled for an FRC with residual tensile strength of 0.6. It was indeed this concrete with a residual strength of 0.6 which resulted in adequate performance according to a numerical analysis of a caisson (Sect. 2.1.1). The numerical modelling of the laboratory tests results in limit curves, visualised in Fig. 2, which are used for comparison with laboratory data. Since these limit curves identify the mechanical performance that should be attained with 95% confidence – i.e. characteristic values – these curves cannot be compared directly with the laboratory observations. For good comparison these later have to be statistically treated.

As can be seen on Fig. 2 the characteristic flexural stress in the notched beams according to the EN 14651 is 4,6 MPa for f_{R1}, f_{R2}, f_{R3} (respectively at crack mouth opening displacements of 0.5; 1.5 and 2.5 mm). The residual force for ASTM C1550 is 23.3; 23.8 and 24.2 kN for a deflection of respectively 0.4; 0.6 and 0.8 mm. These values are considered the minimum performance an FRC should, in these tests, attain with 95% confidence in order to be regarded as adequate for the foreseen application.

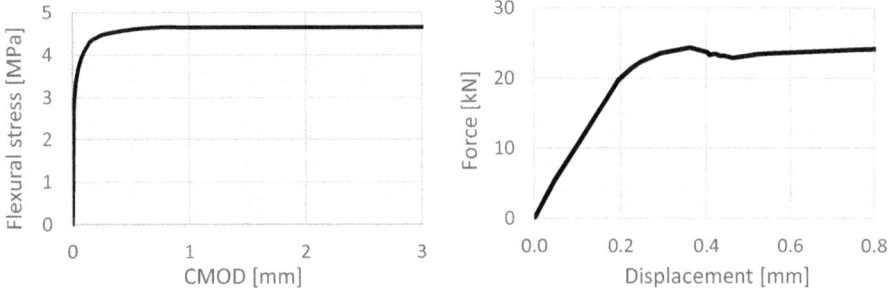

Fig. 2. Limit curve for EN 14651(left) and ASTM C1550 (right)

3.2 Experimental Program – Phase 1

In the first experimental phase three fiber concrete mixtures (with the same basic concrete composition (see Sect. 2.2, Table 3) but different fibers) were assessed with the aim to identify the most suitable fiber among the three types under study for the application. More specifically experiments were conducted with a concrete with 50 kg/m^3 of the 80/30 SL hooked end steel fibers, a second with 50 kg/m^3 of the 65/60 BG hooked end steel fiber and a third concrete with 30 kg/m^3 of the fibra-flex fiber. The concretes are nominated as 3D50, 5D50 and FF30. An attempt was made to produce a concrete with 50 kg/m^3 fibra-flex fiber, but the workability was insufficient for the foreseen application, obligating us to reduce the fiber content to 30 kg/m$^{3.}$ All experiments were conducted at concrete 14 days of age. Although in most studies concrete is assessed at 28 days of age; due to time pressure it was decided to test all, but two (see phase 2), concrete in this study at 14 days age.

The fiber-type clearly has a significant impact on the average performance in both test cases, as can be seen in Table 3. The concretes based on both hooked end fiber had similar average performances and clearly outperform the concrete based on fibra-flex fibers in the EN 14651-test when it comes to average performance. However the data in Table 4 does not take into account the dispersion observed between samples (made with the same concrete batch). Indeed the dispersion observed during execution of these lab trials (see Table 5) has to be taken into account to assess the characteristic performance of these concretes. The fibra-flex fiber has demonstrated an overall low dispersion, for both tests. The 80/20 hooked end fiber demonstrated, mostly, a slightly increase in dispersion. However both these fibers outperform the 65/60 hooked end fiber when it comes to dispersion - and thus characteristic performance -, especially in the ASTM C1550-test.

Table 4. Average performance in lab trials and the limit for characteristic values according to the numerical analysis

	EN 14651 [MPa]			ASTM C1550 [kN]		
	f_{R1}	f_{R2}	f_{R3}	0,4 mm	0,6 mm	0,8 mm
FF30	2.4	1.4	0.9	29	29	26
3D50	7.3	6.4	5.4	29	29	29
5D50	8.2	7.2	6.3	28	25	26
Characteristic limit according to numerical analysis (see Sect. 3.1)	4.6	4.6	4.6	23.3	23.8	24.2
3D55	8.6	7.4	6.3	31	30	30

Table 5. Standard deviation of performance in lab trials

	EN 14651 [MPa]			ASTM C1550 [kN]		
	f_{R1}	f_{R2}	f_{R3}	0.4 mm	0.6 mm	0.8 mm
FF30	0.60	0.26	0.13	1.15	2.00	0.58
3D50	0.88	0.92	0.71	1.53	1.00	1.53
5D50	1.00	1.24	0.92	2.08	3.61	4.04

Based on the information gained in this first laboratory phase all further experiments were conducted with Bekaert's 3D 80/30 hooked end fiber due to its relative good mechanical performance in both tests and its relatively low dispersion compared to the Bekaert's 5D 65/60 hooked end fiber. It might be interesting to know the caissons walls are relatively thin in thickness, being only 12 cm, therefore also for construction feasibility the shorter fibers of 30 mm length might be preferred over the longer ones of 60 mm length.

Additionally, for the 80/30 hooked end fiber it was verified whether it was technically feasible to add more fibers to the concrete and what effect this would have on the mechanical performance. This concrete, with 55 kg/m^3 fibers is denominated as 3D55 and seems to result in an increased mechanical performance, especially in the EN 14651 for f_{R1} as well as f_{R2} and f_{R3}, as can be seen in Table 4. However, no repetitions of this test were executed to verify the repeatability and it still has to be verified whether this observed difference is significant when compared to batch-to-batch variations (see also experimental phase 2).

3.3   Experimental Program – Phase 2

In a consequent lab phase the concrete composition varied, except for the fiber type and content. The aim was to identify for a given fiber type and content, 55 kg/m^3 of Bekaert's 3D the 80/30 SL hooked end fiber, the optimal concrete composition, taking into account industrial constraints which limit the maximal dimension of the granulates to 14 mm and also limits the choice of raw materials.

Eight different concrete compositions with a fiber dosage of 55 kg/m^3 and a D$_{max}$ of 14 mm were produced and the mechanical performance according to EN 14651, i.e. residual bending strengths, were assessed at an age of 14 days. The concrete compositions were similar, but varied slightly among each other to assess the influence of variations in composition. The variations investigated were among others:

- a variation of cement content between 350 and 365 kg/m^3;
- a variation of cement type between CEM I 52.5 N and CEM I 42.5 R;
- a variation in the content of calcareous filler varied between 0 and 50 kg/m^3;
- a variation in the water-to-cement ratio from 0.45 to 0.50;
- a variation in the ratio of granulate caliber 0/4 to granulate caliber 6/14 between 0.8 and 1.2.

The EN 14651 was executed on four or five concrete notched beams for each of these different concrete compositions. Not only were the average residual bending strengths (f_{R1}, f_{R2} and f_{R3}) evaluated but also the coefficients of variation. The coefficients of variation for all three factors f_{Ri} (Table 6) were assessed for normality by means of a Spahiro-Wilk test. The p-values resulting from these tests can be found in Table 6. Since p-values are all 15% or higher, the H0 hypotheses: the distribution of the population is normal with unspecified mean and standard deviation, is not rejected. We thus conclude the coefficients of variation observed during testing of these concretes according to EN 14651 is not affected by the concrete composition itself but merely coincidence. An anova test following a levene's test result in respectively p-values of 66% and 19%. Therefore we cannot reject the H0 hypotheses: the coefficient of variation of the populations are all equal (levene), nor the H0 hypothesis: the mean of the populations are all equal (anova). We thus conclude all measured coefficients of variations are originating from the same normal distributed population, with best estimate for the mean and standard deviation of respectively 0.116 and 0.048. We denominate this coefficient of variations as the overall within-batch coefficient of variation.

Table 6. Within-batch coefficient of variation observed for EN 14651

	Coefficient of variation for f_{Ri} (EN 14651)		
	f_{R1}	f_{R2}	f_{R3}
Test A	0.19	0.17	0.16
Test B	0.17	0.12	0.11
Test C	0.17	0.13	0.13
Test D	0.07	0.06	0.09
Test E	0.06	0.04	0.04
Test F	0.12	0.10	0.12
Test G	0.09	0.07	0.06
Test H	0.18	0.20	0.13
Saphiro-Wilk p-value	**15%**	**74%**	**84%**

To assess the variations one could expect in pure repetitions, in between batches of a same concrete, we used data from two FRCs (concrete A and concrete B) that were both duplicated and tested at 28 days. The data is originating in both cases from FRC with Bekaert's 3D the 80/30 SL hooked end fiber at a dosage of 55 kg/m^3, concrete A and B vary among each other in among others D_{max} (14 mm and 20 mm). Indeed, these concrete compositions do not respect all technical constraints that should be taken into account in the current study, also these two concrete compositions are the only ones being tested at 28 days instead of 14 days age. The data is given in Table 7 and consist of two different concrete compositions in duplicate tested according to EN 14651. The set of all six observations on the coefficient of variation can be tested for a normal distribution with the Spahiro-Wilk test, resulting in a p-value of 25%. We conclude the coefficient of variation in between batches, as observed in these tests are originating from a normal distribution with an estimated mean of 0.147 and a standard deviation of 0.043. We denominate this value as the in-between-batch coefficient of variation.

One can observe the coefficient of variation over the pure repetitions tested at 28 days is greater than those observed over Test A through H at 14 days. The origin of this observation is unclear, but one hypothesis might be that the evolution of the concrete over time, increases variations in-between batches.

Table 7. In-between-batch coefficient of variation and average tensile strength of repetitions of concrete A and concrete B [MPa], both concretes are made with 3D fiber.

| | Average tensile strength [MPa] for f_{Ri} (EN 14651) | | | | | |
| | Concrete A, D_{max} 14 | | | Concrete B, D_{max} 20 | | |
	f_{R1}	f_{R2}	f_{R3}	f_{R1}	f_{R2}	f_{R3}
Batch A	8.03	8.10	7.11	6.68	6.97	6.35
Batch B	6.15	6.29	5.64	8.40	8.23	7.06
Coefficient of variation	**0.19**	**0.18**	**0.16**	**0.16**	**0.12**	**0.08**

Table 8. Average tensile stress observed for EN 14651

| | Average tensile stress [MPa] for f_{Ri} (EN 14651) | | |
	f_{R1}	f_{R2}	f_{R3}
Test A	6.8	7.5	6.7
Test B	7.0	7.1	6.2
Test C	7.2	7.7	7.0
Test D	7.8	7.8	6.8
Test E	8.6	8.3	7.3
Test F	7.3	7.8	6.8
Test G	7.2	7.7	7.0
Test H	7.4	8.1	7.0
Saphiro-Wilk p-value	**14%**	**61%**	**61%**
Overal average	**7.4**	**7.8**	**6.9**

Just like the within-batch coefficient of variation the averages for all three f_{Ri} for Test A through H (Table 8) were assessed for normality by means a Spahirio-Wilk test. Since these averages are considered normal distributed (p-values > 10%) we can conclude that the effect of a change in concrete composition in f_{Ri} is comparable or smaller as the effect of factors beyond our control (e.g. technician, etc.) and can again merely be explained by coincidence. An anova test following a levene' test for each f_{Ri} result in respectively p-values of 0.1% and 47%. We thus conclude, even though the difference of the average performance for the slightly varying concrete compositions is merely affected by coincidence, the performance observed for f_{R1}, f_{R2} and/or f_{R3} are not equal to each other.

To significantly increase the mechanical performance of the FRC according to EN 14651, one can thus not refer to the relatively small changes in concrete composition investigated in this study. The statistical analysis showed that the small deviations observed in the average performance and/or in the scatter for similar, though varying concrete compositions can merely be explained by coincidence.

If the comparison of the mixtures (Test A through Test H) were to be done on the basis of the calculated characteristic performances by means of observed standard deviation (within-batch coefficient of variation) and average performance for each concrete composition separately, excluding the in-between-batch variation, one would probably faulty conclude Test D is the better among the concrete compositions tested. This conclusion would solely and wrongly be based on the relatively higher average performance and low scatter within the batch.

To calculate the characteristic performance of the Test A through H we should estimate the average performance and the coefficient of variation. Higher in this text it was estimated that the average performance might vary in between batches, with a coefficient of variation normal distributed with an estimated mean of 0.147 and a standard deviation of 0.043. Also the coefficient of variation for specimens made within the same batch of concrete was assessed, being normal distributed with a best estimate for the mean and standard deviation of respectively 0.116 and 0.048.

The resulting characteristic value, for which 95% of the population performs equal or better can be calculated according to Eq. 1 and Eq. 2.

$$f_{Rik} = \bar{x}_i + Z\sqrt{\sigma_1{}^2 + \sigma_2{}^2} \tag{1}$$

With

f_{Rik} the characteristic value for respectively f_{R1k}, f_{R2k} or f_{R3k},

$\bar{x}_i$ the average value for respectively f_{R1k}, f_{R2k} or f_{R3k},

σ_1 the estimate for the within-batch standard deviation,

σ_2 the estimate for in-between-batch standard deviation,

Z according to (Eq. 2).

$$Z == t_{0,95;D.F.} \times \sqrt{1 + \frac{1}{n}} \qquad (2)$$

With.

$t_{0,95;DF}$ the t-distribution coefficient for a cumulative frequency of 0.95 in one tail and D.F. degrees of freedom (D.F. = $n-$ 1), being −1.895 for D.F. = 7,

n the number of individual observations.

The group of concretes manufactured in the second phase of this study, being Test A to H, with 55 kg/m^3 of Bekaert's 3D hooked end 80/30 SL fiber has thus a characteristic value for f_{R1k}, f_{R2k} and f_{R3k} of respectively 4.64 MPa, 4.85 MPa and 4.28 MPa.

We must conclude these values, which are representative for all eight compositions, are under the limit of what is acceptable in EN 14651 for f_{R3k}, with a limit value of 4.6 MPa as being calculated in Sect. 2.1 of this text. The concrete with a D$_{max}$ of 14 mm and the 3D-fiber at dosage of 55 kg/m^3 does not perform sufficiently well for the foreseen application and its performance cannot be increased sufficiently by applying only small variations in the composition. Indeed all observed differences in mechanical performance in Test A to Test H can merely be explained by coincidence. It must concluded, only an increased fiber content beyond 55 kg/m^3 of this 80/30 hooked end fiber might increase in sufficient amounts the mechanical performance of these concretes. A variation of the other concrete components has no significant effect. Although it must be admitted alternative solutions, e.g. hybrid solutions with micro-fibers, might exist.

Based on current data it can not be estimated with sufficient confidence what the minimum amount of these 80/30 hooked end fibers should be for the foreseen application. However we observe there is only a gap of 0.32 MPa (or 7.5% relatively speaking) to be bridged. Since this gap is not regarded impossible to be bridged with these fibers. Research will be proceeded and follow-up tasks shall focus first on higher fiber dosages. If an adequate mechanical performance can be demonstrated, the concrete shall be tested again in ASTM C1550 for validation and also at older ages, a minimum of 28 days age.

4   Conclusions

In this paper among three possible commercial fiber types one –Bekaert's 80/30 SL hooked end fiber - seemed appropriate for the intended use i.e. the construction of a caisson for nuclear waste.

We also assessed the performance of different FRC with varying compositions although with a constant fiber content at 55 kg/m^3 and constant fiber type. It is demonstrated the observed differences in terms of performance in flexural tensile strength according to EN 14651 can be explained by coincidence. Small changes in terms of the content of calcareous filler, water-to-cement-ratio, the ratio granulate 0/4 to granulate 6/14 or even cement content or cement type had no important influence on the flexural tensile strength.

It is also demonstrated in this paper concrete with at a dosage of 55 kg/m^3 of 80/30 SL hooked end fiber does not perform sufficiently well for the production of so-called caissons. Although the observed gap between minimum mechanical performance and laboratory performance is very small, the development of an adequate concrete would probably demand for higher fiber loads beyond 55 kg/m^3 or a significant change in other concrete constituents beyond the current investigated ranges. More research is needed to develop such a concrete.

Acknowledgements. The authors would like to acknowledge the great help from Prof. emeritus Marc Coppens from K.U.Leuven, Belgium for his comments and advice on the statistics. Also, we thank Carmen Andrade from the International Center for Numerical Methods in Engineering, Spain for her fruitful comments on draft versions of this manuscript.

References

1. Soetens, T.: State-of-the-art report on Fibre Reinforced Concrete, Magnel Laboratory For Concrete Research (2016)
2. Schaerlaekens, S., Vyncke, J.: Le béton renforcé de fibres d'acier, 2e Partie, Recherches & Etudes, CSTC Magazine, Hiver (2000)
3. American Concrete Institute Committee 544, State-of-the-art report on fiber reinforced concrete, ACI 544.1R-96, (reapproved) (2002)
4. Schiessl, P., Weydert, R.: Corrosion of steel fibres in carbonated cracked Concrete. In: Concrete in the service of mankind – Radial Concrete Technology, pp. 495–500, CRC Press, Boca Raton (1996)
5. Macros Meson, V., Michel, A., Solgaard, A., Fischer, G., Edvardsen, C., Skovhus, T.L.: Corrosion resistance of steel fibre reinforced concrete – a literature review, Published in fib symposium 2016: performance-based approach for concrete structures (2016)
6. Sadeghi-Pouya, H., Ganjian, E., Claisse, P., Muthuramalingam, K. : Corrosion durability of high performance steel fibre reinforced concrete. In: Proceedings of the Third International Conference on Sustainable Construction Materials and Technologies, Kyoto, Japan, August (2013)

Design and Execution of Floors on Ground and Industrial Pavements with Fibre Reinforced Concrete

Roberto Pombo[1], Marcelo G. Altamirano[2], Graciela M. Giaccio[3], and Raúl L. Zerbino[4(✉)]

[1] Bautec S.A., Beccar, Buenos Aires, Argentina
[2] UTN Regional Pacheco - Bautec S.A., Beccar, Buenos Aires, Argentina
[3] LEMIT-CIC, CIC Researcher, Faculty of Engineering UNLP, La Plata, Argentina
[4] CONICET. LEMIT-CIC, Faculty of Engineering UNLP, La Plata, Argentina
zerbino@ing.unlp.edu.ar

Abstract. Steel and synthetic macrofibres are widely used as a replacement for nominal reinforcement in industrial floors and pavements. Some of the properties of the wet and hardened concrete enhanced by the fibres provide superior behaviour to the slabs containing conventional steel bars reinforcement. Fibres control the extent and size of cracks formed due to drying shrinkage and allow the extension of the slabs joint spacing. Fibres improve the post-cracking properties of concrete increasing flexural and ultimate load carrying capacity of the slabs. Finally, the combination of fibres with an expansive agent to obtain a "Shrinkage-Compensating-Fibre-Reinforced Concrete" allows the design of large slabs with almost no cracks and with tight joints widths. Although the same considerations for a good concrete mix design without fibres are also good for a FRC, care should be taken to avoid increasing potential shrinkage and curling using inappropriate materials and proportions. Examples of large floor and pavement works in Argentina using these technologies are presented, mentioning the different fibre types incorporated and explaining the FRC properties considered in the design. Comparison between laboratory and on field results is shown. FRC characterization included the evaluation of workability, free and restrained shrinkage, compressive and flexural residual strengths. In addition, comparisons between polymer and steel FRC behaviour are made.

Keywords: Crack control · Fibre reinforced concrete · Floors on ground · Industrial pavements · Polymer macrofibres

1 Introduction

The benefits of incorporating steel and synthetic macrofibres as a replacement for nominal reinforcement in industrial floors and pavements are widely proved. The behaviour of slabs-on-grade is governed by the interaction between the constructed layer, the supporting layer and the loading. In addition to various improvements in the properties of the fresh and hardened concrete, the use of Fibre Reinforced Concrete

© RILEM 2022
P. Serna et al. (Eds.): BEFIB 2021, RILEM Bookseries 36, pp. 640–651, 2022.
https://doi.org/10.1007/978-3-030-83719-8_55

(FRC) is advantageous as it provides significant rotational capacity to the slab due to its pseudo-ductility enhanced by the fibres. Fibres improve the post-cracking properties of concrete increasing flexural and ultimate load carrying capacity of the slabs.

Fibres tend to reduce segregation and bleeding in the fresh concrete, improving surface quality [1]. When the appearance of some micro or macro cracks takes place, the fibres can keep them much more closed than in the case of plain concrete. In this way, the fibres contribute to minimize the effects of the drying shrinkage allowing increase spacing between joints. Joint spacing depends on the type and dosage of used fibres. For instance, according to authors experience of more than 20 years, incorporating 30 to 40 kg/m^3 of steel fibres, joint spacings of 25 to 30 m can be achieved. FRC with polypropylene fibres at their usual dosage of 3 to 5 kg/m^3 can reach 10 to 15 m of joint spacing. In addition, as the slab tends to curl permanently, microcracks probably occur on the surface, the presence of fibres allowing some relaxation of the slab without the cracks being visible. The consequence is that curling is reduced in some extent [1]. The concrete at the top of the slab has almost the same properties as at the bottom, particularly drying shrinkage. Consequently, the FRC will improve its resistance to impact and abrasion.

The combination of fibres with an expansive agent to obtain a Shrinkage-Compensating-Fibre-Reinforced Concrete (ShCFRC) [2] allows the design of large slabs with almost no cracks and with tight joints widths.

Floor concrete has some specific requirements; both the materials and the mixture proportions should be selected based on those requirements. Although the same considerations for a good concrete mix design without fibres are valid for a FRC, care should be taken to avoid increasing potential shrinkage and curling using inappropriate materials and proportions. Floor execution operations can be carried out in the same way with FRC as with plain concrete. The pour, screeding, the eventual incorporation into the surface of a wear course and the finishing processes are not altered.

This paper shows selected examples of large floor and pavement works made in Argentina using polymer and steel FRC and ShCFRC concretes; the motivations and advantages in structural design and construction issues are discussed.

2   Requirements for Industrial Floors

Concretes for industrial floors have some specific requirements [3–8]. Materials and the mixture proportions, which mean mix design, should be adequately selected. The main requirements are: adequate workability at time of placing and screeding, uniformity of setting time, finishability, economy, abrasion and impact resistance, flexural strength and minimum shrinkage potential.

Regarding the component materials different types of cements can be used in concrete for industrial floors, however cements promoting excessive shrinkage or set retardation must not be used. Diverse mineral additions as ground granulated blast furnace slag (GGBFS) or fly ash can be used but it should be carefully studied before. The incorporation of silica fume is not recommended not only since it is expensive, but also because it can increase water demand and plastic shrinkage, and increments in compressive strength are not essential in concrete floors. The use of GGBFS can increase long term shrinkage [9]. Fly ash incorporation enhances the pumpability but usually extends the setting time (mainly in cold weather) and can affect the final colour. Concerning the aggregates, the maximum size should be limited in the case of FRC in accordance to fibre length. Natural rounded sand (usually with a fineness modulus > 2.5) is preferred and excessive crushed sand contents can affect finishability and surface aspect.

Regarding the mixture proportions [7], the binder content must be optimized, as increasing concrete strength by higher cement or mineral additions contents can negatively affect the elastic modulus and the drying shrinkage, which are closely related with curling process. Reducing the void space between aggregates minimizes water demand, costs, and cracking and curling risks. The w/c ratio must be as low as possible. However, a reduction in mixing water content not necessary reduces shrinkage in the same proportion. The mortar content should be enough for pumping, placing and finishing but excess of mortar can increase shrinkage. The aggregate shape, texture and grading affect the required mortar volume, but this volume depends on the placing and construction method adopted. Potential shrinkage may be affected by many factors as temperature of concrete at discharge, excessive slump, and aggregate maximum-size, among others [10]. Air content must be as low as possible to improve densification of the surface and to avoid delamination.

When dry shakes are used, it may be advantageous to increase the bleeding water. It must not bleed after spreading the dry shake in order to avoid its delamination.

Among the fresh concrete properties, and although placeability and finishability have difficulties to be evaluated, they have similar or even higher importance than other properties as compressive strength, abrasion resistance or durability.

The setting time must be considered without excessive retardations for the environmental conditions expected in the project. Care should be taken for certain chemical admixtures. It is very important to warrant that setting time between batches not differ too much.

Concerning hardened concrete properties, the abrasion resistance has great importance in most applications and the flexural strength has to be known for structural design purposes.

3   Shrinkage Compensating Fibre Reinforced Concrete

Expansive agents, which can be ettringite or CaO based, are added in the case of Shrinkage-Compensating Concrete (ShCC). ShCC is an expansive concrete that, duly restrained by reinforcement or other means, during the wet curing period generates an initial expansion equal to or slightly greater than the expected drying shrinkage. Due to the restraint, during the expansion stage the ShCC will experience some compression, which will then be relieved during the shrinkage stage. Once exposed to the air, there is less shrinkage than in the concrete without an expansive agent. The expected result is that, in the final state of equilibrium, the ShCC remains with zero tension or a slight residual compression stress, eliminating the risk of cracking [11]. ShCC usually is used in combination with conventional steel bars. The reinforcement is initially tensioned during expansion and cause the structure to be compressed [12].

However, the combination of synthetic macrofibres with an expansive agent to obtain a Shrinkage-Compensating-Fibre-Reinforced Concrete is designed to simultaneously take advantage of two properties: the modification of the behavior of the concrete, both in the fresh state (bleeding, segregation, etc.) and in the hardened state (microcracks generation) provided by the fibres and the compensation of the contraction performed by the expansive agent.

4 Selected Cases

4.1 Case 1: Logistic Facility (2017) – Shrinkage-Compensating-Fibre-Reinforced Concrete Floor on a Poor Soil

A logistic warehouse for the storage of large steel coils (of about 15 tons/each one), which implies large concentrated loads (see Fig. 1), was projected on the coast of the Riachuelo river (Dock Sud). The load capacity of the ground at that place is very low, as a consequence of the deposit of contaminated sediments along the years, and foreseeable settlements can be high and variable. It is difficult and very expensive to design a floor that does not deform and/or does not fracture on a ground of these characteristics. Originally a conventional reinforced concrete floor was built and after a short time of use it was quickly out of service as a consequence of the slab cracks originated by the ground settlements.

Fig. 1. Logistic facility (2017), steel coils storage

The solution adopted, steel FRC, gives the concrete certain ductility which allows the slabs to be deformed without wide opening of the cracks that originate. The residual flexural capacity of the SFRC was taken into account in the structural design to reduce the slabs thickness.

The use of steel FRC combined with an expansive agent (ShCFRC) allowed increasing the spacing between joints (about 35 m) and, specially, minimizing the joint openings. The dimensions of the warehouse were 100 m by 30 m, so three 0.15 m-thick slabs of 30 m by 33 m were projected. This means 450 m^3 of ShCFRC. In this case hooked-end steel fibres were used at a dosage of 40 kg/m^3. Figure 2 shows preliminary flexural test results performed in accordance to EN 14651 Standard [13] for

FRC incorporating between 20 to 50 kg/m³ of fibres in a base concrete with a cylindrical compressive strength near 40 MPa. The specified compressive strength at 28 days was 30 MPa and the slump 90 ± 10 mm. Table 1 shows the adopted FRC mixture proportions.

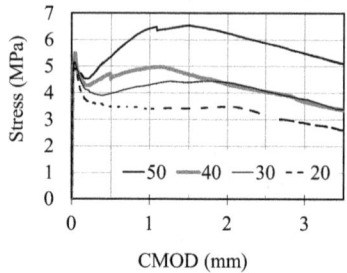

Fig. 2. Stress-CMOD curves from bending tests (EN 14651 Standard) of FRC incorporating 20, 30, 40 or 50 kg/m³ of steel fibres.

Table 1. Case 1: Summary of FRC mixture proportions.

Material	kg/m³
Ordinary portland cement	330
Water	160
Natural sand	810
20 mm crushed stone	1070
Water Reducing Admixture	1.65
Type G (CaO) - Expansive agent	10
Hooked-end steel fibres	40

A naphthalene-based superplasticizer was also added with the aim to achieve the desired slump. The dosage of expansive agent was varied from 5 to 25 kg/m³; resulting in an expansion of 0,075% for 10 kg/m³. For contents higher than 15 kg/m³ there were excessive expansions and multiple cracks.

The successful in choosing this ShCFRC technology was demonstrated with following results:

- When manoeuvring heavy coils on the slabs bending capacity is of great importance. It is well known that steel fibres reinforced concrete has significantly increased this property.
- Considering that the slabs must withstand intense loading and unloading cycles of the coils, the fatigue performance of concrete is also very important. A better fatigue behaviour was expected in ShCFRC where the fibres are distributed throughout the slab thickness with respect to the conventional reinforced concrete solution.

– In this application it was not possible to install a vapour barrier to avoid penetration of aggressive agents because it will be broken by the settlements of the ground. Cracks openings between 0,3 to 0,5 mm and joint widths less than 5 mm were achieved. This means that a floor less susceptible to the penetration of aggressive agents, such as those generated by the contaminated industrial sediments, could be done.

4.2 Case 2: Logistic Facility (2019) - Shrinkage Compensating Fibre Reinforced Concrete Floor in a Robotic Storage System

In the Distribution Centre of whole surface of 80,000 m^2 located in Moreno, Buenos Aires Province, the storage and distribution operations of sector of 7,000 m^2 are performed by a robotic system. The assembly of this system had the following requirements:

The installation of the racks required the placement of countless anchors very close to each other and, consequently the following considerations were important (see Anchor plan in Fig. 3):

– On the floor slab there should be no steel bars of diameters greater than 8 mm, since a larger diameter implies an additional drilling cost.
– The spacing between joints should be as large as possible to minimize the possibility that an anchor is positioned on any of them. It was not allowed to drill in joints areas; minimal distance between joints and an anchor installation had to be 450 mm.
– The assembly precision required the minimum possible joint width and joint movement.
– To ensure sufficient flatness was required.

In addition, the project included a high concentration of columns in a sector due to the presence of an elevated platform.

The adopted solution was a ShCFRC. The concrete mixture design and the characteristics of the fibres and the expansive agent were similar to the previous case. The project was successful because the following floor characteristics were achieved:

– Steel fibres did not interfere the drilling jobs.
– Steel fibres allowed the execution of joints spacing of about 25 m x 30 m.
– ShCFRC provide a better long-term deformation behaviour of the slabs minimizing joint openings and reducing warping/curling effects which ensures the flatness and levelness performances required.
– FRC allowed eliminating the joints that would have to be done in the interception of each of the platform columns.
– Steel fibre reinforced concrete residual flexural strength allowed reducing slab thickness (0.16 m).

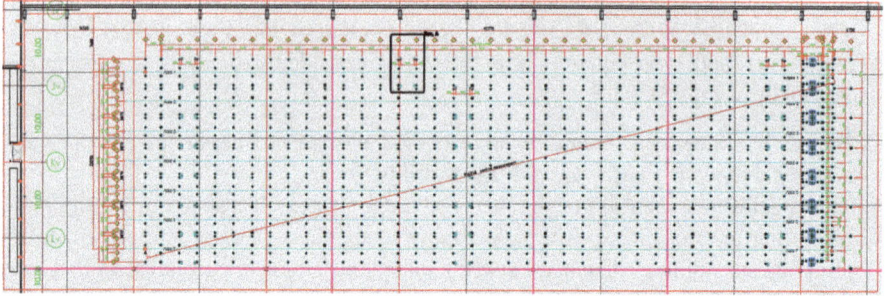

Fig. 3. Moreno distribution centre (2019). Anchor plan (bottom).

4.3   Case 3: Loads Transfer Centre in Buenos Aires City (2015) – Synthetic Fibre Reinforced Concrete Pavement in a Huge Heavy-Duty Industrial and Trucking Facility

The Loads Transfer Centre in Buenos Aires City is a huge construction consisting of 112 warehouses of 900 m^2 with a surrounding pavement of 110,000 m^2. This Centre receives merchandise in large trucks and distributes them to the city in smaller vehicles. The original project of the pavement consisted of saw-cut square slabs, 5 m side, 0.23 m thickness, executed with a 30 MPa compressive strength reinforced concrete, with a steel mesh of 6 mm diameter separated at 15 cm in both directions. The goal was to reduce the thickness of the slabs to make the project economically viable. The FRC technology was selected for this purpose.

In previous tests, three FRC incorporating 4 kg/m^3 of monofilament synthetic macrofibres (polypropylene copolymer 60 mm length, 0.6 mm diameter) were done using cement contents of 310, 350 and 390 kg/m^3. The w/c ratios were 0.48, 0.42 and 0.37 resulting compressive strength at 28 days of 31.8, 38.3 and 45.8 respectively. Natural sand (70%) combined with crushed sand (30%) and 30 mm maximum size crushed stone were used as aggregates. A conventional water reducing admixture combined with a naphthalene-based superplasticizer were used. Figure 4 shows representative stress – deflection curves obtained in bending tests performed in accordance to ASTM C1609 Standard [14]. The formulation of a FRC (cement content 350 kg/m^3)

with an equivalent resistance (R_{e3}) of 33% [15], made possible to reduce the originally thickness from 0.23 m to an average of 0.20 m, conducing to an economy of 3300 m^3 of concrete, which was the main purpose of the project.

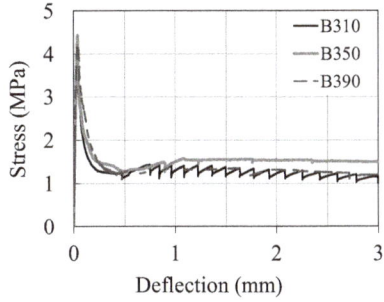

Fig. 4. Stress – deflection curves obtained in bending tests (ASTM C1609).

Another very important additional benefits were: a) Joints spacing were increased from 5 to 7 m, which means saving of 40% in linear meters of joints, b) joint widths were significantly reduced (from 8 mm to 4 mm) c) curling effects were minimized when comparing with the traditional solution in conventional reinforced concrete. As a consequence of these properties the hydraulic project was improved because water seepage through the joints was reduced by 40% minimizing the risk of base contamination and less curling allowed to design lower slopes of a long pavement.

Figure 5 includes different pictures showing the development of the construction and Fig. 6.a presents a lay out of the pavement; the long length (near 500 m) of the pavement can be appreciated. This means an appreciable difference in the levels between the extreme points. A panoramic view of the finished work is presented in Fig. 6.b.

Fig. 5. Construction of the loads transfer centre in Buenos Aires City (2015).

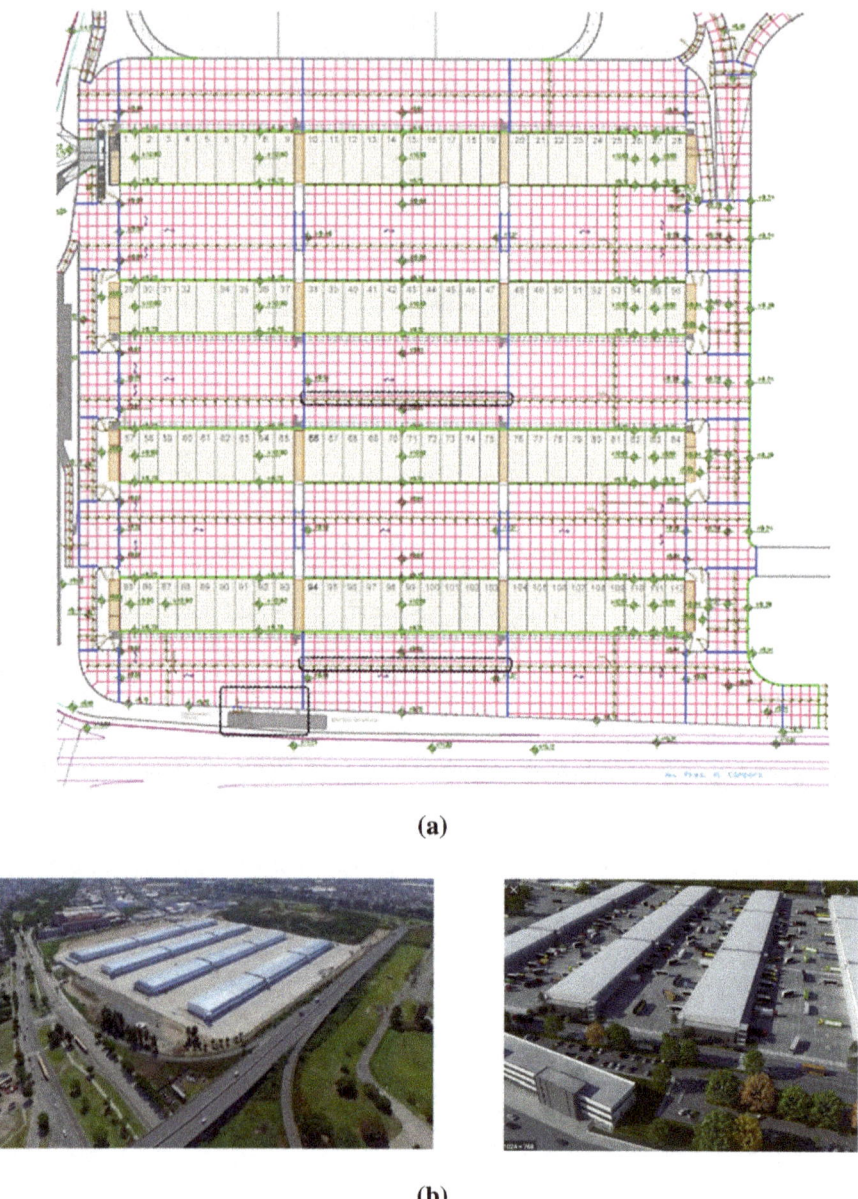

(a)

(b)

Fig. 6. Loads transfer centre in Buenos Aires City (2015): a) Lay out of the pavement; b) Panoramic views.

A survey carried out at the beginning of 2018, more than two years after finishing of the work, indicated that there were only 39 slabs with cracks on a total of near 2500, which represents less than 2%. Then, the durability of the pavement increases. For all these factors the use of a Synthetic Fibre Reinforced Concrete Pavement allows a decrease of the maintenance costs both of the pavement and of the equipment that circulates on the pavement; main problem during operation.

4.4   Case 4: Warehouse in the Coast of the Rio De La Plata River (2019) – Synthetic Fibre Reinforced Concrete Floor with Slabs Gently Deformed by the Loads

Synthetic macrofibres can also be used to improve the performance of industrial floors in poor substrates. In this example the bearing capacity of the soil where the floor was executed has similar characteristics as the Case 1: very low with anticipated high and variable settlements. It was necessary to design a floor that, with sustained loads over time, that is to say with the permanent storage of goods, deform gently without abrupt fractures, to allow a secure displacement of the lifters.

The proposed solution assumes the execution of a floor, which would surely deform with the expected settlements, until they are reduced to a negligible value. The incorporation of synthetic macrofibres transforms the concrete into a more ductile material, reducing the consequences of the deformations, keeping closed any possible cracks that could be generated. In this case, 3 kg/m^3 of polypropylene copolymer fibres 50 mm length and 0,3 mm diameter were used. Adopted joint spacing was 8 m. Obviously, the residual properties were different than those of Case 1.

Initial deformations that took place before the deposit came into service, as a consequence of the expected settlement of the soil. Some cracks are barely visible on the surface. Taking into account the characteristics of the concrete reinforced with synthetic macrofibres, the slabs were initially deformed as shows Fig. 7.a in schematic way. Later as a consequence of loads application, the deformation took place in the manner indicated in Fig. 7.b. Slabs with plain or steel bars reinforced concrete would have deformed as shows Fig. 7.c; in this way, without fibres, the movement of the lifters would have been difficult, if not impossible.

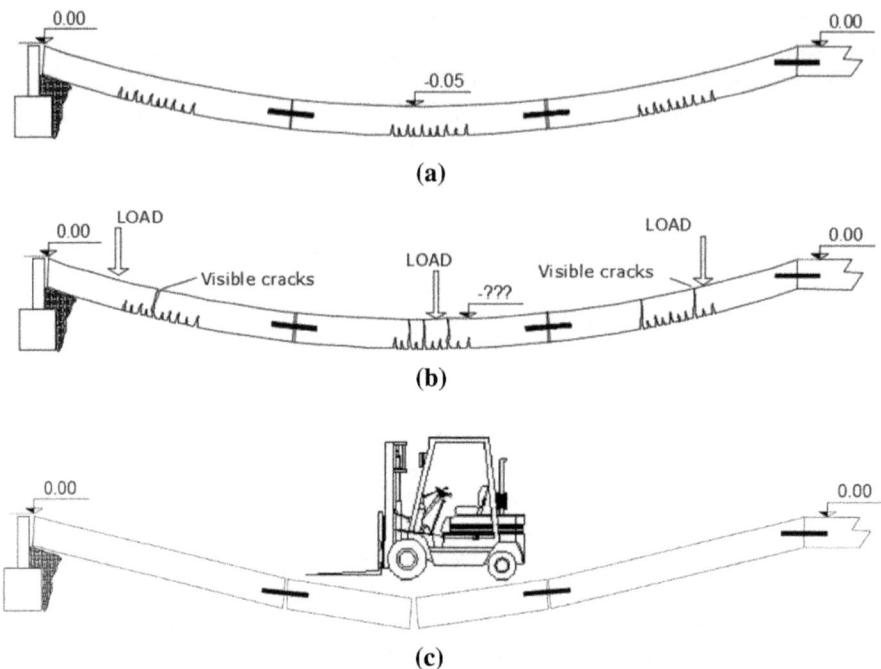

Fig. 7. Warehouse in the coast of the Rio de la Plata river (2019). 7.a (top) Initial deformations; 7.b (centre) Deformations after loading; 7.c (bottom) Slabs without fibres.

5   Concluding Remarks

Selected examples of large floor and pavement works in Argentina have been presented in this paper with the aim of showing technical and economic advantages that can be obtained from the use of polymer and steel FRC and ShCFRC technologies in design and execution of industrial floors. They have shown that:

- When poor soils are present, the use of ShCSFRC provides a superior slabs performance compared to plain concrete. The concrete ductility combined with the expansive effect allows the slabs to be deformed without wide opening of the cracks that originate.
- The control of the generation of cracks, the lower curling effect and the smaller opening of joints that are achieved with the ShCSFRC allow improving the conditions of the assemblies of precision. The slabs are less deformed.
- The contribution of synthetic macrofibres to improve the post-cracking behaviour of concrete increasing flexural and ultimate load carrying capacity of the slabs, allows a reduction of industrial pavement thicknesses. This means a very important economy for the project.
- The incorporation of synthetic macrofibres transforms the concrete into a more ductile material, reducing the consequences of the deformations.

References

1. Holland, J.: Macro Polymeric Fibres for Slabs on Ground Concrete Construction Magazine, 1 June (2008)
2. Paul, B., Polivka, M., Metha, P.: Properties of fiber reinforced shrinkage-compensating concrete. ACI Mat. J. **78**, 488–492 (1981)
3. TR 34–2003: Concrete Industrial Ground floors: A Guide to Design and Construction, The Concrete Society, England (2003)
4. Portland Cement Association, Concrete floors on ground, Fourth edition (2008)
5. Boyd, R.: Designing Floors Slabs-on-Grade, Hanley Wood Inc. (1992)
6. Walker W., Holland J.: Reinforcement for slabs on ground, Concrete Construction, Floors (2007)
7. ACI Committee 302, 'Guide for Concrete Floor and Slab Construction', ACI 302.1R-04 (2004)
8. Holland, J.: Proper Use of Slab Reinforcement Systems: Avoiding Myths & Misconceptions, World of Concrete (2013)
9. Fornasier, G., Zitzer, L., Pombo, J.R.: Some considerations about the use of slag-cements in the formulation of shrinkage compensating concretes. In: Proceedings 6th International Colloquium-Industrial Floors 2007, vol. 2, Technische Akademie Esslingen, Germany (2007)
10. ACI Committee 360, Guide to Design of Slabs-on-Ground, ACI 360R-10 (2010)
11. Fernandez Luco, L., Pombo, J.R., Torrent, R.: Shrinkage Compensating Concrete in Argentina Concr. Inter. **25** (5), (2003)
12. ACI Committee 223, Guide for the Use of Shrinkage-Compensating Concrete, ACI 223R-10 (2010)
13. CEN/TC 229, EN 14651:2005 Test method for metallic fibered concrete - Measuring the flexural tensile strength (limit of proportionality (LOP), residual) Méthode (2005)
14. ASTM C1609 / C1609M–07. Standard test method for flexural performance of fiber-reinforced concrete (using beam with third-point loading), ASTM Standards, vol. 04.02 (2012)
15. Altoubat, S.A., Roesler, J.R., Lange, D.A., Rieder, K.A.: Simplified method for concrete pavement design with discrete structural fibers. Cons. Build. Mat. **22**, 384–393 (2008)

Design Optimization of Fibres Reinforced Concrete Railway Tracks by Using Non-linear Finite Elements Analysis

Jean-Louis Tailhan and Pierre Rossi[✉]

MAST-EMGCU, Univ Gustave Eiffel, IFSTTAR,
F-77447 Marne-la-Vallée, France
pierre.rossi@univ-eiffel.fr

Abstract. Between 2007 and 2014, IFSTTAR, Alstom and other industrials partners have developed a new concept of railways track called New Ballastless Track (NBT). The concept was validated under 10 million fatigue cycles on a real-size mockup at IFSTTAR

A first numerical study, using a non-linear model was performed to evaluate the possibility of replacement of the original reinforced concrete layer of the track slab by a steel fiber reinforced concrete, to simplify the construction of the NBT track and to take advantage of the redistribution of mechanical stresses on a hyper-static structure. This study led to the conclusion that this replacement was very relevant.

This paper is on the optimization of this Fibres Reinforced Concrete Railway Tracks solution by using the same non-linear numerical model. It is shown that this optimization procedure leads to a significant reduction of CO_2 emissions compared with the initial one.

Keywords: Fibre reinforced concrete · Railway tracks · Numerical model · Design optimization · Carbone footprint

1 Introduction

Steel fiber reinforced concrete (SFRC) is increasingly used in structural applications. One of the principal reasons of its gain in popularity is due to the recent emergence of national and international recommendations for the design of structures using this type of material. These recommendations are efficient for designing simply supported structural elements subjected to bending. However, they do not possess a sufficient physical base to propose relevant solutions for more complex structures such as these statically indeterminate. Hence, it is claimed that the most efficient approach for designing such structures with respect to both safety and sustainable development is to use non-linear finite element analysis.

IFSTTAR has been developing, since 1985, a probabilistic discrete cracking model to simulate the cracking process of concrete. Then, this model was extended to analyze reinforced concrete and fibre reinforced concrete structures. All these numerical models have been, today, fully and deeply validated [1–9].

© RILEM 2022
P. Serna et al. (Eds.): BEFIB 2021, RILEM Bookseries 36, pp. 652–665, 2022.
https://doi.org/10.1007/978-3-030-83719-8_56

Between 2007 and 2014, IFSTTAR, Alstom and other industrials partners have developed a new concept of railways track called New Ballastless Track (NBT) [10]. The concept is based on two superimposed independent layers of concrete slabs (i.e. foundation and track slabs) and was designed to achieve a 100 years life span under a mixed high speed and freight traffic. It was first validated through a FEM model and then under 10 million fatigue cycles on a real-size mockup at IFSTTAR [10–12]. A trial section of 1 km was then built by Alstom on the French railways network [10] and open to normal speed traffic, by end of 2013. It was monitored during 2 years and after 5 years of service, it behaves well, confirming the relevance of the concept.

Figure 1 presents the original concept and the geometry of the experimental mock-up. In this figure, the upper track layer is made of reinforced concrete (C35/45) called BC5 while the foundation slab is made of plain concrete (C25/30) called BC3. An elastic layer was used to reproduce the mechanical reaction of the ground simulating the soil bearing capacity.

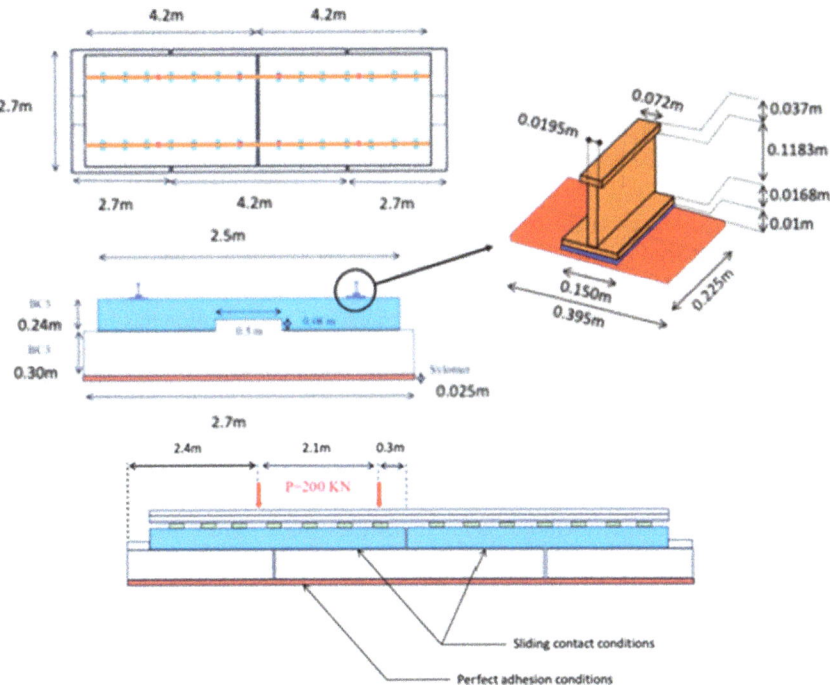

Fig. 1. Example of high quality Enhanced MetaFile (.EMF) format chart.

A first numerical study [13] was performed to evaluate the possibility of replacement of the original reinforced concrete layer BC5 of the track slab by a steel fiber reinforced concrete, to simplify the construction of the NBT track and to take advantage of the redistribution of mechanical stresses on a hyper-static structure. This numerical study was performed by using the probabilistic discrete cracking model evocated above.

The comparison between the two technical solutions was made by analyzing and comparing their respective cracking process.

The principal conclusions related to this first numerical work were the following:

- When in a given concrete structure, *the tensile stresses are localized*, the traditional rebars are more efficient to control cracks than the presence of fibres (the opening cracks are smaller). For that, it is necessary, of course, to get an easy and optimal placement of the rebars in relation with cracks orientation. That is the case in railway tracks.
- On the contrary, when *the tensile stresses are diffused, that is the case in statically indeterminate mechanical situations*, the fibres are significantly more efficient than traditional rebars in regard to this cracks control (more cracks less opened). The parts of the railway track slabs which are located between the zones delimited by the joints in the foundation slab are concerned by this conclusion.

For the practice, it can be argued that, for railway track slabs; the use of 78 kg/m^3 de fibres (it was the case considered in the framework of this first numerical study) is mechanically more efficient than the usual ratio of rebars. On the other hand, if this technical solution is adopted, it is preferable to use, in zones of high concentration of tensile stresses, some local rebars (low quantity) to ensure a necessary level of safety to the structure.

The objective of this second study on the same topic is to optimize the design of the steel fibres reinforced railway track (always by using the probabilistic discrete cracking model).

To achieve this objective, the following strategy of optimization has been followed:

- Step 1: Lower dosage of fibres has been considered (compared with the first study). Indeed, this dosage passes from 78 kg/m^3 to 60 kg/m^3. The local placement of rebars in zones of high concentration of tensile stresses (it means in the upper track layer, B3, at the vertical of the joints of the foundation slab, B5) is also considered. The same sectional percentage of bottom rebars than the one used in the reinforced concrete solution is chosen, but le length of these rebars are a lot of shorter because it passes from 4.2 m to 40 cm.
- Step 2: The same technical solution than the one of the step 1 is considered except for the sectional percentage of local bottom rebars which is divided by two.

2 Effect of the Local Reinforced Rebars

2.1 Finite Elements Mesh

Figure 2 presents the finite elements mesh related to FRCRT with the local reinforced rebars. The total section of local rebars are, respectively, 3.38 10^{-3} m^2 for the step 1 of the optimization procedure and 1.69 10^{-3} m^2 for the second step.

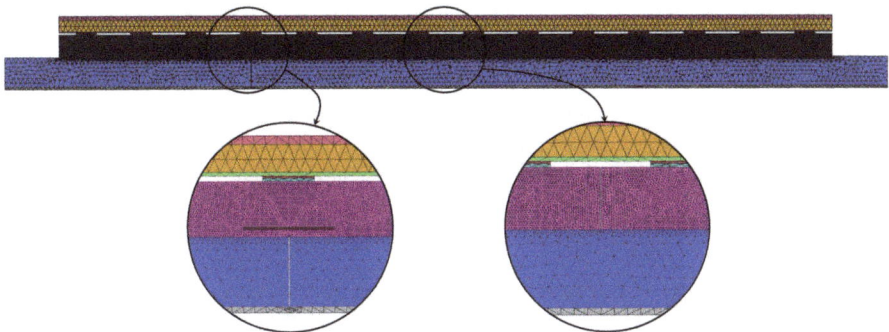

Fig. 2. 2D finite element mesh related to the numerical simulations of the FRCRT with local rebars.

As observed on this figure, the mesh is two-dimensional (2D). Indeed, although the mechanical problem is a three-dimensional (3D) one, it is unreasonable to perform non-linear 3D simulations which should consume too much computational time with the models used (see Sect. 2.2). The numerical simulations are then performed under 2D and plane stress conditions, since these specific models also take into account volume effects (see Sect. 2.2) and the width (i.e. the length measured in the direction perpendicular to the plane of the figure) of the mockup has to be considered.

For the reason mentioned above and related to volume effects (see Sect. 2.2), considering the total width of the mockup will cause the models to produce a poor description of the cracking processes. But, nevertheless, a more detailed description of the cracking processes can be achieved by choosing to perform simulations with a smaller value of this width; since, of course, the simulation also gives acceptable information about the longitudinal distribution of the elastic tensile stresses at the bottom of the track slab (where cracks will be created) compared with 3D simulation.

The width of the cast iron plates under the rail, and equal to *0.395 m* (see Fig. 1), seems to be the smaller one it is reasonable to consider in the present 2D numerical analysis. This choice leads to underestimate the real stresses diffusion under the rails and, so, to underestimate the global stiffness of the mechanical system (all the mockup). To solve this problem, one possible solution (used with success in [13]) is to increase the Young modulus of the materials constituting the mechanical system, especially for the track and foundation slabs and the elastic layer (where the stresses are diffused). To achieve this objective, the following steps of numerical simulations have been performed:

- First step: 3D elastic simulation of the full mockup with the real stiffness of the materials.
- Second step: 2D elastic simulation (plane stresses conditions) with the full width of the track slab mockup and with the real stiffness of the materials.
- Third step: 2D elastic simulations with smaller width (0.395 m) of the mockup and different values of stiffness related to the different materials.

Remark: In all these numerical simulations (3D and 2D), the rebars are not considered.

This search of the best materials stiffness related to the 2D simulation on the reduced mockup width leads to the values summarized in Table 1.

Table 2 presents the values of the others materials properties used in the framework of these elastic numerical simulations. They are independent of the type of linear numerical simulation performed.

Table 1. Values of the materials Young modulus used in the elastic numerical simulations.

Materials	Young modulus (MPa)		
	3D Simulation	2D Simulation full wide	2D Simulation reduced wide
Rails	210000	210000	210000
Pad	7.702	7.702	7.702
Iron Saddles	17200	17200	17200
Track Slab	35000	35000	43750
Foundation Slab	24000	24000	32500
Sylomer	1.2	1.2	1.625

Table 2. Values of the materials young modulus used in the elastic numerical simulations.

Materials	Poisson coefficient	Density
Rails	0.3	7.8
Soles	0.4	1.2
Iron Saddles	0.3	7.8
Track Slab	0.25	2.5
Foundation Slab	0.25	–
Sylomer	0.32	–

2.2   Numerical Model

2.2.1   Probabilistic Explicit Cracking Model of Concrete

The model was first developed at IFSTTAR (formerly LCPC) by Rossi [14–16] and recently improved by Tailhan et al. [17]. It describes the behaviour of concrete via its two major characteristics: heterogeneity and sensitivity to scale effects [18]. The physical basis of the model (presented in detail in [14–16]) can be summarized as follow:

- The heterogeneity of concrete is due to its composition. The local mechanical characteristics (tensile strength f_t, shear strength τ_c) are randomly distributed.
- The scale effects are a consequence of the heterogeneity of the material. The mechanical response directly depends on the volume of material that is stressed.
- The cracking process is controlled by defects in the cement paste, by the heterogeneity of the material, and by the development of tensile stress gradients.

The following points specify how the numerical model accounts for these physical evidences:

- The model is developed in the framework of the finite element method, each element representing a given volume of heterogeneous material.
- The tensile strength is distributed randomly over all elements of the mesh using a Weibull distribution function whose characteristics depend on the ratio: *volume of the finite element/volume of the largest aggregate*, and the compressive strength (as a good indicator of the quality of the cement paste). The volume of the finite element depends on the mesh, while the volume of the largest aggregate is a property of the concrete [14–16].
  Remark: a Weibull distribution function is the best to take into account the rupture in tension of a brittle and heterogeneous material as concrete.
- The shear strength is also distributed randomly over all elements using a distribution function: (1) its mean value is independent of the mesh size and is assumed equal to the half of the average compressive strength of the concrete and (2) its deviation depends on the element size, and is the same (for elements of same size) as that of the tensile strength.

Concerning the cracks representation, two approaches are proposed, depending if 2D or 3D numerical simulations are concerned [17].

In what follows, only aspects related to 2D modelling are presented.

In 2D simulations, the cracks are explicitly represented by 2D non-linear interface elements (quadratic elements) of zero thickness. These elements connect volume elements representing un-cracked plain concrete. Failure criteria of Rankin in tension and Tresca in shear are used. As far as tensile or shear stresses remain lower than their critical values, the interface element ensures the continuity of displacements between the nodes of its two neighboring volume elements. The material cell gathering these two volume elements and the interface element remains therefore elastic. Once one of the preceding failure criteria is reached, the interface element opens and an elementary crack is created. The tensile and shear strengths as well as the normal and tangential stiffness values, related to this interface element, become equal to zero [14–16]. In case of crack re-closure, the interface element recovers its normal stiffness and follows a classical Coulomb's law [14–16]. This numerical model is summarized in Fig. 3.

Note that in this modelling approach, the creation and the propagation of a crack is the result of the creation of elementary failure planes that randomly appear and can coalesce to form the macroscopic cracks).

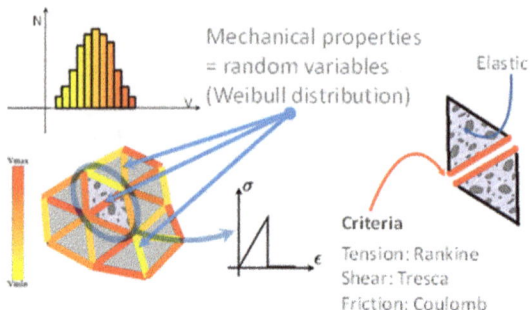

Fig. 3. Probabilistic concrete cracking model – explicit approach

2.2.2   Probabilistic Explicit Cracking Model of Fibre Reinforced Concrete (FRC)

The creation of cracks in the concrete matrix is represented by an elastic perfectly brittle behaviour, whereas the bridging effect of the fibres is described by the following modelling approach.

Normal and tangential stresses in the interface element linearly increase with normal and tangential displacements when a "broken" interface element re-opens to take into account the elastic bridging effect of the fibres inside the crack. Physically speaking, the rigidity of the fibres (inside the cracks) is more important in tension than in shear. Thus, the interface element rigidity is considered different for normal and tangential displacements for the two different failure criteria. In 2D, normal and tangential rigidities of the interface element are K_n' and K_t' respectively. The post-cracking elastic behaviour exists until it reaches a threshold value, ζ_0, related to the normal displacement (Fig. 4). The mechanical behaviour of the interface element changes once this threshold value is reached. The normal stress is considered as linearly decreasing with the normal displacement in order to take into account the damage of the bond between the concrete and the fibre, and fibre pullout. The decreasing evolution is obtained by using a damage model.

Finally, the interface element is considered definitively broken when the normal displacement reaches a threshold value, ζ_c (Fig. 4). This value corresponds to the state where the effect of fibres is considered negligible. It is determined from a uniaxial tensile test and characterised the FRC. At this point, its normal and tangential rigidities are set to zero.

The post-cracking energy dissipated by the bridging effect of the fibres is considered randomly distributed over the mesh elements. The random distribution chosen is a log-normal distribution function with a mean value independent of the mesh elements size [19] and a standard deviation, due to the heterogeneity of the material, increasing as the mesh elements size decreases. The choice of a log-normal distribution function is an arbitrary one. It is convenient to avoid having negative values of the post-cracking energy when the element meshes are very small. To model a given structural element, the distribution function is determined in the following manner:

- The mean value is directly obtained experimentally from a certain number of uniaxial tensile tests on notched specimens, more specifically from the load-crack opening experimental curves (it is the best way of determination). If the uniaxial tensile tests have not been performed, that mean value can be determined also by analysing bending test results using an inverse approach.
- The standard deviation, which depends on the mesh elements size, is determined by an inverse analysis approach that consists of simulating the uniaxial tests with different element mesh sizes. As the mean value of the post-cracking energy is known from the experimental results, several numerical simulations are realized for each mesh size to determine the standard deviation that best fits the experimental results. The inverse analysis approach thus allows finding a relation between the standard deviation and the finite element mesh size. As for the mean value, the standard deviation can be determined by analysing bending test results using an inverse approach.

The threshold parameters ζ_0 and ζ_c are determined by an inverse analysis approach to best fit the simplified triangular stress-displacement curve representing the post-cracking energy (Fig. 4) to the experimental tensile softening curve.

Figure 4 presents the numerical mechanical behaviour adopted to represent the experimental post-cracking behaviour. Only the *normal stress-normal displacement* curve is considered in this figure.

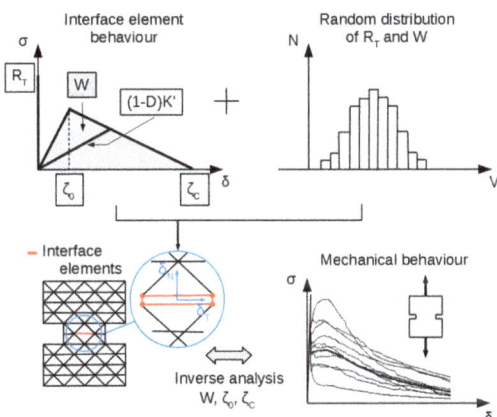

Fig. 4. Probabilistic FRC model

2.2.3 Rebars and Concrete/Steel Bond Modelling

The rebars are considered, in the 2D modelling approach, as equivalent plates mixing concrete and steel rebars. The reinforcement ratio at the top of the track slab being different of the one located at the bottom, two equivalent plates are considered in the simulations. The heights of these plates are taken equal to the greater diameter of the local rebars. The equivalent Young moduli of the plates are determined by using a classical mixture rule (see relation 1).

$$E = (EaAa + EbAb)/(Aa + Ab) = (Ab/Aa + Ab) [Eb + taEa] \quad (1)$$

Where:

E_a and E_b are respectively the steel and the concrete Young moduli;

A_a and A_b are respectively the steel and the concrete sections;

t_a is a local reinforcement ratio (for the considered plate).

Numerically speaking, the equivalent plates are composed of one layer of volume elements. Here again, the volume elements are all interfaced by interface elements. These interface elements allow the crack crossing through the equivalent plates. They are exactly the same as those used to simulate concrete cracking and follow the same opening criteria. In this case, the volume of concrete considered for each opening of interface element is that of the cumulated volume of elements surrounding it (it is, of course, an approximation).

After the opening of interface element, its residual stiffness is calculated considering these two following assumptions:

- The interface element section considered is that of the equivalent plate crossed by the crack.
- A total rupture of adherence along the two volume elements concerned by the interface element is assumed.

Table 3 presents the values related respectively to the height, the width, the Young modulus and the Poisson coefficient of the equivalent plates in the simulations. The two sectional percentages of local rebars are considered.

Table 3. Parameters values related to the equivalent plates in the simulations

2D reinforcement	High (m)	Width (m)	ta (%)	E (MPa)	Poisson coefficient
full ratio of local rebars	0.016	0.395	8.81	50506	0.25
half ratio of local rebars	0.016	0.395	4.4	42687	0.25

2.3 Non-linear numerical simulations

The parameters values (in relation with the size of finite element meshes, see paragraphs 2.1 and 2.2) used in the non-linear simulations are given in Table 4.

The FRC parameters, considered in this study, are determined from the FRC successfully used in the previous numerical study [13]. This FRC, as evocated before, contained 78 kg/m^3 of steel fibres. These fibres are 60 mm long hooked end ones.

In this present work, it has been decided, as evocated before, to consider the same matrix with the same fibres but with a dosage of 60 kg/m^3.

To evaluate this FRC parameters values, a simple assumption is made based on some knowledge concerning FRCs [20]. Thus, it is assumed that, with the type of FRC used in this study (type of matrix and type of steel fibre), the post-cracking dissipated energy is lineary proportional to the fibres dosage when this one is between 50 and 80 kg/m^3.

Table 4. Parameters values related to the equivalent plates in the simulations.

Parameter		Value
RT	Mean value	4.87 [MPa]
	Standard deviation	0.67 [MPa]
	Mean value	$2.56 \ 10^{-3}$ [MPa.mm]
	Standard deviation	$2.6 \ 10^{-3}$ [MPa.mm]
	ζ_0	50 [µm]
	ζ_c	4 [mm]

It is also assumed that the standard deviation related to this post-cracking energy is kept identical. That means that the coefficient of dispersion related to the FRC containing 60 kg/m^3 of fibres is higher than those related to the FRC containing 78 kg/m^3 of fibres. This assumption is reasonable because it is well known that the mechanical degree of heterogeneity decreases when the percentage of fibres increases when this percentage stay below what is called the saturation percentage of fibres [20].

In Fig. 5 are presented cracking patterns of the FRCRT with the full local reinforced rebars. Three different simulations related to different random distributions are concerned.

In Fig. 6 are presented cracking patterns of the FRCRT with half ratio of local reinforced rebars. Three different simulations related to different random distributions are concerned.

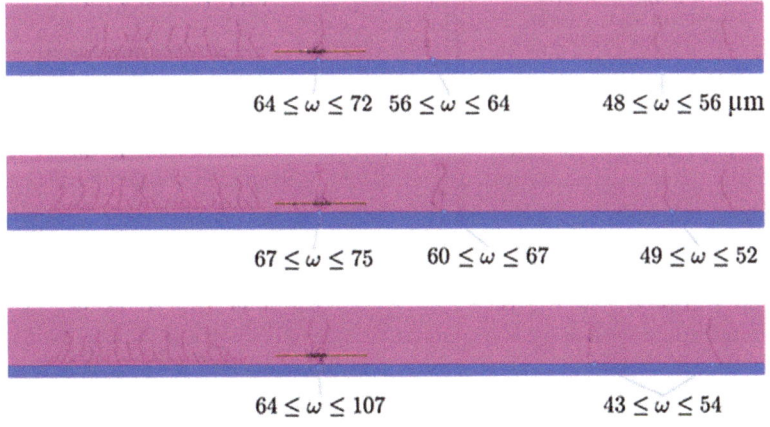

64 ≤ ω ≤ 72   56 ≤ ω ≤ 64   48 ≤ ω ≤ 56 µm

67 ≤ ω ≤ 75   60 ≤ ω ≤ 67   49 ≤ ω ≤ 52

64 ≤ ω ≤ 107   43 ≤ ω ≤ 54

NB: All other cracks having openings lower than 25µm to 30µm

Fig. 5. Cracking patterns related to the FRCRT with full local rebars– 3 simulations.

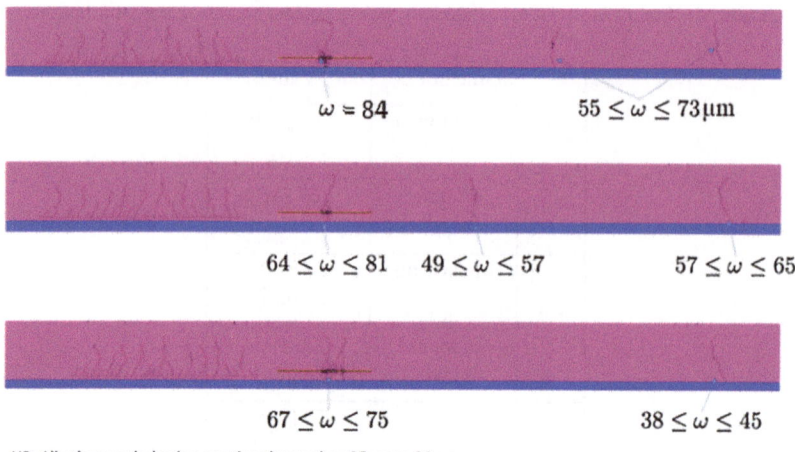

$\omega \simeq 84$ $55 \leq \omega \leq 73\,\mu\mathrm{m}$

$64 \leq \omega \leq 81$ $49 \leq \omega \leq 57$ $57 \leq \omega \leq 65$

$67 \leq \omega \leq 75$ $38 \leq \omega \leq 45$

NB: All other cracks having openings lower than 25μm to 30μm

Fig. 6. Cracking patterns related to the FRCRT with half ratio of local rebars – 3 simulations.

From Figs. 5 and 6, the following comments can be made:

- The cracks opening are very well controlled by mixing fibres and local rebars. All the cracks openings are smaller than 100 μm that is the maximal crack opening allowed by the industrial for this application.
- When cracks opening at the vertical of the joints of the foundation slab are concerned, there is no significant difference between using total section or half section of reinforcement. This can be explained by the fact that the anchorage length of the rebars through the cracks are the same and that the rebars are just more stressed when the section of reinforcement is half.

2.3.1   Carbone Footprint Evaluations

In this paragraph a very simple comparison related to the carbone footprint of two technical solutions of railway track is proposed: the classical solution of railway tracks reinforced only with rebars [8] and that of railway tracks reinforced with a mix of steel fibres and rebars.

This comparison is based on some data from literature [21].

As, in the two technical solutions considered, the thickness of the upper layer track is the same, the comparison will be based only on the reinforcements analysis.

2.4   Classical Solution

The reinforcement ratio at the top of the track slab is different of the one located at the bottom of the track slab [8]. They are, respectively, $1.53 \ 10^{-3} \ \mathrm{m}^2$ (7 rebars of diameter 14 mm + 4 rebars of diameter 12 mm) and $3.38 \ 10^{-3} \ \mathrm{m}^2$ (14 rebars of diameter 16 mm + 5 rebars of diameter 12 mm).

The length of the rebars to be considered in this work is 4.2 m.

So, the total volume of rebars steel related to one upper layer track is **28.622 10^{-3}m^3**.

In consequence the weight of rebars steel related to the classical solution is **223.25 kg**.

In [21], it is found that 1.932 kg of CO_2 is emitted per ton of steel rebars.

So, in the framework of this study, *the quantity of CO_2 emitted by this solution is* **0.43 kg**.

2.5   Mixed Solution

The fibres dosage of the FRC used in this study is 60 kg/m^3. The volume of one top layer railway track being equal to **2.72m^3**, the weight of fibres used is **163.2 kg**.

The length of the local rebars is 0,4 m and their total section is **1.69 10^{-3}m^2**(for the low sectional local reinforcement). So, the total volume of local rebars is **6.76 10^{-4}m^3**and the weight of local rebars is **0.53 kg**.

In [21], it is found that 2.425 kg of CO_2 is emitted per ton of steel fibres.So, in the framework of this study, *the quantity of CO_2 emitted by this solution is* **0.40 kg**.

If this very simplistic analysis is taken into account, it can be considered that ***the mixed solution emits 7% of CO_2 less than the classical solution.***

3   Conclusions

This study concerns the use of numerical models to analyze and compare the cracking process of two types of Railway Tracks: one classical in reinforced concrete and the second one in FRC with local rebars (mixed solution). The models used are Probabilistic Explicit Cracking ones developed by IFSTTAR and fully validated in the framework of previous works.

The main result obtained can be summarized as following:

- The mixed solution is, at least, so mechanically efficient (to control craks opening) than the classical one.
- The mixed solution emits 7% of CO_2 less than the classical one.

In the future, it would be interesting for confirming the present results, to perform experimental tests on a mockup with a slab track made of the FRC (with local rebars) used in that study, as it was previously done with the reinforced concrete during original development of NBT concept. It should be also necessary to confirm or improve the very rough analysis proposed, in the present work, concerning the carbone footprint of the technical solutions studied.

Finally, it should be also interesting, to use the numerical models presented in this work to continue the optimization of the structural design of the SFR track slab (other quality of matrix, of percentage of fibers, of slab thickness…).

As example, if it was mechanically relevant (in relation with the cracks opening control) to decrease about 30% the thickness of the upper track layer by increasing the content of fibres passing from 60 kg/m^3 to 78 kg/m^3 and by conserving the same section of local rebars (1.69 10^{-3} m^2), it would be observed:

- A decrease of 30% of the carbone footprint related to the concrete.
- A decrease of 7.2% of the carbone footprint related to the reinforcement.

At this very rougth analysis, it should be important to add the decrease of material cost and working time.

References

1. Phan, T.S., Tailhan, J.-L., Rossi, P.: 3D numerical modelling of concrete structural element reinforced with ribbed flat steel rebars. Struct. Concr. **14**(4), 378–388 (2013)
2. Tailhan, J.L., Rossi, P., Daviau-Desnoyers, D.: Numerical modelling of cracking in steel fibre reinforced concrete (SFRC) structures. Cement Concr. Compos. **55**, 315–321 (2015)
3. Rossi, P., Daviau-Desnoyers, D., Tailhan, J.L.: Analysis of cracking in steel fibre reinforced concrete (SFRC) structures in bending using probabilistic modelling. Struct. Concr. **16**(3), 381–388 (2015)
4. Rastiello, G., et al.: Macroscopic probabilistic cracking approach for the numerical modelling of fluid leakage in concrete. Ann. Solid Struct. Mech. **7**(1–2), 1–16 (2015)
5. Rossi, P., et al.: Numerical models for designing steel fiber reinforced concrete structures: why and which ones?. In: Massicotte, B., Charron, J.-P., Plizzari, G., Mobasher, B., (eds.) FRC 2014: ACI-fib International Workshop, FIB Bulletin 79 – ACI SP-310, pp. 289–300 (2016)
6. Rossi, P., Tailhan, J.L.: Numerical modelling of the cracking behaviour of steel fibre reinforced concrete (SFRC) beam on grade. Struct. Concr. **18**(4), 571–576 (2017)
7. Rossi, P., Daviau-Desnoyers, D., Tailhan, J.L.: Probabilistic numerical model of ultra-high performance fiber reinforced concrete (UHPRFC) cracking process. Cement Concr. Compos. **90**, 119–125 (2018)
8. Nader, C., Rossi, P., Tailhan, J.L.: Numerical strategy for developing a probabilistic model for elements of reinforced concrete. Struct. Concr. **18**(6), 883–892 (2017)
9. Nader, C., Rossi, P., Tailhan J.L.: Multi-scale strategy for modelling macrocracks propagation in reinforced concrete structures, Cement Concrete Compos. 99, 262-274 htts://doi.org/https://doi.org/10.1016/j.cemconcomp.2018.04.012 (2018)
10. Robertson, I., et al.: Advantages of a new ballastless trackform. Constr. Build. Mater. **92**, 16–22 (2015)
11. Chapeleau, X., et al.: Study of ballastless track structure monitoring by distributed optical fiber sensors on a real-scale mockup in laboratory. Eng. Struct. **56**, 1751–1757 (2013)
12. Sedran, T., et al.: Development of a new concrete slabs track (NBT). In: 12th International Symposium on Concrete Road, Prague, Czech Republic, 23–26 September (2014)
13. Tailhan, J.L., Rossi, P., Sedran, T., Comparison between the cracking process of reinforced concrete and fibres reinforced concrete railway tracks by using non-linear finite elements analysis. In: 10th International Conference on Fracture Mechanics of Concrete and Concrete Structures, FRAMCOS 10, Edited by Pijaudier-Cabot, Grassl and Laborderie, Bayonne - France , 23–26 Juin 2019
14. Rossi, P., Richer, S.: Numerical modelling of concrete cracking based on a stochastic approach. Mater. Struct. **20**, 334–337 (1987)
15. Rossi, P., Wu, X.: Probabilistic model for material behaviour analysis and appraisement of concrete structures. Mag. Concr. Res. **44**(161), 271–280 (1992)
16. Rossi, P., Ulm, F.-J., Hachi, F.: Compressive behaviour of concrete: physical mechanisms and modeling. J. Eng. Mech. **122**(11), 1038–1043 (1996)

17. Tailhan, J.L., Dal Pont, S., Rossi, P.: From local to global probabilistic modelling of concrete cracking. Ann. Solid. Struct. Mech. **1**, 103–115 (2010)
18. Rossi, P., Wu, X., Le Maou, F., Belloc, A.: Scale effect on concrete in tension. Mater. Struct. **27**, 437–444 (1994)
19. Rossi, P.: Experimental study of scaling effect related to post-cracking behaviours of metal fibres reinforced (MFRC). Eur. J. Environ. Civ. Eng. **16**(10), 1261–1268 (2012)
20. Rossi, P.: Les bétons de fibres métallique, Edité par les Presses de l'Ecole Nationale des Ponts et Chaussées (1998). (in french)
21. ITAtech Guidance For Precast Fibre Reinforced Concrete Segments 2016 vol 1: Design Aspects ISBN: 978–2–9701013–2–1.

Elevated Flat Slab of Fibre Reinforced Concrete Non-linear Simulation up to Failure

Alejandro Nogales[(✉)] and Albert de la Fuente

Civil Engineer, Universitat Politècnica de Catalunya, Barcelona, Spain
alejandro.nogales@upc.edu

Abstract. The growing use of fibre reinforced concrete (FRC) on structural concrete has made that several codes and guidelines have included models for design in which the traditional reinforcement has been substituted partially or totally. Among the industrial applications, fibres as reinforcement of elevated flat slabs is gaining interest due the post-break bearing capacity of the material. This technology has already been used for real scale structures. This research contribution is focused on a parametrical analysis of FRC elevated flat slab by means of non-linear finite element simulation. The model is calibrated and compared with the real scale experimental test and with experimental slabs tested up to failure that can be found in the literature. The main goal of this paper is to carry out a parametrical analysis of the slab combining different types of reinforcement: FRC and hybrid reinforcement (fibres + conventional reinforcement) under design loads eventually choosing an amount of fibres for an actual test The results demonstrated that the combination of fibres and rebar improves the structure against failure, reduces deformation and presents a crack pattern better for cracking control.

Keywords: Fibre reinforced concrete · Non-linear analysis · Elevated flat salbs · Hybrid reinforcement

1 Introduction

Over the last decades, the use of fibres for structural concrete reinforcement has noticeable grown. The growing use of FRC has made that several codes and guidelines such as DBV [1], RILEM [2], CNR-DT [3], EHE-08 [4], *fib* Model Code (*fib* MC-2010) [5] and ACI 544 [6] have included models for designing structural elements in which the traditional rebar reinforcement has been partially or totally substituted. In this regard, *fib* MC-2010 has introduced FRC classes in order to classify the post cracking strengths of FRC. The codes are based on state limit methods and gather constitutive equations able to reproduce the post-break response of FRC valid for sectional analysis. The publication of design codes for FRC has boosted the use of this technique in building industry. In this sense, FRC is used for elevated slabs supported on piles or foundations [7, 8], where failure and cracking behaviour has often been a troublesome concern for designers. These structures are likely to be under flexural and

© RILEM 2022
P. Serna et al. (Eds.): BEFIB 2021, RILEM Bookseries 36, pp. 666–677, 2022.
https://doi.org/10.1007/978-3-030-83719-8_57

bending loading scenarios and allow exploiting the post-cracking tensile capacity and strength of FRC, bearing part of the total tensile forces, leading to a conventional reinforcement reduction. Apart from the technical point of view, buildings cast with FRC take profit from time and costs saving during the building process: fibres are added directly at the concrete plant and time needs reduces significantly compared to conventional reinforcement concrete where handing and placing of steel rebar operations are required. Besides, from sustainability point of view, indirect costs, social aspects and environmental issues are factors to take into account and therefore FRC is an attractive solution [9].

Nonetheless, despite of the existence of design codes and guidelines together with the experimental and numerical evidences that prove the suitability of using FRC in pile supported flat slabs, there is still controversy and barriers to its implementation [9]. FRC results are commonly compared with hybrid reinforced concrete or reinforced concrete (RC) and a lack of research is found when comparing between FRC classes. The reported experiences highlight the need of intensifying the research on: (1) the influence of the FRC strength class (flexural residual strength) on ultimate limit state (ULS), even with the combination of traditional reinforcement (hybrid reinforcement) and (2) the safety factors to be used for the design.

In view of this, the aim of this research contribution is focused on a parametrical analysis of an elevated flat slab by means of a non-linear finite element (FE) model capable of reproducing the post-cracking response of FRC and develop the stress redistribution mechanisms that are produced in the structures hyperstatically supported. The model was validated with actual elevated slabs tested up to failure found in the literature and the results are compared with the experimental test. The main goal of this paper is to carry out a parametric analysis of an elevated FRC slab considering design loads and different FRC classes. The analysis was also done from a safety point of view, taking into account the nominal resistance of the structure and the global safety factors.

2   Numerical Modelling of FRC

According to *fib* MC-2010 FRC classes can be classified by its residual strength characteristic values ($f_{Ri,k}$ (i = 1 to 4)) obtained form 3 point bending test notched beams according to EN 14651:2005. Two parameters are used for classifying: f_{R1k}, which represents the residual strength for a crack opening displacement (CMOD) of 0.5 mm and a letter (a, b, c, d or e) that represents the f_{R3k}/f_{R1k} ratio, where f_{R3k} stands for a CMOD of 2.5 mm. In order to stablish the stress-strain (σ-ε) or stress-crack width (σ-w) constitutive relations, the equations proposed by *fib* MC-2010 are adopted, using the variables' mean values for the simulations. The compressive behaviour has been characterised according to the expression proposed by *fib* MC-2010. The implementation of these equations have been done using the software ABAQUS [10] and the "Concrete Damage Plasticity" (CDP) model available. In this regard, the σ-w tensile and σ-ε compression curves proposed by the *fib* MC-2010 are used for the analysis, the characteristic length (L_{ch}), to turn crack opening into strain, implemented by default by the software is the size of the element.

The CDP parameters adopted in this analysis are the default ones proposed in ABAQUS User's Manual [10] for plain concrete and are reported in Table 1, where f_{b0}/f_{c0} and K represent the ratio of biaxial compressive strength to uniaxial compressive strength and the yield surface shape parameter, respectively. The analysis was per-

Table 1. CDP parameters used for the analysis.

Dilation angle	Eccentricity	f_{b0}/f_{c0}	K	Viscosity
30	0.1	1.16	0.667	1e–5

formed using the ABAQUS explicit dynamic algorithm (quasi-static analysis) whose integration method is known as Forward Euler.

3   Experimental Model Validation

A point load experimental test on a suspended elevated slab carried out by Gossla [11] is used for the model validation, this experimental test has previously been used for model validation by Facconi [12]. The slab composed by nine (9) 6.0 × 6.0 m^2 span, 0.2 m height resting on a grid of 16 0.3 × 0.3 m^2 square columns was tested applying a punctual load at the centre of the slab (Fig. 1a). The fibre dosage was 100 kg/m^3 of an undulated commercial steel fibre (1.3 mm in diameter and 50 mm in length). In the actual test, anti-progressive collapse reinforcement was placed to guarantee safety during the test, this reinforcement has no contribution in the structural performance and therefore is not modelled. The average compressive concrete strength (f_{cm}) was 43.7 N/mm^2 and the post-cracking response properties used were obtained by means of an inverse analysis and reported by Soranakom et al. [13] (σ_1 = 2.5, σ_2 = 1.75, σ_3 = 1.06, σ_4 = 0 MPa and w_1 = 0, w_2 = 0.25, w_3 = 1.25, w_4 = 2 mm).

A non-linear 3D model was created (Fig. 1b) to simulate both the geometry and loading conditions presented in Fig. 1a. For the sake of simplicity due to the double-plane symmetry, in both geometry and loading conditions, only a quarter of the slab was modelled. In order to provide symmetry to the model no displacement nor rotations were allowed in both symmetry planes (see Fig. 1b); the columns were modelled as simple supports where vertical displacement is restricted (U_y = 0), detailed in Fig. 1c. The load was applied on a circular steel plate of 20 mm of diameter (Fig. 1d) by means

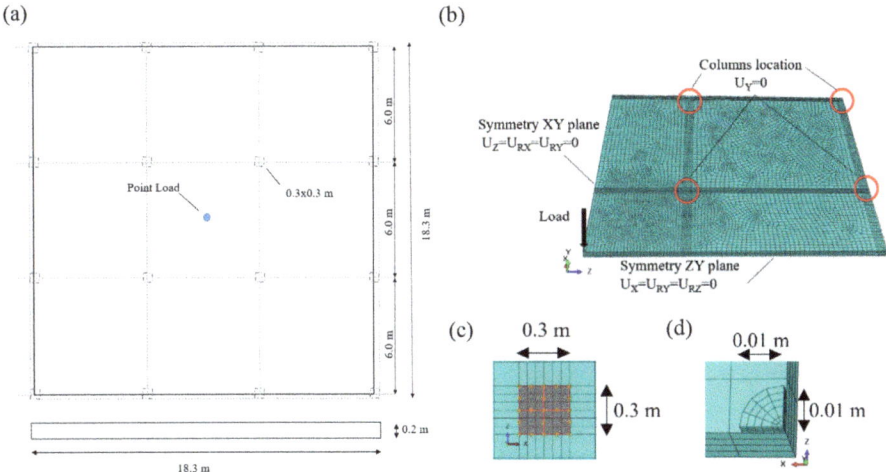

Fig. 1. Elevated slab test (a) Test set up (b) FE model (c) Columns detail (d) Loading plate.

of displacement control in order to guarantee a proper convergence. The mesh consists of 8-noded solid linear hexahedral elements (C3D8R) used for modelling concrete, this leading to an amount of 44720 elements.

Figure 2 presents the numerical and experimental load-deflection relationship. It can be noticed that the model presents a post-cracking response similar to that experimental in terms of load and deflection (with a maximum deviation of 40 kN). At first stages the stiffness are similar for both tests (up to 150 kN) then a ductile response is observed up to failure. The experimental cracking load ($P_{cr,exp}$) in the mid span bottom layer was 200 kN, whereas the numerical cracking load ($P_{cr,num}$) was 185 kN (−7.5%), from the safe side. The failure mode of the numerical model is unclear since the model is not able to reproduce the unloading curve; contrarily the load-deflection curve stabilizes and a constant load is found, even when the stress of the most demanded area (at the bottom face under where the point load is applied) has reached zero value. In this sense, as failure criteria the load level at which this area reach 0.02 strain (ultimate strain ε_{Fu}, according to *fib* MC-2010) is considered as the failure load, 466 kN (see Fig. 2), <1% of difference with respect to the experimental test.

The model does not provide the crack width directly, in this sense, the plastic strain (where tensile stresses have exceed tensile strength) is used instead to compare the crack pattern. The crack patterns measured after the test (up to failure) and those obtained numerically are included in Fig. 3. The first cracks appeared on top centre columns and progressively grew towards the near columns (see Fig. 3a). The second crack appeared on the bottom layer caused by postive bending moments and bifurqued into two directions paralel to the slab edges, plotted in Fig. 3b.

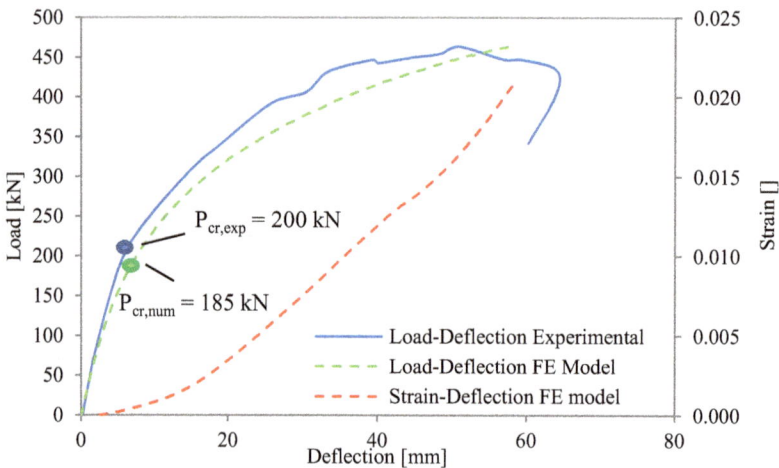

Fig. 2. Experimental and numerical load-displacement curves and strain-displacement.

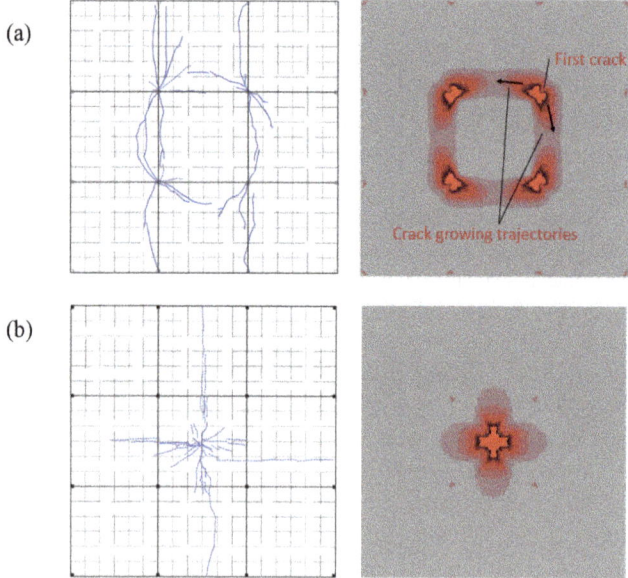

Fig. 3. Crack pattern at maximum load for experimental and numerical test. (a) Top layer and (b) bottom layer.

Therefore, according to the results presented in Fig. 2 and Fig. 3, it can be remarked that the model can reproduce properly the mechanical response of FRC elevated slabs subjected to point loads that generate flexural stresses and capture the ductile mechanism up to failure.

4   Case Study

An extensive experimental program on the characterization of self-compacting FRC composites and full-scale pile supported flat slabs was planned. A parametrical analysis was carried out to determine the fibre content which guaranteed the structural response higher than the design load (F_d). According to the Spanish code CTE [14], the flat slab was designed to resist, besides the self-weight (SW) of 4.8 kN/m^2, a Dead Load (DL) of 2.0 kN/m^2 and a live load (LL) of 3.0 kN/m^2 this leading to a F_d of 14 kN/m^2 taking into account partial safety factors from *fib* MC-2010. Besides using different FRC contents, three simulations combining FRC with conventional reinforcement (hybrid reinforcement) were done.

4.1   Model Description

The geometry of the elevated flat-slab consists in a rectangular slab of 12.00 m 10.00 m in size and 0.2 m in thickness supported by a grid of 9 square columns (0.25 m), forming 4 bays 6.00 × 5.00 m meters each. The supporting columns were 3.00 m high. A quarter of the slab was modelled, due to symmetry, constraining displacements and rotations at the symmetry planes (see Fig. 4a). The columns are modelled as simple supports where the vertical displacement is restricted ($U_y = 0$) at the slab-column contact surface. The mesh is composed by 18000 elements and the concrete was modelled using C3D8R hexahedral solid elements.

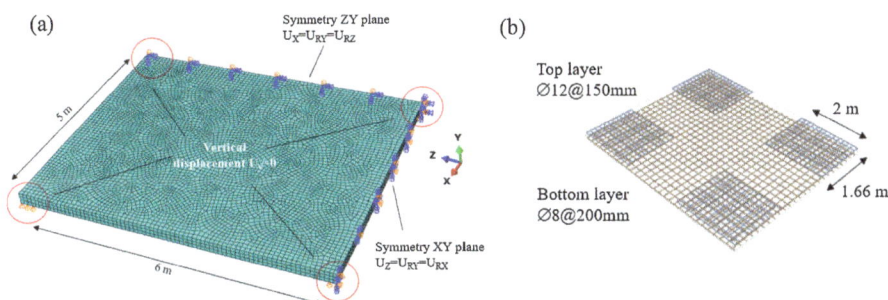

Fig. 4. Study case slab (a) geometry and boundary conditions (b) rebar configuration

Figure 5 depicts three types of FRC commonly used for these structural elements and used for the simulations: FRC class 4c (FRC4), FRC class 6c (FRC6) and FRC class 8c (FRC8) and the constitutive curves using mean values properties (FRC4$_m$, FRC6$_m$ and FRC8$_m$). f_{Fts} and f_{Ftu} stand for serviceability and ultimate residual strength respectively. A commonly used f_{R3m}/f_{R1m} ratio for structural application equal to 1 has been chosen since it has a flat Load-CMOD behaviour after cracking. $f_{ctm} = 4.07$ N/mm^2, $f_{cm} = 58$ N/mm^2 and $E_{cm} = 32900$ N/mm^2 were assumed as concrete mechanical properties for all simulations.

Furthermore, a combination of fibres and steel rebar (hybrid reinforcement) was studied as the slab reinforcement configuration. The three FRC strength classes are combined with a rebar mesh of Ø8 every 200 mm, in both directions, at the bottom layer and Ø12 every 150 mm, in both directions, on the top layer covering a surface equal to a third part of the bay length. A clear cover of 25 mm is considered. The rebar configuration is presented in Fig. 4b. The nomenclature for hybrid solutions from now on is HYB followed by the fibre reinforced concrete strength (4, 6 or 8). A common value of steel yielding stress (f_y) equal to 600 MPa is adopted for the analysis. 2-noded 3D linear truss elements T3D2 are chosen for the rebar, embedded constraint is used for join linear and solid elements which leads to a perfect bond between concrete and steel. It must be highlighted that the reinforcement ratio adopted is considered for a hybrid solution (combining rebar and fibres) and not to withstand the design loads only with the rebar.

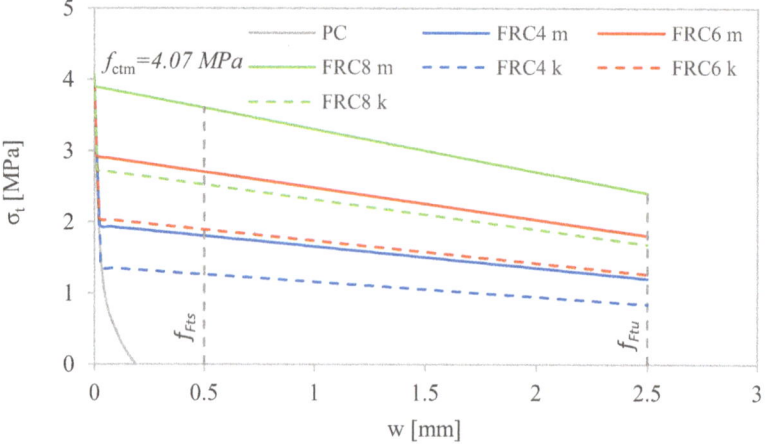

Fig. 5. Tensile constitutive equations of the concrete mixes used for the parametric analysis.

4.2 Results

After being applied the self-weight (4.8 kN/m^2) the slabs were tested up to failure. Figure 6 shows the load-deflection (δ) curve for all reinforcement configurations. In this sense, in order to evaluate the failure load of FRC and hybrid solutions for ULS, two failure criteria have been considered: (1) FRC reaching ε_{Fu} (20‰ of strain) and (2) rebar longitudinal strain (ε_s) equal to 10‰ (in hybrid solutions). In view of this, PC and RC configurations reached the lowest ultimate loads (q_u). These q_u, not including the self-weight, were 10.1 (FRC4), 14.0 (FRC6) and 18.0 (FRC8) kN/m^2 for FRC solutions and 14.9 (HYB4), 18.6 (HYB6) and 23.2 (HYB8) kN/m^2 for hybrid reinforcement. Figure 7 shows q_u versus the FRC residual strength ($f_{R1m} = f_{R3m}$). The results show that adopting a hybrid solution improves the slab strength against failure when compared to FRC and the efficiency decreases when FRC classes are adopted: 47%, 32% and 28% of improvement for 4, 6 and 8 MPa hybrid solutions.

As Fig. 6 shows, the maximum deflection (δ_{max}) for each simulations is reached at different load levels. In order to compare the deformations between different cases, the ductility index ξ is proposed. ξ is set as $\delta_{max}/\delta_{fr1m=fr3m=4}$, where $\delta_{fr1m=fr3m=4}$ stands for the maximum deflection reached by the configuration with less amount of fibres of each type of reinforcement (FRC4 and HYB4). Figure 8 presents ξ – residual strength curve. The results showed that FRC solutions presented higher ductility as post-cracking strengths increased. In contrast, for hybrid reinforcement configurations, increasing the FRC residual strength turned into improving the structure against deflection and presented lower values of ξ, reaching less δ_{max} and increasing stiffness.

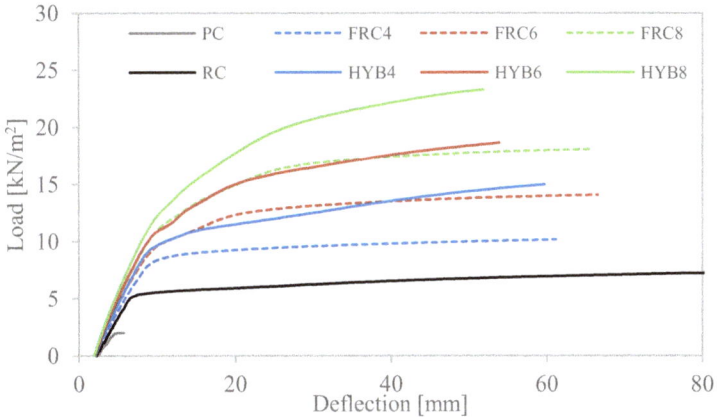

Fig. 6. Simulation results (a) Load – deflection (b) ultimate load – residual strength.

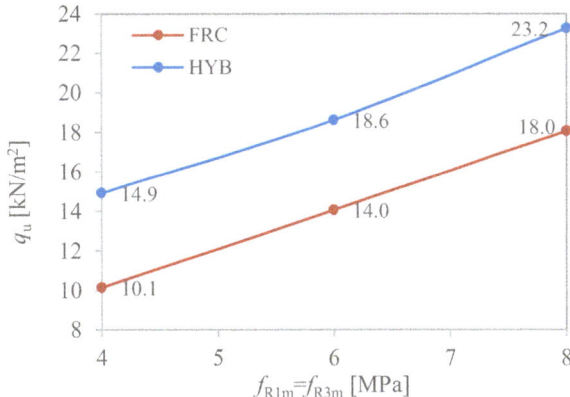

Fig. 7. Simulation results ultimate load – residual strength.

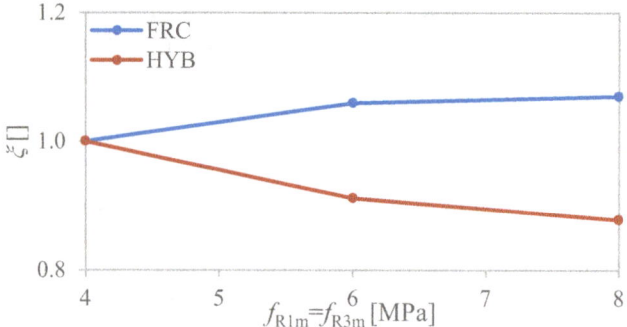

Fig. 8. Ductility index (ξ) – residual strength for FRC and hybrid solutions.

The crack pattern at the bottom layer at failure is shown in Fig. 9a for FRC and in Fig. 9b for hybrid solution, in this case FRC4 and HYB4 respectively. Due to positive bending moments, two crack crossed the bays perpendicular to the larger side of the slab. It is worth noticing that in case of FRC the cracks were concentrated in the centred region where the damage is found, in contrast in hybrid solution the crack pattern showed a distributed shape, being this pattern more favourable in order to control the cracks and reduce the crack width.

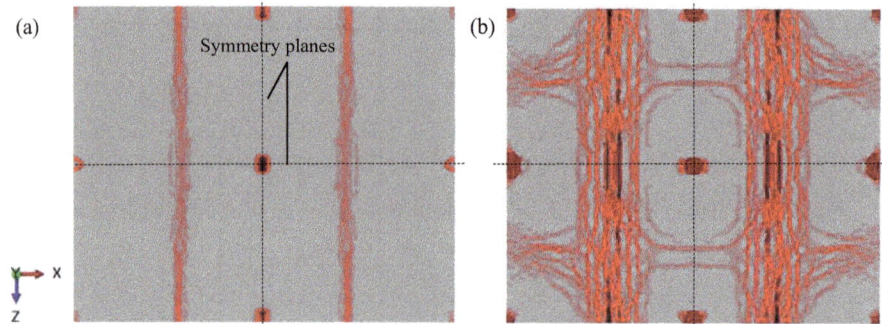

Fig. 9. Crack pattern at the bottom layer at failure for (a) FRC4 and (b) HYB4

4.3   Structural Safety

Aiming to determine the FRC residual strength that would be used for the flat-slab experimental test [15], the structural safety analysis for the FRC solutions is presented in this section. In this sense, the Eurocode 2 [16] proposes a method to aid the assessment of the structural capacity by means of non-linear methods, the main challenge being the definition of the partial safety factor to be applied to both resistance and loading, since the partial safety factor gathered in the codes and guidelines are based on linear elastic methods. In order to fulfil this aspect, *fib* MC-2010 gathers a method that permits to obtain a resistance global safety factor for the whole structure by considering the variability of the mechanical and geometrical properties. The result of this reliability-based model is the safety coefficient to derive the design value of the structural bearing capacity, R_d (Eq. (1)); where R_m stands for the mean value of the structural bearing capacity (considering mean values for mechanical and geometrical properties) and γ_R is the global safety factor. The γ_R coefficient can be calculated as per Eurocode 2 (Eq. (2)), where V_R is the resistance coefficient of variation and α_R and β stands for the resistance sensitivity factor and reliability index respectively, taken as $\alpha_R = 0.8$ and $\beta = 3.8$.

$$R_d = \frac{R_m}{\gamma_R} \tag{1}$$

$$\gamma_R = e^{(\alpha_R \cdot \beta \cdot V_R)} \tag{2}$$

The ECOV method proposed by Cervenka [17] allows estimating V_R by means of a logarithmic ratio between R_m and R_k, the latest being the characteristic resistance value obtained by implementing the non-linear solution with characteristic magnitudes of both mechanical and geometrical properties.

$$V_R = \frac{1}{1,65} ln\left(\frac{R_m}{R_k}\right) \tag{3}$$

For this purpose, three additional FE simulations have been carried out using characteristic values of each of the involved mechanical variables. The σ-w curves for these simulations are plotted in Fig. 5, the f_{Rk}/f_{Rm} ratio equal to 0.7 (CoV = 18.3% assuming a normal distribution) suggested by the *fib* MC-2010 is adopted. Regarding this aspect, the greater the f_{Rm} is, the lower the variability of f_{Rm} is expected (the amount of fibres is higher and the intrinsic scatter tends to be reduced [18]) and, thus, this relation f_{Rk}/f_{Rm} could be >0.7 (CoV < 18.3%). Table 2 gathers the ultimate loads obtained considering mean ($q_{u,m}$) and characteristic ($q_{u,k}$) resistance values as well as V_R, γ_R y R_d. R_m and R_k values are obtained by adding the self-weight (4.8 kN/m^2) to $q_{u,m}$ and $q_{u,k}$ respectively.

Table 2. Mean, characteristic and design structural resistance, variation and global safety factor.

	$q_{u,m}$ [kN/m²]	R_m [kN/m²]	$q_{u,k}$ [kN/m²]	R_k [kN/m²]	V_R	γ_R	R_d
FRC4	10.1	14.9	7.3	12.1	0.12	1.46	10.2
FRC6	14.0	18.8	10.5	15.3	0.12	1.47	12.8
FRC8	18.0	22.8	13.1	17.9	0.14	1.56	14.6

The global safety factors gathered in Table 2 resulted to be 1.46 for FRC4, 1.47 for FRC6 and 1.56 for FRC8 (f_{Rm} = 4, 6 and 8 MPa respectively). In this case study, the flat slab F_d was 14.0 kN/m^2, the ultimate limit state design condition ($R_d \geq F_d$) is therefore fulfilled for FRC solution of f_{Rm} = 8 MPa. It must be remarked that, according to the abovementioned comment about the CoV of f_R, γ_R should decrease as f_R increases; hence, the γ_R obtained could be on the safe side for the higher values of f_R.

The FRC slab of the experimental test was cast with FRC of $f_{R1m} = 8$ and $f_{R3m} = 7.5$ N/mm^2 and was loaded up to 15 kN/m^2 without reaching structural collapse. The FRC8 ($f_{R1m} = f_{R3m} = 8$ N/mm^2) provided 14.6 kN/m^2 for R_d in the numerical simulation, this meaning that the model results using the global safety factors presented in this research contribution are in good agreement with the experimental test, even with a reasonable margin of safety.

5   Conclusions

Simulations of a FRC pile-supported flat slab were carried out by means of a non-linear FE model. Several FRC strengths classes were considered in order to quantify the effect of those on the mechanical response and the structural safety of associated to each FRC strength class, the combination of rebars and fibres was also considered.

Based on the results presented in this research contribution, the following conclusions can be drawn:

- The results of the validation and the experimental real scale test allow confirming that the model is adequate to perform the analysis of this type of structures.
- The hybridization of structural fibres and rebars led to more efficient solutions in terms of bearing capacity. In this sense, the structural efficiency is more noticeable when low FRC strength classes are considered. Likewise, increasing the FRC strength resulted in an improvement of the deformation capacity at ultimate limit state (for the maximum load) while the hybrid solutions, although showing a remarkable ductility, a slight embrittlement respect to the FRC alternatives.
- Hybrid configurations presented a more distributed crack pattern favourable for cracking control and crack width reduction.
- The numerical model together with the safety format presented within the *fib* Model Code for non-linear analysis allowed safely design the real-scale slab constructed (see [15]). It must be highlighted that even the experimental results confirmed a reasonable safety margin respect the numerical results.
- These are preliminary conclusions of a research which is being currently carried out within the context of the PhD thesis of the first author.

Acknowledgements. The first author acknowledges the Spanish Ministry of Science, Innovation and University for providing support through the PhD Industrial Fellowship (DI-17-09390) in collaboration with Smart Engineering Ltd. (UPC's Spin-Off). This research has been possible due to economic funds provided by the Spanish Ministry of Economy, Industry and Competitiveness (MINECO) through the RTC-2016-5263-5 financial support associate to the eFIB Project (building process optimization and structural element design using fibre reinforced concrete instead of conventional rebar) carried out along with SACYR Ingeniería e Infraestructura.

References

1. DBV, Guide to Good Practice: Steel Fibre Concrete. German Society for Concrete and Construction Technology (DBV), Berlin, Germany (2001)
2. RILEM TC 162-TDF, Test and design methods for steel fibre reinforced concrete. Design with σ-ε method. Mater. Struct. 35, 262–278 (2003). https://doi.org/10.1617/13837
3. CNR-DT 204/2006, Guide for the Design and Construction of Fibre-Reinforced Concrete Structures, no. 75. Italian National Research Council (CNR), Rome, Italy (2007). https://doi.org/10.14359/10516
4. EHE-08, Instrucción de Hormigón Estructrural (EHE-08) (2008)
5. FIB, fib Model Code for Concrete Structures 2010. International Federation for Structural Concrete (fib), Lausanne (2013). https://doi.org/10.1002/9783433604090
6. ACI 544, Fibre-Reinforced Concrete: Design and Construction of Steel Fibre- Reinforced Precast Concrete Tunnel Segments (2014)
7. Destrée, X., Mandl, J.: Steel fibre only reinforced concrete in free suspended elevated slabs: case studies, design assisted by testing route, comparison to the latest SFRC standard documents. Tailor Made Concr. Struct. 437–443 (2008). https://doi.org/10.1016/j.jhydrol.2010.04.024
8. di Prisco, M., Martinelli, P., Parmentier, B.: On the reliability of the design approach for FRC structures according to fib Model Code 2010: the case of elevated slabs. Struct. Concr. **17**, 588–602 (2016). https://doi.org/10.1002/suco.201500151
9. de la Fuente, A., Casanovas-Rubio, M.D.M., Pons, O., Armengou, J.: Sustainability of column-supported RC slabs: fiber reinforcement as an alternative. J. Constr. Eng. Manag. **145**, 1–12 (2019). https://doi.org/10.1061/(ASCE)CO.1943-7862.0001667
10. C. Dassault systems Simulia, Abaqus Analysis user's manual 6.12-3, 1137 (2012). http://www.maths.cam.ac.uk/computing/software/abaqus_docs/docs/v6.12/pdf_books/BENCHMARKS.pdf
11. Gossla, U.: Development of SFRC Free Suspended Elevated Flat Slabs, research report for Arcelor Mittal (2005)
12. Facconi, L., Plizzari, G., Minelli, F.: Elevated slabs made of hybrid reinforced concrete: proposal of a new design approach in flexure. Struct. Concr. 1–16 (2018). https://doi.org/10.1002/suco.201700278
13. Soranakom, C., Mobasher, B., Destree, X.: Numerical simulation of FRC round panel tests and full-scale elevated slabs. ACI Special Publication, vol. 248, pp. 31–40 (2007). http://www.concrete.org/Publications/GetArticle.aspx?m=icap&pubID=19008
14. Gobierno de España, DB SE- AE Seguridad Estructural, Acciones en la Edificación., Boletín Of. Del Estado, pp. 1–42 (2009)
15. Aidarov, S., Mena, F., de la Fuente, A.: Self-compacting steel fibre reinforced concrete: material characterization and real scale test up to failure of a pile supported flat slab. In: BEFIB 2020 (2020)
16. DIN EN 1992-1-1:2011-01, Eurocode 2, 2014. https://doi.org/10.2788/35386
17. Cervenka, V.: Global safety formats in fib Model Code 2010 for design of concrete structures. In: 11th International Probabilistic Work, pp. 31–40 (2013)
18. Cavalaro, S.H.P., Aguado, A.: Intrinsic scatter of FRC: an alternative philosophy to estimate characteristic values. Mater. Struct. **48**(11), 3537–3555 (2014). https://doi.org/10.1617/s11527-014-0420-6

TBM Thrust on Fibre Reinforced Concrete Precast Segment Simulation

Alejandro Nogales[✉] and Albert de la Fuente

Civil Engineer, Universitat Politècnica de Catalunya, Barcelona, Spain
alejandro.nogales@upc.edu

Abstract. Fibre reinforced concrete (FRC) is gaining acceptance as a structural material for casting precast segments as this has proven to lead to various advantages respect to the traditional reinforced concrete, especially for improving the crack control during transient loading situations. Concentrated loads induced during the excavation stage by Tunnel Boring Machines (TBMs) is still a matter of discussion into the tunnelling construction field, this having a strong impact from both technical and economic perspectives. In this regard, although specific codes and guidelines have been published and intense research has been carried out, this pointing out the benefits of using FRC as partial or total substitution of conventional reinforcement; however, there is still not a thorough research on the effectiveness of the FRC strength class (as defined into the *fib* Model Code 2010) to control crack widths in tunnel linings during the TBM thrust phase. To this end, the objective of this research contribution consists in carrying out a parametric analysis, considering different FRC strength classes (including hybrid reinforcements), related with the crack patterns due to the TBM-thrust on FRC segments. This was performed by means of a comprehensive non-linear finite element (FE) numerical simulation. The results derived from this research are meant to be useful for those stakeholders involved in the design of precast tunnel linings.

Keywords: Fibre reinforced concrete · Non-linear analysis · Precast segments · TBM · Tunnelling · Concentrated load

1 Introduction

Over the last decades, the use of structural fibres as partial/total replacement of the traditional steel bars has grown [1, 2]. Among the structural applications where FRC is used is tunnelling, precast tunnel segments being suitable elements for this replacement [3]. The use of fibres has been proven as a potential solution and many tunnels were made up with this material in the recent years, among other examples: the "Barcelona metro line 9" [4, 5], "Monte Lirio" in Panamá [6, 7] and the "Prague Metro Line" [8, 9]. Indeed, the recently *fib* Bulletin 83 [10] gather more than 70 experiences of the use of fibres in tunnel linings and provides design provisions for FRC tunnel segments.

During excavation, TBM hydraulic jacks take support against the previously place ring in order to advance forward, these jacks induce high concentrated loads in the segments. Experiences on tunnel design and researches have confirmed that in terms of

© RILEM 2022
P. Serna et al. (Eds.): BEFIB 2021, RILEM Bookseries 36, pp. 678–689, 2022.
https://doi.org/10.1007/978-3-030-83719-8_58

loading and concrete cracking, the most demanding scenario can be during the thrust stage [11, 12].

In this sense, actions in tunnels can be classified in: (1) Primary loads due to soil-structure interactions and water pressure and (2) Secondary loads, which occur during construction called transient phases, which include demoulding, storage, transportation, handling, placing and TBM jack's thrust. Primary loads induce compressive forces combined with low shear forces and bending moments on the ring can be resisted by the concrete matrix and a combination of fibre and conventional reinforcement in linings with large diameters [13] (larger than 6 m). However, tunnels with smaller diameters (metro and hydraulic tunnels) might be subjected to compression in service stage, and it is during transient phases where tensile stresses appear either due to bending or thrust loads. The latter can lead to high concentrated loads and to tensile stresses, which result in splitting and spalling.

These stress patterns caused by concentrated loads are complex and its magnitude and distribution (within a D-region) are both difficult to be assessed. Leonhardt [14] and Iyengar [15] were the first who carried out research in this topic of transfer zones in pre-stressed structures. These studies concluded that concentrated loads induce a tri-axial state of stresses where a principal tensile component of stresses (splitting stresses) acts orthogonally to the paths of compressions. In addition, as a result of compatibility demands with respect to deformed cross section, tensile stresses (spalling stresses) appear in segments.

This state of stresses and the curved shape of the segments makes necessary a complex detailed reinforcement, which may leave uncovered areas where spalling and splitting take place. Alternatively (or complimentary), FRC can deal with this matter due to the randomly distribution nature of the fibres within the whole segment.

FRC for structural applications can be produced with different types of fibres (mainly metallic and synthetic), providing to the concrete matrix an improvement of: ductility; post-crack tensile strength (residual strength); impact resistance and a more convenient crack pattern for serviceability purposes. Due to the growing use of FRC, many guidelines have included design recommendations for its mechanical characterization and design [16–19] and specific documents for FRC precast segments have also been published recently [10, 20].

This topic is of great interest since it influences the design of precast tunnel segments (segment geometry, reinforcement amount and distribution) and cracking during this stage can jeopardise costs, structural durability and the serviceability of the tunnel leading to repairs and maintenance actions. The complex state of stresses generated by the concentrated loads is difficult to predict and this is only possible by means of experimental test and/or non-linear FE analysis [10].

The aim of this contribution is to assess the structural performance of precast FRC tunnel linings subjected to thrust jack forces. To this purpose, different classes of FRC according to *fib* MC-2010 were considered as constituent material of segments and both the bearing and crack control capacities were identified and quantified numerically by means of a model implemented with ABAQUS [21]. The model was previously validated with results derived from experimental tests available in the literature. Both the results and conclusions achieved are found to be of interest in terms of structural and economic optimization in those cases for which the TBM thrust is the design governing phase.

2   Numerical Modelling of FRC

According to *fib* MC-2010 FRC classes can be classified by its characteristic residual strength values (f_{Rk}) obtained from 3 point bending test notched beams according to EN 14651:2005. Two parameters are used for its classification: f_{R1k}, which represents the residual strength for a crack opening displacement of 0.5 mm and a letter (a, b, c, d or e) that represents the f_{R3k}/f_{R1k} ratio, where f_{R3k} stands for a CMOD of 2.5 mm. In order to establish the stress-strain (σ-ε) or stress-crack width (w-ε) constitutive relations, the equations proposed by *fib* MC-2010 are adopted, using mean values of f_{LOP}, f_{R1} and f_{R3} for the simulations. The compressive behaviour was simulated according to the expression proposed by *fib* MC-2010. These equations were implemented in the "Concrete Damage Plasticity" model available in ABAQUS.

3   Experimental Model Validation

3.1   FRC Blocks Subjected to Concentrated Loads

Small-scale experimental tests carried out by Schnütgen and Erdem [22], which consisted of steel FRC anchor blocks subjected to concentrated loads oriented to assess the response against splitting, were used for the validation. To that end, two fibre mixes were used: 35 kg/m³ (FRC-A) and 60 kg/m³ (FRC-B). The concrete compressive strengths (f_{cm}) were 58.2 MPa and 50.2 MPa for FRC-A and FRC-B respectively. Residual strengths were derived from NBN-B-15–238:1992 [23] test: f_{R1} = 5.41 MPa and f_{R3} = 4.81 MPa for FRC-A and 6.49 and 5.96 MPa for FRC-B. The blocks had a square base of 350 mm and 700 mm height and a surface load on top of 150 × 350 mm (Fig. 1a). The FE model was composed by 16800 eight-noded hexahedral elements (C3D8R), no vertical displacement allowed at the base and the load was applied by displacement control (Fig. 1b).

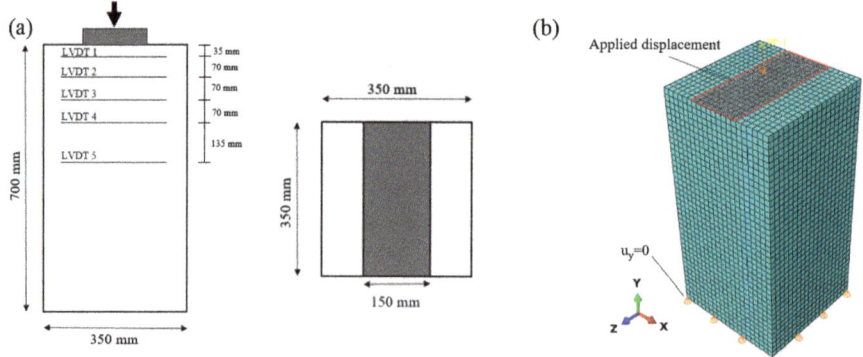

Fig. 1. Splitting test (a) Test set up (b) FE model adopted.

The results of both experimental and numerical tests are presented in Fig. 2.

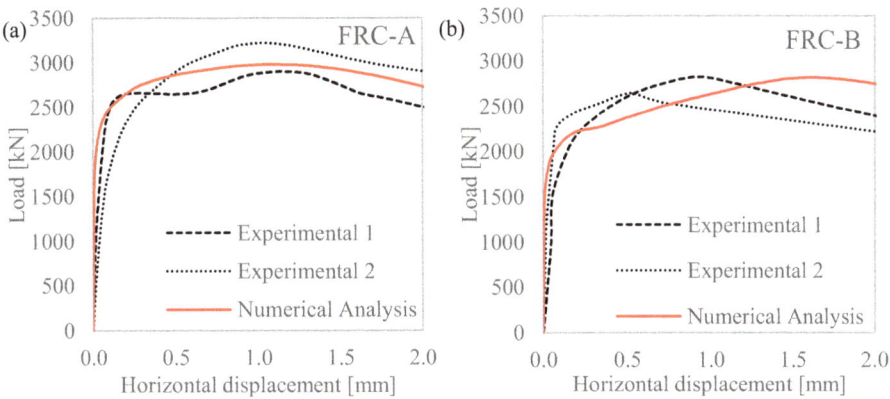

Fig. 2. Load-Displacement chart for experimental and numerical test (a) FRC-A (b) FRC-B.

The horizontal displacement was measured at a distance of 175 mm from the top (LVDT 3) where the maximum tensile stresses appear. The load-displacement results of the numerical simulations are in good agreement with the experimental test, the block cast with FRC-A had a maximum load of 2880 kN, whereas those obtained experimentally were 3200 kN and 2870 kN. Regarding the FRC-B test, the numerical simulation reached 2815 kN whereas the experimental test were 2830 kN and 2650 kN.

3.2   Precast Curved Segment Subjected to Concentrated Loads

In order to check the suitability of the model for curved segments subjected to con-centrated loads, the results of the experimental research carried out by Conforti [24] were considered. This experimental program was meant to analyse the suitability of polypropylene macrofibres for controlling crack width resulting from concentrated loads. To this end, tests on curved segments of 1810 mm length (internal diameter of 3200 mm), 1200 mm height and 250 thick subjected to concentrated loads by means of two loading shoes (500×250 mm^2) (Fig. 3a) were performed. The segment was cast with 10 kg/m^3 polypropylene fibres, had 49.9 N/mm^2 f_{cm} and 2.97 and 4.61 N/mm^2 for f_{R1m} and f_{R3m} respectively, according to EN-14651:2005. The FE model is composed by 8100 C3D8R elements, no vertical displacement allowed at the bottom face and the load was applied by displacement control (Fig. 3b). The operational load (P_{nom}) and accidental load (P_{acc}, considered as the highest operational load) of the TBM were 785 and 1130 kN respectively.

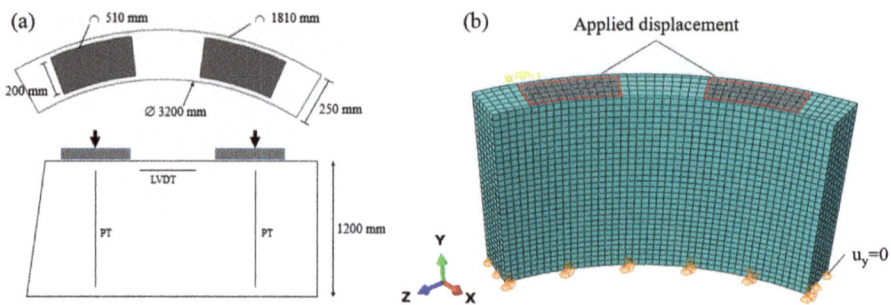

Fig. 3. Segment test (a) Test set up (b) meshed model and boundary conditions.

In Fig. 4a the load-vertical displacement is presented.

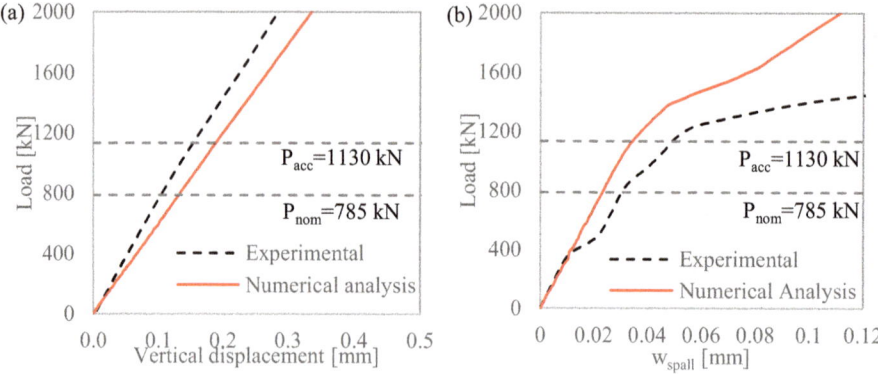

Fig. 4. Experimental and numerical comparison (a) Load-vertical displacement (b) Load-spalling crack opening.

It is worth noticing that the model performs similarly to the experimental test up to 1200 kN, but had a stiffer perfomance. The spalling load (load level that generates the spalling crack, P_{spall}) is 955 kN and 979 kN for both experimental and numerical (1.60%) and the splitting load (P_{sp}) was 1600 kN in the experimental test whereas 1585 kN is registered in the model (0.93%). The spalling crack (w_{spall}) is measured and depicted in Fig. 4b, its location in both experimental and numerical test was between pads, the w_{spall} at P_{acc} is 0.050 mm for the experimental test and in the numerical model is 0.034 mm.

4 Parametric Study

The segment geometry and jack configuration are both taken from precast segment of an actual metro tunnel lining under construction. The inner diameter of the segment is 4075 mm, 1500 mm height and 350 mm thickness (Fig. 5). Four jacks for each segment, the loading surface being 222×500 mm^2 placed 42 mm from the inner edge of the segment.

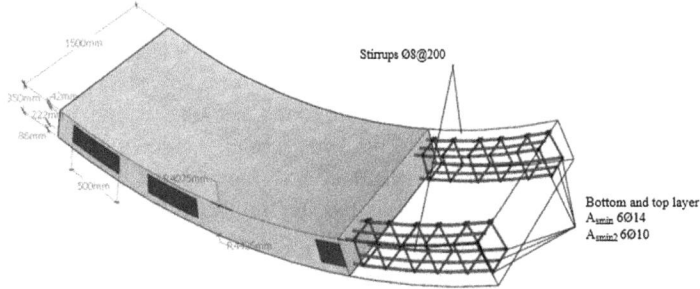

Fig. 5. Geometrical properties of the segment and rebar distribution.

4.1 Materials

A C50/60 concrete has been used for the production of these segments; thus, according to the *fib* MC-2010, the values $f_{ctm} = 4.07$ N/mm^2, f_{cm} 58 N/mm^2 and $E_{cm} = 32900$ were assumed. The FRC strength classes 1, 3 and 5 (related to f_{R1k}) with f_{R3k}/f_{R1k} ratios a, b, c, d and e, these resulting in a total of 16 different concretes (15 FRC + 1 PC). The commonly accepted ratio $f_{Rk}/f_{Rm} = 0.7$ was used to estimate f_{Rm}. In this sense, it must be emphasized that this ratio might be greater as f_{Rm} increase, and viceversa. The constitutive equation according to *fib* MC-2010 is used for assessing the serviceability and ultimate residual strengths, f_{Fts} and f_{Ftu} respectively, calculation and are presented in Table 1.

Table 1. Serviceability and ultimate residual strength for FRC adopted solutions.

| | f_{R1k} | | | | | |
| | 1 | | 3 | | 5 | |
f_{R3k}/f_{R1k}	f_{Fts} [MPa]	f_{Ftu} [MPa]	f_{Fts} [MPa]	f_{Ftu} [MPa]	f_{Fts} [MPa]	f_{Ftu} [MPa]
a	0.64	0.14	1.92	0.42	3.21	0.71
b		0.28		0.85		1.42
c		0.42		1.28		2.14
d		0.57		1.71		2.85
e		0.71		2.14		3.57

Furthermore, aiming at optimising the reinforcement configuration, hybrid rein-forcements (RC + FRC) are also considered. Figure 5 shows the cage configuration, this being composed by two chords made by 2 × 3Ø14 curved rebar with Ø8@200mm stirrups, with a clear cover of 50mm. This amount (per face) corresponds to the minimum amount ($A_{s,min}$) required by *fib* MC-2010 to guarantee the ductile response in flexure. Additionally, $A_{s,min}/2$ (2 × 3Ø10 each face) a ratio of has also been included into the analysis.

4.2   Model Description

In Fig. 6 is presented a general view of the segment model, were is depicted the mesh and the boundary conditions. Simplifications must be taken in order to guarantee a robust model with a non-dependent mesh able to reduce computational calculation time while giving accurate results. To this end, only a single segment was modelled. The interactions with the surrounding segments at both longitudinal and radial joints were not taken into account, these interactions has negligible influence on local behaviour which do not affect the results of splitting and spalling phenomena that are produced under the loading zone and between pads, besides this assumption helps to avoid convergence problems and reduce computational time. For the same reason, in spite of setting a regular mesh, bolt and gasket holes were not modelled.

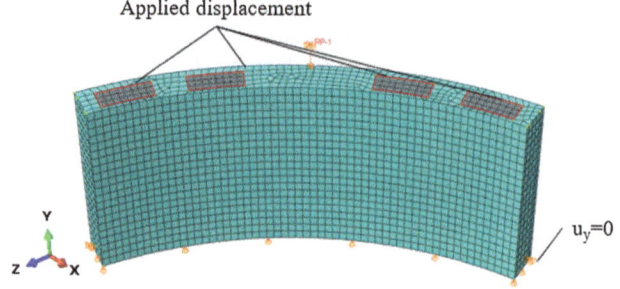

Fig. 6. Meshed model and boundary conditions.

The mesh is set by70 mm size C3D8R elements and the rebar are modelled using linear T3D2 elements embedded in the concrete section, adopting a perfect bond rebar-concrete.

5   Results

5.1   Centred Thrust on FRC Segment

The first crack is produced due to spalling stresses, between centred pads, the cracking load ($P_{cr} = P_{spall}$) is 1975 kN for each jack. The TBM operational load (P_{nom}) must be lower than P_{cr}, taking this into account the assumption $P_{cr} = P_{nom}$ is adopted.

Likewise, P_{acc} was considered as two times the cracking load ($2P_{cr}$). Figure 7 presents the non-dimensional load (P/P_{spall}) versus displacement of the jack.

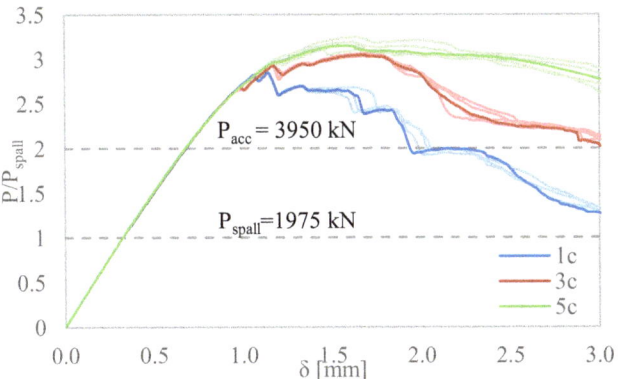

Fig. 7. Non-dimensional load-displacement curve for FRC. Stress pattern for P_{spall} and P_{acc}.

As it can be noticed, the mechanical response of the segments is the same in all cases up to $2.5P_{spall}$. Figure 8a depicts $P/P_{spall} - w_{spall}$ graph, in light colours are plotted "a, b, d and e" FRC classes and in dark colour "c" class. The results allow confirming that f_{R1} has great influence in cracking control whereas, the f_{R3}/f_{R1} ratio do not play a significant role (once f_{R1} is set). Figure 8b shows the non-dimensional load – w_{sp} (splitting crack), this being produced at 2173 kN ($P_{sp} = 1.1P_{spall}$), w_{sp} is minor than 0.05 mm for P_{acc} and no significant differences are noticed between solutions.

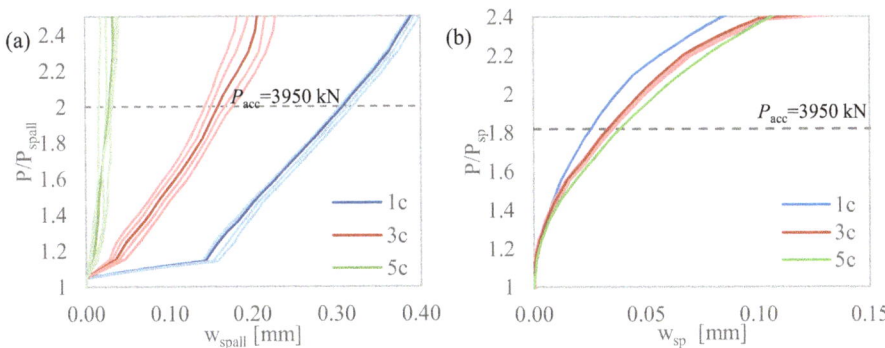

Fig. 8. Non-dimensional load – crack width (a) Spalling (b) Splitting.

5.2   Eccentric Thrust on FRC Segment

During the excavation phase, there may be situations where the thrust exerted by the jacks is not centred and this can generate eccentricities. In order to study this phenomenon, simulations were carried out considering 30 mm eccentricity towards inside the tunnel (e^-) and outside the tunnel (e^+). The eccentric thrust does affect the cracking load as can be seen in Fig. 9, where these loads are presented non-dimensional with respect to P_{spall} for centred thrust (1975 kN).

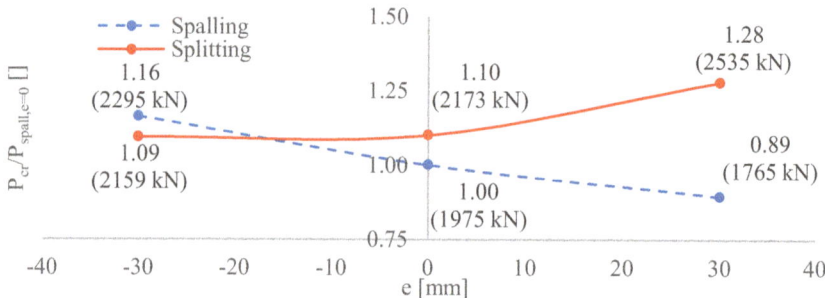

Fig. 9. Cracking load/P_{spall} for $e = 0$ – eccentricity.

For e^+, P_{spall} reduces and P_{sp} increases. In the case of e^-, P_{spall} increases with respect to the centred thrust and P_{sp} reduces. These results have to be taken with precaution since the existence of the gasket could slightly modify these results; likewise, these tendencies are highly dependent on the load area and the curvature of the segment, hence, other results are expected for geometric configurations.

Figures 10a and 10b depict the non-dimensional load – crack width charts for the solution 1c, 3c and 5c. Based on the results presented in Fig. 10a, it can be noticed that as the FRC strength class increases the spalling crack width is reduced drastically. The results shown in Fig. 10b reveal that eccentric thrust has a relevant influence on splitting cracks width whilst the FRC strength class barely affects the response. The FRC 3c strength class seems to be the most suitable for controlling the crack width (while optimizing the amount of fibres) since the crack widths ranges between 0.12 mm ($e = -30$ mm) and 0.18 mm ($e = +30$ mm), which is an acceptable range ($w < 0.20$ mm) for dealing with the posterior service conditions; this considering that the probability of reaching P_{acc} must be, by definition, very low.

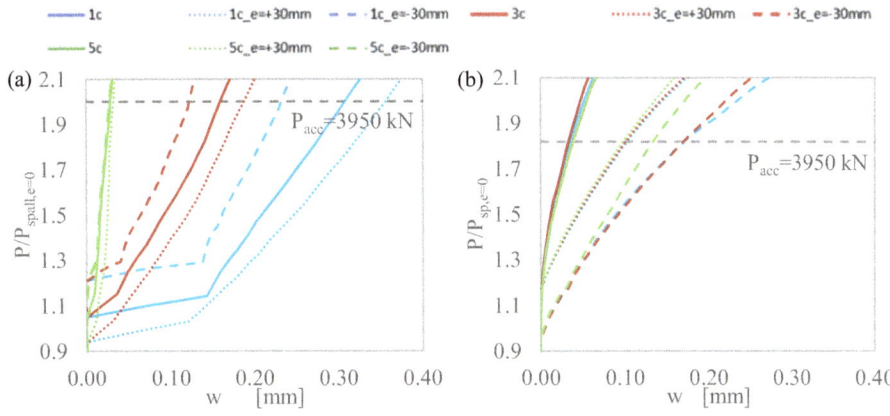

Fig. 10. (a) Cracking load/P_{spall} for $e = 0$ – spalling crack width (b) Cracking load/P_{sp} for $e = 0$ – splitting crack width.

5.3   Thrust on Hybrid Reinforcement Segment

For the studied geometry, the wider cracks are those due to spalling stresses. With the aim of reducing the crack width, a hybrid reinforcement (rebar + fibres) is proposed for the areas where spalling stresses appear. The rebar distribution is plotted in Fig. 5. Simulations were carried out with two rebar configurations ($A_{\text{s,min}}$ and $A_{\text{s,min2}}$) combined with 1c, 3c and 5c for centred and eccentric thrust. Figure 11 plots ξ-eccentric thrust for P_{acc}. Where ξ stands for $w_{\text{spall}}/w_{\text{spall}} = 0.2$ mm (assumed as a well-controlled crack opening, which can be considered different according other premises).

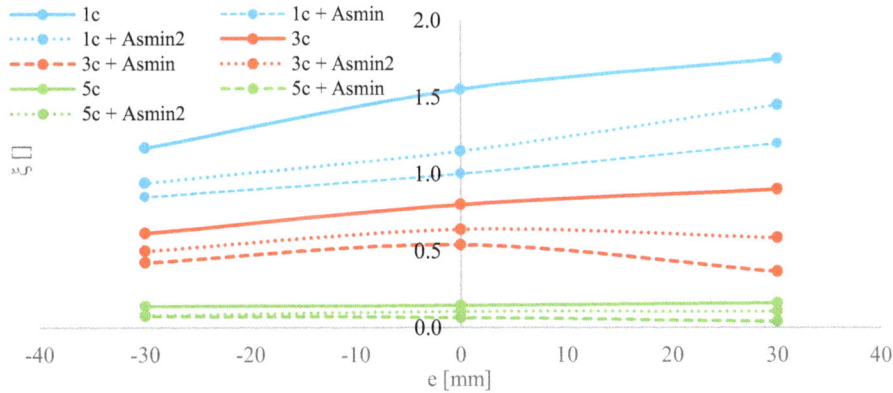

Fig. 11. $w_{\text{spall,acc}}/0.2$ versus eccentricity for accidental load (3950 kN)

The results prove that the hybrid solutions help to reduce the crack width for low resistance classes (namely 1c), however, is less efficient for higher classes (3c and 5c) where the crack width is mainly controlled by the residual strength of FRC ($\xi \leq 1$). Hybrid reinforcement is a suitable solution when either very strict crack width limitations are imposed (e.g., $w \leq 0.15$ mm, $\xi \leq 0.75$) and/or when traditional steel reinforcement is also required to resist bending moments; for which the fibres as unique reinforcement are not a suitable solution from the economic point of view [13].

6   Conclusions

The effect of TBM thrust on precast segments was studied using a non-linear FE analysis and taking into account the residual strength of FRC and hybrid reinforcement. By using this model, spalling and splitting cracks for a range of thrust eccentricities are estimated and compared according to the different reinforcement configurations.

Based on the results presented in this research contribution, the following conclusions can be drawn:

- The residual flexural strength for a crack width of 0.5 mm (f_{R1}) resulted to have a predominant influence in cracking control whilst the ratio f_{R3}/f_{R1} seems not play a significant role.
- The spalling cracks appear to be those governing the FRC post-cracking tensile strength requirements. Eccentric thrust has a relevant influence on cracking performance; in this sense, attention must be paid for both cracking phenomena, specially, for splitting cracks for which cracks up to 6.0 times wider with respect to the centred thrust were measured.
- Hybrid reinforcement leads to crack reduction and this can be considered as a suitable solution when severe crack width limitations are stablished. The combination of rebar and fibres has resulted to increase the efficiency in crack controlling when the FRC strength class decrease and viceversa.

Acknowledgements. The first author acknowledges the Spanish Ministry of Science, Innovation and University for providing support through the PhD Industrial Fellowship (DI-17-09390) in collaboration with Smart Engineering Ltd. (UPC's Spin-Off). This research has been possible due to the economic funds provided by the SAES project (BIA2016-78742-C2-1-R) of the Spanish Ministry of Economy, Industry and Competitiveness (MINECO).

References

1. de la Fuente, A., Blanco, A., Pujadas, P., Aguado, A.: Diseño óptimo de dovelas de hormigón reforzado con fibras para el revestimiento de túneles. Hormigón y Acero. **65**, 267–279 (2015). https://doi.org/10.1016/j.hya.2014.11.002
2. di Prisco, M., Plizzari, G., Vandewalle, L.: Fibre reinforced concrete : new design perspectives. Mater. Struct. 1261–1281 (2009). https://doi.org/10.1617/s11527-009-9529-4
3. de Waal, R.G.A.: Steel fibres reinforced tunnel segments for the application in shield driven tunnel linings, Technische Universiteit Delft (2000)

4. Burgers, R, Walraven, J., Plizzari, G.A., Tiberti, G.: Structural behaviour of SFRC tunnel segments during TBM operations. Undergr. Sp. - 4th Dimens. Metropolises, pp. 1461–1468 (2007)
5. Gettu, R., Ramos, G., Aguado, A., García, T., Barragán, B.: Steel fiber reinforced concrete for the Barcelona metro line 9 tunnel lining. In: BEFIB 2004, Proceedings of the 6th RILEM Symposium FRC. RILEM Symposium, pp. 1–46 (2004)
6. Caratelli, A., Meda, A., Rinaldi, Z., Romualdi, P.: Structural behaviour of precast tunnel segments in fiber reinforced concrete. Tunn. Undergr. Space Technol. **26**, 284–291 (2011). https://doi.org/10.1016/j.tust.2010.10.003
7. Caratelli, A., Meda, A., Rinaldi, Z.: Design according to MC2010 of a fibre-reinforced concrete tunnel in Monte Lirio, Panama. Struct. Concr. **13**, 166–173 (2012). https://doi.org/10.1002/suco.201100034
8. Beňo, J., Hilar, M.: Steel fibre reinforced concrete for tunnel lining - verification by extensive laboratory testing and numerical Modelling. Acta Polytech. **53**, 329–337 (2013)
9. Hilar, M., Vítek, J., Vítek, P., Pukl, R.: Load testing and numerical modelling of SFRC segments (2012)
10. FIB Bulletin 83, Precast Tunnel Segments in Fibre-Reinforced Concrete (2017)
11. Cavalaro, S., Blom, C.B., Walraven, J., Aguado, A.: Structural analysis of contact deficiencies in segmented lining. Packer behaviour under simple and coupled stresses. Tunn. Undergr. Space Technol. **26**, 734–749 (2011)
12. Sugimoto, M.: Causes of shield segment damages during construction. In: International Symposium on Underground Excavation and Tunnelling, pp. 67–74 (2006)
13. Liao, L., de la Fuente, A., Cavalaro, S., Aguado, A.: Design of FRC tunnel segments considering the ductility requirements of the fib Model Code 2010: application to the Barcelona Metro line 9. Tunn. Undergr. Space Technol. **47**, 200–210 (2015)
14. Leonhardt, F., Mönnig, E.: Berlin: Springer-Verlag; 1973 (Italian version:1986) Casi speciali di dimensionamento nlle costruzioni in c.a. e c.a.pConcr. Struct. 66–66 (1973). https://doi.org/10.1201/9781439828410.ch37
15. Iyengar, K.: Two-dimensional theories in anchorage zone stresses in Post-tensioned prestressed beams. Heron **32**, 45–56 (1962)
16. FIB, fib Model Code for Concrete Structures 2010. International Federation for Structural Concrete (fib), Lausanne (2013). https://doi.org/10.1002/9783433604090
17. EHE-08, Instrucción de Hormigón Estructrural (EHE-08) (2008)
18. RILEM TC 162-TDF, Test and design methods for steel fibre reinforced concrete. Design with σ-ε method. Mater. Struct. **35**, 262–278 (2003). https://doi.org/10.1617/13837
19. ACI 544, Fibre-Reinforced Concrete: Design and Construction of Steel Fibre-Reinforced Precast Concrete Tunnel Segments (2014)
20. ACI 544.7R-16, Report on Design and Construction of Fiber-Reinforced Precast Concrete Tunnel Segments. ACI Comm. Rep., pp. 1–36 (2016)
21. C. Dassault systems Simulia, Abaqus Analysis user's manual 6.12-3, 1137 (2012)
22. Schnütgen, B., Erdem, E.: Sub-task 4.4 - Splitting of SFRC induced by local forces - Annex A (2001)
23. NBN-B-15-238:1992. Testing of fiber reinforced concrete. Bending test on prismatic specimens. Belgian Code - In French (1992)
24. Conforti, A., Tiberti, G., Plizzari, G., Caratelli, A., Meda, A.: Precast tunnel segments reinforced by macro-synthetic fibers. Tunn. Undergr. Space Technol. **63**, 1–11 (2016). https://doi.org/10.1016/j.tust.2016.12.005

Design, Specification and Failure Investigation of Fibre Reinforced Concrete Ground Bearing Industrial Floors and Hardstandings and Pile Suspended Industrial Ground Floors

Chris H. Peaston[✉]

Peaston Concrete Consultancy, Nottingham, UK
chris@peastonconcrete.com

Abstract. This paper discusses the significant observations of the Author's involvement in the design and failure investigation of many fibre concrete floors and pavements over the last two decades. Serviceability failures of ground bearing industrial floors are more common than overloading. Nonetheless the extension of the related thickness design methodologies to hardstandings often fails to adequately recognise that the dominant failure mechanism is fatigue; such designs remain empirical but lack comprehensive substantiating trial data. Some methods of suspended fibre concrete floor design misrepresent the true global factor of safety, although full-scale load testing demonstrates that structural performance is not well represented by standard beam tests, and the true global factor of safety is close to that required by typical industry standard guidance.

Keywords: Industrial floors · Hardstandings · Pile suspended slabs · Full-scale tests

1 Introduction

Over the course of the last twenty years the Author has been involved in the design, specification and failure investigation of a significant number of fibre reinforced concrete (FRC) ground bearing industrial floors and hardstandings and pile suspended ground floors. In this paper the Author summarises his experience of design and failure investigation in these common applications against the background of developing design and testing guidance for FRC [1]. This experience includes numerous investigations of ground bearing industrial floors and hardstandings, and a significant number of pile suspended industrial ground slabs. The work has included full-scale load tests in some cases and the application of these results to the assessment of others.

2 Ground Bearing Industrial Floors

Ground bearing concrete industrial floors were the first significant application of FRC, with UK design guidance initially appearing in the Concrete Society's Technical Report TR34 1994 second edition [2]. The guidance was incomplete and bespoke

© RILEM 2022
P. Serna et al. (Eds.): BEFIB 2021, RILEM Bookseries 36, pp. 690–701, 2022.
https://doi.org/10.1007/978-3-030-83719-8_59

detailed design methodologies were developed by the fibre manufacturers. Thickness design was typically governed by ultimate capacity under concentrated loads, such as from rack legs, and determined as a function of the sum of the positive (sagging) and negative (hogging) moment capacities of the slab, M_p and M_n respectively.

TR34 second edition termed this sum the limit moment, M_o, and evaluated it as the elastic bending resistance multiplied by the factor $(1 + R_{e,3})$ in which $R_{e,3}$ is the equivalent flexural strength ratio determined in the JCI test [3]. This factor was referred to as an enhancement factor and, in some cases, mistakenly led to the belief that the addition of fibres significantly increases the flexural strength of the FRC composite.

The thickness design guidance presented in the 2003 third edition of TR34 [2] was significantly more comprehensive and avoided the use of M_o and the enhancement factor. The positive and negative moment capacities were still determined by elastic theory using the flexural tensile strength of the concrete, with the $R_{e,3}$ multiplier used to determine the post crack positive bending capacity of the section at the point of collapse.

The changes introduced in TR34 third edition adopted a more conservative approach to the thickness design, with a consequent increase in thickness for FRC floors. As a result, it was rare to receive a compliant design, even when requested in the project specification, as the fibre manufacturers did not want to lose the commercial advantage of design methods deemed acceptable under the previous edition.

These bespoke methodologies also relied on residual strengths derived from testing conducted by the manufacturers themselves, since the closed-loop net-displacement control standard test requirements were beyond the capability of typical commercial materials testing laboratories. The manufacturers were often reluctant to provide substantiating evidence in the form of standard bending test data and, when it was provided, it was frequently inadequate with respect to the methodology and number of tests conducted.

Many of these issues have been addressed in the 2013 fourth edition of TR34 [2] although, despite these shortcomings, overloading failure of concrete industrial ground bearing floors is uncommon, and such floors are more likely to exhibit serviceability failures related to inadequate detailing with respect to thermal and shrinkage induced movement. Such failures manifest either in the form of cracking, or the failure of joints to accommodate movement while also performing robustly under often aggressive trafficking conditions.

During the last twenty years the Author has been involved in expert investigations of numerous serviceability failures of ground bearing industrial floors, and has helped project teams deliver successful FRC floors for clients including Clarks Shoes and Jaguar Land Rover. Among the lessons learnt is the necessity to consider the full range of temperature-change induced movement in unheated warehouses. In one example, joint repairs involving the reinstatement of a saw-cut joint detail undertaken in the winter did not allow for the expansion of the floor in warmer weather, which lead to the closure of the joint and further arris damage.

3   Hardstandings

3.1   Design Guidance for Unreinforced and bar Reinforced Hardstandings

There has been an increasing trend for the design methodologies developed for FRC industrial ground bearing floors to be applied to hardstandings of the type commonly found around distribution centres, waste-to-energy schemes and the like. This is against the background of there being no universally accepted design methodology for such applications, and a common failure to recognise that the critical limit state with respect to pavement failure is typically fatigue, rather than collapse as a result of a single load application. This may lead to wheel path cracking concurrent with serviceability failures related to the inability to accommodate a range of movement greater than that typical of internal concrete ground bearing floors.

The trafficking of such pavements frequently includes highway-legal heavy goods vehicles (HGVs) but may also include specialised materials handling equipment such as shovel loaders. In the UK, design guidance for concrete hardstandings is provided by Britpave [4] and in TR66 [5]. Both documents are based on the provisions for concrete highway pavements in the UK DMRB [6], which presents empirically derived thickness design equations based on trials conducted in the 1960s as the UK motorway network was developed [7]. As a result, the use of FRC in pavements is not covered and, while the recent guidance documents both allude to this application, the reader is referred to the fibre manufacturers for the necessary detail.

The UK guidance documents [4, 5] limit trafficking to 20 million standard axles (msa) and allow some relaxations with respect to highway pavement design, which would typically be subjected to significantly higher highway-legal traffic loading. These include the use of an unbound subbase and an extrapolation to permit the use of standard welded fabric reinforcement providing 393 mm^2/m, which is less than the minimum 500 mm^2/m requirement of the highway standard.

The guidance provides equations for unreinforced (URC) and jointed reinforced (JRC) options, both with or without a 1 m trafficked edge strip. A minimum thickness of 150 mm or 175 mm is required respectively for bound and unbound subbases and for 20 msa design traffic, with a typical minimum foundation stiffness of 100 MPa, concrete strength class C32/40 and minimum reinforcement leads to a thickness of the order of 200 mm with a difference of around 10% for the URC and JRC options.

3.2   Design Guidance for FRC Hardstandings Based on TR34 Second Edition

As there is no relevant fatigue model, a variety of approaches to pavement design has been adopted by the fibre manufacturers, each based on their bespoke methods developed to meet the requirements of TR34. One such method adopts the enhancement factor identified in Sect. 2 above, which is used to calculate a 'fictitious' flexural strength that significantly exceeds the cracking strength of the composite. The stresses arising from the wheel loads are then calculated using an elastic analysis for the internal, edge and corner loading cases. The resulting elastic stresses are multiplied by a factor of 2 to account for the multiple load applications based on the recommendations of TR550 [8]. An allowance is also made for shrinkage stresses before comparing with the 'fictitious' flexural strength.

This methodology therefore combines elements of elastic and plastic design methods, although this is fundamentally flawed given that the TR550 fatigue factors relate to uncracked concrete. In principle the cracking flexural strength, which is typically unaffected by fibre addition, should thus be halved before it is compared with the calculated elastic stresses. The methodology is not based on any established or justifiable engineering principle and the impression that a minimum factor of safety (FoS) of 2 is achieved is unrealistic.

3.3 Design Guidance for FRC Hardstandings Based on TR34 Third Edition

In an alternative approach adopted by one fibre producer, the thickness design of the SFRC slab is based on the ultimate strength analysis of the slab using the plastic analysis or yield-line theory outlined in TR34 third edition. The yield line theory allows for the development of formulae to determine the slab stresses for the interior, edge and corner load cases. The TR34 equations are used to calculate the ultimate collapse load in each case and the fatigue factor is then worked out as the quotient of the collapse load and wheel load.

It is not made clear what minimum FoS is required, although typical calculations appear to aim for a minimum FoS of 2 for all three load cases based on the recommendations of the Portland Cement Association [9]. A material partial safety factor of 1.5 is also applied to the calculated bending capacity of the FRC section, thus giving the impression that a minimum global FoS of 3 is achieved. The method is therefore also a mixture of plastic analysis, in which the failure mechanism is based on the yield lines caused by a point load, and checked using a fatigue factor relating to allowable stress in the uncracked concrete.

3.4 Design Guidance for FRC Hardstandings Based on Equivalent Bending Capacity

A third method adopted by some fibre manufacturers involves designing a reinforced concrete slab using conventional methods and replacing the bar reinforcement with fibres to give a section with an equivalent moment of resistance in the cracked FRC. The method involves using conventional reinforced concrete analysis to calculate the bending capacity of the bar reinforced section and equates this to the residual bending capacity of the cracked FRC section of the same thickness. The method is used for both singly and doubly reinforced sections.

While it seems reasonable to calculate the equivalent bending capacity of a singly reinforced section in this manner to determine the capacity under a concentrated load, it is not clear how fatigue is dealt with in such comparisons. It is unclear that the bending capacity of the cracked FRC section is the correct basis on which to compare the fatigue performance of a conventionally reinforced pavement under traffic loading.

3.5   Summary Analysis of FRC Pavement Design Methods

There is no single universally adopted design procedure for fibre reinforced concrete pavements and no fatigue model which might be used in such a methodology. In the absence of a fatigue model, design procedures are generally bespoke methodologies developed by the fibre manufacturers based on established methods for internal industrial floor slabs. The addition of fibres at dosages that are practicable for most bulk field applications such as pavements do not have a significant effect on the cracking strength of the concrete.

The addition of fibres at practicable rates only affects the post-cracking behaviour of the FRC. Nonetheless, the design methods adopted by the fibre manufacturers and flooring contractors apply fatigue models developed on uncracked concrete. These methods are used to present calculations which give the impression that a suitable FoS has been adopted, although this is unlikely to be the case. The designs are effectively empirical, but lack the coordinated trial data with which they might be substantiated. FRC pavement design is therefore at best experiential and this has led to some significant failures.

Hardstandings trafficked predominately by highway legal HGVs are typically of the order of 200 mm thick and the difference in thickness for conventionally designed URC and JRC options is around 10%. It is likely that the fatigue performance of an FRC pavement falls somewhere in between that of conventional URC and JRC options, and therefore there is unlikely to be any cost advantage over URC options. Claims that the use of fibres permit increased joint spacing are also empirically based with some suppliers being significantly less conservative than others.

4   Pile Suspended Industrial Ground Floors

4.1   Design Guidance and Methodologies

UK design guidance with respect to pile suspended FRC industrial ground slabs first appeared in the 2003 third edition of TR34. This too was limited, and subsequent designs also therefore relied on the bespoke methodologies and material properties respectively developed and derived on behalf of the fibre manufacturers. This had the potential to lead to the presentation of design calculations which, on inspection, might be found to have a lower FoS than those commonly associated with standard practice [10].

One such example appears in Destrée [11], which illustrates a calculation in standard form that calculates the required moment capacity of a rectangular plate spanning between lines of piles loaded with a typical rack leg load arrangement at its midspan. This leads to the design equation:

$$1.5 * \left(\frac{P * 1.5 * L_N}{8 * b} + \frac{G_{MAX} * 1.35 + Q_I * 1.5}{16} * L_N^2 \right) \leq M_L \tag{1}$$

in which b is the corridor width, P the total load in each corridor, L_N the net span, G_{MAX} the self-weight plus permanent load, Q_I the variable load and M_L the bending capacity of the FRC section. The respective partial factors for point and UDL actions were taken as 1.5 and 1.35 which, along with the material partial factor of 1.5, were presented as complying with the Eurocode EC2 [10]. For a typical high-bay rack leg load, which occurs independently of anything other than self-weight, the UDL contribution is small compared with the rack leg loading and the global FoS is notionally 2.25.

Nonetheless there was little justification for the determination of the net span which was taken as:

$$L_N = L - p - h \tag{2}$$

in which p and h are respectively the pile diameter at the slab soffit and slab thickness, and the moment capacity of the FRC section implies a value of flexural tensile strength significantly greater than is likely to be obtained in a simple beam test [11, 12]. At around the time of the Destrée publications, and along with several erstwhile Arup colleagues, the Author reviewed a design based on this methodology. It was understood that the FRC section moment capacity was back-calculated from bespoke indeterminate round panel tests [12] which, it was considered, were not representative of the discrete yield lines that were likely to form in practice.

At the time a solution based on the Destrée methodology, which was adjusted to use an effective span equal to the pile spacing and therefore compatible with equilibrium considerations, along with a moment capacity derived from a simple standard bend test, could not be demonstrated to have the required FoS. As a result, alternative FRC solutions using either reinforcing bars placed along the pile grid, or fabric reinforcement placed in the top of the section at the pile locations and the bottom of the section at the midspan, were used on several Arup projects.

4.2   Full-Scale Load Test on FRC Pile Suspended Ground Floor

The inability to demonstrate an adequate FoS in pile suspended FRC slabs by conventional code and industry guidance-based calculation [10, 13] has resulted in several disputes, in which the adequacy of the design methodology was questioned, and eventually led to a full-scale load test being undertaken in one such case. The salient details of the Destrée methodology-based design calculations for the relevant governing back to back rack leg load case are reproduced in Table 1 below.

Table 1. Summary design details for load tested slab.

Racking 2.7 × 0.9 m and 250 mm c-c					
Racking aisles at 1.8 m					
Corridor width =1.8 + 0.9 + 0.9 + 0.25 = 3.85 m					
Piles	Pile grid:	$L_1 =$	3.284	(m)	
		$L_2 =$	3.317	(m)	
	Pile head diameter	$p =$	900	(mm)	
Loads	Point loads:	$P =$	360.00	(kN)	(90 kN rack leg load; 4 per pile bay)
	Distance between corridors	$b =$	3.85	(m)	(c-c distance between aisles)
Floor	Thickness:	$h =$	250	(mm)	
Loads	Self-weight floor:	$G_{MAX} =$	6.25	(kN/m²)	
Characteristic 28-day cube strength			40	(N/mm²)	
Fibres	Dosage rate:		45	(kg/m³)	
Yield moment:		$M_L =$	63.18		

The design calculations used the longer of the two net spans calculated using Eq. (2) above, along with Eq. (1) and the input data in Table 1 above, to determine a section capacity of 60.65 kNm/m, which was less than the stated moment capacity of 63.18 kNm/m. This was presented as being consistent with the EC2 requirement for a global FoS of 2.25.

A full-scale load test was conducted on a section of the slab constructed to the above design. The test section was isolated from the rest of the slab by saw cutting through the full slab depth to leave a test section 3 piles wide by 4 piles long. The loads were applied through 8 jacks, located along the centre line in the centre span of the isolated section and arranged to simulate the design rack leg loads (Fig. 1). The test section was sampled to determine depth, concrete strength and fibre content and was loaded to failure, which occurred at 180 kN. The average distance between the hogging line cracks, which did not coincide with the pile centrelines, was measured as 2.7 m.

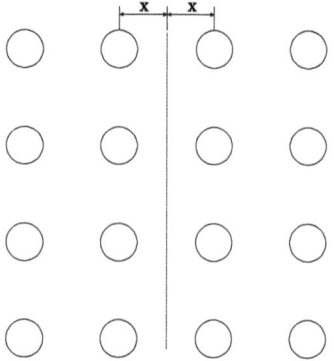

Fig. 1. Indicative pile grid and loading centreline in isolated test area

4.3   Back Calculation of Bending Capacity from Full-Scale Load Test

The salient details of the load test are provided in Table 2 below. The folding plate yield line mechanism developed at a load of 180 kN per jack, which was twice the working rack leg load of 90 kN.

Table 2. Details of full-scale load test

Number of jacks on test panel	$N =$	8	
Width of test panel	$b =$	8.9	m
Concrete density	$c =$	25	kN/m^3
Slab thickness	$t =$	0.255	m
Average distance between hogging moment cracks	$L =$	2.7	m
Dead load udl	$w =$	$c*t = 6.375$	kN/m^2
Line load from jacks	$P =$	$\frac{180*N}{b} = 161.8$	kN/m
Mean bending moment at yield line due to dead load udl	$\bar{w} =$	$\frac{1}{16}*w*L^2 = 2.90$	kNm/m
Mean bending moment at yield line due to live load from jacks	$\bar{P} =$	$\frac{1}{8}*P*L = 54.61$	kNm/m
Total mean bending moment at yield line	$T =$	$\bar{w}+\bar{P} = 57.51$	kNm/m

It should be noted that the average spacing of 2.7 m between the hogging cracks is consistent with the effective span calculated using the provisions of TR34 fourth edition. Nonetheless, the loading arrangement, in which the adjacent spans were unloaded, may have influenced the location at which the yield lines formed.

The unfactored bending resistance of the slab required to take the service load can be calculated from the details in Table 3 below.

Table 3. Details of load tested slab for determination of required service load capacity

Take:	Pile grid as	= 3.317 m in both directions
	Effective pile diameter as	= 0.9 m
	Thickness of slab	= 0.25 m
	Self-weight floor, w,	= 6.25 kPa
	Point load, P,	= 4 * 90 = 360 kN
	Distance between point loads, b,	= 3.85 m
	Effective span	= 3.317 – 0.7 * 0.9 = 2.687 m

Equation (1) can be used with the unfactored loads in Table 3 above to determine the required service load capacity as 34.23 kNm/m. The global FoS implied by the quotient of the back-calculated capacity 57.51 kNm/m (Table 2) to the required capacity is therefore 1.68. This should be regarded as an upper bound value since, had the adjacent spans also been loaded consistent with the likely true load condition, the yield lines may have aligned more closely with the pile centrelines. This would lead to a concurrent increase in the effective span, and reduction in load capacity for identical back-calculated moment capacity.

The global FoS of 1.68 is significantly lower than the global FoS of 2.25 claimed by the design, although it should also be noted that TR34 has consistently adopted a partial load factor of 1.2 for well-defined rack leg loads rather than the factor of 1.5 adopted in the Destrée methodology. The global FoS required by TR34 is therefore 1.8 which, while it is still not met, more favourably compares with the global factor of 1.68 derived from the full-scale load test.

These observations also illustrate that the derivation of FRC material strengths based on standard tests on relatively small prisms, as is required by EN 14651 [14] on which the MC2010 [15] guidance for FRC design is based, is likely to underestimate full-scale FRC performance. This is recognised in MC2010, which permits the use of a redistribution factor to account for such favourable effects.

4.4   Application of Full-Scale Load Test Results to Other Slabs

4.4.1   Design Details

The salient details of the Destrée methodology-based design calculations for the governing back to back leg load case for a comparable design to that for the load tested slab are reproduced in Table 4 below. Once again it was not possible to show by conventional calculation that the design had either the claimed or required global FoS. The results of the previous load test were used to estimate the FoS in the second case using a procedure outlined in the following sections.

Table 4. Summary design details for comparable slab.

Racking (2.8 * 0.9 m and 200 mm c-c)					
Racking aisles at 1.8 m					
Corridor width =1.8 + 0.9 + 0.9 + 0.2 = 3.80 m					
Piles	Pile grid:	$L_1 =$	3.063	(m)	
		$L_2 =$	3.063	(m)	
	Pile head diameter	$p =$	900	(m)	
Loads	Point loads:	$P =$	360.00	(kN)	(90 kN rack leg load; 4 per pile bay)
	Distance between corridors	$b =$	3.80	(m)	(c-c distance between aisles)
Floor	Thickness:	$h =$	270	(mm)	
Loads	Self-weight floor:	$G_{MAX} =$	6.75	(kN/m^2)	
Characteristic 28-day cube strength			35	(N/mm^3)	
Fibres	Dosage rate:		40	(kg/m^2)	
Yield moment:		$M_L =$	66.78		

4.4.2   Required Capacity of Second Slab

The unfactored bending resistance of the second slab required to take the service load was calculated from the details in Table 5 below.

Table 5. Details of second slab for determination of bending resistance

Take:	Pile grid as	= 3.063 m in both directions
	Effective pile diameter as	= 0.9 m
	Thickness of slab	= 0.27 m
	Self-weight floor, w,	= 6.75 kPa
	Point load, P,	= $4 * 90 + 30$ $(MHE) = 390$ kN
	Distance between point loads, b,	= 3.8 m
	Effective span	= $3.063 - 0.7 * 0.9 = 2.433$ m

Equation (1) was used with the unfactored loads in Table 5 above to determine the required service load capacity of the second slab as 31.31 kNm/m.

4.4.3   Second Slab Global Factor of Safety

The mean tensile strength relating to the C32/40 strength class adopted in the load tested slab design was calculated as 3.0 N/mm^2 using the following EC2 relationship.

$$f_{ctm} = 0.3 f_{ck}^{(2/3)} \tag{3}$$

From this the mean flexural strength was derived using the following respective relationships presented in the TR34 third and fourth editions. The relationship given in the fourth edition is consistent with EC2.

$$f_{ctm,fl} = \left[1 + \left(\frac{200}{h}\right)^{0.5}\right] f_{ctm} \tag{4}$$

and

$$f_{ctm,fl} = (1.6 - h/1000) f_{ctm} \tag{5}$$

In the case of Eq. 4 this leads to a multiplying factor of 1.89 for the 255 mm thick load-tested slab and a corresponding mean flexural strength of 5.67 N/mm^2. The moment capacity of the uncracked slab was then calculated as 61.45 kNm/m based on elastic theory using the calculated mean flexural strength.

The effective post-cracking residual strength deriving from the full-scale load test was then calculated as 93.6% from the quotient of the above theoretical bending capacity (61.45 kNm/m) and the back-calculated capacity derived from the full-scale load test (57.51 kNm/m (Table 2)). This ratio was then used to obtain an effective flexural strength of 5.31 N/mm^2 and Eq. 4 used to back-calculate an effective direct tensile strength of 2.8 N/mm$^{2.}$

The full-scale effective tensile strength is of fundamental significance with respect to the full-scale post-cracking performance, and may be used to derive the bending capacity of other slabs of similar geometry and material properties with an identical fibre. Its value might equally have been calculated using the alternate TR34 fourth edition or EC2 Eq. 5 and corresponding ratio of back-calculated to theoretical bending capacities. However Eq. 5 leads to a significantly lower value of mean flexural strength and a bending capacity which is less than the back-calculated value. The above approach based on Eq. 4 therefore appears to more accurately reflect the measured strengths.

Using the back-calculated effective direct tensile strength of 2.8 N/mm^2 derived from the full-scale load test, the TR34 third edition relationship Eq. 4 was used to calculate the mean flexural strength of the 270 mm thick second slab as 5.21 N/mm^2 with a corresponding moment capacity of 63.3 kNm/m. The calculated moment capacity was then adjusted to allow for both the variations in fibre content and strength. The respective adjustments of 0.93 and 0.915 for fibre content and strength were based respectively on a knowledge of the variation of the equivalent flexural strength in a standard JCI bending test and the EC2 relationship given in Eq. 3.

This leads to a post-cracking bending capacity of 53.90 kNm/m for the as designed 270 mm thick slab and a corresponding global FoS of 1.72. This value represents the highest global FoS likely to be obtained should a load test be conducted on the 270 mm thick slab, and is predicated on the assumption that the insitu slab has been constructed in accordance with the design with respect to thickness, concrete strength and fibre content.

5   Conclusions

On the basis on the work discussed above the following conclusions may be drawn:

- The thickness design of ground bearing floor slabs is well understood and rarely leads to under design with respect to load capacity. Such slabs are more likely to suffer serviceability failures due to the effects of trafficking on joints;
- FRC pavements are more likely to fail in fatigue rather than as a result of a single point load application. There are no adequate fatigue models on which to base FRC pavement design, which is therefore empirical but lacks comprehensive trial data on which to base such designs;
- Pile suspended floors designed using the Destrée methodology are likely to significantly overestimate the true global FoS, although full-scale load testing demonstrates that a global FoS roughly consistent with that required by TR34 is achievable;
- Full-scale testing also demonstrates that pile suspended floor designs based on FRC bending capacity derived from standard prismatic bending tests are likely to be conservative, thereby supporting the adoption of redistribution factors in contemporary design guidance, although these should not be extrapolated;

- Compliance with contemporary design standards and related application specific guidance reduces the probability of failure, although many currently serviceable FRC industrial floors and hardstandings are likely to have been designed using compromised methodologies that would not conform to current standard practice.

Acknowledgements. The Author would like to acknowledge both the contributions of various erstwhile Arup colleagues to the work described in this paper, with particular mention to Ian Feltham and Michael Sataya, and the granting of permission to publish this paper based on work undertaken on behalf of the firm.

References

1. Peaston, C.H.: An Introduction to the Characterisation and Design of Fibre Reinforced Concrete. Institute of Concrete Technology Yearbook 2015/16.
2. Concrete Society: Concrete Industrial Ground Floors – A Guide to Design and Construction. Technical Report No 34, Fourth Edition 2013, Third Edition 2003, Second Edition 1994.
3. Japan Concrete Institute, JCI-SF4: Method for test for flexural toughness of fibre reinforced concrete, Standard SF4, JCI Standards for Test Methods for Fiber Reinforced Concrete, pp. 45–51 (1983)
4. Britpave: Concrete Hardstanding Design Handbook – Guidelines for the Design of Concrete Hardstandings. Report BP/17, Second Edition 2007, first published 2005
5. Concrete Society: External in-situ concrete paving. Technical Report 66, Camberley (2007)
6. The Highways Agency: Design Manual for Roads and Bridges. HD 26/06 Pavement Design, London (2012)
7. Mayhew, H.C., Harding H.M.: Thickness design of concrete roads research report 87. The Transport Research Laboratory, Crowthorne (1987)
8. British Cement Association: Design of floors on ground. Technical Report 550. British Cement Association, Crowthorne (1982)
9. Portland Cement Association: New PCA thickness design procedure for concrete highway and street pavements', Portland Cement Association, Skokie, Illinois, USA (1984)
10. British Standards Institution: EN 1992-1-1 Eurocode 2: Design of concrete structures – part 1–1: general rules and rules for buildings
11. Destrée, X.: Structural application of steel fibre as principal reinforcing: conditions – design – examples. RILEM PRO15, Fibre Reinforced Concretes (FRC) BEFIB 2000, pp. 291–301 (2000)
12. Destrée, X.: Structural application of steel fibre as only reinforcing in free suspended elevated slabs: conditions – design – examples. In: 6th RILEM Symposium on Fibre-Reinforced Concrete (FRC) – BEFIB 2004, vol. 2, pp. 1073–1082 (2004)
13. Concrete Society: Guidance for the design of steel-fibre-reinforced concrete. Technical Report No 63 (2007)
14. British Standards Institution: EN 14651, Test method for metallic fibre concrete - measuring the flexural tensile strength (limit of proportionality (LOP), residual)
15. Fédération International du Béton: fib Model code for concrete structures 2010. Ernst & Sohn (2013)

Self-compacting Steel Fibre Reinforced Concrete: Material Characterization and Real Scale Test up to Failure of a Pile Supported Flat Slab

Stanislav Aidarov[1,2(✉)], Francisco Mena[2], and Albert de la Fuente[2]

[1] Smart Engineering Ltd., UPC Spin-Off, Barcelona, Spain
stanislav.aidarov@upc.edu
[2] Polytechnic University of Catalonia, Barcelona, Spain

Abstract. Steel Fibre Reinforced Concrete (SFRC) is increasingly being used in the construction industry providing structural, technological and economic benefits. However, this relatively new material has not demonstrated its full potential due to presence of certain aspects related to both concrete mix and structural design that should be further investigated. Specifically, pile-supported flat slabs is an interesting field of application of the SFRC; however, there are still aspects needing attention and solutions in order to make this structural application more attractive from both economic and technical points of view, these being (among others): the effect of new types of fibres on material properties and more insight regarding the structural capacity at both serviceability and ultimate limit state. Complementing the existing knowledge in aforementioned areas might lead to considerable expansion of SFRC application in elements with high structural responsibility.

To this end, an extensive experimental program was carried out within the industrial-oriented project eFIB that contained characterization of 15 concrete mixes with different fibre types and its content (up to 120 kg/m^3) and construction of 10 × 12 m SFRC flat slab. This prototype was loaded in different stages with permanent and life-loads in order to study both cracking and time deformation responses; the structure was eventually led to failure. The results of the described research are presented and discussed herein.

Keywords: Steel fibre · Self-compacting concrete · Real scale test · Flat slab

1 Introduction

The use of Steel Fibre Reinforced Concrete (SFRC) in the elements with different structural responsibility is already an attractive alternative for the solutions in which, not long ago, the traditional reinforcement was treated as axiomatic. Addition of steel fibres to the cement-based composites demonstrates the remarkable results in terms of residual tensile strength, crack resistance, durability and fatigue [1–4] what allows to apply this technological material in pavements, precast elements and elevated flat slabs [5–8].

© RILEM 2022
P. Serna et al. (Eds.): BEFIB 2021, RILEM Bookseries 36, pp. 702–713, 2022.
https://doi.org/10.1007/978-3-030-83719-8_60

The latter could be named as one the most complex challenges for the aforementioned material due to occurrence of relatively high stresses within the service life-span of the structure. Despite the existence of already executed elevated SFRC flat slabs which have highlighted the technological, economic and environmental positive outcomes, this approach is still of concern to most of engineers due to certain aspects that require further study, such as: maintenance of structural integrity under ultimate loading conditions; the effect of long-term loads in terms of deformation and cracking; potential reduction of fibre content to make the solution more competitive in comparison with those traditional (ex., reinforced concrete with steel bars) and the established criteria of quality control (material and structure). These aspects are among the first to be analysed in order to confirm the technical feasibility of the partial (or even total) substitution of traditional reinforcement in pile supported slabs.

In this context, the industrial-oriented project eFIB (financially supported by Spanish Ministry of Economy, Industry and Competitiveness; reference: RTC-2016-5263-5) was carried out with the aim to increase the knowledge base in the above-described aspects. In turn, the purpose of the research contribution is to report the results obtained from the characterization of SFRC mixes with different fibre type/content and real-scale testing of a SFRC elevated flat slab.

2 Characterization of SFRC

2.1 Development of SFRC Mixes

Development and posterior characterization of produced SFRC mixes was divided into three phases in accordance with Fig. 1. The first phase was concentrated on the study of the properties in fresh and hardened states of six different concrete mixes, in which, two types of fibres were used, varying its content from 60 to 120 kg/m^3 with interval of 30 kg/m^3.

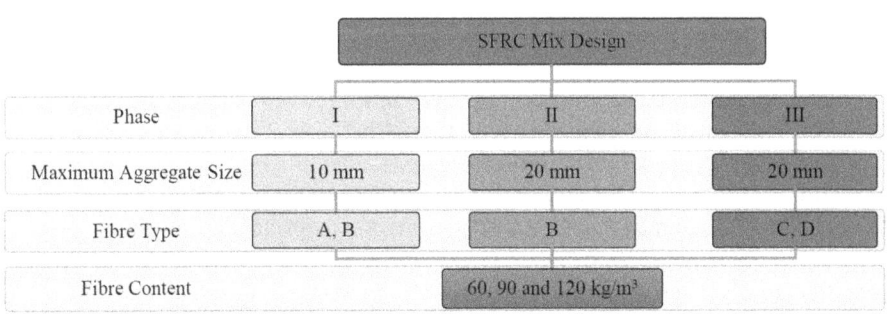

Fig. 1. Phases of SFRCs characterization

The influence of maximum aggregate size on the rheological and mechanical properties of the material was the main goal of the second phase. Fibre B was selected for this phase due to better performance of the latter during the first part of the study.

SFRC mixes with modified granular skeleton and same range of fibre contents (60, 90 and 120 kg/m^3) demonstrated the similar performance in terms of studied properties with respect to previously obtained results. Therefore, the concrete composition of the material with maximum aggregate size of 20 mm was maintained for the characterization of concrete mixes reinforced by fibres C y D – third part of the analysis. The examination of fifteen SFRCs considering four fibre types and three different contents was carried out; the important details to be pointed out are: (1) fibres with the tensile strength ranging from 1500 to 2300 N/mm^2 and with the different type of anchorages were taken into consideration (Table 1) and (2) the principal variables of the concrete granular skeleton did not face any significant modifications for the sake of comparability (Table 2).

2.2 Analysis of SFRC Properties in the Fresh State

Owe to the type of application, self-compactability was found to be convenient. Therefore, each concrete mix was tested for required slump flow in accordance with the Spanish Code on Structural Concrete (EHE-08) [9]. The obtained results demonstrated possible development of the self-compacting concrete mixes even with fibre content of 120 kg/m^3; information on the achieved slump flow values is depicted in Fig. 2. The coding X(Y)-Z, where X, Y and Z being respectively the fibre type, maximum aggregate size and fibre amount was established for referring to each SFRC mix.

Table 1. Properties of studied fibres

Fibre	A	B	C	D
f_{yf} [Mpa]	1800	1900	1500	2300
Length [mm]	50 ± 3	60 ± 3	60	60
Diameter [mm]	1	0.9	0.9	0.9
Aspect ratio	50	67	65	65
Geometrical form	Hooked-end			

Table 2. Granular skeleton (in kg/m^3)

Material	Dosage
Cement	425
Filler	25
Water	200
Fine aggregate	1325
Coarse aggregate	400

Minimum permitted slump flow for concrete to be considered as self-compacting per EHE-08 is equal to 55 cm. Figure 2 shows that only 2 out of 15 SFRCs did not reach this value. However, these mixes contained only 60 kg/m^3, therefore, most likely a slight lack of superplasticizer was the cause of insufficient flowability. Furthermore, the effect of maximum aggregate size increment on compactability of SFRCs could be appreciated on the example of mixes B (10) and B (20); the reduction of slump flows was equal to 10, 7 and 10 cm for the identical fibre contents of 60, 90 and 120 kg/m^3, respectively.

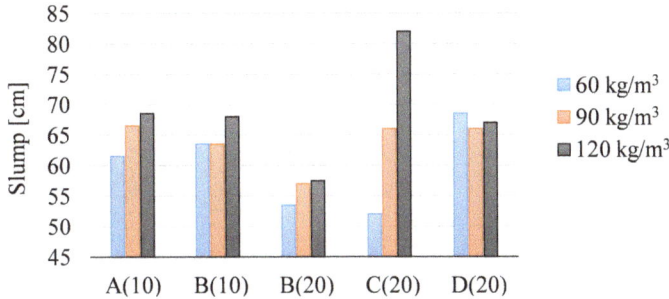

Fig. 2. Slump flows of studied SFRCs.

Apart from slump-flow testing [10], the estimation of density and air content of fresh concrete was carried out [11, 12]. Density ranged between 2.16 and 2.40 g/cm³ – common values for self-compacting concretes, whereas the obtained values of air contents were comprised between 3.0 and 11.0% (Fig. 3). Analysing the results presented in Fig. 3, it was decided to disregard the performance of SFRC B(20)-60 in hardened state due to high value of air content which might be caused by anomalous effect of one of the applied additives.

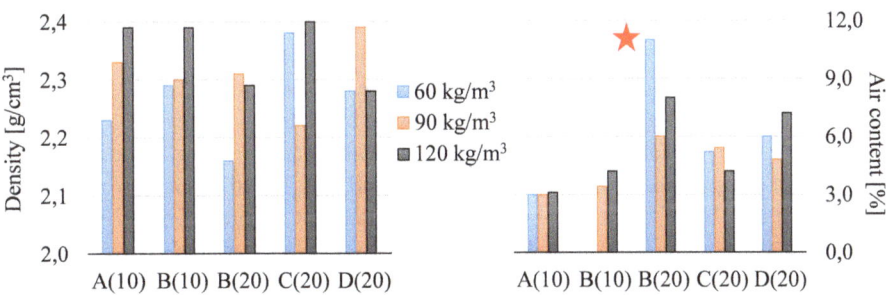

Fig. 3. Density and air content in fresh SFRCs.

2.3   Analysis of SFRC Properties in the Hardened State

2.3.1   Compressive Strength and Young Modulus

Compressive strength of produced SFRCs was evaluated through testing 6 cast cylinders for each mix [13]; three of those were analysed at 7 days and the remaining – at 28 days. Mean values of the compressive strength at 7 days ($f_{cm,7}$) were ranged between 38.3 and 52.8 MPa, whilst the performance at 28 days varied from 50.2 to 67.9 MPa (Fig. 4). The ratios $f_{cm,7}$ to $f_{cm,28}$ of studied materials were within the range of 0.67 to 0.80 – expected values for normal/high strength concretes.

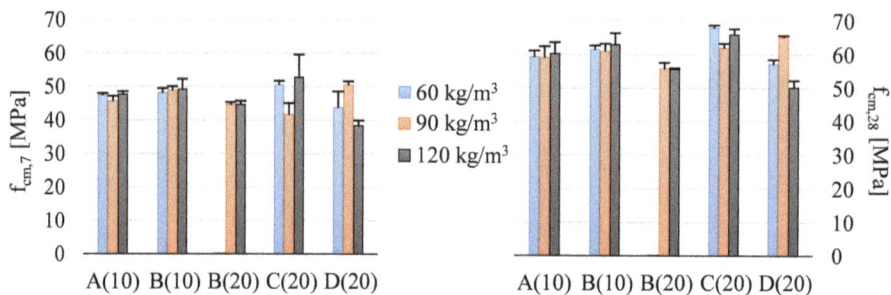

Fig. 4. Mean values of $f_{cm,7}$ and $f_{cm,28}$ with corresponding standard deviations.

The results depicted in Fig. 4 are in line with the previous works which highlighted the absence of any significant influence of the fibre content on compressive strength of the concrete; this parameter mainly depends on the material matrix, i.e. amount of the cement paste and granular skeleton [14, 15]. Considering the above stated, the modification of maximum aggregate size could have had an effect on the studied mechanical properties. However, the amount of coarse aggregate was not significant in the concrete dosages throughout the experimental programme (Table 2) due to established workability requirements. Therefore, the difference in compressive strength of the studied SFRCs can possibly be explained by the variation of moisture content of aggregates and environmental conditions which, along with the incorporated additives, had an impact on the air content in the elaborated concrete mixes.

The Young Modulus was comprised between 30.0 and 38.1 GPa – found to be sufficient for providing stiffness to the slab. However, the presence of considerable amount of fine aggregates in the designed dosages resulted in lower values of Young Modulus in comparison with those suggested by RILEM TC 162-TDF [16] for SFRCs (black line, Fig. 5). In fact, the obtained results better met the prediction of EHE-08 for conventional concrete (blue line, Fig. 5). This observation acknowledges the certain reduction of Modulus of Elasticity due to requirement of increased binder content in order to provide the self-compactability of the material in question [17].

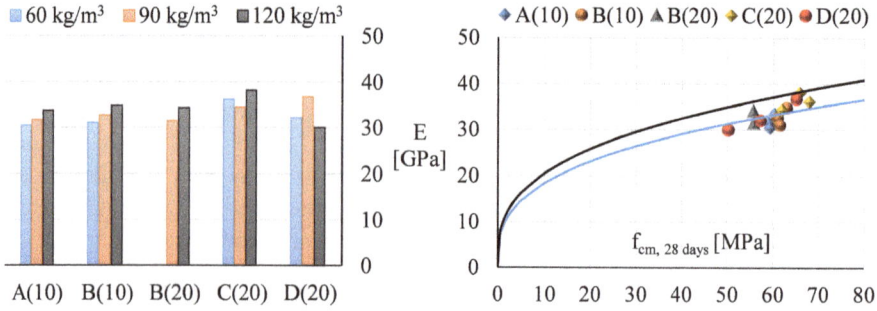

Fig. 5. Young Modulus of SFRCs.

2.3.2 Residual Tensile Strength

Measurements of residual tensile strength were carried out for each SFRC in accordance with EN 14651 [18]. Under a simply-supported three point bending configuration on the notched prisms, the limit of proportionality (f_{LOP}) and flexural residual strengths (f_R) corresponding to certain values of crack mouth opening displacement (CMOD = 0.5, 1.5, 2.5 and 3.5 mm) were found (Table 3).

Table 3. Post-cracking residual strengths (in MPa) and fib MC-2010 classification

SFRC	$f_{LOP,m}$ (CV)	$f_{LOP,k}$	$f_{R1,m}$ (CV)	$f_{R1,k}$	$f_{R3,m}$ (CV)	$f_{R3,k}$	Class
A(10)-60	3.6 (7.8)	3.2	3.9 (13.0)	3.1	4.4 (22.7)	2.8	3b
A(10)-90	4.1 (14.1)	3.1	6.4 (29.1)	3.3	6.1 (22.7)	3.9	3d
A(10)-120	5.0 (14.6)	3.8	8.0 (26.5)	4.5	6.6 (19.7)	4.4	4c
B(10)-60	4.7 (11.4)	3.8	7.5 (25.6)	4.4	7.7 (27.5)	4.3	4c
B(10)-90	5.4 (14.6)	4.1	11.4 (19.0)	7.84	10.4 (13.7)	8.07	7c
B(10)-120	5.6 (12.3)	4.5	12.1 (6.2)	10.9	11.3 (21.3)	7.73	10a
B(20)-60	4.6 (11.3)	3.8	6.8 (23.8)	4.2	7.1 (15.6)	5.3	4d
B(20)-90	5.52 (17.0)	4.0	10.5 (25.6)	6.1	9.0 (29.7)	4.6	6b
B(20)-120	6.3 (7.9)	5.4	11.7 (12.1)	9.4	10.8 (17.3)	7.8	9b
C(20)-60	4.7 (2.2)	4.9	7.3 (14.9)	5.5	7.1 (16.6)	5.2	5c
C(20)-90	5.2 (6.6)	4.8	9.6 (15.6)	7.2	9.8 (15.3)	7.3	7c
C(20)-120	6.6 (14.1)	5.1	14.0 (12.9)	11.0	14.3 (15.3)	10.7	11c
D(20)-60	4.2 (12.4)	3.4	6.6 (25.9)	3.8	7.44 (11.5)	6.0	3e
D(20)-90	4.8 (9.0)	4.1	10.0 (12.9)	7.9	10.4 (11.7)	8.4	7c
D(20)-120	5.9 (20.6)	9.1	13.7 (20.6)	9.1	13.9 (29.2)	7.2	9b

Mean values of f_{R1} ranged between 4.0 and 14.0 MPa, whereas f_{R3} presented values from 4.2 to 14.3 MPa. These results complied with the minimum established requirements to substitute conventional reinforcement at ultimate limit state according to the *fib* Model Code 2010 [17]: $f_{R1k}/f_{LOPk} > 0.4$ and $f_{R3k}/f_{R1k} > 0.5$.

3 Real Scale Testing of SFRC Flat Slab

3.1 Previous Experiences and Construction of SFRC Prototype

The selection of a suitable fibre type and amount for the design and construction of a SFRC flat slab was preceded by the study of already existing cases (Table 4). The concrete mix with fibre content of 100 kg/m^3 was predominantly used in those cases with significant span to depth ratio (L_{max}/d), in which the so called Anti-Progressive Collapse reinforcement (APC) was also included. This reinforcement should be placed in the bottom of the slab in alignment with columns in both directions and this is intended to be capable of providing post-failure resistance of the structural element [19].

Analysing the collected information, it was decided to execute the SFRC pile supported flat slab with span to depth ratio of 30 (the upper limit of those already constructed, see Table 3), consisting of 4 panels of 5 × 6 m² each with a slab thickness of 0.2 m supported onto 0.25 m-square cross-section reinforced concrete columns, see Fig. 6. Residential building loads according to the Spanish Code (self-weight of 4.8 kN/m², dead-load of 2.0 kN/m² and 3.0 kN/m² of live-load) were considered. Additionally, following the recommendations of Canadian Standards, the APC reinforcement was installed.

Table 4. Previous experiences of SFRC flat slabs

Fibre type		C_f	L_{max}/h	Traditional	Building/prototype	Country
Length [mm]	Diam [mm]	[kg/m³]	[m/m]	reinforcement		
50	1,3	100	30	Yes[1]	Prototype	Luxembourg
37	0,5	90	16	Yes[1]	Prototype	Portugal
50	1,3	100	28	Yes[1]	Prototype	Estonia
–	–	100	24	Yes[1]	Building	Lithuania
50	1,3	100	27	Yes[2]	Building	Estonia
50	1,3	100	27	Yes[2]	Building	Spain

NB: Yes[1] – only presence of APC bars; Yes[2] – presence of APC bars + traditional reinforcement in the areas with particularly high stresses

Based on the established parameters of geometry and loads, the design of SFRC prototype was carried out by means of yield line method [20, 21] and non-linear numerical analysis [22]. The report provided by ACI [23] was used for the analytical solution, whereas numerical simulations were performed through the software "ABAQUS" applying different constitutive models in accordance with RILEM [16] and *fib* Model Code [17].

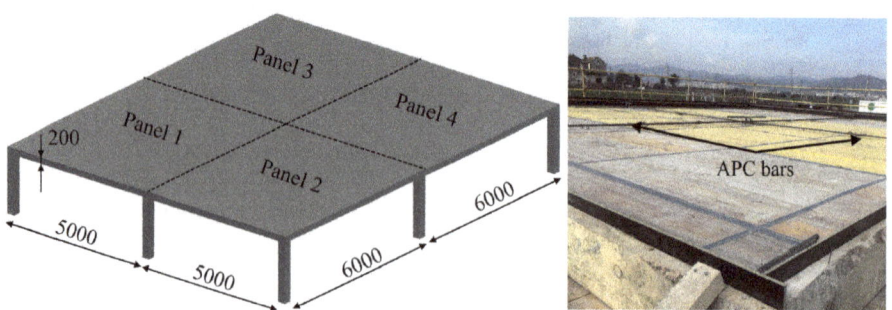

Fig. 6. Geometry of the prototype and installed APC bars

Design output along with the obtained information of studied SFRCs (fibre content – residual tensile strength relationship) permitted to reduce fibre content in the selected concrete mix by 30% in comparison with previous experiences (Table 3), providing the required mechanical properties. Applying the established material, SFRC pile supported flat slab was constructed and, during the execution, the appropriate workability was pointed out: no fibre-balling was detected and the rheological properties of the material did not differ significantly with those achieved throughout experimental program (Fig. 7).

Fig. 7. a) SFRC pumping b) SFRC flat slab in the fresh state

After the construction and during the curing period, the effect of climatic conditions (very low temperatures in Barcelona, including snow) coupled with autogenous shrinkage of concrete was observed – these phenomena led to appearance of certain cracking of upper surface of the slab. However, the observed cracks were controlled by presence of fibres and had no influence on the structural capacity of the SFRC element. The evaluation of shrinkage cracks was followed by formwork removal and installation of prisms in the centres of each field of the SFRC slab in order to monitor the produced deflections during the loading process that is described below.

3.2 Loading of SFRC Prototype

Main objective of the loading process was to evaluate the response of the structure in terms of cracking and deformations under long-term loads of considerable magnitudes. For this purpose, the load was provided by means of concrete cubes of 0.5×0.5 0.6 m ($\approx$350 kg each). It is worth noticing that the structure was loaded gradually, dividing the whole process into four steps. This approach permitted the identification and measurement of both instantaneous and long-term deflections for different values of uniformly distributed loads (Fig. 8).

Fig. 8. a) Second phase of loading; b) Fourth phase of loading

Time-spans between loading phases depended on the response of the structure, i.e. the next phase could only be started when there was no any increment of deflection within the certain time period. Crack patterns on the bottom surface of the slab were measured during the whole loading process. The fourth phase, the total load (self-weight included) was equal to 9.8 kN/m² (which includes both the permanent and variable loads).

After a stabilization process (which took place within 4.5 months), a loading tests was performed in order to bring to failure the structure. To this end, water tanks were placed onto the slab, according to different distributions, and these were filled in phases of 0.1 m (1.0 kN/m² per phase). Firstly, panels 1 and 2 of the prototype (Fig. 9a) were loaded up to reaching a total equivalent load of 14.0 kN/m² (as reference, design load being $q_d = 1.35 \cdot (4.8 + 2.0) + 1.50 \cdot 3.0 = 13.7$ kN/m²). However, SFRC prototype showed barely no increase of cracking respect the fourth phase and still potential deformation capacity, so the loading process continued until reaching a load of 16.0 kN/m². Even at this load level, no loss of integrity or deflections that would have led to predict an imminent failure were observed, so the slab was left fully-loaded until the next day. The following morning, the increment of deflection was equal to 10 mm and still no signs of failure; hence, the structure proved to have a significant strength and ductility capacity.

Fig. 9. SFRC flab under ultimate loading conditions: a) loading of the first half of the slab; b) loading of the rest of the slab

The last load step consisted in transportation of water tanks on the other part of the slab (panels 3 and 4) with posterior repetition of the test (Fig. 9b). Those panels were also loaded up to 16.0 and 14.5 kN/m², respectively, despite the already damaged half of the structure.

The deflections of the structure were monitored by means of prisms and topography. These were measured before and after every load step to quantify the instantaneous component. The time-dependent component of the deflections (owe to creep, shrinkage and temperature variations) was measured 3 days per week. The results derived from the measurements during four phases proved the homogeneity of the material resistant properties and the proper position of the loads as the total deflections measured in the mid-span of the four bays showed no significant differences (see Fig. 10). In this sense, blue, green, orange and red cubes correspond to the I, II, III and IV loading stage (cumulative), respectively.

The same structural response was observed under the ultimate loading conditions: the deformation difference of first and second panels was equal to 2 mm under the load of 16.0 kN/m². Even when the water tanks were emptied and some reduction of the deflections was detected, the mentioned difference kept being in the same range – it reduced to 1 mm.

An analysis of the cracking patterns at different stages of the load was carried out in parallel with the evaluation of produced deformations of the structure. The obtained information was aligned with the expected patterns derived from both the numerical analyses and the Johansen theory (for similar load and support configurations) [7, 22].

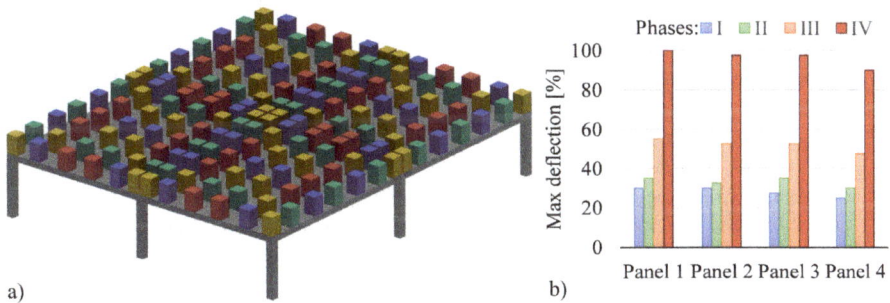

Fig. 10. Measured deflections of SFRC slab

4 Conclusions

An extensive experimental programme on characterizing concrete mixes with various fibre contents and a full-scale test of a SFRC pile supported flat slab was conducted. Suitable workability of self-compacting concrete even with fibre content of 120 kg/m³ was observed, this providing a remarkable residual tensile strength of the material that allowed producing a SFRC flat slab of 30 in aspect ratio and 0.2 m thickness, with a 30% of fibre amount reduction respect to the existing experiences. More detailed conclusions which may be derived based on the results and analysis presented below:

- The examined materials demonstrated flexural hardening behaviour, fulfilling the established requirements to permit the substitution (also partial) of conventional reinforcement according to the *fib* Model Code 2010.
- The 7c strength FRC class (considering mean values), according to the classification of the *fib* Mode Code 2010, proved that the traditional steel bars can be replaced by fibres providing a suitable structural response for loads representative of residential buildings in a column-supported slab with a span to depth ratio of 30.
- The tested prototype presented a greater bearing capacity than that asked for residential buildings according to the Spanish standards demonstrating no loss of integrity even under the load of 16 kN/m^2.

The results presented herein are partial since these are part of an ongoing PhD thesis developed by the first author. The numerical-aided design method proposed for this type of FRC structural elements as well as analyses and provisions for both the crack width and deflections are being developed and these will be published.

Acknowledgements. The authors wish to express their gratitude to the Spanish Ministry of Economy and Competitiveness for the financial support in the scope of the project eFIB (reference: RTC-2016-5236-5) which was carried out along with SACYR Ingeniería e Infraestruturas. The first author also thanks the Department of Enterprise and Education of Catalan Government for providing support through the PhD Industrial Fellowship (2018 DI 77) in collaboration with Smart Engineering Ltd.

References

1. Blanco, A., Pujadas, P., de la Fuente, A., Cavalaro, S.H.P., Aguado, A.: Influence of the type of fiber on the structural response and design of FRC slabs. J. Struct. Eng. (U.S.) **142**, 1–11 (2016)
2. Barros, J.A.O., Taheri, M., Salehian, H.: A model to simulate the moment-rotation and crack width of FRC members reinforced with longitudinal bars. Eng. Struct. **100**, 43–56 (2015)
3. Carlesso, D.M., de la Fuente, A., Cavalaro, S.H.P.: Fatigue of cracked high performance fiber reinforced concrete subjected to bending. Constr. Build. Mater. **220**, 444–455 (2019)
4. Vasanelli, E., Micelli, F., Aiello, M.A., Plizzari, G.: Crack width prediction of FRC beams in short and long term bending condition. Mater. Struct. Constr. **47**, 39–54 (2014)
5. de la Fuente, A., Pujadas, P., Blanco, A., Aguado, A.: Experiences in Barcelona with the use of fibres in segmental linings. Tunn. Undergr. Sp. Technol. **27**, 60–71 (2012)
6. Belletti, B., Cerioni, R., Meda, A., Plizzari, G.: Design aspects on steel fiber-reinforced concrete pavements. J. Mater. Civ. Eng. **1561**, 1–2 (2008)
7. Aidarov, S., Mena Sebastiá, F., de la Fuente, A.: Structural response of a fibre reinforced concrete pile-supported flat slab: full-scale test. Eng. Struct. **239**, 112292 (2021)
8. Destrée, X., Mandl, J.: Steel fibre only reinforced concrete in free suspended elevated slabs: case studies, design assisted by testing route, comparison to the latest SFRC standard documents. Tailor Made Concr. Struct. 437–443 (2008)
9. Ministerio de Fomento: Instrucción de Hormigón Estructural (EHE-08) (2008)
10. CEN 2019 EN 12350-8:2019: Testing fresh concrete. Self-compacting concrete. Slump-flow test
11. CEN 2019 EN 12350-6:2019: Testing fresh concrete. Density

12. CEN 2019 EN 12350-7:2019: Testing fresh concrete. Air content. Pressure methods
13. CEN 2019 EN 12390-3:2019: Testing hardened concrete. Compressive strength of test specimens
14. König, G, Kützing, L.: Modelling the increase of ductility of HPC under compressive forces - a fracture mechanical approach, pp. 251–260. RILEM Publications, Mainz (1999)
15. Sato, Y., Van Mier, J.G.M., Walraven, J.C.: Fifth RILEM Symposium on Fibre-Reinforced Concretes (FRC): Lyon, France, 13–15 September, pp. 791–800. RILEM Publications (2000)
16. RILEM TC162-TDF 2003 RILEM TC162-TDF: Test and design methods for steel fibre reinforced concrete. Final Recomendation. Mater. Struct. 36 560–567
17. fib 2010: fib Model Code for Concrete Structures 2010
18. CEN 2007 EN 14651:2007: Test method for metallic fibre concrete. Measuring the flexural tensile strength (limit of proportionality (LOP), residual)
19. Mitchell, D., Cook, W.D.: Preventing progressive collapse of slab structures. J. Struct. Eng. **110**, 1513–1532 (1984)
20. Johansen, K.W.: Yield-Line Theory. Cement and Concrete Association, London (1962)
21. Johansen, K.W.: Yield-Line Formulae for Slabs. Cement and Concrete Association, London (1972)
22. Nogales, A., de la Fuente, A.: Numerical-aided flexural-based design of fibre reinforced concrete column-supported flat slabs. Eng Struct. **232**, 1–24 (2021)
23. ACI Committee 544, American Concrete Institute: Report on the design and construction of steel fiber-reinforced concrete elevated slabs. American Concrete Institute (2015)

Structural Behavior of Precast Tunnel Segments Reinforced by Macro-synthetic Fibers During Temporary Loading Phases

Ivan Trabucchi[1], Antonio Mudadu[1], Giuseppe Tiberti[1(✉)],
Antonio Conforti[1], Giovanni Plizzari[1], and Ralf Winterberg[2]

[1] Department of Civil, Environmental, Architectural Engineering
and Mathematics, University of Brescia, via Branze n. 43, 25123 Brescia, Italy
giuseppe.tiberti@unibs.it
[2] BarChip Inc., Kanda System Building 7F 7 Kanda, Konya-cho,
Chiyoda-ku, Tokyo, Japan

Abstract. Fibre Reinforced Concrete (FRC) has been increasingly used in precast tunnel linings. FRC, with or without conventional reinforcement could be an adequate solution for fulfilling the structural requirements. The design process of segmental concrete linings in ground conditions generally refers to the final permanent embedded in ground condition as well as to temporary phases. Among these temporary phases, the TBM thrust jack phase is a crucial loading condition during construction, which may noticeably govern the segment design. Tunnel elements having an internal diameter of 3.50 m and a thickness of 0.20 m were studied through full-scale tests to shed new lights on the possible use of macro-synthetic fibres in segmental lining. Precast segments were tested under bending in order to reproduce typical flexural conditions occurring during de-moulding, storage and transportation. Moreover, in order to study lining behaviour during TBM operations, segments were also tested under high concentrated loads for reproducing the excavation phase. Three main reinforcement solutions have been analysed: macro-synthetic fibres only (MSFRC segments), combination of an optimized amount of rebars and macro-synthetic fibres (Hybrid segments) and the reference solution made by conventional bars reinforcement (RC segments).

Keywords: Fibre reinforced concrete · Macro-synthetic fibres · Precast tunnel segments · Flexural test · Point load test

1 Introduction

The growing desire to find innovative and sustainable solutions in the reinforcement configuration for precast tunnel elements are moving the efforts of practitioners and researches to the adoption of Fibre Reinforced Concrete (FRC). The latter could be adopted with or without conventional rebars and it is progressively used in several precast tunnel lining projects [1, 2]. Steel Fibre Reinforced Concrete (SFRC) is usually adopted for tunnelling application [3, 4], even though there is a general growing interest in the scientific community for the use of macro-synthetic fibres. Tiberti et al.

© RILEM 2022
P. Serna et al. (Eds.): BEFIB 2021, RILEM Bookseries 36, pp. 714–726, 2022.
https://doi.org/10.1007/978-3-030-83719-8_61

[5] highlighted how Polypropylene Fibre Reinforced Concrete (PFRC) having performance class 2e, according to *fib* Model Code 2010 classification [6], allows to satisfactorily control local splitting phenomenon due to high concentrated loads. The effectiveness of macro-synthetic fibres in conjunction with conventional rebars (hybrid solution) in precast segments was pointed out by di Prisco et al. [7] and by Conforti et al. [8] for small diameter tunnels (around 3 m). Moreover, Conforti et al. [9] proved the validity of the hybrid solution also for tunnel with higher internal diameter (around 6 m).

The governing load cases in the design process of segmental tunnel linings can be divided in two categories: temporary loading conditions (de-moulding, storage, transportation and TBM thrust jack phase) and permanent embedded loading condition (ground and water pressure, service loads, etc.).

Within this framework, an extensive experimental campaign was developed to shed lights on macro-synthetic fibres application in tunnelling. Ten precast tunnel segments were cast in order to carry out full-scale tests. Five segments were tested under bending, while the remaining segments were loaded by high concentrated loads (point load tests).

2   Experimental Program

The research program included two types of tests: flexural and point load. The former simulates the typical bending actions resulting from the production and transitional stages while the latter reproduces the loads applied by the boring machine during the excavation process, which were investigated for high load-levels. The broad experimental program consists of testing ten full-scale segments making possible to evaluate the scatter of results in terms of structural behaviour of full-scale segments. These aspects contribute to the novelty of this research with respect to previous ones [8, 9].

2.1   Specimen Geometry

It was selected a lining geometry corresponding to the case study of "Scilla tunnel". The latter is one of the main structures that composes a power line between Calabria and Sicilia (Italy). The tunnel, 2800 m long, was excavated by means of a Tunnel Boring Machine (TBM). Each ring is composed by four different precast elements: two trapezoidal and two parallelogrammic segments. The tunnel has an internal diameter (D_i) of 3.50 m and a thickness (t) of 0.20 m (Fig. 1). The trapezoidal shaped precast tunnel segment considered in this experimental campaign, identified as "C1" in Fig. 1, is characterized by an average length (l) of approximately 2906 mm and a width (b_w) of 1100 mm.

According to the original project of "Scilla tunnel" the TBM moves forward pushing itself by means of two loading shoes for each segment. The nominal service load (S.L.) and the exceptional thrust load for each loading shoe were 700 kN and 1000 kN, respectively. Moreover, in case of emergency, the TBM could apply a load equal to 1600 kN (maximum pressure in the hydraulic system, equal to 2.28 times the S.L.) for each loading shoe.

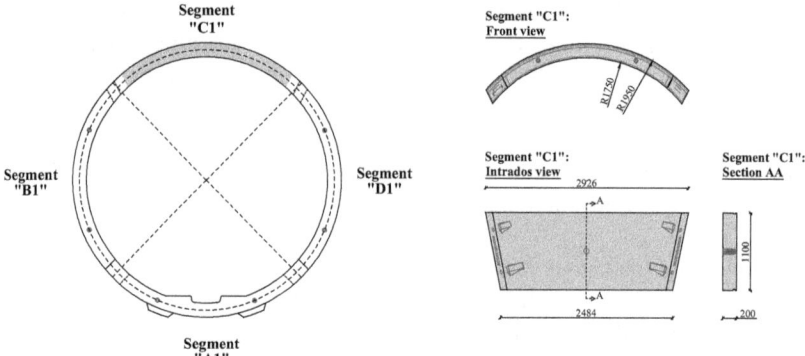

Fig. 1. Tunnel lining cross section and segment C1 details (dimensions in mm).

2.2   Reinforcement Solutions of Precast Tunnel Segments

Three reinforcement solutions were designed and tested in this experimental campaign: 1) macro-synthetic fibres only (MSFRC segments); 2) hybrid solution (Hybrid segments) and 3) traditional steel rebars only (which represents the reference solution, RC segments). However, a preliminary materials mechanical characterization [10], was carried out in order to select a PFRC adequate for precast tunnel elements applications. Continuous embossed polypropylene (PP) fibres 48 mm long with a diameter of 0.70 mm (aspect ratio of 68), tensile strength of 640 MPa, elastic modulus of 12000 MPa and density of about 900 kg/m^3 were adopted. Fibres were added to a concrete matrix with a target mean cube compressive strength of about 60–65 MPa in four different dosage: 4 kg/m^3, 6 kg/m^3, 8 kg/m^3 and 10 kg/m^3. The post-cracking behaviour of PFRCs was evaluated by means of Three-Point Bending Tests (3PBTs) on notched beams according to EN 14651 [11]. After this preliminary phase, PFRC8 with a fibres dosage equal to 8 kg/m^3 was selected for casting full-scale precast tunnel elements. Therefore, four segments reinforced only with macro-synthetic fibres were cast, two were tested under flexure (MSFRC-B1, MSFRC-B2) and two were subjected to point load tests (MSFRC-P1, MSFRC-P2). In these elements fibres withstand the tensile stresses that appear during the whole segment lifetime.

As far as the RC reinforcement solution is concerned, Fig. 2a reports the reinforcement details. The clear concrete cover on curved rebars was 3.5 cm, leading to a segment effective depth (d) of 16.1 cm. According to this reinforcement configuration, two segments were cast, one for each type of test (RC-B and RC-P). The overall steel content was about 110 kg/m^3 and it is made by the following bars:

- 8Ø8 curved rebars both as compression and tension reinforcement (A and B, Fig. 2a) in order to have a double-reinforced section, characterized by typical (for precast tunnel elements) longitudinal reinforcement ratio (ρ_s) of 0.23%. In addition, U-shaped rebars were adopted at bar ends for having a proper anchorage in concrete. These rebars are placed to mainly withstand flexural actions, circumferential splitting forces, as well as to control spalling cracks;

- stirrups Ø8@12 cm with 4 legs (E, Fig. 2a) as minimum shear reinforcement (shear reinforcement ratio $\rho_w = 0.15\%$);
- stirrups Ø8@12 cm with 2 legs (D, Fig. 2a) along all segment sides to withstand radial splitting forces.

By considering the hybrid solution (Fig. 2b), four segments were produced using a combination of rebars and a dosage of 8 kg/m^3 of macro-synthetic fibres (two for bending and two for point load tests, Hybrid-B1, Hybrid-B2 and Hybrid-P1, Hybrid-P2, respectively). The clear concrete cover was the same of the RC segment, i.e. 3.5 cm. The total steel content was 35 kg/m^3, which is significantly smaller (−68%) if compared to RC segments. Moreover, another strong point of this solution is that simplifies the steel cage construction and facilitates its positioning, reducing the overall production time and storage space. In detail, this reinforcement solution was composed by:

- top and bottom chord made by 2Ø8 curved rebars both as compression and tension reinforcement (A and B, Fig. 2b), with stirrups Ø6@20 cm with 2 legs (D, Fig. 2b), leading to a double-reinforced section with $\rho_s = 0.11\%$. The latter was chosen smaller than the minimum amount of flexural reinforcement required by MC2010 [6] for RC members, as similarly assumed in the previous research carried out by Conforti et al. [9]. Chord rebars and fibres are also adopted in order to mutually resist the tensile stresses in the spalling area and for better controlling the corresponding cracking phenomena during the TBM thrust phase;
- macro-synthetic fibres for substituting the minimum amount of shear reinforcement (according to Equation 7.7–14 of MC2010 [6]) and for withstanding the local tensile stresses in the splitting zones (in case of FRC presenting a class "2e" or higher [5]).

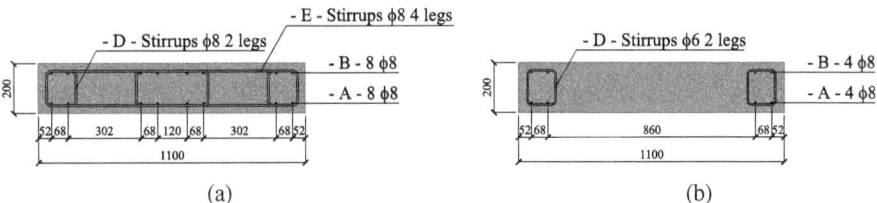

(a)   (b)

Fig. 2. Reinforcement details: (a) traditional solution and (b) hybrid solution (dimensions in mm).

2.3   Material Properties

Ten full-scale tunnel segments were cast. Accordingly, for each batch, at least six small beams (150 × 150 × 600 mm) and ten cubes (150 × 150 × 150 mm) were produced to verify the post-cracking behaviour and the compressive strength, respectively. The segments were de-moulded one day after casting. The full-scale flexural testes were carried out at about 60 days, while point load tests were carried out after 160 days.

Consequently, 3PBTs on small notched beams according to EN14651 [11] as well as compressive tests on cubes were done the same day of full-scale tests. Table 1 reports the cube compressive strength ($f_{c,cube}$), the limit of proportionality (f_L) and the mean values of residual flexure strengths ($f_{R,1}$, $f_{R,2}$, $f_{R,3}$, $f_{R,4}$ according to EN 14651 [11]) for each concrete batch identified with the same name of the corresponding precast tunnel segment tested under flexure and point load tests. The results reported in Table 1 are in agreement with those obtained during the preliminary materials mechanical characterization [10], confirming the consistency and repeatability of casting procedure adopted for PFRC8. In addition, twelve 8 mm bars were tested to evaluate the tensile strength of longitudinal steel bars adopted in the production of traditional and hybrid segments. The average yielding strength was 532 MPa (CoV: 0.04) while the tensile strength was 647 MPa (CoV: 0.03).

Table 1. Mean mechanical properties measured on standard specimens (CoV in brackets).

Concrete	$f_{c,cube}$	f_L	$f_{R,1}$	$f_{R,2}$	$f_{R,3}$	$f_{R,4}$
	[MPa]	[MPa]	[MPa]	[MPa]	[MPa]	[MPa]
RC-B	61.5		–	–	–	–
	(0.06)	–				
MSFRC-B1	68.9	5.19	3.21	4.08	4.58	4.67
	(0.03)	(0.07)	(0.11)	(0.12)	(0.11)	(0.12)
MSFRC-B2	68.8	4.93	2.96	3.77	4.21	4.35
	(0.01)	(0.04)	(0.12)	(0.14)	(0.11)	(0.16)
Hybrid-B1	73.1	6.14	3.18	4.11	4.62	4.59
	(0.03)	(0.05)	(0.18)	(0.20)	(0.21)	(0.20)
Hybrid-B2	75.1	5.90	3.67	4.83	5.49	5.42
	(0.01)	(0.04)	(0.19)	(0.20)	(0.19)	(0.16)
RC-P	78.4		–	–	–	–
	(0.11)	–				
MSFRC-P1	79.9	6.38	3.30	4.29	4.82	4.64
	(0.01)	(0.14)	(0.11)	(0.12)	(0.14)	(0.15)
MSFRC-P2	84.4	6.47	3.59	4.72	5.35	4.98
	(0.02)	(0.07)	(0.16)	(0.17)	(0.17)	(0.19)
Hybrid-P1	69.2	6.83	3.36	4.45	5.00	4.76
	(0.12)	(0.06)	(0.09)	(0.11)	(0.12)	(0.12)
Hybrid-P2	77.6	6.66	3.48	4.50	5.01	4.57
	(0.05)	(0.08)	(0.18)	(0.19)	(0.19)	(0.18)

3   Experimental Results

3.1   Flexural Tests

3.1.1   Test Set-Up and Instrumentation

A three-points bending test procedure (net span 1600 mm) was adopted in order to study the structural response of five trapezoidal segments tested by means of a displacement-controlled procedure. The flexural tests were performed by using a quasi-static loading rate of 0.3 mm/min. Figure 3 shows the test set-up. Two supports were continuous on the entire segment width, while the load was applied at segment extrados by means of two steel plates (150 × 200 mm). A stiff steel girder was placed between the electro-mechanical actuator (loading capacity of 500 kN) and the steel plates, in order to ensure the distribution of the load on the two loading points. Figure 4 exhibits the intrados and front view of segment "C1" highlighting the adopted instrumentation. Three Linear Variable Differential Transducers (LVDTs) were placed on segment intrados to measure the mid-span deflection at segment centre (D3) and sides (D1 and D2). In addition, the supports settlements were measured by four LVDTs placed at the sides of each support (D4. D5, D6, D7, Fig. 4). Therefore, the actual mid-span deflection was calculated by considering the vertical displacement at supports. The flexural crack opening was measured by means of two Potentiometric Transducers (PTs), identified as W1 and W2 in Fig. 4.

Fig. 3. Flexural test: loading system.

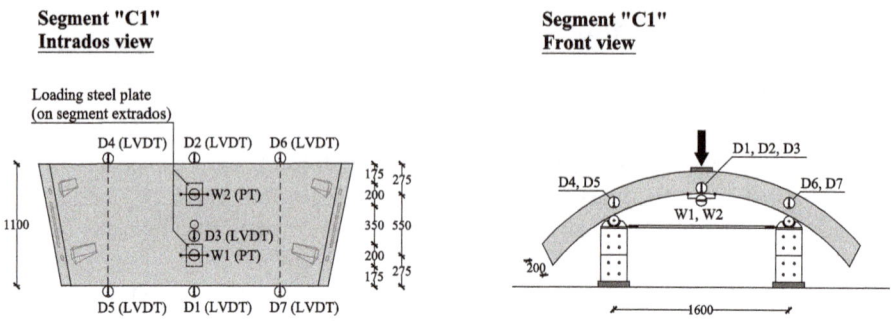

Fig. 4. Flexural test: adopted instrumentation (dimensions in mm).

3.1.2   Experimental Results of Flexural Tests

Figure 5 shows the experimental curves in terms of load vs. mid-span net deflection for the segments tested under flexural test: RC-B, MSFRC-B1, MSFRC-B2, Hybrid-B1 and Hybrid-B2. The mid-span net deflection was obtained as mean value of the deflections measured by LVDTs D1, D2 and D3, by also considering the support settlements, as retrieved by LVDTs D4, D5, D6 and D7. From these graphs, the flexural response of the three different reinforcement solutions can be compared. Moreover, Table 2 summarizes the main experimental results in terms of flexural: cracking load (P_{cr}), initial peak load (P_{peak}), maximum load after initial peak load (P_{max}). Table 2 also shows the corresponding mid-span deflection at: P_{cr} (δ_{cr}), P_{peak} (δ_{peak}) and P_{max} (δ_{max}).

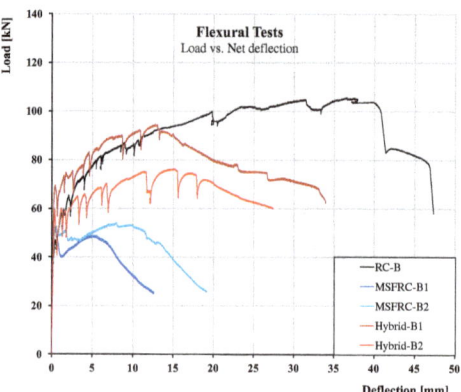

Fig. 5. Flexural tests: experimental curves of load vs. mid-span net deflection for RC-B, Hybrid-B1, Hybrid-B2, MSFRC-B1 and MSFRC-B2 segments.

Table 2. Flexural tests: main experimental results.

Specimen	P_{cr} [kN]	δ_{cr} [mm]	P_{peak} [kN]	δ_{peak} [mm]	P_{max} [kN]	δ_{max} [mm]	$\delta_{max}/\delta_{peak}$ [-]
RC-B	35.1	0.18	46.9	0.38	105.6	36.6	96.3
MSFRC-B1	47.8	0.28	51.4	0.46	48.6	4.90	10.6
MSFRC-B2	43.5	0.23	52.4	0.58	53.9	8.02	13.8
Hybrid-B1	50.6	0.18	69.4	0.40	94.5	12.7	31.8
Hybrid-B2	41.6	0.21	59.6	0.38	76.3	14.9	39.4

The traditional reinforcement solution (black line in Fig. 5) showed the typical behaviour of concrete elements with an adequate amount of longitudinal reinforcement. In fact, after the first crack load (P_{cr}) equal to 35.1 kN, the load increased up to P_{peak} equal to 46.9 kN, due to the typical sectional stresses re-distribution. Afterwards, the load further increased up to 105.6 kN (P_{max}), due to the longitudinal reinforcement. The RC solution exhibited a high ductility, this has been simply quantified by means of the ratio between δ_{max} and δ_{peak} equal to 96.3 (Table 2). In addition, the traditional solution presents a well-distributed crack pattern; the cracks spacing in the central part of the segment intrados is 14.1 cm.

As far as the MSFRC segments are concerned, both segments showed a comparable stiffness up to first cracking load, in fact, both the load and the mid span deflection were comparable (P_{cr} and δ_{cr}, Table 2). Continuing with the tests, both segments reached a similar peak load, equal to 51.4 kN and 52.4 kN. Nevertheless, by further increasing the applied displacement by means of electro-mechanical actuator the two segments showed a slightly different behaviour. MSFRC-B1 segment presented a load drop after the peak load, followed by an increment up to P_{max}. MSFRC-B2 segment exhibited a smaller load decrease and a progressive increase of load up to P_{max} equal to 53.9 kN, which is higher than the corresponding P_{peak} (52.4 kN). In addition, MSFRC-B2 segment had higher ductility than MSFRC-B1 (Table 2). The better flexural behaviour (in terms of bearing capacity and ductility) of MSFRC-B2 is probably related to a more spatial distributed crack pattern, as compared to MSFRC-B1: basically the two flexural cracks exhibited by MSFRC-B2 are developing along the entire width of the segment while in segment MSFRC-B1 only one crack completely developed. Accordingly, MSFRC-B1 tends to localize earlier the deformation in one crack while MSFRC-B2 delayed the crack localization. Moreover, in order to better understand the different post-peak performance, the two MSFRC segments were split into two portions following their main crack plane. Therefore, it was possible to manual counting the effective number of fibres that crossed the main crack plane: in particular, 1750 fibres were crossing the fracture section of MSFRC-B1, corresponding to an average fibre density of 0.80 fibres/cm^2 while MSFRC-B2 segment had 2157 fibres, for a fibre density equal to 0.98 fibres/cm^2. The fibres counting justifies the better performance of MSFRC-B2 segment compared to MSFRC-B1.

Figure 5 also reports the tests results obtained by hybrid segments. It can be observed that Hybrid-B1 segment exhibited a flexural response comparable to the traditional solution while Hybrid-B2 segment had a slightly lower flexural performance (as compared to Hybrid-B1). In particular, Hybrid-B1 and Hybrid-B2 segments reach a maximum bearing capacity, P_{max}, equal to 94.5 kN and 76.3 kN, respectively. Therefore, the second hybrid segment presents a maximum load about 19% lower with respect to the first one, even though the final crack pattern and mean crack spacing were similar. However, even if the flexural response of the two hybrid segments was slightly different, both of them presents a similar trend. In order to better understand the different flexural performance, both segments were split into portions, following the main crack plane, in order to manual counting the effective numbers of fibres that bridge the crack. In particular, in the Hybrid-B1 segment, 2085 fibres were counted, corresponding to an average fibre density of 0.98 fibres/cm^2 while Hybrid-B2 segment had 1802 fibres, corresponding to an average density of 0.83 fibres/cm^2. Therefore, the different flexural performance of hybrid segments could be related to the effective fibre density (-15%) on cracked surface. Both hybrid reinforcement solutions showed also rather significant ductility after segment cracking. The latter was expressed as the ratio between δ_{max} and δ_{peak} as reported in Table 2.

Fig. 6. Point load tests: loading system.

3.2 Point Load Tests

3.2.1 Test Set-Up and Instrumentation

A self-equilibrated system was adopted to simulate the loads applied by the TBM on the lining. It is composed by a stiff concrete beam and two steel reacting frames (with maximum thrust capacity of 6000 kN). Figure 6 reports the testing device and highlights as one single segment was tested instead of the whole ring. It should be noted that the tunnel element was loaded by two loading shoes, having the same geometry of the ones adopted in the original project. The tests were performed in load-control, as it happens during the excavation process. The loading procedure consisted in three load cycles. In fact, after reaching, for each loading shoe, the nominal service load (700 kN), the exceptional load (1600 kN, 2.28 × S.L.) and 2750 kN (3.93 × S.L.), segments were unloaded up to 30 kN (the weight of each steel ring).

Figure 7 shows the intrados of segments with reported the instrumentation details. Referring to the intrados, three Potentiometric Transducers (PTs), identified as SPL-int-mid-1, SPL- int-L and SPL-int-R were adopted between the loading shoes for measuring spalling crack opening. These cracks generally arise as a result of the curved shape assumed locally by the segment. The corresponding spalling crack development along the depth of the segment was investigated with other two PTs, identified as SPL-int-mid-2 and SPL-int-mid-3 (intrados, Fig. 7). As result of the diffusion of the high compressive stresses exerted by the hydraulic jacks within the segment, tensile transverse stresses acting in local tangential direction arise under the loading shoes, leading to possible splitting cracks. The latter were detected with four PTs identified as SPT-INT-L-1 and SPT-INT-L-2 for left loading shoes and SPT-INT-R-1 and SPT-INT-R-2 for right one (intrados, Fig. 7). Two PTs, DSPL-int-L and DSPL-int-R, were used under left and right loading shoes for measuring vertical displacements under the jacks (intrados, Fig. 7). Similar instrumentation (identified with suffix ext) was adopted for evaluating cracking phenomena occurring at segment extrados, with the exception for the spalling crack that has been measured with a unique PT (SPL-ext-mid, Fig. 7).

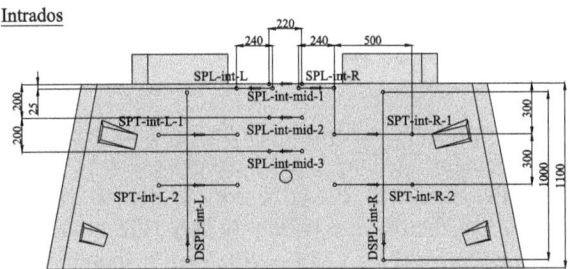

Fig. 7. Point load tests: adopted instrumentation (dimensions in mm).

3.2.2 Experimental Results of Point Load Tests

The global behaviour in terms of load vs. vertical displacement and the local behaviour in terms of crack pattern propagation and crack opening evolution are presented herein. Figure 9a reports the average vertical displacements under left loading shoes (average values between intrados and extrados measurements), the curve of the RC sample is not reported for defective readings of DSPL-ext-L. In terms of bearing capacity, it can be noted that all segments were loaded up to almost four times the service load, exhibiting a noticeable bearing capacity and similar stiffness.

As far as the crack patter is concerned, during the tests two main types of cracks arose: spalling cracks in the region between the loading shoes and splitting cracks underneath the loading shoes. Spalling cracks in RC and Hybrid segments developed as multiple cracking while MSFRC segments developed one single main crack. Cracks were detected neither at service load (700 kN) nor at exceptional load (1000 kN) in all segments. First cracks had always occurred between TBM thrust shoes (spalling cracks) and, subsequently, cracks arose under the TBM thrust shoes, as result of splitting tensile transverse stresses. Figure 8 shows the typical crack pattern of Hybrid-

P1; it can be noticed that a first spalling crack was captured around 1100 kN in the middle intrados region of the segment and then develop along the segment depth up to 2300 kN. In the region between the loading shoes other cracks, inclined along the segment depth (with respect to the loading direction), developed (Fig. 8). The first splitting crack (due to a tangential tensile stresses) arose beneath the right loading shoe for a load of about 1200 kN, with a little inclination with respect to the loading direction.

Intrados

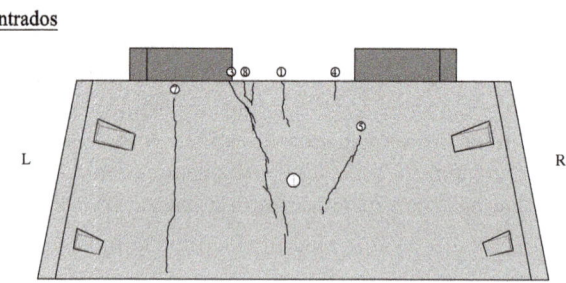

Fig. 8. Point load tests: final crack pattern exhibited by Hybrid-P1 sample (intrados view).

Based on the experimental evidence spalling cracks lead to the widest crack opening. The corresponding spalling cracks development is plotted in Fig. 9b (as provided by gauge SPL-int-mid-1). It can be noticed that all the segments, including MSFRC, presented a satisfactory behaviour. In fact, the relative displacement was smaller than 0.1 mm up to 1500 kN (2.14 × S.L.) but it should be underlined that not all the segments were cracked at this load level. Moreover, for all load levels considered, the Hybrid segments have shown equal or lower relative displacement than reference RC segment. Therefore, in the hybrid solution, the synergy activated between

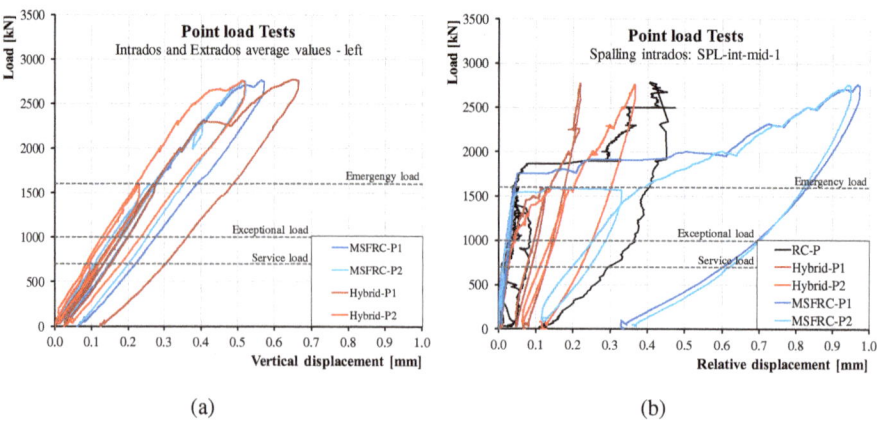

(a)   (b)

Fig. 9. Point load tests: (a) average vertical displacement on left side and (b) load vs. relative displacement curves of spalling phenomenon.

steel rebars and PP fibres allowed to better control spalling cracking phenomena. On the contrary, MSFRC segments (without rebars) were not able to assure a spalling crack control comparable to RC solution, leading to higher relative displacement for the high load levels studied (Fig. 9b).

4 Concluding Remarks

The structural feasibility of macro-synthetic fibres in a precast tunnel lining characterized by a small internal diameter (3.50 m) was evaluated through a broad experimental campaign. The main outcomes of this research can be summarized as follows:

- the combination of PP fibres (8 kg/m^3) and a very low amount of conventional reinforcement (steel content of about 35 kg/m^3, Hybrid segment) improved the flexural behaviour in terms of bearing capacity as well as ductility with respect to MSFRC segments;
- the Hybrid-B1 segment presented similar bearing capacity with respect to traditional RC one. The lightly lower bearing capacity of specimen Hybrid-B2 could be explained with the different fibre density found in cracked surface of segments (after testing);
- point load tests (for simulating TBM thrust phase) evidenced that traditional reinforced and hybrid segments (RC and Hybrid segments, respectively), exhibited a multiple spalling cracking in the region between the loading shoes, while in MSFRC specimens (with fibres only) one single spalling crack appeared between the two loading plates;
- the hybrid reinforcement solution is particularly effective in controlling spalling cracking phenomena occurring between loading shoes, providing similar or better local behaviour with respect to reference RC solution.

Acknowledgements. The Authors are grateful to TETRARENT S.r.L. for the grant of segment formworks. Moreover, the Authors would like to express their gratitude to Engs. A. Piardi, S. Salem, D. Rivetta and S. Treccani for their support during the experimental activities.

References

1. ITA report n. 16: Twenty years of FRC tunnel segments practice: lessons learnt and proposed design principles, p. 71 (2016). ISBN 978-2-970-1013-5-2
2. fib bulletin No. 83: Precast tunnel segments in fibre-reinforced concrete, *fib* WP 1.4.1, p. 168 (2017). ISBN: 978-2-88394-123-6
3. Caratelli, A., Meda, A., Rinaldi, Z., Romualdi, P.: Structural behaviour of precast tunnel segments in fiber reinforced concrete. Tunn. Undergr. Space Technol. **26**, 284–291 (2011). https://doi.org/10.1016/j.tust.2010.10.003
4. Di Carlo, F., Meda, A., Rinaldi, Z.: Design procedure of precast fiber reinforced concrete segments for tunnel lining construction. Struct. Concr. **17**(5), 747–759 (2016)

5. Tiberti, G., Conforti, A., Plizzari, G.A.: Precast segments under TBM hydraulic jacks: experimental investigation on the local splitting behavior. Tunn. Undergr. Space Technol. **50**, 438–450 (2015). https://doi.org/10.1016/j.tust.2015.08.013
6. FIB Model Code for Concrete Structures 2010: Ernst & Sohn, p. 434 (2013). ISBN 978-3-433-03061-5
7. Di Prisco, M., Tomba, S., Bonalumi, P., Meda, A.: On the use of macro synthetic fibres in precast tunnel segments. In: Proceedings of the International Conference Fibre Concrete 2015 (FC2015), Prague, Czech Republic, pp. 478–483 (2015)
8. Conforti, A., Tiberti, G., Plizzari, G.A., Caratelli, A., Meda, A.: Precast tunnel segments reinforced by macro-synthetic fibers. Tunn. Undergr. Space Technol. **63**, 1–11 (2017). https://doi.org/10.1016/j.tust.2016.12.005
9. Conforti, A., Trabucchi, I., Tiberti, G., Plizzari, G.A., Caratelli, A., Meda, A.: Precast tunnel segments for metro tunnel lining: a hybrid reinforcement solution using macro-synthetic fibers. Eng. Struct. **199** (2019). https://doi.org/10.1016/j.engstruct.2019.109628
10. Trabucchi, I., Conforti, A., Tiberti, G., Mudadu, A., Plizzari, G.A., Winterberg, R.: Flexural behaviour of precast tunnel segments reinforced by macro-synthetic fibres. In: Proceedings of the World Tunnel Congress 2019 (WTC2019), Naples, Italy, p. 9 (2019)
11. EN 14651: Test method for metallic fibre concrete–measuring the flexural tensile strength (limit of proportionally (LOP), residual); European Committee for Standardization, p. 18 (2005)

Fibre Reinforced Cement Sheaths for Zonal Isolation in Oil Wells – Quantification and Mitigation of Shrinkage-Induced Cracking

Pablo Alberdi-Pagola[✉], Victor Marcos-Meson, and Gregor Fischer

Civil Engineering Department, Technical University of Denmark (DTU),
Kongens Lyngby, Denmark
ppag@byg.dtu.dk

Abstract. The formation of cracks in the cement sheath and delamination at the cement to steel casing interface due to shrinkage compromises the overall structural stability, imperviousness and durability of oil wells. Particularly, the flow of downhole fluids (e.g. methane or oil) through the cement sheath has become an environmental issue both in offshore and onshore oil wells. This study investigates the impact of fibre reinforcement on the initiation and propagation of cracks in a simulated oil-well section. The study combines a modified setup based on the ASTM-C1581 ring-test with Digital Image Correlation (DIC) in order capture, quantify and measure the initiation and propagation of cracks in the hardening cement and characterize the cracking pattern observed due to autogenous and drying shrinkage deformations and resulting self-induced stresses in the section. The experimental results obtained show the beneficial effect of fibre reinforcement on reducing the extent of cracking by increasing the post-crack ductility of the hardening cement. However, fibre reinforcement had a negligible role in preventing cracking initiation, which is governed by the cement matrix and degree of restraint. This study highlights the benefits of fibre reinforcement as a mitigation measure to reduce shrinkage-induced damage in oil-well cements.

Keywords: Shrinkage · Oil & Gas · Cement sheath · Ring-test · Cracking · Digital Image Correlation (DIC) · PVA fibres

1 Introduction

The Oil-Well Cement Sheath (OWCS) is one of the key components in an oil well: it surrounds the steel casing that connects the surface (e.g. the seabed) and the production formation (i.e. where the oil and gas are extracted). Its main functions are: i) to isolate the reservoir from the surface, avoiding the flow of hydrocarbons from the production zone, ii) provide structural stability to the steel casing, and iii) ensure the durability of the steel casing, protecting it from aggressive conditions (e.g. to prevent corrosion). An inadequate performance of the OWCS entails critical durability and environmental issues (Reddy et al. 2009). In particular, the formation of cracks in the OWCS during service has been related to a critical loss of its isolation function and subsequent hydrocarbon leaks to the environment (Bois et al. 2011; Boukhelifa et al. 2004).

© RILEM 2022
P. Serna et al. (Eds.): BEFIB 2021, RILEM Bookseries 36, pp. 727–738, 2022.
https://doi.org/10.1007/978-3-030-83719-8_62

Cracking of the OWCS during service is primarily attributed to two mechanisms: i) volumetric changes (i.e. shrinkage) of the cement paste during hardening (e.g. during installation) and cooling of the well (e.g. during abandonment) (Reddy et al. 2009), and ii) mechanical loading due to external factors, such as expansion of the steel casing during routine pressure testing (Vralstad et al. 2015). The former is considered the main cause of crack formation in the OWCS (Allena and Newtson 2011; Lura et al. 2009); yet, its study is generally based on numerical modelling and there is limited experimental data to describe and quantify the crack-development process of the OWCS due to shrinkage.

Experimental investigations have generally focused on limiting the extent of cracking due to shrinkage of the hardened OWCS, e.g. using expansive cements and additives (Liu et al. 2016; Parker 1966; Patel et al. 2019); yet not fully addressing key mechanical aspects, which require an enhanced post-crack ductility in the OWCS. Strategies that focus on enhancing the ductility of the OWCS are under discussion (Jafariesfad et al. 2017). Among these, the use of Fibre Reinforced Cement (FRC) composites is regarded as a viable option to control the extent of cracking in the OWCS at low additional cost and potentially with significant performance enhancements.

The addition of polymeric fibre reinforcement in OWCS has been reported to substantially improve the mechanical capacity of the OWCS, such as the ultimate compressive and tensile capacity (Ahmed et al. 2018; Elkatatny et al. 2020; Yang and Deng 2018) and post-crack ductility (Yang and Deng 2018). Additional benefits to the fresh cement slurry properties, such as a reduction of the thickening time (i.e. decreasing the "wait on cement time"), make it a viable alternative to unreinforced cement slurries (Ahmed et al. 2018; Elkatatny et al. 2020).

However, the potential benefits of fibre reinforcement on the performance of the OWCS under shrinkage loads are not well described, and experiments on representative oil-well casing sections are needed. Former studies have investigated the crack development process in scaled OWCS sections (i.e. ring-shaped specimens) combining Digital Image Correlation (DIC) and ring-shaped specimens (Paegle et al. 2019). These studies concluded that DIC is capable of describing the cracking process over time with a high level of detail. Further development on the combination of DIC and existing standards for shrinkage of cement-based materials in comparable (i.e. based on the ASTM-C1581 "Ring-test" method) has focused on providing a detailed description at the micro-scale of the initiation and development of cracks in a simulated OWCS (Alberdi-Pagola et al. 2021).

This paper focuses on quantifying the performance of fibre-reinforced OWCS under restrained shrinkage comprising a low volume fraction of polymeric fibres compared to a plain cement OWCS, using a combination of DIC methods and the ASTM-C1581 standard ring test (Alberdi-Pagola et al. 2021).

2   Methodology

This study investigates the effect of autogenous shrinkage on the crack formation and propagation process of fibre-reinforced cement in a representative oil-well section (using a standard 9–5/8 in. steel casing section). The experiments combined main principles of the ASTM-C1581 standard ring test (with minor modifications as the abovementioned steel ring diameter) and Digital Image Correlation (DIC) following the method described in (Alberdi-Pagola et al. 2021).

Experiments were conducted on two samples with three replicates each: i) a reference sample made of a plain cement slurry, and ii) a sample of fibre-reinforced cement slurry with 0.15%-wt. of Polyvinyl Alcohol (PVA) fibres.

2.1   Materials and Specimen Preparation

Six ring-shaped specimens with the dimensions shown in Fig. 1 were prepared using the same basic cement slurry. The two samples are labelled as reference and FRC. The fibre reinforced cement sample contained 0.15%-wt. of Polyvinyl Alcohol (PVA) fibres. The cement slurry was made of a class-G Portland cement (Dyckerhoff) with a w/c ratio of 0.44; including dispersion and fluid-loss control additives (BASF) (see Table 1). The fibres used were monofilament fibrillated PVA fibres with an average length of 10mm and a diameter of approx. 15 µm.

Table 1. Mixture proportions.

Component	Quantity (g/l)	
	Reference	FRC
Cement	1333	1333
Water	587	587
Dispersant (BASF Liquiment K3F)	0.7	0.7
Fluid-loss control (BASF Polytrol FL 34)	8.0	8.0
PVA fibre	0	1.9

The specimen preparation and material characterization were carried out following the API-RP10b standard (ANSI/API 2010). The results of the material characterization are shown in Table 2. The final setting time was obtained following the EN 196-3 [1] instead of the API-RP10b due to the available equipment.

Table 2. Material Characterization.

Test	Reference	FRC
Compressive strength at 7 days [MPa]	35	36
Mean density [g/cm^3]	1.951	1.950
Final setting time	12 h 20 min	

The specimens were cured fully covered with a PVC formwork for 24 h in a climate-controlled room at 25 ± 1°C with 70 ± 5% RH. After curing, the specimens were demoulded, setting the initial test time (i.e. time "0 h"). The testing room was kept at 25 ± 1 C with 70 ± 5% RH during the testing period. The surfaces of the specimens were painted with an acrylic base white paint to avoid evaporation and a uniform, stochastic pattern of black dots with an average diameter of 0.1 mm was sprayed over the white base.

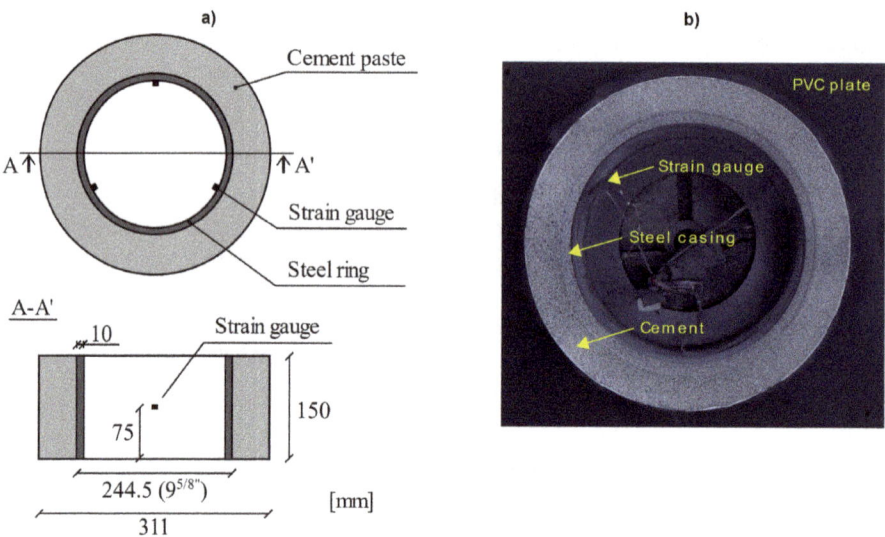

Fig. 1. Test specimens and experimental setup, showing: a) specimen dimensions, and b) experimental setup

2.2   Experimental Setup

The autogenous shrinkage experiments were performed following the ASTM C1581 standard, comprising a few minor modifications: i) outer diameter of the steel ring was 244,5 mm instead of 330 mm, ii) the top face of the ring was sealed using acrylic paint (water loss properties tested according to standard ASTM C309-19) instead of wax or aluminium tape to obtain a visible surface and iii) the drying of the cement from the side surface was avoided by painting it with the same acrylic paint and maintaining a high relative humidity in the room.

The specimens were tested in a tented 1 × 2 m test chamber under controlled temperature and relative humidity. The temperature was maintained at 25 ± 1 °C and the relative humidity was 70 ± 5%-RH. The tests were done following the method described in (Alberdi-Pagola et al. 2021).

The age at cracking was detected measuring the impedance of three strain gauges (with a gage resistance of 120 Ω) connected as a Wheatstone full-bridge with a DC 5 V excitation, installed at the inner part of the steel ring with a logging frequency of 1/30 Hz.

The Digital Image Correlation (DIC) setup consisted on a 35 mm format high-resolution DSLR camera (Canon EOS 5DS R) placed at the top with a 60 mm lens.

The camera was programmed to acquire one image per hour.

2.3   Data Processing and Analysis

The raw data obtained from the strain gauges (voltage differential) was filtered using a 5th order median filter and averaged with a moving average filter with a 5-point box-size. Afterwards, the individual strains (derived from the Wheatstone equation) of the three strain-gages was averaged. The strain values were subsequently interpreted according to the ASTM C1581 Standard to calculate the age at cracking.

The images were pre-processed using MATLAB, following the methodology described in (Alberdi-Pagola et al. 2021); the images were cropped to the region of interest and lens and specimen distortions were corrected using an adaptation procedure. The resulting images were processed using GOM Correlate 2019 to calculate the position matrices of the black dots that compound the speckle pattern at each time-step with a 1 μm resolution. Afterwards, the displacement matrices and derived strain matrices were calculated in MATLAB comparing the position matrices of different time-steps against the reference stage (i.e. time 0).

Cracks were identified afterwards by localizing the local maxima in the strain matrix (i.e. using its derivative) over time. Crack openings below 5 μm were neglected, considering the resolution limit which the pattern size used was capable of showing with minimum noise.

The following items are described herein:

- Crack width, as the distance between edges at each point of the crack perpendicular to the crack direction.
- Cracked area, as the sum of all the crack widths multiplied with their length, where each measured point of the mesh has a length of 1 μm.

3   Results and Discussion

The results from the investigation comprise i) experimental data from the DIC analysis (i.e. displacement and strain data at the specimen surface) that describes the crack pattern and crack width (see Fig. 2) and ii) results from the strain-gauge measurements (i.e. ASTM C1581) that indicate the time to cracking. The latter was used to validate the results from the DIC acquisition method.

For clarity, the results presented herein are shown for a representative replicate of each sample, but discussed for the whole sample, since good agreement was found within the three replicates of each sample. A summary of the mean results for each sample are shown at the end of this section.

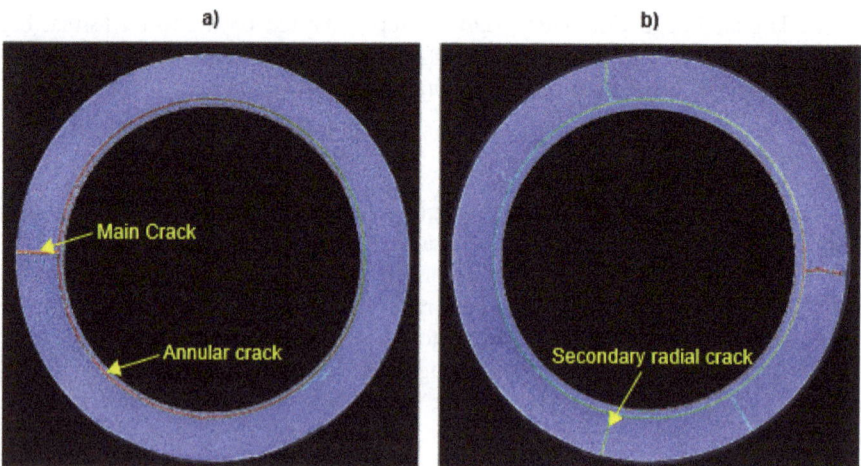

Fig. 2. DIC results showing strain field overlay over the test-specimen surface for: a) reference sample and b) FRC sample. Results are shown for one selected replicate of each sample

3.1   Age at Cracking – Verification of DIC Data

The age at cracking is defined by the ASTM C1581 standard as the time when a drop in the strain measurements is detected. The rest of the information collected by the strain gauges beyond crack formation (i.e. drop in strain) is not relevant for this experiment. The age at cracking (see Fig. 3) was detected at the same time by the strain gauges placed at the inner face of the steel casing (i.e. according to ASTM C1581) and by the analysis of the DIC data, described in this case as the total cracked area over the specimen surface. The results presented show a good agreement between the two measurement techniques. For the strain measurement, note that the compressive strain in the steel ring is shown positive sign.

The measurement of strain development according to the ASTM C1581 standard shows that the addition of fibres did not affect the age at cracking. Therefore, the results show no effect of the fibres in the shrinkage process of the cement paste. However, the cracked area measured with the DIC shows to be significantly reduced by the addition of fibres especially after the initial crack has formed. According to the results, the fibres do not allow a stress-free opening of those cracks, providing ductile properties to the cement sheath and decreasing the cracked area by approximately 80%.

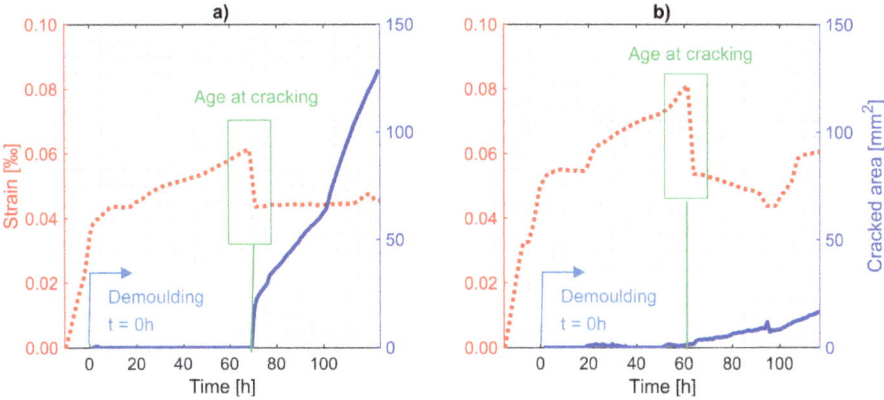

Fig. 3. Age at Cracking calculated from strain gauges and cracked area from DIC, for: a) reference sample and b) FRC sample

3.2 Crack Distribution and Widths

Figure 4 shows the analysed crack pattern over the specimen surface. From both samples can be derived that on average, the number of radial cracks developed in the reference specimens is around 40% lower than the ones developed in the specimens with fibres, but with substantially larger crack widths.

The fibre-reinforced sample (see Fig. 4b) had most radial cracks within a range of 5–0 µm; whereas the reference sample (see Fig. 4a) presented a single main crack significantly larger than the subsequent secondary radial cracks (i.e. approx. 500 µm in the main crack vs 20 to 40 µm in the secondary cracks).

Likewise, the reference sample (Fig. 4a) showed a clear de-bonding of the cement sheath from the steel casing, i.e. the annular crack at the interface between both elements, also known in the oil & gas sector as "microannulus". In the reference sample, the annulus developed a width between 100–300 µm, while the fibre-reinforced sample had a limited de-bonding remaining below 50 µm.

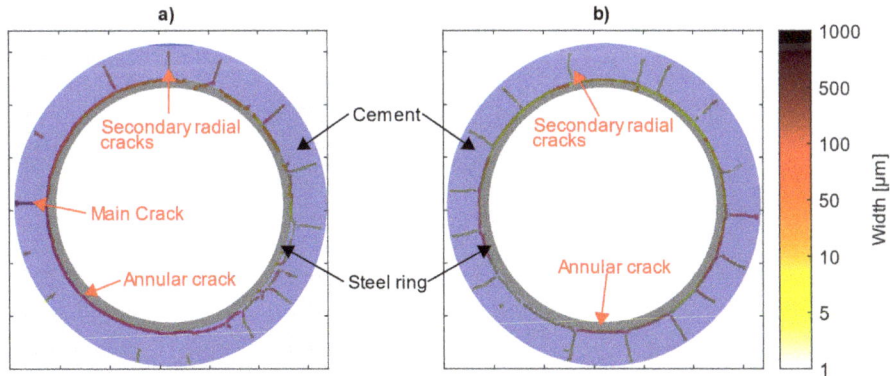

Fig. 4. Crack pattern and width for: a) reference sample, and b) FRC sample

In both samples, the development of the annulus was the immediate and direct result of the formation of radial cracks and, in particular, the formation of the main radial crack at the age of cracking.

Figure 5 presents an overview of the crack development in both samples over time. The effect of the fibres, effectively reducing the crack opening, is clearly visible maintaining the crack widths below 60 μm. Without fibres, however, the crack widths open (and continue opening) up to several hundreds of μm.

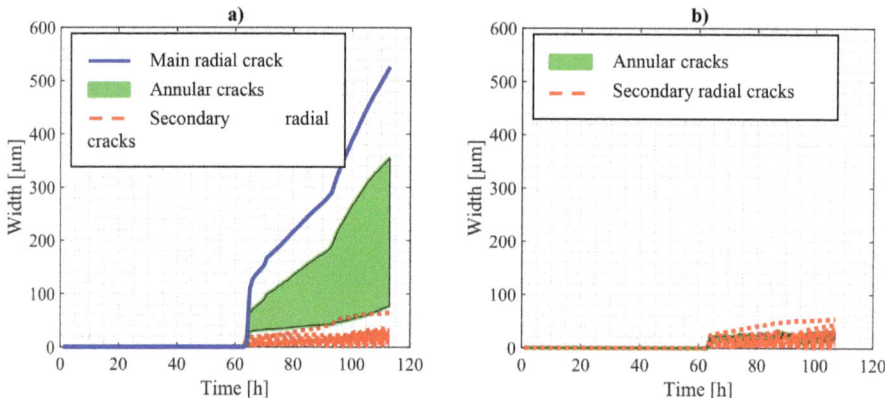

Fig. 5. Crack mean width development over time, for: a) reference sample, and b) FRC sample. Note: in a) annular cracks are shown as a range limited by the maximum and minimum width on the annulus.

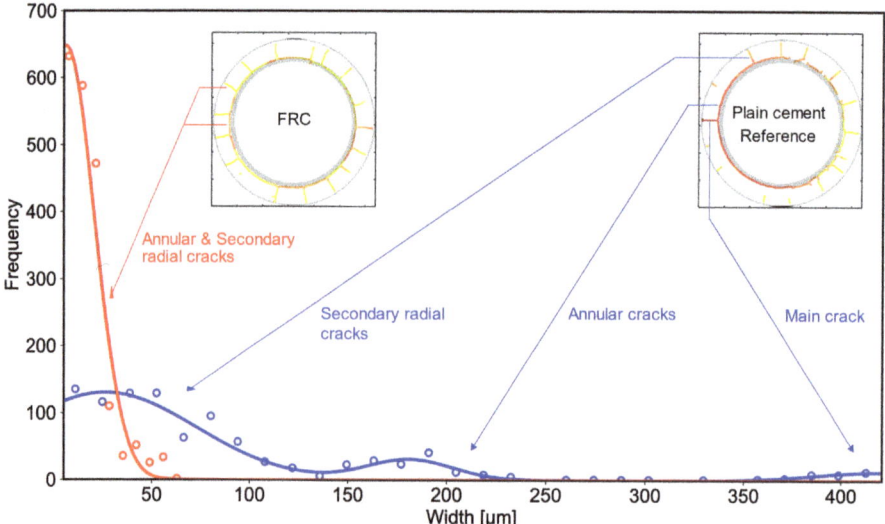

Fig. 6. Crack width distribution at t = 110 h (end of measurements)

Figure 6 shows a more detailed analysis of one single time-step in Fig. 5. For clarity, the width measurements have been approximated with Gaussian distributions. The characterization of the groups of cracks detected in the replicates is shown in Table 3.

The cracks in the reference specimen are clearly distributed in 3 groups:

- Secondary radial cracks
- Annular cracks
- Main crack (stress free) in the reference sample

The cracks developed in the fibre reinforced sample, however, do not differentiate themselves in several groups but are limited in one group. According to their width, radial cracks or annulus cracks are part of the same distribution. Furthermore, the mean width of the detected cracks is smaller than the mean width of the secondary radial cracks of the reference sample while the spread of those values is narrower.

Table 3. Crack mean properties

Test	Reference	FRC
Secondary cracks [μm]	18 ± 7	11 ± 3
Annulus cracks [μm]	146 ± 39	
Main crack 3 [μm]	452 ± 67	–

3.3   Summary and Engineering Implications

A summary of the results obtained during the testing campaign is shown in Table 4, showing the mean values obtained from three replicates tested for each sample (i.e. reference and FRC). The results show an overall better performance of the FRC sample compared to the reference sample, resulting in substantially lower amount of cracking both as mean crack width and total cracked area.

The type and the volume of fibres used in this study were selected based on practical reasons. The selection of PVA fibres was done to avoid the settlement of the fibres in the bottom of the cement sheath, as could happen if denser fibres as steel fibres were used. The selection of the 0.15%-wt fibre content was done to reduce effect of the addition of fibres in the viscosity of the cement slurry. The main purpose was to maintain this viscosity as similar as possible to the one of the reference sample to avoid problems during the pumping process of the cement into the well.

The results showed that, using the fraction of fibres added in this study, the fibres had a limited effect in the shrinkage process and the age at cracking. However, the addition of fibres had a significant beneficial impact on the post-crack behaviour of the FRC sample, reducing substantially the average crack width and ultimately the total cracked area by approx. one-order of magnitude. Further research is needed to find the minimum amount of fibres that can provide the reported benefits regarding shrinkage induced cracking.

The addition of fibres resulted in a larger number of cracks but of a smaller magnitude, resulting in around 80% smaller total cracked area compared to the unreinforced sample. Once activated, the fibres restrained the free opening of the crack. As a consequence, the stresses are redirected to adjacent sections distributing the energy released over a much larger area. Likewise, the addition of fibres reduced substantially the development of the "micro-annulus", by maintaining a residual pressure over the cement ring. Overall, the addition of fibres maintained all cracks (i.e. radial and annular) below the 60 μm range. This is of critical importance for the design of OWCS cement mixes since it may reduce the extent of potential leakages trough the cement sheath and may facilitate healing of such cracks over time.

Finally, the selection of the 9–5/8″ section is explained. The oil wells are known to have different number of sections with different casing diameters depending on the depth, the formation properties and the pressures developed in underground conditions. However, one common practice when designing the oil wells is to reach the reservoir with the 7″ casing, meaning that the previous section (i.e. 9–5/8″) is the last one that is cemented. Therefore, the cement sheath of the 9–5/8″ section is the first barrier against hydrocarbon leaks for many oil wells, which makes relevant the study of the crack development in this particular section.

Table 4. Summary of test results

	Reference	FRC
Age at cracking t [h]	54 ± 11	54 ± 10
Cracked area at end of measurements (t = 110 h) [mm²]	129 ± 15	17 ± 3
Max. crack width at t = 110 h [μm]	463 ± 67	54 ± 9
Max. annulus crack width at t = 110 h [μm]	288 ± 42	42 ± 5
Mean secondary radial crack width at t = 110 h [μm]	18 ± 7	11 ± 3

4   Conclusions

This paper quantifies experimentally the performance of fibre-reinforced OWCS under restrained shrinkage, comprising a low volume fraction of PVA fibres compared to a plain cement OWCS.

The effects of fibre addition were clearly visible: i) the crack widths did not develop above 60 μm, while 500 μm wide cracks have been measured in the reference sample, and ii) the cracked area in FRC specimens was 80% smaller than in the reference specimens. It is expected that the permeability of a cracked cement sheath in an oil well is significantly reduced due to the combined effect of smaller maximum crack width and total cracked area.

The shrinkage-induced cracks were, for the reference sample, distributed in three distinct groups according to the crack width (i.e. secondary radial cracks, annular cracks and main radial crack). For the FRC sample there was only one group of cracks with a very narrow crack width and distribution. Furthermore, the cracks developed in the FRC sample are on average 50% narrower than the secondary radial cracks on the reference specimens.

The addition of a low volume fraction of PVA fibres improves significantly the function of the OWCS in terms of zonal isolation as a result of the observed reduction in maximum crack width and total cracked area. Further research investigating the potential use of fibre reinforced cements to reduce shrinkage-induced cracking in OWCS in field applications is needed.

References

Ahmed, A., Elkatatny, S., Gajbhiye, R., Rahman, M.K., Sarmah, P., Yadav, P.: Effect of polypropylene fibers on oil-well cement properties at HPHT condition. In: Society of Petroleum Engineers - SPE Kingdom of Saudi Arabia Annual Technical Symposium and Exhibition 2018, SATS 2018 (2018)

Alberdi-Pagola, P., Marcos-Meson, V., Paegle, I., Fischer, G.: Quantitative analysis of restrained cracking in cement paste in an oil well sheath using Digital Image Correlation (2021, unpublished)

Allena, S., Newtson, C.M.: State-of-the-art review on early-age shrinkage of concrete. Indian Concr. J. **85**, 14–20 (2011)

ANSI/API: Specification for Cements and Materials for Well Cementing. US (2010)

Bois, A.-P., Garnier, A., Rodot, F., Sain-Marc, J., Aimard, N.: How to prevent loss of zonal isolation through a comprehensive analysis of microannulus formation. SPE Drill. Complet. **26**, 13–31 (2011)

Boukhelifa, L., Moroni, N., James, S.G., Le Roy-Delage, S., Thiercelin, M.J., Lemaire, G.: Evaluation of cement systems for oil and gas well zonal isolation in a full-scale annular geometry. In: IADC/SPE Drilling Conference, pp. 825–839 (2004)

Elkatatny, S., Gajbhiye, R., Ahmed, A., Mahmoud, A.A.: Enhancing the cement quality using polypropylene fiber. J. Petrol. Explor. Prod. Technol. **10**(3), 1097–1107 (2020). https://doi.org/10.1007/s13202-019-00804-4 .

Jafariesfad, N., Geiker, M.R., Gong, Y., Skalle, P., Zhang, Z., He, J.: Cement sheath modification using nanomaterials for long-term zonal isolation of oil wells: review. J. Pet. Sci. Eng. **156**, 662–672 (2017)

Liu, H., Bu, Y., Nazari, A., Sanjayan, J.G., Shen, Z.: Low elastic modulus and expansive well cement system: the application of gypsum microsphere. Constr. Build. Mater. **106**, 27–34 (2016)

Lura, P., Jensen, O.M., Weiss, J.: Cracking in cement paste induced by autogenous shrinkage. Mater. Struct. Constr. **42**, 1089–1099 (2009)

Paegle, I., Marcos-Meson, V., Fischer, G.: Characterization of crack formation and development in the oil&gas well casing cement sheath. In: IOP Conference Series: Materials Science and Engineering, vol. 660, 012063 (2019)

Parker, P.N.: Expanding cement - a new development in well cementing. J. Pet. Technol. **18**, 559–564 (1966)

Patel, H., Salehi, S., Teodoriu, C.: Assessing mechanical integrity of expanding cement. In: Society of Petroleum Engineers - SPE Oklahoma City Oil and Gas Symposium 2019, OKOG 2019. Society of Petroleum Engineers (2019)

Reddy, B.R., Xu, Y., Ravi, K., Gray, D., Pattillo, P.D.: Cement-shrinkage measurement in oilwell cementing - a comparative study of laboratory methods and procedures. SPE Drill. Complet. **24**, 104–114 (2009)

Vralstad, T., Skorpa, R., Opedal, N., De Andrade, J.: Effect of thermal cycling on cement sheath integrity: realistic experimental tests and simulation of resulting leakages. In: Society of Petroleum Engineers - SPE Kingdom of Saudi Arabia Annual Technical Symposium and Exhibition (2015)

Yang, Y., Deng, Y.: Mechanical properties of hybrid short fibers reinforced oil well cement by polyester fiber and calcium carbonate whisker. Constr. Build. Mater. **182**, 258–272 (2018)

Recent FRC Developments in Uruguay: Quality Control, Durability and Three Structural Applications

Luis Segura-Castillo[(⊠)], Nicolás García, Diego Figueredo,
Andrés Clavijo, Enzo González, Bruna Muniz, Iliana Rodriguez,
and Gemma Rodriguez

Facultad de Ingeniería, Universidad de la República, Montevideo, Uruguay
lsegura@fing.edu.uy

Abstract. FRC in Uruguay is usually used in pavements and slabs-on-ground; design based on empirical experience of suppliers and post-cracking strength is not controlled. This paper presents the results of five research projects developed in Uruguay to address FRC use and to show its potential. A) Montevideo test (MVD), a stable test, which requires simple equipment and uses small samples was developed. Correlation between MVD and EN14651 show a qualitative and quantitative equivalence between them. B) The extent to which the cracking improvement given by fibres collaborates with concrete durability, was experimentally studied in beams exposed to a chloride environment. C) A direct experimental assessment of the SFRC orientation factor was performed for an elevated slab. D) A cross-section of Eladio Dieste's famous vaults system is composed by a layer of masonry with steel in between them and a layer of mortar with steel mesh reinforcement. Through an experimental research, an equivalent response was obtained by sprayed FRC as a replacement of the mortar-mesh layer. E) Precast concrete sandwich panels, composed of two concrete layers separated by a layer of insulation, were experimental and numerically studied. Traditional reinforcement in the concrete layers was replaced by fibres, optimising the panel geometry and reinforcement.

Keywords: Fibre · Concrete · Montevideo test · EN14651 · Chloride · Slab · Full-scale · Experimental · Vaults · Sandwich panel

1 Introduction

Fibre Reinforced Concrete (FRC) in Uruguay is mainly used in pavements and slabs on ground and, to a lesser extent, in small precast and slope stabilization. All of them, applications with reduced structural responsibility. Besides a few exceptions, FRC design is based on empirical experience of fibre suppliers and no control of the post-cracking strength is conducted.

This paper presents preliminary results of multiple research projects developed by the *Structural Concrete Group* (GHE) of the *University of the Republic* to address the use of FRC and also show its potential. Efforts were made to produce both "basic" research, aiming at material properties or characterization, but mainly to "applied"

© RILEM 2022
P. Serna et al. (Eds.): BEFIB 2021, RILEM Bookseries 36, pp. 739–748, 2022.
https://doi.org/10.1007/978-3-030-83719-8_63

research, where specific applications were developed to show the main advantages and capabilities of this material.

A specific intention of the research group is to develop its project together with local companies that may use this technology. This objective was fulfilled as in all the projects different actors (concrete suppliers, construction companies, materials suppliers) where involved, in many case with financial support. Further economic support was obtained from the state research and development agency (ANII) through different research projects. In this paper we summarize two of the basic and three of the applied projects.

2   Simple Characterization of FRC: Montevideo Test (MVD)

Montevideo test (MVD), a test to characterize the post-cracking tensile strength of FRC, was developed [1]. The test (Fig. 1c), based on the Wedge Splitting test, is stable, requires simple equipment (no need of close-loop machine), is controlled through the stroke displacement (no clip-gauge needed), and use small samples with simple preparation (Fig. 1a and b). Furthermore, an explicit aim during development was to obtain a test with good correlation to the EN14651 test. It can be seen that the crack pattern obtained through both tests is analogous (Fig. 1d).

Experimental campaigns were carried out with different types of FRC. On each campaign, samples were tested under the MVD and the EN14651 test. Figure 1e and f show the results of the test. P3, S35 and S90 correspond to FRC with 3, 35 and 90 kg/m^3 respectively of Plastic (P) and steel (S) fibres. Point lines correspond to the MVD test and continuous lines to the EN14651 test. CMOD in the MVD test was obtained with a linear transformation of the stroke displacement, as shown [1].

MVD and EN14651 show a qualitative and quantitative equivalence between their load-deflection curves. Furthermore, with a linear transformation of the load (FMVD*1.35, shown in dashed lines in Fig. 1), an excellent correlation can be seen, with errors smaller than 5% for all the residual loads in FRC with softening, and smaller than 10% for the case with hardening. Correlation factor (1.35) was found by a simple trial and error procedure, to reduce the error of the correlation.

Both plastic and steel fibres, FRC with both softening and hardening behaviour, and even different type of matrix show the same correlation factor. Therefore, MVD test results can be directly transformed into EN14651 results, despite type or dosage of FRC used. MVD test can simplify FRC control and, also, be used for research, e.g. fibre orientation factor can be directly evaluated from cores from the structures. A paper showing the correlation between tests with several results from an extensive campaign is under development.

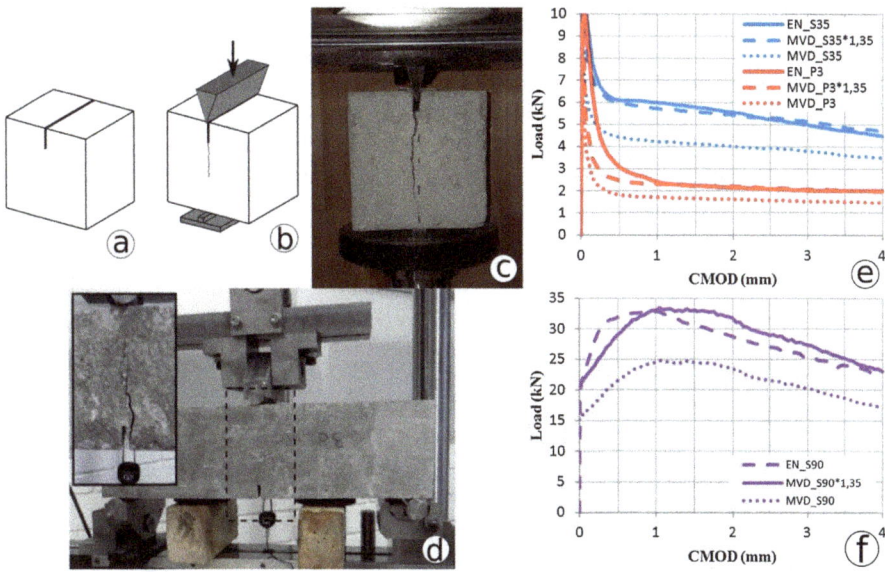

Fig. 1. MVD test: a) typical core geometry and preparation; b) test set-up; c) image of the test; d) EN14651 test; correlation of results with e) softening; and f) hardening behaviour.

3   Reinforcment Corrosion of FRC Under a Chloride Environment

One of the biggest problems in the durability of reinforced concrete (RC) structures is the corrosion of reinforcement by contamination with chlorides in the marine environment. It has been shown that cracking behaviour of traditional RC is improved if fibres are used. To which extent this also collaborates with the improvement in durability is still an open question [2]. In this sense, the objective of this research was to assess, through an experimental program, to what degree the addition of fibres in RC is beneficial for the durability of structures under attack by a chloride environment.

The corrosion initiation time was evaluated by testing in several RC beams (with and without fibres) under induced corrosion. 4 series, 6 beams each, of RC short beams ($15 \times 15 \times 30$ cm^3) were produced. Each beam had a cross section of 150 mm $\times$ 150 mm, and a length of 300 mm. Each beam had 2 ϕ8 mm bars, positioned to have a 25 mm concrete cover (Fig. 2a). Each series comprised 2 plain concrete (RC), 2 steel (SFRC) and 2 plastic (PFRC) fibre reinforced concrete. The amount of fibres used was 0.5% in volume.

After 28 days, the beams were notched and pre-cracked introducing a wedge in the notch (Fig. 2b). Crack with was measured with an optic microscope connected to a PC (Fig. 2d), and fixed in the desired width with a stainless wedge (Fig. 2c). Beams were subjected to continuous immersion in a 5% NaCl solution (Fig. 2e). Half-cell corrosion potential [3] was measured for 180 days. Each series had a different concrete matrix (w/c = water/cement ratio of 0.45 and 0.6) and different crack width (w = 0.2 and

0.5 mm). Note that some values fall below the usual ranges recommended for marine environments.

Results are under study, but preliminary analysis show that corrosion initiation may be delayed by the presence of fibres in the concrete in some cases. Figure 2f show the corrosion initiation time results for the series with w/c = 0.6 and w = 0.2 mm. It can be seen a clear increase in the initiation times for the FRC beams, even with a better performance of the plastic fibres over the steel ones.

The better performance in the response may be produced by the higher tortuosity of the crack morphology caused by the fibres bridging the crack.

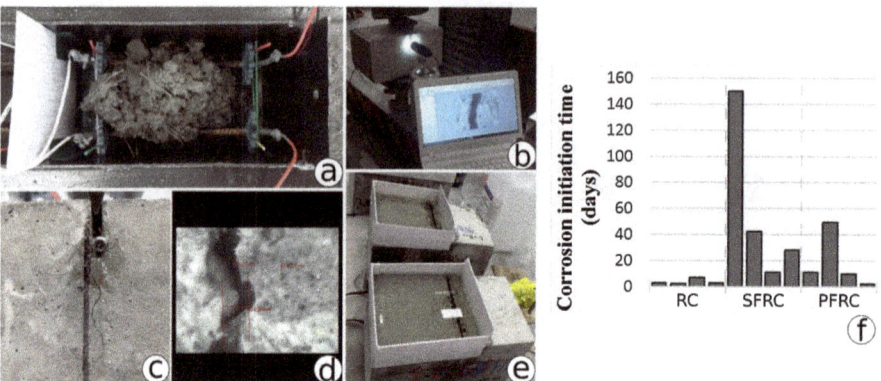

Fig. 2. a) casting of beams; b) pre-cracking; c) crack fixed; d) measure of crack width; e) beams under continuous immersion; f) corrosion initiation times results for the w/c = 0.6, w = 0.2 mm Series.

4   Experimental Evaluation of the Orientation Factor in SFRC Elevated Slab

No universally accepted or complete guideline for the design of elevated FRC slabs exist. A good advancement for the standardization of members with FRC only was made by the Model Code 2010 (MC2010) [4]. It allows the design of FRC structures based on the results of small beams tested under EN14651 [5]. One of the difficulties in this approach to design lies in incorporating the difference in behaviours between the small samples and the real element used. These differences can be caused, for example, by dissimilar fibre orientation. To solve this issue, MC2010 proposed an orientation factor (K) that accounts for this effect. However, no guide is given in how to obtain this factor. Previous studies have assessed it through non-destructive magnetic method (inductive method) or inverse analysis to fit slab tests results [6].

Here we show a direct experimental assessment of the orientation factor of a FRC elevated slab, obtained comparing the mechanical results of beams cast during the slab construction and beams extracted from the slab.

A full-scale (6.2 × 6.2 × 0.13 m^3) self-compacting SFRC (90 kg/m^3) elevated slab was built (Fig. 3a) and tested up to failure in two opposite panels. No conventional reinforcement was used. Preliminary structural results, already reported [7], showed that the ultimate load capacity was between 156 and 211 kN when the slab was punctually loaded in the centre of the panels.

During the slab construction, 5 beams (150 × 150 × 600 mm³) were cast, pouring the self-compacting concrete from one end and letting it flow to complete the mould, as recommended by [8]. These specimens were tested at 28 days in 3 point bending, following EN14651 under central vertical displacement control.

After the load tests of the slab, cores (approximately 125 × 150 × 550 mm³) were extracted from unaltered regions of the slab with a chainsaw (Fig. 3b and Fig. 3c). A somewhat random concrete cast was used for the slab. So no explicit relationship exists between the direction of the cores extracted and the direction of the concrete flow. Extracted cores were tested under three point bending (Fig. 3d), with a 500 mm span. In order to obtain flexion result of cores representative of the slab cross-section structural behaviour, the full depth of the cores (125 mm) was preserved (ie, the extracted cores were tested unnotched). Data of the central displacement and crack opening (CMOD) was obtained using LVDTs. As both extracted cores (full depth) and specimens cast during slab construction (with notched depth) had the same cress-Section (125 mm of height and 150 mm of width) in the failure area, results can be directly compared to obtain an experimental value of the fibre orientation factor K.

Results show (Fig. 3e), for the cast specimens (dark lines, average in orange), an average value of 32,0 kN and 27,0 kN for F_{R1} and F_{R3}, respectively. On the other hand, the extracted cores (grey lines, average in blue) show values of 25,8 kN and 15,1 kN for the same parameters.

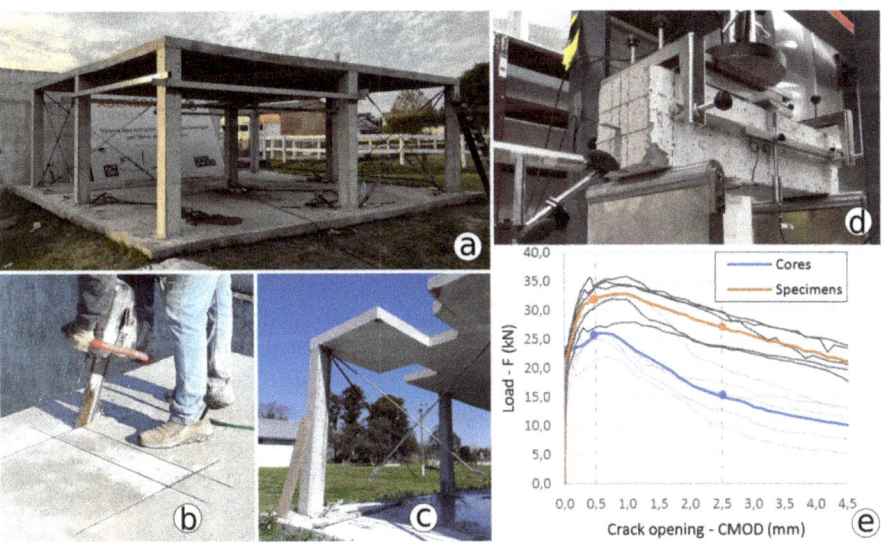

Fig. 3. a) Full scale experimental SFRC elevated slab; b) extraction of cores from the slab; c) slab after extraction of cores; d) flexure test on extracted cores; e) flexure test results.

The small thickness of the slab, connected with the utilization of SCC probably lead to a 2D in plane orientation of fibres. On the other hand, on the cast specimens, as the SCC was allowed to flow in one direction inside the mould, it is more likely that fibres acquired a 1D orientation in the direction of the mould, with the associated higher residual strength. Blanco et al. (2014) [6] suggested that working with different values of K for f_{R1} and f_{R3} can help to improve results when using the constitutive equation proposed in the Model Code 2010. According to those results, an average value of $K_{F1} = 1{,}24$ and $K_{F3} = 1{,}79$ can be estimated.

Although an experimental orientation factor is an important insight towards the development of this type of elements, two drawbacks may be mentioned about the methodology used. First, the differences in the test set up between the test used in cores and in the specimens may introduce error. Secondly, possible specimen damage during sawcut from the elevated slab may reduce the strength of the cores, specially by cutting the fibres next to the surface and reducing its anchorage capacity to the concrete matrix.

5   Sprayed FRC for Masonry Vault Systems

Eladio Dieste (Uruguay, 1917–2000) developed a system of structures that are as efficient as they are architecturally expressive [9, 10]. He used brick to achieve simpler and more economic solutions than those that used concrete [11]. His major creations are characterized by two typologies: barrel vaults (free-standing) and Gaussian vaults (double curvature) (Fig. 4a).

The basic components of a cross-section of Dieste's vaults are: a layer of stacked masonry with steel reinforcement in between them, and on top of these, a layer of mortar with steel mesh reinforcement (Fig. 4b). The purpose of this project is to generate the technical knowledge necessary to be able to include new technologies into Eladio Dieste's traditional technique. The proposed solution consists of substituting the mortar and steel mesh, for a layer of Sprayed Steel Fibre Reinforced Concrete (SSFRC) that will provide the structure with an equivalent structural capacity, while also achieving a cost reduction and increasing the speed of construction. The efficiency of this solution was tested comparing test specimen's behaviour under four points pure flexure.

All test beams were built with hollow bricks. Mortar and hot-rolled steel ribbed bars ($\varphi 4.2$, class S600) was used for the control specimens (Fig. 4b), while SSFRC was used for the rest (Fig. 4c). Two different quantities of steel fibres were tested: 35 and 65 kg/m3. The beams nominal dimensions are: Length = 295 cm; width = 52 cm; height = 15 cm for the control specimens or 17 cm for SSFRC. The main load was applied uniformly across the width of the specimens at two points, using an electro-mechanic press (UTM) and was measured by means of load cells (Fig. 4d).

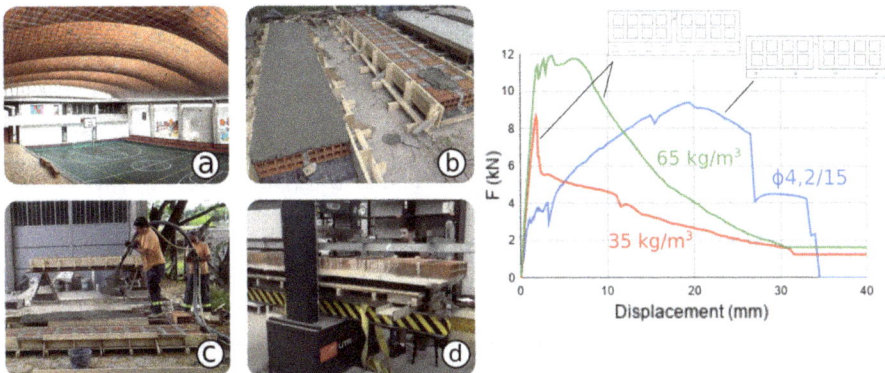

Fig. 4. a) Dieste's Gaussian vaults; casting of test samples with b) Dieste's solution and c) SSFRC solution; d) image during testing of the beam; and e) load-displacement curves for the three types of samples.

Figure 4e shows the relationship between applied force and mid-span displacement (measured by LVDTs) for the three types of beams tested. Both methods are able to achieve the same maximum load capacity, although with a difference in ductility. Control beam showed a lineal behaviour up to mortar cracking, which took place relatively early (for a load at around 3 kN). Then a steady increase in the load up to the maximum load (above 9 kN) and a small decrease until a certain displacement where there is a sudden drop as one of the reinforcement snaps.

Different behaviour was observed for the FRC beams. The maximum load in these was strongly influenced by the amount of fibres present in the concrete. Structural behaviour also showed a lineal behaviour at the beginning of the loading trajectory up until the concrete cracking, which occurred at a higher load (between 8 and 12 kN) than the reference beam. In this point, the beam with low volume of fibres showed a sudden drop in load, of about 3 kN, followed by a gradual reduction of load as the mid displacement increased. Cracking was characterized by a single crack that crossed the beam transversally, and opened gradually while the load decreased.

On the other hand, in the beam with higher amount of fibres the load remained approximately constant for a period after concrete cracked. In this period, multiple cracks were formed in the beam, in the area under pure flexion. Then, it also showed a gradual reduction of the load, coinciding with the opening of one of the previously formed cracks.

As the proposed solution is meant to work in a large area, with a statically undetermined behaviour, both solutions may be applicable. A full structure analysis should be made to assess its potential. Additionally, FRC with hardening behaviour showed improved cracking, with small cracks even for the maximum loads.

6   Precast Concrete Sandwich Panels

Fibre-reinforced concrete (FRC) has been used in numerous types of precast elements around the world, as has been shown that reductions in production costs and time can be obtained. In this project, vertical precast concrete sandwich panels, composed of two concrete layers separated by a layer of insulation (Fig. 5a), were experimental and numerically studied. Traditional steel mesh reinforcement in the concrete layers was replaced by steel (FRCM) and synthetic (FRCS) fibres (Fig. 5b).

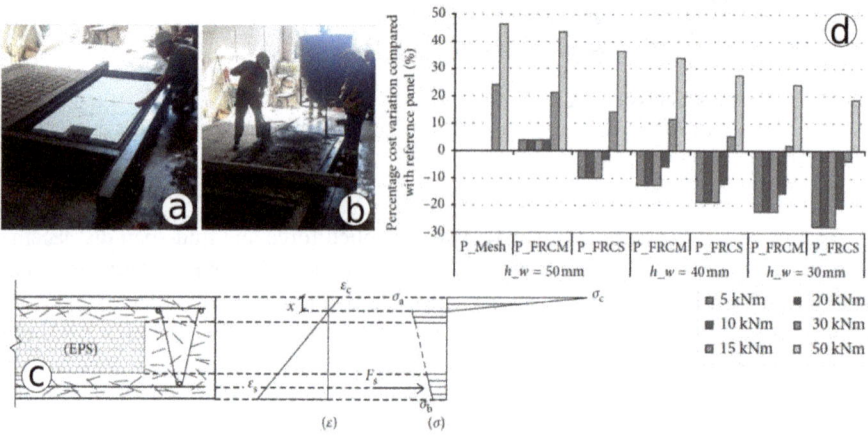

Fig. 5. Precast sandwich panels: a) positioning of insulation layer, b) cast of upper concrete layer, c) strain-stress diagrams for cross-section analysis, d) cost analysis.

FRCM and FRCS were characterized [12], and the panels tested in flexure. To evaluate the structural behaviour of FRC panels, a simple model (Fig. 5c), mainly based on the Spanish Concrete Instruction Guidelines (EHE-08, Annex 14) [13], was developed. The model allowed consideration of the contribution of the fibres to withstand internal tensile forces of the panels and therefore be able to substitute for the steel mesh in the panel wythes.

The results, already published [14], showed that it was possible to optimise panel geometry and reinforcement, reducing layer thickness and adjusting the different types of reinforcement to comply with the design conditions. The cost-efficiency analysis (Fig. 5d) of different panel geometries and amounts of reinforcement showed that it is possible to reduce the cost of production of up to 10% if the reinforcement was modified and of more than 20% when concrete wythes thickness was reduced. The higher reductions are obtained for bending moment smaller than 20 kNm, i.e. for panels with no elevated structural capacity. Besides the reduction in production time, and possible economic advantage, good cracking behaviour was observed when FRC was used.

7 Conclusions

A summary of five research projects developed by the *Structural Concrete Group* (GHE) of the *University of the Republic* in the last years was shown in this paper. Through a transversal analysis of the five projects, it can be concluded that:

- The close contact of the GHE (academy) with the local companies (industry) is encouraging a beneficial synergy, which favours the introduction of new technologies, like fibre reinforced concrete, in our country.
- It must also be highlighted that efforts are made, not only to introduce existing technologies as they are, but also to adapt or combine with local experience, capabilities and traditions.
- For example, incorporating the FRC into Dieste's structural typology would help to make visible it's potential.
- Also, a simplified test, like Montevideo test, is essential in countries with very limited resources, and where there are almost no press with the requirements necessary to perform the reference test (EN14651). It may also reduce quality control costs and reduce industrial waste, and therefore, improve sustainability.

Acknowledgements. The authors would like to thank all the companies (Maccaferri, TEYMA, Dieste & Montañez, Cielo Azul, Sika, Bromberg-Controls, Ferrocement, Astori, Laboratorio de Vialidad, MTOP, Guardia Republicana, Atenko), students (Virginia García, Melissa Eguren, Patricio Cañette, Guzman Rippe, Viviana Sfeir, Diego Gonzalez, Santiago Infante, Mauricio Porcelli, Agustín Silveira, Juan Graña, Emiliano Espinosa), and Department professors (Agustín Spalvier, Cinthia Planchón, Alina Aulet) who were involved in the different projects. Funding was also made available from the *Agencia Nacional de Investigación e Innovación* (ANII), Uruguay, through different Research Projects, and the *Comisión Sectorial de Investigación Científica* (CSIC) of the University. We gratefully acknowledge: Aparicio Daglio and Gonzalo Larrambebere for their support in Dieste's vaults project and specially to Mauricio Montaña for his early support, which boosted the development of the research group.

References

1. Segura-Castillo, L., Monte, R., De Figueiredo, A.D.: Characterisation of the tensile constitutive behaviour of fibre-reinforced concrete : a new configuration for the wedge splitting test. Constr. Build. Mater. **192**, 731–741 (2018)
2. Berrocal, C.G., Lundgren, K., Löfgren, I.: Cement and concrete research corrosion of steel bars embedded in fi bre reinforced concrete under chloride attack : state of the art. Cem. Concr. Res. **80**, 69–85 (2016)
3. Elsener, B., Gulikers, J., Polder, R., Raupach, M.: Half-cell potential measurements – Potential mapping on reinforced concrete structures, vol. 36, pp. 461–471 (2003)
4. FIB: Model Code 2010, Vol 1 & 2. International Federation for Structural Concrete (fib), Lausanne, Switzerland (2010)
5. EN 14651: Test method for metallic fibre concrete—Measuring the flexural tensile strength (limit of proportionality (LOP), residual) (2005)

6. Blanco, A., Pujadas, P., de la Fuente, A., Cavalaro, S., Aguado, A.: Assessment of the fibre orientation factor in SFRC slabs. Compos. Part B **68**, 343–354 (2015)
7. Segura-Castillo, L., Figueredo, D., Rodríguez, I., García, N.: First experimental full-scale elevated FRSCC slab in South America. In: fib/RILEM BEFIB2020, Valencia, Spain (2020)
8. Kasper, T., Stang, H., Mjoernell, P., Thrane, L.N., Reimer, L.: Design guideline for structural applications of Steel Fibre Reinforced concrete. In: SFRC CONSORTIUM (2014)
9. Pedreschi, R.: Eminent structural engineer: eladio dieste — engineer, master builder and architect (1917–2000). Struct. Eng. Int. **24**(2), 301–304 (2014)
10. Anderson, S.: ELADIO DIESTE Innovation in Structural Art. Princeton Architectural Press (2004)
11. Larrambebere, G.: Summer theatre. Constr. Build. Mater. **41**, 918–925 (2013)
12. Rodríguez de Sensale, G., Segura-Castillo, L., Rodríguez Viacava, I., Rolfi Netto, R., Miguez Passada, D., Fernández Iglesias, E.: Fibre reinforced self-compacting concrete for precast. HORMIGÓN Y ACERO **69**(284), 69–75 (2018)
13. CPH, EHE-08: Instrucción del Hormigón Estructural (in Spanish) (2008)
14. Segura-Castillo, L., García, N., Rodriguez Viacava, I., Rodríguez de Sensale, G.: Structural model for fibre-reinforced precast concrete sandwich panels. Adv. Civ. Eng. **2018**, 11 (2018)

Large-Scale Pressure-Swelling Tests on Panels Made of Strain-Hardening Cement-Based Composites with Different Bedding

Steffen Müller[(✉)] and Viktor Mechtcherine

Institute of Construction Material, Technical University Dresden,
Dresden, Germany
steffen.mueller@tu-dresden.de

Abstract. Strain-Hardening Cement-Based Composites (SHCC) are a material class of short-fiber-reinforced concretes which has a strain-hardening behavior due to micromechanical adaptation of the concrete matrix to the short fibers under tensile stress after occurrence of the initial concrete failure. Since its first scientific description in the early 90s, this material has been a constant topic in the research area. In the meantime, extensive knowledge of mechanical properties, cyclic properties, fiber-matrix interactions, modeling and durability aspects is available.

Nevertheless, the number of implemented practical application is relatively small, which is justified on one hand to the barriers from standardization and approval procedures and on the other hand by missing knowledge on the transferability of the results from the laboratory scale on test specimens in the scale 1:1.

In addition to possible applications in building construction, the material seems to have the potential in road construction. It could combine the advantages of the two proven construction methods, asphalt and concrete construction, with the exclusion of some disadvantages, like joints, rods and others. In particular, a joint less and robust concrete construction could be realized by SHCC.

This article presents results from the central loaded square panel tests with different bedding materials. The size of the panel was 2.5 m by 2.5 m and was loaded with up to 4 million cycles. The height of the loading was started on the single wheel load according to German road-building standards and during the experiment increased to 4 times.

The relevant deformation and internal strains were measured and evaluated. After the loading the crack patterns were analyzed and specific cracks were investigated in detail, i.e. small samples from a specific crack were gathered out of the main block and the crack surface were analyzed in the electron microscope. The found failure mechanism were compared with cyclic tests on lab scale specimens to create a missing link between the two scales.

Keywords: SHCC · Strain hardening · Large scale testing · Cyclic loading

© RILEM 2022
P. Serna et al. (Eds.): BEFIB 2021, RILEM Bookseries 36, pp. 749–760, 2022.
https://doi.org/10.1007/978-3-030-83719-8_64

1 Introduction

Highly ductile concrete is a material class of short-fibre reinforced concretes, which exhibits a strain-hardening behaviour due to a micromechanical adaptation of the concrete matrix to the short fibres under tensile stress after the first crack has occurred. Since its first scientific description in the early 1990s [1] this material has been a constant topic in the research field. Meanwhile, a comprehensive knowledge of mechanical properties [2–5], cyclic properties [6–9], fibre-matrix interactions [10], modelling [11] and durability aspects [12] is available. Nevertheless, the number of practical applications implemented is relatively low, which is due to the obstacles from standards and approval procedures as well as the lack of data sets on the transferability of results from laboratory scale to test specimens on a 1:1 scale.

Apart from possible applications in building construction, the material seems to have the potential in road and path construction by combining the advantages of both proven construction methods, asphalt and concrete construction respectively, while excluding some disadvantages. In particular, a jointless concrete construction method could be realized with high-ductile concrete.

In this article results from the centric pressure-swell loading of a square slab (2.5 m × 2.5 m) with different bedding materials are presented. The applied force parameters are based on the design relevant axle loads commonly used in road construction.

2 Material and Experimental Setup

2.1 Material

A matrix composition developed at the TU Dresden was chosen for concreting the large scale slabs, which has been used in this or similar forms in some projects in the meantime. The mechanical properties of the reference material under monotonic and cyclic loading were first investigated by Jun and Mechtcherine [8] and finally by Müller and Mechtcherine [9]. Altmann et al. [12] dealt with the durability of the SHCC, while the first large-scale tensile tests were performed by Mündecke and Mechtcherine [13]. In addition, mechanical properties at elevated temperatures at different strain rates were investigated [14, 15].

Table 1 shows the composition of the mix. The binder was a combination of Portland cement type I 42.5 R and fly ash. In the study, fine-grained quartz sand with a particle size of 0.06 to 0.20 mm was used as aggregate. The limitation of the aggregate size is necessary to achieve a uniform fibre distribution. A superplasticizer and a stabilizer were added to achieve good workability and uniform fibre distribution as well as a reduced blood tendency of the mixture. Furthermore, the investigated highly ductile concrete contained 2% by volume PVA-fibre with a length of 12 mm and a diameter of 40 μm with the trade name Kuralon REC II.

Table 1. SHCC mix design

Component	[kg/m³]
Cement CEM I, 42.5 R	505
Fly Ash	621
Fine-grained sand 0,06–0,2 mm	536
Water	338
Superplasticizer	10
Viscosity agent	4.8
PVA-fibre	26

The mixing was carried out in parallel in two laboratory mixers with a mixing volume of 100l (synchronous mixer, Pemat), and 60l (single-shaft compulsory mixer, Elba). Both mixing volumes were combined and transported to the installation site. Paving was carried out by means of a casting process and manual surface smoothing.

2.2 Experimental Setup

Within the scope of the investigations, 4 different road superstructures on a scale of 1:1 were set up in a stationary test stand with a hydropulse system in the Otto Mohr Laboratory of the Technical University of Dresden. The test field has maximum edge dimensions of 2.5 × 2.5 m. In their basic structure, these variants correspond to a typical road cross-section for high loads. The basis for all test variants was a gravel base course that met the requirements of the German road construction regulations TL SoB [16] and ZTV SoB [17]. In contrast to the recommended concrete base layer thickness of 30 cm, the variants investigated 1 to 3 used 15 cm highly ductile concrete. Option 4 had a further reduced concrete height of 7.5 cm. Variant 1 was built directly onto the gravel base layer, whereas variant 2 was paved on an 8 cm thick asphalt intermediate layer and the 3rd and 4th variants were built on 5 cm soft bedding with edge support. The soft bedding was realised with a layer of low pressure resistant polystyrene. An overview of the variants can be seen in Fig. 1.

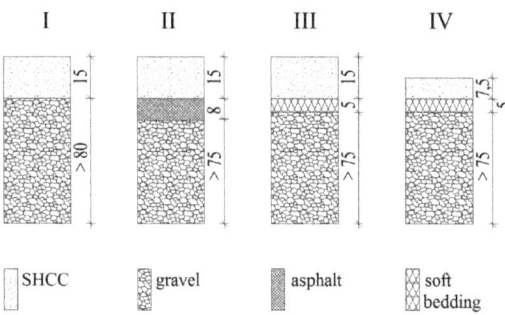

Fig. 1. Pavement variants under investigation.

The pressure force is applied centrally on the surface by a hydraulic ram with a diameter of 30 cm. The load regime consisted of 3 different cyclic loads and one static load. The first load scheme consisted of 1 million load cycles with a normative single wheel load of 57.5 kN at room temperature. In the second load regime the plate temperature was increased to 40 °C and another 1 million cycles with 57.5 kN were applied. The third load scheme was carried out at room temperature with 115 kN and 1 million cycles. A static load of 220 kN was applied at the end of the loading regime. The dynamic load was applied at approx. 5 Hz. In the case of large deformations in the entire system, the 5 Hz could not be maintained due to mechanical reasons. The lower reversal point corresponded in each case to 10% of the maximum load of the respective cycle. An overview about the load regimes is given in Table 2.

Table 2. Loading regime

Load scheme	Type of load	Upper reversal point [kN]	Lower reversal point [kN]	Number of cycles [-]	Temperature [°C]
1	Cyclic	57.5	5.8	1 Mio	20
2	Cyclic	57.5	5.8	1 Mio	40
3	Cyclic	115	11.5	1 Mio	20
4	Quasi static, monoton	220	–	1	20

Each experiment was extensively instrumented. In addition to the typical recording parameters such as force, displacement at load application and test time, another 20 surface deformations, the hall and component temperature as well as internal strain states were recorded.

3   Results and Discussion

3.1   Test Results Plate I and II

After a small number of load cycles, the test of variant 1 showed a permanent deformation of approx. 0.2 mm in the area of the load introduction, decreasing accordingly towards the edge of the plate. These deformations are due to settlement of the gravel base layer according to the load and adjustments in the contact zone of the load introduction plate and the plate made of highly ductile concrete.

The following consideration of the deformations over the number of load cycles shows a discontinuous decrease and increase of the measured deformation values. This phenomenon is not based on a material characteristic, but is linked to the measurement technology. Measured variables such as deformation, strain and stress are electrically measured by small changes in resistance or inductance. The basic assumption is that only the resistance of the specific measuring point changes, whereas all other parameters remain constant. In the present case, the above-mentioned assumption does not apply because the room temperature of the test hall could not be kept constant over the test period of approx. 48 h. Figure 2 shows a significant night-time reduction in temperature. This temperature change couples into the conductor according to the physical principles and changes its resistance, so that a summation of the resistance changes occurs.

One criterion for determining the internal damage is the observation of the deformation amplitude within a cycle. Laboratory-scale investigations have shown that the stiffness decreases with progressive material degradation. Transferred to the large-scale tests, this would mean that an increase in the deformation amplitudes over the number of load cycles can be expected. In the case of variant I, the amplitude remained almost constant within the first two load cycles over all 2 million load cycles. Only when the load was doubled, in load case 3 the measured deformation also doubled. After a short adjustment, approx. 75,000 cycles, which is probably again due to settlement processes of the ballast bed, the deformation amplitude remained constant. Progressive damage could not be detected. After a small number of load cycles, the test of variant 1 showed a permanent deformation of approx. 0.2 mm in the area of the load introduction, decreasing accordingly towards the edge of the plate. These deformations are due to settlement of the gravel base layer according to the load and adjustments in the contact zone of the load introduction plate and the plate made of highly ductile concrete.

The following consideration of the deformations over the number of load cycles shows a discontinuous decrease and increase of the measured deformation values. This phenomenon is not based on a material characteristic, but is linked to the measurement technology. Measured variables such as deformation, strain and stress are electrically measured by small changes in resistance or inductance. The basic assumption is that only the resistance of the specific measuring point changes, whereas all other parameters remain constant. In the present case, the above-mentioned assumption does not apply because the room temperature of the test hall could not be kept constant over the test period of approx. 48 h. Figure 2a shows a significant night-time reduction in temperature. This temperature change couples into the conductor according to the physical principles and changes its resistance, so that a summation of the resistance changes occurs.

One criterion for determining the internal damage is the observation of the deformation amplitude within a cycle. Laboratory-scale investigations have shown that the stiffness decreases with progressive material degradation. Transferred to the large-scale tests, this would mean that an increase in the deformation amplitudes over the number of load cycles can be expected. In the case of variant I, the amplitude remained almost constant within the first two load cycles over all 2 million load cycles (cp. Figure 2b). Only when the load was doubled, in load case 3 the measured deformation also doubled. After a short adjustment, approx. 75,000 cycles, which is probably again due to settlement processes of the ballast bed, the deformation amplitude remained constant. Progressive damage could not be detected.

After removal of the plate of variant I, no cracks became visible on either the upper or lower side. The theory of a non-progressive degradation could therefore be confirmed for this case. Rather, the plate in this structure behaved like a load transfer device from the loading device to the gravel bedding. Due to the relatively high stiffness of the foundation, no sufficiently high loads were applied to create strains within the highly ductile concrete slab that would have caused or promoted crack growth due to bending stress. In this case, the stresses in the case of pure force transmission, without any movement of the bedding, would only be 1.6 N/mm², which cannot lead to any damage to a concrete structure in the compression force range. Damage under the given load levels only occurs due to larger deformations and the associated bending stresses in the slab, as they are caused by softer bedding.

Despite the absence of bending cracks on the underside of the slab, a horizontal crack became visible in the cutted samples taken near the load application point. This is due to the deformation of the slab despite the relatively stiff bedding. If the slab deforms analogous to a bending component, a triangular load distribution between the upper and lower side of the cross-section occurs according to elasticity theory. This stress leads to changes in length, whereby the change in length is largely dependent on the stiffness or Youngth's modulus.

In highly ductile concrete, the tensile and compressive modulus of elasticity is not the same, which means that the change in length cannot be constant. These conditions cause shear stresses which are the reason for horizontal cracks. A corresponding graphic illustration of the processes is given in Fig. 3.

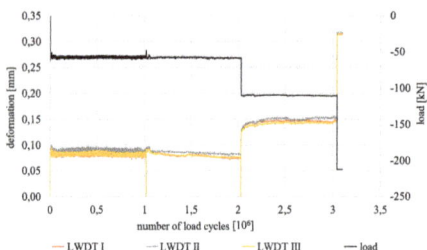

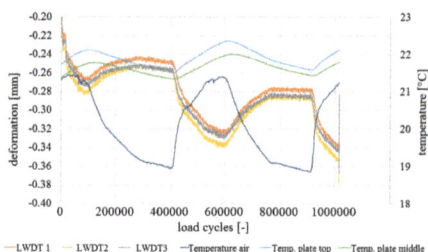

Fig. 2a. Effect of changing room temperature on the deformation measurement values by IWA's.

Fig. 2b. Amplitude of the center deformation variant I.

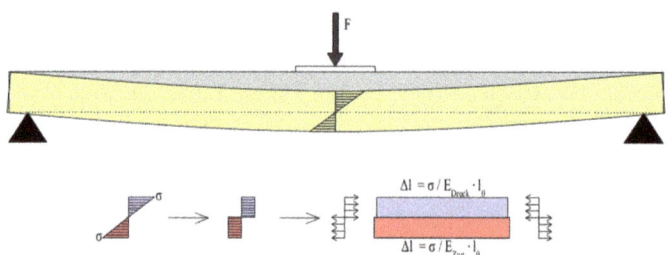

Fig. 3. Scheme for the formation of horizontal cracks.

The behaviour of variant II is very similar to that of variant I. Neither in the area of the real deformations in the middle of the plate nor in the amplitudes are any significant changes in the component behaviour to be recognized. The bedding on asphalt is therefore so stiff that the load is largely activated in compression components and only to a small extent as bending load.

3.2   Test Results Plate III and IV

The loading of the plate according to construction variant III resulted in an initial settling of 0.6 mm within the first cycles, a value approx. 3 times greater than in the previously investigated variants. In the course of the loads of load case 1, this value increases to slightly more than 1 mm centric deflection in the middle of the cycle in the middle of the plate. Within load case 2 the plate deformation increases by a further 0.25 mm. In summary, at the end of load case 2, there are center deflections of 1.25 mm. By doubling the load to 110 kN high tension level, the central deflection also increases by a further 0.5 mm. Over load case 3, a significant increase in the deformations in the middle of the plate by 0.75 mm can be seen, resulting in a final deformation level of 2.5 mm for this load variant.

Also the observation of the cycle amplitude shows the effect of the much softer bedding. If the values are very stable in the first two load cases with an average amplitude of 0.4 mm, this amplitude changes to 0.8 mm in load case 3 and increases to 0.95 mm at the loading point (Fig. 4a). Such an increase is a clear indication of induced damage, which increases further over the load case. Such a behaviour would also be consistent with the prediction of bending stress, which exceeds the theoretical tensile strength at the underside of the panel.

After finishing the loading, the slab could be removed without stability problems at the 4 hooks embedded in the respective corner points. A star-shaped crack pattern appeared on the underside, as is known in building construction in the area of punching through problems. The cracks were very fine. To improve visibility the cracks were wetted and after successful identification they were traced and measured. The calculated mean value of the crack widths on the underside of the slab is 70 µm, with more than half of the examined cracks showing crack widths of less than 40 µm. Due to the soft bedding, structure variant III cannot directly conduct the applied loads through the microstructure as pressure components. In this case, the load transfer must take the form of bending stress on the plate. Dimensioning, material composition and load result in visible damage to the component without leading to material failure.

Design variant 4 pursued the objective of being safely located in the area of crack formation on the underside of the slab even under load case 1. For this purpose, the design was chosen analogous to that of variant 4. Only the slab height of the SHCC slab to be investigated was reduced by 50%.

In this variant, the deformation of the slab centre showed a high value of 5 mm within the consolidation, which increased by a further 3 mm over load case 1. The load in load case 2 showed an almost linear increase in deformations by 2.5 mm due to the heavily damaged slab and the additional temperature influence. A further increase in the load in load case 3 could only be recorded for the first 150,000 cycles, since the inductive displacement transducers then ran out of their measuring range (cp. Figure 4a). After approx. 2.4 million cycles, the test was aborted because the deformations were so high that neither the machine control could provide reproducible results nor measured values about deformations could be recorded (Fig. 4b).

Within load case 3, one edge LWDT changes its deformation increase into a decrease, which is due to a buckling or lifting of the slab corners from the edge support. In the corner areas this was almost 25 cm just before the test shutdown.

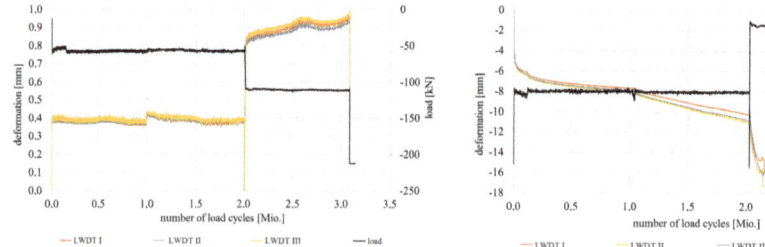

Fig. 4a. Amplitude of the center deformation variant III

Fig. 4b. Center deformation, variant IV, cycle mean-values

After the premature termination of the loading due to the large deformations and clearly forming separation cracks, the slab could be removed from the experimental stand as one element at the 4 attachment points. The developing crack pattern was similar to variant III, only much more pronounced. In a radius of 0.5 m around the centre of the slab, 93 cracks were counted, most of which ran to the edge of the slab. Furthermore, 3 clearly visible, strongly opened cracks on the underside became prominent. For a detailed investigation, both the central and most severely damaged element and a supplementary slab to the edge were taken. In the central part, the plate collapsed along the separation cracks after cutting, leaving 5 individual elements (see Fig. 5). The cracks were determined along the green line in the middle section and along the middle vertical in the edge section. The mean crack width of the edge piece is 55 μm, for single cracks between 10 and 400 μm. In the middle range, values between 20 and 400 μm were measured, with an average value of 93 μm.

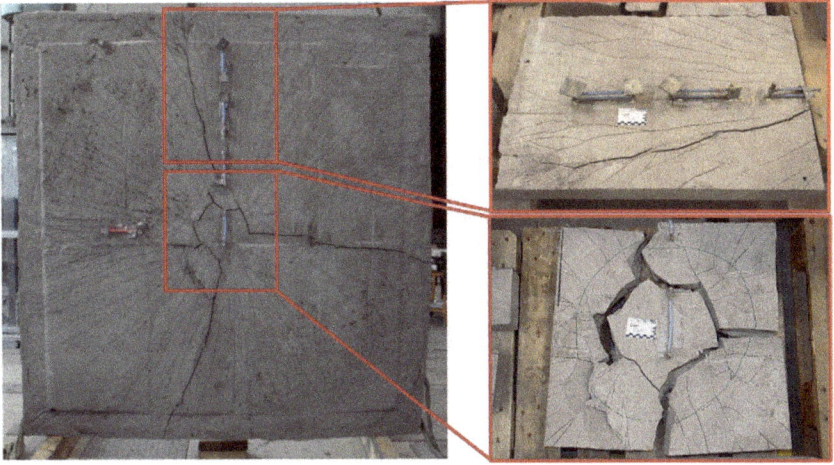

Fig. 5. Crack pattern of construction variant IV with details plate edge and plate centre

The damage in the test set-up variant IV corresponds to that of variant III in terms of damage mechanisms, but the damage is more obvious due to the halved panel height and the resulting 4 times the edge stress on the underside of the panel. The wide open cracks on the underside of the panel clearly show that fibres were not able to transfer the load between the crack flanks over all applied cycles.

3.3   Fracture Surface

For fracture surface examination, a specimen was prepared from the central plate with a widely opened crack (approx. 200 μm). In design variant 3, the crack-bridging effect of the fibres was nevertheless so strong that considerable additional mechanical energy had to be applied to separate the crack flanks. Accordingly, these fracture patterns are only of limited significance, as it is difficult to distinguish between preparation and actual damage caused by the test. Clearly visible in the image are long fibres on the crack flank, with the typical pull-out traces as they also occur in the quasi static test. Furthermore, only a small part of the fibres is broken. The major part of the fracture surface damage is therefore less attributable to cyclic loading than to the post-test separation of the crack flanks. In the area of the fibre exit point from the matrix into the existing crack, damage was found in some fibres, as is typical for tests with many load cycles on small specimens. On the one hand, increased abrasions or tapering's are visible at the exit point and on the other hand, an incipient defibrillation can be observed in places.

The fracture surface investigation for test set-up 4 was carried out on a completely separate crack from the centre of the plate. Nearly all fibres on the fracture surface show different damages and are mostly torn off in the contact zone (see Fig. 6a and 6b). In addition, the fibres show severe damage, largely due to the stress cycling performed under high load (Fig. 6c). Among other things, defibrillation phenomena and notch lines were detected (Fig. 6d), which are characteristic of the damage mode "fibre fatigue" observed on small tensile test specimens. The matrix is in good condition and has only a very small number of structural loosening.

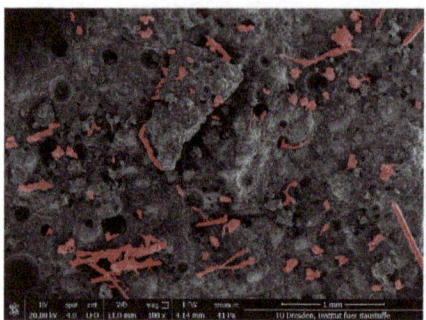

Fig. 6a. Fracture surface variant IV, fibre material marked red.

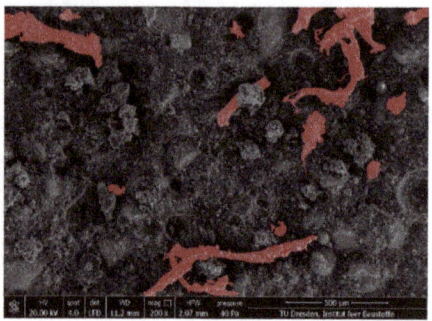

Fig. 6b. Detail of fracture surface variant IV, fibre material marked red

Fig. 6c. Heavily damaged single fibre

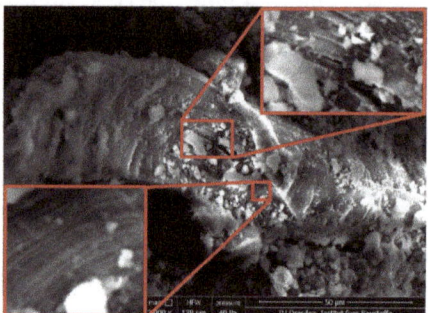

Fig. 6d. Heavily damaged fiber with deffibrilationand rest lines

4   Conclusions

The test results of the large-format laboratory tests with different slab geometries and slab beddings showed clearly different load-bearing behaviour.

- The test setups I–III withstood the doubled loads compared to the design loads in road construction with a reduced overall height at the same time.
- The variants with relatively stiff bedding, such as crushed stone or asphalt, with a slab construction height of 15 cm showed almost no damage. The predominant load transfer mechanism in these variants is the load transfer by compressive stress transfer from the load introduction into the subsoil.

- Construction variants III and IV, with soft bedding, carry the acting forces into the subsoil via the bending stress of the slab. A clear crack pattern, analogous to a push-through problem in slab-column connections, is developed.
- The fracture surfaces of the experimental set-ups 3 and 4 were examined by microscopic images and damage mechanisms were identified. These damage mechanisms are similar to those found in small laboratory test specimens.
- The following damage was found: fibre pull-out, fibre fatigue failure and mechanical fibre degradation. Matrix damage could not be observed due to the missing pressure component in the crack surface.
- Looking at the different damage mechanisms, the large format panels behaved analogously to the smaller laboratory test specimens.

Acknowledgements. The authors would like to thank the German Federal Highway Research Institute (BAST) for the financial support of the work and the institute of urban and pavement engineering for the collaboration in this project. A big thank is given to the Otto-Mohr-Laboratory of the TU Dresden for the provision of the test rig and numerous measurement techniques. Furthermore, we would like to thank the numerous technical staff and student assistants who helped to realize the tests.

References

1. Li, V.C.: From micromechanics to structural engineering-the design of cementitous composites for civil engineering applications (1993)
2. Curosu, I., Liebscher, M., Mechtcherine, V., Bellmann, C., Michel, S.: Tensile behavior of high-strength strain-hardening cement-based composites (HS-SHCC) made with high-performance polyethylene, aramid and PBO fibers. Cem. Concr. Res. **98**, 71–81 (2017)
3. Fischer, G., Li, V.C.: International RILEM Workshop on High Performance Fiber Reinforced Cementitoius Composites (HPFRCC) in Structural Applications. E&Fn Spon (2006)
4. Toledo Filho, R.D., Silva, F.A., Koenders, E.A.B., Fairbairrn, E.M.R.: Strain hardening cementitious composites (SHCC2-Rio). In: RILEM Proceedings PRO, vol. 81 (2011)
5. Schlangen, E., Beltran, M.G.S., Lukovic, M., Ye, G.: SHCC3: 3rd International RILEM Conference on Strain Hardening Cementitious Composites. RILEM Publications SARL, Dordrecht (2014)
6. Fukuyama, H., Suwada, H., Ilseung, Y.: HPFRCC damper for structural control. In: Proceedings, DFRCC Int'l Workshop, pp. 219–228 (2002)
7. Douglas, K.S., Billington, S.L.: Rate dependence in high-performance fiber reinforced cementbased composites for seismic applications. In: Proceedings, HPFRCC-2005 international workshop, Honolulu, Hawaii, USA (2005)
8. Jun, P., Mechtcherine, V.: Behaviour of strain-hardening cement-based composites (SHCC) under monotonic and cyclic tensile loading: part 1–experimental investigations. Cem. Concr. Compos. **32**(10), 801–809 (2010)
9. Müller, S., Mechtcherine, V.: Fatigue behaviour of strain-hardening cement-based composites (SHCC). Cem. Concr. Res. **92**, 75–83 (2017)
10. Ranjbarian, M., Mechtcherine, V.: A novel test setup for the characterization of bridging behaviour of single microfibres embedded in a mineral-based matrix. Cem. Concr. Compos. **92**(January), 92–101 (2018)

11. Jun, P., Mechtcherine, V.: Behaviour of strain-hardening cement-based composites (SHCC) under monotonic and cyclic tensile loading: part 1–Modelling. Cem. Concr. Compos. **32** (10), 810–818 (2010)

12. Altmann, F., Sickert, J.-U., Mechtcherine, V., Kaliske, M.: A fuzzy-probabilistic durability concept for strain-hardening cement-based composites (SHCCs) exposed to chlorides: part 1: concept development. Cem. Concr. Compos. **34**(6), 754–762 (2012)

13. Mündecke, E., Mechtcherine, V.: Untersuchung zum Tragverhalten von zugbeanspruchten Bauteilen aus hochduktilem Beton und Stahlbewehrung. Beton- und Stahlbetonbau **110**(3), 220–227 (2015)

14. Mechtcherine, V., et al.: Behaviour of strain-hardening cement-based composites under high strain rates. J. Adv. Concr. Technol. **9**(1), 51–62 (2011)

15. Mechtcherine, V., de Andrade Silva, F., Müller, S., Jun, P., Toledo Filho, R.D.: Coupled strain rate and temperature effects on the tensile behavior of strain-hardening cement-based composites (SHCC) with PVA fibers. Cem. Concr. Res. **42**(11), 1417–1427 (2012)

16. TL SoB-StB 04: Technische Lieferbedingungen für Baustoffgemische und Böden zur Herstellung von Schichten ohne Bindemittel im Straßenbau, Ausgabe 2004, Forschungsgesellschaft für Straßen- und Verkehrswesen (FGSV) (2007)

17. ZTV SoB-StB 04: Zusätzliche Technische Vertragsbedingungen und Richtlinien für den Bau von Schichten ohne Bindemittel im Straßenbau, Ausgabe 2004, Forschungsgesellschaft für Straßen- und Verkehrswesen (FGSV) (2007)

SFRC Underwater Slab for the Potsdamer Platz (PP) Berlin

Horst Falkner[1(✉)] and Ulf Hinke[2]

[1] University of Brunswick, Brunswick, Germany
horst.falkner@t-online.de
[2] IBF, Stuttgart, Germany

Abstract. Daimler-Benz (DB) the well-known car-producer decided to make the largest immobile investment in Europe, in the heart of Berlin. DB bought a large area in 1988, just near the wall. Nobody knew at that time that the wall would be taken down by the East Berlin people in November 1989. Since our office have made an engineering task in Stuttgart for DB, at a building in 8-m-deep groundwater, for DB it was clear that the same engineer should make the job as well in Berlin. Nobody knew at that time, that the water conditions of those two projects were completely different. In the project we handled before, the groundwater could be lowered by pumps. DB wanted the same concept for the PP. But in Berlin the water conditions are completely different. The depth of the water for the building pit was between 20 to 22 m and the water could not be lowered. The level of the groundwater was not allowed to be touched because the Berlin people get all their drinking water from the perfect filtered groundwater. The idea of a steel fibre concrete slab could be a solution for this project. However, we had no guideline, no code and no experience. Therefore, our procedure to design a safe and stable SFRC underwater slab was based on tests in scale but also on tests in the field with real dimensions. Our main task was to convince the authorities from the Senate of Berlin, DB and the German Railway Authorities. With this concept we got the agreement to build the deepest pit ever constructed before in a depth of 22 m.

Keywords: Steel fibre concrete · Up to 22-m water pressure · Large scale tests at the University of Brunswick · 1 to 1 scale tests at the structural site

1 Introduction

The Potsdamer Platz (PP) covers an area of about 85,000 m^2. The area of the PP railway station has 25,000 m^2. In scale: longitudinal, about 700 m, width about 150 to 300 m (Fig. 1). The city road traffic was planned to be integrated with PP by a double Tunnel System. The PP railway station is open for S- and U-Trains. A special challenge was the foundation of the underwater slab. The water level of the Berlin groundwater - depending on the ground level - will be found 2–3 m below the ground-level. The PP-buildings, however, were founded in 20 to 22 m. Therefore, special investigations and research were required because there had been worldwide no experience on a FRC-slab subjected to such extreme loadings: the SFRC-slab positioned and anchored by piles had to carry 170 to 190 KN/m^2.

© RILEM 2022
P. Serna et al. (Eds.): BEFIB 2021, RILEM Bookseries 36, pp. 761–770, 2022.
https://doi.org/10.1007/978-3-030-83719-8_65

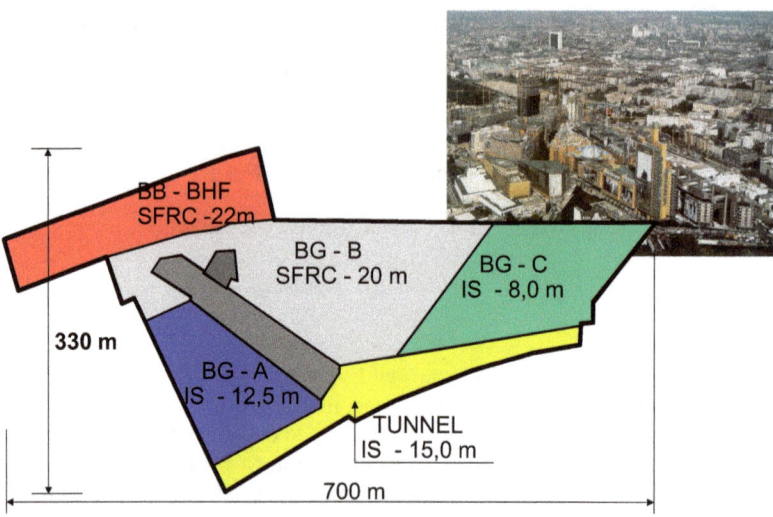

Fig. 1. Layout Potsdamer Platz: 700 m length joint-less building pit.

In a VDI-Congress in Berlin 1995, there were great discussions about the behaviour of SFRC being used for slabs in deep pits. In the PP-pits the SFRC will be brought down through the water using a large pipe, through a height of 18 to 20 m. The main argument in the discussion was that the steel-fibres never will be well distributed over the height of the slab. 90% of the Congress-Participants had the opinion that the fibres would be concentrated at the ground, in a layer of 20 to 30 cm. The slab above would be plain concrete. Such a statically system never would be able to carry the high-water load.

In the opinion of the Congress attendees, we should immediately stop our present procedure and find a new and safe concept. The congress arguments required a long brainstorming weekend. There one of the samples we discussed was the concreting of high reinforced concrete walls. If one considers a high wall to be filled with concrete this is to be done from the top. When additional vibrators are used one would get a perfect concrete surface. If cylinders are drilled from the wall, one will find that the aggregate is well distributed over the height. There would not be heavy stones at the bottom, and a sandy condition at the top. Why should it be different with the steel-fibres? The decision was made to continue with the SFRC concept. However, in order to prove our theoretical model, we decided to ask DB for an additional test in the deep water.

With the knowledge of the 90s it was not possible to analyse the system behaviour of a slab which was anchored on piles in a space of 3.2 × 3.2 m. Mainly the deformation within the slab but also the lifting of the piles due to water pressure, according to our knowledge in the 90s could be done, only by an applied test programme it was possible. We therefore tried to find another method to get the ductile behaviour of the

SFRC slab. It was not possible to think on a real scale test with by a slab of 1.2 m. Finally, we ended up in a scale model in the order of 1:4. Such tests could be made at the laboratory of the University Brunswick, Germany.

With tests in the laboratory, it should be possible, to get the different deformation behaviour using various fibres, but also the contribution of fibres to the ductility.

2   Construction of the Site

In order to realize the project in the deep water-condition, it had to be constructed a back-anchored pit wall. Figure 2 shows the steps, necessary to get finally the dry pit-construction by:

- sheet-pile or diaphragm wall (Fig. 2a)
  dry excavation to the level of the groundwater (2–3 m),
- placing the anchor above the groundwater (Fig. 2a)
- soil-excavation in the groundwater up to the required level −20, or −22 m (Fig. 2a)
- placing the shake piles with a length up to 25 m (Fig. 2b)
- concreting the SFRC-slab with the assistance of 60 divers from all over Europe (Fig. 2b)
- pumping out of the water (Fig. 2c).

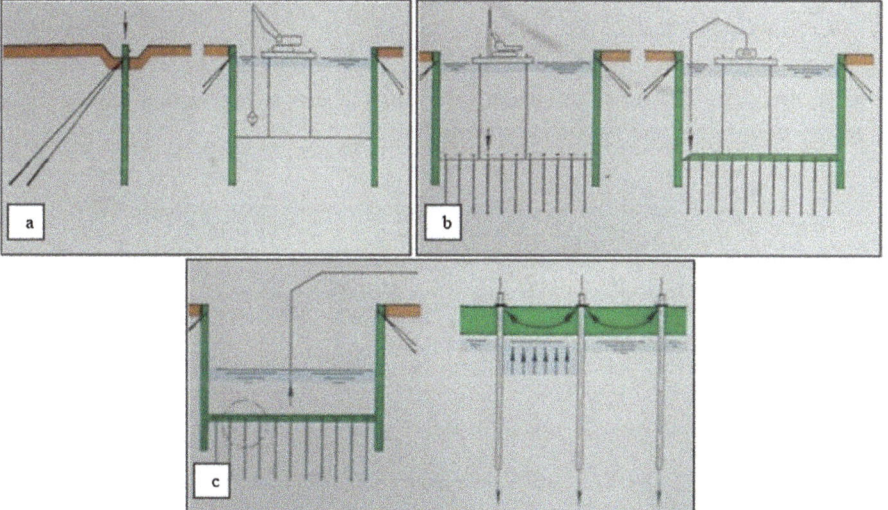

Fig. 2. Building pit construction at the Potsdamer Platz

Their task was by feeling and touching to declare, that everything is ok or might be wrong in the deep and very dirty water condition. After hardening of the SFRC-slab, about 4 weeks, the water was pumped out.

A special congress dealing with the expected uplift took place in Berlin. The uplift was expected between 3 to 5 mm, at the lower bound and 1.50 m at the upper bound. This value we could only comment by saying in case of such a large uplift, we will have only small people and sport-cars in the deepest garage. After pumping out the water, the final uplift of the SFRC-slab was between 3 to 5 mm.

It was always interesting to talk about the amount of fibres being in 1 m³ SFRC. Nobody could tell you how many fibres are necessary, in order to get the desired result in a test. The main reservations against fibres were: "It does not work, too expensive, not allowed, not pumpable, fibre clumps, no compaction, disintegration, pile heads not enclosed or, those few fibres will not help at all."

Figure 3 shows the amount of steel fibres with the characteristic 50/06. The most impressive data is the information that 40 kg of steel fibres represent an amount of 364,000 fibres in 1 m³ of concrete.

Steel fibres 50/0.6

10 fibres = 1,1g
10.000 fibres = 1,1 kg
364.000 fibres = 40 kg
728.000 anchorages !!

Fig. 3. Amount of fibres having 40 kg/m³.

3   Laboratory Tests

For the carrying capacity of the underwater-SFRC the one-dimensional model shows cracking in the field and over the supports Fig. 4.

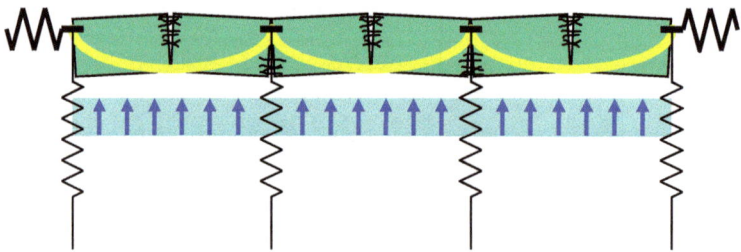

Fig. 4. Underwater SFRC slabs arch model: cracking in the support and field.

It was difficult to find the right test model simulating the water pressure around 200 to 220 kN/m². In the final decision a cork-layer of 6 cm was placed between the slab, which was pre-stressed against the floor in our lab. We found that the cork layer ensures an even and continuous stress distribution under the slab. For the test set-up it was decided to use a slab of 35 cm thickness and outside dimensions of 4 × 4 m. This slab was pre-stressed against a cork-layer of 6 cm as seen in Fig. 5. To react against arguments that the fibres might sink to the bottom of the slab, when the SFRC will be poured from the water level down 20 to 22 m, we decided for the lab-test, to pour the SFRC from the roof of our lab. This is a height of about 17 m. When we checked the concrete after testing by cylinder cubes, we could not find any influence in the distribution of the fibres.

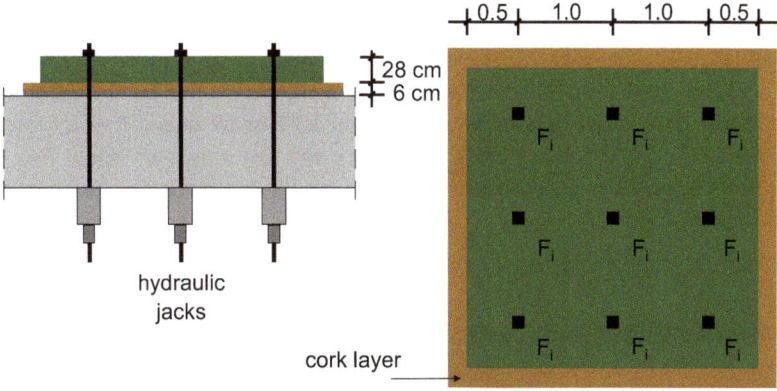

Fig. 5. Underwater SFRC slabs test setup: prestress to simulate pressure.

We have designed 3 test members with the dimensions as shown in Fig. 5: Number 1 was plain concrete in order to get the basic data compared to the following fibre test with fibres of 40 kg 50/06, and 60 kg 60/08 per m^3. The test results are shown in Fig. 6. One can realize that the fibres change the caring capacity dramatically. While the test with plain concrete (black line) suddenly failed, the SFRC-members (red and blue lines) showed a great ductility in the order of 15 to 25 times compared to plain concrete. The caring capacity was more than twice the value of plain concrete. While the unreinforced member was completely broken, SFRC-members remained as a solid slab. Since the test with the 40 kg fibres with 50/06 per m^3 gave the better results compared to the thicker fibres and 60 kg.

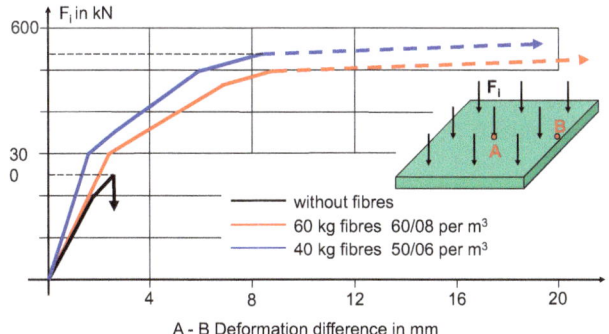

Fig. 6. Test results of the members: plain concrete and SFR-slabs.

Figure 7 shows the test results of a 35 cm thick slab for an unreinforced slab (a) compared to a steel fibre slab (b) having twice as much test load and the deformation capability is about ten times the value obtained for plain concrete. First, we made a secret test without visitors in our lab. Depending on this test results, we were able to invite all important partners to our lab. The partners at the time were: Daimler-Benz and the German Federal Railway because we also built the Railway station PP, the Senate who was responsible to give the permission to build the PP with the new material SFRC, which had never been used in Germany before in a constructive member. Actually, we know, that every test is a singular procedure. A test number 2 could give a result that might be 10 to 30% less compared to number 1. In the presence of our about 35 visitors test number 2 was about 10% better than the secret test No. 1. Definitely, we expected many questions, when the test was over. Only two simple questions were raised and two weeks later; we got the permission to build the SFRC-slab, designed in our office.

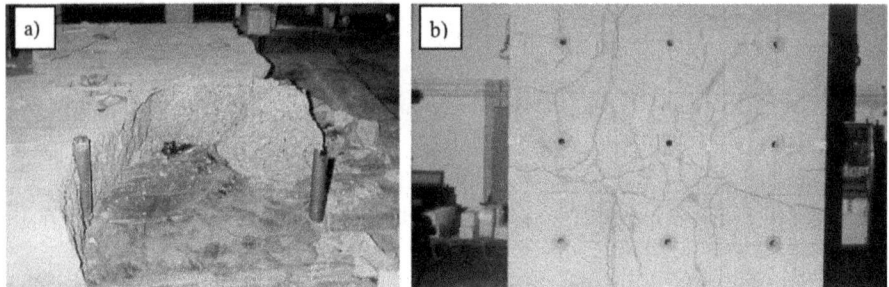

Fig. 7. Test result comparison between plain concrete (a) and SFRC slab (b).

4   Field Test

When taxi drivers drove from the central station to the PP-site, they always commented negatively. They could not understand why over a long period of time one could only see a large lake at the site Fig. 8. In the opinion of the taxi drivers everything went wrong the PP-site. The so-called "Lake" was not only in summer (Fig. 8a) but also in winter, when ice covered the surface as seen in Fig. 8b. There was the hugest dredger ever built to excavate the soil Fig. 9. With one shovel only it could carry 20 m^3 in the boat. People joked: "There will be a new lake built by Daimler-Benz in the Centre of Berlin".

Fig. 8. a) Lake DB in summer; b) Lake DB in winter.

Fig. 9. Water dredger with 20 m³ shovel.

From our laboratory test we have learned the contribution of the fibres with respect to the carrying capacity and the important large deformation compared to plain concrete. However, we still had to find out how SFRC concrete would behave under the influence of water pressure in a depth of 20 to 22 m. After many discussions we decided to make a water test using 3 moulds of the dimension $4.0 \times 1.2 \times 1.2$ m. In these moulds 2 piles with their anchor heads had been welded. The purpose was to find out if SFRC would cover well the anchor-head. The 3 moulds should be brought down to the ground in 20-m-deep water (Fig. 10a). In addition, we wanted to find out how a delay of the hardening process would be influenced by the cold water of about 4 °C. The real time of the hardening process was necessary for the casting of the slab. In the casting process of the underwater slab, we had to place the SFRC in rows of 150 up to 300 m. When one row would had been concreted, the placing of the SFRC would start again where the row started. This process required a delay of about 30 to 80 h. For the 3 test specimen we therefore decided to test the hardening process by using a delayer with 20 h for mould 1, 40 h for mould 2 and 80 h for the third mould. With all together 4 divers, the moulds were brought down in the deep cold 4 °C water. For the total work in the water 60 divers of whole Europe had done the work in the water (Fig. 10b). Their only work was, by touching and feeling to confirm the required task. The divers are mainly not educated structural engineers.

Fig. 10. a) Field test with steel box; b) One of 60 divers; c) Best distribution of fibres.

Since we were very confident about the coming test results, it was proposed to DB to invite the Berlin TV, when the lifting of the moulds 4 weeks after hardening would be lifted back to the air. In a service container it was possible to get the information from the three test-moulds. The first mould with a hardening delay of 20 h did not give any information after 30 to 40 h. New questions raised: Does SFRC not harden in deep cold water? Finally, after 70 to 80 h the temperature showed an increase up to a maximum of 15 °C. The second mould which should harden after 40 h required 80 to 100 h and the temperature raised only up to 12 °C. Crazy questions came upon: Does SFRC not harden in deep cold water? Nobody could give a precise answer. The only answer which could be given: Nobody knew at that time, in which condition the SFRC would be, when it would be raised back on earth. But there was one consequence: The Berlin TV was asked NOT to show up at the site.

Mould 3 gave us the information of hardening after just 180 to 200 h. The increase of temperature was below 10 °C. After 4 weeks the moulds were lifted. A majority of the experts had the opinion that those tests would confirm the arguments of the Congress that SFRC would not work in deep cold water.

When the steel walls had been removed, everything looked very solid. Now the task was, to drill cylinder tubes, as many as possible, and in all directions in order to find out the real distribution of the fibres and the quality of the concrete. The result was better than expected. The concrete quality and the distribution of the fibres in all the tubes were perfect (Fig. 10c). Also, the heads of the piles had been surrounded properly by SFRC. In the laboratory the concrete quality and the distribution of the fibres of the cylinders had been analysed. All that what Congress people had in mind did not happen. The only question which remained was the delay of the hardening process, which could not be explained. Important for us was, that it had no consequence to the strength of the concrete. The test results enabled the progress of the further work. It then could be elaborated the structural behaviour of a SFRC slab under high water pressure.

After all, everybody was nervous when the days of pumping out the water became reality (see Fig. 11). Will the SFRC-slab fulfil the design results? What and how much would be the lifting of the slab? Will the slab be tight enough, so that we get a dry pit? Well, the pumping went on and finally a worker went down in the deep pit of 20 m. He did not know that there was a pressure of 200 kN/m^2 under his feet and 60 Mio tons in the whole pit. Everybody was happy and celebrated this special day. But the work went on.

Fig. 11. a) Pressure under the slab; b) Dry Pit 34.000 m^2.

5   Closing Remarks

In spring 1993 the CEO, at the time Edzard Reuter, declared "Today we begin to build the new city quarter, the Potsdamer Platz. The whole project will be finished by October 28, 1998." When we heard this fixing-date of the opening of the PP, we were shocked, because at that time, nobody knew about the difficulties which would arise not only for the design but also for the transformation at the site. The PP was officially inaugurated on October 28, 1998 (Fig. 12).

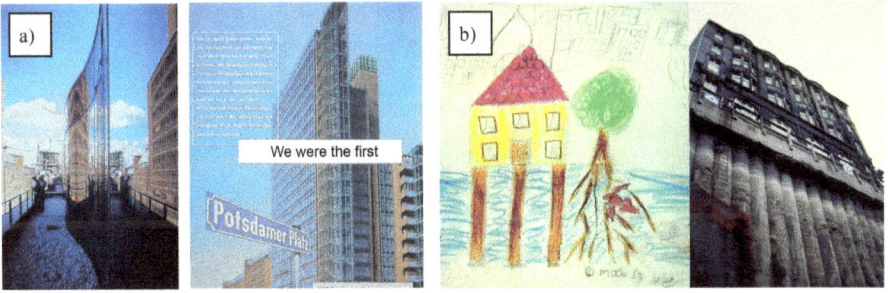

Fig. 12. a) Daimler-Benz: "We were the first"; b) Assistance by children.

References

1. Falkner, H.: Bauwerke in Wechselwirkung mit dem Baugrund. Geotechnik, pp. 121–129 (1994)
2. Brem, G., Wooge, M.: Potsdamer Platz Berlin – Ausführung dichter Baugruben mit rückverankerten Unterwasserbetonsohlen. Bauingenieur **72**, 53–59 (1997)

3. Falkner, H, Henke, V.: Steel fibre concrete for underwater concrete slabs. In: Proceedings (32) of the International RILEM Conference on Production Methods and Workability on Concrete, S. 79
4. Falkner, H.: New technology for the Potsdamer Platz – steel fibre concrete for underwater concrete slabs. In: Tagungsband, 3rd Eurolab Symposium, Berlin, 5–7 Juni 1996
5. Falkner, H.: Meßprogramm für die Baugrube Potsdamer Platz Berlin – Meßtechnische Überprüfung von Lastansätzen und Verformungen. VDI-Berichte Nr. 1196 (1995)
6. Falkner, H., Henke, V., Hinke, U.: Stahlfaserbeton für tiefe Baugruben im Grundwasser. Bauingenieur, pp. S. 47–52 (1997)

Polypropylene Fibre Reinforced Concrete for the Structural Panels of the Pavillions of the Motril Port (Spain)

Elisa Valero Ramos[1(✉)], Juana Sánchez Gómez[2],
Diego Jimenez López[2], Aurora Montalbán[2], and Albert de la Fuente[3]

[1] University of Granada - UGR RNM 909, Granada, Spain
[2] UGR RNM 909, Granada, Spain
[3] Civil and Environmental Engineering Department,
Universitat Politècnica de Catalunya, Barcelona, Spain

Abstract. Two new pavilions for the Motril's Port, finished in October 2020, were designed, and constructed considering polypropylene macro-fibre reinforced concrete (PPFRC) as structural material. The main purpose of this decision was to avoid the use of any type of steel reinforcements in the elements exposed directly to the marine environment (ex., façades) so that the durability and service life were positively benefited altogether with the maintenance costs and the aesthetic requirements. The object of analysis of this scientific contribution are the 2250 × 3480 × 150 mm PPFRC precast panels designed with a C50/60 with PP macrofibres as unique reinforcement. These panels are part of the external envelope and also provide structural support to the roof (and the loads to which this is subjected to). The PPFRC panels resulted from an optimized-oriented design integrating aspects related with material properties, structural behaviour, construction process and production. The solution proposed is, ultimately, oriented to maximize the sustainability of these type of structural elements.

Keywords: Fatigue · Fibre reinforced concrete · Cracked section · Polypropylene macrofiber

1 Introduction

Fibre reinforced concrete (FRC) is a material already accepted for structural purposes [1, 2] in different applications: industrial pavements [3–5], buried pipelines for sewerage [6, 7], tunnel linings [8–10] among others. Steel fibres have been primarily considered for these applications; however, synthetic macrofibres are emerging as an effective concrete reinforcement since the recent enhancements made on these technology and the successful experiences and knowledge gained allow considering this type of fibres in a wide range of structural applications [11–13]. In this line, this contribution is meant to present a novel application of polypropylene macrofibres as unique reinforcement of concrete panels used in a specific real project.

This industrial-research project proposed a novel precast polypropylene macro-fibre reinforced concrete (PPFRC, hereafter) for port building facilities (see Fig. 1).

© RILEM 2022
P. Serna et al. (Eds.): BEFIB 2021, RILEM Bookseries 36, pp. 771–779, 2022.
https://doi.org/10.1007/978-3-030-83719-8_66

In seafront areas, traditional steel-based reinforced concrete solutions are prone to suffer for degradation and deterioration due to the aggressive marine conditions. The challenge is to develop an affordable low embodied carbon construction system, able to open new horizons in the precast concrete construction paradigm. Implementing a strategy of sustainable construction for seafront buildings is of paramount importance both socially and environmentally.

Fig. 1. (a) Placement of the panels and (b) pavilions in its final configuration

2 Project Background

This research on concrete began in 1996 with the restoration of the Manantiales (Fig. 2) by Elisa Valero, one of the most iconic works of the architect Felix Candela (1910–1997). With this first achievement, the former initiated a new attentive path towards structural design as a fundamental factor in the formalisation of architecture by highlighting the possibilities of the industry in the development of new materials that favour rationality and economy of the existing means.

The company Heidelberg, on the Arvision Price, proposed Elisa Valero to exhibit her research on concrete (Fig. 3). For this purpose, self-supporting slender display panels made of concrete were designed and produced. The design was oriented two-fold: to be a homage to Felix and to be stable and resistant with a thickness of 2 cm. This was achieved by using synthetic macro-fibres as reinforcement to avoid the concrete cover (which would lead to an increase of the thickness due to durability aspects) necessary in case of using traditional steel reinforcing bars.

The research context in which this work has grown is given by the research team RNM 909 Sustainable Housing and Urban Recycling from the University of Granada. Conscious of the need to build resilient infrastructures to promote inclusive sustainable industrialization and foster innovation. Since 2019, the team has been developing various research projects for the Port Authority of Motril focusing on how to contribute to mitigate climate change. This project is based on a collaborative approach methodology to synergise the expertise and knowledge of engineers and architects.

Undoubtedly, this collaboration and sharing of expert knowledge will greatly increase the chances of success of complex projects.

Through these collaborations and being aware of the international concern on sustainable construction, the research team have detected a high number of buildings associated to Port facilities, which could profit from an innovative building approach (police control point, markets, points of sale and information, toilets, coffee shops, and other programs servers). Hence, there is a real need to develop new building strategies, which properly address such situations. To change the Port's building facilities paradigm would generate a significant impact.

Fig. 2. Restaurante Manantiales in Xochimilco (México) – designed and constructed by Félix Candela in 1956–1957 and posteriorly restored by Elisa Valero (1996).

Fig. 3. Concrete Exhibition: concrete works by Elisa Valero

3   Structural and PPFRC Design for the Prototype

FRC precast concrete has been successfully used in marine/chloride aggressive environments for more than twenty years (ex., in ground supported floors for logistic operations, tunnels linings in contact with soils contaminated by chlorides). Nevertheless, owe to the strong inertia in the construction industry, these elements were designed and constructed considering the same inherited shapes of the conventional reinforced concrete. The key point of this project was to achieve a holistic design by integrating aspects belonging to material properties, structural behaviour, constructions process and production to develop a structural efficient and sustainable system.

To achieve this goal, it was essential to understand the opportunities provided by PPFRC. In this sense, the avoidance of the conventional steel rebars allowed making the design (in terms of shapes) more flexible. Moreover, the use of polypropylene macrofibres permitted to provide the mechanical strength required at service and ultimate limit states, as well as to guarantee the durability of the material in a chloride aggressive environment.

To fulfil the different requirements established (shape flexibility, material consumption optimization, no chloride sensitivity, flexural hardening response, and others) the PPFRC mix presented in Table 1 was proposed.

Table 1. Mix proportions of the PPFRC

Material	Amount (kg/m^3)
CEM I 52.5	350
Sand 0/2	1.000
Gravel 6/12	800
Superplasticizer	2.8
MasterFiber 400	5
MasterFiber 246	8

This mix presented in Table 1 proved to be flexible in terms of workability for a wide range of the amount of fibres. The Abram's cone settlement averaged 20 cm.

Characterization of the concrete compressive strength was carried out by testing cylinders at 28 days. The characteristic compressive strength ($f_{ck,28}$) achieved was 60 N/mm^2. Likewise, notched-beam tests on 600 $\times$ 150 $\times$ 150 mm beams were carried out according with the EN 14651:2007 to characterize both the pre- and post-cracking flexural response of the PPFRC. The results are presented in Fig. 4 in terms of the ratio f_{Ri}/f_{LOP} for $i = 1$–4.

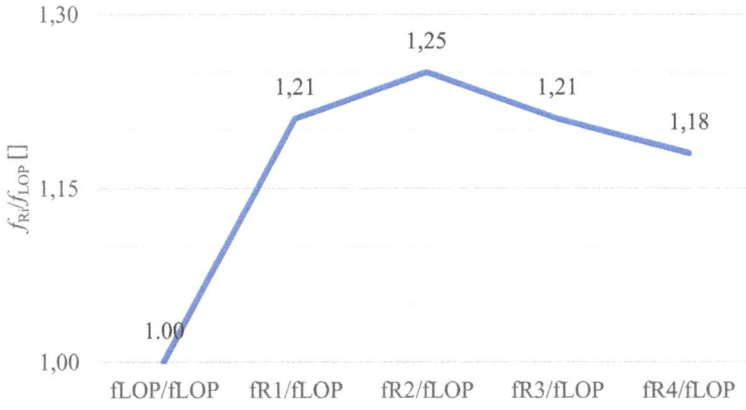

Fig. 4. f_{Ri}/f_{LOP} (mean values) obtained from the notched-beam tests on the designed PPFRC.

The results presented in Fig. 4 allow confirming that the PPFRC composite designed presents a flexural-hardening response as $f_{Ri}/f_{LOP} > 1.0$. Therefore, the material is capable to generate stable cracking patterns and to control the crack with during service conditions while providing bearing capacity and ductility in case of reaching an ultimate limit state. The structural design of the PPFRC panels was carried out by MBCC Group by means of in-house software (previously validated with extensive experimental programs). In this regard, it must be remarked that the flexural strength during transient loading conditions (demoulding, lifting and transport) are those that govern the design of these elements.

4   Prefabrication and Building Process

Both pavilions were built with the same system, using the rectangular geometry for different uses. The first was designed to gather a scanner for the lorries, with dimensions of 19.15 m × 4.40 m, and the second for the offices of freight forwarders of 40.20 m × 4.20 m.

The precast elements were designed for guaranteeing re-configuration in plant of the pavilion and demountability. These conditions led to the necessity of overlapping and coupling joints of the panels with those of the foundations and roofs. In this sense, to minimize the number of joints, the size of the panels was increased. Consequently, the number of pieces was reduced and the transport, handling and assembling resulted cheaper than the traditional solution (proposed in the former project).

In the first panel trials (Fig. 5), an attempt to introduce the thermal insulation inside the panel was made (like a sandwich). Nonetheless, due to the need of installing the thermal bridges at the perimeter contour and connector, for structural purposes, this solution was found incompatible with the requirements established. Finally, it was necessary to implement a continuous insulation to meet the internal comfort and energy efficiency requirements the pavilions as offices.

Fig. 5. Lifting and constructability tests carried out at the precast concrete plant.

It must be mentioned that the manufacturing required to cast the pieces in two thin layers, resulting in a panel with a total thickness of 15 cm.

Loading tests were carried out in the plan facilities (see Fig. 6). The panels showed the required behaviours in terms of cracking (no crack formation for the expected wind load) and ductile response after the cracking occurrence. To this end, panels were tested under a simple supported beam configuration subjected to a line load at the middle of the span. The results evidenced that both mechanical criteria were achieved, these validating the mechanical performance required per project specifications (Fig. 7).

Fig. 6. Cracking and bearing capacity tests of the first trials.

Fig. 7. Construction process: (a) lifting of the panels; (b) connection of the panels to the foundation; (c) temporary support conditions were panels work as cantilevered elements against wind; and (d) final configuration of the panels.

Nowadays, this facility is in service (see Fig. 8) without showing any evidence of cracking or corrosion (not expected due to the absence of steel-based reinforcements). The port authority showed a great satisfaction for the efficiency on the installation (economic), architectural design and appearance (social) and reduction of the environmental impact due to the reduction of material consumption and corrosion free reinforcements (minimization of repair costs and materials consumption), this PPFRC modular construction proving to be a sustainable solution for this conditions.

Fig. 8. Frontal view of one pavilion under service conditions.

5   Conclusions

In this industrial-oriented research project a new polypropylene macrofibre reinforced concrete (PPFRC) panel system was investigated and two large-scale pavilion prototypes were constructed for the Motril's Port (Spain). The main outcomes and conclusions from this project were:

- PPFRC resulted to be a suitable material from the structural and technical perspective. A post-cracking ductile performance was guaranteed by means of designing a flexural-hardening composite, which also allowed to reach to shape and design flexibility required for this panels.
- The constructability and mechanical capacity were tested and validated within the plant facilities, and those were found to match the project requirements. In this sense, higher productivity in all the stages was reported (i.e., fabrication, transport, and installation).

Although very positive outcomes were observed, there is still research to be carried out in this line. In this sense, a comprehensive sustainability analysis should be performed to quantify the benefits from the environmental and social perspectives (economic benefits were already identified) derived from the use of PP fibres as reinforcement for concrete (i.e.., no corrosion problems, lower CO_2 emissions embedded in the production of these fibres, among others).

Acknowledgements. The first author wants to thank the Autoridad Portuaria de Motril for its help. Likewise, the RNM 909 research team wants to thank Jose María Vaquero and his engineering team from MBCC Group for developing the structural design of the panels.

References

1. di Prisco, M., Plizzari, G., Vandewalle, L.: Fibre reinforced concrete: new design perspectives. Mater. Struct. Constr. **42**, 1261–1281 (2009). https://doi.org/10.1617/s11527-009-9529-4

2. Walraven, J.C.: High performance fiber reinforced concrete: progress in knowledge and design codes. Mater. Struct. Constr. **42**, 1247–1260 (2009). https://doi.org/10.1617/s11527-009-9538-3

3. Meda, A., Plizzari, G.A., Riva, P.: Fracture behavior of SFRC slabs on grade. Mater. Struct. Constr. **37**, 405–411 (2004). https://doi.org/10.1617/14093

4. Meda, A., Plizzari, G.A.: New design approach for steel fiber-reinforced concrete slabs-on-ground based on fracture mechanics. ACI Struct. J. **101**, 298–303 (2004). https://doi.org/10.14359/13089

5. Falkner, H., Huang, Z., Teutsch, M.: Comparative study of plain and steel fiber reinforced concrete ground slabs. Concr. Int. **17**, 45–51 (1995)

6. de la Fuente, A., Escariz, R.C., De Figueiredo, A.D., Molins, C., Aguado, A.: A new design method for steel fibre reinforced concrete pipes. Constr. Build. Mater. **30**, 547–555 (2012). https://doi.org/10.1016/j.conbuildmat.2011.12.015

7. Monte, R., de la Fuente, A., De Figueiredo, A.D., Aguado, A.: Barcelona test as an alternative method to control and design fiber-reinforced concrete pipes. ACI Struct. J. **113**, 1175–1184 (2016). https://doi.org/10.14359/51689018

8. Caratelli, A., Meda, A., Rinaldi, Z.: Design according to MC2010 of a fibre-reinforced concrete tunnel in Monte Lirio. Panama. Struct. Concr. **13**, 166–173 (2012). https://doi.org/10.1002/suco.201100034

9. Plizzari, G.A., Tiberti, G.: Steel fibers as reinforcement for precast tunnel segments. Tunn. Undergr. Sp. Technol. (2006). https://doi.org/10.1016/j.tust.2005.12.079

10. Liao, L., de la Fuente, A., Cavalaro, S., Aguado, A.: Design procedure and experimental study on fibre reinforced concrete segmental rings for vertical shafts. Mater. Des. **92**, 590–601 (2016). https://doi.org/10.1016/j.matdes.2015.12.061

11. de la Fuente, A., Escariz, R.C., de Figueiredo, A.D., Aguado, A.: Design of macro-synthetic fibre reinforced concrete pipes. Construct. Build. Mater. **43**, 523–532 (2013)

12. Pujadas, P., Blanco, A., Cavalaro, S., de la Fuente, A., Aguado, A.: Fibre distribution in macro-plastic fibre reinforced concrete slab-panels. Constr. Build. Mater. **64**, 496–503 (2014). https://doi.org/10.1016/j.conbuildmat.2014.04.067

13. Pujadas, P., Blanco, A., Cavalaro, S., Aguado, A.: Plastic fibres as the only reinforcement for flat suspended slabs: experimental investigation and numerical simulation. Constr. Build. Mater. **57**, 92–104 (2014). https://doi.org/10.1016/j.conbuildmat.2014.01.082

Use of Macro Synthetic Fibre in Segmental Tunnels

Ralf Winterberg[⊠]

BarChip Inc., Kurashiki, Okayama, Japan
rwinterberg@barchip.com

Abstract. Fibre reinforced concrete is becoming widely utilized in segmental linings due to the improved mechanical performance, robustness and durability of the segments. Further, significant time and cost savings can be achieved in segment production and by reduced reject or repair rates of segments during temporary loading conditions. The replacement of traditional rebar cages with fibres further allows changing a crack control governed design to a purely structural design with more freedom in detailing. A further benefit of replacing steel rebar cages by fibres is the significantly reduced carbon footprint.

Recently completed research at the University of Brescia on full scale tunnel segments reinforced with macro synthetic fibres (MSF) confirmed the feasibility as well as the fulfilment of the structural requirements of the Model Code 2010. Parallel, field experience has shown that MSF reinforced concrete segments perform robust and dependable even under very difficult conditions.

This paper presents and discusses the key findings from the segment research as well as three successfully completed reference projects using MSF as primary structural reinforcement.

Keywords: Macro synthetic fibre · Fibre reinforced concrete · Segmental lining · TBM · Utility tunnel · Tunnel segment

1 Introduction

Fibre reinforcement for tunnel segments is a relatively young technology. However, recent publications such as the ITAtech "Guidance for precast FRC segments – Volume 1: Design aspects" [1], or the British PAS 8810 "Tunnel design – Design of concrete segmental tunnel linings – Code of practice" [2] have now given more credibility to this reinforcement type and the basis for design. Over the last 20 years, fibre reinforced concrete segmental tunnel linings have been adopted in numerous projects around the world. A comprehensive overview of these projects is given in the ITA WG2 report [3].

Macro synthetic fibre reinforced shotcrete (MSFRS) has reached maturity as an engineered material and is widely used in all forms of sprayed tunnel linings, for temporary as well as for permanent ground support in both, mining and civil tunnel applications [4]. However, the application in precast tunnel segments is relatively new. MSF has distinct advantages over steel reinforcements as it is non-corrosive and doesn't suffer matrix embrittlement and the inherent performance loss with age [5, 6]. Thus, making it an ideal reinforcement for segmental linings, especially in critical environments. In conjunction with the reduced embodied carbon footprint these advantages make MSF an

© RILEM 2022
P. Serna et al. (Eds.): BEFIB 2021, RILEM Bookseries 36, pp. 780–795, 2022.
https://doi.org/10.1007/978-3-030-83719-8_67

and sustainable alternative to steel reinforcements, attaining growing interest amongatt-
design engineers, contractors, and project owners.   ra-

There are no special design standards or guidelines for macro synthetic fibrec-
reinforced concrete (MSFRC). The regular design principles and methodologies fortiv-
steel fibre reinforced concrete (SFRC) apply, such as the *fib* Model Code 2010 [7].e,
Macro synthetic fibre can be used when the project specifications and performancein-
criteria are met: "if concrete reinforced with this fibrous synthetic reinforcementno-
exhibits adequate post-cracking residual strengths then it can be considered suitable forva-
structural purposes as well as SFRC" [3]. In the past, the absence of global standardstiv-
for steel fibres hasn't come in the way of their acceptance and use in various appli-e,
cations. Continuous developments and successfully completed projects show that the
same currently applies to high performance macro synthetic fibres.

Continuous R&D and successful testing along with successful field experience is
key to the acceptance of a relatively new construction material. The reinforcement used
in precast tunnel segments is mainly required to provide them with sufficient robustness
to withstand the loads occurring during transient conditions such as handling, stacking
and storage, as well as those observed during construction, mainly handling, installation and TBM thrust forces. Supporting the investigations from the industry, various
academic studies have been conducted to validate the use of MSFRC in precast concrete tunnel segments when subject to the typical loads expected from the rams of a
tunnel boring machine (TBM), and drive its adoption [8–11].

2 Segment Research Project at the University of Brescia

2.1 Introduction

Comprehensive research had been carried out at the University of Brescia, Italy, on full
scale tunnel segments reinforced with BarChip MSF [12, 13]. A digest with the main
findings will be presented here.

The experimental campaign was conducted to assess the viability of MSF reinforcement for use in precast tunnel segments. The campaign focused on the temporary
load stages and was conducted in three stages:

1) Material characterization of MSFRC in a base concrete mix with a target mean
   compressive strength of 65 MPa on cubes at dose rates of 4, 6, 8 and 10 kg/m^3, to
   identify the optimal fibre dose rate for the next two stages of full-scale segment
   testing:
2) Full scale tunnel segments were subjected to flexural testing, replicating the load
   condition observed during demoulding, storage, and handling (also using a 3-point
   bending setup) and
3) Full scale tunnel segments were exposed to in-plane point loads, replicating the
   highly concentrated loads exerted by the TBM rams during propulsion.

The second two stages considered three different reinforcement configurations, with a
total of 10 full scale tunnel segments being cast across the two stages, i.e. conventionally
reinforced (RC segments), MSFRC and a low percentage of optimised reinforcement
(Hybrid segments), and macro synthetic fibres only (MSFRC segments).

2.2   Material Characterization

All four mixes show a stable post-cracking response characterised by an initial soft-ening behaviour, followed by a hardening branch with a significant increase in residual flexural strength out to 3.5 mm CMOD. Utilizing higher fibre dosages reduces the post-peak drop of the curve and enhances the recovery observed in the hardening branch. It should also be noted that for all dose rates except 4 kg/m^3, the minimum requirements by fib MC 2010 [7] for FRC were met to be considered adequate for structural pur-poses. The residual strength properties were assessed in accordance with EN 14651 [14] and presented in Table 1.

Table 1. Mean mechanical properties of MSFRC mixes (CoV in brackets below) [12]

Mix	$f_{c,cube}$ (MPa)	f_{cm} (MPa)	f_{ct} (MPa)	fL (MPa)	$f_{R,1}$ (MPa)	$f_{R,2}$ (MPa)	$f_{R,3}$ (MPa)	$f_{R,4}$ (MPa)	FRC Class
MSFRC4	65.9 (0.03)	57.3 (0.04)	3.68 (0.13)	5.52 (0.11)	2.07 (0.16)	2.54 (0.19)	2.76 (0.19)	2.48 (0.19)	1.5d
MSFRC6	65.8 (0.06)	54.6 (0.03)	3.81 (0.27)	5.61 (0.07)	2.92 (0.15)	3.75 (0.16)	4.10 (0.17)	3.75 (0.17)	2e
MSFRC8	66.9 (0.06)	59.5 (0.02)	4.61 (0.39)	5.92 (0.11)	3.41 (0.11)	4.45 (0.11)	4.94 (0.10)	4.47 (0.09)	2.5e
MSFRC10	62.6 (0.03)	52.0 (0.07)	4.29 (0.18)	5.82 (0.06)	3.74 (0.10)	4.97 (0.12)	5.53 (0.12)	5.26 (0.12)	3e

2.3   Full Scale Segment Testing

Ten full-scale tunnel segments were cast in total, comprising five segments to be tested in flexure and five segments to be tested under point load conditions to simulate the jacking forces exerted by the TBM rams during propulsion. The moulds for the seg-ments used in this experimental program have been adopted from the segmental lining of the Scilla tunnel in Sicily, Italy. The segmentation as well as the chosen segment

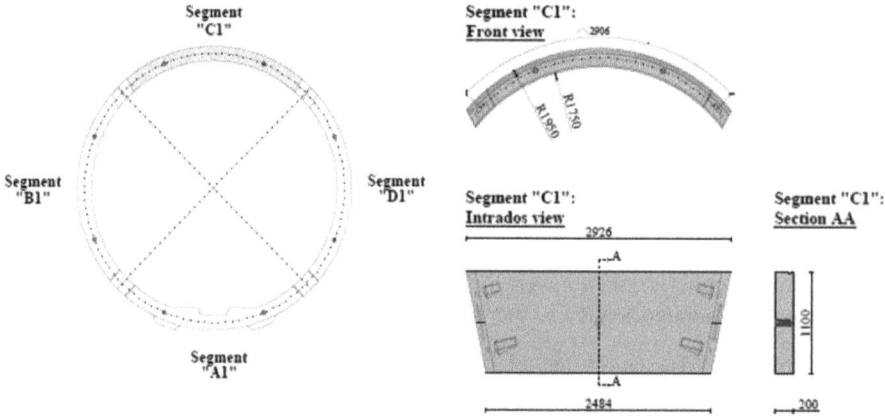

Fig. 1. Scilla precast segmental lining: Segmentation (left) and adopted segment C1 (right) [12]

dimensions, are depicted in Fig. 1.

Based on the results obtained in the material characterization phase, a fibre dosage of 8 kg/m^3 was chosen, yielding an FRC class 2.5e according to MC 2010, and residual strength only slightly lower than the 10 kg/m^3 mix. More significantly, the residual strength exhibited by the MSFRC8 mix satisfies the structural requirements of MC 2010, so that the fibres are used to satisfy the flexural requirements, as minimum shear reinforcement as well as both splitting and spalling reinforcement.

Three different reinforcement configurations were considered; macro-synthetic fibres only (MSFRC segments), a combination of conventional steel reinforcement and macro synthetic fibres designed to provide an optimized solution (Hybrid segments) and a reference segment using a full conventional rebar cage (RC segments). For each of the two testing configurations, two fibre only and two hybrid segments were cast, while only one segment was cast as the reference solution. Each of these specimens is identified using the suffix 'B1' or 'B2' for the flexural tests, and 'P1' or P2' for the point load tests.

RC Reference Segment Reinforcement: The overall steel content was equal to 110 kg/m^3 and a clear concrete cover of 35 mm to the curved reinforcement was utilized, providing an effective depth of 161 mm. This configuration represents a longitudinal reinforcement ratio $\rho_s = 0.23\%$, a typical reinforcement solution adopted in practice.

Hybrid Segment Reinforcement: The Hybrid segments comprised 8 kg/m^3 of fibres and an optimized reinforcement concentrated in chords at the leading and trailing edge. The longitudinal reinforcement ratio ρ_s equals to 0.11%, which is smaller than the minimum amount of flexural reinforcement required by MC 2010 for RC members as similarly assumed in previous research [11]. The reinforcement cover and effective depth of the Hybrid segments were consistent with those adopted in the RC segments. The overall steel rebar content is 35 kg/m^3, which is significantly less (about 70%) than the traditional RC segment.

2.4   Flexural Performance of Full Scale Segments

Five full-scale segments were cast for testing in a flexural loading arrangement to evaluate their flexural behaviour, which is required for temporary construction phases (demoulding, storage, transportation, and positioning) and the in-service stage. A central flexural load was applied to the segment in a three-point bending configuration, including two full width rollers at either end providing continuous support at 1.60 m span (Fig. 2, left).

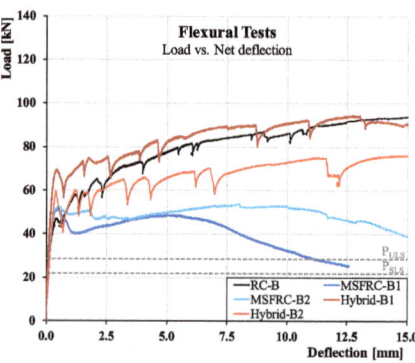

Fig. 2. Hybrid – B1 segment prior to flexural testing (left) and Experimental curves of load vs. mid-span net deflection for the segments tested in flexure up to 15 mm (right) [12]

The flexural response of the three different reinforcement solutions can be observed and compared in Fig. 2 (right), while also being assessed against the two design criteria for SLS and ULS from the original Scilla project, which are plotted as dotted lines. Note that all specimens, especially the fibre reinforced ones, exhibited significantly higher initial peak loads than the ultimate design load of P_{ULS} = 28.6 kN.

The reinforced segment RC-B showed a first cracking load (P_{cr}) at 35.1 kN, after which an increase in load is observed up to the flexural initial peak load (P_{peak} = 46.9 kN). A maximum bearing load of the segment was reached with a load of 105.6 kN, at a deflection of 36.6 mm.

The Hybrid segments exhibit a flexural response similar to the RC segment (which contains about three times the amount of rebars) up to a mid-span deflection of 15 mm, showing significant ductility after initial cracking. The hybrid solution exhibited a greater number of cracks, compared to the traditional RC solution, particularly at the edges of the segment in the areas of the lateral chord reinforcement. Multiple cracking enables the achievement of higher peak loads due to better crack control.

Both MSFRC segments exhibited a very similar load at first crack, which is approximately 30% higher than the RC segment. Both segments then present a further increase in load bearing capacity up to P_{peak}, due to sectional redistribution of stresses and the residual strength contribution of the fibres, which was 10% higher than the referring load for the RC segment.

It should also be noted that the ductility requirements for FRC only structures, according to MC 2010 Sect. 7.7.2 [7], were met with both of the MSFRC segments (i.e. $\delta_u > 20 \, \delta_{SLS}$).

2.5   Point Load Tests on Full Scale Segments

Point load tests were carried out on five segments with the three reinforcement solutions to simulate the jacking loads transferred by the thrust shoes during TBM propulsion, while monitoring the crack development of each segment at increasing load

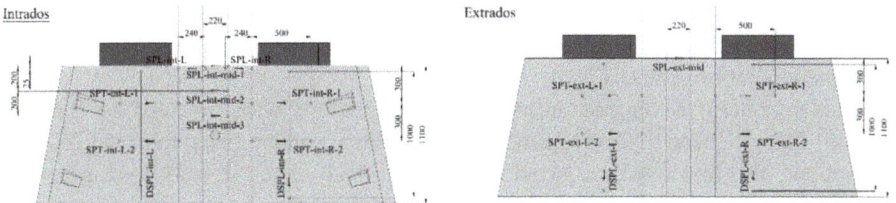

Fig. 3. Instrumentation details utilized in the point load tests (all dimensions in mm) [12]

levels. The two loading shoes herein had the same dimensions as the thrust shoes adopted in the original project (Fig. 3).

The point loads were applied through a self-supporting frame which consisted of a stiff concrete beam and two steel reaction frames. Based on the original construction conditions, three load levels were examined (applied to each loading shoe):

- Nominal service load (SL): 700 kN
- Exceptional thrust load: 1000 kN (1.43 × SL)
- Emergency thrust load: 1600 kN (2.28 × SL, equal to a contact pressure of 23.5 MPa)

These load levels were attained in three cycles under load control and then each time unloaded to assess the residual crack widths. The final maximum load level for each shoe applied in this test was equal to 2750 kN (3.93 × SL), just below the maximum capacity of the hydraulic jacks (3000 kN).

Figure 4 depicts the bearing capacity of the segments as load vs. vertical displacement below the left thrust shoe for all tested specimens. Unfortunately, the displacement of the RC sample could not be reported in Fig. 4 (right) due to faulty readings from the vertical transducer. Nevertheless, it is clear that all reinforcement solutions allowed the segments to withstand the maximum load level of 2750 kN for each loading shoe (3.93 × SL), which obviously does not represent the maximum bearing capacity of these segments.

Analysing the mean vertical displacements (between intrados and extrados) under both loading shoes, it can be observed that the maximum displacement for all tested

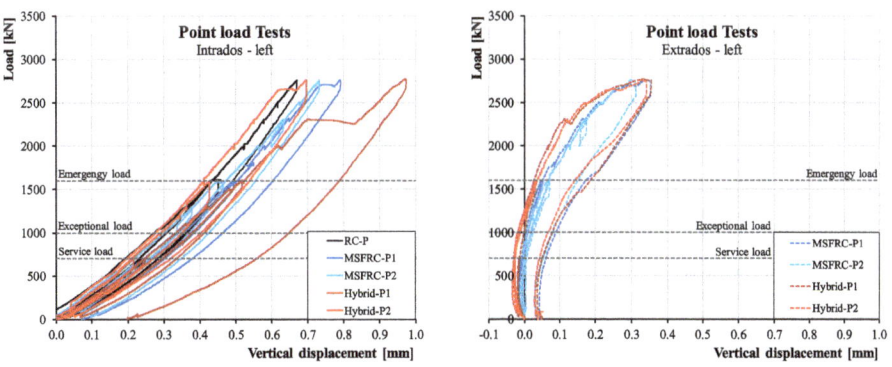

Fig. 4. Load vs. vertical displacement curves; intrados (left) and extrados (right)

specimens was between 0.52–0.66 mm (left shoe) and 0.43–0.54 mm (right shoe). As expected, the specimens demonstrated comparable axial stiffness, independently of the reinforcement solution adopted.

The crack patterns developed during the point load tests were obtained by visual inspection. The RC and hybrid segments presented a similar behaviour; for both configurations, the first crack occurred between the load shoes at a load of around 1000–1100 kN. Progressively increasing the load, splitting cracks appeared in the area beneath the loading shoes, at a load of between 1200 and 1300 kN.

Considering the fibre only segments, only one single spalling crack appears between the two loading plates; however, this crack did not occur until a load of between 1540 and 1750 kN. This equates to a substantially higher load than that observed for the RC and hybrid segments. The crack pattern of the splitting tangential cracks is similar to those experienced by the RC and hybrid segments, however again,

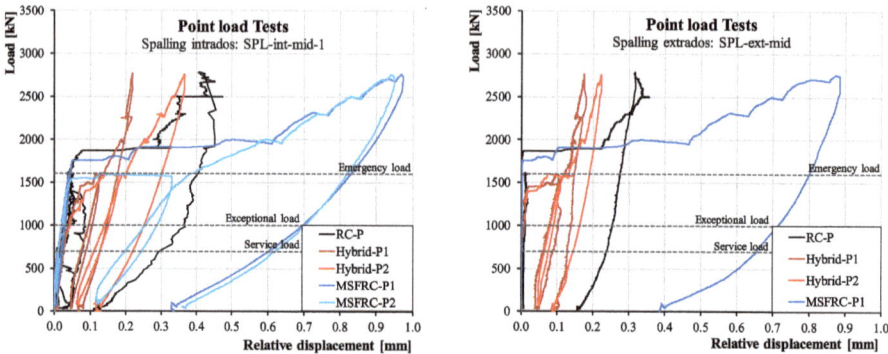

Fig. 5. Load vs relative displacement curves of spalling phenomenon in the mid zone; intrados (SPL-int-mid-1 transducer, left) and extrados (SPL-ext-mid transducer, right). NB: The spalling crack of MSFRC-P2 developed at the edge of the transducer and was hence not detected.

the load required to induce the first tangential splitting crack is much higher than for the traditional and hybrid reinforcement solutions at 1600–1865 kN. A similar trend was observed for the spalling crack phenomena.

Analysing the spalling behaviour, all segments, including the MSFRC segments, presented a satisfactory performance, i.e. very small relative displacement of less than 0.1 mm up to 1500 kN (2.14 × SL), though it should be underlined that not all segments were cracked at this load level. Figure 5 portrays the load vs. spalling relative displacement of the five tested segments. Considering the spalling crack development, for all load levels, the hybrid segments demonstrated equal or lower relative displacements (crack widths) than the reference RC segment. The mutual collaboration of the local longitudinal bars with fibres in the cracked region between the shoes is particularly effective at controlling spalling cracking, as proven by the hybrid segments. It is observed that, for very high loads (>2 × SL) the MSFRC segments show a higher relative displacement in the spalling mid-zones, however, at a load level equal or above the emergency thrust load level.

2.6 Conclusions of the Research Program

- The concrete mix design adopted in this research can be considered adequate for structural purposes in tunnel segments according to fib Model Code 2010 with a minimum of 6 kg/m^3 of BarChip 48 macro synthetic fibres. The fibres can be successfully used to substitute the minimum amount of shear reinforcement and are satisfactory for withstanding the local tensile stresses in the splitting zones.
- MSFRC segments with 8 kg/m^3 of fibres display a higher flexural cracking load than the RC segments utilizing a steel bar content of 110 kg/m^3. Both MSFRC segments were satisfactory regarding the service and ultimate loads prescribed by the designers of the original project. No cracks were present at the ultimate design load.
- The combination of 8 kg/m^3 of fibres and a very low amount of traditional steel reinforcement (35 kg/m^3) in the Hybrid segments, provided a bearing capacity remarkably similar to the RC segments. It must be highlighted that the Hybrid segments contain 70% less conventional steel reinforcement, hence significantly reducing labour time and costs when compared to the RC segment.
- Considering the point load tests, all reinforcement solutions adopted allowed the segments to reach the maximum capacity of the laboratory loading system (i.e. 2750 kN for each loading shoe), which corresponds to 3.93 times the service load, and is also significantly higher (1.72 times) than the maximum design values of the TBM thrust shoes adopted on the original project (1600 kN).
- RC and Hybrid segments displayed multiple spalling cracks in the region between the loading shoes, while the MSFRC segments only exhibited a single crack between the two loading plates.
- In the Hybrid segments, the synergy between local longitudinal reinforcement and the fibres in the cracked region between the loading shoes is especially effective in controlling spalling cracking, since for all load levels investigated the Hybrid segments have shown equal or lower relative displacements than the reference RC segment. This is also true when analysing the spalling crack widths with the digital microscope, where the hybrid solutions obtain a crack width which is 40% smaller (up to 1600 kN) than the reference RC segment.
- The MSFRC segments, which demonstrated a significantly higher first crack load, presented similar residual spalling crack openings when compared with the RC segment, as well as no significant differences in terms of the residual splitting crack widths.
- In the Hybrid segments, the residual splitting crack at the end of the second load cycle (1600 kN) is equal to that observed in the RC segment. Contrarily, the spalling cracks at the end of the second load cycle demonstrated how the combination of bars and macro synthetic fibres allow the hybrid solution to guarantee a lower residual crack width than the traditional RC solution.

3   Project References

3.1   Harefield to Southall Gas Transfer Tunnel, UK (2008)

This project for National Grid comprised the design and construction of an 18 km long and 1220 mm diameter gas transfer pipeline, and connection into the existing network within existing installations at both Harefield and Southall in Middlesex. Construction was in a substantially urban environment requiring over 2.3 km to be constructed in four tunnels to ensure infrastructure was not disturbed. The pipeline became operational in 2009.

Ring layout:
- Internal diameter 2.59 m
- Thickness 180 mm
- Ring length 1.00 m
- Segmentation: 7 + 1
- Ground conditions: London blue clay
- Concrete specifications:
  Concrete class C45/55
  Flexural strength 5.0 MPa
  Residual strength 2.4 MPa
- Reinforcement: 7 kg/m³ BarChip Shogun macro synthetic fibres

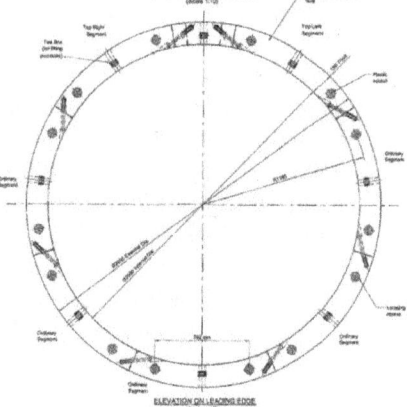

Fig. 6. Ring layout and project specifications (Harefield to Southall tunnel)

One of the tunnels, approx. 1.0 km long with 2.59 m internal diameter was driven by a TBM and segmentally lined using fibre reinforced precast concrete segments. The tunnel, 21 m below the surface, carries the 1.22 m diameter high pressure gas pipeline. The tunnel lining was a wedge block design, see Fig. 6. This was the first ever project

Fig. 7. Segment installation (left) and view into the tunnel (right)

using MSF to replace steel reinforcement in precast concrete tunnel segments (see also project reference in [1]) (Fig. 7).

The use of BarChip fibres in the precast segments proved very effective in meeting all the design requirements as well as ensuring the required robustness for transient load stages such as the mechanical erection and the jacking forces during TBM propulsion. The segments have performed dependable and no reject was reported.

Using MSF has further lowered the overall carbon footprint of the project, negated any corrosion either of the internal steel gas pipe through cathodic protection or of the segment reinforcement itself, and also provided significant manufacturing efficiencies, including reduced labour costs and faster production and as such offered cost savings to the project.

3.2   General Interceptor Collector Santoña-Laredo-Colindres, Spain (2016)

The Santoña–Laredo General Interceptor Collector is a 1.5 km long × 3.50 m ID subsea tunnel in northern Spain. The tunnel is part of the Santoña Marshlands Sanitation Project. It was constructed with a 4.30 m Mixshield TBM. Due to the exposure to corrosive media, such as sewer and subsea conditions, the tunnel lining has high requirements on robustness and durability. The project presented numerous technical challenges, such as umbilical assembly of the TBM at the bottom of a deep shaft and

Ring layout:

- Internal diameter 3.50 m
- Thickness 250 mm
- Ring length 1.20 m
- Segmentation: 5 + 1
- Ground conditions: limestone, sands, clay pockets
- Concrete specifications:

  Concrete class C45/55
  Flexural strength $f_{ctk,fl}$ = 3.60 MPa
  Residual strength f_{R1k} = f_{R3k} = 1.50 MPa

- Reinforcement: 5 kg/m³ BarChip 48 macro synthetic fibres + 16 kg/m³ bursting ladders

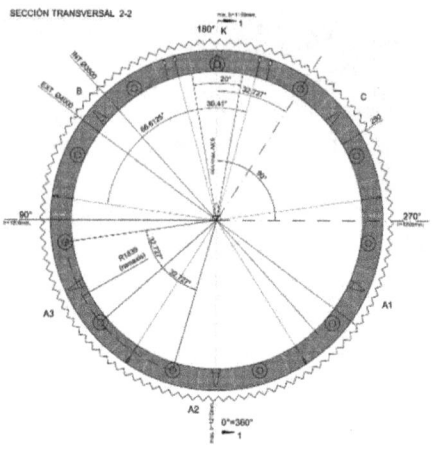

Fig. 8. Ring layout and project specifications (Santoña-Laredo tunnel) [17]

further space constraints on the surface due to the worksite location being in an urban area [15].

The review of the manufacturing process and the marine exposure of the tunnel segments led to the decision to replace conventional rebar cages with macro synthetic fibre. The structural design of the FRC segments is based on the Spanish EHE-08 [16] and the fib Model Code 2010 [7]. To better accommodate the fibre reinforcement solution, smaller segments were adopted to limit the segments' aspect ratio below ten for increased robustness in the various loading stages [17], see Fig. 8.

The Finite Element Analysis yielded maximum spalling stresses during TBM propulsion requiring additional reinforcement at the leading ring joints. This was provided as bursting ladders, consisting of 3 nos. dia. 8 mm connected by a dia. 6 mm continuous stagger bar.

Due to the limited space available in the launching shaft it was necessary to design a temporary reaction structure consisting of half rings assembled and jointed together for the umbilical launch of the TBM (Fig. 9). These half rings were fixed and supported on the reaction concrete block by grouting as the TBM progress required. In addition, the half ring structure was also used as a support for the gantries during the different stages of the backup assembly. Because of the difficulty of driving the TBM under these conditions, the initial advance ratio was very low with an average of 1.0 m per day since every advance cycle required the installation of a new half ring and fixing

Fig. 9. "Half ring" segment structure employed to push the Mixshield during the start of excavation (left) and hoses along the tunnel before positioning of gantries 5, 6 and 7 (right)

operations. Also, eccentric thrust conditions with increased forces needed to be coped with [15].

Switching to BarChip fibre reinforcement eliminated more than 80% of steel reinforcement. The remaining bursting ladders are solely for jacking forces along the leading ring joint. Eliminating the traditional rebar cage and its inherent labour reduced the production cycle times by nearly 50%. A cost assessment including segment manufacture and reduced repair or reject rate, due to significantly improved robustness of the FRC segments, revealed a total cost saving of nearly 40% for the rings, compared to the traditional rebar cage design [17].

The gained experience in these unusual operations of umbilical TBM launch was important to complete the know-how and will help minimizing the difficulties in future projects during the first stages of boring. Launching TBM's in shafts with lack of space is a possible practice, which requires reduced advance ratios, but provides a reduction of shaft structure costs. A complex vertical assembly of the TBM is possible, but it can be a highly demanding process requiring skill, manpower and a dependable robustness of the tunnel segments.

3.3   Blacksnake Creek Stormwater Runoff Tunnel, USA (2019)

The first use of a macro synthetic fibre reinforced segmental lining in North America was installed in the Blacksnake Creek stormwater runoff tunnel in St Joseph, Missouri. The 2 km long × 2.74 m ID segmentally lined tunnel has been driven using a 3.425 m diameter EPB TBM. The tunnel is part of the Blacksnake Creek Storm Water

Ring layout:
- Internal diameter 2.743 m (108 in.)
- Thickness 190.5 mm (7.5 in.)
- Ring length 1.219 m (48 in.)
- Segmentation: 6 trapezoidal segments
- Ground conditions: predominantly shale & claystone
- Concrete specifications:
  Concrete class C45/55 (f'_c = 48 MPa)
  Residual strength f^D_{600} = f^D_{150} = 3.20 MPa
- Reinforcement: 7 kg/m³ BarChip 54 macro synthetic fibres

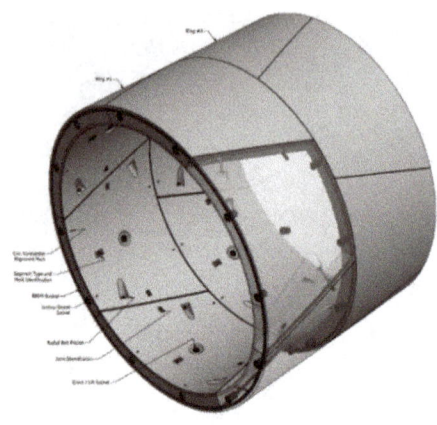

Fig. 10. Ring layout and project specifications (Blacksnake Creek tunnel) [18]

Fig. 11. Half-ring segment installation for TBM propulsion (left) and completed rings (right)

Separation Improvement Project of the City of St. Joseph, Missouri. Technical challenges comprised an umbilical TBM launch in a shaft of limited size, initially with a smaller temporary screw conveyor, and the employment of tunnel segments forming a crescent as a launch cradle, which caused eccentricity of the jacking forces [18].

The segmentation of the universal rings was selected to be six trapezoidal segments (Fig. 10), each having a developed centre arc length of 1.536 m. This yields a segment aspect ratio of 8.1, which is well below the acknowledged limit of 10 to ensure segment robustness for temporary load cases [1].

The launch of the EPB TBM, which had a total length of 78 m, required the construction of a 15 m diameter secant pile shaft. A special electrical and hydraulic umbilical assembly was designed to help launch the TBM. The special umbilicals were used to link the gantries on the surface to the main TBM body in the launching cradle on the bottom of the shaft.

After the TBM was launched and in the ground the short launch screw conveyor was removed and the longer main screw conveyor was installed. Further, the tail shield and the segment transporter belly pan were installed. During this stage, a precast concrete segmental tunnel liner "half-ring" structure was used to push and advance the TBM further into the excavation (Fig. 11 left).

The initial tunnel drive installation has proven that the fibre reinforced segments are robust enough to withstand the temporary load cases, i.e. transportation, hoisting, handling, and installation, without experiencing damage to the segments. In addition, the half-ring segment sections used in the shaft for the launching process (Fig. 11 left) were inspected after the initial push and no damage occurred to the segments.

The biggest challenge for the TBM was controlling steering throughout the alignment in all three distinct geologies due to machine difficulties. The TBM had to be used in exceptionally high pressure mode to overcome steering complications at various times. During these times the fibre reinforced segments remained robust and generally didn't step in, crack, or become damaged due to the high pressures. During recent tunnel inspections though, a few segments with fine longitudinal cracks were observed, which were likely due to the machine pushing off the segments in that period. However, the fibres were able to control developing crack widths to remain within the specified limits of the segmental lining design i.e. 0.008 in or 0.20 mm respectively. No segment had to be rejected.

3.4   Conclusions from the Reference Projects

These successfully completed projects with various challenges, including umbilical TBM launch in confined shafts, TBM steering difficulties or being a World's or America's First, confirm that a strong partnership between all parties involved in these projects yields an innovative, technically and economically leading solution, delivering a quality project to the owner.

Ongoing research and continuous developments on macro synthetic fibres and MSF reinforced concrete have made it today being a modern and cost-efficient construction material. Eliminating durability issues regarding corrosion of the primary reinforcement yields significant advantages for the design being no longer governed by serviceability limit states.

The use of tunnel segments reinforced with BarChip MSF has demonstrated robust, durable, and dependable performance in tunnel projects even under very difficult conditions during the TBM launch phase. The performance and reliability of the MSF reinforced segmental linings has outperformed expectations in these projects.

The presented reference projects build further confidence in MSF reinforced segmental linings for use in future projects. These types of utility tunnelling projects, e.g.

sewage and irrigation, CSO, power, or gas transfer tunnels, are widely existing in the world market and they present a huge opportunity for MSF reinforced concrete linings, benefiting from the proven advantages.

References

1. ITAtech: ITAtech Guidance for Precast Fibre Reinforced Concrete Segments – Vol. 1 Design Aspects. ITAtech Activity Group Support, ITAtech Report No. 7, April (2016). www.ita-aites.org
2. PAS 8810: Tunnel Design – Design of Concrete Segmental Tunnel Linings – Code of Practice. The British Standards Institution, BSI Standards Ltd. (2016)
3. ITA WG2: Twenty Years of FRC Tunnel Segments Practice: Lessons Learnt and Proposed Design Procedure. Report from ITA/AITES Working Group 2 Research (2015). www.ita-aites.org
4. Nitschke, A., Winterberg, R.: Performance of macro synthetic fiber reinforced tunnel linings. In: Proceedings of the World Tunnel Congress 2016, San Francisco, 22–28 April (2016)
5. Bernard, E.S.: Age-dependent changes in post-cracking performance of fibre reinforced concrete for tunnel segments. In: Proceedings of the 15th Australian Tunnelling Conference 2014, Sydney, 17–19 September (2014)
6. Bernard, E.S.: Changes in long-term performance of fibre reinforced shotcrete due to corrosion and embrittlement. Tunn. Undergr. Space Technol. **98**(2020), 103335 (2020). https://doi.org/10.1016/j.tust.2020.103335
7. fib Model Code: Concrete Structures. International Federation for Structural Concrete, Lausanne, Switzerland (2010)
8. Tiberti, G., Conforti, A., Plizzari, G.A.: Precast segments under TBM hydraulic jacks: experimental investigation on the local splitting behavior. Tunn. Undergr. Space Technol. **50**, 438–450 (2015)
9. Di Prisco, M., Tomba, S., Bonalumu, O., Meda, A.: On the use of macro synthetic fibers in precast tunnel segments. In: Proceedings of the International Conference Fiber Concrete 2015 (FC2015), Prague, Czech Republic (2015)
10. Conforti, A., Tiberti, G., Plizzari, G.A., Caratelli, A., Meda, A.: Precast tunnel segments reinforced by macro-synthetic fibers. Tunn. Undergr. Space Technol. **63**, 1–11 (2017)
11. Conforti, A., Trabucchi, I., Tiberti, G., Plizzari, G.A., Caratelli, A., Meda, A.: Precast tunnel segments for metro tunnel lining: A hybrid reinforcement solution using macro-synthetic fibers. Eng. Struct. **199**, 109628 (2019)
12. Plizzari, G.A., Tiberti, G., Conforti, A., Trabucchi, I., Mudadu, A.: Experimental study on the use of macro-synthetic fibers in precast tunnel segments. Technical Report, University of Brescia, Italy, June (2020)
13. Trabucchi, I., Mudadu, A., Tiberti, G., Conforti, A., Plizzari, G.A., Winterberg, R.: Structural behavior of precast tunnel segments reinforced by macro-synthetic fibers during temporary loading phases. In: Proceedings BEFIB 2021 RILEM-fib X International Symposium on Fibre Reinforced Concrete, Valencia, Spain, 20–22 September (2021)
14. EN 14651: Test method for metallic fiber concrete - Measuring the flexural tensile strength (limit of proportionality (LOP), residual). European Committee for Standardisation (CEN) (2005)
15. Winterberg, R., Justa Cámara, R., Sualdea Abad, D.: Santoña–Laredo general interceptor collector – challenges and solutions. In: Proceedings of the FRC 2018, Fibre Reinforced Concrete: From Design to Structural Applications, Desenzano, Italy, 28–30 June (2018)

16. EHE-08: Instrucción de Hormigón Estructural. Anejo 14 "Hormigón con fibras", Spanish Concrete Design Code (2008)
17. Winterberg, R., Mey Rodríguez, L., Justa Cámara, R., Sualdea Abad, D.: Segmental lining design using macro synthetic fibre reinforcement. In: Proceedings of the FRC 2018, Fibre Reinforced Concrete: From Design to Structural Applications, Desenzano, Italy, 28–30 June (2018)
18. Winterberg, R., Garbeth, M.R., Glynn, B.: Innovation in durable segments for CSO tunnels. In: Serna, P., Llano-Torre, A., Martí-Vargas, J.R., Navarro-Gregori, J. (eds.): BEFIB 2020. RB, vol. 30. Springer, Cham (2021). https://doi.org/10.1007/978-3-030-58482-5_67

Design and Verification of Elevated Slabs Made with Hybrid Reinforced Concrete: Case Studies

Luca Facconi[✉], Fausto Minelli, and Giovanni Plizzari

Department of Civil, Environmental, Architectural Engineering and
Mathematics, University of Brescia, Brescia, Italy
luca.facconi@unibs.it

Abstract. The literature has recently reported many real applications with the
use of Fiber Reinforced Concrete (FRC) for the construction of elevated slabs.
Most of these case studies suggests the use of steel fibers as the only reinforcement in addition to the Anti-Progressive Collapse (APC) used to prevent
brittle failure mechanisms. However, this reinforcement arrangement does not
usually represent an optimized design solution as high amounts of fibers
($V_f > 70$ kg/m^3) are often required for reaching the minimum capacity to resist
design loads as well as to the internal actions due to restrained shrinkage.

This paper focuses on the design of FRC elevated slabs by using the most
recent design provisions reported in the fib Model Code 2010. A simplified
design procedure based on a consolidated design practice is described. The use
of a properly designed combination of fibers and conventional reinforcement,
referred to as Hybrid Reinforcement, is proposed. Different case studies taken
from the literature are considered to prove the effectiveness of the proposed
design approach.

Keywords: Fiber Reinforced Concrete · Elevated slab · Hybrid
reinforcement · Top reinforcement · Shrinkage · Design · Verification

1 Introduction

Because of its good mechanical performances in tension, Fiber Reinforced Concrete
(FRC) is a material that may be fruitfully used for structures, such as elevated slabs,
which are characterized by a high internal redundancy. Most of the on-site applications
well described in the literature (Destrée and Mandl, 2008 [1]; Gossla, 2006 [2]; Parmentier et al., 2014 [3]) as well as recent code recommendations (ACI 544.6R-15 [4])
suggest that the use of steel fibers as the only flexural reinforcement represents the most
effective design solution to replace conventional reinforcement in elevated slabs.
Therefore, in order to fulfil the serviceability and safety requirements recommended by
structural codes, slabs (with fibers only) must be reinforced with very high contents of
steel fibers (e.g., $V_f > 70$ kg/m^3) in addition to the Anti-Progressive Collapse
(APC) reinforcement located along the column alignments. A previous work authored
by Facconi et al. (2019) [5] proved that the use of steel fibers in combination with
conventional reinforcement (i.e., Hybrid Reinforcement) may optimize structural

© RILEM 2022
P. Serna et al. (Eds.): BEFIB 2021, RILEM Bookseries 36, pp. 796–808, 2022.
https://doi.org/10.1007/978-3-030-83719-8_68

performance without using very high contents of fibers. Based on the design requirements reported by fib Model Code 2010 (MC2010) [6], a procedure was proposed to perform preliminary design and verification of the slabs containing Hybrid Reinforcement.

In the present paper, the procedure for designing FRC slabs discussed in a previous work [5] will be improved and then used to design the Hybrid Reinforcement of two different elevated slabs reported by literature. Particular attention will be devoted to investigate the influence of top reinforcement on slab behavior at both service and ultimate limit states. Finally, the influence of shrinkage on the structural response will be also investigated.

2 Proposed Design Procedure

The design approach described in the following was presented and validated in a previous work [5]. The method was subdivided in two main stages, namely "Preliminary design" and "Verification" stage. The main novelty here proposed is represented by the approach adopted to determine the slab thickness in preliminary design (see point 2). It is worth remarking that the present procedure allows performing the flexural design and verification of the structure, whereas other mechanisms, such as punching, can be verified separately by using the approaches suggested by literature a well as by structural codes.

2.1 Preliminary Design

Before performing structure verification, designers must carry out preliminary trial design in order to find the structure properties that ensure the achievement of the selected performance targets. The proposed preliminary design procedure is based on the linear elastic analysis commonly used by designers. Unlike the procedure originally proposed, an improved method for dimensioning the slab thickness (i.e., first step of the procedure) is here reported. In more detail, the preliminary stage consists of the following main steps:

1. Choice of mechanical properties of materials (FRC and reinforcing steel).
2. Choice of the slab thickness (t). A preliminary simple calculation of the thickness can be done considering the moment (i.e., $m_{E,SLS}$ resulting from moment combination of bending and torsional moments) acting in the area of the structure exhibiting the maximum deflection. The latter results from a linear elastic analysis implementing the load combination suggested by Eurocode 0 [7] for Serviceability Limit State (SLS) conditions, which reads as follows:

$$E_{SLS} = \begin{cases} \sum_{j \geq 1} G_{k,j} + P + \psi_{1,1} \cdot Q_{k,1} + \sum_{i > 1} \psi_{2,i} \cdot Q_{k,i} & \text{(frequent combination)} \\ \sum_{j \geq 1} G_{k,j} + P + \sum_{i > 1} \psi_{2,i} \cdot Q_{k,i} & \text{(quasi - permanent combination)} \end{cases}$$

$$(1)$$

where G, P and Q represent permanent, pre-stressing and variable actions, respectively; ψ is the factor for the combination of variable actions. Fig. 1a reports a simplified distribution of bending stresses acting over a cross-section at SLS. Concrete in compression is assumed linear elastic so that the maximum compressive stress (i.e., $\varepsilon_c \cdot E_c$) depends on the Young's modulus of concrete (E_c) and on the maximum compressive strain (ε_c). The distribution of tensile stresses follows the linear stress-crack width model proposed by MC2010 [6] for FRC. Thus, if the post-cracking residual strengths f_{R1} and f_{R3} are both known from Three Point Bending tests on notched prisms according EN14651 [8], the characteristic tensile strength $f_{t(w),k}$ results from the following equation:

$$f_{t(w),k} = f_{ftu,k} + \left(f_{Ftu,k} - f_{Fts,k}\right) \cdot \frac{w-2.5}{2} \tag{2}$$

where w is the crack width (in mm); $f_{Fts\ k} = 0.45f_{R1k}$; $f_{Ftu\ k} = 0.5f_{R3k} - 0.2f_{R1k}$; f_{R1k} and f_{R3k} are the characteristic residual strengths corresponding to Crack Mouth Opening Displacement (CMOD) values of 0.5 mm and 2.5 mm, respectively. Crack control is particularly critical in those sections of the slab not containing conventional reinforcement. Therefore, if one considers the schematic of Fig. 1a, the following cubic polynomial is obtained from equilibrium:

$$x^3 - x^2 \cdot \left(3t - \varepsilon_{cr} \cdot w_{max} - \frac{E_c \cdot w_{max}}{2 \cdot f_{t(0.5w_{max}),k}}\right) + x \cdot 3t^2 - t^3 = 0 \tag{3}$$

where ε_{cr} is the cracking strain of concrete, w_{max} is the maximum crack width depending on the exposure environmental class and x is the distance of the neutral axis from the top edge of the cross section. Note that the tensile strength $f_{t(0.5w_{max}),k}$ is obtained from Eq. (2) by assuming a crack width w equal to $0.5 \cdot w_{max}$. As the calculation of x from the Eq. (3) is not always straightforward, the following approximate solution based on the second-degree Tylor polynomials is here proposed:

$$\left(x - \frac{t}{2}\right)^2 \cdot (\alpha - 3t) - t \cdot \left(x - \frac{t}{2}\right) \cdot \left(\alpha + \frac{3}{2}t\right) + \frac{t^2}{4} \cdot (\alpha - t) = 0 \tag{4}$$

where $\alpha = w_{max} \cdot (2 \cdot \varepsilon_{cr} + E_c/f_{t(0.5w_{max}),k})$. In spite of its simpler formulation, Eq. (4) generally provides values of x that differ less than 5% as compared to those resulting from Eq. (3). Once the position of the neutral axis is known, the resisting moment can be calculated as follows:

$$m_{RSLS,FRC} = \frac{1}{2} \cdot f_{t(0.5w_{max}),k} \cdot (t - x - k) \cdot \left(t + k + \frac{x}{3}\right) \geq m_{E,SLS} \tag{5}$$

where $k = (t-x)^2 \cdot \varepsilon_{cr}/w_{max}$. The slab thickness must be chosen so that the resisting moment $m_{RSLS,FRC}$ is not lower than the total moment ($m_{E,SLS}$) acting in the slab under service loading conditions. Typical values of the resisting moment for different FRC classes are shown in the diagram of Fig. 2. The diagram may be used as a design chart able to provide the minimum slab thickness corresponding to the required maximum crack width and bending moment acting at the SLS.

Beside the previous requirement related to the maximum crack width, the structure deformability must also be also checked. To this end, the slab slenderness (t/L = thickness/span length) must fall within the range $1/35 \leq t/L \leq 1/25$.

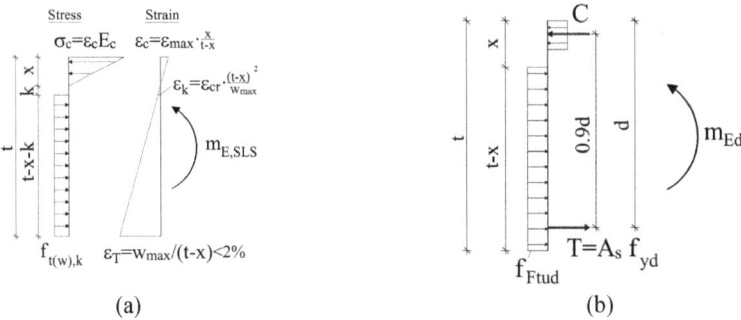

Fig. 1. Simplified bending stress distribution acting over a section at SLS (a) and ULS (b).

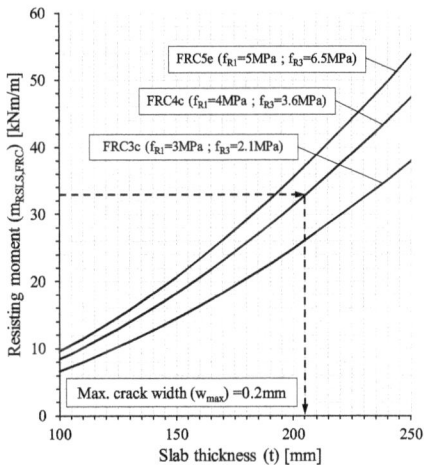

Fig. 2. Resisting moment of the FRC slab cross section (no conventional reinforcement – $w_{max} = 0.2$ mm) at SLS.

3. Determination of the design loads (E_d). As required by the Eurocode 0 [7], the design loads (ULS) have to be combined according to the following relationship:

$$E_d = \sum_{j \geq 1} \gamma_{G,j} \cdot G_{k,j} + \gamma_P \cdot P + \gamma_{Q,1} \cdot Q_{k,1} + \sum_{i > 1} \gamma_{Q,i} \cdot \psi_{0,i} \cdot Q_{k,i} \qquad (6)$$

where the coefficients γ_G, γ_P, γ_Q are the partial factors for actions.

4. Determination of the internal actions through Linear Elastic Finite Element Analysis (LEFEA).
5. Design of conventional reinforcement combined with fibers. The maximum design bending moments ($m_{Ed,x}$, $m_{Ed,y}$) acting in the two orthogonal directions (x and y) can be evaluated as follows:

$$m_{Ed,x} = m_{d,x} \pm \left| m_{d,xy} \right| \; ; \; m_{Ed,y} = m_{d,y} \pm \left| m_{d,xy} \right| \qquad (7)$$

where $m_{d,x}$, $m_{d,y}$ and $m_{d,xy}$ are respectively the bending and the torsional design moments obtained from the LEFEA of the slab.

The simplified cross-section model shown in Fig. 1b allows to determine the design resisting moment due to FRC only ($m_{Rd,FRC}$). The model assumes a constant distribution of tensile stresses below the neutral axis depth (x), which is assumed equal to 0.1 times the thickness (t). Therefore:

$$m_{Rd,FRC} = \frac{1}{2} f_{Ftu,d} \cdot t \cdot (t - x) = 0.45 \cdot f_{Ftu,d} \cdot t^2 \qquad (8)$$

where $f_{Ftu,d} = f_{R3k}/(3 \times \gamma_c)$; $\gamma_c = 1.5$ is the partial safety factor for FRC according MC2010 [6].

Conventional reinforcement must be placed where the design moment (m_{Ed}) is higher than the resisting moment provided by FRC only ($m_{Rd,FRC}$). For instance, if one considers bending moments acting in the x-direction only, additional top and bottom rebars have to be added in some regions of the slab located at columns and midspan, respectively (Fig. 3).

The moment distribution provided by LEFEA may be considered to calculate the total amount of additional reinforcement. As shown in Fig. 3, which reports some of the envelope of design bending moments concerning the slab described at paragraph 3.2, the comparison between the resisting moment ($m_{Rd,FRC}$) and the design moments (m_{Ed}) provides the length L_{int}. Based on L_{int}, the following calculation can be carried out to determine the total area of conventional reinforcement in x- and y- direction:

$$A_{s,x} = \frac{\int_0^{L_{int}} \left(m_{Ed,x} - m_{Rd,FRC} \right) dx}{0.9 \cdot f_{yd} \cdot d} \; ; \; A_{s,y} = \frac{\int_0^{L_{int}} \left(m_{Ed,y} - m_{Rd,FRC} \right) dy}{0.9 \cdot f_{yd} \cdot d} \qquad (9)$$

where $A_{s,x}$ and $A_{s,y}$ are the total reinforcement areas in x- and y-direction (where required), respectively; d is the effective depth of the slab; $f_{yd} = f_{yk}/\gamma_s$, f_{yk} and $\gamma_s = 1.15$ are the design yield strength, the characteristic yield strength and the partial safety factor for steel rebars, respectively.

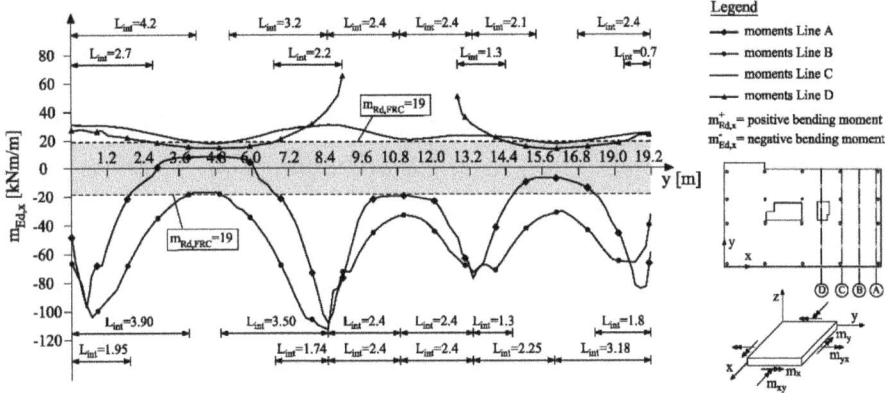

Fig. 3. Distribution of bending moments $m_{Ed,x}$ acting along some of the most critical sections of the FRC slab located in an office building (see paragraph 3.2 of this paper).

For the sake of example, Fig. 4 and Fig. 6 show the additional reinforcement layouts designed for the two case studies discussed below. As one may note, bottom rebars are located close to the column alignments whereas top rebars, which are required to resist negative moments, are placed over the columns. Note that the APC reinforcement must be added to the additional reinforcement designed to resist bending moments.

2.2   Verification

The procedure to verify the FRC slab consists of the following three main steps:

1. Perform Non-Linear Finite Element Analysis (NLFEA) of the slab to determine the global capacity (R_d). The simulation is performed by implementing the constitutive laws of FRC suggested by MC2010 [6] (clause 5.6.4). Safety verification can be performed by assuming the global safety format approach suggested by MC2010, as follows:

$$E_d \leq R_d = \frac{R_m}{\gamma_R^* \cdot \leq \gamma_{Rd}} \tag{10}$$

where R_d and R_m are the design and mean global resistance of the structure, respectively; γ_R^* is the global resistance safety factor; $\gamma_{Rd} = 1.06$ is the model uncertainty factor. According to the method for the estimation of the coefficient of variation of resistance (V_R), γ_R^* is related to V_R as follows:

$$\gamma_R^* = \exp(\alpha_R \cdot \beta \cdot V_R) \quad \text{with} \quad V_R = \frac{1}{1.65} \cdot \leq \ln\left(\frac{R_m}{R_k}\right) \tag{11}$$

where β = 3.8 (50 years) and $α_R$ = 0.8, which corresponds to a failure probability of about 10^{-4}. Note that R_m and R_k are determined by considering the mean and characteristic mechanical properties if materials, respectively.

2. Check the safety and serviceability minimum requirements for FRC structures according MC2010 (see clause 7.7).
3. Check the punching resistance according either MC2010 or Eurocode 2, in case the numerical model is not able to account for it.

3   Case Studies

3.1   Effect of Shrinkage on Design of a FRC Elevated Slab

The design procedure described above was successfully used in a previous study [5] to design the hybrid reinforcement of a slab having the same geometry of that tested by Destrée and Mandl [1]. The slab had three spans in both directions and was supported by 16 columns placed at a distance (axis-to-axis) of 6 m. The total thickness and the effective depth were assumed equal to 200 mm and 170 mm, respectively. Two different FRCs, namely FRC4c and FRC5e-B, were adopted (Table 1). As shown in Table 1, the two materials had the same mechanical properties in compression, which were defined according to Eurocode 2 [9] provisions for concrete C30/37 (f_{ck} = 30 MPa). On the contrary, as two contents (25 and 102 kg/m^3) of steel fibers (fiber tensile strength = 2000 MPa; aspect ratio 80) were adopted, two different residual tensile strengths (f_{Fts}, f_{Ftu}) were calculated according to the linear post-cracking model suggested by MC2010 (clause 5.6.4). Note that the mean tensile strengths were conventionally assumed 30% higher than the corresponding characteristic values.

The details of the Hybrid Reinforcement resulting from the proposed design approach are depicted in the schematic of Fig. 4. The slab made with FRC4c (Fig. 4a) was provided with both bottom reinforcement along the column alignments and top rebars located over the columns. Unlike the slab FRC4c, the slab constructed with concrete FRC5e-B (Fig. 4b) did not contain top reinforcement. The total Steel Content (i.e., SC = (W_{fiber} + W_{rebars})/$V_{concrete}$, where W_{rebars} = total weight of fibers, W_{rebars} = total weight of rebars, and $V_{concrete}$ = total concrete volume) of the slab FRC4c and FRC5e-B resulted equal to 57 kg/m^3 (SC = 25 + 33 = 57 kg/m^3) and 107 kg/m^3 (SC = 102 + 5 = 107 kg/m^3), respectively. It is worth remarking that the length of rebars reported in the schematic does not include anchorages (therefore, the real W_{rebars} is little higher). Moreover, to highlight the actual potential of the Hybrid Reinforcement, the minimum APC rebars were not considered. The additional reinforcement shown in Fig. 4 was designed to withstand a self-weight G_{1k} = 5 kN/m^2 ($γ_{G1}$ = 1–1.35, Eurocode 0), a gravity load G_{2k} = 5 kN/m^2 ($γ_{G2}$ = 0–1.5, Eurocode 0) and a variable load Q_k = 2 kN/m^2 ($γ_Q$ = 0–1.5 Eurocode 0).

Table 1. Mechanical properties used to design and simulate the FRC elevated slab.

Property	Unit	Materials	
Designation		FRC4c	FRC5e-B
Classification (according MC2010)	-	4c	5e
Steel fiber content (W_{fiber})	kg/m^3	25	102
Mean modulus of elasticity (E_{cm})	MPa	32800	
Poisson's coefficient (v)	-	0.2	
Mean cylindrical compressive strength (f_{cm})	MPa	38	
Characteristic cylindrical compressive strength (f_{ck})	MPa	30	
Compressive strain at peak strength (ε_{c1})	‰	2.2	
Ultimate compressive strain (ε_{cu})	‰	3.5	
Mean tensile strength (f_{ctm})	MPa	2.9	
Characteristic tensile strength (f_{ctk})	MPa	2.0	
Mean serviceability residual tensile strength ($f_{Fts,m}$)	MPa	2.3	2.9
Mean ultimate residual tensile strength ($f_{Ftu,m}$)	MPa	1.3	8.0
Characteristic serviceability residual tensile strength ($f_{Fts,k}$)	MPa	1.8	2.3
Characteristic ultimate residual tensile strength ($f_{Ftu,k}$)	MPa	1.0	5.6

Structure verification was performed by simulating the non-linear behavior of the slab with the finite element program Diana 10.3 [10].The compressive behavior of concrete was modelled by the stress-strain parabolic law defined by Eurocode 2 (clause 3.1.5) [9], whereas the tensile constitutive law of FRC followed the linear model reported by MC2010 (clause5.6.4) [6]. An elastic-plastic law with strain hardening was adopted for the constitutive behavior of steel rebars. The latter was characterized by a Young modulus of 210,000 MPa, a characteristic/mean yield strength of 450 MPa/ 482 MPa and by an ultimate characteristic/mean tensile strength of 520 MPa/560 MPa. The maximum steel elongation was assumed equal to 7.5%. In the simulations, the self-weight was initially applied and the loads G_{2k} and Q_k were then monotonically increased until the maximum capacity of the slab was attained.

As already discussed in [5], the reinforcement arrangement of Fig. 4a represents an optimal combination of fibers and rebars able to provide the slab with a sufficient ultimate capacity and a good behavior at SLS. The structural performance of the Hybrid Reinforcement with FRC5e (Fig. 4b) was as much effective as that exhibited by that of FRC4c (Fig. 4a) but the total steel content (SC) was about 95% higher (107 kg/m^3 vs. 55 kg/m^3). The latter was mainly related to the high amount of fibers required due to the lack of top reinforcement over the columns.

The study reported herein focused on the effects of shrinkage on slab behavior. To this end, the slabs depicted in Fig. 4 were re-analyzed by including an initial total shrinkage strain of 300 με, evenly distributed over the slab depth. Shrinkage strain was totally applied before increasing the vertical loads. This assumption allowed to simulate the typical condition that occurs during construction, before removing the molds. According to numerical results, the most significant cracks grew on top surface, especially in the areas located near the perimeter columns. As expected, the slab FRC4c was affected by cracks having the widest width. In fact, as shown by the crack contour

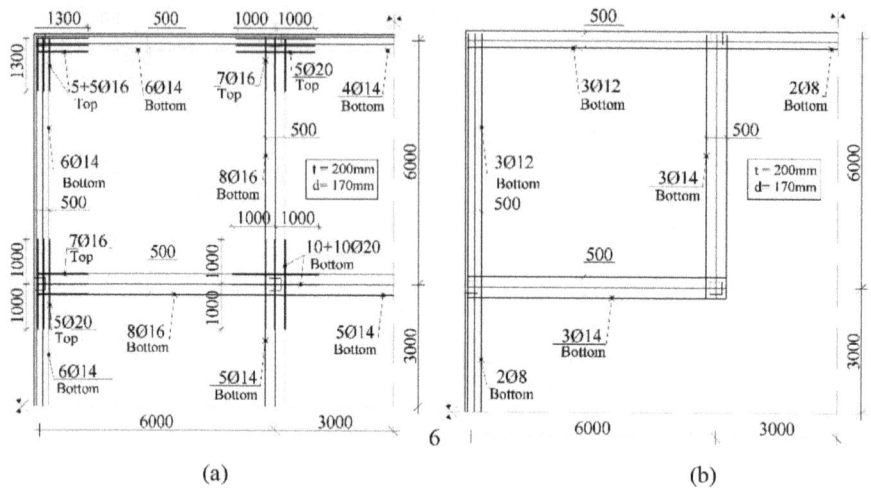

Fig. 4. Schematic of the main details of the Hybrid Reinforcement designed for the slab made with FRC4c (a) and FRC5e-B (b). (dimensions in mm)

obtained from the analysis with the characteristic law of materials (Fig. 5a), the cracks with the maximum width (i.e., from 0.5 mm to 0.8 mm) grew in few sections not strengthened by top rebars. On the contrary, most of cracks presented widths ranging from 0.4 to 0.5 mm. To perform a better control of cracking phenomena without increasing the fiber content, top rebars may be combined with an additional reinforcing mesh located solely in the most critical areas of the slab. As an alternative, glass fibers could be properly combined with steel fibers to get a higher tensile strength right after first cracking [11]. Finally, the use of Shrinkage-Reducing Admixtures (SRA) represents a valid alternative to reduce shrinkage strain.

The effect of shrinkage on the ultimate behavior of the slabs is well highlighted by the total overload ($E' = G_{2k} + Q_k$) – maximum deflection diagram of Fig. 5b. For the sake of brevity, the curves shown in the diagram refers only to the simulations performed by implementing the mean properties of materials and by uniformly spreading the overload on the whole surface of the slab. The horizontal dotted line reported in the diagram represents the minimum value of the overload, i.e. $E'_{min} = 15.6$ kN/m^2, that allows to fulfil the global safety criterion represented by Eq. (8). Therefore, slabs having a maximum capacity lower than that minimum threshold would not be safe. According to numerical results, shrinkage strain reduced the capacity of the slab FRC4c from 20.5 kN/m^2 to 17 kN/m^2. Top reinforcement located over the perimeter columns performed a control of cracking process leading to a limited (17%) reduction of structure resistance. As a result, slab FRC4c was still safe, as its capacity remained higher than the minimum acceptable value (E'_{min}). The lack of top reinforcement in the slab FRC5e-B explains the significant (41%) reduction of the bearing capacity, which reduced from 18.7 kN/m^2 to 11 kN/m^2, when considering shrinkage. As the reduced capacity is considerably lower than E'_{min}, the slab should be re-designed to accomplish the safety requirement. In view of this, the best design choice may consist in increasing

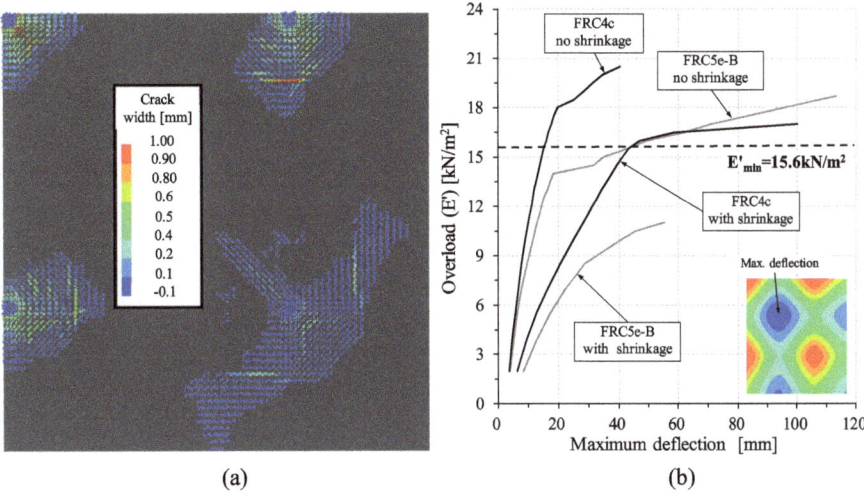

(a)   (b)

Fig. 5. Contour of shrinkage cracks acting on top surface of the FRC4c slab (a). Effect of shrinkage strain on the total overload-deflection mean response of the slabs FRC4c and FRC5e-B (b).

the slab thickness, since the in rease of the tensile performance of FRC would be really difficult to get on the field. It is also worth remarking the important reduction of flexural stiffness due to shrinkage cracking. In fact, within the overload range 2–6 kN/m², the average stiffness of the slab FRC4c varied from 1.6 kN/(m²·mm) to 0.5 kN/(m²·mm), corresponding to 70% stiffness reduction. Regarding the slab FRC5e-B, the stiffness reduction was even higher (150%), as it reduced from initial 0.96 kN/(m²·mm) to 0.38 kN/(m²·mm). Thus, top reinforcement appears quite effective in limiting the stiffness reduction as well as the related increment of deflection.

3.2   Design of a Slab Floor Located in an Office Building

To assess the applicability of the proposed method to the design of real structures, the floor of an office building described by Maturana et al. [12] is considered. As shown in Fig. 6, the slab had dimensions of about 20x33 m² and was supported by a total of 20 columns. The adopted structural grid subdivides the slab in rectangular panels with columns-to-column span lengths of 7.8x8.0 m², 7.8x4.9 m² and 7.8x5.4 m². The structure had a total thickness of 300 mm and was cast with FRC (f_{ck} = 30 MPa) containing 100 kg/m³ of steel fibers having a length of 50 mm, a diameter of 1.3 mm and a tensile strength of 900 MPa. Bending tests on simply supported FRC round panels (diam. 2.1 m – depth = 200 mm) provided a mean residual strength of 2.2 MPa (CoV = 3%) resulting from a simple equation (i.e., $M_{max}/(0.45 \cdot h^2)$) that relates the maximum plastic bending moment (i.e., the moment corresponding to the peak capacity) to the depth (h) of the round panel [12]. Slab reinforcement consisted mainly of fibers and a steel mesh Ø10 mm/150x150 mm (effective depth = 270 mm) laid on the bottom surface of the two 7.8x8.0 m² corner-bays of the floor. Therefore, except for

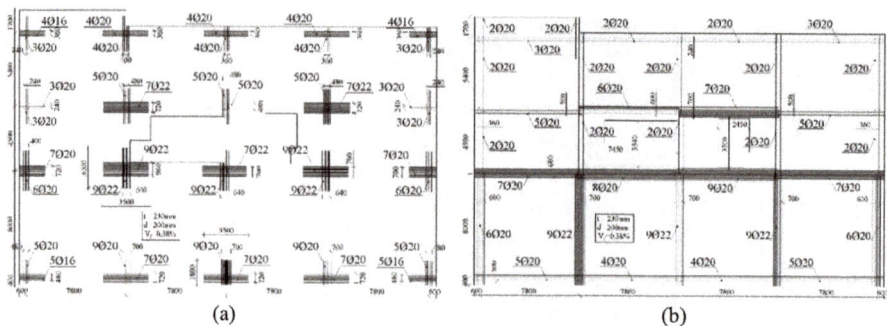

Fig. 6. Floor geometry and details of the Hybrid Reinforcements: top reinforcement (a); bottom reinforcement (b). (Dimensions in mm)

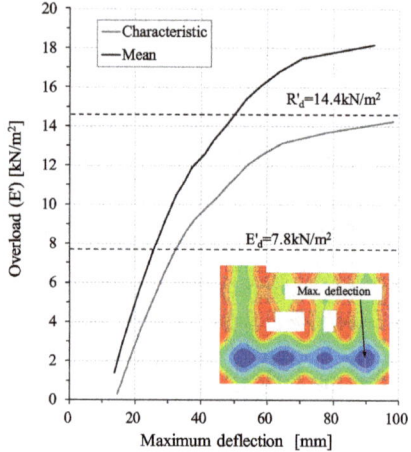

Fig. 7. Verification of the floor by NLFEA.

the APC reinforcement (i.e. 3Ø25 mm along column alignments) and some complementary rebars not considered herein, the total SC (fibers + bottom reinforcement) was approximately equal to 106 kg/m³. This reinforcement arrangement allowed the slab to withstand a maximum design load (E_d) of about 13.8 kN/m², which includes a self-weight (G_{1k}) of 7.5 KN/m², a distributed gravity load (G_{2k}) of 1.2 kN/m² and a variable load (Q_k) of 2 kN/m². In addition to the usual partial factors for actions ($\gamma_{G1} = \gamma_{G2} = 1$–1.35, $\gamma_Q = 0$–1.5), the design value of the variable load was reduced by a combination factor equal to 0.7 that took into account for the large extent (>100 m²) of the floor surface.

The design method proposed herein was used to re-design the slab reinforcement. The new proposal combined the FRC4c (Table 1) with the conventional reinforcement depicted in the schematic of Fig. 6. By considering the anchorage length recommended by Eurocode 2 [9] for rebars, a total SC of 65 kg/m^3 was obtained. Unlike the original project of the floor, the slab thickness and the effective depth were reduced to 230 mm and 200 mm, respectively. To determine the slab thickness, the maximum moment $m_{E,SLS}$ = 33 kNm/m (i.e., maximum moment resulting from the envelope of the load combinations) resulting from the LEFEA was considered. Based on the diagram of Fig. 1a, which assumes a maximum crack width (i.e., w_{max} = 0.2 mm) suitable for an exposure class XC1 according Eurocode 2, the minimum slab thickness is equal to 205 mm. This value can be increased to 230 mm in order to fulfil also the minimum thickness required to control the slab deflection, i.e. t $\geq$ L/35 = 8000/35 = 220 mm.

A finite element model consisting of 14011 four-node linear isoparametric flat shell elements was implemented in Diana 10.3 [10]. The same constitutive laws adopted in the case study described at §3.1 for the compressive and tensile behavior of both FRC and reinforcing steel were adopted. The non-linear simulation was performed by keeping the self-weight constant, whereas the gravity and the variable loads (E′) were both increased up to the maximum capacity of the slab. Unlike the assumptions made by Maturana et al. [12], the partial factor γ_{G2} was assumed equal to 0–1.5 and G_{2k} was increased to 3.2 kN/m^2. Moreover, because of the reduced thickness of the slab, the self-weight is here equal to G_{1k} = 5.5 kN/m^2. The overload (E′) vs. maximum deflection response of the slab is shown in Fig. 7 for the load combination with the overload applied on the whole surface of the slab. As one may observe, the design action is lower than the structural resistance resulting from Eq. (8) (E'_d = 7.8 kN/m^2 < 14.4 kN/m^2 = R'_d). Therefore, the safety requirement of MC2010 is fully satisfied.

4 Concluding Remarks

From the numerical investigations performed in the present paper, the following main concluding remarks can be drawn:

- Hybrid Reinforcement proved to be really effective even in the design of actual elevated slabs characterized by more complex geometries.
- Compared to SFRC slabs mainly reinforced with steel fibers, the use of Hybrid Reinforcement allows to reduce the total steel content up to 50%, confirming the findings of previous papers.
- Long term shrinkage may have a significant detrimental effect on the cracking, deflection control and bearing capacity of slabs reinforced with steel fibers only. A proper combination of bottom and top rebars placed over the columns may allow to smooth the influence of shrinkage strain leading to an improved behavior of the structure at both SLS and ULS conditions.

References

1. Destrée, X, Mandl, J,: Steel fibre only reinforced concrete in free suspended elevated slabs: Case studies, design assisted by testing route, comparison to the latest SFRC standard documents. In: Walraven J, Stoelhorst D. (eds.) Tailor Made Structure, International FIB 2008 Symposium, vol. 111. CRC Press, Amsterdam, Boca Raton, FL (2008)
2. Gossla, U.: Flachdecken aus Stahlfaserbeton, Beton-und Stahlbetonbau 101. Heft. **2**, 94–102 (2006)
3. Parmentier B, Van Itterbeeck P, Skowron A.: The behavior of SFRC flat slabs: The Limelette full-scale experiments for supporting design model codes. In: Charron J.P., Massicotte B., Mobasher B., Plizzari G. eds. FRC 2014 Joint ACI-fib International Workshop - FRC 2014: ACI-fib International Workshop. FIB Bulletin 79 – ACI SP-310 (2016)
4. ACI 544.6R-15 - Report on design and construction of steel fiber-reinforced concrete elevated slabs. Reported by ACI Committee 544. Farmington Hills, MI:American Concrete Institute (2015)
5. Facconi, L., Plizzari, G., Minelli, F.: Elevated slabs made of hybrid reinforced concrete: proposal of a new design approach in flexure. Struct. Concr. **20**(1), 52–67 (2019)
6. Fib Model Code for Concrete Structures 2010. Fédération Internationale du Béton. Ernst & Sohn, Lausanne (2013)
7. EN 1990: Eurocode 0 - Basis of Structural Design. European Committee for Standardization, Brussels (2006)
8. EN 14651–5: Precast Concrete Products – Test Method for Metallic Fibre Concrete – Measuring the Flexural Tensile Strength, European Standard (2005)
9. EN 1992–2: Eurocode 2. Design of Concrete Structures. Part 2: Concrete bridges: design and detailing rules (2005)
10. Diana 10.3: User's manual. Delft, TNO DIANA BV, The Netherlands (2019)
11. Barragán, B., Facconi, L., Laurence, O., Plizzari, G.: Design of glass-fibre-reinforced concrete floors according to the fib Model Code 2010. In: Massicotte, B., Charron, J.-P., Plizzari, G., Mobasher, B. (eds.) Fibre Reinforced Concrete: from Design to Structural Applications - FRC 2014: ACI-fib International Workshop, p. 311–320. FIB Bulletin 79 – ACI SP-310 (2016)
12. Maturana, A., Canales, J., Orbe, A., Cuadrado, J.: Plastic Analysis and Testing of multidirectional SFRC flag slabs. Informes de la Construcción, **66**(535), e031 (2014). https://doi.org/10.3989/ic.13.021

A New Sustainability Assessment Method for Façade Cladding Panels: A Case Study of Fiber/Textile Reinforced Cement Sheets

Payam Sadrolodabaee[1]([✉]), S. M. Amin Hosseini[2], Monica Ardunay[1], Josep Claramunt[1], and Albert de la Fuente[1]

[1] Polytechnic University of Catalonia - BarcelonaTECH, Barcelona, Spain
payam.sadrolodabaee@upc.edu
[2] Resume Tech, Barcelona, Spain

Abstract. As the building sector is one of the leading responsible for energy consumption and CO_2 emissions, criteria of sustainability, availability, and recyclability should be considered for developing materials even in the envelopes. Façade, as the first element against the undesirable external impact, may contribute to building sustainability by reducing the amount of energy consumption and providing indoor environment quality for the inhabitants. The envelope excluding its aesthetic function should fulfill certain requirements such as strength, flexibility, ductility, lightness, thermal and acoustical insulation, durability, and sustainability. Fiber/Textile cement sheets as an interesting architectural material attract great interest during the last decade, especially those reinforced with more sustainable fibers like vegetables or textile wastes. In this sense, this paper presents a novel model to evaluate the sustainability index of façade cladding panel, especially the fiber/textile cement board. To this end, a new model for assessing objectively the façade cladding sustainability was designed and developed based on MIVES according to the value function concept and seminars of experts.

Keywords: Cladding panels · Fiber cement boards · MIVES · Recycled fibers · Sustainability

1 Introduction

Sustainable construction is considered as a way to contribute to sustainable development by protecting the environment. The building sector causes some negative effects on the environment during various phases including materials production, construction on the site, the usage phase and the demolition or end of life [1]. In this sense, based on the statistics, the construction sector is responsible for about 36% of the European Union's total CO_2 emissions, 40% of its final energy consumption and approximately 46% of the total waste [2]. That is why innovations to improve the energy efficiency of the buildings as well as developing more environmentally-friendly materials are then of practical importance.

The building enclosure is one of the dominant parts of each building that plays a pivotal role in sustainability and energy efficiency [3]. Currently, "Ventilated Façade",

© RILEM 2022
P. Serna et al. (Eds.): BEFIB 2021, RILEM Bookseries 36, pp. 809–819, 2022.
https://doi.org/10.1007/978-3-030-83719-8_69

a double-wall construction comprising an external cladding panel and the outdoor side of the external wall, is among the most used constructive solutions of building envelopes. This system has continuity between the thermic and impermeable envelopes, avoiding thermal bridges and consequently avoiding the energy loss and the presence of water vapor condensations [4, 5].

The outer cladding material should satisfy both the architectural and engineering properties to meet the desired needs. Ceramics and natural stones, in addition to wood and aluminum composites, are among the conventional materials used for cladding facades. The former group usually has excessive weight and high stiffness, which limits their size and necessitates a complex supporting structure. Furthermore, the partial breaking of this heavy material which can lead to objects falling on public roads can impose a serious risk for pedestrians. The latter group is more flexible and lightweight but less durable, has a lower hardness, and is more expensive compared to the former [5]. For these reasons the new generation of composite panels, fiber/textile reinforced cement boards (FRC/TRC boards), has been developed for building envelopes.

During the past decades, the incorporation of various types of fibers such as asbestos, steel, glass and polymeric into cement-based materials was proved to enhance the performance in terms of ductility, tensile or flexural strength, toughness, fatigue strength, impact resistance and energy absorption capacity of matrices [6–10]. However, some limitations such as harmful health impacts, expenses, and environmental pollution [11] make researchers discover more sustainable fibers. Vegetable and cellulosic together with textile waste fibers have been recently attracted great interest as a suitable reinforcement in mortars and composites for low-performance structural applications [12–18]. Vegetable fibers as available and biodegradable, along with textile waste as recycled and inexpensive fibers, could allow the development of more sustainable construction materials.

Most of the existing sustainability assessment methods and tools in the building sector only consider the environmental aspect, thus leading to a non-comprehensive sustainability assessment of buildings [19]. The tools and methods which include more than one sustainability requirement, adding economic, social, or even technical and functional to environmental requirements, have recently increased [20]. Furthermore, there is scarce research on sustainability assessment of independent building elements or structures. Regarding the facade cladding panels, a limited number of studies have employed life cycle assessment to evaluate the only environmental performance [21, 22].

In this regard, the objective of this conference paper is to present a new sustainability assessment model for analyzing the sustainability index (SI) of the cladding façade panels by evaluating more than one requirement based on MIVES (from the Spanish Integrated Value Model for the Sustainability Assessment). MIVES is a new comprehensive and integrated Multi-Criteria Decision Making (MCDM) method capable of assessing viable solutions for sustainability while considering economic, environmental, social and even technical requirements [23]. This new sustainability model for the façade cladding panel would be applied to a new Textile Reinforced Mortar (TRM) panel to consider the potentiality of this new board as a sustainable cladding material in a residential building in Barcelona, Spain.

This new cement sheet, with a surface mass of about 16 kg/m^2 and thickness of 10 mm, constituted of cementitious mortar reinforced by textile waste (TW) nonwoven fabrics as explained deeply in [17]. Figure 1 demonstrates the fabrication process of the panel as well as the flexural test configuration. The optimum panel was made up of the Portland cement paste reinforced by the 6 layers of the reinforcement. Table 1 gathers the mean of the flexural strength, toughness and flexural stiffness of the pre-cracked zone for the optimum cement sheet in both unaged and aged conditions. The aged condition consisted of subjecting the plates to 25 dry-wet cycles, after 28 curing days, in order to estimate the durability of the TW nonwoven fabric within the mortar matrix.

2   Model Design

A holistic approach was implemented in this paper to present a general model for the sustainability assessment of cladding panels. To develop the method, data collection and data analysis of cladding panels, particularly oriented to fiber cement sheets, was already done and the model design and model application are in progress. The design of the requirement tree, based on the local characteristics of the case study and its demands, is the first step of the model design. Then, the assignation of weights for indices (requirement, criteria, and indicator) will be evaluated by a group of multi-disciplinary experts through the analytical hierarchy process (AHP). Finally, the value function concept is used to transform the results obtained by each indicator to a non-dimensional magnitude value while minimizing the subjectivity in the assessment [24]. Thus, different alternatives could be evaluated and compared through the model and a sustainability index of each one is obtained. In the following parts, the aforementioned steps of MIVES are defined and designed.

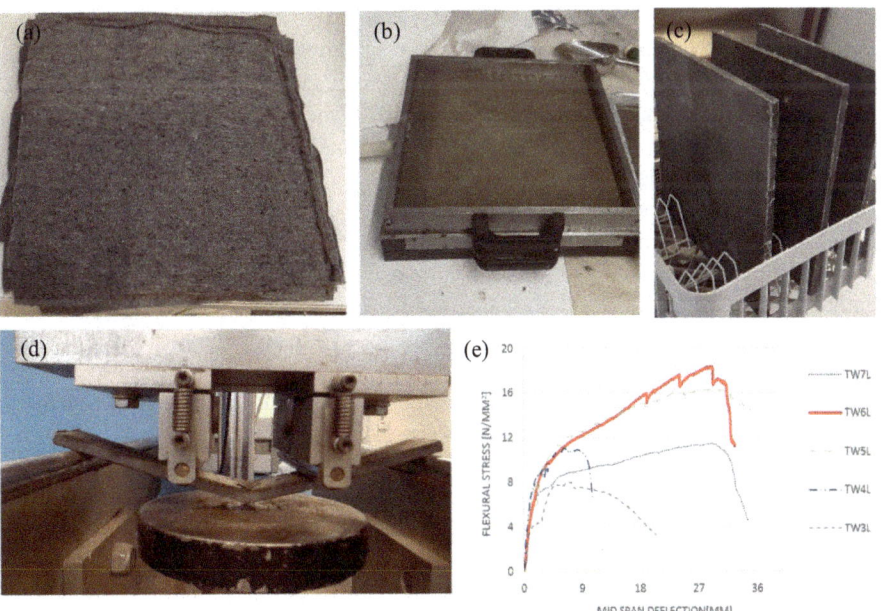

Fig. 1. TW composites: a)TW fabric; b)Casting in the mold; c)TW cement boards; d)Flexural configuration; e)Experimental flexural-deflcetion curve

Table 1. Mechanical characteristics of TW cement board composite.

Mechanical properties	Unaged sample	Aged sample*
Mean 28-day flexural strength [MPa]	15.5	10
Mean toughness index [KJ/m^2]	9.7	6.8
Mean flexural stiffness [GPa]	11.3	12

*aged sample: after exposure to accelerated aging of wet-dry cycles

2.1   Requirements Tree

The requirements tree is a hierarchical diagram in which the various characteristics of the products or processes to be evaluated are organized, at three levels: requirements, criteria and indicators. This requirements' tree was developed based on data collected from an extensive literature review and seminars with multidisciplinary experts in this subject. Based on the MIVES concept, indicators should be independent of each other and considered only once [24], so some indicators should be disregarded due to either their lack of representativeness or due to certain overlapping with other indicators already considered.

The preliminary requirement tree of this research was shown in Table 2. As can be seen, the first level of the tree included 4 requirements (R): economic, environmental, social and technical. The second hierarchical level consisted of 8 criteria (C) and the last level included 14 indicators (I).

The *economic requirement* (R_1) measures the estimated investment for the implementation of cladding façade with the *cost* (C_1) being the determining criterion. The *environmental requirement* (R_2) assesses the environmental effect of the panels during the life cycle through *emission* (C_2), *resource consumption* (C_3) *and waste* (C_4) criteria. The *social requirement* (R_3) considers the health and welfare of people by three criteria: (C_5) *comfort*, (C_6) *safety*, and (C_7) *aesthetic*. Finally, the *technical requirement* (R_4) assesses the mechanical properties of a material by the *reliability* (C_8) criterion.

The *cost* (C_1) includes two indicators: the (I_1) *construction cost* of the external cladding panel including material fabrication, transportation, labor costs and installation. To estimate this indicator, around 40 composite panels including different fiber cement/mortar boards, ceramics, aluminum panels and galvanized steel sheets were analyzed. To this end, the online BEDEC database from the Technological Institute of Catalonia (ITeC) [25] was used. The other indicator is (I_2) *maintenance cost* which estimated expected operations such as cleaning, repairing and replacing during the service life of the panel, usually considered to be 50 years.

The *emission* (C_2) criterion is represented by the indicator of CO_2 *emissions* (I_3) to measure the amount of CO_2 emissions for each cladding panel in the phases of manufacturing and construction. The BEDEC [25] together with The Inventory of Carbon & Energy (ICE) [26] were considered as the reference databases for quantifying this indicator.

The *resource consumption* (C_3) criterion is aimed at quantifying the natural resources consumed, through two indicators. The *energy consumption* (I_4) indicator accounts for the amount of energy consumed during the manufacturing and construction phases. Once more, BEDEC and ICE were considered as a database. The *material consumption* (I_5) considering the amount of raw materials and water used in the fabrication stage, estimated by BEDEC and Environmental Product Declaration (EPD) of materials.

The *waste* (C_4) criterion is assessed by the indicator of *Solid waste management* (I_6) to estimate the amount of waste materials used in the manufacturing phase or the quantity that can be recycled after demolition. A measurable scale of 0 to 20 is used to rate the recyclability of cladding panels through seminars with multidisciplinary engineers and literature reviews.

Comfort (C_5) considers two indicators: (I_7) *acoustic performance*, which evaluates the rate of sound absorption of material by NRC, Noise Reduction Coefficient, a standard rating for how well a material absorbs sound. (I_8) *Thermal performance*, which assesses the thermal conductivity of the different cladding panels, the ability to conduct the heat.

Safety (C_6) is meant to assess the security and safety of persons, including occupants or labors. This includes the indicators of *fire vulnerability* (I_9), which evaluates the durability of the panel against fire, and *risk vulnerability* (I_{10}) which considers the probability of any accidents for persons, the public and labors during the construction or assembly of panels. The former will be assessed in min while the latter on a scale of 0 to 10.

The *aesthetic* (C_7) criterion including *Consistency with the surrounding* indicator (I_{10}) assesses the appearance and architectural style of the façade cladding through seminars with multidisciplinary architects. A measurable scale of 0–10 is used to rate the visual quality of the façade panels.

Finally, the *reliability* (C_8) assesses the flexural or tensile resistance of the panels through the *Strength to weight ratio* (I_{12}) parameter. Moreover, the energy absorption and toughness of the panels are measured by the *Toughness and ductility* (I_{13}) indicator to avoid using the brittle material. *Durability* (I_{14}) considers the resilience of the panels

Table 2. Requirements tree for assessing the façade cladding sustainability.

Requirements	Criteria	Indicators	Units	Function Shape
R_1.Economic	C_1. Cost	I_1.Construction Cost	€/m²	DC_X
		I_2. Maintenance cost	€/m²	DC_X
R_2.Environmental	C_2.Emission	I_3. CO_2 equivalent emission	Kg/m²	DC_X
	C_3. Resource consumption	I_4. Energy consumption	MJ/m²	DC_X
		I_5. Material consumption	Kg/m²	DC_X
	C_4.Waste	I_6. Solid waste management	Points	I_nC_V
R_3.Social	C_5.Comfort	I_7. Acoustic performance	points	I_nS
		I_8. Thermal performance	m² k/w	DS
	C_6.Safety	I_9. Fire vulnerability	min	I_nC_V
		I_{10}. Risk vulnerability	Points	DC_X
	C_7.Aesthetic	I_{11}.Consistency with surrounding	Points	I_nL
R_4.Technical	C_8.Reliability	I_{12}.Strength to weight ratio	Points	I_nC_V
		I_{13}.Toughness and ductility	KJ/m²	I_nC_V
		I_{14}. Durability	Points	I_nC_V

in exposure to extreme weather conditions (wet/dry or freeze/thaw cycles).

2.2   Assignation of Weights for Each Parameter

To express the importance of each parameter (requirement, criteria, and indicator) and prioritize them, weights would be assigned. For instance, the weight of I_1 is considered as $\lambda_1 = 65\%$ while the weight of I_2 is $\lambda_2 = 35\%$, showing the greater importance of the former. The weightings of the tree will be assigned based on previous studies and the knowledge of the professors and experts from architecture and civil engineering faculties, namely Universitat Politècnica de Catalunya, involved in the seminar using the analytical hierarchy process (AHP) [27]. The AHP method enables the most consistent weighting judgments and helps organize the process efficiently, reduce the model complexity and subjectivity, and decrease possible disagreements between the team members [28].

2.3   Establish Value Functions for Each Parameter

To homogenize the indicator's units into dimensionless values ranging from 0.0 to 1.0, the value function concept is used. These values, from 0.0 to 1.0, represent the minimum and maximum degree of satisfaction in terms of sustainability, respectively [23, 29]. To determine the satisfaction value for each indicator, first of all, the tendency (increasing or decreasing) of the value function should be defined. An increasing (I_n) function means the decision maker's satisfaction increases with an increase in the measurement variable. In contrast, a decreasing (D) value function shows that an increase in the measurement unit causes a decrease in satisfaction [23].

Secondly, two points that have a satisfaction value of 0.0 (X_{min}) and 1.0 (X_{max}) should be defined for each indicator. These points are usually established according to existing rules and regulations, experience with previous projects, and the value produced by the different alternatives with respect to the indicator. For instance, X_{min} and X_{max} for the I_1 indicator (construction cost) are determined as 20.5 and 183 €/m2 while for I_9 (fire vulnerability) are specified as 30 and 200 min.

Afterward, the value function shape, concave, convex, linear or S-shaped, should be selected [30] in order to connect the two coordinate points, (X_{min}, 0) and (X_{max}, 1) [23]. A concave-shaped (C_v) is used when satisfaction increases rapidly or decreases slightly, while the convex (C_x) one is more suitable when the satisfaction tendency is contrary to the concave curve. linear (L) function is presented for steadily increases/decreases of satisfaction, while an S-shaped (S) function is used if the satisfaction tendency contains a combination of concave and convex functions (see Fig. 2).

As can be seen in Table 2, seven indicators have a decreasing tendency including six convex shapes and one S-shaped, while five indicators have increasing concaved shapes. Finally, there is one indicator with an increasing S-shaped and another one with an increasing linear function. Figure 3 demonstrates the function shapes of the I_1 and I_9 indicators as examples. As the satisfaction value of the *construction cost* decreases rapidly with increasing the price, so the convex shape was chosen (Fig. 3-a). On the other hand, for *fire vulnerability*, satisfaction increases at first more rapidly when the

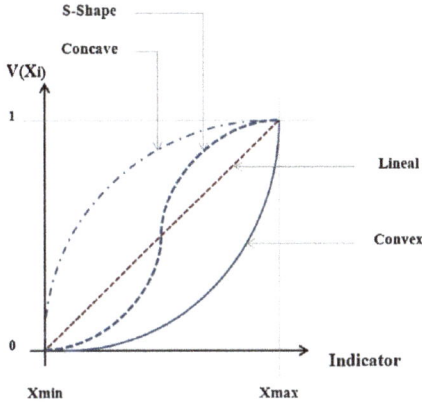

Fig. 2. Different value function shapes.

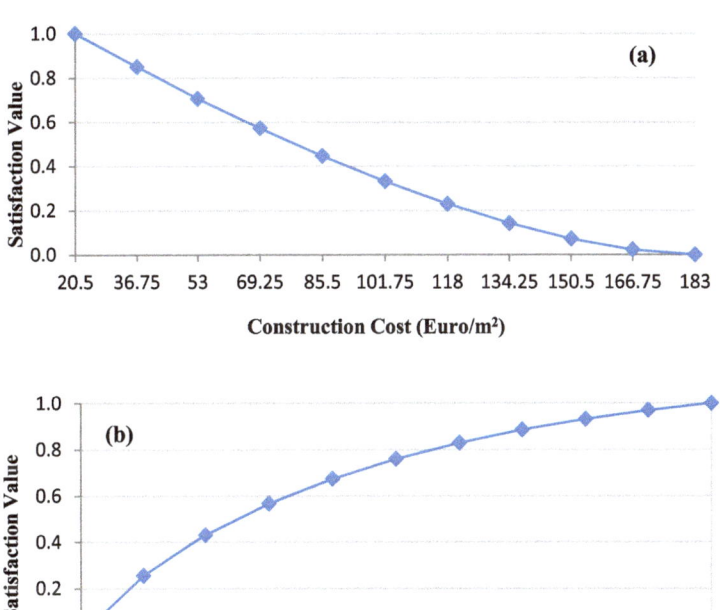

Fig. 3. Value function shapes of indicators: a) construction cost; b) fire vulnerability

fire performance of the cladding improves (Fig. 3-b). Thus, a concave function could be suitable.

Finally, the mathematical expression of MIVES, Eq. (1), for obtaining each indicator value satisfaction, V_i, would be applied. To normalizing the indicators' values between a range of 0.0 and 1.0 as well as achieving factor B for the previous equation, Eq. (2) is used. The constitutive parameters of the value functions were established based upon the scientific literature, existing rules and regulations, experience with previous projects and the background of experts (including professors and experts practitioners who attended the seminars). Thus, the sets of indicator values (Vi(xi)) that are between 0.0 and 1.0 is generated.

$$V_i = A + B \cdot \left[1 - e^{-k_i \cdot \left(\frac{|X_{Alt_i} - X_{min}|}{c_i} \right)^{P_i}} \right] \tag{1}$$

$$B = \left[1 - e^{k_i \cdot \left(\frac{|X_{max} - X_{min}|}{c_i} \right)^{P_i}} \right]^{-1} \tag{2}$$

Where:

V_i: The satisfaction value of the indicator being evaluated.

A: The response value X_{min} (indicator abscissa), generally A = 0;

B: The factor preventing the function from leaving the range (0.0, 1.0).

X_{Alt_i}: The indicator abscissa that generates the value, V_i;

X_{min}: The point of minimum satisfaction, with a value of 0.

X_{max}: The point of maximum satisfaction, with a value of 1.

P_i: A shape factor that determines whether the curve is concave or convex, linear or S-shaped;

(P < 1 the curve is concave; P > 1 the curve is convex or S-shaped; P = 1 it is linear).

C_i: The factor that establishes the value of the abscissa for the inflection point in curves with $P_i > 1$;

K_i: The factor that defines the response value to C_i;

For instance, the construction cost of the textile waste fabric reinforced cement boards (X_1) is estimated as 52 €/m2. Thus, the satisfaction value of this indicator, V_1, is calculated as 0.72.

2.4   Assess the Sustainability Index (SI)

After the assessment of the satisfaction value of the indicators, the formula that is presented in Eq. (3) should be applied to each tree level. In this equation, the indicator value ($V_i(x_i)$) has previously been determined and the weights (λ_i) are assigned to determine the sustainability value of each branch.

$$SI = \sum \lambda i V_i(x_i) \tag{3}$$

Where:

SI = Sustainability Index;

$V_i(x_i)$: The value function of each indicator, criterion or requirement;

λ_i: The associated weight.

Thus, the Sustainability Index (SI) of the cladding panel can be computed. Besides calculating the overall sustainability index, this model also makes it possible to calculate economic, social, environmental and technical satisfaction indexes separately. In this case, the weaknesses and strengths of each alternative can be identified from different points of view. For instance, the satisfaction value of the *economic requirement* (R_1) will be equal to the satisfaction value of the *cost criterion* (C_1) and will be calculated through $\lambda_1 V_1 + \lambda_2 V_2$.

3   Summary and Discussion

A new multi-objective approach for sustainability assessment of the cladding façade panels specifically oriented to the fiber/textile cement board is presented in this research study. This sustainability model, based on the MIVES, uses the value function concept as well as the AHP process to quantify the satisfaction and preferences of the stakeholders and users. The proposed model:

- Considers various indicators in the scope of economic, environmental, social and technical performances that can be applied to any façade cladding. Nonetheless, the function and the location of the building could affect the final identification and filtration of the indicators.
- Helps decision-makers to compare the sustainability value of all alternatives and choose the optimum solution for cladding panels.

This model is conceptually presented herein, while a real case study, applied on the textile waste fabric reinforced cement boards, is being prepared to serve as an example of application within the context of the Ph.D. thesis of the first author.

Acknowledgement. The authors express their gratitude to the Spanish Ministry of Economy, Industry and Competitiveness for the financial support received under the scope of the projects CREEF (PID2019-108978RB-C32) and RECYBUILDMAT (PID2019-108067RB-I00).

References

1. Pons, O., Wadel, G.: Environmental impacts of prefabricated school buildings in Catalonia. Habitat Int. **35**(4), 553–563 (2011)
2. Gálvez-Martos, J.L., Styles, D., Schoenberger, H., Zeschmar-Lahl, B.: Construction and demolition waste best management practice in Europe. Resour. Conserv. Recycl. **136**, 166–178 (2018)
3. Hartkopf, V., Aziz, A., Loftness, V.: Facades facade and Enclosures, Building building for Sustainability. In: Encyclopedia of Sustainability Science and Technology, pp. 3675–3705. Springer, New York (2012). https://doi.org/10.1007/978-1-4419-0851-3
4. Fernández, L.J., Claramunt, J., Llerena, A., Torrens, D., Ardanuy, M., Zamora, J.L.: "First International Conference on Bio-based Building Materials Nonwoven Flax Fiber Mats and White Portland Cement Composites for Building Envelopes (2015)

5. Claramunt, J., Fernández-Carrasco, L.J., Ventura, H., Ardanuy, M.: Natural fiber nonwoven reinforced cement composites as sustainable materials for building envelopes. Constr. Build. Mater. **115**, 230–239 (2016)
6. Ardanuy, M., Claramunt, J., Filho, R.D.T.: Cellulosic fiber reinforced cement-based composites: a review of recent research. Constr. Build. Mater. **79**, 115–128 (2015). https://doi.org/10.1016/j.conbuildmat.2015.01.035
7. Pacheco-Torgal, F., Jalali, S.: Cementitious building materials reinforced with vegetable fibres: a review. Constr. Build. Mater. **25**(2), 575–581 (2011)
8. Brandt, A.M.: Fibre reinforced cement-based (FRC) composites after over 40 years of development in building and civil engineering. Compos. Struct. **86**(1–3), 3–9 (2008)
9. Zani, G., Colombo, M., Di Prisco, M.: High performance cementitious composites for sustainable roofing panels. In: Proceedings 10th fib International PhD Symposium Civil Engineering, pp. 333–338, February 2014
10. Promis, G., Bach, T.Q., Gabor, A., Hamelin, P.: Failure behavior of e-glass fiber- and fabric-reinforced IPC composites under tension and compression loading. Mater. Struct. **47**(4), 631–645 (2013). https://doi.org/10.1617/s11527-013-0085-6
11. Wei, J., Meyer, C.: Improving degradation resistance of sisal fiber in concrete through fiber surface treatment. Appl. Surf. Sci. **289**, 511–523 (2014)
12. Correia, C., Francisco, S., Savastano, H., Moacyr, V.: Utilization of vegetal fibers for production of reinforced cementitious materials (2017)
13. Ramirez, M., Claramunt, J., Ventura, H., Ardanuy, M.: Evaluation of the mechanical performance and durability of binary blended CAC-MK/natural fiber composites. Constr. Build. Mater. (2019)
14. Gonzalez-Lopez, L., Claramunt, J., Hsieh, Y.L., Ventura, H., Ardanuy, M.: Surface modification of flax nonwovens for the development of sustainable, high performance, and durable calcium aluminate cement composites. Compos. Part B Eng. **191**, 107955 (2020)
15. Claramunt, J., Ventura, H., Fernández-Carrasco, L.J., Ardanuy, M.: Tensile and flexural properties of cement composites reinforced with flax nonwoven fabrics. Mater. (Basel) **10**(2), 1–12 (2017)
16. Sadrolodabaee, P., Claramunt, J., Ardanuy, M., de la Fuente, A.: Mechanical and durability characterization of a new textile waste micro-fiber reinforced cement composite for building applications. Case Stud. Constr. Mater. **14**, 00492 (2021)
17. Sadrolodabaee, P., Claramunt, J., Ardanuy, M., de la Fuente, A.: Characterization of a textile waste nonwoven fabric reinforced cement composite for non-structural building components. Constr. Build. Mater. **276**, 122179 (2021)
18. Sadrolodabaee, P., Claramunt, J., Ardanuy, M., de la Fuente, A.: A textile waste fiber-reinforced cement composite: comparison between short random fiber and textile reinforcement. Materials **14**, 3742 (2021). https://doi.org/10.3390/ma14133742
19. Ding, G.K.C.: Sustainable construction-The role of environmental assessment tools. J. Environ. Manage. **86**(3), 451–464 (2008)
20. Salzer, C., Wallbaum, H., Lopez, L.F., Kouyoumji, J.L.: Sustainability of social housing in Asia: a holistic multi-perspective development process for bamboo-based construction in the Philippines. Sustainability **8**(2), 151 (2016)
21. Kvočka, D., et al.: Life cycle assessment of prefabricated geopolymeric façade cladding panels made from large fractions of recycled construction and demolition waste. Mater. (Basel) **13**(18), 3931 (2020)
22. Han, B., Wang, R., Yao, L., Liu, H., Wang, Z.: Life cycle assessment of ceramic façade material and its comparative analysis with three other common façade materials. J. Clean. Prod. **99**, 86–93 (2015)

23. Alarcon, B., et al.: A value function for assessing sustainability: application to industrial buildings. Sustainability **3**(1), 35–50 (2010)
24. Hosseini, S.M.A., De, A., Pons, O.: Multi-criteria decision-making method for assessing the sustainability of post-disaster temporary housing units technologies : a case study in Bam, 2003. Sustain. Cities Soc. **20**, 38–51 (2016)
25. ITeC - Institute of Construction Technology of Catalonia. https://itec.cat/. Accessed: 08 Nov 2020
26. Hammond, G., Jones, C.: Inventory of Carbon and Energy (ICE) (2008)
27. Saaty, T.L.: How to make a decision: the analytic hierarchy process. Eur. J. Oper. Res. **48**(1), 9–26 (1990)
28. De La Fuente, A., Pons, O., Josa, A., Aguado, A.: Multi-criteria decision making in the sustainability assessment of sewerage pipe systems. J. Clean. Prod. **112**, 4762–4770 (2016)
29. Aguado, A., del Caño, A., de la Cruz, M.P., Gómez, D., Josa, A.: Sustainability assessment of concrete structures within the spanish structural concrete code. J. Constr. Eng. Manag. **138**(2), 268–276 (2012)
30. Gilani, G., Pons, O., de la Fuente, A.: Towards the Façades of the Future: A New Sustainability Assessment Approach. IOP Conf. Ser. Earth Environ. Sci. **290**(1), 012075 (2019)

Validation Testing of Precast Tunnel Lining Segments Using Polymeric Fibers

Devansh Patel[1], Chidchanok Pleesudjai[1], Yiming Yao[2],
Steve Schaef[3], and Barzin Mobasher[1(✉)]

[1] School of Sustainable Engineering and Built Environment,
Arizona State University, Tempe, AZ, USA
`barzin@asu.edu`
[2] School of Civil Engineering, Southeast University, Nanjing, China
[3] Development Admixture Systems-BASF North America, Jersey City, USA

Abstract. Fiber reinforcement in precast tunnel segments is quite attractive since it reduces the cost and labor associated with the fabrication process while it improves the post-cracking behavior considerably. Among the benefits, one can address improvements in handling, fatigue, impact resistance and durability while reducing the crack widths significantly. Flexural tests of full-scale precast tunnel segments are conducted using a newly developed testing laboratory for large structural testing. Using the closed-form material properties obtained from the flexural beam specimens, the response of full sections is predicted and the test results are compared with the full-scale test results obtained on the testing facility. Using the proposed methodologies, one can develop proper material and structural models for the accurate design of tunnel segments due to their unique design characteristics.

Keywords: Tunnel lining · Precast concrete · Fiber reinforced concrete · Polymeric fiber · Residual strength · Simulation

1 Introduction

The development of excavation technology using the tunnel boring machines (TBM) has resulted in widespread use and study of precast concrete tunnel linings (PCTLs). In addition to easier handling and installation processes, higher quality and more economical production of PCTL can be achieved since it is fabricated in specialized and controlled precast plants [1]. The use of FRC for tunnel lining can reduce the cost and labor associated with the fabrication process while it improves the post-cracking behavior considerably. The performance of steel fibers has been widely studied in the past few years, even though the corrosion factor still influences the performance of SFRC segments over its service life [2, 3]. The use of polymeric fibers can counter this issue due to its non-corrosive property especially in aggressive soil environment and hydraulic or wastewater tunnels where excessive corrosion may take place.

© RILEM 2022
P. Serna et al. (Eds.): BEFIB 2021, RILEM Bookseries 36, pp. 820–830, 2022.
https://doi.org/10.1007/978-3-030-83719-8_70

The objective of this work is to develop the design, analysis, and verification tools for procedures that would allow for the design of precast tunnel lining using macro synthetic fibers. In addition, the effort determines the mechanical properties of fiber-reinforced sections and proposes design guidelines that allow for replacing the conventional rebars with a certain volume fraction of macro synthetic fibers. It is expected that by using appropriate mix designs and implementing material behavior in the design procedures, sustainability initiatives can be supported and documented in the context of long term performance, ductility, and cost-saving [4–6].

Several tunnel focused research projects highlight the beneficial effect of FRC in the presence of concentrated loads and bursting [1, 7]. Use of fibers increases the fatigue and impact resistance [8] of the segments in mitigation of unintentional impact loads during handling and tunnel construction operations. Segments can be manufactured using FRC only or hybrid solution combination of conventional rebar and fiber reinforcement, i.e. a hybrid system which is very attractive for larger diameter tunnels with high internal forces. A fiber only or hybrid solution depending on performance requirements may therefore present the most ideal solution. Using the current technology with high performance FRC segments, tunnel rings of more than 23 ft (7 m) diameter have been successfully used to include fibers.

Current serviceability based procedures are empirically based extensions of equivalent elastic analysis. Recent adoption of fibers as reinforcement in the Model Code 2010 [9], ACI-318 building code, and ACI 544 Fiber Reinforced Concrete Committee is promoting the use of fibers in structural concrete. Non-linear procedures such as elastic-plastic solutions close the gap between properties, analysis, modeling and service life modeling. Closed-form relationships were developed based on plastic design. The goal of Hybrid Reinforcement Concrete Research (HRC) is to reevaluate traditional reinforcing technologies by analysis of a host of available reinforcing technologies using polymeric macro fibers, and conventional reinforcements. The proposed experimental and analytical approach will lead to enhanced technology development in various areas including geotechnical engineering, transportation, water resources, pavement materials and systems.

2 Research Objectives

The intent of this comprehensive program is to determine the material properties and develop structural design guidelines for macro synthetic fibers used at different volume fractions. The test methodology will develop improved methods to characterize average residual strength in fiber-reinforced concrete mixtures made with a variety of mixtures for applications in tunneling. The goal is to test the materials in terms of optimum mix design and to develop a calibration procedure that allows for design and verification of mechanical properties. Using a system design approach where the analysis tools (e.g. structural analysis) and the design optimization tools (e.g. non-linear programming) are closely linked to design requirements [10].

3   FRC Design Codes, Standards and Recommendations

FRC precast tunnel segments have been designed using constitutive laws recommended by international design codes and standards such as German DBV [11], RILEM TC 162-TDF [12], Italian CNR DT 204/2006 [13], Spanish EHE-08 [14], ACI-544 [15] and fib Model Code 2010 [9]. These codes propose stress-crack width or stress-strain constitutive laws of FRC as a linear post-cracking behavior (hardening or softening) or as a rigid perfectly plastic behavior based on bending test results. A majority of these design guides are related to the adoption of non-linear fracture mechanics parameters to analyze cracking phenomena and obtain allowable stress crack width relationship [16, 17]. This method is used to determine post-crack residual tensile strength from results of standard beam tests such as ASTM C1609 and EN 14651, as one of the key design parameters for FRC tunnel segments [18]. ACI 544.7R report [19] is currently the leading document for design of FRC based tunnel linings for a variety of load cases. It can be used in conjunction with ACI 544.8R report [20] to obtain the required constitutive law for the calculation of resistance moment and axial force-bending moment interaction diagrams for use with FRC precast tunnel segments [21].

4   Materials Characterization Tests

The large-scale testing laboratory at Arizona State University (ASU) uses the newly developed load frame that extends testing Full-scale flexure tests for applications such as tunnel lining. Figure 1 shows the schematic testing frame for large scale tunnel segments. Tests were conducted on fiber-reinforced precast tunnel segments and results were used to characterize the performance of macro polypropylene fiber-reinforced concrete (PFRC) segments with dosages of 14 lb/yd^3 (8.3 kg/m^3) equivalent to 0.91% volume fraction and labeled as PP14. These tests can validate the use of PFRC segments since macro-synthetic PP fibers do not suffer from corrosion that may occur in hydraulic tunnels or tunnels in presence of aggressive soil environments hence ensuring a longer service life [3].

The compressive strength was measured on cylinders in accordance with ASTM C39 cylinder testing. The tensile behavior was deduced from the flexure response through bending standard ASTM C1609 [22] which is a 4 point bend unnotched tests on $150 \times 150 \times 600$ mm specimens (Fig. 2(a)).

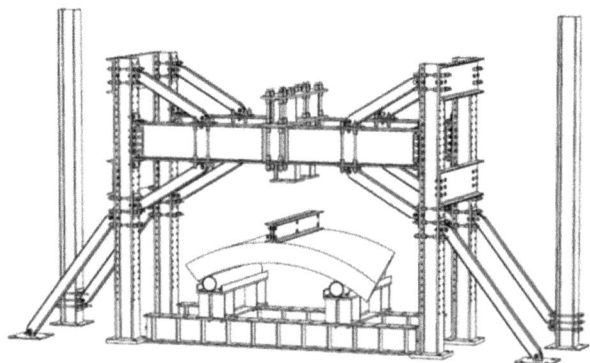

Fig. 1. Load frame setup for precast tunnel segments

5 Experimental Setup and Results

Three replicate precast tunnel segments were tested for 3-point-bending full scale flexure test, prior to which, a standard ASTM C1609 test was conducted on a total of 4 beams on the 28th day of casting tunnel lining specimens to measure the strength parameters, fiber performance, and material properties as per ACI 544.8R report and ASTM C39M test was conducted on three 4″ × 8″ cylinders following each full scale test. The polypropylene fibers used had a nominal length of 54 mm (2.1 in.) and a tensile strength of 585 MPa (85 ksi).

Standard ASTM C1609 beam tests were performed on four beams of size b = 152 mm (6 in.), h = 152 mm (6 in.), and span = 457 mm (18 in.), to study the material properties and fiber response based on the load-deflection response of each beam. Figure 2 (b) shows the load-deflection plots of 4 PFRC beams tested on the 28th day of casting.

The full-scale three-point bending flexural tests were conducted on 3 replicate PP-FRC precast tunnel lining segments to measure the flexure performance in terms of load-deflection. Results were compared with the simulated response using the back-calculated analytical response of ASTM C1609 tests that uses the parameterized stress-strain models. The tests were conducted under load control using a loading beam and steel cylinders on top as shown in Fig. 3, with a total self-weight of 220 lbs. The internal diameter of the tunnel was 2.7 m (8.75 feet) and the outer diameter was 3 m (10 feet). The dimensions of the segment were, span 'L' of 100 cm (39 in.), depth 'd' of 19 cm (7.5 in.) & breadth 'b' of 121 cm (48 in.). The segments were instrumented with 4 LVDTs to measure the deflection, 4 strain gages to measure the strains both in compression and tension zones and 4 string sensors to measure the crack opening at the tension zone. Digital image correlation (DIC) was also set up to measure precise strains, stresses, deflection, and crack width.

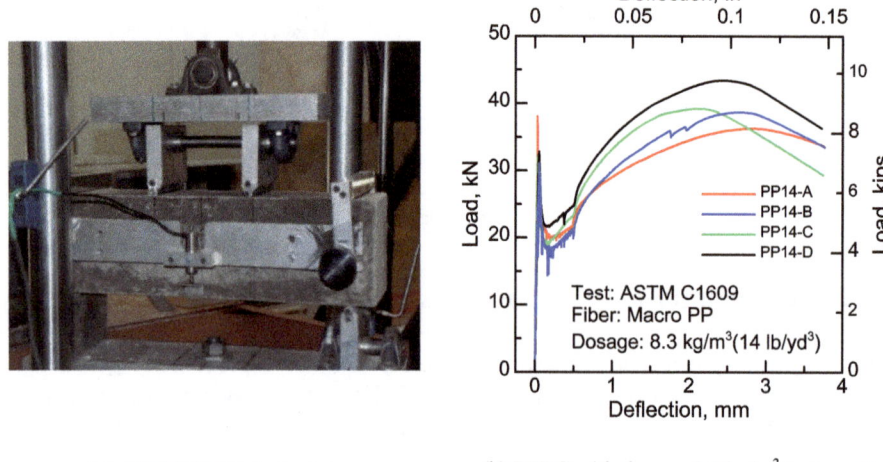

(a) ASTM C1609 test setup   (b) PFRC with dosage 8.3 kg/m³ test results

Fig. 2. ASTM C1609 test results

6 Test Results of PFRC Segments

A plot comparing the load vs. deflection response of the three replicate PFRC segments is shown in Fig. 4. PP14_N4 shows the highest first crack strength and post crack strength. PP14_N1 and PP14_N4 show strain hardening behavior with higher post-peak response compared to the first crack strength. All the specimens reach maximum post-peak strength at around 4 mm deflection. The large-scale test results also correlate with the C1609 tests showing a similar response with a strain-hardening behavior. Figure 5(a) and (d) show the failure of the tunnel segments with the crack opening and fibers straining in the tension zone. Table 1 summarizes the test results of all the tested PP-FRC segments.

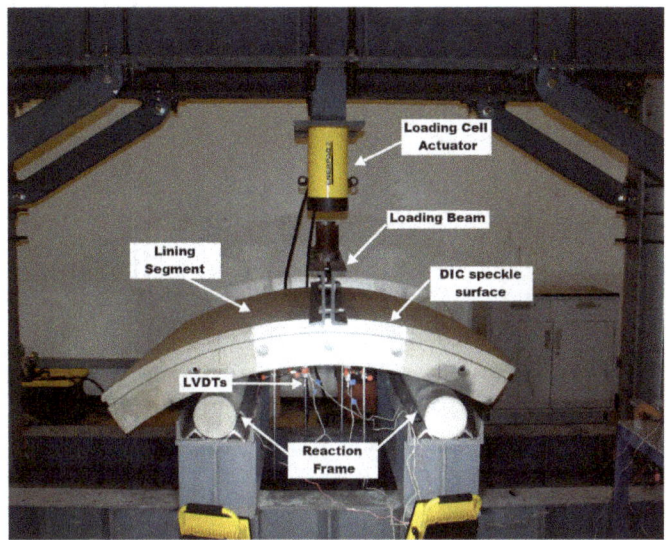

(a)   Flexure Test Setup

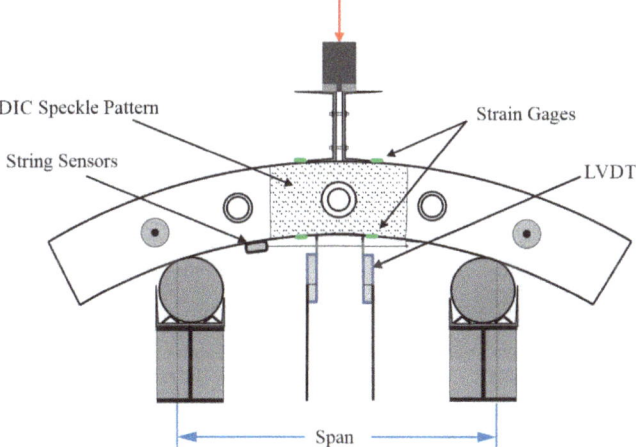

(b)   Schematic diagram of the test setup with the Instrumentations

Fig. 3. Large scale load frame and instrumentations for flexure testing

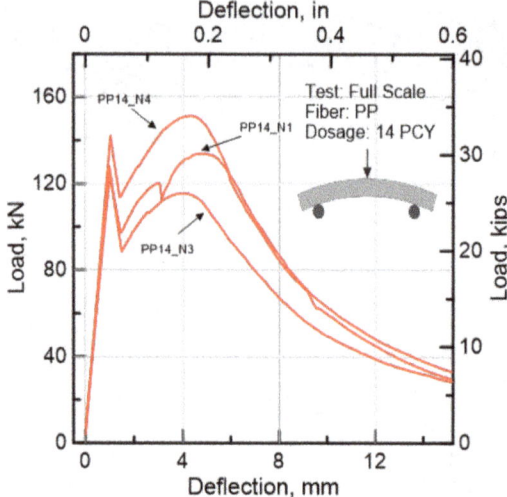

Fig. 4. Load vs. deflection comparison of all three PFRC segments

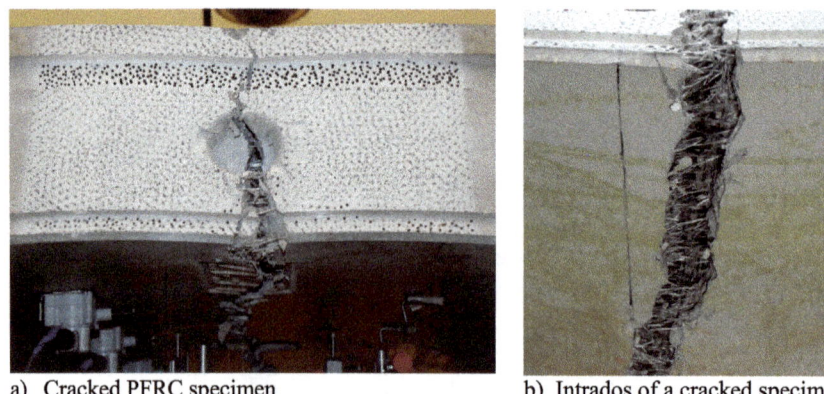

a) Cracked PFRC specimen   b) Intrados of a cracked specimen

Fig. 5. Flexure test failure details

Table 1. Experimental test results of all 3 tunnel lining segments

Fiber Type	SampleID	First Crack Load, $P_{cr,}$ (kN)	First Crack Deflection, δ_{cr} (mm)	Load at L/300 deflection (3.3 mm) (kN)	Load at L/150 deflection (6.6 mm) (kN)	Maximum Load, P_{max} (kN)	Deflection at Max. Load, δ_{max} (mm)	P_{max}/P_{cr}
PP14	PP14_N1	124	1.00	119	111	134	4.83	1.08
	PP14_N3	128	0.99	113	85	128	0.99	1
	PP14_N4	142	1.02	145	108	151	4.32	1.06
	Average	131	1.00	126	102	138	3.38	1.05
	σ_{std}	10	0.01	17	14	12	2.1	0.04
	CV(%)	7	1	14	14	9	62	4

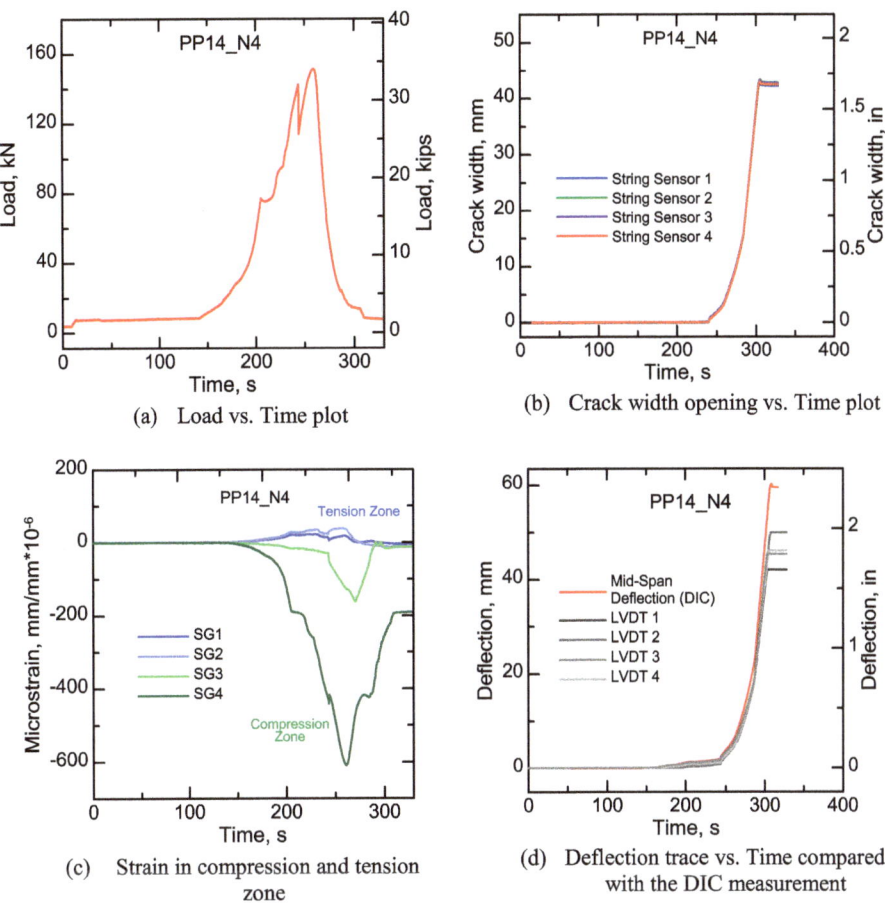

(a) Load vs. Time plot

(b) Crack width opening vs. Time plot

(c) Strain in compression and tension zone

(d) Deflection trace vs. Time compared with the DIC measurement

Fig. 6. Time history plots of PP14_N4 for various instrumentations

Figure 6 shows plots of the data collected from the instrumentations from one of the tests. The DIC method uses a non-contact means of measuring displacements results in precise mid-span deflections compared to LVDTs mounted within 10 cm from the centerline. Figure 6(d) shows the agreement between the deflection measured from LVDTs and DIC. Figure 6(c) shows the strain vs. time plot in which two strain gages in the compression and two strain gages in tension results are presented. A common trend is also observed in all other panels with the compressive strain always higher than the tensile strain, this can be because the strain gages in the tension zone were attached near the grout hole resulting in lower measure of strain.

7 Model Simulation and Correlations

The ASTM C1609 test results can be used to extract the material properties by back-calculating the load-deflection plot and using closed-form solutions as proposed in ACI 544.8R report [10]. A quad-linear tension model can be used for precise back-calculation of the C1609 results based on the complete load-deflection response. The Tensile stress-strain and compression stress-strain models used for back-calculation are shown in Fig. 7. The tensile curve has four linear segments for strain softening/hardening and the compression response has two linear terms for an elastic perfectly plastic model. Two of the ASTM C1609 beam test results were back-calculated using the closed-form equations based on these stress-strain models and simulations were generated as shown in Fig. 8(d). The stress-strain models used to back-calculate the curves are shown in Fig. 8(a–b). The elastic range however shows a higher stiffness compared to the experimental data. This is because the deflections of the panels were measured with respect to the strong floor and not the centerline of the specimen at the support. The observed discrepancies are therefore attributed to the spurious deformations and the settlement of the rollers at the sliding bearings that support the curved-shaped specimen along a thin contact line. While this initial compliance of the entire support frame and the specimen can be adjusted to account for the LVDT that was placed on the ground, the observed differences serve to show that the initial stiffness values are not a direct representation of the elastic modulus of the specimen.

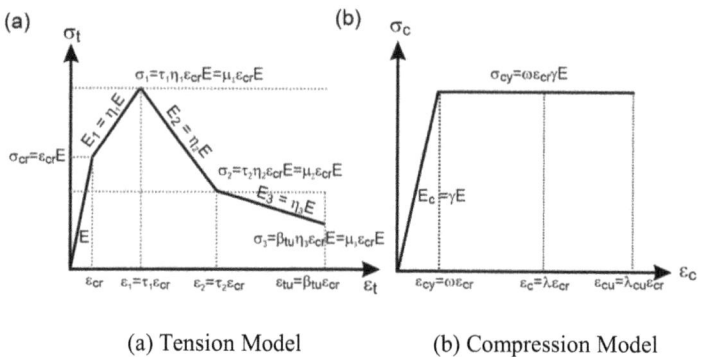

(a) Tension Model   (b) Compression Model

Fig. 7. Stress-strain material models

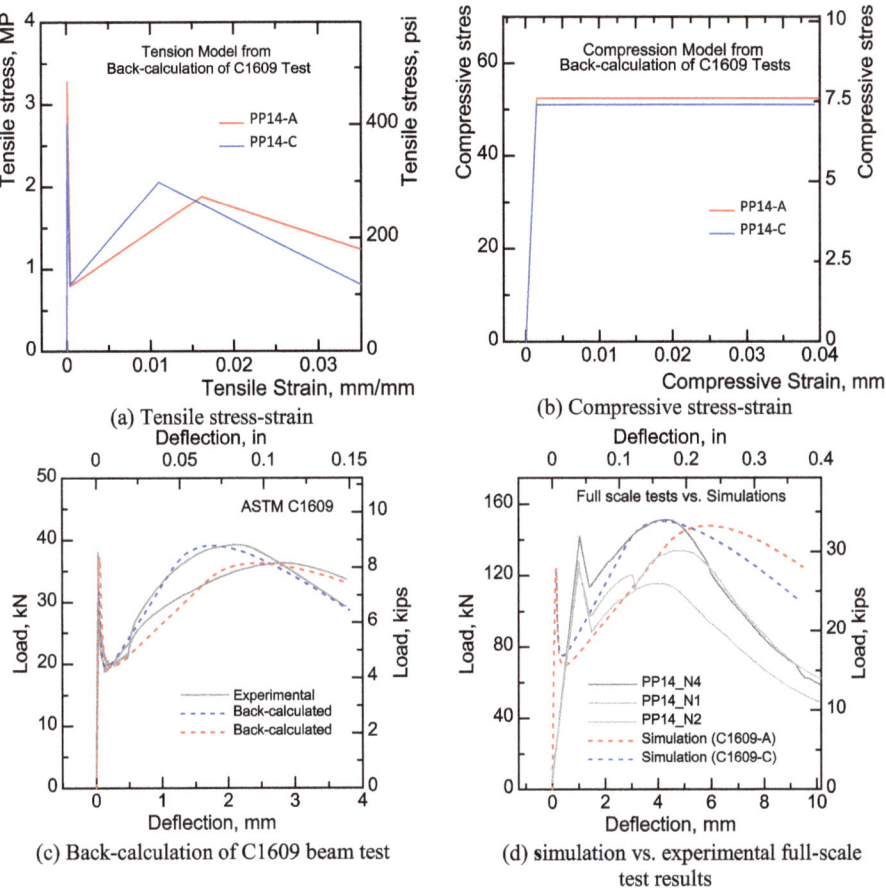

Fig. 8. Material properties extracted from ASTM C1609 tests of PFRC beams

8 Conclusions

Polymeric fibers can be used as alternative reinforcement of tunnel segments as validated by sufficient post-crack strength when only PFRC segments were tested. The study conducted was also able to validate the method used for predicting full-scale responses that use ASTM C1609 tests to back-calculate and extract material parameters which then uses closed-form solutions to predict the full-scale response of the flexure tests. This method of testing and verification using the standard flexural tests can further be used in designing sections with any depth, fiber dosage or hybrid sections.

References

1. de Waal, R.G.A.: Steel Fiber Reinforced Tunnel Segments, Ph.D Dissertation, Delft University of Technology, The Netherlands (1999)

2. Nehdi, M., Abbas, S., Soliman, A.: Exploratory study of ultra-high performance fiber reinforced concrete tunnel lining segments with varying steel fiber lengths and dosages. Eng. Struct. **101**, 733–742 (2015). https://doi.org/10.1016/j.engstruct.2015.07.012

3. Conforti, A., Trabucchi, I., Tiberti, G., Plizzari, G., Caratelli, A., Meda, A.: Precast tunnel segments for metro tunnel lining: a hybrid reinforcement solution using macro-synthetic fibers. Eng. Struct. **199**, 15 (2019)

4. Plizzari G.A.; Tiberti G.: Structural behaviour of SFRC tunnel segments. In: Carpinteri, A., Gambarova, P., Ferro, G., Plizzari, G.A. (eds.) Proceedings of the 6th International Conference on Fracture Mechanics of Concrete and Concrete Structures (FraMCos 2007), Catania, Italy, vol. 3, pp. 1577–1584 (2007)

5. Plizzari, G.A., Tiberti, G.: Steel fibers as reinforcement for precast tunnel segments. Tunn. Undergr. Space Technol. **21**(3–4), 438–439 (2006)

6. De la Fuente, A., Pujadas, P., Blanco, A., Aguado, A.: Experiences in Barcelona with the use of fibres in segmental linings. Tunn. Undergr. Space Technol. **27**(1), 60–71 (2012)

7. Schnütgen, B.: Design of precast steel fiber reinforced tunnel elements. In: Proceedings of the RILEM TC 162-TDF Workshop, Test and Design Methods for Steel Fiber Reinforced Concrete – Background and Experiences, Bochum, Germany, pp.145–152 (2003)

8. di Prisco, M., Felicetti, R.: On fatigue of plain and fibre-reinforced concrete ground slabs. In: 6th International RILEM Symposium, (BEFIB 2004), Varenna, Italy, pp. 1195–1206 (2004)

9. Fédération internationale du béton (2013). Fib model code for concrete structures (2010). http://site.ebrary.com/id/10780745

10. ACI 544.8R-16: Report on Indirect Method to Obtain Stress-Strain Response of Fiber-Reinforced Concrete (FRC), ACI Committee 544 ACI 544.8R (2016)

11. DBV. Design Principles of Steel Fibre Reinforced Concrete for Tunnelling Works, DBV Recommendation, German Concrete Association, pp. 19–29 (1992)

12. RILEM TC 162-TDF. Test and Design Methods for Steel Fibre Reinforced Concrete. Stress–Strain Design Method: Final Recommendation, Materials and Structures, **36**, 262, pp. 560–567 (2003)

13. CNR DT 204/2006, Guidelines for the Design, Construction and Production Control of Fibre Reinforced Concrete Structures, Italian National Research Council – CNR (2006)

14. EHE-08. Code on Structural Concrete, ANNEX 14 – Recommendations for Using Concrete with Fibres, Ministry of Public Works and Transport, Madrid, Spain, pp. 1-55 (2008)

15. ACI 544.7R-16 Report on Design and Construction of Fiber-Reinforced Precast Concrete Tunnel Segments ACI Committee 544 ACI 544.7R (2016)

16. Hillerborg, A., Modker, M., Petersson, P.E.: Analysis of crack formation and crack growth in concrete by means of fracture mechanics and finite elements. Cem. Concr. Res. **6**, 773–782 (1976)

17. Soranakom, C., Mobasher, B.: Flexural design of fiber reinforced concrete. ACI Mater. J. **106**, 461–469 (2009)

18. ACI 544.6R-15 Report on Design and Construction of Steel Fiber-Reinforced Concrete Elevated Slabs, Committee Chair, ACI Committee 544 ACI 544.6R (2015)

19. ACI 566.7R-16 Report on Design and Construction of Fiber-Reinforced Precast Concrete Tunnel Segments ACI Committee 544 ACI 544.7R (2016)

20. 9R-17: Report on Measuring Mechanical Properties of Hardened Fiber-Reinforced Concrete, ACI Committee 544 ACI 544.8R (2017)

21. Yao, Y., Mobasher, B., Soranakom, C.: Closed form solutions for flexural design of hybrid steel fiber reinforced concrete beam. Eng. Struct. **100**, 164–177 (2015)

22. ASTM Standard C1609/C 1609M, "Standard Test Method for Flexural Performance of Fiber-Reinforced Concrete (Using Beam With Third-Point Loading)", ASTM International, West Conshohocken, PA (2007)

Wind Tower FRC Foundations: Research and Design

Marco di Prisco[1(✉)], Claudio di Prisco[1], Giancarlo Fraraccio[2],
Bruno Dal Lago[3], Paolo Martinelli[1], Luca Flessati[1],
Matteo Colombo[1], and Giulio Zani[1]

[1] Department of Civil and Environmental Engineering,
Politecnico di Milano, Milan, Italy
marco.diprisco@polimi.it
[2] Enel Green Power, Engineering and Construction Unit, Rome, Italy
[3] Department of Theoretical and Applied Sciences,
Università degli Studi dell'Insubria, Varese, Italy

Abstract. The design of massive structures, like cast-on-site foundations for wind towers, is not an easy task for design. In fact, this kind of structure is not specifically taken into account by Eurocode 2 standard. The huge concrete volume requires a huge amount of reinforcement, even if only the minimum amount is selected, and it is not clear if the lateral confinement exerted by the truncated cone geometry, mainly loaded along only one radial direction, corresponding to that characterized by the fastest wind direction, really requires the huge transversal reinforcement conventionally introduced in current wind towers. Within the research programme developed in 2019 between Enel Green Power, the Enel Group company dedicated to the development, construction and operation of renewables across the world, and Politecnico di Milano, an experimental investigation, carried out on prototypes characterized by reduced scales (1:15 for the whole structure and 1:4 for the investigation of the only core foundation, where the cylindrical stem is anchored), was conceived to calibrate a reliable modelling with the aim of being extended to full-scale structure, optimizing the required reinforcement. In this perspective, the use of fibre reinforced concrete can significantly improve the construction time by giving a specific toughness to the whole foundation, limiting also the crack opening and the correlated durability problems in the serviceability limit states.

Keywords: Wind tower foundations · Reinforced concrete · Fibre reinforced concrete · Reduced scale prototypes · Finite element modelling

1 Introduction

Wind tower R/C foundations are massive concrete structures that present some design peculiarities strictly correlated to their huge sizes: circular shapes with diameters that may be of more than 20 m and significant depths exceeding 4 m. These structures are usually cast in place and are reinforced with radial bottom and top bars equally spaced, stirrups linked to the radial bars, a post-tensioning anchor-bolt cage fixed by two rings and vertical prestressed bars needed to connect the vertical tubular steel structure long

---

The original version of this chapter was revised: The author name has been amended. The correction to this chapter is available at https://doi.org/10.1007/978-3-030-83719-8_84

P. Serna et al. (Eds.): BEFIB 2021, RILEM Bookseries 36, pp. 831–842, 2022.
https://doi.org/10.1007/978-3-030-83719-8_71

up to more than 120 m with the basis. The main live load is the wind, that flows mainly in one direction and induces load cycles with a frequency that is not close to that of the wind tower first mode. The engineering problem is dominated by the soil-structure interaction. From the construction point of view execution speed is very important and therefore a reduction in the preparation time of the steel cage allowed by fibre use could be very significant.

2 Experimental Programme

The massive structure is usually designed by modelling the foundation as a slab with variable thickness by adopting shells elements and computing both the longitudinal and the transverse bending moment as well as the shear along the load direction. These checks are jointed to the local problem of the bursting acting in proximity of the basket with post-tensioned bars. The research investigation was aimed at clarifying two main design doubts: the first is related to the shear reinforcement, trying to define a reliable design approach for its sizing. The second doubt concerns the soil interaction, because the deformability of the cracked structure and its deformability in the transverse direction cannot be easily modelled by a Winkler model, because the soil very easily reaches non-linearity and the evolution of the soil reaction moves with respect to that predicted by a Winkler model. To better understand the stress evolution correlated to these doubts, two prototypes realized taking into account a standard Wind Turbine shallow foundation, were investigated: the first one reproduces the whole structure in a scale 1:15 and a portion of the soil foundation by means of a tank made of a metal corrugated sheet 1.5 m deep and with a diameter of 3.64 m; the second one reproduces only the central core strip, where shear is expected more critical, simplifying the tapered foundation suitable constraints able to transmit the bending moment. The two prototypes were mainly used to calibrate a Finite Element model that could be instrumental to investigate the real cases underlying the reliability of the current design approach and improving it accordingly to the most sophisticated one. Both the prototypes investigated three solutions: the real R/C solution, a solution without the shear reinforcement and a solution in FRC removing the shear reinforcement and reducing the radial and the circumferential reinforcement of about 50%.

3 Materials

Both the prototypes were cast in a precast plant. The mix design of the materials used is described in Table 1. The average cubic compressive strength was 58 MPa, that correspond to C40/50. It is important to observe that the FRC proposed in the real case should be a 3c one, accordingly with Model Code 2010, that means a fibre reinforced concrete with characteristic residual strengths $f_{R1k} \geq 3$ MPa and $f_{R3k} \geq 2.7$ MPa. Nevertheless, in the 1:15 prototype preparation we were forced to use a large amount of small fibres, because the reduced sizes of the cage forced a high flowability. The

Table 1. Concrete mix design.

Component	Dosage [kg/m³ - l]	Plain	FRC1	FRC2
Cement 42.5 R	300	✓	✓	✓
Limestone filler	100	✓	✓	✓
Sifted sand 0–3 mm	620	✓	✓	✓
Washed sand 0–12 mm	640	✓	✓	✓
Gravel 8–15 mm	510	✓	✓	✓
Superplasticizer	1.5	✓	✓	✓
Water	180	✓	✓	✓
Steel fibres 65/35 4D	50		✓	
HC steel fibres OL 13/16	70			✓

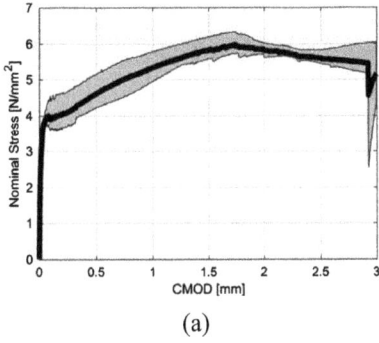
(a)

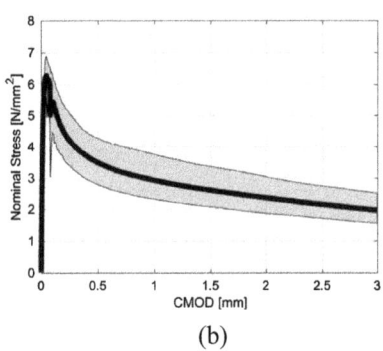
(b)

Fig. 1. Bending behaviour of FRC1 (a) and FRC2 (b) materials according to EN14651 beam test.

relatively low-strength of the matrix produced low-performance of the FRC material, but this result was seen in favour of safety. The bending behaviour identified according to EN14651 tests is briefly described in Fig. 1. FRC1 was used for prototype 1:4 and FRC2 for prototype 1:15.

4   Full Structure Prototype

The 1:15 prototype set-up is described in Fig. 2. It consisted in a truncated-cone foundation, a steel pipe (720 kg) connected by means of a bolted flanged connection to the basket with the post-tensioned vertical bars, the truncated-cone foundation, the tank filled with about 15.5 m³ of sand and four added masses to reproduce the stress pattern computed in the real structure according to the similitude theory.

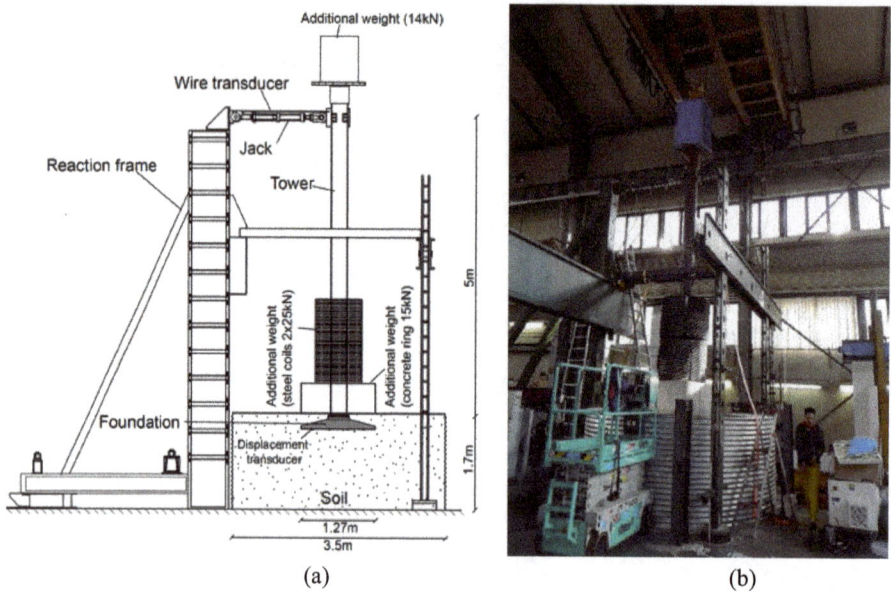

Fig. 2. Test-set-up of the prototype of the whole structure: (a) sketch; (b) lab photo.

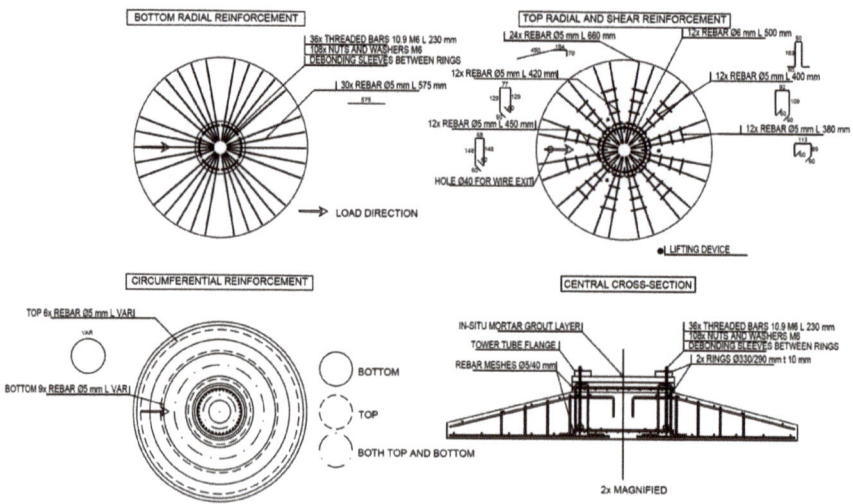

Fig. 3. Reinforcement introduced in the scaled foundation

The top one (1350 kg) reproduces the weight of the shaft, the engine and the wind-tower blades; the two steel coils (2500 kg × 2) and the concrete circular crown (1500 kg) reproduce the weight of the concrete foundation and they are applied directly in contact with the soil. The reinforcement introduced in the prototype of the concrete foundation is specified in Fig. 3, while a picture of the reinforcement cage as well as of

Fig. 4. Concrete R/C foundation: (a) cage; (b) top view.

the concrete foundation are shown in Fig. 4. The soil layers were deposited by means of a sand deposition device (Figs. 5a, 5b) suitably designed in order to guarantee a relative density of about 80%. Four vertical LVDTs were put in the soil to measure the vertical displacements and the base rotation; two LVDTs measured the vertical and the horizontal displacement of the side more solicited by the horizontal jack that reproduces the wind action, while another LVDT measured the top horizontal displacement and a load cell the load applied by the jack. In only one test also pressure sensors were located under the foundation to investigate the soil pressures. Twenty-six strain gauges measured the strains in the radial bars and in the stirrups. Each test was split in several load steps: an accurate description is shown in Table 2.

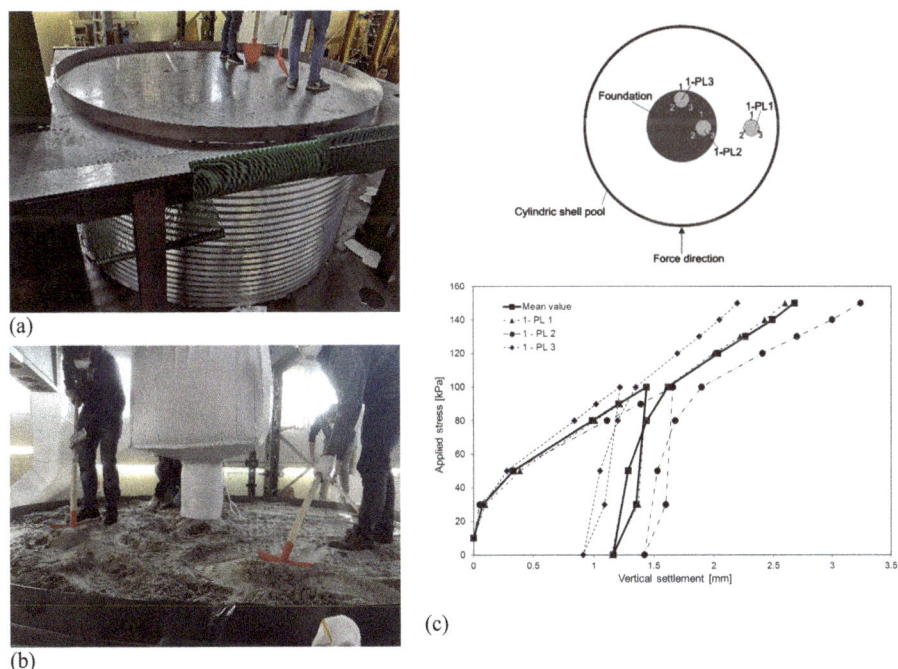

Fig. 5. Soil preparation: (a) sand deposition device; (b) sand dispenser; (c) initial test G

Table 2. Test steps

Phase	Type	Description
G01	Gravity	Placement of the base ring element
G02	Gravity	Placement of the first coil
G03	Gravity	Placement of the second coil
G04	Gravity	Placement of the top cap with steel mass
C01	Cyclic	1 pseudo-static cycle from 0 to 0.62 ULS load (P = 3.0 kN)
C02	Cyclic	1 pseudo-static Single cycle from 0 to 0.62 ULS load and then 1000 cycles from 0.41 to 0.82 ULS load (3.65 kN) f = 0.10–0.22 Hz
C03	Cyclic	Single pseudo-static cycle 0.62 - 0.00 - 1.00 - 0.62 ULS load and repetition of 1000 cycles 0.41 to 0.82 ULS load; f = 0.1–0.22 Hz
C04	Cyclic	200 cycles from 0.06 to 0.67 ULS load; f = 0.10–0.22 Hz
M01	Monotonic	From 0 to maximum jack stroke and unloading (pseudo-static)

The reinforcement ratios of the circumferential and radial reinforcement was carefully reproduced: the comparison between the full-scale and the prototype is shown in Figs. 6a, b. The progressive additions of the masses (G01-G04) has allowed the characterization of the soil stiffness (Fig. 5c). The ULS behaviour for the six tests (two nominal identical tests for each solution investigated) expressed in terms of applied horizontal load vs. horizontal top displacement is proposed in Fig. 6c. It is possible to observe as F03 tests corresponding to Hybrid solution exhibit a response very close to that exhibited by conventional R/C (curves F01) in the last monotonic load step (M01). A careful analysis of the previous load steps G and C, has clarified that the small load jump corresponds to the first bending crack at the intrados (Fig. 6c). The comparison underlines a delaying of the first crack, even if the curves are relatively close each other, with the only exception of the curve F01–1, where the soil was characterized by a lower density, certified by the response in the steps G01–G04.

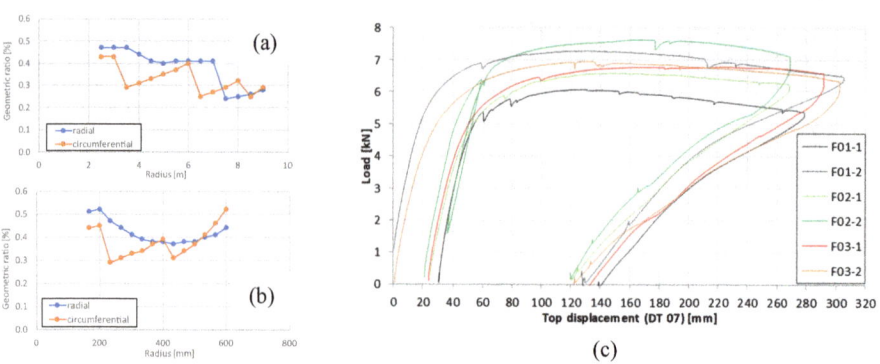

Fig. 6. (a, b) Reinforcement ratios in the real structure and in the prototype; (c) global response

5   Critical Core STRIP Prototype

The prototype 1:4 was investigated to calibrate in a simple test the bending/shear interaction in the central core along the critical strip aligned to the load direction. The load transmitted by the shaft to the foundation is here imposed by means of two jacks controlled by the same hydraulic circuit and a load cell, reproducing the bending component and neglecting the shear transmission. A load cell controlled also the horizontal bottom constraint reaction. The three solutions investigated reproduce the R/C real case (SR; Fig. 7a), the same solution without the shear reinforcement (PC; Fig. 7b) and the Hybrid solution without shear reinforcement, but with random fibres (FRC; Fig. 7c). The test set-up is exemplified in Fig. 7d, while the instrumental equipment adopted is described in Fig. 7e. The load procedure considered 4 phases load controlled: a first phase (A) of 25 cycles at 150 kN; a second phase (B) of 25 cycles at 200 kN; a third phase of 400 cycles (C) at 250 kN and finally a fourth phase (D) monotonic up to failure. Table 3 resumes the failure mode and the final load step for the three solutions investigated by means of two nominally identical tests.

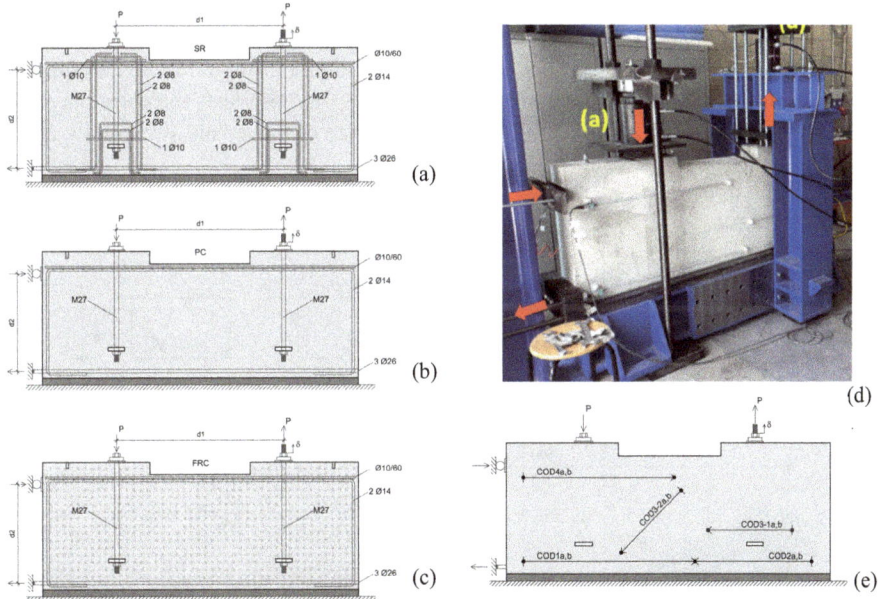

Fig. 7. Critical strip prototype: (a,b) R/C solution with and without shear reinforcement; (c) FRC solution; (d) test-set-up; (e) measure equipment.

Table 3. Test results prototype 1:4.

Test	Failure load (kN)	Load phase	Failure mode
SR-1	250 (124° cicle)	C	Shear
SR-2	250 (14° cicle)	C	Spalling
PC-1	217.5	B-C	Shear
PC-2	200 (19° cicle)	B	Shear
FRC-1	369.6	D	Shear
FRC-2	353.5	D	Yielding Φ26 bars

The Hybrid solution is the only one that was able to reach phase D up to a load that is more than 40% the R/C solution, even without bursting and shear reinforcement.

6 Modelling

After a check carried out by using a 2D Finite element code, as frequently adopted in the real design of standard wind turbine shallow foundations, a non-linear modelling of both the prototypes was performed by means of ABAQUS code and using the concrete damage-plasticity model [1–4]. The tests were first aimed at reproducing carefully the behaviour of the simple R/C beam resumed in Fig. 8. In these tests the roles played by the Finite Element adopted, the bond model, the choice of the tensile strength [5, 6] and

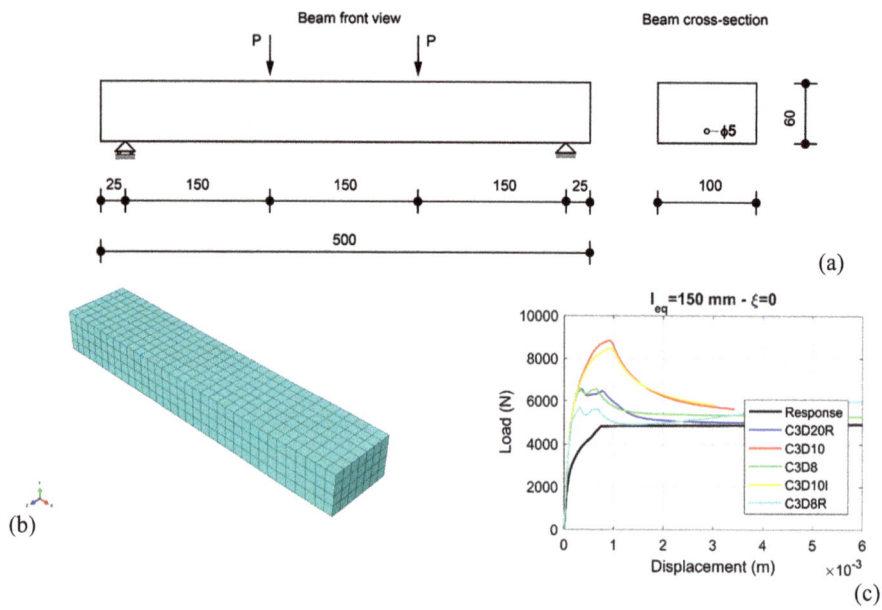

Fig. 8. Unnotched beam specimen: (a) test-set-up; (b) mesh adopted; (c) influence of element choice on the global response

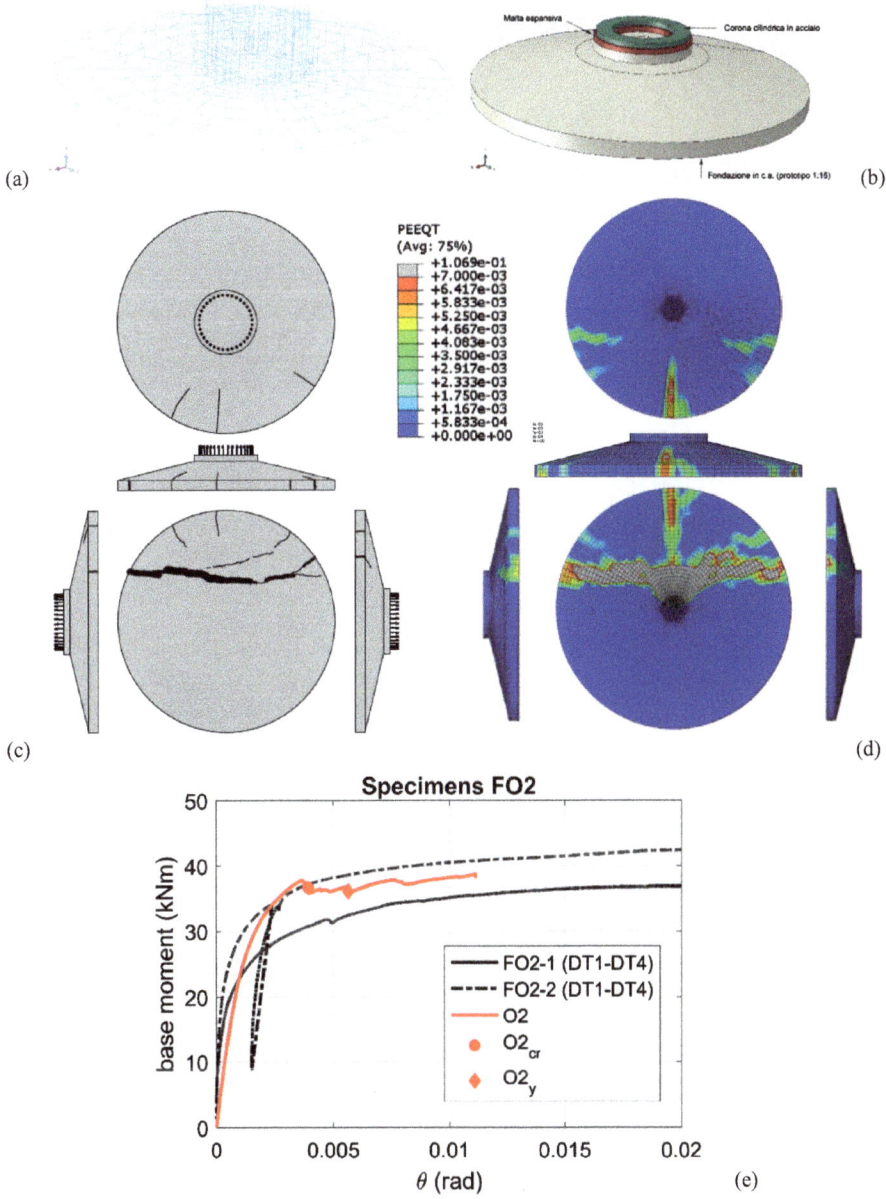

Fig. 9. Prototype of the whole structure: (a) cage mesh; (b) 3D foundation model; (c) experimental crack pattern; (d) F.E. crack pattern; (e) global response.

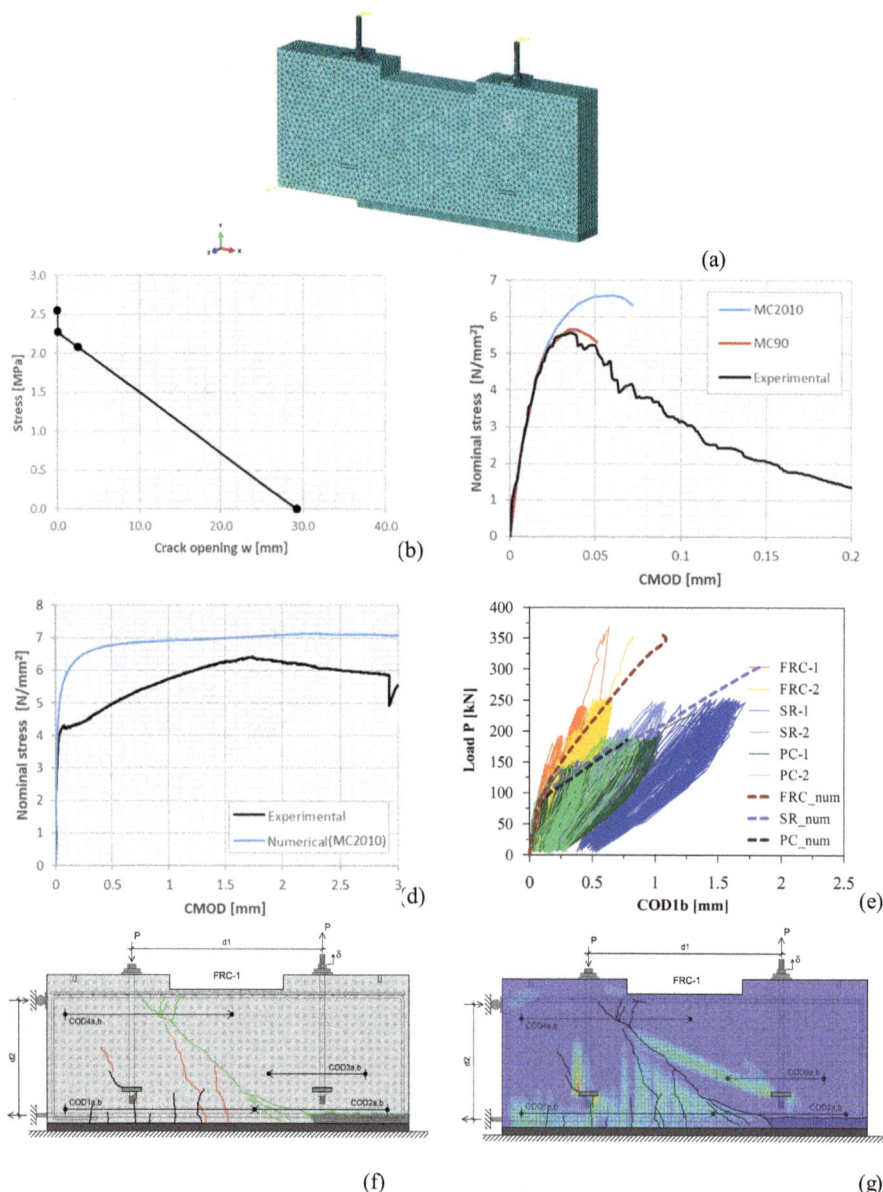

Fig. 10. Core strip model: (a) 3D mesh; (b) tensile uniaxial law; (c) plain beam test; (d) FRC beam test; (e, f) experimental and model crack pattern.

of the creep parameters were analysed. In Fig. 8d the role of the finite element used is shown. The careful numerical calibration, carried out employing an elasto-plastic model based on a not-associated flow rule and a Mohr-Coulomb constitutive model for the soil, allowed to reach a reasonable reproduction of both the prototypes. As an example, in Fig. 9 the reinforcement mesh (Fig. 9a), the solid 3D model (Fig. 9b), the experimental and the simulated crack pattern (Figs. 9c, d), and the mechanical response expressed in terms of bending moment vs. base rotation is shown. The quite good agreement is significantly affected by the soil density, because the non-linear behaviour depends on both soil plasticity and foundation cracking and therefore is a classic example of strong soil-structure interaction. The prototype 1:4 was analysed with the same approach. The mesh optimized is shown in Fig. 10a. The post-peak strength is modelled by means of a softening model in agreement with Model Code 1990 (Fig. 10b, c): in fact the fracture energy proposed in MC2010 for plain concrete seems to overestimate the mechanical response of the FRC plain concrete beam tested according to EN14651 (Fig. 10d). The global test response is shown in Fig. 10d: the computed crack pattern is very close to that experimentally obtained (Fig. 10e, f). The good agreement of the monotonic response and the predicted one for the three solutions is highlighted in Fig. 10e in terms of load vs. COD1b.

7   Conclusions

The research investigation on the wind-tower foundation has highlighted the following remarks:

- The prototype 1:4 has clearly indicated as the hybrid solution can be regarded as an improved solution in relation to the conventional R/C solution, because it is able to better control crack propagation during cycling loads, allowing the reaching of a maximum bending moment in the 2D critical core strip, aligned to the wind direction, about 40% higher than the R/C one.
- The prototype 1:15 exhibited a strong soil-structure interaction: even if the cracking in hybrid solution is delayed, with respect to the conventional R/C solution, the main difference on the maximum bending moment is observed in the two R/C tests, where a significant difference in the soil density was measured.
- The modelling required a careful calibration of the type of element, of the mechanical parameters and of the bond interaction.
- Even though in the prototype 1:15 no significant difference in the global response was detected between the investigated solutions, size effect could amplify the differences between the solutions investigated and, in particular, between that characterized by the full lack of shear reinforcement and fibres. A specific analysis is in progress.

Acknowledgements. The authors thank the MS Eng. Pietro Marveggio, Gabriele Frigerio, Claudio Urso and the technicians Giacomo Vazzana, Daniele Spinelli, Andrea Stefanoni and Daniele Sandrinelli for their precious contribution in the complex execution of the tests.

References

1. ABAQUS: Analysis User's Manual, version 6.14, vol. 2 (2012)
2. Alfarah, B., López-Almansa, F., Oller, S.: New methodology for calculating damage variables evolution in Plastic Damage Model for RC structures. Eng. Struct. **132**, 70–86 (2017)
3. Lee, J., Fenves, G.L.: Plastic-damage model for cyclic loading of concrete structures. Eng. Mech. **124**(8), 892–900 (1998)
4. Lubliner, J., Oliver, J., Oller, S., Oñate, E.: A plastic-damage model for concrete. Solids Struct. **25**(3), 299–326 (1989)
5. CEB-FIP model code 1990. Bull. d'Inform. **213**, 214 (1993)
6. fib Model Code 2010, Fédération Int. du Béton Ernst & Sohn, Lausanne (2013)

Textile-Reinforced Concrete

Development of Polymeric Textile Reinforced Concrete Structural Members

Vikram Dey[1], Anling Li[2], Gozdem Dittel[3], Thomas Gries[3],
Steve Schaef[4], and Barzin Mobasher[5(✉)]

[1] DCI Engineers, Los Angeles, CA, USA
[2] College of Civil Engineering, Hunan University,
Changsha, People's Republic of China
[3] Institut Fuer Textiltechnik (ITA), RWTH-Aachen University,
Aachen, Germany
[4] Development Admixture Systems- BASF North America, Jersey City, USA
[5] School of Sustainable Engineering and Built Environment,
Arizona State University, Tempe, AZ, USA
barzin@asu.edu

Abstract. The life cycle costs of structural systems are a function of raw materials, labor, energy, environmental impact, serviceability, and durability. Textile Reinforced Concrete (TRC) using alternative reinforcing systems are cost competitive and have gained popularity among new construction materials. TRC composites are uniquely lightweight with a very high specific strength, stiffness, and ductility that compete and outperform light gage steel and wood products. TRC's high potential durability and high strength is amenable to continuous production and formability, thus making it highly sustainable. An effective manufacturing technique was developed using automated pultrusion process for efficient production of TRC structural sections. The objective of this study is to develop and characterize the properties of PP based mesh reinforcement. Mechanical properties of textile reinforcement systems are used to justify the use of textiles for use in the production of TRC composite structural sections. Test results of flexural and tension specimens are discussed in terms of closed loop test results as well as Digital Imgae Correlation (DIC) technique. The study indicates that PP-TRC composites with 4% (representing both warp and weft directions) textile volume fraction can reach maximum tensile strength of 11 MPa with strain capacity of 23%. Flexural samples show an apparent flexural strength of 40 MPa and deflection capacity of up to 40 mm for a flexural sample on a span of 254 mm.

1 Introduction

Textile Reinforced Concrete (TRC) composites are a new class of sustainable construction materials with superior tensile strength and ductility. Textiles that provide the distributed reinforcement are made from single or multifilament yarns. The geometric pattern of fabric consists of interconnected longitudinal and transverse fill yarns and the yarn to yarn spacing dictates the structure and density of the weave which in turn affects the penetrability of the cementitious matrix into the mesh. An excellent bond

© RILEM 2022
P. Serna et al. (Eds.): BEFIB 2021, RILEM Bookseries 36, pp. 845–854, 2022.
https://doi.org/10.1007/978-3-030-83719-8_72

develops with cement matrix by mechanical interlock and anchorage of yarns which results in reinforcement of the matrix. A wealth of recent information pertaining to the methodologies, properties, and areas of applications for textile reinforced cement based materials are available [1–3].

Using high performance fibers such as AR-Glass and Carbon, very encouraging tensile strength and ductility responses have been observed by various experimental programs. Tensile strength of the order of 20–25 MPa and strain capacity of 3–5% can be consistently obtained [4, 5]. However the use of polymeric fabrics has not been addressed significantly. TRC laminates could be easily adopted and modified to fabricate composites with structural properties at competitive costs. Furthermore, structural shapes such as channels, angles, hollow sections and plates of any desired length and cross-sections are possible for high volume manufacturing.

A proprietary fibrillated polypropylene microfiber developed by BASF Corporation was used to realize two dimensional warp knitted textile layers. Two different stitching patterns were selected namely pillar and tricot stitch. The textiles were warp knitted in the Composites Division of Institut für Textiltechnik (ITA) of RWTH Aachen University. A composite processing method based on pultrusion technique was developed to manufacture TRC composite laminates for mechanical testing.

The specimens developed from the pultrusion technique were tested under uniaxial tension and four-point bending, and effects of warp knitting technique was evaluated. Mechanical properties of these TRC laminates (two dimensional panels) were further compared to the unidirectional laminated composites systems (panels reinforced in one dimension) developed during an earlier study. It is known that unidirectional composites exhibit excellent in-plane but poor inter-laminar properties. This is due to lack of reinforcements in the thickness direction and leads to poor damage tolerance in the presence of inter-laminar shear stresses. However, tricot warp knitted fabrics are ideal for reinforcing thin sections which undergo biaxial states of stress by the virtue of their balanced ply and improved interlaminar properties.

2   Textile Manufacturing

Textiles were made from multifilament yarns using an open grid approach. The potential penetrability of the cement into the bundle spaces depends on the knitting structure and the density of the grid. The tightening of the joints that connect the warp and weft yarns is affected by the knitting process which uses a specific yarn system for the attachment of orthogonal yarns. Figure 1 shows the different configurations used. Pillar warp knit is a basic one-face warp in which a knit is worked from one fully threaded warp. In the first row the thread forms a stitch on the first needle; in the second row this occurs on the third needle. Stitches are made alternately, first on one side, then on the other. In tricot warp knit one thread forms stitches alternately in two neighboring columns. By the pulling of loops, stitches are made alternately on one side, then on the other [6].

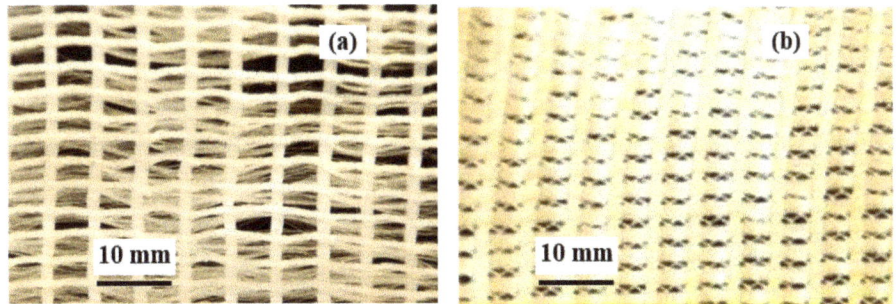

Fig. 1. Textiles warp knitted from MF40 microfibers (a) Pillar warp knit pattern, and (b) tricot warp knit pattern

3   Pultrusion of Cementitious Textile Composites

The automated production process uses a pneumatically controlled, pull-press-release mechanism system. The continuous manufacturing process enables casting of long full-scale sections with optimized cross sections. Multiple pneumatic pressure points allow for shear pressure to aid the impregnation process. The system reduces labor intensive manufacturing cost for prefabricated construction elements. Various shapes of specimens can be manufactured in this process.

The efficient TRC production method by pultrusion is shown in Fig. 2 [7, 8, 9, 10]. During the pultrusion, fabric passes through a water bath, a slurry infiltration chamber, and then through a set of rollers to squeeze the matrix between the fabric openings while simultaneously removing excess matrix. The matrix coated fabric is clamped and pulled through by a set of pneumatic pistons that are controlled by a set of solenoid valves, as shown in Fig. 2(b). The pultruded fabric is placed on a mold wherein multiple fabric layers are stacked and pressed using a pneumatic press. The fabric layers are held at a constant air pressure between 70–140 kPa for 24 h before being demolded and cured in a curing chamber. Proper impregnation of cement paste into the open cell structure of the textile is an operational concern that depends on rheology and processing parameters. Lack of uniform impregnation, can compromise the strength due to improper stress transfer between the matrix and textile. The automated pultrusion system consists of five central stations oriented in succession along a continuous production line. The feed section includes the fiber spool, water and paste baths; guidance sections DC powered rollers and the pneumatic pulling mechanism as shown in Fig. 2. Ultimately the take up section pulls the impregneated textile under a press that applies a hydrostatic load.

4   Experimental Program

A mortar mix design was used based on a blend of Type I/II Portland Cement and fly ash as a partial substitution of 15% of cement. A sand/cement ratio of 0.5, and a water/binder ratio of 0.36 were used High range water reducer (HRWR) developed by

BASF Construction Chemicals was used to ensure workability of this mix which crucial to ensure matrix penetrability within the open spaces of the yarn. Detailed mix design is summarized in Table 1. Samples were cured for 28 days and tested under direct tension and flexure under static loading to characterize the strength, strain capacity, stiffness, and distributed cracking mechanisms.

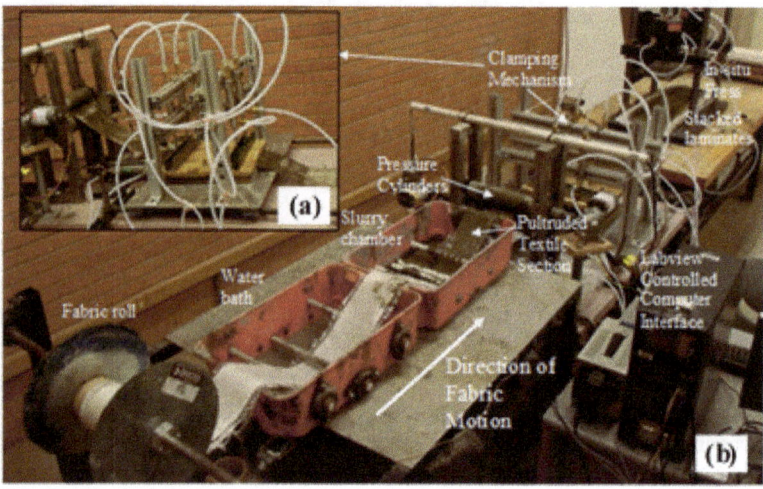

Fig. 2. Pultrusion setup (a) Clamping and pulling mechanism, (b) Pultrusion of textile composites.

Table 1. Mix design for the cementitious matrix

Material	Portland cement	Fly ash (Class F)	Fine silica sand	Water	HRWR (BASF)
Weight fraction (%)	48%	7.2%	24%	20%	0.03%

5   Tension and Flexural Properties of TRC Composites

Experimental setup for tension and flexural tests are shown in Figs. 3a and b respectively. The strain curves of tensile stress based on stroke displacement observed in this study are presented in Fig. 4. Results show that the first crack strength is about 2.8–3.0 MPa. There is distributed cracking phase that starts with the first cracking and is followed by significant multiple cracking until a strain level of 0.03 for pillar warp knitted to a strain level of 0.07 for the tricot warp knitted fabrics. In the multiple cracking phase, the tensile strength of tricot TRC sample is bigger than of the pillar TRC sample. This could be attributed to the tightening and interlocking effect of the stitching pattern contributing to better capacity. Both sets of sample exhibit a strain capacity of up to about 23% at a strength of 11 MPa.

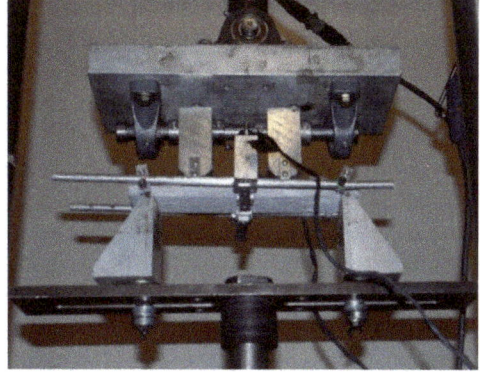

Fig. 3. Experimental setup for (a) Tensile response of PP-TRC specimen with 4% Tricot-Warp Knitted Textile: (a) cracking distribution, (b) flexural test on TRC specimens

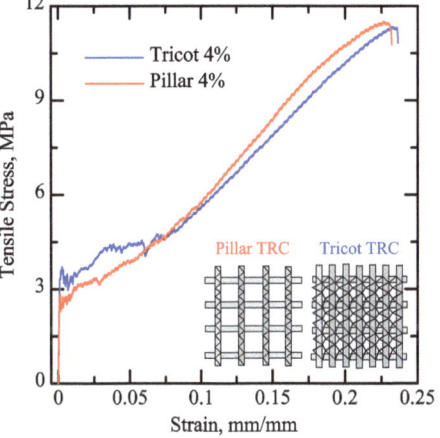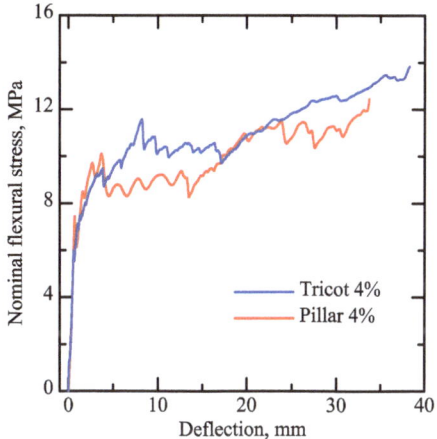

Fig. 4. Tensile properties of pillar warp knitted and tricot warp knitted TRC composites

Fig. 5. Flexural response of pultruded PP-TRC composites

The flexure responses of the two TRC composite types specimen are shown in Fig. 5. The tricot warp knitted TRC composite showed a slightly better response than that of the pillar warp knitted. This is likely due to additional stitching within the tricot textile which provides for a slight increase in load transfer among the yarns and thus increases the ductility. The maximum nominal flexural stress is a slightly bigger than the tensile strength for the TRC materials.

6   Measurement of Crack Spacing Using DIC

Digital Image Correlation (DIC), is a non-contacting optical full field deformation measurement approach that can better address the complex behavior of this class of materials. DIC technique was developed by Sutton et al. [11] and Bruck et al. [12] and has been widely applied for composites, and reinforced concrete sections [13, 14, 15]. In order to perform DIC, an area of interest (AOI) is manually specified and further divided into an evenly spaced virtual grid. The strain fields can be derived by smoothing and differentiating the displacement fields. Distributed cracking mechanism and local strain fields can be easily documented by the DIC method. The technique can also be used for automated determination of crack density, crack spacing, and damage evolution. Different fabric systems have different characteristic responses which correlate the crack spacing and composite stiffness. Statistical measures of evolution of distributed cracking system as a function of applied strain can be also correlated with both the tensile response and the stiffness degradation.

Figure 6 shows the nature of sequential formation of distributed cracking with specific spacing in a tension test. Evolution of distributed cracking and redistribution of local strain fields for a typical specimen is documented in the Fig. 6. The discontinuous distribution of the longitudinal displacement can be used to measure the crack spacing and correlated with experimental strain measured locally with transducers as shown in Figs. 7(a) and (b) which compare the stress-strain and crack spacing vs. strain. Since the onset of first crack, a general decrease in crack spacing is observed until it reaches a steady state defined as saturation crack spacing. Beyond this stage additional imposed strain results in widening of the existing cracks, as no new crack is formed [1].

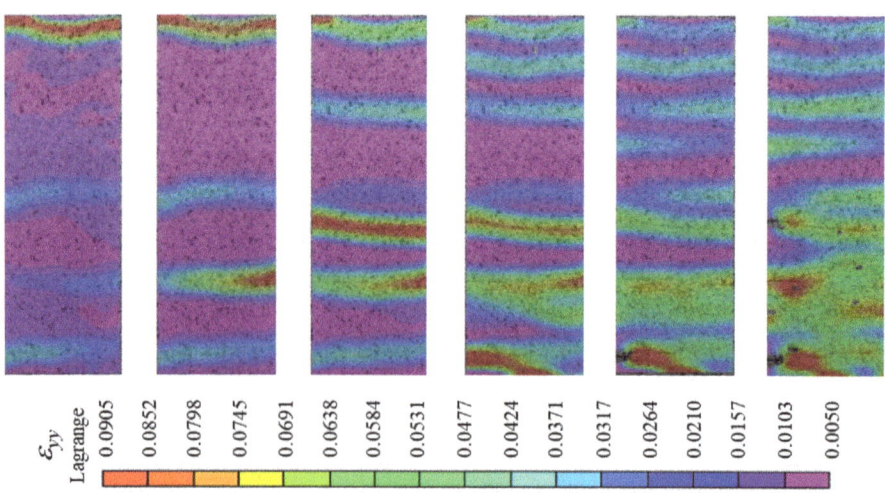

Fig. 6. Distributed cracking observed PP-TRC specimen with 4% Pillar Warp Knitted Textile

Using the DIC software Vic-2D, a vitual extensometers tool could be use to measure the displacement and crack width at any position on the AOI as shown in Fig. 8(a) and (b). The crack width curves of tensile stress based on DIC analysis observed in this study for Tricot TRC sample are presented in Fig. 8(a) as individual crack width measurements are shown. It is shown that the crack width of a crack may decrease during the multi-cracking stage as a new crack forms it its vicinity. The crack width of each crack increases during the crack stabilization stage. The average crack width increases after a crack forms and before the failure in textile rupture.

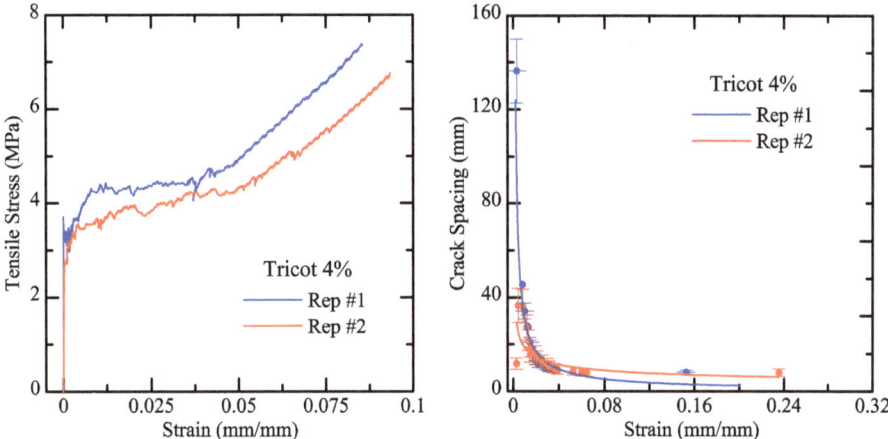

Fig. 7. (a) Stress vs. strain response (b) crack spacing vs. strain response of tricot-warp knitted samples.

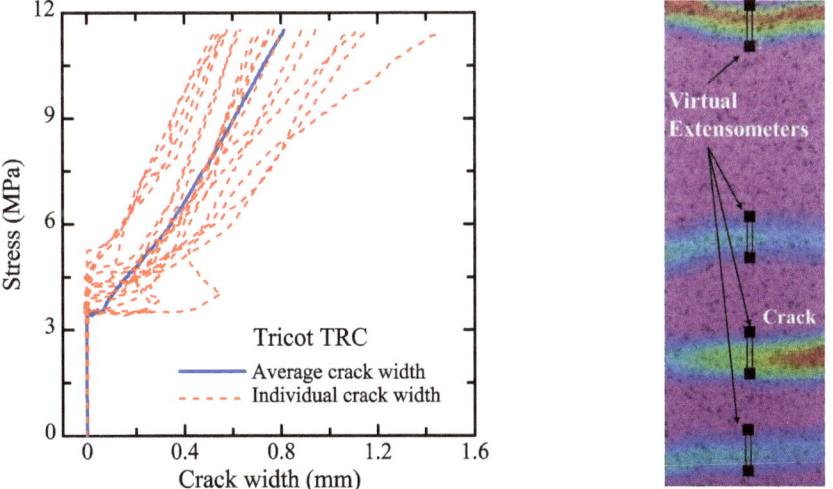

Fig. 8. Stress-crack width and crack spacing-strain response observed in PP-TRC specimen with 4% Tricot-Warp Knitted Textile

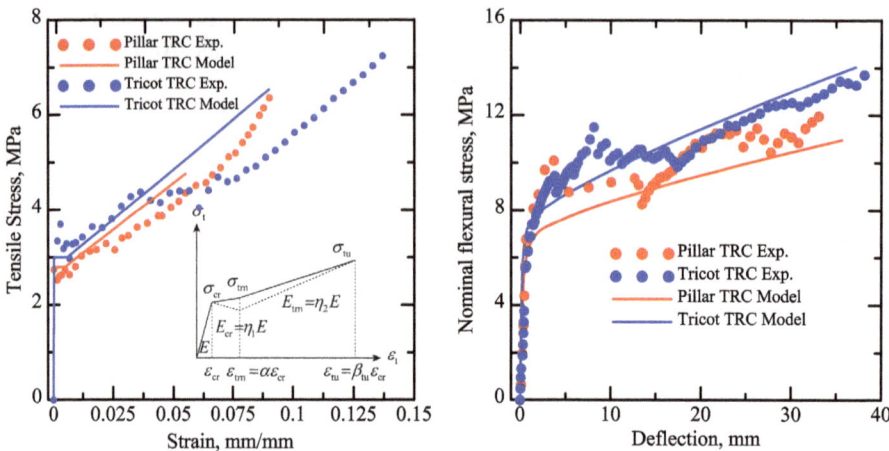

Fig. 9. Nominal flexureal stress versue deflection response along with the backcalcuated constitutive tensile response: (a) tensile stress-strain diagrams (b) nominal flexureal stress-deflection curves simulated using the backcalcuated response

7   Computation of Material Tensile Property Using Inverse Analysis

An inverse analysis procedure is used to obtain the tensile material properties from the experimental load-deflection flexural tests [1]. A tri-linear tension and an elastic-perfectly-plastic compression stress-strain model is used to fit the flexural stress-defection data. The parameters of the tensile model and the curves for the fit representing the TRC material are shown in Fig. 9(a). Figure 9(b) shows the nominal flexural stress-deflection curves simulated using the backcalcuated computed data.

Based on the model of the material shown in the inset of Fig. 9(a), the nominal flexural stress-deflection response of the TRC specimens is calculated and shown that the model could well simulate the flexural response of the TRC material according to the tensile properties obtained from the test, the inverse analysis parameters of the TRC specimens are shown in Table 2. This model could be used to design the flexural performance of structural elements constructed form the TRC materials. Note that when the specimen fails under flexural loading, the tensile stress of the bottom fiber is about 50% of the tensile strength of the material, so the tensile strength of the material is not fully utilized.

Table 2. Inverse analysis parameters of the TRC specimens.

ID	Elastic modulus, E (GPa)	Cracking tensile strain, ε_{cr} ($\mu\varepsilon$)	Cracking strength (MPa)	α, Transition strain/cracking strain	ω, Compressive yield strain	η_1, Modulus ratio	η_2, Modulus ratio	β_{tu}, Normalized tensile strain
Pillar TRC	28	100	2.8	50	16.4	0	0.0014	550
Tricot TRC	30	100	3.0	60	15.3	0	0.0014	900

8   Conclusions

Tensile and flexural properties of polypropylene textile reinforced concrete manufactured using an automated pneumatic pultrusion setup was studied. This computer controlled manufacturing system could be easily adopted and modified to fabricate different structural shapes such as channels, angles, hollow sections and rectangular plates of any desired length and cross-sections. This would enhance the cost effectiveness, quality control and manufacturing efficiency of such cementitious composite materials which could easily replace existing building materials for several structural applications. The effect of two different textile warp knitting style on mechanical properties was evaluated in this study. It was found out that most specimens exhibited strain hardening behavior which further resulted in distributed cracking along the gage length of the specimen. While the effect of warp knitting, did not have much impact on the tensile properties of the composite, the flexural properties especially toughness measured seemed to be have a positive effect to the tricot warp knitting technique due to enhanced interlocking and tightening of the warp knitted filaments. It is to be noted that due to the two-dimensional nature of the biaxial warp knitted textiles, these TRC composites are far more efficient in resisting realistic biaxial and high strain rate loading scenarios such as blast or impact loading, compared to traditional FRC or unidirectional fiber composites.

References

1. Mobasher, B.: Mechanics of Fiber and Textile Reinforced Cement Composites, p. 480. CRC Press, Boca Raton (2011). ISBN 9781439806609
2. Brameshuber, W., Brockmann, T., Hegger, J., Molter, M.: Textilbeton - Betontechnologie und Tragverhalten, Untersuchungen zum Textilbewehrten Beton, Beton 09/2002, Seiten, pp. 424–429 (2002)
3. Peled, A., Mobasher, B., Bentur, A.: Textile reinforced concrete. In: Modern Concrete Technology, vol. 18. Taylor and Francis (2017)
4. Peled, A., Mobasher, B.: Pultruded fabric-cement composites. ACI Mater. J. **102**(1), 15–23 (2005)
5. Reinhardt, H.-W., Krüger, M., Grosse, C.U.: Concrete prestressed with textile fabric. J. Adv Concr. Technol. **1**(3), 231–239 (2003)
6. Weblinks. http://warpknits.vrtxinc.com/product/all-categories/tricot-knit-fabric, http://www.norcostco.com/sharkstooth-scrim-material-20-flame-retardant-white.aspx, http://en.texsite.info/Tricot_weave
7. Yao, Y., Zhu, D., Zhang, H., Li, G., Mobasher, B.: Tensile behaviors of basalt, carbon, glass and aramid fabrics under various strain rates. J. Mater. Civ. Eng. (2016). https://doi.org/10.1061/(ASCE)MT.1943-5533.0001587
8. Rambo, D., Yao, Y., Silva, F.A., Filho, R., Mobasher, B.: Experimental investigation and modelling of the temperature effects on the tensile behavior of textile reinforced refractory concretes. Cem. Concr. Compos. **75**, 51–61 (2016). https://doi.org/10.1016/j.cemconcomp.2016.11.003

9. Nishigaki, T., Suzuki, K., Matuhashi, T., Sasaki, H.: High strength continuous carbon fiber reinforced cement composite (CFRC). In: Brandt, A.M., Marshall, I.H. (eds.) Proceeding of the Third International Symposium on Brittle Matrix Composites, pp. 344–355. Elsevier Applied Science, Warsaw (1991)

10. Mobasher, B., Pivacek, A., Haupt, G.J.: Cement based cross-ply laminates. J. Adv. Cem. Based Mater. **6**, 144–152 (1997)

11. Sutton, M.A., Wolters, W.J., Peters, W.H., Ranson, W.F., McNeil, S.R.: Determination of displacements using an improved digital correlation method. Image Vis. Comput **1**(3), 133–139 (1983). https://doi.org/10.1016/0262-8856(83)90064-1

12. Bruck, H.A., McNeil, S.R., Sutton, M.A., Peters, W.H.: Digital image correlation using Newton-Raphson method of partial differential correction. ExpMech **29**(3), 261–267 (1989). https://doi.org/10.1007/BF02321405

13. Destrebecq, J.-F., Toussaint, E., Ferrier, E.: ExpMech **51**(6), 879–890 (2011)

14. Koerber, H., Xavier, J., Camanho, P.P.: High strain rate characterisation of unidirectional carbon-epoxy IM7-8552 in transverse compression and in-plane shear using digital image correlation. Mech. Mater. **42**, 1004–1019 (2010). https://doi.org/10.1016/j.mechmat.2010.09.003

15. Gao, G., Huang, S., Xia, K., Li, Z.: Application of digital image correlation (DIC) in dynamic notched semi-circular bend (NSCB) tests. Exp. Mech. **55**(1), 95–104 (2014). https://doi.org/10.1007/s11340-014-9863-5

Geopolymers

Mechanical Evaluation of Na-Based Strain-Hardening Geopolymer Composites (SHGC) Reinforced with PVA, UHMWPE, and PBO Fibers

Ana C. C. Trindade[1]($\boxtimes$), Iurie Curosu[2], Marco Liebscher[2],
Viktor Mechtcherine[2], and Flávio de A. Silva[1]

[1] Department of Civil and Environmental Engineering, Pontifícia Universidade
Católica Do Rio de Janeiro (PUC-Rio), Rio de Janeiro, RJ 22451-900, Brazil
ana.trindade@aluno.puc-rio.br
[2] TU Dresden, Institute of Construction Materials, 01062 Dresden, Germany

Abstract. Strain-hardening geopolymer composites (SHGC) show increased deformation capacity due to a multiple cracking tolerance under tensile loading. To evaluate their mechanical performance, a common metakaolin-based mixture was produced. Three types of short fibers were evaluated as disperse reinforcement: polyvinyl alcohol (PVA), ultra-high molecular weight polyethylene (UHMWPE), and poly(p-phenylen-2,6-benzobisoxazole) (PBO). The composites' mechanical features were analyzed in compression, three-point-bending, and tension tests with subsequent Environmental Scanning Electron Microscopy (ESEM) analysis of the fracture surfaces. Digital Image Correlation (DIC) was used to evaluate the extent of multiple cracking and crack widths under uniaxial tension. Additionally, single-fiber pullout tests were performed. PBO-based composites yielded the highest mechanical properties, reaching a 4.8 MPa peak stress in tension at a strain level of 2.3%, with a larger number of cracks. PVA and UHMWPE-based materials, however, demonstrated a lower mechanical performance, because of their larger diameter, lower mechanical properties and fiber-matrix adhesion.

Keywords: Geopolymer · SHGC · PVA · PE · PBO

1 Introduction

Strain-hardening geopolymer-based composites (SHGC) [1, 2] correspond to a class of materials known to result in a pronounced multiple cracking and high strain capacity under tensile loading, obtained by incorporating small amounts ($\sim$ 2% vol.) of synthetic micro-fibers into purposefully designed matrices [2, 3]. Their high inelastic deformability and excellent crack control makes a promising alternative for new constructions and retrofitting applications, also enhancing the performance of structures exposed to severe loading and environmental conditions [2]. The development of geopolymer-based composites occurred due to a demand for new alternative binders capable of enhancing the efficiency of strain-hardening cementitious composites

© RILEM 2022
P. Serna et al. (Eds.): BEFIB 2021, RILEM Bookseries 36, pp. 857–867, 2022.
https://doi.org/10.1007/978-3-030-83719-8_73

(SHCC) [4, 5], both thermally and mechanically. This occurred since SHCC presents a relatively high amount of cement, due to strict limitations regarding aggregates' content [5], which can be highly associated with a significant CO_2 footprint [6, 7] as well as water demand, which is typical for cement-based mixtures [8]. Geopolymers [9] appear as an adequate alternative, due to their varied raw materials availability, mostly based on aluminosilicate binders mixed with alkali solutions [10, 11]. The "geopolymer" name was introduced by Davidovits [9], who established a binder based on alumi-nosilicate materials to enhance the thermal resistance of structural elements. Its empirical formula can be described as $M_2O \cdot Al_2O_3 \cdot xSiO_2 \cdot 11H_2O$ (where M = Na, K, Cs; and x represents the Si/Al ratio used) [9]. The reaction and hardening processes, aside from the fast-setting time [12], enhanced thermal resistance [12], high chemical [13], and long-term durability [14], appear as suitable conditions in improving the long-term performance of the composites.

Various geopolymer-based composites have been investigated containing synthetic [15], mineral [16], and natural fabrics [14], as well as polymer micro-fibers [1–3, 17]. The latter approach was well explored by Nematollahi et al. [18] in studies on the production of fly ash based SHGC through a comprehensive analysis of the matrix and the establishment of optimal volume fractions of PE and PVA reinforcements. Additionally, Trindade et al. [1, 2] presented a complete mechanical evaluation of sand incorporation into KGP and NaGP metakaolin-based binders reinforced with PVA fibers, showing enhanced performance of the latter [1], and a PVA *versus* PE fibers comparison on the SHGC dynamic tensile behavior [2]. However, despite reaching reasonable tensile strength in the order of 4.5 MPa and adequate deformation values, further development can be designed by incorporating new and improved types of reinforcements, such as PBO fibers. Based on previous investigations on cementitious binders (SHCC) [19], this high-performance reinforcement is expected to result in superior mechanical properties [20], also exhibiting an improved behavior in high temperatures demands, which is in accordance with the geopolymer thermal capabilities [11].

Therefore, this study aims at a comprehensive evaluation of a well-established metakaolin-based geopolymer material (NaGP) concerning their use as a SHGC when reinforced with PVA, UHMWPE, and PBO fibers. The resulting mechanical performance characteristics are discussed based on compression, bending, uniaxial tension, pullout tests, and image observations (DIC and ESEM).

2   Experimental Program

2.1   Materials and Processing

The GP mixture was produced through the combination of Metamax metakaolin (MK) and an alkali-based solution (water glass, WG). Sodium hydroxide in pellets (reagent grade > 90%) was dissolved in deionized water, where hydrophilic fumed silica was added and mixed for 24 h using a magnetic stirrer, forming the stable water-glass solution. The WG-to-MK ratio was 1.71. Quartz fine sand (max. ϕ = 0.2 mm) was incorporated in a 50% mass fraction of metakaolin. The MK chemical composition is presented in Table 1.

Table 1. MK chemical composition.

Composition	SiO_2	Al_2O_3	TiO_2	Fe_2O_3	K_2O	MgO	CaO	Na_2O	LOI
wt%	53.0	43.8	1.70	0.43	0.19	0.03	0.02	0.23	0.46

Short fibers 2% in vol. of PVA (Kuraray, Japan), UHMWPE (DSM, The Netherlands), and PBO (Toyobo, Japan) were used as reinforcements, with 12 mm lengths for the first two, and half of this value of the latter. Their physical and mechanical properties are presented in Table 2. A 10-L planetary mixer was used for the preparation of the composites as follows: (i) addition of metakaolin and water glass inside the mixer; (ii) mixing for 3 min at intermediate speed (198 rpm) to ensure adequate homogeneity and reactivity; (iii) addition of aggregates; (iii) mixing for 1 min at intermediate speed (198 rpm); (iv) addition of fibers; (v) mixing for 3 min at higher speed (365 rpm) ensuring a proper fiber distribution. The fresh mix was then poured into the molds, which required a vibration step of 1 min both for consolidation and removal of voids. The molds were sealed in plastic bags for 48 h at room temperature to prevent early dehydration [1, 2]. Subsequently, the samples were removed from the molds and cured inside dry plastic bags for 2 weeks, to prevent early dehydration.

Prismatic specimens of 160 mm × 40 mm × 40 mm were produced for the bending tests. Dumbbell-shaped specimens were produced for uniaxial tension tests, with a nominal cross-section in the gauge region of 40 mm × 24 mm.

Table 2. Physical and mechanical properties of the investigated PVA, UHMWPE, and PBO fibers

Fiber type	PVA (Kuralon® K-II REC15)	UHMWPE (Dyneema® SK62)	PBO-AS (Zylon®)
Length [mm]	12	12	6
Nominal diameter [µm]	40	20	13
Density [g/cm³]	1.26	0.97	1.54
Tensile strength [MPa]	1600	2500	5800
Young's modulus [GPa]	40	80	180
Elongation at break [%]	7	3.5	3.5

The manufacturing process of single-fiber pullout specimens used a special rectangular polymer-made mold, allowing a longitudinal channel in the middle of the mold. The width of the channel corresponded to the fiber embedment length of 2 mm. The fibers were then transversely placed bridging the channel with a 10 mm-spacing between the fibers and carefully fixed with wax at their ends. The mold was subsequently filled with the fresh matrix. Demolding consisted of extracting the beams and cutting between the fibers, resulting in at least 10 single specimens for each parameter variation. All shapes and geometries are presented in Fig. 1.

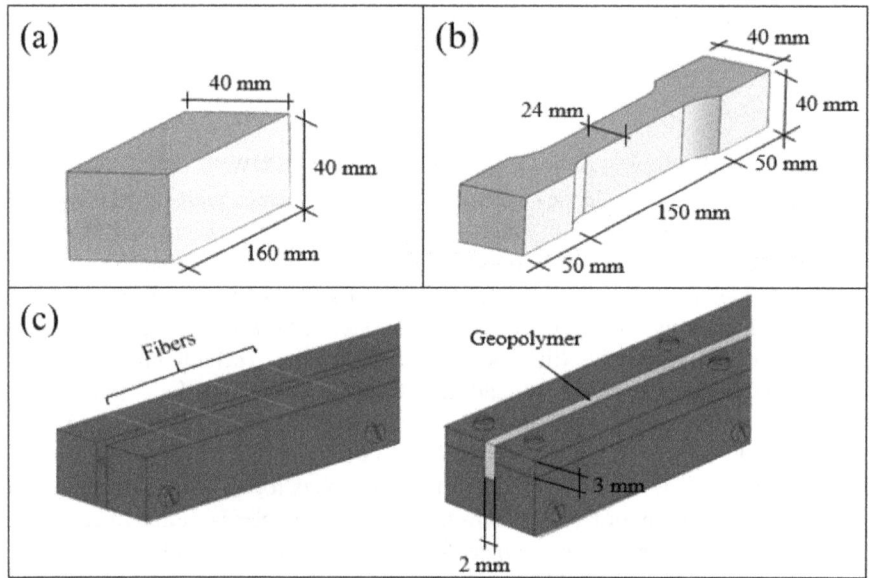

Fig. 1. Shapes and dimensions of (a) Prismatic specimen, (b) dumbbell-shaped specimen, and (c) mold and production of pullout specimens.

2.2   Testing Methods

A servo-hydraulic universal testing system with a load capacity of 200 kN was used to perform three-point bending tests with all material variations presented previously. The tests were carried out based on the BS EN 196-1 [21]; 3 specimens were tested for each variation under a load-controlled rate of 50 N/s. The span between supports was 100 mm.

Their compressive stress-deformation responses were obtained in the same testing system using 40 mm cubic samples, at a 2400 N/s loading rate. Hydraulic testing equipment Instron 8501 with closed-loop control and a load capacity of 100 kN was used to perform the uniaxial tension tests on the SHGC under a displacement rate of 0.04 mm/s. Three dumbbell-shaped specimens were tested for each composite variation. The specimens were glued at their ends in 20 mm-thick steel rings, bolted to the testing machine, ensuring non-rotatable boundary conditions. Two Linear Variable Differential Transducers (LVDTs) were placed on each side of the specimen using a steel frame to measure the deformation in a 100 mm gauge length. Figure 2a presents the testing configuration. During the tension tests, optical measurements were performed, to monitor and quantify the specimen's deformation and crack formation by using Digital Image Correlation (DIC). A black and white speckle pattern was sprayed onto the specimens for this purpose. The optical sampling rate was 1 frame/5 s. The frames were processed with a commercial software ARAMIS 5M. Single-fiber pullout tests were performed for all three fiber-matrix combinations in an electromechanical testing machine Zwick 1445 with a 0.05 mm/s displacement rate using a 10 N capacity load cell. Figure 2b presents the testing configuration, with further details in a previous work [1]. The specimens were glued on a flat aluminum plate, which was screwed to the lower part of the machine. The free fiber end was glued to an upper plate, which was attached to the force sensor. Finally, an ESEM Quanta 250 FEG was used for microscopic analysis of the fracture surfaces of the GP fiber-reinforced composites.

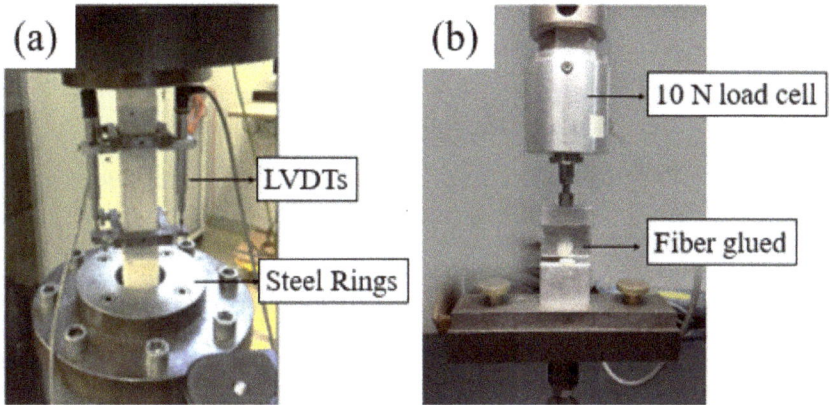

Fig. 2. (a) Uniaxial tension, and (b) pullout test configurations.

3   Results and Discussion

The compressive curves obtained with the plain sodium-based geopolymer (NaGP) material are presented in Fig. 3a, where it is possible to observe an average strength of 56.8 MPa and average Young's modulus of 9.6 GPa. Assuming that the effect of fibers is not substantial, these properties can be attributed to all composite variations since all of them were based on NaGP binders. The composites flexural curves obtained for the three types of SHGC reinforced with PVA, UHMWPE, and PBO fibers are shown in Fig. 3b, while the corresponding average mechanical parameters are summarized in Table 3.

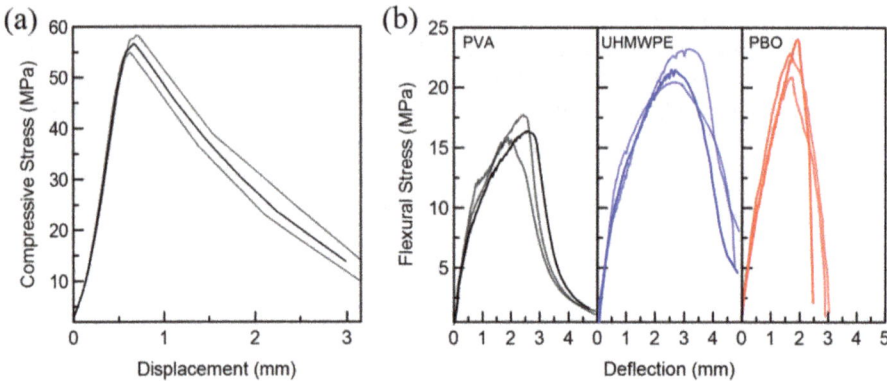

Fig. 3. (a) Compression and (b) three-point-bending tests results.

Table 3. Three-point-bending tests results.

Composite	NaGP$_{PVA}$	NaGP$_{PE}$	NaGP$_{PBO}$
First crack stress [MPa]	9.9 (1.1)	10.8 (0.7)	10.5 (0.5)
Flexural strength [MPa]	17.3 (0.9)	21.3 (1.1)	22.6 (1.4)
Deflection at peak stress [mm]	2.7 (0.2)	2.9 (0.4)	1.9 (0.1)
Work-to-fracture [N/mm]	27.3 (3.2)	40.5 (4.5)	26.5 (2.4)

All composites presented a ductile behavior with multiple cracks formations, due to efficient crack bridging by the micro-fibers. The three types of fibers acted as internal micro-confinements, enhancing the toughness and damage tolerance of the specimens. They all presented similar first crack stresses, reaching average 9.9, 10.8 and 10.5 MPa, for PVA, PE, and PBO reinforcements, respectively. This response is expected since the matrix properties rule the behavior of the composites in such elastic phase of loading (before cracking) [19]. Also, the slightly increased values found for PE and PBO reinforcements can be attributed to increased crack-bridging due to a smaller fiber diameter compared to PVA [1, 2]. Composites based on PBO fibers demonstrated an improved flexural strength of 22.6 MPa, followed by 21.3 and 17.3 MPa found for PE and PVA. This response is a direct result of the PBO fibers higher mechanical properties, and possibly fiber-matrix debonding mechanisms, even with half the length of PVA and PE filaments (6 mm for PBO). However, regarding maximum deflection and fracture energy, both PVA and PE fibers showed higher values, possibly indicating a greater crack opening for PVA and PE. This is also evidenced by the stress drops in the curves (not visible for PBO), characteristic of multiple crack formation, which will be further verified through DIC evaluations. To identify and properly quantify the effects of fiber type on the fiber-matrix interaction, single-fiber pullout tests appear to be instrumental.

The tensile behavior of NaGP composites reinforced with PVA, PE, and PBO fibers are presented in Fig. 4; Table 4 provides the average values obtained from the stress-strain curves.

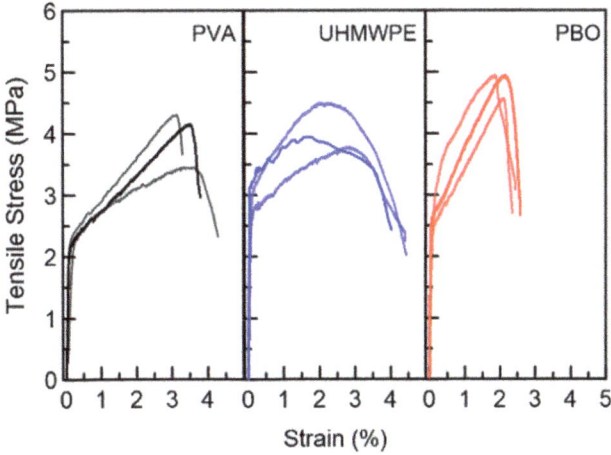

Fig. 4. Uniaxial tension tests results.

A pronounced enhanced first-crack stress can be distinguished for PE and PBO reinforcements when compared to PVA, reaching average 3.3, 3.1 and 2.9 MPa, respectively. This result could be expected in considering the previous flexural findings. Also, PBO-based composites demonstrated a superior mechanical capacity, reaching 4.8 MPa in stress, followed by 4.4 MPa with PE, and 4.2 MPa with PVA. An optimal balance between the cracking strength of the matrix and the crack-bridging capacity of the fibers was attained. The differences in fiber-matrix interactions for all types of matrices will be discussed when presenting the results of pullout tests. Regarding deformation values, it is possible to observe enlarged deformations for PVA and PE-based composites since they reached an average 3.2% of maximum deformation when compared to the 2.3% found for PBO. Again, no stress drops can be seen in the PBO curves, which can be associated with a controlled debonding in a low crack width. Therefore, the cracking parameters will be discussed through DIC evaluations in Fig. 5.

Table 4. Uniaxial tension tests results.

Composite	NaGP_PVA	NaGP_PE	NaGP_PBO
First crack stress [MPa]	2.9 (0.07)	3.3 (0.35)	3.1 (0.20)
Tensile stress [MPa]	4.2 (0.5)	4.4 (0.3)	4.8 (0.2)
Deformation at max. stress [%]	3.2 (0.8)	3.2 (0.7)	2.3 (0.2)
Fracture energy [N/mm]	8.1 (1.8)	8.9 (0.4)	10.7 (0.5)

The multiple cracking and fracture occurrence were evaluated, allowing the determination of the number of cracks, average crack width, and crack spacing. Figure 5 presents a typical analysis of the crack patterns at the final strain stage of loading (2.3%), where most of the cracks have already been formed. A higher crack density was found for PBO-reinforced composites, when compared to the PVA and PE-based materials. The average crack widths recorded for PVA, PE, and PBO-based composites were: 64, 31, and 7 µm, respectively. The cracks found in PBO-based composites were narrower than in PVA and PE. It is interesting to notice the slightly better crack control in comparison to common SHCC, which typically yields increased crack widths and crack spacings in stronger matrices [23, 24].

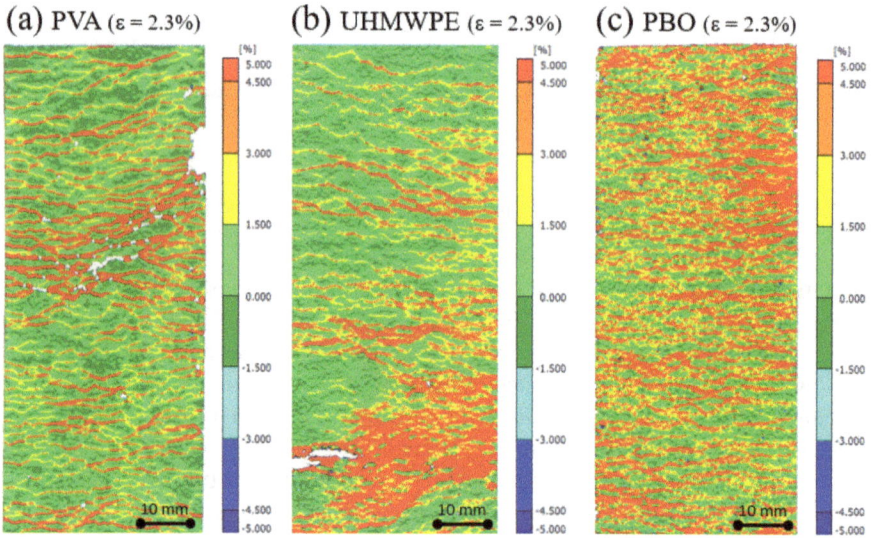

Fig. 5. DIC analysis – cracking evaluation, for representative (a) PVA, (b) UHMWPE, and (c) PBO fiber-reinforced NaGP.

The pullout test results are presented in Fig. 6 and Table 5. All fibers demonstrated a complete fiber pullout, which is indicated by the relatively steady and long descending branches of force-displacement curves. Opposed to the interfacial damage usually recorded in common PVA-SHCC materials [22], the pullout of the PVA fibers from NaGP resulted in a slip-softening behavior, indicating a dramatic reduction of the chemical adhesion through binder substitution. Also, it is interesting to notice a great difference for the bond strength values found for PBO reinforcements, 2.36 MPa, when compared to 0.94 and 0.81 MPa reached with PVA and PE. This enhancement is associated with the fiber's improved mechanical properties in a similar embedment pullout length (2 mm), contributing also to the composites crack bridging, justifying the slightly greater flexural and tensile load-bearing capacities found previously, even with half the length compared to the other two reinforcements. The latter being the main responsible for the small improvements in stress, when compared to the 12 mm lengths of both PVA and PE reinforcements.

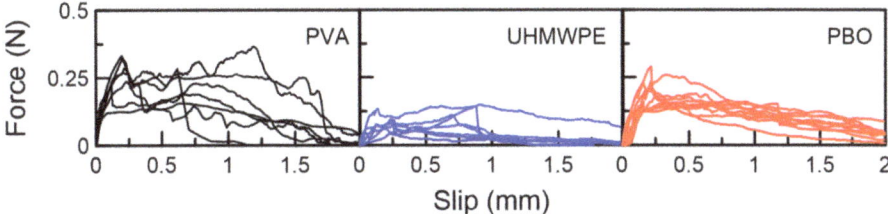

Fig. 6. Pullout tests results for (a) PVA, (b) UHMWPE, and (c) PBO fibers.

Table 5. Pullout tests results.

Fiber	PVA	UHMWPE	PBO
Peak force [N]	0.23 (0.06)	0.10 (0.03)	0.19 (0.08)
Bond strength [MPa]	0.94 (0.22)	0.81 (0.14)	2.36 (0.69)

Figure 7 presents the fracture surfaces of all three composites, showing the surfaces of the partially pulled out fibers after complete failure in the uniaxial tension tests, without any evident fiber surface damage. Micrographs also revealed some regions of fiber agglomerations, especially for PE reinforcements, due to its increased number of filaments, when compared to the same PVA fiber content, evidencing a need for rheological and mixing optimizations. The images clearly show the differences in fiber diameters, and the high homogeneity of the NaGP matrix and its pronounced brittleness, marked by the fine loose fragments.

Fig. 7. Fracture surfaces of (a) PVA, (b) UHMWPE, and (c) PBO fiber reinforced NaGP.

Therefore, these results demonstrate the superior crack-bridging behavior of the PBO fibers, evidenced by the composites' ability to withstand higher loads, showing further developments for the SHGC technology, allowing the use of varied fibers into a well-known geopolymer binder for a varied range of applications, such as retrofit in elements subjected to dynamic loading and possibly requiring long-term durability.

4   Conclusions

The results of the experimental investigations showed that PBO-based composites mixtures exhibited higher tensile and flexural strengths when compared to the same PVA and UHMWPE variations. In general, the fibrous reinforcement was effective in ensuring typical strain-hardening behavior under tensile loading, accompanied by pronounced multiple cracking. PBO-based strain-hardening geopolymer composites (SHGC) yielded strain capacities of 2.3% and tensile stresses of 4.8 MPa, while the same PVA and UHMWPE achieved only 4.4 and 4.2 MPa on average.

Pullout tests made possible a comprehensive analysis of the fiber-matrix interactions, evidencing a stronger fiber anchorage for PBO-based composites, reaching a considerably higher bond strength of 2.36 MPa. In contrast, PVA and UHMWPE-based composites showed weaker bonds, which led to larger crack openings, also enhancing their ultimate deformations. DIC evaluations of crack formation showed that the PBO-based SHGC exhibited average crack widths of 7 μm, while PVA and UHMWPE variations exhibited 64 μm and 7 μm, respectively, because of the stronger bond, higher fiber stiffness and smaller diameter.

In summary, improved strain capacity and smoother shape of the stress-strain curves were observed for SHGCs produced with PBO reinforcements. However, despite the higher stresses, more abrupt failure modes were found, showing that it could be relevant to combine PBO with PE fibers in future works and applications. This indicates further potential for the development of SHGC materials for a wide range of applications, including structural elements subject to dynamic loading.

Acknowledgements. The authors acknowledge the financial support provided by the Brazilian organizations CNPq and CAPES (Probral project number 8887.144079/2017-00) and by the DAAD program (Probral project number 8887.144079/2017-00) in Germany. Also, the use of the facilities at the TU Dresden and the support of the laboratory staff at the Institute of Construction Materials are greatly appreciated.

References

1. Trindade, A.C.C., Curosu, I., Liebscher, M., Mechtcherine, V., de Andrade Silva, F.: On the mechanical performance of K-and Na-based strain-hardening geopolymer composites (SHGC) reinforced with PVA fibers. Constr. Build. Mater. **248**, 118558 (2020)
2. Trindade, A.C.C., Heravi, A.A., Curosu, I., Liebscher, M., de Andrade Silva, F., Mechtcherine, V.: Tensile behavior of strain-hardening geopolymer composites (SHGC) under impact loading. Cem. Concr. Compos. **113**, 103703 (2020)
3. Batista, R.P., Trindade, A.C.C., Ribeiro Borges, P.H., Silva, F.A.: Silica fume as precursor in the development of sustainable and high-performance MK-based alkali-activated materials reinforced with short PVA fibres, Front. Mater. **6**(77) (2019)
4. Mechtcherine, V.: Novel cement-based composites for the strengthening and repair of concrete structures. Constr. Build. Mater. **41**, 365–373 (2013)
5. Choi, W.C., Yun, H.D., Kang, J.W., Kim, S.W.: Development of recycled strain-hardening cement-based composite (SHCC) for sustainable infrastructures. Compos. B Eng. **43**(2), 627–635 (2012)

6. Li, V.: Sustainability of engineered cementitious composites (ECC) infrastructure. In: Engineered Cementitious Composites (ECC) Bendable Concrete for Sustainable and Resilient Infrastructure, pp. 261–312. Springer, Heidelberg (2019). https://doi.org/10.1007/978-3-662-58438-5_8

7. Schneider, M., Romer, M., Tschudin, M., Bolio, H.: Sustainable cement production—present and future. Cem. Concr. Res. **41**(7), 642–650 (2011)

8. Shi, C., Jiménez, A.F., Palomo, A.: New cements for the 21st century: the pursuit of an alternative to Portland cement. Cem. Concr. Res. **41**(7), 750–763 (2011)

9. Davidovits, J.: Geopolymers: inorganic polymeric new materials. J. Therm. Anal. Calorim. **37**(8), 1633–1656 (1991)

10. Purdon, A.O.: The action of alkalis on blast-furnace slag. J. Soc. Chem. Indust. **59**(9), 191–202 (1940)

11. Vickers, L., Van Riessen, A., Rickard, W.D.: Fire-Resistant Geopolymers: Role of Fibres and Fillers to Enhance Thermal Properties. Springer, Singapore (2015). https://doi.org/10.1007/978-981-287-311-8

12. Kriven, W.M., Bell, J.L., Gordon, M.: Microstructure and microchemistry of fully-reacted geopolymers and geopolymer matrix composites. Ceram. Trans. **153**, 227–250 (2003)

13. Provis, J.L., Van Deventer, J.S.J.: Geopolymers: Structures, Processing, Properties and Industrial Applications. Elsevier, Amsterdam (2009)

14. Trindade, A.C.C., Alcamand, H.A., Borges, P.H.R., Silva, F.A.: On the durability behaviour of natural fiber reinforced geopolymers. Ceram. Sci. Proc. **38**(3), 215–228 (2017)

15. Menna, C., et al.: Use of geopolymers for composite external reinforcement of RC members. Compos. B Eng. **45**(1), 1667–1676 (2013)

16. Dias, D.P., Thaumaturgo, C.: Fracture toughness of geopolymeric concretes reinforced with basalt fibers. Cem. Concr. Compos. **27**(1), 49–54 (2005)

17. Nematollahi, B., Sanjayan, J., Ahmed Shaikh, F.U.: Tensile strain hardening behaviour of PVA fiber-reinforced engineered geopolymer composite. J. Mater. Civil Eng. **27**(10), 04015001 (2015)

18. Nematollahi, B., Sanjayan, J., Qiu, J., Yang, E.H.: Micromechanics-based investigation of a sustainable ambient temperature cured one-part strain hardening geopolymer composite. Constr. Build. Mater. **131**, 552–563 (2017)

19. Curosu, I., Liebscher, M., Mechtcherine, V., Bellmann, C., Michel, S.: Tensile behavior of high-strength strain-hardening cement-based composites (HS-SHCC) made with high-performance polyethylene, aramid and PBO fibers. Cem. Concr. Res. **98**, 71–81 (2017)

20. Gong, T., Heravi, A., Alsous, G., Curosu, I., Mechtcherine, V.: The impact-tensile behavior of cementitious composites reinforced with carbon textile and short polymer fibers. Appl. Sci. **9**(19), 4048 (2019)

21. BS EN 196-1: Methods of Testing Cement. Part 1: Determination of Strength (2016)

22. Barbosa, V.F., MacKenzie, K.J.: Thermal behaviour of inorganic geopolymers and composites derived from sodium polysialate. Mater. Res. Bull. **38**(2), 319–331 (2003)

23. Van Zijl, G.P.A.G., Wittmann, F.H.: Durability of Strain-Hardening Fibre-Reinforced Cement-Based Composites (SHCC), vol. 4. Springer, Heidelberg (2010). https://doi.org/10.1007/978-94-007-0338-4

24. Curosu, I., et al.: Influence of fiber type on the tensile behavior of high-strength strain-hardening cement-based composites (SHCC) at elevated temperatures. Mater. Des. **198**, 109397 (2020)

Nano-technologies Related to FRC

Influence of Dispersion Methods of Microcrystalline Cellulose on the Mechanical Behavior of Cement Pastes

Letícia O. de Souza[1(✉)], Lourdes Maria Silva Souza[2],
and Flávio de A. Silva[1]

[1] Department of Civil and Environmental Engineering, Pontifícia Universidade
Católica do Rio de Janeiro (PUC-Rio), Rio de Janeiro, Brazil
[2] Tecgraf Institute, Pontifícia Universidade Católica do Rio de Janeiro
(PUC-Rio), Rio de Janeiro, Brazil

Abstract. The idea of using cellulose-based reinforcement is to explore all the benefits of the cellulose, the main component responsible for the strength of plants. With notable mechanical and physical properties, microcrystalline cellulose (MCC) inclusions are able to reinforce different types of matrices. In cementitious materials, the investigations about microcrystalline cellulose inclusions are quite recent. Moreover, the challenge involving dispersion is one of the limiting factors with regard to the full effectiveness of their properties. In order to assess the effect of the inclusion of MCC on the mechanical properties of cement pastes, compression and four-point bending tests were carried out with several contents of MCC, from 0.05% to 1% in mass of cement. The present work also aims to compare different methods of dispersion of microcrystalline cellulose into water by means of mechanical and chemical approaches, for further addition in cement pastes. The methods include mechanical stirring, superplasticizer addition, surfactant addition, and ultrasonication. Results show an impressive enhancement of the flexural properties of more than 5 and 8 times for strength and modulus, respectively with 0.75% of MCC. The use of ultrasonication as a dispersion method showed to be the most effective to disperse 0.05% of MCC.

Keywords: Microcrystalline cellulose · Cement paste · Dispersion · Mechanical properties

1 Introduction

Cementitious materials are the most used in the construction industry and are used for structural and non-structural applications. Thus, improvements in mechanical properties while considering environmental matter are of interest. One promising alternative of reinforcement less harmful to the environment is the use of cellulosic-based fibers. The use of natural reinforcement is an ancient practice and the techniques involving these materials are improving with time. New technologies enable different scales of natural fibers: from macro textiles, short fibers, and pulp until micro and nanocellulose [1–5]. The hierarchical microstructure of the natural fibers enables the extraction of

© RILEM 2022
P. Serna et al. (Eds.): BEFIB 2021, RILEM Bookseries 36, pp. 871–878, 2022.
https://doi.org/10.1007/978-3-030-83719-8_74

micro- and nano dimension components with superior properties compared to the original source. Thus, the purpose of the addition of micro- and nano-cellulosic particles is to remove parts not as strong and stable as cellulose from their source.

Microcrystalline cellulose (MCC) consists almost exclusively of cellulose crystals and is commercially available. They have a rod-like morphology with variable length, from 2 to 260 μm, and a diameter of around 50 μm [6]. The addition of MCC in cementitious materials alters hydration reactions and products, fresh properties, hardened microstructure, and mechanical response [5–10]. Due to its small size, MCC may fill a gap in the microstructure of the matrix, improving the packing density. Besides, the presence of hydroxyl groups on the surface of MCC also affects the interface between cement particles and products and MCC. The addition of MCC in cementitious materials is quite recent. There are few papers on the subject and the first is dated from 2013 [5]. By that time, Hoyos et al. [5] investigated the effect of MCC on rheology, hydration, microstructure and mechanical properties of cement pastes and mortar. However, the authors did not find positive effects on compressive and flexural properties upon the addition of 3% of MCC [5]. Anju et al. [7] observed an increase in both compressive and flexural strength of 45% and 200%, respectively, with the addition of surface-modified MCC. Parveen et al. [8] investigated the mechanical behaviour of mortars containing MCC by means of flexural and compressive tests, reaching improvements of 31% and 66%, respectively, with 0.5% of MCC. Parveen et al. [6] and Silva et al. [9] also observed significant improvement in mechanical properties, as compressive and flexural strengths, with the addition of different contents of MCC: up to 0.6% [9] and 1% [6].

Despite the advantages of the addition of MCC in cementitious materials, the dispersion of micro and nano materials is still one of the main challenges in the use of such particles. When this type of material is added into a cementitious matrix, they tend to form agglomerates and become a weak spot of the element [11–13]. This behaviour usually limits the amount of addition, which beyond a given content, compromises the mechanical response. The amount of addition that provides the best response is commonly called the optimum content. There are some studies about the dispersion of MCC in cementitious matrix [6–9]. While Anju et al. [7] used a silane coupling agent to modify the surface of MCC, Parveen et al. [8], Parveen et al. [6] and Silva et al. [9] developed MCC suspensions in water for further addition into the cementitious matrix. Parveen et al. [8] combined the addition of surfactant with ultrasonication bath in order to produce the aqueous suspension. Parveen et al. [6] investigated several time periods of ultrasonication bath to ensure MCC dispersion with the optimal time of 30 min. Silva et al. [9] prepared the MCC aqueous suspension using 45 min of magnetic stirring. Even though the aforementioned researches considered the use of dispersion approaches, no comparison with mixtures without dispersion was made.

The present work aims to assess the mechanical properties of cement pastes reinforced with microcrystalline cellulose. Four-point bending tests were performed for composites with 0%, 0.05%, 0.10%, 0.50%, 0.75%, and 1% of MCC, in mass of cement. Compression tests were performed in specimens reinforced with 0% and

0.05% wt. of MCC with no specifv dispersion method applied and in specimens with MCC prepared with dispersion techniques. The dispersion includes mechanical approaches, as mechanical stirring and ultrasonication, and chemical approaches, as superplasticizer and surfactant addition. In an attempt to produce an aqueous suspension, the chemical approaches were combined with mechanical stirring.

2 Materials and Methods

2.1 Materials

The Portland cement used was the Brazilian Type V with low mineral additions, in accordance to NBR 16697. The water/cement ratio was 0.45 and tap water was used. The applied surfactant was Pluronic F-127, from Sigma-Aldrich and the used superplasticizer additive was Glenium 51 MS, with a solid content of 32%, from BASF. The used microcrystalline cellulose was MCC Avicel PH-101, from Sigma-Aldrich.

2.2 Methods

2.2.1 Dispersion Techniques

The dispersion techniques were applied on solutions of MCC and part of the mixing water. The content of MCC corresponds to 0.05% of the cement weight used for the production of three cylinders. Each dispersion methodology is discussed in the following as well as the specific amount of water.

- Mechanical stirring (MecS): The total mixing water was used in the solution with MCC submitted to a mixer revolving at around 900 rpm for 30 min;
- Ultrasonication (US): 20 ml of the mixing water was used in the solution with MCC submitted to 30 min of ultrasonication. A sonicator SONICS Vibra-Cell was used at maximum power with a half-inch tip. At every 10 min, the process was interrupted in order to cool down the solution to room temperature;
- Superplasticizer (Super): Solution composed by the total mixing water, MCC and the superplasticizer corresponding to 1% of the cement mass submitted to 30 min in the mechanical stirrer.
- Pluronic Surfactant (Surf): the total mixing water was used in the solution with MCC and the surfactant. The solution was submitted to a mixer revolving at around 900 rpm for 30 min. The proportion of MCC:surfactant in weight was of 5:1.

2.2.2 Fabrication of the Composites

The fabrication of the specimens of cement paste reinforced with MCC for the bending tests (without dispersion methods) occurred on a planetary mixer of 5l of volume. First, the water was added with half of the cement and the mixture was manually blended. The remaining cement and MCC were added and mixed for 1 min at 136 rpm and

3 min at 281 rpm. The fresh matrix was poured in plate moulds of dimensions of $500 \times 60 \times 10$ mm^3 (length × width × thickness) and stored for 28 days in water.

For the specimens reinforced with 0.05% of MCC with dispersion methods applied, after the dispersion the solution was added and mixed on the planetary mixer. The remaining water, if any, was added and the procedure repeated in the same manner as for non-dispersing specimens. The freshly matrix was poured in cylinders with 50 mm of diameter and 100 mm of height and stored for 28 days in water.

2.2.3   Four-Point Bending Tests

The four-point bending tests were performed with a servo-hydraulic MTS testing machine with a load capacity of 100 kN. The tests were carried on 28-day aged specimens, at a displacement rate of 0.05 mm/min. Two 25-mm displacement transducers were added to the system for accurate axial displacement acquisition. An additional load-cell of 2.5 kN was included to the system for more accurate load acquisition. At least three specimens for each variation were tested.

2.2.4   Compression Tests

The compression tests were performed with a servo-hydraulic MTS testing machine, model 810 with a load capacity of 500 kN. Direct axial compression tests were carried on 28-day aged specimens, at a displacement rate of 0.4 mm/min. Two 5-mm linear variable displacement transducers (LVDTs) were added to the system on an acrylic arrangement for accurate axial displacement acquisition. Three specimens for each variation were tested.

3   Results and Discussion

3.1   Impact of MCC on Flexural Behaviour

Figure 1 shows typical equivalent bending stress versus displacement curves of cement plates reinforced with different contents of MCC. Table 1 lists the mechanical parameters calculated from the experimental data obtained. It is important to point out that the specimens' geometry is close to plates (see Sect. 2.2.2), so this may have impacted the bending results. From the results obtained, it is possible to notice that the addition of any amount of MCC influenced the flexural mechanical properties of the specimens. Even the addition of 0.05% wt. of MCC promoted a considerable enhancement of the strength more than three times and the flexural modulus more than six times. The mechanical parameter that diminished was the deflection capacity. This behaviour had been already observed elsewhere [6, 8, 9] and was attributed to the high crystallinity presented by MCC, which stiffens the material. Besides, the low aspect ratio of MCC and the strong interface between the particles and the hydration products may also lead to a reduced deflection capacity of the composite.

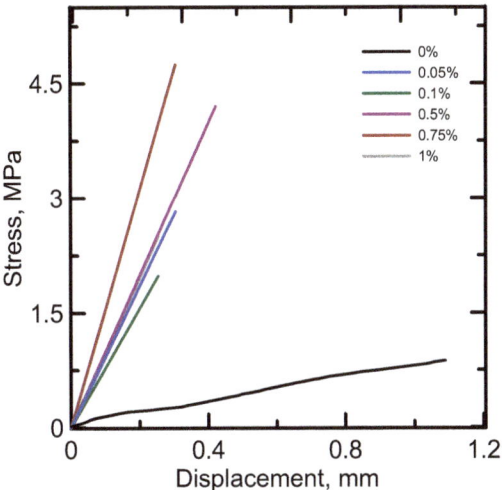

Fig. 1. Equivalent bending stress versus deflection curves of cement paste plates reinforced with 0%, 0.05%, 0.10%, 0.50%, 0.75%, and 1% of MCC, in mass.

Table 1. Flexural strength, modulus od elasticity and maximum deflection calculated from the four-point bending tests of cement paste plates reinforced with 0%, 0.05%, 0.10%, 0.50%, 0.75%, and 1% of MCC, in mass. Standard deviations are included.

MCC content (%)	Strength (MPa)	Modulus (GPa)	Deflection (mm)
0	1.1 ± 0.1	1.6 ± 0.6	1.15 ± 0.34
0.05	3.5 ± 0.5	10.1 ± 1.2	0.33 ± 0.02
0.10	2.9 ± 0.3	8.3 ± 1.4	0.34 ± 0.05
0.50	4.3 ± 1.1	11.4 ± 2.7	0.42 ± 0.02
0.75	5.7 ± 0.8	13.7 ± 3.8	0.39 ± 0.08
1.00	4.0 ± 0.5	11.6 ± 2.6	0.34 ± 0.11

The mechanical parameters increased with the amount of MCC addition up to 0.75%. The addition of 0.75% of MCC promoted the best behaviour regarding the mechanical properties, with an improvement of the strength and modulus of elasticity of more than 500% and 800%, respectively. Beyond the dosage of 0.75%, the mechanical parameters analysed began to decrease.

The trend of improving the parameters until a given amount and subsequent decrease is observed for various nanomaterials inclusions and is related to the agglomeration effect [4, 11, 14–20]. The agglomerates may act as defects or stress concentrators and are formed due to strong inter-particle interactions [11–13]. This phenomenon is magnified when the size of the particles decreases [12, 21]. Thus, one of the ways to overcome this issue is to use dispersion methods that will be discussed in the following section.

3.2   Influence of Dispersion Methods on the Compressive Behaviour

Table 2 shows the mechanical parameters calculated from the data of the compressive tests obtained. Comparing Tables 1 and 2, it is clear that the influence of the MCC was more significant on the flexural behaviour than on the compressive. The addition of 0.05% of MCC resulted in a slightly increase of 7% of the compressive strength and 15% on the modulus, even though, if considering the statistical error, this difference decreased. The strain capacity, similar to the deflection, was negatively affected.

The compressive modulus increased, whether a dispersion method applied or not, for all the types of MCC inclusion. Regarding the dispersion methods, the ultrasonication promoted a noteworthy improvement of the compressive strength, increasing from 7% of the WD specimens to 12%. This result indicates that the ultrasonication was able to create a well-dispersed MCC solution prior to the mixing and they remained to disperse with the addition of the cement. In the case of ultrasonication, the dispersion is carried by the cavitation phenomenon, where the energy in form of bubbles is transferred with the bubbles collapse [22, 23].

With the exception of ultrasonication, none of the other dispersion methods resulted in an increase of the mechanical properties in comparison to the MCC without dispersion (WD). In fact, no dispersion method was able to improve the compressive strength compared to the non-reinforced specimens, besides the ultrasonication.

The difference between the flexural and compressive mechanical behaviour upon the addition of MCC may indicate the tensile stresses are more affected by the MCC than the compressive ones.

Table 2. Compressive strength, modulus of elasticity and strain at calculated from the compressive tests of cement paste specimens reinforced with 0% (Ref) and 0.05% of MCC with no dispersion method applied (WD) and subjected to the dispersion methods: mechanical stirring (MecS), ultrasonication (US), superplasticizer (Super), and surfactant (Surf). Standard deviations are included.

Dispersion method	Strength (MPa)	Modulus (GPa)	Strain at break (%)
Ref (0%)	54.3 ± 1.7	17.0 ± 2.8	0.62 ± 0.08
WD	58.4 ± 1.4	19.6 ± 0.8	0.58 ± 0.03
MecS	51.2 ± 5.0	20.9 ± 0.3	0.37 ± 0.05
Super	46.4 ± 2.5	16.7 ± 0.6	0.47 ± 0.05
Surf	49.4 ± 6.5	19.4 ± 0.4	0.36 ± 0.06
US	61.0 ± 2.7	18.7 ± 1.9	0.56 ± 0.08

4   Conclusions

In the present paper, microcrystalline cellulose (MCC) was used as reinforcement on the micro-scale of cement pastes. The influence of several contents, from 0% to 1% of the micro cellulose was investigated under four-point bending tests. Different dispersion methods were applied on solutions of water and MCC in an attempt to improve the compressive behaviour of composites reinforced with 0.05% of MCC. Based on the experimental results, the following conclusions can be drawn:

- The MCC is capable of reinforcing cement pastes, resulting in important changes in mechanical behaviour. The flexural behaviour was more affected by the micro cellulose: even a small content of MCC, 0.05%wt., enhanced the flexural strength and modulus more than three times and six times, respectively.
- Under four-point bending testing, the optimal content of MCC was 0.75% with an improvement of the strength of more than five times and modulus of eight times. Beyond 0.75%, microcrystalline may form agglomerates that lead to a decrease in the mechanical parameters analysed.
- Regarding the dispersion methods applied, the ultrasonication enabled the enhancement of the compressive strength in comparison with specimens with MCC not submitted to other dispersion methods.
- The specimens had a reduced strain capacity in both compressive and flexural tests after the addition of MCC when compared to the non-reinforced ones. Taking into account the trend of the mechanical parameters analysed, this loss of strain and deflection capacity was probably caused by the strong bond between particles of MCC and hydration products.

Acknowledgements. This work was supported by the Conselho Nacional de Desenvolvimento Científico e Tecnológico – CNPq and Coordenação de aperfeiçoamento de Pessoal de Nível Superior – CAPES (Brazilian National Science Foundations).

References

1. Castoldi, R.S., de Souza, L.M.S., Silva, F.A.: Comparative study on the mechanical behavior and durability of polypropylene and sisal fiber reinforced concretes. Constr. Build. Mater. **211**, 617–628 (2019)
2. Fidelis, M.E.A., Silva, F.A., Filho, R.D.T.: The influence of fiber treatment on the mechanical behavior of jute textile reinforced concrete. Key. Eng. Mater. **600**, 469–474 (2014)
3. Agopyan, V., Savastano, H., Jr., John, V.M., Cincotto, M.A.: Developments on vegetable fibre – cement based materials in São Paulo, Brazil: an overview. Cem. Concr. Compos. **27**, 527–36 (2005)
4. Ardanuy, M., Claramunt, J., Arévalo, R., Parés, F., Aracri, E., Vidal, T.: Nanofibrillated cellulose (Nfc) as a potential reinforcement for high performance cement mortar composites. BioResources **7**(3), 3883–3894 (2012)
5. Gómez, H.C., Cristia, E., Vázquez, A.: Effect of cellulose microcrystalline particles on properties of cement based composites. Mater. Des. **51**, 810–818 (2013)
6. Parveen, S., Rana, S., Ferreira, S., Filho, A., Fangueiro, R.: Ultrasonic dispersion of micro crystalline cellulose for developing cementitious composites with excellent strength and stiffness. Ind. Crops Prod. **122**, 156–165 (2018)
7. Anju, T.R., Ramamurthy, K., Dhamodharan, R.: Surface modified microcrystalline cellulose from cotton as a potential mineral admixture in cement mortar composite. Cem. Concr. Compos. **74**, 147–153 (2016)

8. Parveen, S., Rana, S., Fangueiro, R., Paiva, M.C.: A novel approach of developing micro crystalline cellulose reinforced cementitious composites with enhanced microstructure and mechanical performance. Cem. Concr. Compos. **78**, 146–161 (2017)
9. Silva, L., Parveen, S., Filho, A., Zottis, A., Rana, S., Vanderlei, R., et al.: A facile approach of developing micro crystalline cellulose reinforced cementitious composites with improved microstructure and mechanical performance. Powder Technol. **338**, 654–663 (2018)
10. Peters, S.J., Rushing, T.S., Landis, E.N., Cummins, T.K.: Nanocellulose and microcellulose fibers for concrete. Transp. Res. Rec. J. Transp. Res. Board **2142**, 25–28 (2010)
11. Cao, Y., Zavattieri, P., Youngblood, J., Moon, R., Weiss, J.: The relationship between cellulose nanocrystal dispersion and strength. Constr. Build. Mater. **119**, 71–79 (2016)
12. Dufresne, A.: Nanocellulose: a new ageless bionanomaterial. Mater. Today **16**, 220–227 (2013)
13. Mondal, S.: Preparation, properties and applications of nanocellulosic materials. Carbohydr. Polym. **163**, 301–316 (2017)
14. Mohamed, M.A.S., Ghorbel, E., Wardeh, G.: Valorization of micro-cellulose fibers in selfcompacting concrete. Constr. Build. Mater. **24**, 2473–2480 (2010)
15. Nilsson, J., Sargenius, P.: Effect of microfibrillar cellulose on concrete equivalent mortar fresh and hardened properties (2011)
16. Onuaguluchi, O., Panesar, D.K., Sain, M.: Properties of nanofibre reinforced cement composites. Constr. Build. Mater. **63**, 119–124 (2014)
17. Mejdoub, R., Hammi, H., Suñol, J.J., Khitouni, M., M'nif, A., Boufi, S.: Nanofibrillated cellulose as nanoreinforcement in Portland cement: thermal, mechanical and microstructural properties. J. Compos. Mater. **51**, 2491–503 (2017)
18. Peters, S., Rushing, T., Landis, E., Cummins, T.: Nanocellulose and microcellulose fibers for concrete. Transp. Res. Rec. J. Transp. Res. Board **2142**, 25–28 (2010)
19. Stephenson, K.M.: Characterizing the behavior and properties of nanocellulose reinforced ultra high performance concrete, Maine (2011)
20. Cao, Y., Zavaterri, P., Youngblood, J., Moon, R., Weiss, J.: The influence of cellulose nanocrystal additions on the performance of cement paste. Cem. Concr. Compos. **56**, 73–83 (2015)
21. Eichhorn, S.J., Dufresne, A., Aranguren, M., Marcovich, N.E., Capadona, J.R., Rowan, S.J., et al.: Review: current international research into cellulose nanofibres and nanocomposites. J. Mater. Sci. **45**, 1–33 (2010)
22. Noltingk, B.E., Neppiras, E.A.: Cavitation produced by ultrasonics. Proc. Phys. Soc. Sect. B **63**, 674–685 (1950)
23. Jorge, E., Chartier, T., Boch, P.: Ultrasonic dispersion of ceramic powders. J. Am. Ceram. Soc. **73**(8), 2552–2554 (1990)

Effect of Nano-SiO$_2$ Coating on the Mechanical Recovery of Debonded Fiber-Cement Interface Under Water Curing

Bo Wu$^{(\boxtimes)}$ and Jishen Qiu

Department of Civil and Environmental Engineering, Hong Kong University of Science and Technology, Clear Water Bay, Hong Kong, China
bwuar@connect.ust.hk

Abstract. Recent studies have revealed that interfacial shear strength at debonded fiber/matrix interface can recover under water/dry conditioning cycles which is mainly attributed to the formation of CaCO$_3$ that fills the cracks and enhances the interfacial friction. However, the CO$_3$$^{2-}$ ions (i.e., dissolved CO$_2$) is often scarce at the deep part of a FRCC, significantly limiting the interfacial healing. In this study, SiO$_2$ nanoparticles were successfully coated onto PVA fiber surface via sol-gel method, which is confirmed by SEM-EDS characterizations. The SiO$_2$ nanoparticles coating layer can react with Ca(OH)$_2$ to form C-S-H gel that refines the interfacial zone and potentially promote the healing degree after water conditioning. Single-fiber pull-out and preload-reload results indicate that the SiO$_2$ coating not only enhanced the original interfacial bond, but also greatly helped the recovery at the debonded-and-healed interface. A healing degree estimation way has been proposed in this study and the rationality also has been elucidated.

Keywords: Fiber/cement interface · Self-healing · Nano-SiO$_2$ coating · Mechanical recovery

1 Introduction

The intrinsically brittle cementitious materials are susceptible to cracking due to restrained shrinkage, thermal expansion, and most importantly mechanical loading, which greatly expedites the invasion of external agents, e.g., sulphates, carbonates and chlorides, shortening the service life of R/C structures. Manually repairing the crack is a straightforward measure, but it does not work well to early fine cracks as it is difficult to detect, hence it is important to develop self-healing cementitious materials [1]. Actually, in the presence of water, healing products, e.g., CaCO$_3$ or secondary calcium-silicate-hydrate (C-S-H), can form because of the presence of Ca^{2+} and unhydrated clinkers within hardened cement cracks, only if its width is controlled to approximately 150 µm or less [2]. As a result, microfiber reinforced cementitious composites (FRCCs) are tailored for reducing crack width and improving self-healing capability; a representative example is the engineered cementitious composites (ECCs), which has demonstrated significant recovery of cement matrix by controlling the crack width to around 60 µm with only 2% (by vol.) polymeric microfibers [3, 4].

© RILEM 2022
P. Serna et al. (Eds.): BEFIB 2021, RILEM Bookseries 36, pp. 879–888, 2022.
https://doi.org/10.1007/978-3-030-83719-8_75

The self-healing of FRCC actually happens at two scales. First, in a matrix crack major amount of $CaCO_3$ can form as the dissolved CO_2 (or CO_3^{2-}) reacting with the free Ca^{2+} ions [5–7]; second, recently we have discovered that the same $CaCO_3$ may form between the debonded microfiber and cement matrix [8]. While the first self-healing, i.e., the healing of matrix crack, enhances the durability of FRCC by blocking the external agents [9], it is often the second self-healing, i.e., the fiber/cement interfacial healing (hereafter interfacial healing) that determines the mechanical properties of the fiber-bridging [10]. In a load-bearing structure, the healing of matrix crack may be eliminated if the crack keeps widening because of the poor recovery of fiber-bridging; it is important to study so as to tailor the interfacial healing for robust self-healing in FRCC. Although the formation of $CaCO_3$ has been observed at the debonded interface, this self-healing mechanism will be largely limited due to lack of CO_2, e.g., in a dam or a bridge pillar that is deep in the water, where normally the concentration of CO_2 is about 10 ppm [11]. Thus, we propose to engage a new interfacial healing mechanism by coating a layer of nano-SiO_2 onto the surface of polymeric fiber, based on the following rationale: in FRCC, when a crack-bridging microfiber is debonded from the cement matrix, an interfacial gap will form and expose the nano-SiO_2 to the "water reservoir" in matrix crack; in the presence of water, the nano-SiO_2 will combine free Ca^{2+} ions, mostly from the dissolution of redundant $Ca(OH)_2$ crystals in capillary pores, to form secondary C-S-H (apart from the primary C-S-H from cement hydration); the new C-S-H may refine the fiber/matrix interface and likely to enhance the interfacial friction because of its expansive nature [12, 13].

In this study, we examined the feasibility of this hypothetical self-healing mechanism by demonstrating 1) that a layer of nano-SiO_2 particles can be robustly coated onto the surface of a polyvinyl alcohol (PVA) fiber, the most used fiber type in self-healing ECC; and 2) the effect on nano-SiO_2 coating on the healing-induced recovery of interfacial properties. The sol-gel method was applied to coat nano-SiO_2 on the PVA surface; the single microfiber test, following a preloading-healing-reloading protocol, was carried out to pristine and modified fibers to quantify the interfacial recovery; in addition, the fibers were kept in a simulated cement pore solution to grow healing products, which were characterized as an indirect way to reveal the relevant chemical reaction.

2   Experimental Program

2.1   Fiber Surface Modification

The process of growing nano-SiO_2 on PVA fiber surface is illustrated in Fig. 1a. The current PVA fibers have a length of 18 mm and a diameter of 200 μm; while the typical PVA fibers adopted in ECC have a length of 8–12 mm and a diameter of 40 μm [2, 3]. The large fiber size eliminated the possibility of fiber rupture without altering the surface properties.

Sol-gel method mainly involves two steps [14]: hydrolysis and condensation, as clearly shown in Fig. 1 (a). Specifically, 45 ml ethanol, 4 ml deionized water, 4 ml ammonia, and 2 ml tetraethyl orthosilicate (TEOS) were mixed and then the pristine

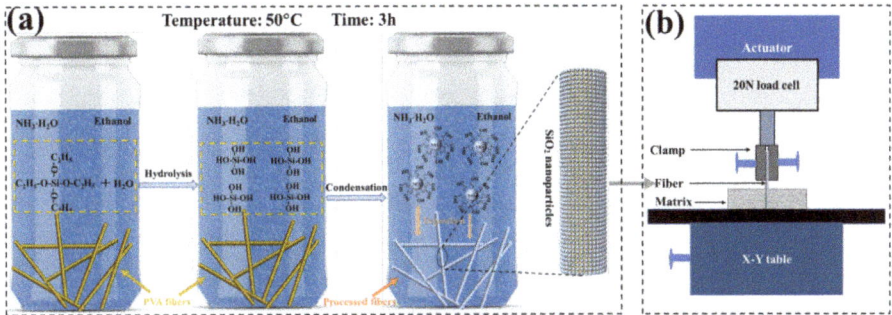

Fig. 1. Illustration of the experiment: (a) coating nano-SiO₂ onto PVA fiber via sol-gel method and (b) single fiber pull-out test set-up

PVA fibers were added, the mixtures was kept at 50 °C for 3 h; finally the fibers were collected by filtration, they were washed with deionized water three times and oven-dried at 70 °C for around 8 h. Both the pristine and modified PVA fibers were examined by scanning electron microscope (SEM, JEOL-6390) and energy dispersive spectrometer (EDS, QUANTAX 4010).

2.2 Single Fiber Pull-Out Test

Single fiber pull-out tests were carried out to determine the fiber/cement interfacial properties. The preparation of single fiber specimen was made by embedding the fibers into a fresh cement paste (52.5R, water/cement ratio of 0.35 by weight); the large multi-fiber specimen was demolded after 24 h of ambient curing, and cured in a moisture room (22 ± 3 °C, relative humidity of 95%) for 27 days, before it was cut into single fiber specimens with a thickness of 2 ± 0.2 mm. All the pull-out tests were conducted on an electrical universal material testing machine (Lyllod, model EZ50) with an actuator displacement rate of 0.5 mm/min, as illustrated in Fig. 1b; in the tensile fixture, there is a X-Y table at the bottom for aiming the fiber tip to the clamper at the top. For both pristine and modified PVA fibers, two pull-out protocols were implemented. In the first protocol, the fiber was directly pulled out by a monotonic loading; the monotonically loaded specimens not only revealed the effect of SiO₂ coating on pristine fiber/cement interface, but also served as controls for healed specimen. In the second protocol, the single fiber was preloaded to fully debond it from the cement matrix; then the whole specimen was removed from the setup and conditioned under tap water at room temperature for 10 days to engage interfacial healing; finally, the healed specimens were re-loaded to induce complete fiber pull-out.

Figure 2 shows typical the tensile load (P) vs. actuator displacement (u) curve of the monotonic fiber pull-out and the preloaded-healed-reloaded fiber pull-out. According to the micromechanics-based fiber pull-out model [4, 15], a sudden and large drop of P (from P_a to P_b) marks the end of interfacial crack propagation, i.e., the fiber being fully debonded from the cement matrix; so the preloading was stopped at this characteristic point.

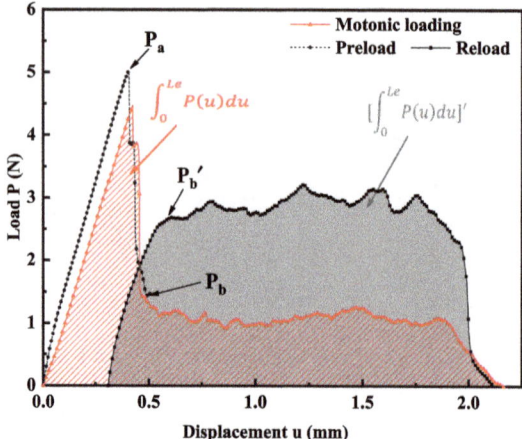

Fig. 2. Representative single fiber pull-out load (P) vs. displacement (u) curve: the monotonically loaded specimen and the preloaded-healed-reloaded specimen.

Interfacial chemical bond (G_d) is the surface energy at the fiber/cement interface, it is an indicator of the strength of chemical link; interfacial frictional bond (τ) represents physical resistance when pulling the fiber out. Both G_d and τ can be calculated based on the P-u curve acquired from single fiber pull-out test (e.g., Fig. 2) [3, 4], using the following equations:

$$G_d = \frac{2(P_a - P_b)^2}{\pi^2 E_f d_f^3} \, (\text{J/m}^2) \tag{1}$$

$$\tau = \frac{P_b}{\pi d_f L_e} \, (\text{N/m}^2) \tag{2}$$

where the E_f, d_f, and L_e, are Young's modulus, diameter, and embedment length of the PVA fiber, respectively. P_a and P_b are the characteristic tensile loads taken right before and right after the moment of full fiber debonding. The total work required to completely pull out a fiber can be determined by the area under the curve, as shown in Fig. 2 (hatched area and coloured area), and the energy absorption of per unit area can be calculated by following equation:

$$G_W = \frac{\int_0^{le} P(u)du}{\pi d_f L_e} \, (\text{J/m}^2) \tag{3}$$

The ratio between the post-healing interfacial properties (x′) and the pristine interfacial properties (x) was calculated to assess the effect of self-healing. The effect of self-healing on frictional bond and energy absorption for fiber pull-out was quantified by

$$\eta = \frac{\tau'}{\tau} \times 100\% \tag{4}$$

$$\xi = \frac{G'_W}{G_W} \times 100\% \tag{5}$$

2.3   Characterization of the Healing Products

It is difficult to characterize the interfacial healing product in situ due to the limited amount of healing product formed in the fine gap and the new compounds may include the same element as bulk cement matrix. As a result, both pristine and coated PVA fibers were kept in saturated $Ca(OH)_2$ solution, which simulates the capillary pore solution in cement paste, to grow healing products. Specifically, the fibers were kept in the solution in a hermetically-sealed glass vial under room temperature for 3 days; after that they were removed, rinsed with deionized water for three times, and oven-dried at 70 °C overnight. The same SEM-EDS was used to characterize the products formed on the fibers surface.

3   Results and Discussion

3.1   Nano-SiO₂ Coating and Its Effect on Interfacial Healing Product

Surface morphology of pristine fiber and modified fiber is shown in Fig. 3. The pristine fiber has a generally smooth surface with longitudinal shallow grooves, which are likely attributed to the uneven shrinkage during the double diffusion of the solidification process, a key step of fiber manufacture [14]; and only the element of C and O are detected on pristine fiber. On the modified fiber, the shallow grooves were covered with a layer of tightly packed round particles ranging from 100 to 320 nm, as shown in Fig. 3d. There were also loose larger particles scattered on this first layer, i.e., the granular texture seen in Fig. 3c, which indicates multiple layers of SiO_2 nanoparticles may grow on PVA surfaces. Apart from the C and O elements detected in pristine fiber surface, the Si element with the content of around 8.6% was detected in modified fiber surface, this contrast indicates that nano-SiO₂ have been successfully coated onto PVA fiber surface.

After immersion in saturated $Ca(OH)_2$ solution, the surface tomography and chemical composition are illustrated in Fig. 4. On the pristine fiber (Fig. 4a and 4b), scattered crystal-like particles were found. Element Ca was detected at different parts; the observed particles are more likely to be $Ca(OH)_2$ instead of $CaCO_3$, as $Ca(OH)_2$ solution was kept in a hermetically-sealed glass vial to avoid contact with CO_2. On the modified fiber (Fig. 3c and 3d), there were also new substance formed. Besides the crystal-like particles, there was also a thin layer of membrane-like substance that covers the original nano-SiO₂ texture. In addition, element Ca deposited on the modified fiber (2.5%) was almost twice that on pristine fiber (1.4%). This can be indicative of that the C-S-H gel can be formed via the reaction between the nano-SiO₂ particles and $Ca(OH)_2$.

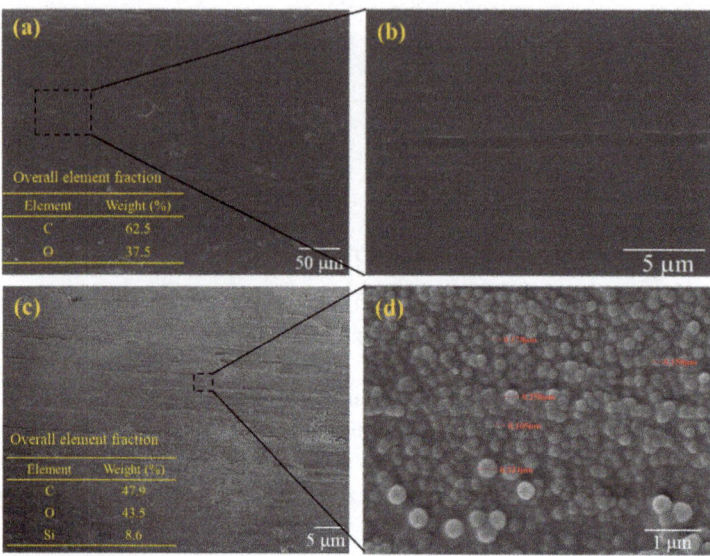

Fig. 3. Surface topography and chemical composition of (a, b) pristine (c, d) modified fiber.

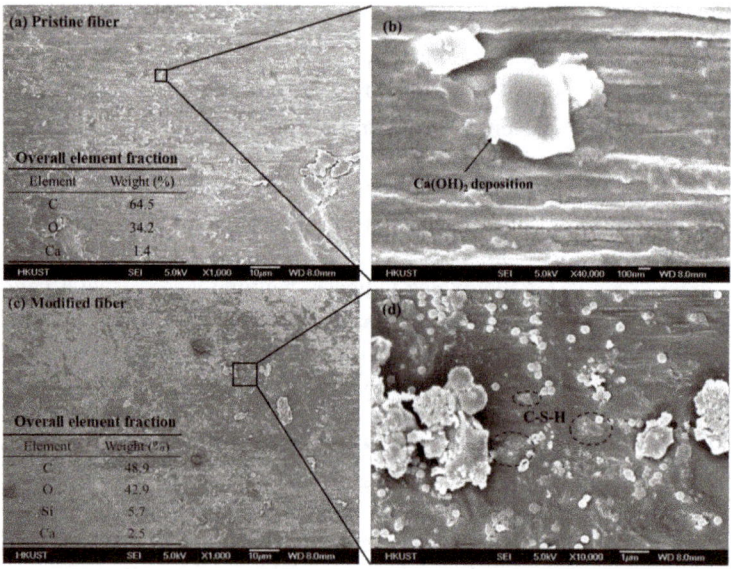

Fig. 4. Surface topography and chemical composition of (a, b) pristine PVA fiber, and (c, d) processed fiber after three-day conditioning in saturated $Ca(OH)_2$ solution.

3.2   Single Fiber Pull-Out Results: Monotonic Loading

Monotonic loading results are shown in Fig. 5a and 5b. The sudden drop from P_a to P_b is observed for both pristine and modified fiber, which indicates the completion of fiber/matrix debonding [3, 4]. In addition, the P_a of the modified fiber was apparently higher than that of pristine fiber, while the P_b of the two groups were comparable. Figure 5c shows that the nano-SiO₂ coating significantly enhanced the G_d (from 4.28 J/m² to 9.55 J/m²), the frictional bond was also improved by about 25% (from 0.88 MPa to 1.10 MPa), implying the formation of C-S-H gel can refine the interface zone and increase the frictional bond.

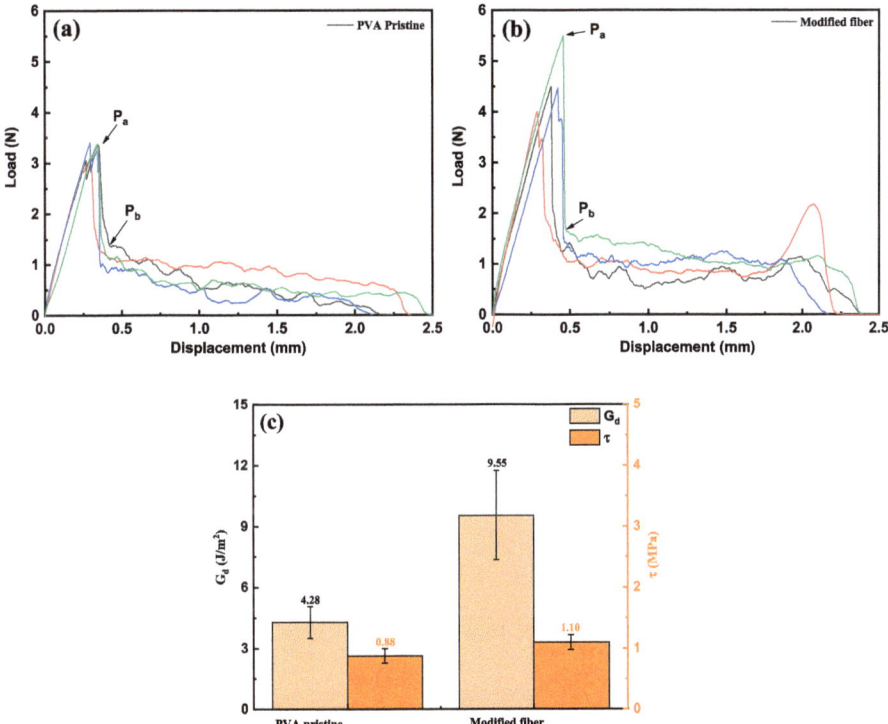

Fig. 5. Single fiber pull out curves of (a) pristine PVA fiber, (b) modified fiber and (c) Chemical bond and frictional bond of pristine PVA fiber and modified PVA fiber.

Figure 6 compares the fiber surface morphology of pristine and modified PVA fiber pulled out from matrix. On the pristine fiber, there are wide spread crystal-like particles, which is similar to the shape of particles in Fig. 4b; they are rich in Ca but free of Si, likely to be Ca(OH)₂. On the modified fiber, there is no obvious scratch, and a different type of hydration product with smaller particle size are observed, similar as the products shape in Fig. 4d; they contain both Ca and Si, likely to be C-S-H. This contrast implies that the loose Ca(OH)₂ particles, which stands for the weak spot at the interface, can be consumed by SiO_2 to form C-S-H gel. Combining the results of mechanical testing and SEM-EDS characterization, the stronger chemical bond from modified fiber may be attributed to the generation of C-S-H gel via reaction of SiO_2 coating and $Ca(OH)_2$, which not only creates stronger chemical link to the –OH group on PVA, but also refines the interfacial zone to enhance the frictional bond.

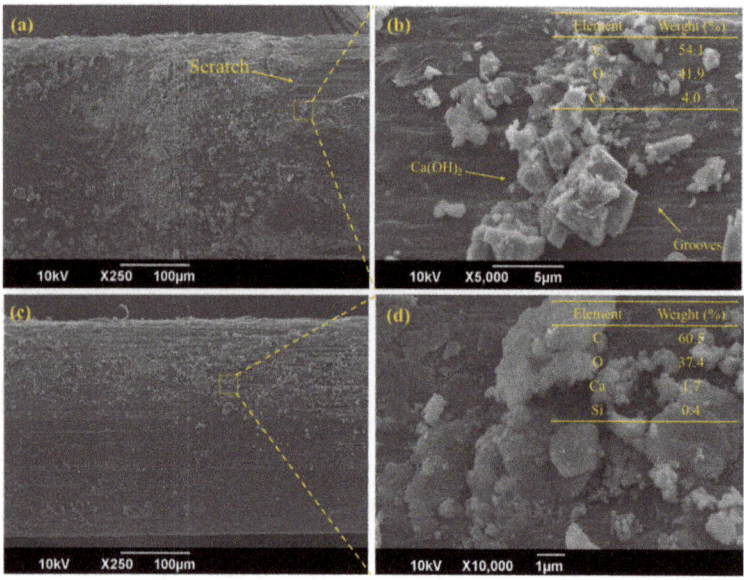

Fig. 6. Surface morphology and elementary composition of (a, b) pristine fiber and (c, d) modified pulled out from matrix.

3.3   Single Fiber Pull-Out Results: Effect of Interfacial Healing

Figure 7a shows the preloading and reloading curves of pristine fibers, while Fig. 7b shows those of the modified fibers. For the pristine fiber, the reloading curve almost follows the preloading curve, demonstrating slip-softening during fiber pull-out; the recorded P_b' is comparable with P_b. These results indicate that the effect of interfacial healing was limited. In previous studies, significant interfacial healing was observed with pristine PVA fibers [8], this is because the preloaded specimen was conditioned under water/dry cycles, which allows CO_2 to penetrate in to form $CaCO_3$. However, in this study the preloaded specimen was constantly cured under water, where is scarce of CO_2. For the modified fiber, the reloading curve noticeably rises from the preloading curve; slip-hardening effect is observed during the fiber pull-out; the P_b' is significantly higher than P_b. The interfacial healing indeed happened despite the absence of CO_2, indicating a different healing mechanism works.

The energy absorption G_w to different fiber displacements (0.5, 1.0, and 1.5 mm) were calculated for both pristine and modified fibers, the results are shown in Fig. 7c. Under either monotonic loading or reloading, the modified fiber induced significantly higher energy absorption at all the studied displacements. It suggests that the effect of SiO$_2$ coating on the pristine interface and healed interface existed until the complete pull-out of the fiber. Figure 7d shows of the effect of interfacial healing by quantifying the healing degree η (Eq. 4) and ξ (Eq. 5). The η and ξ of the pristine fiber were either close to or lower than one, indicating that there was little self-healing at the debonded fiber/cement interface. On the other hand, the η and ξ of the modified fiber were significantly larger than the pristine fiber and of course one, indicating the effect of nano-SiO$_2$ coating on enhancing interfacial healing.

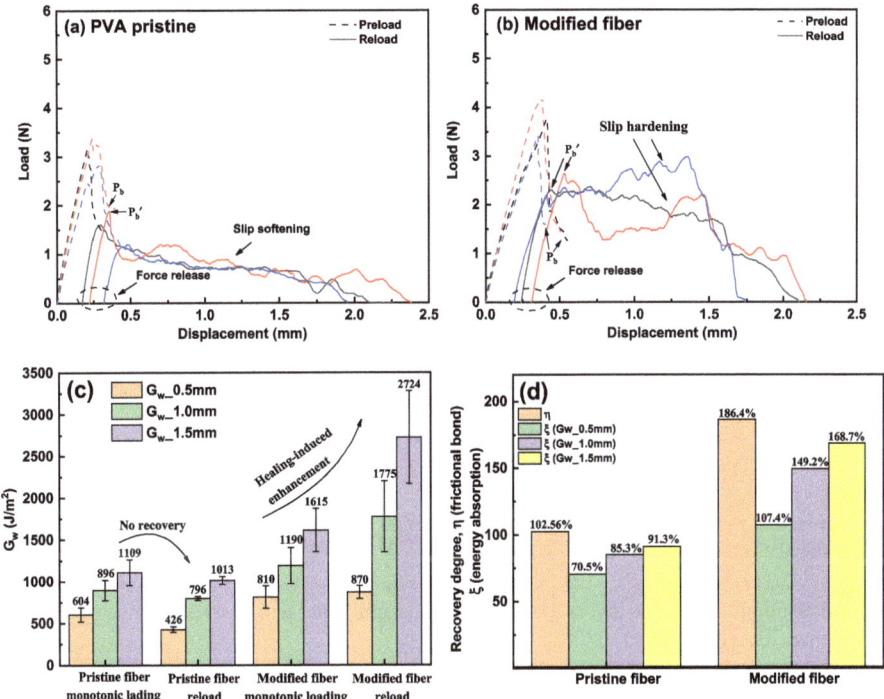

Fig. 7. Preloading and reloading curves of (a) pristine fiber and (b) modified fiber; (c) the energy absorption G$_w$ at different studied displacements (0.5 mm, 1.0 mm, and 1.5 mm); (d) the interfacial healing degree characterized by frictional bond (η, Eq. 4) and energy absorption (ξ, Eq. 5)

4 Conclusions

This work studies the effect of nano-SiO$_2$ particles coated to PVA fiber surface on the fiber/cement interface, especially the interfacial bond after self-healing. The following conclusions are drawn based on the experimental results:

- The nano-SiO$_2$ particles can grow on the PVA fiber surface as a thin layer of coating via sol-gel method, which exhibits higher activity than PVA itself in the mock pore solution of saturated $Ca(OH)_2$ and forms a new product likely to be C-S-H.

- Monotonic single fiber pull-out results suggested that the nano-SiO_2 coating can greatly enhance the both the chemical bond and frictional bond at a pristine fiber/cement interface.
- Reloading of debonded and healed interface indicates that the pristine PVA surface is unlikely to induce interfacial healing under the current water-only conditioning, while the SiO_2 coating induces obvious interfacial healing, as shown by the enhanced frictional bond and the interfacial energy absorption after the same water-only conditioning.

References

1. Li, W., Dong, B., Yang, Z., et al.: Recent advances in intrinsic self-healing cementitious materials. Adv. Mater. **30**, 1705679 (2018)
2. Qiu, J., Tan, H.S., Yang, E.H.: Coupled effects of crack width, slag content, and conditioning alkalinity on autogenous healing of engineered cementitious composites. Cem. Concr. Comp. **73**, 203–212 (2016)
3. Yang, E.H., Wang, S., Yang, Y., et al.: Fiber-bridging constitutive law of engineered cementitious composites. J. Adv. Concr. Technol. **6**, 181–193 (2008)
4. Li, V.C.: Introduction to Engineered Cementitious Composites (ECC), pp. 27–30. Springer, Heidelberg (2019)
5. Palin, D., Wiktor, V., Jonkers, H.M.: Autogenous healing of marine exposed concrete: characterization and quantification through visual crack closure. Cem. Concr. Res. **73**, 17–24 (2015)
6. Maes, M., Snoeck, D., De Belie, N.: Chloride penetration in cracked mortar and the influence of autogenous crack healing. Constr. Build. Mater. **115**, 114–124 (2016)
7. Xue, C., Li, W., Luo, Z., et al.: Effect of chloride ingress on self-healing recovery of smart cementitious composite incorporating crystalline admixture and MgO expansive agent. Cem. Concr. Res. **139**, 106252 (2021)
8. Qiu, J., He, S., Yang, E.H.: Autogenous healing and its enhancement of interface between micro polymeric fiber and hydraulic cement matrix. Cem. Concr. Res. **124**, 105830 (2019)
9. Sahmaran, M., Li, M., Li, V.C.: Transport properties of engineered cementitious composites under chloride exposure. ACI Mater. J. **104**, 604–611 (2007)
10. Colt, J.: Dissolved Gas Concentration in Water: Computation as Functions of Temperature, Salinity and Pressure, 2nd edn. Elsevier, Amsterdam (2012)
11. Pi, Z., Xiao, H., Du, J., et al.: Interfacial microstructure and bond strength of nano-SiO_2-coated steel fibers in cement matrix. Cem. Concr. Comp. **103**, 1–10 (2019)
12. Lu, M., Xiao, H., Liu, M., et al.: Improved interfacial strength of SiO_2 coated carbon fiber in cement matrix. Cem. Concr. Comp. **91**, 21–28 (2018)
13. Yang, X., Zhu, L., Chen, Y., et al.: Controlled hydrophilic/hydrophobic property of silica films by manipulating the hydrolysis and condensation of tetraethoxysilane. Appl. Surf. Sci. **376**, 1–9 (2016)
14. Fujiwara, H., Shibayama, M., Chen, J.H., et al.: Preparation of high-strength poly (vinyl alcohol) fibers by crosslinking wet spinning. J. Appl. Polym. Sci. **37**, 1403–1414 (1989)
15. Redon, C., Li, V.C., Wu, C., Hoshiro, H., Saito, T., Ogawa, A.: Measuring and modifying interface properties of PVA fibers in ECC matrix. J. Mater. Civil. Eng. **13**, 399–406 (2001)

Piezoresistivity of Carbon Black/Cement-Based Sensor Enhanced with Polypropylene Fibre

Wengui Li[1(✉)], Wenkui Dong[1], and Surendra P. Shah[2]

[1] School of Civil and Environmental Engineering, University of Technology Sydney, Sydney, NSW 2007, Australia
wengui.li@uts.edu.au

[2] Center for Advanced Construction Materials, University of Texas at Arlington, Arlington, TX 76019, USA

Abstract. In this study, polypropylene (PP) was added to develop carbon black (CB)/cementitious composites as cement-based sensors. The mechanical properties and piezoresistivity were been experimentally investigated. The compressive strength slightly decreased, while the flexural strength was significantly increased with the increased amount of PP fibres. The improvement is mainly achieved by the reduced CB concentration in cement matrix and the excellent tensile strength of PP fibres. Under the cyclic compression, the piezoresistivity increased by three times for 0.4 wt% PP fibres filled CB/cementitious composite, regardless of the loading rates. The flexural stress sensing efficiency was considerably lower than that of compressive stress sensing, but it increased with the amount of PP fibres. Electrical conductivity increased with the amount of PP fibres, due to the enclosed CB nanoparticles and more conductive passages. Moreover, fitting formulas were proposed and used to evaluate the self-sensing capacity, with the attempts to apply cement-based sensors for structural health monitoring.

Keywords: Cement-based sensor · Carbon black · Polypropylene fibre · Piezoresistivity · Microstructure

1 Introduction

With the increasing demand for structural health monitoring (SHM) of concrete, the cement-based composites possessing piezoresistivity or self-sensing properties have attracted much more attentions. The self-sensing behaviour of the cement-based sensors originates from their improved electrical conductivity brought by the additional conductive fillers, and then the electrical signals can be altered as the loading or deformation changes. Once the relationship between the changed electrical signal and the applied loadings or deformations is calibrated, the cement-based sensor can be used in concrete structures for SHM. Generally, in comparison to the conventional sensors, it has been found that the cement-based sensors not only have better sensing efficiency, low cost and high adaptability, but also cause minimal defects on the mechanical properties and durability of the concrete structures when the sensors are embedded into the detection system [1–4].

© RILEM 2022
P. Serna et al. (Eds.): BEFIB 2021, RILEM Bookseries 36, pp. 889–899, 2022.
https://doi.org/10.1007/978-3-030-83719-8_76

The sensing capacity of cement-based sensors has been summarized with different conductive fillers under various stress and environmental conditions [5, 6]. Generally, the conductive fillers in the forms of fibres have better performance to enhance the electrical conductivity and sensing efficiency of the cement-based sensors. Moreover, the cement-based sensors filled with conductive fibres having high aspect ratio possess higher piezoresistivity than that with conductive fibres with small aspect ratio [7–9]. As for the monitoring efficiency under different stress modes and magnitudes, investigators found that the cement-based sensors have capacity to monitor the health conditions under both static and dynamic loadings [10–12]. However, experimental results also showed different sensing efficiency under various stress magnitudes, with the higher sensitivity at the level of low stress but the relatively lower sensitivity at the high stress level [13]. The uneven distribution of conductive fillers and the microstructural instability of the brittle cementitious composite are responsible for the altered sensitivity.

In terms of the environmental factors, temperature changes can influence the electrical resistivity of the cement-based sensors, while the sensing efficiency of the cement-based sensor is limitedly affected. Nevertheless, both the electrical resistivity and sensing efficiency are greatly fluctuated by the water content of the cement-based sensor. There exists an optimal value of water content to achieve the cement-based sensors with highest piezoresistive sensitivity [14]. In addition to the self-sensing characteristics, investigators proposed that the cement-based sensors filled with conductive nanoparticles can be provided with improved mechanical strengths, durability and ductility [15–18]. For example, Yu et al. [19] and Li et al. [20] respectively observed higher compressive strength for the cementitious composite reinforced with 0.1 wt.% and 0.5 wt.% carbon nanotubes (CNTs) than that without CNTs, and found that the higher content of CNTs could more considerably improve the compressive strength of cementitious composite. However, for the flexural strength of CNTs reinforced cementitious composite, investigators found the optimum content as low as 0.08 wt.%. This demonstrates that the optimal contents of CNTs in the cementitious composites can greatly vary to achieve the highest compressive and flexural strength.

Studies attempted to apply carbon black (CB) nanoparticles mixing with cementitious composite to manufacture the cement-based sensors, because of their lower cost and stable electrical outputs than CNTs. According to the experimental results, the CB filled cement-based sensors can successfully monitor the compressive stress and strain. However, the CB filled cement-based sensors normally have lower sensing efficiency, whose gauge factor (fractional changes of resistivity per unit strain) reaches only a few dozens and accounts for approximately one tenth of the sensitivity of cement-based sensors with CNTs. In particular, mechanical properties of CB filled cementitious composite significantly decrease with the increase of CB content because of CB agglomerates or poor cohesion to cement matrix, which prohibit the further application of CB nanoparticles in the cementitious composites.

To overcome these disadvantages, authors have strived to improve the sensing efficiency of the CB filled cement-based sensor by the addition of conductive rubber fibres or crumbs. The study does improve the gauge factor and the piezoresistive sensitivity of CB filled cement-based sensor, but the mechanical properties significantly decrease with the increase of rubber fibres which restrict the practical application of CB filled cement-based sensors. In addition, the mixture of CB and CNTs have been studied on the mechanical and sensing properties of cementitious composites. Thanks to the improved uniform distribution of nanomaterials, the mechanical properties of the cement-based sensor firstly increase with the increase of concentration of fillers, then decrease as the content gets larger than 1.5 vol.%. The sensing efficiency improvement is unknown because the cement-based sensors merely filled with CB or CNTs are not investigated and compared. Overall, the application potential of CB filled cement-based sensors is still confined either by the relatively low sensing efficiency or the reduced mechanical properties.

Therefore, in this paper, much lower-cost polypropylene (PP) fibres rather than the aforementioned conductive rubber fibres or CNTs have been proposed to assist the CB distribution in the cement-based sensors, in order to achieve sensors with both improved piezoresistivity and mechanical properties. The directional distribution of CB nanoparticles attaching onto the surface of PP fibres promotes the generation of conductive passages and contact points. The effects of PP fibres on the physical and mechanical properties of concrete have been widely explored, but none of these studies investigates the effects of PP fibres in the conductive cementitious composite or cement-based sensors. Authors firstly studied the effects of PP fibres on the mechanical properties and self-sensing behaviours of cement-based sensors. It is expected that the PP fibres can stimulate the usage of CB nanoparticles to manufacture the cement-based sensors at low cost, high sensitivity and practicality. Most importantly, this paper provides a new method to enhance the electrical and piezoresistive properties of specimen by adopting nonconductive fibres possessing high aspect ratio and tensile strength.

2 Experimental Preparation

The commercially available General Purpose cement and silica fume were used for mix design. The silica fume with average size of 360 nm is prepared to partially replace the cement, to improve the microstructures of cementitious composite and improve the uniform dispersion of CB nanoparticles. The substitution rate chosen is 10 wt.% to the weight of cement. The CB with average particle size of 20 nm is used. It has excellent electrical conductivity with the resistivity <0.43 $\Omega \cdot cm$. The physical and electrical properties of CB are listed in Table 1. The micromorphology of CB nanoparticles by scanning electron microscope (SEM) is shown in Fig. 1.

Table 1. Physical properties of carbon black.

Average particle size (nm)	Resistivity ($\Omega \cdot cm$)	Apparent density (g/l)	DBP (ml/100 g)	Surface area (m^2/g)	pH value	Ash content (%)
20	<0.43	0.375	280	254	7.5	<0.3

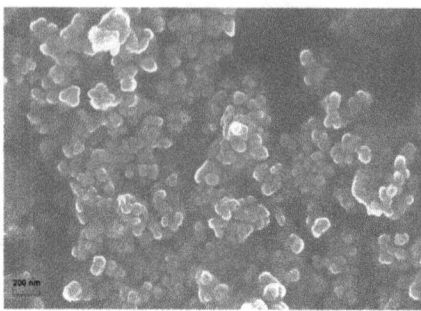

Fig. 1. Micromorphology of as received CB nanoparticles.

The commercially available PP fibres have 9 mm in length and the diameters ranging from 18 to 48 µm. It has been reported that the higher aspect ratio of conductive fibres in the cementitious composite can enhance the electrical conductivity and piezoresistivity. Therefore, the high aspect ratio of PP fibres provides a promising opportunity to be applied in the conductive cement-based sensor. Additional physical and mechanical properties of PP fibres are shown in Table 2. The polycarboxylate based superplasticizer is used to improve the workability of cementitious composites, as well as to stimulate the uniform dispersion of CB nanoparticles.

Table 2. Main physical and mechanical properties of PP fibre.

Appearance	Density (g/cm³)	Length (mm)	Diameter (µm)	Water absorption	Tensile strength (MPa)	Elastic modulus (GPa)	Elongation	Safety
White monofilament	0.91	9	18–48	None	>486	>4.8	>15%	Non-toxic

Generally, there are five groups with different concentrations of PP fibres ranging from 0, 0.1%, 0.2%, 0.3% and 0.4 wt.% by weight of total binder in the CB filled cement-based sensors, and the mix proportions are listed in Table 3. To ensure the even distribution of PP fibres in the cementitious composite, they are firstly added into the weighted tap water for dispersion. The dispersion of PP fibres in the aqueous solution can be conducted through gentle mechanical stirring. In addition, it is noted that the superplasticizer should be added after the mechanical stirring stage, to avoid the mass production of air bubbles. Afterwards, the CB nanoparticles are added into the PP fibres/superplasticizer aqueous solution. Ultrasonic dispersion is applied on the PP fibres/CB aqueous solution after the simple treatment of mechanical stirring. Particularly, the ultrasonic duration is 1 h with the frequency of 40 kHz, and every 10 min the sonication bath is updated to prevent the temperature increase, which probably affects the dispersion efficiency of CB nanoparticles.

After the preparation of PP fibres/CB aqueous solution, the solution is poured into a Hobart mixer, followed by the addition of cement and silica fume. The cementitious composite is mixed at low mixing rate of 140 ± 5 r/min for 1 min, and then mixed at

Table 3. Mix proportions of PP fibres and CB filled cement-based sensors.

Index	Cement	Silica fume	Carbon black (%)	PP fibres (%)	water	Superplasticizer (%)
P0[*]	0.9	0.1	0.5	0	0.4[*]	0.8
P1[*]	0.9	0.1	0.5	0.1	0.4	0.8
P2	0.9	0.1	0.5	0.2	0.4	0.8
P3	0.9	0.1	0.5	0.3	0.4	0.8
P5	0.9	0.1	0.5	0.4	0.4	0.8

Note: P0 and P1 present the CB filled cement-based sensors without and with 0.1% PP fibres, respectively; figures under the cement, silica fume, CB, PP fibres, water and superplasticizer represent their ratios to the weight of binder, e.g. 0.4[*] under water means the water to binder ratio of 0.1%

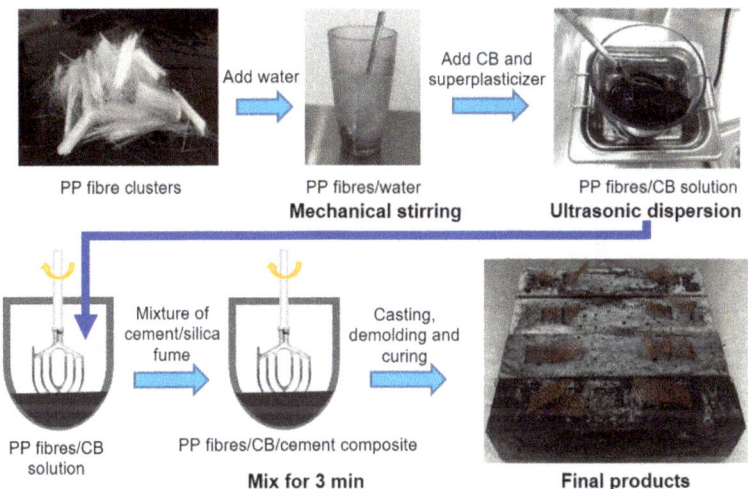

Fig. 2. Schematic diagram of manufacturing procedure of CB/cementitious composites with PP fibre.

high mixing rate of 280 ± 10 r/min for another 3 min. The moulds at the size of $160 \text{ mm} \times 40 \text{ mm} \times 40 \text{ mm}$ are casted and compacted in a vibration table. Meanwhile, four copper meshes are symmetrically embedded into the cementitious composite as electrodes. The specimen are cured in the standard curing chamber for 1 day before demolding, and are further cured in the chamber for another 27 days. The final products of PP fibres/CB cement-based sensor and the manufacturing procedures are illustrated in Fig. 2.

3 Results and Discussion

3.1 Electrical Resistivity

Figure 3 shows the electrical resistivity of the CB filled cementitious composites with different contents of PP fibres. It could be observed that the electrical resistivity dramatically decreased by more than one order of magnitude for the specimens filled with 0.1 wt.% PP fibres. Afterwards, with the increase of PP fibres, the electrical resistivity of the cementitious composite decreased at a slow rate and reached the value of 3.4×10^5 $\Omega \cdot cm$ at the dosage of 0.4 wt.% PP fibres. With the presence of well-dispersed CB nanoparticles, it shows that the PP fibres at a small concentration of 0.4 wt.% to the weight of binder have the capacity to decrease the electrical resistivity of the CB cementitious composite by approximately two orders of magnitude.

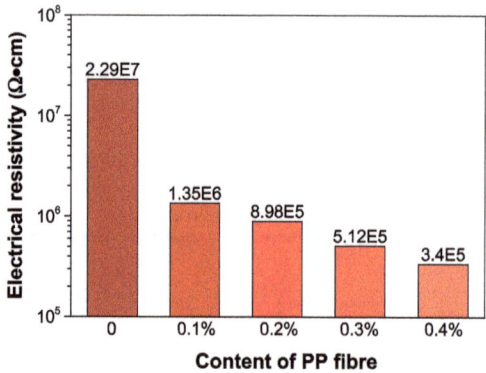

Fig. 3. Electrical resistivity for CB/cementitious composite with various contents of PP fibre.

To explain phenomenon of decreased electrical resistivity, Fig. 4 shows the micromorphology of PP fibres with four different magnifications in the CB filled cementitious composite. Generally, it could be detected that the PP fibres were surrounded by the cement hydration products and CB nanoparticles. The nonconductive hydration products made no differences on the electrical resistivity of PP fibres, while the enclosed CB nanoparticles could generate the conductive passages through their contacts and greatly decrease the electrical resistivity of the PP fibres. Previous studies had demonstrated that the conductive fibrous fillers had better performance to enhance the electrical conductivity of the cementitious composite than the spherical fillers. The high aspect ratio PP fibres with improved conductivity are just similar to the conductive fibrous fillers, which can stimulate the electrical conductivity of the whole cementitious composite by the inner connection among PP fibres. Secondly, in comparison to the density of CB distribution in the PP fibres and cement matrix, it seems that the CB nanoparticles had higher possibility to be absorbed in the surface of PP fibres, rather than spread in the cement matrix.

It has been proposed that the CB preferentially distributed in the epoxy with high polarity. Generally, the CB possesses the hydrophobic groups in the surface such as

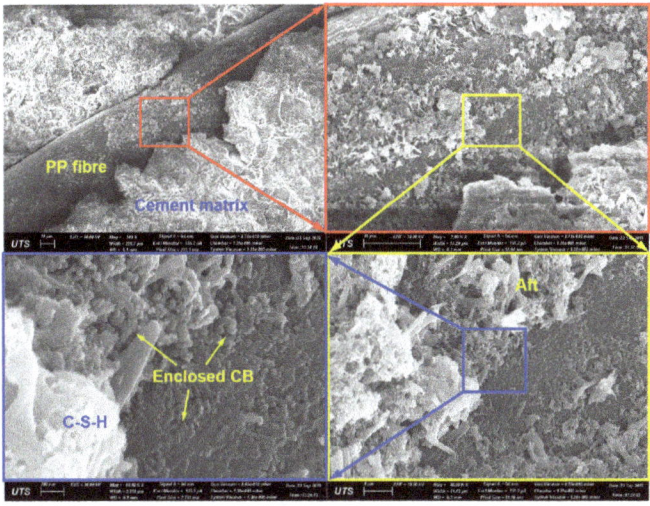

Fig. 4. Microstructure of PP fibres in CB/cement matrix and functional groups in CB surface.

aryl group, ether group and ester group, that is why the CB nanoparticles having the difficulties to disperse in the aqueous solution. However, it seems that the nonpolar PP fibres had much higher lipophilicity than cement matrix and aqueous solution, hence CB nanoparticles were tended to be absorbed by the surface of PP fibres. In other words, the addition of PP fibres led to the directional distribution of CB nanoparticles in the cement matrix, especially along with the PP fibres, which was beneficial for the creation of conductive passages and decreased the electrical resistivity of the cementitious composite. However, the high concentration of CB nanoparticles on the surface of PP fibres represented decreased CB concentration in the cement matrix, which actually had a detrimental effect on the conductivity of cementitious composite. From the above experimental results, the conclusion can be drawn that the positive effect by CB enclosed PP fibres on the electrical conductivity of CB filled cementitious composite overwhelmed the negativity brought by lower CB concentration in cement matrix. More discussions on the reduced CB concentration in cement matrix will be presented in the following section.

3.2   Self-sensing Capacity

Figure 5 illustrates the fractional changes of resistivity (FCR) as a function to compressive stress for the specimens with various contents of PP fibres under various loading rates. For all specimens, the FCR (absolute value, the same below) increased with the increment of compressive stress and decreased when the stress returned to zero, and performed good piezoresistivity. Linear fittings of these results were attached, showing that the relationship between FCR changes and compressive strength was almost linear. Moreover, the overlapped or very close curves in the loading and unloading process indicated that the piezoresistivity of specimen showed good repeatability. Therefore, it could be deduced that the CB filled cementitious composite

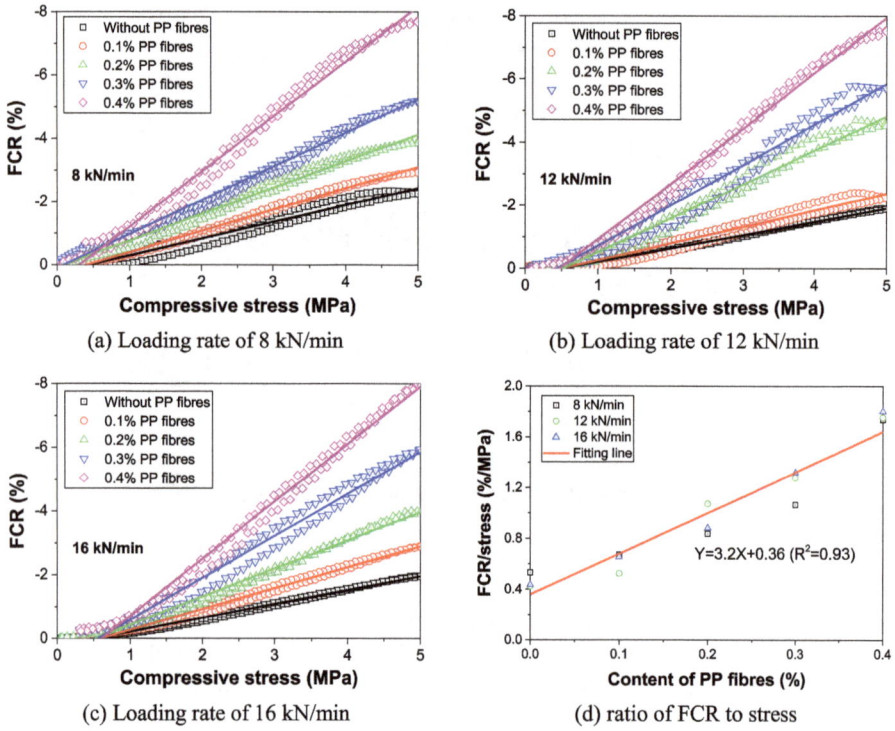

Fig. 5. Fractional changes of resistivity as a function of compressive stress under various loading rates and ratio of FCR to stress with various PP contents

with/without PP fibres were both eligible to be a stress sensor, regardless of the loading rates.

Generally, the tendency of FCR changes was similar under various loading rates, which demonstrated that the loading rates almost possess no impact on the piezoresistivity. Since the loading rates in reality are always unpredictable and multiple, the results lay the foundation of applying the CB/PP fibres filled cementitious composite into the real engineering project. In terms of the piezoresistive sensitivity, the specimen without PP fibres reached the FCR peaks of approximately 2% at the stress magnitude of 5 MPa. However, it seems that the increasing rate of FCR increased with the increased content of PP fibres, and the largest FCR nearly 8% occurred for the specimen filled with 0.4 wt.% PP fibres at the same stress magnitude.

The results showed that the CB cementitious composite filled with 0.4 wt.% PP fibres could improve the piezoresistive sensitivity by three times more than those without PP fibres. The SEM images have showed that the PP fibres were enclosed with conductive CB nanoparticles, and the higher FCR mainly originated from the conductive PP fibres, which were more easily connected to form conductive passages given their high aspect ratio. Howbeit, the CB nanoparticles were much more difficult to contact with each other due to the spherical properties, and that was why the FCR

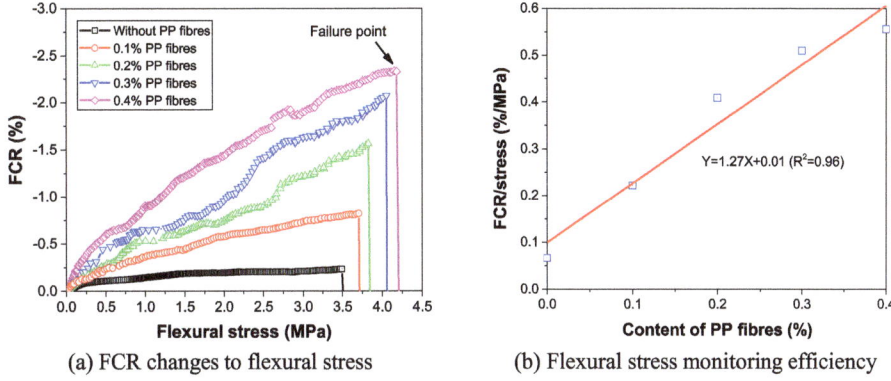

(a) FCR changes to flexural stress   (b) Flexural stress monitoring efficiency

Fig. 6. FCR changes during flexural failure and flexural stress sensitivity at various of PP fibres

changes were lower for the CB cementitious composite without PP fibres when subjected to identical compressive stress.

It illustrates the ratios of FCR to compressive stress as a function to the content of PP fibres. The ratio of FCR changes by unit stress can be a coefficient to evaluate the piezoresistive sensitivity. Regardless of the loading rates applied, the fitting curve showed that the piezoresistive sensitivity of CB cementitious composite possessed linear relationship with the added dosage of PP fibres. Therefore, the fitting curve can be used to predict the compressive stress sensing efficiency for the CB cementitious composite filled with different contents of PP fibres.

The results of FCR as a function to flexural stress by three-point-bend test until failure are displayed in Fig. 6. Similarly, the electrical resistivity of the specimen decreased with the increase of flexural stress, even though the FCR were much smaller than the values under the same stress of uniaxial compression. In addition to the mentioned improved flexural strength of specimen, it was found that the flexural stress sensing ability increased with the increased PP fibres, by the FCR increased from 0.23% to 2.35% when the dosage of PP fibres reached 0.4 wt.%. The more than 10 times higher FCR is mainly attributed to the more contact points induced by PP fibres. As for the failure detection, it could be found that the FCR rapidly increased at the failure point. This is easy to understand because of the brittleness of cementitious composite, the permanent cracks are firstly generated from the central loading point.

The bridging effect of PP fibres maintained the decreased electrical resistivity even there are existing micro-cracks. Still, the cracks can suddenly run through the beam when the stress exceeds flexural strength, thus the PP fibres are failed to maintain the conductive passages and that is why the electrical resistivity rapidly increases. It shows the flexural stress sensing efficiency by plotting the relationship between the ratios of FCR to flexural stress and the content of PP fibres. It seems that the flexural stress sensing efficiency increased with the increase of PP fibres. Similarly, the fitting curve is proposed to assess the piezoresistive sensitivity to the content of PP fibres.

4 Conclusions

The influences of PP fibres on the electrical resistivity, compressive and flexural strength, and the self-sensing properties of functional CB filled cementitious composite were studied in this study. The following conclusions can be drawn up:

- Although the PP fibres were electrically insulative, the enclosed CB nanoparticles on the surface could increase the conductive passages through PP fibres, thus the electrical conductivity of the composite was enhanced with the increase of PP fibres.
- Even though the CB nanoparticles could stimulate the cement hydration, the compressive strength of the composite decreased with the increasing PP fibres under the gradually decreasing rate. This was probably due to the reduced CB dosage in the cement matrix, which released the problem of CB agglomeration.
- The flexural strength of composite considerably improved with the increase of PP fibres. In addition to the released problem of CB agglomeration, the higher flexural strength was mainly owing to the high tensile properties of PP fibres and their bridging effect in the cementitious composite.
- The compressive stress sensing efficiency of composite linearly increased with the increase of PP fibres, regardless of the various loading rates. The higher sensitivity was attributed to more contact points, which was much easier generated by conductive PP fibres rather than spherical CB nanoparticles.
- The flexural stress sensing ability was much weaker than that of compressive stress, and the sensing efficiency dramatically increased by nearly 6 times when the added PP fibres reached 0.4 wt.%. Similarly, the flexural stress as a function to resistivity changes and content of PP fibres was established.

Acknowledgements. All the authors appreciate the financial supports from the Australian Research Council (ARC), University of Technology Sydney Research Academic Program at Tech Lab (UTS RAPT), and University of Technology Sydney Tech Lab Blue Sky Research Scheme.

References

1. Dong, W., Li, W., Tao, Z., Wang, K.: Piezoresistive properties of cement-based sensors: review and perspective. Constr. Build. Mater. **203**, 146–163 (2019)
2. Han, B., Ding, S., Yu, X.: Intrinsic self-sensing concrete and structures: a review. Measurement **59**, 110–128 (2015)
3. Dong, W., Li, W., Wang, K., Luo, Z., Sheng, D.: Self-sensing capabilities of cement-based sensor with layer-distributed conductive rubber fibres. Sens. Actuat. A Phys. **301**, 111763 (2020)
4. Dong, W., Li, W., Wang, K., Vessalas, K., Zhang, S.: Mechanical strength and self-sensing capacity of smart cementitious composite containing conductive rubber crumbs. J. Intel. Mater. Syst. Struct. **31**(10), 1325–1340 (2020)

5. Teomete, E.: The effect of temperature and moisture on electrical resistance, strain sensitivity and crack sensitivity of steel fiber reinforced smart cement composite. Smart Mater. Struct. **25**(7), 075024 (2016)

6. Yıldırım, G., Öztürk, O., Al-Dahawi, A., Ulu, A., Şahmaran, M.: Self-sensing capability of engineered cementitious composites: effects of aging and loading conditions. Constr. Build. Mater. **231**, 117132 (2020)

7. Baeza, F., Galao, O., Zornoza, E., Garcés, P.: Effect of aspect ratio on strain sensing capacity of carbon fiber reinforced cement composites. Mater. Des. **51**, 1085–1094 (2013)

8. Park, J., Jang, J., Wang, Z., Kwon, D., DeVries, K.: Self-sensing of carbon fiber/carbon nanofiber–epoxy composites with two different nanofiber aspect ratios investigated by electrical resistance and wettability measurements. Compos. A Appl. Sci. Manuf. **41**(11), 1702–1711 (2010)

9. Wen, S., Chung, D.: Electrical-resistance-based damage self-sensing in carbon fiber reinforced cement. Carbon **45**(4), 710–716 (2007)

10. Sasmal, S., Ravivarman, N., Sindu, B., Vignesh, K.: Electrical conductivity and piezo-resistive characteristics of CNT and CNF incorporated cementitious nanocomposites under static and dynamic loading. Compos. A Appl. Sci. Manuf. **100**, 227–243 (2017)

11. D'Alessandro, A., et al.: Static and dynamic strain monitoring of reinforced concrete components through embedded carbon nanotube cement-based sensors. Shock Vib. **2017**, 1–11 (2017). https://doi.org/10.1155/2017/3648403

12. Dong, W., Li, W., Shen, L., Sun, Z., Sheng, D.: Piezoresistivity of smart carbon nanotubes (CNTs) reinforced cementitious composite under integrated cyclic compression and impact. Composite Structures **241**, 112106 (2020)

13. Pang, S., Gao, H., Xu, C., Quek, S., Du, H.: Strain and damage self-sensing cement composites with conductive graphene nanoplatelet. In: Sensors and Smart Structures Technologies for Civil, Mechanical, and Aerospace Systems 2014, International Society for Optics and Photonics, p. 906126 (2014)

14. Dong, W., Li, W., Lu, N., Qu, F., Vessalas, K., Sheng, D.: Piezoresistive behaviours of cement-based sensor with carbon black subjected to various temperature and water content. Compos. B Eng. **178**, 107488 (2019)

15. Galao, O., Baeza, F.J., Zornoza, E., Garcés, P.: Strain and damage sensing properties on multifunctional cement composites with CNF admixture. Cem. Concr. Compos. **46**, 90–98 (2014)

16. Al-Dahawi, A., et al.: Electrical percolation threshold of cementitious composites possessing self-sensing functionality incorporating different carbon-based materials. Smart Mater. Struct. **25**(10), 105005 (2016)

17. Danoglidis, P., Konsta-Gdoutos, M., Shah, S.: Relationship between the carbon nanotube dispersion state, electrochemical impedance and capacitance and mechanical properties of percolative nanoreinforced OPC mortars. Carbon **145**, 218–228 (2019)

18. Öztürk, O., Yıldırım, G., Keskin, Ü.S., Siad, H.., Şahmaran, M..: Nano-tailored multi-functional cementitious composites. Compos. B Eng. **182**, 107670 (2020)

19. Yu, X., Kwon, E.: A carbon nanotube/cement composite with piezoresistive properties. Smart Mater. Struct. **18**(5), 055010 (2009)

20. Li, G., Wang, P., Zhao, X.: Mechanical behavior and microstructure of cement composites incorporating surface-treated multi-walled carbon nanotubes. Carbon **43**(6), 1239–1245 (2005)

UHPFRC

Development of Eco-Efficient UHPC and UHPFRC by Recycling Granite Waste Powder (GWP)

David Bouchard, Thomas Sanchez, Luca Sorelli[(⊠)], and David Conciatori

Department of Civil Engineering (CRIB), Laval University, Quebec City, Canada
luca.sorelli@gci.ulaval.ca

Abstract. Ultra-high performance concrete (UHPC) is a class of cement-based composites with enormous potentials for sustainable construction thanks to its outstanding compressive strength and cracking resistance. While the former allows saving material volumes and reducing the structural weight, the latter confers unprecedent durability. However, one of the major limiting factor of the industrial use of UHPC drawbacks it is high cost of raw materials. Moreover, UHPC are often characterized by high cement content which not only remains unreacted for a considerable percentage, but it also contributes to the anthropic carbon emission. To reduce the environmental impact of UHPC and initial cost, granite waste powder (GWP) which is locally available as an industrial by-product of the granite quarry in Quebec was used to partially replace cement. UHPC mix-designs with low cement content (400 kg/m^3) with 200 kg/m^3 of recycled granite powder were produced by assuring satisfactory workability and the compressive strength. The present results compare the developed UHPC and UHPFRC using granite powder with a commercially available UHPC using various indicators.

Keywords: Recycling · Eco-efficient · Granite powder · Mix optimization · UHPC

1 Introduction

Ultra-high performance concrete (UHPC) is a class of cement-based composites with high compressive strength (120–150 MPa) which was developed in the 1990's [1, 2] by enhancing the degree of compaction of matrix. Furthermore, in Ultra-high performance Fiber Reinforced concrete (UHPFRC) micro-steel fibers are added to confer tensile toughness, which assure multiple cracking and impermeability in service conditions. UHPC are renowned for their very high compressive strength, self-compacting behaviour and enhanced durability potential when compared to the conventional or high-performance concretes used in the industry [3]. The development of national codes and design recommendations is today accelerating their industrial applications with several successful pilot projects worldwide [4, 5].

© RILEM 2022
P. Serna et al. (Eds.): BEFIB 2021, RILEM Bookseries 36, pp. 903–914, 2022.
https://doi.org/10.1007/978-3-030-83719-8_77

Some of the major drawback regarding the use of UHPC and UHPFRC are related to their high initial cost and embedded energy (related to the high amount of cement used in its composition [6, 7]). Massive utilization of different supplementary cementitious materials (SCM) or industrial wastes is a promising approach to overcome those drawbacks [8, 9]. Inclusion of granite waste powder, produced in large quantities around the world by the granite industry as an industrial by-products, in conventional concrete had been studied. The general conclusions [10] indicate that granite waste was a viable replacement for cement or natural sand with regards to compressive strength and workability when the replacement ratio is kept within 10 and 40%. Few studies were made regarding the use of granite powder waste in UHPC [7, 9, 11, 12], with limited cement replacement.

This preliminary work aims at developing low-cement/low-cost UHPC and UHPFRC by optimization of the cementitious matrix and utilization of granite powder. The developed mix-designs were finally compared with a commercially available UHPFRC (Ductal®) [13] with various efficiency indicators.

2   Materials and Methodology

2.1   Materials

Two different types of cement were studied, both provided by Ciment Québec. CEM/GU is an all-purpose cement similar to ASTM type I cement and CEM/HE is a high early-strength cement similar to ASTM type III. Supplementary cementitious materials (SCM) include class-F fly ash (FA) provided by LafargeHolcim and an undensified silica fume (SF) provided by Silicium Québec. The particles size distributions (PSD) for all materials are presented in Fig. 1.

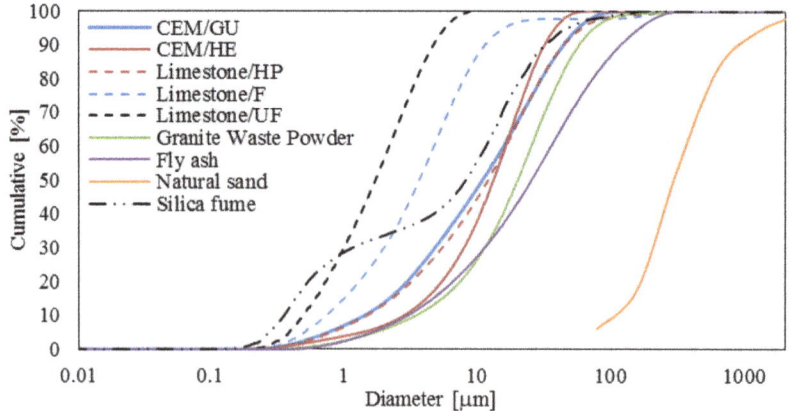

Fig. 1. Particles size distributions (PSD) of materials.

Inert fillers (LF) were used to increase the packing density of the matrix and act as partial replacement for cement. Three grades of crushed limestone fillers were used and all the fillers were provided by OMYA. The grade (PSD) used were medium (LF/HP), fine (LF/F) and ultrafine (LF/UF). Characterization of these materials is provided Fig. 1.

The granite powder was recovered on the dumping site of a Quebec quarry of Polycor (Fig. 2a) and in the settling pond. The granite powder is a very fine powder produced by the cutting and polishing operations of massive granite blocks into thin slabs or others elements. The powder has a very high water retention capacity (Fig. 2b) and the granite mud recovered was let to decant for a period of 24 h. Prior to being used in UHPC mixture, the GP was dried at 110 °C for at least 48 h, until constant mass was achieved. The average mass loss during the whole drying process (decantation and drying) was about 27%. Drying of the GP produced solid agglomerations (Fig. 2c), which were crushed by hand and the material was sieved through a 600 µm sieve to remove any possible contaminants. The dry GP (Fig. 2d) was stored in sealed containers until utilization in UHPC. Extended characterization was made of the granite powder since no information regarding its chemical, mineralogical or physical were provided. Comparison of the chemical content of the granite powder with values obtained from literature [14, 10, 15, 11, 16] is presented in Fig. 3a) and indicated that the chemical content is well within the expected range. SEM images under a magnification of 2000x (Fig. 3b)) and 15000x indicated that the particles were angular and of irregular shape, according with the information found in literature [17, 11, 18, 19]. Analyses of specific surface area indicate that granite powder has a specific surface of 1.749 m²/g.

The sand and aggregates were composed of either fine natural river sand (NRS), coarser silica sand (CSS) or fine aggregates (F.Agg). The fine natural sand was

a)   b)

c)   d)

Fig. 2. Granite waste powder after various steps in the treatment process; a) on site, b) prior to drying, c) after drying, d) sieved and ready to use.

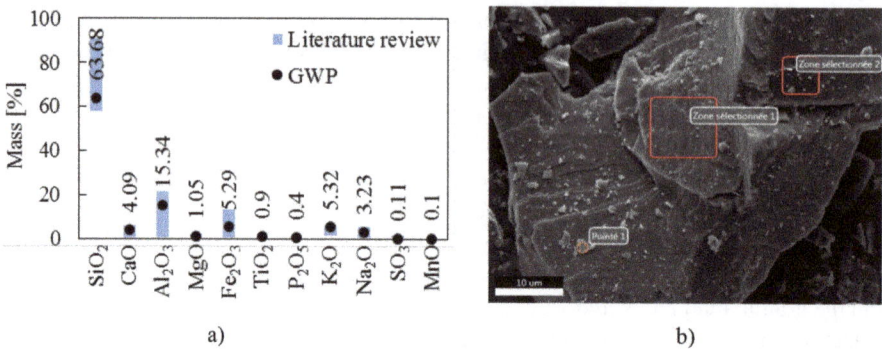

Fig. 3. Characterization of granite powder: a) Chemical composition compared with the expected range from literature, b) SEM images under a 2000x magnification

recovered from the embankment of the Chaudière River, the CSS was provided from Atlantic Silica Sand and the fine aggregates were made from crushed limestone. The PSD of the fine natural sand is provided in Fig. 1. The PSD of CSS and F.Agg are not presented but the materials completely passed through the 2.5 and 4.75 mm meshes and were completely retained on the 0.6 and 2.5 mm meshes for the CSS and F.Agg respectively.

Various polycarboxylate-based superplasticizer (PCE) were used to enhance the fresh and hardened properties of UHPC mixes. The superplasticizers tested were all high-range water reducer admixtures (HRWRa) and were provided by BASF, Chryso, Euclid and Sika. Suitability of these superplasticizers with CEM/GU was checked by measuring the packing density and workability at different dosages (0 to 5%). Specific density of these HRWRa varied from 1.06 to 1.08. For the retained HRWRa, the packing density of CEM/GU was 0.580 ± 0.002, 0.590 ± 0.003, 0.596 ± 0.002, 0.600 ± 0.002 and 0.597 ± 0.002 for a dosage of 1, 2, 3, 4 and 5% respectively. A basic dosage of 3% was selected for cost-efficiency reason and also to allow possible further increase up to 4% (for mixes requiring extra workability for instance) without having a drop in efficiency.

Short, straight steel fibers were used in some mixes to assess the changes in rheological and mechanical properties when mixes were made with and without fibers. The fibers were provided by Bekaert and had a length of 13 mm and a diameter of 0.2 mm. Nominal strength and elastic modulus of the fibers are 2850 MPa and 200 GPa respectively.

The cost, embedded energy (EE) and embodied CO_2 (E-CO_2) were estimated for each material to allow calculation of the total cost and energy of the various mixes. Data were either provided from literature [20, 21, 22, 13, 23, 24, 25, 26] or by the material provider (Table 1). Values for granite powder are not presented in Table 1 since no data were found in literature and the analysis required to obtain those values are beyond the scope of this paper.

Table 1. Cost, embedded energy and embodied CO_2 for the materials used.

Items	Cement	Silica Fume	Fly Ash	Limestone filler	Natural sand	Aggregate	Steel fibers	HRWR
Cost [$/ton]	250	750	60	150	20	20	8000	4000
EE [MJ/kg]	4.5	0	0	0.755	0.048	0.048	32	29.15
E-CO_2 [kg/kg]	0.9	0	0.009	0.019	0.003	0.003	1.5	0.8

2.2 Mix-Designs

The mix-designs developed in this project are presented in Table 2. All mixes were made using a planetary mixer (Hobart N-50) with an estimated final volume of 0.5 L. Preliminary mixes not presented in this paper had been made to optimize the paste composition (cement, SF, FA, GP, LF) using an advanced particles packing model (Compaction-Interaction Packing Model) [27]. Mixes are identified as follow: water to binder ratio – sand/aggregate composition (Fine Sand, Fine+Coarse Sand, Fine Sand +Aggregate) – HRWR (1, 2) – mixing protocol (A, B) – presence of fibers (F).

Table 2. Mixes tested

Material [kg/m³]	0.28-FS-1-A	0.28-F+CS-1-A	0.26-F+CS-1-A	0.26-F+CS-2-A	0.26-F+CS-2-B	0.28-FS-1-A-F	0.28-F+CS-1-A-F	0.25-FS+A-2-B-F
Cement	400	400	400	400	400	400	400	400
Silica fume	80	80	80	80	80	80	80	80
Fly ash	200	200	200	200	200	200	200	200
Granite waste powder	200	200	200	200	200	200	200	200
Limestone	260	260	260	260	260	260	260	260
Natural sand	980	588	588	588	588	980	588	588
Coarse silica sand	0	402	402	402	402	0	402	0
Fine aggregate	0	0	0	0	0	0	0	400
HRWR	32	32	32	32	32	32	32	32
Water	190	190	175	175	175	190	190	170
Steel fibers	0	0	0	0	0	156	156	156
Mixing procedure	A	A	A	A	B	A	A	B

Two different casting procedures (Table 3) were used, as identified in Table 2 as procedure A or procedure B. The cost ($/m^3), embedded energy (MJ/m^3) and embodied CO_2 (kg/m^3) are estimated for each mix and are calculated either for the cementitious matrix only (all material except fibers) or for the complete material. The values for all mixes were estimated at 359 $/m^3, 2976 MJ/m^3 and 395 kg CO_2/m^3 when considering the cementitious matrix only. Cost, EE and E-CO_2 for 1 m^3 were estimated at 1607$, 7968 MJ and 629 kg CO_2 when fibers were used. For Ductal® with 2% steel fibers by volume, values of 1866 $/m^3, 9723 MJ/m^3 and 927 kg/m^3 were determined.

Table 3. Description of the mixing procedures.

A Procedure		B Procedure	
Duration [min]	Action	Duration [min]	Action
2	Dry mixing	2	Dry mixing
1	Add. of water and HRWR	1	Add. of water (≈80%)
5	Mixing	1	Mixing
2	Spread test	1	Add. of water (≈20%) and HRWR
2	Mixing and fiber add	5	Mixing
2	Spread test with fibers	2	Spread test without fiber
		2	Mixing and fibers addition
		2	Spread test with fibers

2.3   Methodology

The fresh property investigated was the workability measured using the mini cone following ASTM C1437 [28] along with the specification for UHPC described in ASTM C1856 [29] . The mini-cone had a top opening of 70 mm and a bottom opening of 120 mm. The diameter of the flow table was 255 mm. Flow diameter were measured both for the static state and the dynamic state (25 drops with the flow table). Two diameters were recorded and averaged for each measurement. For mixes where the flow diameter extended beyond the diameter of the flow table, a value of >250 was recorded as the result.

The hardened properties of interest in this research were the compressive strength measured on cubic samples with sides of 50 mm. The samples were cured at 60 °C 24 h after the casting, for a duration of 6 days. The cure was made by complete immersion of the sample and the temperature was chosen as to accelerate the strength development and allowing faster characterization while minimizing the possibility of increasing the strength beyond the strength expected after 28 days of curing under standard condition [21–23 °C, 95% RH]. 3 samples were used for each mix and the mean values are presented in this paper. The measurements were made using a universal testing machine with a displacement rate of 0.405 mm/min. This rate of loading was in accordance with specification from ASTM C1856 [29].

3 Results

The results for the fresh and hardened properties are presented in Fig. 4 and 5, respectively. For UHPFRC mixes, workability (both under static and dynamic conditions) was determined before and after adding the fibers. Effect of fibers on workability can then be assessed by comparing the different values prior and after fibers addition. Repeatability can also be assessed by comparing the results of corresponding UHPC and UHPFRC (prior to fibers addition) mixes.

- The partial replacement of the fine sand (0.28-FS-1-A) by coarser silica sand (0.28-F+CS-1-A) with the same water and HRWR content led to an increase in workability. This allowed a reduction of water content from 190 L/m^3 (0.28-FS-1-A) to 175 L/m^3 (0.26-F+CS-1-A) to achieve similar workability (175 mm ± 5 and 200 mm ± 5 for static and dynamic values). Comparison between mixes 0.26-F+CS-2-B and 0.25-FS+A-2-B-F (prior to fibers addition) indicate that further water reduction was possible by replacing the coarse silica sand by fine aggregate while maintaining similar or better workability.
- Change of HRWR while maintaining all other parameters constants (0.26-F+CS-X-A) led to a noticeable increase in both static and dynamic spread. Further improvement in workability was possible by delaying the HRWR addition in the mixer, with an increase in static and dynamic spread of ±25 mm and ±20 mm respectively (0.26F+CS-2-X).
- Partially replacing the fine natural river sand by coarser material (F+CS or FS+A) led to a decrease in workability once fibers were added. For 0.28-FS-1-A-F, the difference before and after fiber addition is marginal while workability decreases by ±20 mm with utilization of silica sand (0.28-F+CS-1-A-F) and by about 60 mm with fine aggregate (0.25-FS+A-2-B-F).
- Overall, most mixes presented adequate workability and self-compacting behaviour. Compared with others fillers or SCM (limestone or fly ash), utilization of granite powder slightly decrease workability. Adequate proportioning of the paste composition (through preliminary mixes not presented in this paper) helped minimize this adverse effect.

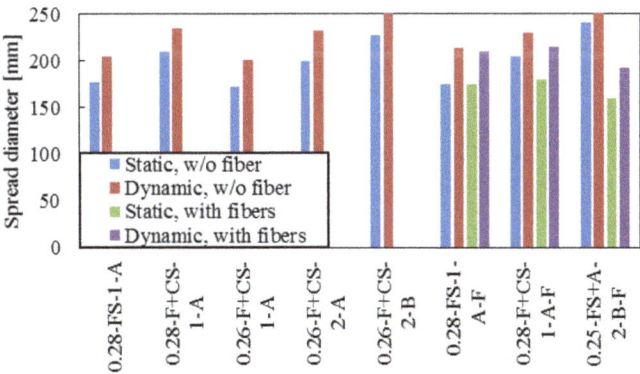

Fig. 4. Workability of mix-designs for both static and dynamic are presented

Figure 5 presents the mean compressive strength achieved on cubic samples. It is worth mentioning that Canadian standard [30] requires a characteristic compressive strength of 120 MPa to qualify the material as UHPC. It should be noted that inclusion of fibres (2%) might has an over-increasing effect on compressive strength due to the small size of the sample (50 mm) compared with the length of the fibres (13 mm). Based on previous works, an addition of 2% steel fibres typically lead to increase in compressive strength in the range of 10 to 20 MPa in 75 by 150 mm cylinders [31]. As such, targeted compressive strength for un-fibres mixes should be between 100 (indicated by the dashed grey line) and 110 while the strength for fibered mixes should be over 120 MPa (indicated by the dashed black line).

- For UHPFRC mixes, compressive strength is at or over 120 MPa. Reduction in water content through the use of coarser particles, change of HRWR and mixing protocol between mix 0.28FS1AF and 0.25-FS+A-2-B-F led to an increase of 8 MPa.
- For UHPC mixes, partially replacing the fine natural sand by the coarse silica sand while maintaining all others parameters constant led to a slight decrease in compressive strength. By reducing the water content to maintain similar workability (0.26F+CS1A), compressive strength was increase up to 95 MPa. Modification in the mixing protocol and HRWR type while maintaining a water/binder ratio of 0.26 led to similar or slightly increased strength.

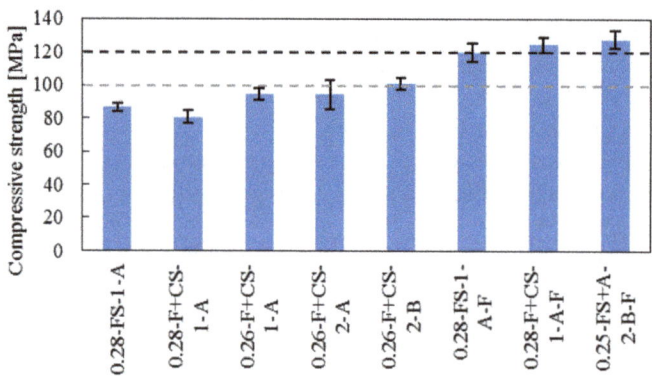

Fig. 5. Compressive strength for cubic samples, mean value presented with standard deviation.

4   Discussion

For all mixes, the cost (a), embedded energy (b) and embodied CO_2 (c) were calculated for the materials only and are presented in Fig. 6. Figure 6d) compares those 3 parameters when considering the total mix (not divided by individual component like Fig. 6a, b and c). The cost relative to the mixing/casting operations and treatment of material were discarded at this point since the mixing time required is mostly affected by the type of mixer used. Moreover, the treatment required for granite waste powder (drying and sieving) is similar to what would be used for natural sand. The results are presented for UHPC and UHPFRC mix-designs. The results (cost, EE, ECO$_2$) were normalized by dividing them by the mean compressive strength. This allows for better comparison between mixes having different compressive strength, as mixes having superior strength would require less volume to support the same loads. The normalized indexes for Ductal® (represented by the dashed lines and based on the values presented at Sect. 2.2) were chosen as the reference values to compare the effectiveness of the developed mixes. A value of 155 MPa was considered as the compressive strength of Ductal® using 2% steel fibers by volume based on experimental results obtained in previous research. Results presented in Fig. 6 indicate that the development of effective mixes using granite powder is possible. When compared with a commercially available UHPC, mixes using fibers presented similar normalized cost. The normalized cost calculated are 13.40, 12.90, 12.57 and 12.04 $/MPa for mixes 0.28-FS-1-A-F, 0.28-F +CS-1-A-F, 0.25-FS+A-2-B-F and Ductal® respectively. Similar or better efficiency

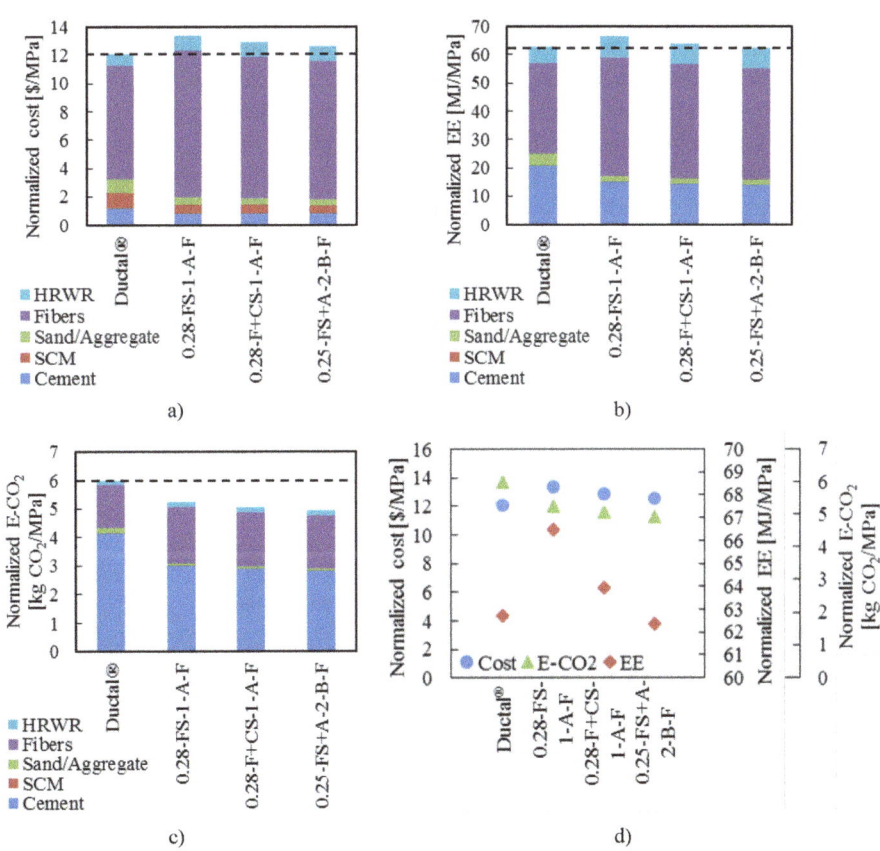

Fig. 6. Comparison of developed UHPFRC mix-design in terms of; a) cost, b) embedded energy, c) embodied CO2, d) comparison for the 3 parameters

was also noted when considering the embedded energy, with values of 66.5, 64, 62.35 and 62.75 MJ/MPa using the same order of comparison. A notable reduction in the normalized CO_2 emission release was possible for all mixes using granite powder and a reduced content of cement. Estimated emissions of CO_2/MPa dropped from 5.98 kg CO_2/MPa for Ductal® down to 4.92 kg CO_2/MPa for mix 0.25-FS+A-2-B-F.

5 Conclusion

This preliminary work presents promising results for developing low-cement, cost-effective UHPC-PG and UHPFRC-PG which recycles granite powder. By partially replacing fine natural sand with either coarse silica sand or fine limestone aggregate, reduction in water was possible while maintaining adequate workability. Delaying the addition of superplasticizer proved to be another effective method to increase workability. A compressive strength greater than 120 MPa was achieved for mixes using a volume content of 2% steel fibers. UHPC with PG mix design achieved a maximal strength of 100 MPa while showing self-compacting behaviour. When compared with a commercially available UHPC, mixes using granite powder showed similar results in terms of normalized cost and embedded energy, while presenting a lower production of CO_2 for each MPa of compressive strength. Further research should be conducted to pursuit the optimization of UHPC using granite powder. In the next work, optimization of the paste content and good proportioning between paste and sand/aggregate content by using particles packing models will be attempted to reduce viscosity as well as further reduce cost, embedded energy and embodied CO_2.

Acknowledgments. We would like to acknowledge NSERC CRD programme and the industrial partner Polycor for financial research support. We further thank Mr. Mathieu Bergeron for the thoughtful research discussion and providing the granite powder samples and Mr. Michel Lessard from Euclid Canada for his time and precious advice. We would finally like to acknowledge BASF, Ciment Québec, Euclid, LafargeHolcim, OMYA, Silicium Québec and Sika for the materials and expertise provided.

References

1. Richard, P., et al.: Composition of reactive powder concretes. Cem. Concr. Res. **25**(7), 1501–1511 (1995). https://doi.org/10.1016/0008-8846(95)00144-2
2. de Larrard, F., Sedran, T.: Optimization of ultra-high-performance concrete by the use of a packing model. Cem. Concr. Res. **24**(6), 997–1009 (1994). https://doi.org/10.1016/0008-8846(94)90022-1
3. Vande Voort, T.L.: Design and field testing of tapered H-shaped ultra high performance concrete piles. Master of Science, Iowa State University, Digital Repository, Ames (2008)
4. Brühwiler, E.: UHPFRC technology to enhance the performance of existing concrete bridges. Struct. Infrastruct. Eng. **16**(1), 94–105 (2020). https://doi.org/10.1080/15732479.2019.1605395

5. Graybeal, B.: Properties and behavior of UHPC-class materials. Office of Infrastructure Research & Development Federal Highway Administration, Final Report FHWA-HRT-18-036, mars (2018)
6. Korpa, A., Kowald, T., Trettin, R.: Phase development in normal and ultra high performance cementitious systems by quantitative X-ray analysis and thermos analytical methods. Cem. Concr. Res. **39**(2), 69–76 (2009). https://doi.org/10.1016/j.cemconres.2008.11.003
7. Zhang, H., Ji, T., He, B., He, L.: Performance of ultra-high performance concrete (UHPC) with cement partially replaced by ground granite powder (GGP) under different curing conditions. Constr. Build. Mater. **213**, 469–482 (2019). https://doi.org/10.1016/j.conbuildmat.2019.04.058
8. Yang, R., et al.: Environmental and economical friendly ultra-high performance-concrete incorporating appropriate quarry-stone powders. J. Clean. Prod. **260**, 121112 (2020). https://doi.org/10.1016/j.jclepro.2020.121112
9. López Boadella, Í., et al.: The influence of granite cutting waste on the properties of ultra-high performance concrete. Materials **12**(4), 634 (2019). https://doi.org/10.3390/ma12040634
10. Singh, S., Nagar, R., Agrawal, V.: A review on Properties of Sustainable Concrete using granite dust as replacement for river sand. J. Clean. Prod. **126**, 74–87 (2016). https://doi.org/10.1016/j.jclepro.2016.03.114
11. de Matos, P.R., Sakata, R.D., Gleize, P.J.P., de Brito, J., Repette, W.L.: Eco-friendly ultra-high performance cement pastes produced with quarry wastes as alternative fillers. J. Cleaner Prod. **269**, 122308 (2020). https://doi.org/10.1016/j.jclepro.2020.122308
12. Vaitkevičius, V., Šerelis, E., Lygutaitė, R.: Production Waste of Granite Rubble Utilisation in Ultra High Performance Concrete. J. Sustain. Archit. Civil Eng. **2**(3), 54–60 (2013). https://doi.org/10.5755/j01.sace.2.3.3873
13. Alsalman, A., Dang, C.N., Martí-Vargas, J.R., Micah Hale, W.: Mixture-proportioning of economical UHPC mixtures. J. Build. Eng. **27**, 100970 (2020). https://doi.org/10.1016/j.jobe.2019.100970
14. Chen, J.J., Li, B.H., Ng, P.L., Kwan, A.K.H.: Adding granite polishing waste as sand replacement to improve packing density, rheology, strength and impermeability of mortar. Powder Technol. **364**, 404–415 (2020). https://doi.org/10.1016/j.powtec.2020.02.012
15. Rashwan, M.A., Al-Basiony, T.M., Mashaly, A.O., Khalil, M.M.: Behaviour of fresh and hardened concrete incorporating marble and granite sludge as cement replacement. J. Build. Eng. **32**, 101697 (2020). https://doi.org/10.1016/j.jobe.2020.101697
16. Ramos, T., Matos, A.M., Schmidt, B., Rio, J., Sousa-Coutinho, J.: Granitic quarry sludge waste in mortar: Effect on strength and durability. Constr. Build. Mater. **47**, 1001–1009 (2013). https://doi.org/10.1016/j.conbuildmat.2013.05.098
17. Adigun, M.A.: Cost effectiveness of replacing sand with crushed granite fine (CGF) in the mixed design of concrete. IOSR-JMCE **10**(1), 01–06 (2013). https://doi.org/10.9790/1684-1010106
18. Mashaly, A.O., Shalaby, B.N., Rashwan, M.A.: Performance of mortar and concrete incorporating granite sludge as cement replacement. Constr. Build. Mater. **169**, 800–818 (2018). https://doi.org/10.1016/j.conbuildmat.2018.03.046
19. Singh, S., Nagar, R., Agrawal, V., Rana, A., Tiwari, A.: Sustainable utilization of granite cutting waste in high strength concrete. J. Clean. Prod. **116**, 223–235 (2016). https://doi.org/10.1016/j.jclepro.2015.12.110
20. Meng, W.: Design and performance of cost-effective ultra-high performance concrete for prefabricated elements. Missouri S&T (2017)
21. Soliman, N.A., Tagnit-Hamou, A.: Using glass sand as an alternative for quartz sand in UHPC. Constr. Build. Mater. **145**, 243–252 (2017). https://doi.org/10.1016/j.conbuildmat.2017.03.187

22. Graybeal, B.: Development of non-proprietary ultra-high performance concrete for use in the highway bridge sector. Federal Highway Administration (FHWA), Technical report FHWA-HRT-13-100, October 2013

23. Chiaia, B., Fantilli, A.P., Guerini, A., Volpatti, G., Zampini, D.: Eco-mechanical index for structural concrete. Constr. Build. Mater. **67**, 386–392 (2014). https://doi.org/10.1016/j.conbuildmat.2013.12.090

24. Gupta, T., Kothari, S., Siddique, S., Sharma, R.K., Chaudhary, S.: Influence of stone processing dust on mechanical, durability and sustainability of concrete. Constr. Build. Mater. **223**, 918–927 (2019). https://doi.org/10.1016/j.conbuildmat.2019.07.188

25. Shi, Y., Long, G., Ma, C., Xie, Y., He, J.: Design and preparation of ultra-high performance concrete with low environmental impact. J. Clean. Prod. **214**, 633–643 (2019). https://doi.org/10.1016/j.jclepro.2018.12.318

26. Müller, H.S., Haist, M., Vogel, M., Moffatt, J.S.: Design approach and properties of a new generation of sustainable structural concretes, p. 17 (2018)

27. Fennis, S.A.A.M.: Design of Ecological Concrete by Particle Packing Optimization. Delft University of Technology (2008)

28. ASTM: Standard Test Method for Flow of Hydraulic Cement Mortar.pdf (2015)

29. ASTM: Standard Practice for Fabricating and Testing Specimens of Ultra-High Performance Concrete. ASTM International (2017)

30. Canadian Standards Association: CSA A23.1:19 Annexe U (2019)

31. Larsen, I.L., Thorstensen, R.T.: The influence of steel fibres on compressive and tensile strength of ultra high performance concrete: A review. Constr. Build. Mater. **256**, 15 (2020). https://doi.org/10.1016/j.conbuildmat.2020.119459

Flexural Behaviour of Ultra High Performance Fiber Reinforced Concrete (UHPFRC) Under Monotonic Loads and Loading-Unloading Cycles

Nicola Generosi[✉], Jacopo Donnini, Giovanni Lancioni, and Valeria Corinaldesi

Università Politecnica delle Marche, Ancona, Italy
n.generosi@pm.univpm.it

Abstract. Ultra High Performance Fiber Reinforced Concrete (UHPFRC) is an innovative material with great mechanical and durability performances, high ductility and toughness. Although the mechanical behaviour of UHPFRC has been extensively studied in the last years, the damage mechanisms and permanent strains of this material when subjected to flexural loads need to be further investigated, in order to quantify and to better predict the performance of UHPFRC structural elements.

This work presents the results of an experimental study on the UHPFRC bending behavior. Both static and cyclic loading-unloading bending tests were performed. The effects of brass-coated steel fibers (diameter of 0.20 mm and length of 13 mm) on the flexural behavior of UHPFRC was investigated, varying the amount of fibers up to 2,5% by volume. Four-point bending tests were performed on prisms with different geometries ($30 \times 70 \times 280$ mm^3 and $70 \times 70 \times 280$ mm^3). Particular attention was paid in the UHPFRC post-cracking behaviour, in order to evaluate the strain-softening and/or strain-hardening phases. Damage progress, number and width of cracks were monitored by means of a Digital Image Correlation (DIC) system on both the frontal and bottom surfaces of the specimens.

Finally, a phase-field model has been implemented in a FE code and numerical simulations have been performed to better understand the effects of different fiber dosages on the mechanical behavior of UHPFRC specimens under cyclic loads. Concrete matrix and fiber reinforcement have been modeled as brittle and elasto-plastic phases of a mixture, whose internal energies are enriched by non-local damage and plasticity contributions.

Keywords: Digital image correlation · UHPFRC · Strain-hardening · Steel fibers · Concrete

1 Introduction

Ultra High Performance Fiber Reinforced Concrete (UHPFRC) is an innovative construction material exhibiting enhanced mechanical and durability properties. The low water-binder ratio and the special selection of fine and ultra-fine particles ensure a high

© RILEM 2022
P. Serna et al. (Eds.): BEFIB 2021, RILEM Bookseries 36, pp. 915–924, 2022.
https://doi.org/10.1007/978-3-030-83719-8_78

compactness and low porosity. The high durability of UHPFRC allows to design long service-life structures with subsequent reduced maintenance costs. On the other hand, the high strength capacity allows to recognize UHPFRCs as suitable materials for structural applications such as tunnel linings, highway or bridge decks overlays, off-shore structures [1] and seismic resistant structures [2]. These structures are subjected to cyclic loads and they are expected to resist millions of cycles during their service life. However, cyclic loads can cause the initiation and development of microcracks inside the concrete structure [3]. The macroscopic consequence of this phenomenon is a modification of the characteristics of the material (strength, stiffness, toughness, durability, etc.) and, in extreme cases, it could lead to fatigue failures.

Some recommendations, technical reports and guidelines regarding the fatigue behaviour of Ultra High Performance Concrete are nowadays available, such as the State-of-art report from the American Federal Highway Administration [4], the Japan Recommendations for Design and Construction of High Performance Fiber Reinforced Cement Composites [5], the *fib* Model Code 2010 [6] and the French standards [7]. However, most of these documents report the fatigue behaviour of concrete under compression, and just a few of them take into consideration the flexural tensile response [8].

The primary goal of this experimental work is to investigate the performance of UHPFRCs under quasi-static and cyclic loading conditions, by varying the amount of steel fibers up to 2.5% by volume. UHPFRC also exhibits a multiple cracking behaviour under flexural tension due to the effective fibers bridging action and load transfer mechanisms (Fig. 1).

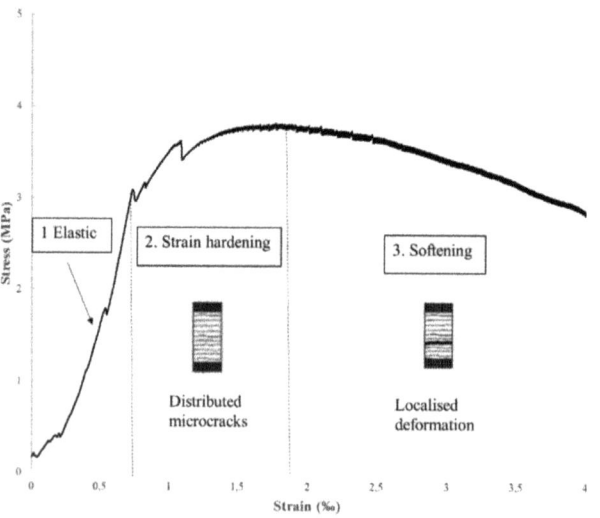

Fig. 1. Schematic representation of flexural response of Ultra High Performance Fiber Reinforced Concrete

To determine these properties, many specimens with same mixes but different fibers content were loaded in four-point bending test until failure. The accuracy of static and cyclic strain measurements using photogrammetry as a function of the measuring rate is investigated. Data from DIC were used to determine micro-crack and local crack strain which were used to investigate micro-crack formation and propagation and their relationship to local crack formation and propagation. This measuring method, also known as digital image correlation (DIC), enables noncontact measurement of deformations on the surface of specimens by means of an applied pattern.

2 Materials and Mix Proportions

2.1 Materials

A Commercial Portland-limestone blended cement (CEM I 52.5 R), in compliance with EN-197/1 [9] was used. The Blaine fineness of cement was 0.48 m^2/g and its specific gravity was 3.15 g/cm^3. As aggregate two different types of quartz sand with particle size 0–1 mm and 0.5–1.1 mm were suitably combined. Silica fume was added at a dosage of 100 kg/m^3. An acrylic-based water-reducing admixture (WRA) was added in powder at dosage of 1.4% by weight of cement with a water to cement ratio (w/c) of 0.33. In addition, an expansive agents (EA) and a shrinkage reducing admixture (SRA) were added at a dosage of 25 kg/m^3 and 9 kg/m^3 respectively. An anti-foaming admixture (AF) was used to limit the content of air bubbles.

The steel microfiber used had length (L$_f$) of 13 mm and diameter (Φ_f) of 0.20 mm, with an aspect ratio (L$_f$/Φ_f) of 65. The volume fractions of microfibers correspond to 1% and 2.5% (78 kg/m^3 and 195 kg/m^3 respectively). The mixture proportions are reported in Table 1.

Table 1. UHPFRC mixtures (kg/m^3)

Specimen	CEM I 52.5R	Water	Sand 0/0.1	Sand 0.5/1.1	Silica fume	WRA	SRA	EA	AF	Fibers
UHPFRC_1	720	240	410	945	100	10	9	25	6	78
UHPFRC_2.5	720	240	410	945	100	11	9	25	6	195

2.2 Specimens Preparation and Curing

Four prismatic specimens with dimensions of 70 × 70 × 280 mm and four of 30 × 70 × 280 mm were prepared for each UHPC mixture (Table 1). All specimens were cast in steel forms, vibrated for 10 s and cured at laboratory conditions (about 20 °C, 65% Relative Humidity) for 24 h, covered with a thin plastic sheet. After that, specimens were demolded and stored in a water tank for 27 days. Both static and cyclic tests were performed at 28 days.

2.3   Test Setup

All specimens were tested under displacement control. During the tests, pictures of the frontal surface of the specimens have been acquired by two digital cameras (model Pixelink® B371F) at 2 frames per second. The cameras were equipped with a lens having a focal length of 16 mm and placed about 1m away from the specimen, in order to reduce the perspective errors due to eventual out-plane motions. The specimen was illuminated using a LED spotlight. A third camera, placed at the bottom, was used to monitor cracks propagation on the lower surface of specimens. Testing apparatus are presented in Fig. 2

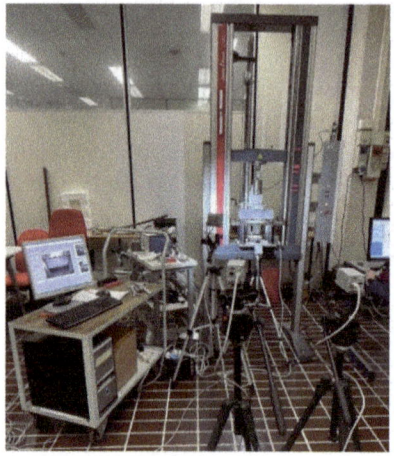

Fig. 2. Four-point cyclic bending test with Digital Image Correlation set-up

3   Experimental Results

3.1   Static Tests

Specimens were tested under monotonic flexural loading until failure. The four-point quasi-static bending tests were performed in displacement control at 1 mm/min. Figure 3 shows the test set-up. Results obtained for the different UHPFRC mixtures are reported in Table 2 as the average value of 2 specimens for each mixture with corresponding Coefficient of Variation (%). Flexural stress-deflection curves are showed in Fig. 4. The flexural stress is calculated as:

$$\sigma = \frac{F\,l}{b\,d^2} \tag{1}$$

where F is the flexural load, l is support span, b and d are respectively width and depth of tested specimens.

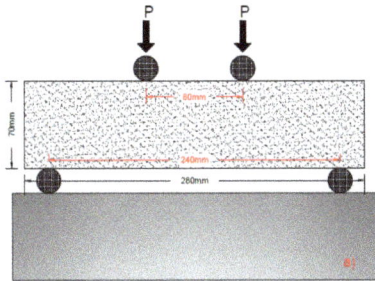 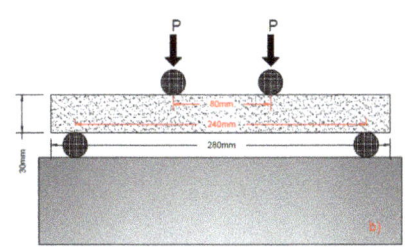

Fig. 3. Four-point bending test set-up: (a) 70 × 70 × 280 mm specimen, (b) 30 × 70 × 280 specimen

The first cracking stress σ_t was defined as the flexural stress corresponding to the first crack formation. The formation of the first crack has been determined by visual inspection (with the help of DIC frames) or looking at the stress-strain curves, in correspondence with a significant slope change in the stress-displacement curve. The average maximum flexural strength σ_{max} and the flexural toughness U_f, for each group of specimens, were also reported in Table 2. Flexural toughness has been calculated by integrating the area under the flexural stress-displacement curves, up to a midspan displacement equal to 3.5 mm. By increasing the amount of fibers, a considerable increase of the flexural toughness was observed. The maximum flexural toughness (about 32 N/mm) was observed for 70 × 70 × 280 mm specimens with 2.5% of fibers, which is almost the double of specimens with 1% of fibers. For the 30 × 70 × 80 specimens instead, similar values of flexural toughness were obtained for both fibers content. From Fig. 4 it can also be noticed that the post-cracking behaviour changes by varying the amount of steel fibers. With 1% dosage of fibers a softening behaviour is observed for all specimens after cracking. Therefore, in this case the first cracking stress σ_t also corresponds to the maximum tensile strength σ_{max}. With a higher dosage of steel fibers, the post-cracking behaviour changes and significant stress-hardening branches can be noticed in the stress-displacement curves (Fig. 4).

Table 2. Experimental results of four-point static bending tests

Dimension (mm)	Specimen		σ_t (MPa)	σ_{max} (MPa)	U_f (N/mm)
70 × 70 × 280	UHPFRC_1%	Average	10.00	10.00	18.79
		CoV(%)	8	8	4
	UHPFRC_2.5%	Average	12.13	13.75	32.01
		CoV(%)	36	17	18
30 × 70 × 280	UHPFRC_1%	Average	9.99	9.99	22.01
		CoV(%)	17	17	2
	UHPFRC_2.5%	Average	8.67	9.87	26.59
		CoV(%)	4	21	27

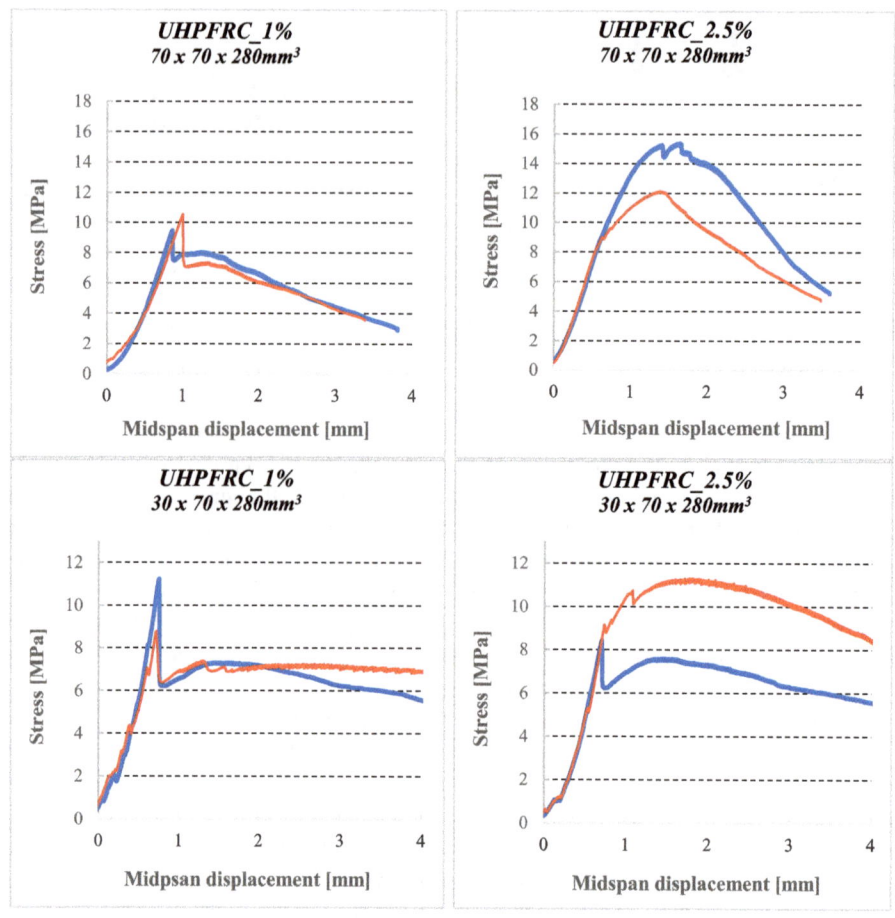

Fig. 4. Flexural stress-displacement curves of four-point static bending tests

3.2   Cyclic Tests

28-days old prisms were tested under cyclic flexural loading until failure. Tests were conducted under displacement control, at 1 mm/min. Unloading-reloading cycles were performed at each 0.5 mm and strains were recorded by DIC. Figure 2 shows the test set-up. Results obtained for the different UHPC mixtures are listed in Table 3 and the stress-deflection curves are showed in Fig. 6.

Table 3. Experimental results of four-point cyclic bending tests

Dimension (mm)	Specimen		Flexural strength (MPa)	Increase of flexural strength (%)	Flexural Toughness U_f (N/mm)	Increase of flexural toughness (%)
70 × 70 × 280	UHPFRC_1%	Average	7.98	–	20.3	–
		CoV(%)	1		4	
	UHPFRC_2.5%	Average	14.29	79	35.39	74
		CoV(%)	2		2	
30 × 70 × 280	UHPFRC_1%	Average	10.9	–	23.21	–
		CoV(%)	3		2	
	UHPFRC_2.5%	Average	14.19	30	38.16	64
		CoV(%)	4		10	

Flexural strength was considerably high for UHPFRC mixture with 2.5% of fiber content thanks to the strain hardening behaviour. Flexural toughness of UHPFRC_2.5% was higher, for both specimen dimensions, in comparison to UHPFRC with 1% of fibers content. The increase in toughness can be attributed to the occurrence of such phenomena associated with cracks-bridging.

The cracking behaviour (crack width, number, spacing) of UHPFRC specimens is investigated. The effect of fibers content on both local crack ad micro-crack formation and propagation was significant. It is clear from Fig. 5 that there is a large variation in the post-cracking behaviour based especially on fibers volume content. Specimens with 1% of fiber content exhibited strain-softening behaviour after cracking and no micro-crack formation was observed in these specimens (Fig. 6a). Specimens with 2.5% fiber showed instead a strain-hardening behaviour with the formation of multiple cracks. This change from unique to multiple-cracking behaviour (Fig. 5), when a critical threshold in fibers dosage is exceeded, is confirmed by some studies present in literature [3, 10].

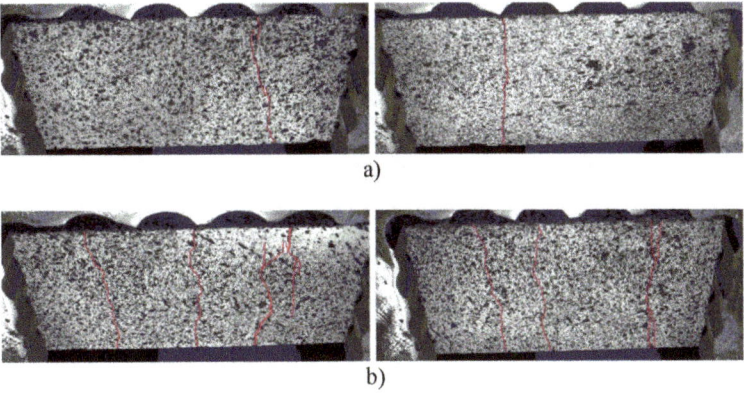

a)

b)

Fig. 5. Visible cracks in the bottom surface of UHPFRC specimens for different fiber dosages: a) UHPFRC_1%, b) UHPFRC_2.5%

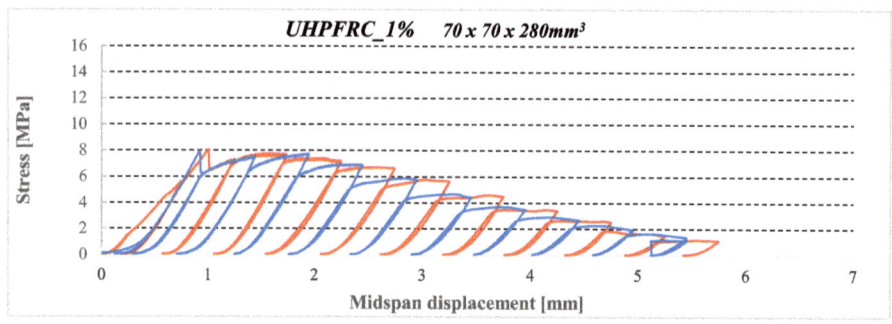

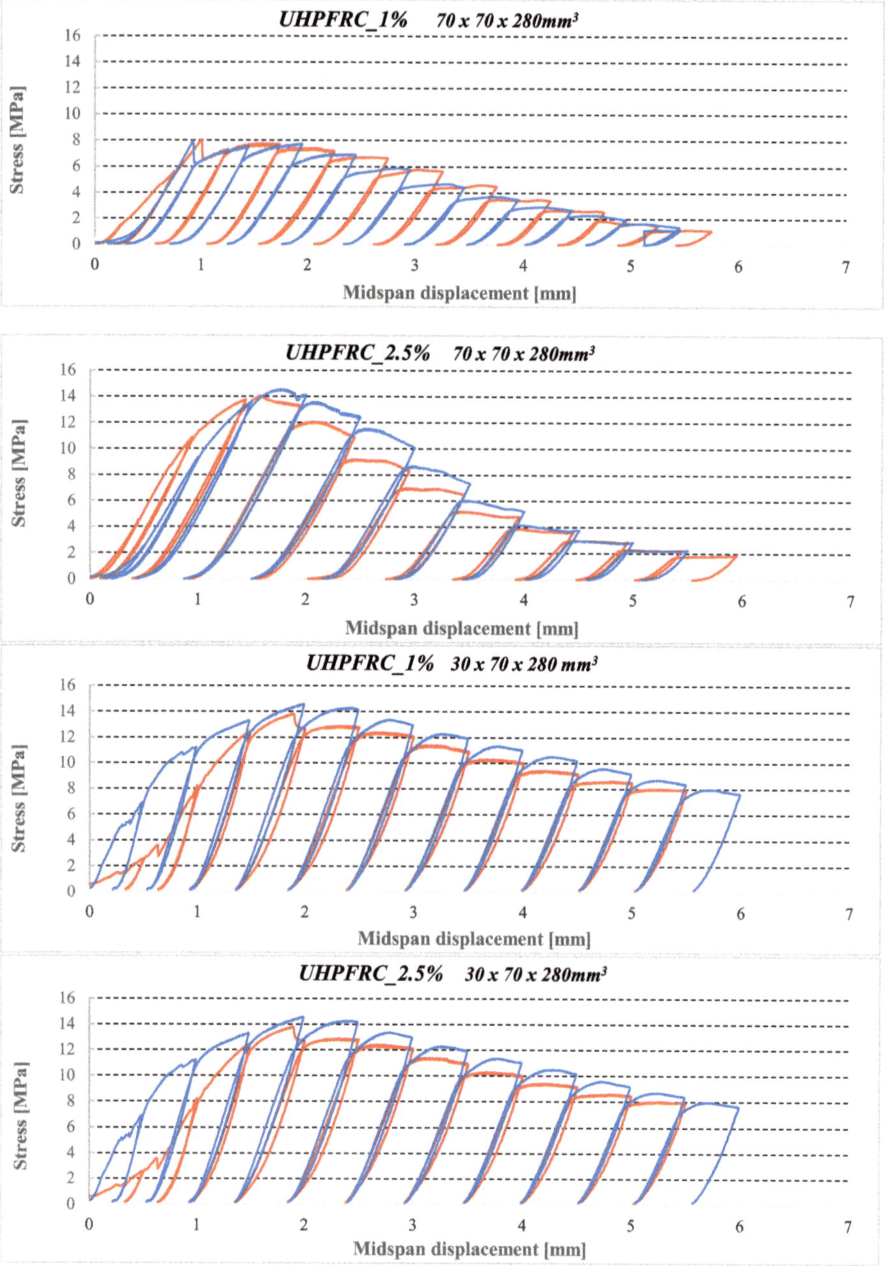

Fig. 6. Stress-displacement curves of four-point cyclic bending tests

4   Numerical Simulations

A phase-field variational model has been developed to better understand the mechanical behavior of the UPFRC mixtures investigated in this study by varying the amount of steel fibers. The UHPFRC is modeled as a mixture of two material phases, describing the cementitious matrix and the reinforcing fibers, whose volume fractions are v_m and v_r. The mechanics of the cementitious matrix and steel fibers is described by two material phases, which account for brittle and ductile elasto-plastic responses. The evolution problem is formulated as an incremental energy minimum problem, where the unknowns are a damage and a plastic strain fields, which correspond to the internal variables associate to the two phases. Analytical estimates and numerical solutions are determined in the two-dimensional setting and implemented in a finite element code, which is used to simulate the behavior of UHPFRC specimens, with different fibers dosage, subjected to four-point bending tests.

A key assumption of the proposed model is that damage and plastic properties of the two phases account for all inelastic phenomena observed in experiments: i. damage properties of matrix allow to describe micro- and multi-cracking; ii. plastic features of reinforcement account for the plastic stretching of fibers, and, mostly, for all inelastic phenomena that occur at the interface level in real UHPFRCs, such as aggregate interlocking, and matrix-fibers frictional mechanisms. Within the model, interface elastic bonds have just the role of transferring stresses between the two phases (Fig. 7).

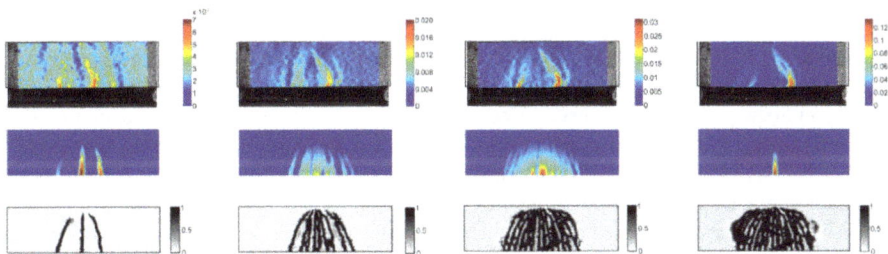

Fig. 7. Frontal strain map of UHPFRC specimens at different steps of the stress-strain curve: experimental maps in line 1, numerical maps in line 2 and damage patterning in line 3.

5   Conclusions

An experimental study was carried out to evaluate the flexural behaviour of UHPFRC under quasi-static and cyclic loads. Four-point bending tests were performed on specimens measuring $30 \times 70 \times 280$ mm^3 and $70 \times 70 \times 280$ mm^3, reinforced with 13-mm steel fibers up to 2.5% by volume.

The main findings are:

- The increase of fiber content allowed to significantly increase the flexural strength of UHPFRCs mixtures for both specimen dimensions.

- Flexural toughness has been greatly raised by increasing the steel fibers volume: with respect to UHPFRC with 1% of fiber content, UHPFRCs with 2.5% of fiber content were able to absorb up to about 2 times higher energy.
- The post-cracking behaviour in bending was strongly influenced by the amount of steel fibers, passing from strain-softening to strain-hardening when 2.5% of fibers was used. Fiber bridging and the toughening mechanisms were responsible for the improvement of the post-cracking behaviour of UHPFRCs.
- The formation of multiple cracks is primarily influenced by the fiber volume content. As the volume content of steel fiber increases from 1% to 2.5%, fiber bridging phenomenon was activated and multiple cracks formed.

As usual for laboratory tests, the results obtained aren't immediately applicable to real-scale structural elements. In order to fill this gap the tests needed to be also performed on prototypes at a scale of 1/2–1/3 in order to be more reliable.

References

1. Jimenez-Martinez, M.: Fatigue of offshore structures: a review of statistical fatigue damage assessment for stochastic loadings. Int. J. Fatigue **132** (2020)
2. Yoo, D.Y., Banthia, N.: Mechanical and structural behaviors of ultra-high-performance fiber-reinforced concrete subjected to impact and blast. Constr. Build. Mater. **149**, 416–431 (2017)
3. Krahl, P.A., Carrazedo, R., El Debs, M.K.: Mechanical damage evolution in UHPFRC: experimental and numerical investigation. Eng. Struct. **170**, 63–77 (2018)
4. Russel, G.H., Graybeal, B.A.: Ultra-high performance concrete : a state-of-the-art report for the bridge community **171** (2013)
5. Japan Society of Civil Engineers: Recommendations for Design and Construction of High Performance Fiber Reinforced Cement Composites with Multiple Fine Cracks (HPFRCC). Concr. Eng. Ser. **82**, 6–10 (2008). Testing Method
6. fib Model Code for Concrete Structures 2010 (2013)
7. AFGC: Documents scientifiques et techniques Bétons fibres à ultra-hautes performances (Ultra High Performance Fibre-Reinforced Concrètes), 152 (2002)
8. Carlesso, D.M., de la Fuente, A., Cavalaro, S.H.P.: Fatigue of cracked high performance fiber reinforced concrete subjected to bending. Constr. Build. Mater. **220**, 444–455 (2019)
9. EN 197-1: Cement - Part 1: Composition, specifications and conformity criteria for common cements (2011)
10. Kim, D.J., Park, S.H., Ryu, G.S., Koh, K.T.: Comparative flexural behavior of hybrid ultra high performance fiber reinforced concrete with different macro fibers. Constr. Build. Mater. **25**, 4144–4155 (2011)

Application of 3D Digital Image Correlation to Capture the Impact Beahviour of UHPFRC Plate

Yuanye He[1,2], Esmaeel Esmaeeli[3(✉)], and Marios N. Soutsos[1]

[1] School of Natural and Built Environment, Queen's University Belfast, Belfast, UK
[2] College of Architecture Engineering, Xinjiang University, Ürümqi, China
[3] Department of Civil and Environmental Engineering, Brunel University London, Uxbridge, UK
`Esmaeel.Esmaeeli@brunel.ac.uk`

Abstract. In this study, a repeated drop-weight test on a 50 mm thick four-side fixed UHPFRC plate was carried out. A high-speed 3D Digital Image Correlation technique (3D DIC) was employed to capture the full-field response of the plate during impact loading. The DIC results were validated against the measurements of physical accelerometer and strain sensors attached on the top surface of the plate. The DIC results in combination with a polynomial surface fitting method was used to obtain the mid-point deflection. Replicability of the plate's dynamic response indicated that only negligible damages occurred in the plate due to repeated drops from 0.5 m. During the subsequent impact from 2.0 m, the overall performance was governed by flexure. Lastly, the plate failed with local failure and flexure after impacted from 3.0 m.

Keywords: UHPFRC plate · Impact behaviour · Fixed support · 3D digital image correlation · Full-field deformation measurements

1 Introduction

Due to its excellent compressive and tensile strength, ductility, and energy absorption capacities, Ultra-High-Performance Fibre-Reinforced Concrete (UHPFRC) is considered a reliable material for construction of protective structures that may experience impact loads or high deformation rates. Several investigations [1, 2] have confirmed that UHPFRC exhibits high impact resistance due to its high compressive strength and excellent resistance to crack development and propagation. Whilst a considerable amount of research [3, 4] has been dedicated to the impact resistance of UHPFRC panels under medium/high velocity impact - velocities in the range of 20 m/s to 500 m/s corresponding to flying debris impacts caused by tornados, airplane collisions or explosion accidents, etc., there are only limited number of experimental studies [5, 6] on the impact resistance and damage assessment of UHPFRC members at low velocities, i.e. velocities less than 10 m/s corresponding to collisions of falling objects, vehicle and vessels, etc.

© RILEM 2022
P. Serna et al. (Eds.): BEFIB 2021, RILEM Bookseries 36, pp. 925–935, 2022.
https://doi.org/10.1007/978-3-030-83719-8_79

A study on the response of reinforced concrete slabs with different supporting conditions to repeated impact loads demonstrated the significant effect of the support type and layout on the number of drops, the damage and crack patterns, and the acceleration response of the specimens [7]. It is also known that plates with fixed boundary condition more realistically represent the protective systems in practice. However, due to the complexity of the impact tests, very limited data on the impact responses of UHPFRC plates are available and they are scarce for plates with fixed boundary conditions.

As impact loading process is more complex compared to static loading, capturing the full-field response of a structure during impact loading is helpful in understanding the transient response of structures and improving the assessment of model predictions. This purpose can be accomplished by using high-speed 3D Digital Image Correlation (DIC).

In this research, an experimental program was carried out to investigate the response of a 50 mm thick UHPFRC plate subjected to repeated low velocity impact loads. With the aim of using 3D image correlation analysis, the stereo-vision system used in this study employed two high-speed cameras to record synchronized images during drop-weight test on the plate. The DIC results were validated against the measurements of physical accelerometer and strain sensors attached to the top surface of the plate. The experimental results in terms of accelerations, strains, mid-point displacement were examined.

2   Experimental Program

2.1   Material and Specimen Preparation

The ingredients and mix proportion of the in-house made UHPFRC used in this study are shown in Table 1. Straight micro steel fibres with a length and diameter of 13 mm and 0.2 mm, respectively, and a tensile strength and Young's modulus of about 2000 MPa and 200 GPa, respectively, were used at a volume content of 2%. The UHPFRC specimens were post-set heat treated for 24 h in 90 °C water after one day of maturity. The 28-day mean compressive strength was determined on three cylinders of 60 mm diameter and 120 mm length as 176 MPa; details of compression test can be found in [8]. Tensile test based on JSCE recommendation [9] was carried out on six dog-bone shaped specimens with the constant cross-section region of 50 mm side square and 414 mm length. According to the results of the tensile tests, the developed UHPFRC had an average tensile cracking and ultimate strength of 8.9 MPa and 13.0 MPa with corresponding strains of 0.018% and 0.421%, respectively.

Table 1. Mix proportion of UHPFRC (kg/m^3)

Silica Sand	CEM I 52.5N	GGBS	Micro Silica Slurry	Water	SP	Steel Fibre
999.4	627.3	399.2	203.7	94.6	32.8	156.8

The square UHPFRC plate was cast with a nominal side of about 700 mm and a thickness of 50 mm. The details of plate are shown in Fig. 1. In order to facilitate fixing the plate to the support rig of the impact test setup and to simulate the fix boundary condition along all sides of the plate, 20 holes of 22 mm diameter were reserved during casting. The casting method adopted was pouring from centre (Fig. 1), as this method results in a high tendency of circumferential alignment of fibres, thus increasing the total number of fibres bridging the radial cracks expected in the slabs [10].

Fig. 1. Details of UHPFRC plate in mm (left) and casting of one plate (right)

2.2 Drop-Weight Impact Test Setup and High-Speed Stereo-Vision System

The schematic of drop-weight impact test setup is shown in Fig. 2. The drop-weight tower is consisted of two steel channel-profiles with internal dimensions of 125×115 mm and a length of 6 m that serves as a guide for impactor. The 40 kg bullet-shape impactor used in this study has a hemispherical head with a diameter of 100 mm. The drop-weight facility is equipped with an electromagnet part enabling to adjust the drop height of the impactor. The steel support frame is composed of a homocentric square-shape structure made of four welded rectangular hollow sections and bolted to a column at each corner. The short rectangular hollow section columns are bolted to a 30 mm thick large rectangular base steel plate. The UHPFRC plate was sandwiched between a 40 mm thick steel clamping frame on its top and the steel support frame underneath. The clamping frame has an overall dimension of 700×700 mm and a rectangle 500×500 mm is cut from its central part to obtain a similar contact area to that of the structure underneath. Twenty holes of 20 mm diameter were perforated on the homocentric square-shape structure at locations concentric with the holes on the UHPFRC plate to facilitate fixing it to the supporting rig using 20 bolts for 20 mm diameter. The bolts were tightened using a torque force of 110 Nm.

In order to facilitate the 3D Digital Image Correlation (DIC) analysis, two Photron FASTCAM SA4 high-speed digital cameras paired in a stereo configuration were mounted vertically to an aluminium frame fixed to the rigid wall. The surface of the UHPFRC plate was first painted with a thin layer of white coating and then marked to generate a random speckle pattern of approximately 3–5 mm in diameter. The stereovision system was first calibrated using a calibration board that has 14 mm grid spacing. Images were recorded at a rate of 10000 frames per second (fps). The field of view (FOV) was 210 × 180 mm.

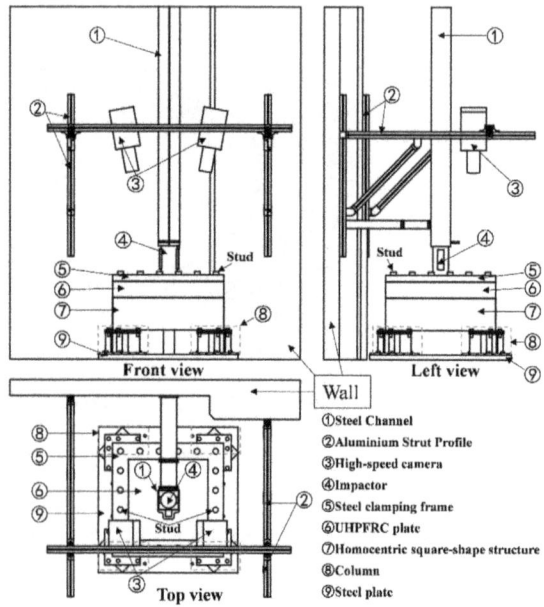

Fig. 2. Drop-weight impact test setup and high-speed stereo-vision system

2.3 Experimental Procedure

An outline of the testing schedule is shown in Table 2 with the aim of comparing the effect of sampling rate and the drop height. Since the sampling rate can influence the results obtained from the experiment, different sampling rates were examined. A constant weight of 40 kg was dropped 8 times on a UHPFRC plate from 0.5 m in order to verify the influence of the sampling rate. Then the impactor was dropped from 2.0 m and 3.0 m. The test was stopped when the plate failure occurred, what was indicated by an almost zero rebound of the impactor [5]. The theoretical impact velocities for drop height of 0.5 mm, 2.0 mm and 3.0 mm are 3.13 m/s, 6.26 m/s and 7.67 m/s, respectively. The arrangement and designation of these sensors are given in Fig. 3. The raw acceleration data corresponding to the first impact of each drop was filtered using a low pass filter with cut-off frequency of 4000 Hz. The accelerations at the same position as the accelerometers were also determined by calculating the second derivative of the displacement obtained by 3D DIC analysis. By comparing the accelerations and strains measured by the physical sensors attached to top surface of the UHPFRC plate with the DIC results, the accuracy of the DIC measurements was examined.

Table 2. Testing schedule

Number of drop	Drop height (m)	Sampling rate Fₛ (kHz)	Accelerometers	Strain gauges
1	0.5	5	/	S1-S6
2	0.5	10	/	S1-S6
3	0.5	20	A2, A4	S1-S6
4	0.5	35	A2, A4	S1-S6
5	0.5	60	A2, A4	S1, S2, S4, S5
6	0.5	120	A2, A4	S2, S5
7	0.5	250	A2, A4	/
8	0.5	25	A1, A2, A3	S1-S6
9	2.0	25	A1, A2, A3	S1-S6
10	3.0	25	A1, A2, A3	S1, S2, S4, S5

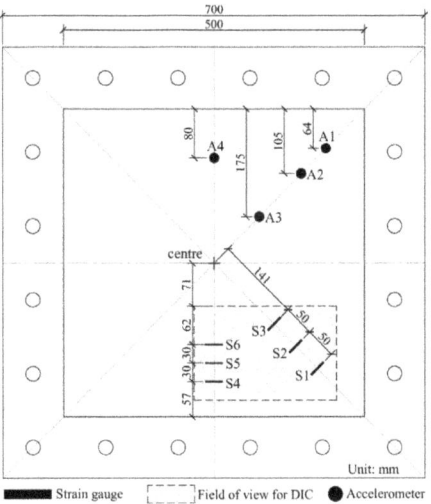

Fig. 3. Arrangements of measurements

3 Test Result and Discussion

3.1 Damage Pattern

In this study, the failure of plate was occurred at the 10th drop. The damage and crack pattern on the top and bottom surface of the UHPFRC at the end of 8th, 9th and 10th drop are shown in Fig. 4. At the end of the first 7 drops, no visible crack was observed.

The 8th drop, however, resulted in a small half-spherical indent on top of the plate at the impactor hit location with only a few micro-cracks at the bottom surface. A possible explanation for this observation is that although the energy imparted in a single 0.5 m drop was insufficient to produce inelastic tensile strains, each drop contributed to damage initiation and progress at the fibres-concrete interface. For the microcracks to appear on the bottom surface of the UHPFRC plate, however, an impact energy equivalent to 8 drops from 0.5 m was needed.

The 2.0 m drop, the 9th drop, caused a notable damage to the top surface of the plate at around the impactor hit region alongside with radial macro-cracks on the opposite face just below the indented area. These radial cracks centred on the impacting point indicate that the behaviour of the plate was governed by a flexural deformation.

At the 10th drop from 3.0 m, a semi-spherical indent damage of 65 mm diameter and 13 mm depth appeared at the top face of the plate while on its opposite face radial and circumferential cracks indicating the flexural and punching shear deformation, respectively, were formed.

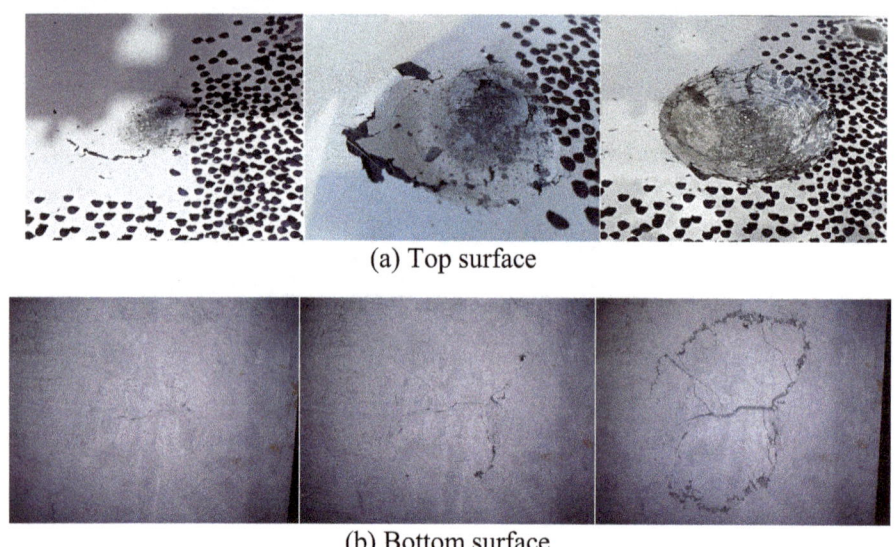

(a) Top surface

(b) Bottom surface

Fig. 4. Plate damage after 8th (left), 9th (middle) and 10th (right) drops.

3.2 Verification of DIC Results

Figure 5 and Fig. 6 show the acceleration time histories obtained from accelerometers and DIC analysis at positions A2 and A4, respectively, for 0.5 m drops. The nomenclature used in the figures refers to the source of signal, location of the sensor and the number of drop, respectively. For example, Acc-A2-3 corresponds to the acceleration obtained from the physical accelerometer at position A2 in the 3rd drop. Since the camera was accidentally disturbed during the 7th and 8th drops, their DIC accelerations are not reported. The graphs representing the measurements of the accelerometers reveal that a higher sampling rate capture more accurately the acceleration response, which manifests in a smoother curve as a result of more data point being available. This explains the difference between the 20 kHz sampling rate curve and the others and that a sampling rate of 25 kHz is sufficient to accurately capture the acceleration response.

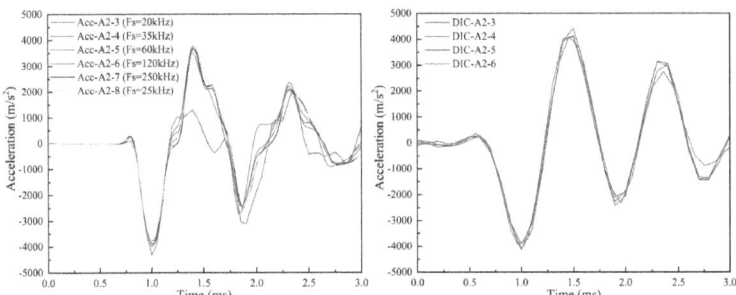

Fig. 5. Acceleration results of accelerometer (left) and DIC (right) at A2 for drop height of 0.5 m

By comparing the accelerometer and DIC results, it can be seen that the acceleration time history of DIC analysis and accelerometer measurements are in good agreement in terms of the peaks and the overall shape for 0.5 m drops. However, since the derivative of the displacement measured in DIC was taken to obtain the acceleration, the wavelength of the DIC curve is slightly longer than accelerometer measurements.

Excluding the results of Acc-A2-3 (Fs = 20 kHz) due to an insufficient sampling rate, it can be seen that the acceleration time history responses of 0.5 m drops are almost identical, indicating no or negligible damage in the plate. This is also confirmed by the fact that no visible crack was observed during the first seven drops.

Figure 7 shows the acceleration obtained from A2 accelerometer and DIC at 2.0 m and 3.0 m drops. Although the overall shape of DIC and accelerometer curves for both drops are in a relatively good agreement, the smaller first negative peak of DIC curve indicates a frame rate higher than 10000 f/s was needed. Since the DIC acceleration is the second derivative of displacement whose sampling rate is determined by frame rate of the camera, in order to capture the acceleration using DIC for large drop height, a larger frame rate is needed compared to smaller drop height as the deformation rates are higher for large drop height than lower ones.

The strain response comparison between strain gauge measurements and DIC results at different drop heights are shown in Fig. 8. The top surface area where the strain gauges were glued was under compression during impact [11]. Since the strain results of 0.5 m drops were identical, only the results of the 2nd drop are presented as an example and they show that regardless of the drop height, the DIC and strain gage

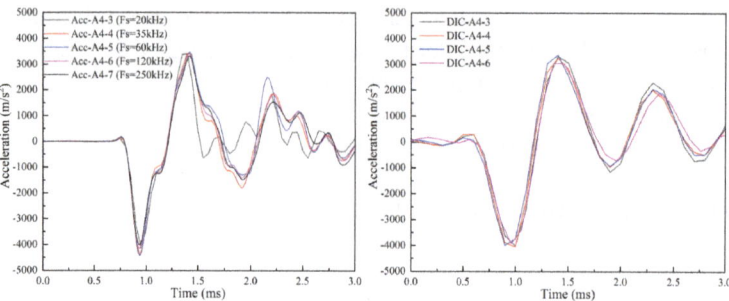

Fig. 6. Acceleration results of accelerometer (left) and DIC (right) at A4 for drop height 0.5 m

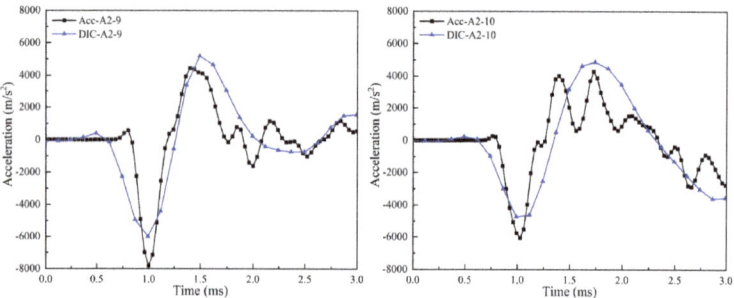

Fig. 7. Acceleration results of accelerometer and DIC at A2 for drop height of 2.0 m (left) and 3.0 m (right)

results are in a good agreement, because the strain and displacement have similar frequency which is lower than the acceleration.

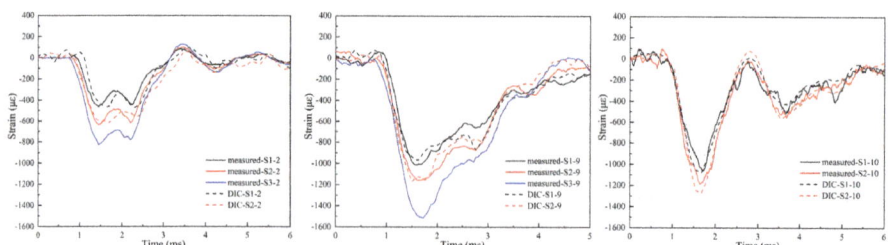

Fig. 8. Comparison of measured strains and DIC

3.3   Mid-point Displacement and Strain Time Histories of First Impact

Due to the limitations of the test setup, the central area of the UHPFRC plate with 140 mm diameter was not within the FOV. Therefore, a 4th order (along both x and y direction) polynomial surface fitting method was employed to extrapolate the displacement of this central part including the plate's mid-point. The mid-point displacement time histories for drop height of 0.5 m, 2.0 m and 3.0 m are shown in Fig. 9. It should be noted that the displacements in these figures are relative to the permanent displacement of previous drop. Since the results of 7th and 8th drops were not reliable, for 0.5 m drops only the result of the 6th drop is presented. It is worth noting that the central deformation obtained from the surface fitting is an estimate based on the assumption of continuous deformation across the plate. Hence, this equivalent deformation omits the localised deformations, e.g., the large central indent in 10th drop, and facilitates the analysis and comparison of the overall deformational performance of the plate in different drops.

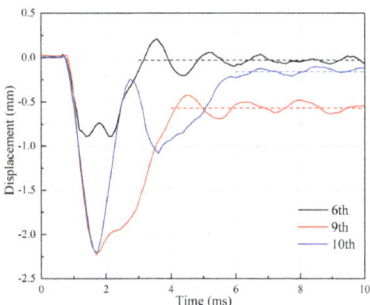

Fig. 9. Mid-point relative displacement time histories of the first hit in drops 8th, 9th and 10th

As shown in Fig. 9, the free vibration of the plate slightly below the zero displacement equilibrium position (the post impact response) for 0.5 m drops confirms the tests observations of negligible damage in the plate. When the drop height increases to 2.0 m, the plate is damaged as reflected by the plate post impact free vibration at a new equilibrium position and the slight decrease in its natural frequency. The new equilibrium position that is known as permanent displacement offset [12] is about 0.57 mm. The plate showed an additional permanent deflation of 0.16 mm as a result of drop from 3.0 m. The plate's post impact natural frequency, however, remains almost the same as for 2.0 m drop.

By comparing the strain measurements of strain gauges bonded on the diagonal (Fig. 10, left) and centre line (Fig. 10, right) of the plate, it is found that the shape of strain time history curve at diagonal line is similar to the shape of mid-point displacement time history curve. This can be used to qualitatively compare and analyse the mid-point displacement of the plate when the mid-point displacement cannot be measured.

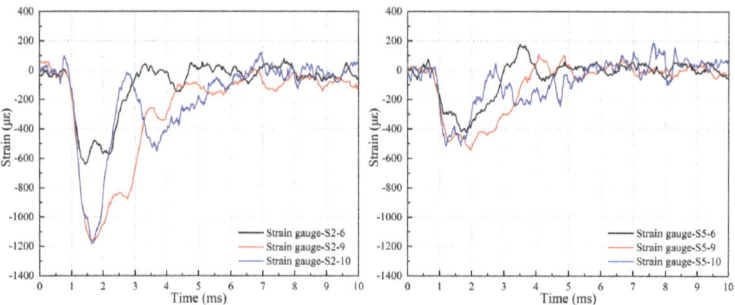

Fig. 10. Strain time histories of first impact for 8th, 9th and 10th drops

4   Conclusions

This paper experimentally investigated the behaviour of a 50 mm thick UHPFRC plate with a fixed boundary condition under repeated low velocity impact load of a 40 kg mass impactor by employing a high-speed 3D DIC system. The study was motivated on the knowledge that a 3D DIC analysis capable of capturing the full-field impact response of a UHPFRC plate with sufficient accuracy can facilitate the development of numerical models.

- For 0.5 m drops, the 3D DIC analysis on images recorded at 10k fps was found sufficient to achieve strain and acceleration results in a good agreement with the measurements of physical sensors. For drops higher than 0.5 m, however, despite a good agreement of strain results, a frame rate higher than 10k is required to obtain accurate acceleration.
- The impact energy imparted in a drop from 0.5 m or at impact velocity of 3.13 m/s was insufficient to produce inelastic tensile strains. However, macrocracks occurred after 8 cumulative drops from 0.5 m.
- The impact energy imparted in a drop from 2.0 m or at impact velocity of 6.26 m/s caused flexure damage. The subsequent impact from 3.0 m caused local failure and flexure. Multiple hits between impactor and plate were observed during one impact when local failure happened.

Acknowledgements. The first author gratefully acknowledges the financial support provided by Xinjiang Uygur Autonomous Region Educational Commission and Xinjiang University from China.

References

1. Yoo, D.Y., Banthia, N.: Mechanical and structural behaviors of UHPFRC concrete subjected to impact and blast. Constr. Build. Mater. **149**, 416–431 (2017)
2. Mao, L., Barnett, S.J.: Investigation of toughness of UHPFRC beam under impact loading. Int. J. Impact Eng. **99**, 26–38 (2017)

3. Beppu, M., Kataoka, S., Ichino, H., Musha, H.: Failure characteristics of UHPFRC panels subjected to projectile impact. Compos. Part B Eng. **182**, 107505 (2020)
4. Sovják, R., Vavřiník, T., Zatloukal, J., Máca, P., Mičunek, T., Frydrýn, M.: Resistance of slim UHPFRC targets to projectile impact using in-service bullets. Int. J. Impact Eng. **76**, 166–177 (2015)
5. Ranade, R., Li, V.C., Heard, W.F., Williams, B.A.: Impact resistance of high strength-high ductility concrete. Cem. Concr. Res. **98**, 24–35 (2017)
6. Othman, H., Marzouk, H.: Dynamic identification of damage control characteristics of ultra-high performance fiber reinforced concrete. Constr. Build. Mater. **157**, 899–908 (2017)
7. Anil, Ö., Kantar, E., Yilmaz, M.C.: Low velocity impact behavior of RC slabs with different support types. Constr. Build. Mater. **93**, 1078–1088 (2015)
8. He, Y., Esmaeeli, E., Soutsos, M.: Effect of casting method and test setup on flexural characterization of UHPFRC. In: Civil Engineering Research in Ireland (CERI2020), pp. 229–233 (2020)
9. JSCE: Recommendations for Design and Construction of High Performance Fiber Reinforced Cement Composites with Multiple Fine Cracks (HPFRCC). Concr. Eng. Ser. **82**, 6–10 (2008). Testing Method
10. Barnett, S.J., Lataste, J.F., Parry, T., Millard, S.G., Soutsos, M.N.: Assessment of fibre orientation in UHPFRC and its effect on flexural strength. Mater. Struct. **43**(7), 1009–1023 (2010)
11. Yi, N.H., Kim, J.H.J., Han, T.S., Cho, Y.G., Lee, J.H.: Blast-resistant characteristics of ultra-high strength concrete and reactive powder concrete. Constr. Build. Mater. **28**(1), 694–707 (2012)
12. Othman, H., Marzouk, H.: Impact response of ultra-high-performance reinforced concrete plates. ACI Struct. J. **113**(6), 1325–1334 (2016)

Experimental Characterization of the Tensile Constitutive Behaviour of Ultra-High Performance Concretes: Effect of Cement and Fibre Type

Francesco Lo Monte[1]([⊠]), Eduardo J. Mezquida-Alcaraz[2],
Juan Navarro-Gregori[2], Pedro Serna[2], and Liberato Ferrara[1]

[1] Politecnico di Milano, Milan, Italy
francesco.lo@polimi.it
[2] Universitat Politècnica de València, Valencia, Spain

Abstract. The Research Project ReSHEALience has been launched in 2018 in the framework of the European Programme Horizon 2020. The project aims at developing a new approach for the design of structures exposed to extremely aggressive environments, based on durability and life cycle analysis. In this regards, the starting point is represented by new advanced Ultra-High Performance Fibre Reinforced Cementitious Composites (UHPFRCCs), called Ultra-High Durability Concretes (UHDC) because of their enhanced durability obtained by means of engineered composition, which should be characterized by strain hardening behaviour under tension in both ordinary and very aggressive conditions. In this context, the first step is to develop an effective approach for identifying the main parameters describing the overall behaviour in tension. In the present study, indirect tension tests have been performed via 4-Point Bending Tests (4PBT). Starting from the test results, a combined experimental-numerical identification procedure has been implemented in order to evaluate the effective material behaviour in direct tension in terms of stress-strain law. Finally, the comparison among three mixes differing for fibre and cement type is reported in terms of tensile response and post-crack localization behaviour.

Keywords: ReSHEALience · Ultra-High Durability Concrete – UHDC · Tensile constitutive behaviour · Identification

1 Introduction

In the present study, the mechanical characterization in tension of three Ultra-High Performance Fibre-Reinforced Cement Composites (UHPFRCC) is reported as a result of the application of a combined experimental-numerical approach aimed at the identification of the constitutive tensile behaviour. This is instrumental for the development of cementitious composites with improved durability, in which the interaction of reduced crack-opening (thanks to fibre-triggered multiple cracking) and engineered self-healing is exploited for significantly increasing the long term performance in aggressive environments.

© RILEM 2022
P. Serna et al. (Eds.): BEFIB 2021, RILEM Bookseries 36, pp. 936–946, 2022.
https://doi.org/10.1007/978-3-030-83719-8_80

This is the first step of the Research Project ReSHEALience, launched in 2018 and involving 14 Partners and 3 linked third parties all around Europe, within the challenging framework of the European Programme Horizon 2020. The main objective is to develop a durability-oriented structural approach for both ordinary and extremely aggressive environments, based on the concepts of Durability Assessment based Design (DAD) and Life Cycle Analysis (LCA).

In parallel to the development of cementitious composites with enhanced durability, a new design approach is under development in which, via the explicit evaluation of key durability parameters, the inherent features and advantages of these advanced materials can be fully exploited [1]. The consistency of the developed approach is now under verification thanks to the survey of 6 full-scale demonstrators, which have been designed and realized according to the above-mentioned Durability-Based Design and are monitored in time in terms of different key performance indicators. Finally, advanced numerical models will help in generalizing the results within a practitioner-friendly framework.

Hence, 3 UHPFRCCs with engineered self-healing capability have been formulated (also called UHDCs in the following [2]) in order to choose the optimal solution. The composition of the mixes is based on the rather established knowledge about UHPFRCC. In particular, fibres are instrumental for obtaining a hardening behaviour in bending and even in direct tension, thanks to the multiple stable crack propagation following the onset of the first crack, till localization of a single unstable propagating crack occurs [3]. This allows for limiting crack opening even for significant tensile deformation by balancing crack-tip toughness and fibre pull-out work [4–6]. In the present case, a combination of cement and slag, small aggregates (maximum size of 2 mm) and structural fibres at dosages higher than 1.5% by volume is adopted.

Self-healing, together with the extensive multiple cracking characterized by very small crack openings, makes it possible to significantly increase the durability of these cementitious composites in the cracked state, which is the ordinary service-life condition of concrete structures [7–13].

In this context, stress-strain law identification in direct tension is of primary importance for the generalization of the results, and a few approaches can be found in the literature [14–20]. This task is, however, not trivial, due to the inherent redundancy of the most common adopted indirect tests such as 4-Point Bending Test - 4PBT (see also [21]).

In the present paper, the approach developed for the characterization of the tensile "constitutive" behaviour of three UHDCs (namely UHPRCCs with improved durability) and for the identification of the main mechanical parameters is described. The approach is based on a combination of 4PBT and inverse analysis. The procedure described allows for the estimation of the constitutive law in direct tension starting from indirect tests.

2 Experimental program

2.1 Concrete Mixes

The reference UHDC mix (XA-CA) contains 600 kg/m^3 of cement (CEM I 52.5), 500 kg/m^3 of slag, 982 kg/m^3 of sand with a maximum aggregate size of 2 mm and 120 kg/m^3 of straight steel fibres (l_f/d_f = 20/0.2 mm). After mixing the solid parts, 200 l/m^3 of water are introduced together with 33 l/m^3 of superplasticizer. Penetron Admix®, in the content of 0.8% by cement mass, is also introduced, whose effect on the overall performance of concrete mixes has been investigated elsewhere [22, 23]. In mix XA-CA-Amf, steel fibres are replaced with 111 kg/m^3 of metallic amorphous fibre ($l_f/w_f/t_f$ = 15/1/0.024 mm), while in mix XA-CA-CEMIII, CEM I is replaced with CEM III.

The constituents' proportions of all mixes are reported in Table 1.

For each mix, six 100 × 100 × 500 mm^3 prismatic beams have been cast. Further studies on the same mixes can be found also in [24, 25].

2.2 Test Setups and Specimen Geometries

The mechanical characterization has been performed via 4-Point Bending Tests (4PBT) on un-notched beams (l·w·h = 500 × 100 × 100 mm^3). 4PBT on un-notched specimens is adopted since it allows to investigate multiple cracking in the central region L$_o$, where bending moment is constant.

During the test, (a) the relative vertical displacement of the mid-span section with respect to the supports and (b) the Crack-Opening Displacement – COD across the central region of the specimen are monitored via two LVDTs each, as represented by red segments in Fig. 1.

Horizontal LVDTs have been fixed to metal platelets, which were glued in the positions reported in the figure, while vertical ones are mounted to an aluminium frame simply supported on the two ends of the specimen. More complex arrangements of transducers could be also employed in order to better determine the crack localization point [20]. In the present study, the tests were stroke-displacement controlled. All tests were performed after at least 90 days from casting, in order to allow slag to develop the maximum possible long term hydration and pozzolanic activity, as compatible with the low water-to-binder ratio employed.

In the curing period, all samples were stored in climate chamber (R.H. = 90%, T = 20 °C).

Table 1. UHDCs' mix compositions.

Constituents [kg/m^3]	XA-CA	XA-CA-Amf	XA-CA-CEMIII
CEM I 52.5	600	600	–
CEM III 52.5	–	–	600
Slag	500	500	500
Water	200	200	200
Steel fibres	120	–	120
Amorphous fibres	–	111	–
Sand (0–2 mm)	982	982	982
Superplasticizer	33	33	33
Crystalline adm	4.8	4.8	4.8

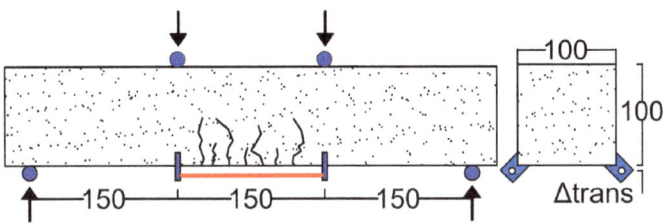

Fig. 1. Scheme for 4PBT on small un-notched beams.

3   Identification of the Mechanical Parameters in tension

The typical behaviour in direct tension and in bending in case of strain-hardening materials is reported schematically in Fig. 2. For both direct tension and bending, there is an initial perfect elastic branch which ends when the first crack occurs (at $\sigma = \sigma_{cr}$). After that, in softening materials, localization occurs and load decreases.

In the case at issue, on the other hand, the amount of fibre allows for increasing the external load after first crack is formed, this being related to the concept of hardening. This is made possible by the bridging-effect of fibres crossing the cracks.

Finally, crack localization occurs (at $\sigma = \sigma_{pk,t}$ in direct tension and at $\sigma = \sigma_{pk,b}$ in bending) and the further tensile deformation concentrates into a single crack with the macroscopic effect that external load starts decreasing. It is worth noting that in the general case $\sigma_{pk,b} \gg \sigma_{pk,t}$.

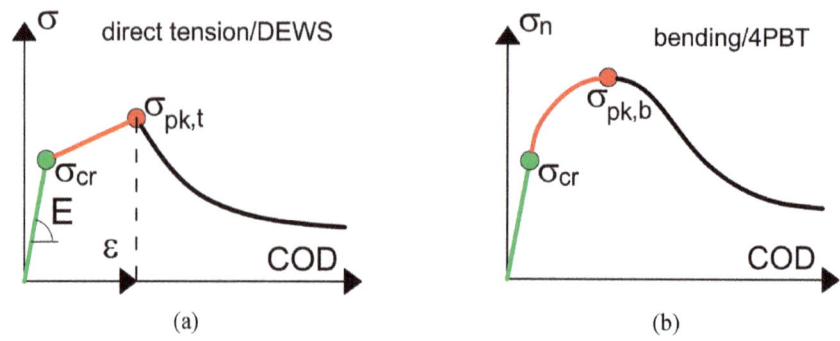

Fig. 2. Schemes of σ_n – COD curves (a) in tension and (b) in bending.

The evaluation of σ_{cr} and $\sigma_{pk,t}$ (and of the correspondent strains) is of primary importance, since they describe the multiple cracking stage. When starting from bending tests, however, such parameters can be identified just by means of a careful back analysis, due to the rather smooth transition among the elastic, the hardening and the softening branches. This is caused by stress redistribution across the specimen depth under bending.

Hence, the estimation of the mechanical response in direct tension needs a combined experimental-numerical approach. In particular, after performing the flexural tests, the inverse analysis is carried out to estimate the mechanical parameters in direct tension. Afterwards, the consistency of the processed mechanical parameters is checked by performing analytical and numerical simulations of the flexural tests, using as input the mechanical parameters calculated via inverse analysis.

Back analysis has been implemented according to the simplified 4-Point Inverse Analysis method (4P-IA) developed at Universitat Politècnica de València [17–20]. The method allows to estimate (1) the elastic modulus, (2) the first cracking stress σ_{cr}, (3) the ultimate strength f_{tu} and the corresponding strain ε_{tu} and (3) the post localization behaviour thanks to the crack opening at $f_{tu}/3$ and at nil stress. Moreover, a softening correction of the f_{tu} parameter obtained from the 4P-IA [18, 26] is adopted in the case of UHPFRCC that exhibits strain-softening tensile behaviour ($f_{tu} < \sigma_{cr}$).

Once estimated the stress-strain relation in direct tension, it is used as input for the numerical/analytical simulation of the bending test. To this purpose both the Hinge Model (analytical approach described in [17–20] and based on closed-form formulations of moment-curvature relationships) and a Non-Linear FEM model described in [27] have been implemented. It is worth remarking that in the Hinge Model, the COD-strain conversion is performed with reference to the LVDT gauge length, this translating in the assumption of non-linear hinge in the multiple-cracking region [18–20].

To define the Non-Linear FEM model of the 4PBT herein used, a discrete cracking approach is considered [26, 27]. The constitutive law for discrete cracking is defined by a traction-separation law. This behaviour is forced only on the mid-span section of the specimen where the macrocrack is set, as shown in Fig. 3. The rest of the specimen is modelled by a smeared cracking approach based on a fixed total strain crack model. The FEM is developed by means of Diana software [28].

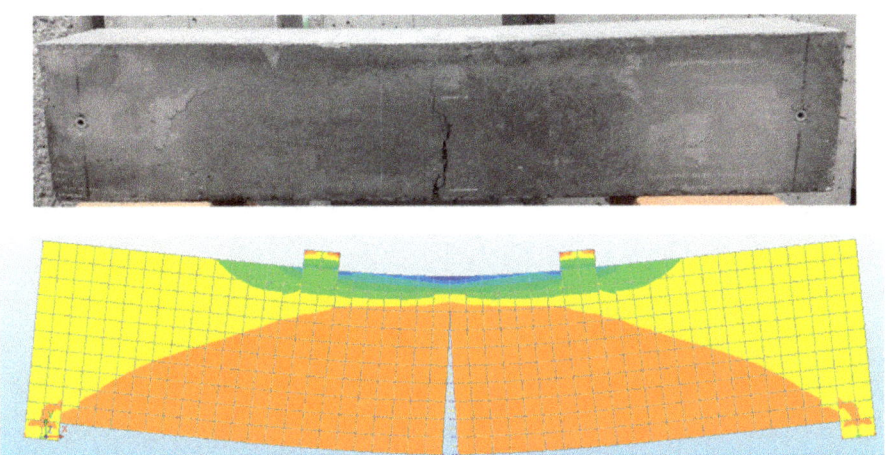

Fig. 3. Non-Linear FEM model of the 4PBT.

4   Results and Discussion

The experimental results on 4PBTs are reported in grey curves in Figs. 4a, b, c in terms of σ – δ curves (where δ is the vertical mid-span displacement), the nominal stress being evaluated as:

$$\sigma = P\,L/\left(b\,h^2\right) \tag{1}$$

where P is the total load applied, L the test span and b, h are the specimen width and thickness, respectively.

The scattering among the curves of a given mix is typical of flexural tests and it is ascribable to the possible differences in terms of fibre arrangement among the specimens and to the inherent heterogeneities of cementitious composites. The average peak stress is equal to 17, 19 and 20 MPa for mixes XA-CA, XA-CA-Amf and XA-CA-CEMIII, respectively. Mixes XA-CA and XA-CA-CEMIII are expected to yield very similar peak stress since the fibre dosage and typology is the same, and this is quite consistent with the results, the lower peak stress of XA-CA being explainable with the higher scattering of the results for this mix. On the other hand, the outcome that also mix XA-CA-Amf provides very similar peak stress to the other two mixes is rather interesting, since the fibre type is very different, consisting in amorphous metallic fibre with a very different aspect ratio.

The main difference among steel and amorphous fibre is observed in the post-localization stage, since the former seems to provide higher ductility with a less pronounced softening branch. These considerations are confirmed by the tensile

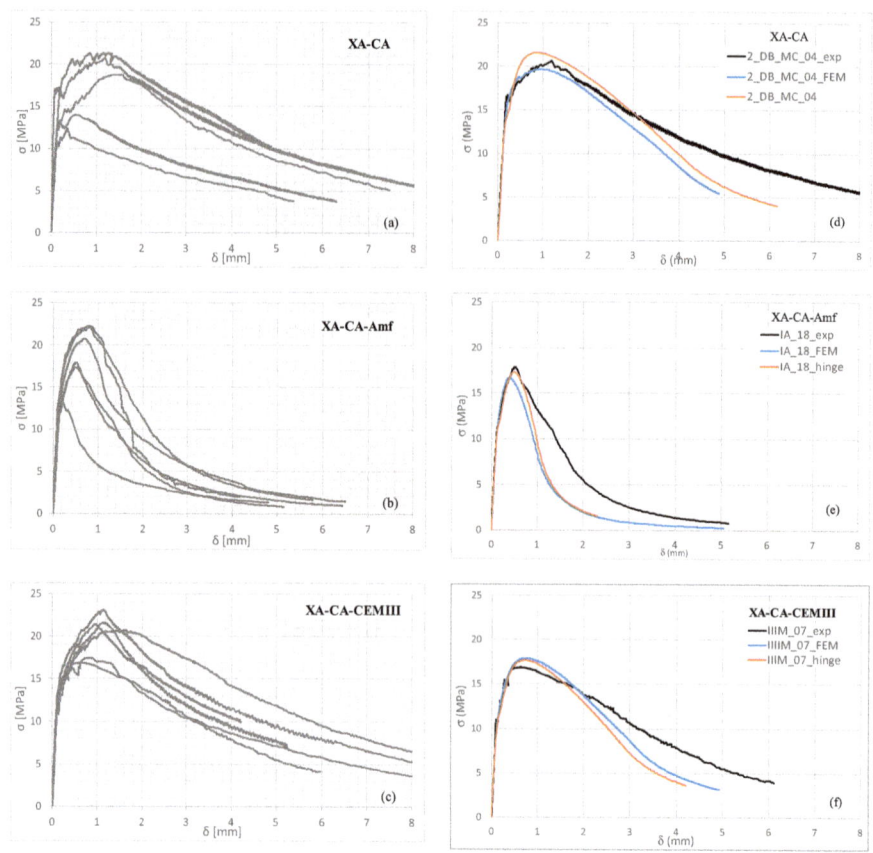

Fig. 4. Plots of σ – δ curves for (a) XA-CA, (b) XA-XA-Amf and (c) XA-CA-CEMIII and (d–f) comparison among one single experimental curve and the correspondent numerical curves obtained via Hinge Model and FEM simulations.

parameters evaluated via 4-Point Inverse Analysis, as shown in Fig. 5, where the direct σ-ε and σ-w curves are reported in grey colour. For each mix, an average curve has been also evaluated (and reported in red colour), this being instrumental for comparing effectively the three mixes (see Fig. 6). It is clear as the three mixes yield an almost perfect plastic behaviour (slightly softening for mix XA-CA and slightly hardening for mixes XA-CA-Amf and XA-CA-CEMIII), with a strain-at-localization around 3–3.5‰ for the mixes with steel fibre and about 1.5‰ for the mix with amorphous fibre (thus corroborating what observed in the bending tests).

As a confirmation of the overall consistency of the results, the constitutive laws obtained via inverse analysis (Fig. 6) have been considered as input for simulating the bending tests via both the Hinge Model and FEM analyses. In Figs. 4d, e, f the comparison among experimental and analytical/numerical results is provided for one specimens per mix, showing a very good agreement.

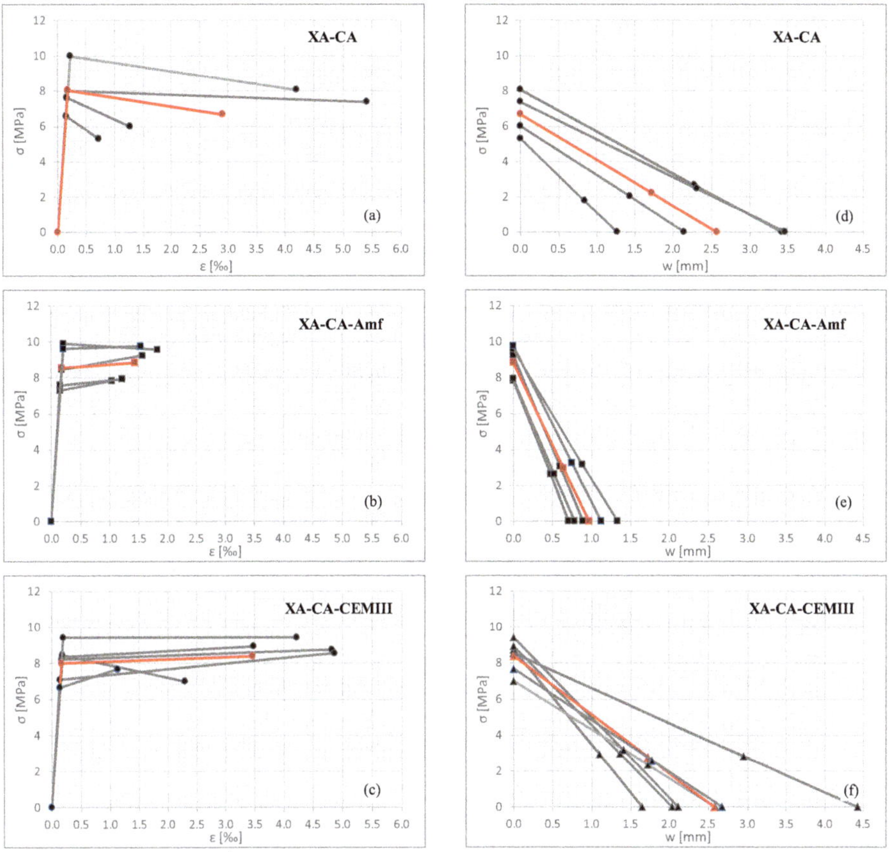

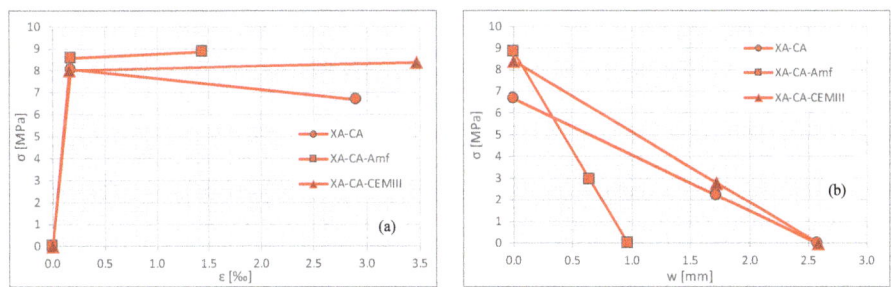

Fig. 5. Plots of σ – ε and σ – δ curves in direct tension as obtained from inverse analysis for (a, d) XA-CA, (b,e) XA-CA-Amf and (c,f) XA-CA-CEMIII.

Fig. 6. Comparison among the three mixes in terms of (a) σ – ε and (b) σ – δ curves in direct tension as obtained from inverse analysis.

5 Concluding remarks

The paper describes the mechanical characterization in tension of three Ultra-High Durability Concretes (UHDC), namely Ultra High-Performance Fibre-Reinforced Cementitious Composites (UHPFRCC) conceived for specific structural applications in extremely aggressive environments.

The tensile characterization is based on an identification procedure which takes advantage of the combination of experimental tests and numerical simulations, aimed at carrying out the main mechanical parameters defining the material behaviour in direct tension. This step is fundamental for moving from the experimental scale to the structural one, thus making it possible the following structural design. The experimental investigation has been performed via 4 Point Bending Tests on 500x100x100 mm^3 small un-notched beams, while back analysis has been implemented via the 4-Point Inverse Analysis (4P-IA) method.

The consistency of the constitutive laws thus evaluated has been assessed by simulating the flexural tests (via both a "traditional" Finite Element Model and the Hinge Model, this latter being an analytical approach based on bending-curvature closed-form formulations), using as input the constitutive laws evaluated by means of inverse analysis. The very good agreement among numerical and experimental curves confirms the overall reliability of the approach.

The identification procedure herein adopted, based on the cross-comparison of experimental results and numerical analysis, proved to be rather easy in implementation and effective in highlighting the overall consistency of the results. This allows to confirm an almost perfect-plastic tensile constitutive response of the investigated UHDC mixes. In particular, mix XA-CA showed a slightly softening behaviour, while mixes XA-CA-Amf and XA-CACEMIII both proved a slightly hardening response after first cracking.

On the other hand, the two mixes with steel fibre yield a higher value of strain-at-localization, this being around 3–3.5‰, while amorphous fibre seems to lead to a more brittle response with a value of strain-at-localization close to 1.5‰. In this regard, it is worth noting as 2‰ is an interesting threshold, this being the yield strain of ordinary steel, which could be closely achieved, or even abundantly overpassed, by all mixes. If also traditional reinforcement is used, in fact, guarantying no crack-localization of concrete at steel-yielding, allows for the optimization of reinforcement amount and durability performance.

On the other hand, the value of cracking and peak stress of the three mixes are rather similar, their average values being comprised in the range 8–9 MPa and 6–8 MPa, respectively. The promising mechanical response in tension provided by the three mixes together with the effectiveness in the identification of the main constitutive parameters via inverse analyses are the bases for the following step, namely the design of the full-scale demonstrators of the Project, these being the final testing ground of the developed procedure.

Acknowledgements. The research activity reported in this paper has been performed in the framework of the ReSHEALience project (Rethinking coastal defence and Green-energy Service infrastructures through enHancEd-durAbiLity high-performance cement-based materials) which has received funding from the European Union's Horizon 2020 research and innovation program under grant agreement No 760824. The information and views set out in this publication do not necessarily reflect the official opinion of the European Commission.

The authors acknowledge the cooperation of MEng. Lorenzo Papa, Stefano Passoni, Angelo Alferi, Nicola Borgioni, Andrea Cervini and Luca Famiani in performing experimental tests, in partial fulfilment of the requirements for the MEng in Civil Engineering and Building Engineering respectively.

The kind collaboration of ReSHEALience partner Penetron Italia (MArch. EnricoMaria Gastaldo Brac) in supplying the crystalline self-healing promoter is also acknowledged.

The authors also thank Mr. Marco Francini (BuzziUnicem) for supplying of cement, Mr. Michele Gadioli and Roberto Rosignoli (Azichem ltd) for supplying of steel fibres, dr Florian Bernard (Saint-Gobain Seva) for supplying of amorphous fibres and Mr. Sandro Moro (BASF Italia) for supplying the superplasticizer employed for casting the investigated UHDC mix.

References

1. Al-Obaidi, S., Bamonte, P., Luchini, M., Mazzantini, I., Ferrara, L.: Durability-based design of structures made with UHP/UHDC in extremely aggressive scenarios: application to a geothermal water basin case study. MDPI Infrastruct. **5**(11), 1–44 (2020). https://doi.org/10.3390/infrastructures5110102
2. Serna, P., et al.: Upgrading the concept of UHPFRC for high durability in the cracked state: the concept of ultra high durability concrete (UHDC) in the approach of the H2020 project reshealience. In: Proceedings of the International Conference on Sustainable Materials Systems and Structures SMSS 2019, Rovinj (Croatia), 20–22 March 2019, pp. 764–771 (2019)
3. Naaman, A.E., Reinhardt, H.W.: Proposed classification of HPFRC composites based on their tensile response. Mats. Structs. **39**(5), 547–555 (2006)
4. Li, V.C., Wu, H.C.: Conditions for pseudo strain-hardening in fiber reinforced brittle matrix composites. Appl. Mechs. Rev. **45**(8), 390–398 (1992)
5. Li, V.C., Stang, H., Krenchel, H.: Micromechanics of crack bridging in fibre-reinforced concrete. Mats. Structs. **26**(8), 486–494 (1993)
6. Li, V.C.: On engineered cementitious composites. A review of the material and its applications. J. Adv. Concr. Tech. **1**(3), 215–230 (2003)
7. Plagué, T., Desmettre, C., Charron, J.-P.: Influence of fiber type and fiber orientation on cracking and permeability of reinforced concrete under tensile loading. Cem. Concr. Res. **94**, 59–70 (2017)
8. Yang, Y., Lepech, M., Yang, E., Li, V.: Autogenous healing of engineered cementitious composites under wet-dry cycles. Cem. Concr. Res. **39**, 382–390 (2009)
9. Zhang, Z., Qian, S., Ma, S.: Investigating mechanical properties and self-healing behavior of micro-cracked ECC with different volume of fly ash. Constr. Build. Mats **52**, 17–23 (2014)
10. Ferrara, L., Ferreira, S.R., Krelani, V., Della Torre, M., Silva, F., Toledo, R.: Natural fibers as promoters of autogenous healing in HPFRCCs: results from on-going Brazil-Italy cooperation (2015)
11. Ferrara, L., Krelani, V., Moretti, F.: Autogenous healing on the recovery of mechanical performance of HPFRCCs: part 2 - correlation between healing of mechanical performance and crack sealing. Cem. Conr. Comp. **73**, 299–315 (2016)

12. Ferrara, L., Krelani, V., Moretti, F., Roig Flores, M., Serna Ros, P.: Effects of autogenous healing on the recovery of mechanical performance of HPFRCCs: part 1. Cem. Concr. Comp. **83**, 76–100 (2017)

13. Snoeck, D., De Belie, N.: Repeated autogenous healing in strain-hardening cementitious composites by using superabsorbent polymers. ASCE J. Mats. Civ. Eng. **28**(1), 1–11 (2016)

14. Baby, F., Graybeal, B., Marchand, P., Toutlemonde, F.: UHPFRC tensile behavior characterization: inverse analysis of four-point bending test results. Mater. Struct. Constr. **46**, 1337–1354 (2013)

15. Qian, S., Li, V.C.: Simplified inverse method for determining the tensile strain capacity of strain hardening cementitious composites. J. Adv. Concr. Technol. **5**, 235–246 (2007)

16. Soranakom, C., Mobasher, B.: Closed-form moment-curvature expressions for homogenized fiber-reinforced concrete. ACI Mater. J. **104**, 351–359 (2007)

17. López, J.Á., Serna, P., Navarro-Gregori, J., Camacho, E.: An inverse analysis method based on deflection to curvature transformation to determine the tensile properties of UHPFRC. Mater. Struct. **48**, 3703–3718 (2015)

18. Mezquida-Alcaraz, E.J., Navarro-Gregori, J., Serna-Ros, P.: Numerical validation of a simplified inverse analysis method to characterize the tensile properties in strain-softening UHPFRC. In: Fibre Concrete 2019, Materials Science and Engineering, vol. 596, p. 012006 (2019). https://doi.org/10.1088/1757-899X/596/1/012006

19. López, J.Á., Serna, P., Navarro-Gregori, J., Camacho, E.: An inverse analysis method based on deflection to curvature transformation to determine the tensile properties of UHPFRC. Mater. Struct. **48**(11), 3703–3718 (2014). https://doi.org/10.1617/s11527-014-0434-0

20. López, J.Á., Serna, P., Navarro-Gregori, J., Coll, H.: A simplified five-point inverse analysis method to determine the tensile properties of UHPFRC from un-notched four-point bending tests. Compos. B **91**, 189–204 (2016)

21. MC 2010, fib Model Code for Concrete Structures 2010, Comité Euro-International du Béton, Lausanne (Switzerland) (2012)

22. Cuenca, E., et al.: Concept of ultra high durability concrete for improved durability in chemical environments: preliminary results. In: Conference on Durable Concrete for Infrastructure under Severe Conditions - Smart admixtures, self-responsiveness and nano-additions, Ghent, Belgium (2019)

23. Cuenca, E., Mezzena, A., Ferrara, L.: Synergy between crystalline admixtures and nano-constituents in enhancing autogenous healing capacity of cementitious composites under cracking and healing cycles in aggressive waters. Constr. Build. Mater. **266**, 121447 (2021). https://doi.org/10.1016/j.conbuildmat.2020.121447

24. Monte, F.L., Ferrara, L.: Tensile behaviour identification in Ultra-High Performance Fibre Reinforced Cementitious Composites: indirect tension tests and back analysis of flexural test results. Mater. Struct. **53**(6), 1–12 (2020)

25. Monte, F.L., Ferrara, L.: Self-healing characterization of UHPFRCC with crystalline admixture: Experimental assessment via multi-test/multi-parameter approach. Constr. Build. Mater. **283**, 122579 (2021)

26. Mezquida-Alcaraz, E.J., Navarro-Gregori, J., Serna-Ros, P.: Direct procedure to characterize the tensile constitutive behavior of strain-softening and strain-hardening UHPFRC. Cement Concrete Compos. **115**, 103854 (2021). https://doi.org/10.1016/j.cemconcomp.2020.103854

27. Mezquida-Alcaraz, E.J., Navarro-Gregori, J., Lopez, J.A., Serna-Ros, P.: Validation of a non-linear hinge model for tensile behavior of UHPFRC using a Finite Element Model. Comput. Concr. **23**, 11–23 (2019)

28. DIANA (Software), User's Manual – Release 10.2, TNO DIANA, Netherlands (2017). https://dianafea.com/manuals/d102/Diana.html

Mechanical and Durability Assessment of Concretes Obtained from Recycled Ultra-High Performance Concretes

Estefania Cuenca[1]([✉]), Marta Roig-Flores[2], Roberto Garofalo[1],
Milena Lozano-Násner[3], Cecilia Ruiz-Muñoz[2], Fabrizio Schillani[1],
Ruben Paul Borg[3], Liberato Ferrara[1], and Pedro Serna[2]

[1] Politecnico di Milano, Milan, Italy
estefania.cuenca@polimi.it
[2] Universitat Politècnica de Valencia, Valencia, Spain
[3] University of Malta, Msida, Malta

Abstract. The aim of this work is to analyse the mechanical and durability properties of Recycled Ultra High Performance Concretes (RUHPC) containing different amounts of recycled fine aggregate obtained from crushing Ultra High Performance Concretes (UHPC). This paper summarizes and compares the results from different experimental campaigns carried out in the framework of the ReSHEALience project (Rethinking coastal defence and Green-energy service infrastructures through enhanced-durAbiLity high-performance cement-based materials) which has received funding from the European Union's Horizon 2020 programme (GA 760824). Mechanical performance was evaluated by means of compressive and flexural tests, whereas durability was evaluated by means of chloride penetration, chloride migration and water absorption capillary tests. The results indicated that replacing 50% or 100% of natural aggregates with recycled aggregates did not significantly affect neither compressive strength nor flexural strength. In the case of high replacement rates, a slight decrease in workability was detected, but the mix retained its self-compacting properties. RUHPC had similar durability performance as UHPC. In conclusion, the results have shown that it is feasible to produce RUHPC; the recycled fine aggregate has shown great potential to be used in the production of new UHPC. Scalability of the recycling procedure to industrial level was also addressed in order to pave the way towards the uptake from the different value chain actors of the construction industry of the innovation potential demonstrated by the research.

Keywords: High Performance Concrete · Recycled aggregates · Durability

1 Introduction

Construction industry is one of the most important consumers of raw materials and hence one of the main responsible of CO_2 emissions. Moreover, it is one of the main producers of waste: as a matter of fact, the Construction and Demolition Waste (CDW) signify 30–40% of solid waste produced in the world [1, 2]. This is due on the

© RILEM 2022
P. Serna et al. (Eds.): BEFIB 2021, RILEM Bookseries 36, pp. 947–957, 2022.
https://doi.org/10.1007/978-3-030-83719-8_81

one hand to the renovation of buildings that are not anymore adequate, neither from the structural nor from the energy point of view, as well as, on the other, to the degradation of materials and reduced structural performance experiences because of aggressiveness and increasing challenges by structural service scenarios [3]. In this framework, concrete represents about a 67% of the total CDW in the US whereas in Europe CDW quantities are linked to the population density and the Gross Domestic Product of each single country [4]. This motivates an increased interest to use CDW through construction and demolition.

Recycled aggregate concrete (RAC) can be used for structural and non-structural purposes and can be classified in recycled concrete aggregate, recycled brick aggregate and recycled mixed aggregates. Recycled aggregates (RA) can be used not only for producing concrete but also for other application such as pavements, roadways and other cementitious composites [5]. Standards indicate minimum requirements for their application, including maximum percentage of crushed concrete, minimum bulk density and maximum absorption capacity [6, 7]. Recycled aggregates have higher water absorption, lower density, lower crushing strength and abrasion resistance compared to natural aggregates. On the other hand, the properties of RAC are affected by the replacement ratio of recycled to natural aggregates, but also by the performance of the concrete from where RAs were obtained. This replacement also influences the mechanical performance of RAC, where generally compressive strength is more influenced than tensile and flexural strength. Reduction in compression can reach up to 30%, for 100% of replacement of aggregates [8].

One of the objectives of ReSHEALience project [9] is to analyse the potential for recycling of UHPC mixes formulated and validated in real applications UHPC mixes, in cradle-to-cradle approach, using the recycled UHPC obtained from a UHPC previously casted and studied in the project. As a matter of fact, UHPC is a relatively new category of construction materials, and the actual possibility for its recycling may look remote since it is also claimed as a longer lasting material because of its higher durability in both the un-cracked and cracked state [10, 11]. Anyway, addressing its recyclability potential has to be meant as a new significance to longer lasting construction materials and structures, moving, also in structure and infrastructure, from an "additive" concept, where new additions are built adjacent to the existing ones to comply with new needs and demands, to a circular concept, where the structure and infrastructures can be self-regenerated and rebuilt to be adapted to new context needs and demands, according to a new concept and using to the largest possible extent its own materials, without any expect loss of performance as if built with "freshly" produced materials employing 100% new raw resources.

2 Objectives

The mechanical and durability properties of Recycled Ultra High Performance Concretes (R-UHPC) containing different amounts of recycled fine aggregate obtained from crushing Ultra High Performance Concretes (UHPC) have been analysed in this paper. This paper summarizes and compares the results from different experimental campaigns carried out in the framework of the ReSHEALience project. One of the campaigns were carried out by University of Malta (UoM) and Politecnico di Milano (PoliMi) whereas the second campaign was carried out by the Universitat Politècnica de Valencia (UPV). Photos of the recycled constituents (aggregates and fibres) obtained from the recycled UHPC are shown in Fig. 1.

3   UOM-Polimi Experimental Campaign

The recycled UHPC concrete (R-UHPC) was obtained from a four-month-old UHPC with an average compressive strength of 150 MPa. The original UHPC was crushed and processed in the laboratory using a jaw crusher (Pascal Engineering) operating at a power of 750 W. Repeated crushing was required in order to reach the finer size of aggregate required to produce UHPC, and to completely separate the steel fibres from the concrete matrix. A magnet was used to remove steel fibres. A sieve analysis was performed on the crushed material up to obtain a grain-size distribution curve similar to that for the natural aggregates used in the original UHPC.

Fig. 1. Photos of the recycled aggregates (left and centre) and fibres (right) obtained from recycling UHPC.

3.1   Mix Design

The UHPC mix composition used as reference has been formulated, validated and produced at the UoM, in the framework of the ReSHEALience project to be employed in the retrofitting of the column support structure of an elevated water tank in the Valletta Grand Harbour, one of the pilots of ReSHEALience [9]. The concrete mixes shown in Table 1 used different replacement percentages of natural aggregates by recycled ones: 0% as reference, 50% R-UHPC and 100% R-UHPC, moreover a mix with 50% of recycled carbonated aggregates was produced. Carbonated aggregates were prepared using an accelerated carbonation chamber set with a temperature of 25 °C and 65% Relative Humidity, under accelerated conditions with a CO_2 concentration of 100%.

Table 1. Mix design proportions for different types of mixture investigated at UoM

Constituents(kg/m³)	Reference	50% R-UHPC R-UHPC-C	100% R-UHPC
Cement 52.5 R	700	700	700
Silica Fume	400	400	400
Superplasticizer (ACE 442)	64	64	64
Water	231	231	231
Natural aggregate 117/F	286	143	0
Natural aggregate 103	409	205	0
Natural aggregate 113	122	61	0
Recycled aggregate	0	409	817
Steel fibres (l_f = 22 mm, d_f = 0.2 mm)	160	160	160
Crystalline admixture (Penetron Admix ®)	5,6	5,6	5,6

3.2   Mechanical Performance of Recycled UHPC (R-UHPC)

The mechanical properties (compressive and flexural strength) of UHPC and R-UHPC were determined at 7, 28, 56, 90 and 120 days. Results are shown in Fig. 2 and Fig. 3. To this purpose three prismatic specimens 40 × 40 × 160 mm were casted for each mix and tested in flexure. Then the six broken halves were tested in compression. From Fig. 2 it can be observed that specimens made with R-UHPC showed higher strength values compared to UHPC, even for larger percentage substitution of natural aggregates. This could be attributed to increased presence of un-reacted cement particles in the finer portion of the recycled aggregates. This also resulted into the evidence that replacement of natural aggregates with recycled ones did not substantially alter the flexural strength capacity and its development over time. The variation is acceptable and it is in accordance with the effects of dispersed fibre reinforcement. Moreover, the concrete with 50% of carbonated recycled aggregates (R-UHPC-C) did not show significant differences as compared to the non-carbonated ones.

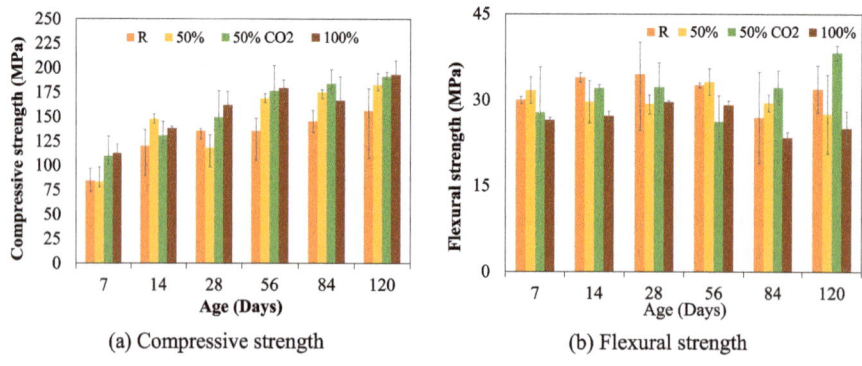

(a) Compressive strength   (b) Flexural strength

Fig. 2. Mechanical properties (compressive and flexural strength values) for UHPC and R-UHPC

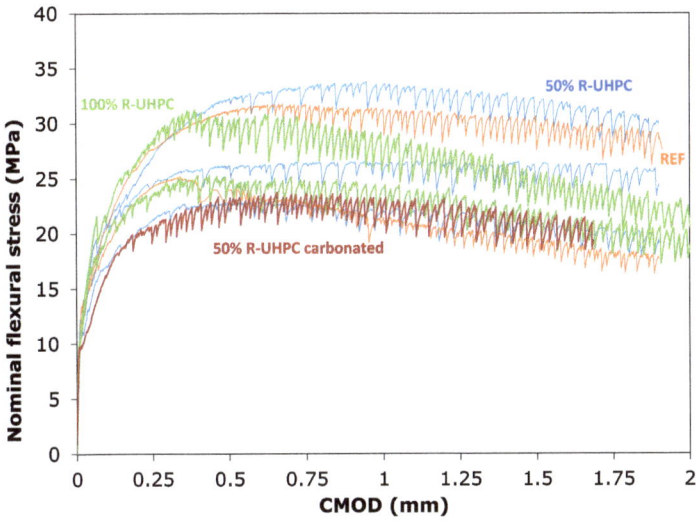

Fig. 3. Flexural stress curves versus CMOD for UHPC (reference) and R-UHPC (50%, 100% and carbonated)

3.3 Durability Properties of Recycled UHPC (R-UHPC)

Durability was evaluated by means of water absorption capillary tests, chloride penetration and chloride migration. Sorptivity tests were carried out at different ages on 40 × 40 × 160 mm prismatic specimens based on EN 13057. To this purpose, the specimens were first dried at 40 °C up to constant weight and then kept in laboratory conditions for 24 h. The lateral surface was waterproofed with silicon up to 20 mm depth from the intrados face and then were immersed in a water basin. The water uptake was measured by weighting the specimens at several intervals for 24 h. The water uptake was plotted versus the square root of time, the sorptivity coefficient [kg/(m^2 h$^{0.5}$)] was determined as the slope of the curve. At 28 days, all concretes showed similar sorptivity coefficients, only the mix with 50% of recycled carbonated aggregates (R-UHPC-C) showed a higher value. Sorptivity coefficients decreased with aging time as shown in Fig. 4 and were very similar for all the analysed mixes.

The Rapid Chloride Ion Penetration test (RCPT) was carried out for all mixes after 28, 56 and 120 days, based on ASTM C1202-97. As shown in Fig. 5a, it is observed a significant improvement with time in the performance of the recycled UHPC specimens as compared to reference ones since, ASTM C1202 considers "very low" chloride ion penetration for values of passing charge between 100 and 1000 coulombs. A continuous decreasing trend for lower charge is observed, obtaining the minimum value at 120 days for all mixes, especially the recycled ones (R-UHPC). Moreover, chloride migration tests were carried out based on NT Build 492. To this purpose, cylindrical

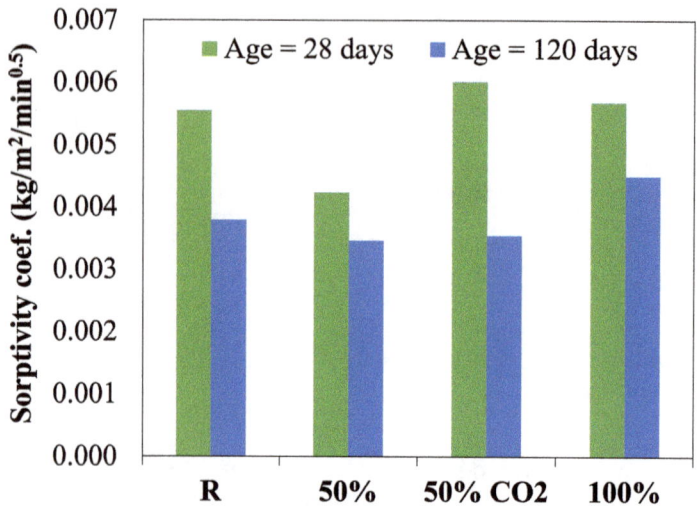

Fig. 4. Sorptivity coefficient for the different investigated mixes at 28 and 120 days

specimens (Ø100 mm, h = 50 mm) were analysed at 28, 56 and 120 days. For each age, the chloride penetration depth was determined and subsequently, the non-steady state chloride migration coefficient (D_{nssm}) was determined (Fig. 5b). Figure 5b shows that D_{nssm} decreases with the age for all concrete mixes. In conclusion, the results from the chloride penetration tests support the potential of the exploitation of recycled UHPC for the production of UHPC.

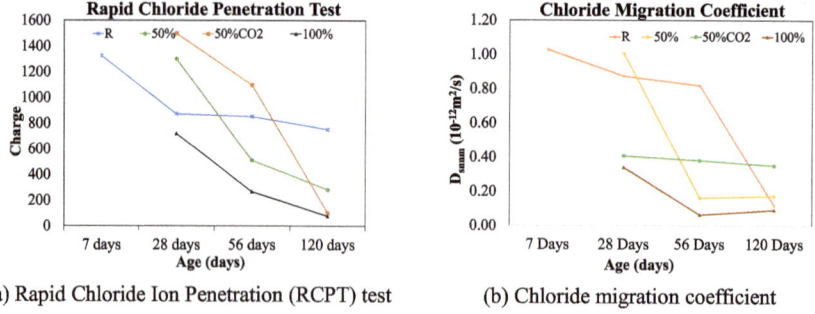

(a) Rapid Chloride Ion Penetration (RCPT) test   (b) Chloride migration coefficient

Fig. 5. Chloride penetration in R-UHPC

4   UPV Experimental Campaign

UPV collaborated with the recycling plant "Gestión y Reciclaje Belcaire" in Moncofa (Castelló) to test the process industrially, which allowed to recycle around 2 tons of UHPC. The process allowed to obtain recycled aggregates of reduced size (<6 mm) and separated most of the steel fibres contained in the original UHPC mixes. At the UPV laboratory, the aggregates were further sieved and classified to obtain fractions with controlled size distribution for obtaining a R-UHPC with a size distribution as similar as possible to the original mixes. Additionally, fibres where further separated using a magnet.

4.1   Mix Design

UPV studied 8 mixes replacing 50% and 100% of two fractions of the aggregates and steel fibres. The replacement rates selected were 50 and 100% to evaluate the effects of high volume replacements to produce UHPC. Recycled fibres had a certain fraction of recycled aggregates strongly attached to them and thus, a correction was made to consider this content in the mixes including recycled fibres. The composition of all the mixes is shown in Table 2.

Table 2. Mix design proportions for different recycled UHPC mixes.

kg / m³	Ref	Fine sand		Medium sand		Fibres		All	
Constituent		50%	100%	50%	100%	50%	100%	50%	100%
Cement 42.5 R-SR	800	800	800	800	800	800	800	800	800
Silica Fume	175	175	175	175	175	175	175	175	175
Water	160	160	160	160	160	160	160	160	160
Silica sand - medium	565	565	565	282.5	-	565	565	282.5	-
Recycled sand - medium	-	-	-	282.5	565	-	-	282.5	565
Silica sand - fine	302	151	-	302	302	302	302	151	-
Recycled sand - fine	-	151	302	-	-	-	-	151	302
Silica sand - flour	225	225	225	225	225	225	225	225	225
Steel microfibres	160	160	160	160	160	80	-	80	-
Recycled steel microfibres	-	-	-	-	-	80	160	80	160
Plasticizer Sika 20 HE	30	30	30	30	30	30	30	30	30

Workability, compressive strength, flexural and tensile strengths were evaluated for these 8 UHDCs to analyse the feasibility of using these recycled materials to produce new UHDC. After casting and demoulding the specimens, they were stored in a standard humidity chamber at 20 °C and 95% RH until testing time.

4.2 Fresh and Hardened State Mechanical Performance of Recycled UHPC (R-UHPC)

Slump flow test was used to evaluate the workability of the mixes. The diameter was measured just after performing the test, and 10 min later. The results in Fig. 6 left show that the mix with the highest workability is the reference mix, and that 50% of replacement of one component is feasible without significant loss of workability. When the replacement ratio increases to 100%, workability is noticeably decreased, especially in the mix "100% all", where three recycled constituents were replaced at the same time.

Compressive strength was tested on 100 mm cubes at the ages of 7 and 28 days. The results (Fig. 6 right) show similar results of compressive strength when replacing the commercial sand and fibres by the recycled materials, with slightly higher variation of results.

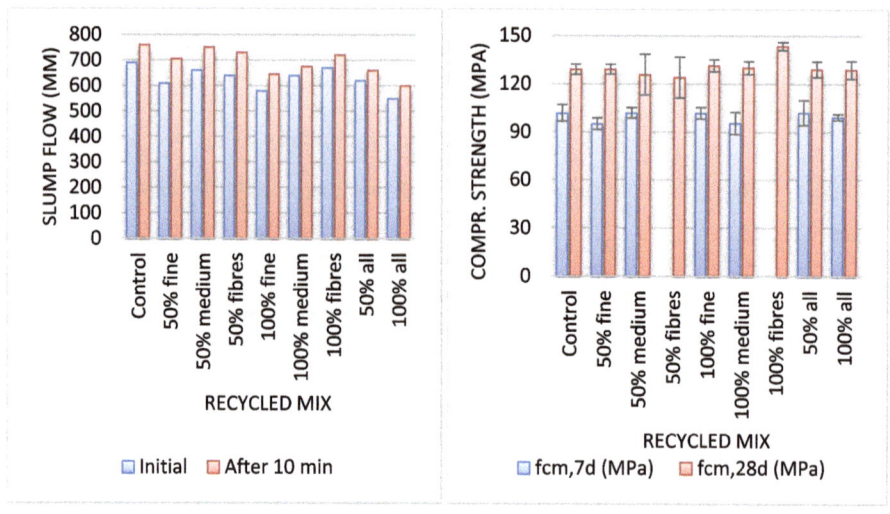

Fig. 6. Slump, initial and after 10 min (left) and compressive strength at 7 and 28 days (right).

In addition to compressive strength, four points flexural tests (4PBT) were performed at the age of 28 days to 100 × 100 × 500 mm^3 prisms to evaluate the flexural behaviour and the tensile strength parameters of all UHPCs mixes employing a simplified inverse analysis (IA). After the flexural test, these curves have been analysed to obtain the tensile behaviour of UHPC following a simplified inverse analysis (IA) method [12]. The constitutive model is defined as a function of five parameters: elastic modulus (E), cracking strength (ft), ultimate cracking strength ($f_{tu} = \gamma \cdot f_t$) and its associated strain (ε_{tu}); crack opening at the intersection of the line that defines the initial slope to the w axis (ω_0). Table 3 shows the tensile parameters obtained for all the mixes.

Table 3. Tensile mechanical properties of the mixes.

Mix	f_t (MPa)		f_{tu} (MPa)		ε_{tu} (‰)		E (MPa)		ω_o (mm)	
	Avg	Std.dev	Avg	Std.dev	Avg	Std.dev	Avg	Std.dev	Avg	Std.dev
Control	9.94	0.16	8.26	0.45	4.71	0.78	50217	1331	2.68	0.42
50% fine	8.96	0.72	7.95	0.69	4.84	1.21	50250	636	3.44	0.28
50% medium	9.50	0.33	7.20	0.51	3.31	0.97	50650	1626	3.56	0.74
50% fibres	9.77	0.17	7.73	0.78	3.47	2.17	52600	2030	2.67	0.79
100% fine	9.61	0.08	9.62	2.56	3.74	3.30	52750	4738	2.67	0.54
100% medium	8.62	0.31	8.48	0.74	5.49	0.62	47500	707	2.72	0.16
100% fibres	8.31	0.53	6.95	0.89	1.55	0.66	51450	5016	1.57	0.50
50% all	6.95	0.95	6.75	1.32	4.71	2.93	48000	2970	2.95	0.55
100% all	7.49	-	5.73	-	0.73	-	46100	-	0.97	-

5 Discussion

Figure 7 shows the results of the fresh properties measured by the slump flow test and compression strength at 28 days for the two experimental campaigns. The results of the slump flow test shows that the aggregates recycled obtained from UHPC are suitable to produce new UHPC since the mixes maintain the self-compacting properties. Similarly, the results show comparable results of compressive strength, reaching typical values of

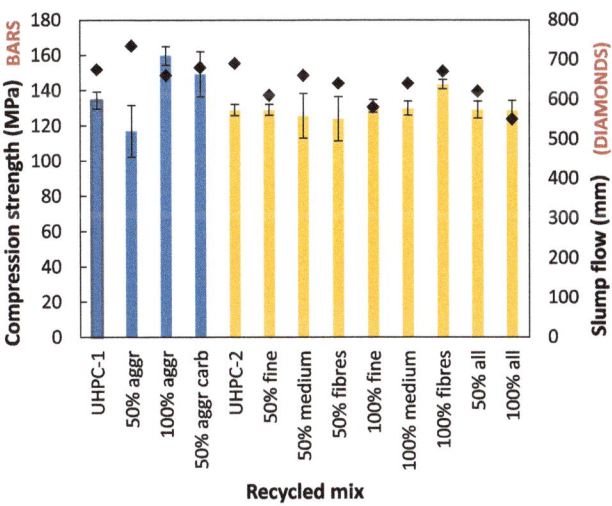

Fig. 7. Compressive strength at 28 days (bars) and slump flow (diamonds) for UHPC and R-UHPC. The outlined bar indicates the reference UHPC in each experimental campaign.

UHPCs, when replacing the commercial sand and fibres by the recycled materials, even when replacing completely some components.

6   Conclusions

The most important finding of these results obtained from the fresh and hardened states is that UHPC can be recycled to be used as aggregate in new UHPC production. This is an important step towards the re-use of waste materials originating from ultra-high durability concrete applications. More specifically, the conclusions of this study can be summarized as follows:

- The replacement of natural aggregates by aggregates recycled from UHPC did not reduce the significantly the self-compacting properties of UHPC. Slump flow slightly reduced when using increasing the replacement of recycled compounds. 50% had no effects on workability.
- Most of the UHPC mixes with recycled aggregates and fibres had a compressive strength similar to their reference UHPC mix, whereas some specimens reached higher compressive strength values probably due to the positive effect that the delayed hydraulic activity of the cement/binder paste layer surrounding the original natural sand particles.
- Flexural strength was not affected by the replacement of natural aggregates by recycled aggregates. Fibre replacement at a 100% replacement, and the complete replacement of aggregates and fibres, however, reduced the flexural strength, especially the post-cracking behaviour.
- The replacement of natural aggregates by recycled aggregates did not affect significantly to the durability properties.

Acknowledgements. The research activity reported in this paper has been performed in the framework of the ReSHEALience project (Rethinking coastal defence and Green-energy Service infrastructures through enHancEd-durAbiLity high-performance cement-based materials) which has received funding from the European Union's Horizon 2020 research and innovation program under grant agreement No 760824. The information and views set out in this publication do not necessarily reflect the official opinion of the European Commission.

References

1. Akhtar, A., Sarmah, A.K.: Construction and demolition waste generation and properties of recycled aggregate concrete: a global perspective. J. Clean. Prod. **186**, 262–281 (2018)
2. López Ruiz, L.A., Roca Ramón, X., Gassó Domingo, S.: The circular economy in the construction and demolition waste sector – a review and an integrative model approach. J. Clean. Prod. **248**, 119238 (2020)
3. Zhang, L., Liu, B., Du, J., Liu, C., Wang, S.: CO2 emission linkage analysis in global construction sectors: alarming trends from . J. Clean. Prod. **221**, 863–877 (2019)
4. Villoria Sáez, P., Osmani, M.: A diagnosis of construction and demolition waste generation and recovery practice in the European Union. J. Clean. Prod. **241**, 118400 (2019)

5. Silva, R.V., de Brito, J., Dhir, R.K.: Use of recycled aggregates arising from construction and demolition waste in new construction applications. J. Clean. Prod. **236**, 117629 (2019)
6. Chen, W., et al.: Adopting recycled aggregates as sustainable construction materials: a review of the scientific literature. Constr. Build. Mater. **218**, 483–496 (2019)
7. Bai, G., Zhu, C., Liu, C., Liu, B.: An evaluation of the recycled aggregate characteristics and the recycled aggregate concrete mechanical properties. Constr. Build. Mater. **240** (2020). Paper 117978
8. Behera, M., Bhattacharyya, S.K., Minocha, A.K., Deoliya, R., Maiti, S.: Recycled aggregate from C&D waste & its use in concrete - a breakthrough towards sustainability in construction sector: a review. Constr. Build. Mater. **68**, 501–516 (2014)
9. Ferrara, L., et al.: An overview on H2020 project reshealience. In: Proceedings of IABSE Symposium, Guimaraes 2019: Towards a Resilient Built Environment Risk and Asset Management, pp. 184–191 (2019)
10. Serna, P., et al.: Upgrading the concept of UHPFRC for high durability in the cracked state: the concept of Ultra High Durability Concrete (UHDC) in the approach of the H2020 project ReSHEALience. In: Proceedings of Sustainable Materials Systems and Structures SMSS 2019, pp. 764–771 (2019)
11. Cuenca, E., Mezzena, A., Ferrara, L.: Synergy between crystalline admixtures and nano-constituents in enhancing autogenous healing capacity of cementitious composites under cracking and healing cycles in aggressive waters. Constr. Build. Mater. **266**, 121447 (2017)
12. López Martínez, J.A.: Characterization of the tensile behaviour of UHPC by means of four-point bending tests. Ph.D. thesis. Universitat Politècnica de València (2017)

An Experimental Evaluation of Direct Tensile Strength for Ultra-high Performance Concrete

An Hoang Le[(✉)]

NTT Hi-Tech Institute, Nguyen Tat Thanh University, Ho Chi Minh City,
Vietnam
lhan@ntt.edu.vn

Abstract. Ultra-high performance concrete (UHPC) has emerged as an advanced material due to its superior mechanical properties and excellent durability. In addition to very high compressive strength exceeding 150 MPa, the tensile strength of UHPC is much larger than that of normal concrete and considered as an important parameter in structural design. To determine the tensile strength, test methods usually include the splitting test, double punch test, flexural test, and direct tension test. Therefore, it is needed to evaluate these methods to get accurate values of tensile strength of UHPC and UHPFRC. The direct tension tests were adopted by many researchers and also considered as a reliable method. In this study, direct tension tests were conducted to evaluate the tensile strength of UHPC. The UHPC was designed to obtain a nominal compressive strength varying between 150 and 200 MPa at the age of 28 days. In addition to plain UHPC, micro steel fibers were added into UHPC mixture with volumes of 1% and 2%. There were two types of prisms for direct tensile tests: (1) Prisms of $40 \times 40 \times 160$ mm without notches; (2) Notched prisms of $40 \times 40 \times 80$ mm. The difference between the direct tension tests on prisms with and without notches was clarified and discussed. Finally, the tensile strengths of UHPC and UHPFRC obtained from this study were compared with those suggested by several guidelines.

Keywords: Ultra-high performance concrete · Direct tensile test · Steel fibers · UHPC · Concrete

1 Introduction

Ultra-high performance concrete (UHPC) is characterized by a very high compressive strength exceeding 150 MPa, possibly attaining 250 MPa and an outstanding durability [1, 2]. The addition of fibers ensures the ductile behavior of UHPC under tension and significantly increases the tensile, flexural and shear strength under different types of structural actions (AFGC-SETRA, 2002) [3]. The superior properties of UHPC in tension and compression can generate many structural advantages, thereby leading to a great attention to many researchers throughout the world. Along with the compressive strength, the tensile strength is one of the most important aspect in the design of structures made of UHPC, especially in the case of UHPC reinforced by fibers (UHPFRC). Generally, there are two common methods to determine the tensile

© RILEM 2022
P. Serna et al. (Eds.): BEFIB 2021, RILEM Bookseries 36, pp. 958–964, 2022.
https://doi.org/10.1007/978-3-030-83719-8_82

strength of concrete: (1) the indirect tension tests including splitting test and flexural test; (2) the direct tension test. The direct tension tests can directly provide material tensile behavior over all stages (elastic, strain hardening, and softening) without backward calculation of the material tensile response, as compared to splitting or flexural tests. However, the direct tension test suffers several difficulties with its performance such as the slippage or the alignment between gripping apparatus and concrete specimens, the stress concentration at the gripping devices, and the effect of specimen size, specimen shape and boundary conditions [4]. Therefore, many researchers have attempted to conduct studies on the direct tension test of UHPC and UHPFRC to get insight into the tensile behavior. Several studies have investigated the difference between the direct tension test on notched and unnotched prisms [6–8]. Moreover, there are currently no testing standards that define the testing procedure, analytical procedure, specimen geometry, test conditions to fully obtain the tensile behavior of UHPC, especially for UHPFRC. Based on this fact, this paper is aimed at presenting an experimental evaluation on the direct tensile strength of UHPC and UHPFRC having compressive strength up to 200 MPa. The adopted method of direct tension test was developed at the University of Kassel – Germany for unnotched prisms of $40 \times 40 \times 160$ mm and notched prisms of $40 \times 40 \times 80$ mm.

2 Experimental Program

UHPC without steel fibers (UHPC) and UHPC with steel fibers of 1% by volume (UHPFRC-SF1%) and 2% by volume (UHPFRC-SF2%) were used in this study, following the recipe of M3Q which was developed at University of Kassel – Germany [1]. The composition of UHPC and UHPFRC in this study can be found in Table 1. The fine-grained UHPC and UHPFRC mix have a water-binder ratio of 0.21 and a maximum grain size of 0.5 mm. The packing density of the M3Q mixture was optimized using the silicafume (Sika ® Silicoll uncompacted) and the superplasticizer was changed to Sika ® ViscoCrete 2810. It should be noted that the M3Q mix was designed to provide a very high self-compacting characteristic and a compressive strength of concrete cylinder at about 200 MPa. Therefore, the necessity for compacting of concrete using external vibration was eliminated. A mixer (Zyklos Gleichlaufmischer ZZ 150 HE) with maximum capacity of 170 L was used for each concrete batch, as seen in Fig. 1. First, all of the dry components (cement, silicafume, quartz sand, ground quartz) were premixed to obtain a good dispersion. Then, water including the superplasticizer was poured into the dry mixture and continued mixing. Once the mixture had enough fluidity and viscosity, the steel fibers were dispersed and mixed together with the mixture. Figure 1 shows the steel fibers and mixer used for this study.

Table 1. Composition of UHPC and UHPFRC mixes

Mix composition	Unit	UHPC	UHPFRC-SF1%	UHPFRC-SF2%
Water	kg/m³	187.98	186.10	184.22
CEM I 52.5R HS-NA	kg/m³	795.40	787.45	779.49
Silica fume	kg/m³	168.60	166.91	165.23
Superplasticizer Sika Viscorete 2810	kg/m³	24.10	23.86	23.62
Ground Quartz W12	kg/m³	198.40	196.42	194.43
Quartz sand 0.125/0.5	kg/m³	971.00	961.29	951.58
Steel fibers	kg/m³	–	78.52	157.04

The steel fibers with a diameter d_f of 0.175 mm and a length l_f of 13 mm were added to the UHPC mix in volume fraction of 1% and 2%, as shown in Fig. 1. When UHPC and UHPFRC mixtures were ready, for each batch of concrete, they were poured into six cylindrical moulds having a diameter of 100 mm and a height of 200 mm and twelve three-gang prism moulds having dimensions of 40 × 40 × 160 mm without any vibrations. The specimens were demolded 48 h after casting and cured at ambient temperature until testing. A total of nine mixes of UHPC and UHPFRC was cast for this study. The compressive strengths f_c and elastic modulus E_c were determined from the compression tests on 3 cylindrical specimens of 100 mm × 200 mm for each batch of concrete in accordance with DIN EN 12390-3:2009-07 and DIN 1048-5, respectively.

a)   Steel fibers   b)   Mixer   c)   Fresh UHPC mixture

Fig. 1. Steel fibers and mixer

The tensile strengths of UHPC and UHPFRC for each batch of concrete were determined through the direct tension tests on six unnotched prisms of 40 × 40 × 160 mm and six notched prisms of 40 × 40 × 80 mm, as described in Fig. 2. A displacement controlled testing machine (Zwick/Roell Z150) with a maximum capacity of 150 kN was used for the direct tension tests on unnotched prisms, as shown in Fig. 2(a). The direct tension tests on notched prisms were performed using tension testing machine (RBO 2000) having a load capacity of 2.0 MN as seen in Fig. 2(b). All specimens were loaded with a speed rate of 0.01 mm/s. When the crack opening exceeds 2 mm, the load rate was increased to 0.05 mm/s to speed up the test.

a) Unnotched prisms of 40x40x160 mm   b) Notched prisms of 40x40x80 mm

Fig. 2. Direct tension test setup

3   Test Results and Discussions

The failure mode of tested prisms was illustrated in Fig. 3. The position of the failure by a dominant crack mainly depends on the homogeneity of the concrete matrix and the distribution of steel fibers. The dominant crack occurred near the middle for the unnotched prisms of UHPFRC, but it positioned far from the middle for the unnotched prisms of UHPC as shown in Fig. 3. In terms of notched prisms, the notches of 5×5 mm in the middle of prisms prevent the failure of the specimens to occur at undesirable locations along the length of the specimens. Therefore the fracture of the notched prisms was induced by cracking in the middle of the specimens as shown in Fig. 3(b).

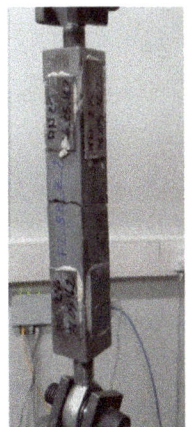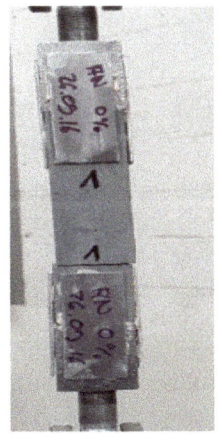

UHPFRC-SF2%   UHPFRC-SF1%   UHPC   b)   Failure of
      a) Failure of unnotched prisms of 40x40x160 mm   notched prisms of
                                                                          40x40x80 mm

Fig. 3. Failure modes of test specimens

Table 2. Test results of compressive strengths and direct tensile strengths

Series	Mix	Steel fiber volume V_f (%)	Average compressive strength f_c (MPa)	Average elastic modulus E_c (MPa)	Average direct tensile strength of unnotched prism $f_{t,un}$ (MPa)	Average direct tensile strength of notched prism $f_{t,n}$ (MPa)
1	Batch 1	0	178.9	48370	7.1	3.7
	Batch 2	1	195.5	49645	7.3	9.1
	Batch 3	2	188.2	48421	8.5	11.3
2	Batch 4	0	198.0	46937	7.7	–
	Batch 5	1	195.5	47881	8.2	8.8
	Batch 6	2	187.8	48580	9.4	11.1
3	Batch 7	0	190.4	46186	8.2	–
	Batch 8	1	195.6	48689	8.4	8.7
	Batch 9	2	192.4	48557	8.7	9.2

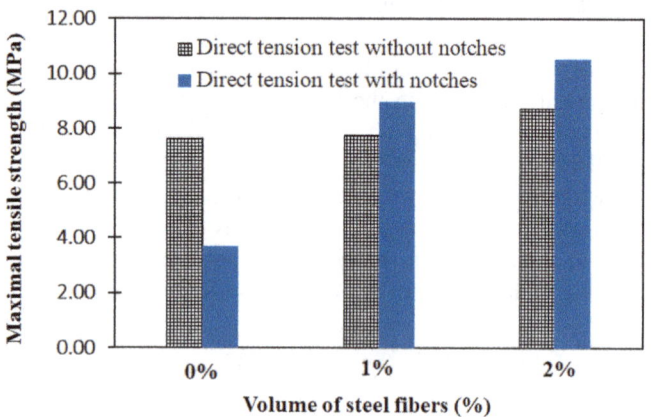

Fig. 4. Comparison between two test methods

Table 2 shows the average tensile strength of unnotched and notched prisms for each concrete batch. It is obviously revealed from Table 1 that for each series, UHPFRC prisms experienced higher values of tensile strength as compared to UHPC prisms. Higher volume of steel fiber resulted in a higher tensile strength for two types of direct tension method. It is observed that for unnotched prisms, the average tensile strengths of UHPFRC-SF2% and UHPFRC-SF1% were increased by 14.29% and 1.46% as compared to UHPC. However, for notched prisms, the average tensile strength of UHPC was found to be 43.82% and 86.33% less than the average tensile strength of UHPFRC-SF1% and UHPFRC-SF2%, respectively.

Figure 4 shows the comparison between direct tension test on notched prisms and unnotched prisms. It can be seen that with the inclusion of steel fibers or increasing fiber volume from 1% to 2%, the level of increase in the tensile strength for notched prisms is much larger than that for unnotched prisms. Based on many previous studies, the measured tensile strengths of notched prisms are found to be more accurate than that of unnotched prisms. Observing the failure mode of unnotched and notched prisms as seen in Fig. 3, it is clearly seen that most of unnotched specimens exhibited a secondary flexure due to the load eccentricity, while proper alignment during the test was ensured for all notched prisms. This phenomenon indicates that the suitable results of direct tension test can be obtained by notched prisms.

Some international guidelines have recommended the values of tensile strength of UHPC and UHPFRC. For example, tensile strengths of UHPC and UHPFRC were found to be 6–10 MPa in FWHA (2013) [6], while JSCE (2004) [5] defined that UHPC and UHPFRC has tensile strength higher than 5 MPa and 9 MPa, respectively. SETRA/AFGC (2002) [3] supposed that tensile strength of UHPC is higher than 6 MPa and UHPFRC has tensile strength varying between 7–15 MPa. Results of this study indicates that the tensile strength of UHPC is lower than that of these recommendations, however, the tensile strength of UHPFRC accords with that of these recommendations.

4   Conclusions

Some aspects of UHPC and UHPFRC can be drawn based on the results of this study, as follows:

- The notched prisms gives a more accurate result of tensile strength as compare to unnotched prisms;
- Higher fiber volume results in higher direct tension strength. For notched prisms in this study, the tensile strength were increased by 43.82% and 86.33% when increasing steel fibers volumns from 0 to 1% and from 0 to 2%.
- The tensile strength of UHPC is lower than that of recommendations in guidelines, however, the tensile strengths of UHPFRC are in line with those suggested by guidelines.
- The effect of various steel fibers and the specimen size on the tensile strength should be further investigated using the direct tension test on notched and unnotched prisms.

Acknowledgements. This research is funded by Vietnam National Foundation for Science and Technology Development (NAFOSTED) under grant number 107.01-2019.325.

References

1. An, L.H., Fehling, E.: Influence of steel fiber content and type on the uniaxial tensile and compressive behavior of UHPC. Constr. Build. Mater. **153**, 790–806 (2017)
2. Le, A.H.: Evaluation of the splitting tensile strength of ultra-high performance concrete. In: Serna, P., Llano-Torre, A., Martí-Vargas, J.R., Navarro-Gregori, J. (eds.) BEFIB 2020. RB, vol. 30, pp. 1149–1160. Springer, Cham (2021). https://doi.org/10.1007/978-3-030-58482-5_101

3. AFGC-SETRA. Ultra high performance fibre-reinforced concretes, Interim recommendations. SETRA, Bagneux, France (2002)
4. Wille, K., El-Tawil, S., Naaman, A.E.: Properties of strain hardening ultra high performance fiber reinforced concrete (UHP-FRC) under direct tensile loading. Cement Concr. Compos. **48**, 53–66 (2014)
5. JSCE. Recommendations for design and construction of ultra-high strength fiber reinforced concrete structures (Draft), Tokyo, Japan, Japan Society of Civil Engineers (2004)
6. Federal Highway Administration (2013). Ultra-High Performance Concrete: A State-of-the-Art Report for the Bridge, Community US Department of Transportation Pub. No. FHWA-HRT-13-060
7. Yuliarti, K., Fehling, E., Mohammed, I., Attitou, A.M.A.: Tensile strength behavior of UHPC and UHPFRC. Procedia Eng. **125**(2015), 1081–1086 (2015)
8. Yuliarti, K., Fehling, E., Hardjasaputra, H., Al-Ani, Y., Attitou, A.M.A.: Axial tensile strength of UHPC and UHPFRC. IOP Conf. Ser. Mater. Sci. Eng. **615**, 012020 (2015)

Finite Element Modelling of UHPFRC Tensile Bars

Eduardo J. Mezquida-Alcaraz[1(✉)], Juan Navarro-Gregori[1],
Majid Khorami[2], and Pedro Serna[1]

[1] Instituto de Ciencia y Tecnología del Hormigón (ICITECH), Universitat
Politècnica de València, Camino de Vera s/n, 46022 València, Spain
edmezal@alumni.upv.es
[2] Facultad de Arquitectura y Urbanismo, Universidad UTE, Calle Rumipamba
S/N y Bourgeois, Quito, Ecuador

Abstract. In the present paper, a numerical modelling study on the uniaxial tensile behaviour of reinforced UHPFRC ties by means of a non-linear finite element model (NLFEM) is carried out. The results obtained from the simulation done by the NLFEM are compared to the results from an experimental programme adopted. These tensile bars (ties) modelled in this work are cast using UHPFRC with 160 kg/m^3 of steel fibres and tested in a direct tensile test. The NLFEM developed for UHPFRC reinforced flexural beams used in previous research is improved and applied in order to validate the mechanical tensile characterisation of UHPFRC when direct tensile reinforced elements are considered. As it happens with the flexural elements, in this case the shrinkage and tension stiffening effects are essential in the model to simulate the reality of the tensile test. After the NLFEM simulation, very accurate results are obtained that lead to consider the reliability of the NLFEM model developed not only for flexural reinforced elements, but also for direct tensile ones.

Keywords: UHPFRC · Direct tensile test · Finite element modelling · Discrete cracking approach · Shrinkage · 3D effects

1 Introduction and Objectives

In order to make ultra-high performance fibre-reinforced concrete (UHPFRC) an efficient and economically competitive material for structural designs and applications, it is necessary to define a complete modelling strategy. This is based on a clear and robust definition of material models and their interaction, a suitable numerical technique that allows complex analyses to be performed and an easy-to-run UHPFRC material characterisation procedure for both strain hardening (SH-UHPFRC) and strain softening (SS-UHPFRC) to accurately establish constitutive material parameters for modelling.

In [1] a numerical 2-dimensional non-linear finite element model (2D-NLFEM) was developed and used to establish a simple reliable direct procedure to characterise tensile UHPFRC behaviour in both cases: SH-UHPFRC and SS-UHPFRC behaviours. Unreinforced four-point bending test (4PBT) to experimentally characterise tensile

© RILEM 2022
P. Serna et al. (Eds.): BEFIB 2021, RILEM Bookseries 36, pp. 965–975, 2022.
https://doi.org/10.1007/978-3-030-83719-8_83

UHPFRC behaviour was modelled and an inverse analysis method to obtain constitutive SH-UHPFRC behaviour developed in [2, 3] was validated in [4] and adapted and improved to be used for SS-UHPFRC in [1, 5].

After establishing and controlling tensile UHPFRC behaviour, a step forward in the structural analysis was taken by modelling the reinforced UHPFRC flexural elements in [6]. The NLFEM was improved by incorporating important effects, such as UHPFRC shrinkage, and taking into account the tension-stiffening behaviour of UHPFRC. Presently, the NLFEM also incorporates reinforcement bars and the bond-slip behaviour between UHPFRC and reinforcement. The results obtained in [6] demonstrated both the reliability of the developed NLFEM and the coherence of the tensile UHPFRC material characterisation process.

The aim of this paper is to model the influence of the reinforcement bars in UHPFRC elements working in direct tension, considering mechanisms like the shrinkage and tension stiffening. Therefore, the results of an experimental programme carried out in a work done by the research group composed of UHPFRC reinforced ties tested in direct tensile test and the specimens to characterise the material properties are adopted to compare to the results obtained from the NLFEM developed. Consequently, the results from the model's application are presented and discussed.

2 Experimental Programme

An experimental programme of UHPFRC reinforced tensile bars was developed in [7–9]. In order to progress with the evolution of the NLFEM and the validation of the characterisation of the tensile behaviour of UHPFRC, the experimental results of the tested tensile bars and the unreinforced 4PBT to characterise the tensile behaviour of UHPFRC cast with 160 kg/m^3 of steel fibres have been adopted. They are compared to the simulated response of the NLFEM developed for the tensile bars. All reinforced concrete tensile bars and 4PBT specimens are tested at 30 days after casting.

2.1 Reinforced Tensile Bars

In Fig. 1 and Fig. 2 the geometry and setup of the direct tensile test carried out in [7–9] are depicted. To develop the NLFEM the same geometry, load and boundary conditions carried out experimentally are considered. The tensile UHPFRC bar is 1 m length and it has a squared section of 60 × 60, 80 × 80 and 100 × 100 mm. It has a reinforcement disposition that can be described in three parts: the beginning and the last 250 mm composed by three reinforcement bars vertically aligned and the central 500 mm where the central reinforcement bar is continued. The diameters of the reinforcement bars considered are ϕ10, ϕ12 and ϕ16 mm.

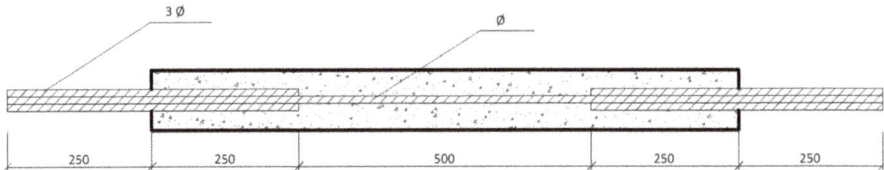

3 Ø   Ø

250   250   500   250   250

Fig. 1. Geometry of the tensile bar specimen [7]

The tensile bar is set in a frame that takes it from the reinforcement in both sides. One side is fixed, that constitutes the reactive side. The other side, considered the active part, is connected to a hydraulic jack that will apply the tensile force to the reinforcement (see Fig. 2). The tensile stress in the rebar is transmitted to the UHPFRC by bond mechanism.

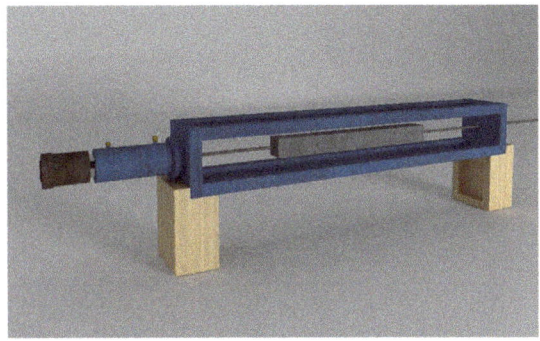

Fig. 2. Tensile bar test setup [7]

In Table 1, the geometric characteristics and the number of tensile test specimens adopted from the experimental programme of [8, 9] cast with UHPFRC with 160 kg/m^3 of steel fibres to be modelled using the NLFEM developed in this paper are summarised.

Table 1. Tensile test specimens from experimental programme from [8, 9]

UHPFRC section (mm)	Rebars φ (mm)	Num. of specimens
60 × 60	10	5
	12	3
80 × 80	10	3
	12	3
	16	3
100 × 100	10	1
	12	3
	16	3

2.2   Material Characterisation

In order to characterise the mechanical behaviour of the UHPFRC used in the tensile bar's batches, a set of 100 mm cubic specimens for the compression test and 100 × 100 × 500 mm unreinforced specimens for the 4PBT are cast together with tensile bars. Figure 3 depicts the stress (σ) - deflection at mid span (δ) experimental curves of the characteristic 50% (average) from the 4PBT of the unreinforced 100 × 100 × 500 mm specimens.

To use bending tests to obtain tensile behaviour, it is necessary to apply inverse analyses. In this case, to obtain the tensile constitutive UHPFRC parameters from 4PBT, a direct procedure to characterise the tensile constitutive behaviour of strain-softening (SS) and strain-hardening (SH) UHPFRC developed by the research group in [1, 5] is applied (see Fig. 4). Table 2 shows the characteristic tensile parameters and the compressive strength for the 50% percentile (average). As it can be observed, the UHPFRC used for casting the tensile bars exhibits SS.

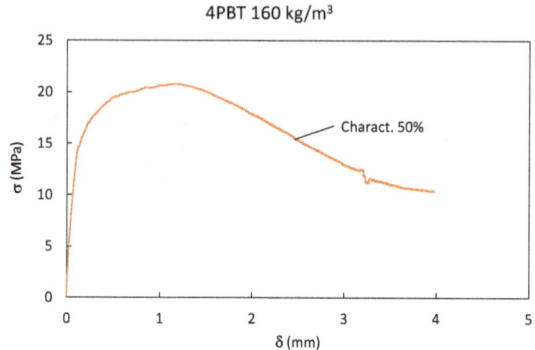

Fig. 3. 4PBT characteristic 5 and 50% σ-δ curves

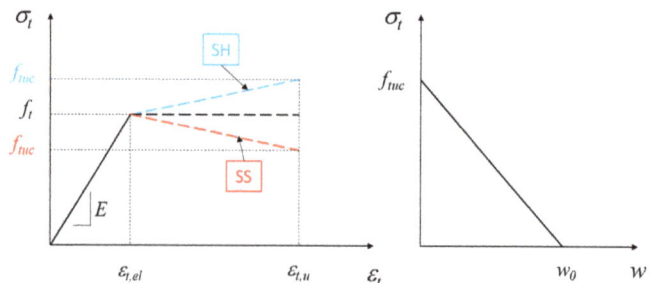

Fig. 4. Tensile constitutive behaviour of SS and SH-UHPFRC

Table 2. Mechanical characterisation for UHPFRC tensile bars

160 kg/m³ of steel fibres						
Charact. σ-δ	f_t (MPa)	f_{tuc} (MPa)	ε_{tu} (‰)	E (MPa)	w_o (mm)	f_c (MPa)
50%	9.62	8.44	3.31	50700	3.24	153.99

3 Numerical Validation

The NLFEM for the UHPFRC tensile bars defined in this work has been developed at two levels: in two dimensions (2D) and in three dimensions (3D). The constitutive model for UHPFRC is based on the discrete cracking model as the interface behaviour. For the 2D-NLFEM this behaviour is forced only at the central specimen section (see Fig. 5). The rest of the specimen is modelled by the smeared cracking approach. The 3D-NLFEM-multicrack consists on the definition of interface behaviour not only at the mid span section but also between all the 3D solid elements in the same vertical, as it is depicted in Fig. 6(left). The solid 3D element is defined by the UHPFRC smeared cracking approach and the boundary interface elements by the stress-crack opening law. In this sense, a *composed finite element* defined by a 3D solid element bounded vertically by two interface elements at each side able to represent the real UHPFRC σ-ε/w constitutive law defined in Fig. 4 is generated. Moreover, in the analysis definition of the 3D-NLFEM-multicrack, a random variation of the FRCFAC factor that multiplies the values of the UHPFRC tension stress in the constitutive behaviour of the NLFEM has been added to better model the heterogeneity of concrete (see Fig. 7). In this way, different behaviours can be expected from different runs of the same algorithm. A variability of 10–11% of the FRCFAC factor is considered (see Fig. 7) to simulate the variability of UHPFRC strength.

The effect of shrinkage in concrete is considered as a material function when the UHPFRC model is defined in NLFEM. Accordingly, the shrinkage function (ε_{cs}) from EN 1992-1-1 Eurocode 2 [10] is used. From it, the obtained values are incremented as a percentage (sh_{inc}) to be adapted to the UHPFRC response. That is, the total shrinkage strain of UHPFRC ($\varepsilon_{csUHPFRC}$) is defined as Eq. (1) for the 2D-NLFEM and as Eq. (2) for the 3D-NLFEM-multicrack. Reinforcement is modelled using Von Mises strain-hardening elasto-plastic behaviour for the steel with a bond-slip behaviour. The results of the UHPFRC characterisation obtained by the direct procedure to characterise the tensile constitutive behaviour from Table 2 are implemented into the constitutive behaviour of the described NLFEM.

$$\varepsilon_{csUHPFRC} = \varepsilon_{cs} \cdot (1 + sh_{inc}/100) \tag{1}$$

$$\varepsilon_{csUHPFRC3D} = \varepsilon_{cs} \cdot (1 + sh_{inc3D}/100) \tag{2}$$

To build the NLFEM the software Diana has been used [11]. For the continuum part of the NLFEM, 10 mm 2D quadratic plane stress elements have been used for the 2D model (see Fig. 5), and 20 mm 3D quadratic structural isoparametric solid brick elements for the 3D model (see Fig. 6(left)). Besides, for the discrete cracking interface to model macrocrack behaviour, quadratic 2D line interface elements are placed at the mid-span section for the 2D model, and quadratic plane 3D interface elements are placed between the solid elements for 3D model. Reinforcement is modelled discretely using truss bond-slip elements (see Fig. 6(right)).

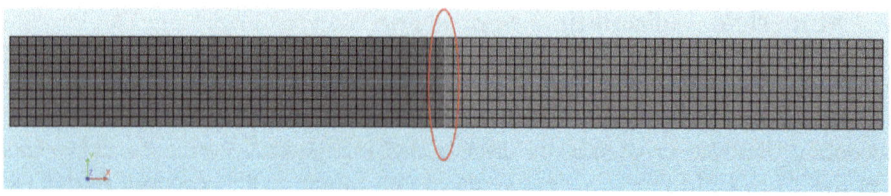

Fig. 5. Mesh for the discrete 2D-NLFEM of tensile bars

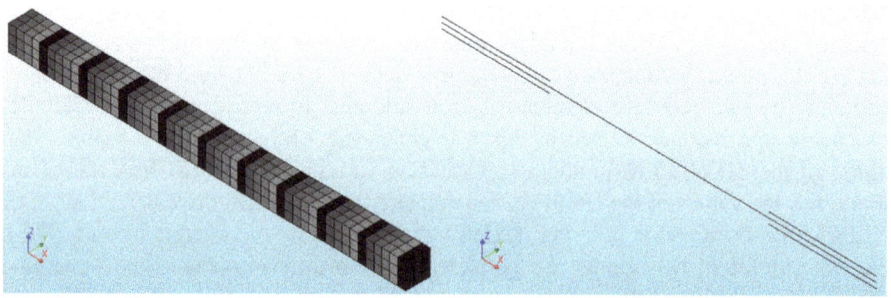

Fig. 6. 3D-NLFEM-multicrack: FEM mesh for UHPFRC (left) and for the reinforcement (right)

Fig. 7. Random distribution of tensile properties (FRCFAC)

4   Results and Discussion

The results from the tensile bar experimental programme are compared to the response of the NLFEM developed.

In Fig. 8, Fig. 9 and Fig. 10 the curves adjusted with the values of sh_{inc3D} necessary for the 3D-NLFEM-multicrack with the random distribution of the FRCFAC for tensile strength are shown for the experimental tensile bars adopted. As it can be observed at the figures, the 3D model adjusts accurately the average experimental response (Experim_50%) with a less value of shrinkage increment (sh_{inc3D}) if it is compared to the shrinkage increment obtained using the 2D model (sh_{inc}). Generally, it can be denoted that the stiffness in the microcrack stabilisation phase (the second slope of the curve) for the 3D model is reduced with respect to the response of the 2D model. Even though this fact generates less accuracy in some cases such as 80×80; $\phi 16$, the 3D model is considered reliable enough.

The difference in the shrinkage increment necessary to adjust de 2D-NLFEM and the 3D-NLFEM-multicrack can be produced due to a *3D effect* that is hidden in the 2D-NLFEM and captured in the 3D-NLFEM-multicrack. In Table 3 the shrinkage increment obtained with the 2D model (sh_{inc}) and from the 3D model (sh_{inc3D}) are compared for the tensile bars considered, and the 3D effect obtained as the difference between the shrinkage increment from the two models.

Using Eq. (1) and Eq. (2), it is possible to obtain the strain due to the shrinkage effect for the case of the 2D-NLFEM ($\varepsilon_{csUHPFRC}$) and for the 3D-NLFEM-multicrack ($\varepsilon_{csUHPFRC3D}$). The strain due to the 3D effects for the case of the 3D-NLFEM-multicrack ($\varepsilon_{3Deffects}$) is obtained as the difference between $\varepsilon_{csUHPFRC3D}$ and $\varepsilon_{csUHPFRC}$. These values are detailed in Table 3.

From Table 3 it can be deduced that, when using the 2D-NLFEM to model the tensile test, the calculated UHPFRC shrinkage strain ($\varepsilon_{csUHPFRC}$) falls within the range of [0.51, 0.96] mm/m. From the other hand, when the 3D-NLFEM-multicrack is used to model the tensile test the calculated UHPFRC 3D shrinkage strain ($\varepsilon_{csUHPFRC3D}$) falls within the range of [0.37, 0.73] mm/m. Therefore, the difference between the two models considered as the strain due to the 3D effects falls within the range of [0.10, 0.23] mm/m. Consequently the $\varepsilon_{csUHPFRC3D}$ range is more in line with [12–15] and the model is able to quantify the 3D effects that affect this particular kind of tensile test.

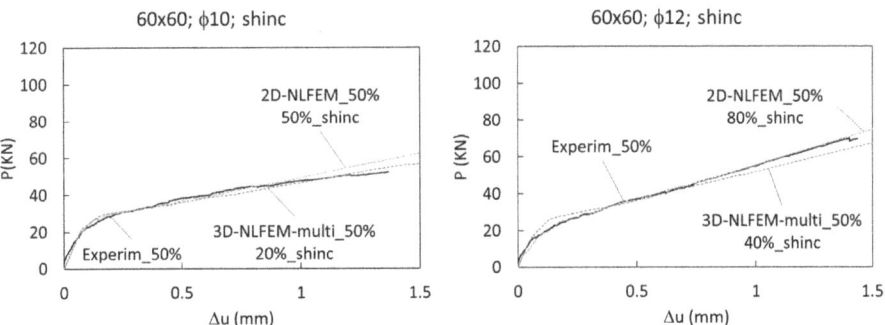

Fig. 8. P-Δu comparison for 60×60 tensile bar

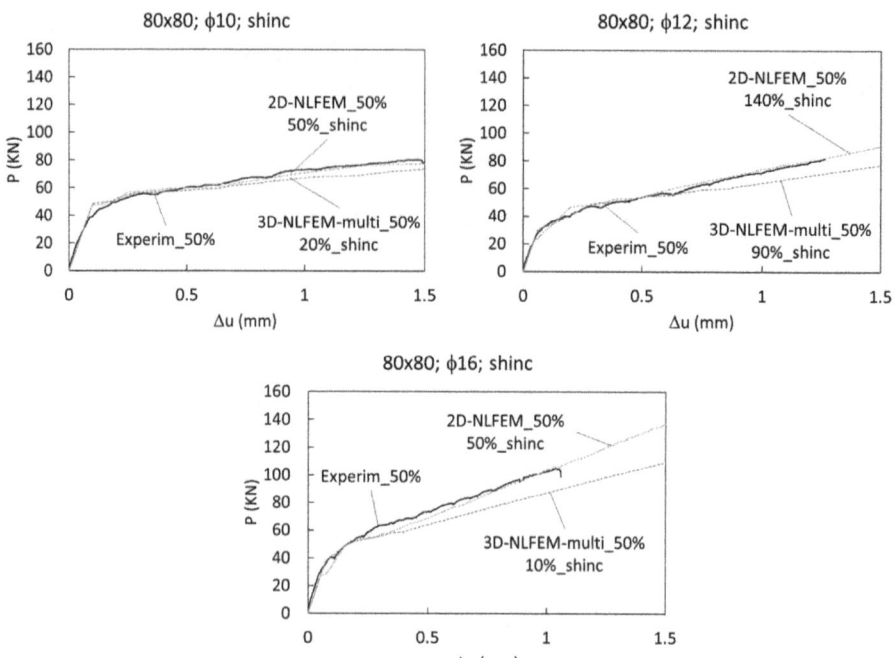

Fig. 9. P-Δu comparison for 80 × 80 tensile bar

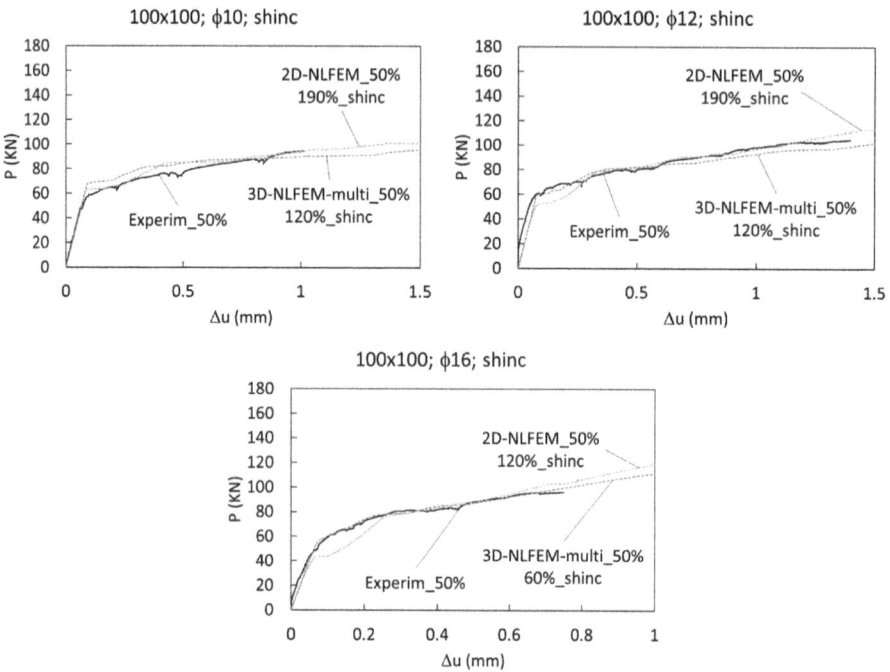

Fig. 10. P-Δu comparison for 100 × 100 tensile bar

Table 3. Shrinkage values for tensile bars from the NLFEM.

Tensile bar	2D-NLFEM			3D-NLFEM-multicrack		
id	ε_{cs} (mm/m)	sh_{inc} (%)	$\varepsilon_{csUHPFRC}$ (mm/m)	sh_{inc3D} (%)	$\varepsilon_{csUHPFRC3D}$ (mm/m)	$\varepsilon_{3Deffects}$ (mm/m)
60×60; $\phi 10$	0.34	50	0.51	20	0.41	0.10
60×60; $\phi 12$	0.34	80	0.61	40	0.48	0.13
80×80; $\phi 10$	0.34	50	0.51	20	0.41	0.10
80×80; $\phi 12$	0.34	140	0.82	90	0.65	0.17
80×80; $\phi 16$	0.34	50	0.51	10	0.37	0.14
100×100; $\phi 10$	0.33	190	0.96	120	0.73	0.23
100×100; $\phi 12$	0.33	190	0.96	120	0.73	0.23
100×100; $\phi 16$	0.33	120	0.73	60	0.53	0.20

5   Conclusions

After applying the NLFEM to the adopted tensile UHPFRC bars, the following con-clusions were drawn:

- The P-Δu response obtained from applying the 2D-NLFEM and the 3D-NLFEM-multicrack when the 50% percentile of the tensile UHPFRC constitutive parameters obtained from 4PBT and the softening correction were used in the model accurately fitted the experimental P-Δu response. This demonstrated not only the reliability of the developed NLFEM, but also the coherence of the process to characterise tensile UHPFRC behaviour in reinforced direct tensile tests.
- When using the 2D-NLFEM to model the tensile test, the calculated UHPFRC shrinkage strain ($\varepsilon_{csUHPFRC}$) fell within the range of [0.51, 0.96] mm/m, while the 3D-NLFEM-multicrack was used to model the tensile test, the calculated UHPFRC 3D shrinkage strain ($\varepsilon_{csUHPFRC3D}$) fell within the range of [0.37, 0.73] mm/m. Therefore, the difference between both models, taken as a strain due to 3D effects, lay within the range of [0.10, 0.23] mm/m. Consequently, the NLFEM developed was able to quantify the 3D effects that affected this particular kind of tensile test.
- It can be concluded that a NLFEM applicable at the 2D and 3D levels was developed to model tensile UHPFRC behaviour for steel-reinforced elements. This model considered shrinkage effects, tension stiffening behaviour and 3D effects due to the particularities of the test, which gave very accurate results compared to the experimental tests. The NLFEM was more accurate because it reproduced real UHPFRC constitutive behaviour with more similarity by means of the composed finite elements defined by a 3D solid element that was bound vertically by two interface elements on each side, and was able to represent the real UHPFRC $\sigma - \varepsilon/w$ constitutive law.
- The flexural and direct tensile tests were reliably modelled using the simple direct procedure to characterise UHPFRC tensile behaviour from 4PBT, which consisted in a 4P-IA and the softening correction for the SS-UHPFRC case. Consequently, this demonstrated not only the reliability of the developed NLFEM, but also the

coherence of the process to characterise UHPFRC tensile behaviour during reinforced flexural and direct tensile tests.

Acknowledgements. This work forms part of project "BIA2016-78460-C3-1-R" supported by the State Research Agency of Spain.

References

1. Mezquida-Alcaraz, E.J., Navarro-Gregori, J., Serna-Ros, P.: Direct procedure to characterize the tensile constitutive behavior of strain-softening and strain-hardening UHPFRC, Cem. Concr. Compos. **115**, 103854 (2021). https://doi.org/10.1016/j.cemconcomp.2020.103854
2. López, J.Á.: Characterisation of the tensile behaviour of UHPFRC by means of four-point bending tests. Ph.D. Thesis, Univ. Politècnica València, València (2017). https://doi.org/10.4995/Thesis/10251/79740
3. López, J.A., Serna, P., Navarro-Gregori, J.: Advances in the development of the first UHPFRC recommendations in Spain: material classification, design and characterization. In: Toutlemonde, F., Resplendino, J.T.Ch. (ed.) UHPFRC 2017 Des. Build. with UHPFRC New Large-Scale Implementations, Recent Tech. Adv. Exp. Stand., RILEM Publications SARL, 2017, pp. 565–574 (2017)
4. Mezquida-Alcaraz, E.J., Navarro-Gregori, J., López, J.A., Serna-Ros, P.: Validation of a non-linear hinge model for tensile behavior of UHPFRC using a Finite Element Model. Comput. Concr. **23**, 11–23 (2019). https://doi.org/10.12989/cac.2019.23.1.011
5. Mezquida-Alcaraz, E.J., Navarro-Gregori, J., Serna-Ros, P.: Numerical validation of a simplified inverse analysis method to characterize the tensile properties in strain-softening UHPFRC. IOP Conf. Ser. Mater. Sci. Eng., p. 12006 (2019)
6. Mezquida-Alcaraz, E.J., Navarro-Gregori, J., Serna, P.: Finite element modelling of UHPFRC flexural-reinforced elements. In: Serna, P., Llano-Torre, A., Martí-Vargas, J.R., Navarro-Gregori, J. (eds.) BEFIB 2020. RB, vol. 30, pp. 639–650. Springer, Cham (2021). https://doi.org/10.1007/978-3-030-58482-5_57
7. Navarro Laguarda, M.Á. : Estudio del comportamiento de tirantes armados de Hormigón de Muy Alta Resistencia (HMAR). Aplicación al diseño de una pasarela sobre el barranco de las ovejas de Alicante. (2018). https://riunet.upv.es:443/handle/10251/114590. Accessed 30 May 2020
8. Khorami, M., Navarro-Gregori, J., Serna, P., Navarro-Laguarda, M.A.: A testing method for studying the serviceability behavior of reinforced UHPFRC tensile ties. IOP Conf. Ser. Mater. Sci. Eng., p. 12022. IOP Publishing (2019)
9. Khorami, M., Navarro-Gregori, J., Serna, P.: Experimental methodology on the serviceability behaviour of reinforced ultra-high performance fibre reinforced concrete tensile éléments. Strain **13** e12361 (2020)
10. The European Union Per Regulation 305/2011, D. 2004/18/EC Directive 98/34/EC, EN 1992-1-1: Eurocode 2: Design of concrete structures, Eur. Stand. (2004)
11. DIANA (Software). User's Manual – Release 10.2, TNO DIANA, Netherlands (2017). https://dianafea.com/manuals/d102/Diana.html
12. Fehling, E., Schmidt, M., Walraven, J., Leutbecher, T., Fröhlich, S.: Ultra-high performance concrete UHPC: Fundamentals, design, examples. Wiley (2014)

13. Xie, T., Fang, C., Mohamad Ali, M.S., Visintin, P.: Characterizations of autogenous and drying shrinkage of ultra-high performance concrete (UHPC): an experimental study, Cem. Concr. Compos. **91**, 156–173 (2018). https://doi.org/10.1016/j.cemconcomp.2018.05.009
14. Fang, C., Ali, M., Xie, T., Visintin, P., Sheikh, A.H.: The influence of steel fibre properties on the shrinkage of ultra-high performance fibre reinforced concrete. Constr. Build. Mater. **242**, 117993 (2020). https://doi.org/10.1016/j.conbuildmat.2019.117993
15. Yoo, D.-Y., Banthia, N., Yoon, Y.-S. : Effectiveness of shrinkage-reducing admixture in reducing autogenous shrinkage stress of ultra-high-performance fiber-reinforced concrete. Cem. Concr. Compos. **64**, 27–36 (2015). https://doi.org/10.1016/j.cemconcomp.2015.09.005

Correction to: Wind Tower FRC Foundations: Research and Design

Marco di Prisco, Claudio di Prisco, Giancarlo Fraraccio,
Bruno Dal Lago, Paolo Martinelli, Luca Flessati, Matteo Colombo,
and Giulio Zani

Correction to:
Chapter "Wind Tower FRC Foundations: Research
and Design" in: P. Serna et al. (Eds.): *Fibre Reinforced*
Concrete: Improvements and Innovations II,
RILEM Bookseries 36,
https://doi.org/10.1007/978-3-030-83719-8_71

In the original version of the book, the following belated correction has been incorporated: The author's name changed from "Giulio Zanis" to "Giulio Zani" in Chapter 71 and in the index. The book and the chapter have been updated with the changes.

The updated version of this chapter can be found at
https://doi.org/10.1007/978-3-030-83719-8_71

© RILEM 2022
P. Serna et al. (Eds.): BEFIB 2021, RILEM Bookseries 36, p. C1, 2022.
https://doi.org/10.1007/978-3-030-83719-8_84

Author Index

A

Abbas, Ali A., 400
Aidarov, Stanislav, 504, 702
Al Marahla, Razan H., 319
Alberdi-Pagola, Pablo, 727
Alberti, Marcos. G., 265
Altamirano, Marcelo G., 640
Altoubat, Salah, 166
Ardunay, Monica, 809

B

Bairán, Jesús Miguel, 552
Bakhshi, Mohammad, 277
Baloch, Hassan, 389
Barros, Joaquim A. O., 277
Bentegri, Imane, 86
Blanco, Ana, 377
Borg, Ruben Paul, 947
Boshoff, William P., 155, 483
Bouchard, David, 903
Boukendakdji, Othmane, 86
Boumakis, Ioannis, 420
Bouteille, Sébastien, 443
Braml, Thomas, 244
Brünig, Harald, 255
Buratti, Nicola, 331, 420

C

Caballero-Jorna, Marta, 220
Cardoso, Camila Vargas, 198
Cardoso, Matheus G., 24
Carlesso, Debora Martinello, 109, 121
Carmona, Sergio, 601
Castellanos, Ramiro, 265
Cavalaro, Sergio H. P., 109, 121, 291
Cavalcante, Ian B., 24

Chan, Ricardo, 13
Charron, Jean-Philippe, 471, 615
Claramunt, Josep, 809
Clavijo, Andrés, 739
Colombo, Matteo, 564, 831
Combrinck, Riaan, 155
Conciatori, David, 903
Conforti, Antonio, 514, 714
Congro, Marcello, 357
Coppens, Erik, 628
Corinaldesi, Valeria, 915
Cuenca, Estefania, 947
Curosu, Iurie, 255, 857

D

de A. Silva, Flávio, 51, 857, 871
de la Fuente, Albert, 109, 121, 342, 377, 504,
    552, 666, 678, 702, 771, 809
de Omena Jucá, Pedro Henrique, 198
De Smedt, Maure, 143
de Souza Castoldi, Raylane, 51
de Souza, Letícia O., 871
de Souza, Lourdes Maria Silva, 51
Debournonville, Guillaume, 628
Del Prete, Clementina, 331, 420
Della Bella, Bruno, 514
Desmettre, Clélia, 471, 615
Dey, Vikram, 845
di Prisco, Claudio, 831
di Prisco, Marco, 539, 564, 831
Dittel, Gozdem, 845
Dong, Wenkui, 889
Donnini, Jacopo, 915
Dooms, Bram, 628
Durigon Cocco, Guilherme, 198

© RILEM 2022
P. Serna et al. (Eds.): BEFIB 2021, RILEM Bookseries 36, pp. 977–979, 2022.
https://doi.org/10.1007/978-3-030-83719-8

E
Enfedaque, Alejandro, 265
Ercegovič, Rok, 208
Esmaeeli, Esmaeel, 925

F
Faccin, Enrico, 456
Facconi, Luca, 456, 525, 579, 796
Falkner, Horst, 761
Fantilli, Alessandro P., 98
Fataar, Humaira, 155
Ferrara, Liberato, 936, 947
Figueredo, Diego, 739
Fischer, Gregor, 727
Fischer, Oliver, 492
Flessati, Luca, 831
Fraraccio, Giancarlo, 831

G
Galeote, Eduardo, 377
Galobardes, Isaac, 13
Gálvez, Jaime C., 265
García, Nicolás, 739
Garcia-Taengua, Emilio, 232, 319
Garofalo, Roberto, 947
Gendron, Frédérick, 471
Generosi, Nicola, 915
Giaccio, Graciela M., 61, 640
González, Enzo, 739
Gries, Thomas, 845
Grünewald, Steffen, 389
Gudžulić, Vladislav, 365

H
Haus, Andreas, 539
He, Yuanye, 925
Heek, Peter, 132
Hinke, Ulf, 761
Hosseini, S. M. Amin, 809
Hunger, Martin, 514

I
Isla, Facundo, 61

J
Jimenez López, Diego, 771

K
Kadri, El-Hadj, 86
Kanstad, Terje, 539
Keuser, Manfred, 244
Khorami, Majid, 965
Koorikkattil, Ajeesh, 433
Kosteski, Luis Eduardo, 198
Kulzer, Felipe Eduardo, 198
Kumar, Dhanendra, 3

L
Lago, Bruno Dal, 831
Lameiras, Rodrigo M., 24
Lancioni, Giovanni, 915
Landler, Josef, 492
Larré de Oliveira, Letícia, 198
Le, An Hoang, 958
Li, Anling, 845
Li, Wengui, 889
Liebscher, Marco, 857
Llano-Torre, Aitor, 291
Look, Katharina, 132, 579
Lozano-Násner, Milena, 947
Luccioni, Bibiana, 61

M
Maalej, Mohamed, 166
Magalhães, Margareth da S., 74
Mantelli, Stefano Giuseppe, 35, 579
Marangon, Ederli, 198
Marchand, Pierre, 443
Marcos-Meson, Victor, 727
Mark, Peter, 132, 579
Martinelli, Paolo, 831
Martí-Vargas, José R., 232
Matthys, Stijn, 389
Mazzotti, Claudio, 331, 420
Mechtcherine, Viktor, 255, 749, 857
Medeghini, Filippo, 579
Mena, Francisco, 702
Meschke, Günther, 365
Mezquida-Alcaraz, Eduardo J., 936, 965
Minelli, Fausto, 35, 456, 525, 539, 579, 796
Mobasher, Barzin, 820, 845
Molins, Climent, 601
Montalbán, Aurora, 771
Monte, Francesco Lo, 936
Moreira Neto, Gerbert P., 74
Moro, Sandro, 514
Moy, Charles K. S., 13
Mudadu, Antonio, 714
Müller, Steffen, 749
Muniz, Bruna, 739

N
Nassif, Nadia, 166
Navarro-Gregori, Juan, 409, 936, 965
Nayar, Sunitha K, 433
Neu, Gerrit E., 365
Nogales, Alejandro, 666, 678
Nunes, Sandra, 174

O
Ortiz-Navas, Francisco, 409, 514

P

Panesar, Daman K., 35
Patel, Devansh, 820
Peaston, Chris H., 690
Piemonti, Alan, 514
Pimentel, Mário, 174
Pleesudjai, Chidchanok, 820
Plizzari, Giovanni, 456, 514, 539, 579, 714, 796
Polanec, David, 208
Pombo, Roberto, 640
Popa, Mihaela-Monica, 255
Pourzarabi, Ali, 564
Příbramský, Vladimír, 307

Q

Qiu, Jishen, 879

R

Ramezansefat, Honeyeh, 277
Ranade, Ravi, 3
Richir, Thomas, 628
Rodriguez, Gemma, 739
Rodriguez, Iliana, 739
Roehl, Deane, 357
Roig-Flores, Marta, 220, 947
Rossi, Pierre, 652
Ruiz-Muñoz, Cecilia, 947

S

Sadrolodabaee, Payam, 809
Saha, Suman, 591
Sánchez Gómez, Juana, 771
Sanchez, Eleazar C. M., 357
Sanchez, Thomas, 903
Schaef, Steve, 514, 820, 845
Scheffler, Christina, 255
Schillani, Fabrizio, 947
Segura-Castillo, Luis, 739
Seltner, Tobias, 244
Serna, Pedro, 220, 232, 291, 409, 936, 947, 965
Shah, Surendra P., 889
Smarzewski, Piotr, 186
Soliman, Amr Ashraf, 3
Sorelli, Luca, 903
Soualhi, Hamza, 86
Sousa, Carlos, 174
Soutsos, Marios N., 925

Souza, Lourdes Maria Silva, 871
Soymi, Olugbenga B., 400
Šušteršič, Jakob,, 208
Sutera, Luca, 504

T

Tailhan, Jean-Louis, 652
Terrade, Benjamin, 443
Tiberti, Giuseppe, 714
Tondolo, Francesco, 98
Torrijos, María Celeste, 61
Tošić, Nikola, 342, 552
Toutlemonde, François, 443
Trabucchi, Ivan, 714
Trindade, Ana C. C., 857
Tung Ngo, Tien, 86

V

Valente, Isabel B., 277
Valerio, Manuela, 504
Valero Ramos, Elisa, 771
Van Itterbeeck, Petra, 628
Vandecruys, Eline, 143
Vandewalle, Lucie, 143
Velasco, Reila V., 74
Venudharan, Veena, 433
Verstrynge, Els, 143
Visser, Divan, 483
Vital, Philipe de O., 74
Vivas, Juan Carlos, 61
Vrijdaghs, Rutger, 143
Vu, Duc-Tam, 443

W

Wan-Wendner, Roman, 420
Winterberg, Ralf, 714, 780
Wu, Bo, 879

Y

Yao, Yiming, 820

Z

Zajc, Andrej, 208
Zani, Giulio, 831
Zerbino, Raúl L., 61, 640
Zohrabyan, Vahan, 244

Lightning Source UK Ltd.
Milton Keynes UK
UKHW020733090922
408595UK00002B/5

9 783030 837211